Practice
"Work the Problems"

NEW

Feature	Description	Benefit	Page
"Assess Your Understanding" contains a variety of problems at the end of each section.			
'Are You Prepared?' Problems	These assess your retention of the prerequisite material you'll need. Answers are given at the end of the section exercises. This feature is related to the Preparing for This Section feature.	Do you always remember what you've learned? Working these problems is the best way to find out. If you get one wrong, you'll know exactly what you need to review and where to review it!	267
Concepts and Vocabulary	These Fill-in-the-Blank and True/False items assess your understanding of key definitions and concepts in the current section.	Learning math is more than memorization – it's about discovering connections. These problems help you understand the 'big ideas' before diving into skill building.	267
Skill Building	Correlated to section examples, these problems provide straightforward practice.	It's important to dig in and develop your skills. These problems provide you with ample practice to do so.	267-269
Applications and Extensions	These problems allow you to apply your skills to real-world problems. These problems also allow you to extend concepts learned in the section.	You will see that the material learned within the section has many uses in everyday life.	270
Discussion and Writing	"Discussion and Writing" problems are colored red. These support class discussion, verbalization of mathematical ideas, and writing and research projects.	To verbalize an idea, or to describe it clearly in writing, shows real understanding. These problems nurture that understanding. They're challenging but you'll get out what you put in.	270
Now Work Problems	Many examples refer you to a related homework problem. These related problems are marked by a pencil and yellow numbers.	If you get stuck while working problems, look for the closest Now Work problem and refer back to the related example to see if it helps.	267

Review
"Study for Quizzes and Tests"

Feature	Description	Benefit	Page
Chapter Reviews at the end of each chapter contain...			
"Things to Know"	A detailed list of important theorems, formulas, and definitions from the chapter.	Review these and you'll know the most important material in the chapter!	345-346
"You Should be Able To..."	Contains a complete list of objectives by section, with corresponding practice exercises.	Do the recommended exercises and you'll have mastery over the key material. If you get something wrong, review the suggested page numbers and try again.	347
Review Exercises	These provide comprehensive review and practice of key skills, matched to the Learning Objectives for each section.	Practice makes perfect. These problems combine exercises from all sections, giving you a comprehensive review in one place.	348-352
Chapter Test	About 15–20 problems that can be taken as a Chapter Test. Be sure to take the Chapter Test under test conditions—no notes!	Be prepared. Take the sample practice test. This will get you ready for your instructor's test.	352
Chapter Projects	The Chapter Project applies what you've learned in the chapter. Additional projects are available on the Instructor's Resource Center (IRC).	The Project gives you an opportunity to apply what you've learned in the chapter to solve a problem related to the opening article. If your instructor allows, these make excellent opportunities to work in a group, which is often the best way of learning math.	353
Cumulative Review	These problem sets appear at the end of each chapter, beginning with Chapter 2. They combine problems from previous chapters, providing an ongoing cumulative review.	These are really important. They will ensure that you are not forgetting anything as you go. These will go a long way toward keeping you constantly primed for quizzes and tests.	353-354

NEW

FOURTH EDITION

PRECALCULUS ESSENTIALS

Enhanced with Graphing Utilities

Michael Sullivan
Chicago State University

Michael Sullivan, III
Joliet Junior College

PEARSON

Prentice
Hall

Upper Saddle River, New Jersey 07458

Library of Congress Cataloging-in-Publication Data

Sullivan, Michael
  Precalculus Essentials: enhanced with graphing utilities / Michael Sullivan, Michael
  Sullivan, III.—4th ed.
    p. cm.
  ISBN 0-13-186668-0
  1. Calculus—Graphic methods—textbooks. I. Sullivan, Michael,
II. Title.
QA303.2.S86 2006
512/.1—dc22   2005048788

Editor-in-Chief: Sally Yagan
Acquisitions Editor: Adam Jaworski
Project Manager: Dawn Murrin
Executive Managing Editor: Kathleen Schiaparelli
Senior Managing Editor: Linda Mihatov Behrens
Production Editor: Bob Walters, Prepress Management, Inc.
Assistant Manufacturing Manager/Buyer: Alan Fischer
Marketing Manager: Halee Dinsey
Marketing Assistant: Joon Moon
Editorial Assistant/Print Supplements Editor: Christopher Truchan
Art Director: Heather Scott
Interior Designer: Judith Matz-Coniglio
Cover Designer: Susan Anderson-Smith
Art Editor: Thomas Benfatti
Director of Creative Services: Paul Belfanti
Director, Image Resource Center: Melinda Reo
Manager, Rights and Permissions: Zina Arabia
Manager, Visual Research: Beth Brenzel
Image Permission Coordinator: Carolyn Gauntt
Cover Image Permission Manager: Karen Sanatar
Photo Researcher: Melinda Alexander
Composition Manager: Allyson Graesser
    Electronic Production Specialists: Karen Stephens and Julie Nazario
Art Studio: Artworks:
    Senior Manager: Patricia Burns
    Production Manager: Ronda Whitson
    Manager, Production Technologies: Matt Haas
    Illustrators: Ryan Currier, Mark Landis
    Art Quality Assurance: Timothy Nguyen, Stacy Smith, Pamela Taylor

© 2006, 2003, 2000, 1996 Pearson Education, Inc.
Pearson Prentice Hall
Pearson Education, Inc.
Upper Saddle River, New Jersey 07458

Printed in the United States of America
10 9 8 7 6 5 4 3 2 1

ISBN 0-13-186668-0

Pearson Education LTD., London
Pearson Education Australia PTY, Limited, Sydney
Pearson Education Singapore, Pte. Ltd
Pearson Education North Asia Ltd. Hong Kong
Pearson Education Canada, Ltd., Toronto
Pearson Educación de Mexico, S.A. de C.V.
Pearson Education – Japan, Tokyo
Pearson Education Malaysia, Pte. Ltd

For the Next Generation
Shannon, Patrick, and Ryan (Murphy)
Michael S., Kevin, and Marissa (Sullivan)
Maeve (Sullivan)
Kaleigh and Billy (O'Hara)

Contents

Chapter 1 Graphs   1

Chapter 2 Functions and Their Graphs   55

Chapter 3 Polynomial and Rational Functions   149

Chapter 4   **Exponential and Logarithmic Functions**   247

Chapter 5   **Trigonometric Functions**   355

Sections in gray are not included in *Precalculus Essentials: Enhanced with Graphing Utilities, 4th Edition and are only available in Precalculus: Enhanced with Graphing Utilities, 4th Edition.*

Three Distinct Series

Students have different goals, learning styles, and levels of preparation. Instructors have different teaching philosophies, styles, and techniques. Rather than write one series to fit all, the Sullivans have written three distinct series. All share the same goal—to develop a high level of mathematical understanding and an appreciation for the way mathematics can describe the world around us. The manner of reaching that goal, however, differs from series to series.

Contemporary Series, Seventh Edition

The Contemporary Series is the most traditional in approach, yet modern in its treatment of precalculus mathematics. Graphing utility coverage is optional and can be included or excluded based on instructor preference.

College Algebra, College Algebra Essentials, Algebra & Trigonometry, Trigonometry, Precalculus, Precalculus Essentials

Enhanced with Graphing Utilities Series, Fourth Edition

This series provides a more thorough integration of graphing utilities into topics, allowing students to explore mathematical concepts and foreshadow ideas usually studied in later courses. Using technology, the approach to solving certain problems differs from the Contemporary series, while the emphasis on understanding concepts and building strong skills does not.

College Algebra, College Algebra Essentials, Algebra & Trigonometry, Trigonometry, Precalculus, Precalculus Essentials

Graphing, Data, and Analysis Series, Third Edition

This series differs from the others, utilizing a functions approach which serves as the organizing principle tying concepts together. Each chapter introduces a new type of function, then develops all concepts pertaining to that particular function. The solutions of equations and inequalities, instead of being developed as stand-alone topics, are developed in the context of the underlying functions. Technology is integrated at the same level as the Enhanced with Graphing Utilities series (described above).

College Algebra, Algebra & Trigonometry, Precalculus

The Essentials Edition

What Is the Goal of the Essentials Edition?

Our goal in publishing this *Essentials* version of Sullivan and Sullivan's **Precalculus Enhanced with Graphing Utilities, 4th Edition**, is to provide a lighter, more tailored, less expensive alternative to the parent textbook.

What's "Essential" About It?

Many one semester courses do not cover the full scope of the parent textbook. So, we selected topics from the parent version that are often not used and removed them, leaving a "core" set of chapters and sections. This "core" is the *Essentials* version.

Is Anything Else Different from the Parent Textbook?

No. The only difference is that some topics have been removed from the parent textbook to create the *Essentials* edition. The chapters and sections that remain which comprise the *Essentials* edition are *identical*, *word-for-word* to those in the parent textbook. Depth of coverage within topics has not been sacrificed in the *Essentials* edition.

What's In and What's Out?

The Table of Contents indicates what is included in this *Essentials* edition ("highlighted"), and what is excluded from this *Essentials* edition (subdued). Why have we left topics in the table of contents that are not included in this *Essentials* edition? Two reasons: (1) we wanted you to be able to determine quickly and easily which book is more appropriate for your course. This is easily done via this format. (2) the parent textbook and the *Essentials* edition share the same supplements. This is made possible by maintaining consistent chapter and section numbering in both books.

What If I Cover Something in My Course That Is Not in the Essentials Edition?

There are two options for you. You can simply ask your Prentice Hall sales representative to send you the parent textbook for review. Or, if you only need two or three additional sections, ask your Prentice Hall representative about custom publishing opportunities, the best way to reach an exact fit.

Preface
to the Instructor

As professors at both an urban university and a community college, Michael Sullivan and Michael Sullivan, III are aware of the varied needs of Precalculus students, ranging from those who have little mathematical background and a fear of mathematics courses, to those having a strong mathematical education and a high level of motivation. For some of your students, this will be their last course in mathematics, whereas others will further their mathematical education. This text is written for both groups.

As a teacher, and as an author of precalculus, engineering calculus, finite mathematics, and business calculus texts, Michael Sullivan understands what students must know if they are to be focused and successful in upper level math courses. However, as a father of four, he also understands the realities of college life. His co-author and son, Michael Sullivan, III, believes in the value of technology as a tool for learning that enhances understanding without sacrificing math skills. Together, both authors have taken great pains to ensure that the text contains solid, student-friendly examples and problems, as well as a clear and seamless writing style.

A tremendous benefit of authoring a successful series is the broad-based feedback we receive from teachers and students. Virtually every change in this edition is the result of their thoughtful comments and suggestions. We are sincerely grateful for this support and hope that we have been able to take these ideas and, building upon a successful third edition, make this series an even better tool for learning and teaching. We continue to encourage you to share with us your experiences teaching from this text.

New Features
in the Fourth Edition

Rather than provide a list of new features here, we have placed that information in the end sheet on the front cover of this book. The new features are easy to spot by the "New" in the left column.

This places the new features in their proper context, as building blocks of an overall learning system that has been carefully crafted over the years to help students get the most out of the time they put into studying. Please take the time to review this, and to discuss it with your students at the beginning of your course. Our experience has been that when students utilize these features, they are more successful in the course.

Organizational Changes
in the Fourth Edition

- The part of former Section 1.4, "Solving Equations," that deals with solving equations algebraically has been moved to the Appendix as Section A.5.

- The balance of former Section 1.4 is now Section 1.3 "Solving Equations Using a Graphing Utility."

- The part of former Section 1.5, "Solving Inequalities," that deals with interval notation and solving inequalities algebraically has been moved to the Appendix as Section A.8.

- The discussion of "Circles," formerly in Section 1.3, "Symmetry; Graphing Key Equations; Circles," now appears as a stand-alone Section 1.5; this allowed for combining the former Section 1.2, "Introduction to Graphing Equations," and Section 1.3 into a single Section 1.2, "Graphs of Equations."

- The discussion of "Composite Functions," former Section 2.6, "Operations on Functions; Composite Functions," now appears as Section 4.1 in Chapter 4, "Exponential and Logarithmic Functions." "Operations on Functions" is now part of Section 2.3, "Properties of Functions."

- Former Section 3.2: "Power Functions and Models" and former Section 3.3, "Polynomials Functions and Models," have been combined into Section 3.2, "Polynomial Functions and Models."

- Former Section 3.4, "Rational Functions I," has been renamed as Section 3.3, "Properties of Rational Functions." Former Section 3.5, Rational Functions II: Analyzing Graphs," has been renamed Section 3.4, "The Graph of a Rational Function; Inverse and Joint Variation."

- Former Section 4.7, "Growth and Decay" and former Section 4.8, Exponential, Logarithmic, and Logistic Models', have been reorganized into Section 4.8, "Exponential Growth and Decay; Newton's Law; Logistic Growth and Decay", and Section 4.9, "Building Exponential, Logarithmic, and Logistic Models from Data".

- Former Section 10.1, "Systems of Linear Equations: Two Equations Containing Two Variables", and former Section 10.2, "Systems of Linear Equations: Three Equations Containing Three Variables", have been combined into one Section 10.1, "Systems of Linear Equations: Substitution and Elimination".

- Former Section A.3, "Polynomials", and Section A.4, "Rational Expressions", have been combined into one Section A.3, "Polynomial and Rational Expressions".

- Former Section A.6, "Equations; Completing the Square; the Quadratic Formula", is now part of Section A.5, "Solving Equations Algebraically".

- Former Section A.8, "Setting up Equations; Applications", is now Section A.7, "Problem Solving: Interest, Mixture, Uniform Motion, and Constant Rate Jobs". The easier applications from former Section A.8 appear in Section A.5.

Using the Fourth Edition Effectively and Efficiently with Your Syllabus

As the chart illustrates, this book has been organized with flexibility of use in mind. Even within a given chapter certain sections are optional and can be omitted without loss of continuity. (See the detail following the flow chart).

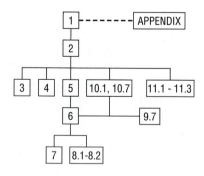

Chapter 1   Graphs

This chapter is a briefer version of the former Chapter 1. A quick coverage of this chapter, which is mainly review material, will enable you to get to Chapter 2, "Functions and Their Graphs", earlier.

Chapter 2   Functions and Their Graphs

Perhaps the most important chapter. Section 2.7 is optional.

Chapter 3   Polynomial and Rational Functions

Topic selection depends on your syllabus.

Chapter 4   Exponential and Logarithmic Functions

Sections 4.1–4.6 follow in sequence. Sections 4.7–4.9 are optional.

Chapter 5   Trigonometric Functions

The sections follow in sequence. Section 5.6 is optional.

Chapter 6   Analytic Trigonometry

Sections 6.2, 6.6, and 6.8 may be omitted in a brief course.

Chapter 7   Applications of Trigonometric Functions

Sections 7.4 and 7.5 are not included in the *Essentials* version of *Precalculus*.

Chapter 8   Polar Coordinates; Vectors

Sections 8.3-8.7 are not included in the *Essentials* version of *Precalculus*.

Chapter 9   Analytic Geometry

Only Section 9.7 is included in the *Essentials* version of *Precalculus*.

Chapter 10   Systems of Equations and Inequalities

Only Sections 10.1 and 10.7 are included in the *Essentials* version of *Precalculus*.

Chapter 11 Sequences; Induction; the Binomial Theorem

Sections 11.4 and 11.5 are not included in the *Essentials* version of *Precalculus*.

Chapter 12 Counting and Probability

This chapter is not included in the *Essentials* version of *Precalculus*.

Chapter 13 A Preview of Calculus: the Limit, Derivative, and Integral of a Function

This chapter is not included in the *Essentials* version of *Precalculus*.

Appendix Review

Sections A.1-A.6 and Sections A.8-A.9 consist of review material, which may be covered at the start of a course or used as a just-in-time review. Specific references to this material occur throughout the text to assist in the review process. Coverage of Section A.7 will depend on your syllabus.

Acknowledgments

Textbooks are written by authors, but evolve from an idea to final form through the efforts of many people. It was Don Dellen who first suggested this series. Don is remembered for his extensive contributions to publishing and mathematics.

We would also like to extend thanks to the following individuals for their important assistance and encouragement in the preparation of this edition: From Prentice Hall: Halee Dinsey for her innovative marketing skills; Adam Jaworski for his substantial contributions, ideas, and enthusiasm; Patrice Jones, who remains a huge fan and supporter; Dawn Murrin for her talent and skill in getting supplements out; Bob Walters, who continues to amaze us with his organizational skill as Production Editor; Sally Yagan for her continued support and genuine interest; and the Prentice Hall Sales Team, for their continued confidence in our books. We also extend out thanks to Teri Lovelace and the entire Laurel Technical Services staff, for their dedication to accuracy in checking manuscript and answers. Special thanks to Celeste Hernandez for her help in preparing chapter projects, and to Kevin Bodden for his work in preparing the chapter tests.

Finally, our grateful thanks to the dedicated users and reviewers of our books, whose collective insights form the backbone of each textbook revision. We apologize for any omissions:

James Africh, College of DuPage
Steve Agronsky, Cal Poly State University
Grant Alexander, Joliet Junior College
Dave Anderson, South Suburban College
Joby Milo Anthony, University of Central Florida
James E. Arnold, University of Wisconsin-Milwaukee
Adel Arshaghi, Center for Educational Merit
Carolyn Autray, University of West Georgia
Agnes Azzolino, Middlesex County College
Wilson P Banks, Illinois State University
Sudeshna Basu, Howard University
Dale R. Bedgood, East Texas State University
Beth Beno, South Suburban College
Carolyn Bernath, Tallahassee Community College
William H. Beyer, University of Akron
Annette Blackwelder, Florida State University
Richelle Blair, Lakeland Community College
Kevin Bodden, Lewis and Clark College
Bob Bradshaw, Ohlone College
Trudy Bratten, Grossmont College
Tim Bremer, Broome Community College
Joanne Brunner, Joliet Junior College
Warren Burch, Brevard Community College
Mary Butler, Lincoln Public Schools
Jim Butterbach, Joliet Junior College
William J. Cable, University of Wisconsin-Stevens Point
Lois Calamia, Brookdale Community College
Jim Campbell, Lincoln Public Schools
Roger Carlsen, Moraine Valley Community College
Elena Catoiu, Joliet Junior College
Mathews Chakkanakuzhi, Palomar College
John Collado, South Suburban College
Nelson Collins, Joliet Junior College

Jim Cooper, Joliet Junior College
Denise Corbett, East Carolina University
Theodore C. Coskey, South Seattle Community College
Donna Costello, Plano Senior High School
Paul Crittenden, University of Nebraska at Lincoln
John Davenport, East Texas State University
Faye Dang, Joliet Junior College
Antonio David, Del Mar College
Duane E. Deal, Ball State University
Timothy Deis, University of Wisconsin-Platteville
Vivian Dennis, Eastfield College
Deborah Dillon, R. L. Turner High School
Guesna Dohrman, Tallahassee Community College
Karen R. Dougan, University of Florida
Louise Dyson, Clark College
Paul D. East, Lexington Community College
Don Edmondson, University of Texas-Austin
Erica Egizio, Joliet Junior College
Christopher Ennis, University of Minnesota
Ralph Esparza, Jr., Richland College
Garret J. Etgen, University of Houston
Pete Falzone, Pensacola Junior College
W.A. Ferguson, University of Illinois-Urbana/Champaign
Iris B. Fetta, Clemson University
Mason Flake, student at Edison Community College
Timothy W. Flood, Pittsburg State University
Merle Friel, Humboldt State University
Richard A. Fritz, Moraine Valley Community College
Dewey Furness, Ricke College
Randy Gallaher, Lewis and Clark College
Tina Garn, University of Arizona
Dawit Getachew, Chicago State University

Wayne Gibson, Rancho Santiago College
Robert Gill, University of Minnesota Duluth
Sudhir Kumar Goel, Valdosta State University
Joan Goliday, Sante Fe Community College
Frederic Gooding, Goucher College
Sue Graupner, Lincoln Public Schools
Jennifer L. Grimsley, University of Charleston
Ken Gurganus, University of North Carolina
James E. Hall, University of Wisconsin-Madison
Judy Hall, West Virginia University
Edward R. Hancock, DeVry Institute of Technology
Julia Hassett, DeVry Institute-Dupage
Christopher Hay-Jahans, University of South Dakota
Michah Heibel, Lincoln Public Schools
LaRae Helliwell, San Jose City College
Brother Herron, Brother Rice High School
Robert Hoburg, Western Connecticut State University
Lynda Hollingsworth, Northwest Missouri State University
Charla Holzbog, Denison High School
Lee Hruby, Naperville North High School
Miles Hubbard, St. Cloud State University
Kim Hughes, California State College-San Bernardino
Ron Jamison, Brigham Young University
Richard A. Jensen, Manatee Community College
Sandra G. Johnson, St. Cloud State University
Tuesday Johnson, New Mexico State University
Moana H. Karsteter, Tallahassee Community College
Donna Katula, Joliet Junior College
Arthur Kaufman, College of Staten Island
Thomas Kearns, North Kentucky University
Shelia Kellenbarger, Lincoln Public Schools

Lynne Kowski, Raritan Valley Community College

Keith Kuchar, Manatee Community College

Tor Kwembe, Chicago State University

Linda J. Kyle, Tarrant Country Jr. College

H.E. Lacey, Texas A & M University

Harriet Lamm, Coastal Bend College

James Lapp, Fort Lewis College

Matt Larson, Lincoln Public Schools

Christopher Lattin, Oakton Community College

Julia Ledet, Lousiana State University

Adele LeGere, Oakton Community College

Kevin Leith, University of Houston

JoAnn Lewin, Edison College

Jeff Lewis, Johnson County Community College

Janice C. Lyon, Tallahassee Community College

Virginia McCarthy, Iowa State University

Jean McArthur, Joliet Junior College

Karla McCavit, Albion College

Tom McCollow, DeVry Institute of Technology

Marilyn McCollum, North Carolina State University

Jill McGowan, Howard University

Will McGowant, Howard University

Angela McNulty, Joliet Junior College

Laurence Maher, North Texas State University

Jay A. Malmstrom, Oklahoma City Community College

Rebecca Mann, Apollo High School

Sherry Martina, Naperville North High School

Alec Matheson, Lamar University

Nancy Matthews, University of Oklahoma

James Maxwell, Oklahoma State University-Stillwater

Marsha May, Midwestern State University

Judy Meckley, Joliet Junior College

David Meel, Bowling Green State University

Carolyn Meitler, Concordia University

Samia Metwali, Erie Community College

Rich Meyers, Joliet Junior College

Eldon Miller, University of Mississippi

James Miller, West Virginia University

Michael Miller, Iowa State University

Kathleen Miranda, SUNY at Old Westbury

Thomas Monaghan, Naperville North High School

Craig Morse, Naperville North High School

Samad Mortabit, Metropolitan State University

Pat Mower, Washburn University

A. Muhundan, Manatee Community College

Jane Murphy, Middlesex Community College

Richard Nadel, Florida International University

Gabriel Nagy, Kansas State University

Bill Naegele, South Suburban College

Karla Neal, Lousiana State University

Lawrence E. Newman, Holyoke Community College

James Nymann, University of Texas-El Paso

Mark Omodt, Anoka-Ramsey Community College

Seth F. Oppenheimer, Mississippi State University

Linda Padilla, Joliet Junior College

E. James Peake, Iowa State University

Kelly Pearson, Murray State University

Philip Pina, Florida Atlantic University

Michael Prophet, University of Northern Iowa

Neal C. Raber, University of Akron

Thomas Radin, San Joaquin Delta College

Ken A. Rager, Metropolitan State College

Kenneth D. Reeves, San Antonio College

Elsi Reinhardt, Truckee Meadows Community College

Jane Ringwald, Iowa State University

Stephen Rodi, Austin Community College

William Rogge, Lincoln Northeast High School

Howard L. Rolf, Baylor University

Mike Rosenthal, Florida International University

Phoebe Rouse, Lousiana State University

Edward Rozema, University of Tennessee at Chattanooga

Dennis C. Runde, Manatee Community College

Alan Saleski, Loyola University of Chicago

Susan Sandmeyer, Jamestown Community College

Linda Schmidt, Greenville Technical College

A.K. Shamma, University of West Florida

Martin Sherry, Lower Columbia College

Tatrana Shubin, San Jose State University

Anita Sikes, Delgado Community College

Timothy Sipka, Alma College

Lori Smellegar, Manatee Community College

Gayle Smith, Loyola Blakefield

John Spellman, Southwest Texas State University

Rajalakshmi Sriram, Okaloosa-Walton Community College

Becky Stamper, Western Kentucky University

Judy Staver, Florida Community College-South

Neil Stephens, Hinsdale South High School

Patrick Stevens, Joliet Junior College

Christopher Terry, Augusta State University

Diane Tesar, South Suburban College

Tommy Thompson, Brookhaven College

Martha K. Tietze, Shawnee Mission Northwest High School

Richard J.Tondra, Iowa State University

Marvel Townsend, University of Florida

Jim Trudnowski, Carroll College

Robert Tuskey, Joliet Junior College

Richard G.Vinson, University of South Alabama

Mary Voxman, University of Idaho

Jennifer Walsh, Daytona Beach Community College

Donna Wandke, Naperville North High School

Darlene Whitkenack, Northern Illinois University

Christine Wilson, West Virginia University

Brad Wind, Florida International University

Mary Wolyniak, Broome Community College

Canton Woods, Auburn University

Tamara S.Worner, Wayne State College

Terri Wright, New Hampshire Community Technical College, Manchester

George Zazi, Chicago State University

Michael Sullivan
Michael Sullivan, III

Student Study-Pak

Everything a student needs to succeed in one place. Free, packaged with the book, or available for purchase standalone. Study-Pak contains:

- *Student Solutions Manual*

  Fully worked solutions to odd-numbered exercises.

- *Pearson Tutor Center*

  Tutors provide one-on-one tutoring for any problem with an answer at the back of the book. Students access the Tutor Center via toll-free phone, fax, or email.

- *CD Lecture Series*

  A comprehensive set of CD-ROMs, tied to the textbook, containing short video clips of an instructor working key book examples.

- *Algebra Review*

  Four chapters of Intermediate Algebra review. Perfect for a slower-paced course or for individual review.

- *Chapter Test Prep Video CD*

  Gives student **step-by-step video solutions to every problem in each Chapter Test.** An instructor works through each chapter test problem providing valuable review to enhance mastery of key chapter content. Easy video navigation allows students to instantly access the solution to any problem.

MathXL®

MathXL® is a powerful online homework, tutorial, and assessment system that accompanies your textbook. Instructors can create, edit, and assign online homework and tests using algorithmically generated exercises correlated at the objective level to the textbook. Student work is tracked in an online gradebook. Students can take chapter tests and receive personalized study plans based on their results. The study plan diagnoses weaknesses and links students to tutorial exercises for objectives they need to study. Students can also access video clips from selected exercises. MathXL® is available to qualified adopters. For more information, visit our website at **www.mathxl.com**, or contact your Prentice Hall sales representative for a demonstration.

MyMathLab

MyMathLab is a text-specific, customizable online course for your textbooks. MyMathLab is powered by CourseCompass™—Pearson Education's online teaching and learning environment—and by MathXL®—our online homework, tutorial, and assessment system. MyMathLab gives you the tools you need to deliver all or a portion of your course online, whether your students are in a lab setting or working from home.

MyMathLab provides a rich and flexible set of course materials, featuring free-response exercises that are algorithmically generated for unlimited practice. Students can use online tools such as video lectures and a multimedia textbook to improve their performance. Instructors can use MyMathLab's homework and test managers to select and assign online exercises correlated to the textbook, and can import TestGen tests for added flexibility. The online gradebook—designed specifically for mathematics—automatically tracks students' homework and test results and gives the instructor control over how to calculate final grades. MyMathLab is available to qualified adopters. For more information, visit our website at **www.mymathlab.com** or contact your Prentice Hall sales representative for a product demonstration.

List of Applications

Photo Credits

TO THE STUDENT

As you begin, you may feel anxious about the number of theorems, definitions, procedures, and equations. You may wonder if you can learn it all in time. Don't worry, your concerns are normal. This textbook was written with you in mind. If you attend class, work hard, and read and study this book, you will build the knowledge and skills you need to be successful. Here's how you can use the book to your benefit.

READ CAREFULLY

When you get busy, it's easy to skip reading and go right to the problems. Don't . . . the book has a large number of examples and clear explanations to help you break down the mathematics into easy-to-understand steps. Reading will provide you with a clearer understanding, beyond simple memorization. Read it before class (not after) so you can ask questions about anything you didn't understand. You'll be amazed at how much more you'll get out of class if you do this.

GETTING THE MOST FROM THIS BOOK

We've provided a brief guide to getting the most from this book (please refer to the "Prepare for Class," "Practice," and "Review" pages on the inside front cover of this book). Spend fifteen minutes reviewing the guide and familiarizing yourself with the features by flipping to the page numbers provided. Then, as you read, use them. This is the best way to make the most of your textbook.

Please do not hesitate to contact us, through Prentice Hall, with any questions, suggestions or comments that would improve this text. We look forward to hearing from you, and best of luck with all your studies.

Best Wishes!

Michael Sullivan
Michael Sullivan, III

Graphs

A LOOK AHEAD In this chapter we make the connection between algebra and geometry through the rectangular coordinate system. The idea of using a system of rectangular coordinates dates back to ancient times, when such a system was used for surveying and city planning. Apollonius of Perga, in 200 BC, used a form of rectangular coordinates in his work on conics, although this use does not stand out as clearly as it does in modern treatments. Sporadic use of rectangular coordinates continued until the 1600s. By that time, algebra had developed sufficiently so that René Descartes (1596–1650) and Pierre de Fermat (1601–1665) could take the crucial step, which was the use of rectangular coordinates to translate geometry problems into algebra problems, and vice versa. This step was important for two reasons. First, it allowed both geometers and algebraists to gain new insights into their subjects, which previously had been regarded as separate, but now were seen to be connected in many important ways. Second, these insights made the development of calculus possible, which greatly enlarged the number of areas in which mathematics could be applied and made possible a much deeper understanding of these areas.

With the advent of modern technology, in particular graphing utilities, now not only are we able to visualize the dual roles of algebra and geometry, but we are also able to solve many problems that required advanced methods before this technology.

30-year mortgage rates back under 6%

WASHINGTON (Reuters)—Interest rates on 30-year mortgages fell under 6% this week, while rates on 15-year and adjustable mortgages also eased, mortgage finance company Freddie Mac said Thursday.

Thirty-year mortgage rates averaged 5.99% in the week ended Aug. 5 compared with 6.08% a week earlier, Freddie Mac said.

Freddie Mac said 15-year mortgages averaged 5.40%, down from 5.49% last week. One-year adjustable rate mortgages dipped to an average of 4.08% from 4.17% last week.

Freddie Mac is a mortgage finance company chartered by Congress that buys mortgages from lenders and packages them into securities for investors or holds them in its own portfolio.

USA Today, August 5, 2004.

—See Chapter Project 1.

OUTLINE

1.1 Rectangular Coordinates; Graphing Utilities

PREPARING FOR THIS SECTION *Before getting started, review the following:*
- Algebra Review (Appendix, Section A.1, pp. 659–667)
- Geometry Review (Appendix, Section A.2, pp. 669–672)

✎ Now work the 'Are You Prepared?' problems on page 8.

OBJECTIVES 1 Use the Distance Formula
2 Use the Midpoint Formula

Figure 1

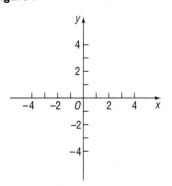

We locate a point on the real number line by assigning it a single real number, called the *coordinate of the point*. For work in a two-dimensional plane, we locate points by using two numbers.

We begin with two real number lines located in the same plane: one horizontal and the other vertical. We call the horizontal line the **x-axis**, the vertical line the **y-axis**, and the point of intersection the **origin O**. See Figure 1. We assign coordinates to every point on these number lines using a convenient scale. In mathematics, we usually use the same scale on each axis; in applications, a different scale is often used on each axis.

The origin O has a value of 0 on both the x-axis and y-axis. Points on the x-axis to the right of O are associated with positive real numbers, and those to the left of O are associated with negative real numbers. Points on the y-axis above O are associated with positive real numbers, and those below O are associated with negative real numbers. In Figure 1, the x-axis and y-axis are labeled as x and y, respectively, and we have used an arrow at the end of each axis to denote the positive direction.

The coordinate system described here is called a **rectangular** or **Cartesian**[*] **coordinate system**. The plane formed by the x-axis and y-axis is sometimes called the **xy-plane**, and the x-axis and y-axis are referred to as the **coordinate axes**.

Figure 2

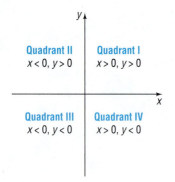

Any point P in the xy-plane can be located by using an **ordered pair** (x, y) of real numbers. Let x denote the signed distance of P from the y-axis (*signed* means that, if P is to the right of the y-axis, then $x > 0$, and if P is to the left of the y-axis, then $x < 0$); and let y denote the signed distance of P from the x-axis. The ordered pair (x, y), also called the **coordinates** of P, then gives us enough information to locate the point P in the plane.

For example, to locate the point whose coordinates are $(-3, 1)$, go 3 units along the x-axis to the left of O and then go straight up 1 unit. We **plot** this point by placing a dot at this location. See Figure 2, in which the points with coordinates $(-3, 1)$, $(-2, -3)$, $(3, -2)$, and $(3, 2)$ are plotted.

The origin has coordinates $(0, 0)$. Any point on the x-axis has coordinates of the form $(x, 0)$, and any point on the y-axis has coordinates of the form $(0, y)$.

If (x, y) are the coordinates of a point P, then x is called the **x-coordinate**, or **abscissa**, of P and y is the **y-coordinate**, or **ordinate**, of P. We identify the point P by its coordinates (x, y) by writing $P = (x, y)$. Usually, we will simply say "the point (x, y)" rather than "the point whose coordinates are (x, y)."

Figure 3

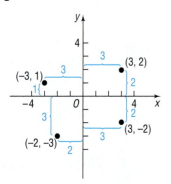

The coordinate axes divide the xy-plane into four sections called **quadrants**, as shown in Figure 3. In quadrant I, both the x-coordinate and the y-coordinate of all points are positive; in quadrant II, x is negative and y is positive; in quadrant III, both x and y are negative; and in quadrant IV, x is positive and y is negative. Points on the coordinate axes belong to no quadrant.

✏️ **NOW WORK PROBLEM 11.**

---
[*]Named after René Descartes (1596–1650), a French mathematician, philosopher, and theologian.

Graphing Utilities

Figure 4
$y = 2x$

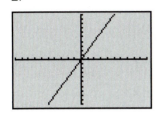

All graphing utilities, that is, all graphing calculators and all computer software graphing packages, graph equations by plotting points on a screen. The screen itself actually consists of small rectangles, called **pixels**. The more pixels the screen has, the better the resolution. Most graphing calculators have 48 pixels per square inch; most computer screens have 32 to 108 pixels per square inch. When a point to be plotted lies inside a pixel, the pixel is turned on (lights up). The graph of an equation is a collection of lighted pixels. Figure 4 shows how the graph of $y = 2x$ looks on a TI-84 Plus graphing calculator.

The screen of a graphing utility will display the coordinate axes of a rectangular coordinate system. However, you must set the scale on each axis. You must also include the smallest and largest values of x and y that you want included in the graph. This is called **setting the viewing rectangle** or **viewing window**. Figure 5 illustrates a typical viewing window.

Figure 5

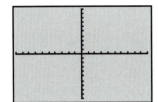

To select the viewing window, we must give values to the following expressions:

Xmin: the smallest value of x shown on the viewing window
Xmax: the largest value of x shown on the viewing window
Xscl: the number of units per tick mark on the x-axis
Ymin: the smallest value of y shown on the viewing window
Ymax: the largest value of y shown on the viewing window
Yscl: the number of units per tick mark on the y-axis

Figure 6 illustrates these settings and their relation to the Cartesian coordinate system.

Figure 6

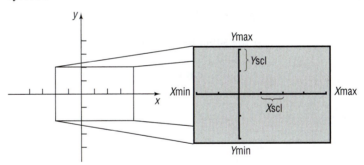

If the scale used on each axis is known, we can determine the minimum and maximum values of x and y shown on the screen by counting the tick marks. Look again at Figure 5. For a scale of 1 on each axis, the minimum and maximum values of x are -10 and 10, respectively; the minimum and maximum values of y are also -10 and 10. If the scale is 2 on each axis, then the minimum and maximum values of x are -20 and 20, respectively; and the minimum and maximum values of y are -20 and 20, respectively.

Conversely, if we know the minimum and maximum values of x and y, we can determine the scales being used by counting the tick marks displayed. We shall follow the practice of showing the minimum and maximum values of x and y in our illustrations so that you will know how the window was set. See Figure 7.

Figure 7

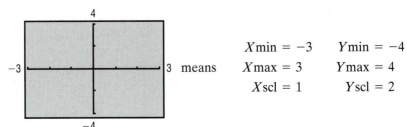

| EXAMPLE 1 | **Finding the Coordinates of a Point Shown on a Graphing Utility Screen** |

Find the coordinates of the point shown in Figure 8. Assume the coordinates are integers.

Figure 8

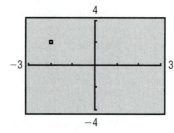

| Solution | First we note that the viewing window used in Figure 8 is |

$$X\text{min} = -3 \qquad Y\text{min} = -4$$
$$X\text{max} = 3 \qquad Y\text{max} = 4$$
$$X\text{scl} = 1 \qquad Y\text{scl} = 2$$

The point shown is 2 tick units to the left on the horizontal axis (scale = 1) and 1 tick up on the vertical scale (scale = 2). The coordinates of the point shown are $(-2, 2)$. ◄

NOW WORK PROBLEMS 15 AND 25.

1 Use the Distance Formula

If the same units of measurement, such as inches, centimeters, and so on, are used for both the x-axis and y-axis, then all distances in the xy-plane can be measured using this unit of measurement.

| EXAMPLE 2 | **Finding the Distance between Two Points** |

Find the distance d between the points $(1, 3)$ and $(5, 6)$.

| Solution | First we plot the points $(1, 3)$ and $(5, 6)$ and connect them with a straight line. See Figure 9(a). We are looking for the length d. We begin by drawing a horizontal line from $(1, 3)$ to $(5, 3)$ and a vertical line from $(5, 3)$ to $(5, 6)$, forming a right triangle, as shown in Figure 9(b). One leg of the triangle is of length 4 (since $|5 - 1| = 4$) and the other is of length 3 (since $|6 - 3| = 3$). By the Pythagorean Theorem, the square of the distance d that we seek is |

$$d^2 = 4^2 + 3^2 = 16 + 9 = 25$$
$$d = \sqrt{25} = 5$$

Figure 9

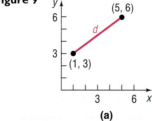

(a)

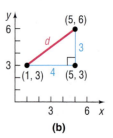

(b)

◄

The **distance formula** provides a straightforward method for computing the distance between two points.

In Words

To compute the distance between two points, find the difference of the x-coordinates, square it, and add this to the square of the difference of the y-coordinates. The square root of this sum is the distance.

Theorem

Distance Formula

The distance between two points $P_1 = (x_1, y_1)$ and $P_2 = (x_2, y_2)$, denoted by $d(P_1, P_2)$, is

$$d(P_1, P_2) = \sqrt{(x_2 - x_1)^2 + (y_2 - y_1)^2} \qquad \textbf{(1)}$$

Proof of the Distance Formula   Let (x_1, y_1) denote the coordinates of point P_1, and let (x_2, y_2) denote the coordinates of point P_2. Assume that the line joining P_1 and P_2 is neither horizontal nor vertical. Refer to Figure 10(a). The coordinates of P_3 are (x_2, y_1). The horizontal distance from P_1 to P_3 is the absolute value of the difference of the x-coordinates, $|x_2 - x_1|$. The vertical distance from P_3 to P_2 is the absolute value of the difference of the y-coordinates, $|y_2 - y_1|$. See Figure 10(b). The distance $d(P_1, P_2)$ that we seek is the length of the hypotenuse of the right triangle, so, by the Pythagorean Theorem, it follows that

$$[d(P_1, P_2)]^2 = |x_2 - x_1|^2 + |y_2 - y_1|^2$$
$$= (x_2 - x_1)^2 + (y_2 - y_1)^2$$
$$d(P_1, P_2) = \sqrt{(x_2 - x_1)^2 + (y_2 - y_1)^2}$$

Figure 10

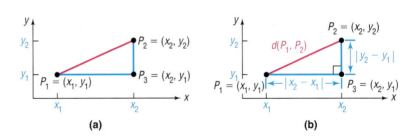

(a)   (b)

Now, if the line joining P_1 and P_2 is horizontal, then the y-coordinate of P_1 equals the y-coordinate of P_2; that is, $y_1 = y_2$. Refer to Figure 11(a). In this case, the distance formula (1) still works, because, for $y_1 = y_2$, it reduces to

$$d(P_1, P_2) = \sqrt{(x_2 - x_1)^2 + 0^2} = \sqrt{(x_2 - x_1)^2} = |x_2 - x_1|$$

Figure 11

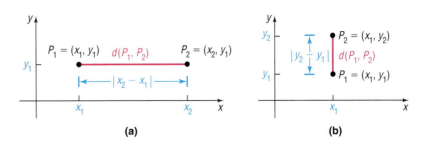

(a)   (b)

A similar argument holds if the line joining P_1 and P_2 is vertical. See Figure 11(b). The distance formula is valid in all cases.   ■

| EXAMPLE 3 | **Finding the Length of a Line Segment** |

Figure 12

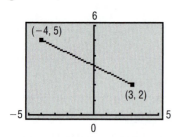

Find the length of the line segment shown in Figure 12.

Solution The length of the line segment is the distance between the points $P_1 = (x_1, y_1) = (-4, 5)$, and $P_2 = (x_2, y_2) = (3, 2)$. Using the distance formula (1) with $x_1 = -4$, $y_1 = 5$, $x_2 = 3$, and $y_2 = 2$, the length d is

$$d = \sqrt{(x_2 - x_1)^2 + (y_2 - y_1)^2} = \sqrt{[3 - (-4)]^2 + (2 - 5)^2}$$
$$= \sqrt{7^2 + (-3)^2} = \sqrt{49 + 9} = \sqrt{58} \approx 7.62$$

◄

NOW WORK PROBLEM 31.

The distance between two points $P_1 = (x_1, y_1)$ and $P_2 = (x_2, y_2)$ is never a negative number. Furthermore, the distance between two points is 0 only when the points are identical, that is, when $x_1 = x_2$ and $y_1 = y_2$. Also, because $(x_2 - x_1)^2 = (x_1 - x_2)^2$ and $(y_2 - y_1)^2 = (y_1 - y_2)^2$, it makes no difference whether the distance is computed from P_1 to P_2 or from P_2 to P_1; that is, $d(P_1, P_2) = d(P_2, P_1)$.

The introduction to this chapter mentioned that rectangular coordinates enable us to translate geometry problems into algebra problems, and vice versa. The next example shows how algebra (the distance formula) can be used to solve geometry problems.

| EXAMPLE 4 | **Using Algebra to Solve Geometry Problems** |

Consider the three points $A = (-2, 1)$, $B = (2, 3)$, and $C = (3, 1)$.

(a) Plot each point and form the triangle ABC.
(b) Find the length of each side of the triangle.
(c) Verify that the triangle is a right triangle.
(d) Find the area of the triangle.

Figure 13

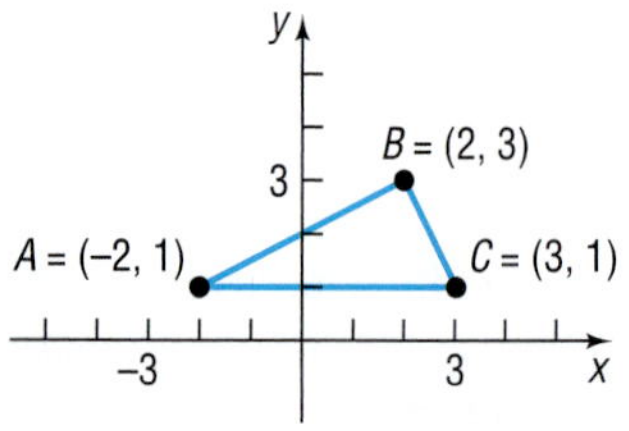

Solution

(a) Points A, B, C and triangle ABC are plotted in Figure 13.

(b) $d(A, B) = \sqrt{[2 - (-2)]^2 + (3 - 1)^2} = \sqrt{16 + 4} = \sqrt{20} = 2\sqrt{5}$
$d(B, C) = \sqrt{(3 - 2)^2 + (1 - 3)^2} = \sqrt{1 + 4} = \sqrt{5}$
$d(A, C) = \sqrt{[3 - (-2)]^2 + (1 - 1)^2} = \sqrt{25 + 0} = 5$

(c) To show that the triangle is a right triangle, we need to show that the sum of the squares of the lengths of two of the sides equals the square of the length of the third side. (Why is this sufficient?) Looking at Figure 13, it seems reasonable to conjecture that the right angle is at vertex B. We shall check to see whether

$$[d(A, B)]^2 + [d(B, C)]^2 = [d(A, C)]^2$$

We find that

$$[d(A, B)]^2 + [d(B, C)]^2 = \left(2\sqrt{5}\right)^2 + \left(\sqrt{5}\right)^2$$
$$= 20 + 5 = 25 = [d(A, C)]^2$$

so it follows from the converse of the Pythagorean Theorem that triangle ABC is a right triangle.

(d) Because the right angle is at vertex B, the sides AB and BC form the base and height of the triangle. Its area is

$$\text{Area} = \frac{1}{2}(\text{Base})(\text{Height}) = \frac{1}{2}\left(2\sqrt{5}\right)\left(\sqrt{5}\right) = 5 \text{ square units}$$

◄

NOW WORK PROBLEM 49.

Figure 14

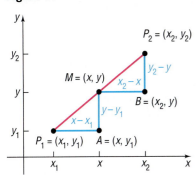

2   Use the Midpoint Formula

We now derive a formula for the coordinates of the **midpoint of a line segment**. Let $P_1 = (x_1, y_1)$ and $P_2 = (x_2, y_2)$ be the endpoints of a line segment, and let $M = (x, y)$ be the point on the line segment that is the same distance from P_1 as it is from P_2. See Figure 14. The triangles P_1AM and MBP_2 are congruent.[*] [Do you see why? Angle $AP_1M = $ angle BMP_2,[†] angle $P_1MA = $ angle MP_2B, and $d(P_1, M) = d(M, P_2)$ is given. So, we have angle–side–angle.] Hence, corresponding sides are equal in length. That is,

$$x - x_1 = x_2 - x \quad \text{and} \quad y - y_1 = y_2 - y$$
$$2x = x_1 + x_2 \qquad\qquad 2y = y_1 + y_2$$
$$x = \frac{x_1 + x_2}{2} \qquad\qquad y = \frac{y_1 + y_2}{2}$$

Theorem

Midpoint Formula

In Words
To find the midpoint of a line segment, average the x-coordinates and average the y-coordinates of the endpoints.

The midpoint $M = (x, y)$ of the line segment from $P_1 = (x_1, y_1)$ to $P_2 = (x_2, y_2)$ is

$$M = (x, y) = \left(\frac{x_1 + x_2}{2}, \frac{y_1 + y_2}{2} \right) \qquad\qquad \textbf{(2)}$$

EXAMPLE 5

Finding the Midpoint of a Line Segment

Find the midpoint of a line segment from $P_1 = (-5, 5)$ to $P_2 = (3, 1)$. Plot the points P_1 and P_2 and their midpoint. Check your answer.

Solution

We apply the midpoint formula (2) using $x_1 = -5$, $y_1 = 5$, $x_2 = 3$, and $y_2 = 1$. Then the coordinates (x, y) of the midpoint M are

$$x = \frac{x_1 + x_2}{2} = \frac{-5 + 3}{2} = -1 \quad \text{and} \quad y = \frac{y_1 + y_2}{2} = \frac{5 + 1}{2} = 3$$

That is, $M = (-1, 3)$. See Figure 15.

Figure 15

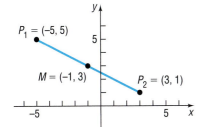

✔ **CHECK:** Because M is the midpoint, we check the answer by verifying that $d(P_1, M) = d(M, P_2)$:

$$d(P_1, M) = \sqrt{[-1 - (-5)]^2 + (3 - 5)^2} = \sqrt{16 + 4} = \sqrt{20}$$
$$d(M, P_2) = \sqrt{[3 - (-1)]^2 + (1 - 3)^2} = \sqrt{16 + 4} = \sqrt{20} \qquad ◀$$

✏ **NOW WORK PROBLEM 55.**

---

[*]The following statement is a postulate from geometry. Two triangles are congruent if their sides are the same length (SSS), or if two sides and the included angle are the same (SAS), or if two angles and the included sides are the same (ASA).
[†]Another postulate from geometry states that the transversal $\overline{P_1P_2}$ forms congruent corresponding angles with the parallel line segments $\overline{P_1A}$ and $\overline{MB}$.

1.1 Assess Your Understanding

'Are You Prepared?'

Answers are given at the end of these exercises. If you get a wrong answer, read the pages listed in red.

1. On the real number line the origin is assigned the number _____. (p. 662)

2. If −3 and 5 are the coordinates of two points on the real number line, the distance between these points is _____. (p. 664)

3. If 3 and 4 are the legs of a right triangle, the hypotenuse is _____. (pp. 669–670)

4. Use the converse of the Pythagorean Theorem to show that a triangle whose sides are of lengths 11, 60, and 61 is a right triangle. (p. 670)

Concepts and Vocabulary

5. If (x, y) are the coordinates of a point P in the xy-plane, then x is called the _____ of P and y is the _____ of P.

6. The coordinate axes divide the xy-plane into four sections called _____.

7. If three distinct points P, Q, and R all lie on a line and if $d(P, Q) = d(Q, R)$, then Q is called the _____ of the line segment from P to R.

8. *True or False:* The distance between two points is sometimes a negative number.

9. *True or False:* The point $(-1, 4)$ lies in quadrant IV of the Cartesian plane.

10. *True or False:* The midpoint of a line segment is found by averaging the x-coordinates and averaging the y-coordinates of the endpoints.

Skill Building

In Problems 11 and 12, plot each point in the xy-plane. Tell in which quadrant or on what coordinate axis each point lies.

11. (a) $A = (-3, 2)$
    (b) $B = (6, 0)$
    (c) $C = (-2, -2)$
    (d) $D = (6, 5)$
    (e) $E = (0, -3)$
    (f) $F = (6, -3)$

12. (a) $A = (1, 4)$
    (b) $B = (-3, -4)$
    (c) $C = (-3, 4)$
    (d) $D = (4, 1)$
    (e) $E = (0, 1)$
    (f) $F = (-3, 0)$

13. Plot the points $(2, 0)$, $(2, -3)$, $(2, 4)$, $(2, 1)$, and $(2, -1)$. Describe the set of all points of the form $(2, y)$, where y is a real number.

14. Plot the points $(0, 3)$, $(1, 3)$, $(-2, 3)$, $(5, 3)$ and $(-4, 3)$. Describe the set of all points of the form $(x, 3)$, where x is a real number.

In Problems 15–18, determine the coordinates of the points shown. Tell in which quadrant each point lies. Assume the coordinates are integers.

15.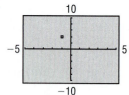

16.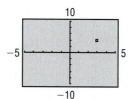

17.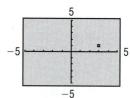

18.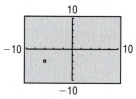

In Problems 19–24, select a setting so that each given point will lie within the viewing window.

19. $(-10, 5)$, $(3, -2)$, $(4, -1)$

20. $(5, 0)$, $(6, 8)$, $(-2, -3)$

21. $(40, 20)$, $(-20, -80)$, $(10, 40)$

22. $(-80, 60)$, $(20, -30)$, $(-20, -40)$

23. $(0, 0)$, $(100, 5)$, $(5, 150)$

24. $(0, -1)$, $(100, 50)$, $(-10, 30)$

In Problems 25–30, determine the viewing window used.

25.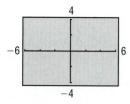

26.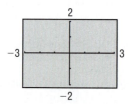

27.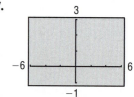

28.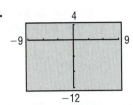

29.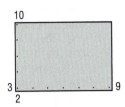

30.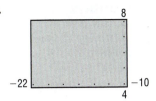

In Problems 31–44, find the distance $d(P_1, P_2)$ between the points P_1 and P_2.

31.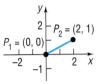

32.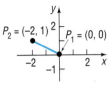

33.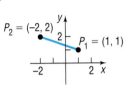

34.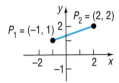

35. $P_1 = (3, -4); \quad P_2 = (5, 4)$

36. $P_1 = (-1, 0); \quad P_2 = (2, 4)$

37. $P_1 = (-3, 2); \quad P_2 = (6, 0)$

38. $P_1 = (2, -3); \quad P_2 = (4, 2)$

39. $P_1 = (4, -3); \quad P_2 = (6, 4)$

40. $P_1 = (-4, -3); \quad P_2 = (6, 2)$

41. $P_1 = (-0.2, 0.3); \quad P_2 = (2.3, 1.1)$

42. $P_1 = (1.2, 2.3); \quad P_2 = (-0.3, 1.1)$

43. $P_1 = (a, b); \quad P_2 = (0, 0)$

44. $P_1 = (a, a); \quad P_2 = (0, 0)$

In Problems 45–48, find the length of the line segment. Assume that the endpoints of each line segment have integer coordinates.

45.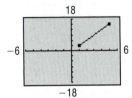

46.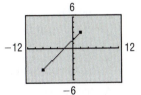

47.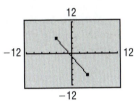

48.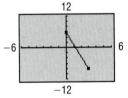

In Problems 49–54, plot each point and form the triangle ABC. Verify that the triangle is a right triangle. Find its area.

49. $A = (-2, 5); \quad B = (1, 3); \quad C = (-1, 0)$

50. $A = (-2, 5); \quad B = (12, 3); \quad C = (10, -11)$

51. $A = (-5, 3); \quad B = (6, 0); \quad C = (5, 5)$

52. $A = (-6, 3); \quad B = (3, -5); \quad C = (-1, 5)$

53. $A = (4, -3); \quad B = (0, -3); \quad C = (4, 2)$

54. $A = (4, -3); \quad B = (4, 1); \quad C = (2, 1)$

In Problems 55–64, find the midpoint of the line segment joining the points P_1 and P_2.

55. $P_1 = (3, -4); \quad P_2 = (5, 4)$

56. $P_1 = (-2, 0); \quad P_2 = (2, 4)$

57. $P_1 = (-3, 2); \quad P_2 = (6, 0)$

58. $P_1 = (2, -3); \quad P_2 = (4, 2)$

59. $P_1 = (4, -3); \quad P_2 = (6, 1)$

60. $P_1 = (-4, -3); \quad P_2 = (2, 2)$

61. $P_1 = (-0.2, 0.3); \quad P_2 = (2.3, 1.1)$

62. $P_1 = (1.2, 2.3); \quad P_2 = (-0.3, 1.1)$

63. $P_1 = (a, b); \quad P_2 = (0, 0)$

64. $P_1 = (a, a); \quad P_2 = (0, 0)$

Applications and Extensions

65. Find all points having an x-coordinate of 2 whose distance from the point $(-2, -1)$ is 5.

66. Find all points having a y-coordinate of -3 whose distance from the point $(1, 2)$ is 13.

67. Find all points on the x-axis that are 5 units from the point $(4, -3)$.

68. Find all points on the y-axis that are 5 units from the point $(4, 4)$.

69. The **medians** of a triangle are the line segments from each vertex to the midpoint of the opposite side (see the figure). Find the lengths of the medians of the triangle with vertices at $A = (0, 0)$, $B = (6, 0)$, and $C = (4, 4)$.

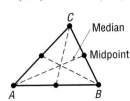

70. An **equilateral triangle** is one in which all three sides are of equal length. If two vertices of an equilateral triangle are $(0, 4)$ and $(0, 0)$, find the third vertex. How many of these triangles are possible?

71. Geometry   Find the midpoint of each diagonal of a square with side of length s. Draw the conclusion that the diagonals of a square intersect at their midpoints.

[**Hint:** Use $(0, 0), (0, s), (s, 0)$, and (s, s) as the vertices of the square.]

72. Geometry   Verify that the points $(0, 0), (a, 0)$ $\left(\dfrac{a}{2}, \dfrac{\sqrt{3}a}{2}\right)$ are the vertices of an equilateral triangle. Then show that the midpoints of the three sides are the vertices of a second equilateral triangle (refer to Problem 70).

In Problems 73–76, find the length of each side of the triangle determined by the three points P_1, P_2, and P_3. State whether the triangle is an isosceles triangle, a right triangle, neither of these, or both. (An **isosceles triangle** *is one in which at least two of the sides are of equal length.)*

73. $P_1 = (2, 1);$ $P_2 = (-4, 1);$ $P_3 = (-4, -3)$

74. $P_1 = (-1, 4);$ $P_2 = (6, 2);$ $P_3 = (4, -5)$

75. $P_1 = (-2, -1);$ $P_2 = (0, 7);$ $P_3 = (3, 2)$

76. $P_1 = (7, 2);$ $P_2 = (-4, 0);$ $P_3 = (4, 6)$

77. Baseball   A major league baseball "diamond" is actually a square, 90 feet on a side (see the figure). What is the distance directly from home plate to second base (the diagonal of the square)?

78. Little League Baseball   The layout of a Little League playing field is a square, 60 feet on a side.* How far is it directly from home plate to second base (the diagonal of the square)?
*SOURCE: *Little League Baseball, Official Regulations and Playing Rules,* 2005.

79. Baseball   Refer to Problem 77. Overlay a rectangular coordinate system on a major league baseball diamond so that the origin is at home plate, the positive x-axis lies in the direction from home plate to first base, and the positive y-axis lies in the direction from home plate to third base.
 (a) What are the coordinates of first base, second base, and third base? Use feet as the unit of measurement.
 (b) If the right fielder is located at $(310, 15)$, how far is it from there to second base?
 (c) If the center fielder is located at $(300, 300)$, how far is it from there to third base?

80. Little League Baseball   Refer to Problem 78. Overlay a rectangular coordinate system on a Little League baseball diamond so that the origin is at home plate, the positive x-axis lies in the direction from home plate to first base, and the positive y-axis lies in the direction from home plate to third base.
 (a) What are the coordinates of first base, second base, and third base? Use feet as the unit of measurement.
 (b) If the right fielder is located at $(180, 20)$, how far is it from there to second base?
 (c) If the center fielder is located at $(220, 220)$, how far is it from there to third base?

81. A Dodge Neon and a Mack truck leave an intersection at the same time. The Neon heads east at an average speed of 30 miles per hour, while the truck heads south at an average speed of 40 miles per hour. Find an expression for their distance apart d (in miles) at the end of t hours.

82. A hot-air balloon, headed due east at an average speed of 15 miles per hour and at a constant altitude of 100 feet, passes over an intersection (see the figure). Find an expression for the distance d (measured in feet) from the balloon to the intersection t seconds later.

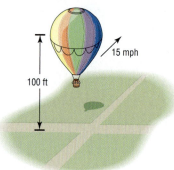

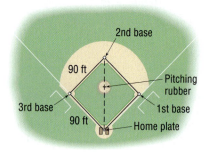

1.2 Graphs of Equations in Two Variables

PREPARING FOR THIS SECTION   *Before getting started, review the following:*

- Linear Equations (Appendix, Section A.5, pp. 694–695)
- Quadratic Equations (Appendix, Section A.5, pp. 696–703)

 Now work the 'Are You Prepared?' problems on page 21.

OBJECTIVES   1 Graph Equations by Hand by Plotting Points
2 Graph Equations Using a Graphing Utility
3 Use a Graphing Utility to Create Tables
4 Find Intercepts from a Graph
5 Find Intercepts from an Equation
6 Use a Graphing Utility to Approximate Intercepts
7 Test an Equation for Symmetry with Respect to the *x*-Axis, the *y*-Axis, and the Origin
8 Know How to Graph Key Equations

1 Graph Equations by Hand by Plotting Points

An **equation in two variables**, say *x* and *y*, is a statement in which two expressions involving *x* and *y* are equal. The expressions are called the **sides** of the equation. Since an equation is a statement, it may be true or false, depending on the value of the variables. Any values of *x* and *y* that result in a true statement are said to **satisfy** the equation.

For example, the following are all equations in two variables *x* and *y*:

$$x^2 + y^2 = 5 \qquad 2x - y = 6 \qquad y = 2x + 5 \qquad x^2 = y$$

The first of these, $x^2 + y^2 = 5$, is satisfied for $x = 1, y = 2$, since $1^2 + 2^2 = 1 + 4 = 5$. Other choices of *x* and *y* also satisfy this equation. It is not satisfied for $x = 2$ and $y = 3$, since $2^2 + 3^2 = 4 + 9 = 13 \neq 5$.

The **graph of an equation in two variables** *x* and *y* consists of the set of points in the *xy*-plane whose coordinates (x, y) satisfy the equation.

EXAMPLE 1   **Determining Whether a Point Is on the Graph of an Equation**

Determine if the following points are on the graph of the equation $2x - y = 6$.

(a) $(2, 3)$ 　　　　　　　　　 (b) $(2, -2)$

Solution   (a) For the point $(2, 3)$, we check to see if $x = 2, y = 3$ satisfies the equation $2x - y = 6$.

$$2x - y = 2(2) - 3 = 4 - 3 = 1 \neq 6$$

The equation is not satisfied, so the point $(2, 3)$ is not on the graph.

(b) For the point $(2, -2)$, we have

$$2x - y = 2(2) - (-2) = 4 + 2 = 6$$

The equation is satisfied, so the point $(2, -2)$ is on the graph.　　◀

NOW WORK PROBLEM 21.

| EXAMPLE 2 | **Graphing an Equation by Hand by Plotting Points** |

Graph the equation: $y = 2x + 5$

Solution

We want to find all points (x, y) that satisfy the equation. To locate some of these points (and thus get an idea of the pattern of the graph), we assign some numbers to x and find corresponding values for y.

Figure 16

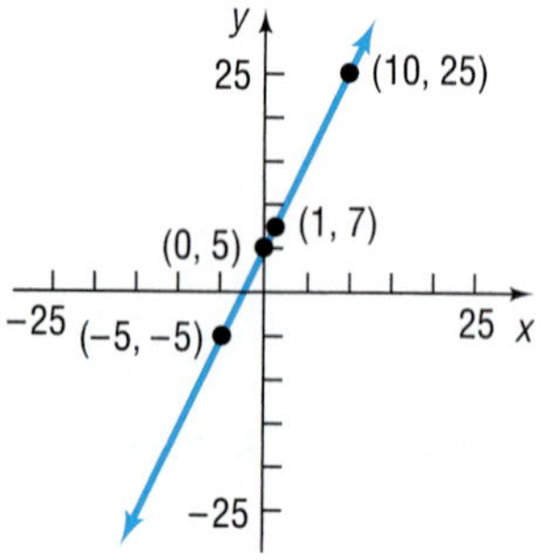

If	Then	Point on Graph
$x = 0$	$y = 2(0) + 5 = 5$	$(0, 5)$
$x = 1$	$y = 2(1) + 5 = 7$	$(1, 7)$
$x = -5$	$y = 2(-5) + 5 = -5$	$(-5, -5)$
$x = 10$	$y = 2(10) + 5 = 25$	$(10, 25)$

By plotting these points and then connecting them, we obtain the graph of the equation (a *line*), as shown in Figure 16. ◀

| EXAMPLE 3 | **Graphing an Equation by Hand by Plotting Points** |

Graph the equation: $y = x^2$

Solution

Table 1 provides several points on the graph. In Figure 17 we plot these points and connect them with a smooth curve to obtain the graph (a *parabola*).

Table 1

x	$y = x^2$	(x, y)
-4	16	$(-4, 16)$
-3	9	$(-3, 9)$
-2	4	$(-2, 4)$
-1	1	$(-1, 1)$
0	0	$(0, 0)$
1	1	$(1, 1)$
2	4	$(2, 4)$
3	9	$(3, 9)$
4	16	$(4, 16)$

Figure 17

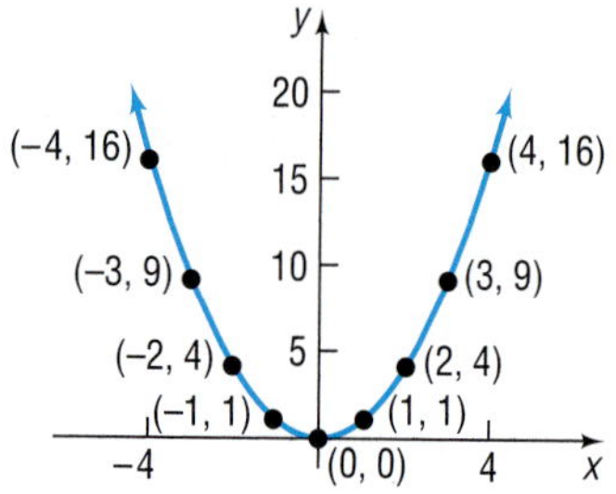

◀

The graphs of the equations shown in Figures 16 and 17 do not show all points. For example, in Figure 16, the point $(20, 45)$ is a part of the graph of $y = 2x + 5$, but it is not shown. Since the graph of $y = 2x + 5$ could be extended out as far as we please, we use arrows to indicate that the pattern shown continues. It is important when illustrating a graph to present enough of the graph so that any viewer of the illustration will "see" the rest of it as an obvious continuation of what is actually there. This is referred to as a **complete graph**.

One way to obtain a complete graph of an equation is to plot a sufficient number of points on the graph until a pattern becomes evident. Then these points are connected with a smooth curve following the suggested pattern. But how many points are sufficient? Sometimes knowledge about the equation tells us. For example, we will learn in Section 1.4 that, if an equation is of the form $y = mx + b$, then its graph is a line. In this case, only two points are needed to obtain the graph.

One purpose of this book is to investigate the properties of equations in order to decide whether a graph is complete. Sometimes we shall graph equations by plotting points. Shortly, we shall investigate various techniques that will enable us to graph an equation without plotting so many points. Other times we shall graph equations using a graphing utility.

2 Graph Equations Using a Graphing Utility

From Examples 2 and 3, we see that a graph can be obtained by plotting points in a rectangular coordinate system and connecting them. Graphing utilities perform these same steps when graphing an equation. For example, the TI-84 Plus determines 95 evenly spaced input values,* uses the equation to determine the output values, plots these points on the screen, and finally (if in the connected mode) draws a line between consecutive points.

To graph an equation in two variables x and y using a graphing utility requires that the equation be written in the form $y = \{$expression in $x\}$. If the original equation is not in this form, replace it by equivalent equations until the form $y = \{$expression in $x\}$ is obtained. Most graphing utilities require the following steps.

Steps for Graphing an Equation Using a Graphing Utility

STEP 1: Solve the equation for y in terms of x.

STEP 2: Get into the graphing mode of your graphing utility. The screen will usually display $Y = \quad$, prompting you to enter the expression involving x that you found in Step 1. (Consult your manual for the correct way to enter the expression; for example, $y = x^2$ might be entered as $x\hat{\ }2$ or as $x*x$ or as xx^Y2).

STEP 3: Select the viewing window. Without prior knowledge about the behavior of the graph of the equation, it is common to select the **standard viewing window**† initially. The viewing window is then adjusted based on the graph that appears. In this text, the standard viewing window will be

$$X\text{min} = -10 \qquad Y\text{min} = -10$$
$$X\text{max} = 10 \qquad Y\text{max} = 10$$
$$X\text{scl} = 1 \qquad Y\text{scl} = 1$$

STEP 4: Graph.

STEP 5: Adjust the viewing window until a complete graph is obtained.

---

*These input values depend on the values of Xmin and Xmax. For example, if Xmin $= -10$ and Xmax $= 10$, then the first input value will be -10 and the next input value will be $-10 + (10 - (-10))/94 = -9.7872$, and so on.
†Some graphing utilities have a ZOOM-STANDARD feature that automatically sets the viewing window to the standard viewing window and graphs the equation.

| EXAMPLE 4 | **Graphing an Equation on a Graphing Utility** |

Graph the equation: $6x^2 + 3y = 36$

STEP 1: We solve for y in terms of x.

$$6x^2 + 3y = 36$$
$$3y = -6x^2 + 36 \quad \text{\color{blue}Subtract } 6x^2 \text{ from both sides of the equation.}$$
$$y = -2x^2 + 12 \quad \text{\color{blue}Divide both sides of the equation by 3 and simplify.}$$

STEP 2: From the graphing mode, enter the expression $-2x^2 + 12$ after the prompt $Y =$. Figure 18 shows the expression entered on a TI-84 Plus.

STEP 3: Set the viewing window to the standard viewing window.

STEP 4: Graph. The screen should look like Figure 19.

STEP 5: The graph of $y = -2x^2 + 12$ is not complete. The value of Ymax must be increased so that the top portion of the graph is visible. After increasing the value of Ymax to 12, we obtain the graph in Figure 20.* The graph is now complete.

Figure 18

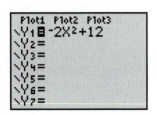

Figure 19

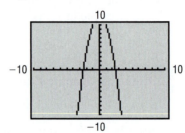

Figure 20

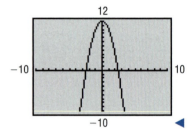

3 Use a Graphing Utility to Create Tables

In addition to graphing equations, graphing utilities can also be used to create a table of values that satisfy the equation. This feature is especially useful in determining an appropriate viewing window when graphing an equation. Many graphing utilities require the following steps to create a table.

Steps for Creating a Table of Values Using a Graphing Utility

STEP 1: Solve the equation for y in terms of x.

STEP 2: Enter the expression in x following the $Y =$ prompt of the graphing utility.

STEP 3: Set up the table. Graphing utilities typically have two modes for creating tables. In the AUTO mode, the user determines a starting point for the table (TblStart) and ΔTbl (pronounced "delta-table"). The ΔTbl feature determines the increment for x. The ASK mode requires the user to enter values of x and then the utility determines the corresponding value of y.

STEP 4: Create the table. The user can scroll within the table if the table was created in AUTO mode.

---

*Some graphing utilities have a ZOOM-FIT feature that determines the appropriate Ymin and Ymax for a given Xmin and Xmax. Consult your owner's manual for the appropriate keystrokes.

| EXAMPLE 5 | **Creating a Table Using a Graphing Utility** |

Create a table that displays the points on the graph of $6x^2 + 3y = 36$ for $x = -3, -2, -1, 0, 1, 2,$ and $3.$

Table 2

Solution

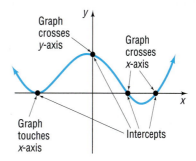

Y₁ = -2X² + 12

STEP 1: We solved the equation for y in Example 4 and obtained $y = -2x^2 + 12.$
STEP 2: Enter the expression in x following the $Y =$ prompt.
STEP 3: We set up the table in the AUTO mode with TblStart $= -3$ and ΔTbl $= 1.$
STEP 4: Create the table. The screen should look like Table 2. ◀

In looking at Table 2, we notice that $y = 12$ when $x = 0.$ This information could have been used to help create the initial viewing window by letting us know that Ymax needs to be at least 12 in order to get a complete graph.

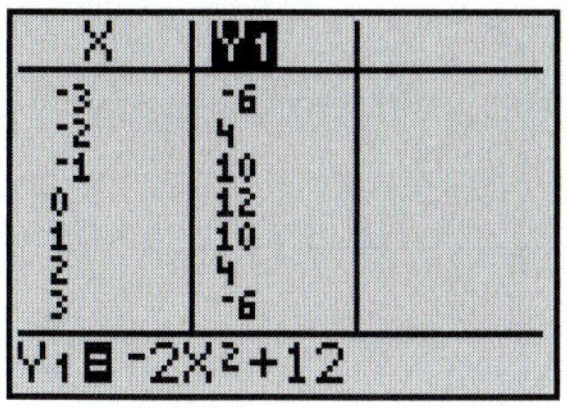

4 Find Intercepts from a Graph

The points, if any, at which a graph crosses or touches the coordinate axes are called the **intercepts**. See Figure 21. The x-coordinate of a point at which the graph crosses or touches the x-axis is an **x-intercept**, and the y-coordinate of a point at which the graph crosses or touches the y-axis is a **y-intercept**. For a graph to be complete, all its intercepts must be displayed.

Figure 21

Graph
crosses
y-axis

Graph
crosses
x-axis

Graph
touches
x-axis

Intercepts

| EXAMPLE 6 | **Finding Intercepts from a Graph** |

Find the intercepts of the graph in Figure 22. What are its x-intercepts? What are its y-intercepts?

Figure 22

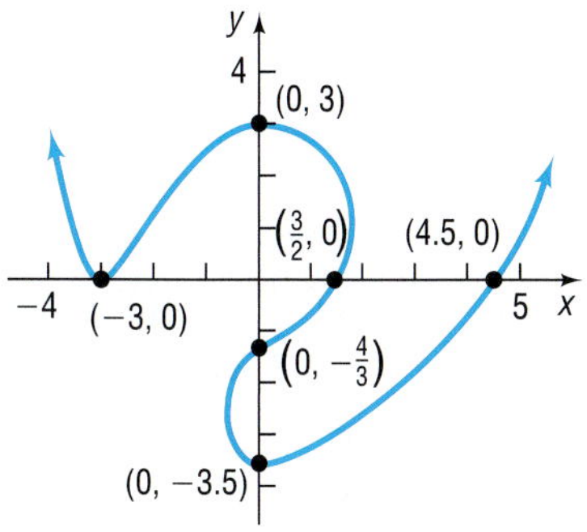

Solution

$(0, 3)$

$\left(\dfrac{3}{2}, 0\right)$ $(4.5, 0)$

$(-3, 0)$

$\left(0, -\dfrac{4}{3}\right)$

$(0, -3.5)$

The intercepts of the graph are the points

$$(-3, 0), \quad (0, 3), \quad \left(\frac{3}{2}, 0\right), \quad \left(0, -\frac{4}{3}\right), \quad (0, -3.5), \quad (4.5, 0)$$

The x-intercepts are $-3, \dfrac{3}{2},$ and 4.5; the y-intercepts are $-3.5, -\dfrac{4}{3},$ and $3.$ ◀

In Example 6, you should notice the following usage: If we do not specify the type of intercept (x- versus y-), then we report the intercept as an ordered pair. However, if we specify the type of intercept, then we only report the coordinate of the intercept. For x-intercepts, we report the x-coordinate of the intercept; for y-intercepts, we report the y-coordinate of the intercept.

✏️ NOW WORK PROBLEM 27(a).

5 Find Intercepts from an Equation

The intercepts of a graph can be found from its equation by using the fact that points on the x-axis have y-coordinates equal to 0 and points on the y-axis have x-coordinates equal to 0.

> **Procedure for Finding Intercepts**
>
> **1.** To find the x-intercept(s), if any, of the graph of an equation, let $y = 0$ in the equation and solve for x.
> **2.** To find the y-intercept(s), if any, of the graph of an equation, let $x = 0$ in the equation and solve for y.

Because the x-intercepts of the graph of an equation are those x-values for which $y = 0$, they are also called the **zeros** (or **roots**) of the equation.

EXAMPLE 7	**Finding Intercepts from an Equation**

Find the x-intercept(s) and the y-intercept(s) of the graph of $y = x^2 - 4$.

Solution To find the x-intercept(s), we let $y = 0$ and obtain the equation

$$x^2 - 4 = 0$$
$$(x + 2)(x - 2) = 0 \qquad \text{Factor.}$$
$$x + 2 = 0 \quad \text{or} \quad x - 2 = 0 \qquad \text{Zero-Product Property}$$
$$x = -2 \quad \text{or} \quad x = 2$$

The equation has two solutions, -2 and 2. The x-intercepts (or zeros) are -2 and 2.

To find the y-intercept(s), we let $x = 0$ in the equation.

$$y = x^2 - 4$$
$$= 0^2 - 4 = -4$$

The y-intercept is -4. ◀

Sometimes the TABLE feature of a graphing utility will reveal the intercepts of an equation. Table 3 shows a table for the equation $y = x^2 - 4$. Can you find the intercepts in the table?

Table 3

X	Y1
-3	5
-2	0
-1	-3
0	-4
1	-3
2	0
3	5

Y1☐X²-4

➤ **NOW WORK PROBLEM 43.**

6 Use a Graphing Utility to Approximate Intercepts

We can use a graphing utility to approximate the intercepts of the graph of an equation, as illustrated in the next example.

EXAMPLE 8	**Finding Intercepts Using a Graphing Utility**

Use a graphing utility to approximate the intercepts of the equation $y = x^3 - 16$.

Solution Figure 23(a) shows the graph of $y = x^3 - 16$.

The eVALUEate feature of a TI-84 Plus graphing calculator accepts as input a value of x and determines the value of y. If we let $x = 0$, we find that the y-intercept is -16. See Figure 23(b).

The ZERO feature of a TI-84 Plus is used to find the x-intercept(s). See Figure 23(c). Rounded to two decimal places, the x-intercept is 2.52. ◄

Figure 23

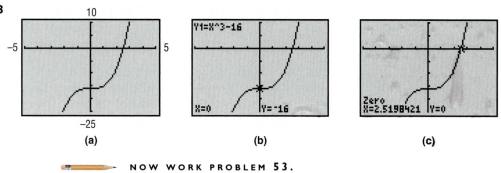

(a)　　　　　　　　(b)　　　　　　　　(c)

NOW WORK PROBLEM 53.

7 **Test an Equation for Symmetry**

We just saw the role that intercepts play in obtaining key points on the graph of an equation. Another helpful tool for graphing equations by hand involves *symmetry*, particularly symmetry with respect to the x-axis, the y-axis, and the origin.

> A graph is said to be **symmetric with respect to the x-axis** if, for every point (x, y) on the graph, the point $(x, -y)$ is also on the graph.
>
> A graph is said to be **symmetric with respect to the y-axis** if, for every point (x, y) on the graph, the point $(-x, y)$ is also on the graph.
>
> A graph is said to be **symmetric with respect to the origin** if, for every point (x, y) on the graph, the point $(-x, -y)$ is also on the graph.

Figure 24 illustrates the definition. Notice that, when a graph is symmetric with respect to the x-axis, the part of the graph above the x-axis is a reflection or mirror image of the part below it, and vice versa. And when a graph is symmetric with respect to the y-axis, the part of the graph to the right of the y-axis is a reflection of the part to the left of it, and vice versa. Symmetry with respect to the origin may be viewed in two ways:

1. As a reflection about the y-axis, followed by a reflection about the x-axis

2. As a projection along a line through the origin so that the distances from the origin are equal

Figure 24

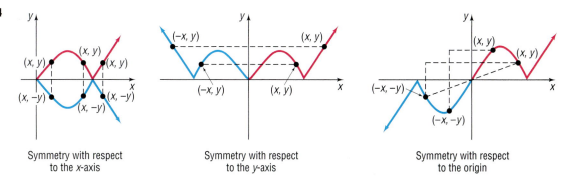

Symmetry with respect to the x-axis

Symmetry with respect to the y-axis

Symmetry with respect to the origin

| EXAMPLE 9 | **Symmetric Points** |

(a) If a graph is symmetric with respect to the x-axis and the point $(4, 2)$ is on the graph, then the point $(4, -2)$ is also on the graph.

(b) If a graph is symmetric with respect to the y-axis and the point $(4, 2)$ is on the graph, then the point $(-4, 2)$ is also on the graph.

(c) If a graph is symmetric with respect to the origin and the point $(4, 2)$ is on the graph, then the point $(-4, -2)$ is also on the graph. ◀

▬▬▬ **NOW WORK PROBLEM 11.**

When the graph of an equation is symmetric with respect to the x-axis, the y-axis, or the origin, the number of points that you need to plot in order to see the pattern is reduced. For example, if the graph of an equation is symmetric with respect to the y-axis, then, once points to the right of the y-axis are plotted, an equal number of points on the graph can be obtained by reflecting them about the y-axis. Because of this, before we graph an equation, we first want to determine whether it has any symmetry. The following tests are used for this purpose.

Tests for Symmetry

To test the graph of an equation for symmetry with respect to the

x-Axis   Replace y by $-y$ in the equation. If an equivalent equation results, the graph of the equation is symmetric with respect to the x-axis.

y-Axis   Replace x by $-x$ in the equation. If an equivalent equation results, the graph of the equation is symmetric with respect to the y-axis.

Origin   Replace x by $-x$ and y by $-y$ in the equation. If an equivalent equation results, the graph of the equation is symmetric with respect to the origin.

| EXAMPLE 10 | **Testing an Equation for Symmetry** |

Test $y = \dfrac{4x^2}{x^2 + 1}$ for symmetry.

Solution   *x-Axis:*   To test for symmetry with respect to the x-axis, replace y by $-y$. Since $-y = \dfrac{4x^2}{x^2 + 1}$ is not equivalent to $y = \dfrac{4x^2}{x^2 + 1}$, the graph of the equation is not symmetric with respect to the x-axis.

y-Axis: To test for symmetry with respect to the y-axis, replace x by $-x$. Since $y = \dfrac{4(-x)^2}{(-x)^2 + 1} = \dfrac{4x^2}{x^2 + 1}$ is equivalent to $y = \dfrac{4x^2}{x^2 + 1}$, the graph of the equation is symmetric with respect to the y-axis.

Origin: To test for symmetry with respect to the origin, replace x by $-x$ and y by $-y$.

$$-y = \frac{4(-x)^2}{(-x)^2 + 1} \qquad \text{Replace } x \text{ by } -x \text{ and } y \text{ by } -y.$$

$$-y = \frac{4x^2}{x^2 + 1} \qquad \text{Simplify.}$$

$$y = -\frac{4x^2}{x^2 + 1} \qquad \text{Multiply both sides by } -1.$$

Since the result is not equivalent to the original equation, the graph of the equation $y = \dfrac{4x^2}{x^2 + 1}$ is not symmetric with respect to the origin. ◀

Figure 25

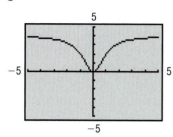

— **Seeing the Concept** —

Figure 25 shows the graph of $y = \dfrac{4x^2}{x^2 + 1}$ using a graphing utility. Do you see the symmetry with respect to the y-axis?

NOW WORK PROBLEMS 27(b) AND 63.

8   Know How to Graph Key Equations

The next three examples use intercepts, symmetry, and point plotting to obtain the graphs of key equations. It is important to know the graphs of these key equations because we use them later. The first of these is $y = x^3$.

EXAMPLE 11   **Graphing the Equation $y = x^3$ by Finding Intercepts and Checking for Symmetry**

Graph the equation $y = x^3$ by hand by plotting points. Find any intercepts and check for symmetry first.

Solution   First, we seek the intercepts. When $x = 0$, then $y = 0$; and when $y = 0$, then $x = 0$. The origin $(0, 0)$ is the only intercept. Now we test for symmetry.

x-*Axis:*   Replace y by $-y$. Since $-y = x^3$ is not equivalent to $y = x^3$, the graph is not symmetric with respect to the x-axis.

y-*Axis:*   Replace x by $-x$. Since $y = (-x)^3 = -x^3$ is not equivalent to $y = x^3$, the graph is not symmetric with respect to the y-axis.

Origin: Replace x by $-x$ and y by $-y$. Since $-y = (-x)^3 = -x^3$ is equivalent to $y = x^3$ (multiply both sides by -1), the graph is symmetric with respect to the origin.

Figure 26

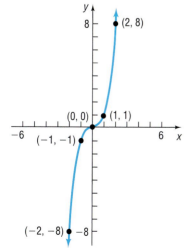

To graph by hand, we use the equation to obtain several points on the graph. Because of the symmetry, we only need to locate points on the graph for which $x \geq 0$. See Table 4. Points on the graph could also be obtained using the TABLE feature on a graphing utility. See Table 5. Do you see the symmetry with respect to the origin from the table? Figure 26 shows the graph.

Table 4

x	$y = x^3$	(x, y)
0	0	$(0, 0)$
1	1	$(1, 1)$
2	8	$(2, 8)$
3	27	$(3, 27)$

Table 5

X	Y1
-3	-27
-2	-8
-1	-1
0	0
1	1
2	8
3	27

Y1 ▪ X^3

◀

| EXAMPLE 12 | Graphing the Equation $x = y^2$ |

Graph the equation $x = y^2$. Find any intercepts and check for symmetry first.

Solution

The lone intercept is $(0, 0)$. The graph is symmetric with respect to the x-axis since $x = (-y)^2$ is equivalent to $x = y^2$. The graph is not symmetric with respect to the y-axis or the origin.

To graph $x = y^2$ by hand, we use the equation to obtain several points on the graph. Because the equation is solved for x, it is easier to assign values to y and use the equation to determine the corresponding values of x. See Table 6. Because of the symmetry, we can restrict ourselves to points whose y-coordinates are positive. We then use the symmetry to find additional points on the graph. For example, since $(1, 1)$ is on the graph, so is $(1, -1)$. Since $(4, 2)$ is on the graph, so is $(4, -2)$, and so on. We plot these points and connect them with a smooth curve to obtain Figure 27.

To graph the equation $x = y^2$ using a graphing utility, we must write the equation in the form $y = \{$expression in $x\}$. We proceed to solve for y.

$$x = y^2$$
$$y^2 = x$$
$$y = \pm\sqrt{x} \qquad \text{\textcolor{blue}{Square Root Method}}$$

To graph $x = y^2$, we need to graph both $Y_1 = \sqrt{x}$ and $Y_2 = -\sqrt{x}$ on the same screen. Figure 28 shows the result. Table 7 shows various values of y for a given value of x when $Y_1 = \sqrt{x}$ and $Y_2 = -\sqrt{x}$. Notice that when $x < 0$ we get an error. Can you explain why?

Table 6

y	$x = y^2$	(x, y)
0	0	(0, 0)
1	1	(1, 1)
2	4	(4, 2)
3	9	(9, 3)

Figure 27

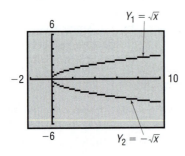

Figure 28

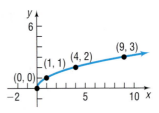

Table 7

X	Y1	Y2
-1	ERROR	ERROR
0	0	0
1	1	-1
2	1.4142	-1.414
3	1.7321	-1.732
4	2	-2
5	2.2361	-2.236

Y1=√(X)

Look again at either Figure 27 or 28. If we restrict y so that $y \geq 0$, the equation $x = y^2$, $y \geq 0$, may be written as $y = \sqrt{x}$. The portion of the graph of $x = y^2$ in quadrant I is the graph of $y = \sqrt{x}$. See Figure 29.

Figure 29

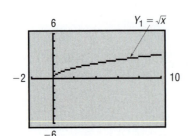

EXAMPLE 13

Graphing the Equation $y = \dfrac{1}{x}$

Graph the equation $y = \dfrac{1}{x}$. Find any intercepts and check for symmetry first.

Solution   We check for intercepts first. If we let $x = 0$, we obtain a 0 in the denominator, which is not defined. We conclude that there is no y-intercept. If we let $y = 0$, we get the equation $\dfrac{1}{x} = 0$, which has no solution. We conclude that there is no x-intercept. The graph of $y = \dfrac{1}{x}$ does not cross or touch the coordinate axes.

Next we check for symmetry.

x-*Axis:*   Replacing y by $-y$ yields $-y = \dfrac{1}{x}$, which is not equivalent to $y = \dfrac{1}{x}$.

y-*Axis:*   Replacing x by $-x$ yields $y = \dfrac{1}{-x} = -\dfrac{1}{x}$, which is not equivalent to $y = \dfrac{1}{x}$.

Origin: Replacing x by $-x$ and y by $-y$ yields $-y = -\dfrac{1}{x}$, which is equivalent to $y = \dfrac{1}{x}$. The graph is symmetric only with respect to the origin.

We can use the equation to form Table 8 and obtain some points on the graph. Because of symmetry, we only find points (x, y) for which x is positive. From Table 8 we infer that, if x is a large and positive number, then $y = \dfrac{1}{x}$ is a positive number close to 0. We also infer that if x is a positive number close to 0, then $y = \dfrac{1}{x}$ is a large and positive number. Armed with this information, we can graph the equation.

Figure 30 illustrates some of these points and the graph of $y = \dfrac{1}{x}$. Observe how the absence of intercepts and the existence of symmetry with respect to the origin were utilized. Figure 31 confirms our algebraic analysis using a TI-84 Plus.

Table 8

x	$y = \dfrac{1}{x}$	(x, y)
$\dfrac{1}{10}$	10	$\left(\dfrac{1}{10}, 10\right)$
$\dfrac{1}{3}$	3	$\left(\dfrac{1}{3}, 3\right)$
$\dfrac{1}{2}$	2	$\left(\dfrac{1}{2}, 2\right)$
1	1	$(1, 1)$
2	$\dfrac{1}{2}$	$\left(2, \dfrac{1}{2}\right)$
3	$\dfrac{1}{3}$	$\left(3, \dfrac{1}{3}\right)$
10	$\dfrac{1}{10}$	$\left(10, \dfrac{1}{10}\right)$

Figure 30

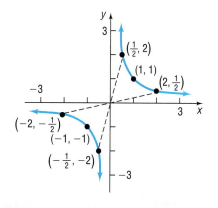

Figure 31

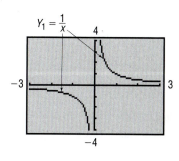

1.2 Assess Your Understanding

'Are You Prepared?'

Answers are given at the end of these exercises. If you get a wrong answer, read the pages listed in red.

1. Solve the equation $2(x + 3) - 1 = -7$. (p. 694)

2. Solve the equation $2x^2 - 9x - 5 = 0$. (pp. 696–703)

Concepts and Vocabulary

3. The points, if any, at which a graph crosses or touches the co-ordinate axes are called _____.

4. Because the *x*-intercepts of the graph of an equation are those *x*-values for which $y = 0$, they are also called _____ or _____.

5. If for every point (x, y) on the graph of an equation the point $(-x, y)$ is also on the graph, then the graph is symmetric with respect to the _____.

6. If the graph of an equation is symmetric with respect to the *y*-axis and -4 is an *x*-intercept of this graph, then _____ is also an *x*-intercept.

7. If the graph of an equation is symmetric with respect to the origin and $(3, -4)$ is a point on the graph, then _____ is also a point on the graph.

8. *True or False:* To find the *y*-intercepts of the graph of an equation, let $x = 0$ and solve for *y*.

9. *True or False:* To graph an equation using a graphing utility, we must first solve for *x* in terms of *y*.

10. *True or False:* If a graph is symmetric with respect to the *x*-axis, then it cannot be symmetric with respect to the *y*-axis.

Skill Building

In Problems 11–20, plot each point. Then plot the point that is symmetric to it with respect to (a) the x-axis; (b) the y-axis; (c) the origin.

11. $(3, 4)$
12. $(5, 3)$
13. $(-2, 1)$
14. $(4, -2)$
15. $(5, -2)$
16. $(-1, -1)$
17. $(-3, -4)$
18. $(4, 0)$
19. $(0, -3)$
20. $(-3, 0)$

In Problems 21–26, tell whether the given points are on the graph of the equation.

21. Equation: $y = x^4 - \sqrt{x}$
    Points: $(0, 0); (1, 1); (-1, 0)$

22. Equation: $y = x^3 - 2\sqrt{x}$
    Points: $(0, 0); (1, 1); (1, -1)$

23. Equation: $y^2 = x^2 + 9$
    Points: $(0, 3); (3, 0); (-3, 0)$

24. Equation: $y^3 = x + 1$
    Points: $(1, 2); (0, 1); (-1, 0)$

25. Equation: $x^2 + y^2 = 4$
    Points: $(0, 2); (-2, 2); \left(\sqrt{2}, \sqrt{2}\right)$

26. Equation: $x^2 + 4y^2 = 4$
    Points: $(0, 1); (2, 0); \left(2, \dfrac{1}{2}\right)$

In Problems 27–34, the graph of an equation is given. (a) Find the intercepts. (b) Indicate whether the graph is symmetric with respect to the x-axis, the y-axis, or the origin.

27.
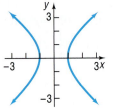

28.
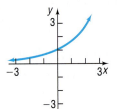

29.
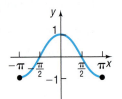

30.
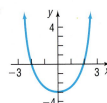

31.
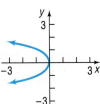

32.
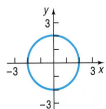

33.
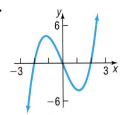

34.
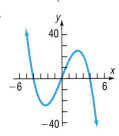

In Problems 35–38, draw a complete graph so that it has the type of symmetry indicated.

35. *y*-axis
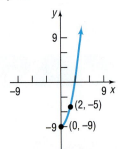

36. *x*-axis
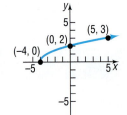

37. Origin
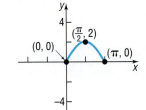

38. *y*-axis
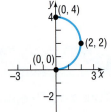

In Problems 39–50, determine the intercepts algebraically and graph each equation by plotting points. Be sure to label the intercepts. Verify your results using a graphing utility.

39. $y = x + 2$

40. $y = x - 6$

41. $y = 2x + 8$

42. $y = 3x - 9$

43. $y = x^2 - 1$

44. $y = x^2 - 9$

45. $y = -x^2 + 4$

46. $y = -x^2 + 1$

47. $2x + 3y = 6$

48. $5x + 2y = 10$

49. $9x^2 + 4y = 36$

50. $4x^2 + y = 4$

In Problems 51–58, graph each equation using a graphing utility. Use a graphing utility to approximate the intercepts rounded to two decimal places. Use the TABLE feature to help to establish the viewing window.

51. $y = 2x - 13$

52. $y = -3x + 14$

53. $y = 2x^2 - 15$

54. $y = -3x^2 + 19$

55. $3x - 2y = 43$

56. $4x + 5y = 82$

57. $5x^2 + 3y = 37$

58. $2x^2 - 3y = 35$

In Problems 59–74, list the intercepts and test for symmetry. Verify your results using a graphing utility.

59. $y^2 = x + 4$

60. $y^2 = x + 9$

61. $y = \sqrt[3]{x}$

62. $y = \sqrt[5]{x}$

63. $x^2 + y - 9 = 0$

64. $y^2 - x - 4 = 0$

65. $9x^2 + 4y^2 = 36$

66. $4x^2 + y^2 = 4$

67. $y = x^3 - 27$

68. $y = x^4 - 1$

69. $y = x^2 - 3x - 4$

70. $y = x^2 + 4$

71. $y = \dfrac{3x}{x^2 + 9}$

72. $y = \dfrac{x^2 - 4}{2x}$

73. $y = \dfrac{-x^3}{x^2 - 9}$

74. $y = \dfrac{x^4 + 1}{2x^5}$

In Problems 75–78, draw a quick sketch of each equation.

75. $y = x^3$

76. $x = y^2$

77. $y = \sqrt{x}$

78. $y = \dfrac{1}{x}$

79. If $(3, b)$ is a point on the graph of $y = 4x + 1$, what is b?

80. If $(-2, b)$ is a point on the graph of $2x + 3y = 2$, what is b?

81. If $(a, 4)$ is a point on the graph of $y = x^2 + 3x$, what is a?

82. If $(a, -5)$ is a point on the graph of $y = x^2 + 6x$, what is a?

Discussion and Writing

In Problem 83, use a graphing utility.

83. (a) Graph $y = \sqrt{x^2}$, $y = x$, $y = |x|$, and $y = (\sqrt{x})^2$, noting which graphs are the same.
(b) Explain why the graphs of $y = \sqrt{x^2}$ and $y = |x|$ are the same.
(c) Explain why the graphs of $y = x$ and $y = (\sqrt{x})^2$ are not the same.
(d) Explain why the graphs of $y = \sqrt{x^2}$ and $y = x$ are not the same.

84. Explain what is meant by a complete graph.

85. What is the standard viewing window?

86. Draw a graph of an equation that contains two x-intercepts; at one the graph crosses the x-axis, and at the other the graph touches the x-axis.

87. Make up an equation with the intercepts $(2, 0)$, $(4, 0)$, and $(0, 1)$. Compare your equation with a friend's equation. Comment on any similarities.

88. Draw a graph that contains the points $(-2, -1)$, $(0, 1)$, $(1, 3)$, and $(3, 5)$. Compare your graph with those of other students. Are most of the graphs almost straight lines? How many are "curved"? Discuss the various ways that these points might be connected.

89. An equation is being tested for symmetry with respect to the x-axis, the y-axis, and the origin. Explain why, if two of these symmetries are present, the remaining one must also be present.

'Are You Prepared?' Answers

1. $\{-6\}$

2. $\left\{-\dfrac{1}{2}, 5\right\}$

1.3 Solving Equations in One Variable Using a Graphing Utility

PREPARING FOR THIS SECTION   *Before getting started, review the following:*

• Linear, Rational, and Quadratic Equations (Appendix, Section A.5, pp. 694–703)

✎ Now work the 'Are You Prepared?' problems on page 26.

OBJECTIVE 1 Solve Equations in One Variable Using a Graphing Utility

1 Solve Equations in One Variable Using a Graphing Utility

In this text, we present two methods for solving equations: algebraic and graphical. We shall see as we proceed through this book that some equations can be solved using algebraic techniques that result in *exact* solutions. For other equations, however, there are no algebraic techniques that lead to an exact solution. For such equations, a graphing utility can often be used to investigate possible solutions. When a graphing utility is used to solve an equation, usually *approximate* solutions are obtained.

One goal of this text is to determine when equations can be solved algebraically. If an algebraic method for solving an equation exists, we shall use it to obtain an exact solution. A graphing utility can then be used to support the algebraic result. However, if an equation must be solved for which no algebraic techniques are available, a graphing utility will be used to obtain approximate solutions. Unless otherwise stated, we shall follow the practice of giving approximate solutions *rounded to two decimal places*.

The ZERO (or ROOT) feature of a graphing utility can be used to find the solutions of an equation when one side of the equation is 0. In using this feature to solve equations, we make use of the fact that the x-intercepts (or zeros) of the graph of an equation are found by letting $y = 0$ and solving the equation for x. Solving an equation for x when one side of the equation is 0 is equivalent to finding where the graph of the corresponding equation crosses or touches the x-axis.

EXAMPLE 1

Using ZERO (or ROOT) to Approximate Solutions of an Equation

Find the solution(s) of the equation $x^3 - x + 1 = 0$. Round answers to two decimal places.

Solution   The solutions of the equation $x^3 - x + 1 = 0$ are the same as the x-intercepts of the graph of $Y_1 = x^3 - x + 1$. We begin by graphing the equation.

Figure 32 shows the graph. From the graph there appears to be one x-intercept (solution to the equation) between -2 and -1.

Using the ZERO (or ROOT) feature of our graphing utility, we determine that the x-intercept, and thus the solution to the equation, is $x = -1.32$, rounded to two decimal places. See Figure 33.

Figure 32

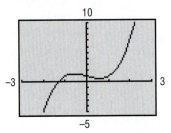

Figure 33

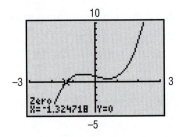

✏ **NOW WORK PROBLEM 5.**

A second method for solving equations using a graphing utility involves the INTERSECT feature of the graphing utility. This feature is used most effectively when one side of the equation is not 0.

EXAMPLE 2   **Using INTERSECT to Approximate Solutions of an Equation**

Find the solution(s) to the equation $4x^4 - 3 = 2x + 1$. Round answers to two decimal places.

Figure 34

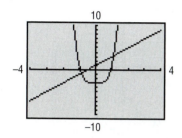

Solution   We begin by graphing each side of the equation as follows: graph $Y_1 = 4x^4 - 3$ and $Y_2 = 2x + 1$. See Figure 34. At the point of intersection of the graphs, the value of the y-coordinate is the same. Thus, the x-coordinate of the point of intersection represents the solution to the equation. Do you see why?

The INTERSECT feature on a graphing utility determines the point of intersection of the graphs. Using this feature, we find that the graphs intersect at $(-0.87, -0.73)$ and $(1.12, 3.23)$, rounded to two decimal places. See Figures 35(a) and (b). The solutions of the equation are $x = -0.87$ and $x = 1.12$, rounded to two decimal places.

Figure 35

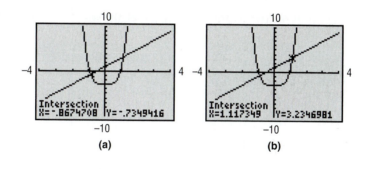

(a)   (b)

◀

NOW WORK PROBLEM 7.

The steps to follow for solving equations graphically are given next.

Steps for Solving Equations Graphically Using ZERO (or ROOT)

STEP 1: Write the equation in the form {expression in x} = 0.

STEP 2: Graph Y_1 = {expression in x}.

STEP 3: Use ZERO (or ROOT) to determine each x-intercept of the graph.

Steps for Solving Equations Graphically Using INTERSECT

STEP 1: Graph Y_1 = {expression in x on the left side of the equation}; Y_2 = {expression in x on the right side of the equation}.

STEP 2: Use INTERSECT to determine the x-coordinate of each point of intersection.

In the next example, we solve an equation both algebraically and graphically.

| EXAMPLE 3 | **Solving an Equation Algebraically and Graphically** |

Solve the equation: $3(x - 2) = 5(x - 1)$

Algebraic Solution

$$3(x - 2) = 5(x - 1)$$
$$3x - 6 = 5x - 5 \qquad \text{Remove the parentheses.}$$
$$3x - 6 - 5x = 5x - 5 - 5x \qquad \text{Subtract 5x from each side.}$$
$$-2x - 6 = -5 \qquad \text{Simplify.}$$
$$-2x - 6 + 6 = -5 + 6 \qquad \text{Add 6 to each side.}$$
$$-2x = 1 \qquad \text{Simplify.}$$
$$\frac{-2x}{-2} = \frac{1}{-2} \qquad \text{Divide each side by } -2.$$
$$x = -\frac{1}{2} \qquad \text{Simplify.}$$

✔ **CHECK:** Let $x = -\dfrac{1}{2}$ in the expression in x on the left side of the equation and simplify. Let $x = -\dfrac{1}{2}$ in the expression in x on the right side of the equation and simplify. If the two expressions are equal, the solution checks.

$$3(x - 2) = 3\left(-\frac{1}{2} - 2\right) = 3\left(-\frac{5}{2}\right) = -\frac{15}{2}$$

$$5(x - 1) = 5\left(-\frac{1}{2} - 1\right) = 5\left(-\frac{3}{2}\right) = -\frac{15}{2}$$

Since the two expressions are equal, the solution $x = -\dfrac{1}{2}$ checks.

Graphing Solution

Graph $Y_1 = 3(x - 2)$ and $Y_2 = 5(x - 1)$. See Figure 36. Using INTERSECT, we find the point of intersection to be $(-0.5, -7.5)$. The solution of the equation is $x = -0.5$.

Figure 36

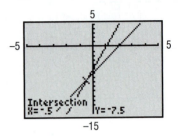

◀

NOW WORK PROBLEM **19**.

1.3 Assess Your Understanding

'Are You Prepared?'

Answers are given at the end of these exercises. If you get a wrong answer, read the pages listed in red.

1. Solve the equation $2x^2 + 5x + 2 = 0$. (pp. 984–997)

2. Solve the equation $2x + 3 = 4(x - 1) + 1$. (p. 986)

Concepts and Vocabulary

3. To solve an equation of the form {expression in x} = 0 using a graphing utility, we graph $Y_1 = $ {expression in x} and use _____ to determine each x-intercept of the graph.

4. *True or False:* In using a graphing utility to solve an equation, exact solutions are always obtained.

Skill Building

In Problems 5–16, use a graphing utility to approximate the real solutions, if any, of each equation rounded to two decimal places. All solutions lie between −10 *and* 10.

5. $x^3 - 4x + 2 = 0$

6. $x^3 - 8x + 1 = 0$

7. $-2x^4 + 5 = 3x - 2$

8. $-x^4 + 1 = 2x^2 - 3$

9. $x^4 - 2x^3 + 3x - 1 = 0$

10. $3x^4 - x^3 + 4x^2 - 5 = 0$

11. $-x^3 - \dfrac{5}{3}x^2 + \dfrac{7}{2}x + 2 = 0$

12. $-x^4 + 3x^3 + \dfrac{7}{3}x^2 - \dfrac{15}{2}x + 2 = 0$

13. $-\dfrac{2}{3}x^4 - 2x^3 + \dfrac{5}{2}x = -\dfrac{2}{3}x^2 + \dfrac{1}{2}$

14. $\frac{1}{4}x^3 - 5x = \frac{1}{5}x^2 - 4$ **15.** $x^4 - 5x^2 + 2x + 11 = 0$ **16.** $-3x^4 + 8x^2 - 2x - 9 = 0$

In Problems 17–36, solve each equation algebraically. Verify your solution using a graphing utility.

17. $2(3 + 2x) = 3(x - 4)$ **18.** $3(2 - x) = 2x - 1$ **19.** $8x - (2x + 1) = 3x - 13$ **20.** $5 - (2x - 1) = 10 - x$

21. $\frac{x+1}{3} + \frac{x+2}{7} = 5$ **22.** $\frac{2x+1}{3} + 16 = 3x$ **23.** $\frac{5}{y} + \frac{4}{y} = 3$ **24.** $\frac{4}{y} - 5 = \frac{18}{2y}$

25. $(x+7)(x-1) = (x+1)^2$ **26.** $(x+2)(x-3) = (x-3)^2$ **27.** $x^2 - 3x - 28 = 0$ **28.** $x^2 - 7x - 18 = 0$

29. $3x^2 = 4x + 4$ **30.** $5x^2 = 13x + 6$ **31.** $x^3 + x^2 - 4x - 4 = 0$ **32.** $x^3 + 2x^2 - 9x - 18 = 0$

33. $\sqrt{x+1} = 4$ **34.** $\sqrt{x-2} = 3$ **35.** $\frac{2}{x+2} + \frac{3}{x-1} = \frac{-8}{5}$ **36.** $\frac{1}{x+1} - \frac{5}{x-4} = \frac{21}{4}$

'Are You Prepared?' Answers

1. $\left\{-2, -\frac{1}{2}\right\}$ **2.** $\{3\}$

1.4 Lines

OBJECTIVES
1. Calculate and Interpret the Slope of a Line
2. Graph Lines Given a Point and the Slope
3. Find the Equation of a Vertical Line
4. Use the Point–Slope Form of a Line; Identify Horizontal Lines
5. Find the Equation of a Line Given Two Points
6. Write the Equation of a Line in Slope–Intercept Form
7. Identify the Slope and y-Intercept of a Line from Its Equation
8. Graph Lines Written in General Form Using Intercepts
9. Find Equations of Parallel Lines
10. Find Equations of Perpendicular Lines

In this section we study a certain type of equation that contains two variables, called a *linear equation*, and its graph, a *line*.

1   Calculate and Interpret the Slope of a Line

Figure 37

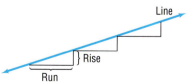

Consider the staircase illustrated in Figure 37. Each step contains exactly the same horizontal **run** and the same vertical **rise**. The ratio of the rise to the run, called the *slope*, is a numerical measure of the steepness of the staircase. For example, if the run is increased and the rise remains the same, the staircase becomes less steep. If the run is kept the same, but the rise is increased, the staircase becomes more steep. This important characteristic of a line is best defined using rectangular coordinates.

Let $P = (x_1, y_1)$ and $Q = (x_2, y_2)$ be two distinct points. If $x_1 \neq x_2$, the **slope m** of the nonvertical line L containing P and Q is defined by the formula

$$m = \frac{y_2 - y_1}{x_2 - x_1}, \qquad x_1 \neq x_2 \qquad \qquad \textbf{(1)}$$

If $x_1 = x_2$, L is a **vertical line** and the slope m of L is **undefined** (since this results in division by 0).

Figure 38(a) provides an illustration of the slope of a nonvertical line; Figure 38(b) illustrates a vertical line.

Figure 38

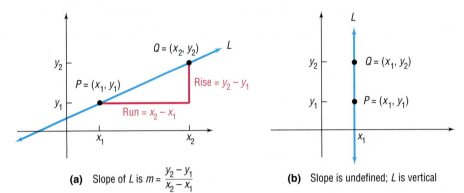

(a) Slope of L is $m = \dfrac{y_2 - y_1}{x_2 - x_1}$

(b) Slope is undefined; L is vertical

As Figure 38(a) illustrates, the slope m of a nonvertical line may be viewed as

$$m = \frac{y_2 - y_1}{x_2 - x_1} = \frac{\text{Rise}}{\text{Run}}$$

We can also express the slope m of a nonvertical line as

$$m = \frac{y_2 - y_1}{x_2 - x_1} = \frac{\text{Change in } y}{\text{Change in } x} = \frac{\Delta y}{\Delta x}$$

That is, the slope m of a nonvertical line L measures the amount that y changes as x changes from x_1 to x_2. This is called the **average rate of change** of y with respect to x.

Two comments about computing the slope of a nonvertical line may prove helpful:

1. Any two distinct points on the line can be used to compute the slope of the line. (See Figure 39 for justification.)

Figure 39
Triangles ABC and PQR are similar (equal angles). Hence, ratios of corresponding sides are proportional so that

Slope using P and Q = $\dfrac{y_2 - y_1}{x_2 - x_1}$

$= \dfrac{d(B, C)}{d(A, C)}$ = Slope using A and B

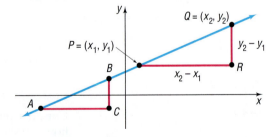

2. The slope of a line may be computed from $P = (x_1, y_1)$ to $Q = (x_2, y_2)$ or from Q to P because

$$\frac{y_2 - y_1}{x_2 - x_1} = \frac{y_1 - y_2}{x_1 - x_2}$$

EXAMPLE 1

Finding and Interpreting the Slope of a Line Containing Two Points

The slope m of the line containing the points $(1, 2)$ and $(5, -3)$ may be computed as

$$m = \frac{-3 - 2}{5 - 1} = \frac{-5}{4} = -\frac{5}{4} \quad \text{or as} \quad m = \frac{2 - (-3)}{1 - 5} = \frac{5}{-4} = -\frac{5}{4}$$

For every 4-unit change in x, y will change by -5 units. That is, if x increases by 4 units, then y will decrease by 5 units. The average rate of change of y with respect to x is $-\dfrac{5}{4}$. ◀

NOW WORK PROBLEMS 7 AND 13.

Square Screens

To get an undistorted view of slope, the same scale must be used on each axis. However, most graphing utilities have a rectangular screen. Because of this, using the same interval for both x and y will result in a distorted view. For example, Figure 40 shows the graph of the line $y = x$ connecting the points $(-4, -4)$ and $(4, 4)$. We expect the line to bisect the first and third quadrants, but it doesn't. We need to adjust the selections for Xmin, Xmax, Ymin, and Ymax so that a **square screen** results. On most graphing utilities, this is accomplished by setting the ratio of x to y at $3:2$.*

Figure 41 shows the graph of the line $y = x$ on a square screen using a TI-84 Plus. Notice that the line now bisects the first and third quadrants. Compare this illustration to Figure 40.

To get a better idea of the meaning of the slope m of a line, consider the following:

Figure 40

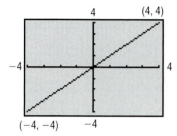

Figure 41

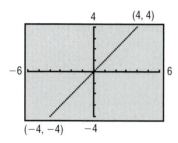

Figure 42

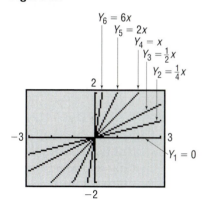

Figure 43

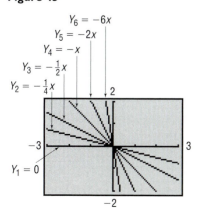

─────── **Seeing the Concept** ───────

On the same square screen, graph the following equations:

$Y_1 = 0$ *Slope of line is 0.*

$Y_2 = \dfrac{1}{4}x$ *Slope of line is $\dfrac{1}{4}$.*

$Y_3 = \dfrac{1}{2}x$ *Slope of line is $\dfrac{1}{2}$.*

$Y_4 = x$ *Slope of line is 1.*

$Y_5 = 2x$ *Slope of line is 2.*

$Y_6 = 6x$ *Slope of line is 6.*

See Figure 42.

─────── **Seeing the Concept** ───────

On the same square screen, graph the following equations:

$Y_1 = 0$ *Slope of line is 0.*

$Y_2 = -\dfrac{1}{4}x$ *Slope of line is $-\dfrac{1}{4}$.*

$Y_3 = -\dfrac{1}{2}x$ *Slope of line is $-\dfrac{1}{2}$.*

$Y_4 = -x$ *Slope of line is -1.*

$Y_5 = -2x$ *Slope of line is -2.*

$Y_6 = -6x$ *Slope of line is -6.*

See Figure 43.

*Most graphing utilities have a feature that automatically squares the viewing window. Consult your owner's manual for the appropriate keystrokes.

Figures 42 and 43 illustrate the following facts:

1. When the slope of a line is positive, the line slants upward from left to right.

2. When the slope of a line is negative, the line slants downward from left to right.

3. When the slope is 0, the line is horizontal.

Figures 42 and 43 also illustrate that the closer the line is to the vertical position, the greater the magnitude of the slope.

2 Graph Lines Given a Point and the Slope

The next example illustrates how the slope of a line can be used to graph the line.

EXAMPLE 2   **Graphing a Line Given a Point and a Slope**

Draw a graph of the line that contains the point $(3, 2)$ and has a slope of:

(a) $\dfrac{3}{4}$ 　　　　　　　　(b) $-\dfrac{4}{5}$

Solution   (a) Slope $= \dfrac{\text{Rise}}{\text{Run}}$. The fact that the slope is $\dfrac{3}{4}$ means that for every horizontal movement (run) of 4 units to the right there will be a vertical movement (rise) of 3 units. If we start at the given point $(3, 2)$ and move 4 units to the right and 3 units up, we reach the point $(7, 5)$. By drawing the line through this point and the point $(3, 2)$, we have the graph. See Figure 44.

(b) The fact that the slope is

$$-\frac{4}{5} = \frac{-4}{5} = \frac{\text{Rise}}{\text{Run}}$$

means that for every horizontal movement of 5 units to the right there will be a corresponding vertical movement of -4 units (a downward movement). If we start at the given point $(3, 2)$ and move 5 units to the right and then 4 units down, we arrive at the point $(8, -2)$. By drawing the line through these points, we have the graph. See Figure 45.

Figure 44

Slope $= \dfrac{3}{4}$

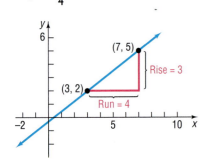

Figure 45

Slope $= -\dfrac{4}{5}$

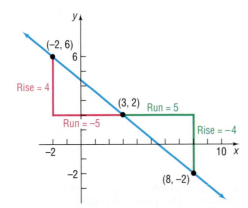

Alternatively, we can set

$$-\frac{4}{5} = \frac{4}{-5} = \frac{\text{Rise}}{\text{Run}}$$

so that for every horizontal movement of -5 units (a movement to the left) there will be a corresponding vertical movement of 4 units (upward). This approach brings us to the point $(-2, 6)$, which is also on the graph shown in Figure 45. ◀

 NOW WORK PROBLEM 19.

3 **Find the Equation of a Vertical Line**

Now that we have discussed the slope of a line, we are ready to derive equations of lines. As we shall see, there are several forms of the equation of a line. Let's start with an example.

EXAMPLE 3 **Graphing a Line**

Graph the equation: $x = 3$

Solution To graph $x = 3$ by hand, recall that we are looking for all points (x, y) in the plane for which $x = 3$. No matter what y-coordinate is used, the corresponding x-coordinate always equals 3. Consequently, the graph of the equation $x = 3$ is a vertical line with x-intercept 3 and undefined slope. See Figure 46(a).

To use a graphing utility, we need to express the equation in the form $y = \{\text{expression in } x\}$. But $x = 3$ cannot be put into this form, so an alternative method must be used. Consult your manual to determine the methodology required to draw vertical lines. Figure 46(b) shows the graph that you should obtain.

Figure 46
$x = 3$

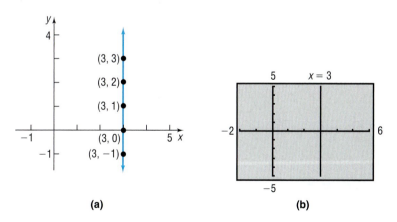

(a)   (b)   ◀

As suggested by Example 3, we have the following result:

Theorem **Equation of a Vertical Line**

A vertical line is given by an equation of the form

$$x = a$$

where a is the x-intercept.

Figure 47

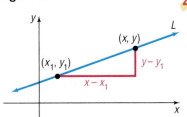

4 Use the Point–Slope Form of a Line; Identify Horizontal Lines

Now let L be a nonvertical line with slope m and containing the point (x_1, y_1). See Figure 47. For any other point (x, y) on L, we have

$$m = \frac{y - y_1}{x - x_1} \quad \text{or} \quad y - y_1 = m(x - x_1)$$

Theorem

Point–Slope Form of an Equation of a Line

An equation of a nonvertical line of slope m that contains the point (x_1, y_1) is

$$y - y_1 = m(x - x_1) \qquad \qquad \textbf{(2)}$$

EXAMPLE 4

Using the Point–Slope Form of a Line

An equation of the line with slope 4 and containing the point $(1, 2)$ can be found by using the point–slope form with $m = 4$, $x_1 = 1$, and $y_1 = 2$.

$$y - y_1 = m(x - x_1)$$
$$y - 2 = 4(x - 1) \qquad m = 4, x_1 = 1, y_1 = 2$$
$$y = 4x - 2$$

See Figure 48 for the graph.

Figure 48

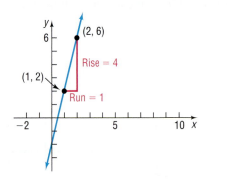

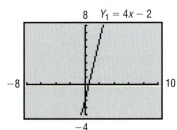

EXAMPLE 5

Finding the Equation of a Horizontal Line

Find an equation of the horizontal line containing the point $(3, 2)$.

Solution

Because all the y-values are equal on a horizontal line, the slope of a horizontal line is 0. To get an equation, we use the point–slope form with $m = 0$, $x_1 = 3$, and $y_1 = 2$.

$$y - y_1 = m(x - x_1)$$
$$y - 2 = 0 \cdot (x - 3) \qquad m = 0, x_1 = 3, \text{ and } y_1 = 2$$
$$y - 2 = 0$$
$$y = 2$$

See Figure 49 for the graph.

Figure 49

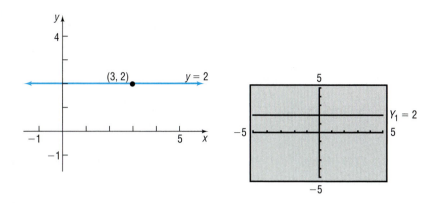

As suggested by Example 5, we have the following result:

Theorem

> **Equation of a Horizontal Line**
>
> A horizontal line is given by an equation of the form
>
> $$y = b$$
>
> where b is the y-intercept.

5 Find the Equation of a Line Given Two Points

We use the slope formula and the point–slope form of a line to find the equation given two points.

EXAMPLE 6   **Finding an Equation of a Line Given Two Points**

Find an equation of the line L containing the points $(2, 3)$ and $(-4, 5)$. Graph the line L.

Solution   We first compute the slope of the line.

$$m = \frac{5 - 3}{-4 - 2} = \frac{2}{-6} = -\frac{1}{3}$$

Figure 50

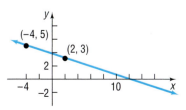

We use the point $(2, 3)$ and the slope $m = -\dfrac{1}{3}$ to get the point–slope form of the equation of the line.

$$y - 3 = -\frac{1}{3}(x - 2)$$

See Figure 50 for the graph.   ◄

In the solution to Example 6, we could have used the other point, $(-4, 5)$, instead of the point $(2, 3)$. The equation that results, although it looks different, is equivalent to the equation that we obtained in the example. (Try it for yourself.)

NOW WORK PROBLEM 33.

6   Write the Equation of a Line in Slope–Intercept Form

Another useful equation of a line is obtained when the slope m and y-intercept b are known. In this event, we know both the slope m of the line and a point $(0, b)$ on the line; then we may use the point–slope form, equation (2), to obtain the following equation:

$$y - b = m(x - 0) \quad \text{or} \quad y = mx + b$$

Theorem

Slope–Intercept Form of an Equation of a Line

An equation of a line L with slope m and y-intercept b is

$$y = mx + b \tag{3}$$

Figure 51 $y = mx + 2$

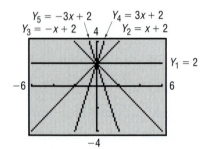

----- **Seeing the Concept** -----

To see the role that the slope m plays, graph the following lines on the same square screen.

$$Y_1 = 2$$
$$Y_2 = x + 2$$
$$Y_3 = -x + 2$$
$$Y_4 = 3x + 2$$
$$Y_5 = -3x + 2$$

See Figure 51. What do you conclude about the lines $y = mx + 2$?

Figure 52 $y = 2x + b$

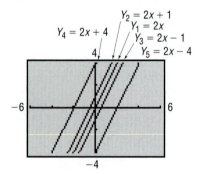

----- **Seeing the Concept** -----

To see the role of the y-intercept b, graph the following lines on the same square screen.

$$Y_1 = 2x$$
$$Y_2 = 2x + 1$$
$$Y_3 = 2x - 1$$
$$Y_4 = 2x + 4$$
$$Y_5 = 2x - 4$$

See Figure 52. What do you conclude about the lines $y = 2x + b$?

7   Identify the Slope and y-Intercept of a Line from Its Equation

When the equation of a line is written in slope–intercept form, it is easy to find the slope m and y-intercept b of the line. For example, suppose that the equation of a line is

$$y = -2x + 3$$

Compare it to $y = mx + b$.

$$y = -2x + 3$$
$$ \uparrow \quad \uparrow$$
$$y = \quad mx + b$$

The slope of this line is -2 and its y-intercept is 3.

NOW WORK PROBLEM 65.

| EXAMPLE 7 | **Finding the Slope and y-Intercept** |

Find the slope m and y-intercept b of the equation $2x + 4y = 8$. Graph the equation.

Solution To obtain the slope and y-intercept, we transform the equation into its slope–intercept form by solving for y.

$$2x + 4y = 8$$
$$4y = -2x + 8$$
$$y = -\frac{1}{2}x + 2 \qquad \textcolor{blue}{y = mx + b}$$

Figure 53
$2x + 4y = 8$

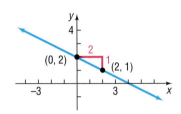

The coefficient of x, $-\dfrac{1}{2}$, is the slope, and the y-intercept is 2. We can graph the line using the fact that the y-intercept is 2 and the slope is $-\dfrac{1}{2}$. Then, starting at the point $(0, 2)$, go to the right 2 units and then down 1 unit to the point $(2, 1)$. See Figure 53.

NOW WORK PROBLEM 71.

8 Graph Lines Written in General Form Using Intercepts

The form of the equation of the line in Example 7, $2x + 4y = 8$, is called the *general form*.

The equation of a line L is in **general form** when it is written as

$$Ax + By = C \tag{4}$$

where A, B, and C are real numbers and A and B are not both 0.

When we want to graph an equation that is written in general form, we can solve the equation for y and write the equation in slope–intercept form as we did in Example 7. Another approach to graphing the equation would be to find its intercepts. Remember, the intercepts of the graph of an equation are the points where the graph crosses or touches a coordinate axis.

| EXAMPLE 8 | **Graphing an Equation in General Form Using Its Intercepts** |

Graph the equation $2x + 4y = 8$ by finding its intercepts.

Solution To obtain the x-intercept, let $y = 0$ in the equation and solve for x.

$$2x + 4y = 8$$
$$2x + 4(0) = 8 \qquad \textcolor{blue}{\text{Let } y = 0.}$$
$$2x = 8$$
$$x = 4 \qquad \textcolor{blue}{\text{Divide both sides by 2.}}$$

The x-intercept is 4 and the point $(4, 0)$ is on the graph of the equation.

Figure 54

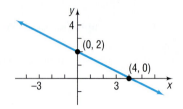

To obtain the y-intercept, let $x = 0$ in the equation and solve for y.

$$2x + 4y = 8$$
$$2(0) + 4y = 8 \qquad \text{Let } x = 0.$$
$$4y = 8$$
$$y = 2 \qquad \text{Divide both sides by 4.}$$

The y-intercept is 2 and the point $(0, 2)$ is on the graph of the equation.

We plot the points $(4, 0)$ and $(0, 2)$ in a Cartesian plane and draw a line through the points. See Figure 54. ◀

◖▭▭▭▭ **N O W   W O R K   P R O B L E M   8 5 .**

Every line has an equation that is equivalent to an equation written in general form. For example, a vertical line whose equation is

$$x = a$$

can be written in the general form

$$1 \cdot x + 0 \cdot y = a \qquad A = 1, B = 0, C = a$$

A horizontal line whose equation is

$$y = b$$

can be written in the general form

$$0 \cdot x + 1 \cdot y = b \qquad A = 0, B = 1, C = b$$

Lines that are neither vertical nor horizontal have general equations of the form

$$Ax + By = C \qquad A \neq 0 \text{ and } B \neq 0$$

Because the equation of every line can be written in general form, any equation equivalent to equation (4) is called a **linear equation**.

9 Find Equations of Parallel Lines

When two lines (in the plane) do not intersect (that is, they have no points in common), they are said to be **parallel**. Look at Figure 55. There we have drawn two lines and have constructed two right triangles by drawing sides parallel to the coordinate axes. These lines are parallel if and only if the right triangles are similar. (Do you see why? Two angles are equal.) And the triangles are similar if and only if the ratios of corresponding sides are equal.

Figure 55
The distinct lines are parallel if and only if their slopes are equal.

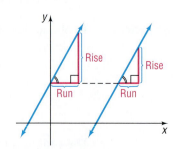

This suggests the following result:

Theorem

> **Criterion for Parallel Lines**
>
> Two nonvertical lines are parallel if and only if their slopes are equal and they have different y-intercepts.

The use of the words "if and only if" in the preceding theorem means that actually two statements are being made, one the converse of the other.

If two nonvertical lines are parallel, then their slopes are equal and they have different y-intercepts.

If two nonvertical lines have equal slopes and they have different y-intercepts, then they are parallel.

EXAMPLE 9   **Showing That Two Lines Are Parallel**

Show that the lines given by the following equations are parallel:

$$L_1:\ 2x + 3y = 6, \qquad L_2:\ 4x + 6y = 0$$

Solution   To determine whether these lines have equal slopes and different y-intercepts, we write each equation in slope–intercept form:

$$L_1:\ 2x + 3y = 6 \qquad\qquad L_2:\ 4x + 6y = 0$$
$$3y = -2x + 6 \qquad\qquad\qquad 6y = -4x$$
$$y = -\frac{2}{3}x + 2 \qquad\qquad\qquad y = -\frac{2}{3}x$$

$$\text{Slope} = -\frac{2}{3};\ y\text{-intercept} = 2 \qquad \text{Slope} = -\frac{2}{3};\ y\text{-intercept} = 0$$

Because these lines have the same slope, $-\dfrac{2}{3}$, but different y-intercepts, the lines are parallel. See Figure 56.

Figure 56
Parallel lines

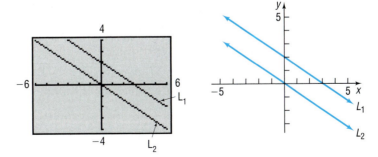

EXAMPLE 10   **Finding a Line That Is Parallel to a Given Line**

Find an equation for the line that contains the point $(2, -3)$ and is parallel to the line $2x + y = 6$.

Solution Since the two lines are to be parallel, the slope of the line that we seek equals the slope of the line $2x + y = 6$. We begin by writing the equation of the line $2x + y = 6$ in slope–intercept form.

$$2x + y = 6$$
$$y = -2x + 6$$

The slope is -2. Since the line that we seek contains the point $(2, -3)$, we use the point–slope form to obtain

$$y - y_1 = m(x - x_1) \qquad \text{Point–slope form}$$
$$y - (-3) = -2(x - 2) \qquad m = -2, x_1 = 2, y_1 = -3$$
$$y + 3 = -2x + 4 \qquad \text{Simplify}$$
$$y = -2x + 1 \qquad \text{Slope–intercept form}$$
$$2x + y = 1 \qquad \text{General form}$$

This line is parallel to the line $2x + y = 6$ and contains the point $(2, -3)$. See Figure 57.

Figure 57

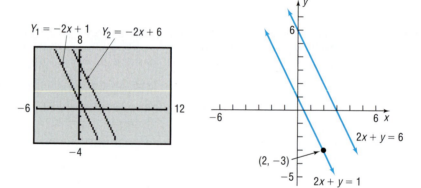

Figure 58
Perpendicular lines

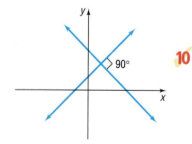

NOW WORK PROBLEM 53.

10 **Find Equations of Perpendicular Lines**

When two lines intersect at a right angle (90°), they are said to be **perpendicular**. See Figure 58.

The following result gives a condition, in terms of their slopes, for two lines to be perpendicular.

Theorem

Criterion for Perpendicular Lines

Two nonvertical lines are perpendicular if and only if the product of their slopes is -1.

Here we shall prove the "only if" part of the statement:

If two nonvertical lines are perpendicular, then the product of their slopes is -1.

You are asked to prove the "if" part of the theorem; that is:

If two nonvertical lines have slopes whose product is -1, then the lines are perpendicular.

Figure 59

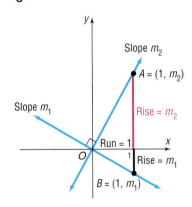

Proof Let m_1 and m_2 denote the slopes of the two lines. There is no loss in generality (that is, neither the angle nor the slopes are affected) if we situate the lines so that they meet at the origin. See Figure 59. The point $A = (1, m_2)$ is on the line having slope m_2, and the point $B = (1, m_1)$ is on the line having slope m_1. (Do you see why this must be true?)

Suppose that the lines are perpendicular. Then triangle OAB is a right triangle. As a result of the Pythagorean Theorem, it follows that

$$[d(O, A)]^2 + [d(O, B)]^2 = [d(A, B)]^2 \qquad \textbf{(5)}$$

By the distance formula, we can write the squares of these distances as

$$[d(O, A)]^2 = (1 - 0)^2 + (m_2 - 0)^2 = 1 + m_2^2$$

$$[d(O, B)]^2 = (1 - 0)^2 + (m_1 - 0)^2 = 1 + m_1^2$$

$$[d(A, B)]^2 = (1 - 1)^2 + (m_2 - m_1)^2 = m_2^2 - 2m_1m_2 + m_1^2$$

Using these facts in equation (5), we get

$$(1 + m_2^2) + (1 + m_1^2) = m_2^2 - 2m_1m_2 + m_1^2$$

which, upon simplification, can be written as

$$m_1m_2 = -1$$

If the lines are perpendicular, the product of their slopes is -1. ■

You may find it easier to remember the condition for two nonvertical lines to be perpendicular by observing that the equality $m_1m_2 = -1$ means that m_1 and m_2 are negative reciprocals of each other; that is, either $m_1 = -\dfrac{1}{m_2}$ or $m_2 = -\dfrac{1}{m_1}$.

EXAMPLE 11   **Finding the Slope of a Line Perpendicular to Another Line**

If a line has slope $\dfrac{3}{2}$, any line having slope $-\dfrac{2}{3}$ is perpendicular to it.   ◄

EXAMPLE 12   **Finding the Equation of a Line Perpendicular to a Given Line**

Find an equation of the line that contains the point $(1, -2)$ and is perpendicular to the line $x + 3y = 6$. Graph the two lines.

Solution   We first write the equation of the given line in slope–intercept form to find its slope.

$$x + 3y = 6$$
$$3y = -x + 6 \qquad \textit{Proceed to solve for y.}$$
$$y = -\frac{1}{3}x + 2 \qquad \textit{Place in the form } y = mx + b.$$

The given line has slope $-\dfrac{1}{3}$. Any line perpendicular to this line will have slope 3.

Because we require the point $(1, -2)$ to be on this line with slope 3, we use the point–slope form of the equation of a line.

$$y - y_1 = m(x - x_1) \qquad \textit{Point–slope form}$$
$$y - (-2) = 3(x - 1) \qquad \textit{m = 3, } x_1 = 1, y_1 = -2.$$

To obtain other forms of the equation, we proceed as follows:

$$y + 2 = 3(x - 1)$$
$$y + 2 = 3x - 3 \qquad \text{Simplify}$$
$$y = 3x - 5 \qquad \text{Slope–intercept form}$$
$$3x - y = 5 \qquad \text{General form}$$

Figure 60 shows the graphs.

Figure 60

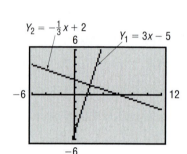

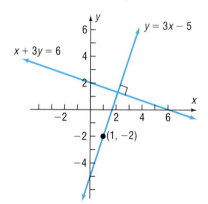

WARNING
Be sure to use a square screen when you graph perpendicular lines. Otherwise, the angle between the two lines will appear distorted. ■

NOW WORK PROBLEM 59.

1.4 Assess Your Understanding

Concepts and Vocabulary

1. The slope of a vertical line is _____; the slope of a horizontal line is _____.

2. Two nonvertical lines have slopes m_1 and m_2, respectively. The lines are parallel if _____ and the _____ are unequal; the lines are perpendicular if _____.

3. A horizontal line is given by an equation of the form _____, where b is the _____.

4. *True or False*: Vertical lines have an undefined slope.

5. *True or False*: The slope of the line $2y = 3x + 5$ is 3.

6. *True or False*: Perpendicular lines have slopes that are reciprocals of one another.

Skill Building

In Problems 7–10, (a) find the slope of the line and (b) interpret the slope.

7.
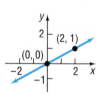

8.

9.
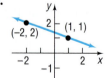

10.
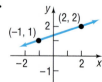

In Problems 11–18, plot each pair of points and determine the slope of the line containing them. Graph the line.

11. $(2, 3)$; $(4, 0)$
12. $(4, 2)$; $(3, 4)$
13. $(-2, 3)$; $(2, 1)$
14. $(-1, 1)$; $(2, 3)$
15. $(-3, -1)$; $(2, -1)$
16. $(4, 2)$; $(-5, 2)$
17. $(-1, 2)$; $(-1, -2)$
18. $(2, 0)$; $(2, 2)$

In Problems 19–26, graph, by hand, the line containing the point P and having slope m.

19. $P = (1, 2)$; $m = 3$
20. $P = (2, 1)$; $m = 4$
21. $P = (2, 4)$; $m = -\dfrac{3}{4}$
22. $P = (1, 3)$; $m = -\dfrac{2}{5}$

23. $P = (-1, 3); m = 0$ **24.** $P = (2, -4); m = 0$ **25.** $P = (0, 3)$; slope undefined   **26.** $P = (-2, 0)$; slope undefined

In Problems 27–32, the slope and a point on a line are given. Use this information to locate three additional points on the line. Answers may vary.

[Hint: *It is not necessary to find the equation of the line. See Example 2.]*

27. Slope 4; point $(1, 2)$

28. Slope 2; point $(-2, 3)$

29. Slope $-\dfrac{3}{2}$; point $(2, -4)$

30. Slope $\dfrac{4}{3}$; point $(-3, 2)$

31. Slope -2; point $(-2, -3)$

32. Slope -1; point $(4, 1)$

In Problems 33–40, find an equation of the line L.

33.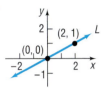

34.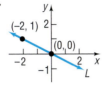

35.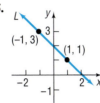

36.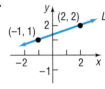

37.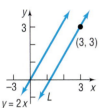
$y = 2x$
L is parallel to $y = 2x$

38.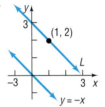
$y = -x$
L is parallel to $y = -x$

39.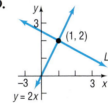
$y = 2x$
L is perpendicular to $y = 2x$

40.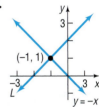
$y = -x$
L is perpendicular to $y = -x$

In Problems 41–64, find an equation for the line with the given properties. Express your answer using either the general form or the slope–intercept form of the equation of a line, whichever you prefer.

41. Slope $= 3$; containing the point $(-2, 3)$

42. Slope $= 2$; containing the point $(4, -3)$

43. Slope $= -\dfrac{2}{3}$; containing the point $(1, -1)$

44. Slope $= \dfrac{1}{2}$; containing the point $(3, 1)$

45. Containing the points $(1, 3)$ and $(-1, 2)$

46. Containing the points $(-3, 4)$ and $(2, 5)$

47. Slope $= -3$; y-intercept $= 3$

48. Slope $= -2$; y-intercept $= -2$

49. x-intercept $= 2$; y-intercept $= -1$

50. x-intercept $= -4$; y-intercept $= 4$

51. Slope undefined; containing the point $(2, 4)$

52. Slope undefined; containing the point $(3, 8)$

53. Parallel to the line $y = 2x$; containing the point $(-1, 2)$

54. Parallel to the line $y = -3x$; containing the point $(-1, 2)$

55. Parallel to the line $2x - y = -2$; containing the point $(0, 0)$

56. Parallel to the line $x - 2y = -5$; containing the point $(0, 0)$

57. Parallel to the line $x = 5$; containing the point $(4, 2)$

58. Parallel to the line $y = 5$; containing the point $(4, 2)$

59. Perpendicular to the line $y = \dfrac{1}{2}x + 4$; containing the point $(1, -2)$

60. Perpendicular to the line $y = 2x - 3$; containing the point $(1, -2)$

61. Perpendicular to the line $2x + y = 2$; containing the point $(-3, 0)$

62. Perpendicular to the line $x - 2y = -5$; containing the point $(0, 4)$

63. Perpendicular to the line $x = 8$; containing the point $(3, 4)$

64. Perpendicular to the line $y = 8$; containing the point $(3, 4)$

In Problems 65–84, find the slope and y-intercept of each line. Graph the line by hand. Check your graph using a graphing utility.

65. $y = 2x + 3$

66. $y = -3x + 4$

67. $\frac{1}{2}y = x - 1$

68. $\frac{1}{3}x + y = 2$

69. $y = \frac{1}{2}x + 2$

70. $y = 2x + \frac{1}{2}$

71. $x + 2y = 4$

72. $-x + 3y = 6$

73. $2x - 3y = 6$

74. $3x + 2y = 6$

75. $x + y = 1$

76. $x - y = 2$

77. $x = -4$

78. $y = -1$

79. $y = 5$

80. $x = 2$

81. $y - x = 0$

82. $x + y = 0$

83. $2y - 3x = 0$

84. $3x + 2y = 0$

In Problems 85–94, (a) find the intercepts of the graph of each equation and (b) graph the equation by hand.

85. $2x + 3y = 6$

86. $3x - 2y = 6$

87. $-4x + 5y = 40$

88. $6x - 4y = 24$

89. $7x + 2y = 21$

90. $5x + 3y = 18$

91. $\frac{1}{2}x + \frac{1}{3}y = 1$

92. $x - \frac{2}{3}y = 4$

93. $0.2x - 0.5y = 1$

94. $-0.3x + 0.4y = 1.2$

95. Find an equation of the x-axis.

96. Find an equation of the y-axis.

In Problems 97–100, match each graph with the correct equation:

(a) $y = x$ (b) $y = 2x$ (c) $y = \dfrac{x}{2}$ (d) $y = 4x$

97.
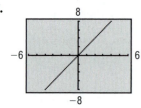

98.
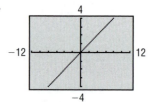

99.
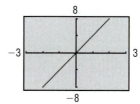

100.
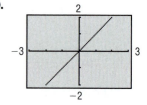

In Problems 101–104, write an equation of each line. Express your answer using either the general form or the slope–intercept form of the equation of a line, whichever you prefer.

101.
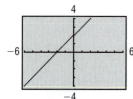

102.
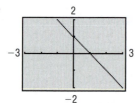

103.
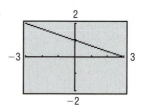

104.
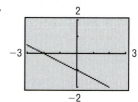

Applications and Extensions

105. Truck Rentals   A truck rental company rents a moving truck for one day by charging $29 plus $0.07 per mile. Write a linear equation that relates the cost C, in dollars, of renting the truck to the number x of miles driven. What is the cost of renting the truck if the truck is driven 110 miles? 230 miles?

106. Cost Equation   The **fixed costs** of operating a business are the costs incurred regardless of the level of production. Fixed costs include rent, fixed salaries, and costs of buying machinery. The **variable costs** of operating a business are the costs that change with the level of output. Variable costs include raw materials, hourly wages, and electricity. Suppose that a manufacturer of jeans has fixed costs of $500 and variable costs of $8 for each pair of jeans manufactured. Write a linear equation that relates the cost C, in dollars, of manufacturing the jeans to the number x pairs of jeans manufactured. What is the cost of manufacturing 400 pairs of jeans? 740 pairs?

107. Cost of Sunday Home Delivery   The cost to the *Chicago Tribune* for Sunday home delivery is approximately $0.53 per newspaper with fixed costs of $1,070,000. Write an equation that relates the cost C and the number x of copies delivered.
SOURCE: *Chicago Tribune*, 2002.

108. Wages of a Car Salesperson   Dan receives $375 per week for selling new and used cars at a car dealership in Oak Lawn, Illinois. In addition, he receives 5% of the profit on any sales that he generates. Write an equation that relates Dan's weekly salary S when he has sales that generate a profit of x dollars.

109. Electricity Rates in Illinois   Commonwealth Edison Company supplies electricity to residential customers for a monthly customer charge of $7.58 plus 8.275 cents per kilowatt-hour for up to 400 kilowatt-hours.

(a) Write an equation that relates the monthly charge C, in dollars, to the number x of kilowatt-hours used in a month, $0 \leq x \leq 400$.
(b) Graph this equation.
(c) What is the monthly charge for using 100 kilowatt-hours?
(d) What is the monthly charge for using 300 kilowatt-hours?
(e) Interpret the slope of the line.
Source: Commonwealth Edison Company, December 2002.

110. Electricity Rates in Florida   Florida Power & Light Company supplies electricity to residential customers for a monthly customer charge of $5.25 plus 6.787 cents per kilowatt-hour for up to 750 kilowatt-hours.
(a) Write an equation that relates the monthly charge C, in dollars, to the number x of kilowatt-hours used in a month, $0 \leq x \leq 750$.
(b) Graph this equation.
(c) What is the monthly charge for using 200 kilowatt-hours?
(d) What is the monthly charge for using 500 kilowatt-hours?
(e) Interpret the slope of the line.
Source: Florida Power & Light Company, January 2003.

111. Measuring Temperature   The relationship between Celsius (°C) and Fahrenheit (°F) degrees of measuring temperature is linear. Find an equation relating °C and °F if 0°C corresponds to 32°F and 100°C corresponds to 212°F. Use the equation to find the Celsius measure of 70°F.

112. Measuring Temperature   The Kelvin (K) scale for measuring temperature is obtained by adding 273 to the Celsius temperature.
(a) Write an equation relating K and °C.
(b) Write an equation relating K and °F (see Problem 111).

113. Product Promotion   A cereal company finds that the number of people who will buy one of its products in the first month that it is introduced is linearly related to the amount of money it spends on advertising. If it spends $40,000 on advertising, then 100,000 boxes of cereal will be sold, and if it spends $60,000, then 200,000 boxes will be sold.
(a) Write an equation describing the relation between the amount A spent on advertising and the number x of boxes sold.
(b) How much advertising is needed to sell 300,000 boxes of cereal?
(c) Interpret the slope.

114. Show that the line containing the points (a, b) and (b, a), $a \neq b$, is perpendicular to the line $y = x$. Also show that the midpoint of (a, b) and (b, a) lies on the line $y = x$.

115. The equation $2x - y = C$ defines a **family of lines**, one line for each value of C. On one set of coordinate axes, graph the members of the family when $C = -4$, $C = 0$, and $C = 2$. Can you draw a conclusion from the graph about each member of the family?

116. Rework Problem 115 for the family of lines $Cx + y = -4$.

Discussion and Writing

117. Which of the following equations might have the graph shown? (More than one answer is possible.)
(a) $2x + 3y = 6$ (e) $x - y = -1$
(b) $-2x + 3y = 6$ (f) $y = 3x - 5$
(c) $3x - 4y = -12$ (g) $y = 2x + 3$
(d) $x - y = 1$ (h) $y = -3x + 3$

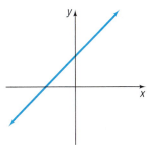

118. Which of the following equations might have the graph shown? (More than one answer is possible.)
(a) $2x + 3y = 6$ (e) $x - y = -1$
(b) $2x - 3y = 6$ (f) $y = -2x - 1$
(c) $3x + 4y = 12$ (g) $y = -\dfrac{1}{2}x + 10$
(d) $x - y = 1$
(h) $y = x + 4$

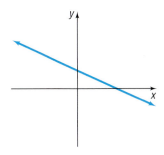

119. The figure shows the graph of two parallel lines. Which of the following pairs of equations might have such a graph?

(a) $x - 2y = 3$
$\quad x + 2y = 7$

(b) $x + y = 2$
$\quad x + y = -1$

(c) $x - y = -2$
$\quad x - y = 1$

(d) $x - y = -2$
$\quad 2x - 2y = -4$

(e) $x + 2y = 2$
$\quad x + 2y = -1$

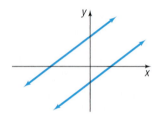

120. The figure shows the graph of two perpendicular lines. Which of the following pairs of equations might have such a graph?

(a) $y - 2x = 2$
$\quad y + 2x = -1$

(b) $y - 2x = 0$
$\quad 2y + x = 0$

(c) $2y - x = 2$
$\quad 2y + x = -2$

(d) $y - 2x = 2$
$\quad x + 2y = -1$

(e) $2x + y = -2$
$\quad 2y + x = -2$

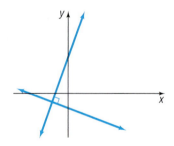

121. The accepted symbol used to denote the slope of a line is the letter *m*. Investigate the origin of this symbolism. Begin by consulting a French dictionary and looking up the French word *monter*. Write a brief essay on your findings.

122. The term *grade* is used to describe the inclination of a road. How does this term relate to the notion of slope of a line? Is a 4% grade very steep? Investigate the grades of some mountainous roads and determine their slopes. Write a brief essay on your findings.

123. Carpentry Carpenters use the term *pitch* to describe the steepness of staircases and roofs. How does pitch relate to slope? Investigate typical pitches used for stairs and for roofs. Write a brief essay on your findings.

124. Can the equation of every line be written in slope–intercept form? Why?

125. Does every line have exactly one *x*-intercept and one *y*-intercept? Are there any lines that have no intercepts?

126. What can you say about two lines that have equal slopes and equal *y*-intercepts?

127. What can you say about two lines with the same *x*-intercept and the same *y*-intercept? Assume that the *x*-intercept is not 0.

128. If two lines have the same slope, but different *x*-intercepts, can they have the same *y*-intercept?

129. If two lines have the same *y*-intercept, but different slopes, can they have the same *x*-intercept?

1.5 Circles

OBJECTIVES 1 Write the Standard Form of the Equation of a Circle
2 Graph a Circle by Hand and by Using a Graphing Utility
3 Work with the General Form of the Equation of a Circle

1 Write the Standard Form of the Equation of a Circle

One advantage of a coordinate system is that it enables us to translate a geometric statement into an algebraic statement, and vice versa. Consider, for example, the following geometric statement that defines a circle.

A **circle** is a set of points in the xy-plane that are a fixed distance r from a fixed point (h, k). The fixed distance r is called the **radius**, and the fixed point (h, k) is called the **center** of the circle.

Figure 61

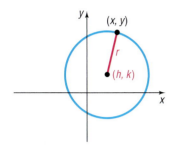

Figure 61 shows the graph of a circle. To find the equation, we let (x, y) represent the coordinates of any point on a circle with radius r and center (h, k). Then the distance between the points (x, y) and (h, k) must always equal r. That is, by the distance formula

$$\sqrt{(x - h)^2 + (y - k)^2} = r$$

or, equivalently,

$$(x - h)^2 + (y - k)^2 = r^2$$

The **standard form of an equation of a circle** with radius r and center (h, k) is

$$(x - h)^2 + (y - k)^2 = r^2 \qquad\qquad \textbf{(1)}$$

The standard form of an equation of a circle of radius r with center at the origin $(0, 0)$ is

$$x^2 + y^2 = r^2$$

If the radius $r = 1$, the circle whose center is at the origin is called the **unit circle** and has the equation

$$x^2 + y^2 = 1$$

See Figure 62.

Figure 62
Unit circle $x^2 + y^2 = 1$

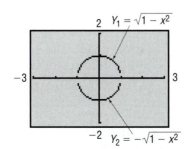
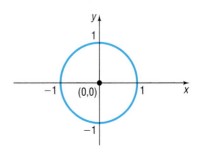

| EXAMPLE 1 | **Writing the Standard Form of the Equation of a Circle** |

Write the standard form of the equation of the circle with radius 5 and center $(-3, 6)$.

Solution Using the form of equation (1) and substituting the values $r = 5, h = -3$, and $k = 6$, we have

$$(x - h)^2 + (y - k)^2 = r^2$$
$$(x + 3)^2 + (y - 6)^2 = 25 \qquad \blacktriangleleft$$

✏ NOW WORK PROBLEM 5.

2 Graph a Circle by Hand and by Using a Graphing Utility

The graph of any equation of the form (1) is that of a circle with radius r and center (h, k).

| EXAMPLE 2 | **Graphing a Circle by Hand and by Using a Graphing Utility** |

Graph the equation: $(x + 3)^2 + (y - 2)^2 = 16$

Solution The graph of the equation is a circle. To graph the equation by hand, we first compare the given equation to the standard form of the equation of a circle. The comparison yields information about the circle.

$$(x + 3)^2 + (y - 2)^2 = 16$$
$$(x - (-3))^2 + (y - 2)^2 = 4^2$$
$$\qquad\quad \uparrow \qquad\qquad\quad \uparrow \qquad \uparrow$$
$$(x - h)^2 + (y - k)^2 = r^2$$

We see that $h = -3, k = 2$, and $r = 4$. The circle has center $(-3, 2)$ and a radius of 4 units. To graph this circle, we first plot the center $(-3, 2)$. Since the radius is 4, we can locate four points on the circle by plotting points 4 units to the left, to the right, up, and down from the center. These four points can then be used as guides to obtain the graph. See Figure 63.

To graph a circle on a graphing utility, we must write the equation in the form $y = \{$expression involving $x\}$.* We must solve for y in the equation

$$(x + 3)^2 + (y - 2)^2 = 16$$
$$(y - 2)^2 = 16 - (x + 3)^2 \qquad \text{Subtract } (x + 3)^2 \text{ from both sides.}$$
$$y - 2 = \pm\sqrt{16 - (x + 3)^2} \qquad \text{Use the Square Root Method.}$$
$$y = 2 \pm \sqrt{16 - (x + 3)^2} \qquad \text{Add 2 to both sides.}$$

Figure 63

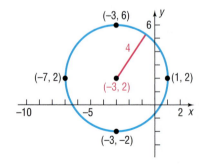

*Some graphing utilities (e.g., TI-83, TI-84, and TI-86) have a CIRCLE function that allows the user to enter only the coordinates of the center of the circle and its radius to graph the circle.

Figure 64

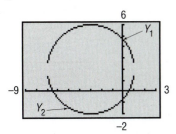

To graph the circle, we graph the top half

$$Y_1 = 2 + \sqrt{16 - (x + 3)^2}$$

and the bottom half

$$Y_2 = 2 - \sqrt{16 - (x + 3)^2}$$

Also, be sure to use a square screen. Otherwise, the circle will appear distorted. Figure 64 shows the graph on a TI-84 Plus.   ◀

NOW WORK PROBLEMS **21(a)** AND **(b)**.

EXAMPLE 3   **Finding the Intercepts of a Circle**

For the circle $(x + 3)^2 + (y - 2)^2 = 16$, find the intercepts, if any, of its graph.

Solution   This is the equation discussed and graphed in Example 2. To find the x-intercepts, if any, let $y = 0$ and solve for x. Then

$$(x + 3)^2 + (y - 2)^2 = 16$$
$$(x + 3)^2 + (0 - 2)^2 = 16 \qquad \text{\color{blue}{y = 0}}$$
$$(x + 3)^2 + 4 = 16 \qquad \text{\color{blue}{Simplify.}}$$
$$(x + 3)^2 = 12 \qquad \text{\color{blue}{Simplify.}}$$
$$x + 3 = \pm\sqrt{12} \qquad \text{\color{blue}{Apply the Square Root Method.}}$$
$$x = -3 \pm 2\sqrt{3} \qquad \text{\color{blue}{Solve for x.}}$$

The x-intercepts are $-3 - 2\sqrt{3} \approx -6.46$ and $-3 + 2\sqrt{3} \approx 0.46$.
   To find the y-intercepts, if any, we let $x = 0$ and solve for y. Then

$$(x + 3)^2 + (y - 2)^2 = 16$$
$$(0 + 3)^2 + (y - 2)^2 = 16 \qquad \text{\color{blue}{x = 0}}$$
$$9 + (y - 2)^2 = 16 \qquad \text{\color{blue}{Simplify.}}$$
$$(y - 2)^2 = 7 \qquad \text{\color{blue}{Simplify.}}$$
$$y - 2 = \pm\sqrt{7} \qquad \text{\color{blue}{Apply the Square Root Method.}}$$
$$y = 2 \pm \sqrt{7} \qquad \text{\color{blue}{Solve for y.}}$$

The y-intercepts are $2 - \sqrt{7} \approx -0.65$ and $2 + \sqrt{7} \approx 4.65$.
   Look back at Figure 63 to verify the approximate locations of the intercepts. ◀

NOW WORK PROBLEM **21(c)**.

3   Work with the General Form of the Equation of a Circle

If we eliminate the parentheses from the standard form of the equation of the circle given in Example 3, we get

$$(x + 3)^2 + (y - 2)^2 = 16$$
$$x^2 + 6x + 9 + y^2 - 4y + 4 = 16$$

which we find, upon simplifying, is equivalent to

$$x^2 + y^2 + 6x - 4y - 3 = 0$$

It can be shown that any equation of the form

$$x^2 + y^2 + ax + by + c = 0$$

has a graph that is a circle, or a point, or has no graph at all. For example, the graph of the equation $x^2 + y^2 = 0$ is the single point $(0, 0)$. The equation $x^2 + y^2 + 5 = 0$, or $x^2 + y^2 = -5$, has no graph, because sums of squares of real numbers are never negative. When its graph is a circle, the equation

$$x^2 + y^2 + ax + by + c = 0$$

is referred to as the **general form of the equation of a circle**.

If an equation of a circle is in the general form, we use the method of completing the square to put the equation in standard form so that we can identify its center and radius.

EXAMPLE 4 **Graphing a Circle Whose Equation Is in General Form**

Graph the equation $x^2 + y^2 + 4x - 6y + 12 = 0$

Solution We complete the square in both x and y to put the equation in standard form. Group the expression involving x, group the expression involving y, and put the constant on the right side of the equation. The result is

$$(x^2 + 4x) + (y^2 - 6y) = -12$$

Next, complete the square of each expression in parentheses. Remember that any number added on the left side of the equation must be added on the right.

$$(x^2 + 4x + 4) + (y^2 - 6y + 9) = -12 + 4 + 9$$

$$\left(\tfrac{4}{2}\right)^2 = 4 \qquad \left(\tfrac{-6}{2}\right)^2 = 9$$

$$(x + 2)^2 + (y - 3)^2 = 1 \quad \text{Factor.}$$

We recognize this equation as the standard form of the equation of a circle with radius 1 and center $(-2, 3)$. To graph the equation by hand, use the center $(-2, 3)$ and the radius 1. See Figure 65 (a).

To graph the equation using a graphing utility, we need to solve for y.

$$(y - 3)^2 = 1 - (x + 2)^2$$

$$y - 3 = \pm\sqrt{1 - (x + 2)^2} \qquad \text{Use the Square Root Method.}$$

$$y = 3 \pm \sqrt{1 - (x + 2)^2} \qquad \text{Add 3 to both sides.}$$

Figure 65(b) illustrates the graph.

Figure 65

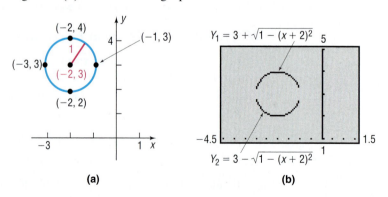

(a) (b)

NOW WORK PROBLEM **25.**

| EXAMPLE 5 | **Finding the General Equation of a Circle** |

Find the general equation of the circle whose center is $(1, -2)$ and whose graph contains the point $(4, -2)$.

Solution   To find the equation of a circle, we need to know its center and its radius. Here, we know that the center is $(1, -2)$. Since the point $(4, -2)$ is on the graph, the radius r will equal the distance from $(4, -2)$ to the center $(1, -2)$. See Figure 66. Thus,

Figure 66

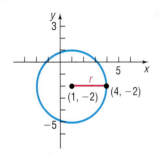

$$r = \sqrt{(4 - 1)^2 + [-2 - (-2)]^2}$$
$$= \sqrt{9} = 3$$

The standard form of the equation of the circle is

$$(x - 1)^2 + (y + 2)^2 = 9$$

Eliminating the parentheses and rearranging terms, we get the general equation

$$x^2 + y^2 - 2x + 4y - 4 = 0$$ ◄

Overview

The discussion in Sections 1.4 and 1.5 about circles and lines dealt with two main types of problems that can be generalized as follows:

1. Given an equation, classify it and graph it.
2. Given a graph, or information about a graph, find its equation.

This text deals with both types of problems. We shall study various equations, classify them, and graph them. Although the second type of problem is usually more difficult to solve than the first, in many instances a graphing utility can be used to solve such problems.

1.5 Assess Your Understanding

Concepts and Vocabulary

1. *True or False:* Every equation of the form
$$x^2 + y^2 + ax + by + c = 0$$
has a circle as its graph.

2. For a circle, the _____ is the distance from the center to any point on the circle.

3. *True or False:* The radius of the circle $x^2 + y^2 = 9$ is 3.

4. *True or False:* The center of the circle
$$(x + 3)^2 + (y - 2)^2 = 13$$
is $(3, -2)$.

Skill Building

In Problems 5–8, find the center and radius of each circle. Write the standard form of the equation.

5.

6.

7.

8.

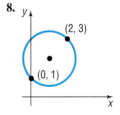

In Problems 9–18, write the standard form of the equation and the general form of the equation of each circle of radius r and center (h, k). By hand, graph each circle.

9. $r = 2$; $(h, k) = (0, 0)$ **10.** $r = 3$; $(h, k) = (0, 0)$ **11.** $r = 2$; $(h, k) = (0, 2)$ **12.** $r = 3$; $(h, k) = (1, 0)$

13. $r = 5$; $(h, k) = (4, -3)$ **14.** $r = 4$; $(h, k) = (2, -3)$ **15.** $r = 4$; $(h, k) = (-2, 1)$ **16.** $r = 7$; $(h, k) = (-5, -2)$

17. $r = \frac{1}{2}$; $(h, k) = \left(\frac{1}{2}, 0\right)$ **18.** $r = \frac{1}{2}$; $(h, k) = \left(0, -\frac{1}{2}\right)$

In Problems 19–30, (a) find the center (h, k) and radius r of each circle; (b) graph each circle; (c) find the intercepts, if any, of the graphs.

19. $x^2 + y^2 = 4$ **20.** $x^2 + (y - 1)^2 = 1$ **21.** $2(x - 3)^2 + 2y^2 = 8$

22. $3(x + 1)^2 + 3(y - 1)^2 = 6$ **23.** $x^2 + y^2 - 2x - 4y - 4 = 0$ **24.** $x^2 + y^2 + 4x + 2y - 20 = 0$

25. $x^2 + y^2 + 4x - 4y - 1 = 0$ **26.** $x^2 + y^2 - 6x + 2y + 9 = 0$ **27.** $x^2 + y^2 - x + 2y + 1 = 0$

28. $x^2 + y^2 + x + y - \frac{1}{2} = 0$ **29.** $2x^2 + 2y^2 - 12x + 8y - 24 = 0$ **30.** $2x^2 + 2y^2 + 8x + 7 = 0$

In Problems 31–36, find the general form of the equation of each circle.

31. Center at the origin and containing the point $(-2, 3)$ **32.** Center $(1, 0)$ and containing the point $(-3, 2)$

33. Center $(2, 3)$ and tangent to the x-axis   **34.** Center $(-3, 1)$ and tangent to the y-axis

35. With endpoints of a diameter at $(1, 4)$ and $(-3, 2)$ **36.** With endpoints of a diameter at $(4, 3)$ and $(0, 1)$

In Problems 37–40, match each graph with the correct equation.
(a) $(x - 3)^2 + (y + 3)^2 = 9$ (b) $(x + 1)^2 + (y - 2)^2 = 4$ (c) $(x - 1)^2 + (y + 2)^2 = 4$ (d) $(x + 3)^2 + (y - 3)^2 = 9$

37. 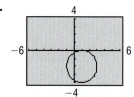   **38.** 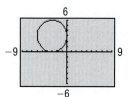   **39.** 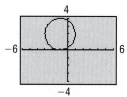   **40.**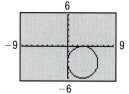

In Problems 41–44, find the standard form of the equation of each circle. Assume that the center has integer coordinates and that the radius is an integer.

41. 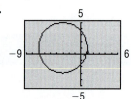   **42.** 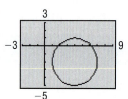   **43.** 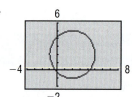   **44.**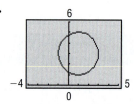

Applications and Extensions

45. Weather Satellites Earth is represented on a map of a portion of the solar system so that its surface is the circle with equation $x^2 + y^2 + 2x + 4y - 4091 = 0$. A weather satellite circles 0.6 unit above Earth with the center of its circular orbit at the center of Earth. Find the equation for the orbit of the satellite on this map.

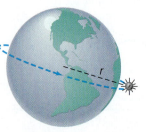

46. The **tangent line** to a circle may be defined as the line that intersects the circle in a single point, called the **point of tangency** (see the figure).

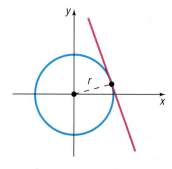

If the equation of the circle is $x^2 + y^2 = r^2$ and the equation of the tangent line is $y = mx + b$, show that:

(a) $r^2(1 + m^2) = b^2$

    [**Hint:** The quadratic equation $x^2 + (mx + b)^2 = r^2$ has exactly one solution.]

(b) The point of tangency is $\left(\dfrac{-r^2 m}{b}, \dfrac{r^2}{b} \right)$.

(c) The tangent line is perpendicular to the line containing the center of the circle and the point of tangency.

47. The Greek method for finding the equation of the tangent line to a circle used the fact that at any point on a circle the lines containing the center and the tangent line are perpendicular (see Problem 46). Use this method to find an equation of the tangent line to the circle $x^2 + y^2 = 9$ at the point $(1, 2\sqrt{2})$.

48. Use the Greek method described in Problem 47 to find an equation of the tangent line to the circle $x^2 + y^2 - 4x + 6y + 4 = 0$ at the point $(3, 2\sqrt{2} - 3)$.

49. Refer to Problem 46. The line $x - 2y + 4 = 0$ is tangent to a circle at $(0, 2)$. The line $y = 2x - 7$ is tangent to the same circle at $(3, -1)$. Find the center of the circle.

50. Find an equation of the line containing the centers of the two circles

$$x^2 + y^2 - 4x + 6y + 4 = 0 \quad \text{and}$$

$$x^2 + y^2 + 6x + 4y + 9 = 0$$

51. If a circle of radius 2 is made to roll along the x-axis, what is an equation for the path of the center of the circle?

Discussion and Writing

52. Which of the following equations might have the graph shown? (More than one answer is possible.)

(a) $(x - 2)^2 + (y + 3)^2 = 13$

(b) $(x - 2)^2 + (y - 2)^2 = 8$

(c) $(x - 2)^2 + (y - 3)^2 = 13$

(d) $(x + 2)^2 + (y - 2)^2 = 8$

(e) $x^2 + y^2 - 4x - 9y = 0$

(f) $x^2 + y^2 + 4x - 2y = 0$

(g) $x^2 + y^2 - 9x - 4y = 0$

(h) $x^2 + y^2 - 4x - 4y = 4$

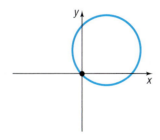

53. Which of the following equations might have the graph shown? (More than one answer is possible.)

(a) $(x - 2)^2 + y^2 = 3$ (e) $x^2 + y^2 + 10x + 16 = 0$

(b) $(x + 2)^2 + y^2 = 3$ (f) $x^2 + y^2 + 10x - 2y = 1$

(c) $x^2 + (y - 2)^2 = 3$ (g) $x^2 + y^2 + 9x + 10 = 0$

(d) $(x + 2)^2 + y^2 = 4$ (h) $x^2 + y^2 - 9x - 10 = 0$

54. Explain how the center and radius of a circle can be used to graph a circle by hand.

55. Explain how the center and radius of a circle can be used to establish an initial viewing window.

56. If the circumference of a circle is 6π, what is its radius?

Chapter Review

Things to Know

Formulas

Distance formula (p. 5) $d = \sqrt{(x_2 - x_1)^2 + (y_2 - y_1)^2}$

Midpoint formula (p. 7) $(x, y) = \left(\dfrac{x_1 + x_2}{2}, \dfrac{y_1 + y_2}{2} \right)$

Slope (p. 27) $m = \dfrac{y_2 - y_1}{x_2 - x_1}$, if $x_1 \neq x_2$; undefined if $x_1 = x_2$

Parallel lines (p. 37)   Equal slopes ($m_1 = m_2$) with different y-intercepts

Perpendicular lines (p. 38)   Product of slopes is -1 ($m_1 \cdot m_2 = -1$)

Equations

Vertical line (p. 31)	$x = a$
Point–slope form of the equation of a line (p. 32)	$y - y_1 = m(x - x_1)$; m is the slope of the line, (x_1, y_1) is a point on the line
Horizontal line (p. 33)	$y = b$
Slope–intercept form of the equation of a line (p. 34)	$y = mx + b$; m is the slope of the line, b is the y-intercept
General form of the equation of a line (p. 35)	$Ax + By = C$; A, B not both 0
Standard form of the equation of a circle (p. 45)	$(x - h)^2 + (y - k)^2 = r^2$; r is the radius of the circle, (h, k) is the center of the circle
Equation of the unit circle (p. 45)	$x^2 + y^2 = 1$
General form of the equation of a circle (p. 48)	$x^2 + y^2 + ax + by + c = 0$

Objectives

Section		You should be able to . . .	Review Exercises
1.1	1	Use the distance formula (p. 4)	1(a)–6(a), 51, 52(a), 54–56
	2	Use the midpoint formula (p. 7)	1(b)–6(b), 55
1.2	1	Graph equations by hand by plotting points (p. 11)	9–14
	2	Graph equations using a graphing utility (p. 13)	8
	3	Use a graphing utility to create tables (p. 14)	8
	4	Find intercepts from a graph (p. 15)	7
	5	Find intercepts from an equation (p. 16)	9–14, 47–50
	6	Use a graphing utility to approximate intercepts (p. 16)	8
	7	Test an equation for symmetry with respect to the x-axis, the y-axis, and the origin (p. 17)	15–22
	8	Know how to graph key equations (p. 19)	23, 24
1.3	1	Solve equations in one variable using a graphing utility (p. 24)	25–28
1.4	1	Calculate and interpret the slope of a line (p. 27)	1(c)–6(c); 1(d)–6(d), 53
	2	Graph lines given a point and the slope (p. 30)	57
	3	Find the equation of a vertical line (p. 31)	31
	4	Use the point–slope form of a line; identify horizontal lines (p. 32)	29, 30
	5	Find the equation of a line given two points (p. 33)	32–34
	6	Write the equation of a line in slope–intercept form (p. 34)	29–38
	7	Identify the slope and y-intercept of a line from its equation (p. 34)	39–42
	8	Graph lines written in general form using intercepts (p. 35)	9, 10
	9	Find equations of parallel lines (p. 36)	35, 36
	10	Find equations of perpendicular lines (p. 38)	37, 38
1.5	1	Write the standard form of the equation of a circle (p. 44)	43–46
	2	Graph a circle by hand and by using a graphing utility (p. 46)	47–50
	3	Work with the general form of the equation of a circle (p. 47)	47–50, 55

Review Exercises

In Problems 1–6, find the following for each pair of points:
(a) *The distance between the points.*
(b) *The midpoint of the line segment connecting the points.*
(c) *The slope of the line containing the points.*
(d) *Then interpret the slope found in part (c).*

1. $(0, 0)$; $(4, 2)$ **2.** $(0, 0)$; $(-4, 6)$ **3.** $(1, -1)$; $(-2, 3)$ **4.** $(-2, 2)$; $(1, 4)$ **5.** $(4, -4)$; $(4, 8)$ **6.** $(-3, 4)$; $(2, 4)$

7. List the intercepts of the following graph.

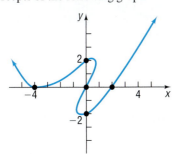

8. Graph $y = -x^2 + 15$ using a graphing utility. Create a table of values to determine a good initial viewing window. Use a graphing utility to approximate the intercepts.

In Problems 9–14, determine the intercepts and graph each equation by hand by plotting points. Verify your results using a graphing utility. Label the intercepts on the graph.

9. $2x - 3y = 6$ **10.** $x + 2y = 4$ **11.** $y = x^2 - 9$ **12.** $y = x^2 + 4$ **13.** $x^2 + 2y = 16$ **14.** $2x^2 - 4y = 24$

In Problems 15–22, test each equation for symmetry with respect to the x-axis, the y-axis, and the origin.

15. $2x = 3y^2$ 　　　　　　　　**16.** $y = 5x$ 　　　　　　　　　　**17.** $x^2 + 4y^2 = 16$ 　　　　　　**18.** $9x^2 - y^2 = 9$

19. $y = x^4 + 2x^2 + 1$ 　　　　**20.** $y = x^3 - x$ 　　　　　　　　**21.** $x^2 + x + y^2 + 2y = 0$ 　　　**22.** $x^2 + 4x + y^2 - 2y = 0$

23. Sketch a graph of $y = x^3$. 　　　　　　　　　　　　　　　　**24.** Sketch a graph of $y = \sqrt{x}$.

In Problems 25–28, use a graphing utility to approximate the solutions of each equation rounded to two decimal places. All solutions lie between −10 and 10.

25. $x^3 - 5x + 3 = 0$ 　　　**26.** $-x^3 + 3x + 1 = 0$ 　　　**27.** $x^4 - 3 = 2x + 1$ 　　　**28.** $-x^4 + 7 = x^2 - 2$

In Problems 29–38, find an equation of the line having the given characteristics. Express your answer using either the general form or the slope–intercept form of the equation of a line, whichever you prefer. Graph the line.

29. Slope $= -2$; containing the point $(3, -1)$ 　　　　　　　　**30.** Slope $= 0$; containing the point $(-5, 4)$

31. Slope undefined; containing the point $(-3, 4)$ 　　　　　　**32.** x-intercept $= 2$; containing the point $(4, -5)$

33. y-intercept $= -2$; containing the point $(5, -3)$ 　　　　　**34.** Containing the points $(3, -4)$ and $(2, 1)$

35. Parallel to the line $2x - 3y = -4$; containing the point $(-5, 3)$ 　**36.** Parallel to the line $x + y = 2$; containing the point $(1, -3)$

37. Perpendicular to the line $x + y = 2$; containing the point $(4, -3)$

38. Perpendicular to the line $3x - y = -4$; containing the point $(-2, 4)$

In Problems 39–42, find the slope and y-intercept of each line.

39. $4x + 6y = 36$ 　　　　　　**40.** $7x - 3y = 42$ 　　　　　　**41.** $\frac{1}{2}x + \frac{5}{2}y = 10$ 　　　　　**42.** $-\frac{2}{3}x + \frac{1}{2}y = 2$

In Problems 43–46, find the standard form of the equation of the circle whose center and radius are given.

43. $(h, k) = (-2, 3); r = 4$ 　　**44.** $(h, k) = (3, 4); r = 4$ 　　**45.** $(h, k) = (-1, -2); r = 1$ 　　**46.** $(h, k) = (2, -4); r = 3$

In Problems 47–50, find the center and radius of each circle. Graph each circle by hand. Determine the intercepts of the graph of each circle.

47. $x^2 + y^2 - 2x + 4y - 4 = 0$ 　**48.** $x^2 + y^2 + 4x - 4y - 1 = 0$ 　**49.** $3x^2 + 3y^2 - 6x + 12y = 0$ 　**50.** $2x^2 + 2y^2 - 4x = 0$

---

51. Show that the points $A = (3, 4), B = (1, 1),$ and $C = (-2, 3)$ are the vertices of an isosceles triangle.

52. Show that the points $A = (-2, 0), B = (-4, 4),$ and $C = (8, 5)$ are the vertices of a right triangle in two ways:
(a) By using the converse of the Pythagorean Theorem
(b) By using the slopes of the lines joining the vertices

53. Show that the points $A = (2, 5), B = (6, 1),$ and $C = (8, -1)$ lie on a straight line by using slopes.

54. Show that the points $A = (1, 5), B = (2, 4),$ and $C = (-3, 5)$ lie on a circle with center $(-1, 2)$. What is the radius of this circle?

55. The endpoints of the diameter of a circle are $(-3, 2)$ and $(5, -6)$. Find the center and radius of the circle. Write the general equation of this circle.

56. Find two numbers y such that the distance from $(-3, 2)$ to $(5, y)$ is 10.

57. Graph the line with slope $\frac{2}{3}$ containing the point $(1, 2)$.

58. Create four problems that you might be asked to do given the two points $(-3, 4)$ and $(6, 1)$. Each problem should involve a different concept. Be sure that your directions are clearly stated.

59. Describe each of the following graphs in the xy-plane. Give justification.
(a) $x = 0$
(b) $y = 0$
(c) $x + y = 0$
(d) $xy = 0$
(e) $x^2 + y^2 = 0$

Chapter Test

1. Suppose the points $(-2, -3)$ and $(4, 5)$ are the endpoints of a diameter of a circle.
   (a) Find the distance between the two points.
   (b) Find the midpoint of the line segment connecting the two points.
   (c) Find the standard equation of the circle containing the two points.
   (d) Graph the circle by hand.

In Problems 2 and 3, graph each equation by hand by plotting points. Determine the intercepts and label them on the graph.

2. $y = x^2 - x - 2$
3. $2x - 7y = 21$

4. Given the linear equation $10x - 6y = 20$, do the following:
   (a) Find the slope of the line.
   (b) Find the intercepts of the graph of the line.
   (c) Graph the equation using the intercepts.
   (d) Find the equation of the line that is perpendicular to the given line that contains the point $(-1, 5)$.
   (e) Find the equation of the line that is parallel to the given line that contains the point $(-1, 5)$.

In Problems 5–7, use a graphing utility to approximate the real solutions of each equation rounded to two decimal places. All solutions lie between -10 and 10.

5. $2x^3 - x^2 - 2x + 1 = 0$
6. $x^4 - 5x^2 - 8 = 0$
7. $-x^3 + 7x - 2 = x^2 + 3x - 3$

8. Test the equation $5x - 2y^2 = 7$ for symmetry with respect to the x-axis, the y-axis, and the origin.

Chapter Projects

1. **Home Mortgages** While you may not be in the market for a home right now, it is probably an event that will occur for you within the next few years. Formula (1) gives the monthly payment P required to pay off a loan L at an annual rate of interest r, expressed as a decimal, but usually given as a percent. The time t, measured in months, is the duration of the loan, so a 15-year mortgage requires $t = 12 \times 15 = 180$ monthly payments.

$$P = L\left[\frac{\dfrac{r}{12}}{1 - \left(1 + \dfrac{r}{12}\right)^{-t}}\right] \quad \textbf{(1)}$$

P = monthly payment
L = loan amount
r = annual rate of interest, expressed as a decimal
t = length of loan, in months

On July 8, 2003, average interest rates were:
(a) 5.52% on 30-year mortgages
(b) 4.85% on 15-year mortgages
   1. For each rate, calculate the monthly payment for a loan of $200,000.
   2. Then compute the total amount paid over the term of each loan.
   3. Calculate the interest paid on each loan.

On August 5, 2004, average interest rates were:
(a) 5.99% on 30-year mortgages
(b) 5.40% on 15-year mortgages
   4. For each rate, calculate the monthly payment for a loan of $200,000.
   5. Then compute the total amount paid over the term of each loan.
   6. Calculate the interest paid on each loan.
   7. Solve equation (1) for L.
   8. If you can afford to pay $1000 per month for a mortgage payment, calculate the amount you can borrow on August 5, 2004.
      (a) On a 30-year mortgage.
      (b) On a 15-year mortgage.
   9. Check with your local lending institution for current rates for 30-year and 15-year mortages. Calculate how much you can borrow with a $1000 per month payment.
   10. Repeat Problem 9 if you can afford a payment of $1300 per month.
   11. Do you think that the interest rate plays an important role in determining how much you can afford to pay for a house?
   12. Comment on the two types of mortgages: 30-year and 15-year. Which would you take? Why?

The following projects are available on the Instructor's Resource Center (IRC):

2. **Project at Motorola** *Mobile Phone Usage*
3. **Economics** *Isocost Lines*

Functions and Their Graphs

2

A LOOK BACK Up to now, our discussion has focused on equations. We have solved equations containing one variable and developed techniques for graphing equations containing two variables.

A LOOK AHEAD In this chapter, we look at a special type of equation involving two variables called a *function*. This chapter deals with what a function is, how to graph functions, properties of functions, and how functions are used in applications. The word function apparently was introduced by René Descartes in 1637. For him, a function simply meant any positive integral power of a variable *x*. Gottfried Wilhelm Leibniz (1646–1716), who always emphasized the geometric side of mathematics, used the word function to denote any quantity associated with a curve, such as the coordinates of a point on the curve. Leonhard Euler (1707–1783) employed the word to mean any equation or formula involving variables and constants. His idea of a function is similar to the one most often seen in courses that precede calculus. Later, the use of functions in investigating heat flow equations led to a very broad definition, due to Lejeune Dirichlet (1805–1859), which describes a function as a rule or correspondence between two sets. It is his definition that we use here.

Portable Phones Power Up

Latest Census Bureau Data Show Rise in Cell Phones

Jan. 24—Cell phone ownership jumped from just over 5.2 million in 1990 to nearly 110 million in 2000—a more than twenty-fold increase in just 10 years, new figures show. Driving the phenomenal growth are several factors—one of which is the decreasing costs. The data note that the average monthly cell phone bill has been almost cut in half—from $81 to just over $45—over the past decade. Survey data, compiled by the Cellular Telecommunications and Internet Association (CTIA) in Washington, D.C., and published in the U.S. Census Bureau's *Statistical Abstract of the United States: 2001* report, point to the rapid adoption for cell phone technology as a key reason behind the growth.

"The cell phone industry has shown remarkable growth over the decade," says Glenn King, chief of the statistical compendia branch of the Commerce Department that produces the abstract. Noting that there were only 55 million users in 1997, "It's doubled in the last three years alone," he says. With cost of service coming down, cell phone service becomes accessible to everyone," says Charles Golvin, a senior analyst with Forrester Research. "People can be in touch anytime and anywhere they want to."

By Paul Eng, abcNEWS.com

—See Chapter Project 1.

OUTLINE

2.1 Functions

PREPARING FOR THIS SECTION *Before getting started, review the following:*
- Intervals (Appendix, Section A.8, pp. 729–730)
- Evaluating Algebraic Expressions, Domain of a Variable (Appendix, Section A.1, pp. 661–662)
- Solving Inequalities (Appendix, Section A.8, pp. 732–734)

✎ Now work the 'Are You Prepared?' problems on page 68.

OBJECTIVES 1 Determine Whether a Relation Represents a Function
2 Find the Value of a Function
3 Find the Domain of a Function
4 Form the Sum, Difference, Product, and Quotient of Two Functions

---

1 **Determine Whether a Relation Represents a Function**

We often see situations where one variable is somehow linked to the value of another variable. For example, an individual's level of education is linked to annual income. Engine size is linked to gas mileage. When the value of one variable is related to the value of a second variable, we have a *relation*. A **relation** is a correspondence between two sets. If x and y are two elements in these sets and if a relation exists between x and y, then we say that x **corresponds** to y or that y **depends on** x, and we write $x \rightarrow y$.

We have a number of ways to express relations between two sets. For example, the equation $y = 3x - 1$ shows a relation between x and y. It says that if we take some number x multiply it by 3 and then subtract 1, we obtain the corresponding value of y. In this sense, x serves as the **input** to the relation and y is the **output** of the relation. We can also express this relation as a graph as shown in Figure 1.

Not only can relations be expressed through an equation or graph, but we can also express relations through a technique called *mapping*. A **map** illustrates a relation by using a set of inputs and drawing arrows to the corresponding element in the set of outputs. Ordered pairs can be used to represent $x \rightarrow y$ as (x, y). We illustrate these two concepts in Example 1.

Figure 1

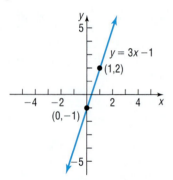

| **EXAMPLE 1** | **Maps and Ordered Pairs as Relations** |

Figure 2 shows a relation between states and the number of representatives each has in the House of Representatives. The relation might be named "number of representatives."

Figure 2

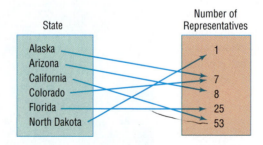

In this relation, Alaska corresponds to 7, Arizona corresponds to 8, and so on. Using ordered pairs, this relation would be expressed as

{(Alaska, 7), (Arizona, 8), (California, 53), (Colorado, 7), (Florida, 25), (North Dakota, 1)} ◀

We now present what is one of the most important concepts in algebra—the *function*. A function is a special type of relation. To understand the idea behind a function, let's revisit the relation presented in Example 1. If we were to ask "How many representatives does Alaska have" you would respond "7." In other words, each input "state" corresponds to a single output "number of representatives."

Let's consider a second relation where we have a correspondence between four people and their phone numbers. See Figure 3. Notice that Maureen has two telephone numbers; therefore, if asked, "What is Maureen's phone number?", you cannot assign a single number to her.

Figure 3

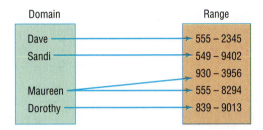

Let's look at one more relation. Figure 4 is a relation that shows a correspondence between "animals" and "life expectancy." If asked to determine the average life expectancy of a dog, we would all respond "11 years." If asked to determine the average life expectancy of a rabbit, we would all respond "7 years."

Figure 4

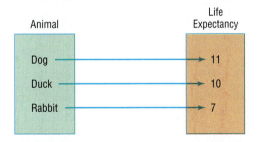

Notice that the relations presented in Figures 2 and 4 have something in common. What is it? The common link between these two relations is that each input corresponds to only one output. This leads to the definition of a *function*.

> Let X and Y be two nonempty sets.[*] A **function** from X into Y is a relation that associates with each element of X exactly one element of Y.

---

[*]The sets X and Y will usually be sets of real numbers, in which case a (real) function results. The two sets can also be sets of complex numbers, and then we have defined a complex function. In the broad definition (due to Lejeune Dirichlet), X and Y can be any two sets.

The set X is called the **domain** of the function. For each element x in X, the corresponding element y in Y is called the **value** of the function at x, or the **image** of x. The set of all images of the elements in the domain is called the **range** of the function. See Figure 5.

Figure 5

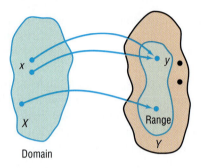

Domain

Since there may be some elements in Y that are not the image of some x in X, it follows that the range of a function may be a subset of Y, as shown in Figure 5.

Not all relations between two sets are functions. The next example shows how to determine whether a relation is a function or not.

| EXAMPLE 2 | **Determining Whether a Relation Represents a Function** |

Determine whether the following relations represent functions. If the relation is a function, then state its domain and range.

(a) See Figure 6. For this relation, the domain represents the level of education and the range represents the unemployment rate.

Figure 6
[SOURCE: *Statistical Abstract of the United States, 2003*]

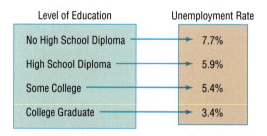

(b) See Figure 7. For this relation, the domain represents the number of calories in a sandwich from a fast-food restaurant and the range represents the fat content (in grams).

Figure 7
[SOURCE: Each company's Web site]

(c) See Figure 8. For this relation, the domain represents the weight of pear-cut diamonds and the range represents their price.

Figure 8
[SOURCE: diamonds.com]

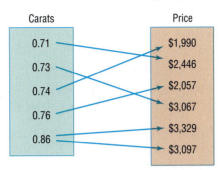

Solution

(a) The relation is a function because each element in the domain corresponds to exactly one element in the range. The domain of the relation is {No High School Diploma, High School Diploma, Some College, College Graduate} and the range of the relation is {7.7%, 5.9%, 5.4%, 3.4%}.

(b) The relation is a function because each element in the domain corresponds to exactly one element in the range. Notice that it is okay for more than one element in the domain to correspond to the same element in the range (McDonald's and Burger King's chicken sandwich both have 23 grams of fat). The domain of the relation is {410, 580, 540, 750, 600, 430}. The range of the relation is {19, 29, 24, 33, 23}.

(c) The relation is not a function because each element in the domain does not correspond to exactly one element in the range. If a 0.86 carat diamond is chosen from the domain, a single price cannot be assigned to it. ◀

NOW WORK PROBLEM **15.**

The idea behind a function is its predictability. If the input is known, we can use the function to determine the output. With "nonfunctions," we don't have this predictability. Look back at Figure 7. The inputs are {410, 580, 540, 750, 600, 430}. The correspondence is "number of fat grams" and the outputs are {19, 29, 24, 33, 23}. If asked "How many fat grams in a 410-calorie sandwich?" we can use the correspondence to answer "19." Now consider Figure 8. If asked "What is the price of a 0.86 carat diamond?" we could not give a single response because two outputs result from the single input "0.86." For this reason, the relation in Figure 8 is not a function.

We may also think of a function as a set of ordered pairs (x, y) in which no two ordered pairs have the same first element, but different second elements. The set of all first elements x is the domain of the function, and the set of all second elements y is its range. Each element x in the domain corresponds to exactly one element y in the range.

In Words

For a function, no input has more than one output.

In Words

For a function, the domain is the set of inputs, and the range is the set of outputs.

EXAMPLE 3   **Determining Whether a Relation Represents a Function**

Determine whether each relation represents a function. If it is a function, state the domain and range.

(a) {(1, 4), (2, 5), (3, 6), (4, 7)}

(b) {(1, 4), (2, 4), (3, 5), (6, 10)}

(c) {(−3, 9), (−2, 4), (0, 0), (1, 1), (−3, 8)}

Solution

(a) This relation is a function because there are no ordered pairs with the same first element and different second elements. The domain of this function is $\{1, 2, 3, 4\}$, and its range is $\{4, 5, 6, 7\}$.

(b) This relation is a function because there are no ordered pairs with the same first element and different second elements. The domain of this function is $\{1, 2, 3, 6\}$, and its range is $\{4, 5, 10\}$.

(c) This relation is not a function because there are two ordered pairs, $(-3, 9)$ and $(-3, 8)$, that have the same first element, but different second elements. ◀

In Example 3(b), notice that 1 and 2 in the domain each have the same image in the range. This does not violate the definition of a function; two different first elements can have the same second element. A violation of the definition occurs when two ordered pairs have the same first element and different second elements, as in Example 3(c).

NOW WORK PROBLEM 19.

Up to now we have shown how to identify when a relation is a function for relations defined by mappings [Example 2] and ordered pairs [Example 3]. We know that relations can also be expressed as equations. We discuss next the circumstances under which equations are functions.

To determine whether an equation, where y depends on x, is a function, it is often easiest to solve the equation for y. If any value of x in the domain corresponds to more than one y, the equation does not define a function; otherwise, it does define a function.

EXAMPLE 4   **Determining Whether an Equation Is a Function**

Determine if the equation $y = 2x - 5$ defines y as a function of x.

Solution

The equation tells us to take an input x, multiply it by 2, and then subtract 5. For any input x, these operations yield only one output y. For example, if $x = 1$, then $y = 2(1) - 5 = -3$. If $x = 3$, then $y = 2(3) - 5 = 1$. For this reason, the equation is a function. ◀

EXAMPLE 5   **Determining Whether an Equation Is a Function**

Determine if the equation $x^2 + y^2 = 1$ defines y as a function of x.

Solution

To determine whether the equation $x^2 + y^2 = 1$, which defines the unit circle, is a function, we need to solve the equation for y.

$$x^2 + y^2 = 1$$

$$y^2 = 1 - x^2$$

$$y = \pm\sqrt{1 - x^2} \qquad \text{Square Root Method}$$

For values of x between -1 and 1, two values of y result. For example, if $x = 0$, then $y = \pm 1$, so two different outputs result from the same input. This means that the equation $x^2 + y^2 = 1$ does not define a function. ◀

NOW WORK PROBLEM 33.

2 Find the Value of a Function

Functions are often denoted by letters such as f, F, g, G, and others. If f is a function, then for each number x in its domain the corresponding image in the range is designated by the symbol $f(x)$, read as "f of x" or as "f at x." We refer to $f(x)$ as the **value of f at the number x**; $f(x)$ is the number that results when x is given and the function f is applied; $f(x)$ does *not* mean "f times x." For example, the function given in Example 4 may be written as $y = f(x) = 2x - 5$. Then $f\left(\dfrac{3}{2}\right) = -2$.

Figure 9 illustrates some other functions. Notice that, in every function illustrated, for each x in the domain there is one value in the range.

Figure 9

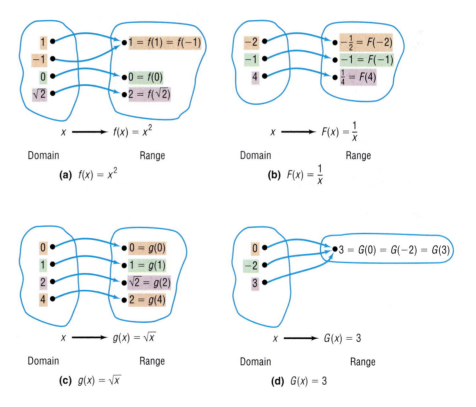

(a) $f(x) = x^2$

(b) $F(x) = \dfrac{1}{x}$

(c) $g(x) = \sqrt{x}$

(d) $G(x) = 3$

Figure 10

Input x

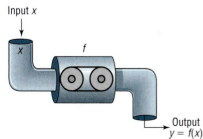

Output $y = f(x)$

Sometimes it is helpful to think of a function f as a machine that receives as input a number from the domain, manipulates it, and outputs the value. See Figure 10.

The restrictions on this input/output machine are as follows:

1. It only accepts numbers from the domain of the function.

2. For each input, there is exactly one output (which may be repeated for different inputs).

For a function $y = f(x)$, the variable x is called the **independent variable**, because it can be assigned any of the permissible numbers from the domain. The variable y is called the **dependent variable**, because its value depends on x.

Any symbol can be used to represent the independent and dependent variables. For example, if f is the *cube function*, then f can be given by $f(x) = x^3$ or $f(t) = t^3$ or $f(z) = z^3$. All three functions are the same. Each tells us to cube the independent variable. In practice, the symbols used for the independent and dependent variables are based on common usage, such as using C for cost in business.

The independent variable is also called the **argument** of the function. Thinking of the independent variable as an argument can sometimes make it easier to find the value of a function. For example, if f is the function defined by $f(x) = x^3$, then f tells us to cube the argument. Thus, $f(2)$ means to cube 2, $f(a)$ means to cube the number a, and $f(x + h)$ means to cube the quantity $x + h$.

EXAMPLE 6　**Finding Values of a Function**

For the function f defined by $f(x) = 2x^2 - 3x$, evaluate

(a) $f(3)$ 　　(b) $f(x) + f(3)$

(c) $f(-x)$ 　　(d) $-f(x)$

(e) $f(x + 3)$ 　(f) $\dfrac{f(x + h) - f(x)}{h}$, $h \neq 0$

Solution　(a) We substitute 3 for x in the equation for f to get

$$f(3) = 2(3)^2 - 3(3) = 18 - 9 = 9$$

(b) $f(x) + f(3) = (2x^2 - 3x) + (9) = 2x^2 - 3x + 9$

(c) We substitute $-x$ for x in the equation for f.

$$f(-x) = 2(-x)^2 - 3(-x) = 2x^2 + 3x$$

(d) $-f(x) = -(2x^2 - 3x) = -2x^2 + 3x$

(e) $f(x + 3) = 2(x + 3)^2 - 3(x + 3)$　　　　Notice the use of parentheses here.

$$= 2(x^2 + 6x + 9) - 3x - 9$$

$$= 2x^2 + 12x + 18 - 3x - 9$$

$$= 2x^2 + 9x + 9$$

(f) $\dfrac{f(x + h) - f(x)}{h} = \dfrac{[2(x + h)^2 - 3(x + h)] - [2x^2 - 3x]}{h}$

↑

$f(x + h) = 2(x + h)^2 - 3(x + h)$

$\quad = \dfrac{2(x^2 + 2xh + h^2) - 3x - 3h - 2x^2 + 3x}{h}$ Simplify.

$\quad = \dfrac{2x^2 + 4xh + 2h^2 - 3h - 2x^2}{h}$ Simplify.

$\quad = \dfrac{4xh + 2h^2 - 3h}{h}$ Simplify.

$\quad = \dfrac{h(4x + 2h - 3)}{h}$ Factor out h.

$\quad = 4x + 2h - 3$ Cancel h.   ◀

Notice in this example that $f(x + 3) \neq f(x) + f(3)$ and $f(-x) \neq -f(x)$. The expression in part (f) is called the **difference quotient** of f, an important expression in calculus.

NOW WORK PROBLEMS 39 AND 73.

Most calculators have special keys that enable you to find the value of certain commonly used functions. For example, you should be able to find the square function $f(x) = x^2$, the square root function $f(x) = \sqrt{x}$, the reciprocal function $f(x) = \dfrac{1}{x} = x^{-1}$, and many others that will be discussed later in this book (such as $\ln x$ and $\log x$). Verify the results of Example 7, which follows, on your calculator.

EXAMPLE 7   **Finding Values of a Function on a Calculator**

(a) $f(x) = x^2$; $f(1.234) = 1.522756$

(b) $F(x) = \dfrac{1}{x}$; $F(1.234) = 0.8103727715$

(c) $g(x) = \sqrt{x}$; $g(1.234) = 1.110855526$ ◀

NOTE   Graphing calculators can be used to evaluate any function that you wish. Figure 11 shows the result obtained in Example 6(a) on a TI-84 Plus graphing calculator with the function to be evaluated, $f(x) = 2x^2 - 3x$, in Y_1.

Figure 11

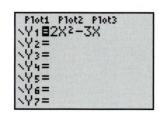

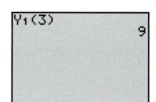

■

Implicit Form of a Function

In general, when a function f is defined by an equation in x and y, we say that the function f is given **implicitly**. If it is possible to solve the equation for y in terms of x, then we write $y = f(x)$ and say that the function is given **explicitly**. For example,

Implicit Form	**Explicit Form**
$3x + y = 5$	$y = f(x) = -3x + 5$
$x^2 - y = 6$	$y = f(x) = x^2 - 6$
$xy = 4$	$y = f(x) = \dfrac{4}{x}$

We list next a summary of some important facts to remember about a function f.

Summary

Important Facts About Functions

(a) For each x in the domain of f, there is exactly one image $f(x)$ in the range; however, an element in the range can result from more than one x in the domain.
(b) f is the symbol that we use to denote the function. It is symbolic of the equation that we use to get from an x in the domain to $f(x)$ in the range.
(c) If $y = f(x)$, then x is called the independent variable or argument of f, and y is called the dependent variable or the value of f at x.

 3 Find the Domain of a Function

Often the domain of a function f is not specified; instead, only the equation defining the function is given. In such cases, we agree that the **domain of f** is the largest set of real numbers for which the value $f(x)$ is a real number. The domain of a function f is the same as the domain of the variable x in the expression $f(x)$.

EXAMPLE 8 **Finding the Domain of a Function**

Find the domain of each of the following functions:

(a) $f(x) = x^2 + 5x$ (b) $g(x) = \dfrac{3x}{x^2 - 4}$ (c) $h(t) = \sqrt{4 - 3t}$

Solution

(a) The function tells us to square a number and then add five times the number. Since these operations can be performed on any real number, we conclude that the domain of f is the set of all real numbers.

(b) The function g tells us to divide $3x$ by $x^2 - 4$. Since division by 0 is not defined, the denominator $x^2 - 4$ can never be 0, so x can never equal -2 or 2. The domain of the function g is $\{x \mid x \neq -2, x \neq 2\}$.

(c) The function h tells us to take the square root of $4 - 3t$. But only nonnegative numbers have real square roots, so the expression under the square root must be nonnegative (greater than or equal to zero). This requires that

$$4 - 3t \geq 0$$
$$-3t \geq -4$$
$$t \leq \frac{4}{3}$$

The domain of h is $\left\{ t \middle| t \leq \dfrac{4}{3} \right\}$ or the interval $\left(-\infty, \dfrac{4}{3} \right]$.   ◄

  NOW WORK PROBLEM 51.

If x is in the domain of a function f, we shall say that **f is defined at x**, or **$f(x)$ exists**. If x is not in the domain of f, we say that **f is not defined at x**, or **$f(x)$ does not exist**. For example, if $f(x) = \dfrac{x}{x^2 - 1}$, then $f(0)$ exists, but $f(1)$ and $f(-1)$ do not exist. (Do you see why?)

We have not said much about finding the range of a function. The reason is that when a function is defined by an equation it is often difficult to find the range.* Therefore, we shall usually be content to find just the domain of a function when only the rule for the function is given. We shall express the domain of a function using inequalities, interval notation, set notation, or words, whichever is most convenient.

When we use functions in applications, the domain may be restricted by physical or geometric considerations. For example, the domain of the function f defined by $f(x) = x^2$ is the set of all real numbers. However, if f is used to obtain the area of a square when the length x of a side is known, then we must restrict the domain of f to the positive real numbers, since the length of a side can never be 0 or negative.

EXAMPLE 9   **Finding the Domain in an Application**

Express the area of a circle as a function of its radius. Find the domain.

Figure 12

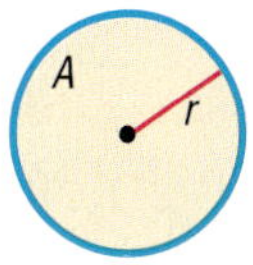

Solution   See Figure 12. We know that the formula for the area A of a circle of radius r is $A = \pi r^2$. If we use r to represent the independent variable and A to represent the dependent variable, the function expressing this relationship is

$$A(r) = \pi r^2$$

In this setting, the domain is $\{r \mid r > 0\}$. (Do you see why?)   ◄

Observe in the solution to Example 9 that we used the symbol A in two ways: It is used to name the function, and it is used to symbolize the dependent variable. This double use is common in applications and should not cause any difficulty.

  NOW WORK PROBLEM 85.

4   **Form the Sum, Difference, Product, and Quotient of Two Functions**

Next we introduce some operations on functions. We shall see that functions, like numbers, can be added, subtracted, multiplied, and divided. For example, if $f(x) = x^2 + 9$ and $g(x) = 3x + 5$, then

$$f(x) + g(x) = (x^2 + 9) + (3x + 5) = x^2 + 3x + 14$$

The new function $y = x^2 + 3x + 14$ is called the *sum function $f + g$*. Similarly,

$$f(x) \cdot g(x) = (x^2 + 9)(3x + 5) = 3x^3 + 5x^2 + 27x + 45$$

The new function $y = 3x^3 + 5x^2 + 27x + 45$ is called the *product function $f \cdot g$*.

---

*In Section 4.2 we discuss a way to find the range for a special class of functions.

The general definitions are given next.

If f and g are functions:

The **sum** $f + g$ is the function defined by

$$(f + g)(x) = f(x) + g(x)$$

The domain of $f + g$ consists of the numbers x that are in the domains of both f and g.

The **difference** $f - g$ is the function defined by

$$(f - g)(x) = f(x) - g(x)$$

The domain of $f - g$ consists of the numbers x that are in the domains of both f and g.

The **product** $f \cdot g$ is the function defined by

$$(f \cdot g)(x) = f(x) \cdot g(x)$$

The domain of $f \cdot g$ consists of the numbers x that are in the domains of both f and g.

The **quotient** $\dfrac{f}{g}$ is the function defined by

$$\left(\frac{f}{g}\right)(x) = \frac{f(x)}{g(x)} \qquad g(x) \neq 0$$

The domain of $\dfrac{f}{g}$ consists of the numbers x for which $g(x) \neq 0$ that are in the domains of both f and g.

EXAMPLE 10 | **Operations on Functions**

Let f and g be two functions defined as

$$f(x) = \frac{1}{x + 2} \quad \text{and} \quad g(x) = \frac{x}{x - 1}$$

Find the following, and determine the domain in each case.

(a) $(f + g)(x)$ (b) $(f - g)(x)$ (c) $(f \cdot g)(x)$ (d) $\left(\dfrac{f}{g}\right)(x)$

Solution   The domain of f is $\{x \mid x \neq -2\}$ and the domain of g is $\{x \mid x \neq 1\}$.

(a) $(f + g)(x) = f(x) + g(x) = \dfrac{1}{x + 2} + \dfrac{x}{x - 1} = \dfrac{x - 1}{(x + 2)(x - 1)} + \dfrac{x(x + 2)}{(x + 2)(x - 1)} = \dfrac{x^2 + 3x - 1}{(x + 2)(x - 1)}$

The domain of $f + g$ consists of those numbers x that are in the domains of both f and g. Therefore, the domain of $f + g$ is $\{x | x \neq -2, x \neq 1\}$.

(b) $(f - g)(x) = f(x) - g(x) = \dfrac{1}{x + 2} - \dfrac{x}{x - 1} = \dfrac{x - 1}{(x + 2)(x - 1)} - \dfrac{x(x + 2)}{(x + 2)(x - 1)} = \dfrac{-(x^2 + x + 1)}{(x + 2)(x - 1)}$

The domain of $f - g$ consists of those numbers x that are in the domains of both f and g. Therefore, the domain of $f - g$ is $\{x | x \neq -2, x \neq 1\}$.

(c) $(f \cdot g)(x) = f(x) \cdot g(x) = \dfrac{1}{x + 2} \cdot \dfrac{x}{x - 1} = \dfrac{x}{(x + 2)(x - 1)}$

The domain of $f \cdot g$ consists of those numbers x that are in the domains of both f and g. Therefore, the domain of $f \cdot g$ is $\{x | x \neq -2, x \neq 1\}$.

(d) $\left(\dfrac{f}{g}\right)(x) = \dfrac{f(x)}{g(x)} = \dfrac{\dfrac{1}{x + 2}}{\dfrac{x}{x - 1}} = \dfrac{1}{x + 2} \cdot \dfrac{x - 1}{x} = \dfrac{x - 1}{x(x + 2)}$

The domain of $\dfrac{f}{g}$ consists of the numbers x for which $g(x) \neq 0$ that are in the domains of both f and g. Since $g(x) = 0$ when $x = 0$, we exclude 0 as well as -2 and 1. The domain of $\dfrac{f}{g}$ is $\{x | x \neq -2, x \neq 0, x \neq 1\}$. ◀

NOW WORK PROBLEM **61.**

 In calculus, it is sometimes helpful to view a complicated function as the sum, difference, product, or quotient of simpler functions. For example,

$F(x) = x^2 + \sqrt{x}$ is the sum of $f(x) = x^2$ and $g(x) = \sqrt{x}$.

$H(x) = \dfrac{x^2 - 1}{x^2 + 1}$ is the quotient of $f(x) = x^2 - 1$ and $g(x) = x^2 + 1$.

Summary

We list here some of the important vocabulary introduced in this section, with a brief description of each term.

Function	A relation between two sets of real numbers so that each number x in the first set, the domain, has corresponding to it exactly one number y in the second set. A set of ordered pairs (x, y) or $(x, f(x))$ in which no first element is paired with two different second elements. The range is the set of y values of the function for the x values in the domain. A function f may be defined implicitly by an equation involving x and y or explicitly by writing $y = f(x)$.
Unspecified domain	If a function f is defined by an equation and no domain is specified, then the domain will be taken to be the largest set of real numbers for which the equation defines a real number.
Function notation	$y = f(x)$ f is a symbol for the function. x is the independent variable or argument. y is the dependent variable. $f(x)$ is the value of the function at x, or the image of x.

2.1 Assess Your Understanding

'Are You Prepared?'

Answers are given at the end of these exercises. If you get a wrong answer, read the pages listed in red.

1. The inequality $-1 < x < 3$ can be written in interval notation as _____. (pp. 729–730)

2. If $x = -2$, the value of the expression $3x^2 - 5x + \dfrac{1}{x}$ is _____. (p. 661)

3. The domain of the variable in the expression $\dfrac{x - 3}{x + 4}$ is _____. (pp. 661–662)

4. Solve the inequality: $3 - 2x > 5$. Graph the solution set. (pp. 732–733)

Concepts and Vocabulary

5. If f is a function defined by the equation $y = f(x)$, then x is called the _____ variable and y is the _____ variable.

6. The set of all images of the elements in the domain of a function is called the _____.

7. If the domain of f is all real numbers in the interval $[0, 7]$ and the domain of g is all real numbers in the interval $[-2, 5]$, the domain of $f + g$ is all real numbers in the interval _____.

8. The domain of $\dfrac{f}{g}$ consists of numbers x for which $g(x)$ _____ 0 that are in the domains of both _____ and _____.

9. If $f(x) = x + 1$ and $g(x) = x^3$, then _____ $= x^3 - (x + 1)$.

10. *True or False:* Every relation is a function.

11. *True or False:* The domain of $(f \cdot g)(x)$ consists of the numbers x that are in the domains of both f and g.

12. *True or False:* The independent variable is sometimes referred to as the argument of the function.

13. *True or False:* If no domain is specified for a function f, then the domain of f is taken to be the set of real numbers.

14. *True or False:* The domain of the function $f(x) = \dfrac{x^2 - 4}{x}$ is $\{x \mid x \neq \pm 2\}$.

Skill Building

In Problems 15–26, determine whether each relation represents a function. For each function, state the domain and range.

15.

16.
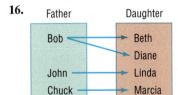

17.
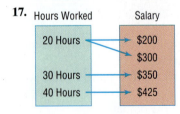

18.
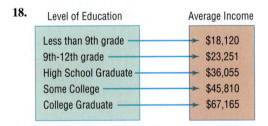

19. $\{(2, 6), (-3, 6), (4, 9), (2, 10)\}$

21. $\{(1, 3), (2, 3), (3, 3), (4, 3)\}$

23. $\{(-2, 4), (-2, 6), (0, 3), (3, 7)\}$

25. $\{(-2, 4), (-1, 1), (0, 0), (1, 1)\}$

20. $\{(-2, 5), (-1, 3), (3, 7), (4, 12)\}$

22. $\{(0, -2), (1, 3), (2, 3), (3, 7)\}$

24. $\{(-4, 4), (-3, 3), (-2, 2), (-1, 1), (-4, 0)\}$

26. $\{(-2, 16), (-1, 4), (0, 3), (1, 4)\}$

In Problems 27–38, determine whether the equation defines y as a function of x.

27. $y = x^2$

28. $y = x^3$

29. $y = \dfrac{1}{x}$

30. $y = |x|$

31. $y^2 = 4 - x^2$

32. $y = \pm\sqrt{1 - 2x}$

33. $x = y^2$

34. $x + y^2 = 1$

35. $y = 2x^2 - 3x + 4$ **36.** $y = \dfrac{3x - 1}{x + 2}$ **37.** $2x^2 + 3y^2 = 1$ **38.** $x^2 - 4y^2 = 1$

In Problems 39–46, find the following values for each function:

(a) $f(0)$ (b) $f(1)$ (c) $f(-1)$ (d) $f(-x)$ (e) $-f(x)$ (f) $f(x + 1)$ (g) $f(2x)$ (h) $f(x + h)$

39. $f(x) = 3x^2 + 2x - 4$ **40.** $f(x) = -2x^2 + x - 1$ **41.** $f(x) = \dfrac{x}{x^2 + 1}$ **42.** $f(x) = \dfrac{x^2 - 1}{x + 4}$

43. $f(x) = |x| + 4$ **44.** $f(x) = \sqrt{x^2 + x}$ **45.** $f(x) = \dfrac{2x + 1}{3x - 5}$ **46.** $f(x) = 1 - \dfrac{1}{(x + 2)^2}$

In Problems 47–60, find the domain of each function.

47. $f(x) = -5x + 4$ **48.** $f(x) = x^2 + 2$ **49.** $f(x) = \dfrac{x}{x^2 + 1}$ **50.** $f(x) = \dfrac{x^2}{x^2 + 1}$

51. $g(x) = \dfrac{x}{x^2 - 16}$ **52.** $h(x) = \dfrac{2x}{x^2 - 4}$ **53.** $F(x) = \dfrac{x - 2}{x^3 + x}$ **54.** $G(x) = \dfrac{x + 4}{x^3 - 4x}$

55. $h(x) = \sqrt{3x - 12}$ **56.** $G(x) = \sqrt{1 - x}$ **57.** $f(x) = \dfrac{4}{\sqrt{x - 9}}$ **58.** $f(x) = \dfrac{x}{\sqrt{x - 4}}$

59. $p(x) = \sqrt{\dfrac{2}{x - 1}}$ **60.** $q(x) = \sqrt{-x - 2}$

In Problems 61–70, for the given functions f and g, find the following functions and state the domain of each.

(a) $f + g$ (b) $f - g$ (c) $f \cdot g$ (d) $\dfrac{f}{g}$

61. $f(x) = 3x + 4;\ \ g(x) = 2x - 3$ **62.** $f(x) = 2x + 1;\ \ g(x) = 3x - 2$

63. $f(x) = x - 1;\ \ g(x) = 2x^2$ **64.** $f(x) = 2x^2 + 3;\ \ g(x) = 4x^3 + 1$

65. $f(x) = \sqrt{x};\ \ g(x) = 3x - 5$ **66.** $f(x) = |x|;\ \ g(x) = x$

67. $f(x) = 1 + \dfrac{1}{x};\ \ g(x) = \dfrac{1}{x}$ **68.** $f(x) = \sqrt{x - 2};\ \ g(x) = \sqrt{4 - x}$

69. $f(x) = \dfrac{2x + 3}{3x - 2};\ \ g(x) = \dfrac{4x}{3x - 2}$ **70.** $f(x) = \sqrt{x + 1};\ \ g(x) = \dfrac{2}{x}$

71. Given $f(x) = 3x + 1$ and $(f + g)(x) = 6 - \dfrac{1}{2}x$, find the function g.

72. Given $f(x) = \dfrac{1}{x}$ and $\left(\dfrac{f}{g}\right)(x) = \dfrac{x + 1}{x^2 - x}$, find the function g.

In Problems 73–78, find the difference quotient of f, that is, find $\dfrac{f(x + h) - f(x)}{h}$, $h \neq 0$, for each function. Be sure to simplify.

73. $f(x) = 4x + 3$ **74.** $f(x) = -3x + 1$ **75.** $f(x) = x^2 - x + 4$

76. $f(x) = x^2 + 5x - 1$ **77.** $f(x) = x^3 - 2$ **78.** $f(x) = \dfrac{1}{x + 3}$

Applications and Extensions

79. If $f(x) = 2x^3 + Ax^2 + 4x - 5$ and $f(2) = 5$, what is the value of A?

80. If $f(x) = 3x^2 - Bx + 4$ and $f(-1) = 12$, what is the value of B?

81. If $f(x) = \dfrac{3x + 8}{2x - A}$ and $f(0) = 2$, what is the value of A?

82. If $f(x) = \dfrac{2x - B}{3x + 4}$ and $f(2) = \dfrac{1}{2}$, what is the value of B?

83. If $f(x) = \dfrac{2x - A}{x - 3}$ and $f(4) = 0$, what is the value of A? Where is f not defined?

84. If $f(x) = \dfrac{x - B}{x - A}$, $f(2) = 0$, and $f(1)$ is undefined, what are the values of A and B?

85. Geometry Express the area A of a rectangle as a function of the length x if the length of the rectangle is twice its width.

86. Geometry Express the area A of an isosceles right triangle as a function of the length x of one of the two equal sides.

87. Constructing Functions Express the gross salary G of a person who earns \$10 per hour as a function of the number x of hours worked.

88. Constructing Functions   Tiffany, a commissioned salesperson, earns $100 base pay plus $10 per item sold. Express her gross salary G as a function of the number x of items sold.

89. Effect of Gravity on Earth   If a rock falls from a height of 20 meters on Earth, the height H (in meters) after x seconds is approximately

$$H(x) = 20 - 4.9x^2$$

(a) What is the height of the rock when $x = 1$ second? $x = 1.1$ seconds? $x = 1.2$ seconds? $x = 1.3$ seconds?

(b) When is the height of the rock 15 meters? When is it 10 meters? When is it 5 meters?

(c) When does the rock strike the ground?

90. Effect of Gravity on Jupiter   If a rock falls from a height of 20 meters on the planet Jupiter, its height H (in meters) after x seconds is approximately

$$H(x) = 20 - 13x^2$$

(a) What is the height of the rock when $x = 1$ second? $x = 1.1$ seconds? $x = 1.2$ seconds?

(b) When is the height of the rock 15 meters? When is it 10 meters? When is it 5 meters?

(c) When does the rock strike the ground?

91. Cost of Trans-Atlantic Travel   A Boeing 747 crosses the Atlantic Ocean (3000 miles) with an airspeed of 500 miles per hour. The cost C (in dollars) per passenger is given by

$$C(x) = 100 + \frac{x}{10} + \frac{36,000}{x}$$

where x is the ground speed (airspeed ± wind).

(a) What is the cost per passenger for quiescent (no wind) conditions?

(b) What is the cost per passenger with a head wind of 50 miles per hour?

(c) What is the cost per passenger with a tail wind of 100 miles per hour?

(d) What is the cost per passenger with a head wind of 100 miles per hour?

92. Cross-sectional Area   The cross-sectional area of a beam cut from a log with radius 1 foot is given by the function $A(x) = 4x\sqrt{1 - x^2}$, where x represents the length, in feet, of half the base of the beam. See the figure. Determine the cross-sectional area of the beam if the length of half the base of the beam is as follows:

(a) One-third of a foot

(b) One-half of a foot

(c) Two-thirds of a foot

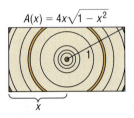

$$A(x) = 4x\sqrt{1 - x^2}$$

93. Economics   The **participation rate** is the number of people in the labor force divided by the civilian population (excludes military). Let $L(x)$ represent the size of the labor force in year x and $P(x)$ represent the civilian population in year x. Determine a function that represents the participation rate R as a function of x.

94. Crimes   Suppose that $V(x)$ represents the number of violent crimes committed in year x and $P(x)$ represents the number of property crimes committed in year x. Determine a function T that represents the combined total of violent crimes and property crimes in year x.

95. Health Care   Suppose that $P(x)$ represents the percentage of income spent on health care in year x and $I(x)$ represents income in year x. Determine a function H that represents total health care expenditures in year x.

96. Income Tax   Suppose that $I(x)$ represents the income of an individual in year x before taxes and $T(x)$ represents the individual's tax bill in year x. Determine a function N that represents the individual's net income (income after taxes) in year x.

97. Some functions f have the property that $f(a + b) = f(a) + f(b)$ for all real numbers a and b. Which of the following functions have this property?

(a) $h(x) = 2x$ (b) $g(x) = x^2$

(c) $F(x) = 5x - 2$ (d) $G(x) = \dfrac{1}{x}$

Discussion and Writing

98. Are the functions $f(x) = x - 1$ and $g(x) = \dfrac{x^2 - 1}{x + 1}$ the same? Explain.

99. Investigate when, historically, the use of the function notation $y = f(x)$ first appeared.

'Are You Prepared?' Answers

1. $(-1, 3)$ **2.** 21.5   **3.** $\{x | x \neq -4\}$ **4.** $\{x | x < -1\}$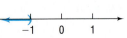

2.2 The Graph of a Function

PREPARING FOR THIS SECTION *Before getting started, review the following:*

- Graphs of Equations (Section 1.2, pp. 11–13)
- Intercepts (Section 1.2, pp. 15–17)

Now work the 'Are You Prepared?' problems on page 75.

OBJECTIVES   1 Identify the Graph of a Function
2 Obtain Information from or about the Graph of a Function

In applications, a graph often demonstrates more clearly the relationship between two variables than, say, an equation or table would. For example, Table 1 shows the average price of gasoline, adjusted for inflation, for the years 1978–2004. If we plot these data and then connect the points, we obtain Figure 13.

Table 1

Year	Price	Year	Price
1978	1.5207	1992	1.3460
1979	1.9263	1993	1.4622
1980	2.4197	1994	1.4166
1981	2.4258	1995	1.4111
1982	2.1136	1996	1.4804
1983	1.8210	1997	1.4697
1984	1.7732	1998	1.2685
1985	1.6998	1999	1.4629
1986	1.3043	2000	1.7469
1987	1.2831	2001	1.6821
1988	1.2674	2002	1.5339
1989	1.3134	2003	1.8308
1990	1.4160	2004	2.0047
1991	1.4490		

Source: Statistical Abstract of the United States.

Figure 13
Price of Gasoline Adjusted
for Inflation, 1978–2004

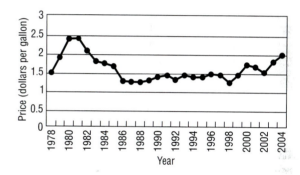

NOTE

When we select a viewing window to graph a function on a calculator, the values of Xmin, Xmax give the domain that we wish to view, while Ymin, Ymax give the range that we wish to view. These settings usually do not represent the actual domain and range of the function. ∎

We can see from the graph that the price of gasoline (adjusted for inflation) rose rapidly from 1978 to 1981 and was falling from 2000 to 2002. The graph also shows that the highest price occurred in 1981. To learn information such as this from an equation requires that some calculations be made.

Look again at Figure 13. The graph shows that for each date on the horizontal axis there is only one price on the vertical axis. The graph represents a function, although the exact rule for getting from date to price is not given.

When a function is defined by an equation in x and y, the **graph of the function** is the graph of the equation, that is, the set of points (x, y) in the xy-plane that satisfies the equation.

1 Identify the Graph of a Function

Not every collection of points in the xy-plane represents the graph of a function. Remember, for a function, each number x in the domain has exactly one image y in the range. This means that the graph of a function cannot contain two points with the same x-coordinate and different y-coordinates. Therefore, the graph of a function must satisfy the following **vertical-line test**.

Theorem

Vertical-line Test

A set of points in the xy-plane is the graph of a function if and only if every vertical line intersects the graph in at most one point.

In other words, if any vertical line intersects a graph at more than one point, the graph is not the graph of a function.

EXAMPLE 1   **Identifying the Graph of a Function**

Which of the graphs in Figure 14 are graphs of functions?

Figure 14

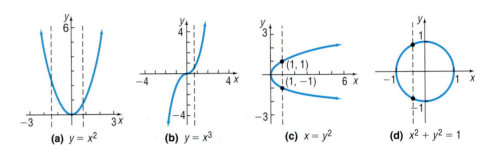

(a) $y = x^2$ (b) $y = x^3$ (c) $x = y^2$ (d) $x^2 + y^2 = 1$

Solution   The graphs in Figures 14(a) and 14(b) are graphs of functions, because every vertical line intersects each graph in at most one point. The graphs in Figures 14(c) and 14(d) are not graphs of functions, because there is a vertical line that intersects each graph in more than one point.   ◀

 NOW WORK PROBLEM 15.

2   **Obtain Information from or about the Graph of a Function**

If (x, y) is a point on the graph of a function f, then y is the value of f at x; that is, $y = f(x)$. The next example illustrates how to obtain information about a function if its graph is given.

EXAMPLE 2   **Obtaining Information from the Graph of a Function**

Let f be the function whose graph is given in Figure 15. (The graph of f might represent the distance that the bob of a pendulum is from its *at-rest* position. Negative values of y mean that the pendulum is to the left of the at-rest position, and positive values of y mean that the pendulum is to the right of the at-rest position.)

Figure 15

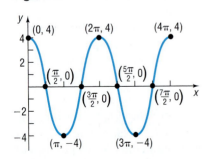

(a) What are $f(0)$, $f\left(\dfrac{3\pi}{2}\right)$, and $f(3\pi)$?

(b) What is the domain of f?

(c) What is the range of f?

(d) List the intercepts. (Recall that these are the points, if any, where the graph crosses or touches the coordinate axes.)

(e) How often does the line $y = 2$ intersect the graph?

(f) For what values of x does $f(x) = -4$?

(g) For what values of x is $f(x) > 0$?

Solution   (a) Since $(0, 4)$ is on the graph of f, the y-coordinate 4 is the value of f at the x-coordinate 0; that is, $f(0) = 4$. In a similar way, we find that when $x = \dfrac{3\pi}{2}$ then $y = 0$, so $f\left(\dfrac{3\pi}{2}\right) = 0$. When $x = 3\pi$, then $y = -4$, so $f(3\pi) = -4$.

(b) To determine the domain of f, we notice that the points on the graph of f will have x-coordinates between 0 and 4π, inclusive; and for each number x between 0 and 4π, there is a point $(x, f(x))$ on the graph. The domain of f is $\{x|0 \leq x \leq 4\pi\}$ or the interval $[0, 4\pi]$.

(c) The points on the graph all have y-coordinates between -4 and 4, inclusive; and for each such number y, there is at least one number x in the domain. The range of f is $\{y|-4 \leq y \leq 4\}$ or the interval $[-4, 4]$.

(d) The intercepts are

$$(0, 4), \left(\frac{\pi}{2}, 0\right), \left(\frac{3\pi}{2}, 0\right), \left(\frac{5\pi}{2}, 0\right), \quad \text{and} \quad \left(\frac{7\pi}{2}, 0\right)$$

(e) If we draw the horizontal line $y = 2$ on the graph in Figure 15, then we find that it intersects the graph four times.

(f) Since $(\pi, -4)$ and $(3\pi, -4)$ are the only points on the graph for which $y = f(x) = -4$, we have $f(x) = -4$ when $x = \pi$ and $x = 3\pi$.

(g) To determine where $f(x) > 0$, we look at Figure 15 and determine the x-values for which the y-coordinate is positive. This occurs on the intervals $\left[0, \dfrac{\pi}{2}\right), \left(\dfrac{3\pi}{2}, \dfrac{5\pi}{2}\right),$ and $\left(\dfrac{7\pi}{2}, 4\pi\right]$. Using inequality notation, $f(x) > 0$ for $0 \leq x < \dfrac{\pi}{2}, \dfrac{3\pi}{2} < x < \dfrac{5\pi}{2},$ and $\dfrac{7\pi}{2} < x \leq 4\pi$. ◄

When the graph of a function is given, its domain may be viewed as the shadow created by the graph on the x-axis by vertical beams of light. Its range can be viewed as the shadow created by the graph on the y-axis by horizontal beams of light. Try this technique with the graph given in Figure 15.

NOW WORK PROBLEMS 9 AND 13.

EXAMPLE 3

Obtaining Information about the Graph of a Function

Consider the function: $f(x) = \dfrac{x}{x + 2}$

(a) Is the point $\left(1, \dfrac{1}{2}\right)$ on the graph of f?

(b) If $x = 2$, what is $f(x)$? What point is on the graph of f?

(c) If $f(x) = 2$, what is x? What point is on the graph of f?

Solution (a) When $x = 1$, then

$$f(x) = \frac{x}{x + 2}$$

$$f(1) = \frac{1}{1 + 2} = \frac{1}{3}$$

The point $\left(1, \frac{1}{3}\right)$ is on the graph of f; the point $\left(1, \frac{1}{2}\right)$ is not.

(b) If $x = 2$, then

$$f(x) = \frac{x}{x + 2}$$

$$f(2) = \frac{2}{2 + 2} = \frac{2}{4} = \frac{1}{2}$$

The point $\left(2, \frac{1}{2}\right)$ is on the graph of f.

(c) If $f(x) = 2$, then

$$f(x) = 2$$

$$\frac{x}{x + 2} = 2$$

$x = 2(x + 2)$ Multiply both sides by x + 2.

$x = 2x + 4$ Remove parentheses.

$x = -4$ Solve for x.

If $f(x) = 2$, then $x = -4$. The point $(-4, 2)$ is on the graph of f. ◀

NOW WORK PROBLEM 25.

EXAMPLE 4   **Average Cost Function**

The average cost $\overline{C}$ of manufacturing x computers per day is given by the function

$$\overline{C}(x) = 0.56x^2 - 34.39x + 1212.57 + \frac{20,000}{x}$$

Determine the average cost of manufacturing:

(a) 30 computers in a day
(b) 40 computers in a day
(c) 50 computers in a day
(d) Graph the function $\overline{C} = \overline{C}(x), 0 < x \leq 80$.
(e) Create a TABLE with TblStart = 1 and ΔTbl = 1. Which value of x minimizes the average cost?

Solution (a) The average cost of manufacturing $x = 30$ computers is

$$\overline{C}(30) = 0.56(30)^2 - 34.39(30) + 1212.57 + \frac{20,000}{30} = \$1351.54$$

(b) The average cost of manufacturing $x = 40$ computers is

$$\overline{C}(40) = 0.56(40)^2 - 34.39(40) + 1212.57 + \frac{20{,}000}{40} = \$1232.97$$

(c) The average cost of manufacturing $x = 50$ computers is

$$\overline{C}(50) = 0.56(50)^2 - 34.39(50) + 1212.57 + \frac{20{,}000}{50} = \$1293.07$$

(d) See Figure 16 for the graph of $\overline{C} = \overline{C}(x)$.

(e) With the function $\overline{C} = \overline{C}(x)$ in Y_1, we create Table 2. We scroll down until we find a value of x for which Y_1 is smallest. Table 3 shows that manufacturing $x = 41$ computers minimizes the average cost at \$1231.74 per computer.

Figure 16

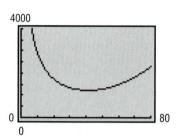

Table 2

X	Y1
1	21179
2	11146
3	7781.1
4	6084
5	5054.6
6	4359.7
7	3856.4

Y1⬛.56X²−34.39X...

Table 3

X	Y1
38	1240.7
39	1235.9
40	1233
41	1231.7
42	1232.2
43	1234.4
44	1238.1

Y1=1231.74487805

NOW WORK PROBLEM 29.

Summary

Graph of a function	The collection of points (x, y) that satisfies the equation $y = f(x)$. A collection of points is the graph of a function provided that every vertical line intersects the graph in at most one point (vertical-line test).

2.2 Assess Your Understanding

'Are You Prepared?'

Answers are given at the end of these exercises. If you get a wrong answer, read the pages listed in red.

1. The intercepts of the equation $x^2 + 4y^2 = 16$ are _____. (pp. 15–17)

2. *True or False:* The point $(-2, -6)$ is on the graph of the equation $x = 2y - 2$. (pp. 11–13)

Concepts and Vocabulary

3. A set of points in the xy-plane is the graph of a function if and only if every _____ line intersects the graph in at most one point.

4. If the point $(5, -3)$ is a point on the graph of f, then $f(\_\_\_) = \_\_\_\_$.

5. Find a so that the point $(-1, 2)$ is on the graph of $f(x) = ax^2 + 4$.

6. *True or False:* A function can have more than one y-intercept.

7. *True or False:* The graph of a function $y = f(x)$ always crosses the y-axis.

8. *True or False:* The y-intercept of the graph of the function $y = f(x)$, whose domain is all real numbers, is $f(0)$.

Skill Building

9. Use the given graph of the function f to answer parts (a)–(n).

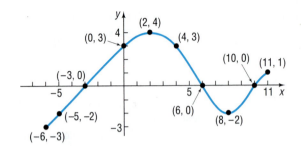

(a) Find $f(0)$ and $f(-6)$.
(b) Find $f(6)$ and $f(11)$.
(c) Is $f(3)$ positive or negative?
(d) Is $f(-4)$ positive or negative?
(e) For what numbers x is $f(x) = 0$?
(f) For what numbers x is $f(x) > 0$?
(g) What is the domain of f?
(h) What is the range of f?
(i) What are the x-intercepts?
(j) What is the y-intercept?

(k) How often does the line $y = \dfrac{1}{2}$ intersect the graph?

(l) How often does the line $x = 5$ intersect the graph?
(m) For what values of x does $f(x) = 3$?
(n) For what values of x does $f(x) = -2$?

10. Use the given graph of the function f to answer parts (a)–(n).

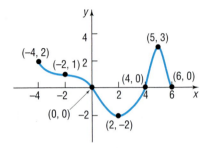

(a) Find $f(0)$ and $f(6)$.
(b) Find $f(2)$ and $f(-2)$.
(c) Is $f(3)$ positive or negative?
(d) Is $f(-1)$ positive or negative?
(e) For what numbers x is $f(x) = 0$?
(f) For what numbers x is $f(x) < 0$?
(g) What is the domain of f?
(h) What is the range of f?
(i) What are the x-intercepts?
(j) What is the y-intercept?
(k) How often does the line $y = -1$ intersect the graph?
(l) How often does the line $x = 1$ intersect the graph?
(m) For what value of x does $f(x) = 3$?
(n) For what value of x does $f(x) = -2$?

In Problems 11–22, determine whether the graph is that of a function by using the vertical-line test. If it is, use the graph to find:
(a) *Its domain and range*
(b) *The intercepts, if any*
(c) *Any symmetry with respect to the x-axis, the y-axis, or the origin*

11.

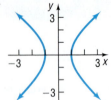

12.

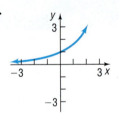

13.

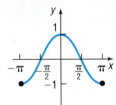

14.

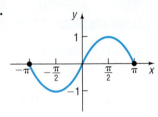

15.

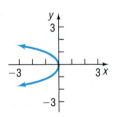

16.

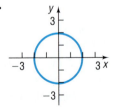

17.

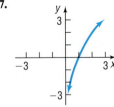

18.

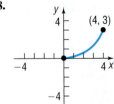

19.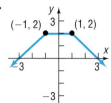

20.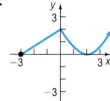

21.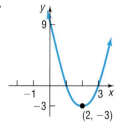

22.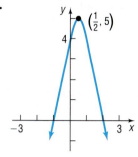

In Problems 23–28, answer the questions about the given function.

23. $f(x) = 2x^2 - x - 1$
(a) Is the point $(-1, 2)$ on the graph of f?
(b) If $x = -2$, what is $f(x)$? What point is on the graph of f?
(c) If $f(x) = -1$, what is x? What point(s) are on the graph of f?
(d) What is the domain of f?
(e) List the x-intercepts, if any, of the graph of f.
(f) List the y-intercept, if there is one, of the graph of f.

24. $f(x) = -3x^2 + 5x$
(a) Is the point $(-1, 2)$ on the graph of f?
(b) If $x = -2$, what is $f(x)$? What point is on the graph of f?
(c) If $f(x) = -2$, what is x? What point(s) are on the graph of f?
(d) What is the domain of f?
(e) List the x-intercepts, if any, of the graph of f.
(f) List the y-intercept, if there is one, of the graph of f.

25. $f(x) = \dfrac{x + 2}{x - 6}$
(a) Is the point $(3, 14)$ on the graph of f?
(b) If $x = 4$, what is $f(x)$? What point is on the graph of f?
(c) If $f(x) = 2$, what is x? What point(s) are on the graph of f?
(d) What is the domain of f?
(e) List the x-intercepts, if any, of the graph of f.
(f) List the y-intercept, if there is one, of the graph of f.

26. $f(x) = \dfrac{x^2 + 2}{x + 4}$
(a) Is the point $\left(1, \dfrac{3}{5}\right)$ on the graph of f?

(b) If $x = 0$, what is $f(x)$? What point is on the graph of f?
(c) If $f(x) = \dfrac{1}{2}$, what is x? What point(s) are on the graph of f?
(d) What is the domain of f?
(e) List the x-intercepts, if any, of the graph of f.
(f) List the y-intercept, if there is one, of the graph of f.

27. $f(x) = \dfrac{2x^2}{x^4 + 1}$
(a) Is the point $(-1, 1)$ on the graph of f?
(b) If $x = 2$, what is $f(x)$? What point is on the graph of f?
(c) If $f(x) = 1$, what is x? What point(s) are on the graph of f?
(d) What is the domain of f?
(e) List the x-intercepts, if any, of the graph of f.
(f) List the y-intercept, if there is one, of the graph of f.

28. $f(x) = \dfrac{2x}{x - 2}$
(a) Is the point $\left(\dfrac{1}{2}, -\dfrac{2}{3}\right)$ on the graph of f?
(b) If $x = 4$, what is $f(x)$? What point is on the graph of f?
(c) If $f(x) = 1$, what is x? What point(s) are on the graph of f?
(d) What is the domain of f?
(e) List the x-intercepts, if any, of the graph of f.
(f) List the y-intercept, if there is one, of the graph of f.

Applications and Extensions

29. Motion of a Golf Ball A golf ball is hit with an initial velocity of 130 feet per second at an inclination of 45° to the horizontal. In physics, it is established that the height h of the golf ball is given by the function

$$h(x) = \frac{-32x^2}{130^2} + x$$

where x is the horizontal distance that the golf ball has traveled. (*continued on page 78*)

(a) Determine the height of the golf ball after it has traveled 100 feet.
(b) What is the height after it has traveled 300 feet?
(c) What is the height after it has traveled 500 feet?
(d) How far was the golf ball hit?
(e) Graph the function $h = h(x)$.
(f) Use a graphing utility to determine the distance that the ball has traveled when the height of the ball is 90 feet.
(g) Create a TABLE with TblStart = 0 and ΔTbl = 25. To the nearest 25 feet, how far does the ball travel before it reaches a maximum height? What is the maximum height?
(h) Adjust the value of ΔTbl until you determine the distance, to within 1 foot, that the ball travels before it reaches a maximum height.

30. Cross-sectional Area The cross-sectional area of a beam cut from a log with radius 1 foot is given by the function $A(x) = 4x\sqrt{1 - x^2}$, where x represents the length, in feet, of half the base of the beam. See the figure.

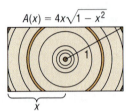

$A(x) = 4x\sqrt{1 - x^2}$

(a) Find the domain of A.
(b) Graph the function $A = A(x)$.

(c) Create a TABLE with TblStart = 0 and ΔTbl = 0.1 for $0 \le x \le 1$. Which value of x maximizes the cross-sectional area? What should be the length of the base of the beam to maximize the cross-sectional area?

31. Cost of Trans-Atlantic Travel A Boeing 747 crosses the Atlantic Ocean (3000 miles) with an airspeed of 500 miles per hour. The cost C (in dollars) per passenger is given by

$$C(x) = 100 + \frac{x}{10} + \frac{36{,}000}{x}$$

where x is the ground speed (airspeed $\pm$ wind).
(a) Graph the function $C = C(x)$.
(b) Create a TABLE with TblStart = 0 and ΔTbl = 50.
(c) To the nearest 50 miles per hour, what ground speed minimizes the cost per passenger?

32. Effect of Elevation on Weight If an object weighs m pounds at sea level, then its weight W (in pounds) at a height of h miles above sea level is given approximately by

$$W(h) = m\left(\frac{4000}{4000 + h}\right)^2$$

(a) If Amy weighs 120 pounds at sea level, how much will she weigh on Pike's Peak, which is 14,110 feet above sea level?
(b) Use a graphing utility to graph the function $W = W(h)$. Use $m = 120$ pounds.
(c) Create a Table with TblStart = 0 and ΔTbl = 0.5 to see how the weight W varies as h changes from 0 to 5 miles.
(d) At what height will Amy weigh 119.95 pounds?
(e) Does your answer to part (d) seem reasonable? Explain.

Discussion and Writing

33. Describe how you would proceed to find the domain and range of a function if you were given its graph. How would your strategy change if you were given the equation defining the function instead of its graph?

34. How many x-intercepts can the graph of a function have?

35. How many y-intercepts can the graph of a function have?

36. Is a graph that consists of a single point the graph of a function? Can you write the equation of such a function?

37. Match each of the following functions with the graphs that best describes the situation.
(a) The cost of building a house as a function of its square footage
(b) The height of an egg dropped from a 300-foot building as a function of time
(c) The height of a human as a function of time
(d) The demand for Big Macs as a function of price
(e) The height of a child on a swing as a function of time

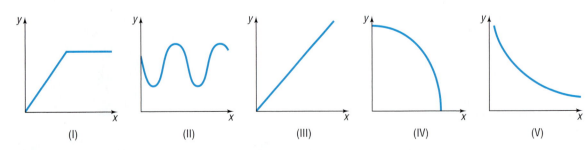

38. Match each of the following functions with the graph that best describes the situation.
  (a) The temperature of a bowl of soup as a function of time
  (b) The number of hours of daylight per day over a 2-year period
  (c) The population of Florida as a function of time
  (d) The distance of a car traveling at a constant velocity as a function of time
  (e) The height of a golf ball hit with a 7-iron as a function of time

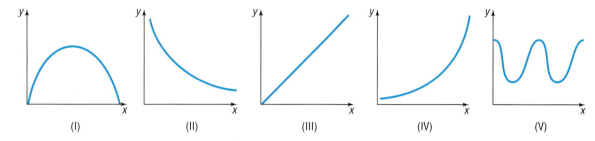

| (I) | (II) | (III) | (IV) | (V) |

---

39. Consider the following scenario: Barbara decides to take a walk. She leaves home, walks 2 blocks in 5 minutes at a constant speed, and realizes that she forgot to lock the door. So Barbara runs home in 1 minute. While at her doorstep, it takes her 1 minute to find her keys and lock the door. Barbara walks 5 blocks in 15 minutes and then decides to jog home. It takes her 7 minutes to get home. Draw a graph of Barbara's distance from home (in blocks) as a function of time.

40. Consider the following scenario: Jayne enjoys riding her bicycle through the woods. At the forest preserve, she gets on her bicycle and rides up a 2000-foot incline in 10 minutes. She then travels down the incline in 3 minutes. The next 5000 feet is level terrain and she covers the distance in 20 minutes. She rests for 15 minutes. Jayne then travels 10,000 feet in 30 minutes. Draw a graph of Jayne's distance traveled (in feet) as a function of time.

41. The following sketch represents the distance d (in miles) that Kevin is from home as a function of time t (in hours). Answer the questions based on the graph. In parts (a)–(g), how many hours elapsed and how far was Kevin from home during this time?

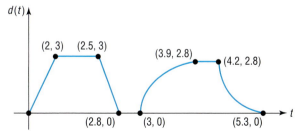

  (a) From $t = 0$ to $t = 2$
  (b) From $t = 2$ to $t = 2.5$
  (c) From $t = 2.5$ to $t = 2.8$

  (d) From $t = 2.8$ to $t = 3$
  (e) From $t = 3$ to $t = 3.9$
  (f) From $t = 3.9$ to $t = 4.2$
  (g) From $t = 4.2$ to $t = 5.3$
  (h) What is the farthest distance that Kevin is from home?
  (i) How many times did Kevin return home?

42. The following sketch represents the speed v (in miles per hour) of Michael's car as a function of time t (in minutes).

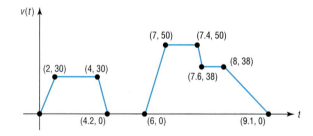

  (a) Over what interval of time is Michael traveling fastest?
  (b) Over what interval(s) of time is Michael's speed zero?
  (c) What is Michael's speed between 0 and 2 minutes?
  (d) What is Michael's speed between 4.2 and 6 minutes?
  (e) What is Michael's speed between 7 and 7.4 minutes?
  (f) When is Michael's speed constant?

43. Draw the graph of a function whose domain is $\{x \mid -3 \leq x \leq 8, \; x \neq 5\}$ and whose range is $\{y \mid -1 \leq y \leq 2, \; y \neq 0\}$. What point(s) in the rectangle $-3 \leq x \leq 8, -1 \leq y \leq 2$ cannot be on the graph? Compare your graph with those of other students. What differences do you see?

44. Is there a function whose graph is symmetric with respect to the x-axis? Explain.

'Are You Prepared?' Answers

  1. $(-4, 0), (4, 0), (0, -2), (0, 2)$ **2.** False

2.3 Properties of Functions

PREPARING FOR THIS SECTION   *Before getting started, review the following:*

• Intervals (Appendix, Section A.8, pp. 729–730)   • Slope of a Line (Section 1.4, pp. 27–29)
• Intercepts (Section 1.2, pp. 15–17)   • Point–Slope Form of a Line (Section 1.4, p. 32)
                                                                       • Symmetry (Section 1.2, pp. 17–19)

 Now work the 'Are You Prepared?' problems on page 88.

OBJECTIVES 1 Determine Even and Odd Functions from a Graph
2 Identify Even and Odd Functions from the Equation
3 Use a Graph to Determine Where a Function Is Increasing, Decreasing, or Constant
4 Use a Graph to Locate Local Maxima and Local Minima
5 Use a Graphing Utility to Approximate Local Maxima and Local Minima and to Determine Where a Function Is Increasing or Decreasing
6 Find the Average Rate of Change of a Function

It is easiest to obtain the graph of a function $y = f(x)$ by knowing certain properties that the function has and the impact of these properties on the way that the graph will look.

1 Determine Even and Odd Functions from a Graph

The words *even* and *odd*, when applied to a function f, describe the symmetry that exists for the graph of the function.

A function f is even if and only if, whenever the point (x, y) is on the graph of f then the point $(-x, y)$ is also on the graph. Using function notation, we define an even function as follows:

A function f is **even** if, for every number x in its domain, the number $-x$ is also in the domain and

$$f(-x) = f(x)$$

A function f is odd if and only if, whenever the point (x, y) is on the graph of f then the point $(-x, -y)$ is also on the graph. Using function notation, we define an odd function as follows:

A function f is **odd** if, for every number x in its domain, the number $-x$ is also in the domain and

$$f(-x) = -f(x)$$

Refer to Section 1.2, where the tests for symmetry are listed. The following results are then evident.

Theorem

A function is even if and only if its graph is symmetric with respect to the y-axis. A function is odd if and only if its graph is symmetric with respect to the origin.

| EXAMPLE 1 | **Determining Even and Odd Functions from the Graph** |

Determine whether each graph given in Figure 17 is the graph of an even function, an odd function, or a function that is neither even nor odd.

Figure 17

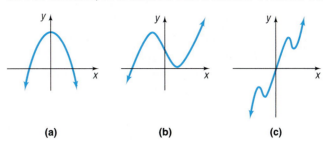

(a)   (b)   (c)

Solution   (a) The graph in Figure 17(a) is that of an even function, because the graph is symmetric with respect to the y-axis.

(b) The function whose graph is given in Figure 17(b) is neither even nor odd, because the graph is neither symmetric with respect to the y-axis nor symmetric with respect to the origin.

(c) The function whose graph is given in Figure 17(c) is odd, because its graph is symmetric with respect to the origin.   ◄

NOW WORK PROBLEMS 21(a), (b), AND (d).

2   Identify Even and Odd Functions from the Equation

A graphing utility can be used to conjecture whether a function is even, odd, or neither. As stated, when the graph of an even function contains the point (x, y), it must also contain the point $(-x, y)$. Therefore, if the graph shows evidence of symmetry with respect to the y-axis, we would conjecture that the function is even. In addition, if the graph shows evidence of symmetry with respect to the origin, we would conjecture that the function is odd.

In the next example, we use a graphing utility to conjecture whether a function is even, odd, or neither. Then we verify our conjecture algebraically.

| EXAMPLE 2 | **Identifying Even and Odd Functions** |

Use a graphing utility to conjecture whether each of the following functions is even, odd, or neither. Verify the conjecture algebraically. Then state whether the graph is symmetric with respect to the y-axis or with respect to the origin.

(a) $f(x) = x^2 - 5$ (b) $g(x) = x^3 - 1$ (c) $h(x) = 5x^3 - x$

Solution   (a) Graph the function. See Figure 18. It appears that the graph is symmetric with respect to the y-axis. We conjecture that the function is even.

Figure 18

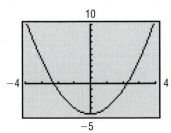

To algebraically verify the conjecture, we replace x by $-x$ in $f(x) = x^2 - 5$. Then

$$f(-x) = (-x)^2 - 5 = x^2 - 5 = f(x)$$

Since $f(-x) = f(x)$, we conclude that f is an even function, and the graph is symmetric with respect to the y-axis.

(b) Graph the function. See Figure 19 on page 82. It appears that there is no symmetry. We conjecture that the function is neither even nor odd.

To algebraically verify that the function is not even, we find $g(-x)$ and compare the result with $g(x)$.

$$g(-x) = (-x)^3 - 1 = -x^3 - 1; \quad g(x) = x^3 - 1$$

Figure 19

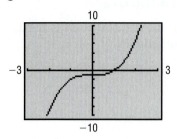

Figure 20

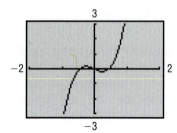

Since $g(-x) \neq g(x)$, the function is not even.

To algebraically verify that the function is not odd, we find $-g(x)$ and compare the result with $g(-x)$.

$$-g(x) = -(x^3 - 1) = -x^3 + 1; \quad g(-x) = -x^3 - 1$$

Since $g(-x) \neq -g(x)$, the function is not odd. The graph is not symmetric with respect to the y-axis nor is it symmetric with respect to the origin.

(c) Graph the function. See Figure 20. It appears that there is symmetry with respect to the origin. We conjecture that the function is odd.

To algebraically verify the conjecture, we replace x by $-x$ in $h(x) = 5x^3 - x$. Then

$$h(-x) = 5(-x)^3 - (-x) = -5x^3 + x = -(5x^3 - x) = -h(x)$$

Since $h(-x) = -h(x)$, h is an odd function, and the graph of h is symmetric with respect to the origin.   ◀

 NOW WORK PROBLEM 33.

3 **Use a Graph to Determine Where a Function Is Increasing, Decreasing, or Constant**

Consider the graph given in Figure 21. If you look from left to right along the graph of the function, you will notice that parts of the graph are rising, parts are falling, and parts are horizontal. In such cases, the function is described as *increasing*, *decreasing*, or *constant*, respectively.

Figure 21

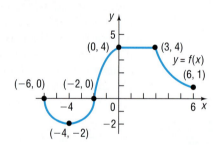

| EXAMPLE 3 | **Determining Where a Function Is Increasing, Decreasing, or Constant from Its Graph** |

Where is the function in Figure 21 increasing? Where is it decreasing? Where is it constant?

Solution   To answer the question of where a function is increasing, where it is decreasing, and where it is constant, we use strict inequalities involving the independent variable x, or we use open intervals* of x-coordinates. The graph in Figure 21 is rising (increasing) from the point $(-4, -2)$ to the point $(0, 4)$, so we conclude that it is increasing on the open interval $(-4, 0)$ or for $-4 < x < 0$. The graph is falling (decreasing) from the point $(-6, 0)$ to the point $(-4, -2)$ and from the point $(3, 4)$ to the point $(6, 1)$. We conclude that the graph is decreasing on the open intervals $(-6, -4)$ and $(3, 6)$ or for $-6 < x < -4$ and $3 < x < 6$. The graph is constant on the open interval $(0, 3)$ or for $0 < x < 3$.   ◀

More precise definitions follow:

*The open interval (a, b) consists of all real numbers x for which $a < x < b$.

WARNING
We describe the behavior of a graph in terms of its x-values. Do not say the graph in Figure 21 is increasing from the point $(-4, -2)$ to $(0, 4)$. Rather, say it is increasing on the interval $(-4, 0)$. ■

A function f is **increasing** on an open interval I if, for any choice of x_1 and x_2 in I, with $x_1 < x_2$, we have $f(x_1) < f(x_2)$.

A function f is **decreasing** on an open interval I if, for any choice of x_1 and x_2 in I, with $x_1 < x_2$, we have $f(x_1) > f(x_2)$.

A function f is **constant** on an interval I if, for all choices of x in I, the values $f(x)$ are equal.

Figure 22 illustrates the definitions. The graph of an increasing function goes up from left to right, the graph of a decreasing function goes down from left to right, and the graph of a constant function remains at a fixed height.

Figure 22

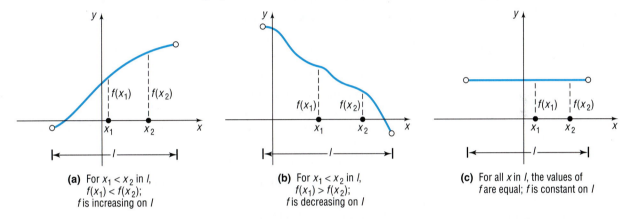

(a) For $x_1 < x_2$ in I, $f(x_1) < f(x_2)$; f is increasing on I

(b) For $x_1 < x_2$ in I, $f(x_1) > f(x_2)$; f is decreasing on I

(c) For all x in I, the values of f are equal; f is constant on I

 NOW WORK PROBLEMS 11, 13, 15, AND 21(c).

4   Use a Graph to Locate Local Maxima and Local Minima

 When the graph of a function is increasing to the left of $x = c$ and decreasing to the right of $x = c$, then at c the value of f is largest. This value is called a *local maximum* of f. See Figure 23(a).

When the graph of a function is decreasing to the left of $x = c$ and is increasing to the right of $x = c$, then at c the value of f is the smallest. This value is called a *local minimum* of f. See Figure 23(b).

Figure 23

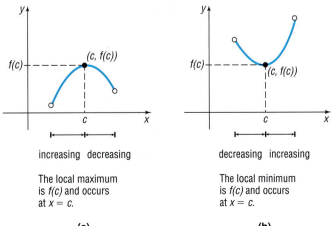

increasing decreasing

The local maximum is $f(c)$ and occurs at $x = c$.

(a)

decreasing increasing

The local minimum is $f(c)$ and occurs at $x = c$.

(b)

> A function f has a **local maximum** at c if there is an open interval I containing c so that, for all x in I, $f(x) \leq f(c)$. We call $f(c)$ a **local maximum of f**.

> A function f has a **local minimum** at c if there is an open interval I containing c so that, for all x in I, $f(x) \geq f(c)$. We call $f(c)$ a **local minimum of f**.

If f has a local maximum at c, then the value of f at c is greater than or equal to the values of f near c. If f has a local minimum at c, then the value of f at c is less than or equal to the values of f near c. The word *local* is used to suggest that it is only near c that the value $f(c)$ is largest or smallest.

EXAMPLE 4

Finding Local Maxima and Local Minima from the Graph of a Function and Determining Where the Function Is Increasing, Decreasing, or Constant

Figure 24

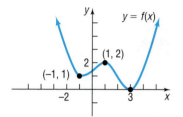

Figure 24 shows the graph of a function f.

(a) At what number(s), if any, does f have a local maximum?
(b) What are the local maxima?
(c) At what number(s), if any, does f have a local minimum?
(d) What are the local minima?
(e) List the intervals on which f is increasing. List the intervals on which f is decreasing.

Solution

The domain of f is the set of real numbers.

(a) f has a local maximum at 1, since for all x close to 1, $x \neq 1$, we have $f(x) \leq f(1)$.
(b) The local maximum is $f(1) = 2$.
(c) f has a local minimum at -1 and at 3.
(d) The local minima are $f(-1) = 1$ and $f(3) = 0$.
(e) The function whose graph is given in Figure 24 is increasing for all values of x between -1 and 1 and for all values of x greater than 3. That is, the function is increasing on the intervals $(-1, 1)$ and $(3, \infty)$ or for $-1 < x < 1$ and $x > 3$. The function is decreasing for all values of x less than -1 and for all values of x between 1 and 3. That is, the function is decreasing on the intervals $(-\infty, -1)$ and $(1, 3)$ or for $x < -1$ and $1 < x < 3$. ◄

WARNING
The y-value is the local maximum or local minimum and it occurs at some x-value. In Figure 24, we say the local maximum is 2 and occurs at x = 1. ∎

 NOW WORK PROBLEMS 17 AND 19.

---

5 ### Use a Graphing Utility to Approximate Local Maxima and Local Minima and to Determine Where a Function Is Increasing or Decreasing

 To locate the exact value at which a function f has a local maximum or a local minimum usually requires calculus. However, a graphing utility may be used to approximate these values by using the MAXIMUM and MINIMUM features.

EXAMPLE 5

Using a Graphing Utility to Approximate Local Maxima and Minima and to Determine Where a Function Is Increasing or Decreasing

(a) Use a graphing utility to graph $f(x) = 6x^3 - 12x + 5$ for $-2 < x < 2$. Approximate where f has a local maximum and where f has a local minimum.
(b) Determine where f is increasing and where it is decreasing.

Solution   (a) Graphing utilities have a feature that finds the maximum or minimum point of a graph within a given interval. Graph the function f for $-2 < x < 2$. Using MAXIMUM, we find that the local maximum is 11.53 and it occurs at $x = -0.82$, rounded to two decimal places. See Figure 25(a). Using MINIMUM, we find that the local minimum is -1.53 and it occurs at $x = 0.82$, rounded to two decimal places. See Figure 25(b).

Figure 25

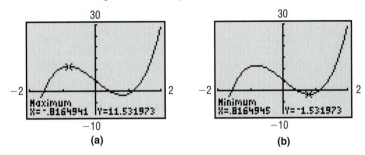

(a)   (b)

(b) Looking at Figures 25(a) and (b), we see that the graph of f is increasing from $x = -2$ to $x = -0.82$ and from $x = 0.82$ to $x = 2$, so f is increasing on the intervals $(-2, -0.82)$ and $(0.82, 2)$ or for $-2 < x < -0.82$ and $0.82 < x < 2$. The graph is decreasing from $x = -0.82$ to $x = 0.82$, so f is decreasing on the interval $(-0.82, 0.82)$ or for $-0.82 < x < 0.82$.   ◄

 NOW WORK PROBLEM 45.

6   Find the Average Rate of Change of a Function

In Section 1.4 we said that the slope of a line could be interpreted as the average rate of change. Often we are interested in the rate at which functions change. To find the average rate of change of a function between any two points on its graph, we calculate the slope of the line containing the two points.

> If c is in the domain of a function $y = f(x)$, the **average rate of change of f** from c to x is defined as
>
> $$\text{Average rate of change} = \frac{\Delta y}{\Delta x} = \frac{f(x) - f(c)}{x - c}, \qquad x \neq c \qquad \textbf{(1)}$$
>
> In calculus, this expression is called the **difference quotient** of f at c.

The symbol Δy in (1) is the "change in y" and Δx is the "change in x." The average rate of change of f is the change in y divided by the change in x.

EXAMPLE 6

Finding the Average Rate of Change

Find the average rate of change of $f(x) = 3x^2$:

(a) From 1 to 3   (b) From 1 to 5   (c) From 1 to 7

Solution   (a) The average rate of change of $f(x) = 3x^2$ from 1 to 3 is

$$\frac{\Delta y}{\Delta x} = \frac{f(3) - f(1)}{3 - 1} = \frac{27 - 3}{3 - 1} = \frac{24}{2} = 12$$

Figure 26

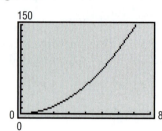

(b) The average rate of change of $f(x) = 3x^2$ from 1 to 5 is

$$\frac{\Delta y}{\Delta x} = \frac{f(5) - f(1)}{5 - 1} = \frac{75 - 3}{5 - 1} = \frac{72}{4} = 18$$

(c) The average rate of change of $f(x) = 3x^2$ from 1 to 7 is

$$\frac{\Delta y}{\Delta x} = \frac{f(7) - f(1)}{7 - 1} = \frac{147 - 3}{7 - 1} = \frac{144}{6} = 24$$ ◀

See Figure 26 for a graph of $f(x) = 3x^2$. The function f is increasing for $x > 0$. The fact that the average rates of change are getting larger indicates that the graph is getting steeper, that is, it is increasing at an increasing rate.

 NOW WORK PROBLEM 53.

EXAMPLE 7 | **Average Rate of Change of a Population**

A strain of E-coli Beu 397-recA441 is placed into a nutrient broth at 30° Celsius and allowed to grow. The data shown in Table 4 are collected. The population is measured in grams and the time in hours. Since population P depends on time t and each input corresponds to exactly one output, we can say that population is a function of time, so $P(t)$ represents the population at time t. For example, $P(2.5) = 0.18$. Figure 27 shows a graph of the function.

(a) Find the average rate of change of the population from 0 to 2.5 hours.
(b) Find the average rate of change of the population from 4.5 to 6 hours.
(c) What is happening to the average rate of change as time passes?

Table 4

Time (hours), x	Population (grams), y
0	0.09
2.5	0.18
3.5	0.26
4.5	0.35
6	0.50

Figure 27

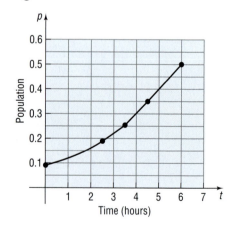

Solution   (a) Average rate of change $= \dfrac{\text{Population at 2.5 hours} - \text{Population at 0 hours}}{2.5 - 0}$

$$= \frac{P(2.5) - P(0)}{2.5 - 0}$$

$$= \frac{0.18 - 0.09}{2.5}$$

$$= 0.036 \text{ gram per hour}$$

On average, the population is increasing at a rate of 0.036 gram per hour from 0 to 2.5 hours.

(b) Average rate of change $= \dfrac{\text{Population at 6 hours } - \text{ Population at 4.5 hours}}{6 - 4.5}$

$$= \dfrac{P(6) - P(4.5)}{6 - 4.5}$$

$$= \dfrac{0.5 - 0.35}{1.5}$$

$$= 0.1 \text{ gram per hour}$$

On average, the population is increasing at a rate of 0.1 gram per hour from 4.5 to 6 hours.

(c) Not only is the size of the population increasing over time, but the rate of increase is also increasing. We say that the population is increasing at an increasing rate. ◀

 NOW WORK PROBLEM 69.

The Secant Line

The average rate of change of a function has an important geometric interpretation. Look at the graph of $y = f(x)$ in Figure 28. We have labeled two points on the graph: $(c, f(c))$ and $(x, f(x))$. The line containing these two points is called the **secant line**; its slope is

$$m_{\text{sec}} = \dfrac{f(x) - f(c)}{x - c}$$

Figure 28

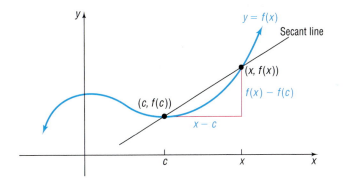

Theorem

Slope of the Secant Line

The average rate of change of a function equals the slope of the secant line containing two points on its graph.

EXAMPLE 8

Finding the Equation of a Secant Line

Suppose that $g(x) = 3x^2 - 2x + 3$.

(a) Find the average rate of change of g from -2 to x.

(b) Use the result of part (a) to find the average rate of change of g from -2 to 1. Interpret this result.

(c) Find an equation of the secant line containing $(-2, g(-2))$ and $(1, g(1))$.

(d) Using a graphing utility, draw the graph of g and the secant line obtained in part (c) on the same screen.

Solution (a) The average rate of change of $g(x) = 3x^2 - 2x + 3$ from -2 to x is

$$\text{Average rate of change} = \frac{g(x) - g(-2)}{x - (-2)}$$

$$= \frac{(3x^2 - 2x + 3) - 19}{x + 2} \qquad g(-2) = 3(-2)^2 - 2(-2) + 3 = 19$$

$$= \frac{3x^2 - 2x - 16}{x + 2}$$

$$= \frac{(3x - 8)(x + 2)}{x + 2} \qquad \textit{Factor.}$$

$$= 3x - 8 \qquad \textit{Cancel like factors.}$$

(b) The average rate of change of g from -2 to x is given by $3x - 8$. Therefore, the average rate of change of g from -2 to 1 is $3(1) - 8 = -5$. The slope of the secant line joining $(-2, g(-2))$ and $(1, g(1))$ is -5.

(c) We use the point–slope form to find an equation of the secant line.

$$y - y_1 = m_{\text{sec}}(x - x_1) \qquad \textit{Point–slope form of the secant line}$$

$$y - 4 = -5(x - 1) \qquad x_1 = 1, y_1 = g(1) = 4, m_{\text{sec}} = -5$$

$$y - 4 = -5x + 5 \qquad \textit{Simplify}$$

$$y = -5x + 9 \qquad \textit{Slope–intercept form of the secant line}$$

(d) Figure 29 shows the graph of g along with the secant line $y = -5x + 9$. ◄

Figure 29

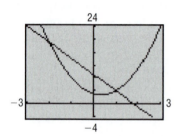

NOW WORK PROBLEM **59.**

2.3 Assess Your Understanding

'Are You Prepared?'

Answers are given at the end of these exercises. If you get a wrong answer, read the pages listed in red.

1. The interval $(2, 5)$ can be written as the inequality _____. (pp. 729–730)

2. The slope of the line containing the points $(-2, 3)$ and $(3, 8)$ is _____. (p. 27)

3. Test the equation $y = 5x^2 - 1$ for symmetry with respect to the x-axis, the y-axis, and the origin. (pp. 17–19)

4. Write the point–slope form of the line with slope 5 containing the point $(3, -2)$. (p. 32)

5. The intercepts of the equation $y = x^2 - 9$ are _____. (pp. 15–17)

Concepts and Vocabulary

6. A function f is _____ on an open interval I if, for any choice of x_1 and x_2 in I, with $x_1 < x_2$, we have $f(x_1) < f(x_2)$.

7. An _____ function f is one for which $f(-x) = f(x)$ for every x in the domain of f; an _____ function f is one for which $f(-x) = -f(x)$ for every x in the domain of f.

8. *True or False:* A function f is decreasing on an open interval I if, for any choice of x_1 and x_2 in I, with $x_1 < x_2$, we have $f(x_1) > f(x_2)$.

9. *True or False:* A function f has a local maximum at c if there is an open interval I containing c so that, for all x in I, $f(x) \le f(c)$.

10. *True or False:* Even functions have graphs that are symmetric with respect to the origin.

Skill Building

In Problems 11–20, use the given graph of the function f.

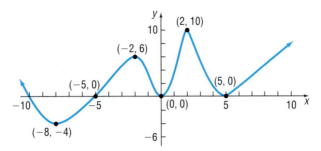

11. Is f increasing on the interval $(-8, -2)$?

12. Is f decreasing on the interval $(-8, -4)$?

13. Is f increasing on the interval $(2, 10)$?

14. Is f decreasing on the interval $(2, 5)$?

15. List the interval(s) on which f is increasing.

16. List the interval(s) on which f is decreasing.

17. Is there a local maximum at 2? If yes, what is it?

18. Is there a local maximum at 5? If yes, what is it?

19. List the numbers at which f has a local maximum. What are these local maxima?

20. List the numbers at which f has a local minimum. What are these local minima?

In Problems 21–28, the graph of a function is given. Use the graph to find:
  (a) *The intercepts, if any*
  (b) *Its domain and range*
  (c) *The intervals on which it is increasing, decreasing, or constant*
  (d) *Whether it is even, odd, or neither*

21.

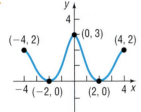

22.

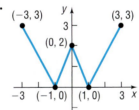

23.

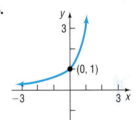

24.

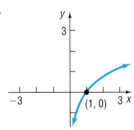

25.

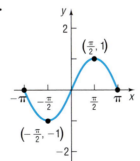

26.

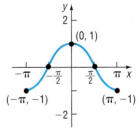

27.

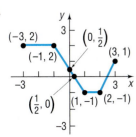

28.

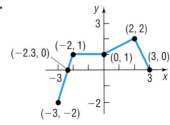

In Problems 29–32, the graph of a function f is given. Use the graph to find:
  (a) The numbers, if any, at which f has a local maximum. What are these local maxima?
  (b) The numbers, if any, at which f has a local minimum. What are these local minima?

29.

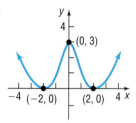

30.

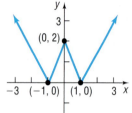

31.

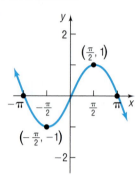

32.

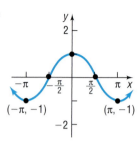

In Problems 33–44, determine algebraically whether each function is even, odd, or neither.

33. $f(x) = 4x^3$

34. $f(x) = 2x^4 - x^2$

35. $g(x) = -3x^2 - 5$

36. $h(x) = 3x^3 + 5$

37. $F(x) = \sqrt[3]{x}$

38. $G(x) = \sqrt{x}$

39. $f(x) = x + |x|$

40. $f(x) = \sqrt[3]{2x^2 + 1}$

41. $g(x) = \dfrac{1}{x^2}$

42. $h(x) = \dfrac{x}{x^2 - 1}$

43. $h(x) = \dfrac{-x^3}{3x^2 - 9}$

44. $F(x) = \dfrac{2x}{|x|}$

In Problems 45–52, use a graphing utility to graph each function over the indicated interval and approximate any local maxima and local minima. Determine where the function is increasing and where it is decreasing. Round answers to two decimal places.

45. $f(x) = x^3 - 3x + 2 \quad (-2, 2)$

46. $f(x) = x^3 - 3x^2 + 5 \quad (-1, 3)$

47. $f(x) = x^5 - x^3 \quad (-2, 2)$

48. $f(x) = x^4 - x^2 \quad (-2, 2)$

49. $f(x) = -0.2x^3 - 0.6x^2 + 4x - 6 \quad (-6, 4)$

50. $f(x) = -0.4x^3 + 0.6x^2 + 3x - 2 \quad (-4, 5)$

51. $f(x) = 0.25x^4 + 0.3x^3 - 0.9x^2 + 3 \quad (-3, 2)$

52. $f(x) = -0.4x^4 - 0.5x^3 + 0.8x^2 - 2 \quad (-3, 2)$

53. Find the average rate of change of $f(x) = -2x^2 + 4$
  (a) From 0 to 2
  (b) From 1 to 3
  (c) From 1 to 4

54. Find the average rate of change of $f(x) = -x^3 + 1$
  (a) From 0 to 2
  (b) From 1 to 3
  (c) From −1 to 1

55. Find the average rate of change of $g(x) = x^3 - 2x + 1$
  (a) From −3 to −2
  (b) From −1 to 1
  (c) From 1 to 3

56. Find the average rate of change of $h(x) = x^2 - 2x + 3$
  (a) From −1 to 1
  (b) From 0 to 2
  (c) From 2 to 5

57. $f(x) = 5x - 2$
  (a) Find the average rate of change from 1 to x.
  (b) Use the result of part (a) to find the average rate of change of f from 1 to 3. Interpret this result.
  (c) Find an equation of the secant line containing $(1, f(1))$ and $(3, f(3))$.
  (d) Using a graphing utility, draw the graph of f and the secant line on the same screen.

58. $f(x) = -4x + 1$
  (a) Find the average rate of change from 2 to x.
  (b) Use the result of part (a) to find the average rate of change of f from 2 to 5. Interpret this result.
  (c) Find an equation of the secant line containing $(2, f(2))$ and $(5, f(5))$.
  (d) Using a graphing utility, draw the graph of f and the secant line on the same screen.

59. $g(x) = x^2 - 2$
  (a) Find the average rate of change from −2 to x.
  (b) Use the result of part (a) to find the average rate of change of g from −2 to 1. Interpret this result.
  (c) Find an equation of the secant line containing $(-2, g(-2))$ and $(1, g(1))$.
  (d) Using a graphing utility, draw the graph of g and the secant line on the same screen.

60. $g(x) = x^2 + 1$
  (a) Find the average rate of change from −1 to x.
  (b) Use the result of part (a) to find the average rate of change of g from −1 to 2. Interpret this result.
  (c) Find an equation of the secant line containing $(-1, g(-1))$ and $(2, g(2))$.
  (d) Using a graphing utility, draw the graph of g and the secant line on the same screen.

61. $h(x) = x^2 - 2x$
  (a) Find the average rate of change from 2 to x.
  (b) Use the result of part (a) to find the average rate of change of h from 2 to 4. Interpret this result.
  (c) Find an equation of the secant line containing $(2, h(2))$ and $(4, h(4))$.
  (d) Using a graphing utility, draw the graph of h and the secant line on the same screen.

62. $h(x) = -2x^2 + x$
  (a) Find the average rate of change from 0 to x.
  (b) Use the result of part (a) to find the average rate of change of h from 0 to 3. Interpret this result.
  (c) Find an equation of the secant line containing $(0, h(0))$ and $(3, h(3))$.
  (d) Using a graphing utility, draw the graph of h and the secant line on the same screen.

Applications and Extensions

63. Constructing an Open Box An open box with a square base is to be made from a square piece of cardboard 24 inches on a side by cutting out a square from each corner and turning up the sides (see the figure).

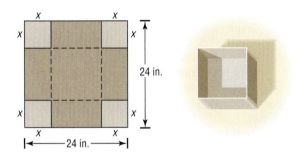

  (a) Express the volume V of the box as a function of the length x of the side of the square cut from each corner.
  (b) What is the volume if a 3-inch square is cut out?
  (c) What is the volume if a 10-inch square is cut out?
  (d) Graph $V = V(x)$. For what value of x is V largest?

64. Constructing an Open Box An open box with a square base is required to have a volume of 10 cubic feet.
  (a) Express the amount A of material used to make such a box as a function of the length x of a side of the square base.
  (b) How much material is required for a base 1 foot by 1 foot?
  (c) How much material is required for a base 2 feet by 2 feet?
  (d) Graph $A = A(x)$. For what value of x is A smallest?

65. Maximum Height of a Ball The height s of a ball (in feet) thrown with an initial velocity of 80 feet per second from an initial height of 6 feet is given as a function of the time t (in seconds) by

$$s(t) = -16t^2 + 80t + 6$$

  (a) Graph $s = s(t)$.
  (b) Determine the time at which height is maximum.
  (c) What is the maximum height?

66. Maximum Height of a Ball On July 1, 2004, the Cassini probe became the first spacecraft to orbit the planet Saturn. Although Saturn is about 764 times the size of Earth, it has a very similar gravitational force. The height s of an object thrown upward from Saturn's surface with an initial velocity of 100 feet per second is given as a function of time t (in seconds) by $s(t) = -17.28t^2 + 100t$.
  (a) Use a graphing utility to graph $s = s(t)$.
  (b) Determine the time at which height is a maximum.
  (c) What is the maximum height?
  (d) The same object thrown from the surface of Earth would have a height given by $s(t) = -16t^2 + 100t$. Determine the maximum height of the object on Earth and compare this to your result from part (c).

67. Minimum Average Cost The average cost of producing x riding lawn mowers per hour is given by

$$\overline{C}(x) = 0.3x^2 + 21x - 251 + \frac{2500}{x}$$

  (a) Graph $\overline{C} = \overline{C}(x)$.
  (b) Determine the number of riding lawn mowers to produce in order to minimize average cost.
  (c) What is the minimum average cost?

68. Medicine Concentration The concentration C of a medication in the bloodstream t hours after being administered is given by

$$C(t) = -0.002x^4 + 0.039t^3 - 0.285t^2 + 0.766t + 0.085.$$

  (a) After how many hours will the concentration be highest?
  (b) A woman nursing a child must wait until the concentration is below 0.5 before she can feed him. After taking the medication, how long must she wait before feeding her child?

69. Revenue from Selling Bikes The following data represent the total revenue that would be received from selling x bicycles at Tunney's Bicycle Shop.

Number of Bicycles, x	Total Revenue, R (Dollars)
0	0
25	28,000
60	45,000
102	53,400
150	59,160
190	62,360
223	64,835
249	66,525

(a) Plot the points (x, R) in a rectangular coordinate system.
(b) Draw a line through the points $(0, 0)$ and $(25, 28000)$ on the graph found in part (a).
(c) Find the average rate of change of revenue from 0 to 25 bicycles.
(d) Interpret the average rate of change found in part (c).
(e) Draw a line through the points $(190, 62360)$ and $(223, 64835)$ on the graph found in part (a).
(f) Find the average rate of change of revenue from 190 to 223 bicycles.
(g) Interpret the average rate of change found in part (f).
(h) What is happening to the average rate of change of revenue as the number of bicycles increases?

70. Cost of Manufacturing Bikes The following data represent the monthly cost of producing bicycles at Tunney's Bicycle Shop.

Number of Bicycles, x	Total Cost of Production, C (Dollars)
0	24,000
25	27,750
60	31,500
102	35,250
150	39,000
190	42,750
223	46,500
249	50,250

(a) Plot the points (x, C) in a rectangular coordinate system.
(b) Draw a line through the points $(0, 24000)$ and $(25, 27750)$ on the graph found in part (a).

(c) Find the average rate of change of the cost from 0 to 25 bicycles.
(d) Interpret the average rate of change found in part (c).
(e) Draw a line through the points $(190, 42750)$ and $(223, 46500)$ on the graph found in part (a).
(f) Find the average rate of change of the cost from 190 to 223 bicycles.
(g) Interpret the average rate of change found in part (f).
(h) What is happening to the average rate of change of cost as the number of bicycles increases?

71. For the function $f(x) = x^2$, compute each average rate of change:
(a) From 0 to 1
(b) From 0 to 0.5
(c) From 0 to 0.1
(d) From 0 to 0.01
(e) From 0 to 0.001
(f) Graph each of the secant lines.
(g) What do you think is happening to the secant lines?
(h) What is happening to the slopes of the secant lines? Is there some number that they are getting closer to? What is that number?

72. For the function $f(x) = x^2$, compute each average rate of change:
(a) From 1 to 2
(b) From 1 to 1.5
(c) From 1 to 1.1
(d) From 1 to 1.01
(e) From 1 to 1.001
(f) Graph each of the secant lines.
(g) What do you think is happening to the secant lines?
(h) What is happening to the slopes of the secant lines? Is there some number that they are getting closer to? What is that number?

Problems 73–80 require the following discussion of a secant line. The slope of the secant line containing the two points $(x, f(x))$ and $(x + h, f(x + h))$ on the graph of a function $y = f(x)$ may be given as

$$m_{\text{sec}} = \frac{f(x + h) - f(x)}{(x + h) - x} = \frac{f(x + h) - f(x)}{h}, \qquad h \neq 0$$

*In calculus, this expression is called the **difference quotient of f**.*
(a) *Express the slope of the secant line of each function in terms of x and h. Be sure to simplify your answer.*
(b) *Find m_{sec} for $h = 0.5, 0.1$, and 0.01 at $x = 1$. What value does m_{sec} approach as h approaches 0?*
(c) *Find the equation for the secant line at $x = 1$ with $h = 0.01$.*
(d) *Graph f and the secant line found in part (c) on the same viewing window.*

73. $f(x) = 2x + 5$ **74.** $f(x) = -3x + 2$ **75.** $f(x) = x^2 + 2x$ **76.** $f(x) = 2x^2 + x$

77. $f(x) = 2x^2 - 3x + 1$ **78.** $f(x) = -x^2 + 3x - 2$ **79.** $f(x) = \dfrac{1}{x}$ **80.** $f(x) = \dfrac{1}{x^2}$

Discussion and Writing

81. Draw the graph of a function that has the following characteristics: domain: all real numbers; range: all real numbers; intercepts: $(0, -3)$ and $(3, 0)$; a local maximum of -2 is at -1; a local minimum of -6 is at 2. Compare your graph with others. Comment on any differences.

82. Redo Problem 81 with the following additional information: increasing on $(-\infty, -1)$, $(2, \infty)$; decreasing on $(-1, 2)$. Again compare your graph with others and comment on any differences.

83. How many *x*-intercepts can a function defined on an interval have if it is increasing on that interval? Explain.

84. Suppose that a friend of yours does not understand the idea of increasing and decreasing functions. Provide an explanation complete with graphs that clarifies the idea.

85. Can a function be both even and odd? Explain.

86. Using a graphing utility, graph $y = 5$ on the interval $(-3, 3)$. Use MAXIMUM to find the local maxima on $(-3, 3)$. Comment on the result provided by the calculator.

'Are You Prepared?' Answers

1. $2 < x < 5$ **2.** 1 **3.** symmetric with respect to the *y*-axis **4.** $y + 2 = 5(x - 3)$ **5.** $(-3, 0), (3, 0), (0, -9)$

2.4 Linear Functions and Models

PREPARING FOR THIS SECTION *Before getting started, review the following:*

• Lines (Section 1.4, pp. 27–36)

• Linear Equations (Appendix, Section A.5, pp. 694–695)

 Now work the 'Are You Prepared?' problems on page 101.

OBJECTIVES 1 Graph Linear Functions
2 Work with Applications of Linear Functions
3 Draw and Interpret Scatter Diagrams
4 Distinguish between Linear and Nonlinear Relations
5 Use a Graphing Utility to Find the Line of Best Fit
6 Construct a Linear Model Using Direct Variation

1 **Graph Linear Functions**

In Section 1.4, we introduced lines. In particular, we discussed the slope–intercept form of the equation of a line $y = mx + b$. When we write the slope–intercept form of a line using function notation, we have a *linear function*.

> A **linear function** is a function of the form
>
> $$f(x) = mx + b$$
>
> The graph of a linear function is a line with slope m and y-intercept b.

EXAMPLE 1 **Graphing a Linear Function**

Figure 30

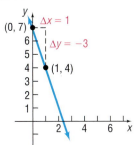

Graph the linear function: $f(x) = -3x + 7$

Solution This is a linear function with slope $m = -3$ and y-intercept 7. To graph this function, we start by plotting the point $(0, 7)$, the y-intercept, and use the slope to find an additional point by moving right 1 unit and down 3 units. See Figure 30. ◄

Alternatively, we could have found an additional point by evaluating the function at some $x \neq 0$. For $x = 1$, we find $f(1) = -3(1) + 7 = 4$ and obtain the point $(1, 4)$ on the graph.

An important property of a linear function is that the average rate of change between any two values is always the same; that is, it is a constant. Moreover, this constant is the slope of the line.

Theorem

> ### Average Rate of Change of Linear Functions
>
> Linear functions have a constant average rate of change. The constant average rate of change of $f(x) = mx + b$ is
>
> $$\frac{\Delta y}{\Delta x} = m$$

Proof The average rate of change of $f(x) = mx + b$ from c to x is

$$\text{Average rate of change} = \frac{f(x) - f(c)}{x - c}$$

$$= \frac{(mx + b) - (mc + b)}{x - c}$$

$$= \frac{mx - mc}{x - c}$$

$$= \frac{m(x - c)}{x - c}$$

$$= m \qquad \blacksquare$$

Based on the theorem just proved, the average rate of change of $f(x) = 3x - 1$ is 3. The average rate of change of $g(x) = -\frac{2}{5}x + 5$ is $-\frac{2}{5}$.

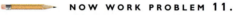 NOW WORK PROBLEM 11.

2 Work with Applications of Linear Functions

There are many applications of linear functions. Let's look at one from accounting.

EXAMPLE 2

Straight-line Depreciation

Book value is the value of an asset that a company uses to create its balance sheet. Some companies depreciate their assets using straight-line depreciation so that the value of the asset declines by a fixed amount each year. The amount of the decline depends on the useful life that the company places on the asset. Suppose that a company just purchased a fleet of new cars for its sales force at a cost of $28,000 per car. The company chooses to depreciate each vehicle using the straight-line method over 7 years. This means that each car will depreciate by $\dfrac{\$28,000}{7} = \$4000$ per year.

(a) Write a linear function that expresses the book value V of each car as a function of its age, x.

(b) Graph the linear function.

(c) What is the book value of each car after 3 years?

(d) Interpret the slope.

(e) When will the book value of each car be $8000?

   [**Hint:** Solve the equation $V(x) = 8000$.]

Solution   (a) If we let $V(x)$ represent the value of each car after x years, then $V(0)$ represents the original value of each car, so $V(0) = \$28{,}000$. The y-intercept of the linear function is \$28,000. Because each car depreciates by \$4000 per year, the slope of the linear function is $-\$4000$. The linear function that represents the value of each car after x years is

$$V(x) = -4000x + 28{,}000$$

(b) Figure 31 shows the graph of $V = V(x)$.

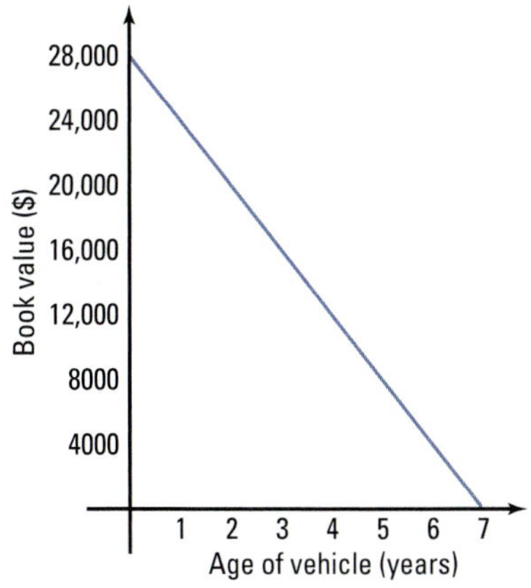

(c) The book value of each car after 3 years is

$$V(3) = -4000(3) + 28{,}000$$
$$= \$16{,}000$$

(d) Since the slope of $V(x) = -4000x + 28{,}000$ is -4000, the average rate of change of book value is $-\$4000$ per year. So, for each additional year that passes, the book value of each car decreases by \$4000.

(e) To find when the book value is \$8000, we solve the equation

$$V(x) = 8000$$
$$-4000x + 28{,}000 = 8000$$
$$-4000x = -20{,}000 \qquad \text{Subtract \$28,000 from each side.}$$
$$x = \frac{-20{,}000}{-4000} = 5 \qquad \text{Divide both sides by } -4000.$$

Each car will have a book value of \$8000 after 5 years. ◀

Figure 31

Book value ($)

28,000
24,000
20,000
16,000
12,000
8000
4000

1 2 3 4 5 6 7
Age of vehicle (years)

 NOW WORK PROBLEM 39.

Next we look at an application from economics.

EXAMPLE 3   **Supply and Demand**

The quantity supplied of a good is the amount of a product that a company is willing to make available for sale at a given price. The quantity demanded of a good is the amount of a product that consumers are willing to purchase at a given price. Suppose that the quantity supplied, S, and quantity demanded, D, of cellular telephones each month are given by the following functions:

$$S(p) = 60p - 900$$
$$D(p) = -15p + 2850$$

where p is the price (in dollars) of the telephone.

(a) The **equilibrium price** of a product is defined as the price at which quantity supplied equals quantity demanded. That is, the equilibrium price is the price at which $S(p) = D(p)$. Find the equilibrium price of cellular telephones. What is the equilibrium quantity, the amount demanded (or supplied), at the equilibrium price?

(b) Determine the prices for which quantity supplied is greater than quantity demanded. That is, solve the inequality $S(p) > D(p)$.

(c) Graph $S = S(p)$, $D = D(p)$ and label the equilibrium price.

Solution

(a) We solve the equation $S(p) = D(p)$.

$$60p - 900 = -15p + 2850 \qquad \textcolor{blue}{S(p) = D(p)}$$
$$60p = -15p + 3750 \qquad \textcolor{blue}{\text{Add 900 to each side.}}$$
$$75p = 3750 \qquad \textcolor{blue}{\text{Add 15p to each side.}}$$
$$p = \$50 \qquad \textcolor{blue}{\text{Divide each side by 75.}}$$

The equilibrium price of cellular phones is $50. To find the equilibrium quantity, we evaluate either $S(p)$ or $D(p)$ at $p = 50$.

$$S(50) = 60(50) - 900 = 2100$$

The equilibrium quantity is 2100 cellular phones. At a price of $50 per phone, the company will sell 2100 phones each month and have no shortages or excess inventory.

(b) We solve the inequality $S(p) > D(p)$.

$$60p - 900 > -15p + 2850 \qquad \textcolor{blue}{S(p) > D(p)}$$
$$60p > -15p + 3750 \qquad \textcolor{blue}{\text{Add 900 to each side.}}$$
$$75p > 3750 \qquad \textcolor{blue}{\text{Add 15p to each side.}}$$
$$p > \$50 \qquad \textcolor{blue}{\text{Divide each side by 75.}}$$

If the company charges more than $50 per phone, then quantity supplied will exceed quantity demanded. In this case the company will have excess phones in inventory.

(c) Figure 32 shows the graphs of $S = S(p)$ and $D = D(p)$.

Figure 32

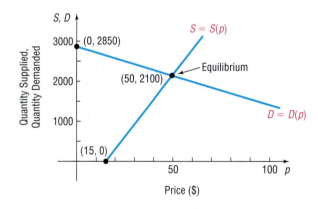

NOW WORK PROBLEM 35.

3 Draw and Interpret Scatter Diagrams

Linear functions can also be created by *fitting* a linear function to data. The first step in finding this relation is to plot the ordered pairs using rectangular coordinates. The resulting graph is called a **scatter diagram**. When drawing a scatter diagram, the independent variable is plotted on the horizontal axis and the dependent variable is plotted on the vertical axis.

| EXAMPLE 4 | **Drawing and Interpreting a Scatter Diagram** |

In baseball, the on-base percentage for a team represents the percentage of time that the players safely reach base. The data given in Table 5 represent the number of runs scored y and the on-base percentage x for teams in the National League during the 2003 baseball season.

Table 5

Team	On-Base Percentage, x	Runs Scored, y	(x, y)
Atlanta	34.9	907	(34.9, 907)
St. Louis	35.0	876	(35.0, 876)
Colorado	34.4	853	(34.4, 853)
Houston	33.6	805	(33.6, 805)
Philadelphia	34.3	791	(34.3, 791)
San Francisco	33.8	755	(33.8, 755)
Pittsburgh	33.8	753	(33.8, 753)
Florida	33.3	751	(33.3, 751)
Chicago Cubs	32.3	724	(32.3, 724)
Arizona	33.0	717	(33.0, 717)
Milwaukee	32.9	714	(32.9, 714)
Montreal	32.6	711	(32.6, 711)
Cincinnati	31.8	694	(31.8, 694)
San Diego	33.3	678	(33.3, 678)
NY Mets	31.4	642	(31.4, 642)
Los Angeles	30.3	574	(30.3, 574)

SOURCE: espn.com

(a) Draw a scatter diagram of the data by hand, treating on-base percentage as the independent variable.

(b) Use a graphing utility to draw a scatter diagram.

(c) Describe what happens as the on-base percentage increases.

Solution (a) To draw a scatter diagram by hand, we plot the ordered pairs listed in Table 5, with the on-base percentage as the x-coordinate and the runs scored as the y-coordinate. See Figure 33(a). Notice that the points in the scatter diagram are not connected.

(b) Figure 33(b) shows a scatter diagram using a TI-84 Plus graphing calculator.

Figure 33

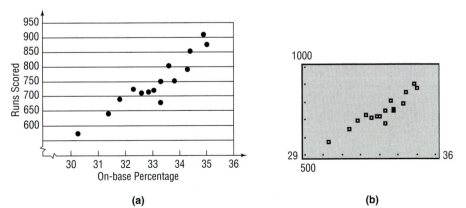

(a)   (b)

(c) We see from the scatter diagrams that, as the on-base percentage increases, the number of runs scored also increases. ◀

NOW WORK PROBLEM 25(a).

4 Distinguish between Linear and Nonlinear Relations

Scatter diagrams are used to help us to see the type of relation that exists between two variables. In this text, we will discuss a variety of different relations that may exist between two variables. For now, we concentrate on distinguishing between linear and nonlinear relations. See Figure 34.

Figure 34

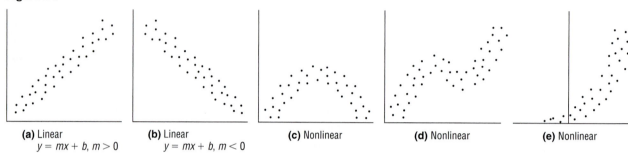

(a) Linear
$y = mx + b, m > 0$

(b) Linear
$y = mx + b, m < 0$

(c) Nonlinear

(d) Nonlinear

(e) Nonlinear

EXAMPLE 5   **Distinguishing between Linear and Nonlinear Relations**

Determine whether the relation between the two variables in Figure 35 is linear or nonlinear.

Figure 35

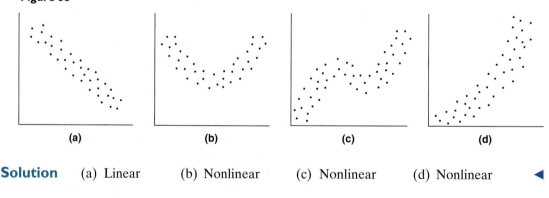

(a)

(b)

(c)

(d)

Solution   (a) Linear   (b) Nonlinear   (c) Nonlinear   (d) Nonlinear   ◀

NOW WORK PROBLEM 19.

In this section we will study data whose scatter diagrams imply that a linear relation exists between the two variables. Nonlinear data will be discussed in later chapters.

5 Use a Graphing Utility to Find the Line of Best Fit

Suppose that the scatter diagram of a set of data appears to be linearly related as in Figure 34(a) or (b). We might wish to find an equation of a line that relates the two variables. One way to obtain an equation for such data is to draw a line through two points on the scatter diagram and determine the equation of the line.

EXAMPLE 6   Finding an Equation for Linearly Related Data

Using the data in Table 5 from Example 4:

(a) Select two points and find an equation of the line containing the points.

(b) Graph the line on the scatter diagram obtained in Example 4(b).

Solution   (a) Select two points, say $(32.6, 711)$ and $(34.9, 907)$. The slope of the line joining the points $(32.6, 711)$ and $(34.9, 907)$ is

$$m = \frac{907 - 711}{34.9 - 32.6} = \frac{196}{2.3} = 85.22$$

rounded to two decimal places. The equation of the line with slope 85.22 and passing through $(32.6, 711)$ is found using the point–slope form with $m = 85.22$, $x_1 = 32.6$, and $y_1 = 711$.

Figure 36

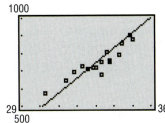

$$y - y_1 = m(x - x_1)$$
$$y - 711 = 85.22(x - 32.6)$$
$$y - 711 = 85.22x - 2778.17$$
$$y = 85.22x - 2067.17$$

(b) Figure 36 shows the scatter diagram with the graph of the line found in part (a). ◄

Select two other points and complete the solution. Add the line obtained to Figure 36.

NOW WORK PROBLEMS **25(b)** AND **(c)**.

The line obtained in Example 6 depends on the selection of points, which will vary from person to person. So the line that we found might be different from the line you found. Although the line we found in Example 6 appears to fit the data well, there may be a line that "fits it better." Do you think your line fits the data better? Is there a line of *best fit*? As it turns out, there is a method for finding the line that best fits linearly related data (called the *line of best fit*).[*]

EXAMPLE 7   Finding the Line of Best Fit

Using the data in Table 5 from Example 4:

(a) Find the line of best fit using a graphing utility.

(b) Graph the line of best fit on the scatter diagram obtained in Example 4(b).

(c) Interpret the slope.

(d) Use the line of best fit to predict the number of runs a team will score if their on-base percentage is 34.1.

---

[*]We shall not discuss the underlying mathematics of lines of best fit in this book.

Solution (a) Graphing utilities contain built-in programs that find the line of best fit for a collection of points in a scatter diagram. Upon executing the LINear REGression program, we obtain the results shown in Figure 37. The output that the utility provides shows us the equation $y = ax + b$, where a is the slope of the line and b is the y-intercept. The line of best fit that relates on-base percentage to runs scored may be expressed as the line $y = 62.17x - 1315.53$.

(b) Figure 38 shows the graph of the line of best fit, along with the scatter diagram.

Figure 37

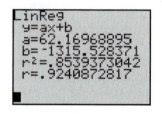

Figure 38

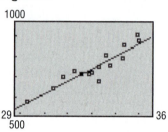

(c) The slope of the line of best fit is 62.17, which means that, for every 1% increase in the on-base percentage, runs scored increase 62.17.

(d) Letting $x = 34.1$ in the equation of the line of best fit, we obtain $y = 62.17(34.1) - 1315.53 \approx 804$ runs. ◄

NOW WORK PROBLEMS 25(d) AND (e).

Does the line of best fit appear to be a good fit? In other words, does the line appear to accurately describe the relation between on-base percentage and runs scored?

And just how "good" is this line of best fit? The answers are given by what is called the *correlation coefficient*. Look again at Figure 37. The last line of output is $r = 0.924$. This number, called the **correlation coefficient,** r, $-1 \leq r \leq 1$, is a measure of the strength of the *linear relation* that exists between two variables. The closer that $|r|$ is to 1, the more perfect the linear relationship is. If r is close to 0, there is little or no *linear* relationship between the variables. A negative value of r, $r < 0$, indicates that as x increases y decreases; a positive value of r, $r > 0$, indicates that as x increases y does also. The data given in Table 5, having a correlation coefficient of 0.924, are indicative of a strong linear relationship with positive slope.

6 Construct a Linear Model Using Direct Variation

When a mathematical model is developed for a real-world problem, it often involves relationships between quantities that are expressed in terms of proportionality.

Force is proportional to acceleration.

For an ideal gas held at a constant temperature, pressure and volume are inversely proportional.

The force of attraction between two heavenly bodies is inversely proportional to the square of the distance between them.

Revenue is directly proportional to sales.

Each of these statements illustrates the idea of **variation**, or how one quantity varies in relation to another quantity. Quantities may vary *directly*, *inversely*, or *jointly*. We discuss direct variation here.

Figure 39
$y = kx, k > 0, x \geq 0$

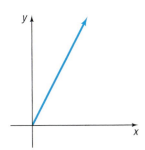

Let x and y denote two quantities. Then y **varies directly** with x, or y is **directly proportional to** x, if there is a nonzero number k such that

$$y = kx$$

The number k is called the **constant of proportionality**.

If y varies directly with x, then y is a linear function of x. The graph in Figure 39 illustrates the relationship between y and x if y varies directly with x and $k > 0, x \geq 0$. Note that the constant of proportionality is, in fact, the slope of the line.

If we know that two quantities vary directly, then knowing the value of each quantity in one instance enables us to write a formula that is true in all cases.

EXAMPLE 8   **Mortgage Payments**

The monthly payment p on a mortgage varies directly with the amount borrowed B. If the monthly payment on a 30-year mortgage is $6.65 for every $1000 borrowed, find a formula that relates the monthly payment p to the amount borrowed B for a mortgage with the same terms. Then find the monthly payment p when the amount borrowed B is $120,000.

Solution   Because p varies directly with B, we know that

$$p = kB$$

for some constant k. Because $p = 6.65$ when $B = 1000$, it follows that

$$6.65 = k(1000)$$
$$k = 0.00665$$

Figure 40

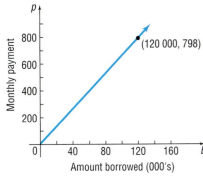

(120 000, 798)

Amount borrowed (000's)

So we have

$$p = 0.00665B$$

We see that p is a linear function of B: $p(B) = 0.00665B$. In particular, when $B = \$120,000$, we find that

$$p(120{,}000) = 0.00665(\$120{,}000) = \$798$$

Figure 40 illustrates the relationship between the monthly payment p and the amount borrowed B. ◀

NOW WORK PROBLEM 45.

2.4 Assess Your Understanding

'Are You Prepared?'

Answers are given at the end of these exercises. If you get a wrong answer, read the pages listed in red.

1. Graph $y = 2x - 3$. (p. 35)

2. Find the slope of the line joining the points $(2, 5)$ and $(-1, 3)$. (p. 27)

3. Find an equation of the line whose slope is -3 and contains the point $(-1, 5)$. (p. 32)

4. Solve $6x - 900 = -15x + 2850$. (p. 694)

Concepts and Vocabulary

5. For the graph of the linear function $f(x) = mx + b$, m is the _____ and b is the _____.

6. A _____ _____ is used to help us to see the type of relation, if any, that may exist between two variables.

7. If x and y are two quantities, then y is directly proportional to x if there is a nonzero number k such that _____.

8. *True or False:* The average rate of change of a linear function $f(x) = mx + b$ is the constant m.

9. *True or False:* The correlation coefficient is a measure of the strength of a linear relation between two variables and must lie between -1 and 1, inclusive.

10. *True or False:* If y varies directly with x, then y is a linear function of x.

Skill Building

For Problems 11–18, graph each linear function by hand. Determine the average rate of change of each function.

11. $f(x) = 2x + 3$

12. $g(x) = 5x - 4$

13. $h(x) = -3x + 4$

14. $p(x) = -x + 6$

15. $f(x) = \dfrac{1}{4}x - 3$

16. $h(x) = -\dfrac{2}{3}x + 4$

17. $F(x) = 4$

18. $G(x) = -2$

In Problems 19–24, examine the scatter diagram and determine whether the type of relation, if any, that may exist is linear or nonlinear.

19.
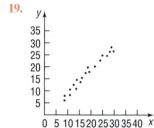

20.
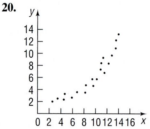

21.
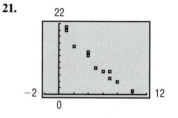

22.
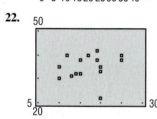

23.
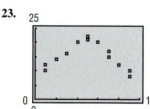

24.
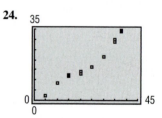

In Problems 25–30:

  (a) *Draw a scatter diagram by hand and using a graphing utility.*

  (b) *Select two points from the scatter diagram and find the equation of the line containing the points selected.*

  (c) *Graph the line found in part (b) on the scatter diagram.*

  (d) *Use a graphing utility to find the line of best fit.*

  (e) *Use a graphing utility to graph the line of best fit on the scatter diagram.*

25.

x	3	4	5	6	7	8	9
y	4	6	7	10	12	14	16

26.

x	3	5	7	9	11	13
y	0	2	3	6	9	11

27.

x	-2	-1	0	1	2
y	-4	0	1	4	5

28.

x	-2	-1	0	1	2
y	7	6	3	2	0

29.

x	-20	-17	-15	-14	-10
y	100	120	118	130	140

30.

x	-30	-27	-25	-20	-14
y	10	12	13	13	18

Applications and Extensions

31. Car Rentals The cost C, in dollars, of renting a moving truck for a day is given by the function $C(x) = 0.25x + 35$, where x is the number of miles driven.

  (a) What is the cost if you drive $x = 40$ miles?

  (b) If the cost of renting the moving truck is $80, how many miles did you drive?

  (c) Suppose that you want to keep the cost below $100. What is the maximum number of miles that you can drive?

32. Phone Charges The monthly cost C, in dollars, of a certain cellular phone plan is given by the function $C(x) = 0.38x + 5$, where x is the number of minutes used on the phone.

  (a) What is the cost if you talk on the phone for $x = 50$ minutes?

  (b) Suppose that your monthly bill is $29.32. How many minutes did you use the phone?

(c) Suppose that you budget yourself $60 per month for the phone. What is the maximum number of minutes that you can talk?

33. Disability Benefits   The average monthly benefit B, in dollars, for individuals on disability is given by the function $B(t) = 19.25t + 585.72$, where t is the number of years since 1990.
  (a) What was the average monthly benefit in 2000 $(t = 10)$?
  (b) In what year will the average monthly benefit be $893.72?
  (c) In what year will the average monthly benefit exceed $1000?

34. Health Expenditures   The total private health expenditures H, in billions of dollars, is given by the function $H(t) = 26t + 411$, where t is the number of years since 1990.
  (a) What was the total private health expenditures in 2000 $(t = 10)$?
  (b) In what year will total private health expenditures be $879 billion?
  (c) In what year will total private health expenditures exceed $1 trillion ($1000 billion)?

35. Supply and Demand   Suppose that the quantity supplied S and quantity demanded D of T-shirts at a concert are given by the following functions:

$$S(p) = -200 + 50p$$
$$D(p) = 1000 - 25p$$

where p is the price in dollars. The equilibrium price of a market is defined as the price at which quantity supplied equals quantity demanded $(S = D)$.
  (a) Find the equilibrium price for T-shirts at this concert. What is the equilibrium quantity?
  (b) Determine the prices for which quantity demanded is greater than quantity supplied.
  (c) What do you think will eventually happen to the price of T-shirts if quantity demanded is greater than quantity supplied?

36. Supply and Demand   Suppose that the quantity supplied S and quantity demanded D of hot dogs at a baseball game are given by the following functions:

$$S(p) = -2000 + 3000p$$
$$D(p) = 10{,}000 - 1000p$$

where p is the price in dollars. The equilibrium price of a market is defined as the price at which quantity supplied equals quantity demanded $(S = D)$.
  (a) Find the equilibrium price for hot dogs at the baseball game. What is the equilibrium quantity?
  (b) Determine the prices for which quantity demanded is less than quantity supplied.
  (c) What do you think will eventually happen to the price of hot dogs if quantity demanded is less than quantity supplied?

---

*The point at which a company's profits equal zero is called the company's **break-even point**. For Problems 37 and 38, let R represent a company's revenue, let C represent the company's costs, and let x represent the number of units produced and sold each day.*
(a) Find the firm's break-even point; that is, find x so that R = C.
(b) Find the values of x such that R(x) > C(x). This represents the number of units that the company must sell to earn a profit.

37. $R(x) = 8x$
  $C(x) = 4.5x + 17{,}500$

38. $R(x) = 12x$
  $C(x) = 10x + 15{,}000$

---

39. Straight-line Depreciation   Suppose that a company has just purchased a new computer for $3000. The company chooses to depreciate the computer using the straight-line method over 3 years.
  (a) Write a linear function that expresses the book value of the computer as a function of its age.
  (b) Graph the linear function.
  (c) What is the book value of the computer after 2 years?
  (d) When will the computer be worth $2000?

40. Straight-line Depreciation   Suppose that a company has just purchased a new machine for its manufacturing facility for $120,000. The company chooses to depreciate the machine using the straight-line method over 10 years.
  (a) Write a linear function that expresses the book value of the machine as a function of its age.

(b) Graph the linear function.
  (c) What is the book value of the machine after 4 years?
  (d) When will the machine be worth $60,000?

41. Cost Function   The simplest cost function is the linear cost function $C(x) = mx + b$, where C is the cost of producing x units. Here the y-intercept b represents the fixed costs of operating a business and the slope m represents the variable cost of producing each unit. Suppose that a small bicycle manufacturer has daily fixed costs of $1800 and each bicycle costs $90 to manufacture.
  (a) Write a linear function that expresses the cost C of manufacturing x bicycles in a day.
  (b) Graph the linear function.
  (c) What is the cost of manufacturing 14 bicycles in a day?
  (d) How many bicycles could be manufactured for $3780?

42. Cost Function Refer to Problem 41. Suppose that the landlord of the building increases the bicycle manufacturer's rent by $100 per month.
  (a) Assuming that the manufacturer is open for business 20 days per month, what are the new daily fixed costs?
  (b) Write a linear function that expresses the cost C of manufacturing x bicycles in a day with the higher rent.
  (c) Graph the linear function.
  (d) What is the cost of manufacturing 14 bicycles in a day?
  (e) How many bicycles could be manufactured for $3780?

43. Truck Rentals A truck rental company rents a truck for one day by charging $29 plus $0.07 per mile.
  (a) Write a linear function that relates the cost C, in dollars, of renting the truck to the number x of miles driven.
  (b) What is the cost of renting the truck if the truck is driven 110 miles? 230 miles?

44. Long Distance A phone company offers a long distance package by charging $5 plus $0.05 per minute.
  (a) Write a linear function that relates the cost C in dollars of talking x minutes.
  (b) What is the cost of talking 105 minutes? 180 minutes?

45. Mortgage Payments The monthly payment p on a mortgage varies directly with the amount borrowed B. If the monthly payment on a 30-year mortgage is $6.49 for every $1000 borrowed, find a linear function that relates the monthly payment p to the amount borrowed B for a mortgage with the same terms. Then find the monthly payment p when the amount borrowed B is $145,000.

46. Mortgage Payments The monthly payment p on a mortgage varies directly with the amount borrowed B. If the monthly payment on a 15-year mortgage is $8.99 for every $1000 borrowed, find a linear function that relates the monthly payment p to the amount borrowed B for a mortgage with the same terms. Then find the monthly payment p when the amount borrowed B is $175,000.

47. Revenue Function At the corner Shell station, the revenue R varies directly with the number g of gallons of gasoline sold. If the revenue is $23.40 when the number of gallons sold is 12, find a linear function that relates revenue R to the number g of gallons of gasoline. Then find the revenue R when the number of gallons of gasoline sold is 10.5.

48. Cost Function The cost C of chocolate-covered almonds varies directly with the number A of pounds of almonds purchased. If the cost is $23.75 when the number of pounds of chocolate-covered almonds purchased is 5, find a linear function that relates the cost C to the number A of pounds of almonds purchased. Then find the cost C when the number of pounds of almonds purchased is 3.5.

49. Sand Castles Researchers at Bournemouth University discovered that the amount of water W varies directly with the amount of sand S that is needed to make the perfect sand castle. If 15 gallons of sand is mixed with 1.875 gallons of water, a perfect sand castle can be built. How many gallons of water should be mixed with 40 gallons of sand to create a perfect sand castle?

50. Physics: Falling Objects The velocity v of a falling object is directly proportional to the time t of the fall. If, after 2 seconds, the velocity of the object is 64 feet per second, what will its velocity be after 3 seconds?

51. Per Capita Disposable Income versus Consumption An economist wishes to estimate a linear function that relates per capita consumption expenditures C and disposable income I. Both C and I are measured in dollars. The following data represent the per capita disposable income (income after taxes) and per capita consumption in the United States for 1995 to 2003.

Year	Per Capita Disposable Income (I)	Per Capita Consumption (C)
1995	20,316	19,061
1996	21,127	19,938
1997	21,871	20,807
1998	22,212	21,385
1999	23,968	22,491
2000	25,472	23,863
2001	26,175	24,690
2002	27,259	25,622
2003	28,227	26,650

SOURCE: U.S. Department of Commerce

Let I represent the independent variable and C the dependent variable.
  (a) Use a graphing utility to draw a scatter diagram.
  (b) Use a graphing utility to find the line of best fit to the data. Express the solution using function notation.
  (c) Interpret the slope. The slope of this line is called the **marginal propensity to consume**.
  (d) Predict the per capita consumption in a year when disposable income is $28,750.
  (e) What is the income in a year when consumption is $26,900?

52. Height versus Head Circumference   A pediatrician wanted to estimate a linear function that relates a child's height, H to their head circumference, C. She randomly selects nine children from her practice, measures their height and head circumference, and obtains the data shown below. Let H represent the independent variable and C the dependent variable.
(a) Use a graphing utility to draw a scatter diagram.
(b) Use a graphing utility to find the line of best fit to the data. Express the solution using function notation.
(c) Interpret the slope.
(d) Predict the head circumference of a child that is 26 inches tall.
(e) What is the height of a child whose head circumference is 17.4 inches?

Height, H (inches)	Head Circumference, C (inches)
25.25	16.4
25.75	16.9
25	16.9
27.75	17.6
26.5	17.3
27	17.5
26.75	17.3
26.75	17.5
27.5	17.5

Source: Denise Slucki, Student at Joliet Junior College

53. Gestation Period versus Life Expectancy   A researcher would like to estimate the linear function relating the gestation period of an animal, G, and its life expectancy, L. She collects the following data.

Animal	Gestation (or incubation) Period, G (days)	Life Expectancy, L (years)
Cat	63	11
Chicken	22	7.5
Dog	63	11
Duck	28	10
Goat	151	12
Lion	108	10
Parakeet	18	8
Pig	115	10
Rabbit	31	7
Squirrel	44	9

Source: Time Almanac 2000

Let G represent the independent variable and L the dependent variable.
(a) Use a graphing utility to draw a scatter diagram.
(b) Use a graphing utility to find the line of best fit to the data. Express the solution using function notation.
(c) Interpret the slope.

(d) Predict the life expectancy of an animal whose gestation period is 89 days.

54. Mortgage Qualification   The amount of money that a lending institution will allow you to borrow mainly depends on the interest rate and your annual income. The following data represent the annual income, I, required by a bank in order to lend L dollars at an interest rate of 7.5% for 30 years.

Annual Income, I ($)	Loan Amount, L ($)
15,000	44,600
20,000	59,500
25,000	74,500
30,000	89,400
35,000	104,300
40,000	119,200
45,000	134,100
50,000	149,000
55,000	163,900
60,000	178,800
65,000	193,700
70,000	208,600

Source: Information Please Almanac, 1999

Let I represent the independent variable and L the dependent variable.
(a) Use a graphing utility to draw a scatter diagram of the data.
(b) Use a graphing utility to find the line of best fit to the data.
(c) Graph the line of best fit on the scatter diagram drawn in part (a).
(d) Interpret the slope of the line of best fit.
(e) Determine the loan amount that an individual would qualify for if her income is $42,000.

55. Demand for Jeans   The marketing manager at Levi-Strauss wishes to find a function that relates the demand D for men's jeans and p, the price of the jeans. The following data were obtained based on a price history of the jeans.

Price ($/Pair), p	Demand (Pairs of Jeans Sold per Day), D
20	60
22	57
23	56
23	53
27	52
29	49
30	44

(a) Does the relation defined by the set of ordered pairs (p, D) represent a function?
(b) Draw a scatter diagram of the data.
(c) Using a graphing utility, find the line of best fit relating price and quantity demanded.

(d) Interpret the slope.
(e) Express the relationship found in part (c) using function notation.
(f) What is the domain of the function?
(g) How many jeans will be demanded if the price is $28 a pair?

56. Advertising and Sales Revenue A marketing firm wishes to find a function that relates the sales S of a product and A,

Advertising Expenditures, A	Sales, S
20	335
22	339
22.5	338
24	343
24	341
27	350
28.3	351

the amount spent on advertising the product. The data are obtained from past experience. Advertising and sales are measured in thousands of dollars.
(a) Does the relation defined by the set of ordered pairs (A, S) represent a function?
(b) Draw a scatter diagram of the data.
(c) Using a graphing utility, find the line of best fit relating advertising expenditures and sales.
(d) Interpret the slope.
(e) Express the relationship found in part (c) using function notation.
(f) What is the domain of the function?
(g) Predict sales if advertising expenditures are $25,000.

Discussion and Writing

57. Maternal Age versus Down Syndrome A biologist would like to know how the age of the mother affects the incidence rate of Down syndrome. The following data represent the age of the mother and the incidence rate of Down syndrome per 1000 pregnancies.

Age of Mother, x	Incidence of Down Syndrome, y
33	2.4
34	3.1
35	4
36	5
37	6.7
38	8.3
39	10
40	13.3
41	16.7
42	22.2
43	28.6
44	33.3
45	50

SOURCE: Hook, E.B., *Journal of the American Medical Association*, 249, 2034-2038, 1983

Draw a scatter diagram treating age of the mother as the independent variable. Explain why it would not make sense to find the line of best fit for these data.

58. Harley Davidson, Inc., Stock The following data represent the closing year-end price of Harley Davidson, Inc., stock for the years 1990 to 2003. We let $x = 1$ represent 1990, $x = 2$ represent 1991, and so on.

Year, x	Closing Price, y
1990 ($x = 1$)	1.1609
1991 ($x = 2$)	2.6988
1992 ($x = 3$)	4.5381
1993 ($x = 4$)	5.3379
1994 ($x = 5$)	6.8032
1995 ($x = 6$)	7.0328
1996 ($x = 7$)	11.5585
1997 ($x = 8$)	13.4799
1998 ($x = 9$)	23.5424
1999 ($x = 10$)	31.9342
2000 ($x = 11$)	39.7277
2001 ($x = 12$)	54.31
2002 ($x = 13$)	46.20
2003 ($x = 14$)	47.53

SOURCE: Yahoo! Finance

Draw a scatter diagram of the data treating the year, x, as the independent variable. Explain why it would not make sense to find the line of best fit for these data.

59. Under what circumstances is a linear function $f(x) = mx + b$ odd? Can a linear function ever be even?

60. Find the line of best fit for the ordered pairs $(1, 5)$ and $(3, 8)$. What is the correlation coefficient for these data? Why is this result reasonable?

61. What does a correlation coefficient of 0 imply?

62. What would it mean if a teacher states that a student's grade in the class is directly proportional to the amount of time that the student studies?

'Are You Prepared?' Answers

1.

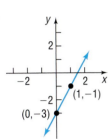

2. $\dfrac{2}{3}$

3. $y = -3x + 2$

4. $\{50\}$

2.5 Library of Functions; Piecewise-defined Functions

PREPARING FOR THIS SECTION   *Before getting started, review the following:*

- Intercepts (Section 1.2, pp. 15–17)

- Graphs of Key Equations (Section 1.2: Example 3, p. 12; Example 11, p. 19; Example 12, p.20; Example 13, p. 21)

Now work the 'Are You Prepared?' problems on page 114.

OBJECTIVES   1 Graph the Functions Listed in the Library of Functions
2 Graph Piecewise-defined Functions

Figure 41

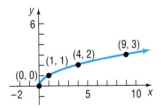

We now introduce a few more functions to add to our list of important functions. We begin with the *square root function*.

In Section 1.2 we graphed the equation $y = \sqrt{x}$. Figure 41 shows a graph of $f(x) = \sqrt{x}$. Based on the graph, we have the following properties:

Properties of $f(x) = \sqrt{x}$

1. The x-intercept of the graph of $f(x) = \sqrt{x}$ is 0. The y-intercept of the graph of $f(x) = \sqrt{x}$ is also 0.

2. The function is neither even nor odd.

3. It is increasing on the interval $(0, \infty)$.

4. It has a minimum value of 0 at $x = 0$.

EXAMPLE 1   **Graphing the Cube Root Function**

(a) Determine whether $f(x) = \sqrt[3]{x}$ is even, odd, or neither. State whether the graph of f is symmetric with respect to the y-axis or symmetric with respect to the origin.

(b) Determine the intercepts, if any, of the graph of $f(x) = \sqrt[3]{x}$.

(c) Graph $f(x) = \sqrt[3]{x}$.

Solution   (a) Because

$$f(-x) = \sqrt[3]{-x} = -\sqrt[3]{x} = -f(x)$$

the function is odd. The graph of f is symmetric with respect to the origin.

(b) The y-intercept is $f(0) = \sqrt[3]{0} = 0$. The x-intercept is found by solving the equation $f(x) = 0$.

$$f(x) = 0$$
$$\sqrt[3]{x} = 0 \qquad f(x) = \sqrt[3]{x}$$
$$x = 0 \qquad \text{Cube both sides of the equation.}$$

The x-intercept is also 0.

(c) We use the function to form Table 6 and obtain some points on the graph. Because of the symmetry with respect to the origin, we only need to find points (x, y) for which $x \geq 0$. Figure 42 shows the graph of $f(x) = \sqrt[3]{x}$.

Table 6

x	$y = f(x) = \sqrt[3]{x}$	(x, y)
0	0	$(0, 0)$
$\frac{1}{8}$	$\frac{1}{2}$	$\left(\frac{1}{8}, \frac{1}{2}\right)$
1	1	$(1, 1)$
2	$\sqrt[3]{2} \approx 1.26$	$\left(2, \sqrt[3]{2}\right)$
8	2	$(8, 2)$

Figure 42

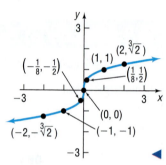

From the results of Example 1 and Figure 42, we have the following properties of the cube root function.

> **Properties of $f(x) = \sqrt[3]{x}$**
>
> 1. The x-intercept of the graph of $f(x) = \sqrt[3]{x}$ is 0. The y-intercept of the graph of $f(x) = \sqrt[3]{x}$ is also 0.
>
> 2. The function is odd.
>
> 3. It is increasing on the interval $(-\infty, \infty)$.
>
> 4. It does not have a local minimum or a local maximum.

EXAMPLE 2   **Graphing the Absolute Value Function**

(a) Determine whether $f(x) = |x|$ is even, odd, or neither. State whether the graph of f is symmetric with respect to the y-axis or symmetric with respect to the origin.

(b) Determine the intercepts, if any, of the graph of $f(x) = |x|$.

(c) Graph $f(x) = |x|$.

Solution   (a) Because

$$f(-x) = |-x| = |x| = f(x)$$

the function is even. The graph of f is symmetric with respect to the y-axis.

(b) The y-intercept is $f(0) = |0| = 0$. The x-intercept is found by solving the equation $f(x) = |x| = 0$. So the x-intercept is 0.

(c) We use the function to form Table 7 and obtain some points on the graph. Because of the symmetry with respect to the y-axis, we only need to find points (x, y) for which $x \geq 0$. Figure 43 shows the graph of $f(x) = |x|$.

Table 7

x	$y = f(x) = |x|$	(x, y)
0	0	(0, 0)
1	1	(1, 1)
2	2	(2, 2)
3	3	(3, 3)

Figure 43

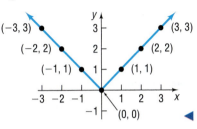

From the results of Example 2 and Figure 43, we have the following properties of the absolute value function.

Properties of $f(x) = |x|$

1. The x-intercept of the graph of $f(x) = |x|$ is 0. The y-intercept of the graph of $f(x) = |x|$ is also 0.

2. The function is even.

3. It is decreasing on the interval $(-\infty, 0)$. It is increasing on the interval $(0, \infty)$.

4. It has a local minimum of 0 at $x = 0$.

───── **Seeing the Concept** ─────

Graph $y = |x|$ on a square screen and compare what you see with Figure 43. Note that some graphing calculators use abs(x) for absolute value.

1 Graph the Functions Listed in the Library of Functions

We now provide a summary of the key functions that we have encountered. In going through this list, pay special attention to the properties of each function, particularly to the shape of each graph. Knowing these graphs will lay the foundation for later graphing techniques.

Linear Function

$$f(x) = mx + b, \qquad m \text{ and } b \text{ are real numbers}$$

Figure 44
Linear Function

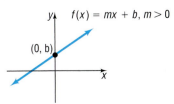

See Figure 44.

The domain of a **linear function** is the set of all real numbers. The graph of this function is a nonvertical line with slope m and y-intercept b. A linear function is increasing if $m > 0$, decreasing if $m < 0$, and constant if $m = 0$.

Constant Function

$$f(x) = b, \qquad b \text{ is a real number}$$

Figure 45
Constant Function

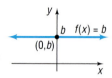

See Figure 45.

A **constant function** is a special linear function ($m = 0$). Its domain is the set of all real numbers; its range is the set consisting of a single number b. Its graph is a horizontal line whose y-intercept is b. The constant function is an even function whose graph is constant over its domain.

Identity Function

$$f(x) = x$$

Figure 46
Identity Function

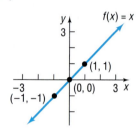

See Figure 46.

The **identity function** is also a special linear function. Its domain and range are the set of all real numbers. Its graph is a line whose slope is $m = 1$ and whose y-intercept is 0. The line consists of all points for which the x-coordinate equals the y-coordinate. The identity function is an odd function that is increasing over its domain. Note that the graph bisects quadrants I and III.

Square Function

$$f(x) = x^2$$

Figure 47
Square Function

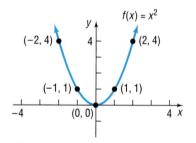

See Figure 47.

The domain of the **square function** f is the set of all real numbers; its range is the set of nonnegative real numbers. The graph of this function is a parabola whose intercept is at $(0, 0)$. The square function is an even function that is decreasing on the interval $(-\infty, 0)$ and increasing on the interval $(0, \infty)$.

Cube Function

$$f(x) = x^3$$

Figure 48
Cube Function

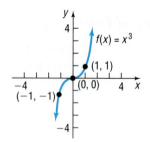

See Figure 48.

The domain and the range of the **cube function** is the set of all real numbers. The intercept of the graph is at $(0, 0)$. The cube function is odd and is increasing on the interval $(-\infty, \infty)$.

Square Root Function

$$f(x) = \sqrt{x}$$

Figure 49
Square Root Function

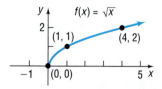

See Figure 49.

The domain and the range of the **square root function** is the set of nonnegative real numbers. The intercept of the graph is at $(0, 0)$. The square root function is neither even nor odd and is increasing on the interval $(0, \infty)$.

Cube Root Function

$$f(x) = \sqrt[3]{x}$$

See Figure 50.

The domain and the range of the **cube root function** is the set of all real numbers. The intercept of the graph is at $(0, 0)$. The cube root function is an odd function that is increasing on the interval $(-\infty, \infty)$.

Reciprocal Function

$$f(x) = \frac{1}{x}$$

Refer to Example 13, page 21, for a discussion of the equation $y = \dfrac{1}{x}$. See Figure 51.

The domain and the range of the **reciprocal function** is the set of all nonzero real numbers. The graph has no intercepts. The reciprocal function is decreasing on the intervals $(-\infty, 0)$ and $(0, \infty)$ and is an odd function.

Absolute Value Function

$$f(x) = |x|$$

See Figure 52.

The domain of the **absolute value function** is the set of all real numbers; its range is the set of nonnegative real numbers. The intercept of the graph is at $(0, 0)$. If $x \geq 0$, then $f(x) = x$, and the graph of f is part of the line $y = x$; if $x < 0$, then $f(x) = -x$, and the graph of f is part of the line $y = -x$. The absolute value function is an even function; it is decreasing on the interval $(-\infty, 0)$ and increasing on the interval $(0, \infty)$.

The notation $\text{int}(x)$ stands for the largest integer less than or equal to x. For example,

$$\text{int}(1) = 1, \quad \text{int}(2.5) = 2, \quad \text{int}\left(\frac{1}{2}\right) = 0, \quad \text{int}\left(-\frac{3}{4}\right) = -1, \quad \text{int}(\pi) = 3$$

This type of correspondence occurs frequently enough in mathematics that we give it a name.

Greatest Integer Function

$$f(x) = \text{int}(x)^* = \text{greatest integer less than or equal to } x$$

*Some books use the notation $f(x) = [\![x]\!]$ instead of $\text{int}(x)$.

Figure 50
Cube Root Function

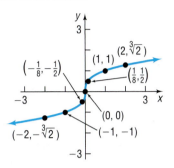

Figure 51
Reciprocal Function

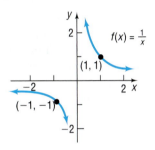

Figure 52
Absolute Value Function

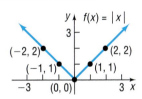

Table 8

x	$y = f(x)$ $= int(x)$	(x, y)
-1	-1	$(-1, -1)$
$-\dfrac{1}{2}$	-1	$\left(-\dfrac{1}{2}, -1\right)$
$-\dfrac{1}{4}$	-1	$\left(-\dfrac{1}{4}, -1\right)$
0	0	$(0, 0)$
$\dfrac{1}{4}$	0	$\left(\dfrac{1}{4}, 0\right)$
$\dfrac{1}{2}$	0	$\left(\dfrac{1}{2}, 0\right)$
$\dfrac{3}{4}$	0	$\left(\dfrac{3}{4}, 0\right)$

We obtain the graph of $f(x) = int(x)$ by plotting several points. See Table 8. For values of x, $-1 \le x < 0$, the value of $f(x) = int(x)$ is -1; for values of x, $0 \le x < 1$, the value of f is 0. See Figure 53 for the graph.

Figure 53
Greatest Integer Function

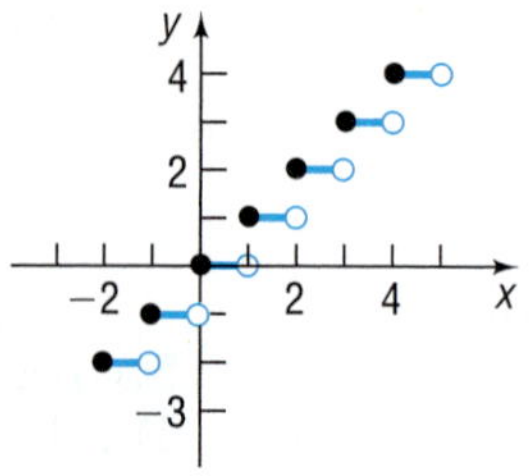

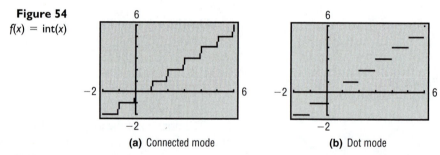

The domain of the **greatest integer function** is the set of all real numbers; its range is the set of integers. The y-intercept of the graph is 0. The x-intercepts lie in the interval $[0, 1)$. The greatest integer function is neither even nor odd. It is constant on every interval of the form $[k, k + 1)$, for k an integer. In Figure 53, we use a solid dot to indicate, for example, that at $x = 1$ the value of f is $f(1) = 1$; we use an open circle to illustrate that the function does not assume the value of 0 at $x = 1$.

From the graph of the greatest integer function, we can see why it also called a **step function**. At $x = 0$, $x = \pm 1$, $x = \pm 2$, and so on, this function exhibits what is called a *discontinuity*; that is, at integer values, the graph suddenly "steps" from one value to another without taking on any of the intermediate values. For example, to the immediate left of $x = 3$, the y-coordinates are 2, and to the immediate right of $x = 3$, the y-coordinates are 3.

Figure 54 shows the graph of $f(x) = int(x)$ on a TI-84 Plus.

Figure 54
$f(x) = int(x)$

(a) Connected mode

(b) Dot mode

The functions that we have discussed so far are basic. Whenever you encounter one of them, you should see a mental picture of its graph. For example, if you encounter the function $f(x) = x^2$, you should see in your mind's eye a picture like Figure 47.

NOTE
When graphing a function using a graphing utility, you can choose either the **connected mode**, in which points plotted on the screen are connected, making the graph appear without any breaks, or the **dot mode**, in which only the points plotted appear. When graphing the greatest integer function with a graphing utility, it is necessary to be in the **dot mode**. This is to prevent the utility from "connecting the dots" when f(x) changes from one integer value to the next. See Figure 54. ■

NOW WORK PROBLEMS 9 THROUGH 16.

2 Graph Piecewise-defined Functions

Sometimes a function is defined differently on different parts of its domain. For example, the absolute value function $f(x) = |x|$ is actually defined by two equations: $f(x) = x$ if $x \ge 0$ and $f(x) = -x$ if $x < 0$. For convenience, we generally combine these equations into one expression as

$$f(x) = |x| = \begin{cases} x & \text{if } x \ge 0 \\ -x & \text{if } x < 0 \end{cases}$$

When functions are defined by more than one equation, they are called **piecewise-defined** functions.

Let's look at another example of a piecewise-defined function.

EXAMPLE 3	**Analyzing a Piecewise-defined Function**

The function f is defined as

$$f(x) = \begin{cases} -x + 1 & \text{if } -1 \le x < 1 \\ 2 & \text{if } x = 1 \\ x^2 & \text{if } x > 1 \end{cases}$$

(a) Find $f(0), f(1)$, and $f(2)$.

(b) Determine the domain of f.

(c) Graph f by hand.

(d) Use the graph to find the range of f.

Solution

(a) To find $f(0)$, we observe that when $x = 0$ the equation for f is given by $f(x) = -x + 1$. So we have

$$f(0) = -0 + 1 = 1$$

When $x = 1$, the equation for f is $f(x) = 2$. Thus,

$$f(1) = 2$$

When $x = 2$, the equation for f is $f(x) = x^2$. So

$$f(2) = 2^2 = 4$$

(b) To find the domain of f, we look at its definition. We conclude that the domain of f is $\{x \mid x \ge -1\}$, or the interval $[-1, \infty)$.

(c) To graph f by hand, we graph "each piece." First we graph the line $y = -x + 1$ and keep only the part for which $-1 \le x < 1$. Then we plot the point $(1, 2)$ because, when $x = 1$, $f(x) = 2$. Finally, we graph the parabola $y = x^2$ and keep only the part for which $x > 1$. See Figure 55.

(d) From the graph, we conclude that the range of f is $\{y \mid y > 0\}$, or the interval $(0, \infty)$.   ◀

Figure 55

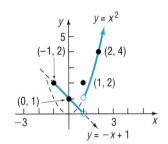

To graph a piecewise-defined function on a graphing calculator, we use the TEST menu to enter inequalities that allow us to restrict the domain function. For example, to graph the function in Example 3 using a TI-84 Plus graphing calculator, we would enter the function in Y_1 as shown in Figure 56(a). We then graph the function and obtain the result in Figure 56(b). When graphing piecewise-defined functions on a graphing calculator, you should use dot mode so that the calculator does not attempt to connect the "pieces" of the function.

Figure 56

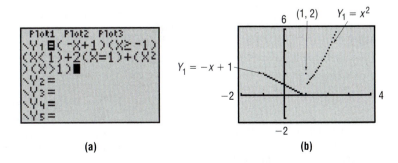

(a)   (b)

🖉 ━━━ **NOW WORK PROBLEM 29.**

| EXAMPLE 4 | Cost of Electricity |

In May 2004, Commonwealth Edison Company supplied electricity to residences for a monthly customer charge of $7.13 plus 8.275¢ per kilowatt-hour (kWhr) for the first 400 kWhr supplied in the month and 6.208¢ per kWhr for all usage over 400 kWhr in the month.

(a) What is the charge for using 300 kWhr in a month?

(b) What is the charge for using 700 kWhr in a month?

(c) If C is the monthly charge for x kWhr, express C as a function of x.

SOURCE: Commonwealth Edison Co., Chicago, Illinois, 2004.

Solution

(a) For 300 kWhr, the charge is $7.13 plus 8.275¢ = $0.08275 per kWhr. That is,

$$\text{Charge} = \$7.13 + \$0.08275(300) = \$31.96$$

(b) For 700 kWhr, the charge is $7.13 plus 8.275¢ per kWhr for the first 400 kWhr plus 6.208¢ per kWhr for the 300 kWhr in excess of 400. That is,

$$\text{Charge} = \$7.13 + \$0.08275(400) + \$0.06208(300) = \$58.85$$

(c) If $0 \le x \le 400$, the monthly charge C (in dollars) can be found by multiplying x times $0.08275 and adding the monthly customer charge of $7.13. So, if $0 \le x \le 400$, then $C(x) = 0.08275x + 7.13$. For $x > 400$, the charge is $0.08275(400) + 7.13 + 0.06208(x - 400)$, since $x - 400$ equals the usage in excess of 400 kWhr, which costs $0.06208 per kWhr. That is, if $x > 400$, then

$$C(x) = 0.08275(400) + 7.13 + 0.06208(x - 400)$$
$$= 40.23 + 0.06208(x - 400)$$
$$= 0.06208x + 15.40$$

The rule for computing C follows two equations:

$$C(x) = \begin{cases} 0.08275x + 7.13 & \text{if } 0 \le x \le 400 \\ 0.06208x + 15.40 & \text{if } x > 400 \end{cases}$$

See Figure 57 for the graph. ◀

Figure 57

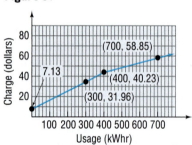

2.5 Assess Your Understanding

'Are You Prepared?'

Answers are given at the end of these exercises. If you get a wrong answer, read the pages listed in red.

1. Sketch the graph of $y = \sqrt{x}$. (p. 20)

2. Sketch the graph of $y = \dfrac{1}{x}$. (p. 21)

3. List the intercepts of the equation $y = x^3 - 8$. (p. 16)

Concepts and Vocabulary

4. The graph of $f(x) = mx + b$ is decreasing if m is _____ than zero.

5. When functions are defined by more than one equation they are called _____ functions.

6. *True or False:* The cube function is odd and is increasing on the interval $(-\infty, \infty)$.

7. *True or False:* The cube root function is odd and is decreasing on the interval $(-\infty, \infty)$.

8. *True or False:* The domain and the range of the reciprocal function are the set of all real numbers.

Skill Building

In Problems 9–16, match each graph to its function.

A. *Constant function* B. *Linear function* C. *Square function*
D. *Cube function* E. *Square root function* F. *Reciprocal function*
G. *Absolute value function* H. *Cube root function*

9.

10.

11.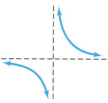

12.

13.

14.

15.

16.

In Problems 17–24, sketch the graph of each function. Be sure to label three points on the graph.

17. $f(x) = x$

18. $f(x) = x^2$

19. $f(x) = x^3$

20. $f(x) = \sqrt{x}$

21. $f(x) = \dfrac{1}{x}$

22. $f(x) = |x|$

23. $f(x) = \sqrt[3]{x}$

24. $f(x) = 3$

25. If $f(x) = \begin{cases} x^2 & \text{if } x < 0 \\ 2 & \text{if } x = 0 \\ 2x + 1 & \text{if } x > 0 \end{cases}$

find: (a) $f(-2)$ (b) $f(0)$ (c) $f(2)$

26. If $f(x) = \begin{cases} x^3 & \text{if } x < 0 \\ 3x + 2 & \text{if } x \geq 0 \end{cases}$

find: (a) $f(-1)$ (b) $f(0)$ (c) $f(1)$

27. If $f(x) = \text{int}(2x)$, find (a) $f(1.2)$, (b) $f(1.6)$, (c) $f(-1.8)$.

28. If $f(x) = \text{int}\left(\dfrac{x}{2}\right)$, find (a) $f(1.2)$, (b) $f(1.6)$, (c) $f(-1.8)$.

In Problems 29–40:

(a) *Find the domain of each function.*
(c) *Graph each function.*

(b) *Locate any intercepts.*
(d) *Based on the graph, find the range.*

29. $f(x) = \begin{cases} 2x & \text{if } x \neq 0 \\ 1 & \text{if } x = 0 \end{cases}$

30. $f(x) = \begin{cases} 3x & \text{if } x \neq 0 \\ 4 & \text{if } x = 0 \end{cases}$

31. $f(x) = \begin{cases} -2x + 3 & x < 1 \\ 3x - 2 & x \geq 1 \end{cases}$

32. $f(x) = \begin{cases} x + 3 & x < -2 \\ -2x - 3 & x \geq -2 \end{cases}$

33. $f(x) = \begin{cases} x + 3 & -2 \leq x < 1 \\ 5 & x = 1 \\ -x + 2 & x > 1 \end{cases}$

34. $f(x) = \begin{cases} 2x + 5 & -3 \leq x < 0 \\ -3 & x = 0 \\ -5x & x > 0 \end{cases}$

35. $f(x) = \begin{cases} 1 + x & \text{if } x < 0 \\ x^2 & \text{if } x \geq 0 \end{cases}$

36. $f(x) = \begin{cases} \dfrac{1}{x} & \text{if } x < 0 \\ \sqrt[3]{x} & \text{if } x \geq 0 \end{cases}$

37. $f(x) = \begin{cases} |x| & \text{if } -2 \leq x < 0 \\ 1 & \text{if } x = 0 \\ x^3 & \text{if } x > 0 \end{cases}$

38. $f(x) = \begin{cases} 3 + x & \text{if } -3 \leq x < 0 \\ 3 & \text{if } x = 0 \\ \sqrt{x} & \text{if } x > 0 \end{cases}$

39. $f(x) = 2\,\text{int}(x)$

40. $f(x) = \text{int}(2x)$

In Problems 41–44, the graph of a piecewise-defined function is given. Write a definition for each function.

41.

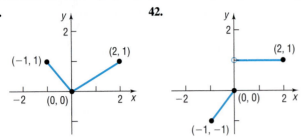

42.

43.

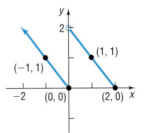

44.

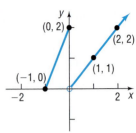

Applications and Extensions

45. Cell Phone Service Sprint PCS offers a monthly cellular phone plan for $35. It includes 300 anytime minutes plus $0.40 per minute for additional minutes. The following function is used to compute the monthly cost for a subscriber:

$$C(x) = \begin{cases} 35 & \text{if } 0 < x \le 300 \\ 0.40x - 85 & \text{if } x > 300 \end{cases}$$

where x is the number of anytime minutes used. Compute the monthly cost of the cellular phone for use of the following anytime minutes:

(a) 200   (b) 365   (c) 301

SOURCE: Sprint PCS.

46. Parking at O'Hare International Airport The short-term parking (no more than 24 hours) fee F (in dollars) for parking x hours at O'Hare International Airport's main parking garage can be modeled by the function

$$F(x) = \begin{cases} 3 & \text{if } 0 < x \le 3 \\ 5\,\text{int}(x + 1) + 1 & \text{if } 3 < x < 9 \\ 50 & \text{if } 9 \le x \le 24 \end{cases}$$

Determine the fee for parking in the short-term parking garage for

(a) 2 hours   (b) 7 hours
(c) 15 hours   (d) 8 hours and 24 minutes

SOURCE: O'Hare International Airport.

47. Cost of Natural Gas In May 2003, the Peoples Gas Company had the following rate schedule for natural gas usage in single-family residences:

Monthly service charge	$9.45
Per therm service charge	
1st 50 therms	$0.36375/therm
Over 50 therms	$0.11445/therm
Gas charge	$0.6338/therm

(a) What is the charge for using 50 therms in a month?
(b) What is the charge for using 500 therms in a month?
(c) Construct a function that relates the monthly charge C for x therms of gas.
(d) Graph this function.

SOURCE: The Peoples Gas Company, Chicago, Illinois, 2003.

48. Cost of Natural Gas In May 2003, Nicor Gas had the following rate schedule for natural gas usage in single-family residences:

Monthly customer charge	$6.45
Distribution charge	
1st 20 therms	$0.2012/therm
Next 30 therms	$0.1117/therm
Over 50 therms	$0.0374/therm
Gas supply charge	$0.7268/therm

(a) What is the charge for using 40 therms in a month?
(b) What is the charge for using 202 therms in a month?
(c) Construct a function that gives the monthly charge C for x therms of gas.
(d) Graph this function.

SOURCE: Nicor Gas, Aurora, Illinois, 2003.

49. Federal Income Tax Two 2004 Tax Rate Schedules are given in the accompanying table. If x equals taxable income and y equals the tax due, construct a function $y = f(x)$ for Schedule X.

REVISED 2004 TAX RATE SCHEDULES										
Schedule X—						**Schedule Y-1—**				
	If Taxable Income		The Tax Is				If Taxable Income		The Tax Is	
	Is over	But Not over	This Amount	Plus This %	Of the Excess over		Is over	But Not over	This Amount	Plus This %	Of the Excess over
Single	$0	$7,150	$0.00	10%	$0.00	**Married**	$0	$14,300	$0.00	10%	$0.00
	$7,150	$29,050	$715.00	15%	$7,150	**Filing**	$14,300	$58,100	$1,430.00	15%	$14,300
	$29,050	$70,350	$4000.00	25%	$29,050	**Jointly**	$58,100	$117,250	$8,000.00	25%	$58,100
	$70,350	$146,750	$14,325.00	28%	$70,350	**or**	$117,250	$178,650	$22,787.50	28%	$117,250
	$146,750	$319,100	$35,717.00	33%	$146,750	**Qualifying**	$178,650	$319,100	$39,979.50	33%	$178,650
	$319,100	—	$92,592.50	35%	$319,100	**Widow(er)**	$319,100	—	$86,328.00	35%	$319,100

SOURCE: Internal Revenue Service

50. Federal Income Tax Refer to the revised 2004 tax rate schedules. If x equals taxable income and y equals the tax due, construct a function $y = f(x)$ for Schedule Y-1.

51. Cost of Transporting Goods A trucking company transports goods between Chicago and New York, a distance of 960 miles. The company's policy is to charge, for each pound, $0.50 per mile for the first 100 miles, $0.40 per mile for the next 300 miles, $0.25 per mile for the next 400 miles, and no charge for the remaining 160 miles.

(a) Graph the relationship between the cost of transportation in dollars and mileage over the entire 960-mile route.
(b) Find the cost as a function of mileage for hauls between 100 and 400 miles from Chicago.
(c) Find the cost as a function of mileage for hauls between 400 and 800 miles from Chicago.

52. Car Rental Costs An economy car rented in Florida from National Car Rental® on a weekly basis costs $95 per week. Extra days cost $24 per day until the day rate exceeds the weekly rate, in which case the weekly rate applies. Also, any part of a day used counts as a full day. Find the cost C of renting an economy car as a piecewise-defined function of the number x of days used, where $7 \leq x \leq 14$. Graph this function.

53. Minimum Payments for Credit Cards Holders of credit cards issued by banks, department stores, oil companies, and so on, receive bills each month that state minimum amounts that must be paid by a certain due date. The minimum due depends on the total amount owed. One such credit card company uses the following rules: For a bill of less than $10, the entire amount is due. For a bill of at least $10 but less than $500, the minimum due is $10. A minimum of $30 is due on a bill of at least $500 but less than $1000, a minimum of $50 is due on a bill of at least $1000 but less than $1500, and a minimum of $70 is due on bills of $1500 or more. Find the function f that describes the minimum payment due on a bill of x dollars. Graph f.

54. Interest Payments for Credit Cards Refer to Problem 53. The card holder may pay any amount between the minimum due and the total owed. The organization issuing the card charges the card holder interest of 1.5% per month for the first $1000 owed and 1% per month on any unpaid balance over $1000. Find the function g that gives the amount of interest charged per month on a balance of x dollars. Graph g.

55. Wind Chill The wind chill factor represents the equivalent air temperature at a standard wind speed that would produce the same heat loss as the given temperature and wind speed. One formula for computing the equivalent temperature is

$$W = \begin{cases} t & 0 \leq v < 1.79 \\ 33 - \dfrac{(10.45 + 10\sqrt{v} - v)(33 - t)}{22.04} & 1.79 \leq v \leq 20 \\ 33 - 1.5958(33 - t) & v > 20 \end{cases}$$

where v represents the wind speed (in meters per second) and t represents the air temperature (°C). Compute the wind chill for the following:

(a) An air temperature of 10°C and a wind speed of 1 meter per second (m/sec)
(b) An air temperature of 10°C and a wind speed of 5 m/sec
(c) An air temperature of 10°C and a wind speed of 15 m/sec
(d) An air temperature of 10°C and a wind speed of 25 m/sec
(e) Explain the physical meaning of the equation corresponding to $0 \leq v < 1.79$.
(f) Explain the physical meaning of the equation corresponding to $v > 20$.

56. Wind Chill Redo Problem 55(a)–(d) for an air temperature of −10°C.

Discussion and Writing

57. Exploration Graph $y = x^2$. Then on the same screen graph $y = x^2 + 2$, followed by $y = x^2 + 4$, followed by $y = x^2 - 2$. What pattern do you observe? Can you predict the graph of $y = x^2 - 4$? Of $y = x^2 + 5$?

58. Exploration Graph $y = x^2$. Then on the same screen graph $y = (x - 2)^2$, followed by $y = (x - 4)^2$, followed by $y = (x + 2)^2$. What pattern do you observe? Can you predict the graph of $y = (x + 4)^2$? Of $y = (x - 5)^2$?

59. Exploration Graph $y = |x|$. Then on the same screen graph $y = 2|x|$, followed by $y = 4|x|$, followed by $y = \frac{1}{2}|x|$.

What pattern do you observe? Can you predict the graph of $y = \frac{1}{4}|x|$? Of $y = 5|x|$?

60. Exploration Graph $y = x^2$. Then on the same screen graph $y = -x^2$. What pattern do you observe? Now try $y = |x|$ and $y = -|x|$. What do you conclude?

61. Exploration Graph $y = \sqrt{x}$. Then on the same screen graph $y = \sqrt{-x}$. What pattern do you observe? Now try $y = 2x + 1$ and $y = 2(-x) + 1$. What do you conclude?

62. Exploration Graph $y = x^3$. Then on the same screen graph $y = (x - 1)^3 + 2$. Could you have predicted the result?

63. Exploration Graph $y = x^2$, $y = x^4$, and $y = x^6$ on the same screen. What do you notice is the same about each graph? What do you notice that is different?

64. Exploration Graph $y = x^3$, $y = x^5$, and $y = x^7$ on the same screen. What do you notice is the same about each graph? What do you notice that is different?

65. Consider the equation

$$y = \begin{cases} 1 & \text{if } x \text{ is rational} \\ 0 & \text{if } x \text{ is irrational} \end{cases}$$

Is this a function? What is its domain? What is its range? What is its y-intercept, if any? What are its x-intercepts, if any? Is it even, odd, or neither? How would you describe its graph?

66. Define some functions that pass through $(0, 0)$ and $(1, 1)$ and are increasing for $x \geq 0$. Begin your list with $y = \sqrt{x}$, $y = x$, and $y = x^2$. Can you propose a general result about such functions?

'Are You Prepared' Answers

1.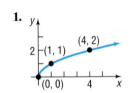

2.

3. $(0, -8), (2, 0)$

2.6 Graphing Techniques: Transformations

OBJECTIVES **1** Graph Functions Using Vertical and Horizontal Shifts
2 Graph Functions Using Compressions and Stretches
3 Graph Functions Using Reflections about the x-Axis or y-Axis

At this stage, if you were asked to graph any of the functions defined by $y = x$, $y = x^2$, $y = x^3$, $y = \sqrt{x}$, $y = \sqrt[3]{x}$, $y = |x|$, or $y = \dfrac{1}{x}$, your response should be, "Yes, I recognize these functions and know the general shapes of their graphs." (If this is not your answer, review the previous section, Figures 46 through 52).

Sometimes we are asked to graph a function that is "almost" like one that we already know how to graph. In this section, we look at some of these functions and develop techniques for graphing them. Collectively, these techniques are referred to as **transformations**.

1 Graph Functions Using Vertical and Horizontal Shifts

———— **Exploration** ————

On the same screen, graph each of the following functions:

$$Y_1 = x^2$$
$$Y_2 = x^2 + 2$$
$$Y_3 = x^2 - 2$$

What do you observe?

Figure 58

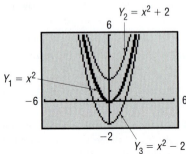

Result Figure 58 illustrates the graphs. You should have observed a general pattern. With $Y_1 = x^2$ on the screen, the graph of $Y_2 = x^2 + 2$ is identical to that of $Y_1 = x^2$, except that it is shifted vertically up 2 units. The graph of $Y_3 = x^2 - 2$ is identical to that of $Y_1 = x^2$, except that it is shifted vertically down 2 units.

We are led to the following conclusion:

> If a real number k is added to the right side of a function $y = f(x)$, the graph of the new function $y = f(x) + k$ is the graph of f **shifted vertically up** k units (if $k > 0$) or **down** $|k|$ units (if $k < 0$).

Let's look at an example.

EXAMPLE 1	**Vertical Shift Down**

Use the graph of $f(x) = x^2$ to obtain the graph of $h(x) = x^2 - 4$.

Solution
Table 9 lists some points on the graphs of $f = Y_1$ and $h = Y_2$. Notice that each y-coordinate of h is 4 units less than the corresponding y-coordinate of f.
The graph of h is identical to that of f, except that it is shifted down 4 units. See Figure 59.

Table 9

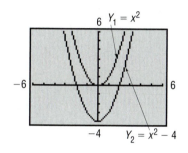

Figure 59

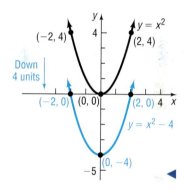

NOW WORK PROBLEM 35.

Figure 60

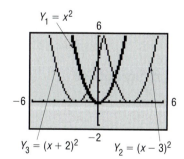

Exploration

On the same screen, graph each of the following functions:

$$Y_1 = x^2$$
$$Y_2 = (x - 3)^2$$
$$Y_3 = (x + 2)^2$$

What do you observe?

Result Figure 60 illustrates the graphs. You should have observed the following pattern. With the graph of $Y_1 = x^2$ on the screen, the graph of $Y_2 = (x - 3)^2$ is identical to that of $Y_1 = x^2$, except it is shifted horizontally to the right 3 units. The graph of $Y_3 = (x + 2)^2$ is identical to that of $Y_1 = x^2$, except it is shifted horizontally to the left 2 units.

We are led to the following conclusion.

> If the argument x of a function f is replaced by $x - h, h > 0$, the graph of the new function $y = f(x - h)$ is the graph of f **shifted horizontally right** h units.
> If the argument x of a function f is repaced by $x + h, h > 0$, the graph of the new function $y = f(x + h)$ is the graph of f **shifted horizontally left** h units.

NOW WORK PROBLEM 39.

Vertical and horizontal shifts are sometimes combined.

EXAMPLE 2 | **Combining Vertical and Horizontal Shifts**

Graph the function $f(x) = (x + 3)^2 - 5$.

Solution | We graph f in steps. First, we note that the rule for f is basically a square function, so we begin with the graph of $y = x^2$ as shown in Figure 61(a). To get the graph of $y = (x + 3)^2$, we shift the graph of $y = x^2$ horizontally 3 units to the left. See Figure 61(b). Finally, to get the graph of $y = (x + 3)^2 - 5$, we shift the graph of $y = (x + 3)^2$ vertically down 5 units. See Figure 61(c). Note the points plotted on each graph. Using key points can be helpful in keeping track of the transformation that has taken place.

Figure 61

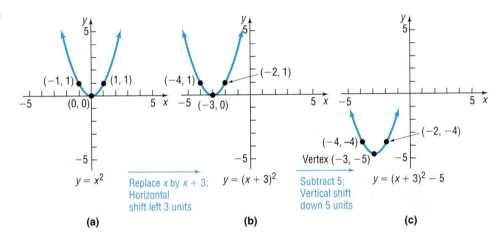

(a) $y = x^2$ Replace x by $x + 3$; Horizontal shift left 3 units

(b) $y = (x + 3)^2$ Subtract 5; Vertical shift down 5 units

(c) $y = (x + 3)^2 - 5$

✔ **CHECK:** Graph $Y_1 = f(x) = (x + 3)^2 - 5$ and compare the graph to Figure 61(c). ◀

In Example 2, if the vertical shift had been done first, followed by the horizontal shift, the final graph would have been the same. (Try it for yourself.)

 NOW WORK PROBLEMS 41 AND 77.

---

2 **Graph Functions Using Compressions and Stretches**

――――― **Exploration** ―――――

On the same screen, graph each of the following functions:

$$Y_1 = |x|$$

$$Y_2 = 2|x|$$

$$Y_3 = \frac{1}{2}|x|$$

Figure 62

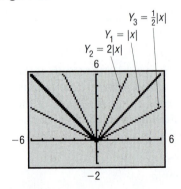

$Y_3 = \frac{1}{2}|x|$
$Y_1 = |x|$
$Y_2 = 2|x|$

Result Figure 62 illustrates the graphs. You should have observed the following pattern. The graph of $Y_2 = 2|x|$ can be obtained from the graph of $Y_1 = |x|$ by multiplying each y-coordinate of $Y_1 = |x|$ by 2. This is sometimes referred to as a vertical *stretch* using a factor of 2.

The graph of $Y_3 = \frac{1}{2}|x|$ can be obtained from the graph of $Y_1 = |x|$ by multiplying each y-coordinate by $\frac{1}{2}$. This is sometimes referred to as a vertical *compression* using a factor of $\frac{1}{2}$.

Look at Tables 10 and 11, where $Y_1 = |x|$, $Y_2 = 2|x|$, and $Y_3 = \dfrac{1}{2}|x|$. Notice that the values for Y_2 in Table 10 are two times the values of Y_1 for each x-value. Therefore, the graph of Y_2 will be vertically *stretched* by a factor of 2. Likewise, the values of Y_3 in Table 11 are half the values of Y_1 for each x-value. Therefore, the graph of Y_3 will be vertically *compressed* by a factor of $\dfrac{1}{2}$.

Table 10

X	Y₁	**Y2**
-2	2	4
-1	1	2
0	0	0
1	1	2
2	2	4
3	3	6
4	4	8

Y₂⬛2abs(X)

Table 11

X	Y₁	**Y3**
-2	2	1
-1	1	.5
0	0	0
1	1	.5
2	2	1
3	3	1.5
4	4	2

Y₃⬛.5abs(X)

> When the right side of a function $y = f(x)$ is multiplied by a positive number a, the graph of the new function $y = af(x)$ is obtained by multiplying each y-coordinate on the graph of $y = f(x)$ by a. The new graph is a **vertically compressed** (if $0 < a < 1$) or a **vertically stretched** (if $a > 1$) version of the graph of $y = f(x)$.

 NOW WORK PROBLEM 43.

What happens if the argument x of a function $y = f(x)$ is multiplied by a positive number a, creating a new function $y = f(ax)$? To find the answer, we look at the following Exploration.

——— **Exploration** ———

On the same screen, graph each of the following functions:

$$Y_1 = f(x) = x^2$$
$$Y_2 = f(4x) = (4x)^2$$
$$Y_3 = f\left(\frac{1}{2}x\right) = \left(\frac{1}{2}x\right)^2$$

Result   You should have obtained the graphs shown in Figure 63. Look at Table 12(a). Notice that (1, 1), (4, 16), and (16, 256) are points on the graph of $Y_1 = x^2$. Also, (0.25, 1), (1, 16), and (4, 256) are points on the graph of $Y_2 = (4x)^2$. For each y-coordinate, the x-coordinate on the graph of Y_2 is $\dfrac{1}{4}$ of the x-coordinate on Y_1. We conclude the graph of $Y_2 = (4x)^2$ is obtained by multiplying the x-coordinate of each point on the graph of $Y_1 = x^2$ by $\dfrac{1}{4}$. The graph of $Y_2 = (4x)^2$ is the graph of $Y_1 = x^2$ compressed horizontally.

Figure 63

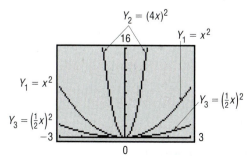

Table 12

X	Y₁	**Y2**
0	0	0
.25	.0625	1
1	1	16
4	16	256
16	256	4096

Y₂⬛(4X)²

X	Y₁	**Y3**
0	0	0
.5	.25	.0625
1	1	.25
2	4	1
4	16	4
8	64	16
16	256	64

Y₃⬛(X/2)²

(a)   (b)

Look at Table 12(b). Notice that (0.5, 0.25), (1, 1), (2, 4), and (4, 16) are points on the graph of $Y_1 = x^2$. Also, (1, 0.25), (2, 1), (4, 4), and (8, 16) are points on the graph of $Y_3 = \left(\dfrac{1}{2}x\right)^2$. For each y-coordinate, the x-coordinate on the graph of Y_3 is 2 times the x-coordinate on Y_1. We conclude the graph of $Y_3 = \left(\dfrac{1}{2}x\right)^2$ is obtained by multiplying the x-coordinate of each point on the graph of $Y_1 = x^2$ by 2. The graph of $Y_3 = \left(\dfrac{1}{2}x\right)^2$ is the graph of $Y_1 = x^2$ stretched horizontally.

> If the argument x of a function $y = f(x)$ is multiplied by a positive number a, the graph of the new function $y = f(ax)$ is obtained by multiplying each x-coordinate of $y = f(x)$ by $\dfrac{1}{a}$. A **horizontal compression** results if $a > 1$, and a **horizontal stretch** occurs if $0 < a < 1$.

Let's look at an example.

EXAMPLE 3   Graphing Using Stretches and Compressions

The graph of $y = f(x)$ is given in Figure 64. Use this graph to find the graphs of

(a) $y = 2f(x)$ (b) $y = f(3x)$

Figure 64

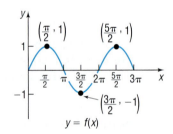

$y = f(x)$

Solution   (a) The graph of $y = 2f(x)$ is obtained by multiplying each y-coordinate of $y = f(x)$ by 2. See Figure 65(a).

(b) The graph of $y = f(3x)$ is obtained from the graph of $y = f(x)$ by multiplying each x-coordinate of $y = f(x)$ by $\dfrac{1}{3}$. See Figure 65(b).

Figure 65

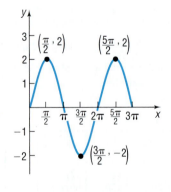

(a) $y = 2f(x)$

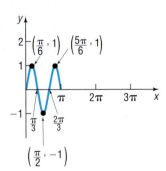

(b) $y = f(3x)$

NOW WORK PROBLEMS 65(e) AND (g).

3 Graph Functions Using Reflections about the *x*-Axis or *y*-Axis

────── **Exploration** ──────────────────────────────

Reflection about the *x*-axis:

(a) Graph $Y_1 = x^2$, followed by $Y_2 = -x^2$.

(b) Graph $Y_1 = |x|$, followed by $Y_2 = -|x|$.

(c) Graph $Y_1 = x^2 - 4$, followed by $Y_2 = -(x^2 - 4) = -x^2 + 4$.

Result   See Tables 13(a), (b), and (c) and Figures 66(a), (b), and (c). In each instance, Y_2 is the reflection about the *x*-axis of Y_1.

Table 13

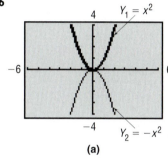   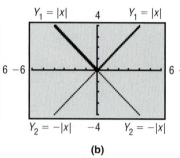 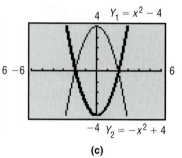

(a) (b) (c)

Figure 66

(Figure 66 graphs)

(a) (b) (c)

> When the right side of the function $y = f(x)$ is multiplied by -1, the graph of the new function $y = -f(x)$ is the **reflection about the *x*-axis** of the graph of the function $y = f(x)$.

   NOW WORK PROBLEM 47.

────── **Exploration** ──────────────────────────────

Reflection about the *y*-axis:

(a) Graph $Y_1 = \sqrt{x}$, followed by $Y_2 = \sqrt{-x}$.

(b) Graph $Y_1 = x + 1$, followed by $Y_2 = -x + 1$.

(c) Graph $Y_1 = x^4 + x$, followed by $Y_2 = (-x)^4 + (-x) = x^4 - x$.

Result   See Tables 14(a), (b), and (c) and Figures 67(a), (b), and (c). In each instance, Y_2 is the reflection about the *y*-axis of Y_1.

Table 14

(Table 14 calculator screens)

(a) (b) (c)

Figure 67

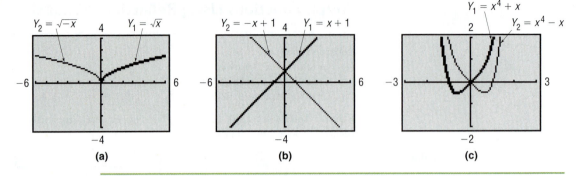

(a)   (b)   (c)

> When the graph of the function $y = f(x)$ is known, the graph of the new function $y = f(-x)$ is the **reflection about the y-axis** of the graph of the function $y = f(x)$.

Summary of Graphing Techniques

Table 15 summarizes the graphing procedures that we have just discussed.

Table 15	To Graph:	Draw the Graph of f and:	Functional Change to $f(x)$
	Vertical shifts		
	$y = f(x) + k, \quad k > 0$	Raise the graph of f by k units.	Add k to $f(x)$.
	$y = f(x) - k, \quad k > 0$	Lower the graph of f by k units.	Subtract k from $f(x)$.
	Horizontal shifts		
	$y = f(x + h), \quad h > 0$	Shift the graph of f to the left h units.	Replace x by $x + h$.
	$y = f(x - h), \quad h > 0$	Shift the graph of f to the right h units.	Replace x by $x - h$.
	Compressing or stretching		
	$y = af(x), \quad a > 0$	Multiply each y-coordinate of $y = f(x)$ by a. Stretch the graph of f vertically if $a > 1$. Compress the graph of f vertically if $0 < a < 1$.	Multiply $f(x)$ by a.
	$y = f(ax), \quad a > 0$	Multiply each x-coordinate of $y = f(x)$ by $\dfrac{1}{a}$. Stretch the graph of f horizontally if $0 < a < 1$. Compress the graph of f horizontally if $a > 1$.	Replace x by ax.
	Reflection about the x-axis		
	$y = -f(x)$	Reflect the graph of f about the x-axis.	Multiply $f(x)$ by -1.
	Reflection about the y-axis		
	$y = f(-x)$	Reflect the graph of f about the y-axis.	Replace x by $-x$.

The examples that follow combine some of the procedures outlined in this section to get the required graph.

EXAMPLE 4   **Determining the Function Obtained from a Series of Transformations**

Find the function that is finally graphed after the following three transformations are applied to the graph of $y = |x|$.

1. Shift left 2 units.

2. Shift up 3 units.

3. Reflect about the *y*-axis.

Solution

1. Shift left 2 units: Replace *x* by *x* + 2. $y = |x + 2|$

2. Shift up 3 units: Add 3. $y = |x + 2| + 3$

3. Reflect about the *y*-axis: Replace *x* by −*x*. $y = |-x + 2| + 3$ ◀

NOW WORK PROBLEM 27.

EXAMPLE 5

Combining Graphing Procedures

Graph the function: $f(x) = \dfrac{3}{x - 2} + 1$

Solution

We use the following steps to obtain the graph of *f*:

STEP 1: $y = \dfrac{1}{x}$ Reciprocal function.

STEP 2: $y = \dfrac{3}{x}$ Multiply by 3; vertical stretch of the graph of $y = \dfrac{1}{x}$ by a factor of 3.

STEP 3: $y = \dfrac{3}{x - 2}$ Replace x by x − 2; horizontal shift to the right 2 units.

STEP 4: $y = \dfrac{3}{x - 2} + 1$ Add 1; vertical shift up 1 unit.

See Figure 68.

Figure 68

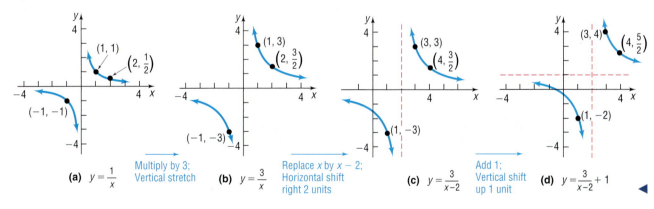

(a) $y = \dfrac{1}{x}$ Multiply by 3; Vertical stretch

(b) $y = \dfrac{3}{x}$ Replace x by x − 2; Horizontal shift right 2 units

(c) $y = \dfrac{3}{x-2}$ Add 1; Vertical shift up 1 unit

(d) $y = \dfrac{3}{x-2} + 1$ ◀

Other orderings of the steps shown in Example 5 would also result in the graph of *f*. For example, try this one:

STEP 1: $y = \dfrac{1}{x}$ Reciprocal function

STEP 2: $y = \dfrac{1}{x - 2}$ Replace x by x − 2; horizontal shift to the right 2 units.

STEP 3: $y = \dfrac{3}{x-2}$ Multiply by 3; vertical stretch of the graph of $y = \dfrac{1}{x-2}$ by a factor of 3.

STEP 4: $y = \dfrac{3}{x-2} + 1$ Add 1; vertical shift up 1 unit.

| EXAMPLE 6 | **Combining Graphing Procedures** |

Graph the function: $f(x) = \sqrt{1-x} + 2$

Solution   We use the following steps to obtain the graph of $y = \sqrt{1-x} + 2$:

STEP 1: $y = \sqrt{x}$ Square root function

STEP 2: $y = \sqrt{x+1}$ Replace x by $x+1$; horizontal shift left 1 unit.

STEP 3: $y = \sqrt{-x+1} = \sqrt{1-x}$ Replace x by $-x$; reflect about y-axis.

STEP 4: $y = \sqrt{1-x} + 2$ Add 2; vertical shift up 2 units.

See Figure 69.

Figure 69

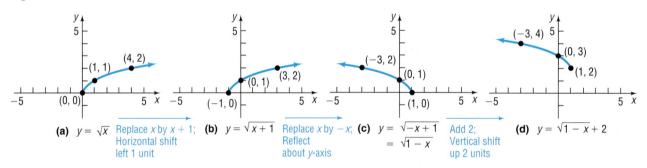

(a) $y = \sqrt{x}$ Replace x by $x+1$;   (b) $y = \sqrt{x+1}$ Replace x by $-x$;   (c) $y = \sqrt{-x+1}$ Add 2;   (d) $y = \sqrt{1-x} + 2$
Horizontal shift   Reflect $= \sqrt{1-x}$ Vertical shift
left 1 unit   about y-axis   up 2 units

✔ **CHECK:** Graph $Y_1 = f(x) = \sqrt{1-x} + 2$ and compare the graph to Figure 69(d).

NOW WORK PROBLEM 57.

2.6 Assess Your Understanding

Concepts and Vocabulary

1. Suppose that the graph of a function f is known. Then the graph of $y = f(x-2)$ may be obtained by a(n) _____ shift of the graph of f to the _____ a distance of 2 units.

2. Suppose that the graph of a function f is known. Then the graph of $y = f(-x)$ may be obtained by a reflection about the _____-axis of the graph of the function $y = f(x)$.

3. Suppose that the x-intercepts of the graph of $y = f(x)$ are $-2, 1$, and 5. The x-intercepts of $y = f(x+3)$ are _____, _____, and _____.

4. *True or False:* The graph of $y = -f(x)$ is the reflection about the x-axis of the graph of $y = f(x)$.

5. *True or False:* To obtain the graph of $y = f(x+2) - 3$, shift the graph of $y = f(x)$ horizontally to the right 2 units and vertically down 3 units.

6. *True or False:* Suppose that the x-intercepts of the graph of $y = f(x)$ are -3 and 2. Then the x-intercepts of the graph of $y = 2f(x)$ are -3 and 2.

Skill Building

In Problems 7–18, match each graph to one of the following functions without using a graphing utility.

A. $y = x^2 + 2$
B. $y = -x^2 + 2$
C. $y = |x| + 2$
D. $y = -|x| + 2$
E. $y = (x - 2)^2$
F. $y = -(x + 2)^2$
G. $y = |x - 2|$
H. $y = -|x + 2|$
I. $y = 2x^2$
J. $y = -2x^2$
K. $y = 2|x|$
L. $y = -2|x|$

7.
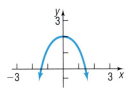

8.
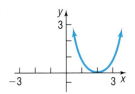

9.
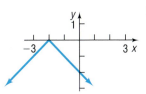

10.
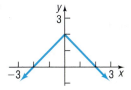

11.
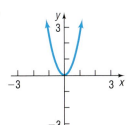

12.
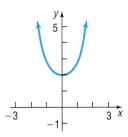

13.
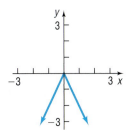

14.
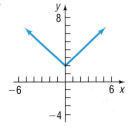

15.
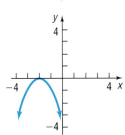

16.
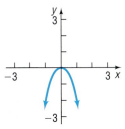

17.
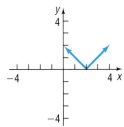

18.
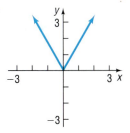

In Problems 19–26, write the function whose graph is the graph of $y = x^3$, but is:

19. Shifted to the right 4 units

20. Shifted to the left 4 units

21. Shifted up 4 units

22. Shifted down 4 units

23. Reflected about the y-axis

24. Reflected about the x-axis

25. Vertically stretched by a factor of 4

26. Horizontally stretched by a factor of 4

In Problems 27–30, find the function that is finally graphed after the following transformations are applied to the graph of $y = \sqrt{x}$.

27. (1) Shift up 2 units
(2) Reflect about the x-axis
(3) Reflect about the y-axis

28. (1) Reflect about the x-axis
(2) Shift right 3 units
(3) Shift down 2 units

29. (1) Reflect about the x-axis
(2) Shift up 2 units
(3) Shift left 3 units

30. (1) Shift up 2 units
(2) Reflect about the y-axis
(3) Shift left 3 units

31. If $(3, 0)$ is a point on the graph of $y = f(x)$, which of the following must be on the graph of $y = -f(x)$?
(a) $(0, 3)$ (b) $(0, -3)$
(c) $(3, 0)$ (d) $(-3, 0)$

32. If $(3, 0)$ is a point on the graph of $y = f(x)$, which of the following must be on the graph of $y = f(-x)$?
(a) $(0, 3)$ (b) $(0, -3)$
(c) $(3, 0)$ (d) $(-3, 0)$

33. If $(0, 3)$ is a point on the graph of $y = f(x)$, which of the following must be on the graph of $y = 2f(x)$?
(a) $(0, 3)$ (b) $(0, 2)$
(c) $(0, 6)$ (d) $(6, 0)$

34. If $(3, 0)$ is a point on the graph of $y = f(x)$, which of the following must be on the graph of $y = \frac{1}{2}f(x)$?
(a) $(3, 0)$ (b) $\left(\frac{3}{2}, 0\right)$
(c) $\left(0, \frac{3}{2}\right)$ (d) $\left(\frac{1}{2}, 0\right)$

In Problems 35–64, graph each function using the techniques of shifting, compressing, stretching, and/or reflecting. Start with the graph of the basic function (for example, $y = x^2$) and show all stages. Verify your results using a graphing utility.

35. $f(x) = x^2 - 1$

36. $f(x) = x^2 + 4$

37. $g(x) = x^3 + 1$

38. $g(x) = x^3 - 1$

39. $h(x) = \sqrt{x - 2}$

40. $h(x) = \sqrt{x + 1}$

41. $f(x) = (x - 1)^3 + 2$

42. $f(x) = (x + 2)^3 - 3$

43. $g(x) = 4\sqrt{x}$

44. $g(x) = \dfrac{1}{2}\sqrt{x}$

45. $h(x) = \dfrac{1}{2x}$

46. $h(x) = 3\sqrt[3]{x}$

47. $f(x) = -\sqrt[3]{x}$

48. $f(x) = -\sqrt{x}$

49. $g(x) = |-x|$

50. $g(x) = \sqrt[3]{-x}$

51. $h(x) = -x^3 + 2$

52. $h(x) = \dfrac{1}{-x} + 2$

53. $f(x) = 2(x + 1)^2 - 3$

54. $f(x) = 3(x - 2)^2 + 1$

55. $g(x) = \sqrt{x - 2} + 1$

56. $g(x) = |x + 1| - 3$

57. $h(x) = \sqrt{-x} - 2$

58. $h(x) = \dfrac{4}{x} + 2$

59. $f(x) = -(x + 1)^3 - 1$

60. $f(x) = -4\sqrt{x - 1}$

61. $g(x) = 2|1 - x|$

62. $g(x) = 4\sqrt{2 - x}$

63. $h(x) = 2\,\text{int}(x - 1)$

64. $h(x) = \text{int}(-x)$

In Problems 65–68, the graph of a function f is illustrated. Use the graph of f as the first step toward graphing each of the following functions:

(a) $F(x) = f(x) + 3$
(b) $G(x) = f(x + 2)$
(c) $P(x) = -f(x)$
(d) $H(x) = f(x + 1) - 2$
(e) $Q(x) = \dfrac{1}{2}f(x)$
(f) $g(x) = f(-x)$
(g) $h(x) = f(2x)$

65.

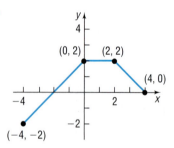

66.

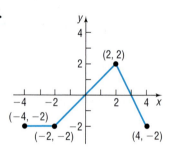

67.

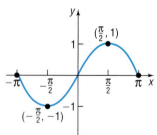

68.

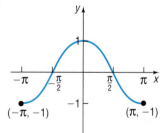

Applications and Extensions

69. Suppose that the x-intercepts of the graph of $y = f(x)$ are -5 and 3.
  (a) What are the x-intercepts of the graph of $y = f(x + 2)$?
  (b) What are the x-intercepts of the graph of $y = f(x - 2)$?
  (c) What are the x-intercepts of the graph of $y = 4f(x)$?
  (d) What are the x-intercepts of the graph of $y = f(-x)$?

70. Suppose that the x-intercepts of the graph of $y = f(x)$ are -8 and 1.
  (a) What are the x-intercepts of the graph of $y = f(x + 4)$?
  (b) What are the x-intercepts of the graph of $y = f(x - 3)$?

  (c) What are the x-intercepts of the graph of $y = 2f(x)$?
  (d) What are the x-intercepts of the graph of $y = f(-x)$?

71. Suppose that the function $y = f(x)$ is increasing on the interval $(-1, 5)$.
  (a) Over what interval is the graph of $y = f(x + 2)$ increasing?
  (b) Over what interval is the graph of $y = f(x - 5)$ increasing?
  (c) What can be said about the graph of $y = -f(x)$?
  (d) What can be said about the graph of $y = f(-x)$?

72. Suppose that the function $y = f(x)$ is decreasing on the interval $(-2, 7)$.
  (a) Over what interval is the graph of $y = f(x + 2)$ decreasing?
  (b) Over what interval is the graph of $y = f(x - 5)$ decreasing?
  (c) What can be said about the graph of $y = -f(x)$?
  (d) What can be said about the graph of $y = f(-x)$?

73. The graph of a function f is illustrated in the figure.
  (a) Draw the graph of $y = |f(x)|$.
  (b) Draw the graph of $y = f(|x|)$.

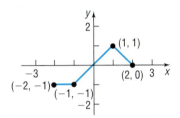

74. The graph of a function f is illustrated in the figure.
  (a) Draw the graph of $y = |f(x)|$.
  (b) Draw the graph of $y = f(|x|)$.

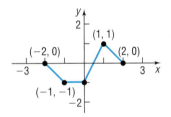

In Problems 75–84, complete the square of each quadratic expression. Then graph each function by hand using the technique of shifting. Verify your results using a graphing utility. (If necessary, refer to the Appendix, Section A.5, to review completing the square.)

75. $f(x) = x^2 + 2x$ **76.** $f(x) = x^2 - 6x$

77. $f(x) = x^2 - 8x + 1$ **78.** $f(x) = x^2 + 4x + 2$

79. $f(x) = x^2 + x + 1$ **80.** $f(x) = x^2 - x + 1$

81. $f(x) = 2x^2 - 12x + 19$

82. $f(x) = 3x^2 + 6x + 1$

83. $f(x) = -3x^2 - 12x - 17$

84. $f(x) = -2x^2 - 12x - 13$

85. The equation $y = (x - c)^2$ defines a *family of parabolas*, one parabola for each value of c. On one set of coordinate axes, graph the members of the family for $c = 0, c = 3$, and $c = -2$.

86. Repeat Problem 85 for the family of parabolas $y = x^2 + c$.

87. Temperature Measurements   The relationship between the Celsius (°C) and Fahrenheit (°F) scales for measuring temperature is given by the equation

$$F = \frac{9}{5}C + 32$$

The relationship between the Celsius (°C) and Kelvin (K) scales is $K = C + 273$. Graph the equation $F = \frac{9}{5}C + 32$ using degrees Fahrenheit on the y-axis and degrees Celsius on the x-axis. Use the techniques introduced in this section to obtain the graph showing the relationship between Kelvin and Fahrenheit temperatures.

88. Period of a Pendulum   The period T (in seconds) of a simple pendulum is a function of its length l (in feet) defined by the equation

$$T = 2\pi\sqrt{\frac{l}{g}}$$

where $g \approx 32.2$ feet per second per second is the acceleration of gravity.
  (a) Use a graphing utility to graph the function $T = T(l)$.
  (b) Now graph the functions $T = T(l + 1), T = T(l + 2)$, and $T = T(l + 3)$.
  (c) Discuss how adding to the length l changes the period T.
  (d) Now graph the functions $T = T(2l), T = T(3l)$, and $T = T(4l)$.
  (e) Discuss how multiplying the length l by factors of 2, 3, and 4 changes the period T.

89. Cigar Company Profits   The daily profits of a cigar company from selling x cigars are given by

$$p(x) = -0.05x^2 + 100x - 2000$$

The government wishes to impose a tax on cigars (sometimes called a *sin tax*) that gives the company the option of either paying a flat tax of $10,000 per day or a tax of 10% on profits. As chief financial officer (CFO) of the company, you need to decide which tax is the better option for the company.
  (a) On the same screen, graph $Y_1 = p(x) - 10{,}000$ and $Y_2 = (1 - 0.10)p(x)$.
  (b) Based on the graph, which option would you select? Why?
  (c) Using the terminology learned in this section, describe each graph in terms of the graph of $p(x)$.
  (d) Suppose that the government offered the options of a flat tax of $4800 or a tax of 10% on profits. Which would you select? Why?

Discussion and Writing

90. Suppose that the graph of a function f is known. Explain how the graph of $y = 4f(x)$ differs from the graph of $y = f(4x)$.

91. Exploration
(a) Use a graphing utility to graph $y = x + 1$ and $y = |x + 1|$.
(b) Graph $y = 4 - x^2$ and $y = |4 - x^2|$.
(c) Graph $y = x^3 + x$ and $y = |x^3 + x|$.

(d) What do you conclude about the relationship between the graphs of $y = f(x)$ and $y = |f(x)|$?

92. Exploration
(a) Use a graphing utility to graph $y = x + 1$ and $y = |x| + 1$.
(b) Graph $y = 4 - x^2$ and $y = 4 - |x|^2$.
(c) Graph $y = x^3 + x$ and $y = |x|^3 + |x|$.
(d) What do you conclude about the relationship between the graphs of $y = f(x)$ and $y = f(|x|)$?

2.7 Mathematical Models: Constructing Functions

OBJECTIVE 1 Construct and Analyze Functions

 Construct and Analyze Functions

Real-world problems often result in mathematical models that involve functions. These functions need to be constructed or built based on the information given. In constructing functions, we must be able to translate the verbal description into the language of mathematics. We do this by assigning symbols to represent the independent and dependent variables and then finding the function or rule that relates these variables.

EXAMPLE 1 | **Area of a Rectangle with Fixed Perimeter**

The perimeter of a rectangle is 50 feet. Express its area A as a function of the length l of a side.

Figure 70

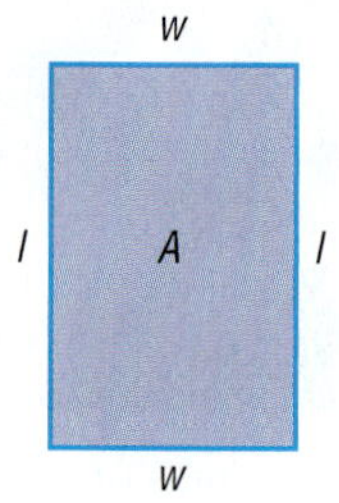

Solution   Consult Figure 70. If the length of the rectangle is l and if w is its width, then the sum of the lengths of the sides is the perimeter, 50.

$$l + w + l + w = 50$$
$$2l + 2w = 50$$
$$l + w = 25$$
$$w = 25 - l$$

The area A is length times width, so

$$A = lw = l(25 - l)$$

The area A as a function of l is

$$A(l) = l(25 - l) \qquad \blacktriangleleft$$

Note that we use the symbol A as the dependent variable and also as the name of the function that relates the length l to the area A. This double usage is common in applications and should cause no difficulties.

EXAMPLE 2 | **Economics: Demand Equations**

In economics, revenue R, in dollars, is defined as the amount of money received from the sale of a product and is equal to the unit selling price p, in dollars, of the product times the number x of units actually sold. That is,

$$R = xp$$

In economics, the Law of Demand states that p and x are related: As one increases, the other decreases. Suppose that p and x are related by the following **demand equation**:

$$p = -\frac{1}{10}x + 20, \qquad 0 \le x \le 200$$

Express the revenue R as a function of the number x of units sold.

Solution   Since $R = xp$ and $p = -\frac{1}{10}x + 20$, it follows that

$$R(x) = xp = x\left(-\frac{1}{10}x + 20\right) = -\frac{1}{10}x^2 + 20x$$

◄

NOW WORK PROBLEM 3.

EXAMPLE 3   **Finding the Distance from the Origin to a Point on a Graph**

Let $P = (x, y)$ be a point on the graph of $y = x^2 - 1$.

(a) Express the distance d from P to the origin O as a function of x.

(b) What is d if $x = 0$?

(c) What is d if $x = 1$?

(d) What is d if $x = \dfrac{\sqrt{2}}{2}$?

(e) Use a graphing utility to graph the function $d = d(x)$, $x \ge 0$. Rounded to two decimal places, find the value(s) of x at which d has a local minimum. [This gives the point(s) on the graph of $y = x^2 - 1$ closest to the origin.]

Figure 71

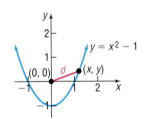

Solution   (a) Figure 71 illustrates the graph. The distance d from P to O is

$$d = \sqrt{(x - 0)^2 + (y - 0)^2} = \sqrt{x^2 + y^2}$$

Since P is a point on the graph of $y = x^2 - 1$, we substitute $x^2 - 1$ for y. Then

$$d(x) = \sqrt{x^2 + (x^2 - 1)^2} = \sqrt{x^4 - x^2 + 1}$$

We have expressed the distance d as a function of x.

(b) If $x = 0$, the distance d is

$$d(0) = \sqrt{1} = 1$$

(c) If $x = 1$, the distance d is

$$d(1) = \sqrt{1 - 1 + 1} = 1$$

(d) If $x = \dfrac{\sqrt{2}}{2}$, the distance d is

Figure 72

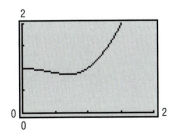

$$d\left(\frac{\sqrt{2}}{2}\right) = \sqrt{\left(\frac{\sqrt{2}}{2}\right)^4 - \left(\frac{\sqrt{2}}{2}\right)^2 + 1} = \sqrt{\frac{1}{4} - \frac{1}{2} + 1} = \frac{\sqrt{3}}{2}$$

(e) Figure 72 shows the graph of $Y_1 = \sqrt{x^4 - x^2 + 1}$. Using the MINIMUM feature on a graphing utility, we find that when $x \approx 0.71$ the value of d is smallest. The local minimum is $d \approx 0.87$ rounded to two decimal places. [By symmetry, it follows that when $x \approx -0.71$ the value of d is also a local minimum.]   ◄

NOW WORK PROBLEM 9.

| EXAMPLE 4 | **Filling a Swimming Pool** |

A rectangular swimming pool 20 meters long and 10 meters wide is 4 meters deep at one end and 1 meter deep at the other. Figure 73 illustrates a cross-sectional view of the pool. Water is being pumped into the pool to a height of 3 meters at the deep end.

Figure 73

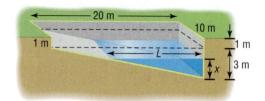

(a) Find a function that expresses the volume V of water in the pool as a function of the height x of the water at the deep end.

(b) Find the volume when the height is 1 meter.

(c) Find the volume when the height is 2 meters.

(d) At what height is the volume 20 cubic meters? 100 cubic meters?

Solution

(a) Let L denote the distance (in meters) measured at water level from the deep end to the short end. Notice that L and x form the sides of a triangle that is similar to the triangle whose sides are 20 meters by 3 meters. This means L and x are related by the equation

$$\frac{L}{x} = \frac{20}{3} \quad \text{or} \quad L = \frac{20x}{3}, \qquad 0 \le x \le 3$$

The volume V of water in the pool at any time is

$$V = \left(\begin{array}{c} \text{cross-sectional} \\ \text{triangular area} \end{array}\right)(\text{width}) = \left(\frac{1}{2}Lx\right)(10) \quad \text{cubic meters}$$

Since $L = \dfrac{20x}{3}$, we have

$$V(x) = \left(\frac{1}{2} \cdot \frac{20x}{3} \cdot x\right)(10) = \frac{100}{3}x^2 \quad \text{cubic meters}$$

(b) When the height x of the water is 1 meter, the volume $V = V(x)$ is

$$V(1) = \frac{100}{3} \cdot 1^2 = 33\frac{1}{3} \quad \text{cubic meters}$$

(c) When the height x of the water is 2 meters, the volume $V = V(x)$ is

$$V(2) = \frac{100}{3} \cdot 2^2 = \frac{400}{3} = 133\frac{1}{3} \quad \text{cubic meters}$$

(d) By solving the equation $\dfrac{100}{3}x^2 = 20$, we find that when $x \approx 0.77$ meter, the volume is 20 cubic meters. By solving the equation $\dfrac{100}{3}x^2 = 100$, we find that when $x \approx 1.73$ meters, the volume is 100 cubic meters. ◀

EXAMPLE 5   **Area of a Rectangle**

Figure 74

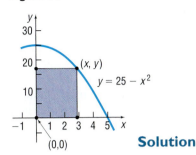

A rectangle has one corner on the graph of $y = 25 - x^2$, another at the origin, a third on the positive y-axis, and the fourth on the positive x-axis. See Figure 74.

(a) Express the area A of the rectangle as a function of x.

(b) What is the domain of A?

(c) Graph $A = A(x)$.

(d) For what value of x is the area largest?

Solution   (a) The area A of the rectangle is $A = xy$, where $y = 25 - x^2$. Substituting this expression for y, we obtain $A(x) = x(25 - x^2) = 25x - x^3$.

(b) Since x represents a side of the rectangle, we have $x > 0$. In addition, the area must be positive, so $y = 25 - x^2 > 0$, which implies that $x^2 < 25$, so $-5 < x < 5$. Combining these restrictions, we have the domain of A as $\{x \mid 0 < x < 5\}$ or $(0, 5)$ using interval notation.

(c) See Figure 75 for the graph of $A = A(x)$.

(d) Using MAXIMUM, we find that the area is a maximum of 48.11 at $x = 2.89$, each rounded to two decimal places. See Figure 76.

Figure 75

Figure 76

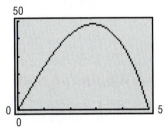

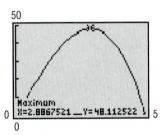

   NOW WORK PROBLEM 15.

EXAMPLE 6   **Making a Playpen** *

A manufacturer of children's playpens makes a square model that can be opened at one corner and attached at right angles to a wall or, perhaps, the side of a house. If each side is 3 feet in length, the open configuration doubles the available area in which the child can play from 9 square feet to 18 square feet. See Figure 77.

Now suppose that we place hinges at the outer corners to allow for a configuration like the one shown in Figure 78.

Figure 77

Figure 78

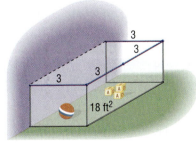

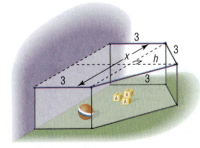

*Adapted from Proceedings, Su*mmer Conference for College Teachers on Applied Mathematics* (University of Missouri, Rolla), 1971.

(a) Express the area A of this configuration as a function of the distance x between the two parallel sides.

(b) Find the domain of A.

(c) Find A if $x = 5$.

(d) Graph $A = A(x)$. For what value of x is the area largest? What is the maximum area?

Solution (a) Refer to Figure 78. The area A that we seek consists of the area of a rectangle (with width 3 and length x) and the area of an isosceles triangle (with base x and two equal sides of length 3). The height h of the triangle may be found using the Pythagorean Theorem.

$$h^2 + \left(\frac{x}{2}\right)^2 = 3^2$$

$$h^2 = 3^2 - \left(\frac{x}{2}\right)^2 = 9 - \frac{x^2}{4} = \frac{36 - x^2}{4}$$

$$h = \frac{1}{2}\sqrt{36 - x^2}$$

The area A enclosed by the playpen is

$$A = \text{area of rectangle} + \text{area of triangle} = 3x + \frac{1}{2}x\left(\frac{1}{2}\sqrt{36 - x^2}\right)$$

$$A(x) = 3x + \frac{x\sqrt{36 - x^2}}{4}$$

Now the area A is expressed as a function of x.

(b) To find the domain of A, we note first that $x > 0$, since x is a length. Also, the expression under the square root must be positive, so

$$36 - x^2 > 0$$
$$x^2 < 36$$
$$-6 < x < 6$$

Figure 79

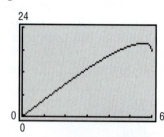

24

0
0   6

Combining these restrictions, we find that the domain of A is $0 < x < 6$, or $(0, 6)$ using interval notation.

(c) If $x = 5$, the area is

$$A(5) = 3(5) + \frac{5}{4}\sqrt{36 - (5)^2} \approx 19.15 \text{ square feet}$$

If the width of the playpen is 5 feet, its area is 19.15 square feet.

(d) See Figure 79. The maximum area is about 19.82 square feet, obtained when x is about 5.58 feet.   ◀

2.7 Assess Your Understanding

Applications and Extensions

1. Volume of a Cylinder The volume V of a right circular cylinder of height h and radius r is $V = \pi r^2 h$. If the height is twice the radius, express the volume V as a function of r.

2. Volume of a Cone The volume V of a right circular cone is $V = \frac{1}{3}\pi r^2 h$. If the height is twice the radius, express the volume V as a function of r.

3. **Demand Equation** The price p, in dollars, and the quantity x sold of a certain product obey the demand equation

$$p = -\frac{1}{6}x + 100, \qquad 0 \le x \le 600$$

(a) Express the revenue R as a function of x. (Remember, $R = xp$.)
(b) What is the revenue if 200 units are sold?
(c) Graph the revenue function using a graphing utility.
(d) What quantity x maximizes revenue? What is the maximum revenue?
(e) What price should the company charge to maximize revenue?

4. **Demand Equation** The price p, in dollars, and the quantity x sold of a certain product obey the demand equation

$$p = -\frac{1}{3}x + 100, \qquad 0 \le x \le 300$$

(a) Express the revenue R as a function of x.
(b) What is the revenue if 100 units are sold?
(c) Graph the revenue function using a graphing utility.
(d) What quantity x maximizes revenue? What is the maximum revenue?
(e) What price should the company charge to maximize revenue?

5. **Demand Equation** The price p, in dollars, and the quantity x sold of a certain product obey the demand equation

$$x = -5p + 100, \qquad 0 \le p \le 20$$

(a) Express the revenue R as a function of x.
(b) What is the revenue if 15 units are sold?
(c) Graph the revenue function using a graphing utility.
(d) What quantity x maximizes revenue? What is the maximum revenue?
(e) What price should the company charge to maximize revenue?

6. **Demand Equation** The price p, in dollars, and the quantity x sold of a certain product obey the demand equation

$$x = -20p + 500, \qquad 0 \le p \le 25$$

(a) Express the revenue R as a function of x.
(b) What is the revenue if 20 units are sold?
(c) Graph the revenue function using a graphing utility.
(d) What quantity x maximizes revenue? What is the maximum revenue?
(e) What price should the company charge to maximize revenue?

7. **Enclosing a Rectangular Field** David has available 400 yards of fencing and wishes to enclose a rectangular area.
(a) Express the area A of the rectangle as a function of the width x of the rectangle.
(b) What is the domain of A?
(c) Graph $A = A(x)$ using a graphing utility. For what value of x is the area largest?

8. **Enclosing a Rectangular Field along a River** Beth has 3000 feet of fencing available to enclose a rectangular field. One side of the field lies along a river, so only three sides require fencing.
(a) Express the area A of the rectangle as a function of x, where x is the length of the side parallel to the river.
(b) Graph $A = A(x)$ using a graphing utility. For what value of x is the area largest?

9. Let $P = (x, y)$ be a point on the graph of $y = x^2 - 8$.
(a) Express the distance d from P to the origin as a function of x.
(b) What is d if $x = 0$?
(c) What is d if $x = 1$?
(d) Use a graphing utility to graph $d = d(x)$.
(e) For what values of x is d smallest?

10. Let $P = (x, y)$ be a point on the graph of $y = x^2 - 8$.
(a) Express the distance d from P to the point $(0, -1)$ as a function of x.
(b) What is d if $x = 0$?
(c) What is d if $x = -1$?
(d) Use a graphing utility to graph $d = d(x)$.
(e) For what values of x is d smallest?

11. Let $P = (x, y)$ be a point on the graph of $y = \sqrt{x}$.
(a) Express the distance d from P to the point $(1, 0)$ as a function of x.
(b) Use a graphing utility to graph $d = d(x)$.
(c) For what values of x is d smallest?

12. Let $P = (x, y)$ be a point on the graph of $y = \dfrac{1}{x}$.
(a) Express the distance d from P to the origin as a function of x.
(b) Use a graphing utility to graph $d = d(x)$.
(c) For what values of x is d smallest?

13. A right triangle has one vertex on the graph of $y = x^3$, $x > 0$, at (x, y), another at the origin, and the third on the positive y-axis at $(0, y)$, as shown in the figure. Express the area A of the triangle as a function of x.

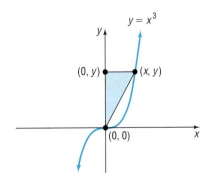

14. A right triangle has one vertex on the graph of $y = 9 - x^2$, $x > 0$, at (x, y), another at the origin, and the third on the positive x-axis at $(x, 0)$. Express the area A of the triangle as a function of x.

15. A rectangle has one corner on the graph of $y = 16 - x^2$, another at the origin, a third on the positive y-axis, and the fourth on the positive x-axis (see the figure).

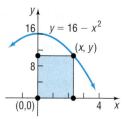

(a) Express the area A of the rectangle as a function of x.
(b) What is the domain of A?
(c) Graph $A = A(x)$. For what value of x is A largest?

16. A rectangle is inscribed in a semicircle of radius 2 (see the figure). Let $P = (x, y)$ be the point in quadrant I that is a vertex of the rectangle and is on the circle.

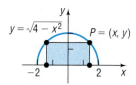

(a) Express the area A of the rectangle as a function of x.
(b) Express the perimeter p of the rectangle as a function of x.
(c) Graph $A = A(x)$. For what value of x is A largest?
(d) Graph $p = p(x)$. For what value of x is p largest?

17. A rectangle is inscribed in a circle of radius 2 (see the figure). Let $P = (x, y)$ be the point in quadrant I that is a vertex of the rectangle and is on the circle.

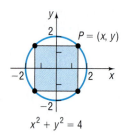

(a) Express the area A of the rectangle as a function of x.
(b) Express the perimeter p of the rectangle as a function of x.
(c) Graph $A = A(x)$. For what value of x is A largest?
(d) Graph $p = p(x)$. For what value of x is p largest?

18. A circle of radius r is inscribed in a square (see the figure).

(a) Express the area A of the square as a function of the radius r of the circle.
(b) Express the perimeter p of the square as a function of r.

19. A wire 10 meters long is to be cut into two pieces. One piece will be shaped as a square, and the other piece will be shaped as a circle (see the figure).

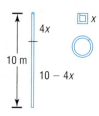

(a) Express the total area A enclosed by the pieces of wire as a function of the length x of a side of the square.
(b) What is the domain of A?
(c) Graph $A = A(x)$. For what value of x is A smallest?

20. A wire 10 meters long is to be cut into two pieces. One piece will be shaped as an equilateral triangle, and the other piece will be shaped as a circle.
(a) Express the total area A enclosed by the pieces of wire as a function of the length x of a side of the equilateral triangle.
(b) What is the domain of A?
(c) Graph $A = A(x)$. For what value of x is A smallest?

21. A wire of length x is bent into the shape of a circle.
(a) Express the circumference of the circle as a function of x.
(b) Express the area of the circle as a function of x.

22. A wire of length x is bent into the shape of a square.
(a) Express the perimeter of the square as a function of x.
(b) Express the area of the square as a function of x.

23. A semicircle of radius r is inscribed in a rectangle so that the diameter of the semicircle is the length of the rectangle (see the figure).

(a) Express the area A of the rectangle as a function of the radius r of the semicircle.
(b) Express the perimeter p of the rectangle as a function of r.

24. An equilateral triangle is inscribed in a circle of radius r. See the figure. Express the circumference C of the circle as a function of the length x of a side of the triangle.

[**Hint:** First show that $r^2 = \dfrac{x^2}{3}$.]

25. An equilateral triangle is inscribed in a circle of radius r. See the figure in Problem 24. Express the area A within the circle, but outside the triangle, as a function of the length x of a side of the triangle.

26. Two cars leave an intersection at the same time. One is headed south at a constant speed of 30 miles per hour, and the other is headed west at a constant speed of 40 miles per hour (see the figure). Express the distance d between the cars as a function of the time t.

[**Hint:** At $t = 0$, the cars leave the intersection.]

27. Two cars are approaching an intersection. One is 2 miles south of the intersection and is moving at a constant speed of 30 miles per hour. At the same time, the other car is 3 miles east of the intersection and is moving at a constant speed of 40 miles per hour.
  (a) Express the distance d between the cars as a function of time t.
     [**Hint:** At $t = 0$, the cars are 2 miles south and 3 miles east of the intersection, respectively.]
  (b) Use a graphing utility to graph $d = d(t)$. For what value of t is d smallest?

28. Inscribing a Cylinder in a Sphere Inscribe a right circular cylinder of height h and radius r in a sphere of fixed radius R. See the illustration. Express the volume V of the cylinder as a function of h.

[**Hint:** $V = \pi r^2 h$. Note also the right triangle.]

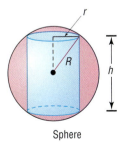

Sphere

29. Inscribing a Cylinder in a Cone Inscribe a right circular cylinder of height h and radius r in a cone of fixed radius R and fixed height H. See the illustration. Express the volume V of the cylinder as a function of r.

[**Hint:** $V = \pi r^2 h$. Note also the similar triangles.]

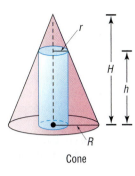

Cone

30. Installing Cable TV MetroMedia Cable is asked to provide service to a customer whose house is located 2 miles from the road along which the cable is buried. The nearest connection box for the cable is located 5 miles down the road (see the figure).

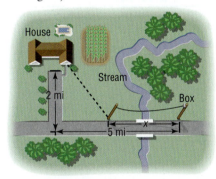

  (a) If the installation cost is \$10 per mile along the road and \$14 per mile off the road, express the total cost C of installation as a function of the distance x (in miles) from the connection box to the point where the cable installation turns off the road. Give the domain.
  (b) Compute the cost if $x = 1$ mile.
  (c) Compute the cost if $x = 3$ miles.
  (d) Graph the function $C = C(x)$. Use TRACE to see how the cost C varies as x changes from 0 to 5.
  (e) What value of x results in the least cost?

31. Time Required to Go from an Island to a Town An island is 2 miles from the nearest point P on a straight shoreline. A town is 12 miles down the shore from P. See the illustration.

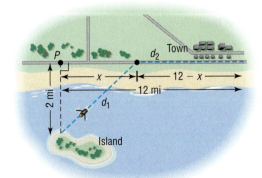

  (a) If a person can row a boat at an average speed of 3 miles per hour and the same person can walk 5 miles per hour, express the time T that it takes to go from the island to town as a function of the distance x from P to where the person lands the boat.

(b) What is the domain of T?

(c) How long will it take to travel from the island to town if the person lands the boat 4 miles from P?

(d) How long will it take if the person lands the boat 8 miles from P?

32. Water is poured into a container in the shape of a right circular cone with radius 4 feet and height 16 feet (see the fig-

ure). Express the volume V of the water in the cone as a function of the height h of the water.

[Hint: The volume V of a cone of radius r and height h is $V = \dfrac{1}{3}\pi r^2 h.$]

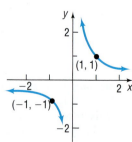

Chapter Review

Library of Functions

Linear function (p. 109)

$f(x) = mx + b$

Graph is a line with slope m and y-intercept b.

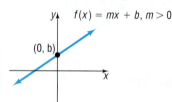

Constant function (p. 110)

$f(x) = b$

Graph is a horizontal line with y-intercept b.

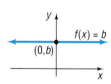

Identity function (p. 110)

$f(x) = x$

Graph is a line with slope 1 and y-intercept 0.

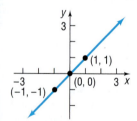

Square function (p. 110)

$f(x) = x^2$

Graph is a parabola with intercept at $(0, 0)$.

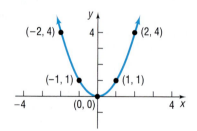

Cube function (p. 110)

$f(x) = x^3$

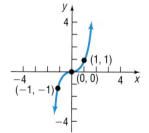

Square root function (p. 110)

$f(x) = \sqrt{x}$

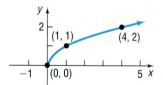

Cube root function (p. 111)

$f(x) = \sqrt[3]{x}$

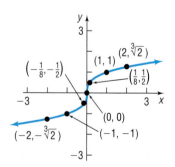

Reciprocal function (p. 112)

$f(x) = \dfrac{1}{x}$

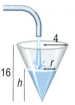

Absolute value function (p. 111)

$f(x) = |x|$

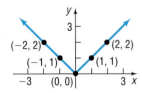

Greatest integer function (pp. 111–112)

$f(x) = \text{int}(x)$

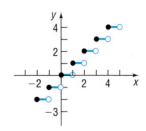

Things to Know

Function (pp. 56–61)	A relation between two sets of real numbers so that each number x in the first set, the domain, has corresponding to it exactly one number y in the second set. The range is the set of y values of the function for the x values in the domain.
	x is the independent variable; y is the dependent variable.
	A function can also be characterized as a set of ordered pairs (x, y) in which no first element is paired with two different second elements.
Function notation (pp. 61–64)	$y = f(x)$
	f is a symbol for the function.
	x is the argument, or independent variable.
	y is the dependent variable.
	$f(x)$ is the value of the function at x, or the image of x.
	A function f may be defined implicitly by an equation involving x and y or explicitly by writing $y = f(x)$.
Difference quotient of f (p. 63 and p. 92)	$\dfrac{f(x + h) - f(x)}{h}, \quad h \neq 0$
Domain (p. 58 and p. 64)	If unspecified, the domain of a function f is the largest set of real numbers for which $f(x)$ is a real number.
Vertical-line test (p. 72)	A set of points in the plane is the graph of a function if and only if every vertical line intersects the graph in at most one point.
Even function f (p. 80)	$f(-x) = f(x)$ for every x in the domain ($-x$ must also be in the domain).
Odd function f (p. 80)	$f(-x) = -f(x)$ for every x in the domain ($-x$ must also be in the domain).
Increasing function (p. 83)	A function f is increasing on an open interval I if, for any choice of x_1 and x_2 in I, with $x_1 < x_2$, we have $f(x_1) < f(x_2)$.
Decreasing function (p. 83)	A function f is decreasing on an open interval I if, for any choice of x_1 and x_2 in I, with $x_1 < x_2$, we have $f(x_1) > f(x_2)$.
Constant function (p. 83)	A function f is constant on an interval I if, for all choices of x in I, the values of $f(x)$ are equal.
Local maximum (p. 84)	A function f has a local maximum at c if there is an open interval I containing c so that, for all x in I, $f(x) \leq f(c)$.
Local minimum (p. 84)	A function f has a local minimum at c if there is an open interval I containing c so that, for all x in I, $f(x) \geq f(c)$.
Average rate of change of a function (p. 85)	The average rate of change of f from c to x is
	$$\dfrac{\Delta y}{\Delta x} = \dfrac{f(x) - f(c)}{x - c}, \qquad x \neq c$$
Linear function (p. 93)	$f(x) = mx + b$
	Graph is a line with slope m and y-intercept b.

Objectives

Review Exercises

In Problems 1 and 2, determine whether each relation represents a function. For each function, state the domain and range.

1. $\{(-1, 0), (2, 3), (4, 0)\}$ **2.** $\{(4, -1), (2, 1), (4, 2)\}$

In Problems 3–8, find the following for each function:

(a) $f(2)$ (b) $f(-2)$ (c) $f(-x)$ (d) $-f(x)$ (e) $f(x - 2)$ (f) $f(2x)$

3. $f(x) = \dfrac{3x}{x^2 - 1}$ **4.** $f(x) = \dfrac{x^2}{x + 1}$ **5.** $f(x) = \sqrt{x^2 - 4}$

6. $f(x) = |x^2 - 4|$ **7.** $f(x) = \dfrac{x^2 - 4}{x^2}$ **8.** $f(x) = \dfrac{x^3}{x^2 - 9}$

In Problems 9–16, find the domain of each function.

9. $f(x) = \dfrac{x}{x^2 - 9}$ **10.** $f(x) = \dfrac{3x^2}{x - 2}$ **11.** $f(x) = \sqrt{2 - x}$ **12.** $f(x) = \sqrt{x + 2}$

13. $h(x) = \dfrac{\sqrt{x}}{|x|}$ **14.** $g(x) = \dfrac{|x|}{x}$ **15.** $f(x) = \dfrac{x}{x^2 + 2x - 3}$ **16.** $F(x) = \dfrac{1}{x^2 - 3x - 4}$

In Problems 17–22, find $f + g, f - g, f \cdot g,$ and $\dfrac{f}{g}$ for each pair of functions. State the domain of each.

17. $f(x) = 2 - x;$ $g(x) = 3x + 1$

18. $f(x) = 2x - 1;$ $g(x) = 2x + 1$

19. $f(x) = 3x^2 + x + 1;$ $g(x) = 3x$

20. $f(x) = 3x;$ $g(x) = 1 + x + x^2$

21. $f(x) = \dfrac{x + 1}{x - 1};$ $g(x) = \dfrac{1}{x}$

22. $f(x) = \dfrac{1}{x - 3};$ $g(x) = \dfrac{3}{x}$

In Problems 23 and 24, find the difference quotient of each function f; that is, find

$$\frac{f(x + h) - f(x)}{h}, \qquad h \neq 0$$

23. $f(x) = -2x^2 + x + 1$

24. $f(x) = 3x^2 - 2x + 4$

25. Using the graph of the function f shown:
 (a) Find the domain and the range of f.
 (b) List the intercepts.
 (c) Find $f(-2)$.
 (d) For what values of x does $f(x) = -3$?
 (e) Solve $f(x) > 0$.
 (f) Graph $y = f(x - 3)$.
 (g) Graph $y = f\left(\dfrac{1}{2}x\right)$.
 (h) Graph $y = -f(x)$.

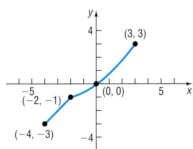

26. Using the graph of the function g shown:
 (a) Find the domain and the range of g.
 (b) Find $g(-1)$.
 (c) List the intercepts.
 (d) For what value of x does $g(x) = -3$?
 (e) Solve $g(x) > 0$.
 (f) Graph $y = g(x - 2)$.
 (g) Graph $y = g(x) + 1$.
 (h) Graph $y = 2g(x)$.

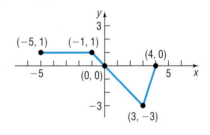

In Problems 27 and 28, use the graph of the function f to find:
(a) *The domain and the range of f*
(b) *The intervals on which f is increasing, decreasing, or constant*
(c) *The local minima and local maxima*
(d) *Whether the graph is symmetric with respect to the x-axis, the y-axis, or the origin*
(e) *Whether the function is even, odd, or neither*
(f) *The intercepts, if any*

27.

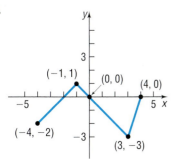

28.

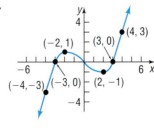

In Problems 29–36, determine (algebraically) whether the given function is even, odd, or neither.

29. $f(x) = x^3 - 4x$

30. $g(x) = \dfrac{4 + x^2}{1 + x^4}$

31. $h(x) = \dfrac{1}{x^4} + \dfrac{1}{x^2} + 1$

32. $F(x) = \sqrt{1 - x^3}$

33. $G(x) = 1 - x + x^3$

34. $H(x) = 1 + x + x^2$

35. $f(x) = \dfrac{x}{1 + x^2}$

36. $g(x) = \dfrac{1 + x^2}{x^3}$

In Problems 37–40, use a graphing utility to graph each function over the indicated interval. Approximate any local maxima and local minima. Determine where the function is increasing and where it is decreasing.

37. $f(x) = 2x^3 - 5x + 1 \quad (-3, 3)$

38. $f(x) = -x^3 + 3x - 5 \quad (-3, 3)$

39. $f(x) = 2x^4 - 5x^3 + 2x + 1 \quad (-2, 3)$

40. $f(x) = -x^4 + 3x^3 - 4x + 3 \quad (-2, 3)$

In Problems 41 and 42, find the average rate of change of f:
(a) *From 1 to 2*
(b) *From 0 to 1*
(c) *From 2 to 4*

41. $f(x) = 8x^2 - x$

42. $f(x) = 2x^3 + x$

In Problems 43–46, find the average rate of change from 2 to x for each function f. Be sure to simplify.

43. $f(x) = 2 - 5x$

44. $f(x) = 2x^2 + 7$

45. $f(x) = 3x - 4x^2$

46. $f(x) = x^2 - 3x + 2$

In Problems 47 and 48, tell which of the following graphs are graphs of functions.

47.

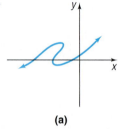

(a)   (b)

48.

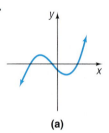   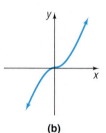

(a)   (b)

In Problems 49–52, graph each linear function.

49. $f(x) = 2x - 5$

50. $g(x) = -4x + 7$

51. $h(x) = \dfrac{4}{5}x - 6$

52. $F(x) = -\dfrac{1}{3}x + 1$

In Problems 53 and 54, sketch the graph of each function. Be sure to label at least three points.

53. $f(x) = |x|$

54. $f(x) = \sqrt[3]{x}$

In Problems 55–66, graph each function using the techniques of shifting, compressing or stretching, and reflections. Identify any intercepts on the graph. State the domain and, based on the graph, find the range.

55. $F(x) = |x| - 4$

56. $f(x) = |x| + 4$

57. $g(x) = -2|x|$

58. $g(x) = \dfrac{1}{2}|x|$

59. $h(x) = \sqrt{x - 1}$

60. $h(x) = \sqrt{x} - 1$

61. $f(x) = \sqrt{1 - x}$

62. $f(x) = -\sqrt{x + 3}$

63. $h(x) = (x - 1)^2 + 2$

64. $h(x) = (x + 2)^2 - 3$

65. $g(x) = 3(x - 1)^3 + 1$

66. $g(x) = -2(x + 2)^3 - 8$

In Problems 67–70,
    (a) *Find the domain of each function.*
    (b) *Locate any intercepts.*
    (c) *Graph each function.*
    (d) *Based on the graph, find the range.*

67. $f(x) = \begin{cases} 3x & -2 < x \le 1 \\ x + 1 & x > 1 \end{cases}$

68. $f(x) = \begin{cases} x - 1 & -3 < x < 0 \\ 3x - 1 & x \ge 0 \end{cases}$

69. $f(x) = \begin{cases} x & -4 \le x < 0 \\ 1 & x = 0 \\ 3x & x > 0 \end{cases}$

70. $f(x) = \begin{cases} x^2 & -2 \le x \le 2 \\ 2x - 1 & x > 2 \end{cases}$

71. Given that f is a linear function, $f(4) = -5$, and $f(0) = 3$, write the equation that defines f.

72. Given that g is a linear function with slope $= -4$ and $g(-2) = 2$, write the equation that defines g.

73. A function f is defined by

$$f(x) = \frac{Ax + 5}{6x - 2}$$

If $f(1) = 4$, find A.

74. A function g is defined by

$$g(x) = \frac{A}{x} + \frac{8}{x^2}$$

If $g(-1) = 0$, find A.

75. Temperature Conversion The temperature T of the air is approximately a linear function of the altitude h for altitudes within 10,000 meters of the surface of Earth. If the surface temperature is 30°C and the temperature at 10,000 meters is 5°C, find the function $T = T(h)$.

76. Speed as a Function of Time The speed v (in feet per second) of a car is a linear function of the time t (in seconds) for $10 \le t \le 30$. If after each second the speed of the car has increased by 5 feet per second and if after 20 seconds the speed is 80 feet per second, how fast is the car going after 30 seconds? Find the function $v = v(t)$.

77. Spheres The volume V of a sphere of radius r is $V = \frac{4}{3}\pi r^3$; the surface area S of this sphere is $S = 4\pi r^2$. If the radius doubles, how does the volume change? How does the surface area change?

78. Page Design A page with dimensions of $8\frac{1}{2}$ inches by 11 inches has a border of uniform width x surrounding the printed matter of the page, as shown in the figure.

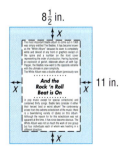

(a) Write a formula for the area A of the printed part of the page as a function of the width x of the border.
(b) Give the domain and the range of A.
(c) Find the area of the printed the page for borders of widths 1 inch, 1.2 inches, and 1.5 inches.
(d) Graph the function $A = A(x)$.

(e) Use TRACE to determine what margin should be used to obtain an area of 70 square inches and of 50 square inches.

79. Strength of a Beam The strength of a rectangular wooden beam is proportional to the product of the width and the cube of its depth (see the figure). If the beam is to be cut from a log in the shape of a cylinder of radius 3 feet, express the strength S of the beam as a function of the width x. What is the domain of S?

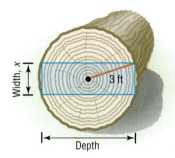

80. Material Needed to Make a Drum A steel drum in the shape of a right circular cylinder is required to have a volume of 100 cubic feet.
(a) Express the amount A of material required to make the drum as a function of the radius r of the cylinder.
(b) How much material is required if the drum is of radius 3 feet?
(c) How much material is required if the drum is of radius 4 feet?
(d) How much material is required if the drum is of radius 5 feet?
(e) Graph $A = A(r)$. For what value of r is A smallest?

81. High School versus College GPA An administrator at Southern Illinois University wants to find a function that relates a student's college grade point average G to the high school grade point average x. She randomly selects eight students and obtains the following data:

High School GPA, x	College GPA, G
2.73	2.43
2.92	2.97
3.45	3.63
3.78	3.81
2.56	2.83
2.98	2.81
3.67	3.45
3.10	2.93

(a) Does the relation defined by the set of ordered pairs (x, G) represent a function?
(b) Draw a scatter diagram of the data. Are the data linear?
(c) Using a graphing utility, find the line of best fit relating high school GPA and college GPA.

(d) Interpret the slope.

(e) Express the relationship found in part (c) using function notation.

(f) What is the domain of the function?

(g) Predict a student's college GPA if her high school GPA is 3.23.

82. Mortgage Payments The monthly payment p on a mortgage varies directly with the amount borrowed B. If the monthly payment on a 30-year mortgage is $854.00 when $130,000 is borrowed, find a linear function that relates the monthly payment p to the amount borrowed B for a mortgage with the same terms. Then find the monthly payment p when the amount borrowed B is $165,000.

83. Revenue Function At the corner Esso station, the revenue R varies directly with the number g of gallons of gasoline sold. If the revenue is $28.89 when the number of gallons sold is 13.5, find a linear function that relates revenue R to the number g of gallons of gasoline. Then find the revenue R when the number of gallons of gasoline sold is 11.2.

84. Landscaping A landscape engineer has 200 feet of border to enclose a rectangular pond. What dimensions will result in the largest pond?

85. Geometry Find the length and width of a rectangle whose perimeter is 20 feet and whose area is 16 square feet.

86. A rectangle has one vertex on the line $y = 10 - x, x > 0$, another at the origin, one on the positive x-axis, and one on the positive y-axis. Find the largest area A that can be enclosed by the rectangle.

87. Minimizing Marginal Cost The marginal cost of a product can be thought of as the cost of producing one additional unit of output. For example, if the marginal cost of producing the 50th product is $6.20, then it cost $6.20 to increase production from 49 to 50 units of output. Callaway Golf Company has determined that the marginal cost C of manufacturing x Big Bertha golf clubs may be expressed by the quadratic function

$$C(x) = 4.9x^2 - 617.4x + 19,600$$

(a) How many clubs should be manufactured to minimize the marginal cost?

(b) At this level of production, what is the marginal cost?

88. Find the point on the line $y = x$ that is closest to the point $(3, 1)$.

89. Find the point on the line $y = x + 1$ that is closest to the point $(4, 1)$.

90. A rectangle has one vertex on the graph of $y = 10 - x^2, x > 0$, another at the origin, one on the positive x-axis, and one on the positive y-axis. Find the largest area A that can be enclosed by the rectangle.

91. Constructing a Closed Box A closed box with a square base is required to have a volume of 10 cubic feet.

(a) Express the amount A of material used to make such a box as a function of the length x of a side of the square base.

(b) How much material is required for a base 1 foot by 1 foot?

(c) How much material is required for a base 2 feet by 2 feet?

(d) Graph $A = A(x)$. For what value of x is A smallest?

92. Cost of a Drum A drum in the shape of a right circular cylinder is required to have a volume of 500 cubic centimeters. The top and bottom are made of material that costs 6¢ per square centimeter; the sides are made of material that costs 4¢ per square centimeter.

500 cc

(a) Express the total cost C of the material as a function of the radius r of the cylinder.

(b) What is the cost if the radius is 4 cm?

(c) What is the cost if the radius is 8 cm?

(d) Graph $C = C(r)$. For what value of r is the cost C least?

Chapter Test

1. Determine whether each relation represents a function. For each function, state the domain and range.
   (a) $\{(2, 5), (4, 6), (6, 7), (8, 8)\}$

   (b) $\{(1, 3), (4, -2), (-3, 5), (1, 7)\}$

   (c)

   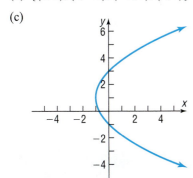

   (d)

   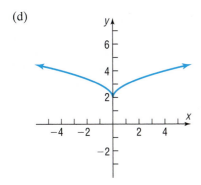

In Problems 2–4, find the domain of each function and evaluate each function at $x = -1$.

2. $f(x) = \sqrt{4 - 5x}$

3. $g(x) = \dfrac{x + 2}{|x + 2|}$

4. $h(x) = \dfrac{x - 4}{x^2 + 5x - 36}$

5. Using the graph of the function f below:
   (a) Find the domain and the range of f.
   (b) List the intercepts.
   (c) Find $f(1)$.
   (d) For what value(s) of x does $f(x) = -3$?
   (e) Solve $f(x) < 0$

   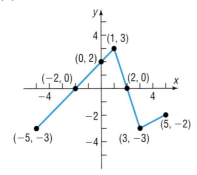

6. Use a graphing utility to graph the function $f(x) = -x^4 + 2x^3 + 4x^2 - 2$ on the interval $(-5, 5)$. Approximate any local maxima and local minima rounded to two decimal places. Determine where the function is increasing and where it is decreasing.

7. Consider the function $g(x) = \begin{cases} 2x + 1 & x < -1 \\ x - 4 & x \geq -1 \end{cases}$.
   (a) Graph the function.
   (b) List the intercepts.
   (c) Find $g(-5)$.
   (d) Find $g(2)$.

8. For the function $f(x) = 3x^2 - 2x + 4$, find the average rate of change from 3 to x.

9. For the functions $f(x) = 2x^2 + 1$ and $g(x) = 3x - 2$, find the following and simplify:
   (a) $f - g$
   (b) $f \cdot g$
   (c) $f(x + h) - f(x)$

10. Graph each function using the techniques of shifting, compressing or stretching, and reflections. Start with the graph of the basic function and show all stages.
    (a) $h(x) = -2(x + 1)^3 + 3$
    (b) $g(x) = |x + 4| + 2$

11. Consider the two data sets:

    Set A:

x	0	1	2	3	4	5	6	7
y	-5	-9	-8	-6	-9	-6	-3	4

    Set B:

x	0	1	2	3	4	5	6	7
y	-4	-5	-1	0	4	4	7	9

    (a) Use a graphing utility to make a scatter diagram for each set of data, and determine which of the two data sets has a stronger linear relationship.
    (b) Use a graphing utility to find the line of best fit for the data set you selected in part (a).

12. The variable interest rate on a student loan changes each July 1 based on the bank prime loan rate. For the years 1992–2004, this rate can be approximated by the model $r(x) = -0.115x^2 + 1.183x + 5.623$, where x is the number of years since 1992 and r is the interest rate as a percent.
    (Source: U.S. Federal Reserve)
    (a) Use a graphing utility to estimate the highest rate during this time period. During which year was the interest rate the highest?
    (b) Use the model to estimate the rate in 2010. Does this value seem reasonable?

13. A community skating rink is in the shape of a rectangle with semicircles attached at the ends. The length of the rectangle is 20 feet less than twice the width. The thickness of the ice is 0.75 inch.

(a) Write the ice volume, V, as a function of the width, x.

(b) How much ice is in the rink if the width is 90 feet?

Chapter Projects

1. **Cell Phone Service** This month when you were paying your bills, you noticed that your original service contract for your cell phone service had expired. Since your cell number is portable to another company now, you decided to investigate different companies as well. Noting that you travel outside your local calling area on a regular basis, you decide to stick with only those plans that have roaming charges included. Here is what you found:

	Anytime minutes included	Charge for each extra minute	Mobile-to-mobile minutes	National Long Distance	Nights and Weekends
Company A:					
Plan A1: $49.99	600	$0.45	Unlimited	Included	Unlimited (after 9 pm)
Plan A2: $59.99	900	$0.35	Unlimited	Included	Unlimited (after 7 pm)
Company B:					
Plan B1: $39.99	450 with rollover	$0.45	Unlimited	Included	3000 min (after 9 pm)
Plan B2: $49.99	600 with rollover	$0.40	Unlimited	Included	unlimited (after 9 pm)
Company C:					
Plan C1: $45.00	300	$0.40	Unlimited	Included	Unlimited (after 9 pm)
Plan C2: $60.00	700	$0.40	Unlimited	Included	Unlimited (after 9 pm)

Each plan requires a two-year contract.

(a) Determine the total cost of each plan for the life of the contract, assuming that you stay within the allotted anytime minutes provided by each contract.

(b) If you expect to use 600 anytime minutes and 2500 night and weekend minutes per month, which plan provides the best deal? If you expect to use 600 anytime minutes and 3500 night and weekend minutes, which plan provides the best deal?

(c) Ignoring any night and weekend usage, if you expect to use 425 anytime minutes each month, which option provides the best deal? What if you use 750 anytime minutes per month?

(d) Each monthly charge includes a specific number of peak time minutes in the monthly fee. Write a function for each option, where C is the monthly cost and x is the number of anytime minutes used.

(e) Graph each of the functions from part (d).

(f) For each of the companies A, B, and C, determine the average price per minute for each plan, based on no extra minutes used. For each company, which plan is better?

(g) Now, looking at the three plans that you found to be the best for Companies A, B, and C, in part (f), which of those three seems to be the best deal?

(h) Based upon your own cell phone usage, which plan would be the best for you?

SOURCE: Based on rates from the websites of the companies: ATTWireless, Cingular, and Sprint PCS, for area code 76201 on October 4, 2004. *(www.attwireless.com, www.cingular.com, www.sprint.com)*

The following projects are available on the Instructor's Resource Center (IRC):

2. **Project at Motorola**   *Pricing Wireless Service*

3. **Cost of Cable**

4. **Oil Spill**

Cumulative Review

In Problems 1–8, find the real solutions of each equation.

1. $-5x + 4 = 0$

2. $x^2 - 7x + 12 = 0$

3. $3x^2 - 5x - 2 = 0$

4. $4x^2 + 4x + 1 = 0$

5. $4x^2 - 2x + 4 = 0$

6. $\sqrt[3]{1 - x} = 2$

7. $\sqrt[5]{1 - x} = 2$

8. $|2 - 3x| = 1$

9. In the complex number system, solve $4x^2 - 2x + 4 = 0$.

10. Solve the inequality $-2 < 3x - 5 < 7$. Graph the solution set.

In Problems 11–14, graph each equation.

11. $-3x + 4y = 12$

12. $y = 3x + 12$

13. $x^2 + y^2 + 2x - 4y + 4 = 0$

14. $y = (x + 1)^2 - 3$

15. For the graph of the function f below:

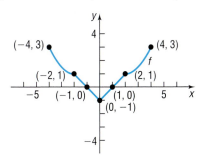

(a) Find the domain and the range of f.

(b) Find the intercepts.

(c) Is the graph of f symmetric with respect to the x-axis, the y-axis, or the origin?

(d) Find $f(2)$.

(e) For what value(s) of x is $f(x) = 3$?

(f) Solve $f(x) < 0$.

(g) Graph $y = f(x) + 2$

(h) Graph $y = f(-x)$.

(i) Graph $y = 2f(x)$.

(j) Is f even, odd, or neither?

(k) Find the interval(s) on which f is increasing.

(l) Find the interval(s) on which f is decreasing.

(m) Find the local maxima and local minima.

(n) Find the average rate of change of f from 1 to 4.

16. Find the distance between the points $P = (-1, 3)$ and $Q = (4, -2)$.

17. Which of the following points are on the graph of $y = x^3 - 3x + 1$?

(a) $(-2, -1)$ (b) $(2, 3)$ (c) $(3, 1)$

18. Determine the intercepts of $y = 3x^2 + 14x - 5$.

19. Use a graphing utility to find the solution(s) of the equation $x^4 - 3x^3 + 4x - 1 = 0$.

Hint: All solutions are between $x = -10$ and $x = 10$.

20. Find the equation of the line perpendicular to the line $y = 2x + 1$ and containing the point $(3, 5)$. Express your answer in slope–intercept form and graph both lines.

21. Determine whether the following relation represents a function: $\{(-3, 8), (1, 3), (2, 5), (3, 8)\}$.

22. For the function f defined by $f(x) = x^2 - 4x + 1$, evaluate:

(a) $f(2)$ (b) $f(x) + f(2)$

(c) $f(-x)$ (d) $-f(x)$

(e) $f(x + 2)$

(f) $\dfrac{f(x + h) - f(x)}{h}$, $h \neq 0$

23. Find the domain of $h(z) = \dfrac{3z - 1}{z^2 - 6z - 7}$.

24. Determine whether the following graph is the graph of a function.

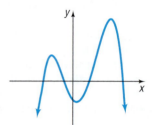

25. Consider the function $f(x) = \dfrac{x}{x + 4}$.

(a) Is the point $\left(1, \dfrac{1}{4}\right)$ on the graph of f?

(b) If $x = -2$, what is $f(x)$? What point is on the graph of f?

(c) If $f(x) = 2$, what is x? What point is on the graph of f?

Polynomial and Rational Functions

3

Ragweed plus bad air will equal fall misery

A hot, dry, smoggy summer is adding up to an insufferable fall for allergy sufferers.

Experts say people with allergies to ragweed, who normally sneeze and wheeze through September, will be dabbing their eyes and nose well into October, even though pollen counts are actually no worse than usual.

The difference this year is that a summer of heavy smog has made the pollen in the air feel much thicker, making it more irritating.

"Smog plays an extremely important role in allergy suffering," says Frances Coates, owner of Ottawa-based Aerobiology Research, which measures pollen levels. "It holds particles in the air a lot longer."

Low rainfall delayed the ragweed season until the end of August. It usually hits early in the month.

SOURCE: Adapted by Louise Surette from *Toronto Star*, September 8, 2001.

See Chapter Project 1.

A LOOK BACK In Chapter 2, we began our discussion of functions. We defined domain and range and independent and dependent variables; we found the value of a function and graphed functions. We continued our study of functions by listing the properties that a function might have, like being even or odd, and we created a library of functions, naming key functions and listing their properties, including their graphs.

A LOOK AHEAD In this chapter, we look at two general classes of functions: polynomial functions and rational functions, and examine their properties. Polynomial functions are arguably the simplest expressions in algebra. For this reason, they are often used to approximate other, more complicated, functions. Rational functions are simply ratios of polynomial functions.

We also introduce methods that can be used to determine the zeros of polynomial functions that are not factored completely. We end the chapter with a discussion of complex zeros.

3.1 Quadratic Functions and Models

PREPARING FOR THIS SECTION *Before getting started, review the following:*

• Intercepts (Section 1.2, pp. 15–17)
• Quadratic Equations (Appendix, Section A.5, pp. 696–703)

• Completing the Square (Appendix, Section A.5, pp. 699)
• Graphing Techniques: Transformations (Section 2.6, pp. 118–126)

 Now work the 'Are You Prepared?' problems on page 163.

OBJECTIVES **1** Graph a Quadratic Function Using Transformations
2 Identify the Vertex and Axis of Symmetry of a Quadratic Function
3 Graph a Quadratic Function Using Its Vertex, Axis, and Intercepts
4 Use the Maximum or Minimum Value of a Quadratic Function to Solve Applied Problems
5 Use a Graphing Utility to Find the Quadratic Function of Best Fit to Data

A *quadratic function* is a function that is defined by a second-degree polynomial in one variable.

> A **quadratic function** is a function of the form
>
> $$f(x) = ax^2 + bx + c \qquad \textbf{(1)}$$
>
> where a, b, and c are real numbers and $a \neq 0$. The domain of a quadratic function is the set of all real numbers.

Many applications require a knowledge of quadratic functions. For example, suppose that Texas Instruments collects the data shown in Table 1, which relate the number of calculators sold at the price p (in dollars) per calculator. Since the price of a product determines the quantity that will be purchased, we treat price as the independent variable. The relationship between the number x of calculators sold and the price p per calculator may be approximated by the linear equation

$$x = 21{,}000 - 150p$$

Table 1

Price per Calculator, p (Dollars)	Number of Calculators, x
60	11,100
65	10,115
70	9,652
75	8,731
80	8,087
85	7,205
90	6,439

Then the revenue R derived from selling x calculators at the price p per calculator is equal to the unit selling price p of the product times the number x of units actually sold. That is,

$$R = xp$$
$$R(p) = (21{,}000 - 150p)p$$
$$= -150p^2 + 21{,}000p$$

So, the revenue R is a quadratic function of the price p. Figure 1 illustrates the graph of this revenue function, whose domain is $0 \le p \le 140$, since both x and p must be non-negative. Later in this section we shall determine the price p that maximizes revenue.

A second situation in which a quadratic function appears involves the motion of a projectile. Based on Newton's second law of motion (force equals mass times acceleration, $F = ma$), it can be shown that, ignoring air resistance, the path of a projectile propelled upward at an inclination to the horizontal is the graph of a quadratic function. See Figure 2 for an illustration. Later in this section we shall analyze the path of a projectile.

Figure 1

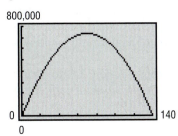

Figure 2
Path of a cannonball

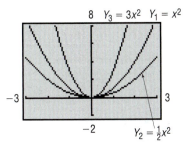

1   Graph a Quadratic Function Using Transformations

We know how to graph the quadratic function $f(x) = x^2$. Figure 3 shows the graph of three functions of the form $f(x) = ax^2, a > 0$, for $a = 1$, $a = \dfrac{1}{2}$, and $a = 3$. Notice that the larger the value of a, the "narrower" the graph is, and the smaller the value of a, the "wider" the graph is.

Figure 3

Figure 4

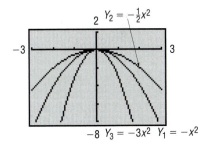

Figure 5
Graphs of a quadratic function,
$f(x) = ax^2 + bx + c, a \neq 0$

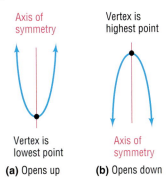

Axis of symmetry

Vertex is highest point

Vertex is lowest point

Axis of symmetry

(a) Opens up
$a > 0$

(b) Opens down
$a < 0$

Figure 4 shows the graphs of $f(x) = ax^2$ for $a < 0$. Notice that these graphs are reflections about the x-axis of the graphs in Figure 3. Based on the results of these two figures, we can draw some general conclusions about the graph of $f(x) = ax^2$. First, as $|a|$ increases, the graph becomes "taller" (a vertical stretch), and as $|a|$ gets closer to zero, the graph gets "shorter" (a vertical compression). Second, if a is positive, then the graph opens "up," and if a is negative, then the graph opens "down."

The graphs in Figures 3 and 4 are typical of the graphs of all quadratic functions, which we call **parabolas**.* Refer to Figure 5, where two parabolas are pictured. The one on the left **opens up** and has a lowest point; the one on the right **opens down** and has a highest point. The lowest or highest point of a parabola is called the **vertex**.

*We shall study parabolas using a geometric definition later in this book.

The vertical line passing through the vertex in each parabola in Figure 5 is called the **axis of symmetry** (usually abbreviated to **axis**) of the parabola. Because the parabola is symmetric about its axis, the axis of symmetry of a parabola can be used to find additional points on the parabola when graphing by hand.

The parabolas shown in Figure 5 are the graphs of a quadratic function $f(x) = ax^2 + bx + c, a \neq 0$. Notice that the coordinate axes are not included in the figure. Depending on the values of a, b, and c, the axes could be placed anywhere. The important fact is that the shape of the graph of a quadratic function will look like one of the parabolas in Figure 5.

In the following example, we use techniques from Section 2.6 to graph a quadratic function $f(x) = ax^2 + bx + c, a \neq 0$. In so doing, we shall complete the square and write the function f in the form $f(x) = a(x - h)^2 + k$.

EXAMPLE 1

Graphing a Quadratic Function Using Transformations

Graph the function $f(x) = 2x^2 + 8x + 5$. Find the vertex and axis of symmetry.

Solution

We begin by completing the square on the right side.

$$f(x) = 2x^2 + 8x + 5$$

$$= 2(x^2 + 4x) + 5 \qquad \text{Factor out the 2 from } 2x^2 + 8x.$$

$$= 2(x^2 + 4x + 4) + 5 - 8 \qquad \text{Complete the square of } 2(x^2 + 4x). \text{ Notice that the factor of 2 requires that 8 be added and subtracted.}$$

$$= 2(x + 2)^2 - 3 \qquad \qquad \qquad \text{(2)}$$

The graph of f can be obtained in three stages, as shown in Figure 6. Now compare this graph to the graph in Figure 5(a). The graph of $f(x) = 2x^2 + 8x + 5$ is a parabola that opens up and has its vertex (lowest point) at $(-2, -3)$. Its axis of symmetry is the line $x = -2$.

Figure 6

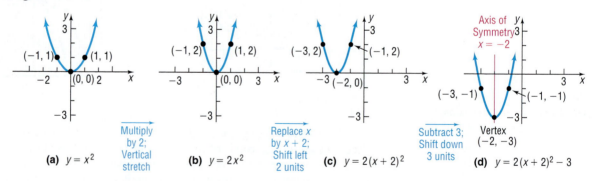

(a) $y = x^2$ Multiply by 2; Vertical stretch   (b) $y = 2x^2$ Replace x by $x + 2$; Shift left 2 units   (c) $y = 2(x + 2)^2$ Subtract 3; Shift down 3 units   (d) $y = 2(x + 2)^2 - 3$

✔ **CHECK:** Use a graphing utility to graph $f(x) = 2x^2 + 8x + 5$ and use the MINIMUM command to locate its vertex.   ◄

NOW WORK PROBLEM 27.

The method used in Example 1 can be used to graph any quadratic function $f(x) = ax^2 + bx + c, a \neq 0$, as follows:

$$f(x) = ax^2 + bx + c$$

$$= a\left(x^2 + \frac{b}{a}x\right) + c \qquad \text{Factor out } a \text{ from } ax^2 + bx.$$

$$= a\left(x^2 + \frac{b}{a}x + \frac{b^2}{4a^2}\right) + c - a\left(\frac{b^2}{4a^2}\right) \qquad \text{Complete the square by adding and subtracting } a\left(\frac{b^2}{4a^2}\right). \text{ Look closely at this step!}$$

$$= a\left(x + \frac{b}{2a}\right)^2 + c - \frac{b^2}{4a} \qquad \text{Factor}$$

$$= a\left(x + \frac{b}{2a}\right)^2 + \frac{4ac - b^2}{4a} \qquad c - \frac{b^2}{4a} = c \cdot \frac{4a}{4a} - \frac{b^2}{4a} = \frac{4ac - b^2}{4a}$$

Based on these results, we conclude the following:

> If $h = -\dfrac{b}{2a}$ and $k = \dfrac{4ac - b^2}{4a}$, then
>
> $$f(x) = ax^2 + bx + c = a(x - h)^2 + k \qquad \text{(3)}$$

The graph of $f(x) = a(x - h)^2 + k$ is the parabola $y = ax^2$ shifted horizontally h units (replace x by $x - h$) and vertically k units (add k). As a result, the vertex is at (h, k), and the graph opens up if $a > 0$ and down if $a < 0$. The axis of symmetry is the vertical line $x = h$.

For example, compare equation (3) with equation (2) of Example 1.

$$f(x) = 2(x + 2)^2 - 3$$
$$= 2(x - (-2))^2 - 3$$
$$= a(x - h)^2 + k$$

We conclude that $a = 2$, so the graph opens up. Also, we find that $h = -2$ and $k = -3$, so its vertex is at $(-2, -3)$.

2 ▸ Identify the Vertex and Axis of Symmetry of a Quadratic Function

We do not need to complete the square to obtain the vertex. In almost every case, it is easier to obtain the vertex of a quadratic function f by remembering that its x-coordinate is $h = -\dfrac{b}{2a}$. The y-coordinate can then be found by evaluating f at $-\dfrac{b}{2a}$ to find $k = f\left(-\dfrac{b}{2a}\right)$.

We summarize these remarks as follows:

> **Properties of the Graph of a Quadratic Function**
>
> $$f(x) = ax^2 + bx + c, \qquad a \neq 0$$
>
> $$\text{Vertex} = \left(-\frac{b}{2a}, f\left(-\frac{b}{2a}\right)\right) \qquad \text{Axis of symmetry: the line } x = -\frac{b}{2a} \qquad \text{(4)}$$
>
> Parabola opens up if $a > 0$; the vertex is a minimum point.
> Parabola opens down if $a < 0$; the vertex is a maximum point.

| **EXAMPLE 2** | **Locating the Vertex without Graphing** |

Without graphing, locate the vertex and axis of symmetry of the parabola defined by $f(x) = -3x^2 + 6x + 1$. Does it open up or down?

Solution   For this quadratic function, $a = -3$, $b = 6$, and $c = 1$. The x-coordinate of the vertex is

$$h = -\frac{b}{2a} = -\frac{6}{-6} = 1$$

The y-coordinate of the vertex is

$$k = f\left(-\frac{b}{2a}\right) = f(1) = -3 + 6 + 1 = 4$$

The vertex is located at the point $(1, 4)$. The axis of symmetry is the line $x = 1$. Because $a = -3 < 0$, the parabola opens down. ◄

3 Graph a Quadratic Function Using Its Vertex, Axis, and Intercepts

The information we gathered in Example 2, together with the location of the intercepts, usually provides enough information to graph $f(x) = ax^2 + bx + c, a \neq 0$, by hand.

The y-intercept is the value of f at $x = 0$; that is, $f(0) = c$.

The x-intercepts, if there are any, are found by solving the quadratic equation

$$ax^2 + bx + c = 0$$

This equation has two, one, or no real solutions, depending on whether the discriminant $b^2 - 4ac$ is positive, 0, or negative. The graph of f has x-intercepts, as follows:

> **The x-intercepts of a Quadratic Function**
>
> **1.** If the discriminant $b^2 - 4ac > 0$, the graph of $f(x) = ax^2 + bx + c$ has two distinct x-intercepts and so will cross the x-axis in two places.
>
> **2.** If the discriminant $b^2 - 4ac = 0$, the graph of $f(x) = ax^2 + bx + c$ has one x-intercept and touches the x-axis at its vertex.
>
> **3.** If the discriminant $b^2 - 4ac < 0$, the graph of $f(x) = ax^2 + bx + c$ has no x-intercept and so will not cross or touch the x-axis.

Figure 7 illustrates these possibilities for parabolas that open up.

Figure 7
$f(x) = ax^2 + bx + c, a > 0$

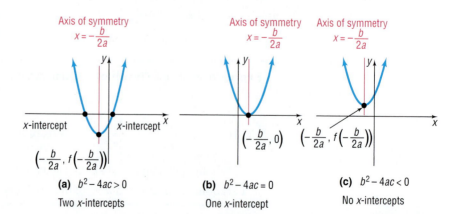

(a) $b^2 - 4ac > 0$
Two x-intercepts

(b) $b^2 - 4ac = 0$
One x-intercept

(c) $b^2 - 4ac < 0$
No x-intercepts

EXAMPLE 3

Graphing a Quadratic Function by Hand Using Its Vertex, Axis, and Intercepts

Solution

Use the information from Example 2 and the locations of the intercepts to graph $f(x) = -3x^2 + 6x + 1$. Determine the domain and the range of f. Determine where f is increasing and where it is decreasing.

In Example 2, we found the vertex to be at $(1, 4)$ and the axis of symmetry to be $x = 1$. The y-intercept is found by letting $x = 0$. The y-intercept is $f(0) = 1$. The x-intercepts are found by solving the equation $f(x) = 0$. This results in the equation

$$-3x^2 + 6x + 1 = 0 \qquad a = -3, b = 6, c = 1$$

The discriminant $b^2 - 4ac = (6)^2 - 4(-3)(1) = 36 + 12 = 48 > 0$, so the equation has two real solutions and the graph has two x-intercepts. Using the quadratic formula, we find that

$$x = \frac{-b + \sqrt{b^2 - 4ac}}{2a} = \frac{-6 + \sqrt{48}}{-6} = \frac{-6 + 4\sqrt{3}}{-6} \approx -0.15$$

and

$$x = \frac{-b - \sqrt{b^2 - 4ac}}{2a} = \frac{-6 - \sqrt{48}}{-6} = \frac{-6 - 4\sqrt{3}}{-6} \approx 2.15$$

The x-intercepts are approximately -0.15 and 2.15.

The graph is illustrated in Figure 8. Notice how we used the y-intercept and the axis of symmetry, $x = 1$, to obtain the additional point $(2, 1)$ on the graph.

The domain of f is the set of all real numbers. Based on the graph, the range of f is the interval $(-\infty, 4]$. The function f is increasing on the interval $(-\infty, 1)$ and decreasing on the interval $(1, \infty)$.

✔ CHECK: Graph $f(x) = -3x^2 + 6x + 1$ using a graphing utility. Use ZERO (or ROOT) to locate the two x-intercepts. Use MAXIMUM to locate the vertex. ◀

Figure 8

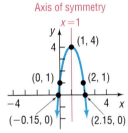

NOW WORK PROBLEM 35.

If the graph of a quadratic function has only one x-intercept or no x-intercepts, it is usually necessary to plot an additional point to obtain the graph by hand.

EXAMPLE 4

Graphing a Quadratic Function by Hand Using Its Vertex, Axis, and Intercepts

Solution

Graph $f(x) = x^2 - 6x + 9$ by determining whether the graph opens up or down and by finding its vertex, axis of symmetry, y-intercept, and x-intercepts, if any. Determine the domain and the range of f. Determine where f is increasing and where it is decreasing.

For $f(x) = x^2 - 6x + 9$, we have $a = 1, b = -6$, and $c = 9$. Since $a = 1 > 0$, the parabola opens up. The x-coordinate of the vertex is

$$h = -\frac{b}{2a} = -\frac{-6}{2(1)} = 3$$

Figure 9

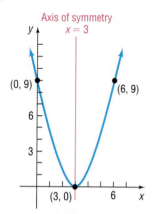

The y-coordinate of the vertex is

$$k = f(3) = (3)^2 - 6(3) + 9 = 0$$

So, the vertex is at $(3, 0)$. The axis of symmetry is the line $x = 3$. The y-intercept is $f(0) = 9$. Since the vertex $(3, 0)$ lies on the x-axis, the graph touches the x-axis at the x-intercept. By using the axis of symmetry and the y-intercept at $(0, 9)$, we can locate the additional point $(6, 9)$ on the graph. See Figure 9.

The domain of f is the set of all real numbers. Based on the graph, the range of f is the interval $[0, \infty)$. The function f is decreasing on the interval $(-\infty, 3)$ and increasing on the interval $(3, \infty)$. ◀

Graph the function in Example 4 by completing the square and using transformations. Which method do you prefer?

NOW WORK PROBLEM 43.

EXAMPLE 5

Graphing a Quadratic Function by Hand Using Its Vertex, Axis, and Intercepts

Graph $f(x) = 2x^2 + x + 1$ by determining whether the graph opens up or down and by finding its vertex, axis of symmetry, y-intercept, and x-intercepts, if any. Determine the domain and the range of f. Determine where f is increasing and where it is decreasing.

Solution

For $f(x) = 2x^2 + x + 1$, we have $a = 2, b = 1$, and $c = 1$. Since $a = 2 > 0$, the parabola opens up. The x-coordinate of the vertex is

$$h = -\frac{b}{2a} = -\frac{1}{4}$$

NOTE

In Example 5, since the vertex is above the x-axis and the parabola opens up, we can conclude the graph of the quadratic function will have no x-intercepts. ∎

The y-coordinate of the vertex is

$$k = f\left(-\frac{1}{4}\right) = 2\left(\frac{1}{16}\right) + \left(-\frac{1}{4}\right) + 1 = \frac{7}{8}$$

So, the vertex is at $\left(-\frac{1}{4}, \frac{7}{8}\right)$. The axis of symmetry is the line $x = -\frac{1}{4}$. The y-intercept is $f(0) = 1$. The x-intercept(s), if any, obey the equation $2x^2 + x + 1 = 0$. Since the discriminant $b^2 - 4ac = (1)^2 - 4(2)(1) = -7 < 0$, this equation has no real solutions, and therefore the graph has no x-intercepts. We use the point $(0, 1)$ and the axis of symmetry $x = -\frac{1}{4}$ to locate the additional point $\left(-\frac{1}{2}, 1\right)$ on the graph. See Figure 10.

The domain of f is the set of all real numbers. Based on the graph, the range of f is the interval $\left[\frac{7}{8}, \infty\right)$. The function f is decreasing on the interval $\left(-\infty, -\frac{1}{4}\right)$ and increasing on the interval $\left(-\frac{1}{4}, \infty\right)$. ◀

NOW WORK PROBLEM 47.

Figure 10

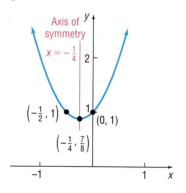

Given the vertex (h, k) and one additional point on the graph of a quadratic function $f(x) = ax^2 + bx + c, a \neq 0$, we can use

$$f(x) = a(x - h)^2 + k \qquad \text{(5)}$$

to obtain the quadratic function.

| EXAMPLE 6 | **Finding the Quadratic Function Given Its Vertex and One Other Point** |

Determine the quadratic function whose vertex is $(1, -5)$ and whose y-intercept is -3. The graph of the parabola is shown in Figure 11.

Figure 11

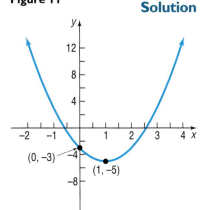

(0, −3)

(1, −5)

Solution The vertex is $(1, -5)$, so $h = 1$ and $k = -5$. Substitute these values into equation (5).

$$f(x) = a(x - h)^2 + k \qquad \text{Equation (5)}$$
$$f(x) = a(x - 1)^2 - 5 \qquad h = 1, k = -5$$

To determine the value of a, we use the fact that $f(0) = -3$ (the y-intercept).

$$f(x) = a(x - 1)^2 - 5$$
$$-3 = a(0 - 1)^2 - 5 \qquad x = 0, y = f(0) = -3$$
$$-3 = a - 5$$
$$a = 2$$

The quadratic function whose graph is shown in Figure 11 is

$$f(x) = a(x - h)^2 + k = 2(x - 1)^2 - 5 = 2x^2 - 4x - 3$$

✔ **CHECK:** Figure 12 shows the graph $f(x) = 2x^2 - 4x - 3$ using a graphing utility.

Figure 12

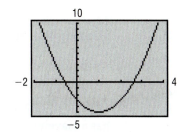

NOW WORK PROBLEM 53.

Summary

Steps for Graphing a Quadratic Function $f(x) = ax^2 + bx + c, a \neq 0$, by Hand.

Option 1

STEP 1: Complete the square in x to write the quadratic function in the form $f(x) = a(x - h)^2 + k$.

STEP 2: Graph the function in stages using transformations.

Option 2

STEP 1: Determine the vertex $\left(-\dfrac{b}{2a}, f\left(-\dfrac{b}{2a}\right)\right)$.

STEP 2: Determine the axis of symmetry, $x = -\dfrac{b}{2a}$.

STEP 3: Determine the y-intercept, $f(0)$.

STEP 4: (a) If $b^2 - 4ac > 0$, then the graph of the quadratic function has two x-intercepts, which are found by solving the equation $ax^2 + bx + c = 0$.

(b) If $b^2 - 4ac = 0$, the vertex is the x-intercept.

(c) If $b^2 - 4ac < 0$, there are no x-intercepts.

STEP 5: Determine an additional point by using the y-intercept and the axis of symmetry.

STEP 6: Plot the points and draw the graph.

4 ## Use the Maximum or Minimum Value of a Quadratic Function to Solve Applied Problems

When a mathematical model leads to a quadratic function, the properties of this quadratic function can provide important information about the model. For example, for a quadratic revenue function, we can find the maximum revenue; for a quadratic cost function, we can find the minimum cost.

To see why, recall that the graph of a quadratic function

$$f(x) = ax^2 + bx + c, \quad a \neq 0$$

is a parabola with vertex at $\left(-\dfrac{b}{2a}, f\left(-\dfrac{b}{2a}\right)\right)$. This vertex is the highest point on the graph if $a < 0$ and the lowest point on the graph if $a > 0$. If the vertex is the highest point $(a < 0)$, then $f\left(-\dfrac{b}{2a}\right)$ is the **maximum value** of f. If the vertex is the lowest point $(a > 0)$, then $f\left(-\dfrac{b}{2a}\right)$ is the **minimum value** of f.

This property of the graph of a quadratic function enables us to answer questions involving optimization (finding maximum or minimum values) in models involving quadratic functions.

| EXAMPLE 7 | **Finding the Maximum or Minimum Value of a Quadratic Function** |

Determine whether the quadratic function

$$f(x) = x^2 - 4x - 5$$

has a maximum or minimum value. Then find the maximum or minimum value.

Solution   We compare $f(x) = x^2 - 4x - 5$ to $f(x) = ax^2 + bx + c$. We conclude that $a = 1, b = -4$, and $c = -5$. Since $a > 0$, the graph of f opens up, so the vertex is a minimum point. The minimum value occurs at

$$x = -\frac{b}{2a} = -\frac{-4}{2(1)} = \frac{4}{2} = 2$$

$a = 1, b = -4$

The minimum value is

$$f\left(-\frac{b}{2a}\right) = f(2) = 2^2 - 4(2) - 5 = 4 - 8 - 5 = -9$$

Table 2

X	Y1
-1	0
0	-5
1	-8
2	-9
3	-8
4	-5
5	0

Y1☐X²−4X−5

✔ **CHECK:** We can support the algebraic solution using the TABLE feature on a graphing utility. Create Table 2 by letting $Y_1 = x^2 - 4x - 5$. From the table we see that the smallest value of y occurs when $x = 2$, leading us to investigate the number 2 further. The symmetry about $x = 2$ confirms that 2 is the x-coordinate of the vertex of the parabola. We conclude that the minimum value is -9 and occurs at $x = 2$. ◄

NOW WORK PROBLEM **61.**

EXAMPLE 8	**Maximizing Revenue**

The marketing department at Texas Instruments has found that, when certain calculators are sold at a price of p dollars per unit, the revenue R (in dollars) as a function of the price p is

$$R(p) = -150p^2 + 21{,}000p$$

What unit price should be established to maximize revenue? If this price is charged, what is the maximum revenue?

Solution The revenue R is

$$R(p) = -150p^2 + 21{,}000p \qquad R(p) = ap^2 + bp + c$$

The function R is a quadratic function with $a = -150$, $b = 21{,}000$, and $c = 0$. Because $a < 0$, the vertex is the highest point on the parabola. The revenue R is therefore a maximum when the price p is

$$p = -\frac{b}{2a} = -\frac{21{,}000}{2(-150)} = -\frac{21{,}000}{-300} = \$70.00$$

$$a = -150, b = 21{,}000$$

The maximum revenue R is

$$R(70) = -150(70)^2 + 21{,}000(70) = \$735{,}000$$

See Figure 13 for an illustration.

Figure 13

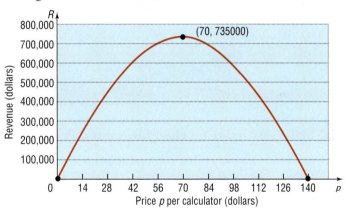

NOW WORK PROBLEM 71.

EXAMPLE 9	**Maximizing the Area Enclosed by a Fence**

A farmer has 2000 yards of fence to enclose a rectangular field. What are the dimensions of the rectangle that encloses the most area?

Figure 14

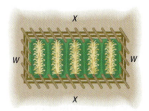

Solution Figure 14 illustrates the situation. The available fence represents the perimeter of the rectangle. If x is the length and w is the width, then

$$2x + 2w = 2000 \tag{6}$$

The area A of the rectangle is

$$A = xw$$

To express A in terms of a single variable, we solve equation (6) for w and substitute the result in $A = xw$. Then A involves only the variable x. [You could also solve equation (6) for x and express A in terms of w alone. Try it!]

$$2x + 2w = 2000 \qquad \text{Equation (6)}$$
$$2w = 2000 - 2x \qquad \text{Solve for } w.$$
$$w = \frac{2000 - 2x}{2} = 1000 - x$$

Then the area A is

$$A = xw = x(1000 - x) = -x^2 + 1000x$$

Now, A is a quadratic function of x.

$$A(x) = -x^2 + 1000x \qquad a = -1, b = 1000, c = 0$$

Figure 15 shows a graph of $A(x) = -x^2 + 1000x$ using a graphing utility. Since $a < 0$, the vertex is a maximum point on the graph of A. The maximum value occurs at

$$x = -\frac{b}{2a} = -\frac{1000}{2(-1)} = 500$$

Figure 15

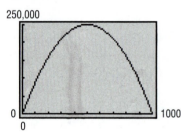

The maximum value of A is

$$A\left(-\frac{b}{2a}\right) = A(500) = -500^2 + 1000(500) = -250{,}000 + 500{,}000 = 250{,}000$$

The largest rectangle that can be enclosed by 2000 yards of fence has an area of 250,000 square yards. Its dimensions are 500 yards by 500 yards.   ◀

⬤▬▬▬▷   **NOW WORK PROBLEM 77.**

EXAMPLE 10	**Analyzing the Motion of a Projectile**

A projectile is fired from a cliff 500 feet above the water at an inclination of 45° to the horizontal, with a muzzle velocity of 400 feet per second. In physics, it is established that the height h of the projectile above the water is given by

$$h(x) = \frac{-32x^2}{(400)^2} + x + 500$$

where x is the horizontal distance of the projectile from the base of the cliff. See Figure 16.

Figure 16

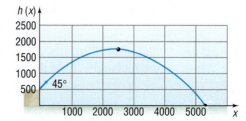

(a) Find the maximum height of the projectile.

(b) How far from the base of the cliff will the projectile strike the water?

Solution   (a) The height of the projectile is given by a quadratic function.

$$h(x) = \frac{-32x^2}{(400)^2} + x + 500 = \frac{-1}{5000}x^2 + x + 500$$

We are looking for the maximum value of h. Since $a < 0$, the maximum value is obtained at the vertex. We compute

$$x = -\frac{b}{2a} = -\frac{1}{2\left(-\dfrac{1}{5000}\right)} = \frac{5000}{2} = 2500$$

The maximum height of the projectile is

$$h(2500) = \frac{-1}{5000}(2500)^2 + 2500 + 500 = -1250 + 2500 + 500 = 1750 \text{ ft}$$

(b) The projectile will strike the water when the height is zero. To find the distance x traveled, we need to solve the equation

$$h(x) = \frac{-1}{5000}x^2 + x + 500 = 0$$

We find the discriminant first.

$$b^2 - 4ac = 1^2 - 4\left(\frac{-1}{5000}\right)(500) = 1.4$$

Then,

$$x = \frac{-b \pm \sqrt{b^2 - 4ac}}{2a} = \frac{-1 \pm \sqrt{1.4}}{2\left(-\dfrac{1}{5000}\right)} \approx \begin{cases} -458 \\ 5458 \end{cases}$$

We discard the negative solution and find that the projectile will strike the water at a distance of about 5458 feet from the base of the cliff. ◀

 **NOW WORK PROBLEM 81.**

— **Seeing the Concept** —

Graph

$$h(x) = \frac{-1}{5000}x^2 + x + 500,$$

$$0 \le x \le 5500$$

Use MAXIMUM to find the maximum height of the projectile, and use ROOT or ZERO to find the distance from the base of the cliff to where it strikes the water. Compare your results with those obtained in Example 10.

EXAMPLE 11

The Golden Gate Bridge

The Golden Gate Bridge, a suspension bridge, spans the entrance to San Francisco Bay. Its 746-foot-tall towers are 4200 feet apart. The bridge is suspended from two huge cables more than 3 feet in diameter; the 90-foot-wide roadway is 220 feet above the water. The cables are parabolic in shape* and touch the road surface at the center of the bridge. Find the height of the cable at a distance of 1000 feet from the center.

Solution   See Figure 17. We begin by choosing the placement of the coordinate axes so that the x-axis coincides with the road surface and the origin coincides with the center of the bridge. As a result, the twin towers will be vertical (height $746 - 220 = 526$ feet above the road) and located 2100 feet from the center. Also, the cable, which has the

Figure 17

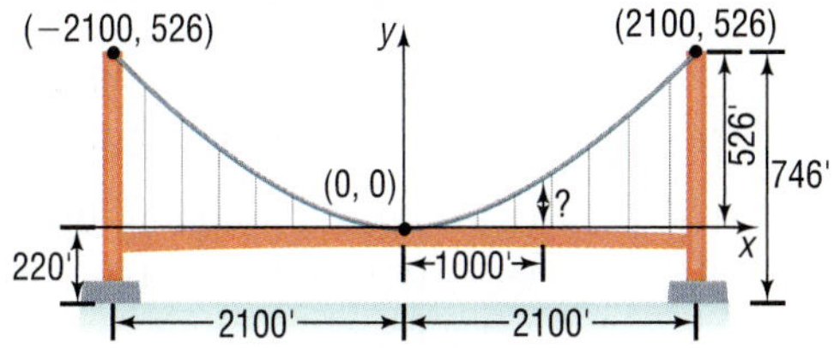

(−2100, 526)   (2100, 526)

526'

(0, 0)

746'

220'   ←1000'→

←——2100'——→←——2100'——→

*A cable suspended from two towers is in the shape of a **catenary**, but when a horizontal roadway is suspended from the cable, the cable takes the shape of a parabola.

shape of a parabola, will extend from the towers, open up, and have its vertex at $(0, 0)$. The choice of placement of the axes enables us to identify the equation of the parabola as $y = ax^2, a > 0$. We can also see that the points $(-2100, 526)$ and $(2100, 526)$ are on the graph.

Based on these facts, we can find the value of a in $y = ax^2$.

$$y = ax^2$$
$$526 = a(2100)^2 \qquad \text{\color{blue}{$x = 2100, y = 526$}}$$
$$a = \frac{526}{(2100)^2}$$

The equation of the parabola is therefore

$$y = \frac{526}{(2100)^2}x^2$$

The height of the cable when $x = 1000$ is

$$y = \frac{526}{(2100)^2}(1000)^2 \approx 119.3 \text{ feet}$$

The cable is 119.3 feet high at a distance of 1000 feet from the center of the bridge. ◀

━━━━ **NOW WORK PROBLEM 83.**

5 ## Use a Graphing Utility to Find the Quadratic Function of Best Fit to Data

In Section 2.4, we found the line of best fit for data that appeared to be linearly related. It was noted that data may also follow a nonlinear relation. Figures 18(a) and (b) show scatter diagrams of data that follow a quadratic relation.

Figure 18

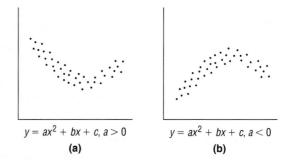

$y = ax^2 + bx + c, a > 0$ $y = ax^2 + bx + c, a < 0$

(a)   (b)

EXAMPLE 12   **Fitting a Quadratic Function to Data**

A farmer collected the data given in Table 3, which shows crop yields Y for various amounts of fertilizer used, x.

(a) With a graphing utility, draw a scatter diagram of the data. Comment on the type of relation that may exist between the two variables.

(b) Use a graphing utility to find the quadratic function of best fit to these data.

(c) Use the function found in part (b) to determine the optimal amount of fertilizer to apply.

Table 3

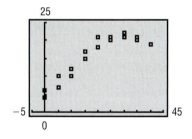

Plot	Fertilizer, x (Pounds/100 ft²)	Yield (Bushels)
1	0	4
2	0	6
3	5	10
4	5	7
5	10	12
6	10	10
7	15	15
8	15	17
9	20	18
10	20	21
11	25	20
12	25	21
13	30	21
14	30	22
15	35	21
16	35	20
17	40	19
18	40	19

(d) Use the function found in part (b) to predict crop yield when the optimal amount of fertilizer is applied.

(e) Draw the quadratic function of best fit on the scatter diagram.

Solution

(a) Figure 19 shows the scatter diagram, from which it appears that the data follow a quadratic relation, with $a < 0$.

(b) Upon executing the QUADratic REGression program, we obtain the results shown in Figure 20. The output that the utility provides shows us the equation $y = ax^2 + bx + c$. The quadratic function of best fit is

$$Y(x) = -0.0171x^2 + 1.0765x + 3.8939$$

where x represents the amount of fertilizer used and Y represents crop yield.

(c) Based on the quadratic function of best fit, the optimal amount of fertilizer to apply is

$$x = -\frac{b}{2a} = -\frac{1.0765}{2(-0.0171)} \approx 31.5 \text{ pounds of fertilizer per 100 square feet}$$

(d) We evaluate the function $Y(x)$ for $x = 31.5$.

$$Y(31.5) = -0.0171(31.5)^2 + 1.0765(31.5) + 3.8939 \approx 20.8 \text{ bushels}$$

If we apply 31.5 pounds of fertilizer per 100 square feet, the crop yield will be 20.8 bushels according to the quadratic function of best fit.

(e) Figure 21 shows the graph of the quadratic function found in part (b) drawn on the scatter diagram.

Figure 19

Figure 20

```
QuadReg
y=ax²+bx+c
a=-.0171212121
b=1.076515152
c=3.893939394
```

Figure 21

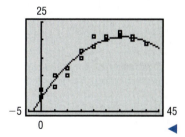

Look again at Figure 20. Notice that the output given by the graphing calculator does not include r, the correlation coefficient. Recall that the correlation coefficient is a measure of the strength of a *linear* relation that exists between two variables. The graphing calculator does not provide an indication of how well the function fits the data in terms of r since a quadratic function cannot be expressed as a linear function.

NOW WORK PROBLEM 101.

3.1 Assess Your Understanding

'Are You Prepared?'

Answers are given at the end of these exercises. If you get a wrong answer, read the pages listed in red.

1. List the intercepts of the equation $y = x^2 - 9$. (pp. 15–17)

2. Solve the equation $2x^2 + 7x - 4 = 0$. (pp. 696–703)

3. To complete the square of $x^2 - 5x$, you add the number _____. (pp. 699)

4. To graph $y = (x - 4)^2$, you shift the graph of $y = x^2$ to the _____ a distance of _____ units. (pp. 118–120)

Concepts and Vocabulary

5. The graph of a quadratic function is called a(n) _____.

6. The vertical line passing through the vertex of a parabola is called the _____.

7. The x-coordinate of the vertex of $f(x) = ax^2 + bx + c$, $a \neq 0$, is _____.

8. *True or False:* The graph of $f(x) = 2x^2 + 3x - 4$ opens up.

9. *True or False:* The y-coordinate of the vertex of $f(x) = -x^2 + 4x + 5$ is $f(2)$.

10. *True or False:* If the discriminant $b^2 - 4ac = 0$, the graph of $f(x) = ax^2 + bx + c, a \neq 0$, will touch the x-axis at its vertex.

Skill Building

In Problems 11–18, match each graph to one the following functions without using a graphing utility.

11. $f(x) = x^2 - 1$

12. $f(x) = -x^2 - 1$

13. $f(x) = x^2 - 2x + 1$

14. $f(x) = x^2 + 2x + 1$

15. $f(x) = x^2 - 2x + 2$

16. $f(x) = x^2 + 2x$

17. $f(x) = x^2 - 2x$

18. $f(x) = x^2 + 2x + 2$

A.
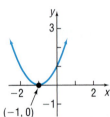

B.
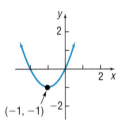

C.
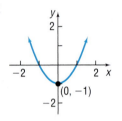

D.
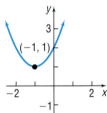

E.
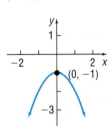

F.
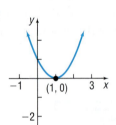

G.
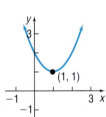

H.
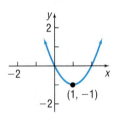

In Problems 19–34, graph the function f by starting with the graph of $y = x^2$ and using transformations (shifting, compressing, stretching, and/or reflection). Verify your results using a graphing utility.

[Hint: *If necessary, write f in the form $f(x) = a(x - h)^2 + k$.*]

19. $f(x) = \dfrac{1}{4}x^2$

20. $f(x) = 2x^2$

21. $f(x) = \dfrac{1}{4}x^2 - 2$

22. $f(x) = 2x^2 - 3$

23. $f(x) = \dfrac{1}{4}x^2 + 2$

24. $f(x) = 2x^2 + 4$

25. $f(x) = \dfrac{1}{4}x^2 + 1$

26. $f(x) = -2x^2 - 2$

27. $f(x) = x^2 + 4x + 2$

28. $f(x) = x^2 - 6x - 1$

29. $f(x) = 2x^2 - 4x + 1$

30. $f(x) = 3x^2 + 6x$

31. $f(x) = -x^2 - 2x$

32. $f(x) = -2x^2 + 6x + 2$

33. $f(x) = \dfrac{1}{2}x^2 + x - 1$

34. $f(x) = \dfrac{2}{3}x^2 + \dfrac{4}{3}x - 1$

In Problems 35–52, graph each quadratic function by determining whether its graph opens up or down and by finding its vertex, axis of symmetry, y-intercept, and x-intercepts, if any. Determine the domain and the range of the function. Determine where the function is increasing and where it is decreasing. Verify your results using a graphing utility.

35. $f(x) = x^2 + 2x$

36. $f(x) = x^2 - 4x$

37. $f(x) = -x^2 - 6x$

38. $f(x) = -x^2 + 4x$

39. $f(x) = 2x^2 - 8x$

40. $f(x) = 3x^2 + 18x$

41. $f(x) = x^2 + 2x - 8$

42. $f(x) = x^2 - 2x - 3$

43. $f(x) = x^2 + 2x + 1$

44. $f(x) = x^2 + 6x + 9$

45. $f(x) = 2x^2 - x + 2$

46. $f(x) = 4x^2 - 2x + 1$

47. $f(x) = -2x^2 + 2x - 3$

48. $f(x) = -3x^2 + 3x - 2$

49. $f(x) = 3x^2 + 6x + 2$

50. $f(x) = 2x^2 + 5x + 3$

51. $f(x) = -4x^2 - 6x + 2$

52. $f(x) = 3x^2 - 8x + 2$

In Problems 53–58, determine the quadratic function whose graph is given.

53.

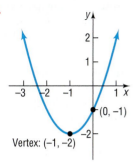

54.

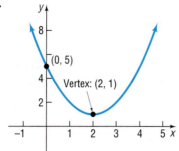

55.

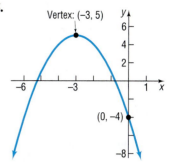

56.

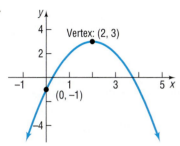

57.

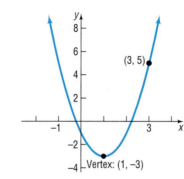

58.

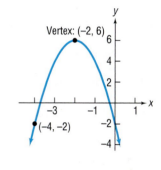

In Problems 59–66, determine, without graphing, whether the given quadratic function has a maximum value or a minimum value and then find the value. Use the TABLE feature on a graphing utility to support your answer.

59. $f(x) = 2x^2 + 12x$

60. $f(x) = -2x^2 + 12x$

61. $f(x) = 2x^2 + 12x - 3$

62. $f(x) = 4x^2 - 8x + 3$

63. $f(x) = -x^2 + 10x - 4$

64. $f(x) = -2x^2 + 8x + 3$

65. $f(x) = -3x^2 + 12x + 1$

66. $f(x) = 4x^2 - 4x$

Applications and Extensions

67. The graph of the function $f(x) = ax^2 + bx + c$ has vertex at $(0, 2)$ and passes through the point $(1, 8)$. Find $a, b,$ and c.

68. The graph of the function $f(x) = ax^2 + bx + c$ has vertex at $(1, 4)$ and passes through the point $(-1, -8)$. Find $a, b,$ and c.

Answer Problems 69 and 70 using the following: A quadratic function of the form $f(x) = ax^2 + bx + c$ with $b^2 - 4ac > 0$ may also be written in the form $f(x) = a(x - r_1)(x - r_2)$, where r_1 and r_2 are the x-intercepts of the graph of the quadratic function.

69. (a) Find a quadratic function whose x-intercepts are -3 and 1 with $a = 1$; $a = 2$; $a = -2$; $a = 5$.

(b) How does the value of a affect the intercepts?

(c) How does the value of a affect the axis of symmetry?

(d) How does the value of a affect the vertex?

(e) Compare the x-coordinate of the vertex with the midpoint of the x-intercepts. What might you conclude?

70. (a) Find a quadratic function whose x-intercepts are -5 and 3 with $a = 1$; $a = 2$; $a = -2$; $a = 5$.

(b) How does the value of a affect the intercepts?

(c) How does the value of a affect the axis of symmetry?

(d) How does the value of a affect the vertex?

(e) Compare the x-coordinate of the vertex with the midpoint of the x-intercepts. What might you conclude?

71. Maximizing Revenue Suppose that the manufacturer of a gas clothes dryer has found that, when the unit price is p dollars, the revenue R (in dollars) is

$$R(p) = -4p^2 + 4000p$$

What unit price for the dryer should be established to maximize revenue? What is the maximum revenue?

72. Maximizing Revenue The John Deere company has found that the revenue from sales of heavy-duty tractors is a function of the unit price p (in dollars) that it charges. If the revenue R is

$$R(p) = -\frac{1}{2}p^2 + 1900p$$

what unit price p (in dollars) should be charged to maximize revenue? What is the maximum revenue?

73. Demand Equation The price p (in dollars) and the quantity x sold of a certain product obey the demand equation

$$p = -\frac{1}{6}x + 100, \qquad 0 \le x \le 600$$

(a) Express the revenue R as a function of x. (Remember, $R = xp$.)

(b) What is the revenue if 200 units are sold?

(c) What quantity x maximizes revenue? What is the maximum revenue?

(d) What price should the company charge to maximize revenue?

74. Demand Equation The price p (in dollars) and the quantity x sold of a certain product obey the demand equation

$$p = -\frac{1}{3}x + 100, \qquad 0 \le x \le 300$$

(a) Express the revenue R as a function of x.
(b) What is the revenue if 100 units are sold?
(c) What quantity x maximizes revenue? What is the maximum revenue?
(d) What price should the company charge to maximize revenue?

75. Demand Equation The price p (in dollars) and the quantity x sold of a certain product obey the demand equation

$$x = -5p + 100, \qquad 0 \le p \le 20$$

(a) Express the revenue R as a function of x.
(b) What is the revenue if 15 units are sold?
(c) What quantity x maximizes revenue? What is the maximum revenue?
(d) What price should the company charge to maximize revenue?

76. Demand Equation The price p (in dollars) and the quantity x sold of a certain product obey the demand equation

$$x = -20p + 500, \qquad 0 \le p \le 25$$

(a) Express the revenue R as a function of x.
(b) What is the revenue if 20 units are sold?
(c) What quantity x maximizes revenue? What is the maximum revenue?
(d) What price should the company charge to maximize revenue?

77. Enclosing a Rectangular Field David has available 400 yards of fencing and wishes to enclose a rectangular area.
(a) Express the area A of the rectangle as a function of the width w of the rectangle.
(b) For what value of w is the area largest?
(c) What is the maximum area?

78. Enclosing a Rectangular Field Beth has 3000 feet of fencing available to enclose a rectangular field.
(a) Express the area A of the rectangle as a function of x where x is the length of the rectangle.
(b) For what value of x is the area largest?
(c) What is the maximum area?

79. Enclosing the Most Area with a Fence A farmer with 4000 meters of fencing wants to enclose a rectangular plot that borders on a river. If the farmer does not fence the side along the river, what is the largest area that can be enclosed? (See the figure.)

$4000 - 2x$

80. Enclosing the Most Area with a Fence A farmer with 2000 meters of fencing wants to enclose a rectangular plot that borders on a straight highway. If the farmer does not fence the side along the highway, what is the largest area that can be enclosed?

81. Analyzing the Motion of a Projectile A projectile is fired from a cliff 200 feet above the water at an inclination of $45°$ to the horizontal, with a muzzle velocity of 50 feet per second. The height h of the projectile above the water is given by

$$h(x) = \frac{-32x^2}{(50)^2} + x + 200$$

where x is the horizontal distance of the projectile from the base of the cliff.
(a) How far from the base of the cliff is the height of the projectile a maximum?
(b) Find the maximum height of the projectile.
(c) How far from the base of the cliff will the projectile strike the water?
(d) Using a graphing utility, graph the function h, $0 \le x \le 200$.
(e) When the height of the projectile is 100 feet above the water, how far is it from the cliff?

82. Analyzing the Motion of a Projectile A projectile is fired at an inclination of $45°$ to the horizontal, with a muzzle velocity of 100 feet per second. The height h of the projectile is given by

$$h(x) = \frac{-32x^2}{(100)^2} + x$$

where x is the horizontal distance of the projectile from the firing point.
(a) How far from the firing point is the height of the projectile a maximum?
(b) Find the maximum height of the projectile.
(c) How far from the firing point will the projectile strike the ground?
(d) Using a graphing utility, graph the function h, $0 \le x \le 350$.
(e) When the height of the projectile is 50 feet above the ground, how far has it traveled horizontally?

83. Suspension Bridge A suspension bridge with weight uniformly distributed along its length has twin towers that extend 75 meters above the road surface and are 400 meters apart. The cables are parabolic in shape and are suspended from the tops of the towers. The cables touch the road surface at the center of the bridge. Find the height of the cables at a point 100 meters from the center. (Assume that the road is level.)

84. Architecture A parabolic arch has a span of 120 feet and a maximum height of 25 feet. Choose suitable rectangular coordinate axes and find the equation of the parabola. Then calculate the height of the arch at points 10 feet, 20 feet, and 40 feet from the center.

85. Constructing Rain Gutters   A rain gutter is to be made of aluminum sheets that are 12 inches wide by turning up the edges 90°. What depth will provide maximum cross-sectional area and hence allow the most water to flow?

86. Norman Windows   A Norman window has the shape of a rectangle surmounted by a semicircle of diameter equal to the width of the rectangle (see the figure). If the perimeter of the window is 20 feet, what dimensions will admit the most light (maximize the area)?

[**Hint:** Circumference of a circle $= 2\pi r$; area of a circle $= \pi r^2$, where r is the radius of the circle.]

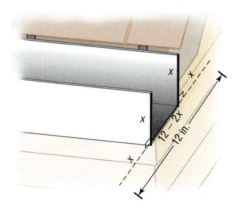

87. Constructing a Stadium   A track and field playing area is in the shape of a rectangle with semicircles at each end (see the figure). The inside perimeter of the track is to be 1500 meters. What should the dimensions of the rectangle be so that the area of the rectangle is a maximum?

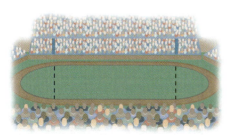

88. Architecture   A special window has the shape of a rectangle surmounted by an equilateral triangle (see the figure). If the perimeter of the window is 16 feet, what dimensions will admit the most light?

[**Hint:** Area of an equilateral triangle $= \left(\dfrac{\sqrt{3}}{4}\right)x^2$, where x is the length of a side of the triangle.]

89. Chemical Reactions   A self-catalytic chemical reaction results in the formation of a compound that causes the formation ratio to increase. If the reaction rate V is given by

$$V(x) = kx(a - x), \qquad 0 \le x \le a$$

where k is a positive constant, a is the initial amount of the compound, and x is the variable amount of the compound, for what value of x is the reaction rate a maximum?

90. Calculus: Simpson's Rule   The figure shows the graph of $y = ax^2 + bx + c$. Suppose that the points $(-h, y_0)$, $(0, y_1)$, and (h, y_2) are on the graph. It can be shown that the area enclosed by the parabola, the x-axis, and the lines $x = -h$ and $x = h$ is

$$\text{Area} = \frac{h}{3}(2ah^2 + 6c)$$

Show that this area may also be given by

$$\text{Area} = \frac{h}{3}(y_0 + 4y_1 + y_2)$$

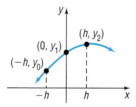

91. Use the result obtained in Problem 90 to find the area enclosed by $f(x) = -5x^2 + 8$, the x-axis, and the lines $x = -1$ and $x = 1$.

92. Use the result obtained in Problem 90 to find the area enclosed by $f(x) = 2x^2 + 8$, the x-axis, and the lines $x = -2$ and $x = 2$.

93. Use the result obtained in Problem 90 to find the area enclosed by $f(x) = x^2 + 3x + 5$, the x-axis, and the lines $x = -4$ and $x = 4$.

94. Use the result obtained in Problem 90 to find the area enclosed by $f(x) = -x^2 + x + 4$, the x-axis, and the lines $x = -1$ and $x = 1$.

95. A rectangle has one vertex on the line $y = 10 - x, x > 0$, another at the origin, one on the positive x-axis, and one on the positive y-axis. Find the largest area A that can be enclosed by the rectangle.

96. Let $f(x) = ax^2 + bx + c$, where a, b, and c are odd integers. If x is an integer, show that $f(x)$ must be an odd integer.

[Hint: x is either an even integer or an odd integer.]

97. Hunting The function $H(x) = -1.01x^2 + 114.3x + 451.0$ models the number of individuals who engage in hunting activities whose annual income is x thousand dollars.

SOURCE: Based on data obtained from the National Sporting Goods Association.

(a) What is the income level for which there are the most hunters? Approximately how many hunters earn this amount?

(b) Using a graphing utility, graph $H = H(x)$. Are the number of hunters increasing or decreasing for individuals earning between $20,000 and $40,000?

98. Advanced Degrees The function $P(x) = -0.008x^2 + 0.868x - 11.884$ models the percentage of the U.S. population in March 2000 whose age is given by x that has earned an advanced degree (more than a bachelor's degree).

SOURCE: Based on data obtained from the U.S. Census Bureau.

(a) What is the age for which the highest percentage of Americans have earned an advanced degree? What is the highest percentage?

(b) Using a graphing utility, graph $P = P(x)$. Is the percentage of Americans that have earned an advanced degree increasing or decreasing for individuals between the ages of 40 and 50?

99. Male Murder Victims The function $M(x) = 0.76x^2 - 107.00x + 3854.18$ models the number of male murder victims who are x years of age $(20 \le x < 90)$.

SOURCE: Based on data obtained from the Federal Bureau of Investigation.

(a) Use the model to approximate the number of male murder victims who are $x = 23$ years of age.

(b) At what age is the number of male murder victims 1456?

(c) Using a graphing utility, graph $M = M(x)$.

(d) Based on the graph drawn in part (c), describe what happens to the number of male murder victims as age increases.

100. Health Care Expenditures The function $H(x) = 0.004x^2 - 0.197x + 5.406$ models the percentage of total income that an individual who is x years of age spends on health care.

SOURCE: Based on data obtained from the Bureau of Labor Statistics.

(a) Use the model to approximate the percentage of total income that an individual who is $x = 45$ years of age spends on health care.

(b) At what age is the percentage of income spent on health care 10%?

(c) Using a graphing utility, graph $H = H(x)$.

(d) Based on the graph drawn in part (c), describe what happens to the percentage of income spent on health care as individuals age.

101. Life Cycle Hypothesis An individual's income varies with his or her age. The following table shows the median income I of individuals of different age groups within the United States for 2001. For each age group, let the class midpoint represent the independent variable, x. For the class "65 years and older," we will assume that the class midpoint is 69.5.

Age	Class Midpoint, x	Median Income, I
15–24 years	19.5	$9,301
25–34 years	29.5	$30,510
35–44 years	39.5	$38,340
45–54 years	49.5	$41,104
55–64 years	59.5	$35,637
65 years and older	69.5	$19,688

SOURCE: U.S. Census Bureau

(a) Draw a scatter diagram of the data. Comment on the type of relation that may exist between the two variables.

(b) Use a graphing utility to find the quadratic function of best fit to these data.

(c) Use the function found in part (b) to determine the age at which an individual can expect to earn the most income.

(d) Use the function found in part (b) to predict the peak income earned.

(e) With a graphing utility, graph the quadratic function of best fit on the scatter diagram.

102. Life Cycle Hypothesis An individual's income varies with his or her age. The following table shows the median income I of individuals of different age groups within the United States for 2000. For each age group, the class midpoint represents the independent variable, x. For the age group "65 years and older," we will assume that the class midpoint is 69.5.

Age	Class Midpoint, x	Median Income, I
15–24 years	19.5	$9,548
25–34 years	29.5	$30,633
35–44 years	39.5	$37,088
45–54 years	49.5	$41,072
55–64 years	59.5	$34,414
65 years and older	69.5	$19,167

SOURCE: U.S. Census Bureau

(a) Draw a scatter diagram of the data. Comment on the type of relation that may exist between the two variables.

(b) Use a graphing utility to find the quadratic function of best fit to these data.

(c) Use the function found in part (b) to determine the age at which an individual can expect to earn the most income.

(d) Use the function found in part (b) to predict the peak income earned.

(e) With a graphing utility, graph the quadratic function of best fit on the scatter diagram.

(f) Compare the results of parts (c) and (d) in 2000 to those of 2001 in Problem 101.

103. Height of a Ball   A shot-putter throws a ball at an inclination of $45°$ to the horizontal. The following data represent the height of the ball h at the instant that it has traveled x feet horizontally.

Distance, x	Height, h
20	25
40	40
60	55
80	65
100	71
120	77
140	77
160	75
180	71
200	64

(a) Draw a scatter diagram of the data. Comment on the type of relation that may exist between the two variables.

(b) Use a graphing utility to find the quadratic function of best fit to these data.

(c) Use the function found in part (b) to determine how far the ball will travel before it reaches its maximum height.

(d) Use the function found in part (b) to find the maximum height of the ball.

(e) With a graphing utility, graph the quadratic function of best fit on the scatter diagram.

104. Miles per Gallon   An engineer collects the following data showing the speed s of a Ford Taurus and its average miles per gallon, M.

Speed, s	Miles per Gallon, M
30	18
35	20
40	23
40	25
45	25
50	28
55	30
60	29
65	26
65	25
70	25

(a) Draw a scatter diagram of the data. Comment on the type of relation that may exist between the two variables.

(b) Use a graphing utility to find the quadratic function of best fit to these data.

(c) Use the function found in part (b) to determine the speed that maximizes miles per gallon.

(d) Use the function found in part (b) to predict miles per gallon for a speed of 63 miles per hour.

(e) With a graphing utility, graph the quadratic function of best fit on the scatter diagram.

Discussion and Writing

105. Make up a quadratic function that opens down and has only one x-intercept. Compare yours with others in the class. What are the similarities? What are the differences?

106. On one set of coordinate axes, graph the family of parabolas $f(x) = x^2 + 2x + c$ for $c = -3, c = 0$, and $c = 1$. Describe the characteristics of a member of this family.

107. On one set of coordinate axes, graph the family of parabolas $f(x) = x^2 + bx + 1$ for $b = -4, b = 0$, and $b = 4$. Describe the general characteristics of this family.

108. State the circumstances that cause the graph of a quadratic function $f(x) = ax^2 + bx + c$ to have no x-intercepts.

109. Why does the graph of a quadratic function open up if $a > 0$ and down if $a < 0$?

110. Refer to Example 8 on page 159. Notice that if the price charged for the calculators is $0 or $140 the revenue is $0. It is easy to explain why revenue would be $0 if the price charged is $0, but how can revenue be $0 if the price charged is $140?

'Are You Prepared?' Answers

1. $(0, -9), (-3, 0), (3, 0)$ **2.** $\left\{-4, \dfrac{1}{2}\right\}$ **3.** $\dfrac{25}{4}$ **4.** right; 4

3.2 Polynomial Functions and Models

PREPARING FOR THIS SECTION *Before getting started, review the following:*
- Polynomials (Appendix, Section A.3, pp. 674–679)
- Using a Graphing Utility to Approximate Local Maxima and Local Minima (Section 2.3, pp. 84–85)
- Graphing Techniques: Transformations (Section 2.6, pp. 118–126)
- Intercepts (Section 1.2, pp. 15–17)

 Now work the 'Are You Prepared?' problems on page 182.

OBJECTIVES
1. Identify Polynomial Functions and Their Degree
2. Graph Polynomial Functions Using Transformations
3. Identify the Zeros of a Polynomial Function and Their Multiplicity
4. Analyze the Graph of a Polynomial Function
5. Find the Cubic Function of Best Fit to Data

 Identify Polynomial Functions and Their Degree

Polynomial functions are among the simplest expressions in algebra. They are easy to evaluate: only addition and repeated multiplication are required. Because of this, they are often used to approximate other, more complicated functions. In this section, we investigate properties of this important class of functions.

A **polynomial function** is a function of the form

$$f(x) = a_n x^n + a_{n-1} x^{n-1} + \cdots + a_1 x + a_0 \qquad \textbf{(1)}$$

where $a_n, a_{n-1}, \ldots, a_1, a_0$ are real numbers and n is a nonnegative integer. The domain is the set of all real numbers.

A polynomial function is a function whose rule is given by a polynomial in one variable. The **degree** of a polynomial function is the degree of the polynomial in one variable, that is, the largest power of x that appears.

EXAMPLE 1   **Identifying Polynomial Functions**

Determine which of the following are polynomial functions. For those that are, state the degree; for those that are not, tell why not.

(a) $f(x) = 2 - 3x^4$ (b) $g(x) = \sqrt{x}$ (c) $h(x) = \dfrac{x^2 - 2}{x^3 - 1}$

(d) $F(x) = 0$ (e) $G(x) = 8$ (f) $H(x) = -2x^3(x-1)^2$

Solution   (a) f is a polynomial function of degree 4.

(b) g is not a polynomial function because $g(x) = \sqrt{x} = x^{\frac{1}{2}}$, so the variable x is raised to the $\dfrac{1}{2}$ power, which is not a nonnegative integer.

(c) h is not a polynomial function. It is the ratio of two polynomials, and the polynomial in the denominator is of positive degree.

(d) F is the zero polynomial function; it is not assigned a degree.

(e) G is a nonzero constant function. It is a polynomial function of degree 0 since $G(x) = 8 = 8x^0$.

(f) $H(x) = -2x^3(x - 1)^2 = -2x^3(x^2 - 2x + 1) = -2x^5 + 4x^4 - 2x^3$. So H is a polynomial function of degree 5. Do you see how to find the degree of H without multiplying out? ◄

■——— **NOW WORK PROBLEMS 11 AND 15.**

We have already discussed in detail polynomial functions of degrees 0, 1, and 2. See Table 4 for a summary of the properties of the graphs of these polynomial functions.

Table 4

Degree	Form	Name	Graph
No degree	$f(x) = 0$	Zero function	The x-axis
0	$f(x) = a_0, \quad a_0 \neq 0$	Constant function	Horizontal line with y-intercept a_0
1	$f(x) = a_1 x + a_0, \quad a_1 \neq 0$	Linear function	Nonvertical, nonhorizontal line with slope a_1 and y-intercept a_0
2	$f(x) = a_2 x^2 + a_1 x + a_0, \quad a_2 \neq 0$	Quadratic function	Parabola: Graph opens up if $a_2 > 0$; graph opens down if $a_2 < 0$

One objective of this section is to analyze the graph of a polynomial function. If you take a course in calculus, you will learn that the graph of every polynomial function is both smooth and continuous. By **smooth**, we mean that the graph contains no sharp corners or cusps; by **continuous**, we mean that the graph has no gaps or holes and can be drawn without lifting pencil from paper. See Figures 22(a) and (b).

Figure 22

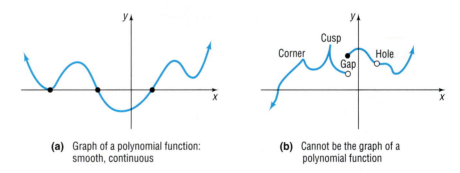

(a) Graph of a polynomial function: smooth, continuous

(b) Cannot be the graph of a polynomial function

2 Graph Polynomial Functions Using Transformations

We begin the analysis of the graph of a polynomial function by discussing *power functions*, a special kind of polynomial function.

In Words
A power function is a function that is defined by a single monomial.

A **power function of degree** n is a function of the form

$$f(x) = ax^n \qquad (2)$$

where a is a real number, $a \neq 0$, and $n > 0$ is an integer.

Examples of power functions are

$$f(x) = 3x \qquad f(x) = -5x^2 \qquad f(x) = 8x^3 \qquad f(x) = -5x^4$$

degree 1   degree 2   degree 3   degree 4

The graph of a power function of degree 1, $f(x) = ax$, is a straight line, with slope a, that passes through the origin. The graph of a power function of degree 2, $f(x) = ax^2$, is a parabola, with vertex at the origin, that opens up if $a > 0$ and down if $a < 0$.

If we know how to graph a power function of the form $f(x) = x^n$, then a compression or stretch and, perhaps, a reflection about the x-axis will enable us to obtain the graph of $g(x) = ax^n$. Consequently, we shall concentrate on graphing power functions of the form $f(x) = x^n$.

We begin with power functions of even degree of the form $f(x) = x^n, n \geq 2$ and n even.

--- **Exploration** ---

Using your graphing utility and the viewing window $-2 \leq x \leq 2, -4 \leq y \leq 16$, graph the function $Y_1 = f(x) = x^4$. On the same screen, graph $Y_2 = g(x) = x^8$. Now, also on the same screen, graph $Y_3 = h(x) = x^{12}$. What do you notice about the graphs as the magnitude of the exponent increases? Repeat this procedure for the viewing window $-1 \leq x \leq 1, 0 \leq y \leq 1$. What do you notice?

Result See Figures 23(a) and (b).

Figure 23

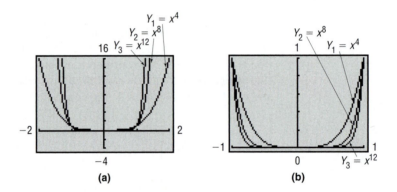

(a)　　　　　　　　(b)

Table 5

X	Y2	Y3
-1	1	1
1	1	1
.5	.00391	2.4E-4
.1	1E-8	1E-12
.01	1E-16	1E-24
.001	1E-24	1E-36
0	0	0

Y2■X^8

The domain of $f(x) = x^n, n \geq 2$ and n even, is the set of all real numbers, and the range is the set of nonnegative real numbers. Such a power function is an even function (do you see why?), so its graph is symmetric with respect to the y-axis. Its graph always contains the origin $(0, 0)$ and the points $(-1, 1)$ and $(1, 1)$.

For large n, it appears that the graph coincides with the x-axis near the origin, but it does not; the graph actually touches the x-axis only at the origin. See Table 5, where $Y_2 = x^8$ and $Y_3 = x^{12}$. For x close to 0, the values of y are positive and close to 0. Also, for large n, it may appear that for $x < -1$ or for $x > 1$ the graph is vertical, but it is not; it is only increasing very rapidly. If you TRACE along one of the graphs, these distinctions will be clear.

To summarize:

NOTE
Don't forget how graphing calculators express scientific notation. In Table 5, $-1E-8$ means -1×10^{-8}. ■

> **Properties of Power Functions, $f(x) = x^n$, n Is an Even Integer**
>
> 1. The graph is symmetric with respect to the y-axis, so f is even.
> 2. The domain is the set of all real numbers. The range is the set of nonnegative real numbers.
> 3. The graph always contains the points $(0, 0)$, $(1, 1)$, and $(-1, 1)$.
> 4. As the exponent n increases in magnitude, the graph becomes more vertical when $x < -1$ or $x > 1$; but for x near the origin, the graph tends to flatten out and lie closer to the x-axis.

Now we consider power functions of odd degree of the form $f(x) = x^n$, *n odd*.

—— **Exploration** ——————————————————————

Using your graphing utility and the viewing window $-2 \leq x \leq 2$, $-16 \leq y \leq 16$, graph the function $Y_1 = f(x) = x^3$. On the same screen, graph $Y_2 = g(x) = x^7$ and $Y_3 = h(x) = x^{11}$. What do you notice about the graphs as the magnitude of the exponent increases? Repeat this procedure for the viewing window $-1 \leq x \leq 1$, $-1 \leq y \leq 1$. What do you notice?

Result   The graphs on your screen should look like Figures 24(a) and (b).

Figure 24

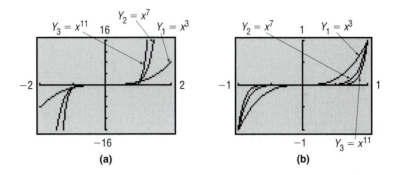

(a)   (b)

The domain and the range of $f(x) = x^n$, $n \geq 3$ and n odd, are the set of real numbers. Such a power function is an odd function (do you see why?), so its graph is symmetric with respect to the origin. Its graph always contains the origin $(0, 0)$ and the points $(-1, -1)$ and $(1, 1)$.

It appears that the graph coincides with the x-axis near the origin, but it does not; the graph actually crosses the x-axis only at the origin. Also, it appears that as x increases the graph is vertical, but it is not; it is increasing very rapidly. TRACE along the graphs to verify these distinctions.

To summarize:

> **Properties of Power Functions, $f(x) = x^n$, n Is an Odd Integer**
>
> **1.** The graph is symmetric with respect to the origin, so f is odd.
>
> **2.** The domain and the range are the set of all real numbers.
>
> **3.** The graph always contains the points $(0, 0)$, $(1, 1)$, and $(-1, -1)$.
>
> **4.** As the exponent n increases in magnitude, the graph becomes more vertical when $x < -1$ or $x > 1$, but for x near the origin, the graph tends to flatten out and lie closer to the x-axis.

The methods of shifting, compression, stretching, and reflection studied in Section 2.6, when used with the facts just presented, will enable us to graph polynomial functions that are transformations of power functions.

EXAMPLE 2   **Graphing Polynomial Functions Using Transformations**

Graph: $f(x) = 1 - x^5$

Solution   Figure 25 shows the required stages.

Figure 25

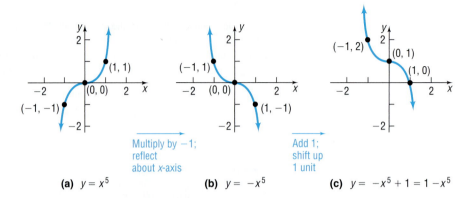

(a) $y = x^5$ **(b)** $y = -x^5$ **(c)** $y = -x^5 + 1 = 1 - x^5$

✔ **CHECK:** Verify the result of Example 2 by graphing $Y_1 = f(x) = 1 - x^5$.   ◀

| **EXAMPLE 3** | **Graphing Polynomial Functions Using Transformations** |

Graph: $f(x) = \dfrac{1}{2}(x - 1)^4$

Solution   Figure 26 shows the required stages.

Figure 26

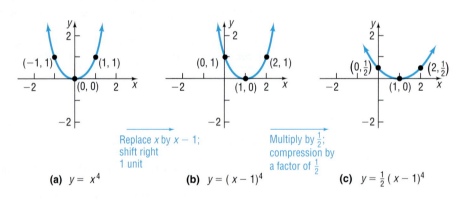

(a) $y = x^4$ **(b)** $y = (x - 1)^4$ **(c)** $y = \frac{1}{2}(x - 1)^4$

✔ **CHECK:** Verify the result of Example 3 by graphing

$$Y_1 = f(x) = \frac{1}{2}(x - 1)^4$$ ◀

✏️ **NOW WORK PROBLEMS 23 AND 27.**

3 **Identify the Zeros of a Polynomial Function
and Their Multiplicity**

Figure 27 shows the graph of a polynomial function with four x-intercepts. Notice that at the x-intercepts the graph must either cross the x-axis or touch the x-axis. Consequently, between consecutive x-intercepts the graph is either above the x-axis or below the x-axis. We will make use of this property of the graph of a polynomial shortly.

Figure 27

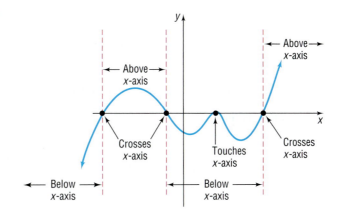

If a polynomial function f is factored completely, it is easy to solve the equation $f(x) = 0$ and locate the x-intercepts of the graph. For example, if $f(x) = (x - 1)^2(x + 3)$, then the solutions of the equation

$$f(x) = (x - 1)^2(x + 3) = 0$$

are identified as 1 and -3. Based on this result, we make the following observations:

> If f is a polynomial function and r is a real number for which $f(r) = 0$, then r is called a (real) **zero of f**, or **root of f**. If r is a (real) zero of f, then
>
> (a) r is an x-intercept of the graph of f.
> (b) $(x - r)$ is a factor of f.

So the real zeros of a function are the x-intercepts of its graph, and they are found by solving the equation $f(x) = 0$.

EXAMPLE 4   **Finding a Polynomial from Its Zeros**

(a) Find a polynomial of degree 3 whose zeros are $-3, 2$, and 5.

(b) Graph the polynomial found in part (a) to verify your result.

Solution   (a) If r is a zero of a polynomial f, then $x - r$ is a factor of f. This means that $x - (-3) = x + 3, x - 2$, and $x - 5$ are factors of f. As a result, any polynomial of the form

$$f(x) = a(x + 3)(x - 2)(x - 5)$$

where a is any nonzero real number, qualifies.

(b) The value of a causes a stretch, compression, or reflection, but does not affect the x-intercepts. We choose to graph f with $a = 1$.

$$f(x) = (x + 3)(x - 2)(x - 5) = x^3 - 4x^2 - 11x + 30$$

Figure 28 shows the graph of f. Notice that the x-intercepts are $-3, 2$, and 5. ◀

Figure 28

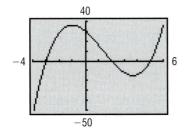

— Seeing the Concept —

Graph the function found in Example 4 for $a = 2$ and $a = -1$. Does the value of a affect the zeros of f? How does the value of a affect the graph of f?

NOW WORK PROBLEM 37.

If the same factor $x - r$ occurs more than once, then r is called a **repeated**, or **multiple, zero of f**. More precisely, we have the following definition.

If $(x - r)^m$ is a factor of a polynomial f and $(x - r)^{m+1}$ is not a factor of f, then r is called a **zero of multiplicity m of f**.

EXAMPLE 5 | **Identifying Zeros and Their Multiplicities**

For the polynomial

$$f(x) = 5(x - 2)(x + 3)^2\left(x - \frac{1}{2}\right)^4$$

2 is a zero of multiplicity 1 because the exponent on the factor $x - 2$ is 1.

-3 is a zero of multiplicity 2 because the exponent on the factor $x + 3$ is 2.

$\dfrac{1}{2}$ is a zero of multiplicity 4 because the exponent on the factor $x - \frac{1}{2}$ is 4. ◀

 NOW WORK PROBLEM 45(a).

In Example 5 notice that, if you add the multiplicities $(1 + 2 + 4 = 7)$, you obtain the degree of the polynomial.

Suppose that it is possible to factor completely a polynomial function and, as a result, locate all the x-intercepts of its graph (the real zeros of the function). The following example illustrates the role that the multiplicity of an x-intercept plays.

EXAMPLE 6 | **Investigating the Role of Multiplicity**

For the polynomial $f(x) = x^2(x - 2)$:

(a) Find the x- and y-intercepts of the graph of f.

(b) Using a graphing utility, graph the polynomial.

(c) For each x-intercept, determine whether it is of odd or even multiplicity.

Solution

Figure 29

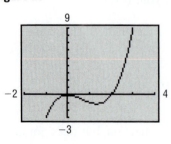

Table 6

(a) The y-intercept is $f(0) = 0^2(0 - 2) = 0$. The x-intercepts satisfy the equation

$$f(x) = x^2(x - 2) = 0$$

from which we find that

$$x^2 = 0 \qquad \text{or} \qquad x - 2 = 0$$
$$x = 0 \qquad \text{or} \qquad x = 2$$

The x-intercepts are 0 and 2.

(b) See Figure 29 for the graph of f.

(c) We can see from the factored form of f that 0 is a zero or root of multiplicity 2, and 2 is a zero or root of multiplicity 1; so 0 is of even multiplicity and 2 is of odd multiplicity. ◀

We can use a TABLE to further analyze the graph. See Table 6. The sign of $f(x)$ is the same on each side of $x = 0$ and the graph of f just touches the x-axis at $x = 0$ (a zero of even multiplicity). The sign of $f(x)$ changes from one side of $x = 2$ to the other and the graph of f crosses the x-axis at $x = 2$ (a zero of odd multiplicity). These observations suggest the following result:

If r Is a Zero of Even Multiplicity

Sign of $f(x)$ does not change from one side of r to the other side of r.

Graph **touches** x-axis at r.

If r Is a Zero of Odd Multiplicity

Sign of $f(x)$ changes from one side of r to the other side of r.

Graph **crosses** x-axis at r.

 NOW WORK PROBLEM 45(b).

Points on the graph where the graph changes from an increasing function to a decreasing function, or vice versa, are called **turning points**.[*]

Figure 30

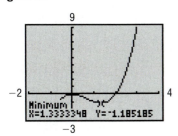

Look at Figure 30. The graph of $f(x) = x^2(x - 2) = x^3 - 2x^2$, has a turning point at $(0, 0)$. After utilizing MINIMUM, we find that the graph also has a turning point at $(1.33, -1.19)$, rounded to two decimal places.

─── **Exploration** ───────────────

Graph $Y_1 = x^3$, $Y_2 = x^3 - x$, and $Y_3 = x^3 + 3x^2 + 4$. How many turning points do you see? How does the number of turning points relate to the degree? Graph $Y_1 = x^4$, $Y_2 = x^4 - \frac{4}{3}x^3$, and $Y_3 = x^4 - 2x^2$. How many turning points do you see? How does the number of turning points compare to the degree?

The following theorem from calculus supplies the answer.

Theorem

If f is a polynomial function of degree n, then f has at most $n - 1$ turning points.

One last remark about Figure 30. Notice that the graph of $f(x) = x^2(x - 2)$ looks somewhat like the graph of $y = x^3$. In fact, for very large values of x, either positive or negative, there is little difference.

─── **Exploration** ───────────────

For each pair of functions Y_1 and Y_2 given in parts (a), (b), and (c), graph Y_1 and Y_2 on the same viewing window. Create a TABLE or TRACE for large positive and large negative values of x. What do you notice about the graphs of Y_1 and Y_2 as x becomes very large and positive or very large and negative?

(a) $Y_1 = x^2(x - 2)$; $Y_2 = x^3$

(b) $Y_1 = x^4 - 3x^3 + 7x - 3$; $Y_2 = x^4$

(c) $Y_1 = -2x^3 + 4x^2 - 8x + 10$; $Y_2 = -2x^3$

[*]Graphing utilities can be used to approximate turning points. Calculus is needed to find the exact location of turning points.

The behavior of the graph of a function for large values of x, either positive or negative, is referred to as its **end behavior**.

Theorem

End Behavior

For large values of x, either positive or negative, that is, for large $|x|$, the graph of the polynomial

$$f(x) = a_n x^n + a_{n-1} x^{n-1} + \cdots + a_1 x + a_0$$

resembles the graph of the power function

$$y = a_n x^n$$

Look back at Figures 23 and 24. Based on this theorem and the previous discussion on power functions, the end behavior of a polynomial can be of only four types. See Figure 31.

Figure 31
End Behavior

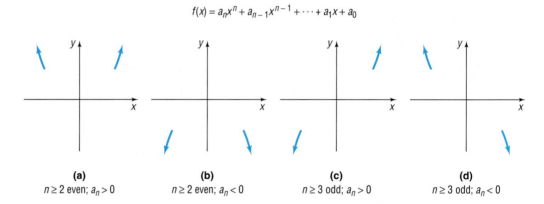

$f(x) = a_n x^n + a_{n-1} x^{n-1} + \cdots + a_1 x + a_0$

(a)
$n \geq 2$ even; $a_n > 0$

(b)
$n \geq 2$ even; $a_n < 0$

(c)
$n \geq 3$ odd; $a_n > 0$

(d)
$n \geq 3$ odd; $a_n < 0$

For example, consider the polynomial function $f(x) = -2x^4 + x^3 + 4x^2 - 7x + 1$. The graph of f will resemble the graph of the power function $y = -2x^4$ for large $|x|$. The graph of f will look like Figure 31(b) for large $|x|$.

✎ **NOW WORK PROBLEM 45(c).**

Summary

Graph of a Polynomial Function $f(x) = a_n x^n + a_{n-1} x^{n-1} + \cdots + a_1 x + a_0, \qquad a_n \neq 0$

Degree of the polynomial f: n

Maximum number of turning points: $n - 1$

At a zero of even multiplicity: The graph of f touches the x-axis.

At a zero of odd multiplicity: The graph of f crosses the x-axis.

Between zeros, the graph of f is either above or below the x-axis.

End behavior: For large $|x|$, the graph of f behaves like the graph of $y = a_n x^n$.

4 Analyze the Graph of a Polynomial Function

| EXAMPLE 7 | **Analyzing the Graph of a Polynomial Function** |

For the polynomial: $f(x) = x^4 - 2x^3 - 8x^2$

(a) Find the degree of the polynomial. Determine the end behavior; that is, find the power function that the graph of f resembles for large values of $|x|$.

(b) Find the x- and y-intercepts of the graph of f.

(c) Determine whether the graph crosses or touches the x-axis at each x-intercept.

(d) Use a graphing utility to graph f.

(e) Determine the number of turning points on the graph of f. Approximate the turning points, if any exist, rounded to two decimal places.

(f) Use the information obtained in parts (a) to (e) to draw a complete graph of f by hand.

(g) Find the domain of f. Use the graph to find the range of f.

(h) Use the graph to determine where f is increasing and where f is decreasing.

Solution

(a) The polynomial is of degree 4. End behavior: the graph of f resembles that of the power function $y = x^4$ for large values of $|x|$.

(b) The y-intercept is $f(0) = 0$. To find the x-intercepts, if any, we first factor f.

$$f(x) = x^4 - 2x^3 - 8x^2 = x^2(x^2 - 2x - 8) = x^2(x - 4)(x + 2)$$

We find the x-intercepts by solving the equation

$$f(x) = x^2(x - 4)(x + 2) = 0$$

So

$$x^2 = 0 \quad \text{or} \quad x - 4 = 0 \quad \text{or} \quad x + 2 = 0$$
$$x = 0 \quad \text{or} \qquad x = 4 \quad \text{or} \qquad x = -2$$

The x-intercepts are -2, 0, and 4.

(c) The x-intercept 0 is a zero of multiplicity 2, so the graph of f will touch the x-axis at 0; -2 and 4 are zeros of multiplicity 1, so the graph of f will cross the x-axis at -2 and 4.

(d) See Figure 32 for the graph of f.

(e) From the graph of f shown in Figure 32, we can see that f has three turning points. Using MAXIMUM, one turning point is at $(0, 0)$. Using MINIMUM, the two remaining turning points are at $(-1.39, -6.35)$ and $(2.89, -45.33)$, rounded to two decimal places.

(f) Figure 33 shows a graph of f using the information obtained in parts (a) to (e).

(g) The domain of f is the set of all real numbers. The range of f is the interval $[-45.33, \infty)$.

(h) Based on the graph, f is decreasing on the intervals $(-\infty, -1.39)$ and $(0, 2.89)$ and is increasing on the intervals $(-1.39, 0)$ and $(2.89, \infty)$. ◀

Figure 32

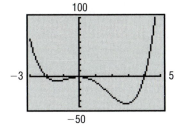

Figure 33

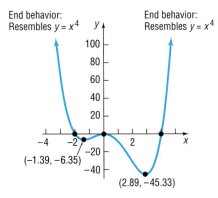

NOW WORK PROBLEM **69**.

For polynomial functions that have noninteger coefficients and for polynomials that are not easily factored, we utilize the graphing utility early in the analysis of the graph. This is because the amount of information that can be obtained from algebraic analysis is limited.

| EXAMPLE 8 | **Using a Graphing Utility to Analyze the Graph of a Polynomial Function** |

For the polynomial $f(x) = x^3 + 2.48x^2 - 4.3155x + 2.484406$:

(a) Find the degree of the polynomial. Determine the end behavior: that is, find the power function that the graph of f resembles for large values of $|x|$.

(b) Graph f using a graphing utility.

(c) Find the x- and y-intercepts of the graph.

(d) Use a TABLE to find points on the graph around each x-intercept.

(e) Determine the local maxima and local minima, if any exist, rounded to two decimal places. That is, locate any turning points.

(f) Use the information obtained in parts (a)–(e) to draw a complete graph of f by hand. Be sure to label the intercepts, turning points, and the points obtained in part (d).

(g) Find the domain of f. Use the graph to find the range of f.

(h) Use the graph to determine where f is increasing and where f is decreasing.

Solution

(a) The degree of the polynomial is 3. End behavior: the graph of f resembles that of the power function $y = x^3$ for large values of $|x|$.

Figure 34

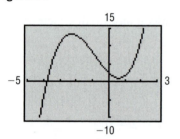

(b) See Figure 34 for the graph of f.

(c) The y-intercept is $f(0) = 2.484406$. In Example 7 we could easily factor $f(x)$ to find the x-intercepts. However, it is not readily apparent how $f(x)$ factors in this example. Therefore, we use a graphing utility's ZERO (or ROOT) feature and find the lone x-intercept to be -3.79, rounded to two decimal places.

(d) Table 7 shows values of x around the x-intercept. The points $(-4, -4.57)$ and $(-2, 13.04)$ are on the graph.

Table 7

X	Y1
-4	-4.574
-2	13.035

Y1■X^3+2.48X²−4...

(e) From the graph we see that it has two turning points: one between -3 and -2 the other between 0 and 1. Rounded to two decimal places, the local maximum is 13.36 and occurs at $x = -2.28$; the local minimum is 1 and occurs at $x = 0.63$. The turning points are $(-2.28, 13.36)$ and $(0.63, 1)$.

(f) Figure 35 shows a graph of f drawn by hand using the information obtained in parts (a) to (e).

Figure 35

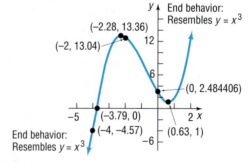

(g) The domain and the range of f are the set of all real numbers.

(h) Based on the graph, f is decreasing on the interval $(-2.28, 0.63)$ and is increasing on the intervals $(-\infty, -2.28)$ and $(0.63, \infty)$. ◀

NOW WORK PROBLEM 81.

Summary

Steps for Graphing a Polynomial by Hand

To analyze the graph of a polynomial function $y = f(x)$, follow these steps:

STEP 1: End behavior: find the power function that the graph of f resembles for large values of x.

STEP 2: (a) Find the x-intercepts, if any, by solving the equation $f(x) = 0$.
(b) Find the y-intercept by letting $x = 0$ and finding the value of $f(0)$.

STEP 3: Determine whether the graph of f crosses or touches the x-axis at each x-intercept.

STEP 4: Use a graphing utility to graph f. Determine the number of turning points on the graph of f. Approximate any turning points rounded to two decimal places.

STEP 5: Use the information obtained in Steps 1 to 4 to draw a complete graph of f by hand.

---

5   Find the Cubic Function of Best Fit to Data

In Section 2.4 we found the line of best fit from data and in Section 3.1 we found the quadratic function of best fit. It is also possible to find polynomial functions of best fit. However, most statisticians do not recommend finding polynomials of best fit of degree higher than 3.

Data that follow a cubic relation should look like Figure 36(a) or (b).

Figure 36

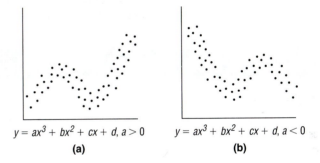

$y = ax^3 + bx^2 + cx + d, a > 0$ $y = ax^3 + bx^2 + cx + d, a < 0$

(a)   (b)

EXAMPLE 9   **A Cubic Function of Best Fit**

The data in Table 8 represent the average miles per gallon of vans, pickups, and sport utility vehicles (SUVs) in the United States for 1991–2001, where 1 represents 1991, 2 represents 1992, and so on.

(a) Draw a scatter diagram of the data using the year x as the independent variable and average miles per gallon M as the dependent variable. Comment on the type of relation that may exist between the two variables x and M.

(b) Using a graphing utility, find the cubic function of best fit $M = M(x)$ to these data.

(c) Graph the cubic function of best fit on your scatter diagram.

(d) Use the function found in part (b) to predict the average miles per gallon in 2002 ($x = 12$).

Table 8

Year, x	Average Miles per Gallon, M
1991, 1	17.0
1992, 2	17.3
1993, 3	17.4
1994, 4	17.3
1995, 5	17.3
1996, 6	17.2
1997, 7	17.2
1998, 8	17.2
1999, 9	17.0
2000, 10	17.4
2001, 11	17.6

SOURCE: United States Federal Highway Administration

Solution   (a) Figure 37 shows the scatter diagram. A cubic relation may exist between the two variables.

(b) Upon executing the CUBIC REGression program, we obtain the results shown in Figure 38. The output the utility provides shows us the equation $y = ax^3 + bx^2 + cx + d$. The cubic function of best fit to the data is $M(x) = 0.0058x^3 - 0.1007x^2 + 0.4950x + 16.6258$.

(c) Figure 39 shows the graph of the cubic function of best fit on the scatter diagram. The function fits the data reasonably well.

Figure 37

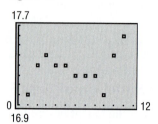

Figure 38

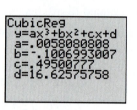

Figure 39

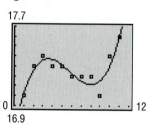

(d) We evaluate the function $M(x)$ at $x = 12$.

$$M(12) = 0.0058(12)^3 - 0.1007(12)^2 + 0.4950(12) + 16.6258 \approx 18.1$$

The model predicts that the average miles per gallon of vans, pickups, and SUVs will be 18.1 miles per gallon in 2002.   ◀

3.2 Assess Your Understanding

'Are You Prepared?'

Answers are given at the end of these exercises. If you get a wrong answer, read the pages listed in red.

1. The intercepts of the equation $9x^2 + 4y = 36$ are _____. (pp. 15–17)

2. *True or False:* The expression $4x^3 - 3.6x^2 - \sqrt{2}$ is a polynomial. (pp. 675–676)

3. To graph $y = x^2 - 4$, you would shift the graph of $y = x^2$ _____ a distance of _____ units. (pp. 118–120)

4. Use a graphing utility to approximate (rounded to two decimal places) the local maxima and local minima of $f(x) = x^3 - 2x^2 - 4x + 5$, for $-10 < x < 10$ (pp. 84–85)

Concepts and Vocabulary

5. The graph of every polynomial function is both _____ and _____.

6. A number r for which $f(r) = 0$ is called a(n) _____ of the function f.

7. If r is a zero of even multiplicity for a function f, the graph of f _____ the x-axis at r.

8. *True or False:* The graph of $f(x) = x^2(x - 3)(x + 4)$ has exactly three x-intercepts.

9. *True or False:* The x-intercepts of the graph of a polynomial function are called turning points.

10. *True or False:* End behavior: the graph of the function $f(x) = 3x^4 - 6x^2 + 2x + 5$ resembles $y = x^4$ for large values of $|x|$.

Skill Building

In Problems 11–22, determine which functions are polynomial functions. For those that are, state the degree. For those that are not, tell why not.

11. $f(x) = 4x + x^3$

12. $f(x) = 5x^2 + 4x^4$

13. $g(x) = \dfrac{1 - x^2}{2}$

14. $h(x) = 3 - \dfrac{1}{2}x$

15. $f(x) = 1 - \dfrac{1}{x}$

16. $f(x) = x(x - 1)$

17. $g(x) = x^{3/2} - x^2 + 2$

18. $h(x) = \sqrt{x}(\sqrt{x} - 1)$

19. $F(x) = 5x^4 - \pi x^3 + \dfrac{1}{2}$

20. $F(x) = \dfrac{x^2 - 5}{x^3}$

21. $G(x) = 2(x - 1)^2(x^2 + 1)$

22. $G(x) = -3x^2(x + 2)^3$

In Problems 23–36, use transformations of the graph of $y = x^4$ or $y = x^5$ to graph each function.

23. $f(x) = (x + 1)^4$

24. $f(x) = (x - 2)^5$

25. $f(x) = x^5 - 3$

26. $f(x) = x^4 + 2$

27. $f(x) = \dfrac{1}{2}x^4$

28. $f(x) = 3x^5$

29. $f(x) = -x^5$

30. $f(x) = -x^4$

31. $f(x) = (x - 1)^5 + 2$

32. $f(x) = (x + 2)^4 - 3$

33. $f(x) = 2(x + 1)^4 + 1$

34. $f(x) = \dfrac{1}{2}(x - 1)^5 - 2$

35. $f(x) = 4 - (x - 2)^5$

36. $f(x) = 3 - (x + 2)^4$

In Problems 37–44, form a polynomial whose zeros and degree are given.

37. Zeros: $-1, 1, 3$;   degree 3

38. Zeros: $-2, 2, 3$;   degree 3

39. Zeros: $-3, 0, 4$;   degree 3

40. Zeros: $-4, 0, 2$;   degree 3

41. Zeros: $-4, -1, 2, 3$;   degree 4

42. Zeros: $-3, -1, 2, 5$;   degree 4

43. Zeros: -1, multiplicity 1;   3, multiplicity 2;   degree 3

44. Zeros: -2, multiplicity 2;   4, multiplicity 1;   degree 3

In Problems 45–56, for each polynomial function: (a) List each real zero and its multiplicity; (b) Determine whether the graph crosses or touches the x-axis at each x-intercept. (c) Find the power function that the graph of f resembles for large values of $|x|$.

45. $f(x) = 3(x - 7)(x + 3)^2$

46. $f(x) = 4(x + 4)(x + 3)^3$

47. $f(x) = 4(x^2 + 1)(x - 2)^3$

48. $f(x) = 2(x - 3)(x + 4)^3$

49. $f(x) = -2\left(x + \dfrac{1}{2}\right)^2(x^2 + 4)^2$

50. $f(x) = \left(x - \dfrac{1}{3}\right)^2(x - 1)^3$

51. $f(x) = (x - 5)^3(x + 4)^2$

52. $f(x) = \left(x + \sqrt{3}\right)^2(x - 2)^4$

53. $f(x) = 3(x^2 + 8)(x^2 + 9)^2$

54. $f(x) = -2(x^2 + 3)^3$

55. $f(x) = -2x^2(x^2 - 2)$

56. $f(x) = 4x(x^2 - 3)$

In Problems 57–80, for each polynomial function f:

(a) *Find the degree of the polynomial. Determine the end behavior: that is, find the power function that the graph of f resembles for large values of $|x|$.*

(b) *Find the x- and y-intercepts of the graph.*

(c) *Determine whether the graph crosses or touches the x-axis at each x-intercept.*

(d) *Graph f using a graphing utility.*

(e) *Determine the local maxima and local minima, if any exist, rounded to two decimal places. That is, locate any turning points.*

(f) *Use the information obtained in parts (a)–(e) to draw a complete graph of f by hand. Be sure to label the intercepts and turning points.*

(g) *Find the domain of f. Use the graph to find the range of f.*

(h) *Use the graph to determine where f is increasing and where f is decreasing.*

57. $f(x) = (x - 1)^2$

58. $f(x) = (x - 2)^3$

59. $f(x) = x^2(x - 3)$

60. $f(x) = x(x + 2)^2$

61. $f(x) = (x + 4)(x - 2)^2$

62. $f(x) = (x - 1)(x + 3)^2$

63. $f(x) = -2(x + 2)(x - 2)^3$

64. $f(x) = -\dfrac{1}{2}(x + 4)(x - 1)^3$

65. $f(x) = (x + 1)(x - 2)(x + 4)$

66. $f(x) = (x - 1)(x + 4)(x - 3)$

67. $f(x) = 4x - x^3$

68. $f(x) = x - x^3$

69. $f(x) = x^2(x - 2)(x + 2)$

70. $f(x) = x^2(x - 3)(x + 4)$

71. $f(x) = (x + 1)^2(x - 2)^2$

72. $f(x) = (x + 1)^3(x - 3)$

73. $f(x) = x^2(x - 3)(x + 1)$

74. $f(x) = x^2(x - 3)(x - 1)$

75. $f(x) = (x + 2)^2(x - 4)^2$

76. $f(x) = (x - 2)^2(x + 2)(x + 4)$

77. $f(x) = x^2(x - 2)(x^2 + 3)$

78. $f(x) = x^2(x^2 + 1)(x + 4)$

79. $f(x) = -x^2(x^2 - 1)(x + 1)$

80. $f(x) = -x^2(x^2 - 4)(x - 5)$

In Problems 81–90, for each polynomial function f:

(a) *Find the degree of the polynomial. Determine the end behavior: that is, find the power function that the graph of f resembles for large values of |x|.*

(b) *Graph f using a graphing utility.*

(c) *Find the x- and y-intercepts of the graph.*

(d) *Use a TABLE to find points on the graph around each x-intercept.*

(e) *Determine the local maxima and local minima, if any exist, rounded to two decimal places. That is, locate any turning points.*

(f) *Use the information obtained in parts (a)–(e) to draw a complete graph of f by hand. Be sure to label the intercepts, turning points, and the points obtained in part (d).*

(g) *Find the domain of f. Use the graph to find the range of f.*

(h) *Use the graph to determine where f is increasing and where f is decreasing.*

81. $f(x) = x^3 + 0.2x^2 - 1.5876x - 0.31752$

82. $f(x) = x^3 - 0.8x^2 - 4.6656x + 3.73248$

83. $f(x) = x^3 + 2.56x^2 - 3.31x + 0.89$

84. $f(x) = x^3 - 2.91x^2 - 7.668x - 3.8151$

85. $f(x) = x^4 - 2.5x^2 + 0.5625$

86. $f(x) = x^4 - 18.5x^2 + 50.2619$

87. $f(x) = 2x^4 - \pi x^3 + \sqrt{5}x - 4$

88. $f(x) = -1.2x^4 + 0.5x^2 - \sqrt{3}x + 2$

89. $f(x) = -2x^5 - \sqrt{2}x^2 - x - \sqrt{2}$

90. $f(x) = \pi x^5 + \pi x^4 + \sqrt{3}x + 1$

Applications and Extensions

In Problems 91–94, construct a polynomial function that might have this graph. (More than one answer may be possible.)

91.
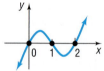

92.

93.
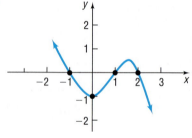

94.
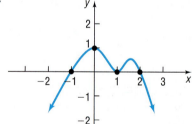

---

95. Cost of Manufacturing The following data represent the cost C (in thousands of dollars) of manufacturing Chevy Cavaliers and the number x of Cavaliers produced.

Number of Cavaliers Produced, x	Cost, C
0	10
1	23
2	31
3	38
4	43
5	50
6	59
7	70
8	85
9	105
10	135

(a) Draw a scatter diagram of the data using x as the independent variable and C as the dependent variable. Comment on the type of relation that may exist between the two variables C and x.

(b) Find the average rate of change of cost from four to five Cavaliers.

(c) What is the average rate of change in cost from eight to nine Cavaliers?

(d) Use a graphing utility to find the cubic function of best fit C = C(x).

(e) Graph the cubic function of best fit on the scatter diagram.

(f) Use the function found in part (d) to predict the cost of manufacturing 11 Cavaliers.

(g) Interpret the y-intercept.

96. Cost of Printing The following data represent the weekly cost C (in thousands of dollars) of printing textbooks and the number x (in thousands of units) of texts printed.

Number of Text Books, x	Cost, C
0	100
5	128.1
10	144
13	153.5
17	161.2
18	162.6
20	166.3
23	178.9
25	190.2
27	221.8

(a) Draw a scatter diagram of the data using x as the independent variable and C as the dependent variable. Comment on the type of relation that may exist between the two variables C and x.

(b) Find the average rate of change in cost from 10,000 to 13,000 textbooks.

(c) What is the average rate of change in the cost of producing from 18,000 to 20,000 textbooks?

(d) Use a graphing utility to find the cubic function of best fit $C = C(x)$.

(e) Graph the cubic function of best fit on the scatter diagram.

(f) Use the function found in part (d) to predict the cost of printing 22,000 texts per week.

(g) Interpret the y-intercept.

97. Motor Vehicle Thefts The following data represent the number T (in thousands) of motor vehicle thefts in the United States for the years 1989–2001, where $x = 1$ represents 1989, $x = 2$ represents 1990, and so on.

Year, x	Motor Vehicle Thefts, T
1989, 1	1565
1990, 2	1636
1991, 3	1662
1992, 4	1611
1993, 5	1563
1994, 6	1539
1995, 7	1472
1996, 8	1394
1997, 9	1354
1998, 10	1243
1999, 11	1152
2000, 12	1160
2001, 13	1226

SOURCE: U.S. Federal Bureau of Investigation

(a) Draw a scatter diagram of the data using x as the independent variable and T as the dependent variable. Comment on the type of relation that may exist between the two variables T and x.

(b) Use a graphing utility to find the cubic function of best fit $T = T(x)$.

(c) Graph the cubic function of best fit on the scatter diagram.

(d) Use the function found in part (b) to predict the number of motor vehicle thefts in 2005.

(e) Do you think the function found in part (b) will be useful in predicting motor vehicle thefts in 2010? Why?

98. Average Annual Miles The following data represent the average number of miles driven (in thousands) annually by vans, pickups, and sport utility vehicles (SUVs) for the years 1993–2001, where $x = 1$ represents 1993, $x = 2$ represents 1994, and so on.

Year, x	Average Miles Driven, M
1993, 1	12.4
1994, 2	12.2
1995, 3	12.0
1996, 4	11.8
1997, 5	12.1
1998, 6	12.2
1999, 7	12.0
2000, 8	11.7
2001, 9	11.1

SOURCE: United States Federal Highway Administration

(a) Draw a scatter diagram of the data using x as the independent variable and M as the dependent variable. Comment on the type of relation that may exist between the two variables M and x.

(b) Use a graphing utility to find the cubic function of best fit $M = M(x)$.

(c) Graph the cubic function of best fit on the scatter diagram.

(d) Use the function found in part (b) to predict the average miles driven by vans, pickups, and SUVs in 2005.

(e) Do you think the function found in part (b) will be useful in predicting the average miles driven by vans, pickups, and SUVs in 2010? Why?

Discussion and Writing

99. Can the graph of a polynomial function have no *y*-intercept? Can it have no *x*-intercepts? Explain.

100. Write a few paragraphs that provide a general strategy for graphing a polynomial function. Be sure to mention the following: degree, intercepts, end behavior, and turning points.

101. Make up a polynomial that has the following characteristics: crosses the *x*-axis at −1 and 4, touches the *x*-axis at 0 and 2, and is above the *x*-axis between 0 and 2. Give your polynomial to a fellow classmate and ask for a written critique of your polynomial.

102. Make up two polynomials, not of the same degree, with the following characteristics: crosses the *x*-axis at −2, touches the *x*-axis at 1, and is above the *x*-axis between −2 and 1. Give your polynomials to a fellow classmate and ask for a written critique of your polynomials.

103. The graph of a polynomial function is always smooth and continuous. Name a function studied earlier that is smooth and not continuous. Name one that is continuous, but not smooth.

104. Which of the following statements are true regarding the graph of the cubic polynomial $f(x) = x^3 + bx^2 + cx + d$? (Give reasons for your conclusions.)
(a) It intersects the *y*-axis in one and only one point.
(b) It intersects the *x*-axis in at most three points.
(c) It intersects the *x*-axis at least once.
(d) For $|x|$ very large, it behaves like the graph of $y = x^3$.
(e) It is symmetric with respect to the origin.
(f) It passes through the origin.

'Are You Prepared?' Answers

1. $(-2, 0), (2, 0), (0, 9)$ **2.** True   **3.** down; 4   **4.** Local maximum of 6.48 at $x = -0.67$; local minimum of -3 at $x = 2$

3.3 Properties of Rational Functions

PREPARING FOR THIS SECTION   *Before getting started, review the following:*

- Rational Expressions (Appendix, Section A.3, pp. 679–683)
- Polynomial Division (Appendix, Section A.4, pp. 685–690)

- Graph of $f(x) = \dfrac{1}{x}$ (Section 1.2, Example 13, p. 21)
- Graphing Techniques: Transformations (Section 2.6, pp. 118–126)

Now work the 'Are You Prepared?' problems on page 195.

OBJECTIVES **1** Find the Domain of a Rational Function
2 Find the Vertical Asymptotes of a Rational Function
3 Find the Horizontal or Oblique Asymptotes of a Rational Function

Ratios of integers are called *rational numbers*. Similarly, ratios of polynomial functions are called *rational functions*.

A **rational function** is a function of the form

$$R(x) = \frac{p(x)}{q(x)}$$

where *p* and *q* are polynomial functions and *q* is not the zero polynomial. The domain is the set of all real numbers except those for which the denominator *q* is 0.

Find the Domain of a Rational Function

| EXAMPLE 1 | **Finding the Domain of a Rational Function** |

(a) The domain of $R(x) = \dfrac{2x^2 - 4}{x + 5}$ is the set of all real numbers x except -5; that is, $\{x \mid x \neq -5\}$.

(b) The domain of $R(x) = \dfrac{1}{x^2 - 4}$ is the set of all real numbers x except -2 and 2, that is, $\{x \mid x \neq -2, x \neq 2\}$.

(c) The domain of $R(x) = \dfrac{x^3}{x^2 + 1}$ is the set of all real numbers.

(d) The domain of $R(x) = \dfrac{-x^2 + 2}{3}$ is the set of all real numbers.

(e) The domain of $R(x) = \dfrac{x^2 - 1}{x - 1}$ is the set of all real numbers x except 1, that is, $\{x \mid x \neq 1\}$.   ◀

It is important to observe that the functions

$$R(x) = \frac{x^2 - 1}{x - 1} \quad \text{and} \quad f(x) = x + 1$$

are not equal, since the domain of R is $\{x \mid x \neq 1\}$ and the domain of f is the set of all real numbers.

 NOW WORK PROBLEM 13.

If $R(x) = \dfrac{p(x)}{q(x)}$ is a rational function and if p and q have no common factors, then the rational function R is said to be in **lowest terms**. For a rational function $R(x) = \dfrac{p(x)}{q(x)}$ in lowest terms, the zeros, if any, of the numerator are the x-intercepts of the graph of R and so will play a major role in the graph of R. The zeros of the denominator of R [that is, the numbers x, if any, for which $q(x) = 0$], although not in the domain of R, also play a major role in the graph of R. We will discuss this role shortly.

We have already discussed the properties of the rational function $f(x) = \dfrac{1}{x}$. (Refer to Example 13, page 21.) The next rational function that we take up is $H(x) = \dfrac{1}{x^2}$.

| EXAMPLE 2 | **Graphing $y = \dfrac{1}{x^2}$** |

Analyze the graph of $H(x) = \dfrac{1}{x^2}$.

Solution The domain of $H(x) = \dfrac{1}{x^2}$ consists of all real numbers x except 0. The graph has no y-intercept, because x can never equal 0. The graph has no x-intercept because the equation $H(x) = 0$ has no solution. Therefore, the graph of H will not cross either coordinate axis.

Because

$$H(-x) = \frac{1}{(-x)^2} = \frac{1}{x^2} = H(x)$$

H is an even function, so its graph is symmetric with respect to the y-axis.

See Figure 40. Notice that the graph confirms the conclusions just reached. But what happens to the graph as the values of x get closer and closer to 0? We use a TABLE to answer the question. See Table 9. The first four rows show that as x approaches 0 the values of $H(x)$ become larger and larger positive numbers. When this happens, we say that $H(x)$ is **unbounded in the positive direction**. We symbolize this by writing $H(x) \to \infty$ [read as "$H(x)$ **approaches infinity**"]. In calculus, **limits** are used to convey these ideas. There we use the symbolism $\lim\limits_{x \to 0} H(x) = \infty$, read as "the limit of $H(x)$ as x approaches 0 is infinity" to mean that $H(x) \to \infty$ as $x \to 0$.

Look at the last three rows of Table 9. As $x \to \infty$, the values of $H(x)$ approach 0 (the end behavior of the graph). This is symbolized in calculus by writing $\lim\limits_{x \to \infty} H(x) = 0$.

Figure 41 shows the graph of $H(x) = \dfrac{1}{x^2}$ drawn by hand.

Figure 40

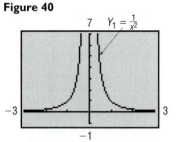

Table 9

X	Y1
.1	100
.01	10000
.001	1E6
1E-4	1E8
10	.01
100	1E-4
1000	1E-6

Y₁◻1/X²

Figure 41

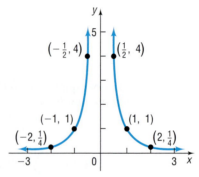

Sometimes transformations (shifting, compressing, stretching, and reflection) can be used to graph a rational function.

EXAMPLE 3   Using Transformations to Graph a Rational Function

Graph the rational function: $R(x) = \dfrac{1}{(x-2)^2} + 1$

Solution   First, notice that the domain of R is the set of all real numbers except $x = 2$. To graph R, we start with the graph of $y = \dfrac{1}{x^2}$. See Figure 42 for the stages.

Figure 42

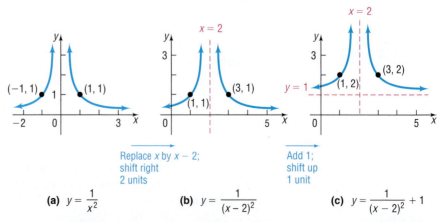

Replace x by $x - 2$; shift right 2 units

Add 1; shift up 1 unit

(a) $y = \dfrac{1}{x^2}$ (b) $y = \dfrac{1}{(x-2)^2}$ (c) $y = \dfrac{1}{(x-2)^2} + 1$

✔ **CHECK:** Graph $Y_1 = \dfrac{1}{(x-2)^2} + 1$ using a graphing utility to verify the graph obtained in Figure 42(c). ◄

 NOW WORK PROBLEM 31.

Asymptotes

Notice that the y-axis in Figure 42(a) is transformed into the vertical line $x = 2$ in Figure 42(c), and the x-axis in Figure 42(a) is transformed into the horizontal line $y = 1$ in Figure 42(c). The **Exploration** that follows will help us analyze the role of these lines.

───── **Exploration** ──

(a) Using a graphing utility and the TABLE feature, evaluate the function $H(x) = \dfrac{1}{(x-2)^2} + 1$ at
$x = 10, 100, 1000,$ and $10{,}000$. What happens to the values of H as x becomes unbounded in
the positive direction, symbolized by $\lim\limits_{x\to\infty} H(x)$?

(b) Evaluate H at $x = -10, -100, -1000,$ and $-10{,}000$. What happens to the values of H as x
becomes unbounded in the negative direction, symbolized by $\lim\limits_{x\to-\infty} H(x)$?

(c) Evaluate H at $x = 1.5, 1.9, 1.99, 1.999,$ and 1.9999. What happens to the values of H as x
approaches 2, $x < 2$, symbolized by $\lim\limits_{x\to 2^-} H(x)$?

(d) Evaluate H at $x = 2.5, 2.1, 2.01, 2.001,$ and 2.0001. What happens to the values of H as x
approaches 2, $x > 2$, symbolized by $\lim\limits_{x\to 2^+} H(x)$?

Result

(a) Table 10 shows the values of $Y_1 = H(x)$ as x approaches ∞. Notice that the values of H are
approaching 1, so $\lim\limits_{x\to\infty} H(x) = 1$.

(b) Table 11 shows the values of $Y_1 = H(x)$ as x approaches $-\infty$. Again the values of H are
approaching 1, so $\lim\limits_{x\to-\infty} H(x) = 1$.

(c) From Table 12 we see that, as x approaches 2, $x < 2$, the values of H are increasing without
bound, so $\lim\limits_{x\to 2^-} H(x) = \infty$.

(d) Finally, Table 13 reveals that, as x approaches 2, $x > 2$, the values of H are increasing without
bound, so $\lim\limits_{x\to 2^+} H(x) = \infty$.

Table 10

X	Y1
10	1.0156
100	1.0001
1000	1
10000	1

Y₁目1/(X−2)²+1

Table 11

X	Y1
−10	1.0069
−100	1.0001
−1000	1
−10000	1

Y₁目1/(X−2)²+1

Table 12

X	Y1
1.5	5
1.9	101
1.99	10001
1.999	1E6
1.9999	1E8

Y₁目1/(X−2)²+1

Table 13

X	Y1
2.5	5
2.1	101
2.01	10001
2.001	1E6
2.0001	1E8

Y₁目1/(X−2)²+1

──

The results of the Exploration reveal an important property of rational functions. The vertical line $x = 2$ and the horizontal line $y = 1$ are called *asymptotes* of the graph of H, which we define as follows:

Let R denote a function:

If, as $x \to -\infty$ or as $x \to \infty$, the values of $R(x)$ approach some fixed number L, then the line $y = L$ is a **horizontal asymptote** of the graph of R.

If, as x approaches some number c, the values $|R(x)| \to \infty$, then the line $x = c$ is a **vertical asymptote** of the graph of R. The graph of R never intersects a vertical asymptote.

Even though the asymptotes of a function are not part of the graph of the function, they provide information about how the graph looks. Figure 43 illustrates some of the possibilities.

Figure 43

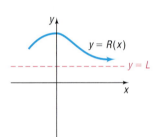

(a) End behavior: As $x \to \infty$, the values of $R(x)$ approach L [$\lim\limits_{x \to \infty} R(x) = L$]. That is, the points on the graph of R are getting closer to the line $y = L$; $y = L$ is a horizontal asymptote.

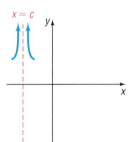

(b) End behavior: As $x \to -\infty$, the values of $R(x)$ approach L [$\lim\limits_{x \to -\infty} R(x) = L$]. That is, the points on the graph of R are getting closer to the line $y = L$; $y = L$ is a horizontal asymptote.

(c) As x approaches c, the values of $|R(x)| \to \infty$ [$\lim\limits_{x \to c^-} R(x) = \infty$; $\lim\limits_{x \to c^+} R(x) = \infty$]. That is, the points on the graph of R are getting closer to the line $x = c$; $x = c$ is a vertical asymptote.

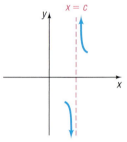

(d) As x approaches c, the values of $|R(x)| \to \infty$ [$\lim\limits_{x \to c^-} R(x) = -\infty$; $\lim\limits_{x \to c^+} R(x) = \infty$]. That is, the points on the graph of R are getting closer to the line $x = c$; $x = c$ is a vertical asymptote.

Figure 44

Oblique asymptote

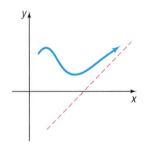

A horizontal asymptote, when it occurs, describes a certain behavior of the graph as $x \to \infty$ or as $x \to -\infty$, that is, its end behavior. **The graph of a function may intersect a horizontal asymptote.**

A vertical asymptote, when it occurs, describes a certain behavior of the graph when x is close to some number c. **The graph of the function will never intersect a vertical asymptote.**

If an asymptote is neither horizontal nor vertical, it is called **oblique**. Figure 44 shows an oblique asymptote. An oblique asymptote, when it occurs, describes the end behavior of the graph. **The graph of a function may intersect an oblique asymptote.**

2 Find the Vertical Asymptotes of a Rational Function

The vertical asymptotes, if any, of a rational function $R(x) = \dfrac{p(x)}{q(x)}$, in lowest terms, are located at the zeros of the denominator of $q(x)$. Suppose that r is a zero, so $x - r$ is a factor. Now, as x approaches r, symbolized as $x \to r$, the values of $x - r$ approach 0, causing the ratio to become unbounded, that is, causing $|R(x)| \to \infty$. Based on the definition, we conclude that the line $x = r$ is a vertical asymptote.

Theorem

Locating Vertical Asymptotes

A rational function $R(x) = \dfrac{p(x)}{q(x)}$, *in lowest terms*, will have a vertical asymptote $x = r$ if r is a real zero of the *denominator* q. That is, if $x - r$ is a factor of the denominator q of a rational function $R(x) = \dfrac{p(x)}{q(x)}$, in lowest terms, then R will have the vertical asymptote $x = r$.

WARNING

If a rational function is not in lowest terms, an application of this theorem will result in an incorrect listing of vertical asymptotes. ■

| **EXAMPLE 4** | **Finding Vertical Asymptotes** |

Find the vertical asymptotes, if any, of the graph of each rational function.

(a) $R(x) = \dfrac{x}{x^2 - 4}$ (b) $F(x) = \dfrac{x + 3}{x - 1}$

(c) $H(x) = \dfrac{x^2}{x^2 + 1}$ (d) $G(x) = \dfrac{x^2 - 9}{x^2 + 4x - 21}$

Solution

(a) R is in lowest terms and the zeros of the denominator $x^2 - 4$ are -2 and 2. Hence, the lines $x = -2$ and $x = 2$ are the vertical asymptotes of the graph of R.

(b) F is in lowest terms and the only zero of the denominator is 1. The line $x = 1$ is the only vertical asymptote of the graph of F.

WARNING

In Example 4(a), the vertical asymptotes are $x = -2$ and $x = 2$. Do not say that the vertical asymptotes are -2 and 2. ■

(c) H is in lowest terms and the denominator has no real zeros because the discriminant of $x^2 + 1$ is $b^2 - 4ac = 0^2 - 4 \cdot 1 \cdot 1 = -4$. The graph of H has no vertical asymptotes.

(d) Factor $G(x)$ to determine if it is in lowest terms.

$$G(x) = \frac{x^2 - 9}{x^2 + 4x - 21} = \frac{(x + 3)(x - 3)}{(x + 7)(x - 3)} = \frac{x + 3}{x + 7}, \qquad x \neq 3$$

The only zero of the denominator of $G(x)$ in lowest terms is -7. Hence, the line $x = -7$ is the only vertical asymptote of the graph of G.   ◄

As Example 4 points out, rational functions can have no vertical asymptotes, one vertical asymptote, or more than one vertical asymptote. However, the graph of a rational function will never intersect any of its vertical asymptotes. (Do you know why?)

───── **Exploration** ──────────────────────────────

Graph each of the following rational functions:

$$R(x) = \frac{1}{x - 1} \quad R(x) = \frac{1}{(x - 1)^2} \quad R(x) = \frac{1}{(x - 1)^3} \quad R(x) = \frac{1}{(x - 1)^4}$$

Each has the vertical asymptote $x = 1$. What happens to the value of $R(x)$ as x approaches 1 from the right side of the vertical asymptote; that is, what is $\lim\limits_{x \to 1^+} R(x)$? What happens to the value of $R(x)$ as x approaches 1 from the left side of the vertical asymptote; that is, what is $\lim\limits_{x \to 1^-} R(x)$? How does the multiplicity of the zero in the denominator affect the graph of R?

──

NOW WORK PROBLEM 45. **(Find the vertical asymptotes, if any.)**

3   Find the Horizontal or Oblique Asymptotes of a Rational Function

The procedure for finding horizontal and oblique asymptotes is somewhat more involved. To find such asymptotes, we need to know how the values of a function behave as $x \to -\infty$ or as $x \to \infty$.

If a rational function $R(x)$ is **proper**, that is, if the degree of the numerator is less than the degree of the denominator, then as $x \to -\infty$ or as $x \to \infty$ the value of $R(x)$ approaches 0. Consequently, the line $y = 0$ (the x-axis) is a horizontal asymptote of the graph.

Theorem	If a rational function is proper, the line $y = 0$ is a horizontal asymptote of its graph.

EXAMPLE 5   **Finding Horizontal Asymptotes**

Find the horizontal asymptotes, if any, of the graph of

$$R(x) = \frac{x - 12}{4x^2 + x + 1}$$

Solution   Since the degree of the numerator, 1, is less than the degree of the denominator, 2, the rational function R is proper. We conclude that the line $y = 0$ is a horizontal asymptote of the graph of R.   ◀

To see why $y = 0$ is a horizontal asymptote of the function R in Example 5, we need to investigate the behavior of R as $x \to -\infty$ and $x \to \infty$. When $|x|$ is un-bounded, the numerator of R, which is $x - 12$, can be approximated by the power function $y = x$, while the denominator of R, which is $4x^2 + x + 1$, can be approximated by the power function $y = 4x^2$. Applying these ideas to $R(x)$, we find

$$R(x) = \frac{x - 12}{4x^2 + x + 1} \underset{\uparrow}{\approx} \frac{x}{4x^2} = \frac{1}{4x} \underset{\uparrow}{\to} 0$$

For $|x|$ unbounded   As $x \to -\infty$ or $x \to \infty$

This shows that the line $y = 0$ is a horizontal asymptote of the graph of R.

We verify these results in Tables 14(a) and (b). Notice, as $x \to -\infty$ [Table 14(a)] or $x \to \infty$ [Table 14(b)], respectively, that the values of $R(x)$ approach 0.

Table 14

X	Y1
-10	-.0563
-100	-.0028
-1000	-3E-4
-10000	-3E-5
-1E5	-3E-6
-1E6	-3E-7
-1E7	-3E-8

Y1⊟(X-12)/(4X²+...

(a)

X	Y1
10	-.0049
100	.00219
1000	2.5E-4
10000	2.5E-5
100000	2.5E-6
1E6	2.5E-7
1E7	2.5E-8

Y1⊟(X-12)/(4X²+...

(b)

If a rational function $R(x) = \dfrac{p(x)}{q(x)}$ is **improper**, that is, if the degree of the numerator is greater than or equal to the degree of the denominator, we must use long division to write the rational function as the sum of a polynomial $f(x)$ plus a proper rational function $\dfrac{r(x)}{q(x)}$. That is, we write

$$R(x) = \frac{p(x)}{q(x)} = f(x) + \frac{r(x)}{q(x)}$$

where $f(x)$ is a polynomial and $\dfrac{r(x)}{q(x)}$ is a proper rational function. Since $\dfrac{r(x)}{q(x)}$ is proper, then $\dfrac{r(x)}{q(x)} \to 0$ as $x \to -\infty$ or as $x \to \infty$. As a result,

$$R(x) = \frac{p(x)}{q(x)} \to f(x), \qquad \text{as } x \to -\infty \text{ or as } x \to \infty$$

The possibilities are listed next.

1. If $f(x) = b$, a constant, then the line $y = b$ is a horizontal asymptote of the graph of R.

2. If $f(x) = ax + b, a \neq 0$, then the line $y = ax + b$ is an oblique asymptote of the graph of R.

3. In all other cases, the graph of R approaches the graph of f, and there are no horizontal or oblique asymptotes.

The following examples demonstrate these conclusions.

EXAMPLE 6

Finding Horizontal or Oblique Asymptotes

Find the horizontal or oblique asymptotes, if any, of the graph of

$$H(x) = \frac{3x^4 - x^2}{x^3 - x^2 + 1}$$

Solution   Since the degree of the numerator, 4, is larger than the degree of the denominator, 3, the rational function H is improper. To find any horizontal or oblique asymptotes, we use long division.

$$
\begin{array}{r}
3x + 3 \\
x^3 - x^2 + 1 \overline{\smash{\big)}\, 3x^4 \quad\quad - x^2} \\
\underline{3x^4 - 3x^3 \quad\quad + 3x} \\
3x^3 - x^2 - 3x \\
\underline{3x^3 - 3x^2 \quad + 3} \\
2x^2 - 3x - 3
\end{array}
$$

As a result,

$$H(x) = \frac{3x^4 - x^2}{x^3 - x^2 + 1} = 3x + 3 + \frac{2x^2 - 3x - 3}{x^3 - x^2 + 1}$$

As $x \to -\infty$ or as $x \to \infty$,

$$\frac{2x^2 - 3x - 3}{x^3 - x^2 + 1} \approx \frac{2x^2}{x^3} = \frac{2}{x} \to 0$$

As $x \to -\infty$ or as $x \to \infty$, we have $H(x) \to 3x + 3$. We conclude that the graph of the rational function H has an oblique asymptote $y = 3x + 3$.

We verify these results in Tables 15(a) and (b) with $Y_1 = H(x)$ and $Y_2 = 3x + 3$. As $x \to -\infty$ or $x \to \infty$, the difference in the values between Y_1 and Y_2 is indistinguishable.

Table 15

X	Y1	Y2
-10	-27.21	-27
-100	-297	-297
-1000	-2997	-2997
-10000	-29997	-29997
-1E5	-3E5	-3E5
-1E6	-3E6	-3E6
-1E7	-3E7	-3E7

Y₁目(3X^4-X²)/(X...

(a)

X	Y1	Y2
10	33.185	33
100	303.02	303
1000	3003	3003
10000	30003	30003
100000	300003	300003
1E6	3E6	3E6
1E7	3E7	3E7

Y₁目(3X^4-X²)/(X...

(b)

◀

| EXAMPLE 7 | **Finding Horizontal or Oblique Asymptotes** |

Find the horizontal or oblique asymptotes, if any, of the graph of

$$R(x) = \frac{8x^2 - x + 2}{4x^2 - 1}$$

Solution   Since the degree of the numerator, 2, equals the degree of the denominator, 2, the rational function R is improper. To find any horizontal or oblique asymptotes, we use long division.

$$
\begin{array}{r}
2 \\
4x^2 - 1 \overline{)8x^2 - x + 2} \\
\underline{8x^2 - 2} \\
- x + 4
\end{array}
$$

As a result,

$$R(x) = \frac{8x^2 - x + 2}{4x^2 - 1} = 2 + \frac{-x + 4}{4x^2 - 1}$$

Then, as $x \rightarrow -\infty$ or as $x \rightarrow \infty$,

$$\frac{-x + 4}{4x^2 - 1} \approx \frac{-x}{4x^2} = \frac{-1}{4x} \rightarrow 0$$

As $x \rightarrow -\infty$ or as $x \rightarrow \infty$, we have $R(x) \rightarrow 2$. We conclude that $y = 2$ is a horizontal asymptote of the graph.

✔ **CHECK:** Verify the results of Example 7 by creating a TABLE with $Y_1 = R(x)$ and $Y_2 = 2$.   ◀

In Example 7, we note that the quotient 2 obtained by long division is the quotient of the leading coefficients of the numerator polynomial and the denominator polynomial $\left(\dfrac{8}{4}\right)$. This means that we can avoid the long division process for rational functions whose numerator and denominator *are of the same degree* and conclude that the quotient of the leading coefficients will give us the horizontal asymptote.

➜ **NOW WORK PROBLEM 41.**

| EXAMPLE 8 | **Finding Horizontal or Oblique Asymptotes** |

Find the horizontal or oblique asymptotes, if any, of the graph of

$$G(x) = \frac{2x^5 - x^3 + 2}{x^3 - 1}$$

Solution   Since the degree of the numerator, 5, is larger than the degree of the denominator, 3, the rational function G is improper. To find any horizontal or oblique asymptotes, we use long division.

$$
\begin{array}{r}
2x^2 - 1 \\
x^3 - 1 \overline{)2x^5 - x^3 + 2} \\
\underline{2x^5 - 2x^2} \\
-x^3 + 2x^2 + 2 \\
\underline{-x^3 + 1} \\
2x^2 + 1
\end{array}
$$

As a result,

$$G(x) = \frac{2x^5 - x^3 + 2}{x^3 - 1} = 2x^2 - 1 + \frac{2x^2 + 1}{x^3 - 1}$$

Then, as $x \to -\infty$ or as $x \to \infty$,

$$\frac{2x^2 + 1}{x^3 - 1} \approx \frac{2x^2}{x^3} = \frac{2}{x} \to 0$$

As $x \to -\infty$ or as $x \to \infty$, we have $G(x) \to 2x^2 - 1$. We conclude that, for large values of $|x|$, the graph of G approaches the graph of $y = 2x^2 - 1$. That is, the graph of G will look like the graph of $y = 2x^2 - 1$ as $x \to -\infty$ or $x \to \infty$. Since $y = 2x^2 - 1$ is not a linear function, G has no horizontal or oblique asymptotes. ◀

We now summarize the procedure for finding horizontal and oblique asymptotes.

Summary

Finding Horizontal and Oblique Asymptotes of a Rational Function R

Consider the rational function

$$R(x) = \frac{p(x)}{q(x)} = \frac{a_n x^n + a_{n-1} x^{n-1} + \cdots + a_1 x + a_0}{b_m x^m + b_{m-1} x^{m-1} + \cdots + b_1 x + b_0}$$

in which the degree of the numerator is n and the degree of the denominator is m.

1. If $n < m$, then R is a proper rational function, and the graph of R will have the horizontal asymptote $y = 0$ (the x-axis).

2. If $n \geq m$, then R is improper. Here long division is used.

   (a) If $n = m$, the quotient obtained will be a number $\dfrac{a_n}{b_m}$, and the line $y = \dfrac{a_n}{b_m}$ is a horizontal asymptote.

   (b) If $n = m + 1$, the quotient obtained is of the form $ax + b$ (a polynomial of degree 1), and the line $y = ax + b$ is an oblique asymptote.

   (c) If $n > m + 1$, the quotient obtained is a polynomial of degree 2 or higher, and R has neither a horizontal nor an oblique asymptote. In this case, for $|x|$ unbounded, the graph of R will behave like the graph of the quotient.

NOTE   The graph of a rational function either has one horizontal or one oblique asymptote or else has no horizontal and no oblique asymptote. ■

3.3 Assess Your Understanding

'Are You Prepared?'

Answers are given at the end of these exercises. If you get a wrong answer, read the pages listed in red.

1. *True or False:* The quotient of two polynomial expressions is a rational expression. (p. 679)

2. What is the quotient and remainder when $3x^3 - 6x^2 + 3x - 4$ is divided by $x^2 - 1$? (pp. 685–687)

3. Graph $y = \dfrac{1}{x}$. (p. 21)

4. Graph $y = 2(x + 1)^2 - 3$ using transformations. (pp. 118–126)

Concepts and Vocabulary

5. The line _____ is a horizontal asymptote of $R(x) = \dfrac{x^3 - 1}{x^3 + 1}$.

6. The line _____ is a vertical asymptote of $R(x) = \dfrac{x - 1}{x + 1}$.

7. For a rational function R, if the degree of the numerator is less than the degree of the denominator, then R is _____.

8. *True or False:* The domain of every rational function is the set of all real numbers.

9. *True or False:* If an asymptote is neither horizontal nor vertical, it is called oblique.

10. *True or False:* If the degree of the numerator of a rational function equals the degree of the denominator, then the ratio of the leading coefficients gives rise to the horizontal asymptote.

Skill Building

In Problems 11–22, find the domain of each rational function.

11. $R(x) = \dfrac{4x}{x - 3}$

12. $R(x) = \dfrac{5x^2}{3 + x}$

13. $H(x) = \dfrac{-4x^2}{(x - 2)(x + 4)}$

14. $G(x) = \dfrac{6}{(x + 3)(4 - x)}$

15. $F(x) = \dfrac{3x(x - 1)}{2x^2 - 5x - 3}$

16. $Q(x) = \dfrac{-x(1 - x)}{3x^2 + 5x - 2}$

17. $R(x) = \dfrac{x}{x^3 - 8}$

18. $R(x) = \dfrac{x}{x^4 - 1}$

19. $H(x) = \dfrac{3x^2 + x}{x^2 + 4}$

20. $G(x) = \dfrac{x - 3}{x^4 + 1}$

21. $R(x) = \dfrac{3(x^2 - x - 6)}{4(x^2 - 9)}$

22. $F(x) = \dfrac{-2(x^2 - 4)}{3(x^2 + 4x + 4)}$

In Problems 23–28, use the graph shown to find:
  (a) *The domain and range of each function*
  (b) *The intercepts, if any*
  (c) *Horizontal asymptotes, if any*
  (d) *Vertical asymptotes, if any*
  (e) *Oblique asymptotes, if any*

23.

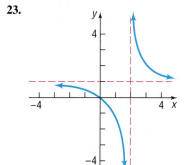

24.

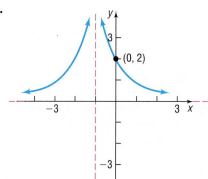

25.

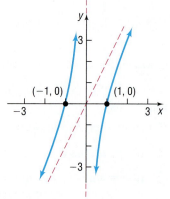

26.

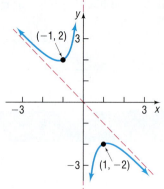

27.

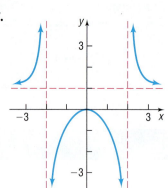

28.

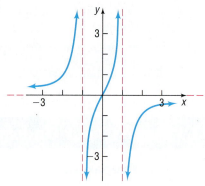

In Problems 29–40, graph each rational function using transformations. Verify your result using a graphing utility.

29. $F(x) = 2 + \dfrac{1}{x}$

30. $Q(x) = 3 + \dfrac{1}{x^2}$

31. $R(x) = \dfrac{1}{(x - 1)^2}$

32. $R(x) = \dfrac{3}{x}$

33. $H(x) = \dfrac{-2}{x + 1}$

34. $G(x) = \dfrac{2}{(x + 2)^2}$

35. $R(x) = \dfrac{-1}{x^2 + 4x + 4}$

36. $R(x) = \dfrac{1}{x - 1} + 1$

37. $G(x) = 1 + \dfrac{2}{(x - 3)^2}$

38. $F(x) = 2 - \dfrac{1}{x + 1}$

39. $R(x) = \dfrac{x^2 - 4}{x^2}$

40. $R(x) = \dfrac{x - 4}{x}$

In Problems 41–52, find the vertical, horizontal, and oblique asymptotes, if any, of each rational function.

41. $R(x) = \dfrac{3x}{x + 4}$

42. $R(x) = \dfrac{3x + 5}{x - 6}$

43. $H(x) = \dfrac{x^3 - 8}{x^2 - 5x + 6}$

44. $G(x) = \dfrac{-x^2 + 1}{x^2 - 5x + 6}$

45. $T(x) = \dfrac{x^3}{x^4 - 1}$

46. $P(x) = \dfrac{4x^5}{x^3 - 1}$

47. $Q(x) = \dfrac{5 - x^2}{3x^4}$

48. $F(x) = \dfrac{-2x^2 + 1}{2x^3 + 4x^2}$

49. $R(x) = \dfrac{3x^4 + 4}{x^3 + 3x}$

50. $R(x) = \dfrac{6x^2 + x + 12}{3x^2 - 5x - 2}$

51. $G(x) = \dfrac{x^3 - 1}{x - x^2}$

52. $F(x) = \dfrac{x - 1}{x - x^3}$

Applications and Extensions

53. Gravity   In physics, it is established that the acceleration due to gravity, g (in meters/sec^2), at a height h meters above sea level is given by

$$g(h) = \frac{3.99 \times 10^{14}}{(6.374 \times 10^6 + h)^2}$$

where 6.374×10^6 is the radius of Earth in meters.
(a) What is the acceleration due to gravity at sea level?
(b) The Sears Tower in Chicago, Illinois, is 443 meters tall. What is the acceleration due to gravity at the top of the Sears Tower?
(c) The peak of Mount Everest is 8848 meters above sea level. What is the acceleration due to gravity on the peak of Mount Everest?
(d) Find the horizontal asymptote of $g(h)$.
(e) Solve $g(h) = 0$. How do you interpret your answer?

54. Population Model   A rare species of insect was discovered in the Amazon Rain Forest. To protect the species, environmentalists declare the insect endangered and transplant the insect into a protected area. The population P of the insect t months after being transplanted is given below.

$$P(t) = \frac{50(1 + 0.5t)}{(2 + 0.01t)}$$

(a) How many insects were discovered? In other words, what was the population when $t = 0$?
(b) What will the population be after 5 years?
(c) Determine the horizontal asymptote of $P(t)$. What is the largest population that the protected area can sustain?

Discussion and Writing

55. If the graph of a rational function R has the vertical asymptote $x = 4$, then the factor $x - 4$ must be present in the denominator of R. Explain why.

56. If the graph of a rational function R has the horizontal asymptote $y = 2$, then the degree of the numerator of R equals the degree of the denominator of R. Explain why.

57. Can the graph of a rational function have both a horizontal and an oblique asymptote? Explain.

58. Make up a rational function that has $y = 2x + 1$ as an oblique asymptote. Explain the methodology that you used.

'Are You Prepared?' Answers

1. True   **2.** Quotient: $3x - 6$; Remainder: $6x - 10$ **3.**

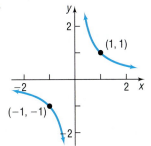

4.

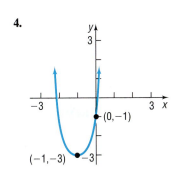

3.4 The Graph of a Rational Function; Inverse and Joint Variation

PREPARING FOR THIS SECTION *Before getting started, review the following:*

• Intercepts (Section 1.2, pp. 15–17)   • Even and Odd Functions (Section 2.3, pp. 80–82)

 Now work the 'Are You Prepared?' problems on page 207.

OBJECTIVES 1 Analyze the Graph of a Rational Function
2 Solve Applied Problems Involving Rational Functions
3 Construct a Model Using Inverse Variation
4 Construct a Model Using Joint or Combined Variation

 Analyze the Graph of a Rational Function

Graphing utilities make the task of graphing rational functions less time consuming. However, the results of algebraic analysis must be taken into account before drawing conclusions based on the graph provided by the utility. We will use the information collected in the last section in conjunction with the graphing utility to analyze the graph of a rational function $R(x) = \dfrac{p(x)}{q(x)}$. The analysis will require the following steps:

Analyzing the Graph of a Rational Function

STEP 1: Find the domain of the rational function.

STEP 2: Write R in lowest terms.

STEP 3: Locate the intercepts of the graph. The x-intercepts, if any, of $R(x) = \dfrac{p(x)}{q(x)}$ in lowest terms, satisfy the equation $p(x) = 0$. The y-intercept, if there is one, is $R(0)$.

STEP 4: Test for symmetry. Replace x by $-x$ in $R(x)$. If $R(-x) = R(x)$, there is symmetry with respect to the y-axis; if $R(-x) = -R(x)$, there is symmetry with respect to the origin.

STEP 5: Locate the vertical asymptotes. The vertical asymptotes, if any, of $R(x) = \dfrac{p(x)}{q(x)}$ in lowest terms are found by identifying the real zeros of $q(x)$. Each zero of the denominator gives rise to a vertical asymptote.

STEP 6: Locate the horizontal or oblique asymptotes, if any, using the procedure given in Section 3.3. Determine points, if any, at which the graph of R intersects these asymptotes.

STEP 7: Graph R using a graphing utility.

STEP 8: Use the results obtained in Steps 1 through 7 to graph R by hand.

EXAMPLE 1 **Analyzing the Graph of a Rational Function**

Analyze the graph of the rational function: $R(x) = \dfrac{x - 1}{x^2 - 4}$

Solution   First, we factor both the numerator and the denominator of R.

$$R(x) = \frac{x - 1}{(x + 2)(x - 2)}$$

STEP 1: The domain of R is $\{x \mid x \neq -2, x \neq 2\}$.

STEP 2: R is in lowest terms.

STEP 3: We locate the x-intercepts by finding the zeros of the numerator. By inspection, 1 is the only x-intercept. The y-intercept is $R(0) = \dfrac{1}{4}$.

STEP 4: Because

$$R(-x) = \frac{-x - 1}{x^2 - 4} = \frac{-(x + 1)}{x^2 - 4}$$

we conclude that R is neither even nor odd. There is no symmetry with respect to the y-axis or the origin.

STEP 5: We locate the vertical asymptotes by finding the zeros of the denominator. Since R is in lowest terms, the graph of R has two vertical asymptotes: the lines $x = -2$ and $x = 2$.

STEP 6: The degree of the numerator is less than the degree of the denominator, so R is proper and the line $y = 0$ (the x-axis) is a horizontal asymptote of the graph. To determine if the graph of R intersects the horizontal asymptote, we solve the equation $R(x) = 0$:

$$\frac{x - 1}{x^2 - 4} = 0$$

$$x - 1 = 0$$

$$x = 1$$

The only solution is $x = 1$, so the graph of R intersects the horizontal asymptote at $(1, 0)$.

STEP 7: The analysis just completed helps us to set the viewing window to obtain a complete graph. Figure 45(a) shows the graph of $R(x) = \dfrac{x - 1}{x^2 - 4}$ in connected mode, and Figure 45(b) shows it in dot mode. Notice in Figure 45(a) that the graph has vertical lines at $x = -2$ and $x = 2$. This is due to the fact that, when the graphing utility is in connected mode, it will "connect the dots" between consecutive pixels. We know that the graph of R does not cross the lines $x = -2$ and $x = 2$, since R is not defined at $x = -2$ or $x = 2$. When graphing rational functions, dot mode should be used if extraneous vertical lines appear. You should confirm that all the algebraic conclusions that we arrived at in Steps 1 through 6 are part of the graph. For example, the graph has vertical asymptotes at $x = -2$ and $x = 2$, and the graph has a horizontal asymptote at $y = 0$. The y-intercept is $\dfrac{1}{4}$ and the x-intercept is 1.

STEP 8: Using the information gathered in Steps 1 through 7, we obtain the graph of R shown in Figure 46. ◄

Figure 45

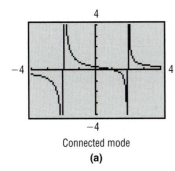

−4 ⎸ 4

Connected mode
(a)

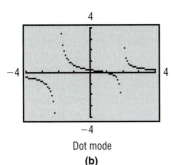

−4 ⎸ 4

Dot mode
(b)

Figure 46

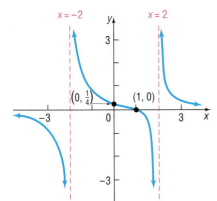

NOW WORK PROBLEM 7.

| EXAMPLE 2 | **Analyzing the Graph of a Rational Function** |

Analyze the graph of the rational function: $R(x) = \dfrac{x^2 - 1}{x}$

Solution

STEP 1: The domain of R is $\{x \mid x \neq 0\}$.

STEP 2: R is in lowest terms.

STEP 3: The graph has two x-intercepts: -1 and 1. There is no y-intercept, since x cannot equal 0.

STEP 4: Since $R(-x) = -R(x)$, the function is odd and the graph is symmetric with respect to the origin.

STEP 5: Since R is in lowest terms, the graph of $R(x)$ has the line $x = 0$ (the y-axis) as a vertical asymptote.

STEP 6: Since the degree of the numerator, 2, is larger than the degree of the denominator, 1, the rational function R is improper. To find any horizontal or oblique asymptotes, we use long division.

$$
\begin{array}{r}
x \\
x{\overline{\smash{\big)}\,x^2 - 1}} \\
\underline{x^2 } \\
-1
\end{array}
$$

NOTE

Because the determinator of the rational function is a monomial, we can also find the oblique asymptote as follows:

$$\frac{x^2 - 1}{x} = \frac{x^2}{x} - \frac{1}{x} = x - \frac{1}{x}$$

Since $\dfrac{1}{x} \to 0$ as $x \to \infty$, $y = x$ is an oblique asymptote. ∎

The quotient is x, so the line $y = x$ is an oblique asymptote of the graph. To determine whether the graph of R intersects the asymptote $y = x$, we solve the equation $R(x) = x$.

$$
\begin{aligned}
R(x) = \frac{x^2 - 1}{x} &= x \\
x^2 - 1 &= x^2 \\
-1 &= 0 \qquad \text{\color{orange}Impossible.}
\end{aligned}
$$

We conclude that the equation $\dfrac{x^2 - 1}{x} = x$ has no solution, so the graph of $R(x)$ does not intersect the line $y = x$.

STEP 7: See Figure 47. We see from the graph that there is no y-intercept and there are two x-intercepts, -1 and 1. The symmetry with respect to the origin is also evident. We can also see that there is a vertical asymptote at $x = 0$. Finally, it is not necessary to graph this function in dot mode since no extraneous vertical lines are present.

STEP 8: Using the information gathered in Steps 1 through 7, we obtain the graph of R shown in Figure 48. Notice how the oblique asymptote is used as a guide in graphing the rational function by hand.

Figure 47

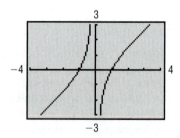

Figure 48

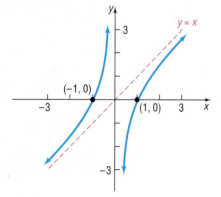

NOW WORK PROBLEM 15.

EXAMPLE 3   **Analyzing the Graph of a Rational Function**

Analyze the graph of the rational function: $R(x) = \dfrac{3x^2 - 3x}{x^2 + x - 12}$

Solution   We factor R to get

$$R(x) = \frac{3x(x - 1)}{(x + 4)(x - 3)}$$

STEP 1: The domain of R is $\{x \mid x \neq -4, x \neq 3\}$.

STEP 2: R is in lowest terms.

STEP 3: The graph has two x-intercepts: 0 and 1. The y-intercept is $R(0) = 0$.

STEP 4: Because

$$R(-x) = \frac{-3x(-x - 1)}{(-x + 4)(-x - 3)} = \frac{3x(x + 1)}{(x - 4)(x + 3)}$$

we conclude that R is neither even nor odd. There is no symmetry with respect to the y-axis or the origin.

STEP 5: Since R is in lowest terms, the graph of R has two vertical asymptotes: $x = -4$ and $x = 3$.

STEP 6: Since the degree of the numerator equals the degree of the denominator, the graph has a horizontal asymptote. To find it, we form the quotient of the leading coefficient of the numerator, 3, and the leading coefficient of the denominator, 1. The graph of R has the horizontal asymptote $y = 3$. To find out whether the graph of R intersects the asymptote, we solve the equation $R(x) = 3$.

$$R(x) = \frac{3x^2 - 3x}{x^2 + x - 12} = 3$$
$$3x^2 - 3x = 3x^2 + 3x - 36$$
$$-6x = -36$$
$$x = 6$$

The graph intersects the line $y = 3$ at $x = 6$, and $(6, 3)$ is a point on the graph of R.

STEP 7: Figure 49(a) shows the graph of R in connected mode. Notice the extraneous vertical lines at $x = -4$ and $x = 3$ (the vertical asymptotes). As a result, we also graph R in dot mode. See Figure 49(b).

STEP 8: Figure 49 does not display the graph between the two x-intercepts, 0 and 1. However, because the zeros in the numerator, 0 and 1, are of odd multiplicity (both are multiplicity 1), we know that the graph of R crosses the x-axis at 0 and 1. Therefore, the graph of R is above the x-axis for $0 < x < 1$. To see this part better, we graph R for $-1 \le x \le 2$ in Figure 50. Using MAXIMUM, we approximate the turning point to be $(0.52, 0.07)$, rounded to two decimal places.

Figure 49

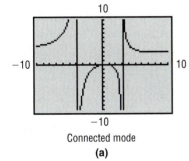

Connected mode
(a)

Dot mode
(b)

Figure 50

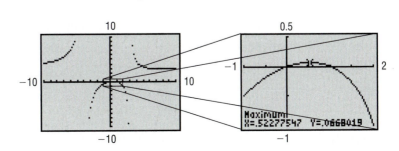

Figure 49 also does not display the graph of R crossing the horizontal asymptote at $(6, 3)$. To see this part better, we graph R for $4 \le x \le 60$ in Figure 51. Using MINIMUM, we approximate the turning point to be $(11.48, 2.75)$, rounded to two decimal places.

Figure 51

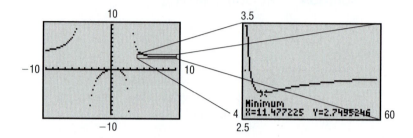

Using this information along with the information gathered in Steps 1 through 7, we obtain the graph of R shown in Figure 52.

Figure 52

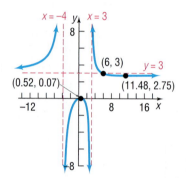

EXAMPLE 4

Analyzing the Graph of a Rational Function with a Hole

Analyze the graph of the rational function: $R(x) = \dfrac{2x^2 - 5x + 2}{x^2 - 4}$

Solution   We factor R and obtain

$$R(x) = \frac{(2x - 1)(x - 2)}{(x + 2)(x - 2)}$$

STEP 1: The domain of R is $\{x \mid x \ne -2, x \ne 2\}$.

STEP 2: In lowest terms,

$$R(x) = \frac{2x - 1}{x + 2}, \qquad x \ne 2$$

STEP 3: The graph has one x-intercept: 0.5. The y-intercept is $R(0) = -0.5$.

STEP 4: Because

$$R(-x) = \frac{2x^2 + 5x + 2}{x^2 - 4}$$

we conclude that R is neither even nor odd. There is no symmetry with respect to the y-axis or the origin.

STEP 5: Since $x + 2$ is the only factor of the denominator of $R(x)$ *in lowest terms* the graph has one vertical asymptote, $x = -2$. However, the rational function is undefined at both $x = 2$ and $x = -2$.

STEP 6: Since the degree of the numerator equals the degree of the denominator, the graph has a horizontal asymptote. To find it, we form the quotient of the leading coefficient of the numerator, 2, and the leading coefficient of the denominator, 1. The graph of R has the horizontal asymptote $y = 2$. To find whether the graph of R intersects the asymptote, we solve the equation $R(x) = 2$.

$$R(x) = \frac{2x - 1}{x + 2} = 2$$

$$2x - 1 = 2(x + 2)$$

$$2x - 1 = 2x + 4$$

$$-1 = 4 \qquad \textcolor{teal}{\textit{Impossible}}$$

The graph does not intersect the line $y = 2$.

Figure 53

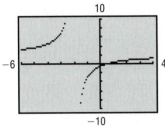

STEP 7: Figure 53 shows the graph of $R(x)$. Notice that the graph has one vertical asymptote at $x = -2$. Also, the function appears to be continuous at $x = 2$.

STEP 8: The analysis presented thus far does not explain the behavior of the graph at $x = 2$. We use the TABLE feature of our graphing utility to determine the behavior of the graph of R as x approaches 2. See Table 16. From the table, we conclude that the value of R approaches 0.75 as x approaches 2. This result is further verified by evaluating R in lowest terms at $x = 2$. We conclude that there is a hole in the graph at $(2, 0.75)$. Using the information gathered in Steps 1 through 7, we obtain the graph of R shown in Figure 54.

Table 16

X	Y1
1.99	.74687
1.999	.74969
1.9999	.74997
2	ERROR
2.0001	.75003
2.001	.75031
2.01	.75312

Y1 ◨ (2X²−5X+2)⁄(...

Figure 54

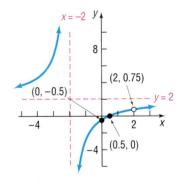

As Example 4 shows, the values excluded from the domain of a rational function give rise to either vertical asymptotes or holes.

NOW WORK PROBLEM 33.

| **EXAMPLE 5** | **Constructing a Rational Function from Its Graph** |

Make up a rational function that might have the graph shown in Figure 55.

Figure 55

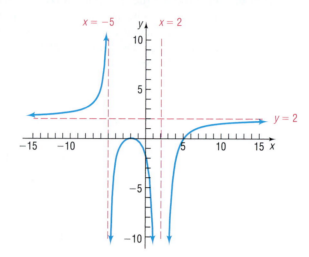

Solution The numerator of a rational function $R(x) = \dfrac{p(x)}{q(x)}$ in lowest terms determines the x-intercepts of its graph. The graph shown in Figure 55 has x-intercepts -2 (even multiplicity; graph touches the x-axis) and 5 (odd multiplicity; graph crosses the x-axis). So one possibility for the numerator is $p(x) = (x + 2)^2(x - 5)$.

The denominator of a rational function in lowest terms determines the vertical asymptotes of its graph. The vertical asymptotes of the graph are $x = -5$ and $x = 2$. Since $R(x)$ approaches ∞ from the left of $x = -5$ and $R(x)$ approaches $-\infty$ from the right of $x = -5$, we know that $(x + 5)$ is a factor of odd multiplicity in $q(x)$. Also, $R(x)$ approaches $-\infty$ from both sides of $x = 2$, so $(x - 2)$ is a factor of even multiplicity in $q(x)$. A possibility for the denominator is $q(x) = (x + 5)(x - 2)^2$.

Figure 56

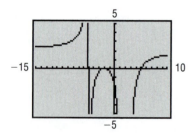

So far we have $R(x) = \dfrac{(x + 2)^2(x - 5)}{(x + 5)(x - 2)^2}$. However, the horizontal asymptote of the graph given in Figure 55, is $y = 2$, so we know that the degree of the numerator must equal the degree in the denominator and the quotient of leading coefficients must be $\dfrac{2}{1}$. This leads to $R(x) = \dfrac{2(x + 2)^2(x - 5)}{(x + 5)(x - 2)^2}$. Figure 56 shows the graph of R drawn on a graphing utility. Since Figure 56 looks similar to Figure 55, we have found a rational function R for the graph in Figure 55. ◄

NOW WORK PROBLEM 51.

2 Solve Applied Problems Involving Rational Functions

| **EXAMPLE 6** | **Finding the Least Cost of a Can** |

Reynolds Metal Company manufactures aluminum cans in the shape of a cylinder with a capacity of 500 cubic centimeters $\left(\dfrac{1}{2} \text{ liter}\right)$. The top and bottom of the can are made of a special aluminum alloy that costs 0.05¢ per square centimeter. The sides of the can are made of material that costs 0.02¢ per square centimeter.

(a) Express the cost of material for the can as a function of the radius r of the can.

(b) Use a graphing utility to graph the function $C = C(r)$.

(c) What value of r will result in the least cost?

(d) What is this least cost?

Figure 57

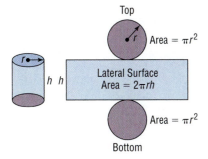

Solution

(a) Figure 57 illustrates the situation. Notice that the material required to produce a cylindrical can of height h and radius r consists of a rectangle of area $2\pi rh$ and two circles, each of area πr^2. The total cost C (in cents) of manufacturing the can is therefore

$$C = \text{Cost of top and bottom} + \text{Cost of side}$$
$$= \underbrace{(2\pi r^2 \text{ cm}^2)}_{\substack{\text{Total area} \\ \text{of top and} \\ \text{bottom}}} \underbrace{(0.05\cancel{c}/\text{cm}^2)}_{\text{Cost/unit area}} + \underbrace{(2\pi rh \text{ cm}^2)}_{\substack{\text{Total area} \\ \text{of side}}} \underbrace{(0.02\cancel{c}/\text{cm}^2)}_{\text{Cost/unit area}}$$
$$= 0.10\pi r^2 + 0.04\pi rh$$

But we have the additional restriction that the height h and radius r must be chosen so that the volume V of the can is 500 cubic centimeters. Since $V = \pi r^2 h$, we have

$$500 = \pi r^2 h \quad \text{or} \quad h = \frac{500}{\pi r^2}$$

Substituting this expression for h, the cost C, in cents, as a function of the radius r, is

Figure 58

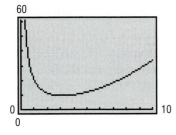

$$C(r) = 0.10\pi r^2 + 0.04\pi r \frac{500}{\pi r^2} = 0.10\pi r^2 + \frac{20}{r} = \frac{0.10\pi r^3 + 20}{r}$$

(b) See Figure 58 for the graph of $C(r)$.

(c) Using the MINIMUM command, the cost is least for a radius of about 3.17 centimeters.

(d) The least cost is $C(3.17) \approx 9.47\cancel{c}$. ◀

NOW WORK PROBLEM **61**.

Figure 59

$y = \dfrac{k}{x}, k > 0, x > 0$

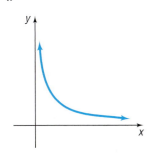

③ Construct a Model Using Inverse Variation

In Section 2.4, we discussed direct variation. We now study **inverse variation**.

> Let x and y denote two quantities. Then y **varies inversely** with x, or y is **inversely proportional** to x, if there is a nonzero constant k such that
>
> $$y = \frac{k}{x}$$

The graph in Figure 59 illustrates the relationship between y and x if y varies inversely with x and $k > 0$, $x > 0$.

| EXAMPLE 7 | **Maximum Weight That Can Be Supported by a Piece of Pine** |

Figure 60

See Figure 60. The maximum weight W that can be safely supported by a 2-inch by 4-inch piece of pine varies inversely with its length l. Experiments indicate that the maximum weight that a 10-foot-long pine 2-by-4 can support is 500 pounds. Write a general formula relating the maximum weight W (in pounds) to the length l (in feet). Find the maximum weight W that can be safely supported by a length of 25 feet.

Solution   Because W varies inversely with l, we know that

$$W = \frac{k}{l}$$

for some constant k. Because $W = 500$ when $l = 10$, we have

$$500 = \frac{k}{10}$$
$$k = 5000$$

Thus, in all cases,

$$W(l) = \frac{5000}{l}$$

Figure 61

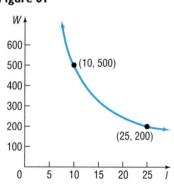

(10, 500)

(25, 200)

In particular, the maximum weight W that can be safely supported by a piece of pine 25 feet in length is

$$W(25) = \frac{5000}{25} = 200 \text{ pounds}$$

Figure 61 illustrates the relationship between the weight W and the length l. ◀

In direct or inverse variation, the quantities that vary may be raised to powers. For example, in the early seventeenth century, Johannes Kepler (1571–1630) discovered that the square of the period of revolution T of a planet around the Sun varies directly with the cube of its mean distance a from the Sun. That is, $T^2 = ka^3$, where k is the constant of proportionality.

 NOW WORK PROBLEM 63.

4   Construct a Model Using Joint or Combined Variation

When a variable quantity Q is proportional to the product of two or more other variables, we say that Q **varies jointly** with these quantities. Finally, combinations of direct and/or inverse variation may occur. This is usually referred to as **combined variation**.

Let's look at an example.

| EXAMPLE 8 | **Loss of Heat Through a Wall** |

The loss of heat through a wall varies jointly with the area of the wall and the difference between the inside and outside temperatures and varies inversely with the thickness of the wall. Write an equation that relates these quantities.

Solution   We begin by assigning symbols to represent the quantities:

$$L = \text{Heat loss} \qquad T = \text{Temperature difference}$$
$$A = \text{Area of wall} \qquad d = \text{Thickness of wall}$$

Then

$$L = k\frac{AT}{d}$$

where k is the constant of proportionality.   ◀

NOW WORK PROBLEM 69.

| **EXAMPLE 9** | **Force of the Wind on a Window** |

Figure 62

See Figure 62. The force F of the wind on a flat surface positioned at a right angle to the direction of the wind varies jointly with the area A of the surface and the square of the speed v of the wind. A wind of 30 miles per hour blowing on a window measuring 4 feet by 5 feet has a force of 150 pounds. What is the force on a window measuring 3 feet by 4 feet caused by a wind of 50 miles per hour?

Solution   Since F varies jointly with A and v^2, we have

$$F = kAv^2$$

where k is the constant of proportionality. We are told that $F = 150$ when $A = 4 \cdot 5 = 20$ and $v = 30$. Then we have

$$150 = k(20)(900) \qquad \textcolor{teal}{F = kAv^2, F = 150, A = 20, v = 30}$$
$$k = \frac{1}{120}$$

The general formula is therefore

$$F = \frac{1}{120}Av^2$$

For a wind of 50 miles per hour blowing on a window whose area is $A = 3 \cdot 4 = 12$ square feet, the force F is

$$F = \frac{1}{120}(12)(2500) = 250 \text{ pounds}$$ ◀

3.4 Assess Your Understanding

'Are You Prepared?'

Answers are given at the end of these exercises. If you get a wrong answer, read the pages listed in red.

1. Find the intercepts of $-3x + 5y = 30$. (pp. 15–17)

2. Determine whether $f(x) = \dfrac{x^2 + 1}{x^4 - 1}$ is even, odd, or neither. Is the graph of f symmetric with respect to the y-axis or origin? (pp. 80–82)

Concepts and Vocabulary

3. If the numerator and the denominator of a rational function have no common factors, the rational function is _____.

4. *True or False:* The graph of a polynomial function sometimes has a hole.

5. *True or False:* The graph of a rational function never intersects a horizontal asymptote.

6. *True or False:* The graph of a rational function sometimes has a hole.

Skill Building

In Problems 7–44, follow Steps 1 through 8 on page 198 to analyze the graph of each function.

7. $R(x) = \dfrac{x+1}{x(x+4)}$

8. $R(x) = \dfrac{x}{(x-1)(x+2)}$

9. $R(x) = \dfrac{3x+3}{2x+4}$

10. $R(x) = \dfrac{2x+4}{x-1}$

11. $R(x) = \dfrac{3}{x^2-4}$

12. $R(x) = \dfrac{6}{x^2-x-6}$

13. $P(x) = \dfrac{x^4+x^2+1}{x^2-1}$

14. $Q(x) = \dfrac{x^4-1}{x^2-4}$

15. $H(x) = \dfrac{x^3-1}{x^2-9}$

16. $G(x) = \dfrac{x^3+1}{x^2+2x}$

17. $R(x) = \dfrac{x^2}{x^2+x-6}$

18. $R(x) = \dfrac{x^2+x-12}{x^2-4}$

19. $G(x) = \dfrac{x}{x^2-4}$

20. $G(x) = \dfrac{3x}{x^2-1}$

21. $R(x) = \dfrac{3}{(x-1)(x^2-4)}$

22. $R(x) = \dfrac{-4}{(x+1)(x^2-9)}$

23. $H(x) = \dfrac{4(x^2-1)}{x^4-16}$

24. $H(x) = \dfrac{x^2+4}{x^4-1}$

25. $F(x) = \dfrac{x^2-3x-4}{x+2}$

26. $F(x) = \dfrac{x^2+3x+2}{x-1}$

27. $R(x) = \dfrac{x^2+x-12}{x-4}$

28. $R(x) = \dfrac{x^2-x-12}{x+5}$

29. $F(x) = \dfrac{x^2+x-12}{x+2}$

30. $G(x) = \dfrac{x^2-x-12}{x+1}$

31. $R(x) = \dfrac{x(x-1)^2}{(x+3)^3}$

32. $R(x) = \dfrac{(x-1)(x+2)(x-3)}{x(x-4)^2}$

33. $R(x) = \dfrac{x^2+x-12}{x^2-x-6}$

34. $R(x) = \dfrac{x^2+3x-10}{x^2+8x+15}$

35. $R(x) = \dfrac{6x^2-7x-3}{2x^2-7x+6}$

36. $R(x) = \dfrac{8x^2+26x+15}{2x^2-x-15}$

37. $R(x) = \dfrac{x^2+5x+6}{x+3}$

38. $R(x) = \dfrac{x^2+x-30}{x+6}$

39. $f(x) = x + \dfrac{1}{x}$

40. $f(x) = 2x + \dfrac{9}{x}$

41. $f(x) = x^2 + \dfrac{1}{x}$

42. $f(x) = 2x^2 + \dfrac{9}{x}$

43. $f(x) = x + \dfrac{1}{x^3}$

44. $f(x) = 2x + \dfrac{9}{x^3}$

In Problems 45–50, graph each function and use MINIMUM to obtain the minimum value, rounded to two decimal places.

45. $f(x) = x + \dfrac{1}{x}, \quad x > 0$

46. $f(x) = 2x + \dfrac{9}{x}, \quad x > 0$

47. $f(x) = x^2 + \dfrac{1}{x}, \quad x > 0$

48. $f(x) = 2x^2 + \dfrac{9}{x}, \quad x > 0$

49. $f(x) = x + \dfrac{1}{x^3}, \quad x > 0$

50. $f(x) = 2x + \dfrac{9}{x^3}, \quad x > 0$

In Problems 51–54, find a rational function that might have this graph. (More than one answer might be possible.)

51.

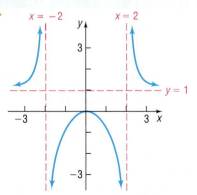

52.

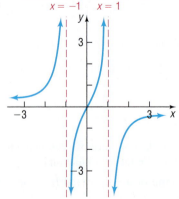

53.

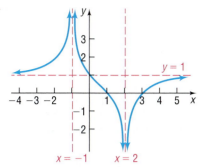

54.

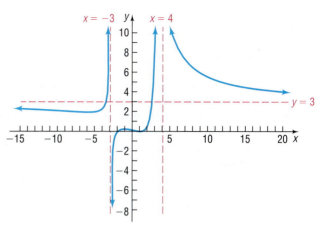

Applications and Extensions

55. Drug Concentration   The concentration C of a certain drug in a patient's bloodstream t hours after injection is given by

$$C(t) = \frac{t}{2t^2 + 1}$$

(a) Find the horizontal asymptote of $C(t)$. What happens to the concentration of the drug as t increases?
(b) Using a graphing utility, graph $C = C(t)$.
(c) Determine the time at which the concentration is highest.

56. Drug Concentration   The concentration C of a certain drug in a patient's bloodstream t minutes after injection is given by

$$C(t) = \frac{50t}{t^2 + 25}$$

(a) Find the horizontal asymptote of $C(t)$. What happens to the concentration of the drug as t increases?
(b) Using a graphing utility, graph $C = C(t)$.
(c) Determine the time at which the concentration is highest.

57. Average Cost   In Problem 95, Exercise 3.2, the cost function C (in thousands of dollars) for manufacturing x Chevy Cavaliers was determined to be

$$C(x) = 0.2x^3 - 2.3x^2 + 14.3x + 10.2$$

Economists define the **average cost function** as

$$\overline{C}(x) = \frac{C(x)}{x}$$

(a) Find the average cost function.
(b) What is the average cost of producing six Cavaliers?
(c) What is the average cost of producing nine Cavaliers?
(d) Using a graphing utility, graph the average cost function.
(e) Using a graphing utility, find the number of Cavaliers that should be produced to minimize the average cost.
(f) What is the minimum average cost?

58. Average Cost   In Problem 96, Exercise 3.2, the cost function C (in thousands of dollars) for printing x textbooks (in thousands of units) was found to be

$$C(x) = 0.015x^3 - 0.595x^2 + 9.15x + 98.43$$

(a) Find the average cost function (refer to Problem 57).
(b) What is the average cost of printing 13,000 textbooks per week?
(c) What is the average cost of printing 25,000 textbooks per week?
(d) Using your graphing utility, graph the average cost function.
(e) Using your graphing utility, find the number of textbooks that should be printed to minimize the average cost.
(f) What is the minimum average cost?

59. Minimizing Surface Area   United Parcel Service has contracted you to design a closed box with a square base that has a volume of 10,000 cubic inches. See the illustration.

(a) Find a function for the surface area of the box.
(b) Using a graphing utility, graph the function found in part (a).
(c) What is the minimum amount of cardboard that can be used to construct the box?
(d) What are the dimensions of the box that minimize the surface area?
(e) Why might UPS be interested in designing a box that minimizes the surface area?

60. Minimizing Surface Area   United Parcel Service has contracted you to design a closed box with a square base that has a volume of 5000 cubic inches. See the illustration.

(a) Find a function for the surface area of the box.
(b) Using a graphing utility, graph the function found in part (a).
(c) What is the minimum amount of cardboard that can be used to construct the box?
(d) What are the dimensions of the box that minimize the surface area?
(e) Why might UPS be interested in designing a box that minimizes the surface area?

61. **Cost of a Can** A can in the shape of a right circular cylinder is required to have a volume of 500 cubic centimeters. The top and bottom are made of material that costs 6¢ per square centimeter, while the sides are made of material that costs 4¢ per square centimeter.
(a) Express the total cost C of the material as a function of the radius r of the cylinder. (Refer to Figure 57.)
(b) Using a graphing utility graph $C = C(r)$. For what value of r is the cost C a minimum?

62. **Material Needed to Make a Drum** A steel drum in the shape of a right circular cylinder is required to have a volume of 100 cubic feet.

(a) Express the amount A of material required to make the drum as a function of the radius r of the cylinder.
(b) How much material is required if the drum's radius is 3 feet?
(c) How much material is required if the drum's radius is 4 feet?
(d) How much material is required if the drum's radius is 5 feet?
(e) Using a graphing utility graph $A = A(r)$. For what value of r is A smallest?

63. **Demand** Suppose that the demand D for candy at the movie theater is inversely related to the price p.
(a) When the price of candy is $2.75 per bag, the theater sells 156 bags of candy on a typical Saturday. Express the demand for candy as a function of its price.
(b) Determine the number of bags of candy that will be sold if the price is raised to $3 a bag.

64. **Driving to School** The time t that it takes to get to school varies inversely with your average speed s.
(a) Suppose that it takes you 40 minutes to get to school when your average speed is 30 miles per hour. Express the driving time to school as a function of average speed.
(b) Suppose that your average speed to school is 40 miles per hour. How long will it take you to get to school?

65. **Pressure** The volume of a gas V held at a constant temperature in a closed container varies inversely with its pressure P. If the volume of a gas is 600 cubic centimeters (cm^3) when the pressure is 150 millimeters of mercury (mm Hg), find the volume when the pressure is 200 mm Hg.

66. **Resistance** The current i in a circuit is inversely proportional to its resistance R measured in ohms. Suppose that when the current in a circuit is 30 amperes the resistance is 8 ohms. Find the current in the same circuit when the resistance is 10 ohms.

67. **Weight** The weight of an object above the surface of Earth varies inversely with the square of the distance from the center of Earth. If Maria weighs 125 pounds when she is on the surface of Earth (3960 miles from the center), determine Maria's weight if she is at the top of Mount McKinley (3.8 miles from the surface of Earth).

68. **Intensity of Light** The intensity I of light (measured in foot-candles) varies inversely with the square of the distance from the bulb. Suppose that the intensity of a 100-watt light bulb at a distance of 2 meters is 0.075 foot-candle. Determine the intensity of the bulb at a distance of 5 meters.

69. **Geometry** The volume V of a right circular cylinder varies jointly with the square of its radius r and its height h. The constant of proportionality is π. (See the figure.) Write an equation for V.

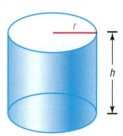

70. **Geometry** The volume V of a right circular cone varies jointly with the square of its radius r and its height h. The constant of proportionality is $\dfrac{\pi}{3}$. (See the figure.) Write an equation for V.

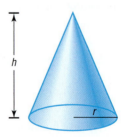

71. **Horsepower** The horsepower (hp) that a shaft can safely transmit varies jointly with its speed (in revolutions per minute, rpm) and the cube of its diameter. If a shaft of a certain material 2 inches in diameter can transmit 36 hp at 75 rpm, what diameter must the shaft have to transmit 45 hp at 125 rpm?

72. Chemistry: Gas Laws The volume V of an ideal gas varies directly with the temperature T and inversely with the pressure P. Write an equation relating V, T, and P using k as the constant of proportionality. If a cylinder contains oxygen at a temperature of 300 K and a pressure of 15 atmospheres in a volume of 100 liters, what is the constant of proportionality k? If a piston is lowered into the cylinder, decreasing the volume occupied by the gas to 80 liters and raising the temperature to 310 K, what is the gas pressure?

73. Physics: Kinetic Energy The kinetic energy K of a moving object varies jointly with its mass m and the square of its velocity v. If an object weighing 25 pounds and moving with a velocity of 100 feet per second has a kinetic energy of 400 foot-pounds, find its kinetic energy when the velocity is 150 feet per second.

74. Electrical Resistance of a Wire The electrical resistance of a wire varies directly with the length of the wire and inversely with the square of the diameter of the wire. If a wire 432 feet long and 4 millimeters in diameter has a resistance

of 1.24 ohms, find the length of a wire of the same material whose resistance is 1.44 ohms and whose diameter is 3 millimeters.

75. Measuring the Stress of Materials The stress in the material of a pipe subject to internal pressure varies jointly with the internal pressure and the internal diameter of the pipe and inversely with the thickness of the pipe. The stress is 100 pounds per square inch when the diameter is 5 inches, the thickness is 0.75 inch, and the internal pressure is 25 pounds per square inch. Find the stress when the internal pressure is 40 pounds per square inch if the diameter is 8 inches and the thickness is 0.50 inch.

76. Safe Load for a Beam The maximum safe load for a horizontal rectangular beam varies jointly with the width of the beam and the square of the thickness of the beam and inversely with its length. If an 8-foot beam will support up to 750 pounds when the beam is 4 inches wide and 2 inches thick, what is the maximum safe load in a similar beam 10 feet long, 6 inches wide, and 2 inches thick?

Discussion and Writing

77. Graph each of the following functions:

$$y = \frac{x^2 - 1}{x - 1} \qquad y = \frac{x^3 - 1}{x - 1}$$

$$y = \frac{x^4 - 1}{x - 1} \qquad y = \frac{x^5 - 1}{x - 1}$$

Is $x = 1$ a vertical asymptote? Why not? What is happening for $x = 1$? What do you conjecture about $y = \frac{x^n - 1}{x - 1}$, $n \geq 1$ an integer, for $x = 1$?

78. Graph each of the following functions:

$$y = \frac{x^2}{x - 1} \qquad y = \frac{x^4}{x - 1} \qquad y = \frac{x^6}{x - 1} \qquad y = \frac{x^8}{x - 1}$$

What similarities do you see? What differences?

79. Write a few paragraphs that provide a general strategy for graphing a rational function. Be sure to mention the following: proper, improper, intercepts, and asymptotes.

80. Create a rational function that has the following characteristics: crosses the x-axis at 2; touches the x-axis at -1; one

vertical asymptote at $x = -5$ and another at $x = 6$; and one horizontal asymptote, $y = 3$. Compare yours to a fellow classmate's. How do they differ? What are their similarities?

81. Create a rational function that has the following characteristics: crosses the x-axis at 3; touches the x-axis at -2; one vertical asymptote, $x = 1$; and one horizontal asymptote, $y = 2$. Give your rational function to a fellow classmate and ask for a written critique of your rational function.

82. The formula on page 206 attributed to Johannes Kepler is one of the famous three Keplerian Laws of Planetary Motion. Go to the library and research these laws. Write a brief paper about these laws and Kepler's place in history.

83. Using a situation that has not been discussed in the text, write a real-world problem that you think involves two variables that vary directly. Exchange your problem with another student's to solve and critique.

84. Using a situation that has not been discussed in the text, write a real-world problem that you think involves two variables that vary inversely. Exchange your problem with another student's to solve and critique.

85. Using a situation that has not been discussed in the text, write a real-world problem that you think involves three variables that vary jointly. Exchange your problem with another student's to solve and critique.

'Are You Prepared?' Answers

1. $(-10, 0)$, $(0, 6)$ **2.** Even; the graph of f is symmetric with respect to the y-axis.

3.5 Polynomial and Rational Inequalities

PREPARING FOR THIS SECTION *Before getting started, review the following:*

• Solving Inequalities (Appendix, Section A.8, pp. 732–733)

Now work the 'Are You Prepared?' problem on page 217.

OBJECTIVES **1** Solve Polynomial Inequalities Algebraically and Graphically
2 Solve Rational Inequalities Algebraically and Graphically

In this section we solve inequalities that involve polynomials of degree 2 and higher, as well as some that involve rational expressions. To solve such inequalities, we use the information obtained in the previous three sections about the graph of polynomial and rational functions. The general idea follows:

Suppose that the polynomial or rational inequality is in one of the forms

$$f(x) < 0 \qquad f(x) > 0 \qquad f(x) \le 0 \qquad f(x) \ge 0$$

Locate the zeros of f if f is a polynomial function, and locate the zeros of the numerator and the denominator if f is a rational function. If we use these zeros to divide the real number line into intervals, then we know that on each interval the graph of f is either above the x-axis $[f(x) > 0]$ or below the x-axis $[f(x) < 0]$. In other words, we have found the solution of the inequality.

The following steps provide more detail.

Steps for Solving Polynomial and Rational Inequalities Algebraically

STEP 1: Write the inequality so that a polynomial or rational expression f is on the left side and zero is on the right side in one of the following forms:

$$f(x) > 0 \qquad f(x) \ge 0 \qquad f(x) < 0 \qquad f(x) \le 0$$

For rational expressions, be sure that the left side is written as a single quotient.

STEP 2: Determine the numbers at which the expression f on the left side equals zero and, if the expression is rational, the numbers at which the expression f on the left side is undefined.

STEP 3: Use the numbers found in Step 2 to separate the real number line into intervals.

STEP 4: Select a number in each interval and evaluate f at the number.
(a) If the value of f is positive, then $f(x) > 0$ for all numbers x in the interval.
(b) If the value of f is negative, then $f(x) < 0$ for all numbers x in the interval.
If the inequality is not strict ($\ge$ or $\le$), include the solutions of $f(x) = 0$ in the solution set, but be careful not to include values of x where the expression is undefined.

✔ Solve Polynomial Inequalities Algebraically and Graphically

EXAMPLE 1	Solving a Polynomial Inequality

Solve the inequality $x^2 \leq 4x + 12$, and graph the solution set.

Algebraic Solution

STEP 1: Rearrange the inequality so that 0 is on the right side.

$$x^2 \leq 4x + 12$$

$$x^2 - 4x - 12 \leq 0 \qquad \text{Subtract } 4x + 12 \text{ from both sides of the inequality.}$$

This inequality is equivalent to the one we wish to solve.

STEP 2: Find the zeros of $f(x) = x^2 - 4x - 12$ by solving the equation $x^2 - 4x - 12 = 0$.

$$x^2 - 4x - 12 = 0$$

$$(x + 2)(x - 6) = 0 \qquad \text{Factor.}$$

$$x = -2 \quad \text{or} \quad x = 6$$

STEP 3: We use the zeros of f to separate the real number line into three intervals.

$$(-\infty, -2) \qquad (-2, 6) \qquad (6, \infty)$$

STEP 4: We select a number in each interval found in Step 3 and evaluate $f(x) = x^2 - 4x - 12$ at each number to determine if $f(x)$ is positive or negative. See Table 17.

Table 17

Interval	$(-\infty, -2)$	$(-2, 6)$	$(6, \infty)$
Number Chosen	-3	0	7
Value of f	$f(-3) = 9$	$f(0) = -12$	$f(7) = 9$
Conclusion	Positive	Negative	Positive

Since we want to know where $f(x)$ is negative, we conclude that $f(x) < 0$ for all x such that $-2 < x < 6$. However, because the original inequality is not strict, numbers x that satisfy the equation $x^2 = 4x + 12$ are also solutions of the inequality $x^2 \leq 4x + 12$. Thus, we include -2 and 6. The solution set of the given inequality is $\{x | -2 \leq x \leq 6\}$ or, using interval notation, $[-2, 6]$.

Graphing Solution

We graph $Y_1 = x^2$ and $Y_2 = 4x + 12$ on the same screen. See Figure 63. Using the IN-TERSECT command, we find that Y_1 and Y_2 intersect at $x = -2$ and at $x = 6$. The graph of Y_1 is below that of Y_2, $Y_1 < Y_2$, between the points of intersection. Since the inequality is not strict, the solution set is $\{x | -2 \leq x \leq 6\}$ or, using interval notation, $[-2, 6]$.

Figure 63

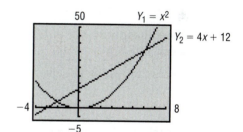

Figure 64

Figure 64 shows the graph of the solution set.

NOW WORK PROBLEMS 5 AND 9.

| EXAMPLE 2 | Solving a Polynomial Inequality |

Solve the inequality $x^4 > x$, and graph the solution set.

Algebraic Solution

STEP 1: Rearrange the inequality so that 0 is on the right side.

$$x^4 > x$$
$$x^4 - x > 0 \qquad \text{Subtract } x \text{ from both sides of the inequality.}$$

This inequality is equivalent to the one we wish to solve.

STEP 2: Find the zeros of $f(x) = x^4 - x$ by solving $x^4 - x = 0$.

$$x^4 - x = 0$$
$$x(x^3 - 1) = 0 \qquad \text{Factor out } x.$$
$$x(x - 1)(x^2 + x + 1) = 0 \qquad \text{Factor the difference of two cubes.}$$
$$x = 0 \quad \text{or} \quad x - 1 = 0 \quad \text{or} \quad x^2 + x + 1 = 0 \qquad \text{Set each factor equal to zero and solve.}$$
$$x = 0 \quad \text{or} \quad x = 1$$

The equation $x^2 + x + 1 = 0$ has no real solutions. (Do you see why?)

STEP 3: We use the zeros to separate the real number line into three intervals:

$$(-\infty, 0) \qquad (0, 1) \qquad (1, \infty)$$

STEP 4: We select a test number in each interval found in Step 3 and evaluate $f(x) = x^4 - x$ at each number to determine if $f(x)$ is positive or negative. See Table 18.

Graphing Solution

We graph $Y_1 = x^4$ and $Y_2 = x$ on the same screen. See Figure 65. Using the INTERSECT command, we find that Y_1 and Y_2 intersect at $x = 0$ and at $x = 1$. The graph of Y_1 is above that of Y_2, $Y_1 > Y_2$, to the left of $x = 0$ and to the right of $x = 1$. Since the inequality is strict, the solution set is $\{x \mid x < 0 \text{ or } x > 1\}$ or, using interval notation, $(-\infty, 0)$ or $(1, \infty)$.

Figure 65

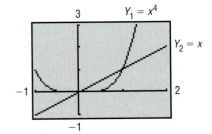

Table 18

Interval	$(-\infty, 0)$	$(0, 1)$	$(1, \infty)$
Number Chosen	-1	$\dfrac{1}{2}$	2
Value of f	$f(-1) = 2$	$f\left(\dfrac{1}{2}\right) = -\dfrac{7}{16}$	$f(2) = 14$
Conclusion	Positive	Negative	Positive

Since we want to know where $f(x)$ is positive, we conclude that $f(x) > 0$ for all numbers x for which $x < 0$ or $x > 1$. Because the original inequality is strict, numbers x that satisfy the equation $x^4 = x$ are not solutions. The solution set of the given inequality is $\{x \mid x < 0 \text{ or } x > 1\}$ or, using interval notation, $(-\infty, 0)$ or $(1, \infty)$.

Figure 66

−2 −1 0 1 2

Figure 66 shows the graph of the solution set.

NOW WORK PROBLEM 21.

2   Solve Rational Inequalities Algebraically and Graphically

EXAMPLE 3	**Solving a Rational Inequality**

Solve the inequality $\dfrac{4x + 5}{x + 2} \geq 3$, and graph the solution set.

Algebraic Solution

STEP 1: We first note that the domain of the variable consists of all real numbers except -2. We rearrange terms so that 0 is on the right side.

$$\dfrac{4x + 5}{x + 2} - 3 \geq 0 \qquad \text{\textcolor{blue}{Subtract 3 from both sides of the inequality.}}$$

STEP 2: For $f(x) = \dfrac{4x + 5}{x + 2} - 3$, find the zeros of f and the values of x at which f is undefined. To find these numbers, we must express f as a single quotient.

$$\begin{aligned} f(x) &= \frac{4x + 5}{x + 2} - 3 && \text{\textcolor{blue}{Least Common Denominator: } } x + 2 \\ &= \frac{4x + 5}{x + 2} - 3 \cdot \frac{x + 2}{x + 2} && \text{\textcolor{blue}{Multiply 3 by } } \tfrac{x + 2}{x + 2} \\ &= \frac{4x + 5 - 3x - 6}{x + 2} && \text{\textcolor{blue}{Write as a single quotient.}} \\ &= \frac{x - 1}{x + 2} && \text{\textcolor{blue}{Combine like terms.}} \end{aligned}$$

The zero of f is 1. Also, f is undefined for $x = -2$.

STEP 3: We use the numbers found in Step 2 to separate the real number line into three intervals.

$$(-\infty, -2) \qquad (-2, 1) \qquad (1, \infty)$$

STEP 4: We select a number in each interval found in Step 3 and evaluate $f(x) = \dfrac{4x + 5}{x + 2} - 3$ at each number to determine if $f(x)$ is positive or negative. See Table 19.

Table 19

	-2			1		$\rightarrow x$

Interval	$(\infty, -2)$	$(-2, 1)$	$(1, \infty)$
Number Chosen	-3	0	2
Value of f	$f(-3) = 4$	$f(0) = -\dfrac{1}{2}$	$f(2) = \dfrac{1}{4}$
Conclusion	Positive	Negative	Positive

We want to know where $f(x)$ is positive. We conclude that $f(x) > 0$ for all x such that $x < -2$ or $x > 1$. Because the original inequality is not strict, numbers x that satisfy the equation $\dfrac{x - 1}{x + 2} = 0$ are also solutions of the inequality. Since $\dfrac{x - 1}{x + 2} = 0$ only if $x = 1$, we conclude that the solution set is $\{x \mid x < -2 \text{ or } x \geq 1\}$ or, using interval notation, $(-\infty, -2)$ or $[1, \infty)$.

Figure 68

Figure 68 shows the graph of the solution set.

Graphing Solution

We first note that the domain of the variable consists of all real numbers except -2. We graph $Y_1 = \dfrac{4x + 5}{x + 2}$ and $Y_2 = 3$ on the same screen. See Figure 67. Using the INTERSECT command, we find that Y_1 and Y_2 intersect at $x = 1$. The graph of Y_1 is above that of Y_2, $Y_1 > Y_2$, to the left of $x = -2$ and to the right of $x = 1$. Since the inequality is not strict, the solution set is $\{x \mid x < -2 \text{ or } x \geq 1\}$. Using interval notation, the solution set is $(-\infty, -2)$ or $[1, \infty)$.

Figure 67

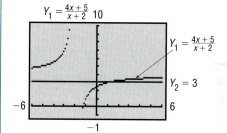

NOW WORK PROBLEM 35.

EXAMPLE 4

Minimum Sales Requirements

Tami is considering leaving her $30,000 a year job and buying a cookie company. According to the financial records of the firm, the relationship between pounds of cookies sold and profit is as exhibited by Table 20.

(a) Draw a scatter diagram of the data in Table 20 with the pounds of cookies sold as the independent variable.

(b) Use a graphing utility to find the quadratic function of best fit.

(c) Use the function found in part (b) to determine the number of pounds of cookies that Tami must sell for the profits to exceed $30,000 a year and therefore make it worthwhile for her to quit her job.

(d) Using the function found in part (b), determine the number of pounds of cookies that Tami should sell to maximize profits.

(e) Using the function found in part (b), determine the maximum profit that Tami can expect to earn.

Table 20

Pounds of Cookies (in Hundreds), x	Profit, P
0	-20,000
50	-5990
75	412
120	10,932
200	26,583
270	36,948
340	44,381
420	49,638
525	49,225
610	44,381
700	34,220

Solution

(a) Figure 69 shows the scatter diagram.

(b) Using a graphing utility, the quadratic function of best fit is

$$P(x) = -0.31x^2 + 295.86x - 20,042.52$$

where P is the profit achieved from selling x pounds (in hundreds) of cookies. See Figure 70.

(c) Since we want profit to exceed $30,000 we need to solve the inequality

$$-0.31x^2 + 295.86x - 20,042.52 > 30,000$$

We graph

$$Y_1 = P(x) = -0.31x^2 + 295.86x - 20,042.52 \quad \text{and} \quad Y_2 = 30,000$$

on the same screen. See Figure 71.

Figure 69

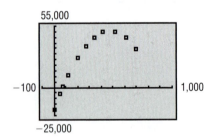

Figure 70

Figure 71

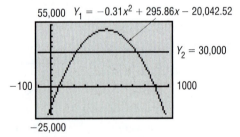

Using the INTERSECT command, we find that Y_1 and Y_2 intersect at $x = 220$ and at $x = 735$. The graph of Y_1 is above that of Y_2, $Y_1 > Y_2$, between the points of intersection. Tami must sell between 220 hundred or 22,000 and 735 hundred or 73,500 pounds of cookies to make it worthwhile to quit her job.

(d) The function $P(x)$ found in part (b) is a quadratic function whose graph opens down. The vertex is therefore the highest point. The x-coordinate of the vertex is

$$x = -\frac{b}{2a} = -\frac{295.86}{2(-0.31)} = 477$$

To maximize profit, 477 hundred (47,700) pounds of cookies must be sold.

(e) The maximum profit is

$$P(477) = -0.31(477)^2 + 295.86(477) - 20,042.52 = \$50,549 \qquad \blacktriangleleft$$

3.5 Assess Your Understanding

'Are You Prepared?'

Answer is given at the end of these exercises. If you get the wrong answer, read the pages listed in red.

1. Solve the inequality: $3 - 4x > 5$. Graph the solution set. (pp. 732–733)

Concepts and Vocabulary

2. *True or False:* A test number for the interval $-5 < x < 1$ is 0.

Skill Building

In Problems 3–56, solve each inequality algebraically. Verify your results using a graphing utility.

3. $(x - 5)(x + 2) < 0$

4. $(x - 5)(x + 2) > 0$

5. $x^2 - 4x > 0$

6. $x^2 + 8x > 0$

7. $x^2 - 9 < 0$

8. $x^2 - 1 < 0$

9. $x^2 + x \geq 12$

10. $x^2 + 7x \leq -12$

11. $2x^2 > 5x + 3$

12. $6x^2 > 6 + 5x$

13. $x(x - 7) < 8$

14. $x(x + 1) < 20$

15. $4x^2 + 9 < 6x$

16. $25x^2 + 16 < 40x$

17. $6(x^2 - 1) \geq 5x$

18. $2(2x^2 - 3x) \geq -9$

19. $(x - 1)(x^2 + x + 1) > 0$

20. $(x + 2)(x^2 - x + 1) > 0$

21. $(x - 1)(x - 2)(x - 3) < 0$

22. $(x + 1)(x + 2)(x + 3) < 0$

23. $x^3 - 2x^2 - 3x \geq 0$

24. $x^3 + 2x^2 - 3x \geq 0$

25. $x^4 > x^2$

26. $x^4 < 4x^2$

27. $x^3 > x^2$

28. $x^3 < 3x^2$

29. $x^4 > 1$

30. $x^3 > 1$

31. $x^2 - 7x - 8 < 0$

32. $x^2 + 12x + 32 \geq 0$

33. $x^4 - 3x^2 - 4 > 0$

34. $x^4 - 5x^2 + 6 < 0$

35. $\dfrac{x + 1}{x - 1} > 0$

36. $\dfrac{x - 3}{x + 1} > 0$

37. $\dfrac{(x - 1)(x + 1)}{x} < 0$

38. $\dfrac{(x - 3)(x + 2)}{x - 1} < 0$

39. $\dfrac{(x - 2)^2}{x^2 - 1} \geq 0$

40. $\dfrac{(x + 5)^2}{x^2 - 4} \geq 0$

41. $6x - 5 < \dfrac{6}{x}$

42. $x + \dfrac{12}{x} < 7$

43. $\dfrac{x + 4}{x - 2} \leq 1$

44. $\dfrac{x + 2}{x - 4} \geq 1$

45. $\dfrac{3x - 5}{x + 2} \leq 2$

46. $\dfrac{x - 4}{2x + 4} \geq 1$

47. $\dfrac{1}{x - 2} < \dfrac{2}{3x - 9}$

48. $\dfrac{5}{x - 3} > \dfrac{3}{x + 1}$

49. $\dfrac{2x + 5}{x + 1} > \dfrac{x + 1}{x - 1}$

50. $\dfrac{1}{x + 2} > \dfrac{3}{x + 1}$

51. $\dfrac{x^2(3 + x)(x + 4)}{(x + 5)(x - 1)} > 0$

52. $\dfrac{x(x^2 + 1)(x - 2)}{(x - 1)(x + 1)} > 0$

53. $\dfrac{2x^2 - x - 1}{x - 4} \leq 0$

54. $\dfrac{3x^2 + 2x - 1}{x + 2} > 0$

55. $\dfrac{x^2 + 3x - 1}{x + 3} > 0$

56. $\dfrac{x^2 - 5x + 3}{x - 5} < 0$

Applications and Extensions

57. For what positive numbers will the cube of a number exceed four times its square?

58. For what positive numbers will the square of a number exceed twice the number?

59. What is the domain of the function $f(x) = \sqrt{x^2 - 16}$?

60. What is the domain of the function $f(x) = \sqrt{x^3 - 3x^2}$?

61. What is the domain of the function $R(x) = \sqrt{\dfrac{x - 2}{x + 4}}$?

62. What is the domain of the function $R(x) = \sqrt{\dfrac{x - 1}{x + 4}}$?

63. Physics A ball is thrown vertically upward with an initial velocity of 80 feet per second. The distance s (in feet) of the ball from the ground after t seconds is $s = 80t - 16t^2$. See the figure.

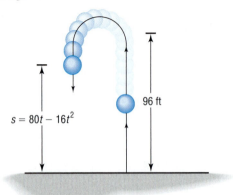

$s = 80t - 16t^2$

96 ft

(a) For what time interval is the ball more than 96 feet above the ground?
(b) Using a graphing utility, graph the relation between s and t.
(c) What is the maximum height of the ball?
(d) After how many seconds does the ball reach the maximum height?

64. Physics A ball is thrown vertically upward with an initial velocity of 96 feet per second. The distance s (in feet) of the ball from the ground after t seconds is $s = 96t - 16t^2$.
(a) For what time interval is the ball more than 112 feet above the ground?
(b) Using a graphing utility, graph the relation between s and t.
(c) What is the maximum height of the ball?
(d) After how many seconds does the ball reach the maximum height?

65. Business The monthly revenue achieved by selling x wristwatches is figured to be $x(40 - 0.2x)$ dollars. The wholesale cost of each watch is $32.
(a) How many watches must be sold each month to achieve a profit (revenue − cost) of at least $50?
(b) Using a graphing utility, graph the revenue function.
(c) What is the maximum revenue that this firm could earn?

(d) How many wristwatches should the firm sell to maximize revenue?
(e) Using a graphing utility, graph the profit function.
(f) What is the maximum profit that this firm can earn?
(g) How many watches should the firm sell to maximize profit?
(h) Provide a reasonable explanation as to why the answers found in parts (d) and (g) differ. Is the shape of the revenue function reasonable in your opinion? Why?

66. Business The monthly revenue achieved by selling x boxes of candy is figured to be $x(5 - 0.05x)$ dollars. The wholesale cost of each box of candy is $1.50.
(a) How many boxes must be sold each month to achieve a profit of at least $60?
(b) Using a graphing utility, graph the revenue function.
(c) What is the maximum revenue that this firm could earn?
(d) How many boxes of candy should the firm sell to maximize revenue?
(e) Using a graphing utility, graph the profit function.
(f) What is the maximum profit that this firm can earn?
(g) How many boxes of candy should the firm sell to maximize profit?
(h) Provide a reasonable explanation as to why the answers found in parts (d) and (g) differ. Is the shape of the revenue function reasonable in your opinion? Why?

67. Cost of Manufacturing In Problem 95 of Section 3.2, a cubic function of best fit relating the cost C of manufacturing x Chevy Cavaliers was found. Budget constraints will not allow Chevy to spend more than $97,000. Determine the number of Cavaliers that could be produced.

68. Cost of Printing In Problem 96 of Section 3.2, a cubic function of best fit relating the cost C of printing x textbooks in a week was found. Budget constraints will not allow the printer to spend more than $170,000 per week. Determine the number of textbooks that could be printed in a week.

69. Prove that, if a, b are real numbers and $a \geq 0, b \geq 0$, then

$$a \leq b \quad \text{is equivalent to} \quad \sqrt{a} \leq \sqrt{b}$$

[Hint: $b - a = \left(\sqrt{b} - \sqrt{a}\right)\left(\sqrt{b} + \sqrt{a}\right).$**]**

Discussion and Writing

70. Make up an inequality that has no solution. Make up one that has exactly one solution.

71. The inequality $x^2 + 1 < -5$ has no solution. Explain why.

'Are You Prepared?' Answer

1. $\left\{x \mid x < -\dfrac{1}{2}\right\}$

−2 −1 −$\frac{1}{2}$ 0 1

3.6 The Real Zeros of a Polynomial Function

PREPARING FOR THIS SECTION   *Before getting started, review the following:*

- Classification of Numbers (Appendix, Section A.1, p. 660)
- Factoring Polynomials (Appendix, Section A.3, pp. 677–679)
- Synthetic Division (Appendix, Section A.4, pp. 687–690)
- Polynomial Division (Appendix, Section A.4, pp. 685–687)
- Quadratic Formula (Appendix, Section A.5, pp. 700–703)

 Now work the 'Are You Prepared?' problems on page 230.

OBJECTIVES   1 Use the Remainder and Factor Theorems
2 Use the Rational Zeros Theorem
3 Find the Real Zeros of a Polynomial Function
4 Solve Polynomial Equations
5 Use the Theorem for Bounds on Zeros
6 Use the Intermediate Value Theorem

In this section, we discuss techniques that can be used to find the real zeros of a polynomial function. Recall that if r is a real zero of a polynomial function f then $f(r) = 0$, r is an x-intercept of the graph of f, and r is a solution of the equation $f(x) = 0$. For polynomial and rational functions, we have seen the importance of the zeros for graphing. In most cases, however, the zeros of a polynomial function are difficult to find using algebraic methods. No nice formulas like the quadratic formula are available to help us find zeros for polynomials of degree 3 or higher. Formulas do exist for solving any third- or fourth-degree polynomial equation, but they are somewhat complicated. No general formulas exist for polynomial equations of degree 5 or higher. Refer to the Historical Feature at the end of this section for more information.

1 Use the Remainder and Factor Theorems

When we divide one polynomial (the dividend) by another (the divisor), we obtain a quotient polynomial and a remainder, the remainder being either the zero polynomial or a polynomial whose degree is less than the degree of the divisor. To check our work, we verify that

$$(\text{Quotient})(\text{Divisor}) + \text{Remainder} = \text{Dividend}$$

This checking routine is the basis for a famous theorem called the **division algorithm[*] for polynomials**, which we now state without proof.

Theorem

> **Division Algorithm for Polynomials**
>
> If $f(x)$ and $g(x)$ denote polynomial functions and if $g(x)$ is a polynomial whose degree is greater than zero, then there are unique polynomial functions $q(x)$ and $r(x)$ such that
>
> $$\frac{f(x)}{g(x)} = q(x) + \frac{r(x)}{g(x)} \quad \text{or} \quad f(x) = q(x)g(x) + r(x) \quad \textbf{(1)}$$
>
> $\quad\quad\quad\quad$ dividend $\quad$ quotient $\quad$ divisor $\quad$ remainder
>
> where $r(x)$ is either the zero polynomial or a polynomial of degree less than that of $g(x)$.

[*]A systematic process in which certain steps are repeated a finite number of times is called an **algorithm**. For example, long division is an algorithm.

In equation (1), $f(x)$ is the **dividend**, $g(x)$ is the **divisor**, $q(x)$ is the **quotient**, and $r(x)$ is the **remainder**.

If the divisor $g(x)$ is a first-degree polynomial of the form

$$g(x) = x - c, \qquad c \text{ a real number}$$

then the remainder $r(x)$ is either the zero polynomial or a polynomial of degree 0. As a result, for such divisors, the remainder is some number, say R, and we may write

$$f(x) = (x - c)q(x) + R \qquad \qquad \text{(2)}$$

This equation is an identity in x and is true for all real numbers x. Suppose that $x = c$. Then equation (2) becomes

$$f(c) = (c - c)q(c) + R$$
$$f(c) = R$$

Substitute $f(c)$ for R in equation (2) to obtain

$$f(x) = (x - c)q(x) + f(c) \qquad \qquad \text{(3)}$$

We have now proved the **Remainder Theorem**.

Remainder Theorem	Let f be a polynomial function. If $f(x)$ is divided by $x - c$, then the remainder is $f(c)$.

EXAMPLE 1 **Using the Remainder Theorem**

Find the remainder if $f(x) = x^3 - 4x^2 - 5$ is divided by

(a) $x - 3$ (b) $x + 2$

Solution (a) We could use long division or synthetic division, but it is easier to use the Remainder Theorem, which says that the remainder is $f(3)$.

$$f(3) = (3)^3 - 4(3)^2 - 5 = 27 - 36 - 5 = -14$$

The remainder is -14.

(b) To find the remainder when $f(x)$ is divided by $x + 2 = x - (-2)$, we evaluate $f(-2)$.

$$f(-2) = (-2)^3 - 4(-2)^2 - 5 = -8 - 16 - 5 = -29$$

The remainder is -29. ◀

Compare the method used in Example 1(a) with the method used in Example 4 on page 981 (synthetic division). Which method do you prefer? Give reasons.

NOTE A graphing utility provides another way to find the value of a function, using the eVALUEate feature. Consult your manual for details. See Figure 72 for the results of Example 1(a). ∎

An important and useful consequence of the Remainder Theorem is the **Factor Theorem**.

Figure 72

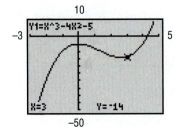

Factor Theorem	Let f be a polynomial function. Then $x - c$ is a factor of $f(x)$ if and only if $f(c) = 0$.

The Factor Theorem actually consists of two separate statements:

> **1.** If $f(c) = 0$, then $x - c$ is a factor of $f(x)$.
>
> **2.** If $x - c$ is a factor of $f(x)$, then $f(c) = 0$.

The proof requires two parts.

Proof

1. Suppose that $f(c) = 0$. Then, by equation (3), we have

$$f(x) = (x - c)q(x) + f(c)$$
$$= (x - c)q(x) + 0$$
$$= (x - c)q(x)$$

   for some polynomial $q(x)$. That is, $x - c$ is a factor of $f(x)$.

2. Suppose that $x - c$ is a factor of $f(x)$. Then there is a polynomial function q such that

$$f(x) = (x - c)q(x)$$

   Replacing x by c, we find that

$$f(c) = (c - c)q(c) = 0 \cdot q(c) = 0$$

   This completes the proof. ∎

One use of the Factor Theorem is to determine whether a polynomial has a particular factor.

EXAMPLE 2 **Using the Factor Theorem**

Use the Factor Theorem to determine whether the function

$$f(x) = 2x^3 - x^2 + 2x - 3$$

has the factor

(a) $x - 1$ (b) $x + 2$

Solution The Factor Theorem states that if $f(c) = 0$ then $x - c$ is a factor.

(a) Because $x - 1$ is of the form $x - c$ with $c = 1$, we find the value of $f(1)$. We choose to use substitution.

$$f(1) = 2(1)^3 - (1)^2 + 2(1) - 3 = 2 - 1 + 2 - 3 = 0$$

See also Figure 73(a). By the Factor Theorem, $x - 1$ is a factor of $f(x)$.

(b) We first need to write $x + 2$ in the form $x - c$. Since $x + 2 = x - (-2)$, we find the value of $f(-2)$. See Figure 73(b). Because $f(-2) = -27 \neq 0$, we conclude from the Factor Theorem that $x - (-2) = x + 2$ is not a factor of $f(x)$.

Figure 73

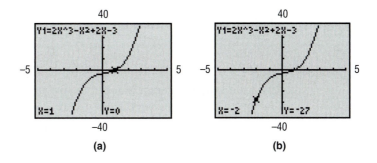

(a)   (b)

In Example 2, we found that $x - 1$ was a factor of f. To write f in factored form, we can use long division or synthetic division. Using synthetic division, we find that

$$
\begin{array}{r|rrrr}
1) & 2 & -1 & 2 & -3 \\
  &   & 2 & 1 & 3 \\
\hline
  & 2 & 1 & 3 & 0
\end{array}
$$

The quotient is $q(x) = 2x^2 + x + 3$ with a remainder of 0, as expected. We can write f in factored form as

$$f(x) = 2x^3 - x^2 + 2x - 3 = (x - 1)(2x^2 + x + 3)$$

NOW WORK PROBLEM 11.

The next theorem concerns the number of real zeros that a polynomial function may have. In counting the zeros of a polynomial, we count each zero as many times as its multiplicity.

Theorem

Number of Real Zeros

A polynomial function of degree n, $n \geq 1$, has at most n real zeros.

Proof The proof is based on the Factor Theorem. If r is a zero of a polynomial function f, then $f(r) = 0$ and, hence, $x - r$ is a factor of $f(x)$. Each zero corresponds to a factor of degree 1. Because f cannot have more first-degree factors than its degree, the result follows. ■

2 **Use the Rational Zeros Theorem**

The next result, called the **Rational Zeros Theorem**, provides information about the rational zeros of a polynomial *with integer coefficients*.

Theorem

Rational Zeros Theorem

Let f be a polynomial function of degree 1 or higher of the form

$$f(x) = a_n x^n + a_{n-1} x^{n-1} + \cdots + a_1 x + a_0, \qquad a_n \neq 0, a_0 \neq 0$$

where each coefficient is an integer. If $\dfrac{p}{q}$, in lowest terms, is a rational zero of f, then p must be a factor of a_0, and q must be a factor of a_n.

EXAMPLE 3 **Listing Potential Rational Zeros**

List the potential rational zeros of

$$f(x) = 2x^3 + 11x^2 - 7x - 6$$

Solution Because f has integer coefficients, we may use the Rational Zeros Theorem. First, we list all the integers p that are factors of $a_0 = -6$ and all the integers q that are factors of the leading coefficient $a_3 = 2$.

$$p: \quad \pm 1, \pm 2, \pm 3, \pm 6$$
$$q: \quad \pm 1, \pm 2$$

Now we form all possible ratios $\dfrac{p}{q}$.

$$\frac{p}{q}: \quad \pm 1, \pm 2, \pm 3, \pm 6, \pm \frac{1}{2}, \pm \frac{3}{2}$$

If f has a rational zero, it will be found in this list, which contains 12 possibilities. ◄

NOW WORK PROBLEM 21.

Be sure that you understand what the Rational Zeros Theorem says: For a polynomial with integer coefficients, *if* there is a rational zero, it is one of those listed. The Rational Zeros Theorem does not say that if a rational number is in the list of potential rational zeros, then it is a zero. It may be the case that the function does not have any rational zeros.

The Rational Zeros Theorem provides a list of potential rational zeros of a function f. If we graph f, we can get a better sense of the location of the x-intercepts and test to see if they are rational. We can also use the potential rational zeros to select our initial viewing window to graph f and then adjust the window based on the results. The graphs shown throughout the text will be those obtained after setting the final viewing window.

3 Find the Real Zeros of a Polynomial Function

EXAMPLE 4	**Finding the Rational Zeros of a Polynomial Function**

Continue working with Example 3 to find the rational zeros of
$$f(x) = 2x^3 + 11x^2 - 7x - 6$$

Solution   We gather all the information that we can about the zeros.

STEP 1: Since f is a polynomial of degree 3, there are at most three real zeros.

STEP 2: We list the potential rational zeros obtained in Example 3:

$$\pm 1, \pm 2, \pm 3, \pm 6, \pm \frac{1}{2}, \pm \frac{3}{2}.$$

Figure 74

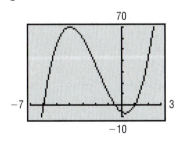

STEP 3: We could, of course, use the Factor Theorem to test each potential rational zero to see if the value of f there is zero. This is not very efficient. The graph of f will tell us approximately where the real zeros are. So we only need to test those rational zeros that are nearby. Figure 74 shows the graph of f. We see that f has three zeros: one near -6, one between -1 and 0, and one near 1. From our original list of potential rational zeros, we will test -6, using eVALUEate.

STEP 4: Because $f(-6) = 0$, we know that -6 is a zero and $x + 6$ is a factor of f. We can use long division or synthetic division to factor f.

$$f(x) = 2x^3 + 11x^2 - 7x - 6$$
$$= (x + 6)(2x^2 - x - 1)$$

Now any solution of the equation $2x^2 - x - 1 = 0$ will be a zero of f. Because of this, we call the equation $2x^2 - x - 1 = 0$ a **depressed equation** of f. Since the degree of the depressed equation of f is less than that of the original polynomial, we work with the depressed equation to find the zeros of f.

The depressed equation $2x^2 - x - 1 = 0$ is a quadratic equation with discriminant $b^2 - 4ac = (-1)^2 - 4(2)(-1) = 9 > 0$. The equation has two real solutions, which can be found by factoring.

$$2x^2 - x - 1 = (2x + 1)(x - 1) = 0$$

$$2x + 1 = 0 \quad \text{or} \quad x - 1 = 0$$

$$x = -\frac{1}{2} \quad \text{or} \quad x = 1$$

The zeros of f are -6, $-\dfrac{1}{2}$, and 1.

Because $f(x) = (x + 6)(2x^2 - x - 1)$, we completely factor f as follows:

$$f(x) = 2x^3 + 11x^2 - 7x - 6 = (x + 6)(2x^2 - x - 1) = (x + 6)(2x + 1)(x - 1)$$

Notice that the three zeros of f are among those given in the list of potential rational zeros in Example 3. ◄

Steps for Finding the Real Zeros of a Polynomial Function

STEP 1: Use the degree of the polynomial to determine the maximum number of zeros.

STEP 2: If the polynomial has integer coefficients, use the Rational Zeros Theorem to identify those rational numbers that potentially can be zeros.

STEP 3: Using a graphing utility, graph the polynomial function.

STEP 4: (a) Use eVALUEate, substitution, synthetic division, or long division to test a potential rational zero based on the graph.
  (b) Each time that a zero (and thus a factor) is found, repeat Step 4 on the depressed equation. In attempting to find the zeros, remember to use (if possible) the factoring techniques that you already know (special products, factoring by grouping, and so on).

| EXAMPLE 5 | **Finding the Real Zeros of a Polynomial Function** |

Find the real zeros of $f(x) = x^5 - x^4 - 4x^3 + 8x^2 - 32x + 48$. Write f in factored form.

Solution   **STEP 1:** There are at most five real zeros.

STEP 2: To obtain the list of potential rational zeros, we write the factors p of $a_0 = 48$ and the factors q of the leading coefficient $a_5 = 1$.

$$p: \quad \pm 1, \pm 2, \pm 3, \pm 4, \pm 6, \pm 8, \pm 12, \pm 16, \pm 24, \pm 48$$
$$q: \quad \pm 1$$

The potential rational zeros consist of all possible quotients $\dfrac{p}{q}$:

$$\frac{p}{q}: \quad \pm 1, \pm 2, \pm 3, \pm 4, \pm 6, \pm 8, \pm 12, \pm 16, \pm 24, \pm 48$$

Figure 75

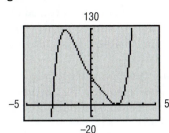

STEP 3: Figure 75 shows the graph of f. The graph has the characteristics that we expect of the given polynomial of degree 5: no more than four turning points, y-intercept 48, and it behaves like $y = x^5$ for large $|x|$.

STEP 4: Since -3 appears to be a zero and -3 is a potential rational zero, we eVALUEate f at -3 and find that $f(-3) = 0$. We use synthetic division to factor f.

$$
\begin{array}{r|rrrrr}
-3) & 1 & -1 & -4 & 8 & -32 & 48 \\
   &   & -3 & 12 & -24 & 48 & -48 \\
\hline
   & 1 & -4 & 8 & -16 & 16 & 0
\end{array}
$$

We can factor f as

$$f(x) = x^5 - x^4 - 4x^3 + 8x^2 - 32x + 48 = (x + 3)(x^4 - 4x^3 + 8x^2 - 16x + 16)$$

We now work with the first depressed equation:

$$q_1(x) = x^4 - 4x^3 + 8x^2 - 16x + 16 = 0$$

Repeat Step 4: In looking back at Figure 75, it appears that 2 might be a zero of even multiplicity. We check the potential rational zero 2 and find that $f(2) = 0$. Using synthetic division,

$$
\begin{array}{r|rrrr}
2) & 1 & -4 & 8 & -16 & 16 \\
   &   & 2 & -4 & 8 & -16 \\
\hline
   & 1 & -2 & 4 & -8 & 0
\end{array}
$$

Now we can factor f as

$$f(x) = (x + 3)(x - 2)(x^3 - 2x^2 + 4x - 8)$$

Repeat Step 4: The depressed equation $q_2(x) = x^3 - 2x^2 + 4x - 8 = 0$ can be factored by grouping.

$$x^3 - 2x^2 + 4x - 8 = (x^3 - 2x^2) + (4x - 8) = x^2(x - 2) + 4(x - 2) = (x - 2)(x^2 + 4) = 0$$
$$x - 2 = 0 \quad \text{or} \quad x^2 + 4 = 0$$
$$x = 2$$

Since $x^2 + 4 = 0$ has no real solutions, the real zeros of f are -3 and 2, with 2 being a zero of multiplicity 2. The factored form of f is

$$f(x) = x^5 - x^4 - 4x^3 + 8x^2 - 32x + 48$$
$$= (x + 3)(x - 2)^2(x^2 + 4) \quad \blacktriangleleft$$

 NOW WORK PROBLEM 39.

4 Solve Polynomial Equations

EXAMPLE 6	**Solving a Polynomial Equation**

Solve the equation: $\quad x^5 - x^4 - 4x^3 + 8x^2 - 32x + 48 = 0$

Solution The solutions of this equation are the zeros of the polynomial function

$$f(x) = x^5 - x^4 - 4x^3 + 8x^2 - 32x + 48$$

Using the result of Example 5, the real zeros are -3 and 2. These are the real solutions of the equation $x^5 - x^4 - 4x^3 + 8x^2 - 32x + 48 = 0$. $\quad \blacktriangleleft$

NOW WORK PROBLEM 63.

In Example 5, the quadratic factor $x^2 + 4$ that appears in the factored form of $f(x)$ is called *irreducible*, because the polynomial $x^2 + 4$ cannot be factored over the real numbers. In general, we say that a quadratic factor $ax^2 + bx + c$ is **irreducible** if it cannot be factored over the real numbers, that is, if it is prime over the real numbers.

Refer back to Examples 4 and 5. The polynomial function of Example 4 has three real zeros, and its factored form contains three linear factors. The polynomial function of Example 5 has two distinct real zeros, and its factored form contains two distinct linear factors and one irreducible quadratic factor.

Theorem

> Every polynomial function (with real coefficients) can be uniquely factored into a product of linear factors and/or irreducible quadratic factors.

We shall prove this result in Section 3.7, and, in fact, we shall draw several additional conclusions about the zeros of a polynomial function. One conclusion is worth noting now. If a polynomial (with real coefficients) is of odd degree, then it must contain at least one linear factor. (Do you see why?) This means that it must have at least one real zero.

COROLLARY

> A polynomial function (with real coefficients) of odd degree has at least one real zero.

5 Use the Theorem for Bounds on Zeros

One challenge in using a graphing utility is to set the viewing window so that a complete graph is obtained. The next theorem is a tool that can be used to find bounds on the zeros. This will assure that the function does not have any zeros outside these bounds. Then using these bounds to set Xmin and Xmax assures that all the x-intercepts appear in the viewing window.

A number M is a **bound** on the zeros of a polynomial if every zero r lies between $-M$ and M, inclusive. That is, M is a bound to the zeros of a polynomial f if

$$-M \leq \text{any zero of } f \leq M$$

Theorem

> **Bounds on Zeros**
>
> Let f denote a polynomial function whose leading coefficient is 1.
>
> $$f(x) = x^n + a_{n-1}x^{n-1} + \cdots + a_1x + a_0$$
>
> A bound M on the zeros of f is the smaller of the two numbers
>
> $$\text{Max}\{1, |a_0| + |a_1| + \cdots + |a_{n-1}|\}, \quad 1 + \text{Max}\{|a_0|, |a_1|, \ldots, |a_{n-1}|\} \quad \textbf{(4)}$$
>
> where Max { } means "choose the largest entry in { }."

An example will help to make the theorem clear.

EXAMPLE 7

Using the Theorem for Finding Bounds on Zeros

Find a bound to the zeros of each polynomial.

(a) $f(x) = x^5 + 3x^3 - 9x^2 + 5$ (b) $g(x) = 4x^5 - 2x^3 + 2x^2 + 1$

Solution   (a) The leading coefficient of f is 1.

$$f(x) = x^5 + 3x^3 - 9x^2 + 5 \qquad a_4 = 0, a_3 = 3, a_2 = -9, a_1 = 0, a_0 = 5$$

We evaluate the expressions in formula (4).

$$\text{Max}\{1, |a_0| + |a_1| + \cdots + |a_{n-1}|\} = \text{Max}\{1, |5| + |0| + |-9| + |3| + |0|\}$$
$$= \text{Max}\{1, 17\} = 17$$
$$1 + \text{Max}\{|a_0|, |a_1|, \ldots, |a_{n-1}|\} = 1 + \text{Max}\{|5|, |0|, |-9|, |3|, |0|\}$$
$$= 1 + 9 = 10$$

The smaller of the two numbers, 10, is the bound. Every zero of f lies between -10 and 10.

(b) First we write g so that its leading coefficient is 1.

$$g(x) = 4x^5 - 2x^3 + 2x^2 + 1 = 4\left(x^5 - \frac{1}{2}x^3 + \frac{1}{2}x^2 + \frac{1}{4}\right)$$

Next we evaluate the two expressions in formula (4) with $a_4 = 0$, $a_3 = -\dfrac{1}{2}$, $a_2 = \dfrac{1}{2}$, $a_1 = 0$, and $a_0 = \dfrac{1}{4}$.

$$\text{Max}\{1, |a_0| + |a_1| + \cdots + |a_{n-1}|\} = \text{Max}\left\{1, \left|\frac{1}{4}\right| + |0| + \left|\frac{1}{2}\right| + \left|-\frac{1}{2}\right| + |0|\right\}$$
$$= \text{Max}\left\{1, \frac{5}{4}\right\} = \frac{5}{4}$$
$$1 + \text{Max}\{|a_0|, |a_1|, \ldots, |a_{n-1}|\} = 1 + \text{Max}\left\{\left|\frac{1}{4}\right|, |0|, \left|\frac{1}{2}\right|, \left|-\frac{1}{2}\right|, |0|\right\}$$
$$= 1 + \frac{1}{2} = \frac{3}{2}$$

The smaller of the two numbers, $\dfrac{5}{4}$, is the bound. Every zero of g lies between $-\dfrac{5}{4}$ and $\dfrac{5}{4}$.   ◄

EXAMPLE 8

Obtaining Graphs Using Bounds on Zeros

Obtain a graph for each polynomial.

(a) $f(x) = x^5 + 3x^3 - 9x^2 + 5$ (b) $g(x) = 4x^5 - 2x^3 + 2x^2 + 1$

Solution

(a) Based on Example 7(a), every zero lies between -10 and 10. Using $X\min = -10$ and $X\max = 10$, we graph $Y_1 = f(x) = x^5 + 3x^3 - 9x^2 + 5$. Figure 76(a) shows the graph obtained using ZOOM-FIT. Figure 76(b) shows the graph after adjusting the viewing window to improve the graph.

Figure 76

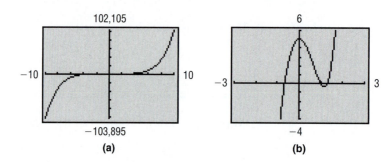

(a)   (b)

Figure 77

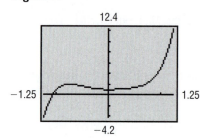

(b) Based on Example 7(b), every zero lies between $-\dfrac{5}{4}$ and $\dfrac{5}{4}$. Using $X\min = -\dfrac{5}{4}$ and $X\max = \dfrac{5}{4}$, we graph $Y_1 = g(x) = 4x^5 - 2x^3 + 2x^2 + 1$. Figure 77 shows the graph after using ZOOM-FIT. Here no adjustment of the viewing window is needed. ◀

🖉━ **NOW WORK PROBLEM 33.**

The next example shows how to proceed when some of the coefficients of the polynomial are not integers.

EXAMPLE 9 **Finding the Zeros of a Polynomial**

Find all the real zeros of the polynomial function

$$f(x) = x^5 - 1.8x^4 - 17.79x^3 + 31.672x^2 + 37.95x - 8.7121$$

Figure 78

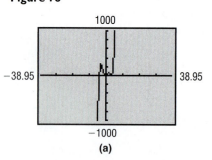

Solution

STEP 1: There are at most five real zeros.

STEP 2: Since there are noninteger coefficients, the Rational Zeros Theorem does not apply.

STEP 3: We determine the bounds of f. The leading coefficient of f is 1 with $a_4 = -1.8$, $a_3 = -17.79$, $a_2 = 31.672$, $a_1 = 37.95$, and $a_0 = -8.7121$. We evaluate the expressions using formula (4).

$$\text{Max}\{1, |-8.7121| + |37.95| + |31.672| + |-17.79| + |-1.8|\} = \text{Max}\{1, 97.9241\}$$
$$= 97.9241$$

$$1 + \text{Max}\{|-8.7121|, |37.95|, |31.672|, |-17.79|, |-1.8|\} = 1 + 37.95$$
$$= 38.95$$

The smaller of the two numbers, 38.95, is the bound. Every real zero of f lies between -38.95 and 38.95. Figure 78(a) shows the graph of f with $X\min = -38.95$ and $X\max = 38.95$. Figure 78(b) shows a graph of f after adjusting the viewing window to improve the graph.

STEP 4: From Figure 78(b), we see that f appears to have four x-intercepts: one near -4, one near -1, one between 0 and 1, and one near 3. The x-intercept near 3 might be a zero of even multiplicity since the graph seems to touch the x-axis at that point.

We use the Factor Theorem to determine if -4 and -1 are zeros. Since $f(-4) = f(-1) = 0$, we know that -4 and -1 are zeros. Using ZERO (or ROOT), we find that the remaining zeros are 0.20 and 3.30, rounded to two decimal places.

There are no real zeros on the graph that have not already been identified. So, either 3.30 is a zero of multiplicity 2 or there are two distinct zeros, each of which is 3.30, rounded to two decimal places. (Example 10 will explain how to determine which is true.) ◄

6 Use the Intermediate Value Theorem

The Intermediate Value Theorem requires that the function be *continuous*. Although calculus is needed to explain the meaning precisely, the *idea* of a continuous function is easy to understand. Very basically, a function f is continuous when its graph can be drawn without lifting pencil from paper, that is, when the graph contains no holes or jumps or gaps. For example, every polynomial function is continuous.

Intermediate Value Theorem

Let f denote a continuous function. If $a < b$ and if $f(a)$ and $f(b)$ are of opposite sign, then f has at least one zero between a and b.

Although the proof of this result requires advanced methods in calculus, it is easy to "see" why the result is true. Look at Figure 79.

Figure 79

If $f(a) < 0$ and $f(b) > 0$ and if f is continuous, there is at least one zero between a and b.

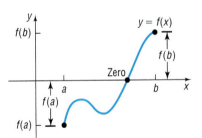

The Intermediate Value Theorem together with the TABLE feature of a graphing utility provides a basis for finding zeros.

| EXAMPLE 10 | **Using the Intermediate Value Theorem and a Graphing Utility to Locate Zeros** |

Continue working with Example 9 to determine whether there is a repeated zero or two distinct zeros near 3.30.

Table 21

X	Y1
3.27	.08567
3.28	.03829
3.29	.00956
3.3	-1E-4
3.31	.0097
3.32	.03936
3.33	.08931

Y1■X^5-1.8X^4-1...

Solution   We use the TABLE feature of a graphing utility. See Table 21. Since $f(3.29) = 0.00956 > 0$ and $f(3.30) = -0.0001 < 0$, by the Intermediate Value Theorem there is a zero between 3.29 and 3.30. Similarly, since $f(3.30) = -0.0001 < 0$ and $f(3.31) = 0.0097 > 0$, there is another zero between 3.30 and 3.31. Now we know that the five zeros of f are distinct. ◄

  NOW WORK PROBLEM 75.

HISTORICAL FEATURE

Tartaglia
(1500–1557)

Formulas for the solution of third- and fourth-degree polynomial equations exist, and, while not very practical, they do have an interesting history.

In the 1500s in Italy, mathematical contests were a popular pastime, and persons possessing methods for solving problems kept them secret. (Solutions that were published were already common knowledge.) Niccolo of Brescia (1500–1557), commonly referred to as Tartaglia ("the stammerer"), had the secret for solving cubic (third-degree) equations, which gave him a decided advantage in the contests. See the Historical Problems.

Girolamo Cardano (1501–1576) found out that Tartaglia had the secret, and, being interested in cubics, he requested it from Tartaglia. The reluctant Tartaglia hesitated for some time, but finally,

swearing Cardano to secrecy with midnight oaths by candlelight, told him the secret. Cardano then published the solution in his book *Ars Magna* (1545), giving Tartaglia the credit but rather compromising the secrecy. Tartaglia exploded into bitter recriminations, and each wrote pamphlets that reflected on the other's mathematics, moral character, and ancestry. See the Historical Problems.

The quartic (fourth-degree) equation was solved by Cardano's student Lodovico Ferrari, and this solution also was included, with credit and this time with permission, in the *Ars Magna*.

Attempts were made to solve the fifth-degree equation in similar ways, all of which failed. In the early 1800s, P. Ruffini, Niels Abel, and Evariste Galois all found ways to show that it is not possible to solve fifth-degree equations by formula, but the proofs required the introduction of new methods. Galois's methods eventually developed into a large part of modern algebra.

Historical Problems

Problems 1–8 develop the Tartaglia–Cardano solution of the cubic equation and show why it is not altogether practical.

1. Show that the general cubic equation $y^3 + by^2 + cy + d = 0$ can be transformed into an equation of the form $x^3 + px + q = 0$ by using the substitution $y = x - \dfrac{b}{3}$.

2. In the equation $x^3 + px + q = 0$, replace x by $H + K$. Let $3HK = -p$, and show that $H^3 + K^3 = -q$.
   [**Hint:** $3H^2K + 3HK^2 = 3HKx$.]

3. Based on Problem 2, we have the two equations
   $$3HK = -p \quad \text{and} \quad H^3 + K^3 = -q$$
   Solve for K in $3HK = -p$ and substitute into $H^3 + K^3 = -q$. Then show that
   $$H = \sqrt[3]{\dfrac{-q}{2} + \sqrt{\dfrac{q^2}{4} + \dfrac{p^3}{27}}}$$
   [**Hint:** Look for an equation that is quadratic in form.]

4. Use the solution for H from Problem 3 and the equation $H^3 + K^3 = -q$ to show that
   $$K = \sqrt[3]{\dfrac{-q}{2} - \sqrt{\dfrac{q^2}{4} + \dfrac{p^3}{27}}}$$

5. Use the results from Problems 2–4 to show that the solution of $x^3 + px + q = 0$ is
   $$x = \sqrt[3]{\dfrac{-q}{2} + \sqrt{\dfrac{q^2}{4} + \dfrac{p^3}{27}}} + \sqrt[3]{\dfrac{-q}{2} - \sqrt{\dfrac{q^2}{4} + \dfrac{p^3}{27}}}$$

6. Use the result of Problem 5 to solve the equation $x^3 - 6x - 9 = 0$.

7. Use a calculator and the result of Problem 5 to solve the equation $x^3 + 3x - 14 = 0$.

8. Use the methods of this chapter to solve the equation $x^3 + 3x - 14 = 0$.

3.6 Assess Your Understanding

'Are You Prepared?'

Answers are given at the end of these exercises. If you get a wrong answer, read the pages listed in red.

1. In the set $\left\{-2, -\sqrt{2}, 0, \dfrac{1}{2}, 4.5, \pi\right\}$, name the numbers that are integers. Which are rational numbers? (p. 660)

2. Factor the expression $6x^2 + x - 2$. (pp. 677–679)

3. Find the quotient and remainder if $3x^4 - 5x^3 + 7x - 4$ is divided by $x - 3$. (pp. 685–687 and pp. 687–690)

4. Solve the equation $x^2 + x - 3 = 0$. (pp. 700–703)

Concepts and Vocabulary

5. In the process of polynomial division, (Divisor)(Quotient) + _____ = _____.

6. When a polynomial function f is divided by $x - c$, the remainder is _____.

7. If a function f, whose domain is all real numbers, is even and if 4 is a zero of f, then _____ is also a zero.

8. *True or False:* Every polynomial function of degree 3 with real coefficients has exactly three real zeros.

9. *True or False:* The only potential rational zeros of $f(x) = 2x^5 - x^3 + x^2 - x + 1$ are $\pm 1, \pm 2$.

10. *True or False:* If f is a polynomial function of degree 4 and if $f(2) = 5$, then

$$\frac{f(x)}{x - 2} = p(x) + \frac{5}{x - 2}$$

where $p(x)$ is a polynomial of degree 3.

Skill Building

In Problems 11–20, use the Factor Theorem to determine whether $x - c$ is a factor of f. If it is, write f in factored form, that is, write f in the form $f(x) = (x - c)$ (quotient).

11. $f(x) = 4x^3 - 3x^2 - 8x + 4; \quad c = 3$

12. $f(x) = -4x^3 + 5x^2 + 8; \quad c = -2$

13. $f(x) = 3x^4 - 6x^3 - 5x + 10; \quad c = 1$

14. $f(x) = 4x^4 - 15x^2 - 4; \quad c = 2$

15. $f(x) = 3x^6 + 2x^3 - 176; \quad c = -2$

16. $f(x) = 2x^6 - 18x^4 + x^2 - 9; \quad c = -3$

17. $f(x) = 4x^6 - 64x^4 + x^2 - 16; \quad c = 4$

18. $f(x) = x^6 - 16x^4 + x^2 - 16; \quad c = -4$

19. $f(x) = 2x^4 - x^3 + 2x - 1; \quad c = -\dfrac{1}{2}$

20. $f(x) = 3x^4 + x^3 - 3x + 1; \quad c = -\dfrac{1}{3}$

In Problems 21–32, tell the maximum number of real zeros that each polynomial function may have. Then list the potential rational zeros of each polynomial function. Do not attempt to find the zeros.

21. $f(x) = 3x^4 - 3x^3 + x^2 - x + 1$

22. $f(x) = x^5 - x^4 + 2x^2 + 3$

23. $f(x) = x^5 - 6x^2 + 9x - 3$

24. $f(x) = 2x^5 - x^4 - x^2 + 1$

25. $f(x) = -4x^3 - x^2 + x + 2$

26. $f(x) = 6x^4 - x^2 + 2$

27. $f(x) = 3x^4 - x^2 + 2$

28. $f(x) = -4x^3 + x^2 + x + 2$

29. $f(x) = 2x^5 - x^3 + 2x^2 + 4$

30. $f(x) = 3x^5 - x^2 + 2x + 3$

31. $f(x) = 6x^4 + 2x^3 - x^2 + 2$

32. $f(x) = -6x^3 - x^2 + x + 3$

In Problems 33–38, find the bounds to the zeros of each polynomial function. Use the bounds to obtain a complete graph of f.

33. $f(x) = 2x^3 + x^2 - 1$

34. $f(x) = 3x^3 - 2x^2 + x + 4$

35. $f(x) = x^3 - 5x^2 - 11x + 11$

36. $f(x) = 2x^3 - x^2 - 11x - 6$

37. $f(x) = x^4 + 3x^3 - 5x^2 + 9$

38. $f(x) = 4x^4 - 12x^3 + 27x^2 - 54x + 81$

In Problems 39–56, find the real zeros of f. Use the real zeros to factor f.

39. $f(x) = x^3 + 2x^2 - 5x - 6$

40. $f(x) = x^3 + 8x^2 + 11x - 20$

41. $f(x) = 2x^3 - 13x^2 + 24x - 9$

42. $f(x) = 2x^3 - 5x^2 - 4x + 12$

43. $f(x) = 3x^3 + 4x^2 + 4x + 1$

44. $f(x) = 3x^3 - 7x^2 + 12x - 28$

45. $f(x) = x^3 - 8x^2 + 17x - 6$

46. $f(x) = x^3 + 6x^2 + 6x - 4$

47. $f(x) = x^4 + x^3 - 3x^2 - x + 2$

48. $f(x) = x^4 - x^3 - 6x^2 + 4x + 8$

49. $f(x) = 2x^4 + 17x^3 + 35x^2 - 9x - 45$

50. $f(x) = 4x^4 - 15x^3 - 8x^2 + 15x + 4$

51. $f(x) = 2x^4 - 3x^3 - 21x^2 - 2x + 24$

52. $f(x) = 2x^4 + 11x^3 - 5x^2 - 43x + 35$

53. $f(x) = 4x^4 + 7x^2 - 2$

54. $f(x) = 4x^4 + 15x^2 - 4$

55. $f(x) = 4x^5 - 8x^4 - x + 2$

56. $f(x) = 4x^5 + 12x^4 - x - 3$

In Problems 57–62, find the real zeros of f rounded to two decimal places.

57. $f(x) = x^3 + 3.2x^2 - 16.83x - 5.31$

58. $f(x) = x^3 + 3.2x^2 - 7.25x - 6.3$

59. $f(x) = x^4 - 1.4x^3 - 33.71x^2 + 23.94x + 292.41$

60. $f(x) = x^4 + 1.2x^3 - 7.46x^2 - 4.692x + 15.2881$

61. $f(x) = x^3 + 19.5x^2 - 1021x + 1000.5$

62. $f(x) = x^3 + 42.2x^2 - 664.8x + 1490.4$

In Problems 63–72, find the real solutions of each equation.

63. $x^4 - x^3 + 2x^2 - 4x - 8 = 0$

64. $2x^3 + 3x^2 + 2x + 3 = 0$

65. $3x^3 + 4x^2 - 7x + 2 = 0$

66. $2x^3 - 3x^2 - 3x - 5 = 0$

67. $3x^3 - x^2 - 15x + 5 = 0$

68. $2x^3 - 11x^2 + 10x + 8 = 0$

69. $x^4 + 4x^3 + 2x^2 - x + 6 = 0$

70. $x^4 - 2x^3 + 10x^2 - 18x + 9 = 0$

71. $x^3 - \dfrac{2}{3}x^2 + \dfrac{8}{3}x + 1 = 0$

72. $x^3 - \dfrac{2}{3}x^2 + 3x - 2 = 0$

In Problems 73–78, use the Intermediate Value Theorem to show that each function has a zero in the given interval. Approximate the zero rounded to two decimal places.

73. $f(x) = 8x^4 - 2x^2 + 5x - 1; \quad [0, 1]$

74. $f(x) = x^4 + 8x^3 - x^2 + 2; \quad [-1, 0]$

75. $f(x) = 2x^3 + 6x^2 - 8x + 2; \quad [-5, -4]$

76. $f(x) = 3x^3 - 10x + 9; \quad [-3, -2]$

77. $f(x) = x^5 - x^4 + 7x^3 - 7x^2 - 18x + 18; \quad [1.4, 1.5]$

78. $f(x) = x^5 - 3x^4 - 2x^3 + 6x^2 + x + 2; \quad [1.7, 1.8]$

Applications and Extensions

79. Cost of Manufacturing In Problem 95 of Section 3.2, you found the cost function for manufacturing Chevy Cavaliers. Use the methods learned in this section to determine how many Cavaliers can be manufactured at a total cost of $3,000,000. In other words, solve the equation $C(x) = 3000$.

80. Cost of Printing In Problem 96 of Section 3.2, you found the cost function for printing textbooks. Use the methods learned in this section to determine how many textbooks can be printed at a total cost of $200,000. In other words, solve the equation $C(x) = 200$.

81. Find k such that $f(x) = x^3 - kx^2 + kx + 2$ has the factor $x - 2$.

82. Find k such that $f(x) = x^4 - kx^3 + kx^2 + 1$ has the factor $x + 2$.

83. What is the remainder when $f(x) = 2x^{20} - 8x^{10} + x - 2$ is divided by $x - 1$?

84. What is the remainder when $f(x) = -3x^{17} + x^9 - x^5 + 2x$ is divided by $x + 1$?

85. One solution of the equation $x^3 - 8x^2 + 16x - 3 = 0$ is 3. Find the sum of the remaining solutions.

86. One solution of the equation $x^3 + 5x^2 + 5x - 2 = 0$ is -2. Find the sum of the remaining solutions.

87. What is the length of the edge of a cube if, after a slice 1 inch thick is cut from one side, the volume remaining is 294 cubic inches?

88. What is the length of the edge of a cube if its volume could be doubled by an increase of 6 centimeters in one edge, an increase of 12 centimeters in a second edge, and a decrease of 4 centimeters in the third edge?

89. Use the Factor Theorem to prove that $x - c$ is a factor of $x^n - c^n$ for any positive integer n.

90. Use the Factor Theorem to prove that $x + c$ is a factor of $x^n + c^n$ if $n \geq 1$ is an odd integer.

91. Let $f(x)$ be a polynomial function whose coefficients are integers. Suppose that r is a real zero of f and that the leading coefficient of f is 1. Use the Rational Zeros Theorem to show that r is either an integer or an irrational number.

92. Prove the Rational Zeros Theorem.

[**Hint:** Let $\dfrac{p}{q}$, where p and q have no common factors except 1 and -1, be a zero of the polynomial function $f(x) = a_n x^n + a_{n-1} x^{n-1} + \cdots + a_1 x + a_0$, whose coefficients are all integers. Show that $a_n p^n + a_{n-1} p^{n-1} q + \cdots + a_1 p q^{n-1} + a_0 q^n = 0$. Now, because p is a factor of the first n terms of this equation, p must also be a factor of the term $a_0 q^n$. Since p is not a factor of q (why?), p must be a factor of a_0. Similarly, q must be a factor of a_n.]

Discussion and Writing

93. Is $\dfrac{1}{3}$ a zero of $f(x) = 2x^3 + 3x^2 - 6x + 7$? Explain.

94. Is $\dfrac{1}{3}$ a zero of $f(x) = 4x^3 - 5x^2 - 3x + 1$? Explain.

95. Is $\dfrac{3}{5}$ a zero of $f(x) = 2x^6 - 5x^4 + x^3 - x + 1$? Explain.

96. Is $\dfrac{2}{3}$ a zero of $f(x) = x^7 + 6x^5 - x^4 + x + 2$? Explain.

'Are You Prepared?' Answers

1. Integers: $\{-2, 0\}$; Rational $\left\{-2, 0, \dfrac{1}{2}, 4.5\right\}$

2. $(3x + 2)(2x - 1)$

3. Quotient: $3x^3 + 4x^2 + 12x + 43$; Remainder: 125

4. $\left\{\dfrac{-1 - \sqrt{13}}{2}, \dfrac{-1 + \sqrt{13}}{2}\right\}$

3.7 Complex Zeros; Fundamental Theorem of Algebra

PREPARING FOR THIS SECTION   *Before getting started, review the following:*

- Complex Numbers (Appendix, Section A.6, pp. 708–713)
- Quadratic Equations with a Negative Discriminant (Appendix, Section A.6, pp. 713–715)

✎ Now work the 'Are You Prepared?' problems on page 237.

OBJECTIVES 1 Use the Conjugate Pairs Theorem
  2 Find a Polynomial Function with Specified Zeros
  3 Find the Complex Zeros of a Polynomial

In Section 3.6 we found the **real** zeros of a polynomial function. In this section we will find the **complex** zeros of a polynomial function. Finding the complex zeros of a function requires finding all zeros of the form $a + bi$. These zeros will be real if $b = 0$.

A variable in the complex number system is referred to as a **complex variable**.

A **complex polynomial function** f of degree n is a function of the form

$$f(x) = a_n x^n + a_{n-1} x^{n-1} + \cdots + a_1 x + a_0 \qquad (1)$$

where $a_n, a_{n-1}, \ldots, a_1, a_0$ are complex numbers, $a_n \neq 0$, n is a nonnegative integer, and x is a complex variable. As before, a_n is called the **leading coefficient** of f. A complex number r is called a **(complex) zero** of f if $f(r) = 0$.

We have learned that some quadratic equations have no real solutions, but that in the complex number system every quadratic equation has a solution, either real or complex. The next result, proved by Karl Friedrich Gauss (1777–1855) when he was 22 years old,[*] gives an extension to complex polynomials. In fact, this result is so important and useful that it has become known as the **Fundamental Theorem of Algebra**.

Fundamental Theorem of Algebra

Every complex polynomial function $f(x)$ of degree $n \geq 1$ has at least one complex zero.

We shall not prove this result, as the proof is beyond the scope of this book. However, using the Fundamental Theorem of Algebra and the Factor Theorem, we can prove the following result:

Theorem

Every complex polynomial function $f(x)$ of degree $n \geq 1$ can be factored into n linear factors (not necessarily distinct) of the form

$$f(x) = a_n(x - r_1)(x - r_2) \cdots \cdots (x - r_n) \qquad (2)$$

where $a_n, r_1, r_2, \ldots, r_n$ are complex numbers. That is, every complex polynomial function of degree $n \geq 1$ has exactly n (not necessarily distinct) zeros.

---

[*]In all, Gauss gave four different proofs of this theorem, the first one in 1799 being the subject of his doctoral dissertation.

Proof Let

$$f(x) = a_n x^n + a_{n-1} x^{n-1} + \cdots + a_1 x + a_0$$

By the Fundamental Theorem of Algebra, f has at least one zero, say r_1. Then, by the Factor Theorem, $x - r_1$ is a factor, and

$$f(x) = (x - r_1)q_1(x)$$

where $q_1(x)$ is a complex polynomial of degree $n - 1$ whose leading coefficient is a_n. Again by the Fundamental Theorem of Algebra, the complex polynomial $q_1(x)$ has at least one zero, say r_2. By the Factor Theorem, $q_1(x)$ has the factor $x - r_2$, so

$$q_1(x) = (x - r_2)q_2(x)$$

where $q_2(x)$ is a complex polynomial of degree $n - 2$ whose leading coefficient is a_n. Consequently,

$$f(x) = (x - r_1)(x - r_2)q_2(x)$$

Repeating this argument n times, we finally arrive at

$$f(x) = (x - r_1)(x - r_2) \cdots \cdot (x - r_n)q_n(x)$$

where $q_n(x)$ is a complex polynomial of degree $n - n = 0$ whose leading coefficient is a_n. Thus, $q_n(x) = a_n x^0 = a_n$, and so

$$f(x) = a_n(x - r_1)(x - r_2) \cdots \cdot (x - r_n)$$

We conclude that every complex polynomial function $f(x)$ of degree $n \geq 1$ has exactly n (not necessarily distinct) zeros. ∎

1 Use the Conjugate Pairs Theorem

We can use the Fundamental Theorem of Algebra to obtain valuable information about the complex zeros of polynomials whose coefficients are real numbers.

Conjugate Pairs Theorem

Let $f(x)$ be a polynomial whose coefficients are real numbers. If $r = a + bi$ is a zero of f, then the complex conjugate $\bar{r} = a - bi$ is also a zero of f.

In other words, for polynomials whose coefficients are real numbers, the zeros occur in conjugate pairs.

Proof Let

$$f(x) = a_n x^n + a_{n-1} x^{n-1} + \cdots + a_1 x + a_0$$

where $a_n, a_{n-1}, \ldots, a_1, a_0$ are real numbers and $a_n \neq 0$. If $r = a + bi$ is a zero of f, then $f(r) = f(a + bi) = 0$, so

$$a_n r^n + a_{n-1} r^{n-1} + \cdots + a_1 r + a_0 = 0$$

We take the conjugate of both sides to get

$$\overline{a_n r^n + a_{n-1} r^{n-1} + \cdots + a_1 r + a_0} = \bar{0}$$

$$\overline{a_n r^n} + \overline{a_{n-1} r^{n-1}} + \cdots + \overline{a_1 r} + \overline{a_0} = \bar{0} \qquad \text{\textit{The conjugate of a sum equals the sum of the conjugates (see the Appendix, Section A.6).}}$$

$$\overline{a_n}(\bar{r})^n + \overline{a_{n-1}}(\bar{r})^{n-1} + \cdots + \overline{a_1}\bar{r} + \overline{a_0} = \bar{0} \qquad \text{\textit{The conjugate of a product equals the product of the conjugates.}}$$

$$a_n(\bar{r})^n + a_{n-1}(\bar{r})^{n-1} + \cdots + a_1\bar{r} + a_0 = 0 \qquad \text{\textit{The conjugate of a real number equals the real number.}}$$

This last equation states that $f(\bar{r}) = 0$; that is, $\bar{r} = a - bi$ is a zero of f. ∎

The value of this result should be clear. If we know that, $3 + 4i$ is a zero of a polynomial with real coefficients, then we know that $3 - 4i$ is also a zero. This result has an important corollary.

COROLLARY

A polynomial f of odd degree with real coefficients has at least one real zero.

Proof Because complex zeros occur as conjugate pairs in a polynomial with real coefficients, there will always be an even number of zeros that are not real numbers. Consequently, since f is of odd degree, one of its zeros has to be a real number. ■

For example, the polynomial $f(x) = x^5 - 3x^4 + 4x^3 - 5$ has at least one zero that is a real number, since f is of degree 5 (odd) and has real coefficients.

EXAMPLE 1

Using the Conjugate Pairs Theorem

A polynomial f of degree 5 whose coefficients are real numbers has the zeros $1, 5i$, and $1 + i$. Find the remaining two zeros.

Solution Since complex zeros appear as conjugate pairs, it follows that $-5i$, the conjugate of $5i$, and $1 - i$, the conjugate of $1 + i$, are the two remaining zeros. ◄

 NOW WORK PROBLEM 7.

2 Find a Polynomial Function with Specified Zeros

EXAMPLE 2

Finding a Polynomial Function Whose Zeros Are Given

(a) Find a polynomial f of degree 4 whose coefficients are real numbers and that has the zeros $1, 1$, and $-4 + i$.

(b) Graph the polynomial found in part (a) to verify your result.

Solution (a) Since $-4 + i$ is a zero, by the Conjugate Pairs Theorem, $-4 - i$ must also be a zero of f. Because of the Factor Theorem, if $f(c) = 0$, then $x - c$ is a factor of $f(x)$. So we can now write f as

$$f(x) = a(x - 1)(x - 1)[x - (-4 + i)][x - (-4 - i)]$$

where a is any real number. If we let $a = 1$, we obtain

$$\begin{aligned}
f(x) &= (x - 1)(x - 1)[x - (-4 + i)][x - (-4 - i)] \\
&= (x^2 - 2x + 1)[x^2 - (-4 + i)x - (-4 - i)x + (-4 + i)(-4 - i)] \\
&= (x^2 - 2x + 1)(x^2 + 4x - ix + 4x + ix + 16 + 4i - 4i - i^2) \\
&= (x^2 - 2x + 1)(x^2 + 8x + 17) \\
&= x^4 + 8x^3 + 17x^2 - 2x^3 - 16x^2 - 34x + x^2 + 8x + 17 \\
&= x^4 + 6x^3 + 2x^2 - 26x + 17
\end{aligned}$$

Figure 80

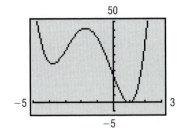

(b) A quick analysis of the polynomial f tells us what to expect:

At most three turning points.

For large $|x|$, the graph will behave like $y = x^4$.

A repeated real zero at 1 so that the graph will touch the x-axis at 1.

The only x-intercept is at 1.

Figure 80 shows the complete graph. (Do you see why? The graph has exactly three turning points and the degree of the polynomial is 4.) ◄

Exploration

Graph the function found in Example 2 for $a = 2$ and $a = -1$. Does the value of a affect the zeros of f? How does the value of a affect the graph of f?

Now we can prove the theorem we conjectured earlier in Section 3.6.

Theorem

> Every polynomial function with real coefficients can be uniquely factored over the real numbers into a product of linear factors and/or irreducible quadratic factors.

Proof Every complex polynomial f of degree n has exactly n zeros and can be factored into a product of n linear factors. If its coefficients are real, then those zeros that are complex numbers will always occur as conjugate pairs. As a result, if $r = a + bi$ is a complex zero, then so is $\bar{r} = a - bi$. Consequently, when the linear factors $x - r$ and $x - \bar{r}$ of f are multiplied, we have

$$(x - r)(x - \bar{r}) = x^2 - (r + \bar{r})x + r\bar{r} = x^2 - 2ax + a^2 + b^2$$

This second-degree polynomial has real coefficients and is irreducible (over the real numbers). Thus, the factors of f are either linear or irreducible quadratic factors. ∎

3 Find the Complex Zeros of a Polynomial

EXAMPLE 3 **Finding the Complex Zeros of a Polynomial**

Find the complex zeros of the polynomial function

$$f(x) = 3x^4 + 5x^3 + 25x^2 + 45x - 18$$

Solution **STEP 1:** The degree of f is 4. So f will have four complex zeros.

STEP 2: The Rational Zeros Theorem provides information about the potential rational zeros of polynomials with integer coefficients. For this polynomial (which has integer coefficients), the potential rational zeros are

$$\pm\frac{1}{3}, \pm\frac{2}{3}, \pm1, \pm2, \pm3, \pm6, \pm9, \pm18$$

Figure 81

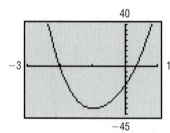

STEP 3: Figure 81 shows the graph of f. The graph has the characteristics that we expect of this polynomial of degree 4: It behaves like $y = 3x^4$ for large $|x|$ and has y-intercept -18. There are x-intercepts near -2 and between 0 and 1.

STEP 4: Because $f(-2) = 0$, we known that -2 is a zero and $x + 2$ is a factor of f. We can use long division or synthetic division to factor f:

$$f(x) = (x + 2)(3x^3 - x^2 + 27x - 9)$$

From the graph of f and the list of potential rational zeros, it appears that $\frac{1}{3}$ may also be a zero of f. Since $f\left(\frac{1}{3}\right) = 0$, we know that $\frac{1}{3}$ is a zero of f. We use synthetic division on the depressed equation of f to factor.

$$\frac{1}{3})\overline{\smash{)}3 \quad -1 \quad 27 \quad -9}$$

$$\begin{array}{r} \underline{ 1 \quad0 \quad9} \\ 3 \quad0 \quad 27 \quad0 \end{array}$$

Using the bottom row of the synthetic division, we find

$$f(x) = (x + 2)\left(x - \frac{1}{3}\right)(3x^2 + 27) = 3(x + 2)\left(x - \frac{1}{3}\right)(x^2 + 9)$$

The factor $x^2 + 9$ does not have any real zeros; its complex zeros are $\pm 3i$. The complex zeros of $f(x) = 3x^4 + 5x^3 + 25x^2 + 45x - 18$ are

$$-2, \frac{1}{3}, 3i, -3i.$$

◀

NOW WORK PROBLEM 33.

3.7 Assess Your Understanding

'Are You Prepared?'

Answers are given at the end of these exercises. If you get a wrong answer, read the pages listed in red.

1. Find the sum and the product of the complex numbers $3 - 2i$ and $-3 + 5i$. (pp. 708–710)

2. In the complex number system, solve the equation $x^2 + 2x + 2 = 0$. (pp. 713–714)

Concepts and Vocabulary

3. Every polynomial function of odd degree with real coefficients will have at least _____ real zero(s).

4. If $3 + 4i$ is a zero of a polynomial function of degree 5 with real coefficients, then so is _____.

5. *True or False:* A polynomial function of degree n with real coefficients has exactly n complex zeros. At most n of them are real zeros.

6. *True or False:* A polynomial function of degree 4 with real coefficients could have $-3, 2 + i, 2 - i$, and $-3 + 5i$ as its zeros.

Skill Building

In Problems 7–16, information is given about a polynomial $f(x)$ whose coefficients are real numbers. Find the remaining zeros of f.

7. Degree 3;   zeros: $3, 4 - i$

8. Degree 3;   zeros: $4, 3 + i$

9. Degree 4;   zeros: $i, 1 + i$

10. Degree 4;   zeros: $1, 2, 2 + i$

11. Degree 5;   zeros: $1, i, 2i$

12. Degree 5;   zeros: $0, 1, 2, i$

13. Degree 4;   zeros: $i, 2, -2$

14. Degree 4;   zeros: $2 - i, -i$

15. Degree 6;   zeros: $2, 2 + i, -3 - i, 0$

16. Degree 6;   zeros: $i, 3 - 2i, -2 + i$

In Problems 17–22, form a polynomial $f(x)$ with real coefficients having the given degree and zeros.

17. Degree 4;   zeros: $3 + 2i$; 4, multiplicity 2

18. Degree 4;   zeros: $i, 1 + 2i$

19. Degree 5;   zeros: $2; -i; 1 + i$

20. Degree 6;   zeros: $i, 4 - i; 2 + i$

21. Degree 4;   zeros: 3, multiplicity 2; $-i$

22. Degree 5;   zeros: 1, multiplicity 3; $1 + i$

In Problems 23–30, use the given zero to find the remaining zeros of each function.

23. $f(x) = x^3 - 4x^2 + 4x - 16$;   zero: $2i$

24. $g(x) = x^3 + 3x^2 + 25x + 75$;   zero: $-5i$

25. $f(x) = 2x^4 + 5x^3 + 5x^2 + 20x - 12$;   zero: $-2i$

26. $h(x) = 3x^4 + 5x^3 + 25x^2 + 45x - 18$;   zero: $3i$

27. $h(x) = x^4 - 9x^3 + 21x^2 + 21x - 130$;   zero: $3 - 2i$

28. $f(x) = x^4 - 7x^3 + 14x^2 - 38x - 60$;   zero: $1 + 3i$

29. $h(x) = 3x^5 + 2x^4 + 15x^3 + 10x^2 - 528x - 352$;   zero: $-4i$

30. $g(x) = 2x^5 - 3x^4 - 5x^3 - 15x^2 - 207x + 108$;   zero: $3i$

In Problems 31–40, find the complex zeros of each polynomial function. Write f in factored form.

31. $f(x) = x^3 - 1$

32. $f(x) = x^4 - 1$

33. $f(x) = x^3 - 8x^2 + 25x - 26$

34. $f(x) = x^3 + 13x^2 + 57x + 85$

35. $f(x) = x^4 + 5x^2 + 4$

36. $f(x) = x^4 + 13x^2 + 36$

37. $f(x) = x^4 + 2x^3 + 22x^2 + 50x - 75$

38. $f(x) = x^4 + 3x^3 - 19x^2 + 27x - 252$

39. $f(x) = 3x^4 - x^3 - 9x^2 + 159x - 52$

40. $f(x) = 2x^4 + x^3 - 35x^2 - 113x + 65$

Discussion and Writing

In Problems 41 and 42, explain why the facts given are contradictory.

41. $f(x)$ is a polynomial of degree 3 whose coefficients are real numbers; its zeros are $4 + i, 4 - i$, and $2 + i$.

42. $f(x)$ is a polynomial of degree 3 whose coefficients are real numbers; its zeros are $2, i$, and $3 + i$.

43. $f(x)$ is a polynomial of degree 4 whose coefficients are real numbers; three of its zeros are $2, 1 + 2i$, and $1 - 2i$. Explain why the remaining zero must be a real number.

44. $f(x)$ is a polynomial of degree 4 whose coefficients are real numbers; two of its zeros are -3 and $4 - i$. Explain why one of the remaining zeros must be a real number. Write down one of the missing zeros.

'Are You Prepared?' Answers

1. Sum: $3i$; product: $1 + 21i$ **2.** $\{-1 - i, -1 + i\}$

Chapter Review

Things to Know

Quadratic function (pp. 150–157)

$f(x) = ax^2 + bx + c$

Graph is a parabola that opens up if $a > 0$ and opens down if $a < 0$.

Vertex: $\left(-\dfrac{b}{2a}, f\left(-\dfrac{b}{2a} \right) \right)$

Axis of symmetry: $x = -\dfrac{b}{2a}$

y-intercept: $f(0)$

x-intercept(s): If any, found by finding the real solutions of the equation $ax^2 + bx + c = 0$.

Power function (pp. 171–174)

$f(x) = x^n, \quad n \geq 2$ even

Domain: all real numbers   Range: nonnegative real numbers

Passes through $(-1, 1), (0, 0), (1, 1)$

Even function

Decreasing on $(-\infty, 0)$, increasing on $(0, \infty)$

$f(x) = x^n, \quad n \geq 3$ odd

Domain: all real numbers   Range: all real numbers

Passes through $(-1, -1), (0, 0), (1, 1)$

Odd function

Increasing on $(-\infty, \infty)$

Polynomial function (p. 170 and pp. 174–178)

$f(x) = a_n x^n + a_{n-1} x^{n-1} + \cdots$
$\quad + a_1 x + a_0, \quad a_n \neq 0$

Domain: all real numbers

At most $n - 1$ turning points

End behavior: Behaves like $y = a_n x^n$ for large $|x|$

Zeros of a polynomial function f (p. 175)

Numbers for which $f(x) = 0$; the real zeros of f are the x-intercepts of the graph of f.

Rational function (p. 186)

$$R(x) = \frac{p(x)}{q(x)}$$

p, q are polynomial functions.

Domain: $\{x \mid q(x) \neq 0\}$

Vertical asymptotes: With $R(x)$ in lowest terms, if $q(r) = 0$, then $x = r$ is a vertical asymptote.

Horizontal or oblique asymptotes: See the summary on page 198.

Inverse Variation (p. 205)

Let x and y denote two quantities. Then y varies inversely with x, or y is inversely proportional to x, if there is a nonzero constant k such that $y = \dfrac{k}{x}$.

Remainder Theorem (p. 220)

If a polynomial $f(x)$ is divided by $x - c$, then the remainder is $f(c)$.

Factor Theorem (p. 220)

$x - c$ is a factor of a polynomial $f(x)$ if and only if $f(c) = 0$.

Rational Zeros Theorem (p. 222)

Let f be a polynomial function of degree 1 or higher of the form

$$f(x) = a_n x^n + a_{n-1} x^{n-1} + \cdots + a_1 x + a_0, \quad a_n \neq 0, a_0 \neq 0$$

where each coefficient is an integer. If $\dfrac{p}{q}$, in lowest terms, is a rational zero of f, then p must be a factor of a_0 and q must be a factor of a_n.

Intermediate Value Theorem (p. 229)

Let f denote a continuous function. If $a < b$ and $f(a)$ and $f(b)$ are of opposite sign, then f has at least one zero between a and b.

Fundamental Theorem of Algebra (p. 233)

Every complex polynomial function $f(x)$ of degree $n \geq 1$ has at least one complex zero.

Conjugate Pairs Theorem (p. 234)

Let $f(x)$ be a polynomial whose coefficients are real numbers. If $r = a + bi$ is a zero of f, then its complex conjugate $\bar{r} = a - bi$ is also a zero of f.

Objectives

Section		You should be able to …	Review Exercises
3.1	1	Graph a quadratic function using transformations (p. 151)	1–6
	2	Identify the vertex and axis of symmetry of a quadratic function (p. 153)	7–16
	3	Graph a quadratic function using its vertex, axis, and intercepts (p. 154)	7–16
	4	Use the maximum or minimum value of a quadratic function to solve applied problems (p. 158)	115–122
	5	Use a graphing utility to find the quadratic function of best fit to data (p. 162)	125
3.2	1	Identify polynomial functions and their degree (p. 170)	23–26
	2	Graph polynomial functions using transformations (p. 171)	1–6, 27–32
	3	Identify the zeros of a polynomial function and their multiplicity (p. 174)	33–40
	4	Analyze the graph of a polynomial function (p. 179)	33–40
	5	Find the cubic function of best fit to data (p. 181)	126
3.3	1	Find the domain of a rational function (p. 187)	41–44
	2	Find the vertical asymptotes of a rational function (p. 190)	41–44
	3	Find the horizontal or oblique asymptotes of a rational function (p. 191)	41–44
3.4	1	Analyze the graph of a rational function (p. 198)	45–56
	2	Solve applied problems involving rational functions (p. 204)	127
	3	Construct a model using inverse variation (p. 205)	123

Review Exercises

In Problems 1–6, graph each function using transformations (shifting, compressing, stretching, and reflection). Verify your result using a graphing utility.

1. $f(x) = (x - 2)^2 + 2$ 　　　　**2.** $f(x) = (x + 1)^2 - 4$ 　　　　**3.** $f(x) = -(x - 4)^2$

4. $f(x) = (x - 1)^2 - 3$ 　　　　**5.** $f(x) = 2(x + 1)^2 + 4$ 　　　　**6.** $f(x) = -3(x + 2)^2 + 1$

In Problems 7–16, graph each quadratic function by determining whether its graph opens up or down and by finding its vertex, axis of symmetry, y-intercept, and x-intercepts, if any.

7. $f(x) = (x - 2)^2 + 2$ 　　**8.** $f(x) = (x + 1)^2 - 4$ 　　**9.** $f(x) = \dfrac{1}{4}x^2 - 16$ 　　**10.** $f(x) = -\dfrac{1}{2}x^2 + 2$

11. $f(x) = -4x^2 + 4x$ 　　**12.** $f(x) = 9x^2 - 6x + 3$ 　　**13.** $f(x) = \dfrac{9}{2}x^2 + 3x + 1$ 　　**14.** $f(x) = -x^2 + x + \dfrac{1}{2}$

15. $f(x) = 3x^2 + 4x - 1$ 　　**16.** $f(x) = -2x^2 - x + 4$

In Problems 17–22, determine whether the given quadratic function has a maximum value or a minimum value, and then find the value.

17. $f(x) = 3x^2 - 6x + 4$ 　　　　**18.** $f(x) = 2x^2 + 8x + 5$ 　　　　**19.** $f(x) = -x^2 + 8x - 4$

20. $f(x) = -x^2 - 10x - 3$ 　　　　**21.** $f(x) = -3x^2 + 12x + 4$ 　　　　**22.** $f(x) = -2x^2 + 4$

In Problems 23–26, determine which functions are polynomial functions. For those that are, state the degree. For those that are not, tell why not.

23. $f(x) = 4x^5 - 3x^2 + 5x - 2$ 　　**24.** $f(x) = \dfrac{3x^5}{2x + 1}$ 　　**25.** $f(x) = 3x^2 + 5x^{1/2} - 1$ 　　**26.** $f(x) = 3$

In Problems 27–32, graph each function using transformations (shifting, compressing, stretching, and reflection). Show all the stages. Verify your result using a graphing utility.

27. $f(x) = (x + 2)^3$ 　　　　**28.** $f(x) = -x^3 + 3$ 　　　　**29.** $f(x) = -(x - 1)^4$

30. $f(x) = (x - 1)^4 - 2$ 　　　　**31.** $f(x) = 2(x + 1)^4 + 2$ 　　　　**32.** $f(x) = (1 - x)^3$

In Problems 33–40, for each polynomial function f:
 (a) *Find the x- and y-intercepts of the graph of f.*
 (b) *Determine whether the graph crosses or touches the x-axis at each x-intercept.*
 (c) *End behavior: Find the power function that the graph of f resembles for large values of $|x|$.*
 (d) *Use a graphing utility to graph f.*
 (e) *Determine the number of turning points on the graph of f. Approximate the turning points if any exist, rounded to two decimal places.*
 (f) *Use the information obtained in parts (a) to (e) to draw a complete graph of f by hand.*
 (g) *Find the domain of f. Use the graph to find the range of f.*
 (h) *Use the graph to determine where f is increasing and where f is decreasing.*

33. $f(x) = x(x + 2)(x + 4)$ 　　　　**34.** $f(x) = x(x - 2)(x - 4)$ 　　　　**35.** $f(x) = (x - 2)^2(x + 4)$

36. $f(x) = (x - 2)(x + 4)^2$ **37.** $f(x) = x^3 - 4x^2$ **38.** $f(x) = x^3 + 4x$

39. $f(x) = (x - 1)^2(x + 3)(x + 1)$ **40.** $f(x) = (x - 4)(x + 2)^2(x - 2)$

In Problems 41–44, find the domain of each rational function. Find any horizontal, vertical, or oblique asymptotes.

41. $R(x) = \dfrac{x + 2}{x^2 - 9}$ **42.** $R(x) = \dfrac{x^2 + 4}{x - 2}$ **43.** $R(x) = \dfrac{x^2 + 3x + 2}{(x + 2)^2}$ **44.** $R(x) = \dfrac{x^3}{x^3 - 1}$

In Problems 45–56, discuss each rational function following the eight steps on page 198.

45. $R(x) = \dfrac{2x - 6}{x}$ **46.** $R(x) = \dfrac{4 - x}{x}$ **47.** $H(x) = \dfrac{x + 2}{x(x - 2)}$ **48.** $H(x) = \dfrac{x}{x^2 - 1}$

49. $R(x) = \dfrac{x^2 + x - 6}{x^2 - x - 6}$ **50.** $R(x) = \dfrac{x^2 - 6x + 9}{x^2}$ **51.** $F(x) = \dfrac{x^3}{x^2 - 4}$ **52.** $F(x) = \dfrac{3x^3}{(x - 1)^2}$

53. $R(x) = \dfrac{2x^4}{(x - 1)^2}$ **54.** $R(x) = \dfrac{x^4}{x^2 - 9}$ **55.** $G(x) = \dfrac{x^2 - 4}{x^2 - x - 2}$ **56.** $F(x) = \dfrac{(x - 1)^2}{x^2 - 1}$

In Problems 57–66, solve each inequality (a) algebraically and (b) graphically.

57. $2x^2 + 5x - 12 < 0$ **58.** $3x^2 - 2x - 1 \geq 0$ **59.** $\dfrac{6}{x + 3} \geq 1$ **60.** $\dfrac{-2}{1 - 3x} < 1$

61. $\dfrac{2x - 6}{1 - x} < 2$ **62.** $\dfrac{3 - 2x}{2x + 5} \geq 2$ **63.** $\dfrac{(x - 2)(x - 1)}{x - 3} > 0$ **64.** $\dfrac{x + 1}{x(x - 5)} \leq 0$

65. $\dfrac{x^2 - 8x + 12}{x^2 - 16} > 0$ **66.** $\dfrac{x(x^2 + x - 2)}{x^2 + 9x + 20} \leq 0$

In Problems 67–70, find the remainder R when $f(x)$ is divided by $g(x)$. Is g a factor of f?

67. $f(x) = 8x^3 - 3x^2 + x + 4$; $g(x) = x - 1$ **68.** $f(x) = 2x^3 + 8x^2 - 5x + 5$; $g(x) = x - 2$

69. $f(x) = x^4 - 2x^3 + 15x - 2$; $g(x) = x + 2$ **70.** $f(x) = x^4 - x^2 + 2x + 2$; $g(x) = x + 1$

71. Find the value of $f(x) = 12x^6 - 8x^4 + 1$ at $x = 4$.   **72.** Find the value of $f(x) = -16x^3 + 18x^2 - x + 2$ at $x = -2$.

In Problems 73 and 74, tell the maximum number of real zeros that each polynomial function may have. Then list the potential rational zeros of each polynomial function. Do not attempt to find the zeros.

73. $f(x) = 2x^8 - x^7 + 8x^4 - 2x^3 + x + 3$ **74.** $f(x) = -6x^5 + x^4 + 5x^3 + x + 1$

In Problems 75–80, find all the real zeros of each polynomial function.

75. $f(x) = x^3 - 3x^2 - 6x + 8$ **76.** $f(x) = x^3 - x^2 - 10x - 8$

77. $f(x) = 4x^3 + 4x^2 - 7x + 2$ **78.** $f(x) = 4x^3 - 4x^2 - 7x - 2$

79. $f(x) = x^4 - 4x^3 + 9x^2 - 20x + 20$ **80.** $f(x) = x^4 + 6x^3 + 11x^2 + 12x + 18$

In Problems 81–84, determine the real zeros of the polynomial function. Approximate all irrational zeros rounded to two decimal places.

81. $f(x) = 2x^3 - 11.84x^2 - 9.116x + 82.46$ **82.** $f(x) = 12x^3 + 39.8x^2 - 4.4x - 3.4$

83. $g(x) = 15x^4 - 21.5x^3 - 1718.3x^2 + 5308x + 3796.8$ **84.** $g(x) = 3x^4 + 67.93x^3 + 486.265x^2 + 1121.32x + 412.195$

In Problems 85–88, find the real solutions of each equation.

85. $2x^4 + 2x^3 - 11x^2 + x - 6 = 0$ **86.** $3x^4 + 3x^3 - 17x^2 + x - 6 = 0$

87. $2x^4 + 7x^3 + x^2 - 7x - 3 = 0$ **88.** $2x^4 + 7x^3 - 5x^2 - 28x - 12 = 0$

In Problems 89–92, find bounds to the zeros of each polynomial function. Obtain a complete graph of f.

89. $f(x) = x^3 - x^2 - 4x + 2$ **90.** $f(x) = x^3 + x^2 - 10x - 5$

91. $f(x) = 2x^3 - 7x^2 - 10x + 35$ **92.** $f(x) = 3x^3 - 7x^2 - 6x + 14$

In Problems 93–96, use the Intermediate Value Theorem to show that each polynomial has a zero in the given interval. Approximate the zero rounded to two decimal places.

93. $f(x) = 3x^3 - x - 1;$ $[0, 1]$

94. $f(x) = 2x^3 - x^2 - 3;$ $[1, 2]$

95. $f(x) = 8x^4 - 4x^3 - 2x - 1;$ $[0, 1]$

96. $f(x) = 3x^4 + 4x^3 - 8x - 2;$ $[1, 2]$

In Problems 97–100, information is given about a complex polynomial $f(x)$ whose coefficients are real numbers. Find the remaining zeros of f. Write a polynomial function whose zeros are given.

97. Degree 3;   zeros: $4 + i, 6$

98. Degree 3;   zeros: $3 + 4i, 5$

99. Degree 4;   zeros: $i, 1 + i$

100. Degree 4;   zeros: $1, 2, 1 + i$

In Problems 101–114, solve each equation in the complex number system.

101. $x^2 + x + 1 = 0$

102. $x^2 - x + 1 = 0$

103. $2x^2 + x - 2 = 0$

104. $3x^2 - 2x - 1 = 0$

105. $x^2 + 3 = x$

106. $2x^2 + 1 = 2x$

107. $x(1 - x) = 6$

108. $x(1 + x) = 2$

109. $x^4 + 2x^2 - 8 = 0$

110. $x^4 + 8x^2 - 9 = 0$

111. $x^3 - x^2 - 8x + 12 = 0$

112. $x^3 - 3x^2 - 4x + 12 = 0$

113. $3x^4 - 4x^3 + 4x^2 - 4x + 1 = 0$

114. $x^4 + 4x^3 + 2x^2 - 8x - 8 = 0$

---

115. Find the point on the line $y = x$ that is closest to the point $(3, 1)$.

[**Hint:** Find the minimum value of the function $f(x) = d^2$, where d is the distance from $(3, 1)$ to a point on the line.]

116. Landscaping   A landscape engineer has 200 feet of border to enclose a rectangular pond. What dimensions will result in the largest pond?

117. Enclosing the Most Area with a Fence   A farmer with 10,000 meters of fencing wants to enclose a rectangular field and then divide it into two plots with a fence parallel to one of the sides (see the figure). What is the largest area that can be enclosed?

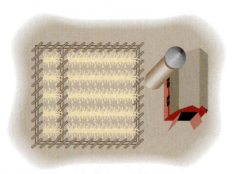

118. A rectangle has one vertex on the line $y = 8 - 2x, x > 0$, another at the origin, one on the positive x-axis, and one on the positive y-axis. Find the largest area A that can be enclosed by the rectangle.

119. Architecture   A special window in the shape of a rectangle with semicircles at each end is to be constructed so that the outside dimensions are 100 feet in length. See the illustration. Find the dimensions that maximizes the area of the rectangle.

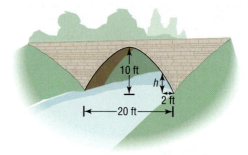

120. Parabolic Arch Bridges   A horizontal bridge is in the shape of a parabolic arch. Given the information shown in the figure, what is the height h of the arch 2 feet from shore?

121. Minimizing Marginal Cost   The marginal cost of a product can be thought of as the cost of producing one additional unit of output. For example, if the marginal cost of producing the 50th product is $6.20, then it cost $6.20 to increase production from 49 to 50 units of output. Callaway Golf Company has determined that the marginal cost C of manufacturing x Big Bertha golf clubs may be expressed by the quadratic function

$$C(x) = 4.9x^2 - 617.4x + 19,600$$

(a) How many clubs should be manufactured to minimize the marginal cost?

(b) At this level of production, what is the marginal cost?

122. Violent Crimes   The function

$$V(t) = -10.0t^2 + 39.2t + 1862.6$$

models the number V (in thousands) of violent crimes committed in the United States t years after 1990. So $t = 0$ represents 1990, $t = 1$ represents 1991, and so on.
(a) Determine the year in which the most violent crimes were committed.
(b) Approximately how many violent crimes were committed during this year?
(c) Using a graphing utility, graph $V = V(t)$. Were the number of violent crimes increasing or decreasing during the years 1994 to 1998?

SOURCE: Based on data obtained from the Federal Bureau of Investigation.

123. **Weight of a Body** The weight of a body varies inversely with the square of its distance from the center of Earth. Assuming that the radius of Earth is 3960 miles, how much would a man weigh at an altitude of 1 mile above Earth's surface if he weighs 200 pounds on Earth's surface?

124. **Resistance due to a Conductor** The resistance (in ohms) of a circular conductor varies directly with the length of the conductor and inversely with the square of the radius of the conductor. If 50 feet of wire with a radius of 6×10^{-3} inch has a resistance of 10 ohms, what would be the resistance of 100 feet of the same wire if the radius is increased to 7×10^{-3} inch?

125. **Advertising** A small manufacturing firm collected the following data on advertising expenditures A (in thousands of dollars) and total revenue R (in thousands of dollars).

Advertising	Total Revenue
20	$6101
22	$6222
25	$6350
25	$6378
27	$6453
28	$6423
29	$6360
31	$6231

(a) Draw a scatter diagram of the data. Comment on the type of relation that may exist between the two variables.
(b) Use a graphing utility to find the quadratic function of best fit to these data.
(c) Use the function found in part (b) to determine the optimal level of advertising for this firm.
(d) Use the function found in part (b) to find the revenue that the firm can expect if it uses the optimal level of advertising.
(e) With a graphing utility, graph the quadratic function of best fit on the scatter diagram.

126. **AIDS Cases in the United States** The following data represent the cumulative number of reported AIDS cases in the United States for 1990–1997.

Year, t	Number of AIDS Cases, A
1990, 1	193,878
1991, 2	251,638
1992, 3	326,648
1993, 4	399,613
1994, 5	457,280
1995, 6	528,215
1996, 7	594,760
1997, 8	653,253

SOURCE: U.S. Center for Disease Control and Prevention

(a) Draw a scatter diagram of the data.
(b) The cubic function of best fit to these data is

$$A(t) = -212t^3 + 2429t^2 + 59{,}569t + 130{,}003$$

Use this function to predict the cumulative number of AIDS cases reported in the United States in 2000.
(c) Use a graphing utility to verify that the function given in part (b) is the cubic function of best fit.
(d) With a graphing utility, draw a scatter diagram of the data and then graph the cubic function of best fit on the scatter diagram.
(e) Do you think the function given in part (b) will be useful in predicting the number of AIDS cases in 2005?

127. **Making a Can** A can in the shape of a right circular cylinder is required to have a volume of 250 cubic centimeters.
(a) Express the amount A of material to make the can as a function of the radius r of the cylinder.
(b) How much material is required if the can is of radius 3 centimeters?
(c) How much material is required if the can is of radius 5 centimeters?
(d) Graph $A = A(r)$. For what value of r is A smallest?

128. Design a polynomial function with the following characteristics: degree 6; four real zeros, one of multiplicity 3; y-intercept 3; behaves like $y = -5x^6$ for large values of $|x|$. Is this polynomial unique? Compare your polynomial with those of other students. What terms will be the same as everyone else's? Add some more characteristics, such as symmetry or naming the real zeros. How does this modify the polynomial?

129. Design a rational function with the following characteristics: three real zeros, one of multiplicity 2; y-intercept 1; vertical asymptotes $x = -2$ and $x = 3$; oblique asymptote $y = 2x + 1$. Is this rational function unique? Compare yours with those of other students. What will be the same as everyone else's? Add some more characteristics, such as symmetry or naming the real zeros. How does this modify the rational function?

130. The illustration shows the graph of a polynomial function.
  (a) Is the degree of the polynomial even or odd?
  (b) Is the leading coefficient positive or negative?
  (c) Is the function even, odd, or neither?
  (d) Why is x^2 necessarily a factor of the polynomial?
  (e) What is the minimum degree of the polynomial?
  (f) Formulate five different polynomials whose graphs could look like the one shown. Compare yours to those of other students. What similarities do you see? What differences?

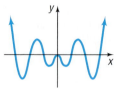

Chapter Test

1. Graph $f(x) = (x - 3)^4 - 2$.

2. $f(x) = 3x^2 - 12x + 4$
  (a) Determine if the function has a maximum or a minimum and explain how you know.
  (b) Algebraically, determine the vertex of the graph.
  (c) Determine the axis of symmetry of the graph.
  (d) Algebraically, determine the intercepts of the graph.
  (e) Graph the function by hand.

3. For the polynomial function $g(x) = 2x^3 + 5x^2 - 28x - 15$,
  (a) Determine the maximum number of real zeros that the function may have.
  (b) Find bounds to the zeros of the function.
  (c) List the potential rational zeros.
  (d) Determine the real zeros of g. Factor g over the reals.

4. Find the complex zeros of $f(x) = x^3 - 4x^2 + 25x - 100$.

5. Solve $3x^3 + 2x - 1 = 8x^2 - 4$ in the complex number system.

In Problems 6 and 7, find the domain of each function. Find any horizontal, vertical, or oblique asymptotes.

6. $g(x) = \dfrac{2x^2 - 14x + 24}{x^2 + 6x - 40}$

7. $r(x) = \dfrac{x^2 + 2x - 3}{x + 1}$

8. Sketch the graph of the function in Problem 7. Label all intercepts, vertical asymptotes, horizontal asymptotes, and oblique asymptotes. Verify your results using a graphing utility.

In Problems 9 and 10, write a function that meets the given conditions.

9. Fourth-degree polynomial with real coefficients; zeros: $-2, 0, 3 + i$

10. Rational function; asymptotes: $y = 2, x = 4$; domain: $\{x \mid x \neq 4, x \neq 9\}$

11. Use the Intermediate Value Theorem to show that the function $f(x) = -2x^2 - 3x + 8$ has at least one real zero on the interval $[0, 4]$.

In Problems 12 and 13, solve each inequality algebraically. Write the solution set in interval notation. Verify your results using a graphing utility.

12. $3x^2 - x - 4 \geq x^2 - 3x + 8$

13. $\dfrac{x + 2}{x - 3} < 2$

14. Given the polynomial function $f(x) = -2(x - 1)^2(x + 2)$, do the following:
  (a) Find the x- and y-intercepts of the graph of f.
  (b) Determine whether the graph crosses or touches the x-axis at each x-intercept.
  (c) Find the power function that the graph of f resembles for large values of $|x|$.
  (d) Using results of parts (a)–(c), describe what the graph of f should look like. Verify this with a graphing utility.
  (e) Approximate the turning points of f rounded to two decimal places.
  (f) By hand, use the information in parts (a)–(e) to draw a complete graph of f.

15. The table gives the average home game attendance A for the St. Louis Cardinals (in thousands) during the years 1995–2003, where x is the number of years since 1995.

SOURCE: www.baseball-almanac.com

x	0	1	2	3	4	5	6	7	8
A	24.570	32.774	32.519	39.453	40.197	41.191	38.390	37.182	35.930

(a) Use a graphing utility to find the quadratic function of best fit to the data. Round coefficients to three decimal places.
(b) Use the function found in part (a) to estimate the average home game attendance for the St. Louis Cardinals in 2004.

Chapter Projects

Date	Grains per cubic meter	Date	Grains per cubic meter
8/4	18	9/8	45
8/9	14	9/10	93
8/12	12	9/13	14
8/13	16	9/15	10
8/17	37	9/17	11
8/18	30	9/20	5
8/23	53	9/22	3
8/24	22	9/27	4
8/25	47	9/29	6
8/26	7	10/1	4
8/31	214	10/4	4
9/2	31	10/5	7
9/3	44	10/6	7
9/6	50		

SOURCE: National Allergy Bureau at the American Academy of Allergy, Asthma and Immunology website (*www.aaaai.org*)

1. **Weed Pollen**   Given in the table below is the Weed pollen count for Lexington, Kentucky, from August 4, 2004, to October 6, 2004, the prime season for weed pollen.

   (a) Let the independent variable D represent the date, where $D = 1$ on 8/4, $D = 6$ on 8/9, $D = 9$ on 8/12… and $D = 64$ on 10/6. Let the dependent variable P represent the number of grains per cubic meter of weed pollen. Draw a scatter diagram of the data using your graphing utility, and by hand on graph paper.

   (b) In the hand-drawn scatter diagram, sketch a smooth curve that fits the data. Try to have the curve pass through as many of the points as closely as possible. How many turning points are there? What does this tell you about the degree of the polynomial that could be fit to the data? If the ends of the graph on the left and on the right are extended downward, how many real zeros are indicated by the graph? How many complex? Why?

   (c) Use a graphing utility to determine the quartic function of the best fit. Graph it on your graphing utility. Does it look like what you expected from your hand-drawn graph? Explain.

   (d) Use a graphing utility to determine the cubic function of best fit. Graph it on your graphing utility. Does it look like what you expected from your hand-drawn graph? Explain.

   (e) Use a graphing utility to determine the quadratic function of best fit. Graph it on your graphing utility. Does it look like what you expected from your hand-drawn graph? Explain.

   (f) Which of the three functions seems to fit the best? Explain your reasoning.

   (g) Investigate weed pollen counts at other times of the year. (Go to the National Allergy Bureau at www.aaaai.org.) Would the polynomial you found in part (c) work for other 2-month periods? Why or why not? What about for the same period of time in other years?

The following projects are available at the Instructor's Resource Center (IRC):

2. **Project at Motorola**   *How Many Cellphones Can I Make?*

3. **First and Second Differences**

4. **Cannons**

5. **Maclaurin Series**

6. **Theory of Equations**

7. **CBL Experiment**

Cumulative Review

1. Find the distance between the points $P = (1, 3)$ and $Q = (-4, 2)$.

2. Solve the inequality $x^2 \geq x$ and graph the solution set.

3. Solve the inequality $x^2 - 3x < 4$ and graph the solution set.

4. Find a linear function with slope -3 that contains the point $(-1, 4)$. Graph the function.

5. Find the equation of the line parallel to the line $y = 2x + 1$ and containing the point $(3, 5)$. Express your answer in slope–intercept form and graph the line.

6. Graph the equation $y = x^3$.

7. Does the relation $\{(3, 6), (1, 3), (2, 5), (3, 8)\}$ represent a function? Why or why not?

8. Solve the equation $x^3 - 6x^2 + 8x = 0$.

9. Solve the inequality $3x + 2 \leq 5x - 1$ and graph the solution set.

10. Find the center and radius of the circle $x^2 + 4x + y^2 - 2y - 4 = 0$. Graph the circle.

11. For the equation $y = x^3 - 9x$, determine the intercepts and test for symmetry.

12. Find an equation of the line perpendicular to $3x - 2y = 7$ that contains the point $(1, 5)$.

13. Is the following graph the graph of a function? Why or why not?

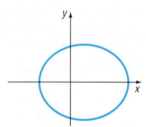

14. For the function $f(x) = x^2 + 5x - 2$, find
    (a) $f(3)$
    (b) $f(-x)$
    (c) $-f(x)$
    (d) $f(3x)$
    (e) $\dfrac{f(x + h) - f(x)}{h}, h \neq 0$

15. Answer the following questions regarding the function $f(x) = \dfrac{x + 5}{x - 1}$.
    (a) What is the domain of f?
    (b) Is the point $(2, 6)$ on the graph of f?
    (c) If $x = 3$, what is $f(x)$? What point is on the graph of f?
    (d) If $f(x) = 9$, what is x? What point is on the graph of f?

16. Graph the function $f(x) = -3x + 7$.

17. Graph $f(x) = 2x^2 - 4x + 1$ by determining whether its graph opens up or down and by finding its vertex, axis of symmetry, y-intercept, and x-intercepts, if any.

18. Find the average rate of change of $f(x) = x^2 + 3x + 1$ from 1 to x. Use this result to find the slope of the secant line containing $(1, f(1))$ and $(2, f(2))$.

19. In parts (a) to (f) use the following graph.

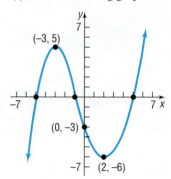

    (a) Determine the intercepts.
    (b) Based on the graph, tell whether the graph is symmetric with respect to the x-axis, the y-axis, and/or the origin.
    (c) Based on the graph, tell whether the function is even, odd, or neither.
    (d) List the intervals on which f is increasing. List the intervals on which f is decreasing.
    (e) List the numbers, if any, at which f has a local maximum. What are these local maxima?
    (f) List the numbers, if any, at which f has a local minimum. What are these local minima?

20. Determine algebraically whether the function
$$f(x) = \frac{5x}{x^2 - 9}$$
is even, odd, or neither.

21. For the function $f(x) = \begin{cases} 2x + 1 & \text{if } -3 < x < 2 \\ -3x + 4 & \text{if } x \geq 2 \end{cases}$
    (a) Find the domain of f.
    (b) Locate any intercepts.
    (c) Graph the function.
    (d) Based on the graph, find the range.

22. Graph the function $f(x) = -3(x + 1)^2 + 5$ using transformations.

23. Suppose that $f(x) = x^2 - 5x + 1$ and $g(x) = -4x - 7$.
    (a) Find $f + g$ and state its domain.
    (b) Find $\dfrac{f}{g}$ and state its domain.

24. **Demand Equation**   The price p (in dollars) and the quantity x sold of a certain product obey the demand equation
$$p = -\frac{1}{10}x + 150, \qquad 0 \leq x \leq 1500$$
    (a) Express the revenue R as a function of x.
    (b) What is the revenue if 100 units are sold?
    (c) What quantity x maximizes revenue? What is the maximum revenue?
    (d) What price should the company charge to maximize revenue?

Exponential and Logarithmic Functions

4

The McDonald's Scalding Coffee Case

April 3, 1996

There is a lot of hype about the McDonald's scalding coffee case. No one is in favor of frivolous cases or outlandish results; however, it is important to understand some points that were not reported in most of the stories about the case. McDonald's coffee was not only hot, it was scalding, capable of almost instantaneous destruction of skin, flesh, and muscle.

Plaintiff's expert, a scholar in thermodynamics applied to human skin burns, testified that liquids at 180 degrees will cause a full thickness burn to human skin in two to seven seconds. Other testimony showed that, as the temperature decreases toward 155 degrees, the extent of the burn relative to that temperature decreases exponentially. Thus, if (the) spill had involved coffee at 155 degrees, the liquid would have cooled and given her time to avoid a serious burn.

ATLA fact sheet © 1995, 1996 Consumer Attorneys of CA. Used with permission of the Association of Trial Lawyers of America.

—See Chapter Project 1.

A LOOK BACK Until now, our study of functions has concentrated on polynomial and rational functions. These functions belong to the class of **algebraic functions**, that is, functions that can be expressed in terms of sums, differences, products, quotients, powers, or roots of polynomials. Functions that are not algebraic are termed **transcendental** (they transcend, or go beyond, algebraic functions).

A LOOK AHEAD In this chapter, we study two transcendental functions: the exponential function and the logarithmic function. These functions occur frequently in a wide variety of applications, such as biology, chemistry, economics, and psychology.

The chapter begins with a discussion of composite functions.

OUTLINE

4.1 Composite Functions

PREPARING FOR THIS SECTION   *Before getting started, review the following:*

• Finding Values of a Function (Section 2.1, pp. 61–63)   • Domain of a Function (Section 2.1, pp. 57–60 and 64–65)

 Now work the 'Are You Prepared?' problems on page 253.

OBJECTIVES   **1** Form a Composite Function
                2 Find the Domain of a Composite Function

 Form a Composite Function

Suppose that an oil tanker is leaking oil and we want to be able to determine the area of the circular oil patch around the ship. See Figure 1. It is determined that the oil is leaking from the tanker in such a way that the radius of the circular oil patch around the ship is increasing at a rate of 3 feet per minute. Therefore, the radius r of the oil patch at any time t, in minutes, is given by $r(t) = 3t$. So after 20 minutes, the radius of the oil patch is $r(20) = 3(20) = 60$ feet. The area A of a circle as a function of the radius r is given by $A(r) = \pi r^2$. The area of the circular patch of oil after 20 minutes is $A(60) = \pi(60)^2 = 3600\pi$ square feet.

Notice that $60 = r(20)$, so $A(60) = A(r(20))$. The argument of the function A is itself a function! In general, we can find the area of the oil patch as a function of time t by evaluating $A(r(t))$. The function $A(r(t))$ is a special type of function called a *composite function*.

Figure 1

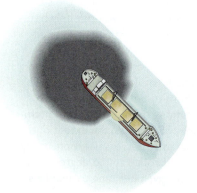

As another example, consider the function $y = (2x + 3)^2$. If we write $y = f(u) = u^2$ and $u = g(x) = 2x + 3$, then, by a substitution process, we can obtain the original function: $y = f(u) = f(g(x)) = (2x + 3)^2$.

In general, suppose that f and g are two functions and that x is a number in the domain of g. By evaluating g at x, we get $g(x)$. If $g(x)$ is in the domain of f, then we may evaluate f at $g(x)$ and thereby obtain the expression $f(g(x))$. The correspondence from x to $f(g(x))$ is called a *composite function* $f \circ g$.

Given two functions f and g, the **composite function**, denoted by $f \circ g$ (read as "f composed with g"), is defined by

$$(f \circ g)(x) = f(g(x))$$

The domain of $f \circ g$ is the set of all numbers x in the domain of g such that $g(x)$ is in the domain of f.

Look carefully at Figure 2. Only those x's in the domain of g for which $g(x)$ is in the domain of f can be in the domain of $f \circ g$. The reason is that if $g(x)$ is not in the domain of f then $f(g(x))$ is not defined. Because of this, the domain of $f \circ g$ is a subset of the domain of g; the range of $f \circ g$ is a subset of the range of f.

Figure 2

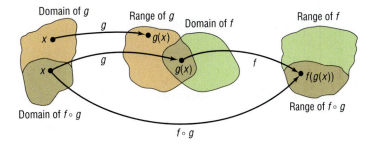

Figure 3 provides a second illustration of the definition. Here, x is the input to the function g, yielding $g(x)$. Then $g(x)$ is the input to the function f, yielding $f(g(x))$. Notice that the "inside" function g in $f(g(x))$ is done first.

Figure 3

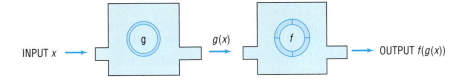

Let's look at some examples.

| **EXAMPLE 1** | **Evaluating a Composite Function** |

Suppose that $f(x) = 2x^2 - 3$ and $g(x) = 4x$. Find:

(a) $(f \circ g)(1)$ (b) $(g \circ f)(1)$ (c) $(f \circ f)(-2)$ (d) $(g \circ g)(-1)$

Solution   (a) $\ (f \circ g)(1) = f(g(1)) = f(4) = 2 \cdot 4^2 - 3 = 29$

$$g(x) = 4x \qquad f(x) = 2x^2 - 3$$
$$g(1) = 4$$

(b) $(g \circ f)(1) = g(f(1)) = g(-1) = 4 \cdot (-1) = -4$

$$\underset{\uparrow}{} \qquad \underset{\uparrow}{}$$

$f(x) = 2x^2 - 3 \quad g(x) = 4x$
$f(1) = -1$

(c) $(f \circ f)(-2) = f(f(-2)) = f(5) = 2 \cdot 5^2 - 3 = 47$

$$\underset{\uparrow}{}$$

$f(-2) = 2(-2)^2 - 3 = 5$

(d) $(g \circ g)(-1) = g(g(-1)) = g(-4) = 4 \cdot (-4) = -16$

$$\underset{\uparrow}{}$$

$g(-1) = -4$ ◀

Figure 4

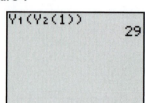

Graphing calculators can be used to evaluate composite functions.* Let $Y_1 = f(x) = 2x^2 - 3$ and $Y_2 = g(x) = 4x$. Then, using a TI-84 Plus graphing calculator, $(f \circ g)(1)$ would be found as shown in Figure 4. Notice that this is the result obtained in Example 1(a).

✏ **NOW WORK PROBLEM 11.**

2 Find the Domain of a Composite Function

EXAMPLE 2 **Finding a Composite Function and Its Domain**

Suppose that $f(x) = x^2 + 3x - 1$ and $g(x) = 2x + 3$.

Find: (a) $f \circ g$ (b) $g \circ f$

Then find the domain of each composite function.

Solution The domain of f and the domain of g are all real numbers.

(a) $(f \circ g)(x) = f(g(x)) = f(2x + 3) = (2x + 3)^2 + 3(2x + 3) - 1$

$$\underset{\uparrow}{}$$

$f(x) = x^2 + 3x - 1$

$= 4x^2 + 12x + 9 + 6x + 9 - 1 = 4x^2 + 18x + 17$

Since the domains of both f and g are all real numbers, the domain of $f \circ g$ is all real numbers.

(b) $(g \circ f)(x) = g(f(x)) = g(x^2 + 3x - 1) = 2(x^2 + 3x - 1) + 3$

$$\underset{\uparrow}{}$$

$g(x) = 2x + 3$

$= 2x^2 + 6x - 2 + 3 = 2x^2 + 6x + 1$

Since the domains of both f and g are all real numbers, the domain of $f \circ g$ is all real numbers.   ◀

Look back at Figure 2. In determining the domain of the composite function $(f \circ g)(x) = f(g(x))$, keep the following two thoughts in mind about the input x.

1. $g(x)$ must be defined so any x not in the domain of g must be excluded.

2. $f(g(x))$ must be defined so any x for which $g(x)$ is not in the domain of f must be excluded.

*Consult your owner's manual for the appropriate keystrokes.

| EXAMPLE 3 | **Finding the Domain of $f \circ g$** |

Find the domain of $(f \circ g)(x)$ if $f(x) = \dfrac{1}{x + 2}$ and $g(x) = \dfrac{4}{x - 1}$

Solution For $(f \circ g)(x) = f(g(x))$, we first note that the domain of g is $\{x \mid x \neq 1\}$, so we exclude 1 from the domain of $f \circ g$. Next, we note that the domain of f is $\{x \mid x \neq -2\}$, which means that $g(x)$ cannot equal -2. We solve the equation $g(x) = -2$ to determine what values of x to exclude.

$$\frac{4}{x - 1} = -2 \qquad\qquad g(x) = -2$$
$$4 = -2(x - 1)$$
$$4 = -2x + 2$$
$$2x = -2$$
$$x = -1$$

We also exclude -1 from the domain of $f \circ g$. The domain of $f \circ g$ is $\{x \mid x \neq -1, x \neq 1\}$.

✔ **CHECK:** For $x = 1$, $g(x) = \dfrac{4}{x - 1}$ is not defined, so $(f \circ g)(x) = f(g(x))$ is not defined.

For $x = -1$, $g(-1) = \dfrac{4}{-2} = -2$, and $(f \circ g)(-1) = f(g(-1)) = f(-2)$ is not defined. ◀

 NOW WORK PROBLEM 21.

| EXAMPLE 4 | **Finding a Composite Function and Its Domain** |

Suppose that $f(x) = \dfrac{1}{x + 2}$ and $g(x) = \dfrac{4}{x - 1}$

Find: (a) $f \circ g$ (b) $f \circ f$

Then find the domain of each composite function.

Solution The domain of f is $\{x \mid x \neq -2\}$ and the domain of g is $\{x \mid x \neq 1\}$.

(a) $(f \circ g)(x) = f(g(x)) = f\left(\dfrac{4}{x - 1}\right) = \dfrac{1}{\dfrac{4}{x - 1} + 2} = \dfrac{x - 1}{4 + 2(x - 1)} = \dfrac{x - 1}{2x + 2} = \dfrac{x - 1}{2(x + 1)}$

$\qquad\qquad \uparrow \qquad\qquad\qquad\qquad \uparrow \qquad\qquad\qquad\qquad \uparrow$

$\qquad f(x) = \dfrac{1}{x + 2} \qquad\qquad$ Multiply by $\dfrac{x - 1}{x - 1}$.

In Example 3, we found the domain of $f \circ g$ to be $\{x \mid x \neq -1, x \neq 1\}$.

We could also find the domain of $f \circ g$ by first looking at the domain of g: $\{x \mid x \neq 1\}$. We exclude 1 from the domain of $f \circ g$ as a result. Then we look at $f \circ g$ and notice that x cannot equal -1, since $x = -1$ results in division by 0. So we also exclude -1 from the domain of $f \circ g$. Therefore, the domain of $f \circ g$ is $\{x \mid x \neq -1, x \neq 1\}$.

(b) $(f \circ f)(x) = f(f(x)) = f\left(\dfrac{1}{x + 2}\right) = \dfrac{1}{\dfrac{1}{x + 2} + 2} = \dfrac{x + 2}{1 + 2(x + 2)} = \dfrac{x + 2}{2x + 5}$

$\qquad\qquad\qquad\qquad\qquad\qquad \uparrow \qquad\qquad \uparrow$

$\qquad\qquad f(x) = \dfrac{1}{x + 2} \qquad\qquad$ Multiply by $\dfrac{x + 2}{x + 2}$.

The domain of $f \circ f$ consists of those x in the domain of f, $\{x | x \neq -2\}$, for which

$$f(x) = \frac{1}{x + 2} \neq -2 \qquad \frac{1}{x + 2} = -2$$

$$1 = -2(x + 2)$$
$$1 = -2x - 4$$
$$2x = -5$$
$$x = -\frac{5}{2}$$

or, equivalently,

$$x \neq -\frac{5}{2}$$

The domain of $f \circ f$ is $\left\{ x | x \neq -\dfrac{5}{2}, x \neq -2 \right\}$.

We could also find the domain of $f \circ f$ by recognizing that -2 is not in the domain of f and so should be excluded from the domain of $f \circ f$. Then, looking at $f \circ f$, we see that x cannot equal $-\dfrac{5}{2}$. Do you see why? Therefore, the domain of $f \circ f$ is $\left\{ x | x \neq -\dfrac{5}{2}, x \neq -2 \right\}$. ◀

NOW WORK PROBLEMS 33 AND 35.

Look back at Example 2, which illustrates that, in general, $f \circ g \neq g \circ f$. Sometimes $f \circ g$ does equal $g \circ f$, as shown in the next example.

EXAMPLE 5	**Showing That Two Composite Functions Are Equal**

If $f(x) = 3x - 4$ and $g(x) = \dfrac{1}{3}(x + 4)$, show that

$$(f \circ g)(x) = (g \circ f)(x) = x$$

for every x in the domain of $f \circ g$ and $g \circ f$.

Solution

$$(f \circ g)(x) = f(g(x))$$

$$= f\left(\frac{x + 4}{3}\right) \qquad g(x) = \frac{1}{3}(x + 4) = \frac{x + 4}{3}$$

$$= 3\left(\frac{x + 4}{3}\right) - 4 \qquad \text{Substitute } g(x) \text{ into the rule for } f, \, f(x) = 3x - 4.$$

$$= x + 4 - 4 = x$$

$$(g \circ f)(x) = g(f(x))$$

$$= g(3x - 4) \qquad f(x) = 3x - 4$$

$$= \frac{1}{3}[(3x - 4) + 4] \qquad \text{Substitute } f(x) \text{ into the rule for } g, \, g(x) = \frac{1}{3}(x + 4).$$

$$= \frac{1}{3}(3x) = x$$

Thus, $(f \circ g)(x) = (g \circ f)(x) = x$. ◀

— **Seeing the Concept** —

Using a graphing calculator, let $Y_1 = f(x) = 3x - 4$, $Y_2 = g(x) = \dfrac{1}{3}(x + 4)$, $Y_3 = f \circ g$, and $Y_4 = g \circ f$. Using the viewing window $-3 \leq x \leq 3, -2 \leq y \leq 2$, graph only Y_3 and Y_4. What do you see? TRACE to verify that $Y_3 = Y_4$.

In Section 4.2, we shall see that there is an important relationship between functions f and g for which $(f \circ g)(x) = (g \circ f)(x) = x$.

NOW WORK PROBLEM 45.

Calculus Application

Some techniques in calculus require that we be able to determine the components of a composite function. For example, the function $H(x) = \sqrt{x + 1}$ is the composition of the functions f and g, where $f(x) = \sqrt{x}$ and $g(x) = x + 1$, because $H(x) = (f \circ g)(x) = f(g(x)) = f(x + 1) = \sqrt{x + 1}$.

EXAMPLE 6	**Finding the Components of a Composite Function**

Find functions f and g such that $f \circ g = H$ if $H(x) = (x^2 + 1)^{50}$.

Solution

The function H takes $x^2 + 1$ and raises it to the power 50. A natural way to decompose H is to raise the function $g(x) = x^2 + 1$ to the power 50. If we let $f(x) = x^{50}$ and $g(x) = x^2 + 1$, then

$$(f \circ g)(x) = f(g(x))$$
$$= f(x^2 + 1)$$
$$= (x^2 + 1)^{50} = H(x)$$

Figure 5

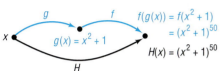

See Figure 5. ◀

Other functions f and g may be found for which $f \circ g = H$ in Example 6. For example, if $f(x) = x^2$ and $g(x) = (x^2 + 1)^{25}$, then

$$(f \circ g)(x) = f(g(x)) = f((x^2 + 1)^{25}) = [(x^2 + 1)^{25}]^2 = (x^2 + 1)^{50}$$

Although the functions f and g found as a solution to Example 6 are not unique, there is usually a "natural" selection for f and g that comes to mind first.

EXAMPLE 7	**Finding the Components of a Composite Function**

Find functions f and g such that $f \circ g = H$ if $H(x) = \dfrac{1}{x + 1}$.

Solution

Here H is the reciprocal of $g(x) = x + 1$. If we let $f(x) = \dfrac{1}{x}$ and $g(x) = x + 1$, we find that

$$(f \circ g)(x) = f(g(x)) = f(x + 1) = \frac{1}{x + 1} = H(x)$$

◀

NOW WORK PROBLEM 53.

4.1 Assess Your Understanding

'Are You Prepared?'

Answers are given at the end of these exercises. If you get a wrong answer, read the pages listed in red.

1. Find $f(3)$ if $f(x) = -4x^2 + 5x$. (p. 62)

2. Find $f(3x)$ if $f(x) = 4 - 2x^2$. (p.62)

3. Find the domain of the function $f(x) = \dfrac{x^2 - 1}{x^2 - 4}$. (pp. 64–65)

Concepts and Vocabulary

4. If $f(x) = x + 1$ and $g(x) = x^3$, then _____ $= (x + 1)^3$.

5. *True or False:* $f(g(x)) = f(x) \cdot g(x)$.

6. *True or False:* The domain of the composite function $(f \circ g)(x)$ is the same as the domain of $g(x)$.

Skill Building

In Problems 7 and 8, evaluate each expression using the values given in the table.

7.

x	−3	−2	−1	0	1	2	3
f(x)	−7	−5	−3	−1	3	5	5
g(x)	8	3	0	−1	0	3	8

    (a) $(f \circ g)(1)$ (b) $(f \circ g)(-1)$ (c) $(g \circ f)(-1)$ (d) $(g \circ f)(0)$ (e) $(g \circ g)(-2)$ (f) $(f \circ f)(-1)$

8.

x	−3	−2	−1	0	1	2	3
f(x)	11	9	7	5	3	1	−1
g(x)	−8	−3	0	1	0	−3	−8

    (a) $(f \circ g)(1)$ (b) $(f \circ g)(2)$ (c) $(g \circ f)(2)$ (d) $(g \circ f)(3)$ (e) $(g \circ g)(1)$ (f) $(f \circ f)(3)$

In Problems 9 and 10, evaluate each expression using the graphs of $y = f(x)$ and $y = g(x)$ shown below.

9. (a) $g(f(-1))$ (b) $g(f(0))$
    (c) $f(g(-1))$ (d) $f(g(4))$

10. (a) $g(f(1))$ (b) $g(f(5))$
    (c) $f(g(0))$ (d) $f(g(2))$

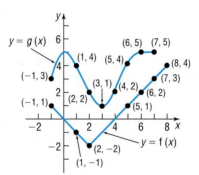

In Problems 11–20, for the given functions f and g, find:

    (a) $(f \circ g)(4)$ (b) $(g \circ f)(2)$ (c) $(f \circ f)(1)$ (d) $(g \circ g)(0)$

11. $f(x) = 2x;\ \ g(x) = 3x^2 + 1$ **12.** $f(x) = 3x + 2;\ \ g(x) = 2x^2 - 1$

13. $f(x) = 4x^2 - 3;\ \ g(x) = 3 - \dfrac{1}{2}x^2$ **14.** $f(x) = 2x^2;\ \ g(x) = 1 - 3x^2$

15. $f(x) = \sqrt{x};\ \ g(x) = 2x$ **16.** $f(x) = \sqrt{x + 1};\ \ g(x) = 3x$

17. $f(x) = |x|;\ \ g(x) = \dfrac{1}{x^2 + 1}$ **18.** $f(x) = |x - 2|;\ \ g(x) = \dfrac{3}{x^2 + 2}$

19. $f(x) = \dfrac{3}{x + 1};\ \ g(x) = \sqrt[3]{x}$ **20.** $f(x) = x^{3/2};\ \ g(x) = \dfrac{2}{x + 1}$

In Problems 21–28, find the domain of the composite function $f \circ g$.

21. $f(x) = \dfrac{3}{x - 1};\ \ g(x) = \dfrac{2}{x}$ **22.** $f(x) = \dfrac{1}{x + 3};\ \ g(x) = \dfrac{-2}{x}$

23. $f(x) = \dfrac{x}{x - 1};\ \ g(x) = \dfrac{-4}{x}$ **24.** $f(x) = \dfrac{x}{x + 3};\ \ g(x) = \dfrac{2}{x}$

25. $f(x) = \sqrt{x};\ \ g(x) = 2x + 3$ **26.** $f(x) = x - 2;\ \ g(x) = \sqrt{1 - x}$

27. $f(x) = x^2 + 1;\ \ g(x) = \sqrt{x - 1}$ **28.** $f(x) = x^2 + 4;\ \ g(x) = \sqrt{x - 2}$

In Problems 29–44, for the given functions f and g, find:

(a) $f \circ g$ (b) $g \circ f$ (c) $f \circ f$ (d) $g \circ g$

State the domain of each composite function.

29. $f(x) = 2x + 3$; $g(x) = 3x$

30. $f(x) = -x$; $g(x) = 2x - 4$

31. $f(x) = 3x + 1$; $g(x) = x^2$

32. $f(x) = x + 1$; $g(x) = x^2 + 4$

33. $f(x) = x^2$; $g(x) = x^2 + 4$

34. $f(x) = x^2 + 1$; $g(x) = 2x^2 + 3$

35. $f(x) = \dfrac{3}{x - 1}$; $g(x) = \dfrac{2}{x}$

36. $f(x) = \dfrac{1}{x + 3}$; $g(x) = \dfrac{-2}{x}$

37. $f(x) = \dfrac{x}{x - 1}$; $g(x) = \dfrac{-4}{x}$

38. $f(x) = \dfrac{x}{x + 3}$; $g(x) = \dfrac{2}{x}$

39. $f(x) = \sqrt{x}$; $g(x) = 2x + 3$

40. $f(x) = \sqrt{x - 2}$; $g(x) = 1 - 2x$

41. $f(x) = x^2 + 1$; $g(x) = \sqrt{x - 1}$

42. $f(x) = x^2 + 4$; $g(x) = \sqrt{x - 2}$

43. $f(x) = ax + b$; $g(x) = cx + d$

44. $f(x) = \dfrac{ax + b}{cx + d}$; $g(x) = mx$

In Problems 45–52, show that $(f \circ g)(x) = (g \circ f)(x) = x$.

45. $f(x) = 2x$; $g(x) = \dfrac{1}{2}x$

46. $f(x) = 4x$; $g(x) = \dfrac{1}{4}x$

47. $f(x) = x^3$; $g(x) = \sqrt[3]{x}$

48. $f(x) = x + 5$; $g(x) = x - 5$

49. $f(x) = 2x - 6$; $g(x) = \dfrac{1}{2}(x + 6)$

50. $f(x) = 4 - 3x$; $g(x) = \dfrac{1}{3}(4 - x)$

51. $f(x) = ax + b$; $g(x) = \dfrac{1}{a}(x - b)$, $a \neq 0$

52. $f(x) = \dfrac{1}{x}$; $g(x) = \dfrac{1}{x}$

In Problems 53–58, find functions f and g so that $f \circ g = H$.

53. $H(x) = (2x + 3)^4$

54. $H(x) = (1 + x^2)^3$

55. $H(x) = \sqrt{x^2 + 1}$

56. $H(x) = \sqrt{1 - x^2}$

57. $H(x) = |2x + 1|$

58. $H(x) = |2x^2 + 3|$

Applications and Extensions

59. If $f(x) = 2x^3 - 3x^2 + 4x - 1$ and $g(x) = 2$, find $(f \circ g)(x)$ and $(g \circ f)(x)$.

60. If $f(x) = \dfrac{x + 1}{x - 1}$, find $(f \circ f)(x)$.

61. If $f(x) = 2x^2 + 5$ and $g(x) = 3x + a$, find a so that the graph of $f \circ g$ crosses the y-axis at 23.

62. If $f(x) = 3x^2 - 7$ and $g(x) = 2x + a$, find a so that the graph of $f \circ g$ crosses the y-axis at 68.

63. Surface Area of a Balloon   The surface area S (in square meters) of a hot-air balloon is given by

$$S(r) = 4\pi r^2$$

where r is the radius of the balloon (in meters). If the radius r is increasing with time t (in seconds) according to the formula $r(t) = \dfrac{2}{3}t^3$, $t \geq 0$, find the surface area S of the balloon as a function of the time t.

64. Volume of a Balloon   The volume V (in cubic meters) of the hot-air balloon described in Problem 63 is given by $V(r) = \dfrac{4}{3}\pi r^3$. If the radius r is the same function of t as in Problem 63, find the volume V as a function of the time t.

65. Automobile Production   The number N of cars produced at a certain factory in 1 day after t hours of operation is given by $N(t) = 100t - 5t^2, 0 \leq t \leq 10$. If the cost C (in dollars) of producing N cars is $C(N) = 15{,}000 + 8000N$, find the cost C as a function of the time t of operation of the factory.

66. Environmental Concerns   The spread of oil leaking from a tanker is in the shape of a circle. If the radius r (in feet) of the spread after t hours is $r(t) = 200\sqrt{t}$, find the area A of the oil slick as a function of the time t.

67. Production Cost   The price p, in dollars, of a certain product and the quantity x sold obey the demand equation

$$p = -\frac{1}{4}x + 100, \qquad 0 \le x \le 400$$

Suppose that the cost C, in dollars, of producing x units is

$$C = \frac{\sqrt{x}}{25} + 600$$

Assuming that all items produced are sold, find the cost C as a function of the price p.

[**Hint:** Solve for x in the demand equation and then form the composite.]

68. Cost of a Commodity   The price p, in dollars, of a certain commodity and the quantity x sold obey the demand equation

$$p = -\frac{1}{5}x + 200, \qquad 0 \le x \le 1000$$

Suppose that the cost C, in dollars, of producing x units is

$$C = \frac{\sqrt{x}}{10} + 400$$

Assuming that all items produced are sold, find the cost C as a function of the price p.

69. Volume of a Cylinder   The volume V of a right circular cylinder of height h and radius r is $V = \pi r^2 h$. If the height is twice the radius, express the volume V as a function of r.

70. Volume of a Cone   The volume V of a right circular cone is $V = \frac{1}{3}\pi r^2 h$. If the height is twice the radius, express the volume V as a function of r.

71. Foreign Exchange   Traders often buy foreign currency in hope of making money when the currency's value changes. For example, on October 20, 2003, one U.S. dollar could purchase 0.857118 Euros and one Euro could purchase 128.6054 yen. Let $f(x)$ represent the number of Euros you can buy with x dollars and let $g(x)$ represent the number of yen you can buy with x Euros.
(a) Find a function that relates dollars to Euros.
(b) Find a function that relates Euros to yen.
(c) Use the results of parts (a) and (b) to find a function that relates dollars to yen. That is, find $g(f(x))$.
(d) What is $g(f(1000))$?

72. If f and g are odd functions, show that the composite function $f \circ g$ is also odd.

73. If f is an odd function and g is an even function, show that the composite functions $f \circ g$ and $g \circ f$ are also even.

'Are You Prepared?' Answers

1. -21 **2.** $4 - 18x^2$ **3.** $\{x \mid x \ne -2, x \ne 2\}$

4.2 One-to-One Functions; Inverse Functions

PREPARING FOR THIS SECTION   *Before getting started, review the following:*

• Functions (Section 2.1, pp. 56–67)   • Increasing/Decreasing Functions (Section 2.3, pp. 82–85)

  Now work the 'Are You Prepared?' problems on page 267.

OBJECTIVES **1** Determine Whether a Function Is One-to-One
                2 Determine the Inverse of a Function Defined by a Map or an Ordered Pair
                3 Obtain the Graph of the Inverse Function from the Graph of the Function
                4 Find the Inverse of a Function Defined by an Equation

 Determine Whether a Function Is One-to-One

In Section 2.1, we presented four different ways to represent a function: as (1) a map, (2) a set of ordered pairs, (3) a graph, and (4) an equation. For example, Figures 6 and 7 illustrate two different functions represented as mappings. The function in

Figure 6

Figure 7

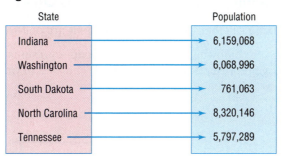

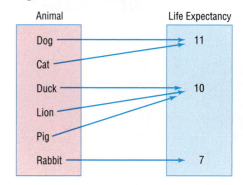

Figure 6 shows the correspondence between states and their population. The function in Figure 7 shows a correspondence between animals and life expectancy.

Suppose we asked a group of people to name the state that has a population of 761,063 based on the function in Figure 6. Everyone in the group would respond South Dakota. Now, if we asked the same group of people to name the animal whose life expectancy is 11 years based on the function in Figure 7, some would respond dog, while others would respond cat. What is the difference between the functions in Figures 6 and 7? In Figure 6, we can see that each element in the domain corresponds to exactly one element in the range. In Figure 7, this is not the case: two different elements in the domain correspond to the same element in the range. We give functions such as the one in Figure 6 a special name.

In Words

A function is not one-to-one if two different inputs correspond to the same output.

A function is **one-to-one** if any two different inputs in the domain correspond to two different outputs in the range. That is, if x_1 and x_2 are two different inputs of a function f, then $f(x_1) \neq f(x_2)$.

Put another way, a function f is one-to-one if no y in the range is the image of more than one x in the domain. A function is not one-to-one if two different elements in the domain correspond to the same element in the range. So, the function in Figure 7 is not one-to-one because two different elements in the domain, dog and cat, both correspond to 11. Figure 8 illustrates the distinction of one-to-one functions, non–one-to-one functions, and nonfunctions.

Figure 8

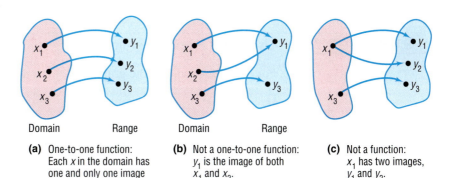

(a) One-to-one function: Each x in the domain has one and only one image in the range

(b) Not a one-to-one function: y_1 is the image of both x_1 and x_2.

(c) Not a function: x_1 has two images, y_1 and y_2.

EXAMPLE 1	**Determining Whether a Function Is One-to-One**

Determine whether the following functions are one-to-one.

(a) For the following function, the domain represents the age of five males and the range represents their HDL (good) cholesterol (mg/dL).

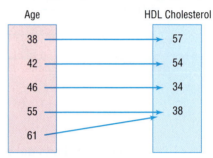

(b) $\{(-2, 6), (-1, 3), (0, 2), (1, 5), (2, 8)\}$

Solution (a) The function is not one-to-one because there are two different inputs, 55 and 61, that correspond to the same output, 38.

(b) The function is one-to-one because there are no two distinct inputs that correspond to the same output. ◀

NOW WORK PROBLEMS 9 AND 13.

If the graph of a function f is known, there is a simple test, called the **horizontal-line test**, to determine whether f is one-to-one.

Figure 9
$f(x_1) = f(x_2) = h$ and
$x_1 \neq x_2; f$ is not a
one-to-one function.

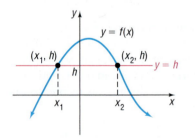

Theorem

Horizontal-line Test

If every horizontal line intersects the graph of a function f in at most one point, then f is one-to-one.

The reason that this test works can be seen in Figure 9, where the horizontal line $y = h$ intersects the graph at two distinct points, (x_1, h) and (x_2, h). Since h is the image of both x_1 and x_2, $x_1 \neq x_2$, f is not one-to-one. Based on Figure 9, we can state the horizontal-line test in another way: If the graph of any horizontal line intersects the graph of a function f at more than one point, then f is not one-to-one.

EXAMPLE 2	**Using the Horizontal-line Test**

For each function, use the graph to determine whether the function is one-to-one.

(a) $f(x) = x^2$ (b) $g(x) = x^3$

Solution (a) Figure 10(a) illustrates the horizontal-line test for $f(x) = x^2$. The horizontal line $y = 1$ intersects the graph of f twice, at $(1, 1)$ and at $(-1, 1)$, so f is not one-to-one.

(b) Figure 10(b) illustrates the horizontal-line test for $g(x) = x^3$. Because every horizontal line will intersect the graph of g exactly once, it follows that g is one-to-one.

Figure 10

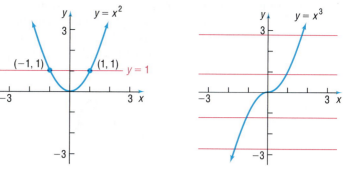

(a) A horizontal line intersects the graph twice; f is not one-to-one

(b) Every horizontal line intersects the graph exactly once; g is one-to-one

◄

 **NOW WORK PROBLEM 17.**

Let's look more closely at the one-to-one function $g(x) = x^3$. This function is an increasing function. Because an increasing (or decreasing) function will always have different y-values for unequal x-values, it follows that a function that is increasing (or decreasing) over its domain is also a one-to-one function.

Theorem

> A function that is increasing on an interval I is a one-to-one function on I.
>
> A function that is decreasing on an interval I is a one-to-one function on I.

2 **Determine the Inverse of a Function Defined by a Map or an Ordered Pair**

In Section 2.1, we said that a function f can be thought of as a machine that receives an input, say x, from the domain, manipulates it, and outputs the value $f(x)$. The **inverse function of f** receives as input $f(x)$, manipulates it, and outputs x. For example, if $f(x) = 2x + 5$, then $f(3) = 11$. Here, the input is 3 and the output is 11. The inverse of f would "undo" what f does and receive as input 11 and provide as output 3. So, if a function takes inputs and multiplies them by 2, the inverse would take inputs and divide them by 2. For the inverse of a function f to itself be a function, f must be one-to-one.

Recall that we have a variety of ways of representing functions. We will discuss how to find inverses for all four representations of functions: (1) maps, (2) sets of ordered pairs (3) graphs, and (4) equations. We begin with finding inverses of functions represented by maps or sets of ordered pairs.

EXAMPLE 3 **Finding the Inverse of a Function Defined by a Map**

Find the inverse of the following function. Let the domain of the function represent certain states, and let the range represent the state's population. State the domain and the range of the inverse function.

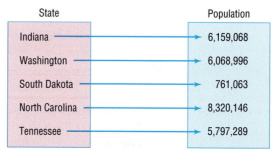

Solution   The function is one-to-one, so the inverse will be a function. To find the inverse function, we interchange the elements in the domain with the elements in the range. For example, the function receives as input Indiana and outputs 6,159,068. So, the inverse receives as input 6,159,068 and outputs Indiana. The inverse function is shown next.

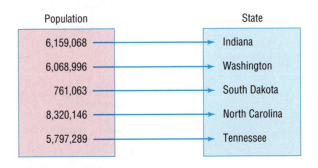

Population   State

The domain of the inverse function is {6 159 068, 6 068 996, 761 063, 8 320 146, 5 797 289}. The range of the inverse function is {Indiana, Washington, South Dakota, North Carolina, Tennessee}.   ◀

If the function f is a set of ordered pairs (x, y), then the inverse of f is the set of ordered pairs (y, x).

EXAMPLE 4   **Finding the Inverse of a Function Defined By a Set of Ordered Pairs**

Find the inverse of the following one-to-one function:

$$\{(-3, -27), (-2, -8), (-1, -1), (0, 0), (1, 1), (2, 8), (3, 27)\}$$

Solution   The inverse of the given function is found by interchanging the entries in each ordered pair and so is given by

$$\{(-27, -3), (-8, -2), (-1, -1), (0, 0), (1, 1), (8, 2), (27, 3)\} \quad ◀$$

✏ **NOW WORK PROBLEMS 23 AND 27.**

Remember, if f is a one-to-one function, its inverse is a function. Then, to each x in the domain of f, there is exactly one y in the range (because f is a function); and to each y in the range of f, there is exactly one x in the domain (because f is one-to-one). The correspondence from the range of f back to the domain of f is called the inverse function of f and is denoted by the symbol f^{-1}. Figure 11 illustrates this definition.

Based on Figure 11, two facts are now apparent about a function f and its inverse f^{-1}.

Figure 11

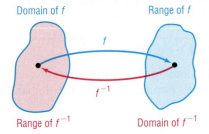

Domain of f Range of f

f

f^{-1}

Range of f^{-1} Domain of f^{-1}

Domain of f = Range of f^{-1} Range of f = Domain of f^{-1}

WARNING
Be careful! f^{-1} is a symbol for the inverse function of f. The -1 used in f^{-1} is not an exponent. That is, f^{-1} does not mean the reciprocal of f; $f^{-1}(x)$ is not equal to $\dfrac{1}{f(x)}$. ▪

Look again at Figure 11 to visualize the relationship. If we start with x, apply f, and then apply f^{-1}, we get x back again. If we start with x, apply f^{-1}, and then apply f, we get the number x back again. To put it simply, what f does, f^{-1} undoes, and vice versa. See the illustration that follows.

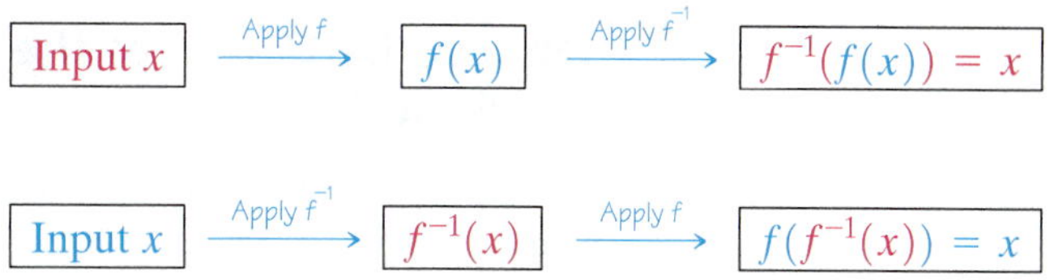

$$\boxed{\text{Input } x} \xrightarrow{\text{Apply } f} \boxed{f(x)} \xrightarrow{\text{Apply } f^{-1}} \boxed{f^{-1}(f(x)) = x}$$

$$\boxed{\text{Input } x} \xrightarrow{\text{Apply } f^{-1}} \boxed{f^{-1}(x)} \xrightarrow{\text{Apply } f} \boxed{f(f^{-1}(x)) = x}$$

In other words,

$$f^{-1}(f(x)) = x \quad \text{where } x \text{ is in the domain of } f$$
$$f(f^{-1}(x)) = x \quad \text{where } x \text{ is in the domain of } f^{-1}$$

Consider the function $f(x) = 2x$, which multiplies the argument x by 2. The inverse function f^{-1} undoes whatever f does. So the inverse function of f is $f^{-1}(x) = \dfrac{1}{2}x$, which divides the argument by 2. For example, $f(3) = 2(3) = 6$ and $f^{-1}(6) = \dfrac{1}{2}(6) = 3$, so f^{-1} undoes what f did. We can verify this by showing that

$$f^{-1}(f(x)) = f^{-1}(2x) = \frac{1}{2}(2x) = x \quad \text{and} \quad f(f^{-1}(x)) = f\left(\frac{1}{2}x\right) = 2\left(\frac{1}{2}x\right) = x$$

See Figure 12.

Figure 12

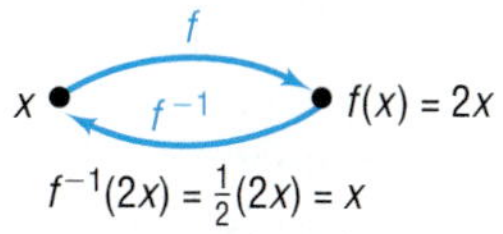

$f^{-1}(2x) = \frac{1}{2}(2x) = x$

EXAMPLE 5

Verifying Inverse Functions

— **Exploration** —
Simultaneously graph $Y_1 = x$, $Y_2 = x^3$, and $Y_3 = \sqrt[3]{x}$ on a square screen with $-3 \le x \le 3$. What do you observe about the graphs of $Y_2 = x^3$, its inverse $Y_3 = \sqrt[3]{x}$, and the line $Y_1 = x$?

Repeat this experiment by simultaneously graphing $Y_1 = x$, $Y_2 = 2x + 3$, and $Y_3 = \dfrac{1}{2}(x - 3)$ on a square screen with $-6 \le x \le 3$. Do you see the symmetry of the graph of Y_2 and its inverse Y_3 with respect to the line $Y_1 = x$?

(a) We verify that the inverse of $g(x) = x^3$ is $g^{-1}(x) = \sqrt[3]{x}$ by showing that

$$g^{-1}(g(x)) = g^{-1}(x^3) = \sqrt[3]{x^3} = x \quad \text{for all } x \text{ in the domain of } g$$
$$g(g^{-1}(x)) = g(\sqrt[3]{x}) = (\sqrt[3]{x})^3 = x \quad \text{for all } x \text{ in the domain of } g^{-1}.$$

(b) We verify that the inverse of $f(x) = 2x + 3$ is $f^{-1}(x) = \dfrac{1}{2}(x - 3)$ by showing that

$$f^{-1}(f(x)) = f^{-1}(2x + 3) = \frac{1}{2}[(2x + 3) - 3] = \frac{1}{2}(2x) = x \quad \text{for all } x \text{ in the domain of } f$$

$$f(f^{-1}(x)) = f\left(\frac{1}{2}(x - 3)\right) = 2\left[\frac{1}{2}(x - 3)\right] + 3 = (x - 3) + 3 = x \quad \text{for all } x \text{ in the domain of } f^{-1}. \blacktriangleleft$$

EXAMPLE 6

Verifying Inverse Functions

Verify that the inverse of $f(x) = \dfrac{1}{x - 1}$ is $f^{-1}(x) = \dfrac{1}{x} + 1$. For what values of x is $f^{-1}(f(x)) = x$? For what values of x is $f(f^{-1}(x)) = x$?

Solution

The domain of f is $\{x | x \ne 1\}$ and the domain of f^{-1} is $\{x | x \ne 0\}$ Now,

$$f^{-1}(f(x)) = f^{-1}\left(\frac{1}{x - 1}\right) = \frac{1}{\dfrac{1}{x - 1}} + 1 = x - 1 + 1 = x \quad \text{provided } x \ne 1.$$

$$f(f^{-1}(x)) = f\left(\frac{1}{x} + 1\right) = \frac{1}{\dfrac{1}{x} + 1 - 1} = \frac{1}{\dfrac{1}{x}} = x \quad \text{provided } x \ne 0 \blacktriangleleft$$

NOW WORK PROBLEM **37.**

Obtain the Graph of the Inverse Function from the Graph of the Function

For the functions in Example 5(b), we list points on the graph of $f = Y_1$ and on the graph of $f^{-1} = Y_2$ in Table 1.

We notice that whenever (a, b) is on the graph of f then (b, a) is on the graph of f^{-1}. Figure 13 shows these points plotted. Also shown is the graph of $y = x$, which you should observe is a line of symmetry of the points.

Figure 13

Table 1

X	Y1
-5	-7
-4	-5
-3	-3
-2	-1
-1	1
0	3
1	5

$Y_1 \boxminus 2X+3$

X	Y2
-7	-5
-5	-4
-3	-3
-1	-2
1	-1
3	0
5	1

$Y_2 \boxminus (1/2)(X-3)$

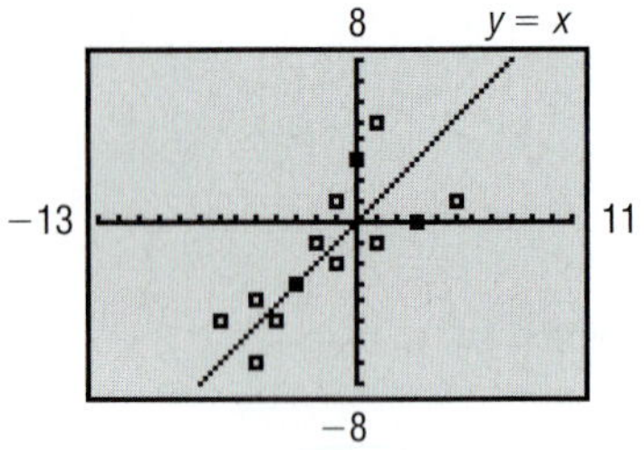

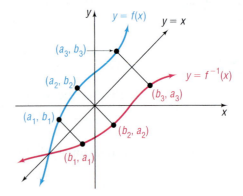

Figure 14

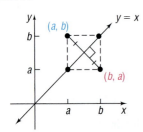

Suppose that (a, b) is a point on the graph of a one-to-one function f defined by $y = f(x)$. Then $b = f(a)$. This means that $a = f^{-1}(b)$, so (b, a) is a point on the graph of the inverse function f^{-1}. The relationship between the point (a, b) on f and the point (b, a) on f^{-1} is shown in Figure 14. The line segment containing (a, b) and (b, a) is perpendicular to the line $y = x$ and is bisected by the line $y = x$. (Do you see why?) It follows that the point (b, a) on f^{-1} is the reflection about the line $y = x$ of the point (a, b) on f.

Theorem

The graph of a function f and the graph of its inverse f^{-1} are symmetric with respect to the line $y = x$.

Figure 15 illustrates this result. Notice that, once the graph of f is known, the graph of f^{-1} may be obtained by reflecting the graph of f about the line $y = x$.

Figure 15

EXAMPLE 7 **Graphing the Inverse Function**

The graph in Figure 16(a) is that of a one-to-one function $y = f(x)$. Draw the graph of its inverse.

Solution   We begin by adding the graph of $y = x$ to Figure 16(a). Since the points $(-2, -1), (-1, 0)$, and $(2, 1)$ are on the graph of f, we know that the points $(-1, -2), (0, -1)$, and $(1, 2)$ must be on the graph of f^{-1}. Keeping in mind that the graph of f^{-1} is the reflection about the line $y = x$ of the graph of f, we can draw f^{-1}. See Figure 16(b).

Figure 16

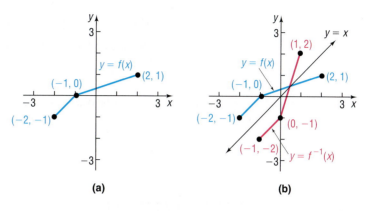

(a)   (b)

◀

NOW WORK PROBLEM 31.

4   Find the Inverse of a Function Defined by an Equation

The fact that the graphs of a one-to-one function f and its inverse function f^{-1} are symmetric with respect to the line $y = x$ tells us more. It says that we can obtain f^{-1} by interchanging the roles of x and y in f. Look again at Figure 15. If f is defined by the equation

$$y = f(x)$$

then f^{-1} is defined by the equation

$$x = f(y)$$

The equation $x = f(y)$ defines f^{-1} *implicitly*. If we can solve this equation for y, we will have the *explicit* form of f^{-1}, that is,

$$y = f^{-1}(x)$$

Let's use this procedure to find the inverse of $f(x) = 2x + 3$. (Since f is a linear function and is increasing, we know that f is one-to-one and so has an inverse function.)

EXAMPLE 8   **Finding the Inverse Function**

Find the inverse of $f(x) = 2x + 3$. Also find the domain and range of f and f^{-1}. Graph f and f^{-1} on the same coordinate axes.

Solution   In the equation $y = 2x + 3$, interchange the variables x and y. The result,

$$x = 2y + 3$$

is an equation that defines the inverse f^{-1} implicitly. To find the explicit form, we solve for y.

$$2y + 3 = x$$
$$2y = x - 3$$
$$y = \frac{1}{2}(x - 3)$$

The explicit form of the inverse f^{-1} is therefore

$$f^{-1}(x) = \frac{1}{2}(x - 3)$$

which we verified in Example 5(c).

Next we find

$$\text{Domain of } f = \text{Range of } f^{-1} = (-\infty, \infty)$$
$$\text{Range of } f = \text{Domain of } f^{-1} = (-\infty, \infty)$$

The graphs of $Y_1 = f(x) = 2x + 3$ and its inverse $Y_2 = f^{-1}(x) = \frac{1}{2}(x - 3)$

are shown in Figure 17. Note the symmetry of the graphs with respect to the line $Y_3 = x$.

Figure 17

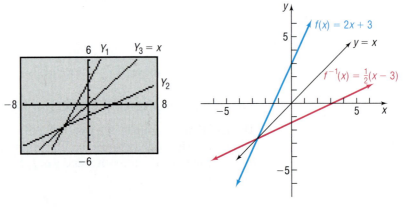

We now outline the steps to follow for finding the inverse of a one-to-one function.

Procedure for Finding the Inverse of a One-to-One Function

STEP 1: In $y = f(x)$, interchange the variables x and y to obtain

$$x = f(y)$$

This equation defines the inverse function f^{-1} implicitly.

STEP 2: If possible, solve the implicit equation for y in terms of x to obtain the explicit form of f^{-1}

$$y = f^{-1}(x)$$

STEP 3: Check the result by showing that

$$f^{-1}(f(x)) = x \quad \text{and} \quad f(f^{-1}(x)) = x$$

EXAMPLE 9   **Finding the Inverse Function**

The function

$$f(x) = \frac{2x + 1}{x - 1}, \qquad x \neq 1$$

is one-to-one. Find its inverse and check the result.

Solution   **STEP 1:** Interchange the variables x and y in

$$y = \frac{2x + 1}{x - 1}$$

to obtain

$$x = \frac{2y + 1}{y - 1}$$

STEP 2: Solve for y.

$$x = \frac{2y + 1}{y - 1}$$

$$x(y - 1) = 2y + 1 \qquad \text{Multiply both sides by } y - 1.$$

$$xy - x = 2y + 1 \qquad \text{Apply the Distributive Property.}$$

$$xy - 2y = x + 1 \qquad \text{Subtract } 2y \text{ from both sides; add } x \text{ to both sides.}$$

$$(x - 2)y = x + 1 \qquad \text{Factor.}$$

$$y = \frac{x + 1}{x - 2} \qquad \text{Divide by } x - 2.$$

The inverse is

$$f^{-1}(x) = \frac{x + 1}{x - 2}, \qquad x \neq 2 \qquad \text{Replace } y \text{ by } f^{-1}(x).$$

STEP 3: ✔ **CHECK:**

$$f^{-1}(f(x)) = f^{-1}\left(\frac{2x + 1}{x - 1}\right) = \frac{\dfrac{2x + 1}{x - 1} + 1}{\dfrac{2x + 1}{x - 1} - 2} = \frac{2x + 1 + x - 1}{2x + 1 - 2(x - 1)} = \frac{3x}{3} = x, \quad x \neq 1$$

$$f(f^{-1}(x)) = f\left(\frac{x + 1}{x - 2}\right) = \frac{2\left(\dfrac{x + 1}{x - 2}\right) + 1}{\dfrac{x + 1}{x - 2} - 1} = \frac{2(x + 1) + x - 2}{x + 1 - (x - 2)} = \frac{3x}{3} = x, \quad x \neq 2$$

◀

───── **Exploration** ─────────────────────────────────

In Example 9, we found that, if $f(x) = \dfrac{2x + 1}{x - 1}$, then $f^{-1}(x) = \dfrac{x + 1}{x - 2}$. Compare the vertical and horizontal asymptotes of f and f^{-1}. What did you find? Are you surprised?

Result   You should have determined that the vertical asymptote of f is $x = 1$ and the horizontal asymptote is $y = 2$. The vertical asymptote of f^{-1} is $x = 2$, and the horizontal asymptote is $y = 1$.

───

NOW WORK PROBLEM 49.

We said in Chapter 2 that finding the range of a function f is not easy. However, if f is one-to-one, we can find its range by finding the domain of the inverse function f^{-1}.

EXAMPLE 10	**Finding the Range of a Function**

Find the domain and range of

$$f(x) = \frac{2x + 1}{x - 1}$$

Solution   The domain of f is $\{x \mid x \neq 1\}$. To find the range of f, we first find the inverse f^{-1}. Based on Example 9, we have

$$f^{-1}(x) = \frac{x + 1}{x - 2}$$

The domain of f^{-1} is $\{x \mid x \neq 2\}$, so the range of f is $\{y \mid y \neq 2\}$.   ◀

NOW WORK PROBLEM 63.

If a function is not one-to-one, then its inverse is not a function. Sometimes, though, an appropriate restriction on the domain of such a function will yield a new function that is one-to-one. Let's look at an example of this common practice.

EXAMPLE 11	**Finding the Inverse of a Domain-restricted Function**

Find the inverse of $y = f(x) = x^2$ if $x \geq 0$.

Solution   The function $y = x^2$ is not one-to-one. [Refer to Example 2(a).] However, if we restrict the domain of this function to $x \geq 0$, as indicated, we have a new function that is increasing and therefore is one-to-one. As a result, the function defined by $y = f(x) = x^2$, $x \geq 0$, has an inverse function, f^{-1}.
    We follow the steps given previously to find f^{-1}.

STEP 1:   In the equation $y = x^2$, $x \geq 0$, interchange the variables x and y. The result is

$$x = y^2, \qquad y \geq 0$$

This equation defines (implicitly) the inverse function.

STEP 2:   We solve for y to get the explicit form of the inverse. Since $y \geq 0$, only one solution for y is obtained, namely, $y = \sqrt{x}$. So $f^{-1}(x) = \sqrt{x}$.

STEP 3:   ✔ **CHECK:** $f^{-1}(f(x)) = f^{-1}(x^2) = \sqrt{x^2} = |x| = x$, since $x \geq 0$
$$f(f^{-1}(x)) = f(\sqrt{x}) = (\sqrt{x})^2 = x.$$ ◀

Figure 18 illustrates the graphs of $Y_1 = f(x) = x^2$, $x \geq 0$, and $Y_2 = f^{-1}(x) = \sqrt{x}$.

Figure 18

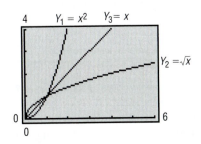

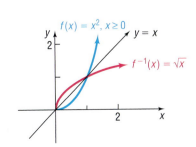

Summary

1. If a function f is one-to-one, then it has an inverse function f^{-1}.
2. Domain of f = Range of f^{-1}; Range of f = Domain of f^{-1}.
3. To verify that f^{-1} is the inverse of f, show that $f^{-1}(f(x)) = x$ for every x in the domain of f and $f(f^{-1}(x)) = x$ for every x in the domain of f^{-1}.
4. The graphs of f and f^{-1} are symmetric with respect to the line $y = x$.
5. To find the range of a one-to-one function f, find the domain of the inverse function f^{-1}.

4.2 Assess Your Understanding

'Are You Prepared?'

Answers are given at the end of these exercises. If you get a wrong answer, read the pages listed in red.

1. Is the set of ordered pairs $\{(1, 3), (2, 3), (-1, 2)\}$ a function? Why or why not? (pp. 56–60)
2. Where is the function $f(x) = x^2$ increasing? Where is it decreasing? (pp. 82–85)
3. What is the domain of $f(x) = \dfrac{x + 5}{x^2 + 3x - 18}$? (pp. 64–65)

Concepts and Vocabulary

4. If every horizontal line intersects the graph of a function f at no more than one point, then f is a(n) _____ function.
5. If f^{-1} denotes the inverse of a function f, then the graphs of f and f^{-1} are symmetric with respect to the line _____.
6. If the domain of a one-to-one function f is $[4, \infty)$, then the range of its inverse, f^{-1}, is _____.
7. *True or False:* If f and g are inverse functions, then the domain of f is the same as the domain of g.
8. *True or False:* If f and g are inverse functions, then their graphs are symmetric with respect to the line $y = x$.

Skill Building

In Problems 9–16, determine whether the function is one-to-one.

9.

Domain	Range
20 Hours	$200
25 Hours	$300
30 Hours	$350
40 Hours	$425

10.

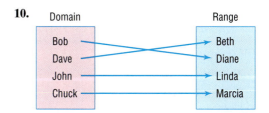

11.

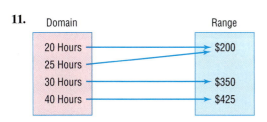

12.

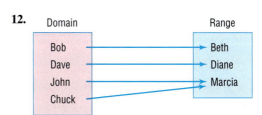

13. $\{(2, 6), (-3, 6), (4, 9), (1, 10)\}$

14. $\{(-2, 5), (-1, 3), (3, 7), (4, 12)\}$

15. $\{(0, 0), (1, 1), (2, 16), (3, 81)\}$

16. $\{(1, 2), (2, 8), (3, 18), (4, 32)\}$

In Problems 17–22, the graph of a function f is given. Use the horizontal-line test to determine whether f is one-to-one.

17.

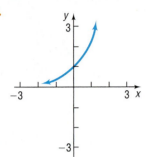

18.

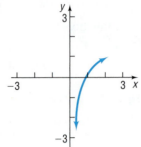

19.

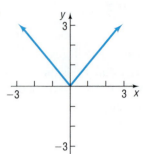

20.

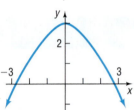

21.

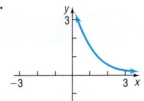

22.

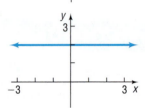

In Problems 23–30, find the inverse of each one-to-one function. State the domain and the range of each inverse function.

23.

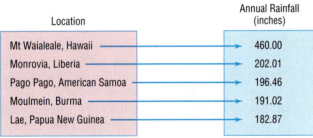

SOURCE: *Information Please Almanac*

24.

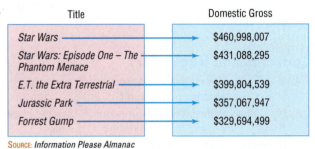

SOURCE: *Information Please Almanac*

25.

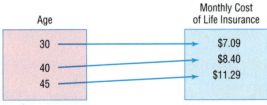

SOURCE: *eterm.com*

26.

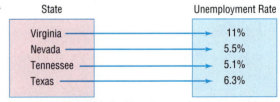

SOURCE: *United States Statistical Abstract*

27. $\{(-3, 5), (-2, 9), (-1, 2), (0, 11), (1, -5)\}$

28. $\{(-2, 2), (-1, 6), (0, 8), (1, -3), (2, 9)\}$

29. $\{(-2, 1), (-3, 2), (-10, 0), (1, 9), (2, 4)\}$

30. $\{(-2, -8), (-1, -1), (0, 0), (1, 1), (2, 8)\}$

In Problems 31–36, the graph of a one-to-one function f is given. Draw the graph of the inverse function f^{-1}. For convenience (and as a hint), the graph of y = x is also given.

31.

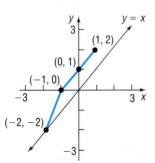

32.

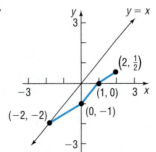

33.

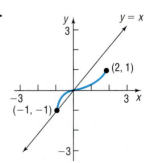

34.

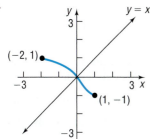

35.

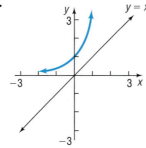

36.

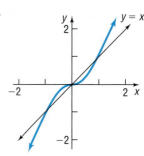

In Problems 37–46, verify that the functions f and g are inverses of each other by showing that $f(g(x)) = x$ and $g(f(x)) = x$. Give any values of x that need to be excluded.

37. $f(x) = 3x + 4;\quad g(x) = \dfrac{1}{3}(x - 4)$

38. $f(x) = 3 - 2x;\quad g(x) = -\dfrac{1}{2}(x - 3)$

39. $f(x) = 4x - 8;\quad g(x) = \dfrac{x}{4} + 2$

40. $f(x) = 2x + 6;\quad g(x) = \dfrac{1}{2}x - 3$

41. $f(x) = x^3 - 8;\quad g(x) = \sqrt[3]{x + 8}$

42. $f(x) = (x - 2)^2, x \ge 2;\quad g(x) = \sqrt{x} + 2$

43. $f(x) = \dfrac{1}{x};\quad g(x) = \dfrac{1}{x}$

44. $f(x) = x;\quad g(x) = x$

45. $f(x) = \dfrac{2x + 3}{x + 4};\quad g(x) = \dfrac{4x - 3}{2 - x}$

46. $f(x) = \dfrac{x - 5}{2x + 3};\quad g(x) = \dfrac{3x + 5}{1 - 2x}$

In Problems 47–58, the function f is one-to-one. Find its inverse and check your answer. State the domain and the range of f and f^{-1}. Graph f, f^{-1}, and y = x on the same coordinate axes.

47. $f(x) = 3x$

48. $f(x) = -4x$

49. $f(x) = 4x + 2$

50. $f(x) = 1 - 3x$

51. $f(x) = x^3 - 1$

52. $f(x) = x^3 + 1$

53. $f(x) = x^2 + 4, \quad x \ge 0$

54. $f(x) = x^2 + 9, \quad x \ge 0$

55. $f(x) = \dfrac{4}{x}$

56. $f(x) = -\dfrac{3}{x}$

57. $f(x) = \dfrac{1}{x - 2}$

58. $f(x) = \dfrac{4}{x + 2}$

In Problems 59–70, the function f is one-to-one. Find its inverse and check your answer. State the domain of f and find its range using f^{-1}.

59. $f(x) = \dfrac{2}{3 + x}$

60. $f(x) = \dfrac{4}{2 - x}$

61. $f(x) = \dfrac{3x}{x + 2}$

62. $f(x) = \dfrac{-2x}{x - 1}$

63. $f(x) = \dfrac{2x}{3x - 1}$

64. $f(x) = \dfrac{3x + 1}{-x}$

65. $f(x) = \dfrac{3x + 4}{2x - 3}$

66. $f(x) = \dfrac{2x - 3}{x + 4}$

67. $f(x) = \dfrac{2x + 3}{x + 2}$

68. $f(x) = \dfrac{-3x - 4}{x - 2}$

69. $f(x) = \dfrac{x^2 - 4}{2x^2}, \quad x > 0$

70. $f(x) = \dfrac{x^2 + 3}{3x^2}, \quad x > 0$

Applications and Extensions

71. Use the graph of $y = f(x)$ given in Problem 31 to evaluate the following:
(a) $f(-1)$ (b) $f(1)$ (c) $f^{-1}(1)$ (d) $f^{-1}(2)$

72. Use the graph of $y = f(x)$ given in Problem 32 to evaluate the following:
(a) $f(2)$ (b) $f(1)$ (c) $f^{-1}(0)$ (d) $f^{-1}(-1)$

73. Find the inverse of the linear function
$$f(x) = mx + b, \quad m \neq 0$$

74. Find the inverse of the function
$$f(x) = \sqrt{r^2 - x^2}, \quad 0 \leq x \leq r$$

75. A function f has an inverse function. If the graph of f lies in quadrant I, in which quadrant does the graph of f^{-1} lie?

76. A function f has an inverse function. If the graph of f lies in quadrant II, in which quadrant does the graph of f^{-1} lie?

77. The function $f(x) = |x|$ is not one-to-one. Find a suitable restriction on the domain of f so that the new function that results is one-to-one. Then find the inverse of f.

78. The function $f(x) = x^4$ is not one-to-one. Find a suitable restriction on the domain of f so that the new function that results is one-to-one. Then find the inverse of f.

79. Height versus Head Circumference The head circumference C of a child is related to the height H of the child (both in inches) through the function
$$H(C) = 2.15C - 10.53$$
(a) Express the head circumference C as a function of height H.

(b) Predict the head circumference of a child who is 26 inches tall.

80. Temperature Conversion To convert from x degrees Celsius to y degrees Fahrenheit, we use the formula $y = f(x) = \frac{9}{5}x + 32$. To convert from x degrees Fahrenheit to y degrees Celsius, we use the formula $y = g(x) = \frac{5}{9}(x - 32)$. Show that f and g are inverse functions.

81. Demand for Corn The demand for corn obeys the equation $p(x) = 300 - 50x$, where p is the price per bushel (in dollars) and x is the number of bushels produced (in millions). Express the production amount x as a function of the price p.

82. Period of a Pendulum The period T (in seconds) of a simple pendulum is a function of its length l (in feet), given by $T(l) = 2\pi\sqrt{\dfrac{l}{g}}$, where $g \approx 32.2$ feet per second per second is the acceleration of gravity. Express the length l as a function of the period T.

83. Given
$$f(x) = \frac{ax + b}{cx + d}$$
find $f^{-1}(x)$. If $c \neq 0$, under what conditions on $a, b, c,$ and d is $f = f^{-1}$?

Discussion and Writing

84. Can a one-to-one function and its inverse be equal? What must be true about the graph of f for this to happen? Give some examples to support your conclusion.

85. Draw the graph of a one-to-one function that contains the points $(-2, -3)$, $(0, 0)$, and $(1, 5)$. Now draw the graph of its inverse. Compare your graph to those of other students. Discuss any similarities. What differences do you see?

86. Give an example of a function whose domain is the set of real numbers and that is neither increasing nor decreasing on its domain, but is one-to-one.

[**Hint:** Use a piecewise-defined function.]

87. If a function f is even, can it be one-to-one? Explain.

88. Is every odd function one-to-one? Explain.

89. If the graph of a function and its inverse intersect, where must this necessarily occur? Can they intersect anywhere else? Must they intersect?

90. Suppose $C(g)$ represents the cost C of manufacturing g cars. Explain what $C^{-1}(800,000)$ represents.

'Are You Prepared?' Answers

1. Yes; for each input x there is one output y.   **2.** Increasing on $(0, \infty)$; decreasing on $(-\infty, 0)$.   **3.** $\{x | x \neq -6, x \neq 3\}$

4.3 Exponential Functions

PREPARING FOR THIS SECTION *Before getting started, review the following:*

- Exponents (Appendix, Section A.1, pp. 664–665, and Section A.9, pp. 743–744)
- Graphing Techniques: Transformations (Section 2.6, pp. 118–126)

- Average Rate of Change (Section 2.3, pp. 85–87)
- Solving Equations (Appendix, Section A.5, pp. 692–703)
- Horizontal Asymptotes (Section 3.3, pp. 189–195)

 Now work the 'Are You Prepared?' problems on page 282.

OBJECTIVES 1 Evaluate Exponential Functions
2 Graph Exponential Functions
3 Define the Number e
4 Solve Exponential Equations

---

1 Evaluate Exponential Functions

In the Appendix, Section A.9, we give a definition for raising a real number a to a rational power. Based on that discussion, we gave meaning to expressions of the form

$$a^r$$

where the base a is a positive real number and the exponent r is a rational number.

  But what is the meaning of a^x, where the base a is a positive real number and the exponent x is an irrational number? Although a rigorous definition requires methods discussed in calculus, the basis for the definition is easy to follow: Select a rational number r that is formed by truncating (removing) all but a finite number of digits from the irrational number x. Then it is reasonable to expect that

$$a^x \approx a^r$$

For example, take the irrational number $\pi = 3.14159\ldots$. Then an approximation to a^π is

$$a^\pi \approx a^{3.14}$$

where the digits after the hundredths position have been removed from the value for π. A better approximation would be

$$a^\pi \approx a^{3.14159}$$

where the digits after the hundred-thousandths position have been removed. Continuing in this way, we can obtain approximations to a^π to any desired degree of accuracy.

Most calculators have an $\boxed{x^y}$ key or a caret key $\boxed{\wedge}$ for working with exponents. To evaluate expressions of the form a^x, enter the base a, then press the $\boxed{x^y}$ key (or the $\boxed{\wedge}$ key), enter the exponent x, and press $\boxed{=}$ (or $\boxed{\text{enter}}$).

EXAMPLE 1   **Using a Calculator to Evaluate Powers of 2**

Using a calculator, evaluate:

(a) $2^{1.4}$ (b) $2^{1.41}$ (c) $2^{1.414}$ (d) $2^{1.4142}$ (e) $2^{\sqrt{2}}$

Figure 19

```
2^1.4
          2.639015822
```

Solution

Figure 19 shows the solution to part (a) using a TI-84 Plus graphing calculator.

(a) $2^{1.4} \approx 2.639015822$

(b) $2^{1.41} \approx 2.657371628$

(c) $2^{1.414} \approx 2.66474965$

(d) $2^{1.4142} \approx 2.665119089$

(e) $2^{\sqrt{2}} \approx 2.665144143$

◀

NOW WORK PROBLEM 11.

It can be shown that the familiar laws for rational exponents hold for real exponents.

Theorem

Laws of Exponents

If s, t, a, and b are real numbers with $a > 0$ and $b > 0$, then

$$a^s \cdot a^t = a^{s+t} \qquad (a^s)^t = a^{st} \qquad (ab)^s = a^s \cdot b^s$$

$$1^s = 1 \qquad a^{-s} = \frac{1}{a^s} = \left(\frac{1}{a}\right)^s \qquad a^0 = 1 \qquad \textbf{(1)}$$

We are now ready for the following definition:

An **exponential function** is a function of the form

$$f(x) = a^x$$

where a is a positive real number ($a > 0$) and $a \neq 1$. The domain of f is the set of all real numbers.

CAUTION It is important to distinguish a power function, $g(x) = x^n$, $n \geq 2$, an integer, from an exponential function, $f(x) = a^x$, $a \neq 1$, a real. In a power function, the base is a variable and the exponent is a constant. In an exponential function, the base is a constant and the exponent is a variable. ■

We exclude the base $a = 1$ because this function is simply the constant function $f(x) = 1^x = 1$. We also need to exclude bases that are negative; otherwise, we would have to exclude many values of x from the domain, such as $x = \frac{1}{2}$ and $x = \frac{3}{4}$.

[Recall that $(-2)^{1/2} = \sqrt{-2}$, $(-3)^{3/4} = \sqrt[4]{(-3)^3} = \sqrt[4]{-27}$, and so on, are not defined in the set of real numbers.]

Some examples of exponential functions are

$$f(x) = 2^x, \qquad F(x) = \left(\frac{1}{3}\right)^x$$

Notice that in each example the base is a constant and the exponent is a variable.

You may wonder what role the base a plays in the exponential function $f(x) = a^x$. We use the following Exploration to find out.

───── **Exploration** ─────

(a) Evaluate $f(x) = 2^x$ at $x = -2, -1, 0, 1, 2$, and 3.
(b) Evaluate $g(x) = 3x + 2$ at $x = -2, -1, 0, 1, 2$, and 3.
(c) Comment on the pattern that exists in the values of f and g.

Result (a) Table 2 shows the values of $f(x) = 2^x$ for $x = -2, -1, 0, 1, 2$, and 3.
(b) Table 3 shows the values of $g(x) = 3x + 2$ for $x = -2, -1, 0, 1, 2$, and 3.

Table 2

x	$f(x) = 2^x$
-2	$f(-2) = 2^{-2} = \dfrac{1}{2^2} = \dfrac{1}{4}$
-1	$\dfrac{1}{2}$
0	1
1	2
2	4
3	8

Table 3

x	$g(x) = 3x + 2$
-2	$g(-2) = 3(-2) + 2 = -4$
-1	-1
0	2
1	5
2	8
3	11

(c) In Table 2 we notice that each value of the exponential function $f(x) = a^x = 2^x$ could be found by multiplying the previous value of the function by the base, $a = 2$. For example,

$$f(-1) = 2 \cdot f(-2) = 2 \cdot \frac{1}{4} = \frac{1}{2}, \quad f(0) = 2 \cdot f(-1) = 2 \cdot \frac{1}{2} = 1, \quad f(1) = 2 \cdot f(0) = 2 \cdot 1 = 2$$

and so on.

Put another way, we see that the ratio of consecutive outputs is constant for unit increases in the input. The constant equals the value of the base a of the exponential function. For example, for the function $f(x) = 2^x$, we notice that

$$\frac{f(-1)}{f(-2)} = \frac{\frac{1}{2}}{\frac{1}{4}} = 2, \quad \frac{f(1)}{f(0)} = \frac{2}{1} = 2, \quad \frac{f(x+1)}{f(x)} = \frac{2^{x+1}}{2^x} = 2$$

and so on.

From Table 3 we see that the function $g(x) = 3x + 2$ does not have the ratio of consecutive outputs that are constant because it is not exponential. For example,

$$\frac{g(-1)}{g(-2)} = \frac{-1}{-4} = \frac{1}{4} \neq \frac{g(1)}{g(0)} = \frac{5}{2}$$

Instead, because $g(x) = 3x + 2$ is a linear function, for unit increases in the input, the outputs increase by a fixed amount equal to the value of the slope, 3.

The results of the Exploration lead to the following result.

Theorem

For an exponential function $f(x) = a^x, a > 0, a \neq 1$, if x is any real number, then

$$\frac{f(x + 1)}{f(x)} = a$$

In Words

For an exponential function $f(x) = a^x$, for 1-unit changes in the input x, the ratio of consecutive outputs is the constant a.

Proof

$$\frac{f(x + 1)}{f(x)} = \frac{a^{x+1}}{a^x} = a^{x+1-x} = a^1 = a \qquad \blacksquare$$

NOW WORK PROBLEM 21.

2 Graph Exponential Functions

First, we graph the exponential function $f(x) = 2^x$.

EXAMPLE 2 | **Graphing an Exponential Function**

Graph the exponential function: $f(x) = 2^x$

Table 4

X	Y1
-10	9.8E-4
-3	.125
-1	.5
0	1
1	2
3	8
10	1024

Y1 ⊟ 2^X

Solution The domain of $f(x) = 2^x$ is the set of all real numbers. We begin by locating some points on the graph of $f(x) = 2^x$, as listed in Table 4.

Since $2^x > 0$ for all x, the range of f is $(0, \infty)$. From this, we conclude that the graph has no x-intercepts, and, in fact, the graph will lie above the x-axis for all x. As Table 4 indicates, the y-intercept is 1. Table 4 also indicates that as $x \to -\infty$ the value of $f(x) = 2^x$ gets closer and closer to 0. We conclude that the x-axis ($y = 0$) is a horizontal asymptote to the graph as $x \to -\infty$. This gives us the end behavior for x large and negative.

To determine the end behavior for x large and positive, look again at Table 4. As $x \to \infty$, $f(x) = 2^x$ grows very quickly, causing the graph of $f(x) = 2^x$ to rise very rapidly. It is apparent that f is an increasing function and hence is one-to-one.

Figure 20 shows the graph of $f(x) = 2^x$. Notice that all the conclusions given earlier are confirmed by the graph.

Figure 20

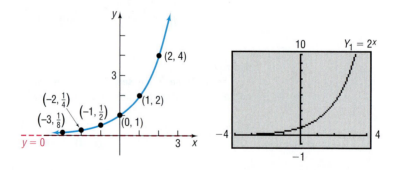

As we shall see, graphs that look like the one in Figure 20 occur very frequently in a variety of situations. For example, look at the graph in Figure 21, which illus-

Figure 21

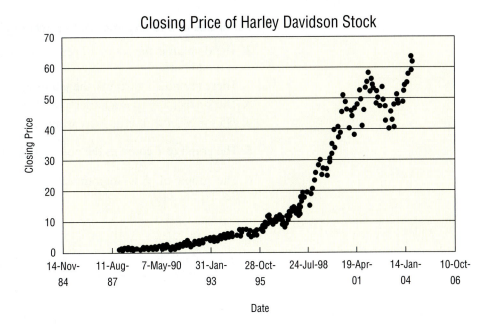

Closing Price of Harley Davidson Stock

trates the closing price of a share of Harley Davidson stock. Investors might conclude from this graph that the price of Harley Davidson stock is *behaving exponentially*; that is, the graph exhibits rapid, or exponential, growth.

We shall have more to say about situations that lead to exponential growth later in this chapter. For now, we continue to seek properties of the exponential functions.

The graph of $f(x) = 2^x$ in Figure 20 is typical of all exponential functions that have a base larger than 1. Such functions are increasing functions and hence are one-to-one. Their graphs lie above the x-axis, pass through the point $(0, 1)$, and thereafter rise rapidly as $x \to \infty$. As $x \to -\infty$, the x-axis $(y = 0)$ is a horizontal asymptote. There are no vertical asymptotes. Finally, the graphs are smooth and continuous, with no corners or gaps.

Figure 22 illustrates the graphs of two more exponential functions whose bases are larger than 1. Notice that for the larger base the graph is steeper when $x > 0$. Figure 23 shows that when $x < 0$ the graph of the equation with the larger base is closer to the x-axis.

Figure 22

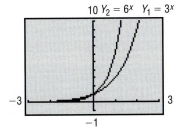

Figure 23

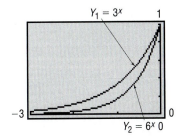

The following display summarizes the information that we have about $f(x) = a^x, a > 1$.

> **Properties of the Exponential Function** $f(x) = a^x$, $a > 1$
>
> **1.** The domain is the set of all real numbers; the range is the set of positive real numbers.
> **2.** There are no x-intercepts; the y-intercept is 1.
> **3.** The x-axis ($y = 0$) is a horizontal asymptote as $x \to -\infty$.
> **4.** $f(x) = a^x$, $a > 1$, is an increasing function and is one-to-one.
> **5.** The graph of f contains the points $(0, 1)$, $(1, a)$, and $\left(-1, \dfrac{1}{a}\right)$.
> **6.** The graph of f is smooth and continuous, with no corners or gaps. See Figure 24.

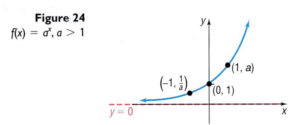

Figure 24
$f(x) = a^x$, $a > 1$

Now we consider $f(x) = a^x$ when $0 < a < 1$.

EXAMPLE 3 **Graphing an Exponential Function**

Graph the exponential function: $f(x) = \left(\dfrac{1}{2}\right)^x$

Solution The domain of $f(x) = \left(\dfrac{1}{2}\right)^x$ consists of all real numbers. As before, we locate some points on the graph, as listed in Table 5. Since $\left(\dfrac{1}{2}\right)^x > 0$ for all x, the range of f is the interval $(0, \infty)$. The graph lies above the x-axis and so has no x-intercepts. The y-intercept is 1. As $x \to -\infty$, $f(x) = \left(\dfrac{1}{2}\right)^x$ grows very quickly. As $x \to \infty$, the values of $f(x)$ approach 0. The x-axis ($y = 0$) is a horizontal asymptote as $x \to \infty$. It is apparent that f is a decreasing function and hence is one-to-one. Figure 25 illustrates the graph.

Table 5

X	Y1
-10	1024
-3	8
-1	2
0	1
1	.5
3	.125
10	9.8E-4

Y1🔲(1/2)^X

Figure 25

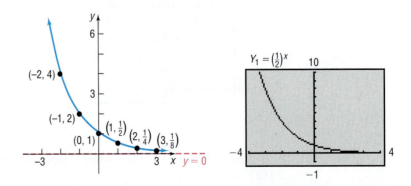

We could have obtained the graph of $y = \left(\dfrac{1}{2}\right)^x$ from the graph of $y = 2^x$ using transformations. The graph of $y = \left(\dfrac{1}{2}\right)^x = 2^{-x}$ is a reflection about the y-axis of the graph of $y = 2^x$. Compare Figures 20 and 25.

The graph of $f(x) = \left(\dfrac{1}{2}\right)^x$ in Figure 25 is typical of all exponential functions that have a base between 0 and 1. Such functions are decreasing and one-to-one. Their graphs lie above the x-axis and pass through the point $(0, 1)$. The graphs rise rapidly as $x \to -\infty$. As $x \to \infty$, the x-axis $(y = 0)$ is a horizontal asymptote. There are no vertical asymptotes. Finally, the graphs are smooth and continuous, with no corners or gaps.

Figure 26 illustrates the graphs of two more exponential functions whose bases are between 0 and 1. Notice that the smaller base results in a graph that is steeper when $x < 0$. Figure 27 shows that when $x > 0$ the graph of the equation with the smaller base is closer to the x-axis.

Figure 26

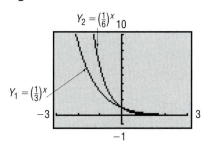

Figure 27

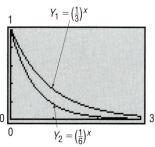

The following display summarizes the information that we have about the function $f(x) = a^x, 0 < a < 1$.

Properties of the Exponential Function $f(x) = a^x, 0 < a < 1$

1. The domain is the set of all real numbers; the range is the set of positive real numbers.
2. There are no x-intercepts; the y-intercept is 1.
3. The x-axis $(y = 0)$ is a horizontal asymptote as $x \to \infty$.
4. $f(x) = a^x, 0 < a < 1$, is a decreasing function and is one-to-one.
5. The graph of f contains the points $(0, 1)$, $(1, a)$, and $\left(-1, \dfrac{1}{a}\right)$.
6. The graph of f is smooth and continuous, with no corners or gaps. See Figure 28.

Figure 28

$f(x) = a^x, 0 < a < 1$

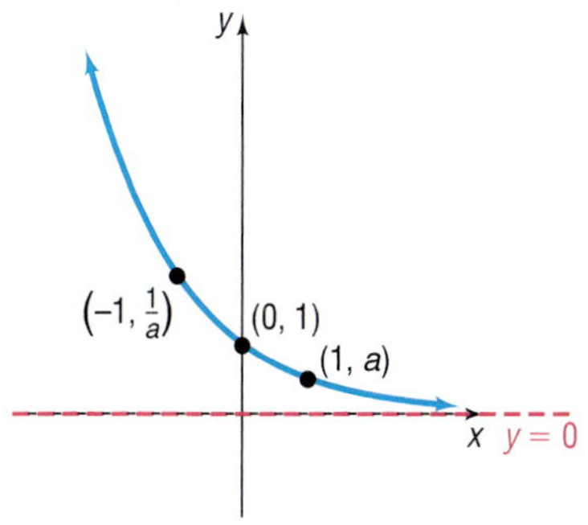

$\left(-1, \dfrac{1}{a}\right)$ $(0, 1)$ $(1, a)$ $y = 0$

| EXAMPLE 4 | **Graphing Exponential Functions Using Transformations** |

Graph $f(x) = 2^{-x} - 3$ and determine the domain, range, and horizontal asymptote of f.

Solution We begin with the graph of $y = 2^x$. Figure 29 shows the stages.

Figure 29

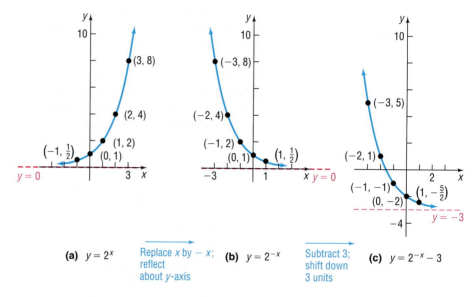

(a) $y = 2^x$ — Replace x by $-x$; reflect about y-axis — (b) $y = 2^{-x}$ — Subtract 3; shift down 3 units — (c) $y = 2^{-x} - 3$

As Figure 29(c) illustrates, the domain of $f(x) = 2^{-x} - 3$ is the interval $(-\infty, \infty)$ and the range is the interval $(-3, \infty)$. The horizontal asymptote of f is the line $y = -3$.

✔ **CHECK:** Graph $Y_1 = 2^{-x} - 3$ to verify the graph obtained in Figure 29(c). ◄

 NOW WORK PROBLEM 37.

3 Define the Number e

As we shall see shortly, many problems that occur in nature require the use of an exponential function whose base is a certain irrational number, symbolized by the letter e.

Let's look at one way of arriving at this important number e.

The **number e** is defined as the number that the expression

$$\left(1 + \frac{1}{n}\right)^n \qquad (2)$$

approaches as $n \to \infty$. In calculus, this is expressed using limit notation as

$$e = \lim_{n \to \infty} \left(1 + \frac{1}{n}\right)^n$$

Figure 32

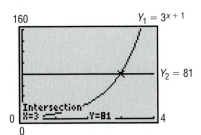

Now we have the same base, 3, on each side, so we can apply property (3) to obtain

$$x + 1 = 4$$
$$x = 3$$

✔ **CHECK:** We verify the solution by graphing $Y_1 = 3^{x+1}$ and $Y_2 = 81$ to determine where the graphs intersect. See Figure 32. The graphs intersect at $x = 3$. The solution set is $\{3\}$.   ◀

NOW WORK PROBLEM 53.

EXAMPLE 7   **Solving an Exponential Equation**

Solve: $e^{-x^2} = (e^x)^2 \cdot \dfrac{1}{e^3}$

Solution   We use the Laws of Exponents first to get the base e on the right side.

$$(e^x)^2 \cdot \frac{1}{e^3} = e^{2x} \cdot e^{-3} = e^{2x-3}$$

As a result,

$$e^{-x^2} = e^{2x-3}$$
$$-x^2 = 2x - 3 \qquad \text{Apply Property (3).}$$
$$x^2 + 2x - 3 = 0 \qquad \text{Place the quadratic equation in standard form.}$$
$$(x + 3)(x - 1) = 0 \qquad \text{Factor.}$$
$$x = -3 \quad \text{or} \quad x = 1 \qquad \text{Use the Zero-Product Property.}$$

The solution set is $\{-3, 1\}$.
You should verify these solutions using a graphing utility.   ◀

Many applications involve the exponential functions. Let's look at one.

EXAMPLE 8   **Exponential Probability**

Between 9:00 PM and 10:00 PM cars arrive at Burger King's drive-thru at the rate of 12 cars per hour (0.2 car per minute). The following formula from statistics can be used to determine the probability that a car will arrive within t minutes of 9:00 PM.

$$F(t) = 1 - e^{-0.2t}$$

(a) Determine the probability that a car will arrive within 5 minutes of 9 PM (that is, before 9:05 PM).

(b) Determine the probability that a car will arrive within 30 minutes of 9 PM (before 9:30 PM).

(c) Graph F using your graphing utility.

(d) What value does F approach as t becomes unbounded in the positive direction?

Figure 33

Solution

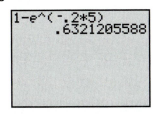

(a) The probability that a car will arrive within 5 minutes is found by evaluating $F(t)$ at $t = 5$.

$$F(5) = 1 - e^{-0.2(5)}$$

We evaluate this expression in Figure 33. We conclude that there is a 63% probability that a car will arrive within 5 minutes.

(b) The probability that a car will arrive within 30 minutes is found by evaluating $F(t)$ at $t = 30$.

Figure 34

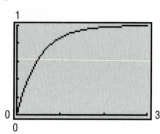

$$F(30) = 1 - e^{-0.2(30)} \approx 0.9975$$

There is a 99.75% probability that a car will arrive within 30 minutes.

(c) See Figure 34 for the graph of F.

(d) As time passes, the probability that a car will arrive increases. The value that F approaches can be found by letting $t \to \infty$. Since $e^{-0.2t} = \dfrac{1}{e^{0.2t}}$, it follows that $e^{-0.2t} \to 0$ as $t \to \infty$. Thus, F approaches 1 as t gets large.   ◀

NOW WORK PROBLEM 81.

Summary

Properties of the Exponential Function

$f(x) = a^x, \quad a > 1$	Domain: the interval $(-\infty, \infty)$; Range: the interval $(0, \infty)$ x-intercepts: none; y-intercept: 1 Horizontal asymptote: x-axis ($y = 0$) as $x \to -\infty$ Increasing; one-to-one; smooth; continuous See Figure 24 for a typical graph.
$f(x) = a^x, \quad 0 < a < 1$	Domain: the interval $(-\infty, \infty)$; Range: the interval $(0, \infty)$. x-intercepts: none; y-intercept: 1 Horizontal asymptote: x-axis ($y = 0$) as $x \to \infty$ Decreasing; one-to-one; smooth; continuous See Figure 28 for a typical graph.

If $a^u = a^v$, then $u = v$.

4.3 Assess Your Understanding

'Are You Prepared?'

Answers are given at the end of these exercises. If you get a wrong answer, read the pages listed in red.

1. $4^3 =$ _____; $8^{2/3} =$ _____; $3^{-2} =$ _____. (pp. 664–665 and pp. 743–744)

2. Solve: $3x^2 + 5x - 2 = 0$. (pp. 696–703)

3. *True or False:* To graph $y = (x - 2)^3$, shift the graph of $y = x^3$ to the left 2 units. (pp. 118–120)

4. Find the average rate of change of $f(x) = 3x - 5$ from $x = 0$ to $x = c$. (pp. 85–88)

5. *True or False:* The function $f(x) = \dfrac{2x}{x - 3}$ has $y = 2$ as a horizontal asymptote. (pp. 189–195)

Concepts and Vocabulary

6. The graph of every exponential function $f(x) = a^x, a > 0, a \neq 1$, passes through three points: _____, _____, and _____.

7. If the graph of the exponential function $f(x) = a^x, a > 0, a \neq 1$, is decreasing, then a must be less than _____.

8. If $3^x = 3^4$, then $x =$ _____.

9. *True or False:* The graphs of $y = 3^x$ and $y = \left(\dfrac{1}{3}\right)^x$ are identical.

10. *True or False:* The range of the exponential function $f(x) = a^x, a > 0, a \neq 1$, is the set of all real numbers.

Skill Building

In Problems 11–20, approximate each number using a calculator. Express your answer rounded to three decimal places.

11. (a) $3^{2.2}$ (b) $3^{2.23}$ (c) $3^{2.236}$ (d) $3^{\sqrt{5}}$ **12.** (a) $5^{1.7}$ (b) $5^{1.73}$ (c) $5^{1.732}$ (d) $5^{\sqrt{3}}$

13. (a) $2^{3.14}$ (b) $2^{3.141}$ (c) $2^{3.1415}$ (d) 2^{π} **14.** (a) $2^{2.7}$ (b) $2^{2.71}$ (c) $2^{2.718}$ (d) 2^{e}

15. (a) $3.1^{2.7}$ (b) $3.14^{2.71}$ (c) $3.141^{2.718}$ (d) π^{e} **16.** (a) $2.7^{3.1}$ (b) $2.71^{3.14}$ (c) $2.718^{3.141}$ (d) e^{π}

17. $e^{1.2}$ **18.** $e^{-1.3}$ **19.** $e^{-0.85}$ **20.** $e^{2.1}$

In Problems 21–28, determine whether the given function is exponential or not. For those that are exponential functions, identify the value of the base a. [**Hint:** Look at the ratio of consecutive values.]

21.

x	f(x)
−1	3
0	6
1	12
2	18
3	30

22.

x	g(x)
−1	2
0	5
1	8
2	11
3	14

23.

x	H(x)
−1	$\frac{1}{4}$
0	1
1	4
2	16
3	64

24.

x	F(x)
−1	$\frac{2}{3}$
0	1
1	$\frac{3}{2}$
2	$\frac{9}{4}$
3	$\frac{27}{8}$

25.

x	f(x)
−1	$\frac{3}{2}$
0	3
1	6
2	12
3	24

26.

x	g(x)
−1	6
0	1
1	0
2	3
3	10

27.

x	H(x)
−1	2
0	4
1	6
2	8
3	10

28.

x	F(x)
−1	$\frac{1}{2}$
0	$\frac{1}{4}$
1	$\frac{1}{8}$
2	$\frac{1}{16}$
3	$\frac{1}{32}$

In Problems 29–36, the graph of an exponential function is given. Match each graph to one of the following functions.

   A. $y = 3^{x}$ B. $y = 3^{-x}$ C. $y = -3^{x}$ D. $y = -3^{-x}$

   E. $y = 3^{x} - 1$ F. $y = 3^{x-1}$ G. $y = 3^{1-x}$ H. $y = 1 - 3^{x}$

29.

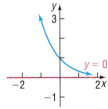

30.

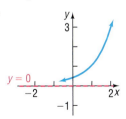

31.

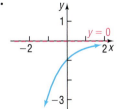

32.

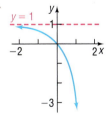

33.

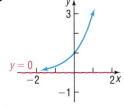

34.

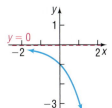

35.

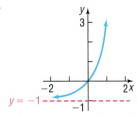

36.

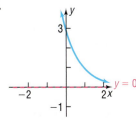

In Problems 37–44, use transformations to graph each function. Determine the domain, range, and horizontal asymptote of each function.

37. $f(x) = 2^x + 1$ **38.** $f(x) = 2^{x+2}$ **39.** $f(x) = 3^{-x} - 2$ **40.** $f(x) = -3^x + 1$

41. $f(x) = 2 + 3(4^x)$ **42.** $f(x) = 1 - 3(2^x)$ **43.** $f(x) = 2 + 3^{x/2}$ **44.** $f(x) = 1 - 2^{-x/3}$

In Problems 45–52, begin with the graph of $y = e^x$ (Figure 30 (a)) and use transformations to graph each function. Determine the domain, range, and horizontal asymptote of each function.

45. $f(x) = e^{-x}$ **46.** $f(x) = -e^x$ **47.** $f(x) = e^{x+2}$ **48.** $f(x) = e^x - 1$

49. $f(x) = 5 - e^{-x}$ **50.** $f(x) = 9 - 3e^{-x}$ **51.** $f(x) = 2 - e^{-x/2}$ **52.** $f(x) = 7 - 3e^{2x}$

In Problems 53–66, solve each equation.

53. $2^{2x+1} = 4$ **54.** $5^{1-2x} = \dfrac{1}{5}$ **55.** $3^{x^3} = 9^x$ **56.** $4^{x^2} = 2^x$ **57.** $8^{x^2-2x} = \dfrac{1}{2}$

58. $9^{-x} = \dfrac{1}{3}$ **59.** $2^x \cdot 8^{-x} = 4^x$ **60.** $\left(\dfrac{1}{2}\right)^{1-x} = 4$ **61.** $\left(\dfrac{1}{5}\right)^{2-x} = 25$ **62.** $4^x - 2^x = 0$

63. $4^x = 8$ **64.** $9^{2x} = 27$ **65.** $e^{x^2} = (e^{3x}) \cdot \dfrac{1}{e^2}$ **66.** $(e^4)^x \cdot e^{x^2} = e^{12}$

67. If $4^x = 7$, what does 4^{-2x} equal?   **68.** If $2^x = 3$, what does 4^{-x} equal?

69. If $3^{-x} = 2$, what does 3^{2x} equal?   **70.** If $5^{-x} = 3$, what does 5^{3x} equal?

In Problems 71–74, determine the exponential function whose graph is given.

71.

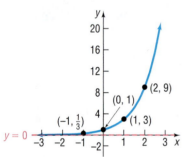

72.

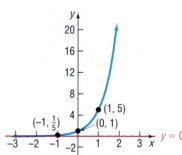

73.

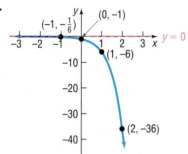

74.

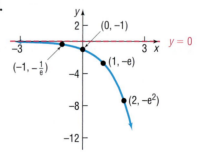

Applications and Extensions

75. Optics If a single pane of glass obliterates 3% of the light passing through it, then the percent p of light that passes through n successive panes is given approximately by the function

$$p(n) = 100(0.97)^n$$

(a) What percent of light will pass through 10 panes?
(b) What percent of light will pass through 25 panes?

76. Atmospheric Pressure The atmospheric pressure p on a balloon or plane decreases with increasing height. This pressure, measured in millimeters of mercury, is related to the height h (in kilometers) above sea level by the function

$$p(h) = 760e^{-0.145h}$$

(a) Find the atmospheric pressure at a height of 2 kilometers (over a mile).
(b) What is it at a height of 10 kilometers (over 30,000 feet)?

77. Depreciation The price p of a Honda Civic DX Sedan that is x years old is given by

$$p(x) = 16{,}630(0.90)^x$$

(a) How much does a 3-year-old Civic DX Sedan cost?
(b) How much does a 9-year-old Civic DX Sedan cost?

78. Healing of Wounds The normal healing of wounds can be modeled by an exponential function. If A_0 represents the original area of the wound and if A equals the area of the wound, then the function

$$A(n) = A_0 e^{-0.35n}$$

describes the area of a wound after n days following an injury when no infection is present to retard the healing. Suppose that a wound initially had an area of 100 square millimeters.

(a) If healing is taking place, how large will the area of the wound be after 3 days?
(b) How large will it be after 10 days?

79. Drug Medication The function

$$D(h) = 5e^{-0.4h}$$

can be used to find the number of milligrams D of a certain drug that is in a patient's bloodstream h hours after the drug has been administered. How many milligrams will be present after 1 hour? After 6 hours?

80. Spreading of Rumors A model for the number N of people in a college community who have heard a certain rumor is

$$N = P(1 - e^{-0.15d})$$

where P is the total population of the community and d is the number of days that have elapsed since the rumor began. In a community of 1000 students, how many students will have heard the rumor after 3 days?

81. Exponential Probability Between 12:00 PM and 1:00 PM, cars arrive at Citibank's drive-thru at the rate of 6 cars per hour (0.1 car per minute). The following formula from probability can be used to determine the probability that a car will arrive within t minutes of 12:00 PM:

$$F(t) = 1 - e^{-0.1t}$$

(a) Determine the probability that a car will arrive within 10 minutes of 12:00 PM (that is, before 12:10 PM).
(b) Determine the probability that a car will arrive within 40 minutes of 12:00 PM (before 12:40 PM).
(c) What value does F approach as t becomes unbounded in the positive direction?
(d) Graph F using a graphing utility.
(e) Using TRACE, determine how many minutes are needed for the probability to reach 50%.

82. Exponential Probability Between 5:00 PM and 6:00 PM, cars arrive at Jiffy Lube at the rate of 9 cars per hour (0.15 car per minute). The following formula from probability can be used to determine the probability that a car will arrive within t minutes of 5:00 PM:

$$F(t) = 1 - e^{-0.15t}$$

(a) Determine the probability that a car will arrive within 15 minutes of 5:00 PM (that is, before 5:15 PM).
(b) Determine the probability that a car will arrive within 30 minutes of 5:00 PM (before 5:30 PM).
(c) What value does F approach as t becomes unbounded in the positive direction?
(d) Graph F using a graphing utility.
(e) Using TRACE, determine how many minutes are needed for the probability to reach 60%.

83. Poisson Probability Between 5:00 PM and 6:00 PM, cars arrive at McDonald's drive-thru at the rate of 20 cars per hour. The following formula from probability can be used to determine the probability that x cars will arrive between 5:00 PM and 6:00 PM.

$$P(x) = \frac{20^x e^{-20}}{x!}$$

where

$$x! = x \cdot (x - 1) \cdot (x - 2) \cdots \cdots 3 \cdot 2 \cdot 1$$

(a) Determine the probability that $x = 15$ cars will arrive between 5:00 PM and 6:00 PM.
(b) Determine the probability that $x = 20$ cars will arrive between 5:00 PM and 6:00 PM.

84. Poisson Probability People enter a line for the *Demon Roller Coaster* at the rate of 4 per minute. The following formula from probability can be used to determine the probability that x people will arrive within the next minute.

$$P(x) = \frac{4^x e^{-4}}{x!}$$

where

$$x! = x \cdot (x - 1) \cdot (x - 2) \cdots \cdots 3 \cdot 2 \cdot 1$$

(a) Determine the probability that $x = 5$ people will arrive within the next minute.
(b) Determine the probability that $x = 8$ people will arrive within the next minute.

85. Relative Humidity The relative humidity is the ratio (expressed as a percent) of the amount of water vapor in the air to the maximum amount that it can hold at a specific temperature. The relative humidity, R, is found using the following formula:

$$R = 10^{\left(\frac{4221}{T+459.4} - \frac{4221}{D+459.4} + 2\right)}$$

where T is the air temperature (in °F) and D is the dew point temperature (in °F).

(a) Determine the relative humidity if the air temperature is 50° Fahrenheit and the dew point temperature is 41° Fahrenheit.
(b) Determine the relative humidity if the air temperature is 68° Fahrenheit and the dew point temperature is 59° Fahrenheit.
(c) What is the relative humidity if the air temperature and the dew point temperature are the same?

86. Learning Curve Suppose that a student has 500 vocabulary words to learn. If the student learns 15 words after 5 minutes, the function

$$L(t) = 500(1 - e^{-0.0061t})$$

approximates the number of words L that the student will learn after t minutes.

(a) How many words will the student learn after 30 minutes?

(b) How many words will the student learn after 60 minutes?

87. Current in a *RL* Circuit The equation governing the amount of current I (in amperes) after time t (in seconds) in a single *RL* circuit consisting of a resistance R (in ohms), an inductance L (in henrys), and an electromotive force E (in volts) is

$$I = \frac{E}{R}[1 - e^{-(R/L)t}]$$

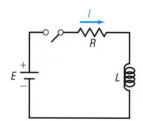

(a) If $E = 120$ volts, $R = 10$ ohms, and $L = 5$ henrys, how much current I_1 is flowing after 0.3 second? After 0.5 second? After 1 second?

(b) What is the maximum current?

(c) Graph this function $I = I_1(t)$, measuring I along the y-axis and t along the x-axis.

(d) If $E = 120$ volts, $R = 5$ ohms, and $L = 10$ henrys, how much current I_2 is flowing after 0.3 second? After 0.5 second? After 1 second?

(e) What is the maximum current?

(f) Graph this function $I = I_2(t)$ on the same coordinate axes as $I_1(t)$.

88. Current in a *RC* Circuit The equation governing the amount of current I (in amperes) after time t (in microseconds) in a single *RC* circuit consisting of a resistance R (in ohms), a capacitance C (in microfarads), and an electromotive force E (in volts) is

$$I = \frac{E}{R}e^{-t/(RC)}$$

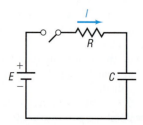

(a) If $E = 120$ volts, $R = 2000$ ohms, and $C = 1.0$ microfarad, how much current I_1 is flowing initially $(t = 0)$? After 1000 microseconds? After 3000 microseconds?

(b) What is the maximum current?

(c) Graph this function $I = I_1(t)$, measuring I along the y-axis and t along the x-axis.

(d) If $E = 120$ volts, $R = 1000$ ohms, and $C = 2.0$ microfarads, how much current I_2 is flowing initially? After 1000 microseconds? After 3000 microseconds?

(e) What is the maximum current?

(f) Graph this function $I = I_2(t)$ on the same coordinate axes as $I_1(t)$.

89. Another Formula for *e* Use a calculator to compute the values of

$$2 + \frac{1}{2!} + \frac{1}{3!} + \cdots + \frac{1}{n!}$$

for $n = 4, 6, 8,$ and 10. Compare each result with e.

[**Hint:** $1! = 1, 2! = 2 \cdot 1, 3! = 3 \cdot 2 \cdot 1,$
$n! = n(n - 1) \cdot \cdots \cdot (3)(2)(1).$]

90. Another Formula for *e* Use a calculator to compute the various values of the expression. Compare the values to e.

$$2 + \cfrac{1}{1 + \cfrac{1}{2 + \cfrac{2}{3 + \cfrac{3}{4 + \cfrac{4}{\text{etc.}}}}}}$$

91. Difference Quotient If $f(x) = a^x$, show that

$$\frac{f(x + h) - f(x)}{h} = a^x \cdot \frac{a^h - 1}{h}$$

92. If $f(x) = a^x$, show that $f(A + B) = f(A) \cdot f(B)$.

93. If $f(x) = a^x$, show that $f(-x) = \dfrac{1}{f(x)}$.

94. If $f(x) = a^x$, show that $f(\alpha x) = [f(x)]^\alpha$.

Problems 95 and 96 provide definitions for two other transcendental functions.

95. The **hyperbolic sine function**, designated by sinh x, is defined as

$$\sinh x = \frac{1}{2}(e^x - e^{-x})$$

(a) Show that $f(x) = \sinh x$ is an odd function.

(b) Graph $f(x) = \sinh x$ using a graphing utility.

96. The **hyperbolic cosine function**, designated by cosh x, is defined as

$$\cosh x = \frac{1}{2}(e^x + e^{-x})$$

(a) Show that $f(x) = \cosh x$ is an even function.

(b) Graph $f(x) = \cosh x$ using a graphing utility.

(c) Refer to Problem 95. Show that, for every x,
$$(\cosh x)^2 - (\sinh x)^2 = 1$$

97. Historical Problem Pierre de Fermat (1601–1665) conjectured that the function
$$f(x) = 2^{(2^x)} + 1$$

for $x = 1, 2, 3, \ldots$, would always have a value equal to a prime number. But Leonhard Euler (1707–1783) showed that this formula fails for $x = 5$. Use a calculator to determine the prime numbers produced by f for $x = 1, 2, 3, 4$. Then show that $f(5) = 641 \times 6{,}700{,}417$, which is not prime.

Discussion and Writing

98. The bacteria in a 4-liter container double every minute. After 60 minutes the container is full. How long did it take to fill half the container?

99. Explain in your own words what the number e is. Provide at least two applications that require the use of this number.

100. Do you think that there is a power function that increases more rapidly than an exponential function whose base is greater than 1? Explain.

101. As the base a of an exponential function $f(x) = a^x, a > 1$, increases, what happens to the behavior of its graph for $x > 0$? What happens to the behavior of its graph for $x < 0$?

102. The graphs of $y = a^{-x}$ and $y = \left(\dfrac{1}{a}\right)^x$ are identical. Why?

'Are You Prepared?' Answers

1. $64; 4; \dfrac{1}{9}$ **2.** $\left\{-2, \dfrac{1}{3}\right\}$ **3.** False   **4.** 3   **5.** True

4.4 Logarithmic Functions

PREPARING FOR THIS SECTION *Before getting started, review the following:*

- Solving Inequalities (Appendix, Section A.8, pp. 732–733)
- Polynomial and Rational Inequalities (Section 3.5, pp. 212–215)

Now work the 'Are You Prepared?' problems on page 296.

OBJECTIVES   1 Change Exponential Expressions to Logarithmic Expressions and Logarithmic Expressions to Exponential Expressions
2 Evaluate Logarithmic Expressions
3 Determine the Domain of a Logarithmic Function
4 Graph Logarithmic Functions
5 Solve Logarithmic Equations

Recall that a one-to-one function $y = f(x)$ has an inverse function that is defined (implicitly) by the equation $x = f(y)$. In particular, the exponential function $y = f(x) = a^x, a > 0, a \neq 1$, is one-to-one and hence has an inverse function that is defined implicitly by the equation
$$x = a^y, \qquad a > 0, \qquad a \neq 1$$

This inverse function is so important that it is given a name, the *logarithmic function*.

> The **logarithmic function to the base a**, where $a > 0$ and $a \neq 1$, is denoted by $y = \log_a x$ (read as "y is the logarithm to the base a of x") and is defined by
>
> $$y = \log_a x \quad \text{if and only if} \quad x = a^y$$
>
> The domain of the logarithmic function $y = \log_a x$ is $x > 0$.

A *logarithm* is merely a name for a certain exponent.

| EXAMPLE 1 | **Relating Logarithms to Exponents** |

(a) If $y = \log_3 x$, then $x = 3^y$. For example, $4 = \log_3 81$ is equivalent to $81 = 3^4$.

(b) If $y = \log_5 x$, then $x = 5^y$. For example, $-1 = \log_5\left(\dfrac{1}{5}\right)$ is equivalent to $\dfrac{1}{5} = 5^{-1}$.

◀

1 **Change Exponential Expressions to Logarithmic Expressions and Logarithmic Expressions to Exponential Expressions**

We can use the definition of a logarithm to convert from exponential form to logarithmic form, and vice versa, as the following two examples illustrate.

| EXAMPLE 2 | **Changing Exponential Expressions to Logarithmic Expressions** |

Change each exponential expression to an equivalent expression involving a logarithm.

(a) $1.2^3 = m$ (b) $e^b = 9$ (c) $a^4 = 24$

Solution We use the fact that $y = \log_a x$ and $x = a^y$, $a > 0$, $a \neq 1$, are equivalent.

(a) If $1.2^3 = m$, then $3 = \log_{1.2} m$.
(b) If $e^b = 9$, then $b = \log_e 9$.
(c) If $a^4 = 24$, then $4 = \log_a 24$.

◀

🖉 **NOW WORK PROBLEM 9.**

| EXAMPLE 3 | **Changing Logarithmic Expressions to Exponential Expressions** |

Change each logarithmic expression to an equivalent expression involving an exponent.

(a) $\log_a 4 = 5$ (b) $\log_e b = -3$ (c) $\log_3 5 = c$

Solution (a) If $\log_a 4 = 5$, then $a^5 = 4$.
(b) If $\log_e b = -3$, then $e^{-3} = b$.
(c) If $\log_3 5 = c$, then $3^c = 5$.

◀

🖉 **NOW WORK PROBLEM 21.**

2 **Evaluate Logarithmic Expressions**

To find the exact value of a logarithm, we write the logarithm in exponential notation and use the fact that if $a^u = a^v$ then $u = v$.

| EXAMPLE 4 | **Finding the Exact Value of a Logarithmic Expression** |

Find the exact value of:

(a) $\log_2 16$ (b) $\log_3 \dfrac{1}{27}$

Solution   (a) $y = \log_2 16$

$$2^y = 16 \qquad \textit{Change to exponential form.}$$
$$2^y = 2^4 \qquad \textit{16} = 2^4$$
$$y = 4 \qquad \textit{Equate exponents.}$$

Therefore, $\log_2 16 = 4$.

(b) $y = \log_3 \dfrac{1}{27}$

$$3^y = \frac{1}{27} \qquad \textit{Change to exponential form.}$$

$$3^y = 3^{-3} \qquad \frac{1}{27} = \frac{1}{3^3} = 3^{-3}$$

$$y = -3 \qquad \textit{Equate exponents.}$$

Therefore, $\log_3 \dfrac{1}{27} = -3$.   ◀

NOW WORK PROBLEM 33.

3 Determine the Domain of a Logarithmic Function

The logarithmic function $y = \log_a x$ has been defined as the inverse of the exponential function $y = a^x$. That is, if $f(x) = a^x$, then $f^{-1}(x) = \log_a x$. Based on the discussion given in Section 4.2 on inverse functions, for a function f and its inverse f^{-1}, we have

$$\text{Domain of } f^{-1} = \text{Range of } f \quad \text{and} \quad \text{Range of } f^{-1} = \text{Domain of } f$$

Consequently, it follows that

> Domain of the logarithmic function = Range of the exponential function = $(0, \infty)$
>
> Range of the logarithmic function = Domain of the exponential function = $(-\infty, \infty)$

In the next box, we summarize some properties of the logarithmic function:

> $y = \log_a x$ (defining equation: $x = a^y$)
>
> Domain: $0 < x < \infty$ Range: $-\infty < y < \infty$

The domain of a logarithmic function consists of the *positive* real numbers, so the argument of a logarithmic function must be greater than zero.

EXAMPLE 5   **Finding the Domain of a Logarithmic Function**

Find the domain of each logarithmic function.

(a) $F(x) = \log_2(x + 3)$ (b) $g(x) = \log_5\left(\dfrac{1 + x}{1 - x}\right)$

(c) $h(x) = \log_{1/2}|x|$

Solution   (a) The domain of F consists of all x for which $x + 3 > 0$, that is, $x > -3$. Using interval notation, the domain of f is $(-3, \infty)$.

(b) The domain of g is restricted to

$$\frac{1 + x}{1 - x} > 0$$

Solving this inequality, we find that the domain of g consists of all x between -1 and 1, that is, $-1 < x < 1$ or, using interval notation, $(-1, 1)$.

(c) Since $|x| > 0$, provided that $x \neq 0$, the domain of h consists of all real numbers except zero or, using interval notation, $(-\infty, 0)$ or $(0, \infty)$. ◄

NOW WORK PROBLEMS 47 AND 53.

4 Graph Logarithmic Functions

Since exponential functions and logarithmic functions are inverses of each other, the graph of the logarithmic function $y = \log_a x$ is the reflection about the line $y = x$ of the graph of the exponential function $y = a^x$, as shown in Figure 35.

Figure 35

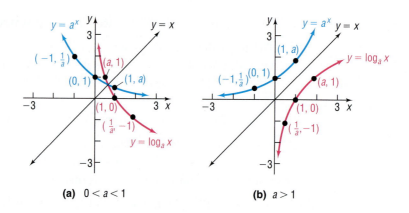

(a) $0 < a < 1$ **(b)** $a > 1$

For example, to graph $y = \log_2 x$, graph $y = 2^x$ and reflect it about the line $y = x$. See Figure 36. To graph $y = \log_{1/3} x$, graph $y = \left(\frac{1}{3}\right)^x$ and reflect it about the line $y = x$. See Figure 37.

Figure 36

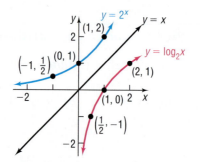

Figure 37

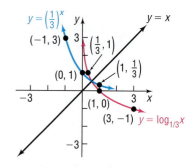

NOW WORK PROBLEM 63.

Properties of the Logarithmic Function $f(x) = \log_a x$

1. The domain is the set of positive real numbers; the range is the set of all real numbers.
2. The x-intercept of the graph is 1. There is no y-intercept.
3. The y-axis ($x = 0$) is a vertical asymptote of the graph.
4. A logarithmic function is decreasing if $0 < a < 1$ and increasing if $a > 1$.
5. The graph of f contains the points $(1, 0)$, $(a, 1)$, and $\left(\dfrac{1}{a}, -1 \right)$.
6. The graph is smooth and continuous, with no corners or gaps.

If the base of a logarithmic function is the number e, then we have the **natural logarithm function**. This function occurs so frequently in applications that it is given a special symbol, **ln** (from the Latin, *logarithmus naturalis*). That is,

$$y = \ln x \quad \text{if and only if} \quad x = e^y \qquad \textbf{(1)}$$

Since $y = \ln x$ and the exponential function $y = e^x$ are inverse functions, we can obtain the graph of $y = \ln x$ by reflecting the graph of $y = e^x$ about the line $y = x$. See Figure 38.

Figure 38

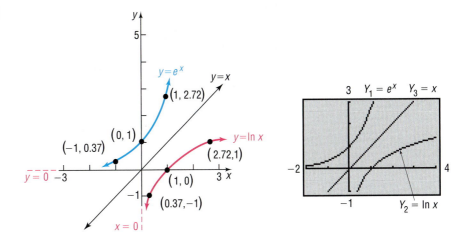

Table 8 displays other points on the graph of $f(x) = \ln x$. Notice for $x \leq 0$ that we obtain an error message. Do you recall why?

Table 8

X	Y₂
-1	ERROR
0.5	-.6931
1	0
2	.69315
2.7183	1
3	1.0986

Y₂ ◼ln(X)

EXAMPLE 6	**Graphing Logarithmic Functions Using Transformations**

Graph $f(x) = -\ln(x + 2)$ by starting with the graph of $y = \ln x$. Determine the domain, range, and vertical asymptote.

Solution   The domain consists of all x for which

$$x + 2 > 0 \quad \text{or} \quad x > -2$$

To obtain the graph of $y = -\ln(x + 2)$, we use the steps illustrated in Figure 39.

Figure 39

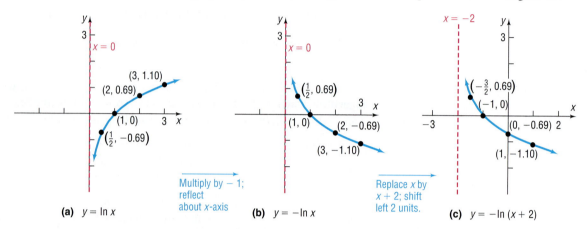

(a) $y = \ln x$

Multiply by -1; reflect about x-axis $\longrightarrow$

(b) $y = -\ln x$

Replace x by $x + 2$; shift left 2 units. $\longrightarrow$

(c) $y = -\ln(x + 2)$

The range of $f(x) = -\ln(x + 2)$ is the interval $(-\infty, \infty)$, and the vertical asymptote is $x = -2$. [Do you see why? The original asymptote $(x = 0)$ is shifted to the left 2 units.]

✔ **CHECK:** Graph $Y_1 = -\ln(x + 2)$ using a graphing utility to verify Figure 39(c).   ◀

   NOW WORK PROBLEM 75.

If the base of a logarithmic function is the number 10, then we have the **common logarithm function**. If the base a of the logarithmic function is not indicated, it is understood to be 10. That is,

$$y = \log x \quad \text{if and only if} \quad x = 10^y$$

Since $y = \log x$ and the exponential function $y = 10^x$ are inverse functions, we can obtain the graph of $y = \log x$ by reflecting the graph of $y = 10^x$ about the line $y = x$. See Figure 40.

Figure 40

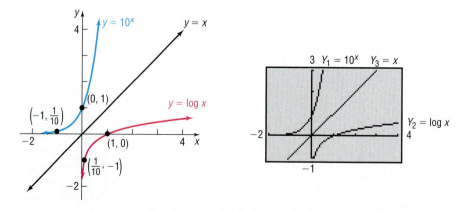

| **EXAMPLE 7** | **Graphing Logarithmic Functions Using Transformations** |

Graph $f(x) = 3 \log(x - 1)$. Determine the domain, range, and vertical asymptote of f.

Solution   The domain consists of all x for which

$$x - 1 > 0 \quad \text{or} \quad x > 1$$

To obtain the graph of $y = 3 \log(x - 1)$, we use the steps illustrated in Figure 41.

Figure 41

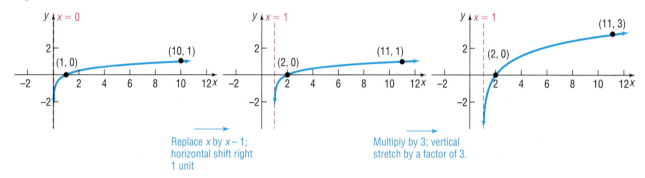

(a) $y = \log x$

Replace x by $x - 1$; horizontal shift right 1 unit

(b) $y = \log (x - 1)$

Multiply by 3; vertical stretch by a factor of 3.

(c) $y = 3 \log (x - 1)$

The range of $f(x) = 3 \log(x - 1)$ is the interval $(-\infty, \infty)$, and the vertical asymptote is $x = 1$.

✔ **CHECK:** Graph $Y_1 = 3 \log(x - 1)$ using a graphing utility to verify Figure 41(c).   ◀

   NOW WORK PROBLEM 85.

5   Solve Logarithmic Equations

Equations that contain logarithms are called **logarithmic equations**. Care must be taken when solving logarithmic equations algebraically. Be sure to check each apparent solution in the original equation and discard any that are extraneous. In the expression $\log_a M$, remember that a and M are positive and $a \neq 1$.

Some logarithmic equations can be solved by changing from a logarithmic expression to an exponential expression.

| **EXAMPLE 8** | **Solving a Logarithmic Equation** |

Solve: (a) $\log_3(4x - 7) = 2$ (b) $\log_x 64 = 2$

Solution   (a) We can obtain an exact solution by changing the logarithm to exponential form.

$$\log_3(4x - 7) = 2$$
$$4x - 7 = 3^2 \qquad \textit{Change to exponential form.}$$
$$4x - 7 = 9$$
$$4x = 16$$
$$x = 4$$

✔ **CHECK:** $\log_3(4x - 7) = \log_3(16 - 7) = \log_3 9 = 2$ $3^2 = 9$

(b) We can obtain an exact solution by changing the logarithm to exponential form.

$$\log_x 64 = 2$$

$$x^2 = 64 \qquad \text{Change to exponential form.}$$

$$x = \pm\sqrt{64} = \pm 8 \qquad \text{Square Root Method.}$$

The base of a logarithm is always positive. As a result, we discard -8; the only solution is 8.

✔ **CHECK:** $\log_8 64 = 2$ $8^2 = 64$ ◀

EXAMPLE 9	**Using Logarithms to Solve Exponential Equations**

Solve: $e^{2x} = 5$

Solution   We can obtain an exact solution by changing the exponential equation to logarithmic form.

$$e^{2x} = 5$$

$$\ln 5 = 2x \qquad \text{Change to a logarithmic expression using (1).}$$

$$x = \frac{\ln 5}{2} \qquad \text{Exact solution.}$$

$$\approx 0.805 \qquad \text{Approximate solution.}$$ ◀

 **NOW WORK PROBLEMS 91 AND 103.**

EXAMPLE 10	**Alcohol and Driving**

The concentration of alcohol in a person's blood is measurable. Recent medical research suggests that the risk R (given as a percent) of having an accident while driving a car can be modeled by the equation

$$R = 6e^{kx}$$

where x is the variable concentration of alcohol in the blood and k is a constant.

(a) Suppose that a concentration of alcohol in the blood of 0.04 results in a 10% risk $(R = 10)$ of an accident. Find the constant k in the equation. Graph $R = 6e^{kx}$ using this value of k.

(b) Using this value of k, what is the risk if the concentration is 0.17?

(c) Using the same value of k, what concentration of alcohol corresponds to a risk of 100%?

(d) If the law asserts that anyone with a risk of having an accident of 20% or more should not have driving privileges, at what concentration of alcohol in the blood should a driver be arrested and charged with a DUI (Driving Under the Influence)?

Solution (a) For a concentration of alcohol in the blood of 0.04 and a risk of 10%, we let $x = 0.04$ and $R = 10$ in the equation and solve for k.

$$R = 6e^{kx}$$

$$10 = 6e^{k(0.04)} \qquad \text{\textcolor{blue}{R = 10; x = 0.04}}$$

$$\frac{10}{6} = e^{0.04k} \qquad \text{\textcolor{blue}{Divide both sides by 6.}}$$

$$0.04k = \ln\frac{10}{6} = 0.5108256 \qquad \text{\textcolor{blue}{Change to a logarithmic expression.}}$$

$$k = 12.77 \qquad \text{\textcolor{blue}{Solve for k.}}$$

See Figure 42 for the graph of $R = 6e^{12.77x}$.

Figure 42

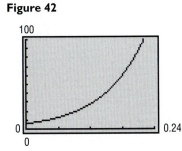

(b) Using $k = 12.77$ and $x = 0.17$ in the equation, we find the risk R to be

$$R = 6e^{kx} = 6e^{(12.77)(0.17)} = 52.6$$

For a concentration of alcohol in the blood of 0.17, the risk of an accident is about 52.6%.

(c) Using $k = 12.77$ and $R = 100$ in the equation, we find the concentration x of alcohol in the blood to be

$$R = 6e^{kx}$$

$$100 = 6e^{12.77x} \qquad \text{\textcolor{blue}{R = 100; k = 12.77}}$$

$$\frac{100}{6} = e^{12.77x} \qquad \text{\textcolor{blue}{Divide both sides by 6.}}$$

$$12.77x = \ln\frac{100}{6} = 2.8134 \qquad \text{\textcolor{blue}{Change to a logarithmic expression.}}$$

$$x = 0.22 \qquad \text{\textcolor{blue}{Solve for x.}}$$

For a concentration of alcohol in the blood of 0.22, the risk of an accident is 100%.

(d) Using $k = 12.77$ and $R = 20$ in the equation, we find the concentration x of alcohol in the blood to be

$$R = 6e^{kx}$$

$$20 = 6e^{12.77x}$$

$$\frac{20}{6} = e^{12.77x}$$

$$12.77x = \ln\frac{20}{6} = 1.204$$

$$x = 0.094$$

A driver with a concentration of alcohol in the blood of 0.094 or more (9.4%) should be arrested and charged with DUI. ◄

NOTE Most states use 0.08 or 0.10 as the blood alcohol content at which a DUI citation is given. ■

Summary

Properties of the Logarithmic Function

$f(x) = \log_a x, \quad a > 1$
$(y = \log_a x \text{ means } x = a^y)$

Domain: the interval $(0, \infty)$; Range: the interval $(-\infty, \infty)$
x-intercept: 1; y-intercept: none; vertical asymptote: $x = 0$ (y-axis); increasing; one-to-one
See Figure 43(a) for a typical graph.

$f(x) = \log_a x, \quad 0 < a < 1$
$(y = \log_a x \text{ means } x = a^y)$

Domain: the interval $(0, \infty)$; Range: the interval $(-\infty, \infty)$
x-intercept: 1; y-intercept: none; vertical asymptote: $x = 0$ (y-axis); decreasing; one-to-one
See Figure 43(b) for a typical graph.

Figure 43

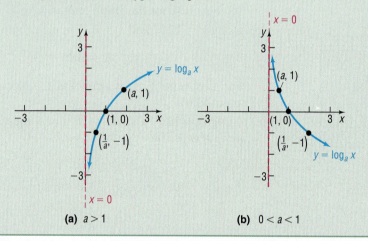

(a) $a > 1$ **(b)** $0 < a < 1$

4.4 Assess Your Understanding

'Are You Prepared?'

Answers are given at the end of these exercises. If you get a wrong answer, read the pages listed in red.

1. Solve the inequality: $3x - 7 \le 8 - 2x$ (pp. 732–733)
2. Solve the inequality: $x^2 - x - 6 > 0$ (pp. 212–214)
3. Solve the inequality: $\dfrac{x-1}{x+4} > 0$ (p. 215)

Concepts and Vocabulary

4. The domain of the logarithmic function $f(x) = \log_a x$ is _____.

5. The graph of every logarithmic function $f(x) = \log_a x$, $a > 0, a \ne 1$, passes through three points: _____, _____, and _____.

6. If the graph of a logarithmic function $f(x) = \log_a x$, $a > 0, a \ne 1$, is increasing, then its base must be larger than _____.

7. *True or False:* If $y = \log_a x$, then $y = a^x$.

8. *True or False:* The graph of $f(x) = \log_a x, a > 0, a \ne 1$, has an x-intercept equal to 1 and no y-intercept.

Skill Building

In Problems 9–20, change each exponential expression to an equivalent expression involving a logarithm.

9. $9 = 3^2$
10. $16 = 4^2$
11. $a^2 = 1.6$
12. $a^3 = 2.1$
13. $1.1^2 = M$
14. $2.2^3 = N$
15. $2^x = 7.2$
16. $3^x = 4.6$
17. $x^{\sqrt{2}} = \pi$
18. $x^\pi = e$
19. $e^x = 8$
20. $e^{2.2} = M$

In Problems 21–32, change each logarithmic expression to an equivalent expression involving an exponent.

21. $\log_2 8 = 3$

22. $\log_3\left(\dfrac{1}{9}\right) = -2$

23. $\log_a 3 = 6$

24. $\log_b 4 = 2$

25. $\log_3 2 = x$

26. $\log_2 6 = x$

27. $\log_2 M = 1.3$

28. $\log_3 N = 2.1$

29. $\log_{\sqrt{2}} \pi = x$

30. $\log_\pi x = \dfrac{1}{2}$

31. $\ln 4 = x$

32. $\ln x = 4$

In Problems 33–44, find the exact value of each logarithm without using a calculator.

33. $\log_2 1$

34. $\log_8 8$

35. $\log_5 25$

36. $\log_3\left(\dfrac{1}{9}\right)$

37. $\log_{1/2} 16$

38. $\log_{1/3} 9$

39. $\log_{10} \sqrt{10}$

40. $\log_5 \sqrt[3]{25}$

41. $\log_{\sqrt{2}} 4$

42. $\log_{\sqrt{3}} 9$

43. $\ln \sqrt{e}$

44. $\ln e^3$

In Problems 45–56, find the domain of each function.

45. $f(x) = \ln(x - 3)$

46. $g(x) = \ln(x - 1)$

47. $F(x) = \log_2 x^2$

48. $H(x) = \log_5 x^3$

49. $f(x) = 3 - 2\log_4 \dfrac{x}{2}$

50. $g(x) = 8 + 5\ln(2x)$

51. $f(x) = \ln\left(\dfrac{1}{x + 1}\right)$

52. $g(x) = \ln\left(\dfrac{1}{x - 5}\right)$

53. $g(x) = \log_5\left(\dfrac{x + 1}{x}\right)$

54. $h(x) = \log_3\left(\dfrac{x}{x - 1}\right)$

55. $f(x) = \sqrt{\ln x}$

56. $g(x) = \dfrac{1}{\ln x}$

In Problems 57–60, use a calculator to evaluate each expression. Round your answer to three decimal places.

57. $\ln \dfrac{5}{3}$

58. $\dfrac{\ln 5}{3}$

59. $\dfrac{\ln \dfrac{10}{3}}{0.04}$

60. $\dfrac{\ln \dfrac{2}{3}}{-0.1}$

61. Find a so that the graph of $f(x) = \log_a x$ contains the point $(2, 2)$.

62. Find a so that the graph of $f(x) = \log_a x$ contains the point $\left(\dfrac{1}{2}, -4\right)$.

In Problems 63–66, graph each logarithmic function by hand.

63. $y = \log_3 x$

64. $y = \log_{1/2} x$

65. $y = \log_{1/4} x$

66. $y = \log_5 x$

In Problems 67–74, the graph of a logarithmic function is given. Match each graph to one of the following functions:

A. $y = \log_3 x$

B. $y = \log_3(-x)$

C. $y = -\log_3 x$

D. $y = -\log_3(-x)$

E. $y = \log_3 x - 1$

F. $y = \log_3(x - 1)$

G. $y = \log_3(1 - x)$

H. $y = 1 - \log_3 x$

67.
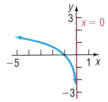

68.
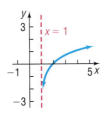

69.
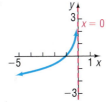

70.
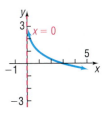

71.
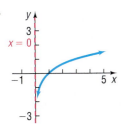

72.
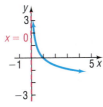

73.
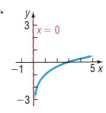

74.
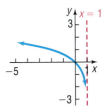

In Problems 75–90, use transformations to graph each function. Determine the domain, range, and vertical asymptote of each function.

75. $f(x) = \ln(x + 4)$

76. $f(x) = \ln(x - 3)$

77. $f(x) = 2 + \ln x$

78. $f(x) = -\ln(-x)$

79. $g(x) = \ln(2x)$

80. $h(x) = \ln\left(\dfrac{1}{2}x\right)$

81. $f(x) = 3 \ln x$

82. $f(x) = -2 \ln x$

83. $f(x) = \log(x - 4)$

84. $f(x) = \log(x + 5)$

85. $h(x) = 4 \log x$

86. $g(x) = -3 \log x$

87. $F(x) = \log(2x)$

88. $G(x) = \log(5x)$

89. $h(x) = 3 + \log(x + 2)$

90. $g(x) = 2 - \log(x + 1)$

In Problems 91–110, solve each equation.

91. $\log_3 x = 2$

92. $\log_5 x = 3$

93. $\log_2(2x + 1) = 3$

94. $\log_3(3x - 2) = 2$

95. $\log_x 4 = 2$

96. $\log_x\left(\dfrac{1}{8}\right) = 3$

97. $\ln e^x = 5$

98. $\ln e^{-2x} = 8$

99. $\log_4 64 = x$

100. $\log_5 625 = x$

101. $\log_3 243 = 2x + 1$

102. $\log_6 36 = 5x + 3$

103. $e^{3x} = 10$

104. $e^{-2x} = \dfrac{1}{3}$

105. $e^{2x+5} = 8$

106. $e^{-2x+1} = 13$

107. $\log_3(x^2 + 1) = 2$

108. $\log_5(x^2 + x + 4) = 2$

109. $\log_2 8^x = -3$

110. $\log_3 3^x = -1$

Applications and Extensions

In Problems 111–114, (a) graph each function and state its domain, range, and asymptote; (b) determine the inverse function; (c) use the graph obtained in (a) to graph the inverse and state its domain, range, and asymptote.

111. $f(x) = 2^x$

112. $f(x) = 5^x$

113. $f(x) = 2^{x+3}$

114. $f(x) = 5^x - 2$

---

115. Chemistry The pH of a chemical solution is given by the formula

$$pH = -\log_{10}[H^+]$$

where $[H^+]$ is the concentration of hydrogen ions in moles per liter. Values of pH range from 0 (acidic) to 14 (alkaline).
(a) What is the pH of a solution for which $[H^+]$ is 0.1?
(b) What is the pH of a solution for which $[H^+]$ is 0.01?
(c) What is the pH of a solution for which $[H^+]$ is 0.001?
(d) What happens to pH as the hydrogen ion concentration decreases?
(e) Determine the hydrogen ion concentration of an orange (pH = 3.5).
(f) Determine the hydrogen ion concentration of human blood (pH = 7.4).

116. Diversity Index Shannon's diversity index is a measure of the diversity of a population. The diversity index is given by the formula

$$H = -(p_1 \log p_1 + p_2 \log p_2 + \cdots + p_n \log p_n)$$

where p_1 is the proportion of the population that is species 1, p_2 is the proportion of the population that is species 2, and so on.
(a) According to the U.S. Census Bureau, the distribution of race in the United States in 2000 was as follows:

Race	Proportion
American Indian or Native	0.014
Alaskan	
Asian	0.041
Black or African American	0.128
Hispanic	0.124
Native Hawaiian or Pacific Islander	0.003
White	0.690.

SOURCE: U.S. Census Bureau

Compute the diversity index of the United States in 2000.
(b) The largest value of the diversity index is given by $H_{max} = \log(S)$, where S is the number of categories of race. Compute H_{max}.
(c) The evenness ratio is given by $E_H = \dfrac{H}{H_{max}}$, where $0 \le E_H \le 1$. If $E_H = 1$, there is complete evenness. Compute the evenness ratio for the United States.
(d) Obtain the distribution of race for the United States in 1990 from the Census Bureau. Compute Shannon's diversity index. Is the United States becoming more diverse? Why?

117. **Atmospheric Pressure**   The atmospheric pressure p on a balloon or an aircraft decreases with increasing height. This pressure, measured in millimeters of mercury, is related to the height h (in kilometers) above sea level by the formula

$$p = 760e^{-0.145h}$$

(a) Find the height of an aircraft if the atmospheric pressure is 320 millimeters of mercury.
(b) Find the height of a mountain if the atmospheric pressure is 667 millimeters of mercury.

118. **Healing of Wounds**   The normal healing of wounds can be modeled by an exponential function. If A_0 represents the original area of the wound and if A equals the area of the wound, then the formula

$$A = A_0e^{-0.35n}$$

describes the area of a wound after n days following an injury when no infection is present to retard the healing. Suppose that a wound initially had an area of 100 square millimeters.

(a) If healing is taking place, after how many days will the wound be one-half its original size?
(b) How long before the wound is 10% of its original size?

119. **Exponential Probability**   Between 12:00 PM and 1:00 PM, cars arrive at Citibank's drive-thru at the rate of 6 cars per hour (0.1 car per minute). The following formula from statistics can be used to determine the probability that a car will arrive within t minutes of 12:00 PM.

$$F(t) = 1 - e^{-0.1t}$$

(a) Determine how many minutes are needed for the probability to reach 50%.
(b) Determine how many minutes are needed for the probability to reach 80%.
(c) Is it possible for the probability to equal 100%? Explain.

120. **Exponential Probability**   Between 5:00 PM and 6:00 PM, cars arrive at Jiffy Lube at the rate of 9 cars per hour (0.15 car per minute). The following formula from statistics can be used to determine the probability that a car will arrive within t minutes of 5:00 PM.

$$F(t) = 1 - e^{-0.15t}$$

(a) Determine how many minutes are needed for the probability to reach 50%.

(b) Determine how many minutes are needed for the probability to reach 80%.

121. **Drug Medication**   The formula

$$D = 5e^{-0.4h}$$

can be used to find the number of milligrams D of a certain drug that is in a patient's bloodstream h hours after the drug has been administered. When the number of milligrams reaches 2, the drug is to be administered again. What is the time between injections?

122. **Spreading of Rumors**   A model for the number N of people in a college community who have heard a certain rumor is

$$N = P(1 - e^{-0.15d})$$

where P is the total population of the community and d is the number of days that have elapsed since the rumor began. In a community of 1000 students, how many days will elapse before 450 students have heard the rumor?

123. **Current in a *RL* Circuit**   The equation governing the amount of current I (in amperes) after time t (in seconds) in a simple RL circuit consisting of a resistance R (in ohms), an inductance L (in henrys), and an electromotive force E (in volts) is

$$I = \frac{E}{R}[1 - e^{-(R/L)t}]$$

If $E = 12$ volts, $R = 10$ ohms, and $L = 5$ henrys, how long does it take to obtain a current of 0.5 ampere? Of 1.0 ampere? Graph the equation.

124. **Learning Curve**   Psychologists sometimes use the function

$$L(t) = A(1 - e^{-kt})$$

to measure the amount L learned at time t. The number A represents the amount to be learned, and the number k measures the rate of learning. Suppose that a student has an amount A of 200 vocabulary words to learn. A psychologist determines that the student learned 20 vocabulary words after 5 minutes.

(a) Determine the rate of learning k.
(b) Approximately how many words will the student have learned after 10 minutes?
(c) After 15 minutes?
(d) How long does it take for the student to learn 180 words?

Loudness of Sound   *Problems 125–128 use the following discussion: The **loudness** $L(x)$, measured in decibels, of a sound of intensity x, measured in watts per square meter, is defined as $L(x) = 10 \log \dfrac{x}{I_0}$, where $I_0 = 10^{-12}$ watt per square meter is the least intense sound that a human ear can detect. Determine the loudness, in decibels, of each of the following sounds.*

125. Normal conversation: intensity of $x = 10^{-7}$ watt per square meter.

126. Heavy city traffic: intensity of $x = 10^{-3}$ watt per square meter.

127. Amplified rock music: intensity of 10^{-1} watt per square meter.

128. Diesel truck traveling 40 miles per hour 50 feet away: intensity 10 times that of a passenger car traveling 50 miles per hour 50 feet away whose loudness is 70 decibels.

*Problems 129 and 130 use the following discussion: The **Richter scale** is one way of converting seismographic readings into numbers that provide an easy reference for measuring the magnitude M of an earthquake. All earthquakes are compared to a **zero-level earthquake** whose seismographic reading measures 0.001 millimeter at a distance of 100 kilometers from the epicenter. An earthquake whose seismographic reading measures x millimeters has **magnitude** M(x), given by*

$$M(x) = \log\left(\frac{x}{x_0}\right)$$

where $x_0 = 10^{-3}$ is the reading of a zero-level earthquake the same distance from its epicenter. Determine the magnitude of the following earthquakes.

129. Magnitude of an Earthquake Mexico City in 1985: seismographic reading of 125,892 millimeters 100 kilometers from the center.

130. Magnitude of an Earthquake San Francisco in 1906: seismographic reading of 7943 millimeters 100 kilometers from the center.

131. Alcohol and Driving The concentration of alcohol in a person's blood is measurable. Suppose that the risk R (given as a percent) of having an accident while driving a car can be modeled by the equation

$$R = 3e^{kx}$$

where x is the variable concentration of alcohol in the blood and k is a constant.

(a) Suppose that a concentration of alcohol in the blood of 0.06 results in a 10% risk ($R = 10$) of an accident. Find the constant k in the equation.

(b) Using this value of k, what is the risk if the concentration is 0.17?

(c) Using the same value of k, what concentration of alcohol corresponds to a risk of 100%?

(d) If the law asserts that anyone with a risk of having an accident of 15% or more should not have driving privileges, at what concentration of alcohol in the blood should a driver be arrested and charged with a DUI?

(e) Compare this situation with that of Example 10. If you were a lawmaker, which situation would you support? Give your reasons.

Discussion and Writing

132. Is there any function of the form $y = x^\alpha, 0 < \alpha < 1$, that increases more slowly than a logarithmic function whose base is greater than 1? Explain.

133. In the definition of the logarithmic function, the base a is not allowed to equal 1. Why?

134. Critical Thinking In buying a new car, one consideration might be how well the price of the car holds up over time. Different makes of cars have different depreciation rates. One way to compute a depreciation rate for a car is given here. Suppose that the current prices of a certain Mercedes automobile are as follows:

		Age in Years			
New	**1**	**2**	**3**	**4**	**5**
$38,000	$36,600	$32,400	$28,750	$25,400	$21,200

Use the formula New = Old(e^{Rt}) to find R, the annual depreciation rate, for a specific time t. When might be the best time to trade in the car? Consult the NADA ("blue") book and compare two like models that you are interested in. Which has the better depreciation rate?

'Are You Prepared?' Answers

1. $x \leq 3$ **2.** $x < -2$ or $x > 3$ **3.** $x < -4$ or $x > 1$

4.5 Properties of Logarithms

OBJECTIVES
1 Work with the Properties of Logarithms
2 Write a Logarithmic Expression as a Sum or Difference of Logarithms
3 Write a Logarithmic Expression as a Single Logarithm
4 Evaluate Logarithms Whose Base Is Neither 10 nor e
5 Graph Logarithmic Functions Whose Base Is Neither 10 nor e

✓ 1 Work with the Properties of Logarithms

Logarithms have some very useful properties that can be derived directly from the definition and the laws of exponents.

| EXAMPLE 1 | **Establishing Properties of Logarithms** |

(a) Show that $\log_a 1 = 0$. (b) Show that $\log_a a = 1$.

Solution (a) This fact was established when we graphed $y = \log_a x$ (see Figure 35). To show the result algebraically, let $y = \log_a 1$. Then

$$y = \log_a 1$$
$$a^y = 1 \qquad \text{\textit{Change to an exponent.}}$$
$$a^y = a^0 \qquad \text{\textit{$a^0 = 1$}}$$
$$y = 0 \qquad \text{\textit{Solve for y.}}$$
$$\log_a 1 = 0 \qquad \text{\textit{$y = \log_a 1$}}$$

(b) Let $y = \log_a a$. Then

$$y = \log_a a$$
$$a^y = a \qquad \text{\textit{Change to an exponent.}}$$
$$a^y = a^1 \qquad \text{\textit{$a^1 = a$}}$$
$$y = 1 \qquad \text{\textit{Solve for y.}}$$
$$\log_a a = 1 \qquad \text{\textit{$y = \log_a a$}}$$ ◀

To summarize:

$$\log_a 1 = 0 \qquad \log_a a = 1$$

Theorem **Properties of Logarithms**

In the properties given next, M and a are positive real numbers, with $a \neq 1$, and r is any real number.

The number $\log_a M$ is the exponent to which a must be raised to obtain M. That is,

$$a^{\log_a M} = M \qquad \qquad \text{(1)}$$

The logarithm to the base a of a raised to a power equals that power. That is,

$$\log_a a^r = r \qquad \qquad \text{(2)}$$

The proof uses the fact that $y = a^x$ and $y = \log_a x$ are inverses.

Proof of Property (1) For inverse functions,

$$f(f^{-1}(x)) = x \quad \text{for every } x \text{ in the domain of } f^{-1}$$

Using $f(x) = a^x$ and $f^{-1}(x) = \log_a x$, we find

$$f(f^{-1}(x)) = a^{\log_a x} = x \quad \text{for} \quad x > 0$$

Now let $x = M$ to obtain $a^{\log_a M} = M$, where $M > 0$. ∎

Proof of Property (2) For inverse functions,

$$f^{-1}(f(x)) = x \quad \text{for all } x \text{ in the domain of } f.$$

Using $f(x) = a^x$ and $f^{-1}(x) = \log_a x$, we find

$$f^{-1}(f(x)) = \log_a a^x = x \quad \text{for all real numbers } x.$$

Now let $x = r$ to obtain $\log_a a^r = r$, where r is any real number. ∎

EXAMPLE 2

Using Properties (1) and (2)

(a) $2^{\log_2 \pi} = \pi$ \quad (b) $\log_{0.2} 0.2^{-\sqrt{2}} = -\sqrt{2}$ \quad (c) $\ln e^{kt} = kt$ ◀

NOW WORK PROBLEM 9.

Other useful properties of logarithms are given next.

Theorem

Properties of Logarithms

In the following properties, M, N, and a are positive real numbers, with $a \neq 1$, and r is any real number.

The Log of a Product Equals the Sum of the Logs

$$\log_a(MN) = \log_a M + \log_a N \tag{3}$$

The Log of a Quotient Equals the Difference of the Logs

$$\log_a\left(\frac{M}{N}\right) = \log_a M - \log_a N \tag{4}$$

The Log of a Power Equals the Product of the Power and the Log

$$\log_a M^r = r \log_a M \tag{5}$$

We shall derive properties (3) and (5) and leave the derivation of property (4) as an exercise (see Problem 101).

Proof of Property (3) Let $A = \log_a M$ and let $B = \log_a N$. These expressions are equivalent to the exponential expressions

$$a^A = M \quad \text{and} \quad a^B = N$$

Now

$$\log_a(MN) = \log_a(a^A a^B) = \log_a a^{A+B} \qquad \text{Law of Exponents}$$
$$= A + B \qquad \text{Property (2) of logarithms}$$
$$= \log_a M + \log_a N \qquad ∎$$

Proof of Property (5) Let $A = \log_a M$. This expression is equivalent to

$$a^A = M$$

Now
$$\log_a M^r = \log_a(a^A)^r = \log_a a^{rA} \qquad \text{Law of Exponents}$$
$$= rA \qquad\qquad \text{Property (2) of logarithms}$$
$$= r \log_a M$$

∎

NOW WORK PROBLEM 13.

2 Write a Logarithmic Expression as a Sum or Difference of Logarithms

Logarithms can be used to transform products into sums, quotients into differences, and powers into factors. Such transformations prove useful in certain types of calculus problems.

EXAMPLE 3 | **Writing a Logarithmic Expression as a Sum of Logarithms**

Write $\log_a\left(x\sqrt{x^2 + 1}\right), x > 0$, as a sum of logarithms. Express all powers as factors.

Solution
$$\log_a\left(x\sqrt{x^2 + 1}\right) = \log_a x + \log_a \sqrt{x^2 + 1} \qquad \text{Property (3)}$$
$$= \log_a x + \log_a(x^2 + 1)^{1/2}$$
$$= \log_a x + \frac{1}{2}\log_a(x^2 + 1) \qquad \text{Property (5)}$$

◀

EXAMPLE 4 | **Writing a Logarithmic Expression as a Difference of Logarithms**

Write
$$\ln\frac{x^2}{(x - 1)^3}, \qquad x > 1$$
as a difference of logarithms. Express all powers as factors.

Solution
$$\ln\frac{x^2}{(x - 1)^3} = \ln x^2 - \ln(x - 1)^3 = 2\ln x - 3\ln(x - 1)$$

↑ Property (4) ↑ Property (5)

◀

EXAMPLE 5 | **Writing a Logarithmic Expression as a Sum and Difference of Logarithms**

Write
$$\log_a\frac{\sqrt{x^2 + 1}}{x^3(x + 1)^4}, \qquad x > 0$$
as a sum and difference of logarithms. Express all powers as factors.

CAUTION
In using properties (3) through (5), be careful about the values that the variable may assume. For example, the domain of the variable for $\log_a x$ is $x > 0$ and for $\log_a(x - 1)$ it is $x > 1$. If we add these functions, the domain is $x > 1$. That is, the equality

$$\log_a x + \log_a(x - 1) = \log_a[x(x - 1)]$$

is true only for $x > 1$.

■

Solution
$$\log_a\frac{\sqrt{x^2 + 1}}{x^3(x + 1)^4} = \log_a \sqrt{x^2 + 1} - \log_a[x^3(x + 1)^4] \qquad \text{Property (4)}$$

$$= \log_a \sqrt{x^2 + 1} - [\log_a x^3 + \log_a(x + 1)^4] \qquad \text{Property (3)}$$

$$= \log_a(x^2 + 1)^{1/2} - \log_a x^3 - \log_a(x + 1)^4$$

$$= \frac{1}{2}\log_a(x^2 + 1) - 3\log_a x - 4\log_a(x + 1) \qquad \text{Property (5)}$$

◀

NOW WORK PROBLEM 45.

3 Write a Logarithmic Expression as a Single Logarithm

Another use of properties (3) through (5) is to write sums and/or differences of logarithms with the same base as a single logarithm. This skill will be needed to solve certain logarithmic equations discussed in the next section.

EXAMPLE 6   **Writing Expressions as a Single Logarithm**

Write each of the following as a single logarithm.

(a) $\log_a 7 + 4 \log_a 3$ (b) $\dfrac{2}{3} \ln 8 - \ln(3^4 - 8)$

(c) $\log_a x + \log_a 9 + \log_a(x^2 + 1) - \log_a 5$

Solution   (a) $\begin{aligned}\log_a 7 + 4 \log_a 3 &= \log_a 7 + \log_a 3^4 \quad &\text{Property (5)}\\ &= \log_a 7 + \log_a 81 \\ &= \log_a(7 \cdot 81) \quad &\text{Property (3)}\\ &= \log_a 567\end{aligned}$

(b) $\begin{aligned}\dfrac{2}{3}\ln 8 - \ln(3^4 - 8) &= \ln 8^{2/3} - \ln(81 - 8) \quad &\text{Property (5)}\\ &= \ln 4 - \ln 73 \\ &= \ln\left(\dfrac{4}{73}\right) \quad &\text{Property (4)}\end{aligned}$

(c) $\begin{aligned}\log_a x + \log_a 9 + \log_a(x^2+1) - \log_a 5 &= \log_a(9x) + \log_a(x^2+1) - \log_a 5 \\ &= \log_a[9x(x^2+1)] - \log_a 5 \\ &= \log_a\left[\dfrac{9x(x^2+1)}{5}\right]\end{aligned}$ ◀

> **WARNING**   A common error made by some students is to express the logarithm of a sum as the sum of logarithms.
>
> $$\log_a(M+N) \quad \textit{is not equal to} \quad \log_a M + \log_a N$$
>
> *Correct statement* $\log_a(MN) = \log_a M + \log_a N$ Property (3)
>
> Another common error is to express the difference of logarithms as the quotient of logarithms.
>
> $$\log_a M - \log_a N \quad \textit{is not equal to} \quad \dfrac{\log_a M}{\log_a N}$$
>
> *Correct statement* $\log_a M - \log_a N = \log_a\left(\dfrac{M}{N}\right)$ Property (4)
>
> A third common error is to express a logarithm raised to a power as the product of the power times the logarithm.
>
> $$(\log_a M)^r \quad \textit{is not equal to} \quad r \log_a M$$
>
> *Correct statement* $\log_a M^r = r \log_a M$ Property (5) ■

 NOW WORK PROBLEM 51.

Two other properties of logarithms that we need to know are consequences of the fact that the logarithmic function $y = \log_a x$ is one-to-one.

Theorem   **Properties of Logarithms**

In the following properties, M, N, and a are positive real numbers, with $a \neq 1$.

$$\text{If } M = N, \text{ then } \log_a M = \log_a N. \tag{6}$$

$$\text{If } \log_a M = \log_a N, \text{ then } M = N. \tag{7}$$

When property (6) is used, we start with the equation $M = N$ and say "take the logarithm of both sides" to obtain $\log_a M = \log_a N$.

Properties (6) and (7) are useful for solving *exponential and logarithmic equations*, a topic discussed in the next section.

4 Evaluate Logarithms Whose Base Is Neither 10 nor e

Logarithms to the base 10, common logarithms, were used to facilitate arithmetic computations before the widespread use of calculators. (See the Historical Feature at the end of this section.) Natural logarithms, that is, logarithms whose base is the number e, remain very important because they arise frequently in the study of natural phenomena.

Common logarithms are usually abbreviated by writing **log**, with the base understood to be 10, just as natural logarithms are abbreviated by **ln**, with the base understood to be e.

Most calculators have both $\boxed{\log}$ and $\boxed{\ln}$ keys to calculate the common logarithm and natural logarithm of a number. Let's look at an example to see how to approximate logarithms having a base other than 10 or e.

EXAMPLE 7   **Approximating Logarithms Whose Base Is Neither 10 nor e**

Approximate $\log_2 7$. Round the answer to four decimal places.

Solution   Let $y = \log_2 7$. Then $2^y = 7$, so

$$2^y = 7$$
$$\ln 2^y = \ln 7 \qquad \text{Property (6)}$$
$$y \ln 2 = \ln 7 \qquad \text{Property (5)}$$
$$y = \frac{\ln 7}{\ln 2} \qquad \text{Exact solution}$$
$$y \approx 2.8074 \qquad \text{Approximate solution rounded to four decimal places}$$

Example 7 shows how to approximate a logarithm whose base is 2 by changing to logarithms involving the base e. In general, we use the **Change-of-Base Formula**.

Theorem   **Change-of-Base Formula**

If $a \neq 1$, $b \neq 1$, and M are positive real numbers, then

$$\log_a M = \frac{\log_b M}{\log_b a} \tag{8}$$

Proof   We derive this formula as follows: Let $y = \log_a M$. Then

$$a^y = M$$

$$\log_b a^y = \log_b M \qquad \text{Property (6)}$$

$$y \log_b a = \log_b M \qquad \text{Property (5)}$$

$$y = \frac{\log_b M}{\log_b a} \qquad \text{Solve for } y.$$

$$\log_a M = \frac{\log_b M}{\log_b a} \qquad y = \log_a M \qquad ■$$

Since calculators have keys only for $\boxed{\log}$ and $\boxed{\ln}$, in practice, the Change-of-Base Formula uses either $b = 10$ or $b = e$. That is,

$$\log_a M = \frac{\log M}{\log a} \quad \text{and} \quad \log_a M = \frac{\ln M}{\ln a} \qquad \text{(9)}$$

EXAMPLE 8

Using the Change-of-Base Formula

Approximate: (a) $\log_5 89$ (b) $\log_{\sqrt{2}} \sqrt{5}$
Round answers to four decimal places.

Solution   (a) $\log_5 89 = \dfrac{\log 89}{\log 5} \approx \dfrac{1.949390007}{0.6989700043} \approx 2.7889$

or

$$\log_5 89 = \frac{\ln 89}{\ln 5} \approx \frac{4.48863637}{1.609437912} \approx 2.7889$$

(b) $\log_{\sqrt{2}} \sqrt{5} = \dfrac{\log \sqrt{5}}{\log \sqrt{2}} = \dfrac{\frac{1}{2} \log 5}{\frac{1}{2} \log 2} \approx 2.3219$

or

$$\log_{\sqrt{2}} \sqrt{5} = \frac{\ln \sqrt{5}}{\ln \sqrt{2}} = \frac{\frac{1}{2} \ln 5}{\frac{1}{2} \ln 2} \approx 2.3219 \qquad ◀$$

NOW WORK PROBLEMS 17 AND 65.

5   Graph Logarithmic Functions Whose Base Is Neither 10 nor e

We also use the Change-of-Base Formula to graph logarithmic functions whose base is neither 10 nor e.

EXAMPLE 9 **Graphing a Logarithmic Function Whose Base Is Neither 10 nor e**

Use a graphing utility to graph $y = \log_2 x$.

Figure 44

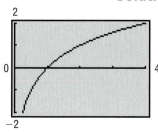

Solution   Since graphing utilities only have logarithms with the base 10 or the base e, we need to use the Change-of-Base Formula to express $y = \log_2 x$ in terms of logarithms with base 10 or base e. We can graph either $y = \dfrac{\ln x}{\ln 2}$ or $y = \dfrac{\log x}{\log 2}$ to obtain the graph of $y = \log_2 x$. See Figure 44.

✔ **CHECK:** Verify that $y = \dfrac{\ln x}{\ln 2}$ and $y = \dfrac{\log x}{\log 2}$ result in the same graph by graphing each on the same screen. ◀

NOW WORK PROBLEM 73.

Summary

Properties of Logarithms

In the list that follows, $a > 0$, $a \neq 1$, and $b > 0$, $b \neq 1$; also, $M > 0$ and $N > 0$.

Definition $y = \log_a x$ means $x = a^y$

Properties of logarithms

$\log_a 1 = 0$; $\log_a a = 1$

$a^{\log_a M} = M$; $\log_a a^r = r$

$\log_a(MN) = \log_a M + \log_a N$

$\log_a\left(\dfrac{M}{N}\right) = \log_a M - \log_a N$

$\log_a M^r = r \log_a M$

If $M = N$, then $\log_a M = \log_a N$.

If $\log_a M = \log_a N$, then $M = N$.

Change-of-Base Formula $\log_a M = \dfrac{\log_b M}{\log_b a}$

HISTORICAL FEATURE

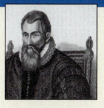

John Napier
(1550–1617)

Logarithms were invented about 1590 by John Napier (1550–1617) and Joost Bürgi (1552–1632), working independently. Napier, whose work had the greater influence, was a Scottish lord, a secretive man whose neighbors were inclined to believe him to be in league with the devil. His approach to logarithms was very different from ours; it was based on the relationship between arithmetic and geometric sequences, discussed in a later chapter, and not on the inverse function relationship of logarithms to exponential functions (described in Section 4.4). Napier's tables, pub-lished in 1614, listed what would now be called *natural logarithms* of sines and were rather difficult to use. A London professor, Henry Briggs, became interested in the tables and visited Napier. In their conversations, they developed the idea of common logarithms, which were published in 1617. Their importance for calculation was immediately recognized, and by 1650 they were being printed as far away as China. They remained an important calculation tool until the advent of the inexpensive handheld calculator about 1972, which has decreased their calculational, but not their theoretical, importance.

A side effect of the invention of logarithms was the popular-ization of the decimal system of notation for real numbers.

4.5 Assess Your Understanding

Concepts and Vocabulary

1. The logarithm of a product equals the _____ of the logarithms.

2. If $\log_8 M = \dfrac{\log_5 7}{\log_5 8}$, then $M =$ _____.

3. $\log_a M^r =$ _____.

4. *True or False:* $\ln(x + 3) - \ln(2x) = \dfrac{\ln(x + 3)}{\ln(2x)}$

5. *True or False:* $\log_2(3x^4) = 4\log_2(3x)$

6. *True or False:* $\log_2 16 = \dfrac{\ln 16}{\ln 2}$

Skill Building

In Problems 7–22, use properties of logarithms to find the exact value of each expression. Do not use a calculator.

7. $\log_3 3^{71}$ **8.** $\log_2 2^{-13}$ **9.** $\ln e^{-4}$ **10.** $\ln e^{\sqrt{2}}$

11. $2^{\log_2 7}$ **12.** $e^{\ln 8}$ **13.** $\log_8 2 + \log_8 4$ **14.** $\log_6 9 + \log_6 4$

15. $\log_6 18 - \log_6 3$ **16.** $\log_8 16 - \log_8 2$ **17.** $\log_2 6 \cdot \log_6 4$ **18.** $\log_3 8 \cdot \log_8 9$

19. $3^{\log_3 5 - \log_3 4}$ **20.** $5^{\log_5 6 + \log_5 7}$ **21.** $e^{\log_{e^2} 16}$ **22.** $e^{\log_{e^2} 9}$

In Problems 23–30, suppose that $\ln 2 = a$ and $\ln 3 = b$. Use properties of logarithms to write each logarithm in terms of a and b.

23. $\ln 6$ **24.** $\ln \dfrac{2}{3}$ **25.** $\ln 1.5$ **26.** $\ln 0.5$

27. $\ln 8$ **28.** $\ln 27$ **29.** $\ln \sqrt[5]{6}$ **30.** $\ln \sqrt[4]{\dfrac{2}{3}}$

In Problems 31–50, write each expression as a sum and/or difference of logarithms. Express powers as factors.

31. $\log_5(25x)$ **32.** $\log_3 \dfrac{x}{9}$ **33.** $\log_2 z^3$ **34.** $\log_7(x^5)$

35. $\ln(ex)$ **36.** $\ln \dfrac{e}{x}$ **37.** $\ln(xe^x)$ **38.** $\ln \dfrac{x}{e^x}$

39. $\log_a(u^2 v^3)$, $u > 0, v > 0$ **40.** $\log_2 \left(\dfrac{a}{b^2}\right)$, $a > 0, b > 0$ **41.** $\ln\left(x^2\sqrt{1-x}\right)$, $0 < x < 1$ **42.** $\ln\left(x\sqrt{1+x^2}\right)$, $x > 0$

43. $\log_2\left(\dfrac{x^3}{x-3}\right)$, $x > 3$ **44.** $\log_5\left(\dfrac{\sqrt[3]{x^2+1}}{x^2-1}\right)$, $x > 1$ **45.** $\log\left[\dfrac{x(x+2)}{(x+3)^2}\right]$, $x > 0$ **46.** $\log\left[\dfrac{x^3\sqrt{x+1}}{(x-2)^2}\right]$, $x > 2$

47. $\ln\left[\dfrac{x^2-x-2}{(x+4)^2}\right]^{1/3}$, $x > 2$ **48.** $\ln\left[\dfrac{(x-4)^2}{x^2-1}\right]^{2/3}$, $x > 4$

49. $\ln \dfrac{5x\sqrt{1+3x}}{(x-4)^3}$, $x > 4$ **50.** $\ln\left[\dfrac{5x^2\sqrt[3]{1-x}}{4(x+1)^2}\right]$, $0 < x < 1$

In Problems 51–64, write each expression as a single logarithm.

51. $3\log_5 u + 4\log_5 v$ **52.** $2\log_3 u - \log_3 v$ **53.** $\log_3 \sqrt{x} - \log_3 x^3$

54. $\log_2\left(\dfrac{1}{x}\right) + \log_2\left(\dfrac{1}{x^2}\right)$ **55.** $\log_4(x^2-1) - 5\log_4(x+1)$ **56.** $\log(x^2+3x+2) - 2\log(x+1)$

57. $\ln\left(\dfrac{x}{x-1}\right) + \ln\left(\dfrac{x+1}{x}\right) - \ln(x^2-1)$ **58.** $\log\left(\dfrac{x^2+2x-3}{x^2-4}\right) - \log\left(\dfrac{x^2+7x+6}{x+2}\right)$ **59.** $8\log_2 \sqrt{3x-2} - \log_2\left(\dfrac{4}{x}\right) + \log_2 4$

60. $21\log_3 \sqrt[3]{x} + \log_3(9x^2) - \log_3 9$ **61.** $2\log_a(5x^3) - \dfrac{1}{2}\log_a(2x+3)$ **62.** $\dfrac{1}{3}\log(x^3+1) + \dfrac{1}{2}\log(x^2+1)$

63. $2\log_2(x+1) - \log_2(x+3) - \log_2(x-1)$ **64.** $3\log_5(3x+1) - 2\log_5(2x-1) - \log_5 x$

In Problems 65–72, use the Change-of-Base Formula and a calculator to evaluate each logarithm. Round your answer to three decimal places.

65. $\log_3 21$ **66.** $\log_5 18$ **67.** $\log_{1/3} 71$ **68.** $\log_{1/2} 15$

69. $\log_{\sqrt{2}} 7$ **70.** $\log_{\sqrt{5}} 8$ **71.** $\log_\pi e$ **72.** $\log_\pi \sqrt{2}$

In Problems 73–78, graph each function using a graphing utility and the Change-of-Base Formula.

73. $y = \log_4 x$ **74.** $y = \log_5 x$ **75.** $y = \log_2(x+2)$ **76.** $y = \log_4(x-3)$

77. $y = \log_{x-1}(x+1)$ **78.** $y = \log_{x+2}(x-2)$

Applications and Extensions

In Problems 79–88, express y as a function of x. The constant C is a positive number.

79. $\ln y = \ln x + \ln C$ **80.** $\ln y = \ln(x + C)$

81. $\ln y = \ln x + \ln(x + 1) + \ln C$

82. $\ln y = 2 \ln x - \ln(x + 1) + \ln C$

83. $\ln y = 3x + \ln C$

84. $\ln y = -2x + \ln C$

85. $\ln(y - 3) = -4x + \ln C$

86. $\ln(y + 4) = 5x + \ln C$

87. $3 \ln y = \frac{1}{2} \ln(2x + 1) - \frac{1}{3} \ln(x + 4) + \ln C$

88. $2 \ln y = -\frac{1}{2} \ln x + \frac{1}{3} \ln(x^2 + 1) + \ln C$

89. Find the value of $\log_2 3 \cdot \log_3 4 \cdot \log_4 5 \cdot \log_5 6 \cdot \log_6 7 \cdot \log_7 8$.

90. Find the value of $\log_2 4 \cdot \log_4 6 \cdot \log_6 8$.

91. Find the value of $\log_2 3 \cdot \log_3 4 \cdot \cdots \cdot \log_n(n + 1) \cdot \log_{n+1} 2$.

92. Find the value of $\log_2 2 \cdot \log_2 4 \cdot \cdots \cdot \log_2 2^n$.

93. Show that $\log_a\left(x + \sqrt{x^2 - 1}\right) + \log_a\left(x - \sqrt{x^2 - 1}\right) = 0$.

94. Show that $\log_a\left(\sqrt{x} + \sqrt{x - 1}\right) + \log_a\left(\sqrt{x} - \sqrt{x - 1}\right) = 0$.

95. Show that $\ln(1 + e^{2x}) = 2x + \ln(1 + e^{-2x})$.

96. Difference Quotient   If $f(x) = \log_a x$, show that $\dfrac{f(x + h) - f(x)}{h} = \log_a\left(1 + \dfrac{h}{x}\right)^{1/h}$, $h \neq 0$.

97. If $f(x) = \log_a x$, show that $-f(x) = \log_{1/a} x$.

98. If $f(x) = \log_a x$, show that $f(AB) = f(A) + f(B)$.

99. If $f(x) = \log_a x$, show that $f\left(\dfrac{1}{x}\right) = -f(x)$.

100. If $f(x) = \log_a x$, show that $f(x^\alpha) = \alpha f(x)$.

101. Show that $\log_a\left(\dfrac{M}{N}\right) = \log_a M - \log_a N$, where a, M, and N are positive real numbers, with $a \neq 1$.

102. Show that $\log_a\left(\dfrac{1}{N}\right) = -\log_a N$, where a and N are positive real numbers, with $a \neq 1$.

Discussion and Writing

103. Graph $Y_1 = \log(x^2)$ and $Y_2 = 2 \log(x)$ using a graphing utility. Are they equivalent? What might account for any differences in the two functions?

104. Write an example that illustrates why $\log_2(x + y) \neq \log_2 x + \log_2 y$.

105. Write an example that illustrates why $(\log_a x)^r \neq r \log_a x$.

106. Does $3^{\log_3 -5} = -5$? Why or Why not?

4.6 Logarithmic and Exponential Equations

PREPARING FOR THIS SECTION   *Before getting started, review the following:*

• Solving Equations Using a Graphing Utility (Section 1.3, pp. 24–26)

• Solving Quadratic Equations (Appendix, Section A.5, pp. 696–704)

Now work the 'Are You Prepared?' problems on page 313.

OBJECTIVES   **1** Solve Logarithmic Equations Using the Properties of Logarithms

2 Solve Exponential Equations

3 Solve Logarithmic and Exponential Equations Using a Graphing Utility

1 Solve Logarithmic Equations Using the Properties of Logarithms

In Section 4.4 we solved logarithmic equations by changing a logarithm to exponential form. Often, however, some manipulation of the equation (usually using the properties of logarithms) is required before we can change to exponential form.

Our practice will be to solve equations, whenever possible, by finding exact solutions using algebraic methods and exact or approximate solutions using a graphing utility. When algebraic methods cannot be used, approximate solutions will be obtained using a graphing utility. The reader is encouraged to pay particular attention to the form of equations for which exact solutions are possible.

| EXAMPLE 1 | Solving a Logarithmic Equation |

Solve: $2 \log_5 x = \log_5 9$

Algebraic Solution

Because each logarithm is to the same base, 5, we can obtain an exact solution as follows:

$$2 \log_5 x = \log_5 9$$
$$\log_5 x^2 = \log_5 9 \qquad \text{\small\color{teal}$\log_a M^r = r \log_a M$}$$
$$x^2 = 9 \qquad \text{\small\color{teal}If $\log_a M = \log_a N$, then $M = N$.}$$
$$x = 3 \quad \text{or} \quad \cancel{x = -3} \qquad \text{\small\color{teal}Recall that logarithms of negative numbers are not defined, so, in the expression $2 \log_5 x$, x must be positive. Therefore, -3 is extraneous and we discard it.}$$

✔ CHECK: $2 \log_5 3 \overset{?}{=} \log_5 9$

$\qquad\qquad \log_5 3^2 \overset{?}{=} \log_5 9 \qquad \text{\small\color{teal}$r \log_a M = \log_a M^r$}$

$\qquad\qquad \log_5 9 = \log_5 9$

The solution set is $\{3\}$.

Graphing Solution

To solve the equation using a graphing utility, graph $Y_1 = 2 \log_5 x = \dfrac{2 \log x}{\log 5}$ and

$Y_2 = \log_5 9 = \dfrac{\log 9}{\log 5}$, and determine the point of intersection. See Figure 45.

Figure 45

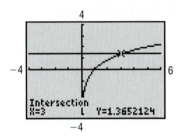

NOW WORK PROBLEM 9.

| EXAMPLE 2 | Solving a Logarithmic Equation |

Solve: $\log_4(x + 3) + \log_4(2 - x) = 1$

Algebraic Solution

To obtain an exact solution, we need to express the left side as a single logarithm. Then we will change the expression to exponential form.

$$\log_4(x + 3) + \log_4(2 - x) = 1$$
$$\log_4[(x + 3)(2 - x)] = 1 \qquad \text{\small\color{teal}$\log_a M + \log_a N = \log_a(MN)$}$$
$$(x + 3)(2 - x) = 4^1 = 4 \qquad \text{\small\color{teal}Change to an exponential expression.}$$
$$-x^2 - x + 6 = 4 \qquad \text{\small\color{teal}Simplify.}$$
$$x^2 + x - 2 = 0 \qquad \text{\small\color{teal}Place the quadratic equation in standard form.}$$
$$(x + 2)(x - 1) = 0 \qquad \text{\small\color{teal}Factor.}$$
$$x = -2 \quad \text{or} \quad x = 1 \qquad \text{\small\color{teal}Zero-Product Property}$$

Since the arguments of each logarithmic expression in the equation are positive for both $x = -2$ and $x = 1$, neither is extraneous. We leave the check to you.

Graphing Solution

Graph $Y_1 = \log_4(x + 3) + \log_4(2 - x) = $

$\dfrac{\log(x + 3)}{\log 4} + \dfrac{\log(2 - x)}{\log 4}$ and $Y_2 = 1$ and

determine the points of intersection. See Figure 46.

Figure 46

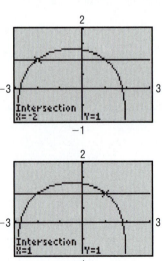

The solution set is $\{-2, 1\}$.

NOW WORK PROBLEM 13.

2 Solve Exponential Equations

In Sections 4.3 and 4.4, we solved certain exponential equations algebraically by expressing each side of the equation with the same base. However, many exponential equations cannot be rewritten so each side has the same base. In such cases, properties of logarithms along with algebraic techniques can sometimes be used to obtain a solution.

EXAMPLE 3   **Solving an Exponential Equation**

Solve: $4^x - 2^x - 12 = 0$

Algebraic Solution

We note that $4^x = (2^2)^x = 2^{2x} = (2^x)^2$, so the equation is actually quadratic in form, and we can rewrite it as

$$(2^x)^2 - 2^x - 12 = 0 \qquad \text{Let } u = 2^x; \text{ then } u^2 - u - 12 = 0.$$

Now we can factor as usual.

$$(2^x - 4)(2^x + 3) = 0 \qquad (u - 4)(u + 3) = 0$$
$$2^x - 4 = 0 \quad \text{or} \quad 2^x + 3 = 0 \qquad u - 4 = 0 \quad \text{or} \quad u + 3 = 0$$
$$2^x = 4 \qquad\qquad 2^x = -3 \qquad u = 2^x = 4 \qquad u = 2^x = -3$$

The equation on the left has the solution $x = 2$, since $2^x = 4 = 2^2$; the equation on the right has no solution, since $2^x > 0$ for all x. The only solution is 2.

The solution set is $\{2\}$.

Graphing Solution

Graph $Y_1 = 4^x - 2^x - 12$ and determine the x-intercept. See Figure 47.

Figure 47

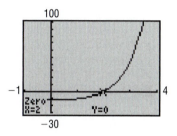

In the algebraic solution of the previous example, we were able to write the exponential expression using the same base after utilizing some algebra, obtaining an exact solution to the equation. When this is not possible, logarithms can sometimes be used to obtain the solution.

EXAMPLE 4   **Solving an Exponential Equation**

Solve: $2^x = 5$

Algebraic Solution

Since 5 cannot be written as an integral power of 2, we write the exponential equation as the equivalent logarithmic equation.

$$2^x = 5$$
$$x = \log_2 5 = \frac{\ln 5}{\ln 2}$$

Change-of-Base Formula (9), Section 4.5

Alternatively, we can solve the equation $2^x = 5$ by taking the natural logarithm (or common logarithm) of each side. Taking the natural logarithm,

$$2^x = 5$$
$$\ln 2^x = \ln 5 \qquad \text{If } M = N, \text{ then } \ln M = \ln N.$$
$$x \ln 2 = \ln 5 \qquad \ln M^r = r \ln M$$
$$x = \frac{\ln 5}{\ln 2} \qquad \text{Exact solution}$$
$$\approx 2.322 \qquad \text{Approximate solution}$$

Graphing Solution

Graph $Y_1 = 2^x$ and $Y_2 = 5$ and determine the x-coordinate of the point of intersection. See Figure 48.

Figure 48

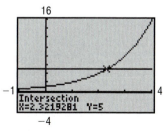

The approximate solution, rounded to three decimal places, is 2.322.

NOW WORK PROBLEM 21.

EXAMPLE 5   Solving an Exponential Equation

Solve: $8 \cdot 3^x = 5$

Algebraic Solution

$$8 \cdot 3^x = 5$$

$$3^x = \frac{5}{8} \qquad \text{Solve for } 3^x.$$

$$x = \log_3\left(\frac{5}{8}\right) = \frac{\ln\left(\frac{5}{8}\right)}{\ln 3} \qquad \text{Exact solution}$$

$$\approx -0.428 \qquad \text{Approximate solution} \quad \blacktriangleleft$$

Graphing Solution

Graph $Y_1 = 8 \cdot 3^x$ and $Y_2 = 5$ and determine the x-coordinate of the point of intersection. See Figure 49.

Figure 49

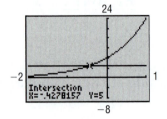

The approximate solution, rounded to three decimal places, is –0.428. $\blacktriangleleft$

EXAMPLE 6   Solving an Exponential Equation

Solve: $5^{x-2} = 3^{3x+2}$

Algebraic Solution

Because the bases are different, we first apply Property (6), Section 4.5 (take the natural logarithm of each side), and then use appropriate properties of logarithms. The result is an equation in x that we can solve.

$$5^{x-2} = 3^{3x+2}$$

$$\ln 5^{x-2} = \ln 3^{3x+2} \qquad \text{If } M = N, \ln M = \ln N.$$

$$(x - 2)\ln 5 = (3x + 2)\ln 3 \qquad \ln M^r = r \ln M$$

$$(\ln 5)x - 2\ln 5 = (3\ln 3)x + 2\ln 3 \qquad \text{Distribute.}$$

$$(\ln 5)x - (3\ln 3)x = 2\ln 3 + 2\ln 5 \qquad \text{Place terms involving } x \text{ on the left.}$$

$$(\ln 5 - 3\ln 3)x = 2(\ln 3 + \ln 5) \qquad \text{Factor.}$$

$$x = \frac{2(\ln 3 + \ln 5)}{\ln 5 - 3\ln 3} \qquad \text{Exact solution}$$

$$\approx -3.212 \qquad \text{Approximate solution} \quad \blacktriangleleft$$

Graphing Solution

Graph $Y_1 = 5^{x-2}$ and $Y_2 = 3^{3x+2}$ and determine the x-coordinate of the point of intersection. See Figure 50.

Figure 50

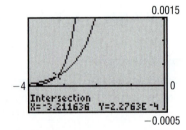

The approximate solution, rounded to three decimal places, is –3.212. $\blacktriangleleft$

 NOW WORK PROBLEM 29.

3 Solve Logarithmic and Exponential Equations Using a Graphing Utility

The algebraic techniques introduced in this section to obtain exact solutions apply only to certain types of logarithmic and exponential equations. Solutions for other types are usually studied in calculus, using numerical methods. For such types, we can use a graphing utility to approximate the solution.

| EXAMPLE 7 | **Solving Equations Using a Graphing Utility** |

Figure 51

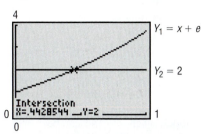

Solve: $x + e^x = 2$

Express the solution(s) rounded to two decimal places.

Solution The solution is found by graphing $Y_1 = x + e^x$ and $Y_2 = 2$. Y_1 is an increasing function (do you know why?), and so there is only one point of intersection for Y_1 and Y_2. Figure 51 shows the graphs of Y_1 and Y_2. Using the INTERSECT command, the solution is 0.44 rounded to two decimal places. ◀

━━━ **NOW WORK PROBLEM 53.**

4.6 Assess Your Understanding

'Are You Prepared?'

Answers are given at the end of these exercises. If you get a wrong answer, read the pages listed in red.

1. Solve $x^2 - 7x - 30 = 0$. (pp. 696–703)

2. Solve $(x + 3)^2 - 4(x + 3) + 3 = 0$. (pp. 703–704)

3. Approximate the solution(s) to $x^3 = x^2 - 5$ using a graphing utility. (pp. 24–26)

4. Approximate the solution(s) to $x^3 - 2x + 2 = 0$ using a graphing utility. (pp. 24–26)

Skill Building

In Problems 5–50, solve each equation. Verify your solution using a graphing utility.

5. $\log_4(x + 2) = \log_4 8$

6. $\log_5(2x + 3) = \log_5 3$

7. $\dfrac{1}{2}\log_3 x = 2\log_3 2$

8. $-2\log_4 x = \log_4 9$

9. $2\log_5 x = 3\log_5 4$

10. $3\log_2 x = -\log_2 27$

11. $3\log_2(x - 1) + \log_2 4 = 5$

12. $2\log_3(x + 4) - \log_3 9 = 2$

13. $\log x + \log(x + 15) = 2$

14. $\log_4 x + \log_4(x - 3) = 1$

15. $\ln x + \ln(x + 2) = 4$

16. $\ln(x + 1) - \ln x = 2$

17. $2^{2x} + 2^x - 12 = 0$

18. $3^{2x} + 3^x - 2 = 0$

19. $3^{2x} + 3^{x+1} - 4 = 0$

20. $2^{2x} + 2^{x+2} - 12 = 0$

21. $2^x = 10$

22. $3^x = 14$

23. $8^{-x} = 1.2$

24. $2^{-x} = 1.5$

25. $3^{1-2x} = 4^x$

26. $2^{x+1} = 5^{1-2x}$

27. $\left(\dfrac{3}{5}\right)^x = 7^{1-x}$

28. $\left(\dfrac{4}{3}\right)^{1-x} = 5^x$

29. $1.2^x = (0.5)^{-x}$

30. $0.3^{1+x} = 1.7^{2x-1}$

31. $\pi^{1-x} = e^x$

32. $e^{x+3} = \pi^x$

33. $5(2^{3x}) = 8$

34. $0.3(4^{0.2x}) = 0.2$

35. $\log_a(x - 1) - \log_a(x + 6) = \log_a(x - 2) - \log_a(x + 3)$

36. $\log_a x + \log_a(x - 2) = \log_a(x + 4)$

37. $\log_{1/3}(x^2 + x) - \log_{1/3}(x^2 - x) = -1$

38. $\log_4(x^2 - 9) - \log_4(x + 3) = 3$

39. $\log_2(x + 1) - \log_4 x = 1$
[**Hint:** Change $\log_4 x$ to base 2.]

40. $\log_2(3x + 2) - \log_4 x = 3$

41. $\log_{16} x + \log_4 x + \log_2 x = 7$

42. $\log_9 x + 3\log_3 x = 14$

43. $\left(\sqrt[3]{2}\right)^{2-x} = 2^{x^2}$

44. $\log_2 x^{\log_2 x} = 4$

45. $\dfrac{e^x + e^{-x}}{2} = 1$

46. $\dfrac{e^x + e^{-x}}{2} = 3$

47. $\dfrac{e^x - e^{-x}}{2} = 2$

48. $\dfrac{e^x - e^{-x}}{2} = -2$

[**Hint:** Multiply each side by e^x.]

49. $\log_5 x + \log_3 x = 1$
[**Hint:** Use the Change-of-Base Formula and factor $\log x$.]

50. $\log_2 x + \log_6 x = 3$

In Problems 51–64, use a graphing utility to solve each equation. Express your answer rounded to two decimal places.

51. $\log_5(x + 1) - \log_4(x - 2) = 1$

52. $\log_2(x - 1) - \log_6(x + 2) = 2$

53. $e^x = -x$

54. $e^{2x} = x + 2$

55. $e^x = x^2$

56. $e^x = x^3$

57. $\ln x = -x$

58. $\ln(2x) = -x + 2$

59. $\ln x = x^3 - 1$

60. $\ln x = -x^2$

61. $e^x + \ln x = 4$

62. $e^x - \ln x = 4$

63. $e^{-x} = \ln x$

64. $e^{-x} = -\ln x$

Applications and Extensions

65. A Population Model The population of the United States in 2000 was 282 million people. In addition, the population of the United States was growing at a rate of 1.1% per year. Assuming that this growth rate continues, the model $P(t) = 282(1.011)^{t-2000}$ represents the population P (in millions of people) in year t.

(a) According to this model, when will the population of the United States be 303 million people?

(b) According to this model, when will the population of the United States be 355 million people?

SOURCE: *Statistical Abstract of the United States*

66. A Population Model The population of the world in 2000 was 6.08 billion people. In addition, the population of the world was growing at a rate of 1.26% per year. Assuming that this growth rate continues, the model $P(t) = 6.08(1.0126)^{t-2000}$ represents the population P (in millions of people) in year t.

(a) According to this model, when will the population of the world be 9.17 billion people?

(b) According to this model, when will the population of the world be 11.55 billion people?

SOURCE: *Statistical Abstract of the United States*

67. Depreciation The value V of a Chevy Cavalier Coupe that is t years old can be modeled by $V(t) = 14{,}512(0.82)^t$.

(a) According to the model, when will the car be worth $9000?

(b) According to the model, when will the car be worth $4000?

(c) According to the model, when will the car be worth $2000?

SOURCE: *Kelley Blue Book*

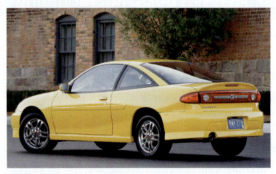

68. Depreciation The value V of a Dodge Stratus that is t years old can be modeled by $V(t) = 19{,}282(0.84)^t$.

(a) According to the model, when will the car be worth $15,000?

(b) According to the model, when will the car be worth $8000?

(c) According to the model, when will the car be worth $2000?

SOURCE: *Kelley Blue Book*

Discussion and Writing

69. Fill in reasons for each step in the following two solutions.

Solve: $\log_3(x - 1)^2 = 2$

Solution A

$\log_3(x - 1)^2 = 2$

$(x - 1)^2 = 3^2 = 9$ _____

$(x - 1) = \pm 3$ _____

$x - 1 = -3$ or $x - 1 = 3$ _____

$x = -2$ or $x = 4$ _____

Solution B

$\log_3(x - 1)^2 = 2$

$2\log_3(x - 1) = 2$ _____

$\log_3(x - 1) = 1$ _____

$x - 1 = 3^1 = 3$ _____

$x = 4$ _____

Both solutions given in Solution A check. Explain what caused the solution $x = -2$ to be lost in Solution B.

'Are You Prepared?' Answers

1. $\{-3, 10\}$ **2.** $\{-2, 0\}$ **3.** $\{-1.43\}$ **4.** $\{-1.77\}$

4.7 Compound Interest

PREPARING FOR THIS SECTION   *Before getting started, review the following:*

• Simple Interest (Appendix, Section A.7, pp. 718–719)

✎ Now work the 'Are You Prepared?' problems on page 322.

OBJECTIVES 1 Determine the Future Value of a Lump Sum of Money
2 Calculate Effective Rates of Return
3 Determine the Present Value of a Lump Sum of Money
4 Determine the Time Required to Double or Triple a Lump Sum of Money

1 Determine the Future Value of a Lump Sum of Money

Interest is money paid for the use of money. The total amount borrowed (whether by an individual from a bank in the form of a loan or by a bank from an individual in the form of a savings account) is called the **principal**. The **rate of interest**, expressed as a percent, is the amount charged for the use of the principal for a given period of time, usually on a yearly (that is, per annum) basis.

> **Simple Interest Formula**
>
> If a principal of P dollars is borrowed for a period of t years at a per annum interest rate r, expressed as a decimal, the interest I charged is
>
> $$I = Prt \qquad \textbf{(1)}$$
>
> Interest charged according to formula (1) is called **simple interest**.

In working with problems involving interest, we define the term **payment period** as follows:

Annually:	Once per year	**Monthly:**	12 times per year
Semiannually:	Twice per year	**Daily:**	365 times per year[*]
Quarterly:	Four times per year		

When the interest due at the end of a payment period is added to the principal so that the interest computed at the end of the next payment period is based on this new principal amount (old principal + interest), the interest is said to have been **compounded**. **Compound interest** is interest paid on principal and previously earned interest.

EXAMPLE 1   **Computing Compound Interest**

A credit union pays interest of 8% per annum compounded quarterly on a certain savings plan. If $1000 is deposited in such a plan and the interest is left to accumulate, how much is in the account after 1 year?

---

[*]Most banks use a 360-day "year." Why do you think they do?

Solution   We use the simple interest formula, $I = Prt$. The principal P is \$1000 and the rate of interest is $8\% = 0.08$. After the first quarter of a year, the time t is $\frac{1}{4}$ year, so the interest earned is

$$I = Prt = (\$1000)(0.08)\left(\frac{1}{4}\right) = \$20$$

The new principal is $P + I = \$1000 + \$20 = \$1020$. At the end of the second quarter, the interest on this principal is

$$I = (\$1020)(0.08)\left(\frac{1}{4}\right) = \$20.40$$

At the end of the third quarter, the interest on the new principal of $\$1020 + \$20.40 = \$1040.40$ is

$$I = (\$1040.40)(0.08)\left(\frac{1}{4}\right) = \$20.81$$

Finally, after the fourth quarter, the interest is

$$I = (\$1061.21)(0.08)\left(\frac{1}{4}\right) = \$21.22$$

After 1 year the account contains $\$1061.21 + \$21.22 = \$1082.43$.   ◀

The pattern of the calculations performed in Example 1 leads to a general formula for compound interest. To fix our ideas, let P represent the principal to be invested at a per annum interest rate r that is compounded n times per year, so the time of each compounding period is $\frac{1}{n}$ years. (For computing purposes, r is expressed as a decimal.) The interest earned after each compounding period is given by formula (1).

$$\text{Interest} = \text{principal} \times \text{rate} \times \text{time} = P \cdot r \cdot \frac{1}{n} = P \cdot \left(\frac{r}{n}\right)$$

The amount A after one compounding period is

$$A = P + I = P + P \cdot \left(\frac{r}{n}\right) = P \cdot \left(1 + \frac{r}{n}\right)$$

After two compounding periods, the amount A, based on the new principal $P \cdot \left(1 + \frac{r}{n}\right)$, is

$$A = \underbrace{P \cdot \left(1 + \frac{r}{n}\right)}_{\substack{\text{New} \\ \text{principal}}} + \underbrace{P \cdot \left(1 + \frac{r}{n}\right)\left(\frac{r}{n}\right)}_{\substack{\text{Interest on} \\ \text{new principal}}} = P \cdot \left(1 + \frac{r}{n}\right)\left(1 + \frac{r}{n}\right) = P \cdot \left(1 + \frac{r}{n}\right)^2$$

After three compounding periods, the amount A is

$$A = P \cdot \left(1 + \frac{r}{n}\right)^2 + P \cdot \left(1 + \frac{r}{n}\right)^2\left(\frac{r}{n}\right) = P \cdot \left(1 + \frac{r}{n}\right)^2 \cdot \left(1 + \frac{r}{n}\right) = P \cdot \left(1 + \frac{r}{n}\right)^3$$

Continuing this way, after n compounding periods (1 year), the amount A is

$$A = P \cdot \left(1 + \frac{r}{n}\right)^n$$

Because t years will contain $n \cdot t$ compounding periods, after t years we have

$$A = P \cdot \left(1 + \frac{r}{n}\right)^{nt}$$

Theorem

Compound Interest Formula

The amount A after t years due to a principal P invested at an annual interest rate r compounded n times per year is

$$A = P \cdot \left(1 + \frac{r}{n}\right)^{nt} \qquad\qquad (2)$$

— **Exploration** ——————

To see the effects of compounding interest monthly on an initial deposit of $1, graph $Y_1 = \left(1 + \dfrac{r}{12}\right)^{12x}$ with $r = 0.06$ and $r = 0.12$ for $0 \le x \le 30$. What is the future value of $1 in 30 years when the interest rate per annum is $r = 0.06$ (6%)? What is the future value of $1 in 30 years when the interest rate per annum is $r = 0.12$ (12%)? Does doubling the interest rate double the future value?

For example, to rework Example 1, we would use $P = \$1000, r = 0.08, n = 4$ (quarterly compounding), and $t = 1$ year to obtain

$$A = P \cdot \left(1 + \frac{r}{n}\right)^{nt} = 1000\left(1 + \frac{0.08}{4}\right)^{4 \cdot 1} = \$1082.43$$

In equation (2), the amount A is typically referred to as the **future value** of the account, while P is called the **present value**.

 NOW WORK PROBLEM 3.

EXAMPLE 2

Comparing Investments Using Different Compounding Periods

Investing $1000 at an annual rate of 10% compounded annually, semiannually, quarterly, monthly, and daily will yield the following amounts after 1 year:

Annual compounding ($n = 1$): $\quad A = P \cdot (1 + r)$
$$= (\$1000)(1 + 0.10) = \$1100.00$$

Semiannual compounding ($n = 2$): $\quad A = P \cdot \left(1 + \frac{r}{2}\right)^2$
$$= (\$1000)(1 + 0.05)^2 = \$1102.50$$

Quarterly compounding ($n = 4$): $\quad A = P \cdot \left(1 + \frac{r}{4}\right)^4$
$$= (\$1000)(1 + 0.025)^4 = \$1103.81$$

Monthly compounding ($n = 12$): $\quad A = P \cdot \left(1 + \frac{r}{12}\right)^{12}$
$$= (\$1000)(1 + 0.00833)^{12} = \$1104.71$$

Daily compounding ($n = 365$): $\quad A = P \cdot \left(1 + \frac{r}{365}\right)^{365}$
$$= (\$1000)(1 + 0.000274)^{365} = \$1105.16 \blacktriangleleft$$

From Example 2 we can see that the effect of compounding more frequently is that the amount after 1 year is higher: $1000 compounded 4 times a year at 10% results in $1103.81; $1000 compounded 12 times a year at 10% results in $1104.71; and $1000 compounded 365 times a year at 10% results in $1105.16. This leads to the following question: What would happen to the amount after 1 year if the number of times that the interest is compounded were increased without bound?

Let's find the answer. Suppose that P is the principal, r is the per annum interest rate, and n is the number of times that the interest is compounded each year. The amount after 1 year is

$$A = P \cdot \left(1 + \frac{r}{n}\right)^n$$

Rewrite this expression as follows:

$$A = P \cdot \left(1 + \frac{r}{n}\right)^n = P \cdot \left(1 + \frac{1}{\frac{n}{r}}\right)^n = P \cdot \left[\left(1 + \frac{1}{\frac{n}{r}}\right)^{n/r}\right]^r = P \cdot \left[\left(1 + \frac{1}{h}\right)^h\right]^r \quad \text{(3)}$$

$$h = \frac{n}{r}$$

Now suppose that the number n of times that the interest is compounded per year gets larger and larger; that is, suppose that $n \to \infty$. Then $h = \dfrac{n}{r} \to \infty$, and the expression in brackets equals e. [Refer to equation (2), page 317.] That is, $A \to Pe^r$.

Table 9 compares $\left(1 + \dfrac{r}{n}\right)^n$, for large values of n, to e^r for $r = 0.05$, $r = 0.10$, $r = 0.15$, and $r = 1$. The larger that n gets, the closer $\left(1 + \dfrac{r}{n}\right)^n$ gets to e^r.

No matter how frequent the compounding, the amount after 1 year has the definite ceiling Pe^r.

Table 9

	$\left(1 + \frac{r}{n}\right)^n$			
	$n = 100$	$n = 1000$	$n = 10,000$	e^r
$r = 0.05$	1.0512580	1.0512698	1.051271	1.0512711
$r = 0.10$	1.1051157	1.1051654	1.1051704	1.1051709
$r = 0.15$	1.1617037	1.1618212	1.1618329	1.1618342
$r = 1$	2.7048138	2.7169239	2.7181459	2.7182818

When interest is compounded so that the amount after 1 year is Pe^r, we say the interest is **compounded continuously**.

Theorem

Continuous Compounding

The amount A after t years due to a principal P invested at an annual interest rate r compounded continuously is

$$A = Pe^{rt} \quad \text{(4)}$$

EXAMPLE 3   **Using Continuous Compounding**

The amount A that results from investing a principal P of $1000 at an annual rate r of 10% compounded continuously for a time t of 1 year is

$$A = \$1000e^{0.10} = (\$1000)(1.10517) = \$1105.17$$

◀

NOW WORK PROBLEM 11.

2 Calculate Effective Rates of Return

The **effective rate of interest** is the equivalent annual simple rate of interest that would yield the same amount as compounding after 1 year. For example, based on Example 3, a principal of $1000 will result in $1105.17 at a rate of 10% compounded continuously. To get this same amount using a simple rate of interest would require that interest of $1105.17 − $1000.00 = $105.17 be earned on the principal. Since $105.17 is 10.517% of $1000, a simple rate of interest of 10.517% is needed to equal 10% compounded continuously. The effective rate of interest of 10% compounded continuously is 10.517%.

Based on the results of Examples 2 and 3, we find the following comparisons:

	Annual Rate	Effective Rate
Annual compounding	10%	10%
Semiannual compounding	10%	10.25%
Quarterly compounding	10%	10.381%
Monthly compounding	10%	10.471%
Daily compounding	10%	10.516%
Continuous compounding	10%	10.517%

EXAMPLE 4	**Computing the Effective Rate of Interest**

On January 2, 2004, $2000 is placed in an Individual Retirement Account (IRA) that will pay interest of 7% per annum compounded continuously.

(a) What will the IRA be worth on January 1, 2024?

(b) What is the effective rate of interest?

Solution   (a) The amount A after 20 years is

$$A = Pe^{rt} = \$2000e^{(0.07)(20)} = \$8110.40$$

(b) First, we compute the interest earned on $2000 at $r = 7\%$ compounded continuously for 1 year.

$$A = \$2000e^{0.07(1)}$$
$$= \$2145.02$$

So the interest earned is $2145.02 − $2000.00 = $145.02. Use the simple interest formula $I = Prt$, with $I = \$145.02$, $P = \$2000$, and $t = 1$, and solve for r, the effective rate of interest.

$$\$145.02 = \$2000 \cdot r \cdot 1$$

$$r = \frac{\$145.02}{\$2000} = 0.07251$$

The effective rate of interest is 7.251%.   ◄

— Exploration —————

For the IRA described in Example 4, how long will it be until $A = \$4000$? $6000?

[**Hint:** Graph $Y_1 = 2000e^{0.07x}$ and $Y_2 = 4000$. Use INTERSECT to find x.]

▬▬▬- **NOW WORK PROBLEM 23.**

 ## Determine the Present Value of a Lump Sum of Money

When people engaged in finance speak of the "time value of money," they are usually referring to the *present value* of money. The **present value** of A dollars to be received at a future date is the principal that you would need to invest now so that it will grow to A dollars in the specified time period. The present value of money to be received at a future date is always less than the amount to be received, since the amount to be received will equal the present value (money invested now) *plus* the interest accrued over the time period.

We use the compound interest formula (2) to get a formula for present value. If P is the present value of A dollars to be received after t years at a per annum interest rate r compounded n times per year, then, by formula (2),

$$A = P \cdot \left(1 + \frac{r}{n}\right)^{nt}$$

To solve for P, we divide both sides by $\left(1 + \dfrac{r}{n}\right)^{nt}$. The result is

$$\frac{A}{\left(1 + \dfrac{r}{n}\right)^{nt}} = P \quad \text{or} \quad P = A \cdot \left(1 + \frac{r}{n}\right)^{-nt}$$

Theorem

Present Value Formulas

The present value P of A dollars to be received after t years, assuming a per annum interest rate r compounded n times per year, is

$$P = A \cdot \left(1 + \frac{r}{n}\right)^{-nt} \tag{5}$$

If the interest is compounded continuously, then

$$P = Ae^{-rt} \tag{6}$$

To prove (6), solve formula (4) for P.

EXAMPLE 5 **Computing the Value of a Zero-Coupon Bond**

A zero-coupon (noninterest-bearing) bond can be redeemed in 10 years for $1000. How much should you be willing to pay for it now if you want a return of

(a) 8% compounded monthly?

(b) 7% compounded continuously?

Solution (a) We are seeking the present value of $1000. We use formula (5) with $A = \$1000$, $n = 12$, $r = 0.08$, and $t = 10$.

$$P = A \cdot \left(1 + \frac{r}{n}\right)^{-nt} = \$1000\left(1 + \frac{0.08}{12}\right)^{-12(10)} = \$450.52$$

For a return of 8% compounded monthly, you should pay $450.52 for the bond.

(b) Here we use formula (6) with $A = \$1000$, $r = 0.07$, and $t = 10$.

$$P = Ae^{-rt} = \$1000e^{-(0.07)(10)} = \$496.59$$

For a return of 7% compounded continuously, you should pay $496.59 for the bond. ◀

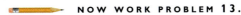 **NOW WORK PROBLEM 13.**

EXAMPLE 6 | **Rate of Interest Required to Double an Investment**

What annual rate of interest compounded annually should you seek if you want to double your investment in 5 years?

Solution   If P is the principal and we want P to double, the amount A will be $2P$. We use the compound interest formula with $n = 1$ and $t = 5$ to find r.

$$A = P \cdot \left(1 + \frac{r}{n}\right)^{nt}$$

$2P = P \cdot (1 + r)^5$ $A = 2P, n = 1, t = 5$

$2 = (1 + r)^5$ Cancel the P's.

$1 + r = \sqrt[5]{2}$ Take the fifth root of each side.

$r = \sqrt[5]{2} - 1 \approx 1.148698 - 1 = 0.148698$

The annual rate of interest needed to double the principal in 5 years is 14.87%. ◀

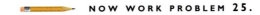 **NOW WORK PROBLEM 25.**

---

4 **Determine the Time Required to Double or Triple a Lump Sum of Money**

EXAMPLE 7 | **Doubling and Tripling Time for an Investment**

(a) How long will it take for an investment to double in value if it earns 5% compounded continuously?

(b) How long will it take to triple at this rate?

Solution   (a) If P is the initial investment and we want P to double, the amount A will be $2P$. We use formula (4) for continuously compounded interest with $r = 0.05$. Then

$$A = Pe^{rt}$$

$2P = Pe^{0.05t}$ $A = 2P, r = 0.05$

$2 = e^{0.05t}$ Cancel the P's.

$0.05t = \ln 2$ Rewrite as a logarithm.

$t = \dfrac{\ln 2}{0.05} \approx 13.86$ Solve for t.

It will take about 14 years to double the investment.

(b) To triple the investment, we set $A = 3P$ in formula (4).

$$A = Pe^{rt}$$
$$3P = Pe^{0.05t} \qquad A = 3P, r = 0.05$$
$$3 = e^{0.05t} \qquad \text{Cancel the P's.}$$
$$0.05t = \ln 3 \qquad \text{Rewrite as a logarithm.}$$
$$t = \frac{\ln 3}{0.05} \approx 21.97 \qquad \text{Solve for t.}$$

It will take about 22 years to triple the investment. ◀

NOW WORK PROBLEM 31.

4.7 Assess Your Understanding

'Are You Prepared?'

Answers are given at the end of these exercises. If you get a wrong answer, read the pages listed in red.

1. What is the interest due if $500 is borrowed for 6 months at a simple interest rate of 6% per annum? (pp. 718–719)

2. If you borrow $5000 and, after 9 months, pay off the loan in the amount of $5500, what per annum rate of interest was charged? (pp. 718–719)

Skill Building

In Problems 3–12, find the amount that results from each investment.

3. $100 invested at 4% compounded quarterly after a period of 2 years

4. $50 invested at 6% compounded monthly after a period of 3 years

5. $500 invested at 8% compounded quarterly after a period of $2\frac{1}{2}$ years

6. $300 invested at 12% compounded monthly after a period of $1\frac{1}{2}$ years

7. $600 invested at 5% compounded daily after a period of 3 years

8. $700 invested at 6% compounded daily after a period of 2 years

9. $10 invested at 11% compounded continuously after a period of 2 years

10. $40 invested at 7% compounded continuously after a period of 3 years

11. $100 invested at 10% compounded continuously after a period of $2\frac{1}{4}$ years

12. $100 invested at 12% compounded continuously after a period of $3\frac{3}{4}$ years

In Problems 13–22, find the principal needed now to get each amount; that is, find the present value.

13. To get $100 after 2 years at 6% compounded monthly

14. To get $75 after 3 years at 8% compounded quarterly

15. To get $1000 after $2\frac{1}{2}$ years at 6% compounded daily

16. To get $800 after $3\frac{1}{2}$ years at 7% compounded monthly

17. To get $600 after 2 years at 4% compounded quarterly

18. To get $300 after 4 years at 3% compounded daily

19. To get $80 after $3\frac{1}{4}$ years at 9% compounded continuously

20. To get $800 after $2\frac{1}{2}$ years at 8% compounded continuously

21. To get $400 after 1 year at 10% compounded continuously

22. To get $1000 after 1 year at 12% compounded continuously

23. Find the effective rate of interest for $5\frac{1}{4}$% compounded quarterly.

24. What interest rate compounded quarterly will give an effective interest rate of 7%?

25. What rate of interest compounded annually is required to double an investment in 3 years?

26. What rate of interest compounded annually is required to double an investment in 10 years?

In Problems 27–30, which of the two rates would yield the larger amount in 1 year?

[**Hint:** *Start with a principal of $10,000 in each instance.*]

27. 6% compounded quarterly or $6\frac{1}{4}$% compounded annually

28. 9% compounded quarterly or $9\frac{1}{4}$% compounded annually

29. 9% compounded monthly or 8.8% compounded daily

31. How long does it take for an investment to double in value if it is invested at 8% per annum compounded monthly? Compounded continuously?

30. 8% compounded semiannually or 7.9% compounded daily

32. How long does it take for an investment to double in value if it is invested at 10% per annum compounded monthly? Compounded continuously?

Applications and Extensions

33. If Tanisha has $100 to invest at 8% per annum compounded monthly, how long will it be before she has $150? If the compounding is continuous, how long will it be?

34. If Angela has $100 to invest at 10% per annum compounded monthly, how long will it be before she has $175? If the compounding is continuous, how long will it be?

35. How many years will it take for an initial investment of $10,000 to grow to $25,000? Assume a rate of interest of 6% compounded continuously.

36. How many years will it take for an initial investment of $25,000 to grow to $80,000? Assume a rate of interest of 7% compounded continuously.

37. What will a $90,000 house cost 5 years from now if the inflation rate over that period averages 3% compounded annually?

38. Sears charges 1.25% per month on the unpaid balance for customers with charge accounts (interest is compounded monthly). A customer charges $200 and does not pay her bill for 6 months. What is the bill at that time?

39. Jerome will be buying a used car for $15,000 in 3 years. How much money should he ask his parents for now so that, if he invests it at 5% compounded continuously, he will have enough to buy the car?

40. John will require $3000 in 6 months to pay off a loan that has no prepayment privileges. If he has the $3000 now, how much of it should he save in an account paying 3% compounded monthly so that in 6 months he will have exactly $3000?

41. George is contemplating the purchase of 100 shares of a stock selling for $15 per share. The stock pays no dividends. The history of the stock indicates that it should grow at an annual rate of 15% per year. How much will the 100 shares of stock be worth in 5 years?

42. Tracy is contemplating the purchase of 100 shares of a stock selling for $15 per share. The stock pays no dividends. Her broker says that the stock will be worth $20 per share in 2 years. What is the annual rate of return on this investment?

43. A business purchased for $650,000 in 1994 is sold in 1997 for $850,000. What is the annual rate of return for this investment?

44. Tanya has just inherited a diamond ring appraised at $5000. If diamonds have appreciated in value at an annual rate of 8%, what was the value of the ring 10 years ago when the ring was purchased?

45. Jim places $1000 in a bank account that pays 5.6% compounded continuously. After 1 year, will he have enough money to buy a computer system that costs $1060? If another bank will pay Jim 5.9% compounded monthly, is this a better deal?

46. On January 1, Kim places $1000 in a certificate of deposit that pays 6.8% compounded continuously and matures in 3 months. Then Kim places the $1000 and the interest in a passbook account that pays 5.25% compounded monthly. How much does Kim have in the passbook account on May 1?

47. Will invests $2000 in a bond trust that pays 9% interest compounded semiannually. His friend Henry invests $2000 in a certificate of deposit that pays $8\frac{1}{2}\%$ compounded continuously. Who has more money after 20 years, Will or Henry?

48. Suppose that April has access to an investment that will pay 10% interest compounded continuously. Which is better: To be given $1000 now so that she can take advantage of this investment opportunity or to be given $1325 after 3 years?

49. Colleen and Bill have just purchased a house for $150,000, with the seller holding a second mortgage of $50,000. They promise to pay the seller $50,000 plus all accrued interest 5 years from now. The seller offers them three interest options on the second mortgage:
(a) Simple interest at 12% per annum
(b) $11\frac{1}{2}\%$ interest compounded monthly
(c) $11\frac{1}{4}\%$ interest compounded continuously
Which option is best; that is, which results in the least interest on the loan?

50. The First National Bank advertises that it pays interest on savings accounts at the rate of 4.25% compounded daily. Find the effective rate if the bank uses (a) 360 days or (b) 365 days in determining the daily rate.

Problems 51–54 involve zero-coupon bonds. A zero-coupon bond is a bond that is sold now at a discount and will pay its face value at the time when it matures; no interest payments are made.

51. A zero-coupon bond can be redeemed in 20 years for $10,000. How much should you be willing to pay for it now if you want a return of:
(a) 10% compounded monthly?
(b) 10% compounded continuously?

52. A child's grandparents are considering buying a $40,000 face value zero-coupon bond at birth so that she will have enough money for her college education 17 years later. If they want a rate of return of 8% compounded annually, what should they pay for the bond?

53. How much should a $10,000 face value zero-coupon bond, maturing in 10 years, be sold for now if its rate of return is to be 8% compounded annually?

54. If Pat pays $12,485.52 for a $25,000 face value zero-coupon bond that matures in 8 years, what is his annual rate of return?

55. **Time to Double or Triple an Investment** The formula

$$t = \frac{\ln m}{n \ln\left(1 + \dfrac{r}{n}\right)}$$

can be used to find the number of years t required to multiply an investment m times when r is the per annum interest rate compounded n times a year.
(a) How many years will it take to double the value of an IRA that compounds annually at the rate of 12%?

(b) How many years will it take to triple the value of a savings account that compounds quarterly at an annual rate of 6%?
(c) Give a derivation of this formula.

56. **Time to Reach an Investment Goal** The formula

$$t = \frac{\ln A - \ln P}{r}$$

can be used to find the number of years t required for an investment P to grow to a value A when compounded continuously at an annual rate r.
(a) How long will it take to increase an initial investment of $1000 to $8000 at an annual rate of 10%?
(b) What annual rate is required to increase the value of a $2000 IRA to $30,000 in 35 years?
(c) Give a derivation of this formula.

Discussion and Writing

57. Explain in your own words what the term *compound interest* means. What does *continuous compounding* mean?

58. Explain in your own words the meaning of *present value*.

59. **Critical Thinking** You have just contracted to buy a house and will seek financing in the amount of $100,000. You go to several banks. Bank 1 will lend you $100,000 at the rate of 8.75% amortized over 30 years with a loan origination fee of 1.75%. Bank 2 will lend you $100,000 at the rate of 8.375% amortized over 15 years with a loan origination fee of 1.5%. Bank 3 will lend you $100,000 at the rate of 9.125% amortized over 30 years with no loan origination fee. Bank 4 will lend you $100,000 at the rate of 8.625% amortized over 15 years with no loan origination fee. Which loan would you take? Why? Be sure to have sound reasons for your choice.

Use the information in the table to assist you. If the amount of the monthly payment does not matter to you, which loan would you take? Again, have sound reasons for your choice. Compare your final decision with others in the class. Discuss.

	Monthly Payment	Loan Origination Fee
Bank 1	$786.70	$1,750.00
Bank 2	$977.42	$1,500.00
Bank 3	$813.63	$0.00
Bank 4	$990.68	$0.00

'Are You Prepared?' Answers

1. $15   2. 13.33%

4.8 Exponential Growth and Decay; Newton's Law; Logistic Growth and Decay

OBJECTIVES 1 Find Equations of Populations That Obey the Law of Uninhibited Growth
2 Find Equations of Populations That Obey the Law of Decay
3 Use Newton's Law of Cooling
4 Use Logistic Models

1 Find Equations of Populations That Obey the Law of Uninhibited Growth

Many natural phenomena have been found to follow the law that an amount A varies with time t according to

$$A = A_0 e^{kt} \tag{1}$$

Here A_0 is the original amount ($t = 0$) and $k \neq 0$ is a constant.

If $k > 0$, then equation (1) states that the amount A is increasing over time; if $k < 0$, the amount A is decreasing over time. In either case, when an amount A varies over time according to equation (1), it is said to follow the **exponential law** or the **law of uninhibited growth** ($k > 0$) **or decay** ($k < 0$). See Figure 52.

Figure 52

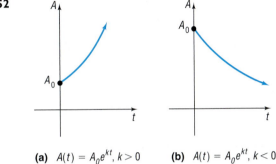

(a) $A(t) = A_0 e^{kt}, k > 0$ **(b)** $A(t) = A_0 e^{kt}, k < 0$

For example, we saw in Section 4.7 that continuously compounded interest follows the law of uninhibited growth. In this section we shall look at three additional phenomena that follow the exponential law.

Cell division is the growth process of many living organisms, such as amoebas, plants, and human skin cells. Based on an ideal situation in which no cells die and no by-products are produced, the number of cells present at a given time follows the law of uninhibited growth. Actually, however, after enough time has passed, growth at an exponential rate will cease due to the influence of factors such as lack of living space and dwindling food supply. The law of uninhibited growth accurately reflects only the early stages of the cell division process.

The cell division process begins with a culture containing N_0 cells. Each cell in the culture grows for a certain period of time and then divides into two identical cells. We assume that the time needed for each cell to divide in two is constant and does not change as the number of cells increases. These new cells then grow, and eventually each divides in two, and so on.

Uninhibited Growth of Cells

A model that gives the number N of cells in a culture after a time t has passed (in the early stages of growth) is

$$N(t) = N_0 e^{kt}, \quad k > 0 \tag{2}$$

where N_0 is the initial number of cells and k is a positive constant that represents the growth rate of the cells.

In using formula (2) to model the growth of cells, we are using a function that yields positive real numbers, even though we are counting the number of cells, which must be an integer. This is a common practice in many applications.

| EXAMPLE 1 | **Bacterial Growth** |

A colony of bacteria grows according to the law of uninhibited growth according to the function $N(t) = 100e^{0.045t}$, where N is measured in grams and t is measured in days.

(a) Determine the initial amount of bacteria.

(b) What is the growth rate of the bacteria?

(c) Graph the function using a graphing utility.

(d) What is the population after 5 days?

(e) How long will it take for the population to reach 140 grams?

(f) What is the doubling time for the population?

Solution (a) The initial amount of bacteria, N_0, is obtained when $t = 0$, so

$$N_0 = N(0) = 100e^{0.045(0)} = 100 \text{ grams.}$$

(b) Compare $N(t) = 100e^{0.045t}$ to $N(t) = 100e^{kt}$. The value of k, 0.045, indicates a growth rate of 4.5%.

Figure 53

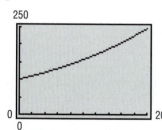

(c) Figure 53 shows the graph of $N(t) = 100e^{0.045t}$.

(d) The population after 5 days is $N(5) = 100e^{0.045(5)} = 125.2$ grams.

(e) To find how long it takes for the population to reach 140 grams, we solve the equation $N(t) = 140$.

$$100e^{0.045t} = 140$$

$e^{0.045t} = 1.4$ *Divide both sides of the equation by 100.*

$0.045t = \ln 1.4$ *Rewrite as a logarithm.*

$t = \dfrac{\ln 1.4}{0.045}$ *Divide both sides of the equation by 0.045.*

≈ 7.5 days

(f) The population doubles when $N(t) = 200$ grams, so we find the doubling time by solving the equation $200 = 100e^{0.045t}$ for t.

$$200 = 100e^{0.045t}$$

$2 = e^{0.045t}$ *Divide both sides of the equation by 100.*

$\ln 2 = 0.045t$ *Rewrite as a logarithm.*

$t = \dfrac{\ln 2}{0.045}$ *Divide both sides of the equation by 0.045.*

≈ 15.4 days

The population doubles approximately every 15.4 days. ◀

 NOW WORK PROBLEM 1.

| EXAMPLE 2 | **Bacterial Growth** |

A colony of bacteria increases according to the law of uninhibited growth.

(a) If the number of bacteria doubles in 3 hours, find the function that gives the number of cells in the culture.

(b) How long will it take for the size of the colony to triple?

(c) How long will it take for the population to double a second time (that is, increase four times)?

Solution   (a) Using formula (2), the number N of cells at a time t is

$$N(t) = N_0 e^{kt}$$

where N_0 is the initial number of bacteria present and k is a positive number. We first seek the number k. The number of cells doubles in 3 hours, so we have

$$N(3) = 2N_0$$

But $N(3) = N_0 e^{k(3)}$, so

$$N_0 e^{k(3)} = 2N_0$$
$$e^{3k} = 2 \qquad \text{\color{blue}{Divide both sides by } } N_0.$$
$$3k = \ln 2 \qquad \text{\color{blue}{Write the exponential equation as a logarithm.}}$$
$$k = \frac{1}{3} \ln 2$$

Formula (2) for this growth process is therefore

$$N(t) = N_0 e^{\left(\frac{1}{3}\ln 2\right)t}$$

(b) The time t needed for the size of the colony to triple requires that $N = 3N_0$. We substitute $3N_0$ for N to get

$$3N_0 = N_0 e^{\left(\frac{1}{3}\ln 2\right)t}$$
$$3 = e^{\left(\frac{1}{3}\ln 2\right)t}$$
$$\left(\frac{1}{3}\ln 2\right)t = \ln 3$$
$$t = \frac{3\ln 3}{\ln 2} \approx 4.755 \text{ hours}$$

It will take about 4.755 hours or 4 hours, 45 minutes for the size of the colony to triple.

(c) If a population doubles in 3 hours, it will double a second time in 3 more hours, for a total time of 6 hours.   ◀

2   Find Equations of Populations That Obey the Law of Decay

Radioactive materials follow the law of uninhibited decay.

Uninhibited Radioactive Decay

The amount A of a radioactive material present at time t is given by

$$A(t) = A_0 e^{kt}, \qquad k < 0 \qquad \qquad \textbf{(3)}$$

where A_0 is the original amount of radioactive material and k is a negative number that represents the rate of decay.

All radioactive substances have a specific **half-life**, which is the time required for half of the radioactive substance to decay. In **carbon dating**, we use the fact that all living organisms contain two kinds of carbon, carbon 12 (a stable carbon) and carbon 14 (a radioactive carbon with a half-life of 5600 years). While an organism is living, the ratio of carbon 12 to carbon 14 is constant. But when an organism dies, the original amount of carbon 12 present remains unchanged, whereas the amount of carbon 14 begins to decrease. This change in the amount of carbon 14 present relative to the amount of carbon 12 present makes it possible to calculate when an organism died.

EXAMPLE 3   **Estimating the Age of Ancient Tools**

Traces of burned wood along with ancient stone tools in an archeological dig in Chile were found to contain approximately 1.67% of the original amount of carbon 14.

(a) If the half-life of carbon 14 is 5600 years, approximately when was the tree cut and burned?

(b) Using a graphing utility, graph the relation between the percentage of carbon 14 remaining and time.

(c) Determine the time that elapses until half of the carbon 14 remains. This answer should equal the half-life of carbon 14.

(d) Use a graphing utility to verify the answer found in part (a).

Solution   (a) Using formula (3), the amount A of carbon 14 present at time t is

$$A(t) = A_0 e^{kt}$$

where A_0 is the original amount of carbon 14 present and k is a negative number. We first seek the number k. To find it, we use the fact that after 5600 years half of the original amount of carbon 14 remains, so $A(5600) = \dfrac{1}{2} A_0$. Then,

$$\frac{1}{2} A_0 = A_0 e^{k(5600)}$$

$$\frac{1}{2} = e^{5600k} \qquad \text{\textcolor{teal}{Divide both sides of the equation by } } A_0.$$

$$5600k = \ln \frac{1}{2} \qquad \text{\textcolor{teal}{Rewrite as a logarithm.}}$$

$$k = \frac{1}{5600} \ln \frac{1}{2} \approx -0.000124$$

Formula (3) therefore becomes

$$A(t) = A_0 e^{\frac{\ln \frac{1}{2}}{5600} t}$$

If the amount A of carbon 14 now present is 1.67% of the original amount, it follows that

$$0.0167 A_0 = A_0 e^{\frac{\ln \frac{1}{2}}{5600} t}$$

$$0.0167 = e^{\frac{\ln \frac{1}{2}}{5600} t} \qquad \text{\textcolor{teal}{Divide both sides of the equation by } } A_0.$$

$$\frac{\ln \frac{1}{2}}{5600} t = \ln 0.0167 \qquad \text{\textcolor{teal}{Rewrite as a logarithm.}}$$

$$t = \frac{5600}{\ln \frac{1}{2}} \ln 0.0167 \approx 33{,}062 \text{ years}$$

The tree was cut and burned about 33,062 years ago. Some archeologists use this conclusion to argue that humans lived in the Americas 33,000 years ago, much earlier than is generally accepted.

(b) Figure 54 shows the graph of $y = e^{\frac{\ln\frac{1}{2}}{5600}x}$, where y is the fraction of carbon 14 present and x is the time.

(c) By graphing $Y_1 = 0.5$ and $Y_2 = e^{\frac{\ln\frac{1}{2}}{5600}x}$, where x is time, and using INTERSECT, we find that it takes 5600 years until half the carbon 14 remains. The half-life of carbon 14 is 5600 years.

(d) By graphing $Y_1 = 0.0167$ and $Y_2 = e^{\frac{\ln\frac{1}{2}}{5600}x}$, where x is time, and using INTERSECT, we find that it takes 33,062 years until 1.67% of the carbon 14 remains. ◄

Figure 54

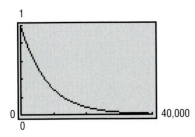

NOW WORK PROBLEM 3.

3 Use Newton's Law of Cooling

Newton's Law of Cooling[*] states that the temperature of a heated object decreases exponentially over time toward the temperature of the surrounding medium.

> **Newton's Law of Cooling**
>
> The temperature u of a heated object at a given time t can be modeled by the following function:
>
> $$u(t) = T + (u_0 - T)e^{kt}, \quad k < 0 \qquad (4)$$
>
> where T is the constant temperature of the surrounding medium, u_0 is the initial temperature of the heated object, and k is a negative constant.

EXAMPLE 4   **Using Newton's Law of Cooling**

An object is heated to 100°C (degrees Celsius) and is then allowed to cool in a room whose air temperature is 30°C.

(a) If the temperature of the object is 80°C after 5 minutes, when will its temperature be 50°C?

(b) Using a graphing utility, graph the relation found between the temperature and time.

(c) Using a graphing utility, verify that after 18.6 minutes the temperature is 50°C.

(d) Using a graphing utility, determine the elapsed time before the object is 35°C.

(e) What do you notice about the temperature as time passes?

Solution   (a) Using formula (4) with $T = 30$ and $u_0 = 100$, the temperature (in degrees Celsius) of the object at time t (in minutes) is

$$u(t) = 30 + (100 - 30)e^{kt} = 30 + 70e^{kt}$$

---

[*]Named after Sir Isaac Newton (1642–1727), one of the cofounders of calculus.

where k is a negative constant. To find k, we use the fact that $u = 80$ when $t = 5$. Then

$$80 = 30 + 70e^{k(5)}$$
$$50 = 70e^{5k}$$
$$e^{5k} = \frac{50}{70}$$
$$5k = \ln \frac{5}{7}$$
$$k = \frac{1}{5} \ln \frac{5}{7} \approx -0.0673$$

Formula (4) therefore becomes

$$u(t) = 30 + 70e^{\frac{\ln \frac{5}{7}}{5} t}$$

We want to find t when $u = 50°C$, so

$$50 = 30 + 70e^{\frac{\ln \frac{5}{7}}{5} t}$$
$$20 = 70e^{\frac{\ln \frac{5}{7}}{5} t}$$
$$e^{\frac{\ln \frac{5}{7}}{5} t} = \frac{20}{70}$$
$$\frac{\ln \frac{5}{7}}{5} t = \ln \frac{2}{7}$$
$$t = \frac{5}{\ln \frac{5}{7}} \ln \frac{2}{7} \approx 18.6 \text{ minutes}$$

The temperature of the object will be 50°C after about 18.6 minutes or 18 minutes, 37 seconds.

(b) Figure 55 shows the graph of $y = 30 + 70e^{\frac{\ln \frac{5}{7}}{5} x}$, where y is the temperature and x is the time.

Figure 55

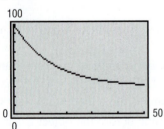

(c) By graphing $Y_1 = 50$ and $Y_2 = 30 + 70e^{\frac{\ln \frac{5}{7}}{5} x}$, where x is time, and using INTERSECT, we find that it takes $x = 18.6$ minutes (18 minutes, 37 seconds) for the temperature to cool to 50°C.

(d) By graphing $Y_1 = 35$ and $Y_2 = 30 + 70e^{\frac{\ln \frac{5}{7}}{5} x}$, where x is time, and using INTERSECT, we find that it takes $x = 39.22$ minutes (39 minutes, 13 seconds) for the temperature to cool to 35°C.

(e) As x increases, the value of $e^{\frac{\ln \frac{5}{7}}{5} x}$ approaches zero, so the value of y, the temperature of the object, approaches 30°C, the air temperature of the room. ◀

NOW WORK PROBLEM **13**.

4 Use Logistic Models

The exponential growth model $A(t) = A_0 e^{kt}$, $k > 0$, assumes uninhibited growth, meaning that the value of the function grows without limit. Recall that we stated that cell division could be modeled using this function, assuming that no cells die and no by-products are produced. However, cell division would eventually be limited by factors such as living space and food supply. The **logistic model** can describe situations where the growth or decay of the dependent variable is limited. The logistic model is given next.

Logistic Model

In a logistic growth model, the population P after time t obeys the equation

$$P(t) = \frac{c}{1 + ae^{-bt}} \qquad \text{(5)}$$

where a, b, and c are constants with $c > 0$. The model is a growth model if $b > 0$; the model is a decay model if $b < 0$.

The number c is called the **carrying capacity** (for growth models) because the value $P(t)$ approaches c as t approaches infinity; that is, $\lim\limits_{t \to \infty} P(t) = c$. The number $|b|$ is the growth rate for $b > 0$ and the decay rate for $b < 0$. Figure 56(a) shows the graph of a typical logistic growth function, and Figure 56(b) shows the graph of a typical logistic decay function.

Figure 56

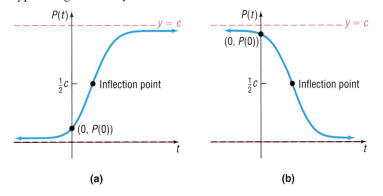

(a)   (b)

Based on the figures, we have the following properties of logistic growth functions.

Properties of the Logistic Growth Function, Equation (5)

1. The domain is the set of all real numbers. The range is the interval $(0, c)$, where c is the carrying capacity.

2. There are no x-intercepts; the y-intercept is $P(0)$.

3. There are two horizontal asymptotes: $y = 0$ and $y = c$.

4. $P(t)$ is an increasing function if $b > 0$ and a decreasing function if $b < 0$.

5. There is an **inflection point** where $P(t)$ equals $\frac{1}{2}$ of the carrying capacity. The inflection point is the point on the graph where the graph changes from being curved upward to curved downward for growth functions and the point where the graph changes from being curved downward to curved upward for decay functions.

6. The graph is smooth and continuous, with no corners or gaps.

| **EXAMPLE 5** | **Fruit Fly Population** |

Fruit flies are placed in a half-pint milk bottle with a banana (for food) and yeast plants (for food and to provide a stimulus to lay eggs). Suppose that the fruit fly population after t days is given by

$$P(t) = \frac{230}{1 + 56.5e^{-0.37t}}$$

(a) State the carrying capacity and the growth rate.

(b) Determine the initial population.

(c) Use a graphing utility to graph $P(t)$.

(d) What is the population after 5 days?

(e) How long does it take for the population to reach 180?

(f) How long does it take for the population to reach one-half of the carrying capacity?

Solution

(a) As $t \to \infty$, $e^{-0.37t} \to 0$ and $P(t) \to 230/1$. The carrying capacity of the half-pint bottle is 230 fruit flies. The growth rate is $|b| = |0.37| = 37\%$.

(b) To find the initial number of fruit flies in the half-pint bottle, we evaluate $P(0)$.

$$P(0) = \frac{230}{1 + 56.5e^{-0.37(0)}}$$

$$= \frac{230}{1 + 56.5}$$

$$= 4$$

So initially there were 4 fruit flies in the half-pint bottle.

(c) See Figure 57 for the graph of $P(t)$.

Figure 57

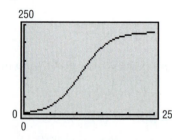

(d) To find the number of fruit flies in the half-pint bottle after 5 days, we evaluate $P(5)$.

$$P(5) = \frac{230}{1 + 56.5e^{-0.37(5)}} \approx 23 \text{ fruit flies}$$

After 5 days, there are approximately 23 fruit flies in the bottle.

(e) To determine when the population of fruit flies will be 180, we solve the equation

$$\frac{230}{1 + 56.5e^{-0.37t}} = 180$$

$$230 = 180(1 + 56.5e^{-0.37t})$$

$$1.2778 = 1 + 56.5e^{-0.37t} \qquad \text{Divide both sides by 180.}$$

$$0.2778 = 56.5e^{-0.37t} \qquad \text{Subtract 1 from both sides.}$$

$$0.0049 = e^{-0.37t} \qquad \text{Divide both sides by 56.5.}$$

$$\ln(0.0049) = -0.37t \qquad \text{Rewrite as a logarithmic expression.}$$

$$t \approx 14.4 \text{ days} \qquad \text{Divide both sides by } -0.37.$$

It will take approximately 14.4 days (14 days, 9 hours) for the population to reach 180 fruit flies.

We could also solve this problem by graphing $Y_1 = \dfrac{230}{1 + 56.5e^{-0.37t}}$ and $Y_2 = 180$ and using INTERSECT. See Figure 58.

(f) One-half of the carrying capacity is 115 fruit flies. We solve $P(t) = 115$ by graphing $Y_1 = \dfrac{230}{1 + 56.5e^{-0.37t}}$ and $Y_2 = 115$ and using INTERSECT. See Figure 59. The population will reach one-half of the carrying capacity in about 10.9 days (10 days, 22 hours).

Figure 58

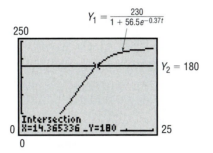

Figure 59

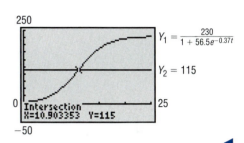

Exploration

On the same viewing rectangle, graph

$$Y_1 = \frac{500}{1 + 24e^{-0.03t}} \text{ and } Y_2 = \frac{500}{1 + 24e^{-0.08t}}.$$

What effect does the growth rate $|b|$ have on the logistic growth function?

Look back at Figure 59. Notice the point where the graph reaches 115 fruit flies (one-half of the carrying capacity): the graph changes from being curved upward to being curved downward. Using the language of calculus, we say the graph changes from increasing at an increasing rate to increasing at a decreasing rate. For any logistic growth function, when the population reaches one-half the carrying capacity, the population growth starts to slow down.

NOW WORK PROBLEM 21.

| **EXAMPLE 6** | **Wood Products** |

The EFISCEN wood product model classifies wood products according to their life-span. There are four classifications; short (1 year), medium short (4 years), medium long (16 years), and long (50 years). Based on data obtained from the European Forest Institute, the percentage of remaining wood products after t years for wood products with long life-spans (such as those used in the building industry) is given by

$$P(t) = \frac{100.3952}{1 + 0.0316e^{0.0581t}}$$

(a) What is the decay rate?

(b) Use a graphing utility to graph $P(t)$.

(c) What is the percentage of remaining wood products after 10 years?

(d) How long does it take for the percentage of remaining wood products to reach 50 percent?

(e) Explain why the numerator given in the model is reasonable.

Solution

(a) The decay rate is $|b| = |-0.0581| = 5.81\%$.

(b) The graph of $P(t)$ is given in Figure 60.

Figure 60

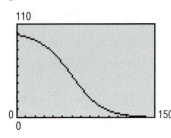

(c) Evaluate $P(10)$.

$$P(10) = \frac{100.3952}{1 + 0.0316e^{0.0581(10)}} \approx 95.0$$

So 95% of wood products remain after 10 years.

(d) Solve the equation $P(t) = 50$.

$$\frac{100.3952}{1 + 0.0316e^{0.0581t}} = 50$$

$100.3952 = 50(1 + 0.0316e^{0.0581t})$	
$2.0079 = 1 + 0.0316e^{0.0581t}$	Divide both sides by 50.
$1.0079 = 0.0316e^{0.0581t}$	Subtract 1 from both sides.
$31.8956 = e^{0.0581t}$	Divide both sides by 0.0316.
$\ln(31.8956) = 0.0581t$	Rewrite as a logarithmic expression.
$t \approx 59.6$ years	Divide both sides by 0.0581.

It will take approximately 59.6 years for the percentage of wood products remaining to reach 50%.

(e) The numerator of 100.3952 is reasonable because the maximum percentage of wood products remaining that is possible is 100%. ◄

 NOW WORK PROBLEM 27.

4.8 Assess Your Understanding

Applications and Extensions

1. **Growth of an Insect Population** The size P of a certain insect population at time t (in days) obeys the function $P(t) = 500e^{0.02t}$.
   (a) Determine the number of insects at $t = 0$ days.
   (b) What is the growth rate of the insect population?
   (c) Graph the function using a graphing utility.
   (d) What is the population after 10 days?
   (e) When will the insect population reach 800?
   (f) When will the insect population double?

2. **Growth of Bacteria** The number N of bacteria present in a culture at time t (in hours) obeys the law of uninhibited growth $N(t) = 1000e^{0.01t}$.
   (a) Determine the number of bacteria at $t = 0$ hours.
   (b) What is the growth rate of the bacteria?
   (c) Graph the function using a graphing utility.
   (d) What is the population after 4 hours?
   (e) When will the number of bacteria reach 1700?
   (f) When will the number of bacteria double?

3. **Radioactive Decay** Strontium 90 is a radioactive material that decays according to the function $A(t) = A_0e^{-0.0244t}$, where A_0 is the initial amount present and A is the amount present at time t (in years). Assume that a scientist has a sample of 500 grams of strontium 90.
   (a) What is the decay rate of strontium 90?

(b) Graph the function using a graphing utility.
(c) How much strontium 90 is left after 10 years?
(d) When will 400 grams of strontium 90 be left?
(e) What is the half-life of strontium 90?

4. **Radioactive Decay** Iodine 131 is a radioactive material that decays according to the function $A(t) = A_0e^{-0.087t}$, where A_0 is the initial amount present and A is the amount present at time t (in days). Assume that a scientist has a sample of 100 grams of iodine 131.
   (a) What is the decay rate of iodine 131?
   (b) Graph the function using a graphing utility.
   (c) How much iodine 131 is left after 9 days?
   (d) When will 70 grams of iodine 131 be left?
   (e) What is the half-life of iodine 131?

5. **Growth of a Colony of Mosquitoes** The population of a colony of mosquitoes obeys the law of uninhibited growth. If there are 1000 mosquitoes initially and there are 1800 after 1 day, what is the size of the colony after 3 days? How long is it until there are 10,000 mosquitoes?

6. **Bacterial Growth** A culture of bacteria obeys the law of uninhibited growth. If 500 bacteria are present initially and there are 800 after 1 hour, how many will be present in the culture after 5 hours? How long is it until there are 20,000 bacteria?

7. **Population Growth**   The population of a southern city follows the exponential law. If the population doubled in size over an 18-month period and the current population is 10,000, what will the population be 2 years from now?

8. **Population Decline**   The population of a midwestern city follows the exponential law. If the population decreased from 900,000 to 800,000 from 2003 to 2005, what will the population be in 2007?

9. **Radioactive Decay**   The half-life of radium is 1690 years. If 10 grams are present now, how much will be present in 50 years?

10. **Radioactive Decay**   The half-life of radioactive potassium is 1.3 billion years. If 10 grams are present now, how much will be present in 100 years? In 1000 years?

11. **Estimating the Age of a Tree**   A piece of charcoal is found to contain 30% of the carbon 14 that it originally had.
    (a) When did the tree from which the charcoal came die? Use 5600 years as the half-life of carbon 14.
    (b) Using a graphing utility, graph the relation between the percentage of carbon 14 remaining and time.
    (c) Using INTERSECT, determine the time that elapses until half of the carbon 14 remains.
    (d) Verify the answer found in part (a).

12. **Estimating the Age of a Fossil**   A fossilized leaf contains 70% of its normal amount of carbon 14.
    (a) How old is the fossil?
    (b) Using a graphing utility, graph the relation between the percentage of carbon 14 remaining and time.
    (c) Using INTERSECT, determine the time that elapses until half of the carbon 14 remains.
    (d) Verify the answer found in part (a).

13. **Cooling Time of a Pizza**   A pizza baked at 450°F is removed from the oven at 5:00 PM into a room that is a constant 70°F. After 5 minutes, the pizza is at 300°F.
    (a) At what time can you begin eating the pizza if you want its temperature to be 135°F?
    (b) Using a graphing utility, graph the relation between temperature and time.
    (c) Using INTERSECT, determine the time that needs to elapse before the pizza is 160°F.
    (d) TRACE the function for large values of time. What do you notice about y, the temperature?

14. **Newton's Law of Cooling**   A thermometer reading 72°F is placed in a refrigerator where the temperature is a constant 38°F.
    (a) If the thermometer reads 60°F after 2 minutes, what will it read after 7 minutes?

    (b) How long will it take before the thermometer reads 39°F?
    (c) Using a graphing utility, graph the relation between temperature and time.
    (d) Using INTERSECT, determine the time needed to elapse before the thermometer reads 45°F.
    (e) TRACE the function for large values of time. What do you notice about y, the temperature?

15. **Newton's Law of Heating**   A thermometer reading 8°C is brought into a room with a constant temperature of 35°C.
    (a) If the thermometer reads 15°C after 3 minutes, what will it read after being in the room for 5 minutes? For 10 minutes?
    (b) Graph the relation between temperature and time. TRACE to verify that your answers are correct.

    [**Hint:** You need to construct a formula similar to equation (4).]

16. **Thawing Time of a Steak**   A frozen steak has a temperature of 28°F. It is placed in a room with a constant temperature of 70°F. After 10 minutes, the temperature of the steak has risen to 35°F. What will the temperature of the steak be after 30 minutes? How long will it take the steak to thaw to a temperature of 45°F? [See the hint given for Problem 15.] Graph the relation between temperature and time. TRACE to verify that your answer is correct.

17. **Decomposition of Salt in Water**   Salt (NaCl) decomposes in water into sodium (NA^+) and chloride (Cl^-) ions according to the law of uninhibited decay. If the initial amount of salt is 25 kilograms and, after 10 hours, 15 kilograms of salt is left, how much salt is left after 1 day? How long does it take until $\frac{1}{2}$ kilogram of salt is left?

18. **Voltage of a Conductor**   The voltage of a certain conductor decreases over time according to the law of uninhibited decay. If the initial voltage is 40 volts, and 2 seconds later it is 10 volts, what is the voltage after 5 seconds?

19. **Radioactivity from Chernobyl**   After the release of radioactive material into the atmosphere from a nuclear power plant at Chernobyl (Ukraine) in 1986, the hay in Austria was contaminated by iodine 131 (half-life 8 days). If it is all right to feed the hay to cows when 10% of the iodine 131 remains, how long do the farmers need to wait to use this hay?

20. **Pig Roasts**   The hotel Bora-Bora is having a pig roast. At noon, the chef put the pig in a large earthen oven. The pig's original temperature was 75°F. At 2:00 PM the chef checked the pig's temperature and was upset because it had reached only 100°F. If the oven's temperature remains a constant 325°F, at what time may the hotel serve its guests, assuming that pork is done when it reaches 175°F?

21. Proportion of the Population That Owns a VCR The logistic growth model

$$P(t) = \frac{0.9}{1 + 6e^{-0.32t}}$$

relates the proportion of U.S. households that own a VCR to the year. Let $t = 0$ represent 1984, $t = 1$ represent 1985, and so on.

(a) Determine the maximum proportion of households that will own a VCR.

(b) What proportion of households owned a VCR in 1984 ($t = 0$)?

(c) Use a graphing utility to graph $P(t)$.

(d) What proportion of households owned a VCR in 1999 ($t = 15$)?

(e) When will 0.8 (80%) of U.S. households own a VCR?

(f) How long will it be before 0.45 (45%) of the population owns a VCR?

22. Market Penetration of Intel's Coprocessor The logistic growth model

$$P(t) = \frac{0.90}{1 + 3.5e^{-0.339t}}$$

relates the proportion of new personal computers (PCs) sold at Best Buy that have Intel's latest coprocessor t months after it has been introduced.

(a) Determine the maximum proportion of PCs sold at Best Buy that will have Intel's latest coprocessor.

(b) What proportion of computers sold at Best Buy will have Intel's latest coprocessor when it is first introduced ($t = 0$)?

(c) Use a graphing utility to graph $P(t)$.

(d) What proportion of PCs sold will have Intel's latest coprocessor $t = 4$ months after it is introduced?

(e) When will 0.75 (75%) of PCs sold at Best Buy have Intel's latest coprocessor?

(f) How long will it be before 0.45 (45%) of the PCs sold by Best Buy have Intel's latest coprocessor?

23. Population of a Bacteria Culture The logistic growth model

$$P(t) = \frac{1000}{1 + 32.33e^{-0.439t}}$$

represents the population (in grams) of a bacterium after t hours.

(a) Determine the carrying capacity of the environment.

(b) What is the growth rate of the bacteria?

(c) Determine the initial population size.

(d) Use a graphing utility to graph $P(t)$.

(e) What is the population after 9 hours?

(f) When will the population be 700 grams?

(g) How long does it take for the population to reach one-half of the carrying capacity?

24. Population of an Endangered Species Often environmentalists capture an endangered species and transport the species to a controlled environment where the species can produce offspring and regenerate its population. Suppose that six American bald eagles are captured, transported to Montana, and set free. Based on experience, the environmentalists expect the population to grow according to the model

$$P(t) = \frac{500}{1 + 83.33e^{-0.162t}}$$

where t is measured in years.

(a) Determine the carrying capacity of the environment.

(b) What is the growth rate of the bald eagle?

(c) Use a graphing utility to graph $P(t)$.

(d) What is the population after 3 years?

(e) When will the population be 300 eagles?

(f) How long does it take for the population to reach one-half of the carrying capacity?

25. The *Challenger* Disaster After the *Challenger* disaster in 1986, a study was made of the 23 launches that preceded the fatal flight. A mathematical model was developed involving the relationship between the Fahrenheit temperature x around the O-rings and the number y of eroded or leaky primary O-rings. The model stated that

$$y = \frac{6}{1 + e^{-(5.085 - 0.1156x)}}$$

where the number 6 indicates the 6 primary O-rings on the spacecraft.

(a) What is the predicted number of eroded or leaky primary O-rings at a temperature of 100°F?

(b) What is the predicted number of eroded or leaky primary O-rings at a temperature of 60°F?

(c) What is the predicted number of eroded or leaky primary O-rings at a temperature of 30°F?

(d) Graph the equation. At what temperature is the predicted number of eroded or leaky O-rings 1? 3? 5?

SOURCE: Linda Tappin, "Analyzing Data Relating to the *Challenger* Disaster," *Mathematics Teacher*, Vol. 87, No. 6, September 1994, pp. 423–426.

26. Word Users According to a survey by Olsten Staffing Services, the percentage of companies reporting usage of Microsoft Word t years since 1984 is given by

$$P(t) = \frac{99.744}{1 + 3.014e^{-0.799t}}$$

(a) What is the growth rate in the percentage of Microsoft Word users?
(b) Use a graphing utility to graph $P(t)$.
(c) What was the percentage of Microsoft Word users in 1990?
(d) During what year did the percentage of Microsoft Word users reach 90%?
(e) Explain why the numerator given in the model is reasonable? What does it imply?

27. Home Computers The logistic model

$$P(t) = \frac{95.4993}{1 + 0.0405e^{0.1968t}}$$

represents the percentage of households that do not own a personal computer t years since 1984.

(a) Evaluate and interpret $P(0)$.
(b) Use a graphing utility to graph $P(t)$.
(c) What percentage of households did not own a personal computer in 1995?
(d) In what year will the percentage of households that do not own a personal computer reach 20%?

SOURCE: U.S. Department of Commerce

28. Farmers The logistic model

$$W(t) = \frac{14,656.248}{1 + 0.059e^{0.057t}}$$

represents the number of farm workers in the United States t years after 1910.

(a) Evaluate and interpret $W(0)$.
(b) Use a graphing utility to graph $W(t)$.
(c) How many farm workers were there in the United States in 1990?

(d) When did the number of farm workers in the United States reach 10,000?
(e) According to this model, what happens to the number of farm workers in the United States as t approaches ∞? Based on this result, do you think that it is reasonable to use this model to predict the number of farm workers in the United States in 2060? Why?

SOURCE: U.S. Department of Agriculture

29. Birthdays The logistic model

$$P(n) = \frac{113.3198}{1 + 0.115e^{0.0912n}}$$

models the probability that, in a room of n people, no two people share the same birthday.

(a) Use a graphing utility to graph $P(n)$.
(b) In a room of $n = 15$ people, what is the probability that no two share the same birthday?
(c) How many people must be in a room before the probability that no two people share the same birthday is 10%?
(d) What happens to the probability as n increases? Explain what this result means.

30. Bank Failures The logistic model

$$F(t) = \frac{130.118}{1 + 0.00022e^{3.042t}}$$

models the number of bank failures in the United States t years since 1991.

(a) Evaluate and interpret $F(0)$.
(b) Use a graphing utility to graph $F(t)$.
(c) How many bank failures were there in 1995 ($t = 4$)?
(d) In what year did the number of bank failures reach 10?
(e) What does the model imply will happen to the number of bank failures as time passes?

SOURCE: Federal Deposit Insurance Corporation

4.9 Building Exponential, Logarithmic, and Logistic Models from Data

PREPARING FOR THIS SECTION *Before getting started, review the following:*

• Scatter Diagrams; Linear Curve Fitting (Section 2.4, pp. 96–100)
• Quadratic Functions of Best Fit (Section 3.1, pp. 162–163)

OBJECTIVES 1 Use a Graphing Utility to Fit an Exponential Function to Data
2 Use a Graphing Utility to Fit a Logarithmic Function to Data
3 Use a Graphing Utility to Fit a Logistic Function to Data

In Section 2.4 we discussed how to find the linear function of best fit ($y = ax + b$), and in Section 3.1 we discussed how to find the quadratic function of best fit ($y = ax^2 + bx + c$).

In this section we will discuss how to use a graphing utility to find equations of best fit that describe the relation between two variables when the relation is thought to be exponential $(y = ab^x)$, logarithmic $(y = a + b \ln x)$, or logistic $\left(y = \dfrac{c}{1 + ae^{-bx}}\right)$. As before, we draw a scatter diagram of the data to help to determine the appropriate model to use.

Figure 61 shows scatter diagrams that will typically be observed for the three models. Below each scatter diagram are any restrictions on the values of the parameters.

Figure 61

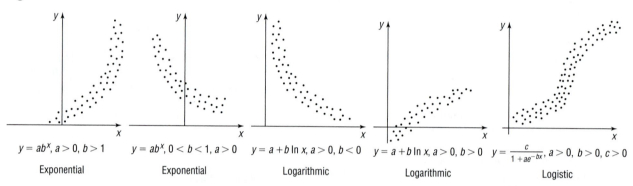

$y = ab^x, a > 0, b > 1$	$y = ab^x, 0 < b < 1, a > 0$	$y = a + b \ln x, a > 0, b < 0$	$y = a + b \ln x, a > 0, b > 0$	$y = \dfrac{c}{1 + ae^{-bx}}, a > 0, b > 0, c > 0$
Exponential	Exponential	Logarithmic	Logarithmic	Logistic

Most graphing utilities have REGression options that fit data to a specific type of curve. Once the data have been entered and a scatter diagram obtained, the type of curve that you want to fit to the data is selected. Then that REGression option is used to obtain the curve of *best fit* of the type selected.

The correlation coefficient r will appear only if the model can be written as a linear expression. As it turns out, r will appear for the linear, power, exponential, and logarithmic models, since these models can be written as a linear expression. Remember, the closer $|r|$ is to 1, the better the fit.

Let's look at some examples.

1 Use a Graphing Utility to Fit an Exponential Function to Data

We saw in Section 4.7 that the future value of money behaves exponentially, and we saw in Section 4.8 that growth and decay models also behave exponentially. The next example shows how data can lead to an exponential model.

EXAMPLE 1 | **Fitting an Exponential Function to Data**

Beth is interested in finding a function that explains the closing price of Harley Davidson stock at the end of each year. She obtains the data shown in Table 10.

(a) Using a graphing utility, draw a scatter diagram with year as the independent variable.

(b) Using a graphing utility, fit an exponential function to the data.

(c) Express the function found in part (b) in the form $A = A_0 e^{kt}$.

(d) Graph the exponential function found in part (b) or (c) on the scatter diagram.

(e) Using the solution to part (b) or (c), predict the closing price of Harley Davidson stock at the end of 2004.

(f) Interpret the value of k found in part (c).

Table 10

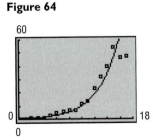

Year, x	Closing Price, y
1987 (x = 1)	0.392
1988 (x = 2)	0.7652
1989 (x = 3)	1.1835
1990 (x = 4)	1.1609
1991 (x = 5)	2.6988
1992 (x = 6)	4.5381
1993 (x = 7)	5.3379
1994 (x = 8)	6.8032
1995 (x = 9)	7.0328
1996 (x = 10)	11.5585
1997 (x = 11)	13.4799
1998 (x = 12)	23.5424
1999 (x = 13)	31.9342
2000 (x = 14)	39.7277
2001 (x = 15)	54.31
2002 (x = 16)	46.20
2003 (x = 17)	47.53

SOURCE: *http://finance.yahoo.com*

Solution

(a) Enter the data into the graphing utility, letting 1 represent 1987, 2 represent 1988, and so on. We obtain the scatter diagram shown in Figure 62.

(b) A graphing utility fits the data in Figure 62 to an exponential function of the form $y = ab^x$ by using the EXPonential REGression option. From Figure 63 we find $y = ab^x = 0.47547(1.3582)^x$.

Figure 62

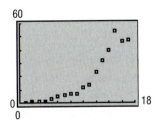

Figure 63

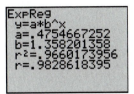

(c) To express $y = ab^x$ in the form $A = A_0 e^{kt}$, where $x = t$ and $y = A$, we proceed as follows:

$$ab^x = A_0 e^{kt}, \qquad x = t$$

When $x = t = 0$, we find $a = A_0$. This leads to

$$a = A_0, \qquad b^x = e^{kt}$$
$$b^x = (e^k)^t$$
$$b = e^k \qquad x = t$$

Since $y = ab^x = 0.47547(1.3582)^x$, we find that $a = 0.47547$ and $b = 1.3582$:

$$a = A_0 = 0.47547 \quad \text{and} \quad b = 1.3582 = e^k$$

We want to find k, so we rewrite $1.3582 = e^k$ as a logarithm and obtain

$$k = \ln(1.3582) \approx 0.3062$$

As a result, $A = A_0 e^{kt} = 0.47547 e^{0.3062t}$.

(d) See Figure 64 for the graph of the exponential function of best fit.

(e) Let $t = 18$ (end of 2004) in the function found in part (c). The predicted closing price of Harley Davidson stock at the end of 2004 is

$$A = 0.47547 e^{0.3062(18)} = \$117.70$$

(f) The value of k represents the annual interest rate compounded continuously.

$$A = A_0 e^{kt} = 0.47547 e^{0.3062t}$$
$$= Pe^{rt} \qquad \text{Equation (4), Section 4.7}$$

The price of Harley Davidson stock has grown at an annual rate of 30.62% (compounded continuously) between 1987 and 2003. ◀

Figure 64

60

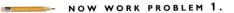

0   18
0

NOW WORK PROBLEM **1**.

2 Use a Graphing Utility to Fit a Logarithmic Function to Data

Many relations between variables do not follow an exponential model; instead, the independent variable is related to the dependent variable using a logarithmic model.

EXAMPLE 2	**Fitting a Logarithmic Function to Data**

Jodi, a meteorologist, is interested in finding a function that explains the relation between the height of a weather balloon (in kilometers) and the atmospheric pressure (measured in millimeters of mercury) on the balloon. She collects the data shown in Table 11.

Table 11

Atmospheric Pressure, p	Height, h
760	0
740	0.184
725	0.328
700	0.565
650	1.079
630	1.291
600	1.634
580	1.862
550	2.235

(a) Using a graphing utility, draw a scatter diagram of the data with atmospheric pressure as the independent variable.

(b) It is known that the relation between atmospheric pressure and height follows a logarithmic model. Using a graphing utility, fit a logarithmic function to the data.

(c) Draw the logarithmic function found in part (b) on the scatter diagram.

(d) Use the function found in part (b) to predict the height of the weather balloon if the atmospheric pressure is 560 millimeters of mercury.

Solution

(a) After entering the data into the graphing utility, we obtain the scatter diagram shown in Figure 65.

(b) A graphing utility fits the data in Figure 65 to a logarithmic function of the form $y = a + b \ln x$ by using the Logarithm REGression option. See Figure 66. The logarithmic function of best fit to the data is

$$h(p) = 45.7863 - 6.9025 \ln p$$

where h is the height of the weather balloon and p is the atmospheric pressure. Notice that $|r|$ is close to 1, indicating a good fit.

(c) Figure 67 shows the graph of $h(p) = 45.7863 - 6.9025 \ln p$ on the scatter diagram.

Figure 65

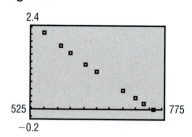

Figure 66

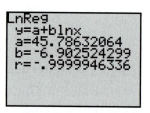

Figure 67

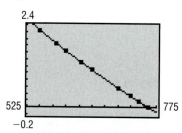

(d) Using the function found in part (b), Jodi predicts the height of the weather balloon when the atmospheric pressure is 560 to be

$$h(560) = 45.7863 - 6.9025 \ln 560$$
$$\approx 2.108 \text{ kilometers} \qquad \blacktriangleleft$$

 NOW WORK PROBLEM 7.

3 Use a Graphing Utility to Fit a Logistic Function to Data

Logistic growth models can be used to model situations for which the value of the dependent variable is limited. Many real-world situations conform to this scenario. For example, the population of the human race is limited by the availability of natural resources such as food and shelter. When the value of the dependent variable is limited, a logistic growth model is often appropriate.

| EXAMPLE 3 | **Fitting a Logistic Function to Data** |

The data in Table 12 represent the amount of yeast biomass in a culture after t hours.

Table 12

Time (in hours)	Yeast Biomass	Time (in hours)	Yeast Biomass
0	9.6	10	513.3
1	18.3	11	559.7
2	29.0	12	594.8
3	47.2	13	629.4
4	71.1	14	640.8
5	119.1	15	651.1
6	174.6	16	655.9
7	257.3	17	659.6
8	350.7	18	661.8
9	441.0		

SOURCE: Tor Carlson (Über Geschwindigkeit und Grösse der Hefevermehrung in Würze, Biochemische Zeitschrift, Bd. 57, pp. 313–334, 1913)

(a) Using a graphing utility, draw a scatter diagram of the data with time as the independent variable.
(b) Using a graphing utility, fit a logistic function to the data.
(c) Using a graphing utility, graph the function found in part (b) on the scatter diagram.
(d) What is the predicted carrying capacity of the culture?
(e) Use the function found in part (b) to predict the population of the culture at $t = 19$ hours.

Solution
(a) See Figure 68 for a scatter diagram of the data.
(b) A graphing utility fits a logistic growth model of the form $y = \dfrac{c}{1 + ae^{-bx}}$ by using the LOGISTIC regression option. See Figure 69. The logistic function of best fit to the data is

$$y = \frac{663.0}{1 + 71.6e^{-0.5470x}}$$

where y is the amount of yeast biomass in the culture and x is the time.

Figure 68

Figure 69

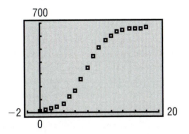

Figure 70

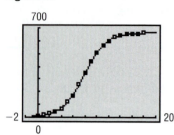

(c) See Figure 70 for the graph of the logistic function of best fit.

(d) Based on the logistic growth function found in part (b), the carrying capacity of the culture is 663.

(e) Using the logistic growth function found in part (b), the predicted amount of yeast biomass at $t = 19$ hours is

$$y = \frac{663.0}{1 + 71.6e^{-0.5470(19)}} = 661.5$$

◄

NOW WORK PROBLEM 9.

4.9 Assess Your Understanding

Applications and Extensions

1. **Biology**   A strain of E-coli Beu 397-recA441 is placed into a nutrient broth at 30°Celsius and allowed to grow. The following data are collected. Theory states that the number of bacteria in the petri dish will initially grow according to the law of uninhibited growth. The population is measured using an optical device in which the amount of light that passes through the petri dish is measured.

Time (hours), x	Population, y
0	0.09
2.5	0.18
3.5	0.26
4.5	0.35
6	0.50

SOURCE: Dr. Polly Lavery, Joliet Junior College

(a) Draw a scatter diagram treating time as the predictor variable.

(b) Using a graphing utility, fit an exponential function to the data.

(c) Express the function found in part (b) in the form $N(t) = N_0 e^{kt}$.

(d) Graph the exponential function found in part (b) or (c) on the scatter diagram.

(e) Use the exponential function from part (b) or (c) to predict the population at $x = 7$ hours.

(f) Use the exponential function from part (b) or (c) to predict when the population will reach 0.75.

2. **Biology**   A strain of E-coli SC18del-recA718 is placed into a nutrient broth at 30°Celsius and allowed to grow. The following data are collected. Theory states that the number of bacteria in the petri dish will initially grow according to the law of uninhibited growth. The population is measured using an optical device in which the amount of light that passes through the petri dish is measured.

Time (hours), x	Population, y
2.5	0.175
3.5	0.38
4.5	0.63
4.75	0.76
5.25	1.20

SOURCE: Dr. Polly Lavery, Joliet Junior College

(a) Draw a scatter diagram treating time as the predictor variable.

(b) Using a graphing utility, fit an exponential function to the data.

(c) Express the function found in part (b) in the form $N(t) = N_0 e^{kt}$.

(d) Graph the exponential function found in part (b) or (c) on the scatter diagram.

(e) Use the exponential function from part (b) or (c) to predict the population at $x = 6$ hours.

(f) Use the exponential function from part (b) or (c) to predict when the population will reach 2.1.

3. Chemistry   A chemist has a 100-gram sample of a radioactive material. He records the amount of radioactive material every week for 6 weeks and obtains the following data:

Week	Weight (in Grams)
0	100.0
1	88.3
2	75.9
3	69.4
4	59.1
5	51.8
6	45.5

(a) Using a graphing utility, draw a scatter diagram with week as the independent variable.
(b) Using a graphing utility, fit an exponential function to the data.
(c) Express the function found in part (b) in the form $A(t) = A_0e^{kt}$.
(d) Graph the exponential function found in part (b) or (c) on the scatter diagram.
(e) From the result found in part (b), determine the half-life of the radioactive material.
(f) How much radioactive material will be left after 50 weeks?
(g) When will there be 20 grams of radioactive material?

4. Chemistry   A chemist has a 1000-gram sample of a radioactive material. She records the amount of radioactive material remaining in the sample every day for a week and obtains the following data:

Day	Weight (in Grams)
0	1000.0
1	897.1
2	802.5
3	719.8
4	651.1
5	583.4
6	521.7
7	468.3

(a) Using a graphing utility, draw a scatter diagram with day as the independent variable.
(b) Using a graphing utility, fit an exponential function to the data.
(c) Express the function found in part (b) in the form $A(t) = A_0e^{kt}$.
(d) Graph the exponential function found in part (b) or (c) on the scatter diagram.

(e) From the result found in part (b), find the half-life of the radioactive material.
(f) How much radioactive material will be left after 20 days?
(g) When will there be 200 grams of radioactive material?

5. Finance   The following data represent the amount of money an investor has in an investment account each year for 10 years. She wishes to determine the average annual rate of return on her investment.

Year	Value of Account
1994	$10,000
1995	$10,573
1996	$11,260
1997	$11,733
1998	$12,424
1999	$13,269
2000	$13,968
2001	$14,823
2002	$15,297
2003	$16,539

(a) Using a graphing utility, draw a scatter diagram with time as the independent variable and the value of the account as the dependent variable.
(b) Using a graphing utility, fit an exponential function to the data.
(c) Based on the answer in part (b), what was the average annual rate of return from this account over the past 10 years?
(d) If the investor plans on retiring in 2021, what will the predicted value of this account be?
(e) When will the account be worth $50,000?

6. Finance   The following data show the amount of money an investor has in an investment account each year for 7 years. He wishes to determine the average annual rate of return on his investment.

Year	Value of Account
1997	$20,000
1998	$21,516
1999	$23,355
2000	$24,885
2001	$27,434
2002	$30,053
2003	$32,622

(a) Using a graphing utility, draw a scatter diagram with time as the independent variable and the value of the account as the dependent variable.

(b) Using a graphing utility, fit an exponential function to the data.

(c) Based on the answer to part (b), what was the average annual rate of return from this account over the past 7 years?

(d) If the investor plans on retiring in 2020, what will the predicted value of his account be?

(e) When will the account be worth $80,000?

7. Economics and Marketing The following data represent the price and quantity demanded in 2005 for IBM personal computers.

Price ($/Computer)	Quantity Demanded
2300	152
2000	159
1700	164
1500	171
1300	176
1200	180
1000	189

(a) Using a graphing utility, draw a scatter diagram of the data with price as the dependent variable.

(b) Using a graphing utility, fit a logarithmic function to the data.

(c) Using a graphing utility, draw the logarithmic function found in part (b) on the scatter diagram.

(d) Use the function found in part (b) to predict the number of IBM personal computers that would be demanded if the price were $1650.

8. Economics and Marketing The following data represent the price and quantity supplied in 2005 for IBM personal computers.

Price ($/Computer)	Quantity Supplied
2300	180
2000	173
1700	160
1500	150
1300	137
1200	130
1000	113

(a) Using a graphing utility, draw a scatter diagram of the data with price as the dependent variable.

(b) Using a graphing utility, fit a logarithmic function to the data.

(c) Using a graphing utility, draw the logarithmic function found in part (b) on the scatter diagram.

(d) Use the function found in part (b) to predict the number of IBM personal computers that would be supplied if the price were $1650.

9. Population Model The following data represent the population of the United States. An ecologist is interested in finding a function that describes the population of the United States.

Year	Population
1900	76,212,168
1910	92,228,496
1920	106,021,537
1930	123,202,624
1940	132,164,569
1950	151,325,798
1960	179,323,175
1970	203,302,031
1980	226,542,203
1990	248,709,873
2000	281,421,906

SOURCE: U.S. Census Bureau

(a) Using a graphing utility, draw a scatter diagram of the data using the year as the independent variable and population as the dependent variable.

(b) Using a graphing utility, fit a logistic function to the data.

(c) Using a graphing utility, draw the function found in part (b) on the scatter diagram.

(d) Based on the function found in part (b), what is the carrying capacity of the United States?

(e) Use the function found in part (b) to predict the population of the United States in 2004.

(f) When will the United States population be 300,000,000?

(g) Compare actual U.S. Census figures to the prediction found in part (e).

10. Population Model The following data represent the world population. An ecologist is interested in finding a function that describes the world population.

Year	Population (in Billions)
1993	5.531
1994	5.611
1995	5.691
1996	5.769
1997	5.847
1998	5.925
1999	6.003
2000	6.080
2001	6.157

SOURCE: U.S. Census Bureau

(a) Using a graphing utility, draw a scatter diagram of the data using year as the independent variable and population as the dependent variable.
(b) Using a graphing utility, fit a logistic function to the data.
(c) Using a graphing utility, draw the function found in part (b) on the scatter diagram.
(d) Based on the function found in part (b), what is the carrying capacity of the world?
(e) Use the function found in part (b) to predict the population of the world in 2004.
(f) When will world population be 7 billion?
(g) Compare actual U.S. Census figures to the prediction found in part (e).

11. Population Model The following data represent the population of Illinois. An urban economist is interested in finding a model that describes the population of Illinois.

Year	Population
1900	4,821,550
1910	5,638,591
1920	6,485,280
1930	7,630,654
1940	7,897,241
1950	8,712,176
1960	10,081,158
1970	11,110,285
1980	11,427,409
1990	11,430,602
2000	12,419,293

SOURCE: U.S. Census Bureau

(a) Using a graphing utility, draw a scatter diagram of the data using year as the independent variable and population as the dependent variable.

(b) Using a graphing utility, fit a logistic function to the data.
(c) Using a graphing utility, draw the function found in part (b) on the scatter diagram.
(d) Based on the function found in part (b), what is the carrying capacity of Illinois?
(e) Use the function found in part (b) to predict the population of Illinois in 2010.

12. Population Model The following data represent the population of Pennsylvania. An urban economist is interested in finding a model that describes the population of Pennsylvania.

(a) Using a graphing utility, draw a scatter diagram of the data using year as the independent variable and population as the dependent variable.
(b) Using a graphing utility, fit a logistic function to the data.
(c) Using a graphing utility, draw the function found in part (b) on the scatter diagram.
(d) Based on the function found in part (b), what is the carrying capacity of Pennsylvania?
(e) Use the function found in part (b) to predict the population of Pennsylvania in 2010.

Year	Population
1900	6,302,115
1910	7,665,111
1920	8,720,017
1930	9,631,350
1940	9,900,180
1950	10,498,012
1960	11,319,366
1970	11,800,766
1980	11,864,720
1990	11,881,643
2000	12,281,054

SOURCE: U.S. Census Bureau

Chapter Review

Things to Know

Composite Function (p. 249) $(f \circ g)(x) = f(g(x))$

One-to-one function f (p. 257)   A function whose inverse is also a function
For any choice of elements x_1, x_2 in the domain of f, if $x_1 \neq x_2$, then $f(x_1) \neq f(x_2)$.

Horizontal-line test (p. 258)   If every horizontal line intersects the graph of a function f in at most one point, then f is one-to-one.

Inverse function f^{-1} of f (pp. 259–263)   Domain of f = Range of f^{-1}; Range of f = Domain of f^{-1}
$f^{-1}(f(x)) = x$ for all x in the domain of f and $f(f^{-1}(x)) = x$ for all x in the domain of f^{-1}.
Graphs of f and f^{-1} are symmetric with respect to the line $y = x$.

Properties of the exponential function (pp. 276 and 277)

$f(x) = a^x, \quad a > 1$ Domain: the interval $(-\infty, \infty)$

Range: the interval $(0, \infty)$

x-intercepts: none; y-intercept: 1

Horizontal asymptote: x-axis ($y = 0$) as $x \to -\infty$

Increasing; one-to-one; smooth; continuous

See Figure 24 for a typical graph.

$f(x) = a^x, \quad 0 < a < 1$ Domain: the interval $(-\infty, \infty)$

Range: the interval $(0, \infty)$

x-intercepts: none; y-intercept: 1

Horizontal asymptote: x-axis ($y = 0$) as $x \to \infty$

Decreasing; one-to-one; smooth; continuous

See Figure 28 for a typical graph.

Number e (p. 278)

Value approached by the expression $\left(1 + \dfrac{1}{n}\right)^n$ as $n \to \infty$; that is, $\displaystyle\lim_{n \to \infty}\left(1 + \dfrac{1}{n}\right)^n = e$

Property of exponents (p. 280)

If $a^u = a^v$, then $u = v$.

Properties of the logarithmic functions (pp. 287–291)

$f(x) = \log_a x, \quad a > 1$ Domain: the interval $(0, \infty)$

($y = \log_a x$ means $x = a^y$)   Range: the interval $(-\infty, \infty)$

x-intercept: 1; y-intercept; none

Vertical asymptote: $x = 0$ (y-axis)

Increasing; one-to-one; smooth; continuous

See Figure 35(b) for a typical graph.

$f(x) = \log_a x, \quad 0 < a < 1$ Domain: the interval $(0, \infty)$

($y = \log_a x$ means $x = a^y$)   Range: the interval $(-\infty, \infty)$

x-intercept: 1; y-intercept; none

Vertical asymptote: $x = 0$ (y-axis)

Decreasing; one-to-one; smooth; continuous

See Figure 35(a) for a typical graph.

Natural logarithm (p. 291)

$y = \ln x$ means $x = e^y$.

Properties of logarithms (pp. 301–302, 305)

$\log_a 1 = 0 \qquad \log_a a = 1 \qquad a^{\log_a M} = M \qquad \log_a a^r = r$

$\log_a(MN) = \log_a M + \log_a N \qquad \log_a\!\left(\dfrac{M}{N}\right) = \log_a M - \log_a N$

$\log_a M^r = r \log_a M$

If $M = N$, then $\log_a M = \log_a N$.

If $\log_a M = \log_a N$, then $M = N$.

Formulas

Change-of-Base Formula (p. 306)

$\log_a M = \dfrac{\log_b M}{\log_b a}$

Compound Interest Formula (p. 317)

$A = P \cdot \left(1 + \dfrac{r}{n}\right)^{nt}$

Continuous compounding (p. 318)

$A = Pe^{rt}$

Present Value Formulas (p. 320)

$P = A \cdot \left(1 + \dfrac{r}{n}\right)^{-nt}$ or $P = Ae^{-rt}$

Growth and decay (p. 325) $A(t) = A_0e^{kt}$

Newton's Law of Cooling (p. 329) $u(t) = T + (u_0 - T)e^{kt}, \quad k < 0$

Logistic model (p. 331) $P(t) = \dfrac{c}{1 + ae^{-bt}}$

Objectives

Section		You should be able to . . .	Review Exercises
4.1	1	Form a composite function (p. 248)	1–12
	2	Find the domain of a composite function (p. 250)	7–12
4.2	1	Determine whether a function is one-to-one (p. 256)	13(a), 14(a), 15, 16
	2	Determine the inverse of a function defined by a map on or an ordered pair (p. 259)	13(b), 14(b)
	3	Obtain the graph of the inverse function from the graph of the function (p. 262)	15, 16
	4	Find the inverse of a function defined by an equation (p. 263)	17–22
4.3	1	Evaluate exponential functions (p. 271)	23(a), (c); 24(a), (c), 87(a)
	2	Graph exponential functions (p. 274)	55–59, 62, 63
	3	Define the number e (p. 278)	59, 62, 63
	4	Solve exponential equations (p. 280)	65–68, 73, 74, 76, 77, 78
4.4	1	Change exponential expressions to logarithmic expressions and logarithmic expressions to exponential expressions (p. 288)	25–28
	2	Evaluate logarithmic expressions (p. 288)	23(b), (d), 24(b), (d), 33–34, 85, 86, 88(a), 89
	3	Determine the domain of a logarithmic function (p. 289)	29–32
	4	Graph logarithmic functions (p. 290)	60, 61, 64
	5	Solve logarithmic equations (p. 293)	69, 70, 75
4.5	1	Work with the properties of logarithms (p. 300)	35–38
	2	Write a logarithmic expression as a sum or difference of logarithms (p. 303)	39–44
	3	Write a logarithmic expression as a single logarithm (p. 304)	45–50
	4	Evaluate logarithms whose base is neither 10 nor e (p. 305)	51, 52
	5	Graph logarithmic functions whose base is neither 10 nor e (p. 306)	53, 54
4.6	1	Solve logarithmic equations using the properties of logarithms (p. 309)	79, 80
	2	Solve exponential equations (p. 311)	71, 72, 81–84
	3	Solve logarithmic and exponential equations using a graphing utility (p. 312)	65–84
4.7	1	Determine the future value of a lump sum of money (p. 315)	90
	2	Calculate effective rates of return (p. 319)	90
	3	Determine the present value of a lump sum of money (p. 320)	91
	4	Determine the time required to double or triple a lump sum of money (p. 321)	90
4.8	1	Find equations of populations that obey the law of uninhibited growth (p. 324)	95
	2	Find equations of populations that obey the law of decay (p. 327)	93, 96
	3	Use Newton's Law of Cooling (p. 329)	94
	4	Use logistic models (p. 330)	97
4.9	1	Use a graphing utility to fit an exponential function to data (p. 338)	98
	2	Use a graphing utility to fit a logarithmic function to data (p. 340)	99
	3	Use a graphing utility to fit a logistic function to data (p. 340)	100

Review Exercises

In Problems 1–6, for the given functions f and g find:

(a) $(f \circ g)(2)$ (b) $(g \circ f)(-2)$ (c) $(f \circ f)(4)$ (d) $(g \circ g)(-1)$

1. $f(x) = 3x - 5; \quad g(x) = 1 - 2x^2$

2. $f(x) = 4 - x; \quad g(x) = 1 + x^2$

3. $f(x) = \sqrt{x + 2}; \quad g(x) = 2x^2 + 1$

4. $f(x) = 1 - 3x^2; \quad g(x) = \sqrt{4 - x}$

5. $f(x) = e^x; \quad g(x) = 3x - 2$

6. $f(x) = \dfrac{2}{1 + 2x^2}; \quad g(x) = 3x$

In Problems 7–12, find $f \circ g$, $g \circ f$, $f \circ f$, and $g \circ g$ for each pair of functions. State the domain of each composite function.

7. $f(x) = 2 - x; \quad g(x) = 3x + 1$

8. $f(x) = 2x - 1; \quad g(x) = 2x + 1$

9. $f(x) = 3x^2 + x + 1; \quad g(x) = |3x|$

10. $f(x) = \sqrt{3x}; \quad g(x) = 1 + x + x^2$

11. $f(x) = \dfrac{x + 1}{x - 1}; \quad g(x) = \dfrac{1}{x}$

12. $f(x) = \sqrt{x - 3}; \quad g(x) = \dfrac{3}{x}$

In Problems 13 and 14, (a) verify that the function is one-to-one, and (b) find the inverse of the given function.

13. $\{(1, 2), (3, 5), (5, 8), (6, 10)\}$

14. $\{(-1, 4), (0, 2), (1, 5), (3, 7)\}$

In Problems 15 and 16, state why the graph of the function is one-to-one. Then draw the graph of the inverse function f^{-1}. For convenience (and as a hint), the graph of $y = x$ is also given.

15.

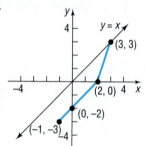

16.

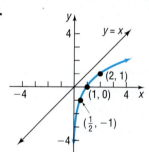

In Problems 17–22, the function f is one-to-one. Find the inverse of each function and check your answer. Find the domain and the range of f and f^{-1}.

17. $f(x) = \dfrac{2x + 3}{5x - 2}$

18. $f(x) = \dfrac{2 - x}{3 + x}$

19. $f(x) = \dfrac{1}{x - 1}$

20. $f(x) = \sqrt{x - 2}$

21. $f(x) = \dfrac{3}{x^{1/3}}$

22. $f(x) = x^{1/3} + 1$

In Problems 23 and 24, $f(x) = 3^x$ and $g(x) = \log_3 x$.

23. Evaluate: (a) $f(4)$ (b) $g(9)$ (c) $f(-2)$ (d) $g\left(\dfrac{1}{27}\right)$

24. Evaluate: (a) $f(1)$ (b) $g(81)$ (c) $f(-4)$ (d) $g\left(\dfrac{1}{243}\right)$

In Problems 25 and 26, convert each exponential expression to an equivalent expression involving a logarithm. In Problems 27 and 28, convert each logarithmic expression to an equivalent expression involving an exponent.

25. $5^2 = z$

26. $a^5 = m$

27. $\log_5 u = 13$

28. $\log_a 4 = 3$

In Problems 29–32, find the domain of each logarithmic function.

29. $f(x) = \log(3x - 2)$ **30.** $F(x) = \log_5(2x + 1)$ **31.** $H(x) = \log_2(x^2 - 3x + 2)$ **32.** $F(x) = \ln(x^2 - 9)$

In Problems 33–38, evaluate each expression. Do not use a calculator.

33. $\log_2\left(\dfrac{1}{8}\right)$ **34.** $\log_3 81$ **35.** $\ln e^{\sqrt{2}}$

36. $e^{\ln 0.1}$ **37.** $2^{\log_2 0.4}$ **38.** $\log_2 2^{\sqrt{3}}$

In Problems 39–44, write each expression as the sum and/or difference of logarithms. Express powers as factors.

39. $\log_3\left(\dfrac{uv^2}{w}\right)$, $u > 0, v > 0, w > 0$ **40.** $\log_2\left(a^2\sqrt{b}\right)^4$, $a > 0, b > 0$ **41.** $\log\left(x^2\sqrt{x^3 + 1}\right)$, $x > 0$

42. $\log_5\left(\dfrac{x^2 + 2x + 1}{x^2}\right)$, $x > 0$ **43.** $\ln\left(\dfrac{x\sqrt[3]{x^2 + 1}}{x - 3}\right)$, $x > 3$ **44.** $\ln\left(\dfrac{2x + 3}{x^2 - 3x + 2}\right)^2$, $x > 2$

In Problems 45–50, write each expression as a single logarithm.

45. $3\log_4 x^2 + \dfrac{1}{2}\log_4 \sqrt{x}$ **46.** $-2\log_3\left(\dfrac{1}{x}\right) + \dfrac{1}{3}\log_3 \sqrt{x}$

47. $\ln\left(\dfrac{x - 1}{x}\right) + \ln\left(\dfrac{x}{x + 1}\right) - \ln(x^2 - 1)$ **48.** $\log(x^2 - 9) - \log(x^2 + 7x + 12)$

49. $2\log 2 + 3\log x - \dfrac{1}{2}[\log(x + 3) + \log(x - 2)]$ **50.** $\dfrac{1}{2}\ln(x^2 + 1) - 4\ln\dfrac{1}{2} - \dfrac{1}{2}[\ln(x - 4) + \ln x]$

In Problems 51 and 52, use the Change-of-Base Formula and a calculator to evaluate each logarithm. Round your answer to three decimal places.

51. $\log_4 19$ **52.** $\log_2 21$

In Problems 53 and 54, graph each function using a graphing utility and the Change-of-Base Formula.

53. $y = \log_3 x$ **54.** $y = \log_7 x$

In Problems 55–64, use transformations to graph each function. Determine the domain, range, and any asymptotes.

55. $f(x) = 2^{x-3}$ **56.** $f(x) = -2^x + 3$ **57.** $f(x) = \dfrac{1}{2}(3^{-x})$ **58.** $f(x) = 1 + 3^{2x}$

59. $f(x) = 1 - e^x$ **60.** $f(x) = 3 + \ln x$ **61.** $f(x) = \dfrac{1}{2}\ln x$ **62.** $f(x) = 3e^x$

63. $f(x) = 3 - e^{-x}$ **64.** $f(x) = 4 - \ln(-x)$

In Problems 65–84, solve each equation.

65. $4^{1-2x} = 2$ **66.** $8^{6+3x} = 4$ **67.** $3^{x^2+x} = \sqrt{3}$ **68.** $4^{x-x^2} = \dfrac{1}{2}$

69. $\log_x 64 = -3$ **70.** $\log_{\sqrt{2}} x = -6$ **71.** $5^x = 3^{x+2}$ **72.** $5^{x+2} = 7^{x-2}$

73. $9^{2x} = 27^{3x-4}$ **74.** $25^{2x} = 5^{x^2-12}$ **75.** $\log_3 \sqrt{x - 2} = 2$ **76.** $2^{x+1} \cdot 8^{-x} = 4$

77. $8 = 4^{x^2} \cdot 2^{5x}$ **78.** $2^x \cdot 5 = 10^x$ **79.** $\log_6(x + 3) + \log_6(x + 4) = 1$

80. $\log(7x - 12) = 2\log x$ **81.** $e^{1-x} = 5$ **82.** $e^{1-2x} = 4$

83. $2^{3x} = 3^{2x+1}$ **84.** $2^{x^3} = 3^{x^2}$

In Problems 85 and 86, use the following result: If x is the atmospheric pressure (measured in millimeters of mercury), then the formula for the altitude $h(x)$ (measured in meters above sea level) is

$$h(x) = (30T + 8000) \log\left(\frac{P_0}{x}\right)$$

where T is the temperature (in degrees Celsius) and P_0 is the atmospheric pressure at sea level, which is approximately 760 millimeters of mercury.

85. Finding the Altitude of an Airplane At what height is a Piper Cub whose instruments record an outside temperature of 0°C and a barometric pressure of 300 millimeters of mercury?

86. Finding the Height of a Mountain How high is a mountain if instruments placed on its peak record a temperature of 5°C and a barometric pressure of 500 millimeters of mercury?

---

87. Amplifying Sound An amplifier's power output P (in watts) is related to its decibel voltage gain d by the formula $P = 25e^{0.1d}$.

(a) Find the power output for a decibel voltage gain of 4 decibels.
(b) For a power output of 50 watts, what is the decibel voltage gain?

88. Limiting Magnitude of a Telescope A telescope is limited in its usefulness by the brightness of the star that it is aimed at and by the diameter of its lens. One measure of a star's brightness is its *magnitude*; the dimmer the star, the larger its magnitude. A formula for the limiting magnitude L of a telescope, that is, the magnitude of the dimmest star that it can be used to view, is given by

$$L = 9 + 5.1 \log d$$

where d is the diameter (in inches) of the lens.
(a) What is the limiting magnitude of a 3.5-inch telescope?
(b) What diameter is required to view a star of magnitude 14?

89. Salvage Value The number of years n for a piece of machinery to depreciate to a known salvage value can be found using the formula

$$n = \frac{\log s - \log i}{\log(1 - d)}$$

where s is the salvage value of the machinery, i is its initial value, and d is the annual rate of depreciation.
(a) How many years will it take for a piece of machinery to decline in value from $90,000 to $10,000 if the annual rate of depreciation is 0.20 (20%)?
(b) How many years will it take for a piece of machinery to lose half of its value if the annual rate of depreciation is 15%?

90. Funding a College Education A child's grandparents purchase a $10,000 bond fund that matures in 18 years to be used for her college education. The bond fund pays 4% interest compounded semiannually. How much will the bond fund be worth at maturity? What is the effective rate of interest? How long would it take the bond to double in value under these terms?

91. Funding a College Education A child's grandparents wish to purchase a bond that matures in 18 years to be used for her college education. The bond pays 4% interest compounded semiannually. How much should they pay so that the bond will be worth $85,000 at maturity?

92. Funding an IRA First Colonial Bankshares Corporation advertised the following IRA investment plans.

Target IRA Plans

For each $5000 Maturity Value Desired	
Deposit:	At a Term of:
$620.17	20 Years
$1045.02	15 Years
$1760.92	10 Years
$2967.26	5 Years

(a) Assuming continuous compounding, what was the annual rate of interest that they offered?
(b) First Colonial Bankshares claims that $4000 invested today will have a value of over $32,000 in 20 years. Use the answer found in part (a) to find the actual value of $4000 in 20 years. Assume continuous compounding.

93. Estimating the Date That a Prehistoric Man Died The bones of a prehistoric man found in the desert of New Mexico contain approximately 5% of the original amount of carbon 14. If the half-life of carbon 14 is 5600 years, approximately how long ago did the man die?

94. Temperature of a Skillet A skillet is removed from an oven whose temperature is 450°F and placed in a room whose temperature is 70°F. After 5 minutes, the temperature of the skillet is 400°F. How long will it be until its temperature is 150°F?

95. World Population The growth rate of the world's population in 2003 was $k = 1.16\% = 0.0116$. The population of the world in 2003 was 6,302,486,693. Letting $t = 0$ represent 2003, use the uninhibited growth model to predict the world's population in the year 2010.

SOURCE: U.S. Census Bureau.

96. Radioactive Decay The half-life of radioactive cobalt is 5.27 years. If 100 grams of radioactive cobalt is present now, how much will be present in 20 years? In 40 years?

97. Logistic Growth The logistic growth model

$$P(t) = \frac{0.8}{1 + 1.67e^{-0.16t}}$$

represents the proportion of new cars with a Global Positioning System (GPS). Let $t = 0$ represent 2003, $t = 1$ represent 2004, and so on.
(a) What proportion of new cars in 2003 had a GPS?
(b) Determine the maximum proportion of new cars that have a GPS.
(c) Using a graphing utility, graph $P(t)$.
(d) When will 75% of new cars have a GPS?

98. CBL Experiment The following data were collected by placing a temperature probe in a portable heater, removing the probe, and then recording temperature over time.

Time (sec.)	Temperature (°F)
0	165.07
1	164.77
2	163.99
3	163.22
4	162.82
5	161.96
6	161.20
7	160.45
8	159.35
9	158.61
10	157.89
11	156.83
12	156.11
13	155.08
14	154.40
15	153.72

According to Newton's Law of Cooling, these data should follow an exponential model.
(a) Using a graphing utility, draw a scatter diagram for the data.
(b) Using a graphing utility, fit an exponential function to the data.

(c) Graph the exponential function found in part (b) on the scatter diagram.
(d) Predict how long it will take for the probe to reach a temperature of 110°F.

99. Wind Chill Factor The following data represent the wind speed (mph) and wind chill factor at an air temperature of 15°F.

Wind Speed (mph)	Wind Chill Factor (°F)
5	7
10	3
15	0
20	−2
25	−4
30	−5
35	−7

SOURCE: U.S. National Weather Service

(a) Using a graphing utility, draw a scatter diagram with wind speed as the independent variable.
(b) Using a graphing utility, fit a logarithmic function to the data.
(c) Using a graphing utility, draw the logarithmic function found in part (b) on the scatter diagram.
(d) Use the function found in part (b) to predict the wind chill factor if the air temperature is 15°F and the wind speed is 23 mph.

100. Spreading of a Disease Jack and Diane live in a small town of 50 people. Unfortunately, Jack and Diane both have a cold. Those who come in contact with someone who has this cold will themselves catch the cold. The following data represent the number of people in the small town who have caught the cold after t days.

Days, t	Number of People with Cold, C
0	2
1	4
2	8
3	14
4	22
5	30
6	37
7	42
8	44

(a) Using a graphing utility, draw a scatter diagram of the data. Comment on the type of relation that appears to exist between the days and number of people with a cold.

(b) Using a graphing utility, fit a logistic function to the data.

(c) Graph the function found in part (b) on the scatter diagram.

(d) According to the function found in part (b), what is the maximum number of people who will catch the cold? In reality, what is the maximum number of people who could catch the cold?

(e) Sometime between the second and third day, 10 people in the town had a cold. According to the model found in part (b), when did 10 people have a cold?

(f) How long will it take for 46 people to catch the cold?

Chapter Test

1. Given $f(x) = \dfrac{x + 2}{x - 2}$ and $g(x) = 2x + 5$, find:

   (a) $f \circ g$ and state its domain

   (b) $(g \circ f)(-2)$

   (c) $(f \circ g)(-2)$

2. Determine whether the function is one-to-one.

   (a) $y = 4x^2 + 3$

   (b) $y = \sqrt{x + 3} - 5$

3. Find the inverse of $f(x) = \dfrac{2}{3x - 5}$ and check your answer. State the domain and range of f and f^{-1}.

4. If the point $(3, -5)$ is on the graph of a one-to-one function f, what point must be on the graph of f^{-1}?

In Problems 5–7, find the unknown value without using a calculator.

5. $3^x = 243$

6. $\log_b 16 = 2$

7. $\log_5 x = 4$

In Problems 8–11, use a calculator to evaluate each expression. Round your answer to three decimal places.

8. $e^3 + 2$

9. $\log 20$

10. $\log_3 21$

11. $\ln 133$

In Problems 12 and 13, use transformations to graph each function. Determine the domain, range, and any asymptotes.

12. $f(x) = 4^{x+1} - 2$

13. $g(x) = 1 - \log_5(x - 2)$

In Problems 14–19, solve each equation.

14. $5^{x+2} = 125$

15. $\log(x + 9) = 2$

16. $8 - 2e^{-x} = 4$

17. $\log(x^2 + 3) = \log(x + 6)$

18. $7^{x+3} = e^x$

19. $\log_2(x - 4) + \log_2(x + 4) = 3$

20. Write $\log_2\left(\dfrac{4x^3}{x^2 - 3x - 18}\right)$ as the sum and/or difference of logarithms. Express powers as factors.

21. A 50-mg sample of a radioactive substance decays to 34 mg after 30 days. How long will it take for there to be 2 mg remaining?

22. The average cost of college at 4-year private colleges was $19,710 in 2003–2004. This was a 6% increase from the previous year.

   (a) If the cost of college increases by 6% each year, what will be the average cost of college at 4-year private colleges in 2013–2014?

   (b) College savings plans allow individuals to put money aside now to help pay for college later. If one such plan offers a rate of 5% compounded continuously, how much would Angie have needed to put in a college savings plan in 2003 in order to pay for 1 year of the cost of college at a 4-year private college in 2013?

23. The decibel level, D, of sound is given by the equation $D = 10 \log\left(\dfrac{I}{I_0}\right)$, where I is the intensity of the sound and $I_0 = 10^{-12}$ watts per square meter.

   (a) If the shout of a single person measures 80 decibels, how loud will the sound be if two people shout at the same time? That is, how loud would the sound be if the intensity doubled?

   (b) The pain threshold for sound is 125 decibels. If the Athens Olympic Stadium 2004 (Olympiako Stadio Athinas 'Spyros Louis') can seat 74,400 people, how many people in the crowd need to shout at the same time in order for the resulting sound level to meet or exceed the pain threshold? (Ignore any possible sound dampening.)

24. The table shows the estimated number of U.S. cell phone subscribers (in millions), y, from 1985 to 2003. Use a graphing utility to make a scatter diagram of the data. Fit a logistic model to the data and use the model to predict the number of U.S. cell phone subscribers in 2007. Let $x =$ the number of years since 1985.

   SOURCE: Cellular Telecommunications & Internet Association

x	0	2	4	6	8	10	12	14	16	18
y	0.34	1.23	3.51	7.56	16.01	33.76	55.31	86.05	128.37	158.72

Chapter Projects

1. **Hot Coffee** A fast-food restaurant wants a special container to hold coffee. The restaurant wishes the container to quickly cool the coffee from 200° to 130°F and keep the liquid between 110° and 130°F as long as possible. The restaurant has three containers to select from.

   1. The CentiKeeper Company has a container that reduces the temperature of a liquid from 200°F to 100°F in 30 minutes by maintaining a constant temperature of 70°F.

   2. The TempControl Company has a container that reduces the temperature of a liquid from 200°F to 110°F in 25 minutes by maintaining a constant temperature of 60°F.

   3. The Hot'n'Cold Company has a container that reduces the temperature of a liquid from 200°F to 120°F in 20 minutes by maintaining a constant temperature of 65°F.

   You need to recommend which container the restaurant should purchase.

   (a) Use Newton's Law of Cooling to find a function relating the temperature of the liquid over time for each container.

   (b) How long does it take each container to lower the coffee temperature from 200° to 130°F?

   (c) How long will the coffee temperature remain between 110° and 130°F? This temperature is considered the optimal drinking temperature.

   (d) Graph each function using a graphing utility.

   (e) Which company would you recommend to the restaurant? Why?

   (f) How might the cost of the container affect your decision?

The following projects are available on the Instructor's Resource Center (IRC):

2. **Project at Motorola** *Thermal Fatigue of Solder Connections*

3. **Depreciation of a New Car**

4. **CBL Experiment**

Cumulative Review

1. Is the following graph the graph of a function? If it is, is the function one-to-one?

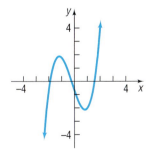

2. For the function $f(x) = 2x^2 - 3x + 1$, find the following:
   (a) $f(3)$ (b) $f(-x)$ (c) $f(x + h)$

3. Determine which of the following points are on the graph of $x^2 + y^2 = 1$.

   (a) $\left(\dfrac{1}{2}, \dfrac{1}{2}\right)$ (b) $\left(\dfrac{1}{2}, \dfrac{\sqrt{3}}{2}\right)$

4. Solve the equation $3(x - 2) = 4(x + 5)$.

5. Graph the line $2x - 4y = 16$.

6. (a) Graph the quadratic function $f(x) = -x^2 + 2x - 3$ by determining whether its graph opens up or down and by finding its vertex, axis of symmetry, y-intercept, and x-intercept(s), if any.
   (b) Solve $f(x) \leq 0$.

7. Determine the quadratic function whose graph is given in the figure.

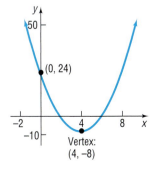

8. Graph $f(x) = 3(x + 1)^3 - 2$ using transformations.

9. Given that $f(x) = x^2 + 2$ and $g(x) = \dfrac{2}{x - 3}$, find $f(g(x))$ and state its domain. What is $f(g(5))$?

10. For the polynomial function $f(x) = 4x^3 + 9x^2 - 30x - 8$:
    (a) Find the real zeros of f.
    (b) Determine the intercepts of the graph of f.
    (c) Use a graphing utility to approximate the local maxima and local minima.
    (d) By hand, draw a complete graph of f. Be sure to label the intercepts and turning points.

11. For the function $g(x) = 3^x + 2$:
    (a) Graph g using transformations. State the domain, range, and horizontal asymptote of g.
    (b) Determine the inverse of g. State the domain, range, and vertical asymptote of g^{-1}.
    (c) On the same graph as g, graph g^{-1}.

12. Solve the equation $4^{x-3} = 8^{2x}$.

13. Solve the equation: $\log_3(x + 1) + \log_3(2x - 3) = \log_9 9$

14. Suppose that $f(x) = \log_3(x + 2)$. Solve:
    (a) $f(x) = 0$.
    (b) $f(x) > 0$.
    (c) $f(x) = 3$.

15. **Data Analysis** The following data represent the percent of all drivers that have been stopped by the police for any reason within the past year by age. The median age represents the midpoint of the upper and lower limit for the age range.

Age Range	Median Age, x	Percentage Stopped, y
16–19	17.5	18.2
20–29	24.5	16.8
30–39	34.5	11.3
40–49	44.5	9.4
50–59	54.5	7.7
≥60	69.5	3.8

(a) Using your graphing utility, draw a scatter diagram of the data treating median age, x, as the independent variable.
(b) Determine a model that you feel best describes the relation between median age and percentage stopped. You may choose from among linear, quadratic, cubic, power, exponential, logarithmic, or logistic models.
(c) Provide a justification for the model that you selected in part (b).

Trigonometric Functions

5

A LOOK BACK In Chapter 2, we began our discussion of functions. We defined domain and range and independent and dependent variables; we found the value of a function and graphed functions. We continued our study of functions by listing properties that a function might have, like being even or odd, and we created a library of functions, naming key functions and listing their properties, including the graph.

A LOOK AHEAD In this chapter we define the trigonometric functions, six functions that have a wide application. We shall talk about their domain and range, see how to find values and graph them, and develop a list of their properties.

There are two widely accepted approaches to the development of the trigonometric functions: one uses right triangles; the other uses circles, especially the unit circle. In this book, we develop the trigonometric functions using the unit circle. In Chapter 7, we present right triangle trigonometry.

Tidal Coastline and Pots of Water

In Florida, they post times of tides coming in and going out very precisely, like 11:23 A.M. How can they be so precise? There is more to tides, the rise and fall of ocean waters, than the gravitational pull of the Moon and Sun.

These are the chief factors, of course. And because the movements of Earth, Sun and Moon in relationship to each other are known precisely, the rhythm of tides rising and falling along coastlines is easy to predict.

Yet the time and heights of high and low tide may vary along different stretches of the same coast, though they are reacting to similar driving forces and pressures.

Historic observation makes possible the exact timing of high tides and low tides along a particular section of coast for a month, a year, or far into the future.

The reason for the difference is oscillation. Think of pots and pans filled with varying levels of water on a table, says Charles O' Reilly, Chief of Tidal Analysis for the Geological Survey of Canada's hydrographic service in Dartmouth, Nova Scotia. Then kick the table.

"You'll notice the water in the pots and pans will slosh differently. That's their natural oscillation," he said. "If you kick the table rhythmically, you'll find each pot continues to slosh differently because it has its own rhythm.

"Now, if you join those pots and pans together, that's sort of like a coastal ocean. They're all feeling the same 'kick,' but they are all responding differently. In order to predict a tide, you have to measure them for some period of time."

SOURCE: *Toronto Star*, June 13, 2001, p. GT02. Reprinted with permission—Torstar Syndication Services.

—See Chapter Project 1.

OUTLINE

5.1 Angles and Their Measure

PREPARING FOR THIS SECTION *Before getting started, review the following:*

• Circumference and Area of a Circle (Appendix, Section A.2, p. 671)

Now work the 'Are You Prepared?' problems on page 366.

OBJECTIVES
1 Convert between Degrees, Minutes, Seconds, and Decimal Forms for Angles
2 Find the Arc Length of a Circle
3 Convert from Degrees to Radians and from Radians to Degrees
4 Find the Area of a Sector of a Circle
5 Find the Linear Speed of an Object Traveling in Circular Motion

A **ray**, or **half-line**, is that portion of a line that starts at a point V on the line and extends indefinitely in one direction. The starting point V of a ray is called its **vertex**. See Figure 1.

Figure 1

Line
V Ray

If two rays are drawn with a common vertex, they form an **angle**. We call one of the rays of an angle the **initial side** and the other the **terminal side**. The angle formed is identified by showing the direction and amount of rotation from the initial side to the terminal side. If the rotation is in the counterclockwise direction, the angle is **positive**; if the rotation is clockwise, the angle is **negative**. See Figure 2. Lowercase Greek letters, such as α (alpha), β (beta), γ (gamma), and θ (theta), will be used to denote angles. Notice in Figure 2(a) that the angle α is positive because the direction of the rotation from the initial side to the terminal side is counterclockwise. The angle β in Figure 2(b) is negative because the rotation is clockwise. The angle γ in Figure 2(c) is positive. Notice that the angle α in Figure 2(a) and the angle γ in Figure 2(c) have the same initial side and the same terminal side. However, α and γ are unequal, because the amount of rotation required to go from the initial side to the terminal side is greater for angle γ than for angle α.

Figure 2

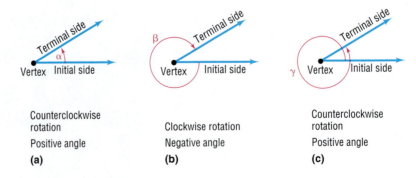

(a) Counterclockwise rotation — Positive angle

(b) Clockwise rotation — Negative angle

(c) Counterclockwise rotation — Positive angle

An angle θ is said to be in **standard position** if its vertex is at the origin of a rectangular coordinate system and its initial side coincides with the positive x-axis. See Figure 3.

Figure 3

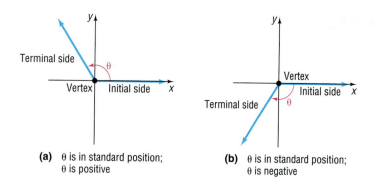

(a) θ is in standard position; θ is positive

(b) θ is in standard position; θ is negative

When an angle θ is in standard position, the terminal side will lie either in a quadrant, in which case we say that θ **lies in that quadrant**, or θ will lie on the *x*-axis or the *y*-axis, in which case we say that θ is a **quadrantal angle**. For example, the angle θ in Figure 4(a) lies in quadrant II, the angle θ in Figure 4(b) lies in quadrant IV, and the angle θ in Figure 4(c) is a quadrantal angle.

Figure 4

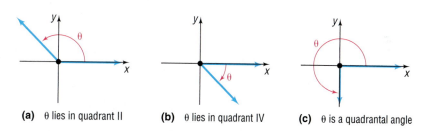

(a) θ lies in quadrant II

(b) θ lies in quadrant IV

(c) θ is a quadrantal angle

We measure angles by determining the amount of rotation needed for the initial side to become coincident with the terminal side. The two commonly used measures for angles are *degrees* and *radians*.

Degrees

The angle formed by rotating the initial side exactly once in the counterclockwise direction until it coincides with itself (1 revolution) is said to measure 360 degrees, abbreviated 360°. **One degree, 1°,** is $\dfrac{1}{360}$ revolution. A **right angle** is an angle that measures 90°, or $\dfrac{1}{4}$ revolution; a **straight angle** is an angle that measures 180°, or $\dfrac{1}{2}$ revolution. See Figure 5. As Figure 5(b) shows, it is customary to indicate a right angle by using the symbol $\llcorner$.

Figure 5

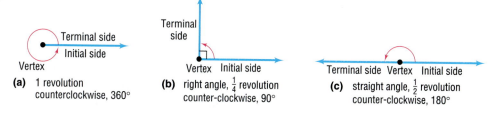

(a) 1 revolution counterclockwise, 360°

(b) right angle, $\frac{1}{4}$ revolution counter-clockwise, 90°

(c) straight angle, $\frac{1}{2}$ revolution counter-clockwise, 180°

It is also customary to refer to an angle that measures θ degrees as an angle *of* θ degrees.

EXAMPLE 1	Drawing an Angle

Draw each angle.

(a) 45°   (b) −90°   (c) 225°   (d) 405°

Solution (a) An angle of 45° is $\frac{1}{2}$ of a right angle. See Figure 6.

(b) An angle of −90° is $\frac{1}{4}$ revolution in the clockwise direction. See Figure 7.

Figure 6

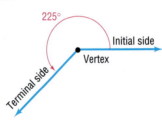

Figure 7

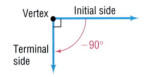

(c) An angle of 225° consists of a rotation through 180° followed by a rotation through 45°. See Figure 8.

(d) An angle of 405° consists of 1 revolution (360°) followed by a rotation through 45°. See Figure 9.

Figure 8

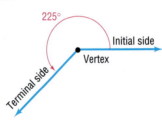

Figure 9

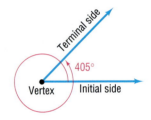

✏ NOW WORK PROBLEM 11.

1 ✔ Convert between Degrees, Minutes, Seconds, and Decimal Forms for Angles

Although subdivisions of a degree may be obtained by using decimals, we also may use the notion of *minutes* and *seconds*. **One minute**, denoted by **1′**, is defined as $\frac{1}{60}$ degree. **One second**, denoted by **1″**, is defined as $\frac{1}{60}$ minute, or equivalently, $\frac{1}{3600}$ degree. An angle of, say, 30 degrees, 40 minutes, 10 seconds is written compactly as 30°40′10″. To summarize:

> 1 counterclockwise revolution = 360°
>
> 1° = 60′   1′ = 60″   **(1)**

It is sometimes necessary to convert from the degree, minute, second notation (D°M′S″) to a decimal form, and vice versa. Check your calculator; it should be capable of doing the conversion for you.

Before getting started, though, you must set the mode to degrees because there are two common ways to measure angles: degree mode and radian mode. (We will define radians shortly.) Usually, a menu is used to change from one mode to another. Check your owner's manual to find out how your particular calculator works.

Now let's see how to convert from the degree, minute, second notation (D°M′S″) to a decimal form, and vice versa, by looking at some examples:

$$15°30' = 15.5° \quad \text{because} \quad 30' = 30 \cdot \left(\frac{1}{60}\right)° = 0.5°$$

$$1' = \left(\frac{1}{60}\right)°$$

$$32.25° = 32°15' \quad \text{because} \quad 0.25° = \left(\frac{1}{4}\right)° = \frac{1}{4}(60') = 15'$$

$$1° = 60'$$

| EXAMPLE 2 | **Converting between Degrees, Minutes, Seconds, and Decimal Forms** |

(a) Convert 50°6′21″ to a decimal in degrees.
(b) Convert 21.256° to the D°M′S″ form.

Algebraic Solution

(a) Because $1' = \left(\frac{1}{60}\right)°$ and $1'' = \left(\frac{1}{60}\right)' = \left(\frac{1}{60} \cdot \frac{1}{60}\right)°$, we convert as follows:

$$50°6'21'' = 50° + 6' + 21''$$

$$= 50° + 6 \cdot \left(\frac{1}{60}\right)° + 21 \cdot \left(\frac{1}{60} \cdot \frac{1}{60}\right)°$$

$$\approx 50° + 0.1° + 0.005833°$$

$$= 50.105833°$$

(b) We proceed as follows:

$$21.256° = 21° + 0.256°$$

$$= 21° + (0.256)(60') \quad \textit{Convert fraction of degree to minutes; } 1° = 60'.$$

$$= 21° + 15.36'$$

$$= 21° + 15' + 0.36'$$

$$= 21° + 15' + (0.36)(60'') \quad \textit{Convert fraction of minute to seconds; } 1' = 60''.$$

$$= 21° + 15' + 21.6''$$

$$\approx 21°15'22''$$

◄

Graphing Solution

(a) Figure 10 shows the solution using a TI-84 Plus graphing calculator.

Figure 10

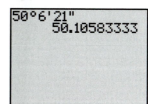

(b) Figure 11 shows the solution using a TI-84 Plus graphing calculator.

Figure 11

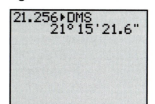

◄

NOW WORK PROBLEMS 23 AND 29.

In many applications, such as describing the exact location of a star or the precise position of a boat at sea, angles measured in degrees, minutes, and even seconds are used. For calculation purposes, these are transformed to decimal form. In other applications, especially those in calculus, angles are measured using *radians*.

Radians

A **central angle** is an angle whose vertex is at the center of a circle. The rays of a central angle subtend (intersect) an arc on the circle. If the radius of the circle is r and the length of the arc subtended by the central angle is also r, then the measure of the angle is **1 radian**. See Figure 12(a).

Figure 12

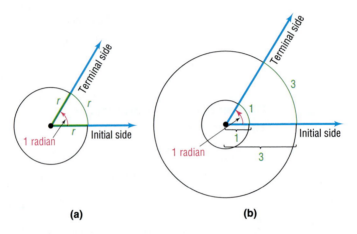

(a)　　　　　　　　(b)

For a circle of radius 1, the rays of a central angle with measure 1 radian would subtend an arc of length 1. For a circle of radius 3, the rays of a central angle with measure 1 radian would subtend an arc of length 3. See Figure 12(b).

2 Find the Arc Length of a Circle

Figure 13

$$\frac{\theta}{\theta_1} = \frac{s}{s_1}$$

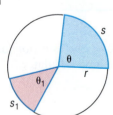

Now consider a circle of radius r and two central angles, θ and θ_1, measured in radians. Suppose that these central angles subtend arcs of lengths s and s_1, respectively, as shown in Figure 13. From geometry, we know that the ratio of the measures of the angles equals the ratio of the corresponding lengths of the arcs subtended by these angles; that is,

$$\frac{\theta}{\theta_1} = \frac{s}{s_1} \tag{2}$$

Suppose that $\theta_1 = 1$ radian. Refer again to Figure 12(a). The amount of arc s_1 subtended by the central angle $\theta_1 = 1$ radian equals the radius r of the circle. Then $s_1 = r$, so equation (2) reduces to

$$\frac{\theta}{1} = \frac{s}{r} \quad \text{or} \quad s = r\theta \tag{3}$$

Theorem

> ### Arc Length
>
> For a circle of radius r, a central angle of θ radians subtends an arc whose length s is
>
> $$s = r\theta \tag{4}$$

NOTE   Formulas must be consistent with regard to the units used. In equation (4), we write

$$s = r\theta$$

To see the units, however, we must go back to equation (3) and write

$$\frac{\theta \text{ radians}}{1 \text{ radian}} = \frac{s \text{ length units}}{r \text{ length units}}$$

$$s \text{ length units} = r \text{ length units} \frac{\theta \text{ radians}}{1 \text{ radian}}$$

Since the radians cancel, we are left with

$$s \text{ length units} = (r \text{ length units})\theta \qquad s = r\theta$$

where θ appears to be "dimensionless" but, in fact, is measured in radians. So, in using the formula $s = r\theta$, the dimension for θ is radians, and any convenient unit of length (such as inches or meters) may be used for s and r. ■

EXAMPLE 3	**Finding the Length of an Arc of a Circle**

Find the length of the arc of a circle of radius 2 meters subtended by a central angle of 0.25 radian.

Solution   We use equation (4) with $r = 2$ meters and $\theta = 0.25$. The length s of the arc is

$$s = r\theta = 2(0.25) = 0.5 \text{ meter} \qquad ◀$$

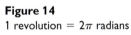 **NOW WORK PROBLEM 71.**

3   **Convert from Degrees to Radians and from Radians to Degrees**

Figure 14
1 revolution = 2π radians

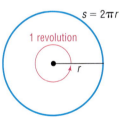
$s = 2\pi r$
1 revolution

Consider a circle of radius r. A central angle of 1 revolution will subtend an arc equal to the circumference of the circle (Figure 14). Because the circumference of a circle equals $2\pi r$, we use $s = 2\pi r$ in equation (4) to find that, for an angle θ of 1 revolution,

$$s = r\theta$$
$$2\pi r = r\theta \qquad \theta = 1 \text{ revolution; } s = 2\pi r$$
$$\theta = 2\pi \text{ radians} \qquad \text{Solve for } \theta.$$

From this we have,

$$1 \text{ revolution} = 2\pi \text{ radians} \qquad \text{(5)}$$

Since 1 revolution = 360°, we have

$$360° = 2\pi \text{ radians}$$

or

$$180° = \pi \text{ radians} \qquad \text{(6)}$$

Divide both sides of equation (6) by 180. Then

$$1 \text{ degree} = \frac{\pi}{180} \text{ radian}$$

Divide both sides of (6) by π. Then

$$\frac{180}{\pi} \text{ degrees} = 1 \text{ radian}$$

We have the following two conversion formulas:

$$1 \text{ degree} = \frac{\pi}{180} \text{ radian} \qquad 1 \text{ radian} = \frac{180}{\pi} \text{ degrees} \qquad \text{(7)}$$

EXAMPLE 4	**Converting from Degrees to Radians**

Convert each angle in degrees to radians.

(a) $60°$ (b) $150°$ (c) $-45°$ (d) $90°$ (e) $107°$

Solution   (a) $60° = 60 \cdot 1 \text{ degree} = 60 \cdot \dfrac{\pi}{180} \text{ radian} = \dfrac{\pi}{3} \text{ radians}$

(b) $150° = 150 \cdot \dfrac{\pi}{180} \text{ radian} = \dfrac{5\pi}{6} \text{ radians}$

(c) $-45° = -45 \cdot \dfrac{\pi}{180} \text{ radian} = -\dfrac{\pi}{4} \text{ radian}$

(d) $90° = 90 \cdot \dfrac{\pi}{180} \text{ radian} = \dfrac{\pi}{2} \text{ radians}$

(e) $107° = 107 \cdot \dfrac{\pi}{180} \text{ radian} \approx 1.868 \text{ radians}$ ◄

Example 4 illustrates that angles that are fractions of a revolution, (a)–(d), are expressed in radian measure as fractional multiples of π, rather than as decimals. For example, a right angle, as in Example 4(d), is left in the form $\dfrac{\pi}{2}$ radians, which is exact, rather than using the approximation $\dfrac{\pi}{2} \approx \dfrac{3.1416}{2} = 1.5708$ radians.

NOW WORK PROBLEMS 35 AND 61.

EXAMPLE 5	**Converting Radians to Degrees**

Convert each angle in radians to degrees.

(a) $\dfrac{\pi}{6} \text{ radian}$ (b) $\dfrac{3\pi}{2} \text{ radians}$ (c) $-\dfrac{3\pi}{4} \text{ radians}$

(d) $\dfrac{7\pi}{3} \text{ radians}$ (e) 3 radians

Solution   (a) $\dfrac{\pi}{6} \text{ radian} = \dfrac{\pi}{6} \cdot 1 \text{ radian} = \dfrac{\pi}{6} \cdot \dfrac{180}{\pi} \text{ degrees} = 30°$

(b) $\dfrac{3\pi}{2} \text{ radians} = \dfrac{3\pi}{2} \cdot \dfrac{180}{\pi} \text{ degrees} = 270°$

(c) $-\dfrac{3\pi}{4}$ radians $= -\dfrac{3\pi}{4} \cdot \dfrac{180}{\pi}$ degrees $= -135°$

(d) $\dfrac{7\pi}{3}$ radians $= \dfrac{7\pi}{3} \cdot \dfrac{180}{\pi}$ degrees $= 420°$

(e) 3 radians $= 3 \cdot \dfrac{180}{\pi}$ degrees $\approx 171.89°$

◄

NOW WORK PROBLEM **47**.

Table 1 lists the degree and radian measures of some commonly encountered angles. You should learn to feel equally comfortable using degree or radian measure for these angles.

Table 1

Degrees	0°	30°	45°	60°	90°	120°	135°	150°	180°
Radians	0	$\dfrac{\pi}{6}$	$\dfrac{\pi}{4}$	$\dfrac{\pi}{3}$	$\dfrac{\pi}{2}$	$\dfrac{2\pi}{3}$	$\dfrac{3\pi}{4}$	$\dfrac{5\pi}{6}$	π
Degrees		210°	225°	240°	270°	300°	315°	330°	360°
Radians		$\dfrac{7\pi}{6}$	$\dfrac{5\pi}{4}$	$\dfrac{4\pi}{3}$	$\dfrac{3\pi}{2}$	$\dfrac{5\pi}{3}$	$\dfrac{7\pi}{4}$	$\dfrac{11\pi}{6}$	2π

EXAMPLE 6   **Finding the Distance between Two Cities**

See Figure 15(a). The latitude of a location L is the angle formed by a ray drawn from the center of Earth to the Equator and a ray drawn from the center of Earth to L. See Figure 15(b). Glasgow, Montana, is due north of Albuquerque, New Mexico. Find the distance between Glasgow (48°9′ north latitude) and Albuquerque (35°5′ north latitude). Assume that the radius of Earth is 3960 miles.

Figure 15

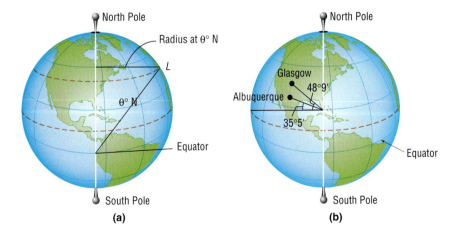

(a)   (b)

Solution   The measure of the central angle between the two cities is $48°9′ - 35°5′ = 13°4′$. We use equation (4), $s = r\theta$, but first we must convert the angle of $13°4′$ to radians.

$$\theta = 13°4′ \approx 13.0667° = 13.0667 \cdot \dfrac{\pi}{180} \text{ radian} \approx 0.228 \text{ radian}$$

We use $\theta = 0.228$ radian and $r = 3960$ miles in equation (4). The distance between the two cities is

$$s = r\theta = 3960 \cdot 0.228 \approx 903 \text{ miles}$$ ◄

When an angle is measured in degrees, the degree symbol will always be shown. However, when an angle is measured in radians, we will follow the usual practice and omit the word *radians*. So, if the measure of an angle is given as $\dfrac{\pi}{6}$, it is understood to mean $\dfrac{\pi}{6}$ radian.

NOW WORK PROBLEM 101.

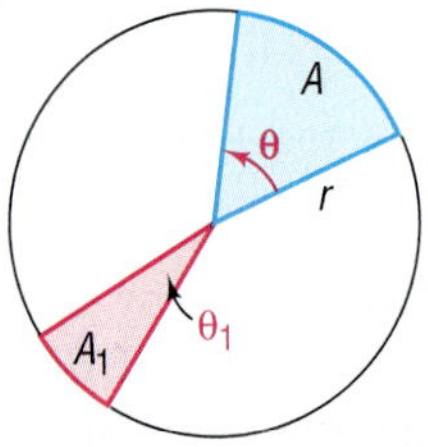

4 **Find the Area of a Sector of a Circle**

Figure 16

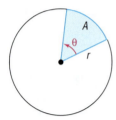

Consider a circle of radius r. Suppose that θ, measured in radians, is a central angle of this circle. See Figure 16. We seek a formula for the area A of the sector formed by the angle θ (shown in blue).

Now consider a circle of radius r and two central angles θ and θ_1, both measured in radians. See Figure 17. From geometry, we know the ratio of the measures of the angles equals the ratio of the corresponding areas of the sectors formed by these angles. That is,

Figure 17

$$\frac{\theta}{\theta_1} = \frac{A}{A_1}$$

$$\frac{\theta}{\theta_1} = \frac{A}{A_1}$$

Suppose that $\theta_1 = 2\pi$ radians. Then $A_1 =$ area of the circle $= \pi r^2$. Solving for A, we find

$$A = A_1 \frac{\theta}{\theta_1} = \pi r^2 \frac{\theta}{2\pi} = \frac{1}{2} r^2 \theta$$

Theorem

Area of a Sector

The area A of the sector of a circle of radius r formed by a central angle of θ radians is

$$A = \frac{1}{2} r^2 \theta \qquad\qquad (8)$$

EXAMPLE 7 **Finding the Area of a Sector of a Circle**

Find the area of the sector of a circle of radius 2 feet formed by an angle of 30°. Round the answer to two decimal places.

Solution We use equation (8) with $r = 2$ feet and $\theta = 30° = \dfrac{\pi}{6}$ radians. [Remember, in equation (8), θ must be in radians.] The area A of the sector is

$$A = \frac{1}{2} r^2 \theta = \frac{1}{2} (2)^2 \frac{\pi}{6} = \frac{\pi}{3} \text{ square feet} \approx 1.05 \text{ square feet}$$

rounded to two decimal places. ◄

NOW WORK PROBLEM 79.

5 Find the Linear Speed of an Object Traveling in Circular Motion

We have already defined the average speed of an object as the distance traveled divided by the elapsed time. Suppose that an object moves around a circle of radius r at a constant speed. If s is the distance traveled in time t around this circle, then the **linear speed** v of the object is defined as

$$v = \frac{s}{t} \tag{9}$$

Figure 18

$$v = \frac{s}{t} \qquad \omega = \frac{\theta}{t}$$

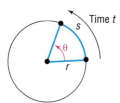

As this object travels around the circle, suppose that θ (measured in radians) is the central angle swept out in time t. See Figure 18. Then the **angular speed** ω (the Greek letter omega) of this object is the angle (measured in radians) swept out, divided by the elapsed time; that is,

$$\omega = \frac{\theta}{t} \tag{10}$$

Angular speed is the way the turning rate of an engine is described. For example, an engine idling at 900 rpm (revolutions per minute) is one that rotates at an angular speed of

$$900\,\frac{\text{revolutions}}{\text{minute}} = 900\,\frac{\text{revolutions}}{\text{minute}} \cdot 2\pi\,\frac{\text{radians}}{\text{revolution}} = 1800\pi\,\frac{\text{radians}}{\text{minute}}$$

There is an important relationship between linear speed and angular speed:

$$\text{linear speed} = v = \underset{\underset{(9)}{\uparrow}}{\frac{s}{t}} = \underset{\underset{s\,=\,r\theta}{\uparrow}}{\frac{r\theta}{t}} = r\left(\frac{\theta}{t}\right)$$

Then, using equation (10), we obtain

$$v = r\omega \tag{11}$$

where ω is measured in radians per unit time.

When using equation (11), remember that $v = \dfrac{s}{t}$ (the linear speed) has the dimensions of length per unit of time (such as feet per second or miles per hour), r (the radius of the circular motion) has the same length dimension as s, and ω (the angular speed) has the dimensions of radians per unit of time. If the angular speed is given in terms of *revolutions* per unit of time (as is often the case), be sure to convert it to *radians* per unit of time before attempting to use equation (11).

EXAMPLE 8   **Finding Linear Speed**

Figure 19

A child is spinning a rock at the end of a 2-foot rope at the rate of 180 revolutions per minute (rpm). Find the linear speed of the rock when it is released.

Solution   Look at Figure 19. The rock is moving around a circle of radius $r = 2$ feet. The angular speed ω of the rock is

$$\omega = 180\,\frac{\text{revolutions}}{\text{minute}} = 180\,\frac{\text{revolutions}}{\text{minute}} \cdot 2\pi\,\frac{\text{radians}}{\text{revolution}} = 360\pi\,\frac{\text{radians}}{\text{minute}}$$

From equation (11), the linear speed v of the rock is

$$v = r\omega = 2 \text{ feet} \cdot 360\pi \frac{\text{radians}}{\text{minute}} = 720\pi \frac{\text{feet}}{\text{minute}} \approx 2262 \frac{\text{feet}}{\text{minute}}$$

The linear speed of the rock when it is released is 2262 ft/min $\approx$ 25.7 mi/hr. ◀

NOW WORK PROBLEM 97.

HISTORICAL FEATURE

Trigonometry was developed by Greek astronomers, who regarded the sky as the inside of a sphere, so it was natural that triangles on a sphere were investigated early (by Menelaus of Alexandria about AD 100) and that triangles in the plane were studied much later. The first book containing a systematic treatment of plane and spherical trigonometry was written by the Persian astronomer Nasir Eddin (about AD 1250).

Regiomontanus (1436–1476) is the person most responsible for moving trigonometry from astronomy into mathematics. His work was improved by Copernicus (1473–1543) and Copernicus's student Rhaeticus (1514–1576). Rhaeticus's book was the first to define the six trigonometric functions as ratios of sides of triangles, although he did not give the functions their present names. Credit for this is due to Thomas Finck (1583), but Finck's notation was by no means universally accepted at the time. The notation was finally stabilized by the textbooks of Leonhard Euler (1707–1783).

Trigonometry has since evolved from its use by surveyors, navigators, and engineers to present applications involving ocean tides, the rise and fall of food supplies in certain ecologies, brain wave patterns, and many other phenomena.

5.1 Assess Your Understanding

'Are You Prepared?'

Answers are given at the end of these exercises. If you get a wrong answer, read the pages listed in red.

1. What is the formula for the circumference C of a circle of radius r? (p. 671)

2. What is the formula for the area A of a circle of radius r? (p. 671)

Concepts and Vocabulary

3. An angle θ is in _____ _____ if its vertex is at the origin of a rectangular coordinate system and its initial side coincides with the positive x-axis.

4. On a circle of radius r, a central angle of θ radians subtends an arc of length $s =$ _____; the area of the sector formed by this angle θ is $A =$ _____.

5. An object travels around a circle of radius r with constant speed. If s is the distance traveled in time t around the circle and θ is the central angle (in radians) swept out in time t, then the linear speed of the object is $v =$ _____ and the angular speed of the object is $\omega =$ _____.

6. *True or False:* $\pi = 180$.

7. *True or False:* $180° = \pi$ radians.

8. *True or False:* On the unit circle, if s is the length of the arc subtended by a central angle θ, measured in radians, then $s = \theta$.

9. *True or False:* The area A of the sector of a circle of radius r formed by a central angle of θ degrees is $A = \dfrac{1}{2}r^2\theta$.

10. *True or False:* For circular motion on a circle of radius r, linear speed equals angular speed divided by r.

Skill Building

In Problems 11–22, draw each angle.

11. $30°$

12. $60°$

13. $135°$

14. $-120°$

15. $450°$

16. $540°$

17. $\dfrac{3\pi}{4}$

18. $\dfrac{4\pi}{3}$

19. $-\dfrac{\pi}{6}$

20. $-\dfrac{2\pi}{3}$

21. $\dfrac{16\pi}{3}$

22. $\dfrac{21\pi}{4}$

In Problems 23–28, convert each angle to a decimal in degrees. Round your answer to two decimal places. Verify your results using a graphing utility.

23. 40°10′25″   **24.** 61°42′21″   **25.** 1°2′3″   **26.** 73°40′40″   **27.** 9°9′9″   **28.** 98°22′45″

In Problems 29–34, convert each angle to D°M′S″ form. Round your answer to the nearest second. Verify your results using a graphing utility.

29. 40.32°   **30.** 61.24°   **31.** 18.255°   **32.** 29.411°   **33.** 19.99°   **34.** 44.01°

In Problems 35–46, convert each angle in degrees to radians. Express your answer as a multiple of π.

35. 30°   **36.** 120°   **37.** 240°   **38.** 330°   **39.** −60°   **40.** −30°

41. 180°   **42.** 270°   **43.** −135°   **44.** −225°   **45.** −90°   **46.** −180°

In Problems 47–58, convert each angle in radians to degrees.

47. $\dfrac{\pi}{3}$ **48.** $\dfrac{5\pi}{6}$ **49.** $-\dfrac{5\pi}{4}$ **50.** $-\dfrac{2\pi}{3}$ **51.** $\dfrac{\pi}{2}$ **52.** 4π

53. $\dfrac{\pi}{12}$ **54.** $\dfrac{5\pi}{12}$ **55.** $-\dfrac{\pi}{2}$ **56.** $-\pi$ **57.** $-\dfrac{\pi}{6}$ **58.** $-\dfrac{3\pi}{4}$

In Problems 59–64, convert each angle in degrees to radians. Express your answer in decimal form, rounded to two decimal places.

59. 17°   **60.** 73°   **61.** −40°   **62.** −51°   **63.** 125°   **64.** 350°

In Problems 65–70, convert each angle in radians to degrees. Express your answer in decimal form, rounded to two decimal places.

65. 3.14   **66.** 0.75   **67.** 2   **68.** 3   **69.** 6.32   **70.** $\sqrt{2}$

In Problems 71–78, s denotes the length of the arc of a circle of radius r subtended by the central angle θ. Find the missing quantity. Round answers to three decimal places.

71. $r = 10$ meters, $\theta = \dfrac{1}{2}$ radian, $s = ?$ **72.** $r = 6$ feet, $\theta = 2$ radians, $s = ?$

73. $\theta = \dfrac{1}{3}$ radian, $s = 2$ feet, $r = ?$ **74.** $\theta = \dfrac{1}{4}$ radian, $s = 6$ centimeters, $r = ?$

75. $r = 5$ miles, $s = 3$ miles, $\theta = ?$ **76.** $r = 6$ meters, $s = 8$ meters, $\theta = ?$

77. $r = 2$ inches, $\theta = 30°$, $s = ?$ **78.** $r = 3$ meters, $\theta = 120°$, $s = ?$

In Problems 79–86, A denotes the area of the sector of a circle of radius r formed by the central angle θ. Find the missing quantity. Round answers to three decimal places.

79. $r = 10$ meters, $\theta = \dfrac{1}{2}$ radian, $A = ?$ **80.** $r = 6$ feet, $\theta = 2$ radians, $A = ?$

81. $\theta = \dfrac{1}{3}$ radian, $A = 2$ square feet, $r = ?$ **82.** $\theta = \dfrac{1}{4}$ radian, $A = 6$ square centimeters, $r = ?$

83. $r = 5$ miles, $A = 3$ square miles, $\theta = ?$ **84.** $r = 6$ meters, $A = 8$ square meters, $\theta = ?$

85. $r = 2$ inches, $\theta = 30°$, $A = ?$ **86.** $r = 3$ meters, $\theta = 120°$, $A = ?$

In Problems 87–90, find the length s and area A. Round answers to three decimal places.

87.   **88.**   **89.**   **90.**

Applications and Extensions

91. Minute Hand of a Clock The minute hand of a clock is 6 inches long. How far does the tip of the minute hand move in 15 minutes? How far does it move in 25 minutes?

92. Movement of a Pendulum A pendulum swings through an angle of 20° each second. If the pendulum is 40 inches long, how far does its tip move each second?

93. Area of a Sector Find the area of the sector of a circle of radius 4 meters formed by an angle of 45°. Round the answer to two decimal places.

94. Area of a Sector Find the area of the sector of a circle of radius 3 centimeters formed by an angle of 60°. Round the answer to two decimal places.

95. Watering a Lawn A water sprinkler sprays water over a distance of 30 feet while rotating through an angle of 135°. What area of lawn receives water?

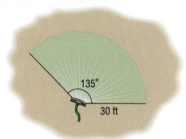

135°

30 ft

96. Designing a Water Sprinkler An engineer is asked to design a water sprinkler that will cover a field of 100 square yards that is in the shape of a sector of a circle of radius 50 yards. Through what angle should the sprinkler rotate?

97. Motion on a Circle An object is traveling around a circle with a radius of 5 centimeters. If in 20 seconds a central angle of $\frac{1}{3}$ radian is swept out, what is the angular speed of the object? What is its linear speed?

98. Motion on a Circle An object is traveling around a circle with a radius of 2 meters. If in 20 seconds the object travels 5 meters, what is its angular speed? What is its linear speed?

99. Bicycle Wheels The diameter of each wheel of a bicycle is 26 inches. If you are traveling at a speed of 35 miles per hour on this bicycle, through how many revolutions per minute are the wheels turning?

100. Car Wheels The radius of each wheel of a car is 15 inches. If the wheels are turning at the rate of 3 revolutions per second, how fast is the car moving? Express your answer in inches per second and in miles per hour.

In Problems 101–104, the latitude of a location L is the angle formed by a ray drawn from the center of Earth to the Equator and a ray drawn from the center of Earth to L. See the figure.

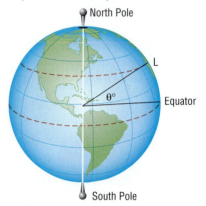

North Pole

L

θ°

Equator

South Pole

101. Distance between Cities Memphis, Tennessee, is due north of New Orleans, Louisiana. Find the distance between Memphis (35°9′ north latitude) and New Orleans (29°57′ north latitude). Assume that the radius of Earth is 3960 miles.

102. Distance between Cities Charleston, West Virginia, is due north of Jacksonville, Florida. Find the distance between Charleston (38°21′ north latitude) and Jacksonville (30°20′ north latitude). Assume that the radius of Earth is 3960 miles.

103. Linear Speed on Earth Earth rotates on an axis through its poles. The distance from the axis to a location on Earth 30° north latitude is about 3429.5 miles. Therefore, a location on Earth at 30° north latitude is spinning on a circle of radius 3429.5 miles. Compute the linear speed on the surface of Earth at 30° north latitude.

104. Linear Speed on Earth Earth rotates on an axis through its poles. The distance from the axis to a location on Earth 40° north latitude is about 3033.5 miles. Therefore, a location on Earth at 40° north latitude is spinning on a circle of radius 3033.5 miles. Compute the linear speed on the surface of Earth at 40° north latitude.

105. Speed of the Moon The mean distance of the Moon from Earth is 2.39×10^5 miles. Assuming that the orbit of the Moon around Earth is circular and that 1 revolution takes 27.3 days, find the linear speed of the Moon. Express your answer in miles per hour.

106. Speed of Earth The mean distance of Earth from the Sun is 9.29×10^7 miles. Assuming that the orbit of Earth around the Sun is circular and that 1 revolution takes 365 days, find the linear speed of Earth. Express your answer in miles per hour.

107. Pulleys Two pulleys, one with radius 2 inches and the other with radius 8 inches, are connected by a belt. (See the figure.) If the 2-inch pulley is caused to rotate at 3 revolutions per minute, determine the revolutions per minute of the 8-inch pulley.

[**Hint:** The linear speeds of the pulleys are the same; both equal the speed of the belt.]

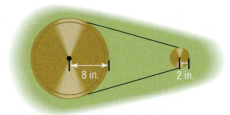

108. **Ferris Wheels**   A neighborhood carnival has a Ferris wheel whose radius is 30 feet. You measure the time it takes for one revolution to be 70 seconds. What is the linear speed (in feet per second) of this Ferris wheel? What is the angular speed in radians per second?

109. **Computing the Speed of a River Current**   To approximate the speed of the current of a river, a circular paddle wheel with radius 4 feet is lowered into the water. If the current causes the wheel to rotate at a speed of 10 revolutions per minute, what is the speed of the current? Express your answer in miles per hour.

110. **Spin Balancing Tires**   A spin balancer rotates the wheel of a car at 480 revolutions per minute. If the diameter of the wheel is 26 inches, what road speed is being tested? Express your answer in miles per hour. At how many revolutions per minute should the balancer be set to test a road speed of 80 miles per hour?

111. **The Cable Cars of San Francisco**   At the Cable Car Museum you can see the four cable lines that are used to pull cable cars up and down the hills of San Francisco. Each cable travels at a speed of 9.55 miles per hour, caused by a rotating wheel whose diameter is 8.5 feet. How fast is the wheel rotating? Express your answer in revolutions per minute.

112. **Difference in Time of Sunrise**   Naples, Florida, is approximately 90 miles due west of Ft. Lauderdale. How much sooner would a person in Ft. Lauderdale first see the rising Sun than a person in Naples?

[**Hint:** Consult the figure. When a person at Q sees the first rays of the Sun, a person at P is still in the dark. The person at P sees the first rays after Earth has rotated so that P is at the location Q. Now use the fact that at the latitude of Ft. Lauderdale in 24 hours a length of arc of $2\pi(3559)$ miles is subtended.]

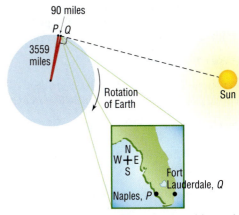

113. **Keeping Up with the Sun**   How fast would you have to travel on the surface of Earth at the equator to keep up with the Sun (that is, so that the Sun would appear to remain in the same position in the sky)?

114. **Nautical Miles**   A **nautical mile** equals the length of arc subtended by a central angle of 1 minute on a great circle[*] on the surface of Earth. (See the figure.) If the radius of Earth is taken as 3960 miles, express 1 nautical mile in terms of ordinary, or **statute**, miles.

115. **Approximating the Circumference of Earth**   Eratosthenes of Cyrene (276–194 BC) was a Greek scholar who lived and worked in Cyrene and Alexandria. One day while visiting in Syene he noticed that the Sun's rays shone directly down a well. On this date 1 year later, in Alexandria, which is 500 miles due north of Syene he measured the angle of the Sun to be about 7.2 degrees. See the figure. Use this information to approximate the radius and circumference of Earth.

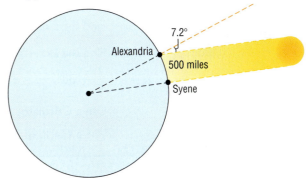

[*]Any circle drawn on the surface of Earth that divides Earth into two equal hemispheres.

116. Pulleys Two pulleys, one with radius r_1 and the other with radius r_2, are connected by a belt. The pulley with radius r_1 rotates at ω_1 revolutions per minute, whereas the pulley with radius r_2 rotates at ω_2 revolutions per minute. Show that $\dfrac{r_1}{r_2} = \dfrac{\omega_2}{\omega_1}$.

Discussion and Writing

117. Do you prefer to measure angles using degrees or radians? Provide justification and a rationale for your choice.

118. What is 1 radian?

119. Which angle has the larger measure: 1 degree or 1 radian? Or are they equal?

120. Explain the difference between linear speed and angular speed.

121. For a circle of radius r, a central angle of θ degrees subtends an arc whose length s is $s = \dfrac{\pi}{180} r\theta$. Discuss whether this is a true or false statement. Give reasons to defend your position.

122. Discuss why ships and airplanes use nautical miles to measure distance. Explain the difference between a nautical mile and a statute mile.

123. Investigate the way that speed bicycles work. In particular, explain the differences and similarities between 5-speed and 9-speed derailleurs. Be sure to include a discussion of linear speed and angular speed.

124. In Example 6, we found that the distance between Albuquerque, NM and Glasgow, MT is approximately 903 miles. According to mapquest.com, the distance is approximately 1300 miles. What might account for the difference?

'Are You Prepared?' Answers

1. $C = 2\pi r$ **2.** $A = \pi r^2$

5.2 Trigonometric Functions: Unit Circle Approach

PREPARING FOR THIS SECTION   *Before getting started, review the following:*

- Pythagorean Theorem (Appendix, Section A.2, pp. 669–671)
- Unit Circle (Section 1.5, p. 45)
- Symmetry (Section 1.2, pp. 17–19)
- Functions (Section 2.1, pp. 56–65)

Now work the 'Are You Prepared?' problems on page 384.

OBJECTIVES   1   Find the Exact Values of the Trigonometric Functions Using a Point on the Unit Circle

    2   Find the Exact Values of the Trigonometric Functions of Quadrantal Angles

    3   Find the Exact Values of the Trigonometric Functions of $\dfrac{\pi}{4} = 45°$

    4   Find the Exact Values of the Trigonometric Functions of $\dfrac{\pi}{6} = 30°$ and $\dfrac{\pi}{3} = 60°$

    5   Find the Exact Values of the Trigonometric Functions for Integer Multiples of $\dfrac{\pi}{6} = 30°$, $\dfrac{\pi}{4} = 45°$, and $\dfrac{\pi}{3} = 60°$

    6   Use a Calculator to Approximate the Value of a Trigonometric Function

    7   Use a Circle of Radius r to Evaluate the Trigonometric Functions

We are now ready to introduce trigonometric functions. The approach that we take uses the unit circle.

The Unit Circle

Recall that the unit circle is a circle whose radius is 1 and whose center is at the origin of a rectangular coordinate system. Also recall that any circle of radius r has circumference of length $2\pi r$. Therefore, the unit circle (radius = 1) has a circumference of length 2π. In other words, for 1 revolution around the unit circle the length of the arc is 2π units.

The following discussion sets the stage for defining the trigonometric functions using the unit circle.

Let t be any real number. We position the t-axis so it is vertical with the positive direction up. We place this t-axis in the xy-plane, so that $t = 0$ is located at the point $(1, 0)$ in the xy-plane.

If $t \geq 0$, let s be the distance from the origin to t on the t-axis. See the red portion of Figure 20(a).

Now look at the unit circle in Figure 20(a). Beginning at the point $(1, 0)$ on the unit circle, travel $s = t$ units in the counterclockwise direction along the circle, to arrive at the point $P = (x, y)$. In this sense, the length $s = t$ units is being **wrapped** around the unit circle.

If $t < 0$, we begin at the point $(1, 0)$ on the unit circle and travel $s = |t|$ units in the clockwise direction to arrive at the point $P = (x, y)$. See Figure 20(b).

Figure 20

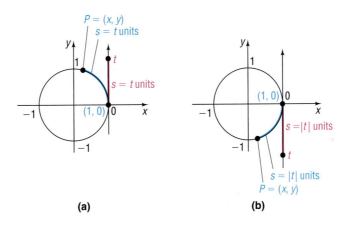

(a)　　　　　　(b)

If $t > 2\pi$ or if $t < -2\pi$, it will be necessary to travel around the unit circle more than once before arriving at the point P. Do you see why?

Let's describe this process another way. Picture a string of length $s = |t|$ units being wrapped around a circle of radius 1 unit. We start wrapping the string around the circle at the point $(1, 0)$. If $t \geq 0$, we wrap the string in the counterclockwise direction; if $t < 0$, we wrap the string in the clockwise direction. The point $P = (x, y)$ is the point where the string ends.

This discussion tells us that, for any real number t, we can locate a unique point $P = (x, y)$ on the unit circle. We call P **the point on the unit circle that corresponds to t.** This is the important idea here. No matter what real number t is chosen, there is a unique point P on the unit circle corresponding to it. We use the coordinates of the point $P = (x, y)$ on the unit circle corresponding to the real number t to define the **six trigonometric functions of t.**

Let t be a real number and let $P = (x, y)$ be the point on the unit circle that corresponds to t.

The **sine function** associates with t the y-coordinate of P and is denoted by

$$\sin t = y$$

The **cosine function** associates with t the x-coordinate of P and is denoted by

$$\cos t = x$$

If $x \neq 0$, the **tangent function** is defined as

$$\tan t = \frac{y}{x}$$

If $y \neq 0$, the **cosecant function** is defined as

$$\csc t = \frac{1}{y}$$

If $x \neq 0$, the **secant function** is defined as

$$\sec t = \frac{1}{x}$$

If $y \neq 0$, the **cotangent function** is defined as

$$\cot t = \frac{x}{y}$$

Notice in these definitions that if $x = 0$, that is, if the point P is on the y-axis, then the tangent function and the secant function are undefined. Also, if $y = 0$, that is, if the point P is on the x-axis, then the cosecant function and the cotangent function are undefined.

Because we use the unit circle in these definitions of the trigonometric functions, they are also sometimes referred to as **circular functions.**

1 Find the Exact Values of the Trigonometric Functions Using a Point on the Unit Circle

| EXAMPLE 1 | **Finding the Values of the Six Trigonometric Functions Using a Point on the Unit Circle** |

Let t be a real number and let $P = \left(-\frac{1}{2}, \frac{\sqrt{3}}{2} \right)$ be the point on the unit circle that corresponds to t. Find the values of $\sin t$, $\cos t$, $\tan t$, $\csc t$, $\sec t$, and $\cot t$.

Solution See Figure 21. We follow the definition of the six trigonometric functions, using

$$P = \left(-\frac{1}{2}, \frac{\sqrt{3}}{2}\right) = (x, y).$$ Then, with $x = -\frac{1}{2}, y = \frac{\sqrt{3}}{2}$, we have

Figure 21

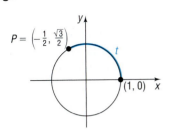

$$\sin t = y = \frac{\sqrt{3}}{2} \qquad \cos t = x = -\frac{1}{2} \qquad \tan t = \frac{y}{x} = \frac{\frac{\sqrt{3}}{2}}{-\frac{1}{2}} = -\sqrt{3}$$

$$\csc t = \frac{1}{y} = \frac{1}{\frac{\sqrt{3}}{2}} = \frac{2\sqrt{3}}{3} \qquad \sec t = \frac{1}{x} = \frac{1}{-\frac{1}{2}} = -2 \qquad \cot t = \frac{x}{y} = \frac{-\frac{1}{2}}{\frac{\sqrt{3}}{2}} = -\frac{\sqrt{3}}{3} \qquad \blacktriangleleft$$

NOW WORK PROBLEM 11.

Trigonometric Functions of Angles

Let $P = (x, y)$ be the point on the unit circle corresponding to the real number t. See Figure 22(a). Let θ be the angle in standard position, measured in radians, whose terminal side is the ray from the origin through P. See Figure 22(b). Since the unit circle has radius 1 unit, from the formula for arc length, $s = r\theta$, we find that

$$s = r\theta = \theta$$
$$\uparrow$$
$$r = 1$$

So, if $s = |t|$ units, then $\theta = t$ radians. See Figures 22(c) and (d).

Figure 22

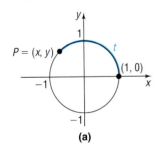
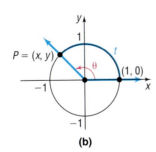
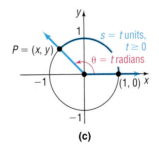
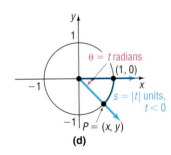

| (a) | (b) | (c) | (d) |

The point $P = (x, y)$ on the unit circle that corresponds to the real number t is the point P on the terminal side of the angle $\theta = t$ radians. As a result, we can say that

$$\sin t = \sin \theta$$
$$\uparrow \qquad\qquad \uparrow$$
$$\text{Real number} \qquad \theta = t \text{ radians}$$

and so on. We can now define the trigonometric functions of the angle θ.

If $\theta = t$ radians, the **six trigonometric functions of the angle θ** are defined as

$$\sin \theta = \sin t \qquad \cos \theta = \cos t \qquad \tan \theta = \tan t$$
$$\csc \theta = \csc t \qquad \sec \theta = \sec t \qquad \cot \theta = \cot t$$

Even though the distinction between trigonometric functions of real numbers and trigonometric functions of angles is important, it is customary to refer to trigonometric functions of real numbers and trigonometric functions of angles collectively as the *trigonometric functions*. We shall follow this practice from now on.

If an angle θ is measured in degrees, we shall use the degree symbol when writing a trigonometric function of θ, as, for example, in $\sin 30°$ and $\tan 45°$. If an angle θ is measured in radians, then no symbol is used when writing a trigonometric function of θ, as, for example, in $\cos \pi$ and $\sec \dfrac{\pi}{3}$.

Finally, since the values of the trigonometric functions of an angle θ are determined by the coordinates of the point $P = (x, y)$ on the unit circle corresponding to θ, the units used to measure the angle θ are irrelevant. For example, it does not matter whether we write $\theta = \dfrac{\pi}{2}$ radians or $\theta = 90°$. The point on the unit circle corresponding to this angle is $P = (0, 1)$. As a result,

$$\sin \frac{\pi}{2} = \sin 90° = 1 \quad \text{and} \quad \cos \frac{\pi}{2} = \cos 90° = 0$$

2 Find the Exact Values of the Trigonometric Functions of Quadrantal Angles

To find the exact value of a trigonometric function of an angle θ or a real number t requires that we locate the point $P = (x, y)$ on the unit circle that corresponds to t. This is not always easy to do. In the examples that follow, we will evaluate the trigonometric functions of certain angles or real numbers for which this process is relatively easy. A calculator will be used to evaluate the trigonometric functions of most other angles.

EXAMPLE 2

Finding the Exact Values of the Six Trigonometric Functions of Quadrantal Angles

Find the exact values of the six trigonometric functions of:

(a) $\theta = 0 = 0°$ (b) $\theta = \dfrac{\pi}{2} = 90°$

(c) $\theta = \pi = 180°$ (d) $\theta = \dfrac{3\pi}{2} = 270°$

Figure 23

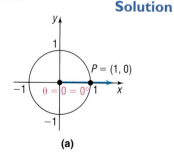

(a)

Solution (a) The point on the unit circle that corresponds to $\theta = 0 = 0°$ is $P = (1, 0)$. See Figure 23(a). Then

$$\sin 0 = \sin 0° = y = 0 \qquad \cos 0 = \cos 0° = x = 1$$

$$\tan 0 = \tan 0° = \frac{y}{x} = 0 \qquad \sec 0 = \sec 0° = \frac{1}{x} = 1$$

Since the y-coordinate of P is 0, $\csc 0$ and $\cot 0$ are not defined.

Figure 23

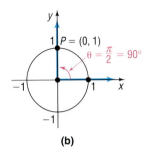

(b)

Figure 23

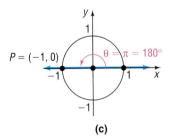

(c)

Figure 23

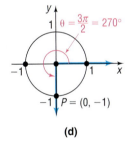

(d)

(b) The point on the unit circle that corresponds to $\theta = \dfrac{\pi}{2} = 90°$ is $P = (0, 1)$. See Figure 23(b). Then

$$\sin\frac{\pi}{2} = \sin 90° = y = 1 \qquad \cos\frac{\pi}{2} = \cos 90° = x = 0$$

$$\csc\frac{\pi}{2} = \csc 90° = \frac{1}{y} = 1 \qquad \cot\frac{\pi}{2} = \cot 90° = \frac{x}{y} = 0$$

Since the x-coordinate of P is 0, $\tan\dfrac{\pi}{2}$ and $\sec\dfrac{\pi}{2}$ are not defined.

(c) The point on the unit circle that corresponds to $\theta = \pi = 180°$ is $P = (-1, 0)$. See Figure 23(c). Then

$$\sin\pi = \sin 180° = y = 0 \qquad \cos\pi = \cos 180° = x = -1$$

$$\tan\pi = \tan 180° = \frac{y}{x} = 0 \qquad \sec\pi = \sec 180° = \frac{1}{x} = -1$$

Since the y-coordinate of P is 0, $\csc\pi$ and $\cot\pi$ are not defined.

(d) The point on the unit circle that corresponds to $\theta = \dfrac{3\pi}{2} = 270°$ is $P = (0, -1)$. See Figure 23(d). Then

$$\sin\frac{3\pi}{2} = \sin 270° = y = -1 \qquad \cos\frac{3\pi}{2} = \cos 270° = x = 0$$

$$\csc\frac{3\pi}{2} = \csc 270° = \frac{1}{y} = -1 \qquad \cot\frac{3\pi}{2} = \cot 270° = \frac{x}{y} = 0$$

Since the x-coordinate of P is 0, $\tan\dfrac{3\pi}{2}$ and $\sec\dfrac{3\pi}{2}$ are not defined. ◀

Table 2 summarizes the values of the trigonometric functions found in Example 2.

Table 2

		Quadrantal Angles					
θ (Radians)	θ (Degrees)	$\sin\theta$	$\cos\theta$	$\tan\theta$	$\csc\theta$	$\sec\theta$	$\cot\theta$
0	0°	0	1	0	Not defined	1	Not defined
$\dfrac{\pi}{2}$	90°	1	0	Not defined	1	Not defined	0
π	180°	0	-1	0	Not defined	-1	Not defined
$\dfrac{3\pi}{2}$	270°	-1	0	Not defined	-1	Not defined	0

There is no need to memorize Table 2. To find the value of a trigonometric function of a quadrantal angle, draw the angle and apply the definition, as we did in Example 2.

| EXAMPLE 3 | **Finding Exact Values of the Trigonometric Functions of Angles That Are Integer Multiples of Quadrantal Angles** |

Find the exact value of:

(a) $\sin(3\pi)$

(b) $\cos(-270°)$

Solution

(a) See Figure 24. The point P on the unit circle that corresponds to $\theta = 3\pi$ is $P = (-1, 0)$, so $\sin(3\pi) = 0$.

(b) See Figure 25. The point P on the unit circle that corresponds to $\theta = -270°$ is $P = (0, 1)$, so $\cos(-270°) = 0$.

Figure 24

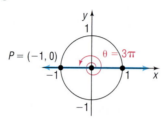

Figure 25

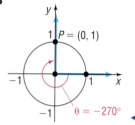

NOW WORK PROBLEMS 19 AND 63.

3 **Find the Exact Values of the Trigonometric Functions of $\dfrac{\pi}{4} = 45°$**

| EXAMPLE 4 | **Finding the Exact Values of the Trigonometric Functions of $\dfrac{\pi}{4} = 45°$** |

Find the exact values of the six trigonometric functions of $\dfrac{\pi}{4} = 45°$.

Solution We seek the coordinates of the point $P = (x, y)$ on the unit circle that corresponds to $\theta = \dfrac{\pi}{4} = 45°$. See Figure 26. First, we observe that P lies on the line $y = x$. (Do you see why? Since $\theta = 45° = \dfrac{1}{2} \cdot 90°$, P must lie on the line that bisects quadrant I.) Since $P = (x, y)$ also lies on the unit circle, $x^2 + y^2 = 1$, it follows that

Figure 26

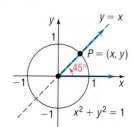

$$x^2 + y^2 = 1$$

$$x^2 + x^2 = 1 \qquad \text{\textcolor{blue}{$y = x, x > 0, y > 0$}}$$

$$2x^2 = 1$$

$$x = \dfrac{1}{\sqrt{2}} = \dfrac{\sqrt{2}}{2}, \qquad y = \dfrac{\sqrt{2}}{2}$$

Then

$$\sin\frac{\pi}{4} = \sin 45° = \frac{\sqrt{2}}{2} \qquad \cos\frac{\pi}{4} = \cos 45° = \frac{\sqrt{2}}{2} \qquad \tan\frac{\pi}{4} = \tan 45° = \frac{\frac{\sqrt{2}}{2}}{\frac{\sqrt{2}}{2}} = 1$$

$$\csc\frac{\pi}{4} = \csc 45° = \frac{1}{\frac{\sqrt{2}}{2}} = \sqrt{2} \qquad \sec\frac{\pi}{4} = \sec 45° = \frac{1}{\frac{\sqrt{2}}{2}} = \sqrt{2} \qquad \cot\frac{\pi}{4} = \cot 45° = \frac{\frac{\sqrt{2}}{2}}{\frac{\sqrt{2}}{2}} = 1 \qquad ◄$$

EXAMPLE 5   **Finding the Exact Value of a Trigonometric Expression**

Find the exact value of each expression.

(a) $\sin 45° \cos 180°$ (b) $\tan\dfrac{\pi}{4} - \sin\dfrac{3\pi}{2}$ (c) $\left(\sec\dfrac{\pi}{4}\right)^2 + \csc\dfrac{\pi}{2}$

Solution   (a) $\sin 45° \cos 180° = \dfrac{\sqrt{2}}{2}\cdot(-1) = -\dfrac{\sqrt{2}}{2}$

                ↑   ↑
       From Example 4   From Table 2

   (b) $\tan\dfrac{\pi}{4} - \sin\dfrac{3\pi}{2} = 1 - (-1) = 2$

                ↑   ↑
       From Example 4   From Table 2

   (c) $\left(\sec\dfrac{\pi}{4}\right)^2 + \csc\dfrac{\pi}{2} = \left(\sqrt{2}\right)^2 + 1 = 2 + 1 = 3$ ◄

✎ **NOW WORK PROBLEM 33.**

---

4   **Find the Exact Values of the Trigonometric Functions of $\dfrac{\pi}{6} = 30°$ and $\dfrac{\pi}{3} = 60°$**

Consider a right triangle in which one of the angles is $\dfrac{\pi}{6} = 30°$. It then follows that the third angle is $\dfrac{\pi}{3} = 60°$. Figure 27(a) illustrates such a triangle with hypotenuse of length 1. Our problem is to determine a and b.

     We begin by placing next to this triangle another triangle congruent to the first, as shown in Figure 27(b). Notice that we now have a triangle whose angles are each 60°. This triangle is therefore equilateral, so each side is of length 1. In particular,

Figure 27

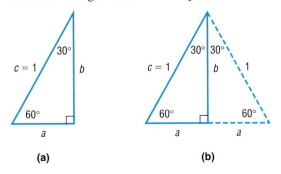

(a)   (b)

the base is $2a = 1$, and so $a = \dfrac{1}{2}$. By the Pythagorean Theorem, b satisfies the equation $a^2 + b^2 = c^2$, so we have

Figure 27

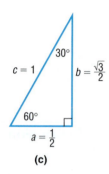

(c)

$$a^2 + b^2 = c^2$$

$$\frac{1}{4} + b^2 = 1 \qquad\qquad a = \frac{1}{2}, c = 1$$

$$b^2 = 1 - \frac{1}{4} = \frac{3}{4}$$

$$b = \frac{\sqrt{3}}{2}$$

This results in Figure 27(c).

EXAMPLE 6 **Finding the Exact Values of the Trigonometric Functions of $\dfrac{\pi}{3} = 60°$**

Find the exact values of the six trigonometric functions of $\dfrac{\pi}{3} = 60°$.

Solution Position the triangle in Figure 27(c) so that the 60° angle is in the standard position. See Figure 28.

Figure 28

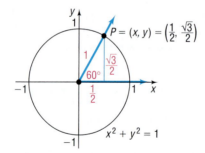

The point on the unit circle that corresponds to $\theta = \dfrac{\pi}{3} = 60°$ is $P = \left(\dfrac{1}{2}, \dfrac{\sqrt{3}}{2}\right)$. Then

$$\sin\frac{\pi}{3} = \sin 60° = \frac{\sqrt{3}}{2} \qquad\qquad \cos\frac{\pi}{3} = \cos 60° = \frac{1}{2}$$

$$\csc\frac{\pi}{3} = \csc 60° = \frac{1}{\frac{\sqrt{3}}{2}} = \frac{2}{\sqrt{3}} = \frac{2\sqrt{3}}{3} \qquad \sec\frac{\pi}{3} = \sec 60° = \frac{1}{\frac{1}{2}} = 2$$

$$\tan\frac{\pi}{3} = \tan 60° = \frac{\frac{\sqrt{3}}{2}}{\frac{1}{2}} = \sqrt{3} \qquad\qquad \cot\frac{\pi}{3} = \cot 60° = \frac{\frac{1}{2}}{\frac{\sqrt{3}}{2}} = \frac{1}{\sqrt{3}} = \frac{\sqrt{3}}{3}$$

◀

EXAMPLE 7

Finding the Exact Values of the Trigonometric Functions of $\frac{\pi}{6} = 30°$

Find the exact values of the trigonometric functions of $\frac{\pi}{6} = 30°$.

Solution

Position the triangle in Figure 27(c) so that the 30° angle is in the standard position. See Figure 29.

The point on the unit circle that corresponds to $\theta = \frac{\pi}{6} = 30°$ is $P = \left(\frac{\sqrt{3}}{2}, \frac{1}{2}\right)$. Then

Figure 29

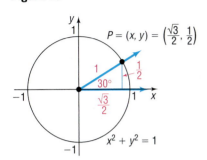

$$\sin\frac{\pi}{6} = \sin 30° = \frac{1}{2} \qquad\qquad \cos\frac{\pi}{6} = \cos 30° = \frac{\sqrt{3}}{2}$$

$$\csc\frac{\pi}{6} = \csc 30° = \frac{1}{\frac{1}{2}} = 2 \qquad \sec\frac{\pi}{6} = \sec 30° = \frac{1}{\frac{\sqrt{3}}{2}} = \frac{2}{\sqrt{3}} = \frac{2\sqrt{3}}{3}$$

$$\tan\frac{\pi}{6} = \tan 30° = \frac{\frac{1}{2}}{\frac{\sqrt{3}}{2}} = \frac{1}{\sqrt{3}} = \frac{\sqrt{3}}{3} \qquad \cot\frac{\pi}{6} = \cot 30° = \frac{\frac{\sqrt{3}}{2}}{\frac{1}{2}} = \sqrt{3}$$

◀

Table 3 summarizes the information just derived for $\frac{\pi}{6} = 30°, \frac{\pi}{4} = 45°,$ and $\frac{\pi}{3} = 60°$. Until you memorize the entries in Table 3, you should draw an appropriate diagram to determine the values given in the table.

Table 3

θ (Radians)	θ (Degrees)	$\sin\theta$	$\cos\theta$	$\tan\theta$	$\csc\theta$	$\sec\theta$	$\cot\theta$
$\frac{\pi}{6}$	30°	$\frac{1}{2}$	$\frac{\sqrt{3}}{2}$	$\frac{\sqrt{3}}{3}$	2	$\frac{2\sqrt{3}}{3}$	$\sqrt{3}$
$\frac{\pi}{4}$	45°	$\frac{\sqrt{2}}{2}$	$\frac{\sqrt{2}}{2}$	1	$\sqrt{2}$	$\sqrt{2}$	1
$\frac{\pi}{3}$	60°	$\frac{\sqrt{3}}{2}$	$\frac{1}{2}$	$\sqrt{3}$	$\frac{2\sqrt{3}}{3}$	2	$\frac{\sqrt{3}}{3}$

 NOW WORK PROBLEM 39.

EXAMPLE 8

Constructing a Rain Gutter

A rain gutter is to be constructed of aluminum sheets 12 inches wide. After marking off a length of 4 inches from each edge, this length is bent up at an angle θ. See Figure 30. The area A of the opening may be expressed as a function of θ as

$$A(\theta) = 16\sin\theta(\cos\theta + 1)$$

Find the area A of the opening for $\theta = 30°$, $\theta = 45°$, and $\theta = 60°$.

Figure 30

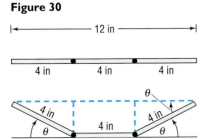

Solution For $\theta = 30°$: $A(30°) = 16\sin 30°(\cos 30° + 1)$

$$= 16\left(\frac{1}{2}\right)\left(\frac{\sqrt{3}}{2} + 1\right) = 4\sqrt{3} + 8 \approx 14.9$$

The area of the opening for $\theta = 30°$ is about 14.9 square inches.

For $\theta = 45°$: $A(45°) = 16 \sin 45°(\cos 45° + 1)$

$$= 16\left(\frac{\sqrt{2}}{2}\right)\left(\frac{\sqrt{2}}{2} + 1\right) = 8 + 8\sqrt{2} \approx 19.3$$

The area of the opening for $\theta = 45°$ is about 19.3 square inches.

For $\theta = 60°$: $A(60°) = 16 \sin 60°(\cos 60° + 1)$

$$= 16\left(\frac{\sqrt{3}}{2}\right)\left(\frac{1}{2} + 1\right) = 12\sqrt{3} \approx 20.8$$

The area of the opening for $\theta = 60°$ is about 20.8 square inches. ◀

5 **Find the Exact Values of the Trigonometric Functions for Integer Multiples of $\dfrac{\pi}{6} = 30°, \dfrac{\pi}{4} = 45°$, and $\dfrac{\pi}{3} = 60°$**

We know the exact values of the trigonometric functions of $\dfrac{\pi}{4} = 45°$. Using symmetry, we can find the exact values of the trigonometric functions of $\dfrac{3\pi}{4} = 135°, \dfrac{5\pi}{4} = 225°$, and $\dfrac{7\pi}{4} = 315°$. Figure 31 shows how.

As Figure 31 shows, using symmetry with respect to the y-axis, the point $\left(-\dfrac{\sqrt{2}}{2}, \dfrac{\sqrt{2}}{2}\right)$ is the point on the unit circle that corresponds to the angle $\dfrac{3\pi}{4} = 135°$.

Similarly, using symmetry with respect to the origin, the point $\left(-\dfrac{\sqrt{2}}{2}, -\dfrac{\sqrt{2}}{2}\right)$ is the point on the unit circle that corresponds to the angle $\dfrac{5\pi}{4} = 225°$. Finally, using symmetry with respect to the x-axis, the point $\left(\dfrac{\sqrt{2}}{2}, -\dfrac{\sqrt{2}}{2}\right)$ is the point on the unit circle that corresponds to the angle $\dfrac{7\pi}{4} = 315°$.

Figure 31

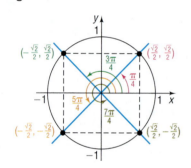

| **EXAMPLE 9** | **Finding Exact Values for Multiples of $\dfrac{\pi}{4} = 45°$** |

Based on Figure 31, we see that

(a) $\sin 135° = \dfrac{\sqrt{2}}{2}$ (b) $\cos \dfrac{5\pi}{4} = -\dfrac{\sqrt{2}}{2}$ (c) $\tan 315° = \dfrac{-\dfrac{\sqrt{2}}{2}}{\dfrac{\sqrt{2}}{2}} = -1$ ◀

Figure 31 can also be used to find exact values for other multiples of $\dfrac{\pi}{4} = 45°$. For example, the point $\left(\dfrac{\sqrt{2}}{2}, -\dfrac{\sqrt{2}}{2}\right)$ is the point on the unit circle that corresponds to the angle $-\dfrac{\pi}{4} = -45°$; the point $\left(\dfrac{\sqrt{2}}{2}, \dfrac{\sqrt{2}}{2}\right)$ is the point on the unit circle that corresponds to the angle $\dfrac{9\pi}{4} = 405°$.

NOW WORK PROBLEMS 53 AND 57.

The use of symmetry also provides information about certain integer multiples of the angles $\dfrac{\pi}{6} = 30°$ and $\dfrac{\pi}{3} = 60°$. See Figures 32 and 33.

Figure 32

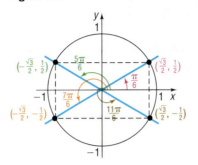

Figure 33

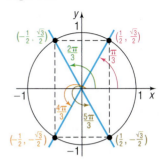

EXAMPLE 10   **Using Figures 32 and 33**

Based on Figures 32 and 33, we see that

(a) $\cos 210° = -\dfrac{\sqrt{3}}{2}$ (b) $\sin(-60°) = -\dfrac{\sqrt{3}}{2}$ (c) $\tan \dfrac{5\pi}{3} = \dfrac{-\dfrac{\sqrt{3}}{2}}{\dfrac{1}{2}} = -\sqrt{3}$ ◀

 NOW WORK PROBLEM 49.

6   Use a Calculator to Approximate the Value of a Trigonometric Function

Before getting started, you must first decide whether to enter the angle in the calculator using radians or degrees and then set the calculator to the correct MODE. Check your instruction manual to find out how your calculator handles degrees and radians. Your calculator has keys marked $\boxed{\sin}$, $\boxed{\cos}$, and $\boxed{\tan}$. To find the values of the remaining three trigonometric functions, secant, cosecant, and cotangent, we use the fact that, if $P = (x, y)$ is a point on the unit circle on the terminal side of θ, then

$$\sec \theta = \frac{1}{x} = \frac{1}{\cos \theta} \qquad \csc \theta = \frac{1}{y} = \frac{1}{\sin \theta} \qquad \cot \theta = \frac{x}{y} = \frac{1}{\dfrac{y}{x}} = \frac{1}{\tan \theta}$$

EXAMPLE 11   **Using a Calculator to Approximate the Value of a Trigonometric Function**

Use a calculator to find the approximate value of:

(a) $\cos 48°$ (b) $\csc 21°$ (c) $\tan \dfrac{\pi}{12}$

Express your answers rounded to two decimal places.

Solution   (a) First we set the MODE to receive degrees. See Figure 34(a). Figure 34(b) shows the solution using a TI-84 Plus graphing calculator. Rounded to two decimal places,

$$\cos 48° = 0.66991306 \approx 0.67$$

Figure 34

(a)

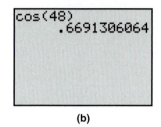

(b)

Figure 35

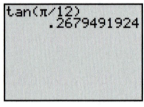

(b) Most calculators do not have a csc key. The manufacturers assume that the user knows some trigonometry. To find the value of csc 21°, use the fact that $\csc 21° = \dfrac{1}{\sin 21°}$. Figure 35 shows the solution using a TI-84 Plus graphing calculator. Rounded to two decimal places,

$$\csc 21° \approx 2.79$$

(c) Set the MODE to receive radians. Figure 36 shows the solution using a TI-84 Plus graphing calculator. Rounded to two decimal places,

$$\tan \frac{\pi}{12} \approx 0.27$$ ◀

Figure 36

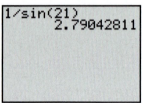

NOW WORK PROBLEM 67.

7 Use a Circle of Radius r to Evaluate the Trigonometric Functions

Until now, to find the exact value of a trigonometric function of an angle θ required that we locate the corresponding point $P = (x, y)$ on the unit circle. In fact, though, any circle whose center is at the origin can be used.

Let θ be any nonquadrantal angle placed in standard position. Let $P = (x, y)$ be the point on the circle $x^2 + y^2 = r^2$ that corresponds to θ, and let $P^* = (x^*, y^*)$ be the point on the unit circle that corresponds to θ. See Figure 37.

Notice that the triangles OA^*P^* and OAP are similar; as a result, the ratios of corresponding sides are equal.

Figure 37

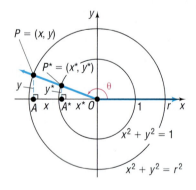

$$\frac{y^*}{1} = \frac{y}{r} \qquad \frac{x^*}{1} = \frac{x}{r} \qquad \frac{y^*}{x^*} = \frac{y}{x}$$

$$\frac{1}{y^*} = \frac{r}{y} \qquad \frac{1}{x^*} = \frac{r}{x} \qquad \frac{x^*}{y^*} = \frac{x}{y}$$

These results lead us to formulate the following theorem:

Theorem

For an angle θ in standard position, let $P = (x, y)$ be the point on the terminal side of θ that is also on the circle $x^2 + y^2 = r^2$. Then

$$\sin \theta = \frac{y}{r} \qquad\qquad \cos \theta = \frac{x}{r} \qquad\qquad \tan \theta = \frac{y}{x}, \quad x \neq 0$$

$$\csc \theta = \frac{r}{y}, \quad y \neq 0 \qquad \sec \theta = \frac{r}{x}, \quad x \neq 0 \qquad \cot \theta = \frac{x}{y}, \quad y \neq 0$$

EXAMPLE 12	**Finding the Exact Values of the Six Trigonometric Functions**

Find the exact values of each of the six trigonometric functions of an angle θ if $(4, -3)$ is a point on its terminal side.

Solution See Figure 38. The point $(4, -3)$ is on a circle of radius $r = \sqrt{4^2 + (-3)^2} = \sqrt{16 + 9} = \sqrt{25} = 5$ with the center at the origin.

Figure 38

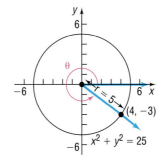

For the point $(x, y) = (4, -3)$, we have $x = 4$ and $y = -3$. Since $r = 5$, we find

$$\sin \theta = \frac{y}{r} = -\frac{3}{5} \qquad \cos \theta = \frac{x}{r} = \frac{4}{5} \qquad \tan \theta = \frac{y}{x} = -\frac{3}{4}$$

$$\csc \theta = \frac{r}{y} = -\frac{5}{3} \qquad \sec \theta = \frac{r}{x} = \frac{5}{4} \qquad \cot \theta = \frac{x}{y} = -\frac{4}{3}$$

◀

   NOW WORK PROBLEM 83.

HISTORICAL FEATURE

The name *sine* for the sine function is due to a medieval confusion. The name comes from the Sanskrit word *jiva* (meaning chord), first used in India by Araybhata the Elder (AD 510). He really meant half-chord, but abbreviated it. This was brought into Arabic as *jiba*, which was meaningless. Because the proper Arabic word *jaib* would be written the same way (short vowels are not written out in Arabic), *jiba* was pronounced as *jaib*, which meant bosom or hollow, and *jiba* remains as the Arabic word for sine to this day. Scholars translating the Arabic works into Latin found that the word *sinus* also meant bosom or hollow, and from *sinus* we get the word *sine*.

The name *tangent*, due to Thomas Finck (1583), can be understood by looking at Figure 39. The line segment $\overline{DC}$ is tangent to the circle at C. If $d(O, B) = d(O, C) = 1$, then the length of the line segment $\overline{DC}$ is

$$d(D, C) = \frac{d(D, C)}{1} = \frac{d(D, C)}{d(O, C)} = \tan \alpha$$

The old name for the tangent is *umbra versa* (meaning turned shadow), referring to the use of the tangent in solving height problems with shadows.

The names of the remaining functions came about as follows. If α and β are complementary angles, then $\cos \alpha = \sin \beta$. Because β is the complement of α, it was natural to write the cosine of α as *sin co α*. Probably for reasons involving ease of pronunciation, the *co* migrated to the front, and then cosine received a three-letter abbreviation to match sin, sec, and tan. The two other cofunctions were similarly treated, except that the long forms *cotan* and *cosec* survive to this day in some countries.

Figure 39

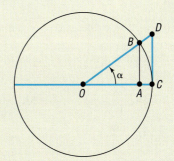

5.2 Assess Your Understanding

'Are You Prepared?'

Answers are given at the end of these exercises. If you get a wrong answer, read the pages listed in red.

1. In a right triangle, with legs a and b and hypotenuse c, the Pythagorean Theorem states that _____. (p. 670)

2. The value of the function $f(x) = 3x - 7$ at 5 is _____. (pp. 61–63)

3. *True or False:* For a function $y = f(x)$, for each x in the domain, there is exactly one element y in the range. (pp. 55–61)

4. What is the equation of the unit circle? (p. 45)

5. What point is symmetric with respect to the y-axis to the point $\left(\frac{1}{2}, \frac{\sqrt{3}}{2}\right)$? (pp. 17–19)

6. If (x, y) is a point on the unit circle in quadrant IV and if $x = \frac{\sqrt{3}}{2}$, what is y? (p. 45)

Concepts and Vocabulary

7. $\tan \dfrac{\pi}{4} + \sin 30° = $ _____.

8. Using a calculator, $\sin 2 = $ _____, rounded to two decimal places.

9. *True or False:* Exact values can be found for the trigonometric functions of 60°.

10. *True or False:* Exact values can be found for the sine of any angle.

Skill Building

In Problems 11–18, t is a real number and P = (x, y) is the point on the unit circle that corresponds to t. Find the exact values of the six trigonometric functions of t.

11. $\left(\dfrac{\sqrt{3}}{2}, \dfrac{1}{2}\right)$

12. $\left(\dfrac{1}{2}, -\dfrac{\sqrt{3}}{2}\right)$

13. $\left(-\dfrac{2}{5}, \dfrac{\sqrt{21}}{5}\right)$

14. $\left(-\dfrac{1}{5}, \dfrac{2\sqrt{6}}{5}\right)$

15. $\left(-\dfrac{\sqrt{2}}{2}, \dfrac{\sqrt{2}}{2}\right)$

16. $\left(\dfrac{\sqrt{2}}{2}, \dfrac{\sqrt{2}}{2}\right)$

17. $\left(\dfrac{2\sqrt{2}}{3}, -\dfrac{1}{3}\right)$

18. $\left(-\dfrac{\sqrt{5}}{3}, -\dfrac{2}{3}\right)$

In Problems 19–28, find the exact value. Do not use a calculator.

19. $\sin \dfrac{11\pi}{2}$

20. $\cos(7\pi)$

21. $\tan(6\pi)$

22. $\cot \dfrac{7\pi}{2}$

23. $\csc \dfrac{11\pi}{2}$

24. $\sec(8\pi)$

25. $\cos\left(-\dfrac{3\pi}{2}\right)$

26. $\sin(-3\pi)$

27. $\sec(-\pi)$

28. $\tan(-3\pi)$

In Problems 29–48, find the exact value of each expression. Do not use a calculator.

29. $\sin 45° + \cos 60°$

30. $\sin 30° - \cos 45°$

31. $\sin 90° + \tan 45°$

32. $\cos 180° - \sin 180°$

33. $\sin 45° \cos 45°$

34. $\tan 45° \cos 30°$

35. $\csc 45° \tan 60°$

36. $\sec 30° \cot 45°$

37. $4 \sin 90° - 3 \tan 180°$

38. $5 \cos 90° - 8 \sin 270°$

39. $2 \sin \dfrac{\pi}{3} - 3 \tan \dfrac{\pi}{6}$

40. $2 \sin \dfrac{\pi}{4} + 3 \tan \dfrac{\pi}{4}$

41. $\sin \dfrac{\pi}{4} - \cos \dfrac{\pi}{4}$

42. $\tan \dfrac{\pi}{3} + \cos \dfrac{\pi}{3}$

43. $2 \sec \dfrac{\pi}{4} + 4 \cot \dfrac{\pi}{3}$

44. $3 \csc \dfrac{\pi}{3} + \cot \dfrac{\pi}{4}$

45. $\tan \pi - \cos 0$

46. $\sin \dfrac{3\pi}{2} + \tan \pi$

47. $\csc \dfrac{\pi}{2} + \cot \dfrac{\pi}{2}$

48. $\sec \pi - \csc \dfrac{\pi}{2}$

In Problems 49–66, find the exact values of the six trigonometric functions of the given angle. If any are not defined, say "not defined." Do not use a calculator.

49. $\dfrac{2\pi}{3}$

50. $\dfrac{5\pi}{6}$

51. $210°$

52. $240°$

53. $\dfrac{3\pi}{4}$

54. $\dfrac{11\pi}{4}$

55. $\dfrac{8\pi}{3}$

56. $\dfrac{13\pi}{6}$

57. $405°$

58. $390°$

59. $-\dfrac{\pi}{6}$

60. $-\dfrac{\pi}{3}$

61. $-45°$

62. $-60°$

63. $\dfrac{5\pi}{2}$

64. 5π

65. $720°$

66. $630°$

In Problems 67–82, use a calculator to find the approximate value of each expression rounded to two decimal places.

67. sin 28° **68.** cos 14° **69.** tan 21° **70.** cot 70°

71. sec 41° **72.** csc 55° **73.** $\sin \dfrac{\pi}{10}$ **74.** $\cos \dfrac{\pi}{8}$

75. $\tan \dfrac{5\pi}{12}$ **76.** $\cot \dfrac{\pi}{18}$ **77.** $\sec \dfrac{\pi}{12}$ **78.** $\csc \dfrac{5\pi}{13}$

79. sin 1 **80.** tan 1 **81.** sin 1° **82.** tan 1°

In Problems 83–92, a point on the terminal side of an angle θ is given. Find the exact values of the six trigonometric functions of θ.

83. $(-3, 4)$ **84.** $(5, -12)$ **85.** $(2, -3)$ **86.** $(-1, -2)$ **87.** $(-2, -2)$

88. $(1, -1)$ **89.** $(-3, -2)$ **90.** $(2, 2)$ **91.** $\left(\dfrac{1}{3}, -\dfrac{1}{4}\right)$ **92.** $(-0.3, -0.4)$

93. Find the exact value of:
$$\sin 45° + \sin 135° + \sin 225° + \sin 315°$$

94. Find the exact value of:
$$\tan 60° + \tan 150°$$

95. If $\sin \theta = 0.1$, find $\sin(\theta + \pi)$.

96. If $\cos \theta = 0.3$, find $\cos(\theta + \pi)$.

97. If $\tan \theta = 3$, find $\tan(\theta + \pi)$.

98. If $\cot \theta = -2$, find $\cot(\theta + \pi)$.

99. If $\sin \theta = \dfrac{1}{5}$, find $\csc \theta$.

100. If $\cos \theta = \dfrac{2}{3}$, find $\sec \theta$.

In Problems 101–112, $f(\theta) = \sin \theta$ and $g(\theta) = \cos \theta$. Find the exact value of each function below if $\theta = 60°$. Do not use a calculator.

101. $f(\theta)$ **102.** $g(\theta)$ **103.** $f\left(\dfrac{\theta}{2}\right)$ **104.** $g\left(\dfrac{\theta}{2}\right)$ **105.** $[f(\theta)]^2$ **106.** $[g(\theta)]^2$

107. $f(2\theta)$ **108.** $g(2\theta)$ **109.** $2f(\theta)$ **110.** $2g(\theta)$ **111.** $f(-\theta)$ **112.** $g(-\theta)$

113. Use a calculator in radian mode to complete the following table.

What can you conclude about the value of $\dfrac{\sin \theta}{\theta}$ as θ approaches 0?

θ	0.5	0.4	0.2	0.1	0.01	0.001	0.0001	0.00001
$\sin \theta$								
$\dfrac{\sin \theta}{\theta}$								

114. Use a calculator in radian mode to complete the following table.

What can you conclude about the value of $\dfrac{\cos \theta - 1}{\theta}$ as θ approaches 0?

θ	0.5	0.4	0.2	0.1	0.01	0.001	0.0001	0.00001
$\cos \theta - 1$								
$\dfrac{\cos \theta - 1}{\theta}$								

Projectile Motion *The path of a projectile fired at an inclination θ to the horizontal with initial speed v_0 is a parabola (see the figure).*

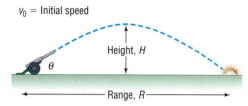

v_0 = Initial speed

Height, *H*

θ

Range, *R*

The range R of the projectile, that is, the horizontal distance that the projectile travels, is found by using the formula

$$R = \frac{v_0^2 \sin(2\theta)}{g}$$

where g ≈ 32.2 feet per second per second ≈ 9.8 meters per second per second is the acceleration due to gravity. The maximum height H of the projectile is

$$H = \frac{v_0^2 (\sin \theta)^2}{2g}$$

In Problems 115–118, find the range R and maximum height H.

115. The projectile is fired at an angle of 45° to the horizontal with an initial speed of 100 feet per second.

116. The projectile is fired at an angle of 30° to the horizontal with an initial speed of 150 meters per second.

117. The projectile is fired at an angle of 25° to the horizontal with an initial speed of 500 meters per second.

118. The projectile is fired at an angle of 50° to the horizontal with an initial speed of 200 feet per second.

119. Inclined Plane If friction is ignored, the time *t* (in seconds) required for a block to slide down an inclined plane (see the figure) is given by the formula

$$t = \sqrt{\frac{2a}{g \sin \theta \cos \theta}}$$

where *a* is the length (in feet) of the base and *g* ≈ 32 feet per second per second is the acceleration due to gravity. How long does it take a block to slide down an inclined plane with base *a* = 10 feet when:
(a) θ = 30°?   (b) θ = 45°?   (c) θ = 60°?

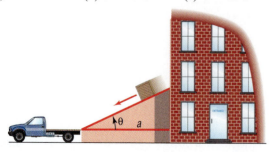

120. Piston Engines In a certain piston engine, the distance *x* (in centimeters) from the center of the drive shaft to the head of the piston is given by

$$x = \cos \theta + \sqrt{16 + 0.5 \cos(2\theta)}$$

where θ is the angle between the crank and the path of the piston head (see the figure). Find *x* when θ = 30° and when θ = 45°.

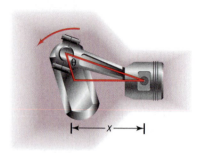

121. Calculating the Time of a Trip Two oceanfront homes are located 8 miles apart on a straight stretch of beach, each a distance of 1 mile from a paved road that parallels the ocean. Sally can jog 8 miles per hour along the paved road, but only 3 miles per hour in the sand on the beach. Because of a river directly between the two houses, it is necessary to jog in the sand to the road, continue on the road, and then jog directly back in the sand to get from one house to the other. See the illustration. The time *T* to get from one house to the other as a function of the angle θ shown in the illustration is

$$T(\theta) = 1 + \frac{2}{3 \sin \theta} - \frac{1}{4 \tan \theta}, \qquad 0° < \theta < 90°$$

(a) Calculate the time *T* for θ = 30°. How long is Sally on the paved road?
(b) Calculate the time *T* for θ = 45°. How long is Sally on the paved road?
(c) Calculate the time *T* for θ = 60°. How long is Sally on the paved road?
(d) Calculate the time *T* for θ = 90°. Describe the path taken. Why can't the formula for *T* be used?

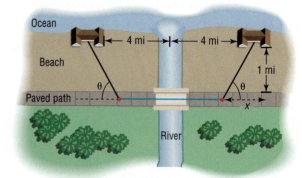

122. Designing Fine Decorative Pieces A designer of decorative art plans to market solid gold spheres encased in clear crystal cones. Each sphere is of fixed radius *R* and will be enclosed in a cone of height *h* and radius *r*. See the illustration. Many cones can be used to enclose the sphere, each having a different slant angle θ. The volume *V* of the cone can be expressed as a function of the slant angle θ of the cone as

$$V(\theta) = \frac{1}{3}\pi R^3 \frac{(1 + \sec \theta)^3}{(\tan \theta)^2}, \qquad 0° < \theta < 90°$$

What volume V is required to enclose a sphere of radius 2 centimeters in a cone whose slant angle θ is 30°? 45°? 60°?

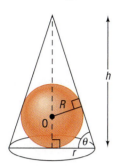

123. Projectile Motion An object is propelled upward at an angle θ, $45° < \theta < 90°$, to the horizontal with an initial velocity of v_0 feet per second from the base of an inclined plane that makes an angle of 45° with the horizontal. See the illustration. If air resistance is ignored, the distance R that it travels up the inclined plane is given by

$$R = \frac{v_0^2 \sqrt{2}}{32} \left[\sin(2\theta) - \cos(2\theta) - 1\right]$$

(a) Find the distance R that the object travels along the inclined plane if the initial velocity is 32 feet per second and $\theta = 60°$.
(b) Graph $R = R(\theta)$ if the initial velocity is 32 feet per second.
(c) What value of θ makes R largest?

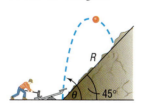

124. If θ, $0 < \theta < \pi$, is the angle between the positive x-axis and a nonhorizontal, nonvertical line L, show that the slope m of L equals $\tan \theta$. The angle θ is called the **inclination** of L.

[**Hint:** See the illustration, where we have drawn the line M parallel to L and passing through the origin. Use the fact that M intersects the unit circle at the point $(\cos \theta, \sin \theta)$.]

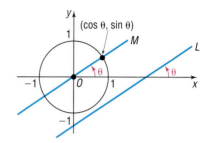

In Problems 125–128, use the figure to approximate the value of the six trigonometric functions at t to the nearest tenth. Then use a calculator to approximate each of the six trigonometric functions at t.

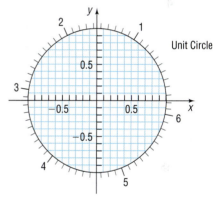

125. (a) $t = 1$ (b) $t = 5.1$ (c) $t = 2.4$
126. (a) $t = 2$ (b) $t = 4$ (c) $t = 5.9$
127. (a) $t = 1.5$ (b) $t = 4.3$ (c) $t = 5.3$
128. (a) $t = 2.7$ (b) $t = 3.9$ (c) $t = 6.1$

Discussion and Writing

129. Write a brief paragraph that explains how to quickly compute the trigonometric functions of 30°, 45°, and 60°.

130. Write a brief paragraph that explains how to quickly compute the trigonometric functions of 0°, 90°, 180°, and 270°.

131. How would you explain the meaning of the sine function to a fellow student who has just completed college algebra?

'Are You Prepared?' Answers

1. $c^2 = a^2 + b^2$ **2.** 8 **3.** True **4.** $x^2 + y^2 = 1$ **5.** $\left(-\dfrac{1}{2}, \dfrac{\sqrt{3}}{2}\right)$ **6.** $-\dfrac{1}{2}$

5.3 Properties of the Trigonometric Functions

PREPARING FOR THIS SECTION   *Before getting started, review the following:*

• Functions (Section 2.1, pp. 56–65)

• Even and Odd Functions (Section 2.3, pp. 80–82)

• Identity (Appendix, Section A.5, p. 692)

 Now work the 'Are You Prepared?' problems on page 399.

OBJECTIVES   1 Determine the Domain and the Range of the Trigonometric Functions
2 Determine the Period of the Trigonometric Functions
3 Determine the Signs of the Trigonometric Functions in a Given Quadrant
4 Find the Values of the Trigonometric Functions Using Fundamental Identities
5 Find the Exact Values of the Trigonometric Functions of an Angle Given One of the Functions and the Quadrant of the Angle
6 Use Even–Odd Properties to Find the Exact Values of the Trigonometric Functions

1   Determine the Domain and the Range of the Trigonometric Functions

Figure 40

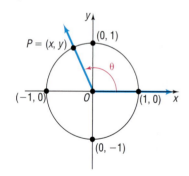

Let θ be an angle in standard position, and let $P = (x, y)$ be the point on the unit circle that corresponds to θ. See Figure 40. Then, by definition,

$$\sin \theta = y \qquad \cos \theta = x \qquad \tan \theta = \frac{y}{x}, \quad x \neq 0$$

$$\csc \theta = \frac{1}{y}, \quad y \neq 0 \qquad \sec \theta = \frac{1}{x}, \quad x \neq 0 \qquad \cot \theta = \frac{x}{y}, \quad y \neq 0$$

For $\sin \theta$ and $\cos \theta$, θ can be any angle, so it follows that the domain of the sine function and cosine function is the set of all real numbers.

> The domain of the sine function is the set of all real numbers.
>
> The domain of the cosine function is the set of all real numbers.

If $x = 0$, then the tangent function and the secant function are not defined. That is, for the tangent function and secant function, the x-coordinate of $P = (x, y)$ cannot be 0. On the unit circle, there are two such points $(0, 1)$ and $(0, -1)$. These two points correspond to the angles $\frac{\pi}{2}(90°)$ and $\frac{3\pi}{2}(270°)$ or, more generally, to any angle that is an odd integer multiple of $\frac{\pi}{2}(90°)$, such as $\frac{\pi}{2}(90°)$, $\frac{3\pi}{2}(270°)$, $\frac{5\pi}{2}(450°)$, $-\frac{\pi}{2}(-90°)$, $-\frac{3\pi}{2}(-270°)$, and so on. Such angles must therefore be excluded from the domain of the tangent function and secant function.

> The domain of the tangent function is the set of all real numbers, except odd integer multiples of $\dfrac{\pi}{2} (90°)$.
>
> The domain of the secant function is the set of all real numbers, except odd integer multiples of $\dfrac{\pi}{2} (90°)$.

If $y = 0$, then the cotangent function and the cosecant function are not defined. For the cotangent function and cosecant function, the y-coordinate of $P = (x, y)$ cannot be 0. On the unit circle, there are two such points, $(1, 0)$ and $(-1, 0)$. These two points correspond to the angles $0(0°)$ and $\pi(180°)$ or, more generally, to any angle that is an integer multiple of $\pi(180°)$, such as $0(0°)$, $\pi(180°)$, $2\pi(360°)$, $3\pi(540°)$, $-\pi(-180°)$, and so on. Such angles must therefore be excluded from the domain of the cotangent function and cosecant function.

> The domain of the cotangent function is the set of all real numbers, except integer multiples of $\pi(180°)$.
>
> The domain of the cosecant function is the set of all real numbers, except integer multiples of $\pi(180°)$.

Next we determine the range of each of the six trigonometric functions. Refer again to Figure 40. Let $P = (x, y)$ be the point on the unit circle that corresponds to the angle θ. It follows that $-1 \le x \le 1$ and $-1 \le y \le 1$. Consequently, since $\sin \theta = y$ and $\cos \theta = x$, we have

$$-1 \le \sin \theta \le 1 \qquad -1 \le \cos \theta \le 1$$

The range of both the sine function and the cosine function consists of all real numbers between -1 and 1, inclusive. Using absolute value notation, we have $|\sin \theta| \le 1$ and $|\cos \theta| \le 1$.

If θ is not an integer multiple of $\pi(180°)$, then $\csc \theta = \dfrac{1}{y}$. Since $y = \sin \theta$ and $|y| = |\sin \theta| \le 1$, it follows that $|\csc \theta| = \dfrac{1}{|\sin \theta|} = \dfrac{1}{|y|} \ge 1$. The range of the cosecant function consists of all real numbers less than or equal to -1 or greater than or equal to 1. That is,

$$\csc \theta \le -1 \quad \text{or} \quad \csc \theta \ge 1$$

If θ is not an odd integer multiple of $\dfrac{\pi}{2} (90°)$, then, by definition, $\sec \theta = \dfrac{1}{x}$. Since $x = \cos \theta$ and $|x| = |\cos \theta| \le 1$, it follows that $|\sec \theta| = \dfrac{1}{|\cos \theta|} = \dfrac{1}{|x|} \ge 1$. The range of the secant function consists of all real numbers less than or equal to -1 or greater than or equal to 1. That is,

$$\sec \theta \le -1 \quad \text{or} \quad \sec \theta \ge 1$$

The range of both the tangent function and the cotangent function is the set of all real numbers. You are asked to prove this in Problems 121 and 122.

$$-\infty < \tan \theta < \infty \qquad -\infty < \cot \theta < \infty$$

Table 4 summarizes these results.

Table 4

Function	Symbol	Domain	Range
sine	$f(\theta) = \sin \theta$	All real numbers	All real numbers from -1 to 1, inclusive
cosine	$f(\theta) = \cos \theta$	All real numbers	All real numbers from -1 to 1, inclusive
tangent	$f(\theta) = \tan \theta$	All real numbers, except odd integer multiples of $\dfrac{\pi}{2}$ (90°)	All real numbers
cosecant	$f(\theta) = \csc \theta$	All real numbers, except integer multiples of π (180°)	All real numbers greater than or equal to 1 or less than or equal to -1
secant	$f(\theta) = \sec \theta$	All real numbers, except odd integer multiples of $\dfrac{\pi}{2}$ (90°)	All real numbers greater than or equal to 1 or less than or equal to -1
cotangent	$f(\theta) = \cot \theta$	All real numbers, except integer multiples of π (180°)	All real numbers

NOW WORK PROBLEM 97.

2 Determine the Period of the Trigonometric Functions

Figure 41

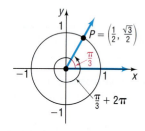

Look at Figure 41. This figure shows that for an angle of $\dfrac{\pi}{3}$ radians the corresponding point P on the unit circle is $\left(\dfrac{1}{2}, \dfrac{\sqrt{3}}{2}\right)$. Notice that, for an angle of $\dfrac{\pi}{3} + 2\pi$ radians, the corresponding point P on the unit circle is also $\left(\dfrac{1}{2}, \dfrac{\sqrt{3}}{2}\right)$. Then

$$\sin \frac{\pi}{3} = \frac{\sqrt{3}}{2} \quad \text{and} \quad \sin\left(\frac{\pi}{3} + 2\pi\right) = \frac{\sqrt{3}}{2}$$

$$\cos \frac{\pi}{3} = \frac{1}{2} \quad \text{and} \quad \cos\left(\frac{\pi}{3} + 2\pi\right) = \frac{1}{2}$$

Figure 42

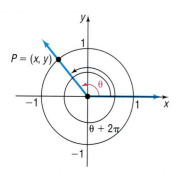

This example illustrates a more general situation. For a given angle θ, measured in radians, suppose that we know the corresponding point $P = (x, y)$ on the unit circle. Now add 2π to θ. The point on the unit circle corresponding to $\theta + 2\pi$ is identical to the point P on the unit circle corresponding to θ. See Figure 42. The values of the trigonometric functions of $\theta + 2\pi$ are equal to the values of the corresponding trigonometric functions of θ.

If we add (or subtract) integer multiples of 2π to θ, the trigonometric values remain unchanged. That is, for all θ.

$$\sin(\theta + 2\pi k) = \sin \theta \qquad \cos(\theta + 2\pi k) = \cos \theta \tag{1}$$
$$\text{where } k \text{ is any integer.}$$

Functions that exhibit this kind of behavior are called *periodic functions*.

A function f is called **periodic** if there is a positive number p such that, whenever θ is in the domain of f, so is $\theta + p$, and

$$f(\theta + p) = f(\theta)$$

If there is a smallest such number p, this smallest value is called the **(fundamental) period** of f.

Based on equation (1), the sine and cosine functions are periodic. In fact, the sine and cosine functions have period 2π. You are asked to prove this fact in Problems 123 and 124. The secant and cosecant functions are also periodic with period 2π, and the tangent and cotangent functions are periodic with period π. You are asked to prove these statements in Problems 125 through 128.

These facts are summarized as follows:

In Words

Tangent and cotangent have period π; the others have period 2π.

Periodic Properties

$$\sin(\theta + 2\pi) = \sin\theta \quad \cos(\theta + 2\pi) = \cos\theta \quad \tan(\theta + \pi) = \tan\theta$$
$$\csc(\theta + 2\pi) = \csc\theta \quad \sec(\theta + 2\pi) = \sec\theta \quad \cot(\theta + \pi) = \cot\theta$$

Because the sine, cosine, secant, and cosecant functions have period 2π, once we know their values for $0 \leq \theta < 2\pi$, we know all their values; similarly, since the tangent and cotangent functions have period π, once we know their values for $0 \leq \theta < \pi$, we know all their values.

EXAMPLE 1 | **Finding Exact Values Using Periodic Properties**

Find the exact value of:

(a) $\sin\dfrac{17\pi}{4}$ (b) $\cos(5\pi)$ (c) $\tan\dfrac{5\pi}{4}$

Solution (a) It is best to sketch the angle first, as shown in Figure 43(a). Since the period of the sine function is 2π, each full revolution can be ignored. This leaves the angle $\dfrac{\pi}{4}$. Then

$$\sin\frac{17\pi}{4} = \sin\left(\frac{\pi}{4} + 4\pi\right) = \sin\frac{\pi}{4} = \frac{\sqrt{2}}{2}$$

Figure 43

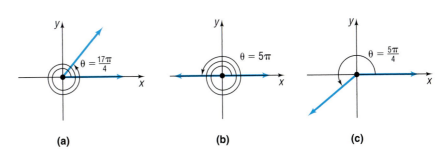

(a)   (b)   (c)

(b) See Figure 43(b). Since the period of the cosine function is 2π, each full revolution can be ignored. This leaves the angle π. Then

$$\cos(5\pi) = \cos(\pi + 4\pi) = \cos \pi = -1$$

(c) See Figure 43(c). Since the period of the tangent function is π, each half-revolution can be ignored. This leaves the angle $\frac{\pi}{4}$. Then

$$\tan \frac{5\pi}{4} = \tan\left(\frac{\pi}{4} + \pi\right) = \tan \frac{\pi}{4} = 1 \qquad \blacktriangleleft$$

The periodic properties of the trigonometric functions will be very helpful to us when we study their graphs later in the chapter.

NOW WORK PROBLEM 11.

3 Determine the Signs of the Trigonometric Functions in a Given Quadrant

Let $P = (x, y)$ be the point on the unit circle that corresponds to the angle θ. If we know in which quadrant the point P lies, then we can determine the signs of the trigonometric functions of θ. For example, if $P = (x, y)$ lies in quadrant IV, as shown in Figure 44, then we know that $x > 0$ and $y < 0$. Consequently,

Figure 44

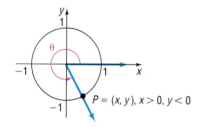

$$\sin \theta = y < 0 \qquad \cos \theta = x > 0 \qquad \tan \theta = \frac{y}{x} < 0$$

$$\csc \theta = \frac{1}{y} < 0 \qquad \sec \theta = \frac{1}{x} > 0 \qquad \cot \theta = \frac{x}{y} < 0$$

Table 5 lists the signs of the six trigonometric functions for each quadrant. See also Figure 45.

Table 5

Quadrant of P	$\sin \theta, \csc \theta$	$\cos \theta, \sec \theta$	$\tan \theta, \cot \theta$
I	Positive	Positive	Positive
II	Positive	Negative	Negative
III	Negative	Negative	Positive
IV	Negative	Positive	Negative

Figure 45

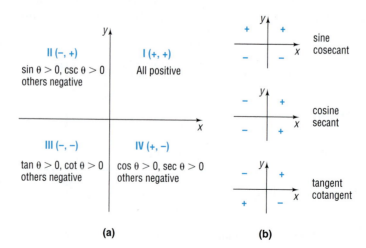

(a)   (b)

EXAMPLE 2	**Finding the Quadrant in Which an Angle θ Lies**

If $\sin\theta < 0$ and $\cos\theta < 0$, name the quadrant in which the angle θ lies.

Solution Let $P = (x, y)$ be the point on the unit circle corresponding to θ. Then $\sin\theta = y < 0$ and $\cos\theta = x < 0$. The point $P = (x, y)$ must be in quadrant III, so θ lies in quadrant III. ◀

 NOW WORK PROBLEM **27**.

4 Find the Values of the Trigonometric Functions Using Fundamental Identities

If $P = (x, y)$ is the point on the unit circle corresponding to θ, then

$$\sin\theta = y \qquad \cos\theta = x \qquad \tan\theta = \frac{y}{x}, \quad \text{if } x \neq 0$$

$$\csc\theta = \frac{1}{y}, \quad \text{if } y \neq 0 \qquad \sec\theta = \frac{1}{x}, \quad \text{if } x \neq 0 \qquad \cot\theta = \frac{x}{y}, \quad \text{if } y \neq 0$$

Based on these definitions, we have the **reciprocal identities:**

> **Reciprocal Identities**
>
> $$\csc\theta = \frac{1}{\sin\theta} \qquad \sec\theta = \frac{1}{\cos\theta} \qquad \cot\theta = \frac{1}{\tan\theta} \qquad \text{(2)}$$

Two other fundamental identities are the **quotient identities.**

> **Quotient Identities**
>
> $$\tan\theta = \frac{\sin\theta}{\cos\theta} \qquad \cot\theta = \frac{\cos\theta}{\sin\theta} \qquad \text{(3)}$$

The proofs of identities (2) and (3) follow from the definitions of the trigonometric functions. (See Problems 129 and 130.)

If $\sin\theta$ and $\cos\theta$ are known, identities (2) and (3) make it easy to find the values of the remaining trigonometric functions.

EXAMPLE 3	**Finding Exact Values Using Identities When Sine and Cosine Are Given**

Given $\sin\theta = \dfrac{\sqrt{5}}{5}$ and $\cos\theta = \dfrac{2\sqrt{5}}{5}$, find the exact values of the four remaining trigonometric functions of θ using identities.

Solution Based on a quotient identity from (3), we have

$$\tan\theta = \frac{\sin\theta}{\cos\theta} = \frac{\dfrac{\sqrt{5}}{5}}{\dfrac{2\sqrt{5}}{5}} = \frac{1}{2}$$

Then we use the reciprocal identities from (2) to get

$$\csc\theta = \frac{1}{\sin\theta} = \frac{1}{\frac{\sqrt{5}}{5}} = \frac{5}{\sqrt{5}} = \sqrt{5} \qquad \sec\theta = \frac{1}{\cos\theta} = \frac{1}{\frac{2\sqrt{5}}{5}} = \frac{5}{2\sqrt{5}} = \frac{\sqrt{5}}{2} \qquad \cot\theta = \frac{1}{\tan\theta} = \frac{1}{\frac{1}{2}} = 2 \quad \blacktriangleleft$$

NOW WORK PROBLEM 35.

The equation of the unit circle is $x^2 + y^2 = 1$, or equivalently, $y^2 + x^2 = 1$. If $P = (x, y)$ is the point on the unit circle that corresponds to the angle θ, then

$$y^2 + x^2 = 1$$

But $y = \sin\theta$ and $x = \cos\theta$, so

$$(\sin\theta)^2 + (\cos\theta)^2 = 1 \tag{4}$$

It is customary to write $\sin^2\theta$ instead of $(\sin\theta)^2$, $\cos^2\theta$ instead of $(\cos\theta)^2$, and so on. With this notation, we can rewrite equation (4) as

$$\sin^2\theta + \cos^2\theta = 1 \tag{5}$$

If $\cos\theta \neq 0$, we can divide each side of equation (5) by $\cos^2\theta$.

$$\frac{\sin^2\theta}{\cos^2\theta} + \frac{\cos^2\theta}{\cos^2\theta} = \frac{1}{\cos^2\theta}$$

$$\left(\frac{\sin\theta}{\cos\theta}\right)^2 + 1 = \left(\frac{1}{\cos\theta}\right)^2$$

Now use identities (2) and (3) to get

$$\tan^2\theta + 1 = \sec^2\theta \tag{6}$$

Similarly, if $\sin\theta \neq 0$, we can divide equation (5) by $\sin^2\theta$ and use identities (2) and (3) to get $1 + \cot^2\theta = \csc^2\theta$, which we write as

$$\cot^2\theta + 1 = \csc^2\theta \tag{7}$$

Collectively, the identities in (5), (6), and (7) are referred to as the **Pythagorean identities.**

Let's pause here to summarize the fundamental identities.

Fundamental Identities

$$\tan\theta = \frac{\sin\theta}{\cos\theta} \qquad \cot\theta = \frac{\cos\theta}{\sin\theta}$$

$$\csc\theta = \frac{1}{\sin\theta} \qquad \sec\theta = \frac{1}{\cos\theta} \qquad \cot\theta = \frac{1}{\tan\theta}$$

$$\sin^2\theta + \cos^2\theta = 1 \qquad \tan^2\theta + 1 = \sec^2\theta \qquad \cot^2\theta + 1 = \csc^2\theta$$

| **EXAMPLE 4** | **Finding the Exact Value of a Trigonometric Expression Using Identities** |

Find the exact value of each expression. Do not use a calculator.

(a) $\tan 20° - \dfrac{\sin 20°}{\cos 20°}$ (b) $\sin^2 \dfrac{\pi}{12} + \dfrac{1}{\sec^2 \dfrac{\pi}{12}}$

Solution   (a) $\tan 20° - \underset{\underset{\dfrac{\sin \theta}{\cos \theta} = \tan \theta}{\uparrow}}{\dfrac{\sin 20°}{\cos 20°}} = \tan 20° - \tan 20° = 0$

(b) $\sin^2 \dfrac{\pi}{12} + \underset{\underset{\cos \theta = \dfrac{1}{\sec \theta}}{\uparrow}}{\dfrac{1}{\sec^2 \dfrac{\pi}{12}}} = \sin^2 \dfrac{\pi}{12} + \underset{\underset{\sin^2 \theta + \cos^2 \theta = 1}{\uparrow}}{\cos^2 \dfrac{\pi}{12}} = 1$ ◀

✏ **NOW WORK PROBLEM 79.**

5   Find the Exact Values of the Trigonometric Functions of an Angle Given One of the Functions and the Quadrant of the Angle

Many problems require finding the exact values of the remaining trigonometric functions when the value of one of them is known and the quadrant in which θ lies can be found. There are two approaches to solving such problems. One approach uses a circle of radius r, the other uses identities.

When using identities, sometimes a rearrangement is required. For example, the Pythagorean identity

$$\sin^2 \theta + \cos^2 \theta = 1$$

can be solved for $\sin \theta$ in terms of $\cos \theta$ (or vice versa) as follows:

$$\sin^2 \theta = 1 - \cos^2 \theta$$
$$\sin \theta = \pm\sqrt{1 - \cos^2 \theta}$$

where the $+$ sign is used if $\sin \theta > 0$ and the $-$ sign is used if $\sin \theta < 0$. Similarly, in $\tan^2 \theta + 1 = \sec^2 \theta$, we can solve for $\tan \theta$ (or $\sec \theta$), and in $\cot^2 \theta + 1 = \csc^2 \theta$, we can solve for $\cot \theta$ (or $\csc \theta$).

| **EXAMPLE 5** | **Finding Exact Values Given One Value and the Sign of Another** |

Given that $\sin \theta = \dfrac{1}{3}$ and $\cos \theta < 0$, find the exact values of each of the remaining five trigonometric functions.

Solution 1
Using a Circle

Figure 46

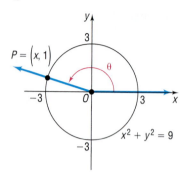

Suppose that $P = (x, y)$ is the point on a circle that corresponds to θ. Since $\sin \theta = \dfrac{1}{3} > 0$ and $\cos \theta < 0$, the point $P = (x, y)$ is in quadrant II. Because $\sin \theta = \dfrac{1}{3} = \dfrac{y}{r}$, we let $y = 1$ and $r = 3$. The point $P = (x, y)$ that corresponds to θ lies on the circle of radius 3, namely $x^2 + y^2 = 9$. See Figure 46.

To find x, we use the fact that $x^2 + y^2 = 9$, $y = 1$, and P is in quadrant II.

$$x^2 + y^2 = 9$$
$$x^2 + 1^2 = 9 \qquad y = 1$$
$$x^2 = 8$$
$$x = -2\sqrt{2} \qquad x < 0$$

Since $x = -2\sqrt{2}$, $y = 1$, and $r = 3$, we find that

$$\cos \theta = \frac{x}{r} = -\frac{2\sqrt{2}}{3} \qquad \tan \theta = \frac{y}{x} = \frac{1}{-2\sqrt{2}} = -\frac{\sqrt{2}}{4}$$

$$\csc \theta = \frac{r}{y} = \frac{3}{1} = 3 \qquad \sec \theta = \frac{r}{x} = \frac{3}{-2\sqrt{2}} = -\frac{3\sqrt{2}}{4} \qquad \cot \theta = \frac{x}{y} = \frac{-2\sqrt{2}}{1} = -2\sqrt{2}$$

◀

Solution 2
Using Identities

First, we solve equation (5) for $\cos \theta$.

$$\sin^2 \theta + \cos^2 \theta = 1$$
$$\cos^2 \theta = 1 - \sin^2 \theta$$
$$\cos \theta = \pm\sqrt{1 - \sin^2 \theta}$$

Because $\cos \theta < 0$, we choose the minus sign.

$$\cos \theta = -\sqrt{1 - \sin^2 \theta} = -\sqrt{1 - \frac{1}{9}} = -\sqrt{\frac{8}{9}} = -\frac{2\sqrt{2}}{3}$$
$$\uparrow$$
$$\sin \theta = \frac{1}{3}$$

Now we know the values of $\sin \theta$ and $\cos \theta$, so we can use identities (2) and (3) to get

$$\tan \theta = \frac{\sin \theta}{\cos \theta} = \frac{\dfrac{1}{3}}{\dfrac{-2\sqrt{2}}{3}} = \frac{1}{-2\sqrt{2}} = -\frac{\sqrt{2}}{4} \qquad \cot \theta = \frac{1}{\tan \theta} = -2\sqrt{2}$$

$$\sec \theta = \frac{1}{\cos \theta} = \frac{1}{\dfrac{-2\sqrt{2}}{3}} = \frac{-3}{2\sqrt{2}} = -\frac{3\sqrt{2}}{4} \qquad \csc \theta = \frac{1}{\sin \theta} = \frac{1}{\dfrac{1}{3}} = 3$$

◀

> **Finding the Values of the Trigonometric Functions When One Is Known**
>
> Given the value of one trigonometric function and the quadrant in which θ lies, the exact value of each of the remaining five trigonometric functions can be found in either of two ways.
>
> **Method 1 Using a Circle of Radius r**
>
> **STEP 1:** Draw a circle showing the location of the angle θ and the point $P = (x, y)$ that corresponds to θ. The radius of the circle is $r = \sqrt{x^2 + y^2}$.
>
> **STEP 2:** Assign a value to two of the three variables x, y, r based on the value of the given trigonometric function and the location of P.
>
> **STEP 3:** Use the fact that P lies on the circle $x^2 + y^2 = r^2$ to find the value of the missing variable.
>
> **STEP 4:** Apply the theorem on page 382 to find the values of the remaining trigonometric functions.
>
> **Method 2 Using Identities**
>
> Use appropriately selected identities to find the value of each of the remaining trigonometric functions.

EXAMPLE 6   **Given One Value of a Trigonometric Function, Find the Remaining Ones**

Given that $\tan \theta = \dfrac{1}{2}$ and $\sin \theta < 0$, find the exact value of each of the remaining five trigonometric functions of θ.

Solution 1 Using a Circle

STEP 1: Since $\tan \theta = \dfrac{1}{2} > 0$ and $\sin \theta < 0$, the point $P = (x, y)$ that corresponds to θ lies in quadrant III. See Figure 47.

STEP 2: Since $\tan \theta = \dfrac{1}{2} = \dfrac{y}{x}$ and θ lies in quadrant III, we let $x = -2$ and $y = -1$.

STEP 3: Then $r = \sqrt{x^2 + y^2} = \sqrt{(-2)^2 + (-1)^2} = \sqrt{5}$, and P lies on the circle $x^2 + y^2 = 5$.

STEP 4: Now apply the theorem on p. 382 using $x = -2$, $y = -1$, $r = \sqrt{5}$.

Figure 47

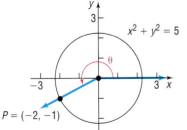

$$\sin \theta = \frac{y}{r} = \frac{-1}{\sqrt{5}} = -\frac{\sqrt{5}}{5} \qquad \cos \theta = \frac{x}{r} = \frac{-2}{\sqrt{5}} = -\frac{2\sqrt{5}}{5}$$

$$\csc \theta = \frac{r}{y} = \frac{\sqrt{5}}{-1} = -\sqrt{5} \qquad \sec \theta = \frac{r}{x} = \frac{\sqrt{5}}{-2} = -\frac{\sqrt{5}}{2} \qquad \cot \theta = \frac{x}{y} = \frac{-2}{-1} = 2 \quad ◀$$

Solution 2
Using Identities

We use the Pythagorean identity that involves $\tan \theta$, that is, $\tan^2 \theta + 1 = \sec^2 \theta$. Since $\tan \theta = \dfrac{1}{2} > 0$ and $\sin \theta < 0$, then θ lies in quadrant III, where $\sec \theta < 0$.

$$\tan^2 \theta + 1 = \sec^2 \theta \qquad \text{Pythagorean identity}$$

$$\left(\frac{1}{2}\right)^2 + 1 = \sec^2 \theta \qquad \tan \theta = \frac{1}{2}$$

$$\sec^2 \theta = \frac{1}{4} + 1 = \frac{5}{4} \qquad \text{Proceed to solve for } \sec \theta.$$

$$\sec \theta = -\frac{\sqrt{5}}{2} \qquad \sec \theta < 0$$

Then

$$\cos \theta = \frac{1}{\sec \theta} = \frac{1}{-\dfrac{\sqrt{5}}{2}} = -\frac{2}{\sqrt{5}} = -\frac{2\sqrt{5}}{5}$$

$$\tan \theta = \frac{\sin \theta}{\cos \theta} \quad \text{so} \quad \sin \theta = \tan \theta \cdot \cos \theta = \left(\frac{1}{2}\right) \cdot \left(-\frac{2\sqrt{5}}{5}\right) = -\frac{\sqrt{5}}{5}$$

$$\csc \theta = \frac{1}{\sin \theta} = \frac{1}{-\dfrac{\sqrt{5}}{5}} = -\frac{5}{\sqrt{5}} = -\sqrt{5}$$

$$\cot \theta = \frac{1}{\tan \theta} = \frac{1}{\dfrac{1}{2}} = 2$$

◀

✎ **NOW WORK PROBLEM 43.**

6 Use Even–Odd Properties to Find the Exact Values of the Trigonometric Functions

Recall that a function f is even if $f(-\theta) = f(\theta)$ for all θ in the domain of f; a function f is odd if $f(-\theta) = -f(\theta)$ for all θ in the domain of f. We will now show that the trigonometric functions sine, tangent, cotangent, and cosecant are odd functions and the functions cosine and secant are even functions.

Theorem

In Words
Cosine and secant are even functions; the others are odd functions.

Even–Odd Properties

$$\sin(-\theta) = -\sin \theta \qquad \cos(-\theta) = \cos \theta \qquad \tan(-\theta) = -\tan \theta$$
$$\csc(-\theta) = -\csc \theta \qquad \sec(-\theta) = \sec \theta \qquad \cot(-\theta) = -\cot \theta$$

Proof Let $P = (x, y)$ be the point on the unit circle that corresponds to the angle θ. See Figure 48. The point Q on the unit circle that corresponds to the angle $-\theta$ will have coordinates $(x, -y)$. Using the definition of the trigonometric functions, we have

$$\sin \theta = y \qquad \sin(-\theta) = -y \qquad \cos \theta = x \qquad \cos(-\theta) = x$$

Figure 48

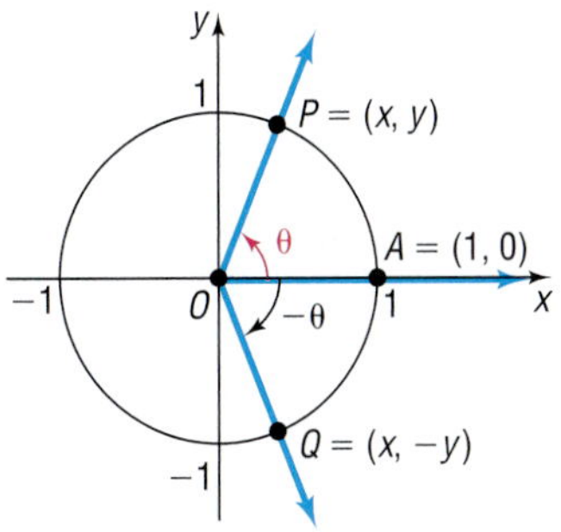

so

$$\sin(-\theta) = -y = -\sin\theta \qquad \cos(-\theta) = x = \cos\theta$$

Now, using these results and some of the fundamental identities, we have

$$\tan(-\theta) = \frac{\sin(-\theta)}{\cos(-\theta)} = \frac{-\sin\theta}{\cos\theta} = -\tan\theta \qquad \cot(-\theta) = \frac{1}{\tan(-\theta)} = \frac{1}{-\tan\theta} = -\cot\theta$$

$$\sec(-\theta) = \frac{1}{\cos(-\theta)} = \frac{1}{\cos\theta} = \sec\theta \qquad \csc(-\theta) = \frac{1}{\sin(-\theta)} = \frac{1}{-\sin\theta} = -\csc\theta \qquad\blacksquare$$

| EXAMPLE 7 | **Finding Exact Values Using Even–Odd Properties** |

Find the exact value of:

(a) $\sin(-45°)$ (b) $\cos(-\pi)$ (c) $\cot\left(-\dfrac{3\pi}{2}\right)$ (d) $\tan\left(-\dfrac{37\pi}{4}\right)$

Solution   (a) $\sin(-45°) = \underset{\uparrow}{-\sin 45°} = -\dfrac{\sqrt{2}}{2}$ (b) $\cos(-\pi) = \underset{\uparrow}{\cos\pi} = -1$

   Odd function   *Even function*

(c) $\cot\left(-\dfrac{3\pi}{2}\right) = \underset{\uparrow}{-\cot\dfrac{3\pi}{2}} = 0$

   Odd function

(d) $\tan\left(-\dfrac{37\pi}{4}\right) = \underset{\uparrow}{-\tan\dfrac{37\pi}{4}} = -\tan\left(\dfrac{\pi}{4} + 9\pi\right) = \underset{\uparrow}{-\tan\dfrac{\pi}{4}} = -1$

   Odd function   *Period is π.*   ◄

 NOW WORK PROBLEM 59.

5.3 Assess Your Understanding

'Are You Prepared?'

Answers are given at the end of these exercises. If you get a wrong answer, read the pages listed in red.

1. The domain of the function $g(x) = \dfrac{x+1}{2x+1}$ is _____.
   (pp. 64–65)

2. A function for which $f(x) = f(-x)$ for all x in the domain of f is called an _____ function. (pp. 80–82)

3. *True or False:* The function $f(x) = \sqrt{x}$ is even.
   (pp. 80–82)

4. *True or False:* The equation $x^2 + 2x = (x+1)^2 - 1$ is an identity. (p. 692)

Concepts and Vocabulary

5. The sine, cosine, cosecant, and secant functions have period _____; the tangent and cotangent functions have period _____.

6. The domain of the tangent function is _____.

7. The range of the sine function is _____.

8. *True or False:* The only even trigonometric functions are the cosine and secant functions.

9. *True or False:* All the trigonometric functions are periodic, with period 2π.

10. *True or False:* The range of the secant function is the set of positive real numbers.

Skill Building

In Problems 11–26, use the fact that the trigonometric functions are periodic to find the exact value of each expression. Do not use a calculator.

11. $\sin 405°$
12. $\cos 420°$
13. $\tan 405°$
14. $\sin 390°$
15. $\csc 450°$
16. $\sec 540°$

17. $\cot 390°$
18. $\sec 420°$
19. $\cos \dfrac{33\pi}{4}$
20. $\sin \dfrac{9\pi}{4}$
21. $\tan(21\pi)$
22. $\csc \dfrac{9\pi}{2}$

23. $\sec \dfrac{17\pi}{4}$
24. $\cot \dfrac{17\pi}{4}$
25. $\tan \dfrac{19\pi}{6}$
26. $\sec \dfrac{25\pi}{6}$

In Problems 27–34, name the quadrant in which the angle θ lies.

27. $\sin\theta > 0, \quad \cos\theta < 0$
28. $\sin\theta < 0, \quad \cos\theta > 0$
29. $\sin\theta < 0, \quad \tan\theta < 0$

30. $\cos\theta > 0, \quad \tan\theta > 0$
31. $\cos\theta > 0, \quad \tan\theta < 0$
32. $\cos\theta < 0, \quad \tan\theta > 0$

33. $\sec\theta < 0, \quad \sin\theta > 0$
34. $\csc\theta > 0, \quad \cos\theta < 0$

In Problems 35–42, sin θ and cos θ are given. Find the exact value of each of the four remaining trigonometric functions.

35. $\sin\theta = -\dfrac{3}{5}, \quad \cos\theta = \dfrac{4}{5}$
36. $\sin\theta = \dfrac{4}{5}, \quad \cos\theta = -\dfrac{3}{5}$
37. $\sin\theta = \dfrac{2\sqrt{5}}{5}, \quad \cos\theta = \dfrac{\sqrt{5}}{5}$

38. $\sin\theta = -\dfrac{\sqrt{5}}{5}, \quad \cos\theta = -\dfrac{2\sqrt{5}}{5}$
39. $\sin\theta = \dfrac{1}{2}, \quad \cos\theta = \dfrac{\sqrt{3}}{2}$
40. $\sin\theta = \dfrac{\sqrt{3}}{2}, \quad \cos\theta = \dfrac{1}{2}$

41. $\sin\theta = -\dfrac{1}{3}, \quad \cos\theta = \dfrac{2\sqrt{2}}{3}$
42. $\sin\theta = \dfrac{2\sqrt{2}}{3}, \quad \cos\theta = -\dfrac{1}{3}$

In Problems 43–58, find the exact value of each of the remaining trigonometric functions of θ.

43. $\sin\theta = \dfrac{12}{13}, \quad \theta$ in quadrant II
44. $\cos\theta = \dfrac{3}{5}, \quad \theta$ in quadrant IV
45. $\cos\theta = -\dfrac{4}{5}, \quad \theta$ in quadrant III

46. $\sin\theta = -\dfrac{5}{13}, \quad \theta$ in quadrant III
47. $\sin\theta = \dfrac{5}{13}, \quad 90° < \theta < 180°$
48. $\cos\theta = \dfrac{4}{5}, \quad 270° < \theta < 360°$

49. $\cos\theta = -\dfrac{1}{3}, \quad \dfrac{\pi}{2} < \theta < \pi$
50. $\sin\theta = -\dfrac{2}{3}, \quad \pi < \theta < \dfrac{3\pi}{2}$
51. $\sin\theta = \dfrac{2}{3}, \quad \tan\theta < 0$

52. $\cos\theta = -\dfrac{1}{4}, \quad \tan\theta > 0$
53. $\sec\theta = 2, \quad \sin\theta < 0$
54. $\csc\theta = 3, \quad \cot\theta < 0$

55. $\tan\theta = \dfrac{3}{4}, \quad \sin\theta < 0$
56. $\cot\theta = \dfrac{4}{3}, \quad \cos\theta < 0$
57. $\tan\theta = -\dfrac{1}{3}, \quad \sin\theta > 0$

58. $\sec\theta = -2, \quad \tan\theta > 0$

In Problems 59–76, use the even–odd properties to find the exact value of each expression. Do not use a calculator.

59. $\sin(-60°)$
60. $\cos(-30°)$
61. $\tan(-30°)$
62. $\sin(-135°)$
63. $\sec(-60°)$
64. $\csc(-30°)$

65. $\sin(-90°)$
66. $\cos(-270°)$
67. $\tan\left(-\dfrac{\pi}{4}\right)$
68. $\sin(-\pi)$
69. $\cos\left(-\dfrac{\pi}{4}\right)$
70. $\sin\left(-\dfrac{\pi}{3}\right)$

71. $\tan(-\pi)$
72. $\sin\left(-\dfrac{3\pi}{2}\right)$
73. $\csc\left(-\dfrac{\pi}{4}\right)$
74. $\sec(-\pi)$
75. $\sec\left(-\dfrac{\pi}{6}\right)$
76. $\csc\left(-\dfrac{\pi}{3}\right)$

In Problems 77–88, use properties of the trigonometric functions to find the exact value of each expression. Do not use a calculator.

77. $\sin^2 40° + \cos^2 40°$
78. $\sec^2 18° - \tan^2 18°$
79. $\sin 80° \csc 80°$
80. $\tan 10° \cot 10°$

81. $\tan 40° - \dfrac{\sin 40°}{\cos 40°}$
82. $\cot 20° - \dfrac{\cos 20°}{\sin 20°}$
83. $\cos 400° \cdot \sec 40°$
84. $\tan 200° \cdot \cot 20°$

85. $\sin\left(-\dfrac{\pi}{12}\right)\csc \dfrac{25\pi}{12}$
86. $\sec\left(-\dfrac{\pi}{18}\right) \cdot \cos \dfrac{37\pi}{18}$
87. $\dfrac{\sin(-20°)}{\cos 380°} + \tan 200°$
88. $\dfrac{\sin 70°}{\cos(-430°)} + \tan(-70°)$

89. If $\sin \theta = 0.3$, find the value of:

$$\sin \theta + \sin(\theta + 2\pi) + \sin(\theta + 4\pi)$$

90. If $\cos \theta = 0.2$, find the value of:

$$\cos \theta + \cos(\theta + 2\pi) + \cos(\theta + 4\pi)$$

91. If $\tan \theta = 3$, find the value of:

$$\tan \theta + \tan(\theta + \pi) + \tan(\theta + 2\pi)$$

92. If $\cot \theta = -2$, find the value of:

$$\cot \theta + \cot(\theta - \pi) + \cot(\theta - 2\pi)$$

93. Find the exact value of:
$\sin 1° + \sin 2° + \sin 3° + \cdots + \sin 358° + \sin 359°$

94. Find the exact value of:
$\cos 1° + \cos 2° + \cos 3° + \cdots + \cos 358° + \cos 359°$

95. What is the domain of the sine function?

96. What is the domain of the cosine function?

97. For what numbers θ is $f(\theta) = \tan \theta$ not defined?

98. For what numbers θ is $f(\theta) = \cot \theta$ not defined?

99. For what numbers θ is $f(\theta) = \sec \theta$ not defined?

100. For what numbers θ is $f(\theta) = \csc \theta$ not defined?

101. What is the range of the sine function?

102. What is the range of the cosine function?

103. What is the range of the tangent function?

104. What is the range of the cotangent function?

105. What is the range of the secant function?

106. What is the range of the cosecant function?

107. Is the sine function even, odd, or neither? Is its graph symmetric? With respect to what?

108. Is the cosine function even, odd, or neither? Is its graph symmetric? With respect to what?

109. Is the tangent function even, odd, or neither? Is its graph symmetric? With respect to what?

110. Is the cotangent function even, odd, or neither? Is its graph symmetric? With respect to what?

111. Is the secant function even, odd, or neither? Is its graph symmetric? With respect to what?

112. Is the cosecant function even, odd, or neither? Is its graph symmetric? With respect to what?

In Problems 113–118, use the periodic and even–odd properties.

113. If $f(\theta) = \sin \theta$ and $f(a) = \dfrac{1}{3}$, find the exact value of:
  (a) $f(-a)$ (b) $f(a) + f(a + 2\pi) + f(a + 4\pi)$

114. If $f(\theta) = \cos \theta$ and $f(a) = \dfrac{1}{4}$, find the exact value of:
  (a) $f(-a)$ (b) $f(a) + f(a + 2\pi) + f(a - 2\pi)$

115. If $f(\theta) = \tan \theta$ and $f(a) = 2$, find the exact value of:
  (a) $f(-a)$ (b) $f(a) + f(a + \pi) + f(a + 2\pi)$

116. If $f(\theta) = \cot \theta$ and $f(a) = -3$, find the exact value of:
  (a) $f(-a)$ (b) $f(a) + f(a + \pi) + f(a + 4\pi)$

117. If $f(\theta) = \sec \theta$ and $f(a) = -4$, find the exact value of:
  (a) $f(-a)$ (b) $f(a) + f(a + 2\pi) + f(a + 4\pi)$

118. If $f(\theta) = \csc \theta$ and $f(a) = 2$, find the exact value of:
  (a) $f(-a)$ (b) $f(a) + f(a + 2\pi) + f(a + 4\pi)$

119. Calculating the Time of a Trip   From a parking lot, you want to walk to a house on the ocean. The house is located 1500 feet down a paved path that parallels the ocean, which is 500 feet away. See the illustration. Along the path you can walk 300 feet per minute, but in the sand on the beach you can only walk 100 feet per minute.

    The time T to get from the parking lot to the beach-house can be expressed as a function of the angle θ shown in the illustration and is

$$T(\theta) = 5 - \frac{5}{3 \tan \theta} + \frac{5}{\sin \theta}, \qquad 0 < \theta < \frac{\pi}{2}$$

Calculate the time T if you walk directly from the parking lot to the house.

[Hint: $\tan \theta = \dfrac{500}{1500}$.**]**

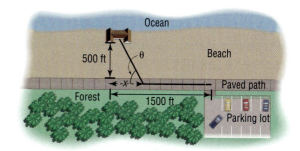

120. Calculating the Time of a Trip   Two oceanfront homes are located 8 miles apart on a straight stretch of beach, each a distance of 1 mile from a paved road that parallels the ocean. Sally can jog 8 miles per hour along the paved road, but only 3 miles per hour in the sand on the beach. Because of a river directly between the two houses, it is necessary to jog in the sand to the road, continue on the road, and then jog directly back in the sand to get from one house to the other. See the illustration. The time T to get from one house to the other as a function of the angle θ shown in the illustration is

$$T(\theta) = 1 + \frac{2}{3 \sin \theta} - \frac{1}{4 \tan \theta} \qquad 0 < \theta < \frac{\pi}{2}$$

(a) Calculate the time T for $\tan \theta = \frac{1}{4}$.
(b) Describe the path taken.
(c) Explain why θ must be larger than 14°.

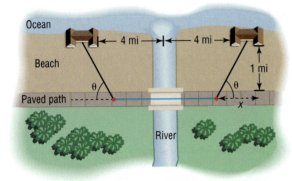

121. Show that the range of the tangent function is the set of all real numbers.

122. Show that the range of the cotangent function is the set of all real numbers.

123. Show that the period of $f(\theta) = \sin \theta$ is 2π.

[**Hint:** Assume that $0 < p < 2\pi$ exists so that $\sin(\theta + p) = \sin \theta$ for all θ. Let $\theta = 0$ to find p. Then let $\theta = \dfrac{\pi}{2}$ to obtain a contradiction.]

124. Show that the period of $f(\theta) = \cos \theta$ is 2π.

125. Show that the period of $f(\theta) = \sec \theta$ is 2π.

126. Show that the period of $f(\theta) = \csc \theta$ is 2π.

127. Show that the period of $f(\theta) = \tan \theta$ is π.

128. Show that the period of $f(\theta) = \cot \theta$ is π.

129. Prove the reciprocal identities given in formula (2).

130. Prove the quotient identities given in formula (3).

131. Establish the identity:
$$(\sin \theta \cos \phi)^2 + (\sin \theta \sin \phi)^2 + \cos^2 \theta = 1$$

Discussion and Writing

132. Write down five properties of the tangent function. Explain the meaning of each.

133. Describe your understanding of the meaning of a periodic function.

134. Explain how to find the value of $\sin 390°$ using periodic properties.

135. Explain how to find the value of $\cos(-45°)$ using even–odd properties.

'Are You Prepared?' Answers

1. $\left\{ x \mid x \neq -\dfrac{1}{2} \right\}$ **2.** Even **3.** False **4.** True

---

5.4 Graphs of the Sine and Cosine Functions*

PREPARING FOR THIS SECTION *Before getting started, review the following:*

• Graphing Techniques: Transformations (Section 2.6, pp. 118–126)

✎ Now work the 'Are You Prepared?' problems on page 414.

OBJECTIVES 1 Graph Transformations of the Sine Function
2 Graph Transformations of the Cosine Function
3 Determine the Amplitude and Period of Sinusoidal Functions
4 Graph Sinusoidal Functions Using Key Points
5 Find an Equation for a Sinusoidal Graph

---

*For those who wish to include phase shifts here, Section 5.6 can be covered immediately after Section 5.4 without loss of continuity.

Since we want to graph the trigonometric functions in the xy-plane, we shall use the traditional symbols x for the independent variable (or argument) and y for the dependent variable (or value at x) for each function. So we write the six trigonometric functions as

$$y = f(x) = \sin x \qquad y = f(x) = \cos x \qquad y = f(x) = \tan x$$
$$y = f(x) = \csc x \qquad y = f(x) = \sec x \qquad y = f(x) = \cot x$$

 Here the independent variable x represents an angle, measured in radians. In calculus, x will usually be treated as a real number. As we said earlier, these are equivalent ways of viewing x.

Graph Transformations of the Sine Function

Since the sine function has period 2π, we need to graph $y = \sin x$ only on the interval $[0, 2\pi]$. The remainder of the graph will consist of repetitions of this portion of the graph.

We begin by constructing Table 6, which lists some points on the graph of $y = \sin x$, $0 \le x \le 2\pi$. As the table shows, the graph of $y = \sin x$, $0 \le x \le 2\pi$, begins at the origin. As x increases from 0 to $\dfrac{\pi}{2}$, the value of $y = \sin x$ increases from 0 to 1; as x increases from $\dfrac{\pi}{2}$ to π to $\dfrac{3\pi}{2}$, the value of y decreases from 1 to 0 to -1; as x increases from $\dfrac{3\pi}{2}$ to 2π, the value of y increases from -1 to 0. If we plot the points listed in Table 6 and connect them with a smooth curve, we obtain the graph shown in Figure 49.

Table 6

x	$y = \sin x$	(x, y)
0	0	$(0, 0)$
$\dfrac{\pi}{6}$	$\dfrac{1}{2}$	$\left(\dfrac{\pi}{6}, \dfrac{1}{2}\right)$
$\dfrac{\pi}{2}$	1	$\left(\dfrac{\pi}{2}, 1\right)$
$\dfrac{5\pi}{6}$	$\dfrac{1}{2}$	$\left(\dfrac{5\pi}{6}, \dfrac{1}{2}\right)$
π	0	$(\pi, 0)$
$\dfrac{7\pi}{6}$	$-\dfrac{1}{2}$	$\left(\dfrac{7\pi}{6}, -\dfrac{1}{2}\right)$
$\dfrac{3\pi}{2}$	-1	$\left(\dfrac{3\pi}{2}, -1\right)$
$\dfrac{11\pi}{6}$	$-\dfrac{1}{2}$	$\left(\dfrac{11\pi}{6}, -\dfrac{1}{2}\right)$
2π	0	$(2\pi, 0)$

Figure 49
$y = \sin x$, $0 \le x \le 2\pi$

The graph in Figure 49 is one period, or **cycle**, of the graph of $y = \sin x$. To obtain a more complete graph of $y = \sin x$, we repeat this period in each direction, as shown in Figure 50(a). Figure 50(b) shows the graph on a TI-84 Plus graphing calculator.

Figure 50
$y = \sin x$, $-\infty < x < \infty$

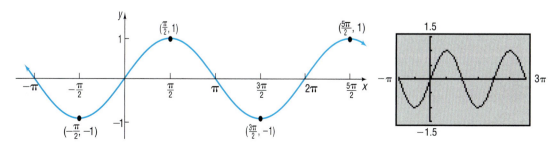

(a)   (b)

The graph of $y = \sin x$ illustrates some of the facts that we already know about the sine function.

Properties of the Sine Function

1. The domain is the set of all real numbers.
2. The range consists of all real numbers from -1 to 1, inclusive.
3. The sine function is an odd function, as the symmetry of the graph with respect to the origin indicates.
4. The sine function is periodic, with period 2π.
5. The x-intercepts are $\ldots, -2\pi, -\pi, 0, \pi, 2\pi, 3\pi, \ldots$; the y-intercept is 0.
6. The maximum value is 1 and occurs at $x = \ldots, -\dfrac{3\pi}{2}, \dfrac{\pi}{2}, \dfrac{5\pi}{2}, \dfrac{9\pi}{2}, \ldots$;

   the minimum value is -1 and occurs at $x = \ldots, -\dfrac{\pi}{2}, \dfrac{3\pi}{2}, \dfrac{7\pi}{2}, \dfrac{11\pi}{2}, \ldots$.

$\hspace{2cm}$ **NOW WORK PROBLEMS 9, 11, AND 13.**

The graphing techniques introduced in Chapter 2, Section 2.6, may be used to graph functions that are transformations of the sine function.

| **EXAMPLE 1** | **Graphing Variations of $y = \sin x$ Using Transformations** |

Use the graph of $y = \sin x$ to graph $y = \sin\left(x - \dfrac{\pi}{4}\right)$.

Solution   Figure 51 illustrates the steps.

Figure 51

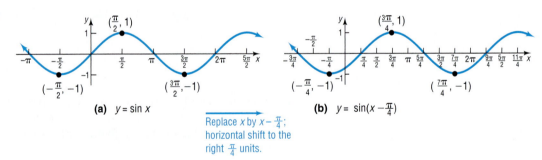

(a) $y = \sin x$

Replace x by $x - \frac{\pi}{4}$;
horizontal shift to the
right $\frac{\pi}{4}$ units.

(b) $y = \sin\left(x - \frac{\pi}{4}\right)$

✔ **CHECK:** Figure 52 shows the graph of $Y_1 = \sin\left(x - \dfrac{\pi}{4}\right)$.

Figure 52

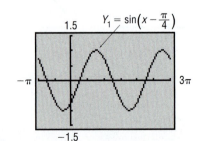

$Y_1 = \sin\left(x - \frac{\pi}{4}\right)$

EXAMPLE 2

Graphing Variations of $y = \sin x$ Using Transformations

Use the graph of $y = \sin x$ to graph $y = -\sin x + 2$.

Solution   Figure 53 illustrates the steps.

Figure 53

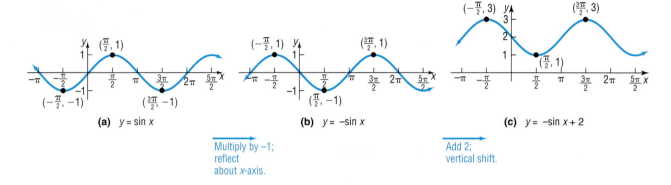

(a) $y = \sin x$ (b) $y = -\sin x$ (c) $y = -\sin x + 2$

Multiply by –1;
reflect
about x-axis.

Add 2;
vertical shift.

✔ **CHECK:** Figure 54 shows the graph of $Y_1 = -\sin x + 2$.

Figure 54 $Y_1 = -\sin x + 2$

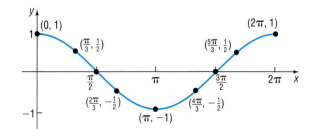

NOW WORK PROBLEM 25.

Table 7

x	$y = \cos x$	(x, y)
0	1	$(0, 1)$
$\dfrac{\pi}{3}$	$\dfrac{1}{2}$	$\left(\dfrac{\pi}{3}, \dfrac{1}{2}\right)$
$\dfrac{\pi}{2}$	0	$\left(\dfrac{\pi}{2}, 0\right)$
$\dfrac{2\pi}{3}$	$-\dfrac{1}{2}$	$\left(\dfrac{2\pi}{3}, -\dfrac{1}{2}\right)$
π	-1	$(\pi, -1)$
$\dfrac{4\pi}{3}$	$-\dfrac{1}{2}$	$\left(\dfrac{4\pi}{3}, -\dfrac{1}{2}\right)$
$\dfrac{3\pi}{2}$	0	$\left(\dfrac{3\pi}{2}, 0\right)$
$\dfrac{5\pi}{3}$	$\dfrac{1}{2}$	$\left(\dfrac{5\pi}{3}, \dfrac{1}{2}\right)$
2π	1	$(2\pi, 1)$

2 ### Graph Transformations of the Cosine Function

The cosine function also has period 2π. We proceed as we did with the sine function by constructing Table 7, which lists some points on the graph of $y = \cos x, 0 \le x \le 2\pi$. As the table shows, the graph of $y = \cos x, 0 \le x \le 2\pi$, begins at the point $(0, 1)$. As x increases from 0 to $\dfrac{\pi}{2}$ to π, the value of y decreases from 1 to 0 to -1; as x increases from π to $\dfrac{3\pi}{2}$ to 2π, the value of y increases from -1 to 0 to 1. As before, we plot the points in Table 7 to get one period or cycle of the graph. See Figure 55.

Figure 55
$y = \cos x, 0 \le x \le 2\pi$

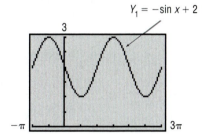

A more complete graph of $y = \cos x$ is obtained by repeating this period in each direction, as shown in Figure 56(a). Figure 56(b) shows the graph on a TI-84 Plus graphing calculator.

Figure 56
$y = \cos x$, $-\infty < x < \infty$

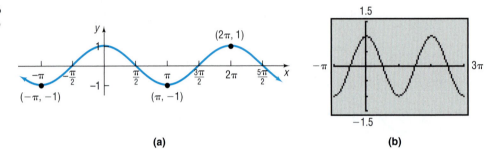

(a)

(b)

The graph of $y = \cos x$ illustrates some of the facts that we already know about the cosine function.

Properties of the Cosine Function

1. The domain is the set of all real numbers.

2. The range consists of all real numbers from -1 to 1, inclusive.

3. The cosine function is an even function, as the symmetry of the graph with respect to the y-axis indicates.

4. The cosine function is periodic, with period 2π.

5. The x-intercepts are $\ldots, -\dfrac{3\pi}{2}, -\dfrac{\pi}{2}, \dfrac{\pi}{2}, \dfrac{3\pi}{2}, \dfrac{5\pi}{2}, \ldots$; the y-intercept is 1.

6. The maximum value is 1 and occurs at $x = \ldots, -2\pi, 0, 2\pi, 4\pi, 6\pi, \ldots$; the minimum value is -1 and occurs at $x = \ldots, -\pi, \pi, 3\pi, 5\pi, \ldots$.

Again, the graphing techniques from Chapter 2 may be used to graph transformations of the cosine function.

EXAMPLE 3 **Graphing Variations of $y = \cos x$ Using Transformations**

Use the graph of $y = \cos x$ to graph $y = 2 \cos x$.

Solution Figure 57 illustrates the steps.

Figure 57

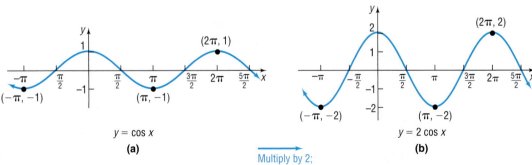

$y = \cos x$

(a)

Multiply by 2; vertical stretch by a factor of 2.

$y = 2 \cos x$

(b)

✔ **CHECK:** Figure 58 shows the graph $Y_1 = 2 \cos x$.

Figure 58

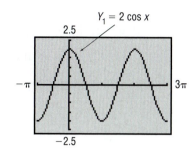

◀

EXAMPLE 4   **Graphing Variations of $y = \cos x$ Using Transformations**

Use the graph of $y = \cos x$ to graph $y = \cos(3x)$.

Solution   Figure 59 illustrates the graph, which is a horizontal compression of the graph of $y = \cos x$. (Multiply each x-coordinate by $\frac{1}{3}$.) Notice that, due to this compression, the period of $y = \cos(3x)$ is $\frac{2\pi}{3}$, whereas the period of $y = \cos x$ is 2π.

Figure 59

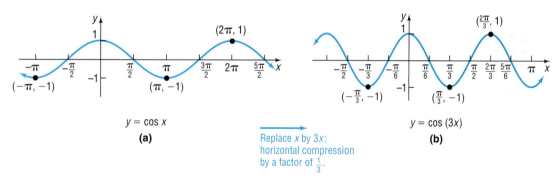

✔ **CHECK:** Figure 60 shows the graph of $Y_1 = \cos(3x)$.

— Seeing the Concept —

Graph $Y_1 = \cos(3x)$ with $X\min = 0$, $X\max = \frac{2\pi}{3}$, and $X\text{scl} = \frac{\pi}{6}$ to verify that the period is $\frac{2\pi}{3}$.

Figure 60

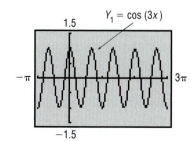

◀

NOW WORK PROBLEM **33.**

Sinusoidal Graphs

Shift the graph of $y = \cos x$ to the right $\dfrac{\pi}{2}$ units to obtain the graph of $y = \cos\left(x - \dfrac{\pi}{2}\right)$. See Figure 61(a). Now look at the graph of $y = \sin x$ in Figure 61(b). We see that the graph of $y = \sin x$ is the same as the graph of $y = \cos\left(x - \dfrac{\pi}{2}\right)$.

Figure 61

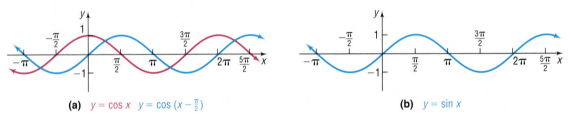

(a) $y = \cos x$ $y = \cos\left(x - \frac{\pi}{2}\right)$ **(b)** $y = \sin x$

Based on Figure 61, we conjecture that

$$\sin x = \cos\left(x - \frac{\pi}{2}\right)$$

— **Seeing the Concept** —

Graph $Y_1 = \sin x$ and $Y_2 = \cos\left(x - \dfrac{\pi}{2}\right)$.

How many graphs do you see?

(We shall prove this fact in Chapter 6.) Because of this similarity, the graphs of sine functions and cosine functions are referred to as **sinusoidal graphs**.

Let's look at some general properties of sinusoidal graphs.

3 ## Determine the Amplitude and Period of Sinusoidal Functions

In Example 3 we obtained the graph of $y = 2 \cos x$, which we reproduce in Figure 62. Notice that the values of $y = 2 \cos x$ lie between -2 and 2, inclusive.

Figure 62
$y = 2 \cos x$

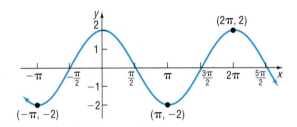

In general, the values of the functions $y = A \sin x$ and $y = A \cos x$, where $A \neq 0$, will always satisfy the inequalities

$$-|A| \leq A \sin x \leq |A| \quad \text{and} \quad -|A| \leq A \cos x \leq |A|$$

respectively. The number $|A|$ is called the **amplitude** of $y = A \sin x$ or $y = A \cos x$. See Figure 63.

Figure 63

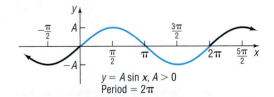

$y = A \sin x, A > 0$
Period $= 2\pi$

In Example 4, we obtained the graph of $y = \cos(3x)$, which we reproduce in Figure 64. Notice that the period of this function is $\dfrac{2\pi}{3}$.

Figure 64
$y = \cos(3x)$

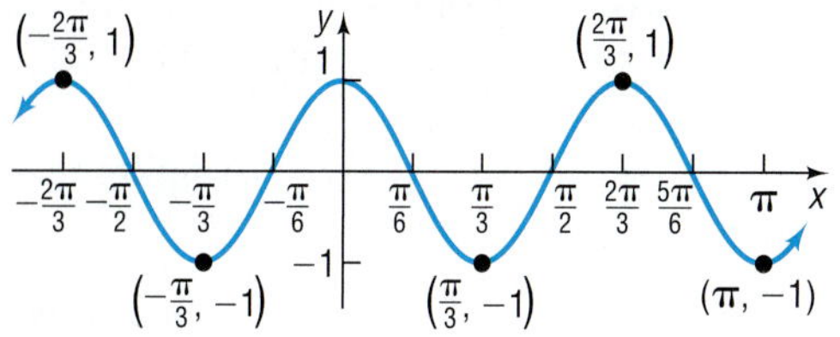

In general, if $\omega > 0$, the functions $y = \sin(\omega x)$ and $y = \cos(\omega x)$ will have period $T = \dfrac{2\pi}{\omega}$. To see why, recall that the graph of $y = \sin(\omega x)$ is obtained from the graph of $y = \sin x$ by performing a horizontal compression or stretch by a factor $\dfrac{1}{\omega}$. This horizontal compression replaces the interval $[0, 2\pi]$, which contains one period of the graph of $y = \sin x$, by the interval $\left[0, \dfrac{2\pi}{\omega}\right]$, which contains one period of the graph of $y = \sin(\omega x)$. The period of the functions $y = \sin(\omega x)$ and $y = \cos(\omega x)$, $\omega > 0$, is $\dfrac{2\pi}{\omega}$.

For example, for the function $y = \cos(3x)$, graphed in Figure 64, $\omega = 3$, so the period is $\dfrac{2\pi}{\omega} = \dfrac{2\pi}{3}$.

One period of the graph of $y = \sin(\omega x)$ or $y = \cos(\omega x)$ is called a **cycle**. Figure 65 illustrates the general situation. The blue portion of the graph is one cycle.

Figure 65

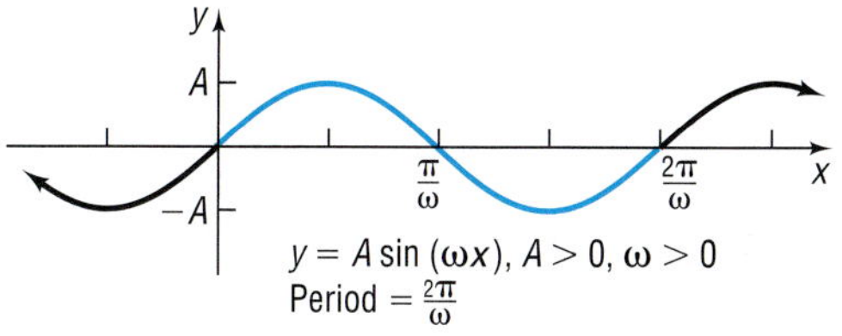

$y = A \sin(\omega x), A > 0, \omega > 0$
Period $= \frac{2\pi}{\omega}$

If $\omega < 0$ in $y = \sin(\omega x)$ or $y = \cos(\omega x)$, we use the Even–Odd Properties of the sine and cosine functions as follows:

$$\sin(-\omega x) = -\sin(\omega x) \quad \text{and} \quad \cos(-\omega x) = \cos(\omega x)$$

This gives us an equivalent form in which the coefficient of x is positive. For example,

$$\sin(-2x) = -\sin(2x) \quad \text{and} \quad \cos(-\pi x) = \cos(\pi x)$$

Theorem

If $\omega > 0$, the amplitude and period of $y = A \sin(\omega x)$ and $y = A \cos(\omega x)$ are given by

$$\text{Amplitude} = |A| \qquad \text{Period} = T = \frac{2\pi}{\omega} \qquad \textbf{(1)}$$

EXAMPLE 5	**Finding the Amplitude and Period of a Sinusoidal Function**

Determine the amplitude and period of $y = 3 \sin(4x)$.

Solution   Comparing $y = 3 \sin(4x)$ to $y = A \sin(\omega x)$, we find that $A = 3$ and $\omega = 4$. From equation (1),

$$\text{Amplitude} = |A| = 3 \qquad \text{Period} = T = \frac{2\pi}{\omega} = \frac{2\pi}{4} = \frac{\pi}{2} \qquad \blacktriangleleft$$

 NOW WORK PROBLEM 41.

4 Graph Sinusoidal Functions Using Key Points

Earlier, we graphed sine and cosine functions using tranformations. We now introduce another method that can be used to graph these functions.

Figure 66 shows one cycle of the graphs of $y = \sin x$ and $y = \cos x$ on the interval $[0, 2\pi]$. Notice that each graph consists of four parts corresponding to the four subintervals:

$$\left[0, \frac{\pi}{2}\right], \quad \left[\frac{\pi}{2}, \pi\right], \quad \left[\pi, \frac{3\pi}{2}\right], \quad \left[\frac{3\pi}{2}, 2\pi\right]$$

Each subinterval is of length $\dfrac{\pi}{2}$ (the period 2π divided by 4, the number of parts), and the endpoints of these intervals give rise to five key points on the graph:

$$\text{For } y = \sin x: (0, 0), \left(\frac{\pi}{2}, 1\right), (\pi, 0), \left(\frac{3\pi}{2}, -1\right), (2\pi, 0)$$

$$\text{For } y = \cos x: (0, 1), \left(\frac{\pi}{2}, 0\right), (\pi, -1), \left(\frac{3\pi}{2}, 0\right), (2\pi, 1)$$

See Figure 66.

Figure 66

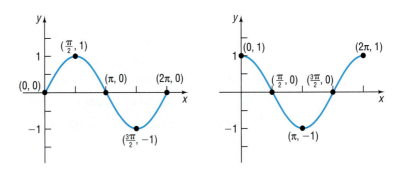

When graphing a sinusoidal function of the form $y = A \sin(\omega x)$ or $y = A \cos(\omega x)$ we use the amplitude to determine the maximum and minimum values of the function. The period is used to divide the x-axis into four subintervals. The endpoints of the subintervals give rise to five key points on the graph, which are used to sketch one cycle. Finally, extend the graph in either direction to make it complete.

Let's look at an example.

EXAMPLE 6   **Graphing a Sinusoidal Function Using Key Points**

Graph: $y = 3 \sin(4x)$

Solution   From Example 5, the amplitude is 3 and the period is $\dfrac{\pi}{2}$. Because the amplitude is 3, the graph of $y = 3 \sin(4x)$ will lie between -3 and 3 on the y-axis. Because the period is $\dfrac{\pi}{2}$, one cycle will begin at $x = 0$ and end at $x = \dfrac{\pi}{2}$.

We divide the interval $\left[0, \dfrac{\pi}{2}\right]$ into four subintervals, each of length $\dfrac{\pi}{2} \div 4 = \dfrac{\pi}{8}$:

$$\left[0, \frac{\pi}{8}\right], \quad \left[\frac{\pi}{8}, \frac{\pi}{4}\right], \quad \left[\frac{\pi}{4}, \frac{3\pi}{8}\right], \quad \left[\frac{3\pi}{8}, \frac{\pi}{2}\right]$$

NOTE

We could also obtain the five key points by evaluating $y = 3 \sin(4x)$ at the endpoints of each subinterval. ∎

The endpoints of these intervals give rise to the x-coordinates of the five key points on the graph. To obtain the y-coordinates of the five key points, multiply the y-coordinates of the five key points in $y = \sin x$ by $A = 3$. Refer to Figure 66. The five key points are

$$(0, 0), \quad \left(\frac{\pi}{8}, 3\right), \quad \left(\frac{\pi}{4}, 0\right), \quad \left(\frac{3\pi}{8}, -3\right), \quad \left(\frac{\pi}{2}, 0\right)$$

We plot these five points and fill in the graph of the sine curve as shown in Figure 67(a). We extend the graph in either direction to obtain the complete graph shown in Figure 67(b).

To graph a sinusoidal function using a graphing utility, we use the amplitude to set Ymin and Ymax and use the period to set Xmin and Xmax. Figure 67(c) shows the graph using a graphing utility.

Figure 67

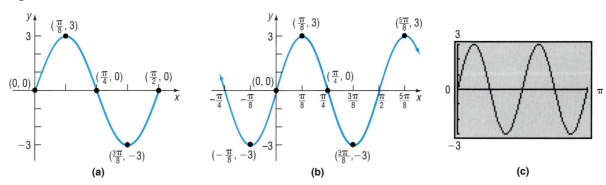

(a)   (b)   (c)

✔ **CHECK:** Graph $y = 3 \sin(4x)$ by hand using transformations. Which graphing method do you prefer?   ◄

NOW WORK PROBLEM **47.**

| EXAMPLE 7 | **Finding the Amplitude and Period of a Sinusoidal Function and Graphing It Using Key Points** |

Determine the amplitude and period of $y = -4\cos(\pi x)$, and graph the function.

Solution Comparing $y = -4\cos(\pi x)$ with $y = A\cos(\omega x)$, we find that $A = -4$ and $\omega = \pi$. The amplitude is $|A| = |-4| = 4$, and the period is $T = \dfrac{2\pi}{\omega} = \dfrac{2\pi}{\pi} = 2$.

The graph of $y = -4\cos(\pi x)$ will lie between -4 and 4 on the y-axis. One cycle will begin at $x = 0$ and end at $x = 2$. We divide the interval $[0, 2]$ into four subintervals, each of length $2 \div 4 = \dfrac{1}{2}$:

$$\left[0, \frac{1}{2}\right], \quad \left[\frac{1}{2}, 1\right], \quad \left[1, \frac{3}{2}\right], \quad \left[\frac{3}{2}, 2\right].$$

The five key points on the graph are

$$(0, -4), \quad \left(\frac{1}{2}, 0\right), \quad (1, 4), \quad \left(\frac{3}{2}, 0\right), \quad (2, -4).$$

We plot these five points and fill in the graph of the cosine function as shown in Figure 68(a). Extending the graph in either direction, we obtain Figure 68(b). Figure 68(c) shows the graph using a graphing utility.

Figure 68

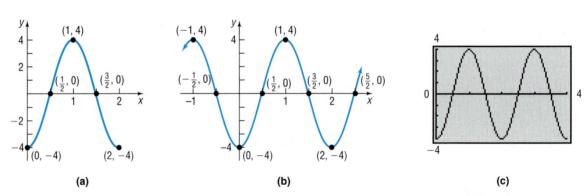

(a)   (b)   (c)

✔ **CHECK:** Graph $y = -4\cos(\pi x)$ by hand using transformations. Which graphing method do you prefer?   ◀

| EXAMPLE 8 | **Finding the Amplitude and Period of a Sinusoidal Function and Graphing It Using Key Points** |

Determine the amplitude and period of $y = 2\sin\left(-\dfrac{\pi}{2}x\right)$, and graph the function.

Solution Since the sine function is odd, we can use the equivalent form:

$$y = -2\sin\left(\frac{\pi}{2}x\right)$$

Comparing $y = -2\sin\left(\dfrac{\pi}{2}x\right)$ to $y = A\sin(\omega x)$, we find that $A = -2$ and $\omega = \dfrac{\pi}{2}$.

The amplitude is $|A| = 2$, and the period is $T = \dfrac{2\pi}{\omega} = \dfrac{2\pi}{\dfrac{\pi}{2}} = 4$.

The graph of $y = -2 \sin\left(\dfrac{\pi}{2}x\right)$ will lie between -2 and 2 on the y-axis. One cycle will begin at $x = 0$ and end at $x = 4$. We divide the interval $[0, 4]$ into four subintervals, each of length $4 \div 4 = 1$:

$$[0, 1], \quad [1, 2], \quad [2, 3], \quad [3, 4]$$

The five key points on the graph are

$$(0, 0), \quad (1, -2), \quad (2, 0), \quad (3, 2), \quad (4, 0)$$

We plot these five points and fill in the graph of the sine function as shown in Figure 69(a). Extending the graph in either direction, we obtain Figure 69(b). Figure 69(c) shows the graph using a graphing utility.

Figure 69

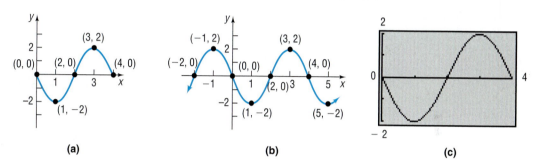

(a)   (b)   (c)

✔ **CHECK:** Graph $y = 2 \sin\left(-\dfrac{\pi}{2}x\right)$ by hand using transformations. Which graphing method do you prefer? ◀

   NOW WORK PROBLEM 61.

5   Find an Equation for a Sinusoidal Graph

We can also use the ideas of amplitude and period to identify a sinusoidal function when its graph is given.

EXAMPLE 9   **Finding an Equation for a Sinusoidal Graph**

Find an equation for the graph shown in Figure 70.

Figure 70

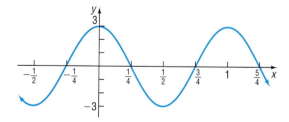

Solution   The graph has the characteristics of a cosine function. Do you see why? So we view the equation as a cosine function $y = A \cos(\omega x)$ with $A = 3$ and period $T = 1$. Then $\dfrac{2\pi}{\omega} = 1$, so $\omega = 2\pi$. The cosine function whose graph is given in Figure 70 is

$$y = A \cos(\omega x) = 3 \cos(2\pi x)$$

✔ **CHECK:** Graph $Y_1 = 3 \cos(2\pi x)$ and compare the result with Figure 70. ◀

EXAMPLE 10	**Finding an Equation for a Sinusoidal Graph**

Find an equation for the graph shown in Figure 71.

Solution The graph is sinusoidal, with amplitude $|A| = 2$. The period is 4, so $\dfrac{2\pi}{\omega} = 4$ or $\omega = \dfrac{\pi}{2}$. Since the graph passes through the origin, it is easiest to view the equation as a sine function,[*] but notice that the graph is actually the reflection of a sine function about the x-axis (since the graph is decreasing near the origin). Thus, we have

$$y = -A \sin(\omega x) = -2 \sin\left(\frac{\pi}{2}x\right)$$

Figure 71

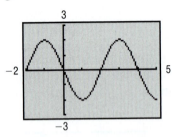

✔ **CHECK:** Graph $Y_1 = -2 \sin\left(\dfrac{\pi}{2}x\right)$ and compare the result with Figure 71. ◀

 N O W   W O R K   P R O B L E M S   7 1   A N D   7 5 .

[*]The equation could also be viewed as a cosine function with a horizontal shift, but viewing it as a sine function is easier.

5.4 Assess Your Understanding

'Are You Prepared?'

Answers are given at the end of these exercises. If you get a wrong answer, read the pages listed in red.

1. Use transformations to graph $y = 3x^2$ (pp. 120–122)

2. Use transformations to graph $y = -x^2$. (pp. 123–124)

Concepts and Vocabulary

3. The maximum value of $y = \sin x$ is _____ and occurs at $x = $ _____.

4. The function $y = A \sin(\omega x)$, $A > 0$, has amplitude 3 and period 2; then $A = $ _____ and $\omega = $ _____.

5. The function $y = 3 \cos(6x)$ has amplitude _____ and period _____.

6. *True or False:* The graphs of $y = \sin x$ and $y = \cos x$ are identical except for a horizontal shift.

7. *True or False:* For $y = 2 \sin(\pi x)$, the amplitude is 2 and the period is $\dfrac{\pi}{2}$.

8. *True or False:* The graph of the sine function has infinitely many x-intercepts.

Skill Building

In Problems 9–18, if necessary, refer to the graphs to answer each question.

9. What is the y-intercept of $y = \sin x$?

10. What is the y-intercept of $y = \cos x$?

11. For what numbers x, $-\pi \le x \le \pi$, is the graph of $y = \sin x$ increasing?

12. For what numbers x, $-\pi \le x \le \pi$, is the graph of $y = \cos x$ decreasing?

13. What is the largest value of $y = \sin x$?

14. What is the smallest value of $y = \cos x$?

15. For what numbers $x, 0 \le x \le 2\pi$, does $\sin x = 0$?

16. For what numbers $x, 0 \le x \le 2\pi$, does $\cos x = 0$?

17. For what numbers $x, -2\pi \le x \le 2\pi$, does $\sin x = 1$? What about $\sin x = -1$?

18. For what numbers $x, -2\pi \le x \le 2\pi$, does $\cos x = 1$? What about $\cos x = -1$?

In Problems 19 and 20, match the graph to a function. Three answers are possible.

A. $y = -\sin x$

B. $y = -\cos x$

C. $y = \sin\left(x - \dfrac{\pi}{2}\right)$

D. $y = -\cos\left(x - \dfrac{\pi}{2}\right)$

E. $y = \sin(x + \pi)$

F. $y = \cos(x + \pi)$

19.

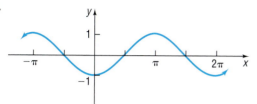

20.

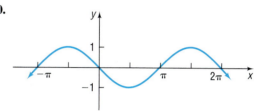

In Problems 21–36, use transformations to graph each function. Verify your results using a graphing utility.

21. $y = 3 \sin x$

22. $y = 4 \cos x$

23. $y = -\cos x$

24. $y = -\sin x$

25. $y = \sin x - 1$

26. $y = \cos x + 1$

27. $y = \sin(x - \pi)$

28. $y = \cos(x + \pi)$

29. $y = \sin(\pi x)$

30. $y = \cos\left(\dfrac{\pi}{2}x\right)$

31. $y = 2 \sin x + 2$

32. $y = 3 \cos x + 3$

33. $y = 4 \cos(2x)$

34. $y = 3 \sin\left(\dfrac{1}{2}x\right)$

35. $y = -2 \sin x + 2$

36. $y = -3 \cos x - 2$

In Problems 37–46, determine the amplitude and period of each function without graphing.

37. $y = 2 \sin x$

38. $y = 3 \cos x$

39. $y = -4 \cos(2x)$

40. $y = -\sin\left(\dfrac{1}{2}x\right)$

41. $y = 6 \sin(\pi x)$

42. $y = -3 \cos(3x)$

43. $y = -\dfrac{1}{2}\cos\left(\dfrac{3}{2}x\right)$

44. $y = \dfrac{4}{3}\sin\left(\dfrac{2}{3}x\right)$

45. $y = \dfrac{5}{3}\sin\left(-\dfrac{2\pi}{3}x\right)$

46. $y = \dfrac{9}{5}\cos\left(-\dfrac{3\pi}{2}x\right)$

In Problems 47–56, match the given function to one of the graphs (A)–(J).

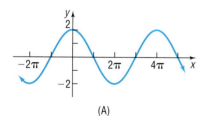

(A)

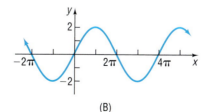

(B)

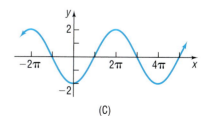

(C)

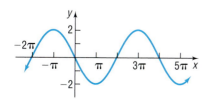

(D)

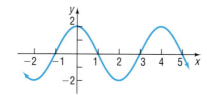

(E)

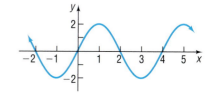

(F)

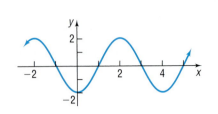

(G)

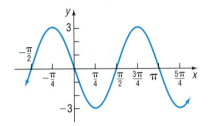

(H)

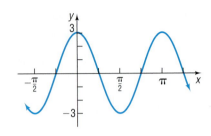

(I)

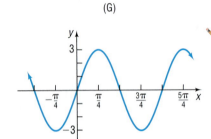

(J)

47. $y = 2 \sin\left(\dfrac{\pi}{2}x\right)$

48. $y = 2 \cos\left(\dfrac{\pi}{2}x\right)$

49. $y = 2 \cos\left(\dfrac{1}{2}x\right)$

50. $y = 3 \cos(2x)$

51. $y = -3 \sin(2x)$

52. $y = 2 \sin\left(\dfrac{1}{2}x\right)$

53. $y = -2 \cos\left(\dfrac{1}{2}x\right)$

54. $y = -2 \cos\left(\dfrac{\pi}{2}x\right)$

55. $y = 3 \sin(2x)$

56. $y = -2 \sin\left(\dfrac{1}{2}x\right)$

In Problems 57–60, match the given function to one of the graphs (A)–(D).

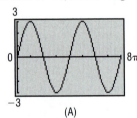

(A)

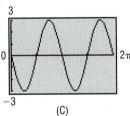

(B)

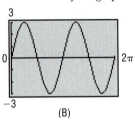

(C)

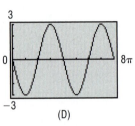

(D)

57. $y = 3 \sin\left(\dfrac{1}{2}x\right)$

58. $y = -3 \sin(2x)$

59. $y = 3 \sin(2x)$

60. $y = -3 \sin\left(\dfrac{1}{2}x\right)$

In Problems 61–70, graph each sinusoidal function.

61. $y = 5 \sin(4x)$

62. $y = 4 \cos(6x)$

63. $y = 5 \cos(\pi x)$

64. $y = 2 \sin(\pi x)$

65. $y = -2 \cos(2\pi x) + 1$

66. $y = -5 \cos(2\pi x) - 2$

67. $y = -4 \sin\left(\dfrac{1}{2}x\right)$

68. $y = -2 \cos\left(\dfrac{1}{2}x\right)$

69. $y = \dfrac{3}{2} \sin\left(-\dfrac{2}{3}x\right)$

70. $y = \dfrac{4}{3} \cos\left(-\dfrac{1}{3}x\right)$

In Problems 71–74, write the equation of a sine function that has the given characteristics.

71. Amplitude: 3
Period: π

72. Amplitude: 2
Period: 4π

73. Amplitude: 3
Period: 2

74. Amplitude: 4
Period: 1

In Problems 75–88, find an equation for each graph.

75.

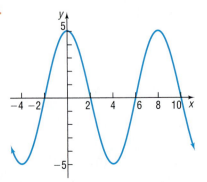

76.

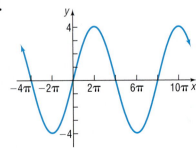

77.

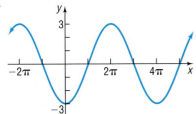

78.

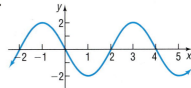

79.

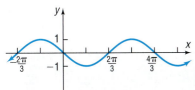

80.

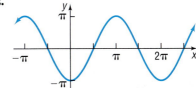

81.

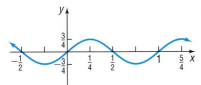

82.

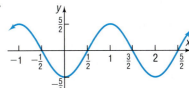

83.

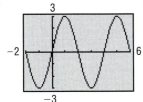

84.

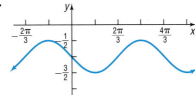

85.

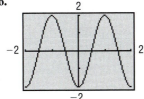

86.

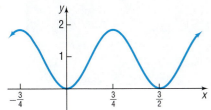

87.

88.

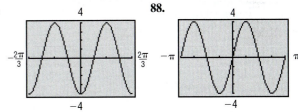

Applications and Extensions

89. Alternating Current (ac) Circuits The current I, in amperes, flowing through an ac (alternating current) circuit at time t is

$$I = 220 \sin(60\pi t), \qquad t \geq 0$$

What is the period? What is the amplitude? Graph this function over two periods.

90. Alternating Current (ac) Circuits The current I, in amperes, flowing through an ac (alternating current) circuit at time t is

$$I = 120 \sin(30\pi t), \qquad t \geq 0$$

What is the period? What is the amplitude? Graph this function over two periods.

91. Alternating Current (ac) Generators The voltage V produced by an ac generator is

$$V = 220 \sin(120\pi t)$$

(a) What is the amplitude? What is the period?
(b) Graph V over two periods, beginning at $t = 0$.
(c) If a resistance of $R = 10$ ohms is present, what is the current I?
[**Hint:** Use Ohm's Law, $V = IR$.]
(d) What is the amplitude and period of the current I?
(e) Graph I over two periods, beginning at $t = 0$.

92. Alternating Current (ac) Generators The voltage V produced by an ac generator is

$$V = 120 \sin(120\pi t)$$

(a) What is the amplitude? What is the period?

(b) Graph V over two periods, beginning at $t = 0$.

(c) If a resistance of $R = 20$ ohms is present, what is the current I?

[**Hint:** Use Ohm's Law, $V = IR$.]

(d) What is the amplitude and period of the current I?

(e) Graph I over two periods, beginning at $t = 0$.

93. Alternating Current (ac) Generators The voltage V produced by an ac generator is sinusoidal. As a function of time, the voltage V is

$$V = V_0 \sin(2\pi ft)$$

where f is the **frequency**, the number of complete oscillations (cycles) per second. [In the United States and Canada, f is 60 hertz (Hz).] The **power** P delivered to a resistance R at any time t is defined as

$$P = \frac{V^2}{R}$$

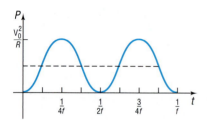

Power in an ac generator

(a) Show that $P = \dfrac{V_0^2}{R} \sin^2(2\pi ft)$.

(b) The graph of P is shown in the figure. Express P as a sinusoidal function.

(c) Deduce that

$$\sin^2(2\pi ft) = \frac{1}{2}[1 - \cos(4\pi ft)]$$

94. Biorhythms In the theory of biorhythms, a sine function of the form

$$P = 50 \sin(\omega t) + 50$$

is used to measure the percent P of a person's potential at time t, where t is measured in days and $t = 0$ is the person's birthday. Three characteristics are commonly measured:

Physical potential: period of 23 days
Emotional potential: period of 28 days
Intellectual potential: period of 33 days

(a) Find ω for each characteristic.

(b) Graph all three functions.

(c) Is there a time t when all three characteristics have 100% potential? When is it?

(d) Suppose that you are 20 years old today ($t = 7305$ days). Describe your physical, emotional, and intellectual potential for the next 30 days.

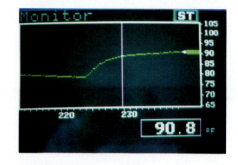

95. Graph $y = |\cos x|$, $-2\pi \le x \le 2\pi$.

96. Graph $y = |\sin x|$, $-2\pi \le x \le 2\pi$.

97. Draw a quick sketch of $y = \sin x$. Be sure to label at least five points.

Discussion and Writing

98. Explain how you would scale the x-axis and y-axis before graphing $y = 3 \cos(\pi x)$.

99. Explain the term *amplitude* as it relates to the graph of a sinusoidal function.

100. Explain how the amplitude and period of a sinusoidal graph are used to establish the scale on each coordinate axis.

101. Find an application in your major field that leads to a sinusoidal graph. Write a paper about your findings.

'Are You Prepared?' Answers

1. Vertical stretch by a factor of 3

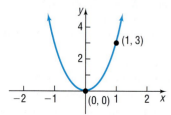

2. Reflection about the x-axis

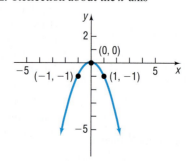

5.5 Graphs of the Tangent, Cotangent, Cosecant, and Secant Functions

PREPARING FOR THIS SECTION   *Before getting started, review the following:*

• Vertical Asymptotes (Section 3.3 pp. 189–191)

✎ Now work the 'Are You Prepared?' problems on page 423.

OBJECTIVES   1   Graph Transformations of the Tangent Function and Cotangent Function
   2   Graph Transformations of the Cosecant Function and Secant Function

1   Graphs Transformations of the Tangent Function and Cotangent Function

Because the tangent function has period π, we only need to determine the graph over some interval of length π. The rest of the graph will consist of repetitions of that graph. Because the tangent function is not defined at $\ldots, -\dfrac{3\pi}{2}, -\dfrac{\pi}{2}, \dfrac{\pi}{2}, \dfrac{3\pi}{2}, \ldots$, we will concentrate on the interval $\left(-\dfrac{\pi}{2}, \dfrac{\pi}{2}\right)$, of length π, and construct Table 8, which lists some points on the graph of $y = \tan x$, $-\dfrac{\pi}{2} < x < \dfrac{\pi}{2}$. We plot the points in the table and connect them with a smooth curve. See Figure 72 for a partial graph of $y = \tan x$, where $-\dfrac{\pi}{3} \le x \le \dfrac{\pi}{3}$.

Table 8

x	$y = \tan x$	(x, y)
$-\dfrac{\pi}{3}$	$-\sqrt{3} \approx -1.73$	$\left(-\dfrac{\pi}{3}, -\sqrt{3}\right)$
$-\dfrac{\pi}{4}$	-1	$\left(-\dfrac{\pi}{4}, -1\right)$
$-\dfrac{\pi}{6}$	$-\dfrac{\sqrt{3}}{3} \approx -0.58$	$\left(-\dfrac{\pi}{6}, -\dfrac{\sqrt{3}}{3}\right)$
0	0	$(0, 0)$
$\dfrac{\pi}{6}$	$\dfrac{\sqrt{3}}{3} \approx 0.58$	$\left(\dfrac{\pi}{6}, \dfrac{\sqrt{3}}{3}\right)$
$\dfrac{\pi}{4}$	1	$\left(\dfrac{\pi}{4}, 1\right)$
$\dfrac{\pi}{3}$	$\sqrt{3} \approx 1.73$	$\left(\dfrac{\pi}{3}, \sqrt{3}\right)$

Figure 72

$y = \tan x, \ -\dfrac{\pi}{3} \le x \le \dfrac{\pi}{3}$

To complete one period of the graph of $y = \tan x$, we need to investigate the behavior of the function as x approaches $-\dfrac{\pi}{2}$ and $\dfrac{\pi}{2}$. We must be careful, though, because $y = \tan x$ is not defined at these numbers. To determine this behavior, we use the table feature on a graphing utility.

Table 9

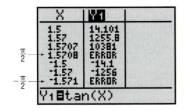

See Table 9. As x gets closer to $\dfrac{\pi}{2} \approx 1.5708$, but remains less than $\dfrac{\pi}{2}$, the values of $\tan x$ are positive and getting larger, so $\tan x$ approaches ∞ $\left(\lim\limits_{x \to \frac{\pi}{2}^-} \tan x = \infty\right)$.

In other words, the vertical line $x = \dfrac{\pi}{2}$ is a vertical asymptote to the graph of $y = \tan x$.

As x gets closer to $-\dfrac{\pi}{2}$, but remains greater than $-\dfrac{\pi}{2}$, the values of $\tan x$ are negative and getting larger in magnitude, so $\tan x$ approaches $-\infty$ $\left(\lim\limits_{x \to -\frac{\pi}{2}^{+}} \tan x = -\infty \right)$. In other words, the vertical line $x = -\dfrac{\pi}{2}$ is also a vertical asymptote to the graph.

With these observations, we can complete one period of the graph. We obtain the complete graph of $y = \tan x$ by repeating this period, as shown in Figure 73(a).

Figure 73(b) shows the graph of $y = \tan x$, $-\infty < x < \infty$, using a graphing utility. Notice we used dot mode when graphing $y = \tan x$. Do you know why?

Figure 73

$y = \tan x$, $-\infty < x < \infty$, x not equal to odd multiples of $\dfrac{\pi}{2}$

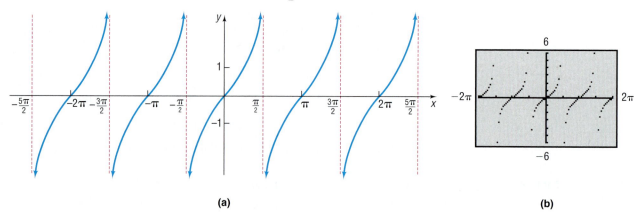

(a)   (b)

The graph of $y = \tan x$ illustrates some facts that we already know about the tangent function.

Properties of the Tangent Function

1. The domain is the set of all real numbers, except odd multiples of $\dfrac{\pi}{2}$.
2. The range is the set of all real numbers.
3. The tangent function is an odd function, as the symmetry of the graph with respect to the origin indicates.
4. The tangent function is periodic, with period π.
5. The x-intercepts are $\ldots, -2\pi, -\pi, 0, \pi, 2\pi, 3\pi, \ldots$; the y-intercept is 0.
6. Vertical asymptotes occur at $x = \ldots, -\dfrac{3\pi}{2}, -\dfrac{\pi}{2}, \dfrac{\pi}{2}, \dfrac{3\pi}{2}, \ldots$.

NOW WORK PROBLEMS 7 AND 15.

EXAMPLE 1   **Graphing Variations of $y = \tan x$ Using Transformations**

Graph: $y = 2 \tan x$

Solution   We start with the graph of $y = \tan x$ and vertically stretch it by a factor of 2. See Figure 74.

Figure 74

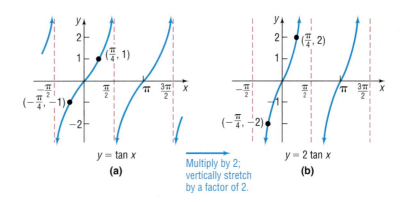

$y = \tan x$
(a)

Multiply by 2;
vertically stretch
by a factor of 2.

$y = 2 \tan x$
(b)

✔ **CHECK:** Figure 75 shows the graph using a graphing utility.

Figure 75

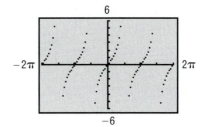

EXAMPLE 2　　**Graphing Variations of $y = \tan x$ Using Transformations**

Graph: $\quad y = -\tan\left(x + \dfrac{\pi}{4}\right)$

Solution　　We start with the graph of $y = \tan x$. See Figure 76.

Figure 76

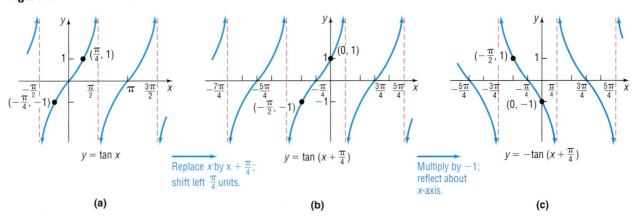

$y = \tan x$

(a)

Replace x by $x + \frac{\pi}{4}$;
shift left $\frac{\pi}{4}$ units.

$y = \tan\left(x + \frac{\pi}{4}\right)$

(b)

Multiply by -1;
reflect about
x-axis.

$y = -\tan\left(x + \frac{\pi}{4}\right)$

(c)

✔ **CHECK:** Graph $Y_1 = -\tan\left(x + \dfrac{\pi}{4}\right)$ and compare the result to Figure 76(c).

NOW WORK PROBLEM 25.

Table 10

x	$y = \cot x$	(x, y)
$\dfrac{\pi}{6}$	$\sqrt{3}$	$\left(\dfrac{\pi}{6}, \sqrt{3}\right)$
$\dfrac{\pi}{4}$	1	$\left(\dfrac{\pi}{4}, 1\right)$
$\dfrac{\pi}{3}$	$\dfrac{\sqrt{3}}{3}$	$\left(\dfrac{\pi}{3}, \dfrac{\sqrt{3}}{3}\right)$
$\dfrac{\pi}{2}$	0	$\left(\dfrac{\pi}{2}, 0\right)$
$\dfrac{2\pi}{3}$	$-\dfrac{\sqrt{3}}{3}$	$\left(\dfrac{2\pi}{3}, -\dfrac{\sqrt{3}}{3}\right)$
$\dfrac{3\pi}{4}$	-1	$\left(\dfrac{3\pi}{4}, -1\right)$
$\dfrac{5\pi}{6}$	$-\sqrt{3}$	$\left(\dfrac{5\pi}{6}, -\sqrt{3}\right)$

Table 11

```
  X        Y1
 .5      1.8305
 .1      9.9666
 .01     99.997
 .001    1000
 3       -7.015
 3.1     -24.03
 3.14    -627.9
Y1■1/tan(X)
```

We obtain the graph of $y = \cot x$ as we did the graph of $y = \tan x$. The period of $y = \cot x$ is π. Because the cotangent function is not defined for integer multiples of π, we will concentrate on the interval $(0, \pi)$. Table 10 lists some points on the graph of $y = \cot x, 0 < x < \pi$.

See Table 11. As x approaches 0, but remains greater than 0, the values of $\cot x$ will be positive and large; so as x approaches 0, with $x > 0$, $\cot x$ approaches ∞ $\left(\lim\limits_{x \to 0^+} \cot x = \infty\right)$. Similarly, as x approaches π, but remains less than π, the values of $\cot x$ will be negative and will approach $-\infty$ $\left(\lim\limits_{x \to \pi^-} \cot x = -\infty\right)$. Figure 77 shows the graph.

Figure 77
$y = \cot x, -\infty < x < \infty$, x not equal to integer multiples of π, $-\infty < y < \infty$

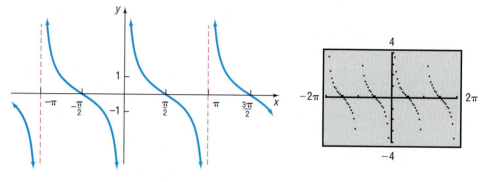

NOW WORK PROBLEM 31.

2 Graph Transformations of the Cosecant Function and Secant Function

The cosecant and secant functions, sometimes referred to as **reciprocal functions**, are graphed by making use of the reciprocal identities

$$\csc x = \frac{1}{\sin x} \quad \text{and} \quad \sec x = \frac{1}{\cos x}$$

For example, the value of the cosecant function $y = \csc x$ at a given number x equals the reciprocal of the corresponding value of the sine function, provided that the value of the sine function is not 0. If the value of $\sin x$ is 0, then x is an integer multiple of π. At such numbers, the cosecant function is not defined. In fact, the graph of the cosecant function has vertical asymptotes at integer multiples of π. Figure 78 shows the graph.

Figure 78
$y = \csc x, -\infty < x < \infty$, x not equal to integer multiples of π, $|y| \geq 1$

EXAMPLE 3 **Graphing Variations of $y = \csc x$ Using Transformations**

Graph: $y = 2 \csc x - 1$

Solution Figure 79 shows the required steps.

Figure 79

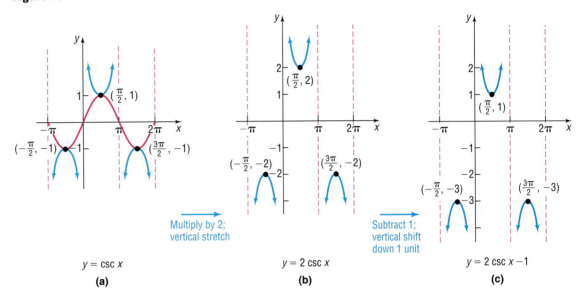

Multiply by 2; vertical stretch

Subtract 1; vertical shift down 1 unit

$y = \csc x$ $y = 2 \csc x$ $y = 2 \csc x - 1$

(a)   (b)   (c)

✔ **CHECK:** Graph $Y_1 = 2 \csc x - 1$ and compare the result with Figure 79. ◀

NOW WORK PROBLEM 37.

Using the idea of reciprocals, we can similarly obtain the graph of $y = \sec x$. See Figure 80.

Figure 80

$y = \sec x, -\infty < x < \infty, x$ not equal to odd multiples of $\dfrac{\pi}{2}, |y| \geq 1$

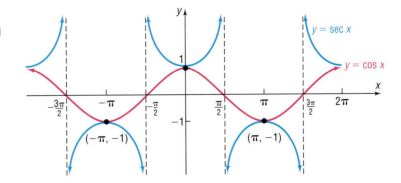

5.5 Assess Your Understanding

'Are You Prepared?'

Answers are given at the end of these exercises. If you get a wrong answer, read the pages listed in red.

1. The graph of $y = \dfrac{3x - 6}{x - 4}$ has a vertical asymptote. What is it? (pp. 189–191)

2. *True or False:* If a function f has the vertical asymptote $x = c$, then $f(c)$ is not defined. (pp. 189–191)

Concepts and Vocabulary

3. The graph of $y = \tan x$ is symmetric with respect to the _____ and has vertical asymptotes at _____.

4. The graph of $y = \sec x$ is symmetric with respect to the _____ and has vertical asymptotes at _____.

5. It is easiest to graph $y = \sec x$ by first sketching the graph of _____.

6. *True or False:* The graphs of $y = \tan x$, $y = \cot x$, $y = \sec x$, and $y = \csc x$ each have infinitely many vertical asymptotes.

Skill Building

In Problems 7–16, if necessary, refer to the graphs to answer each question.

7. What is the y-intercept of $y = \tan x$?

8. What is the y-intercept of $y = \cot x$?

9. What is the y-intercept of $y = \sec x$?

10. What is the y-intercept of $y = \csc x$?

11. For what numbers x, $-2\pi \le x \le 2\pi$, does $\sec x = 1$? What about $\sec x = -1$?

12. For what numbers x, $-2\pi \le x \le 2\pi$, does $\csc x = 1$? What about $\csc x = -1$?

13. For what numbers x, $-2\pi \le x \le 2\pi$, does the graph of $y = \sec x$ have vertical asymptotes?

14. For what numbers x, $-2\pi \le x \le 2\pi$, does the graph of $y = \csc x$ have vertical asymptotes?

15. For what numbers x, $-2\pi \le x \le 2\pi$, does the graph of $y = \tan x$ have vertical asymptotes?

16. For what numbers x, $-2\pi \le x \le 2\pi$, does the graph of $y = \cot x$ have vertical asymptotes?

In Problems 17–20, match each function to its graph.

A. $y = -\tan x$

B. $y = \tan\left(x + \dfrac{\pi}{2}\right)$

C. $y = \tan(x + \pi)$

D. $y = -\tan\left(x - \dfrac{\pi}{2}\right)$

17.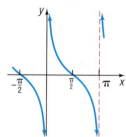

18.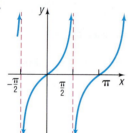

19.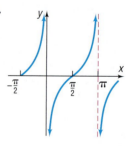

20.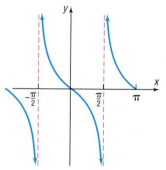

In Problems 21–44, use tranformations to graph each function. Verify your results using a graphing utility.

21. $y = -\sec x$

22. $y = -\cot x$

23. $y = \sec\left(x - \dfrac{\pi}{2}\right)$

24. $y = \csc(x - \pi)$

25. $y = \tan(x - \pi)$

26. $y = \cot(x - \pi)$

27. $y = 3\tan(2x)$

28. $y = 4\tan\left(\dfrac{1}{2}x\right)$

29. $y = \sec(2x)$

30. $y = \csc\left(\dfrac{1}{2}x\right)$

31. $y = \cot(\pi x)$

32. $y = \cot(2x)$

33. $y = -3\tan(4x)$

34. $y = -3\tan(2x)$

35. $y = 2\sec\left(\dfrac{1}{2}x\right)$

36. $y = 2\sec(3x)$

37. $y = -3\csc\left(x + \dfrac{\pi}{4}\right)$

38. $y = -2\tan\left(x + \dfrac{\pi}{4}\right)$

39. $y = \dfrac{1}{2}\cot\left(x - \dfrac{\pi}{4}\right)$

40. $y = 3\sec\left(x + \dfrac{\pi}{2}\right)$

41. $y = \tan x + 2$

42. $y = \cot x - 1$

43. $y = \sec\left(x + \dfrac{\pi}{2}\right) - 1$

44. $y = \csc\left(x - \dfrac{\pi}{4}\right) + 1$

Applications and Extensions

45. Carrying a Ladder around a Corner   Two hallways, one of width 3 feet, the other of width 4 feet, meet at a right angle. See the illustration.

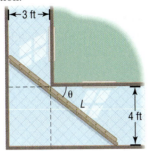

(a) Show that the length L of the line segment shown as a function of the angle θ is
$$L(\theta) = 3 \sec \theta + 4 \csc \theta$$

(b) Graph $L = L(\theta)$, $0 < \theta < \dfrac{\pi}{2}$.

(c) For what value of θ is L the least?

(d) What is the length of the longest ladder that can be carried around the corner? Why is this also the least value of L?

46. A Rotating Beacon   Suppose that a fire truck is parked in front of a building as shown in the figure.

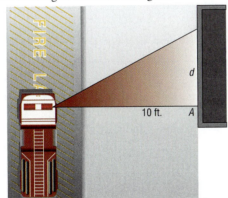

The beacon light on top of the fire truck is located 10 feet from the wall and has a light on each side. If the beacon light rotates 1 revolution every 2 seconds, then a model for determining the distance d that the beacon of light is from point A on the wall after t seconds is given by

$$d = 10 \tan(\pi t)$$

(a) Graph $d = 10 \tan(\pi t)$ for $0 \le t \le 2$.

(b) For what values of t is the function undefined? Explain what this means in terms of the beam of light on the wall.

(c) Fill in the following table.

t	$d = 10 \tan(\pi t)$
0	
0.1	
0.2	
0.3	
0.4	

(d) Compute $\dfrac{d(0.1) - d(0)}{0.1 - 0}$, $\dfrac{d(0.2) - d(0.1)}{0.2 - 0.1}$, and so on, for each consecutive value of t. These are called **first differences**.

(e) Interpret the first differences found in part (d). What is happening to the speed of the beam of light as d increases?

47. Exploration   Graph

$$y = \tan x \quad \text{and} \quad y = -\cot\left(x + \frac{\pi}{2}\right)$$

Do you think that $\tan x = -\cot\left(x + \dfrac{\pi}{2}\right)$?

'Are You Prepared?' Answers

1. $x = 4$ **2.** True

5.6 Phase Shift; Sinusoidal Curve Fitting

> **OBJECTIVES**   **1** Graph Sinusoidal Functions of the Form $y = A \sin(\omega x - \phi)$, Using the Amplitude, Period, and Phase Shift
>
>   **2** Find a Sinusoidal Function from Data

1   **Graph Sinusoidal Functions of the Form y = A sin(ωx − φ), Using the Amplitude, Period, and Phase Shift**

Figure 81
One cycle
$y = A \sin(\omega x)$, $A > 0$, $\omega > 0$

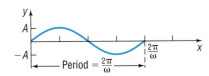

We have seen that the graph of $y = A \sin(\omega x)$, $\omega > 0$, has amplitude $|A|$ and period $T = \dfrac{2\pi}{\omega}$. One cycle can be drawn as x varies from 0 to $\dfrac{2\pi}{\omega}$ or, equivalently, as ωx varies from 0 to 2π. See Figure 81.

We now want to discuss the graph of

$$y = A\sin(\omega x - \phi) = A\sin\left[\omega\left(x - \frac{\phi}{\omega}\right)\right]$$

where $\omega > 0$ and ϕ (the Greek letter phi) are real numbers. The graph will be a sine curve with amplitude $|A|$. As $\omega x - \phi$ varies from 0 to 2π, one period will be traced out. This period will begin when

$$\omega x - \phi = 0 \quad \text{or} \quad x = \frac{\phi}{\omega}$$

and will end when

$$\omega x - \phi = 2\pi \quad \text{or} \quad x = \frac{2\pi}{\omega} + \frac{\phi}{\omega}$$

Figure 82
One cycle $y = A\sin(\omega x - \phi)$, $A > 0$, $\omega > 0$, $\phi > 0$

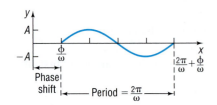

See Figure 82.

We see that the graph of $y = A\sin(\omega x - \phi) = A\sin\left[\omega\left(x - \frac{\phi}{\omega}\right)\right]$ is the same

as the graph of $y = A\sin(\omega x)$, except that it has been shifted $\frac{\phi}{\omega}$ units (to the right if $\phi > 0$ and to the left if $\phi < 0$). This number $\frac{\phi}{\omega}$ is called the **phase shift** of the graph of $y = A\sin(\omega x - \phi)$.

For the graphs of $y = A\sin(\omega x - \phi)$ or $y = A\cos(\omega x - \phi)$, $\omega > 0$,

$$\text{Amplitude} = |A| \qquad \text{Period} = T = \frac{2\pi}{\omega} \qquad \text{Phase shift} = \frac{\phi}{\omega}$$

The phase shift is to the left if $\phi < 0$ and to the right if $\phi > 0$.

EXAMPLE 1

Finding the Amplitude, Period, and Phase Shift of a Sinusoidal Function and Graphing It

Find the amplitude, period, and phase shift of $y = 3\sin(2x - \pi)$, and graph the function.

Solution Comparing

$$y = 3\sin(2x - \pi) = 3\sin\left[2\left(x - \frac{\pi}{2}\right)\right]$$

to

$$y = A\sin(\omega x - \phi) = A\sin\left[\omega\left(x - \frac{\phi}{\omega}\right)\right]$$

we find that $A = 3$, $\omega = 2$, and $\phi = \pi$. The graph is a sine curve with amplitude $|A| = 3$, period $T = \dfrac{2\pi}{\omega} = \dfrac{2\pi}{2} = \pi$, and phase shift $= \dfrac{\phi}{\omega} = \dfrac{\pi}{2}$.

The graph of $y = 3\sin(2x - \pi)$ will lie between -3 and 3 on the y-axis. One cycle will begin at $x = \dfrac{\phi}{\omega} = \dfrac{\pi}{2}$ and end at $x = \dfrac{2\pi}{\omega} + \dfrac{\phi}{\omega} = \pi + \dfrac{\pi}{2} = \dfrac{3\pi}{2}$.

We divide the interval $\left[\dfrac{\pi}{2}, \dfrac{3\pi}{2}\right]$ into four subintervals, each of length $\pi \div 4 = \dfrac{\pi}{4}$:

$$\left[\dfrac{\pi}{2}, \dfrac{3\pi}{4}\right], \quad \left[\dfrac{3\pi}{4}, \pi\right], \quad \left[\pi, \dfrac{5\pi}{4}\right], \quad \left[\dfrac{5\pi}{4}, \dfrac{3\pi}{2}\right]$$

The end points of these subintervals give rise to the following five key points on the graph:

$$\left(\dfrac{\pi}{2}, 0\right), \quad \left(\dfrac{3\pi}{4}, 3\right), \quad (\pi, 0), \quad \left(\dfrac{5\pi}{4}, -3\right), \quad \left(\dfrac{3\pi}{2}, 0\right)$$

We plot these five points and fill in the graph of the sine function as shown in Figure 83(a). Extending the graph in either direction, we obtain Figure 83(b).

Figure 83

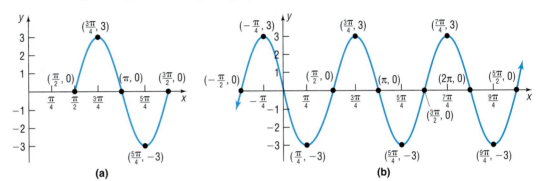

(a)   (b)

The graph of $y = 3\sin(2x - \pi) = 3\sin\left[2\left(x - \dfrac{\pi}{2}\right)\right]$ may also be obtained using transformations. See Figure 84.

Figure 84

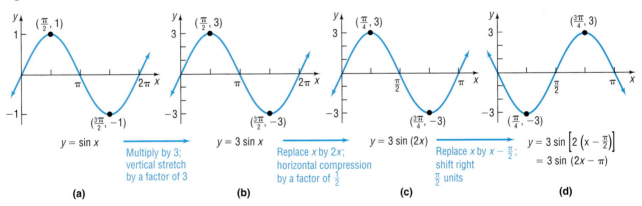

$y = \sin x$ — Multiply by 3; vertical stretch by a factor of 3 → $y = 3\sin x$ — Replace x by $2x$; horizontal compression by a factor of $\frac{1}{2}$ → $y = 3\sin(2x)$ — Replace x by $x - \frac{\pi}{2}$; shift right $\frac{\pi}{2}$ units → $y = 3\sin\left[2\left(x - \frac{\pi}{2}\right)\right]$ $= 3\sin(2x - \pi)$

(a)   (b)   (c)   (d)

✔ **CHECK:** Figure 85 shows the graph of $Y_1 = 3\sin(2x - \pi)$ using a graphing utility.

Figure 85

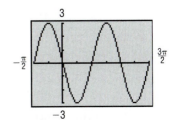

EXAMPLE 2	**Finding the Amplitude, Period, and Phase Shift of a Sinusoidal Function and Graphing It**

Find the amplitude, period, and phase shift of $y = 2 \cos(4x + 3\pi)$, and graph the function.

Solution Comparing

$$y = 2 \cos(4x + 3\pi) = 2 \cos\left[4\left(x + \frac{3\pi}{4}\right)\right]$$

to

$$y = A \cos(\omega x - \phi) = A \cos\left[\omega\left(x - \frac{\phi}{\omega}\right)\right]$$

we see that $A = 2$, $\omega = 4$, and $\phi = -3\pi$. The graph is a cosine curve with amplitude $|A| = 2$, period $T = \dfrac{2\pi}{\omega} = \dfrac{2\pi}{4} = \dfrac{\pi}{2}$, and phase shift $= \dfrac{\phi}{\omega} = -\dfrac{3\pi}{4}$.

The graph of $y = 2 \cos(4x + 3\pi)$ will lie between -2 and 2 on the y-axis. One cycle will begin at $x = \dfrac{\phi}{\omega} = -\dfrac{3\pi}{4}$ and end at $x = \dfrac{2\pi}{\omega} + \dfrac{\phi}{\omega} = \dfrac{\pi}{2} + \left(-\dfrac{3\pi}{4}\right) = -\dfrac{\pi}{4}$.

We divide the interval $\left[-\dfrac{3\pi}{4}, -\dfrac{\pi}{4}\right]$ into four subintervals, each of the length $\dfrac{\pi}{2} \div 4 = \dfrac{\pi}{8}$:

$$\left[-\frac{3\pi}{4}, -\frac{5\pi}{8}\right], \quad \left[-\frac{5\pi}{8}, -\frac{\pi}{2}\right], \quad \left[-\frac{\pi}{2}, -\frac{3\pi}{8}\right], \quad \left[-\frac{3\pi}{8}, -\frac{\pi}{4}\right]$$

The five key points on the graph are

$$\left(-\frac{3\pi}{4}, 2\right), \quad \left(-\frac{5\pi}{8}, 0\right), \quad \left(-\frac{\pi}{2}, -2\right), \quad \left(-\frac{3\pi}{8}, 0\right), \quad \left(-\frac{\pi}{4}, 2\right)$$

We plot these five points and fill in the graph of the cosine function as shown in Figure 86(a). Extending the graph in either direction, we obtain Figure 86(b).

Figure 86

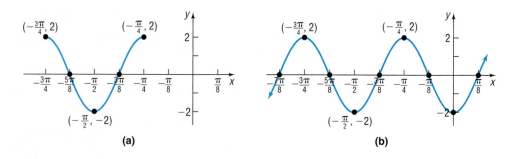

(a)   (b)

The graph of $y = 2 \cos(4x + 3\pi) = 2 \cos\left[4\left(x + \dfrac{3\pi}{4}\right)\right]$ may also be obtained using transformations. See Figure 87.

Figure 87

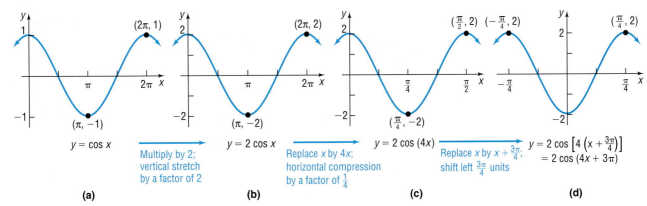

✔ **CHECK:** Graph $Y_1 = 2\cos(4x + 3\pi)$ using a graphing utility. ◀

NOW WORK PROBLEM 3.

Summary

Steps for Graphing Sinusoidal Functions $y = A\sin(\omega x - \phi)$ or $y = A\cos(\omega x - \phi)$

STEP 1: Determine the amplitude $|A|$ and period $T = \dfrac{2\pi}{\omega}$.

STEP 2: Determine the starting point of one cycle of the graph, $\dfrac{\phi}{\omega}$.

STEP 3: Determine the ending point of one cycle of the graph, $\dfrac{2\pi}{\omega} + \dfrac{\phi}{\omega}$.

STEP 4: Divide the interval $\left[\dfrac{\phi}{\omega}, \dfrac{2\pi}{\omega} + \dfrac{\phi}{\omega}\right]$ into four subintervals, each of length $\dfrac{2\pi}{\omega} \div 4$.

STEP 5: Use the endpoints of the subintervals to find the five key points on the graph.

STEP 6: Fill in one cycle of the graph.

STEP 7: Extend the graph in each direction to make it complete.

2   Find a Sinusoidal Function from Data

Scatter diagrams of data sometimes take the form of a sinusoidal function. Let's look at an example.

The data given in Table 12 represent the average monthly temperatures in Denver, Colorado. Since the data represent *average* monthly temperatures collected over many years, the data will not vary much from year to year and so will essentially repeat each year. In other words, the data are periodic. Figure 88 shows the scatter diagram of these data repeated over 2 years, where $x = 1$ represents January, $x = 2$ represents February, and so on.

Table 12

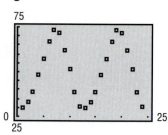

Month, x	Average Monthly Temperature, °F
January, 1	29.7
February, 2	33.4
March, 3	39.0
April, 4	48.2
May, 5	57.2
June, 6	66.9
July, 7	73.5
August, 8	71.4
September, 9	62.3
October, 10	51.4
November, 11	39.0
December, 12	31.0

SOURCE: U.S. National Oceanic and Atmospheric Administration

Figure 88

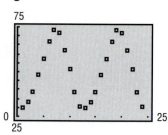

Notice that the scatter diagram looks like the graph of a sinusoidal function. We choose to fit the data to a sine function of the form

$$y = A \sin(\omega x - \phi) + B$$

where A, B, ω, and ϕ are constants.

EXAMPLE 3 **Finding a Sinusoidal Function from Temperature Data**

Fit a sine function to the data in Table 12.

Solution We begin with a scatter diagram of the data for 1 year. See Figure 89. The data will be fitted to a sine function of the form

$$y = A \sin(\omega x - \phi) + B$$

Figure 89

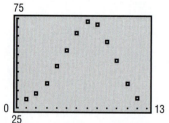

STEP 1: To find the amplitude A, we compute

$$\text{Amplitude} = \frac{\text{largest data value} - \text{smallest data value}}{2}$$

$$= \frac{73.5 - 29.7}{2} = 21.9$$

To see the remaining steps in this process, we superimpose the graph of the function $y = 21.9 \sin x$, where x represents months, on the scatter diagram. Figure 90 shows the two graphs.

To fit the data, the graph needs to be shifted vertically, shifted horizontally, and stretched horizontally.

Figure 90

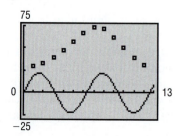

STEP 2: We determine the vertical shift by finding the average of the highest and lowest data value.

$$\text{Vertical shift} = \frac{73.5 + 29.7}{2} = 51.6$$

Now we superimpose the graph of $y = 21.9 \sin x + 51.6$ on the scatter diagram. See Figure 91.

We see that the graph needs to be shifted horizontally and stretched horizontally.

STEP 3: It is easier to find the horizontal stretch factor first. Since the temperatures repeat every 12 months, the period of the function is $T = 12$. Since

$$T = \frac{2\pi}{\omega} = 12, \text{ we have}$$

$$\omega = \frac{2\pi}{12} = \frac{\pi}{6}$$

Now we superimpose the graph of $y = 21.9 \sin\left(\frac{\pi}{6}x\right) + 51.6$ on the scatter diagram. See Figure 92.

We see that the graph still needs to be shifted horizontally.

STEP 4: To determine the horizontal shift, we use the period $T = 12$ and divide the interval $[0, 12]$ into four subintervals of length $12 \div 4 = 3$:

$$[0, 3], \quad [3, 6], \quad [6, 9], \quad [9, 12]$$

The sine curve is increasing on the interval $(0, 3)$, and is decreasing on the interval $(3, 9)$, so a local maximum occurs at $x = 3$. The data indicate that a maximum occurs at $x = 7$ (corresponding to July's temperature), so we must shift the graph of the function 4 units to the right by replacing x by $x - 4$. Doing this, we obtain

$$y = 21.9 \sin\left(\frac{\pi}{6}(x - 4)\right) + 51.6$$

Multiplying out, we find that a sine function of the form $y = A \sin(\omega x - \phi) + B$ that fits the data is

$$y = 21.9 \sin\left(\frac{\pi}{6}x - \frac{2\pi}{3}\right) + 51.6$$

The graph of $y = 21.9 \sin\left(\frac{\pi}{6}x - \frac{2\pi}{3}\right) + 51.6$ and the scatter diagram of the data are shown in Figure 93.

Figure 91

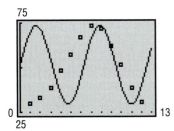

Figure 92

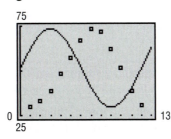

Figure 93

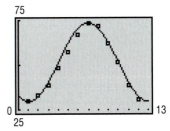

The steps to fit a sine function

$$y = A \sin(\omega x - \phi) + B$$

to sinusoidal data follow:

Steps for Fitting Data to a Sine Function $y = A \sin(\omega x - \phi) + B$

STEP 1: Determine A, the amplitude of the function.

$$\text{Amplitude} = \frac{\text{largest data value} - \text{smallest data value}}{2}$$

STEP 2: Determine B, the vertical shift of the function.

$$\text{Vertical shift} = \frac{\text{largest data value} + \text{smallest data value}}{2}$$

STEP 3: Determine ω. Since the period T, the time it takes for the data to repeat, is $T = \dfrac{2\pi}{\omega}$, we have

$$\omega = \frac{2\pi}{T}$$

STEP 4: Determine the horizontal shift of the function by using the period of the data. Divide the period into four subintervals of equal length. Determine the x-coordinate for the maximum of the sine function and the x-coordinate for the maximum value of the data. Use this information to determine the value of the phase shift, $\dfrac{\phi}{\omega}$.

✏— **NOW WORK PROBLEMS 21(a)–(c).**

Let's look at another example. Since the number of hours of sunlight in a day cycles annually, the number of hours of sunlight in a day for a given location can be modeled by a sinusoidal function.

The longest day of the year (in terms of hours of sunlight) occurs on the day of the summer solstice. The summer solstice is the time when the sun is farthest north. In 2005, the summer solstice occurred on June 21 (the 172nd day of the year) at 2:46 AM EDT. The shortest day of the year occurs on the day of the winter solstice. The winter solstice is the time when the Sun is farthest south (again, for locations in the northern hemisphere). In 2005, the winter solstice occurred on December 21 (the 355th day of the year) at 1:35 PM (EST).

EXAMPLE 4   **Finding a Sinusoidal Function for Hours of Daylight**

According to the *Old Farmer's Almanac*, the number of hours of sunlight in Boston on the summer solstice is 15.283 and the number of hours of sunlight on the winter solstice is 9.067.

(a) Find a sinusoidal function of the form $y = A \sin(\omega x - \phi) + B$ that fits the data.

(b) Use the function found in part (a) to predict the number of hours of sunlight on April 1, the 91st day of the year.

(c) Draw a graph of the function found in part (a).

(d) Look up the number of hours of sunlight for April 1 in the *Old Farmer's Almanac* and compare the actual hours of daylight to the results found in part (b).

Solution   (a) **STEP 1:** Amplitude $= \dfrac{\text{largest data value } - \text{ smallest data value}}{2}$

$$= \frac{15.283 - 9.067}{2} = 3.108$$

STEP 2: Vertical shift $= \dfrac{\text{largest data value } + \text{ smallest data value}}{2}$

$$= \frac{15.283 + 9.067}{2} = 12.175$$

STEP 3: The data repeat every 365 days. Since $T = \dfrac{2\pi}{\omega} = 365$, we find

$$\omega = \frac{2\pi}{365}$$

So far, we have $y = 3.108 \sin\left(\dfrac{2\pi}{365}x - \phi\right) + 12.175$.

STEP 4: To determine the horizontal shift, we use the period $T = 365$ and divide the interval $[0, 365]$ into four subintervals of length $365 \div 4 = 91.25$:

$$[0, 91.25], \quad [91.25, 182.5], \quad [182.5, 273.75], \quad [273.75, 365]$$

The sine curve is increasing on the interval $(0, 91.25)$ and is decreasing on the interval $(91.25, 273.75)$, so a local maximum occurs at $x = 91.25$. Since the maximum occurs on the summer solstice at $x = 172$, we must shift the graph of the function $172 - 91.25 = 80.75$ units to the right by replacing x by $x - 80.75$. Doing this, we obtain

$$y = 3.108 \sin\left(\frac{2\pi}{365}(x - 80.75)\right) + 12.175$$

Multiplying out, we find that a sine function of the form $y = A\sin(\omega x - \phi) + B$ that fits the data is

$$y = 3.108 \sin\left(\frac{2\pi}{365}x - \frac{323\pi}{730}\right) + 12.175$$

(b) To predict the number of hours of daylight on April 1, we let $x = 91$ in the function found in part (a) and obtain

$$y = 3.108 \sin\left(\frac{2\pi}{365} \cdot 91 - \frac{323}{730}\pi\right) + 12.175$$

$$\approx 12.72$$

So we predict that there will be about 12.72 hours $= 12$ hours, 43 minutes of sunlight on April 1 in Boston.

(c) The graph of the function found in part (a) is given in Figure 94.

(d) According to the *Old Farmer's Almanac*, there will be 12 hours 43 minutes of sunlight on April 1 in Boston. Our results agree with the Old Farmer's Almanac! ◀

Figure 94

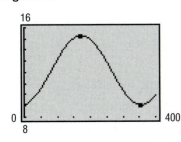

NOW WORK PROBLEM **27.**

Certain graphing utilities (such as a TI-83, TI-84 Plus, and TI-86) have the capability of finding the sine function of best fit for sinusoidal data. At least four data points are required for this process.

EXAMPLE 5 **Finding the Sine Function of Best Fit**

Use a graphing utility to find the sine function of best fit for the data in Table 12. Graph this function with the scatter diagram of the data.

Solution Enter the data from Table 12 and execute the SINe REGression program. The result is shown in Figure 95.

The output that the utility provides shows the equation
$$y = a\sin(bx + c) + d$$

The sinusoidal function of best fit is
$$y = 21.15\sin(0.55x - 2.35) + 51.19$$

where x represents the month and y represents the average temperature.

Figure 96 shows the graph of the sinusoidal function of best fit on the scatter diagram.

Figure 95

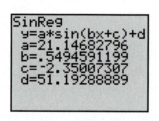

Figure 96

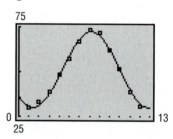

NOW WORK PROBLEMS 21(d) AND (e).

5.6 Assess Your Understanding

Concepts and Vocabulary

1. For the graph of $y = A\sin(\omega x - \phi)$, the number $\dfrac{\phi}{\omega}$ is called the _____.

2. *True or False:* Only two data points are required by a graphing utility to find the sine function of best fit.

Skill Building

In Problems 3–14, find the amplitude, period, and phase shift of each function. Graph each function. Show at least one period. Verify the result using a graphing utility.

3. $y = 4\sin(2x - \pi)$

4. $y = 3\sin(3x - \pi)$

5. $y = 2\cos\left(3x + \dfrac{\pi}{2}\right)$

6. $y = 3\cos(2x + \pi)$

7. $y = -3\sin\left(2x + \dfrac{\pi}{2}\right)$

8. $y = -2\cos\left(2x - \dfrac{\pi}{2}\right)$

9. $y = 4\sin(\pi x + 2)$

10. $y = 2\cos(2\pi x + 4)$

11. $y = 3\cos(\pi x - 2)$

12. $y = 2\cos(2\pi x - 4)$

13. $y = 3\sin\left(-2x + \dfrac{\pi}{2}\right)$

14. $y = 3\cos\left(-2x + \dfrac{\pi}{2}\right)$

In Problems 15–18, write the equation of a sine function that has the given characteristics.

15. Amplitude: 2
Period: π
Phase shift: $\dfrac{1}{2}$

16. Amplitude: 3
Period: $\dfrac{\pi}{2}$
Phase shift: 2

17. Amplitude: 3
Period: 3π
Phase shift: $-\dfrac{1}{3}$

18. Amplitude: 2
Period: π
Phase shift: -2

Applications and Extensions

19. Alternating Current (ac) Circuits   The current I, in amperes, flowing through an ac (alternating current) circuit at time t is

$$I = 120 \sin\left(30\pi t - \frac{\pi}{3}\right), \qquad t \geq 0$$

What is the period? What is the amplitude? What is the phase shift? Graph this function over two periods.

20. Alternating Current (ac) Circuits   The current I, in amperes, flowing through an ac (alternating current) circuit at time t is

$$I = 220 \sin\left(60\pi t - \frac{\pi}{6}\right), \qquad t \geq 0$$

What is the period? What is the amplitude? What is the phase shift? Graph this function over two periods.

21. Monthly Temperature   The following data represent the average monthly temperatures for Juneau, Alaska.

Month, x	Average Monthly Temperature, °F
January, 1	24.2
February, 2	28.4
March, 3	32.7
April, 4	39.7
May, 5	47.0
June, 6	53.0
July, 7	56.0
August, 8	55.0
September, 9	49.4
October, 10	42.2
November, 11	32.0
December, 12	27.1

Source: U.S. National Oceanic and Atmospheric Administration

(a) Use a graphing utility to draw a scatter diagram of the data for one period.
(b) By hand, find a sinusoidal function of the form $y = A \sin(\omega x - \phi) + B$ that fits the data.
(c) Draw the sinusoidal function found in part (b) on the scatter diagram.
(d) Use a graphing utility to find the sinusoidal function of best fit.
(e) Draw the sinusoidal function of best fit on the scatter diagram.

22. Monthly Temperature   The following data represent the average monthly temperatures for Washington, D.C.

Month, x	Average Monthly Temperature, °F
January, 1	34.6
February, 2	37.5
March, 3	47.2
April, 4	56.5
May, 5	66.4
June, 6	75.6
July, 7	80.0
August, 8	78.5
September, 9	71.3
October, 10	59.7
November, 11	49.8
December, 12	39.4

Source: U.S. National Oceanic and Atmospheric Administration

(a) Use a graphing utility to draw a scatter diagram of the data for one period.
(b) By hand, find a sinusoidal function of the form $y = A \sin(\omega x - \phi) + B$ that fits the data.
(c) Draw the sinusoidal function found in part (b) on the scatter diagram.
(d) Use a graphing utility to find the sinusoidal function of best fit.
(e) Graph the sinusoidal function of best fit on the scatter diagram.

23. Monthly Temperature   The following data represent the average monthly temperatures for Indianapolis, Indiana.

Month, x	Average Monthly Temperature, °F
January, 1	25.5
February, 2	29.6
March, 3	41.4
April, 4	52.4
May, 5	62.8
June, 6	71.9
July, 7	75.4
August, 8	73.2
September, 9	66.6
October, 10	54.7
November, 11	43.0
December, 12	30.9

Source: U.S. National Oceanic and Atmospheric Administration

(a) Use a graphing utility to draw a scatter diagram of the data for one period.
(b) By hand, find a sinusoidal function of the form $y = A \sin(\omega x - \phi) + B$ that fits the data.
(c) Draw the sinusoidal function found in part (b) on the scatter diagram.
(d) Use a graphing utility to find the sinusoidal function of best fit.
(e) Graph the sinusoidal function of best fit on the scatter diagram.

24. **Monthly Temperature** The following data represent the average monthly temperatures for Baltimore, Maryland.

Month, x	Average Monthly Temperature, °F
January, 1	31.8
February, 2	34.8
March, 3	44.1
April, 4	53.4
May, 5	63.4
June, 6	72.5
July, 7	77.0
August, 8	75.6
September, 9	68.5
October, 10	56.6
November, 11	46.8
December, 12	36.7

Source: U.S. National Oceanic and Atmospheric Administration

(a) Use a graphing utility to draw a scatter diagram of the data for one period.
(b) By hand, find a sinusoidal function of the form $y = A \sin(\omega x - \phi) + B$ that fits the data.
(c) Draw the sinusoidal function found in part (b) on the scatter diagram.
(d) Use a graphing utility to find the sinusoidal function of best fit.
(e) Graph the sinusoidal function of best fit on the scatter diagram.

25. **Tides** Suppose that the length of time between consecutive high tides is approximately 12.5 hours. According to the National Oceanic and Atmospheric Administration, on Saturday, August 7, 2004, in Savannah, Georgia, high tide occurred at 3:38 AM (3.6333 hours) and low tide occurred at 10:08 AM (10.1333 hours). Water heights are measured as the amounts above or below the mean lower low water. The height of the water at high tide was 8.2 feet, and the height of the water at low tide was −0.6 foot.
(a) Approximately when will the next high tide occur?
(b) Find a sinusoidal function of the form $y = A \sin(\omega x - \phi) + B$ that fits the data.
(c) Draw a graph of the function found in part (b).
(d) Use the function found in part (b) to predict the height of the water at the next high tide.

26. **Tides** Suppose that the length of time between consecutive high tides is approximately 12.5 hours. According to the National Oceanic and Atmospheric Administration, on Saturday, August 7, 2004, in Juneau, Alaska, high tide occurred at 8:11 AM (8.1833 hours) and low tide occurred at 2:14 PM (14.2333 hours). Water heights are measured as the amounts above or below the mean lower low water. The height of the water at high tide was 13.2 feet, and the height of the water at low tide was 2.2 feet.
(a) Approximately when will the next high tide occur?
(b) Find a sinusoidal function of the form $y = A \sin(\omega x - \phi) + B$ that fits the data.
(c) Draw a graph of the function found in part (b).
(d) Use the function found in part (b) to predict the height of the water at the next high tide.

27. **Hours of Daylight** According to the *Old Farmer's Almanac*, in Miami, Florida, the number of hours of sunlight on the summer solstice is 12.75 and the number of hours of sunlight on the winter solstice is 10.583.
(a) Find a sinusoidal function of the form $y = A \sin(\omega x - \phi) + B$ that fits the data.
(b) Use the function found in part (a) to predict the number of hours of sunlight on April 1, the 91st day of the year.
(c) Draw a graph of the function found in part (a).
(d) Look up the number of hours of sunlight for April 1 in the *Old Farmer's Almanac*, and compare the actual hours of daylight to the results found in part (c).

28. **Hours of Daylight** According to the *Old Farmer's Almanac*, in Detroit, Michigan, the number of hours of sunlight on the summer solstice is 13.65 and the number of hours of sunlight on the winter solstice is 9.067.
(a) Find a sinusoidal function of the form $y = A \sin(\omega x - \phi) + B$ that fits the data.
(b) Use the function found in part (a) to predict the number of hours of sunlight on April 1, the 91st day of the year.
(c) Draw a graph of the function found in part (a).
(d) Look up the number of hours of sunlight for April 1 in the *Old Farmer's Almanac*, and compare the actual hours of daylight to the results found in part (c).

29. **Hours of Daylight** According to the *Old Farmer's Almanac*, in Anchorage, Alaska, the number of hours of sunlight on the summer solstice is 16.233 and the number of hours of sunlight on the winter solstice is 5.45.
(a) Find a sinusoidal function of the form $y = A \sin(\omega x - \phi) + B$ that fits the data.
(b) Use the function found in part (a) to predict the number of hours of sunlight on April 1, the 91st day of the year.
(c) Draw a graph of the function found in part (a).
(d) Look up the number of hours of sunlight for April 1 in the *Old Farmer's Almanac*, and compare the actual hours of daylight to the results found in part (c).

30. **Hours of Daylight** According to the *Old Farmer's Almanac*, in Honolulu, Hawaii, the number of hours of sunlight on the summer solstice is 12.767 and the number of hours of sunlight on the winter solstice is 10.783.

(a) Find a sinusoidal function of the form $y = A \sin(\omega x - \phi) + B$ that fits the data.
(b) Use the function found in part (a) to predict the number of hours of sunlight on April 1, the 91st day of the year.
(c) Draw a graph of the function found in part (a).
(d) Look up the number of hours of sunlight for April 1 in the *Old Farmer's Almanac*, and compare the actual hours of daylight to the results found in part (c).

Discussion and Writing

31. Explain how the amplitude and period of a sinusoidal graph are used to establish the scale on each coordinate axis.

32. Find an application in your major field that leads to a sinusoidal graph. Write a paper about your findings.

Chapter Review

Things to Know

Definitions

Angle in standard position (p. 356)	Vertex is at the origin; initial side is along the positive x-axis
1 Degree ($1°$) (p. 357)	$1° = \dfrac{1}{360}$ revolution
1 Radian (p. 360)	The measure of a central angle of a circle whose rays subtend an arc whose length is the radius of the circle

Trigonometric functions (pp. 371–372)

$P = (x, y)$ is the point on the unit circle corresponding to $\theta = t$ radians.

$$\sin t = \sin \theta = y \qquad \cos t = \cos \theta = x \qquad \tan t = \tan \theta = \frac{y}{x}, \quad x \neq 0$$

$$\csc t = \csc \theta = \frac{1}{y}, \quad y \neq 0 \qquad \sec t = \sec \theta = \frac{1}{x}, \quad x \neq 0 \qquad \cot t = \cot \theta = \frac{x}{y}, \quad y \neq 0$$

Trigonometric functions using a circle of radius r (pp. 382–383)

For an angle θ in standard position $P = (x, y)$ is the point on the terminal side of θ that is also on the circle $x^2 + y^2 = r^2$.

$$\sin \theta = \frac{y}{r} \qquad \cos \theta = \frac{x}{r} \qquad \tan \theta = \frac{y}{x}, x \neq 0$$

$$\csc \theta = \frac{r}{y}, \quad y \neq 0 \qquad \sec \theta = \frac{r}{x}, \quad x \neq 0 \qquad \cot \theta = \frac{x}{y}, \quad y \neq 0$$

Periodic function (p. 391)

$f(\theta + p) = f(\theta)$, for all θ, $p > 0$, where the smallest such p is the fundamental period

Formulas

1 revolution $= 360°$ (p. 358)

$\qquad\qquad\quad = 2\pi$ radians (p. 361)

$s = r\theta$ (p. 360)

θ is measured in radians; s is the length of arc subtended by the central angle θ of the circle of radius r; A is the area of the sector.

$A = \dfrac{1}{2}r^2\theta$ (p. 364)

$v = r\omega$ (p. 365)

v is the linear speed along the circle of radius r; ω is the angular speed (measured in radians per unit time).

	TABLE OF VALUES						
θ (Radians)	θ (Degrees)	$\sin\theta$	$\cos\theta$	$\tan\theta$	$\csc\theta$	$\sec\theta$	$\cot\theta$
0	0°	0	1	0	Not defined	1	Not defined
$\dfrac{\pi}{6}$	30°	$\dfrac{1}{2}$	$\dfrac{\sqrt{3}}{2}$	$\dfrac{\sqrt{3}}{3}$	2	$\dfrac{2\sqrt{3}}{3}$	$\sqrt{3}$
$\dfrac{\pi}{4}$	45°	$\dfrac{\sqrt{2}}{2}$	$\dfrac{\sqrt{2}}{2}$	1	$\sqrt{2}$	$\sqrt{2}$	1
$\dfrac{\pi}{3}$	60°	$\dfrac{\sqrt{3}}{2}$	$\dfrac{1}{2}$	$\sqrt{3}$	$\dfrac{2\sqrt{3}}{3}$	2	$\dfrac{\sqrt{3}}{3}$
$\dfrac{\pi}{2}$	90°	1	0	Not defined	1	Not defined	0
π	180°	0	−1	0	Not defined	−1	Not defined
$\dfrac{3\pi}{2}$	270°	−1	0	Not defined	−1	Not defined	0

Fundamental Identities (p. 394)

$$\tan\theta = \frac{\sin\theta}{\cos\theta}, \quad \cot\theta = \frac{\cos\theta}{\sin\theta}$$

$$\csc\theta = \frac{1}{\sin\theta}, \quad \sec\theta = \frac{1}{\cos\theta}, \quad \cot\theta = \frac{1}{\tan\theta}$$

$$\sin^2\theta + \cos^2\theta = 1, \quad \tan^2\theta + 1 = \sec^2\theta, \quad 1 + \cot^2\theta = \csc^2\theta$$

Properties of the Trigonometric Functions

$y = \sin x$ Domain: $-\infty < x < \infty$
(p. 404) Range: $-1 \le y \le 1$
Periodic: period $= 2\pi\,(360°)$
Odd function

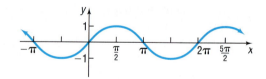

$y = \cos x$ Domain: $-\infty < x < \infty$
(p. 406) Range: $-1 \le y \le 1$
Periodic: period $= 2\pi\,(360°)$
Even function

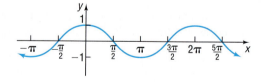

$y = \tan x$ Domain: $-\infty < x < \infty$, except odd multiples of $\dfrac{\pi}{2}\,(90°)$
(p. 420) Range: $-\infty < y < \infty$
Periodic: period $= \pi\,(180°)$
Odd function

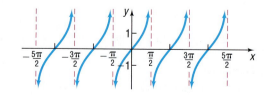

$y = \cot x$ Domain: $-\infty < x < \infty$, except integer multiples of $\pi\,(180°)$
(p. 422) Range: $-\infty < y < \infty$
Periodic: period $= \pi\,(180°)$
Odd function

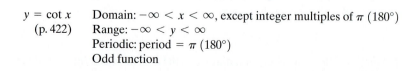
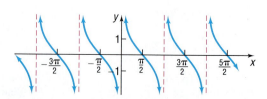

$y = \csc x$ Domain: $-\infty < x < \infty$, except integer multiples of π (180°)
(p. 422)   Range: $|y| \geq 1$
            Periodic: period $= 2\pi$ (360°)
            Odd function

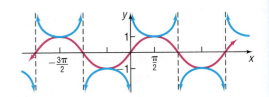

$y = \sec x$ Domain: $-\infty < x < \infty$, except odd multiples of $\dfrac{\pi}{2}$ (90°)
(p. 423)   Range: $|y| \geq 1$
            Periodic: period $= 2\pi$ (360°)
            Even function

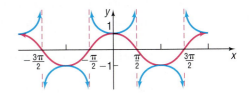

Sinusoidal graphs (pp. 409 and 426)

$y = A \sin(\omega x), \quad \omega > 0$ Period $= \dfrac{2\pi}{\omega}$

$y = A \cos(\omega x), \quad \omega > 0$ Amplitude $= |A|$

$y = A \sin(\omega x - \phi) = A \sin\left[\omega\left(x - \dfrac{\phi}{\omega}\right)\right]$ Phase shift $= \dfrac{\phi}{\omega}$

$y = A \cos(\omega x - \phi) = A \cos\left[\omega\left(x - \dfrac{\phi}{\omega}\right)\right]$

Objectives

Section		You should be able to . . .	Review Exercises
5.1	1	Convert between degrees, minutes, seconds, and decimal forms for angles (p. 358)	82
	2	Find the arc length of a circle (p. 360)	83, 84
	3	Convert from degrees to radians and from radians to degrees (p. 361)	1–8
	4	Find the area of a sector of a circle (p. 364)	83
	5	Find the linear speed of an object traveling in circular motion (p. 365)	85–88
5.2	1	Find the exact values of the trigonometric functions using a point on the unit circle (p. 372)	79
	2	Find the exact values of the trigonometric functions of quadrantal angles (p. 374)	17, 18, 20
	3	Find the exact values of the trigonometric functions of $\dfrac{\pi}{4} = 45°$ (p. 376)	9, 11, 13, 15, 16, 19
	4	Find the exact values of the trigonometric functions of $\dfrac{\pi}{6} = 30°$ and $\dfrac{\pi}{3} = 60°$ (p. 377)	9–15
	5	Find the exact values of the trigonometric functions for integer multiples of $\dfrac{\pi}{6} = 30°$, $\dfrac{\pi}{4} = 45°$, and $\dfrac{\pi}{3} = 60°$ (p. 380)	13–16, 94
	6	Use a calculator to approximate the value of a trigonometric function (p. 381)	75, 76
	7	Use circle of radius r to evaluate the trigonometric functions (p. 388)	80
5.3	1	Determine the domain and the range of the trigonometric functions (p. 388)	81
	2	Determine the period of the trigonometric functions (p. 390)	81
	3	Determine the signs of the trigonometric functions in a given quadrant (p. 392)	77–78
	4	Find the values of the trigonometric functions using fundamental identities (p. 393)	21–30
	5	Find the exact values of the trigonometric functions of an angle given one of the functions and the quadrant of the angle (p. 395)	31–46

Review Exercises

In Problems 1–4, convert each angle in degrees to radians. Express your answer as a multiple of π.

1. $135°$ **2.** $210°$ **3.** $18°$ **4.** $15°$

In Problems 5–8, convert each angle in radians to degrees.

5. $\dfrac{3\pi}{4}$ **6.** $\dfrac{2\pi}{3}$ **7.** $-\dfrac{5\pi}{2}$ **8.** $-\dfrac{3\pi}{2}$

In Problems 9–30, find the exact value of each expression. Do not use a calculator.

9. $\tan \dfrac{\pi}{4} - \sin \dfrac{\pi}{6}$

10. $\cos \dfrac{\pi}{3} + \sin \dfrac{\pi}{2}$

11. $3 \sin 45° - 4 \tan \dfrac{\pi}{6}$

12. $4 \cos 60° + 3 \tan \dfrac{\pi}{3}$

13. $6 \cos \dfrac{3\pi}{4} + 2 \tan\left(-\dfrac{\pi}{3}\right)$

14. $3 \sin \dfrac{2\pi}{3} - 4 \cos \dfrac{5\pi}{2}$

15. $\sec\left(-\dfrac{\pi}{3}\right) - \cot\left(-\dfrac{5\pi}{4}\right)$

16. $4 \csc \dfrac{3\pi}{4} - \cot\left(-\dfrac{\pi}{4}\right)$

17. $\tan \pi + \sin \pi$

18. $\cos \dfrac{\pi}{2} - \csc\left(-\dfrac{\pi}{2}\right)$

19. $\cos 540° - \tan(-45°)$

20. $\sin 630° + \cos(-180°)$

21. $\sin^2 20° + \dfrac{1}{\sec^2 20°}$

22. $\dfrac{1}{\cos^2 40°} - \dfrac{1}{\cot^2 40°}$

23. $\sec 50° \cos 50°$

24. $\tan 10° \cot 10°$

25. $\sec^2 20° - \tan^2 20°$

26. $\dfrac{1}{\sec^2 40°} + \dfrac{1}{\csc^2 40°}$

27. $\sin(-40°) \csc 40°$ **28.** $\tan(-20°) \cot 20°$ **29.** $\cos 410° \sec(-50°)$ **30.** $\cot 200° \tan(-20°)$

In Problems 31–46, find the exact value of each of the remaining trigonometric functions.

31. $\sin \theta = \dfrac{4}{5}$, θ is acute

32. $\cos \theta = \dfrac{3}{5}$, θ is acute

33. $\tan \theta = \dfrac{12}{5}$, $\sin \theta < 0$

34. $\cot \theta = \dfrac{12}{5}$, $\cos \theta < 0$

35. $\sec \theta = -\dfrac{5}{4}$, $\tan \theta < 0$

36. $\csc \theta = -\dfrac{5}{3}$, $\cot \theta < 0$

37. $\sin \theta = \dfrac{12}{13}$, θ in quadrant II

38. $\cos \theta = -\dfrac{3}{5}$, θ in quadrant III

39. $\sin \theta = -\dfrac{5}{13}$, $\dfrac{3\pi}{2} < \theta < 2\pi$

40. $\cos \theta = \dfrac{12}{13}$, $\dfrac{3\pi}{2} < \theta < 2\pi$

41. $\tan \theta = \dfrac{1}{3}$, $180° < \theta < 270°$

42. $\tan \theta = -\dfrac{2}{3}$, $90° < \theta < 180°$

43. $\sec \theta = 3$, $\dfrac{3\pi}{2} < \theta < 2\pi$

44. $\csc \theta = -4$, $\pi < \theta < \dfrac{3\pi}{2}$

45. $\cot \theta = -2$, $\dfrac{\pi}{2} < \theta < \pi$

46. $\tan \theta = -2$, $\dfrac{3\pi}{2} < \theta < 2\pi$

In Problems 47–58, graph each function. Each graph should contain at least one period.

47. $y = 2 \sin(4x)$

48. $y = -3 \cos(2x)$

49. $y = -2 \cos\left(x + \dfrac{\pi}{2}\right)$

50. $y = 3 \sin(x - \pi)$

51. $y = \tan(x + \pi)$

52. $y = -\tan\left(x - \dfrac{\pi}{2}\right)$

53. $y = -2 \tan(3x)$

54. $y = 4 \tan(2x)$

55. $y = \cot\left(x + \dfrac{\pi}{4}\right)$

56. $y = -4 \cot(2x)$

57. $y = \sec\left(x - \dfrac{\pi}{4}\right)$

58. $y = \csc\left(x + \dfrac{\pi}{4}\right)$

In Problems 59–62, determine the amplitude and period of each function without graphing.

59. $y = 4 \cos x$

60. $y = \sin(2x)$

61. $y = -8 \sin\left(\dfrac{\pi}{2}x\right)$

62. $y = -2 \cos(3\pi x)$

In Problems 63–70, find the amplitude, period, and phase shift of each function. Graph each function. Show at least one period.

63. $y = 4 \sin(3x)$

64. $y = 2 \cos\left(\dfrac{1}{3}x\right)$

65. $y = 2 \sin(2x - \pi)$

66. $y = -\cos\left(\dfrac{1}{2}x + \dfrac{\pi}{2}\right)$

67. $y = \dfrac{1}{2} \sin\left(\dfrac{3}{2}x - \pi\right)$

68. $y = \dfrac{3}{2} \cos(6x + 3\pi)$

69. $y = -\dfrac{2}{3} \cos(\pi x - 6)$

70. $y = -7 \sin\left(\dfrac{\pi}{3}x + \dfrac{4}{3}\right)$

In Problems 71–74, find a function whose graph is given.

71.

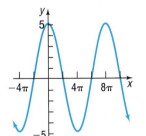

72.

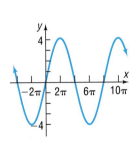

73.

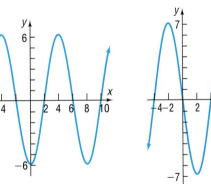

74.

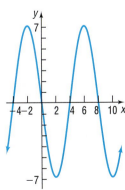

75. Use a calculator to approximate $\sin \dfrac{\pi}{8}$. Round the answer to two decimal places.

76. Use a calculator to approximate $\sec 10°$. Round the answer to two decimal places.

77. Determine the signs of the six trigonometric functions of an angle θ whose terminal side is in quadrant III.

78. Name the quadrant θ lies in if $\cos \theta > 0$ and $\tan \theta < 0$.

79. Find the exact values of the six trigonometric functions if $P = \left(-\dfrac{1}{3}, \dfrac{2\sqrt{2}}{3}\right)$ is the point on the unit circle that corresponds to t.

80. Find the exact value of $\sin t$, $\cos t$, and $\tan t$ if $P = (-2, 5)$ is the point on the circle that corresponds to t.

81. What is the domain and the range of the secant function? What is the period?

82. (a) Convert the angle $32°20'35''$ to a decimal in degrees. Round the answer to two decimal places.
(b) Convert the angle $63.18°$ to D°M'S'' form. Express the answer to the nearest second.

83. Find the length of the arc subtended by a central angle of $30°$ on a circle of radius 2 feet. What is the area of the sector?

84. The minute hand of a clock is 8 inches long. How far does the tip of the minute hand move in 30 minutes? How far does it move in 20 minutes?

85. Angular Speed of a Race Car A race car is driven around a circular track at a constant speed of 180 miles per hour. If the diameter of the track is $\frac{1}{2}$ mile, what is the angular speed of the car? Express your answer in revolutions per hour (which is equivalent to laps per hour).

86. Merry-Go-Rounds A neighborhood carnival has a merry-go-round whose radius is 25 feet. If the time for one revolution is 30 seconds, how fast is the merry-go-round going?

87. Lighthouse Beacons The Montauk Point Lighthouse on Long Island has dual beams (two light sources opposite each other). Ships at sea observe a blinking light every 5 seconds. What rotation speed is required to do this?

88. Spin Balancing Tires The radius of each wheel of a car is 16 inches. At how many revolutions per minute should a spin balancer be set to balance the tires at a speed of 90 miles per hour? Is the setting different for a wheel of radius 14 inches? If so, what is this setting?

89. Alternating Voltage The electromotive force E, in volts, in a certain ac (alternating circuit) circuit obeys the equation

$$E = 120 \sin(120\pi t), \quad t \geq 0$$

where t is measured in seconds.
(a) What is the maximum value of E?
(b) What is the period?
(c) Graph this function over two periods.

90. Alternating Current The current I, in amperes, flowing through an ac (alternating current) circuit at time t is

$$I = 220 \sin\left(30\pi t + \frac{\pi}{6}\right), \quad t \geq 0$$

(a) What is the period?
(b) What is the amplitude?
(c) What is the phase shift?
(d) Graph this function over two periods.

91. Monthly Temperature The following data represent the average monthly temperatures for Phoenix, Arizona.

Month, m	Average Monthly Temperature, T
January, 1	51
February, 2	55
March, 3	63
April, 4	67
May, 5	77
June, 6	86
July, 7	90
August, 8	90
September, 9	84
October, 10	71
November, 11	59
December, 12	52

Source: U.S. National Oceanic and Atmospheric Administration

(a) Use a graphing utility to draw a scatter diagram of the data for one period.
(b) By hand, find a sinusoidal function of the form $y = A \sin(\omega x - \phi) + B$ that fits the data.
(c) Draw the sinusoidal function found in part (b) on the scatter diagram.
(d) Use a graphing utility to find the sinusoidal function of best fit.
(e) Graph the sinusoidal function of best fit on the scatter diagram.

92. Monthly Temperature The following data represent the average monthly temperatures for Chicago, Illinois.

Month, m	Average Monthly Temperature, T
January, 1	25
February, 2	28
March, 3	36
April, 4	48
May, 5	61
June, 6	72
July, 7	74
August, 8	75
September, 9	66
October, 10	55
November, 11	39
December, 12	28

Source: U.S. National Oceanic and Atmospheric Administration

(a) Use a graphing utility to draw a scatter diagram of the data for one period.
(b) By hand, find a sinusoidal function of the form $y = A \sin(\omega x - \phi) + B$ that fits the data.

(c) Draw the sinusoidal function found in part (b) on the scatter diagram.

(d) Use a graphing utility to find the sinusoidal function of best fit.

(e) Graph the sinusoidal function of best fit on the scatter diagram.

93. Hours of Daylight According to the *Old Farmer's Almanac*, in Las Vegas, Nevada, the number of hours of sunlight on the summer solstice is 13.367 and the number of hours of sunlight on the winter solstice is 9.667.

(a) Find a sinusoidal function of the form

$$y = A \sin(\omega x - \phi) + B$$

that fits the data.

(b) Draw a graph of the function found in part (a).

(c) Use the function found in part (a) to predict the number of hours of sunlight on April 1, the 91st day of the year.

(d) Look up the number of hours of sunlight for April 1 in the *Old Farmer's Almanac* and compare the actual hours of daylight to the results found in part (c).

94. Unit Circle Fill in the angles (in degrees and radians) and terminal points P of each angle on the unit circle shown.

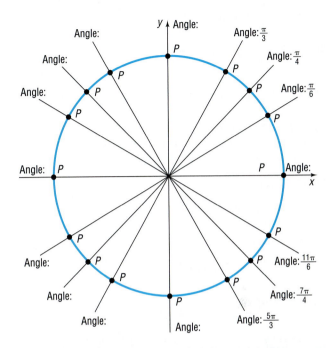

Chapter Test

In Problems 1–3, convert each angle in degrees to radians. Express your answer as a multiple of π.

1. $260°$

2. $-400°$

3. $13°$

In Problems 4–6 convert each angle in radians to degrees.

4. $-\dfrac{\pi}{8}$

5. $\dfrac{9\pi}{2}$

6. $\dfrac{3\pi}{4}$

In Problems 7–12, find the exact value of each expression.

7. $\sin \dfrac{\pi}{6}$

8. $\cos\left(-\dfrac{5\pi}{4}\right) - \cos \dfrac{3\pi}{4}$

9. $\cos(-120°)$

10. $\tan 330°$

11. $\sin \dfrac{\pi}{2} - \tan \dfrac{19\pi}{4}$

12. $2 \sin^2 60° - 3 \cos 45°$

In Problems 13–16, use a calculator to evaluate each expression. Round your answers to three decimal places.

13. $\sin 17°$

14. $\cos \dfrac{2\pi}{5}$

15. $\sec 229°$

16. $\cot \dfrac{28\pi}{9}$

17. Fill in each table entry with the sign of each function.

	sin θ	cos θ	tan θ	sec θ	csc θ	cot θ
θ in QI						
θ in QII						
θ in QIII						
θ in QIV						

18. If $f(x) = \sin x$ and $f(a) = \dfrac{3}{5}$, find $f(-a)$.

In Problems 19–21 find the value of the remaining five trigonometric functions of θ.

19. $\sin \theta = \dfrac{5}{7}$, θ in quadrant II

20. $\cos \theta = \dfrac{2}{3}$, $\dfrac{3\pi}{2} < \theta < 2\pi$

21. $\tan \theta = -\dfrac{12}{5}$, $\dfrac{\pi}{2} < \theta < \pi$

In Problems 22–24, the point (x, y) is on the terminal side of angle θ in standard position. Find the exact value of the given trigonometric function.

22. $(2, 7)$, $\sin \theta$

23. $(-5, 11)$, $\cos \theta$

24. $(6, -3)$, $\tan \theta$

In Problems 25 and 26, graph the function by hand.

25. $y = 2 \sin\left(x - \dfrac{\pi}{6}\right)$

26. $y = \tan\left(-x + \dfrac{\pi}{4}\right) + 2$

27. Write an equation for a sinusoidal graph with the following properties:

$$A = -3 \qquad \text{period} = \frac{2\pi}{3} \qquad \text{phase shift} = -\frac{\pi}{4}$$

28. Logan has a garden in the shape of a sector of a circle; the outer rim of the garden is 25 ft long and the central angle of the sector is 50°. She wants to add a 3 ft wide walk to the outer rim; how many square feet of paving blocks will she need to build the walk?

29. Hungarian Adrian Annus won the gold medal for the hammer throw at the 2004 Olympics in Athens with a winning distance of 83.19 meters.[*] The event consists of swinging a 16 lb weight—attached to a wire 190 cm long—in a circle, then releasing it. Assuming his release is at a 45° angle to the ground, the hammer will travel a distance of $\dfrac{v_0^2}{g}$ meters, where $g = 9.8$ m/sec² and v_0 is the linear speed of the hammer when released. At what rate (rpm) was he swinging the hammer upon release?

30. The Freedom Tower is to be the centerpiece of the rebuilding of the World Trade Center in New York City. The tower will be 1776 feet tall (not including a broadcast antenna). The angle of elevation from the base of an office building to the top of the tower is 34°. The angle of elevation from the helipad on the roof of the office building to the top of the tower is 20°.

(a) How far away is the office building from the Freedom Tower (assume the side of the tower is vertical)? Round to the nearest foot.

(b) How tall is the office building? Round to the nearest foot.

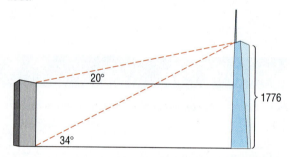

[*]Annus was stripped of his medal after refusing post-medal drug testing.

Chapter Projects

1. **Tides** A partial tide table for September 2001 for Sabine Pass along the Texas Gulf Coast is given in the table.
   (a) On September 15, when was the tide high? This is called *high tide*. On September 19, when was the tide low? This is called *low tide*. Most days will have two low tides and two high tides.
   (b) Why do you think there is a negative height for the low tide on September 14? What is the tide height measured against?

   (c) On your graphing utility, draw a scatter diagram for the data in the table. Let T (time) be the independent variable, with $T = 0$ being 12:00 AM on September 1, $T = 24$ being 12:00 AM on September 2, and so on. Remember that there are 60 minutes in an hour. Let H be the height in feet. Also, make sure that your graphing utility is in radian mode.
   (d) What shape does the data take? What is the period of the data? What is the amplitude? Is the amplitude constant? Explain.
   (e) Using Steps 1–4 given on page 432, fit a sine curve to the data. Let the amplitude be the average of the amplitudes that you found in part (c), unless the amplitude was constant. Is there a vertical shift? Is there a phase shift?
   (f) Using your graphing utility, find the sinusoidal function of best fit. How does it compare to your equation?
   (g) Using the equation found in part (e) and the sinusoidal equation of best fit found in part (f), predict the high tides and the low tides on September 21.
   (h) Looking at the times of day that the low tides occur, what do you think causes the low tides to vary so much each day? Explain. Does this seem to have the same type of effect on the high tides? Explain.

Sept	High Tide Time	Ht (ft)	High Tide Time	Ht (ft)	Low Tide Time	Ht (ft)	Low Tide Time	Ht (ft)	Sun/Moon phase Rise/Set
14	03:08a	2.4	11:12a	2.2	08:14a	2.0	07:19p	−0.1	7:00a/7:23p
15	03:33a	2.4	12:56p	2.2	08:15a	1.9	08:13p	0.0	7:00a/7:22p
16	03:57a	2.3	02:17p	2.3	08:45a	1.6	09:05p	0.3	7:01a/7:20p
17	04:20a	2.2	03:33p	2.3	09:24a	1.4	09:54p	0.5	7:01a/7:19p
18	04:41a	2.2	04:47p	2.3	10:08a	1.0	10:43p	1.0	7:02a/7:08p
19	05:01a	2.0	06:04p	2.3	10:54a	0.7	11:32p	1.4	7:02a/7:17p
20	05:20a	2.0	07:27p	2.3	11:44a	0.4			7:03a/7:15p

SOURCE: www.harbortides.com

The following projects are available at the Instructor's Resource Center (IRC):

2. **Project at Motorola** *Digital Transmission Over the Air*
3. **Identifying Mountain Peaks in Hawaii**
4. **CBL Experiment**

Cumulative Review

1. Find the real solutions, if any, of the equation $2x^2 + x - 1 = 0$.

2. Find an equation for the line with slope -3 containing the point $(-2, 5)$.

3. Find an equation for a circle of radius 4 and center at the point $(0, -2)$.

4. Discuss the equation $2x - 3y = 12$. Graph it.

5. Discuss the equation $x^2 + y^2 - 2x + 4y - 4 = 0$. Graph it.

6. Use transformations to graph the function $y = (x - 3)^2 + 2$.

7. Sketch a graph of each of the following functions. Label at least three points on each graph.
   - (a) $y = x^2$
   - (b) $y = x^3$
   - (c) $y = e^x$
   - (d) $y = \ln x$
   - (e) $y = \sin x$
   - (f) $y = \tan x$

8. Find the inverse function of $f(x) = 3x - 2$.

9. Find the exact value of $(\sin 14°)^2 + (\cos 14°)^2 - 3$.

10. Graph $y = 3 \sin(2x)$.

11. Find the exact value of $\tan \dfrac{\pi}{4} - 3 \cos \dfrac{\pi}{6} + \csc \dfrac{\pi}{6}$.

12. Find an exponential function for the following graph. Express your answer in the form $y = Ab^x$.

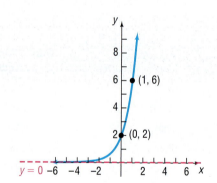

13. Find a sinusoidal function for the following graph.

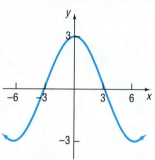

14. (a) Find a linear function that contains the points $(-2, 3)$ and $(1, -6)$. What is the slope? What are the intercepts of the function? Graph the function. Be sure to label the intercepts.

    (b) Find a quadratic function that contains the point $(-2, 3)$ with vertex $(1, -6)$. What are the intercepts of the function? Graph the function.

    (c) Show that there is no exponential function of the form $f(x) = ae^x$ that contains the points $(-2, 3)$ and $(1, -6)$.

15. (a) Find a polynomial function of degree 3 whose y-intercept is 5 and whose x-intercepts are -2, 3, and 5. Graph the function. Label the local minima and local maxima.

    (b) Find a rational function whose y-intercept is 5 and whose x-intercepts are -2, 3, and 5 that has the line $x = 2$ as a vertical asymptote. Graph the function. Answers may vary.

Analytic Trigonometry

6

A LOOK BACK In Chapter 4, we defined inverse functions and developed their properties, particularly the relationship between the domain and range of a function and its inverse. We learned that the graph of a function and its inverse are symmetric with respect to the line $y = x$.

We continued in Chapter 4 by defining the exponential function and the inverse of the exponential function, the logarithmic function.

A LOOK AHEAD In the first two sections of this chapter, we define the six inverse trigonometric functions and investigate their properties. In Sections 6.3 through 6.6 of this chapter, we continue the derivation of identities. These identities play an important role in calculus, the physical and life sciences, and economics, where they are used to simplify complicated expressions. The last two sections of this chapter deal with equations that contain trigonometric functions.

Tremor Brought First Hint of Doom

Underwater earthquakes that triggered the devastating tsunami in PNG would have been felt by villagers about 30 minutes before the waves struck, scientists said yesterday. "The tremor was felt by coastal residents who may not have realized its significance or did not have time to retreat," said associate Professor Ted Bryant, a geoscientist at the University of Wollongong.

The tsunami would have sounded like a fleet of bombers as it crashed into a 30-kilometer stretch of coast. "The tsunami would have been caused by a rapid uplift or drop of the sea floor," said an applied mathematician and cosmologist from Monash University, Professor Joe Monaghan, who is one of Australia's leading experts on tsunamis.

Tsunamis, ridges of water hundreds of kilometers long and stretching from front to back for several kilometers, line up parallel to the beach. "We are talking about a huge volume of water moving very fast—300 kilometers per hour would be typical," Professor Monaghan said.

SOURCE: Peter Spinks, *The Age*, Tuesday, July 21, 1998.

—See Chapter Project 1.

OUTLINE

6.1 The Inverse Sine, Cosine, and Tangent Functions

PREPARING FOR THIS SECTION *Before getting started, review the following:*

- Inverse Functions (Section 4.2, pp. 256–267)
- Values of the Trigonometric Functions (Section 5.2, pp. 374–382)

- Properties of the Sine, Cosine, and Tangent Functions (Section 5.3, pp. 388–393)
- Graphs of the Sine, Cosine, and Tangent Functions (Section 5.4, pp. 402–407, and Section 5.5, pp. 419–422)

Now work the 'Are You Prepared?' problems on page 457.

OBJECTIVES 1 Find the Exact Value of the Inverse Sine, Cosine, and Tangent Functions
2 Find an Approximate Value of the Inverse Sine, Cosine, and Tangent Functions

In Section 4.2 we discussed inverse functions, and we noted that if a function is one-to-one it will have an inverse function. We also observed that if a function is not one-to-one it may be possible to restrict its domain in some suitable manner so that the restricted function is one-to-one.

Next, we review some properties of a one-to-one function f and its inverse function f^{-1}.

1. $f^{-1}(f(x)) = x$ for every x in the domain of f and $f(f^{-1}(x)) = x$ for every x in the domain of f^{-1}.
2. Domain of f = range of f^{-1}, and range of f = domain of f^{-1}.
3. The graph of f and the graph of f^{-1} are symmetric with respect to the line $y = x$.
4. If a function $y = f(x)$ has an inverse function, the equation of the inverse function is $x = f(y)$. The solution of this equation is $y = f^{-1}(x)$.

The Inverse Sine Function

In Figure 1, we reproduce the graph of $y = \sin x$. Because every horizontal line $y = b$, where b is between -1 and 1, intersects the graph of $y = \sin x$ infinitely many times, it follows from the horizontal-line test that the function $y = \sin x$ is not one-to-one.

Figure 1
$y = \sin x, -\infty < x < \infty, -1 \le y \le 1$

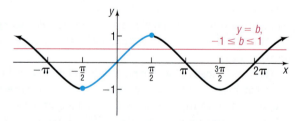

Figure 2
$y = \sin x, -\dfrac{\pi}{2} \le x \le \dfrac{\pi}{2}, -1 \le y \le 1$

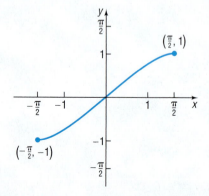

However, if we restrict the domain of $y = \sin x$ to the interval $\left[-\dfrac{\pi}{2}, \dfrac{\pi}{2}\right]$, the restricted function

$$y = \sin x, \qquad -\dfrac{\pi}{2} \le x \le \dfrac{\pi}{2}$$

is one-to-one and so will have an inverse function.* See Figure 2.

---

*Although there are many other ways to restrict the domain and obtain a one-to-one function, mathematicians have agreed to use the interval $\left[-\dfrac{\pi}{2}, \dfrac{\pi}{2}\right]$ to define the inverse of $y = \sin x$.

An equation for the inverse of $y = f(x) = \sin x$ is obtained by interchanging x and y. The implicit form of the inverse function is $x = \sin y$, $-\dfrac{\pi}{2} \le y \le \dfrac{\pi}{2}$. The explicit form is called the **inverse sine** of x and is symbolized by $y = f^{-1}(x) = \sin^{-1} x$.

$$y = \sin^{-1} x \quad \text{means} \quad x = \sin y$$

$$\text{where} \quad -1 \le x \le 1 \quad \text{and} \quad -\frac{\pi}{2} \le y \le \frac{\pi}{2} \tag{1}$$

Because $y = \sin^{-1} x$ means $x = \sin y$, we read $y = \sin^{-1} x$ as "y is the angle or real number whose sine equals x." Alternatively, we can say that "y is the inverse sine of x." Be careful about the notation used. The superscript -1 that appears in $y = \sin^{-1} x$ is not an exponent, but is reminiscent of the symbolism f^{-1} used to denote the inverse function of f. (To avoid this notation, some books use the notation $y = \text{Arcsin } x$ instead of $y = \sin^{-1} x$.)

The inverse of a function f receives as input an element from the range of f and returns as output an element in the domain of f. The restricted sine function, $y = f(x) = \sin x$, receives as input an angle or real number x in the interval $\left[-\dfrac{\pi}{2}, \dfrac{\pi}{2}\right]$ and outputs a real number in the interval $[-1, 1]$. Therefore, the inverse sine function $y = \sin^{-1} x$ receives as input a real number in the interval $[-1, 1]$ or $-1 \le x \le 1$, its domain, and outputs an angle or real number in the interval $\left[-\dfrac{\pi}{2}, \dfrac{\pi}{2}\right]$ or $-\dfrac{\pi}{2} \le y \le \dfrac{\pi}{2}$, its range.

The graph of the inverse sine function can be obtained by reflecting the restricted portion of the graph of $y = f(x) = \sin x$ about the line $y = x$, as shown in Figure 3(a). Figure 3(b) shows the graph using a graphing utility.

Figure 3

$y = \sin^{-1} x$, $-1 \le x \le 1$, $-\dfrac{\pi}{2} \le y \le \dfrac{\pi}{2}$

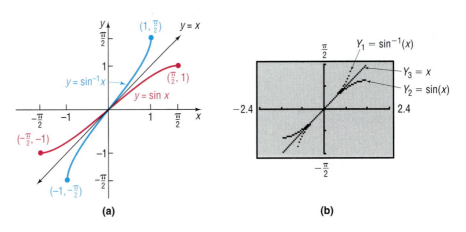

(a) (b)

1 **Find the Exact Value of the Inverse Sine Function**

For some numbers x, it is possible to find the exact value of $y = \sin^{-1} x$.

EXAMPLE 1	**Finding the Exact Value of a Composite Function**

Find the exact value of: $\sin^{-1} 1$

Solution   Let $\theta = \sin^{-1} 1$. We seek the angle, θ, $-\dfrac{\pi}{2} \leq \theta \leq \dfrac{\pi}{2}$, whose sine equals 1.

$$\theta = \sin^{-1} 1, \qquad -\frac{\pi}{2} \leq \theta \leq \frac{\pi}{2}$$

$$\sin \theta = 1, \qquad -\frac{\pi}{2} \leq \theta \leq \frac{\pi}{2} \qquad \textit{By definition of } y = \sin^{-1} x$$

Now look at Table 1 and Figure 4.

Table 1

θ	$\sin \theta$
$-\dfrac{\pi}{2}$	-1
$-\dfrac{\pi}{3}$	$-\dfrac{\sqrt{3}}{2}$
$-\dfrac{\pi}{4}$	$-\dfrac{\sqrt{2}}{2}$
$-\dfrac{\pi}{6}$	$-\dfrac{1}{2}$
0	0
$\dfrac{\pi}{6}$	$\dfrac{1}{2}$
$\dfrac{\pi}{4}$	$\dfrac{\sqrt{2}}{2}$
$\dfrac{\pi}{3}$	$\dfrac{\sqrt{3}}{2}$
$\dfrac{\pi}{2}$	1

Figure 4

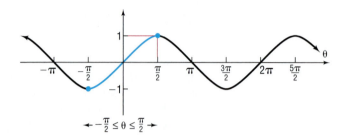

We see that the only angle θ within the interval $\left[-\dfrac{\pi}{2}, \dfrac{\pi}{2}\right]$ whose sine is 1 is $\dfrac{\pi}{2}$. (Note that $\sin \dfrac{5\pi}{2}$ also equals 1, but $\dfrac{5\pi}{2}$ lies outside the interval $\left[-\dfrac{\pi}{2}, \dfrac{\pi}{2}\right]$ and hence is not admissible.) So, since $\sin \dfrac{\pi}{2} = 1$ and $\dfrac{\pi}{2}$ is in $\left[-\dfrac{\pi}{2}, \dfrac{\pi}{2}\right]$, we conclude that

$$\sin^{-1} 1 = \frac{\pi}{2}$$

Figure 5

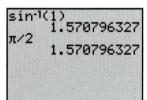

✔ **CHECK:** We can verify the solution by evaluating $\sin^{-1} 1$ with our graphing calculator in radian mode. See Figure 5.   ◀

For the remainder of the section, the reader is encouraged to verify the solutions obtained using a graphing utility.

✎ **NOW WORK PROBLEM 13.**

EXAMPLE 2   **Finding the Exact Value of an Inverse Sine Function**

Find the exact value of: $\sin^{-1}\left(-\dfrac{1}{2}\right)$

Solution   Let $\theta = \sin^{-1}\left(-\dfrac{1}{2}\right)$. We seek the angle θ, $-\dfrac{\pi}{2} \leq \theta \leq \dfrac{\pi}{2}$, whose sine equals $-\dfrac{1}{2}$.

$$\theta = \sin^{-1}\left(-\frac{1}{2}\right), \qquad -\frac{\pi}{2} \leq \theta \leq \frac{\pi}{2}$$

$$\sin \theta = -\frac{1}{2}, \qquad -\frac{\pi}{2} \leq \theta \leq \frac{\pi}{2}$$

(Refer to Table 1 and Figure 4, if necessary.) The only angle within the interval $\left[-\dfrac{\pi}{2}, \dfrac{\pi}{2}\right]$ whose sine is $-\dfrac{1}{2}$ is $-\dfrac{\pi}{6}$. So, since $\sin\left(-\dfrac{\pi}{6}\right) = -\dfrac{1}{2}$ and $-\dfrac{\pi}{6}$ is in the interval $\left[-\dfrac{\pi}{2}, \dfrac{\pi}{2}\right]$, we conclude that

$$\sin^{-1}\left(-\dfrac{1}{2}\right) = -\dfrac{\pi}{6}$$

◄

NOW WORK PROBLEM 19.

2   **Find an Approximate Value of the Inverse Sine Function**

For most numbers x, the value $y = \sin^{-1} x$ must be approximated.

EXAMPLE 3

Finding an Approximate Value of an Inverse Sine Function

Find an approximate value of:

(a) $\sin^{-1} \dfrac{1}{3}$

(b) $\sin^{-1}\left(-\dfrac{1}{4}\right)$

Express the answer in radians rounded to two decimal places.

Solution   Because we want the angle measured in radians, we first set the mode to radians.

(a) Figure 6(a) shows the solution
using a TI-84 Plus graphing
calculator.

(b) Figure 6(b) shows the solution
using a TI-84 Plus graphing
calculator.

Figure 6a

Figure 6b

```
sin⁻¹(1/3)
          .3398369095
```

```
sin⁻¹(-1/4)
          -.2526802551
```

We have $\sin^{-1} \dfrac{1}{3} = 0.34$,
rounded to two decimal places.

We have $\sin^{-1}\left(-\dfrac{1}{4}\right) = -0.25$,
rounded to two decimal places.   ◄

NOW WORK PROBLEM 25.

When we discussed functions and their inverses in Section 4.2, we found that $f^{-1}(f(x)) = x$ for all x in the domain of f and $f(f^{-1}(x)) = x$ for all x in the domain of f^{-1}. In terms of the sine function and its inverse, these properties are of the form

$$f^{-1}(f(x)) = \sin^{-1}(\sin x) = x, \qquad \text{where } -\dfrac{\pi}{2} \le x \le \dfrac{\pi}{2} \qquad \text{(2a)}$$

$$f(f^{-1}(x)) = \sin(\sin^{-1} x) = x, \qquad \text{where } -1 \le x \le 1 \qquad \text{(2b)}$$

For example, because $\dfrac{\pi}{8}$ lies in the interval $\left[-\dfrac{\pi}{2}, \dfrac{\pi}{2}\right]$, the restricted domain of the sine function, we can apply (2a) to get

$$\sin^{-1}\left[\sin\left(\dfrac{\pi}{8}\right)\right] = \dfrac{\pi}{8}$$

Also, because 0.8 lies in the interval $[-1, 1]$, the domain of the inverse sine function, we can apply (2b) to get

$$\sin[\sin^{-1}(0.8)] = 0.8$$

See Figure 7 for these calculations on a graphing calculator.

Figure 7

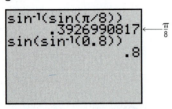

See Figure 8. Because $\dfrac{5\pi}{8}$ is not in the interval $\left[-\dfrac{\pi}{2}, \dfrac{\pi}{2}\right]$,

$$\sin^{-1}\left[\sin\left(\dfrac{5\pi}{8}\right)\right] \neq \dfrac{5\pi}{8}$$

Figure 8

To find $\sin^{-1}\left(\sin\dfrac{5\pi}{8}\right)$, we use the fact that $\sin\dfrac{5\pi}{8} = \sin\dfrac{3\pi}{8}$. Since $\dfrac{3\pi}{8}$ lies in the interval $\left[-\dfrac{\pi}{2}, \dfrac{\pi}{2}\right]$, we can apply (2a) to get

$$\sin^{-1}\left(\sin\dfrac{5\pi}{8}\right) = \sin^{-1}\left(\sin\dfrac{3\pi}{8}\right) = \dfrac{3\pi}{8} \approx 1.178097245$$

Also, because 1.8 is not in the interval $[-1, 1]$,

$$\sin[\sin^{-1}(1.8)] \neq 1.8$$

See Figure 9. Can you explain why the error appears?

Figure 9

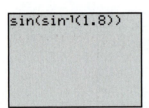

▬▬▬ **NOW WORK PROBLEM 37.**

The Inverse Cosine Function

In Figure 10 we reproduce the graph of $y = \cos x$. Because every horizontal line $y = b$, where b is between -1 and 1, intersects the graph of $y = \cos x$ infinitely many times, it follows that the cosine function is not one-to-one.

Figure 10

$y = \cos x, \ -\infty < x < \infty,$

$-1 \leq y \leq 1$

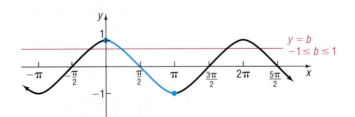

However, if we restrict the domain of $y = \cos x$ to the interval $[0, \pi]$, the restricted function

$$y = \cos x, \qquad 0 \leq x \leq \pi$$

is one-to-one and hence will have an inverse function.[*] See Figure 11.

Figure 11
$y = \cos x, 0 \leq x \leq \pi, -1 \leq y \leq 1$

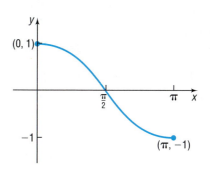

An equation for the inverse of $y = f(x) = \cos x$ is obtained by interchanging x and y. The implicit form of the inverse function is $x = \cos y, 0 \leq y \leq \pi$. The explicit form is called the **inverse cosine** of x and is symbolized by $y = f^{-1}(x) = \cos^{-1} x$ (or by $y = \text{Arccos } x$).

$$y = \cos^{-1} x \quad \text{means} \quad x = \cos y$$
$$\text{where} \quad -1 \leq x \leq 1 \quad \text{and} \quad 0 \leq y \leq \pi \tag{3}$$

Here y is the angle whose cosine is x. Because the range of the cosine function, $y = \cos x$, is $-1 \leq y \leq 1$, the domain of the inverse function $y = \cos^{-1} x$ is $-1 \leq x \leq 1$. Because the restricted domain of the cosine function, $y = \cos x$, is $0 \leq x \leq \pi$, the range of the inverse function $y = \cos^{-1} x$ is $0 \leq y \leq \pi$.

The graph of $y = \cos^{-1} x$ can be obtained by reflecting the restricted portion of the graph of $y = \cos x$ about the line $y = x$, as shown in Figure 12.

Figure 12
$y = \cos^{-1} x, -1 \leq x \leq 1, 0 \leq y \leq \pi$

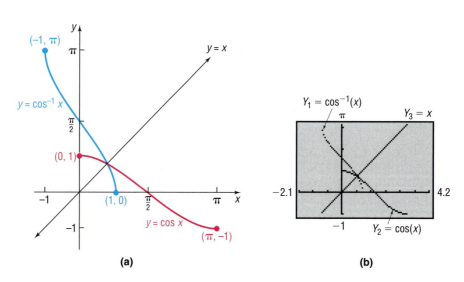

(a)

(b)

[*]This is the generally accepted restriction to define the inverse cosine function.

| EXAMPLE 4 | **Finding the Exact Value of an Inverse Cosine Function** |

Find the exact value of: $\cos^{-1} 0$

Solution Let $\theta = \cos^{-1} 0$. We seek the angle $\theta, 0 \le \theta \le \pi$, whose cosine equals 0.

$$\theta = \cos^{-1} 0, \qquad 0 \le \theta \le \pi$$
$$\cos \theta = 0, \qquad 0 \le \theta \le \pi$$

Look at Table 2 and Figure 13.

Table 2

θ	$\cos \theta$
0	1
$\dfrac{\pi}{6}$	$\dfrac{\sqrt{3}}{2}$
$\dfrac{\pi}{4}$	$\dfrac{\sqrt{2}}{2}$
$\dfrac{\pi}{3}$	$\dfrac{1}{2}$
$\dfrac{\pi}{2}$	0
$\dfrac{2\pi}{3}$	$-\dfrac{1}{2}$
$\dfrac{3\pi}{4}$	$-\dfrac{\sqrt{2}}{2}$
$\dfrac{5\pi}{6}$	$-\dfrac{\sqrt{3}}{2}$
π	-1

Figure 13

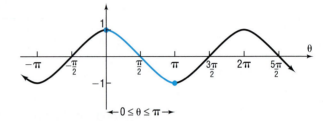

We see that the only angle θ within the interval $[0, \pi]$ whose cosine is 0 is $\dfrac{\pi}{2}$. (Note that $\cos \dfrac{3\pi}{2}$ also equals 0, but $\dfrac{3\pi}{2}$ lies outside the interval $[0, \pi]$ and hence is not admissible.) So, since $\cos \dfrac{\pi}{2} = 0$ and $\dfrac{\pi}{2}$ is in the interval $[0, \pi]$, we conclude that

$$\cos^{-1} 0 = \frac{\pi}{2} \qquad \blacktriangleleft$$

| EXAMPLE 5 | **Finding the Exact Value of an Inverse Cosine Function** |

Find the exact value of: $\cos^{-1}\left(-\dfrac{\sqrt{2}}{2}\right)$

Solution Let $\theta = \cos^{-1}\left(-\dfrac{\sqrt{2}}{2}\right)$. We seek the angle $\theta, 0 \le \theta \le \pi$, whose cosine equals $-\dfrac{\sqrt{2}}{2}$.

$$\theta = \cos^{-1}\left(-\frac{\sqrt{2}}{2}\right), \qquad 0 \le \theta \le \pi$$
$$\cos \theta = -\frac{\sqrt{2}}{2}, \qquad 0 \le \theta \le \pi$$

Look at Table 2 and Figure 14.

Figure 14

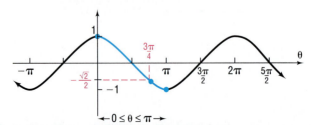

We see that the only angle θ within the interval $[0, \pi]$, whose cosine is $-\dfrac{\sqrt{2}}{2}$ is $\dfrac{3\pi}{4}$. So, since $\cos\dfrac{3\pi}{4} = -\dfrac{\sqrt{2}}{2}$ and $\dfrac{3\pi}{4}$ is in the interval $[0, \pi]$, we conclude that

$$\cos^{-1}\left(-\frac{\sqrt{2}}{2}\right) = \frac{3\pi}{4}$$

◀

🖉 **NOW WORK PROBLEM 23.**

For the cosine function and its inverse, the following properties hold:

$$f^{-1}(f(x)) = \cos^{-1}(\cos x) = x, \qquad \text{where } 0 \le x \le \pi \qquad \textbf{(4a)}$$
$$f(f^{-1}(x)) = \cos(\cos^{-1} x) = x, \qquad \text{where } -1 \le x \le 1 \qquad \textbf{(4b)}$$

EXAMPLE 6   **Finding the Exact Value of an Inverse Cosine Function**

Find the exact value of: (a) $\cos^{-1}\left[\cos\left(\dfrac{\pi}{12}\right)\right]$ (b) $\cos[\cos^{-1}(-0.4)]$

Solution   (a) $\cos^{-1}\left[\cos\left(\dfrac{\pi}{12}\right)\right] = \dfrac{\pi}{12}$ By Property (4a)

(b) $\cos[\cos^{-1}(-0.4)] = -0.4$ By Property (4b)   ◀

🖉 **NOW WORK PROBLEM 39.**

The Inverse Tangent Function

In Figure 15 we reproduce the graph of $y = \tan x$. Because every horizontal line intersects the graph infinitely many times, it follows that the tangent function is not one-to-one.

Figure 15

$y = \tan x$, $-\infty < x < \infty$, x not equal to odd multiples of $\dfrac{\pi}{2}$, $-\infty < y < \infty$

Figure 16

$y = \tan x$, $-\dfrac{\pi}{2} < x < \dfrac{\pi}{2}$, $-\infty < y < \infty$

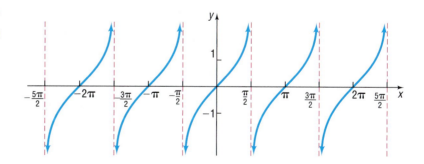

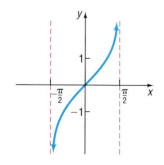

However, if we restrict the domain of $y = \tan x$ to the interval $\left(-\dfrac{\pi}{2}, \dfrac{\pi}{2}\right)$, the restricted function

$$y = \tan x, \qquad -\frac{\pi}{2} < x < \frac{\pi}{2}$$

is one-to-one and hence has an inverse function.* See Figure 16.

*This is the generally accepted restriction.

An equation for the inverse of $y = f(x) = \tan x$ is obtained by interchanging x and y. The implicit form of the inverse function is $x = \tan y, -\dfrac{\pi}{2} < y < \dfrac{\pi}{2}$. The explicit form is called the **inverse tangent** of x and is symbolized by $y = f^{-1}(x) = \tan^{-1} x$ (or by $y = \text{Arctan } x$).

$$y = \tan^{-1} x \quad \text{means} \quad x = \tan y$$

$$\text{where} \quad -\infty < x < \infty \quad \text{and} \quad -\frac{\pi}{2} < y < \frac{\pi}{2} \tag{5}$$

Here y is the angle whose tangent is x. The domain of the function $y = \tan^{-1} x$ is $-\infty < x < \infty$, and its range is $-\dfrac{\pi}{2} < y < \dfrac{\pi}{2}$. The graph of $y = \tan^{-1} x$ can be obtained by reflecting the restricted portion of the graph of $y = \tan x$ about the line $y = x$, as shown in Figure 17.

Figure 17

$y = \tan^{-1} x$,

$-\infty < x < \infty$,

$-\dfrac{\pi}{2} < y < \dfrac{\pi}{2}$

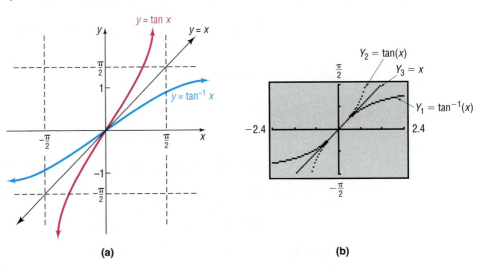

(a)　　　　　　(b)

Table 3

θ	$\tan \theta$
$-\dfrac{\pi}{2}$	Undefined
$-\dfrac{\pi}{3}$	$-\sqrt{3}$
$-\dfrac{\pi}{4}$	-1
$-\dfrac{\pi}{6}$	$-\dfrac{\sqrt{3}}{3}$
0	0
$\dfrac{\pi}{6}$	$\dfrac{\sqrt{3}}{3}$
$\dfrac{\pi}{4}$	1
$\dfrac{\pi}{3}$	$\sqrt{3}$
$\dfrac{\pi}{2}$	Undefined

EXAMPLE 7

Finding the Exact Value of an Inverse Tangent Function

Find the exact value of: $\tan^{-1} 1$

Solution Let $\theta = \tan^{-1} 1$. We seek the angle $\theta, -\dfrac{\pi}{2} < \theta < \dfrac{\pi}{2}$, whose tangent equals 1.

$$\theta = \tan^{-1} 1, \qquad -\frac{\pi}{2} < \theta < \frac{\pi}{2}$$

$$\tan \theta = 1, \qquad -\frac{\pi}{2} < \theta < \frac{\pi}{2}$$

Look at Table 3. The only angle θ within the interval $\left(-\dfrac{\pi}{2}, \dfrac{\pi}{2}\right)$ whose tangent is 1 is $\dfrac{\pi}{4}$. So, since $\tan \dfrac{\pi}{4} = 1$ and $\dfrac{\pi}{4}$ is in the interval $\left(-\dfrac{\pi}{2}, \dfrac{\pi}{2}\right)$, we conclude that

$$\tan^{-1} 1 = \frac{\pi}{4}$$

◀

NOW WORK PROBLEM 17.

EXAMPLE 8   **Finding the Exact Value of an Inverse Tangent Function**

Find the exact value of: $\tan^{-1}\left(-\sqrt{3}\right)$

Solution   Let $\theta = \tan^{-1}\left(-\sqrt{3}\right)$. We seek the angle θ, $-\dfrac{\pi}{2} < \theta < \dfrac{\pi}{2}$, whose tangent equals $-\sqrt{3}$.

$$\theta = \tan^{-1}\left(-\sqrt{3}\right), \qquad -\frac{\pi}{2} < \theta < \frac{\pi}{2}$$

$$\tan\theta = -\sqrt{3}, \qquad -\frac{\pi}{2} < \theta < \frac{\pi}{2}$$

Look at Table 3 or Figure 16 if necessary. The only angle θ within the interval $\left(-\dfrac{\pi}{2}, \dfrac{\pi}{2}\right)$ whose tangent is $-\sqrt{3}$ is $-\dfrac{\pi}{3}$. So, since $\tan\left(-\dfrac{\pi}{3}\right) = -\sqrt{3}$ and $-\dfrac{\pi}{3}$ is in the interval $\left(-\dfrac{\pi}{2}, \dfrac{\pi}{2}\right)$, we conclude that

$$\tan^{-1}\left(-\sqrt{3}\right) = -\frac{\pi}{3}$$ ◀

For the tangent function and its inverse, the following properties hold:

$$f^{-1}(f(x)) = \tan^{-1}(\tan x) = x, \qquad \text{where } -\frac{\pi}{2} < x < \frac{\pi}{2}$$

$$f(f^{-1}(x)) = \tan(\tan^{-1} x) = x, \qquad \text{where } -\infty < x < \infty$$

6.1 Assess Your Understanding

'Are You Prepared?'

Answers are given at the end of these exercises. If you get a wrong answer, read the pages listed in red.

1. What is the domain and the range of $y = \sin x$? (pp. 388–389)

2. A suitable restriction on the domain of the function $f(x) = (x - 1)^2$ to make it one-to-one would be _____. (p. 266)

3. If the domain of a one-to-one function is $[3, \infty)$, then the range of its inverse is _____. (pp. 259–261)

4. *True or False:* The graph of $y = \cos x$ is decreasing on the interval $[0, \pi]$. (p. 406)

5. $\tan\dfrac{\pi}{4} =$ _____; $\sin\dfrac{\pi}{3} =$ _____ (p. 379)

6. $\sin\left(-\dfrac{\pi}{6}\right) =$ _____; $\cos\pi =$ _____ (pp. 374–381)

Concepts and Vocabulary

7. $y = \sin^{-1} x$ means _____, where $-1 \le x \le 1$ and $-\dfrac{\pi}{2} \le y \le \dfrac{\pi}{2}$.

8. The value of $\sin^{-1}\left[\sin\dfrac{\pi}{2}\right]$ is _____.

9. $\cos^{-1}\left[\cos\dfrac{\pi}{5}\right] =$ _____.

10. *True or False:* The domain of $y = \sin^{-1} x$ is $-\dfrac{\pi}{2} \le x \le \dfrac{\pi}{2}$.

11. *True or False:* $\sin(\sin^{-1} 0) = 0$ and $\cos(\cos^{-1} 0) = 0$.

12. *True or False:* $y = \tan^{-1} x$ means $x = \tan y$, where $-\infty < x < \infty$ and $-\dfrac{\pi}{2} < y < \dfrac{\pi}{2}$.

Skill Building

In Problems 13–24, find the exact value of each expression. Verify your results using a graphing utility.

13. $\sin^{-1} 0$

14. $\cos^{-1} 1$

15. $\sin^{-1}(-1)$

16. $\cos^{-1}(-1)$

17. $\tan^{-1} 0$

18. $\tan^{-1}(-1)$

19. $\sin^{-1} \dfrac{\sqrt{2}}{2}$

20. $\tan^{-1} \dfrac{\sqrt{3}}{3}$

21. $\tan^{-1} \sqrt{3}$

22. $\sin^{-1}\left(-\dfrac{\sqrt{3}}{2}\right)$

23. $\cos^{-1}\left(-\dfrac{\sqrt{3}}{2}\right)$

24. $\sin^{-1}\left(-\dfrac{\sqrt{2}}{2}\right)$

In Problems 25–36, use a calculator to find the value of each expression rounded to two decimal places.

25. $\sin^{-1} 0.1$

26. $\cos^{-1} 0.6$

27. $\tan^{-1} 5$

28. $\tan^{-1} 0.2$

29. $\cos^{-1} \dfrac{7}{8}$

30. $\sin^{-1} \dfrac{1}{8}$

31. $\tan^{-1}(-0.4)$

32. $\tan^{-1}(-3)$

33. $\sin^{-1}(-0.12)$

34. $\cos^{-1}(-0.44)$

35. $\cos^{-1} \dfrac{\sqrt{2}}{3}$

36. $\sin^{-1} \dfrac{\sqrt{3}}{5}$

In Problems 37–44, find the exact value of each expression. Do not use a calculator.

37. $\sin[\sin^{-1}(0.54)]$

38. $\tan[\tan^{-1}(7.4)]$

39. $\cos^{-1}\left[\cos\left(\dfrac{4\pi}{5}\right)\right]$

40. $\sin^{-1}\left[\sin\left(-\dfrac{\pi}{10}\right)\right]$

41. $\tan[\tan^{-1}(-3.5)]$

42. $\cos[\cos^{-1}(-0.05)]$

43. $\sin^{-1}\left[\sin\left(-\dfrac{3\pi}{7}\right)\right]$

44. $\tan^{-1}\left[\tan\left(\dfrac{2\pi}{5}\right)\right]$

In Problems 45–56, do not use a calculator. For your answer, also say why or why not.

45. Does $\sin^{-1}\left[\sin\left(-\dfrac{\pi}{6}\right)\right] = -\dfrac{\pi}{6}$?

46. Does $\sin^{-1}\left[\sin\left(\dfrac{2\pi}{3}\right)\right] = \dfrac{2\pi}{3}$?

47. Does $\sin[\sin^{-1}(2)] = 2$?

48. Does $\sin\left[\sin^{-1}\left(-\dfrac{1}{2}\right)\right] = -\dfrac{1}{2}$?

49. Does $\cos^{-1}\left[\cos\left(-\dfrac{\pi}{6}\right)\right] = -\dfrac{\pi}{6}$?

50. Does $\cos^{-1}\left[\cos\left(\dfrac{2\pi}{3}\right)\right] = \dfrac{2\pi}{3}$?

51. Does $\cos\left[\cos^{-1}\left(-\dfrac{1}{2}\right)\right] = -\dfrac{1}{2}$?

52. Does $\cos[\cos^{-1}(2)] = 2$?

53. Does $\tan^{-1}\left[\tan\left(-\dfrac{\pi}{3}\right)\right] = -\dfrac{\pi}{3}$?

54. Does $\tan^{-1}\left[\tan\left(\dfrac{2\pi}{3}\right)\right] = \dfrac{2\pi}{3}$?

55. Does $\tan[\tan^{-1}(2)] = 2$?

56. Does $\tan\left[\tan^{-1}\left(-\dfrac{1}{2}\right)\right] = -\dfrac{1}{2}$?

Applications and Extensions

In Problems 57–62, use the following: The formula

$$D = 24\left[1 - \frac{\cos^{-1}(\tan i \tan \theta)}{\pi}\right]$$

can be used to approximate the number of hours of daylight when the declination of the Sun is $i°$ at a location $\theta°$ north latitude for any date between the vernal equinox and autumnal equinox. The declination of the Sun is defined as the angle i between the equatorial plane and any ray of light from the Sun. The latitude of a location is the angle θ between the Equator and the location on the surface of Earth, with the vertex of the angle located at the center of Earth. See the figure. To use the formula, $\cos^{-1}(\tan i \tan \theta)$ must be expressed in radians.

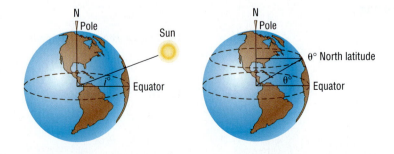

57. Approximate the number of hours of daylight in Houston, Texas (29°45′ north latitude), for the following dates:
(a) Summer solstice ($i = 23.5°$)
(b) Vernal equinox ($i = 0°$)
(c) July 4 ($i = 22°48′$)

58. Approximate the number of hours of daylight in New York, New York (40°45′ north latitude), for the following dates:
(a) Summer solstice ($i = 23.5°$)
(b) Vernal equinox ($i = 0°$)
(c) July 4 ($i = 22°48′$)

59. Approximate the number of hours of daylight in Honolulu, Hawaii (21°18′ north latitude), for the following dates:
(a) Summer solstice ($i = 23.5°$)
(b) Vernal equinox ($i = 0°$)
(c) July 4 ($i = 22°48′$)

60. Approximate the number of hours of daylight in Anchorage, Alaska (61°10′ north latitude), for the following dates:
(a) Summer solstice ($i = 23.5°$)
(b) Vernal equinox ($i = 0°$)
(c) July 4 ($i = 22°48′$)

61. Approximate the number of hours of daylight at the Equator (0° north latitude) for the following dates:
(a) Summer solstice ($i = 23.5°$)
(b) Vernal equinox ($i = 0°$)
(c) July 4 ($i = 22°48′$)
(d) What do you conclude about the number of hours of daylight throughout the year for a location at the Equator?

62. Approximate the number of hours of daylight for any location that is 66°30′ north latitude for the following dates:
(a) Summer solstice ($i = 23.5°$)
(b) Vernal equinox ($i = 0°$)
(c) July 4 ($i = 22°48′$)
(d) The number of hours of daylight on the winter solstice may be found by computing the number of hours of daylight on the summer solstice and subtracting this result from 24 hours, due to the symmetry of the orbital path of Earth around the Sun. Compute the number of hours of daylight for this location on the winter solstice. What do you conclude about daylight for a location at 66°30′ north latitude?

63. Being the First to See the Rising Sun   Cadillac Mountain, elevation 1530 feet, is located in Acadia National Park, Maine, and is the highest peak on the east coast of the United States. It is said that a person standing on the summit will be the first person in the United States to see the rays of the rising Sun. How much sooner would a person atop Cadillac Mountain see the first rays than a person standing below, at sea level?

[**Hint:** Consult the figure. When the person at D sees the first rays of the Sun, the person at P does not. The person at P sees the first rays of the Sun only after Earth has rotated so that P is at location Q. Compute the length of the arc subtended by the central angle θ. Then use the fact that, at the

latitude of Cadillac Mountain, in 24 hours a length of 2π (2710) miles is subtended, and find the time that it takes to subtend this length.]

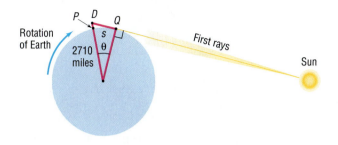

64. The Movie Theater   Suppose that a movie theater has a screen that is 28 feet tall. When you sit down, the bottom of the screen is 6 feet above your eye level. The angle formed by drawing a line from your eye to the bottom of the screen and your eye and the top of the screen is called the **viewing angle**. In the figure θ is the viewing angle. Suppose that you sit x feet from the screen. The viewing angle θ is given by the equation

$$\theta = \tan^{-1}\left(\frac{34}{x}\right) - \tan^{-1}\left(\frac{6}{x}\right).$$

(a) What is your viewing angle if you sit 10 feet from the screen? 15 feet? 20 feet?
(b) Using a graphing utility, graph

$$\theta = \tan^{-1}\left(\frac{34}{x}\right) - \tan^{-1}\left(\frac{6}{x}\right).$$

What value of x results in maximizing the viewing angle?
(c) If there is 5 feet between the screen and the first row of seats and there is 3 feet between each row, which row results in the optimal viewing angle?

65. Area under a Curve   The area under the graph of $y = \dfrac{1}{1 + x^2}$ and above the x-axis between $x = a$ and $x = b$ is given by

$$\tan^{-1} b - \tan^{-1} a$$

See the figure.

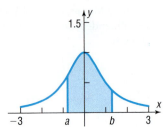

(a) Find the exact area under the graph of $y = \dfrac{1}{1 + x^2}$ and above the x-axis between $x = 0$ and $x = \sqrt{3}$.

(b) Find the exact area under the graph of $y = \dfrac{1}{1 + x^2}$ and above the x-axis between $x = -\dfrac{\sqrt{3}}{3}$ and $x = 1$.

66. **Area under a Curve** The area under the graph of $y = \dfrac{1}{\sqrt{1 - x^2}}$ and above the x-axis between $x = a$ and $x = b$ is given by

$$\sin^{-1} b - \sin^{-1} a$$

See the figure.

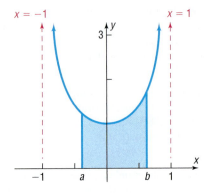

(a) Find the exact area under the graph of $y = \dfrac{1}{\sqrt{1 - x^2}}$ and above the x-axis between $x = 0$ and $x = \dfrac{\sqrt{3}}{2}$.

(b) Find the exact area under the graph of $y = \dfrac{1}{\sqrt{1 - x^2}}$ and above the x-axis between $x = -\dfrac{1}{2}$ and $x = \dfrac{1}{2}$.

'Are You Prepared?' Answers

1. domain: $-\infty < x < \infty$; range: $-1 \le y \le 1$ 2. Two answers are possible: $x \le -1$ or $x \ge 1$ 3. $[3, \infty)$ 4. True

5. $1; \dfrac{\sqrt{3}}{2}$ 6. $-\dfrac{1}{2}; -1$

6.2 The Inverse Trigonometric Functions (Continued)

PREPARING FOR THIS SECTION   *Before getting started, review the following concepts:*

• Finding Exact Values Given the Value of a Trigonometric Function and the Quadrant of the Angle (Section 5.3, pp. 395–398)

• Graphs of the Secant, Cosecant, and Cotangent Functions (Section 5.5, pp. 422–423)

• Domain and Range of the Secant, Cosecant, and Cotangent Functions (Section 5.3, p. 388–390)

Now work the 'Are You Prepared?' problems on page 464.

OBJECTIVES   1 Find the Exact Value of Expressions Involving the Inverse Sine, Cosine, and Tangent Functions

2 Know the Definition of the Inverse Secant, Cosecant, and Cotangent Functions

3 Use a Calculator to Evaluate $\sec^{-1} x$, $\csc^{-1} x$, and $\cot^{-1} x$

1   Find the Exact Value of Expressions Involving the Inverse Sine, Cosine, and Tangent Functions

EXAMPLE 1   **Finding the Exact Value of Expressions Involving Inverse Trigonometric Functions**

Find the exact value of: $\sin^{-1}\left(\sin\dfrac{5\pi}{4}\right)$

Solution $\sin^{-1}\left(\sin\dfrac{5\pi}{4}\right) = \sin^{-1}\left(-\dfrac{\sqrt{2}}{2}\right) = -\dfrac{\pi}{4}$

✔ **CHECK:** We can verify the solution by evaluating $\sin^{-1}\left(\sin\frac{5\pi}{4}\right)$ with our graphing calculator in radian mode. See Figure 18.   ◄

Notice in the solution to Example 1 that we did not use Property (2a), page 451. This is because the argument of the sine function is not in the interval $\left[-\frac{\pi}{2}, \frac{\pi}{2}\right]$, as required. If we use the fact that

$$\sin\frac{5\pi}{4} = \sin\left(-\frac{\pi}{4}\right)$$

then we can use Property (2a):

$$\sin^{-1}\left(\sin\frac{5\pi}{4}\right) = \sin^{-1}\left[\sin\left(-\frac{\pi}{4}\right)\right] = -\frac{\pi}{4}$$

$$\underset{\text{Property (2a)}}{\uparrow}$$

For the remainder of the section, the reader is encouraged to verify the solutions obtained using a graphing utility.

NOW WORK PROBLEM 23.

Figure 18

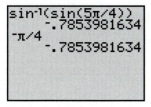

```
sin⁻¹(sin(5π/4))
          -.7853981634
⁻π/4
          -.7853981634
```

EXAMPLE 2	**Finding the Exact Value of Expressions Involving Inverse Trigonometric Functions**

Find the exact value of: $\sin\left(\tan^{-1}\frac{1}{2}\right)$

Figure 19

$\tan\theta = \frac{1}{2}$

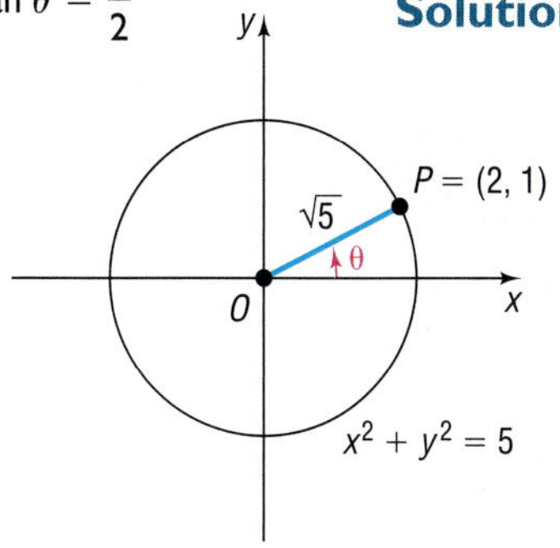

$P = (2, 1)$

$\sqrt{5}$

$x^2 + y^2 = 5$

Solution Let $\theta = \tan^{-1}\frac{1}{2}$. Then $\tan\theta = \frac{1}{2}$, where $-\frac{\pi}{2} < \theta < \frac{\pi}{2}$. We seek $\sin\theta$. Because $\tan\theta > 0$, it follows that $0 < \theta < \frac{\pi}{2}$, so θ lies in quadrant I. Since $\tan\theta = \frac{1}{2} = \frac{y}{x}$, we let $x = 2$ and $y = 1$. Since $r = d(O, P) = \sqrt{2^2 + 1^2} = \sqrt{5}$, the point $P = (x, y) = (2, 1)$ is on the circle $x^2 + y^2 = 5$. See Figure 19. Then, with $x = 2, y = 1$, and $r = \sqrt{5}$, we have

$$\sin\left(\tan^{-1}\frac{1}{2}\right) = \sin\theta = \frac{1}{\sqrt{5}} = \frac{\sqrt{5}}{5}$$

$$\underset{\sin\theta = \frac{y}{r}}{\uparrow}$$

◄

EXAMPLE 3	**Finding the Exact Value of Expressions Involving Inverse Trigonometric Functions**

Find the exact value of: $\cos\left[\sin^{-1}\left(-\frac{1}{3}\right)\right]$

Figure 20

$\sin\theta = -\frac{1}{3}$

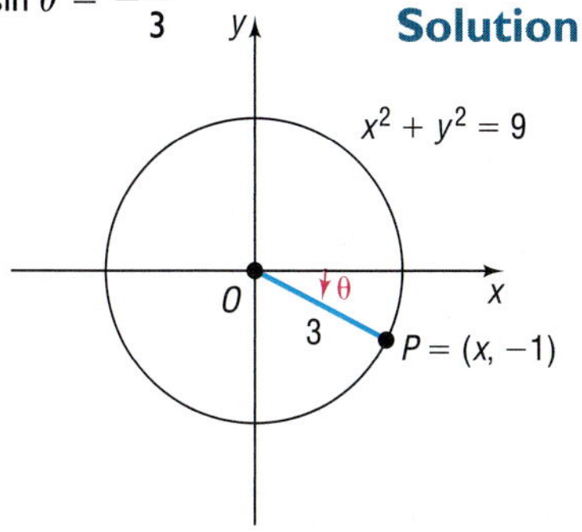

$x^2 + y^2 = 9$

$P = (x, -1)$

Solution Let $\theta = \sin^{-1}\left(-\frac{1}{3}\right)$. Then $\sin\theta = -\frac{1}{3}$ and $-\frac{\pi}{2} \leq \theta \leq \frac{\pi}{2}$. We seek $\cos\theta$. Because $\sin\theta < 0$, it follows that $-\frac{\pi}{2} \leq \theta < 0$, so θ lies in quadrant IV. Since $\sin\theta = \frac{-1}{3} = \frac{y}{r}$, we let $y = -1$ and $r = 3$. The point $P = (x, y) = (x, -1), x > 0$, is on a circle of radius 3, namely, $x^2 + y^2 = 9$. See Figure 20. Then,

$$x^2 + y^2 = 9$$
$$x^2 + (-1)^2 = 9 \qquad \textcolor{orange}{y = -1}$$
$$x^2 = 8$$
$$x = 2\sqrt{2} \qquad \textcolor{orange}{x > 0}$$

Then we have $x = 2\sqrt{2}$, $y = -1$, $r = 3$, so that

$$\cos\left[\sin^{-1}\left(-\frac{1}{3}\right)\right] = \cos\theta = \frac{2\sqrt{2}}{3}$$

$$\uparrow\atop{\cos\theta = \frac{x}{r}}$$

◀

EXAMPLE 4	**Finding the Exact Value of Expressions Involving Inverse Trigonometric Functions**

Find the exact value of: $\tan\left[\cos^{-1}\left(-\frac{1}{3}\right)\right]$

Solution Let $\theta = \cos^{-1}\left(-\frac{1}{3}\right)$. Then $\cos\theta = -\frac{1}{3}$ and $0 \le \theta \le \pi$. We seek $\tan\theta$. Because $\cos\theta < 0$, it follows that $\frac{\pi}{2} < \theta \le \pi$, so θ lies in quadrant II. Since $\cos\theta = \frac{-1}{3} = \frac{x}{r}$, we let $x = -1$ and $r = 3$. The point $P = (x, y) = (-1, y)$, $y > 0$, is on a circle of radius $r = 3$, namely, $x^2 + y^2 = 9$. See Figure 21. Then,

$$x^2 + y^2 = 9$$
$$(-1)^2 + y^2 = 9 \qquad x = -1$$
$$y^2 = 8$$
$$y = 2\sqrt{2} \quad y > 0$$

Then, we have $x = -1$, $y = 2\sqrt{2}$, and $r = 3$, so that

$$\tan\left[\cos^{-1}\left(-\frac{1}{3}\right)\right] = \tan\theta = \frac{2\sqrt{2}}{-1} = -2\sqrt{2}$$

$$\uparrow\atop{\tan\theta = \frac{y}{x}}$$

◀

Figure 21

$\cos\theta = -\frac{1}{3}$

$P = (-1, y)$

$x^2 + y^2 = 9$

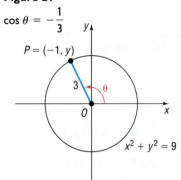

✏ **NOW WORK PROBLEMS 9 AND 27.**

Look at Example 1 and Examples 2–4. What differences do you see?

2 **Know the Definition of the Inverse Secant, Cosecant, and Cotangent Functions**

The inverse secant, inverse cosecant, and inverse cotangent functions are defined as follows:

$$y = \sec^{-1} x \quad \text{means} \quad x = \sec y \tag{1}$$
$$\text{where} \quad |x| \ge 1 \quad \text{and} \quad 0 \le y \le \pi, \quad y \ne \frac{\pi}{2}^*$$

$$y = \csc^{-1} x \quad \text{means} \quad x = \csc y \tag{2}$$
$$\text{where} \quad |x| \ge 1 \quad \text{and} \quad -\frac{\pi}{2} \le y \le \frac{\pi}{2}, \quad y \ne 0^\dagger$$

$$y = \cot^{-1} x \quad \text{means} \quad x = \cot y \tag{3}$$
$$\text{where} \quad -\infty < x < \infty \quad \text{and} \quad 0 < y < \pi$$

---

*Most books use this definition. A few use the restriction $0 \le y < \frac{\pi}{2}, \pi \le y < \frac{3\pi}{2}$.

†Most books use this definition. A few use the restriction $-\pi < y \le -\frac{\pi}{2}, 0 < y \le \frac{\pi}{2}$.

You are encouraged to review the graphs of the cotangent, cosecant, and secant functions in Figures 77, 78, and 80 in Section 5.5 to help you to see the basis for these definitions.

EXAMPLE 5   **Finding the Exact Value of an Inverse Cosecant Function**

Find the exact value of: $\csc^{-1} 2$

Solution   Let $\theta = \csc^{-1} 2$. We seek the angle θ, $-\dfrac{\pi}{2} \le \theta \le \dfrac{\pi}{2}, \theta \ne 0$, whose cosecant equals 2 $\left(\text{or, equivalently, whose sine equals } \dfrac{1}{2}\right)$.

$$\theta = \csc^{-1} 2, \qquad -\frac{\pi}{2} \le \theta \le \frac{\pi}{2}, \quad \theta \ne 0$$
$$\csc \theta = 2, \qquad -\frac{\pi}{2} \le \theta \le \frac{\pi}{2}, \quad \theta \ne 0$$

The only angle θ in the interval $-\dfrac{\pi}{2} \le \theta \le \dfrac{\pi}{2}, \theta \ne 0$, whose cosecant is 2, is $\dfrac{\pi}{6}$, so $\csc^{-1} 2 = \dfrac{\pi}{6}$.   ◀

NOW WORK PROBLEM 39.

3 **Use a Calculator to Evaluate $\sec^{-1} x$, $\csc^{-1} x$, and $\cot^{-1} x$**

Most calculators do not have keys for evaluating the inverse cotangent, cosecant, and secant functions. The easiest way to evaluate them is to convert to an inverse trigonometric function whose range is the same as the one to be evaluated. In this regard, notice that $y = \cot^{-1} x$ and $y = \sec^{-1} x$, except where undefined, each have the same range as $y = \cos^{-1} x$; $y = \csc^{-1} x$, except where undefined, has the same range as $y = \sin^{-1} x$.

EXAMPLE 6   **Approximating the Value of Inverse Trigonometric Functions**

Use a calculator to approximate each expression in radians rounded to two decimal places.

(a) $\sec^{-1} 3$ (b) $\csc^{-1}(-4)$ (c) $\cot^{-1} \dfrac{1}{2}$ (d) $\cot^{-1}(-2)$

Solution   First, set your calculator to radian mode.

(a) Let $\theta = \sec^{-1} 3$. Then $\sec \theta = 3$ and $0 \le \theta \le \pi, \theta \ne \dfrac{\pi}{2}$. We seek $\cos \theta$ because $y = \cos^{-1} x$ has the same range as $y = \sec^{-1} x$, except where undefined. Since $\sec \theta = \dfrac{1}{\cos \theta} = 3$, we have $\cos \theta = \dfrac{1}{3}$. Then $\theta = \cos^{-1} \dfrac{1}{3}$ and

$$\sec^{-1} 3 = \theta = \cos^{-1} \frac{1}{3} \approx 1.23$$

↑
Use a calculator.

(b) Let $\theta = \csc^{-1}(-4)$. Then $\csc \theta = -4$, $-\dfrac{\pi}{2} \le \theta \le \dfrac{\pi}{2}, \theta \ne 0$. We seek $\sin \theta$ because $y = \sin^{-1} x$ has the same range as $y = \csc^{-1} x$, except where undefined. Since $\csc \theta = \dfrac{1}{\sin \theta} = -4$, we have $\sin \theta = -\dfrac{1}{4}$. Then $\theta = \sin^{-1}\left(-\dfrac{1}{4}\right)$, and

$$\csc^{-1}(-4) = \theta = \sin^{-1}\left(-\frac{1}{4}\right) \approx -0.25$$

Figure 22 $\cot\theta = \dfrac{1}{2}, 0 < \theta < \pi$

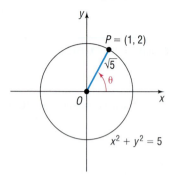

Figure 23
$\cot\theta = -2, 0 < \theta < \pi$

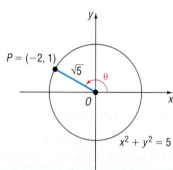

(c) There is no inverse cotangent key on the calculator. We proceed as before. Let $\theta = \cot^{-1}\dfrac{1}{2}$. Then $\cot\theta = \dfrac{1}{2}, 0 < \theta < \pi$. From these facts we know that θ lies in quadrant I. We seek $\cos\theta$ because $y = \cos^{-1}x$ has the same range as $y = \cot^{-1}x$, except where undefined. To find $\cos\theta$ we use Figure 22. Then

$$\cos\theta = \frac{1}{\sqrt{5}}, \ 0 < \theta < \frac{\pi}{2}, \ \theta = \cos^{-1}\!\left(\frac{1}{\sqrt{5}}\right), \text{ and}$$

$$\cot^{-1}\frac{1}{2} = \theta = \cos^{-1}\!\left(\frac{1}{\sqrt{5}}\right) \approx 1.11$$

(d) Let $\theta = \cot^{-1}(-2)$. Then $\cot\theta = -2, 0 < \theta < \pi$. From these facts we know that θ lies in quadrant II. We seek $\cos\theta$. To find it we use Figure 23. Then

$$\cos\theta = -\frac{2}{\sqrt{5}}, \ \frac{\pi}{2} < \theta < \pi, \ \theta = \cos^{-1}\!\left(-\frac{2}{\sqrt{5}}\right), \text{ and}$$

$$\cot^{-1}(-2) = \theta = \cos^{-1}\!\left(-\frac{2}{\sqrt{5}}\right) \approx 2.68$$

NOW WORK PROBLEM 45.

6.2 Assess Your Understanding

'Are You Prepared?'

Answers are given at the end of these exercises. If you get a wrong answer, read the pages listed in red.

1. What is the domain and the range of $y = \sec x$? (pp. 388–390)

2. *True or False:* The graph of $y = \sec x$ is increasing on the interval $\left[0, \dfrac{\pi}{2}\right)$ and on the interval $\left(\dfrac{\pi}{2}, \pi\right]$. (p. 423)

3. If $\cot\theta = -2$ and $0 < \theta < \pi$, then $\cos\theta = $ _____. (pp. 395–398)

Concepts and Vocabulary

4. $y = \sec^{-1}x$ means _____, where $|x|$ _____ and _____ $\le y \le$ _____, $y \ne \dfrac{\pi}{2}$.

5. $\cos(\tan^{-1}1) = $ _____.

6. *True or False:* It is impossible to obtain exact values for the inverse secant function.

7. *True or False:* $\csc^{-1}0.5$ is not defined.

8. *True or False:* The domain of the inverse cotangent function is the set of real numbers.

Skill Building

In Problems 9–36, find the exact value of each expression. Verify your results using a graphing utility.

9. $\cos\!\left(\sin^{-1}\dfrac{\sqrt{2}}{2}\right)$

10. $\sin\!\left(\cos^{-1}\dfrac{1}{2}\right)$

11. $\tan\!\left[\cos^{-1}\!\left(-\dfrac{\sqrt{3}}{2}\right)\right]$

12. $\tan\!\left[\sin^{-1}\!\left(-\dfrac{1}{2}\right)\right]$

13. $\sec\!\left(\cos^{-1}\dfrac{1}{2}\right)$

14. $\cot\!\left[\sin^{-1}\!\left(-\dfrac{1}{2}\right)\right]$

15. $\csc(\tan^{-1}1)$

16. $\sec(\tan^{-1}\sqrt{3})$

17. $\sin[\tan^{-1}(-1)]$

18. $\cos\!\left[\sin^{-1}\!\left(-\dfrac{\sqrt{3}}{2}\right)\right]$

19. $\sec\!\left[\sin^{-1}\!\left(-\dfrac{1}{2}\right)\right]$

20. $\csc\!\left[\cos^{-1}\!\left(-\dfrac{\sqrt{3}}{2}\right)\right]$

21. $\cos^{-1}\!\left(\cos\dfrac{5\pi}{4}\right)$

22. $\tan^{-1}\!\left(\tan\dfrac{2\pi}{3}\right)$

23. $\sin^{-1}\!\left[\sin\!\left(-\dfrac{7\pi}{6}\right)\right]$

24. $\cos^{-1}\!\left[\cos\!\left(-\dfrac{\pi}{3}\right)\right]$

25. $\tan\left(\sin^{-1}\dfrac{1}{3}\right)$ **26.** $\tan\left(\cos^{-1}\dfrac{1}{3}\right)$ **27.** $\sec\left(\tan^{-1}\dfrac{1}{2}\right)$ **28.** $\cos\left(\sin^{-1}\dfrac{\sqrt{2}}{3}\right)$

29. $\cot\left[\sin^{-1}\left(-\dfrac{\sqrt{2}}{3}\right)\right]$ **30.** $\csc[\tan^{-1}(-2)]$ **31.** $\sin[\tan^{-1}(-3)]$ **32.** $\cot\left[\cos^{-1}\left(-\dfrac{\sqrt{3}}{3}\right)\right]$

33. $\sec\left(\sin^{-1}\dfrac{2\sqrt{5}}{5}\right)$ **34.** $\csc\left(\tan^{-1}\dfrac{1}{2}\right)$ **35.** $\sin^{-1}\left(\cos\dfrac{3\pi}{4}\right)$ **36.** $\cos^{-1}\left(\sin\dfrac{7\pi}{6}\right)$

In Problems 37–44, find the exact value of each expression. Verify your results using a graphing utility.

37. $\cot^{-1}\sqrt{3}$ **38.** $\cot^{-1}1$ **39.** $\csc^{-1}(-1)$ **40.** $\csc^{-1}\sqrt{2}$

41. $\sec^{-1}\dfrac{2\sqrt{3}}{3}$ **42.** $\sec^{-1}(-2)$ **43.** $\cot^{-1}\left(-\dfrac{\sqrt{3}}{3}\right)$ **44.** $\csc^{-1}\left(-\dfrac{2\sqrt{3}}{3}\right)$

In Problems 45–56, use a calculator to find the value of each expression rounded to two decimal places.

45. $\sec^{-1}4$ **46.** $\csc^{-1}5$ **47.** $\cot^{-1}2$ **48.** $\sec^{-1}(-3)$

49. $\csc^{-1}(-3)$ **50.** $\cot^{-1}\left(-\dfrac{1}{2}\right)$ **51.** $\cot^{-1}(-\sqrt{5})$ **52.** $\cot^{-1}(-8.1)$

53. $\csc^{-1}\left(-\dfrac{3}{2}\right)$ **54.** $\sec^{-1}\left(-\dfrac{4}{3}\right)$ **55.** $\cot^{-1}\left(-\dfrac{3}{2}\right)$ **56.** $\cot^{-1}(-\sqrt{10})$

---

57. Using a graphing utility, graph $y = \cot^{-1}x$.   **59.** Using a graphing utility, graph $y = \csc^{-1}x$.

58. Using a graphing utility, graph $y = \sec^{-1}x$.

Discussion and Writing

60. Explain in your own words how you would use your calculator to find the value of $\cot^{-1}10$.

61. Consult three books on calculus and write down the definition in each of $y = \sec^{-1}x$ and $y = \csc^{-1}x$. Compare these with the definition given in this book.

'Are You Prepared?' Answers

1. Domain: $\left\{x \mid x \neq \text{ odd multiples of } \dfrac{\pi}{2}\right\}$; Range: $\{y \leq -1 \text{ or } y \geq 1\}$ **2.** True   **3.** $\dfrac{-2\sqrt{5}}{5}$

6.3 Trigonometric Identities

PREPARING FOR THIS SECTION   *Before getting started, review the following:*

• Fundamental Identities (Section 5.3, p. 394)

Now work the 'Are You Prepared?' problems on page 471.

OBJECTIVES **1** Use Algebra to Simplify Trigonometric Expressions
   2 Establish Identities

We saw in the previous chapter that the trigonometric functions lend themselves to a wide variety of identities. Before establishing some additional identities, let's review the definition of an *identity*.

Two functions f and g are said to be **identically equal** if

$$f(x) = g(x)$$

for every value of x for which both functions are defined. Such an equation is referred to as an **identity**. An equation that is not an identity is called a **conditional equation**.

For example, the following are identities:

$$(x + 1)^2 = x^2 + 2x + 1 \qquad \sin^2 x + \cos^2 x = 1 \qquad \csc x = \frac{1}{\sin x}$$

The following are conditional equations:

$$2x + 5 = 0 \qquad \text{True only if } x = -\frac{5}{2}$$

$$\sin x = 0 \qquad \text{True only if } x = k\pi, k \text{ an integer}$$

$$\sin x = \cos x \qquad \text{True only if } x = \frac{\pi}{4} + 2k\pi \text{ or } x = \frac{5\pi}{4} + 2k\pi, k \text{ an integer}$$

The following boxes summarize the trigonometric identities that we have established thus far.

Quotient Identities

$$\tan \theta = \frac{\sin \theta}{\cos \theta} \qquad \cot \theta = \frac{\cos \theta}{\sin \theta}$$

Reciprocal Identities

$$\csc \theta = \frac{1}{\sin \theta} \qquad \sec \theta = \frac{1}{\cos \theta} \qquad \cot \theta = \frac{1}{\tan \theta}$$

Pythagorean Identities

$$\sin^2 \theta + \cos^2 \theta = 1 \qquad \tan^2 \theta + 1 = \sec^2 \theta$$

$$\cot^2 \theta + 1 = \csc^2 \theta$$

Even-Odd Identities

$$\sin(-\theta) = -\sin \theta \qquad \cos(-\theta) = \cos \theta \qquad \tan(-\theta) = -\tan \theta$$

$$\csc(-\theta) = -\csc \theta \qquad \sec(-\theta) = \sec \theta \qquad \cot(-\theta) = -\cot \theta$$

This list of identities comprises what we shall refer to as the **basic trigonometric identities**. These identities should not merely be memorized, but should be *known* (just as you know your name rather than have it memorized). In fact, minor variations of a basic identity are often used. For example, we might want to use

$$\sin^2 \theta = 1 - \cos^2 \theta \quad \text{or} \quad \cos^2 \theta = 1 - \sin^2 \theta$$

instead of $\sin^2 \theta + \cos^2 \theta = 1$. For this reason, among others, you need to know these relationships and be quite comfortable with variations of them.

1 Use Algebra to Simplify Trigonometric Expressions

The ability to use algebra to manipulate trigonometric expressions is a key skill that one must have to establish identities. Some of the techniques that are used in establishing identities are multiplying by a "well-chosen 1," writing a trigonometric expression over a common denominator, rewriting a trigonometric expression in terms of sine and cosine only, and factoring.

| **EXAMPLE 1** | **Using Algebraic Techniques to Simplify Trigonometric Expressions** |

(a) Simplify $\dfrac{\cot\theta}{\csc\theta}$ by rewriting each trigonometric function in terms of sine and cosine.

(b) Show that $\dfrac{\cos\theta}{1+\sin\theta}=\dfrac{1-\sin\theta}{\cos\theta}$ by multiplying the numerator and denominator by $1-\sin\theta$.

(c) Simplify $\dfrac{1+\sin\theta}{\sin\theta}+\dfrac{\cot\theta-\cos\theta}{\cos\theta}$ by rewriting the expression over a common denominator.

(d) Simplify $\dfrac{\sin^2\theta-1}{\tan\theta\sin\theta-\tan\theta}$ by factoring.

Solution

(a) $\dfrac{\cot\theta}{\csc\theta}=\dfrac{\dfrac{\cos\theta}{\sin\theta}}{\dfrac{1}{\sin\theta}}=\dfrac{\cos\theta}{\sin\theta}\cdot\dfrac{\sin\theta}{1}=\cos\theta$

(b) $\dfrac{\cos\theta}{1+\sin\theta}=\dfrac{\cos\theta}{1+\sin\theta}\cdot\dfrac{1-\sin\theta}{1-\sin\theta}=\dfrac{\cos\theta(1-\sin\theta)}{1-\sin^2\theta}=\dfrac{\cos\theta(1-\sin\theta)}{\cos^2\theta}=\dfrac{1-\sin\theta}{\cos\theta}$

$\quad$ Well-chosen 1: $\dfrac{1-\sin\theta}{1-\sin\theta}$

(c) $\dfrac{1+\sin\theta}{\sin\theta}+\dfrac{\cot\theta-\cos\theta}{\cos\theta}=\dfrac{1+\sin\theta}{\sin\theta}\cdot\dfrac{\cos\theta}{\cos\theta}+\dfrac{\cot\theta-\cos\theta}{\cos\theta}\cdot\dfrac{\sin\theta}{\sin\theta}$

$\quad=\dfrac{\cos\theta+\sin\theta\cos\theta+\cot\theta\sin\theta-\cos\theta\sin\theta}{\sin\theta\cos\theta}=\dfrac{\cos\theta+\dfrac{\cos\theta}{\sin\theta}\cdot\sin\theta}{\sin\theta\cos\theta}$

$\qquad\qquad\qquad\qquad\qquad\qquad\qquad\qquad\qquad\qquad\qquad\cot\theta=\dfrac{\cos\theta}{\sin\theta}$

$\quad=\dfrac{\cos\theta+\cos\theta}{\sin\theta\cos\theta}=\dfrac{2\cos\theta}{\sin\theta\cos\theta}=\dfrac{2}{\sin\theta}$

(d) $\dfrac{\sin^2\theta-1}{\tan\theta\sin\theta-\tan\theta}=\dfrac{(\sin\theta+1)(\sin\theta-1)}{\tan\theta(\sin\theta-1)}=\dfrac{\sin\theta+1}{\tan\theta}$ ◄

NOW WORK PROBLEMS 9, 11, AND 13.

2 ## Establish Identities

In the examples that follow, the directions will read "Establish the identity. ... " As you will see, this is accomplished by starting with one side of the given equation (usually the one containing the more complicated expression) and, using appropriate basic identities and algebraic manipulations, arriving at the other side. The selection of appropriate basic identities to obtain the desired result is learned only through experience and lots of practice.

| **EXAMPLE 2** | **Establishing an Identity** |

Establish the identity: $\csc\theta\cdot\tan\theta=\sec\theta$

Solution We start with the left side, because it contains the more complicated expression, and apply a reciprocal identity and a quotient identity.

$$\csc\theta\cdot\tan\theta=\dfrac{1}{\sin\theta}\cdot\dfrac{\sin\theta}{\cos\theta}=\dfrac{1}{\cos\theta}=\sec\theta$$

Having arrived at the right side, the identity is established. ◄

NOTE   A graphing utility can be used to provide evidence of an identity. For example, if we graph $Y_1 = \csc\theta \cdot \tan\theta$ and $Y_2 = \sec\theta$, the graphs appear to be the same. This provides evidence that $Y_1 = Y_2$. However, it does not prove their equality. A graphing utility *cannot* be used to *establish* an identity—identities must be established algebraically.   ■

 **NOW WORK PROBLEM 19.**

EXAMPLE 3	**Establishing an Identity**

Establish the identity: $\sin^2(-\theta) + \cos^2(-\theta) = 1$

Solution   We begin with the left side and apply Even–Odd Identities.

$$
\begin{aligned}
\sin^2(-\theta) + \cos^2(-\theta) &= [\sin(-\theta)]^2 + [\cos(-\theta)]^2 \\
&= (-\sin\theta)^2 + (\cos\theta)^2 \qquad \text{Even–Odd Identities} \\
&= (\sin\theta)^2 + (\cos\theta)^2 \\
&= 1 \qquad\qquad\qquad\qquad \text{Pythagorean Identity} \qquad \blacktriangleleft
\end{aligned}
$$

EXAMPLE 4	**Establishing an Identity**

Establish the identity: $\dfrac{\sin^2(-\theta) - \cos^2(-\theta)}{\sin(-\theta) - \cos(-\theta)} = \cos\theta - \sin\theta$

Solution   We begin with two observations: The left side contains the more complicated expression. Also, the left side contains expressions with the argument $-\theta$, whereas the right side contains expressions with the argument θ. We decide, therefore, to start with the left side and apply Even–Odd Identities.

$$
\begin{aligned}
\frac{\sin^2(-\theta) - \cos^2(-\theta)}{\sin(-\theta) - \cos(-\theta)} &= \frac{[\sin(-\theta)]^2 - [\cos(-\theta)]^2}{\sin(-\theta) - \cos(-\theta)} \\[2mm]
&= \frac{(-\sin\theta)^2 - (\cos\theta)^2}{-\sin\theta - \cos\theta} \qquad \text{Even–Odd Identities} \\[2mm]
&= \frac{(\sin\theta)^2 - (\cos\theta)^2}{-\sin\theta - \cos\theta} \qquad \text{Simplify.} \\[2mm]
&= \frac{(\sin\theta - \cos\theta)\cancel{(\sin\theta + \cos\theta)}}{-\cancel{(\sin\theta + \cos\theta)}} \qquad \text{Factor.} \\[2mm]
&= \cos\theta - \sin\theta \qquad\qquad\qquad \text{Cancel and simplify.} \qquad \blacktriangleleft
\end{aligned}
$$

EXAMPLE 5	**Establishing an Identity**

Establish the identity: $\dfrac{1 + \tan\theta}{1 + \cot\theta} = \tan\theta$

Solution $\dfrac{1 + \tan\theta}{1 + \cot\theta} = \dfrac{1 + \tan\theta}{1 + \dfrac{1}{\tan\theta}} = \dfrac{1 + \tan\theta}{\dfrac{\tan\theta + 1}{\tan\theta}} = \dfrac{\tan\theta\cancel{(1 + \tan\theta)}}{\cancel{\tan\theta + 1}} = \tan\theta$ $\blacktriangleleft$

 **NOW WORK PROBLEM 27.**

When sums or differences of quotients appear, it is usually best to rewrite them as a single quotient, especially if the other side of the identity consists of only one term.

EXAMPLE 6	**Establishing an Identity**

Establish the identity: $\dfrac{\sin \theta}{1 + \cos \theta} + \dfrac{1 + \cos \theta}{\sin \theta} = 2 \csc \theta$

Solution The left side is more complicated, so we start with it and proceed to add.

$$\frac{\sin \theta}{1 + \cos \theta} + \frac{1 + \cos \theta}{\sin \theta} = \frac{\sin^2 \theta + (1 + \cos \theta)^2}{(1 + \cos \theta)(\sin \theta)} \qquad \text{\small\it Add the quotients.}$$

$$= \frac{\sin^2 \theta + 1 + 2 \cos \theta + \cos^2 \theta}{(1 + \cos \theta)(\sin \theta)} \qquad \text{\small\it Remove parentheses in the numerator.}$$

$$= \frac{(\sin^2 \theta + \cos^2 \theta) + 1 + 2 \cos \theta}{(1 + \cos \theta)(\sin \theta)} \qquad \text{\small\it Regroup.}$$

$$= \frac{2 + 2 \cos \theta}{(1 + \cos \theta)(\sin \theta)} \qquad \text{\small\it Pythagorean identity}$$

$$= \frac{2(1 + \cos \theta)}{(1 + \cos \theta)(\sin \theta)} \qquad \text{\small\it Factor and cancel.}$$

$$= \frac{2}{\sin \theta}$$

$$= 2 \csc \theta \qquad \text{\small\it Reciprocal Identity} \qquad \blacktriangleleft$$

Sometimes it helps to write one side in terms of sines and cosines only.

EXAMPLE 7	**Establishing an Identity**

Establish the identity: $\dfrac{\tan \theta + \cot \theta}{\sec \theta \csc \theta} = 1$

Solution $\dfrac{\tan \theta + \cot \theta}{\sec \theta \csc \theta} = \dfrac{\dfrac{\sin \theta}{\cos \theta} + \dfrac{\cos \theta}{\sin \theta}}{\dfrac{1}{\cos \theta} \cdot \dfrac{1}{\sin \theta}} = \dfrac{\dfrac{\sin^2 \theta + \cos^2 \theta}{\cos \theta \sin \theta}}{\dfrac{1}{\cos \theta \sin \theta}}$

$\uparrow$ *Change to sines and cosines.* $\quad\uparrow$ *Add the quotients in the numerator.*

$$= \frac{1}{\cos \theta \sin \theta} \cdot \frac{\cos \theta \sin \theta}{1} = 1$$

$\uparrow$ *Divide the quotients;* $\sin^2 \theta + \cos^2 \theta = 1.$ $\qquad \blacktriangleleft$

✏ **NOW WORK PROBLEM 69.**

Sometimes, multiplying the numerator and denominator by an appropriate factor will result in a simplification.

EXAMPLE 8	**Establishing an Identity**

Establish the identity: $\dfrac{1 - \sin\theta}{\cos\theta} = \dfrac{\cos\theta}{1 + \sin\theta}$

Solution We start with the left side and multiply the numerator and the denominator by $1 + \sin\theta$. (Alternatively, we could multiply the numerator and denominator of the right side by $1 - \sin\theta$.)

$$\dfrac{1 - \sin\theta}{\cos\theta} = \dfrac{1 - \sin\theta}{\cos\theta} \cdot \dfrac{1 + \sin\theta}{1 + \sin\theta} \qquad \textit{Multiply the numerator and denominator by } 1 + \sin\theta.$$

$$= \dfrac{1 - \sin^2\theta}{\cos\theta(1 + \sin\theta)}$$

$$= \dfrac{\cos^2\theta}{\cos\theta(1 + \sin\theta)} \qquad 1 - \sin^2\theta = \cos^2\theta$$

$$= \dfrac{\cos\theta}{1 + \sin\theta} \qquad \textit{Cancel.}$$

◀

NOW WORK PROBLEM 53.

EXAMPLE 9	**Establishing an Identity Involving Inverse Trigonometric Functions**

Show that $\sin(\tan^{-1} v) = \dfrac{v}{\sqrt{1 + v^2}}$.

Solution Let $\theta = \tan^{-1} v$ so that $\tan\theta = v$, $-\dfrac{\pi}{2} < \theta < \dfrac{\pi}{2}$. As a result, we know that $\sec\theta > 0$.

$$\sin(\tan^{-1} v) = \sin\theta = \sin\theta \cdot \dfrac{\cos\theta}{\cos\theta} = \tan\theta\cos\theta = \dfrac{\tan\theta}{\sec\theta} = \dfrac{\tan\theta}{\sqrt{1 + \tan^2\theta}} = \dfrac{v}{\sqrt{1 + v^2}}$$

Multiply by 1: $\dfrac{\cos\theta}{\cos\theta}$. $\quad$ $\dfrac{\sin\theta}{\cos\theta} = \tan\theta$ $\quad$ $\sec^2\theta = 1 + \tan^2\theta$

$\sec\theta > 0$

◀

NOW WORK PROBLEM 99.

WARNING
Be careful not to handle identities to be established as if they were conditional equations. You cannot establish an identity by such methods as adding the same expression to each side and obtaining a true statement. This practice is not allowed, because the original statement is precisely the one that you are trying to establish. You do not know until it has been established that it is, in fact, true. ■

Although a lot of practice is the only real way to learn how to establish identities, the following guidelines should prove helpful.

Guidelines for Establishing Identities

1. It is almost always preferable to start with the side containing the more complicated expression.

2. Rewrite sums or differences of quotients as a single quotient.

3. Sometimes rewriting one side in terms of sines and cosines only will help.

4. Always keep your goal in mind. As you manipulate one side of the expression, you must keep in mind the form of the expression on the other side.

6.3 Assess Your Understanding

'Are You Prepared?'

Answers are given at the end of these exercises. If you get a wrong answer, read the pages listed in red.

1. *True or False:* $\sin^2 \theta = 1 - \cos^2 \theta.$ (p. 394)

2. *True or False:* $\sin(-\theta) + \cos(-\theta) = \cos \theta - \sin \theta.$ (p. 398)

Concepts and Vocabulary

3. Suppose that f and g are two functions with the same domain. If $f(x) = g(x)$ for every x in the domain, the equation is called a(n) _____. Otherwise, it is called a(n) _____ equation.

4. $\tan^2 \theta - \sec^2 \theta = $ _____.

5. $\cos(-\theta) - \cos \theta = $ _____.

6. *True or False:* $\sin(-\theta) + \sin \theta = 0$ for any value of θ.

7. *True or False:* In establishing an identity, it is often easiest to just multiply both sides by a well-chosen nonzero expression involving the variable.

8. *True or False:* $\tan \theta \cdot \cos \theta = \sin \theta$ for any $\theta \neq (2k + 1)\dfrac{\pi}{2}$.

Skill Building

In Problems 9–18, simplify each trigonometric expression by following the indicated direction.

9. Rewrite in terms of sine and cosine: $\tan \theta \cdot \csc \theta$.

10. Rewrite in terms of sine and cosine: $\cot \theta \cdot \sec \theta$.

11. Multiply $\dfrac{\cos \theta}{1 - \sin \theta}$ by $\dfrac{1 + \sin \theta}{1 + \sin \theta}$.

12. Multiply $\dfrac{\sin \theta}{1 + \cos \theta}$ by $\dfrac{1 - \cos \theta}{1 - \cos \theta}$.

13. Rewrite over a common denominator:

$$\frac{\sin \theta + \cos \theta}{\cos \theta} + \frac{\cos \theta - \sin \theta}{\sin \theta}$$

14. Rewrite over a common denominator:

$$\frac{1}{1 - \cos \theta} + \frac{1}{1 + \cos \theta}$$

15. Multiply and simplify: $\dfrac{(\sin \theta + \cos \theta)(\sin \theta + \cos \theta) - 1}{\sin \theta \cos \theta}$

16. Multiply and simplify: $\dfrac{(\tan \theta + 1)(\tan \theta + 1) - \sec^2 \theta}{\tan \theta}$

17. Factor and simplify: $\dfrac{3 \sin^2 \theta + 4 \sin \theta + 1}{\sin^2 \theta + 2 \sin \theta + 1}$

18. Factor and simplify: $\dfrac{\cos^2 \theta - 1}{\cos^2 \theta - \cos \theta}$

In Problems 19–98, establish each identity.

19. $\csc \theta \cdot \cos \theta = \cot \theta$

20. $\sec \theta \cdot \sin \theta = \tan \theta$

21. $1 + \tan^2(-\theta) = \sec^2 \theta$

22. $1 + \cot^2(-\theta) = \csc^2 \theta$

23. $\cos \theta(\tan \theta + \cot \theta) = \csc \theta$

24. $\sin \theta(\cot \theta + \tan \theta) = \sec \theta$

25. $\tan \theta \cot \theta - \cos^2 \theta = \sin^2 \theta$

26. $\sin \theta \csc \theta - \cos^2 \theta = \sin^2 \theta$

27. $(\sec \theta - 1)(\sec \theta + 1) = \tan^2 \theta$

28. $(\csc \theta - 1)(\csc \theta + 1) = \cot^2 \theta$

29. $(\sec \theta + \tan \theta)(\sec \theta - \tan \theta) = 1$

30. $(\csc \theta + \cot \theta)(\csc \theta - \cot \theta) = 1$

31. $\cos^2 \theta(1 + \tan^2 \theta) = 1$

32. $(1 - \cos^2 \theta)(1 + \cot^2 \theta) = 1$

33. $(\sin \theta + \cos \theta)^2 + (\sin \theta - \cos \theta)^2 = 2$

34. $\tan^2 \theta \cos^2 \theta + \cot^2 \theta \sin^2 \theta = 1$

35. $\sec^4 \theta - \sec^2 \theta = \tan^4 \theta + \tan^2 \theta$

36. $\csc^4 \theta - \csc^2 \theta = \cot^4 \theta + \cot^2 \theta$

37. $\sec \theta - \tan \theta = \dfrac{\cos \theta}{1 + \sin \theta}$

38. $\csc \theta - \cot \theta = \dfrac{\sin \theta}{1 + \cos \theta}$

39. $3 \sin^2 \theta + 4 \cos^2 \theta = 3 + \cos^2 \theta$

40. $9 \sec^2 \theta - 5 \tan^2 \theta = 5 + 4 \sec^2 \theta$

41. $1 - \dfrac{\cos^2 \theta}{1 + \sin \theta} = \sin \theta$

42. $1 - \dfrac{\sin^2 \theta}{1 - \cos \theta} = -\cos \theta$

43. $\dfrac{1 + \tan \theta}{1 - \tan \theta} = \dfrac{\cot \theta + 1}{\cot \theta - 1}$

44. $\dfrac{\csc \theta - 1}{\csc \theta + 1} = \dfrac{1 - \sin \theta}{1 + \sin \theta}$

45. $\dfrac{\sec \theta}{\csc \theta} + \dfrac{\sin \theta}{\cos \theta} = 2 \tan \theta$

46. $\dfrac{\csc \theta - 1}{\cot \theta} = \dfrac{\cot \theta}{\csc \theta + 1}$

47. $\dfrac{1 + \sin \theta}{1 - \sin \theta} = \dfrac{\csc \theta + 1}{\csc \theta - 1}$

48. $\dfrac{\cos \theta + 1}{\cos \theta - 1} = \dfrac{1 + \sec \theta}{1 - \sec \theta}$

49. $\dfrac{1 - \sin\theta}{\cos\theta} + \dfrac{\cos\theta}{1 - \sin\theta} = 2\sec\theta$

50. $\dfrac{\cos\theta}{1 + \sin\theta} + \dfrac{1 + \sin\theta}{\cos\theta} = 2\sec\theta$

51. $\dfrac{\sin\theta}{\sin\theta - \cos\theta} = \dfrac{1}{1 - \cot\theta}$

52. $1 - \dfrac{\sin^2\theta}{1 + \cos\theta} = \cos\theta$

53. $\dfrac{1 - \sin\theta}{1 + \sin\theta} = (\sec\theta - \tan\theta)^2$

54. $\dfrac{1 - \cos\theta}{1 + \cos\theta} = (\csc\theta - \cot\theta)^2$

55. $\dfrac{\cos\theta}{1 - \tan\theta} + \dfrac{\sin\theta}{1 - \cot\theta} = \sin\theta + \cos\theta$

56. $\dfrac{\cot\theta}{1 - \tan\theta} + \dfrac{\tan\theta}{1 - \cot\theta} = 1 + \tan\theta + \cot\theta$

57. $\tan\theta + \dfrac{\cos\theta}{1 + \sin\theta} = \sec\theta$

58. $\dfrac{\sin\theta\cos\theta}{\cos^2\theta - \sin^2\theta} = \dfrac{\tan\theta}{1 - \tan^2\theta}$

59. $\dfrac{\tan\theta + \sec\theta - 1}{\tan\theta - \sec\theta + 1} = \tan\theta + \sec\theta$

60. $\dfrac{\sin\theta - \cos\theta + 1}{\sin\theta + \cos\theta - 1} = \dfrac{\sin\theta + 1}{\cos\theta}$

61. $\dfrac{\tan\theta - \cot\theta}{\tan\theta + \cot\theta} = \sin^2\theta - \cos^2\theta$

62. $\dfrac{\sec\theta - \cos\theta}{\sec\theta + \cos\theta} = \dfrac{\sin^2\theta}{1 + \cos^2\theta}$

63. $\dfrac{\tan\theta - \cot\theta}{\tan\theta + \cot\theta} + 1 = 2\sin^2\theta$

64. $\dfrac{\tan\theta - \cot\theta}{\tan\theta + \cot\theta} + 2\cos^2\theta = 1$

65. $\dfrac{\sec\theta + \tan\theta}{\cot\theta + \cos\theta} = \tan\theta\sec\theta$

66. $\dfrac{\sec\theta}{1 + \sec\theta} = \dfrac{1 - \cos\theta}{\sin^2\theta}$

67. $\dfrac{1 - \tan^2\theta}{1 + \tan^2\theta} + 1 = 2\cos^2\theta$

68. $\dfrac{1 - \cot^2\theta}{1 + \cot^2\theta} + 2\cos^2\theta = 1$

69. $\dfrac{\sec\theta - \csc\theta}{\sec\theta\csc\theta} = \sin\theta - \cos\theta$

70. $\dfrac{\sin^2\theta - \tan\theta}{\cos^2\theta - \cot\theta} = \tan^2\theta$

71. $\sec\theta - \cos\theta = \sin\theta\tan\theta$

72. $\tan\theta + \cot\theta = \sec\theta\csc\theta$

73. $\dfrac{1}{1 - \sin\theta} + \dfrac{1}{1 + \sin\theta} = 2\sec^2\theta$

74. $\dfrac{1 + \sin\theta}{1 - \sin\theta} - \dfrac{1 - \sin\theta}{1 + \sin\theta} = 4\tan\theta\sec\theta$

75. $\dfrac{\sec\theta}{1 - \sin\theta} = \dfrac{1 + \sin\theta}{\cos^3\theta}$

76. $\dfrac{1 - \sin\theta}{1 + \sin\theta} = (\sec\theta - \tan\theta)^2$

77. $\dfrac{(\sec\theta - \tan\theta)^2 + 1}{\csc\theta(\sec\theta - \tan\theta)} = 2\tan\theta$

78. $\dfrac{\sec^2\theta - \tan^2\theta + \tan\theta}{\sec\theta} = \sin\theta + \cos\theta$

79. $\dfrac{\sin\theta + \cos\theta}{\cos\theta} - \dfrac{\sin\theta - \cos\theta}{\sin\theta} = \sec\theta\csc\theta$

80. $\dfrac{\sin\theta + \cos\theta}{\sin\theta} - \dfrac{\cos\theta - \sin\theta}{\cos\theta} = \sec\theta\csc\theta$

81. $\dfrac{\sin^3\theta + \cos^3\theta}{\sin\theta + \cos\theta} = 1 - \sin\theta\cos\theta$

82. $\dfrac{\sin^3\theta + \cos^3\theta}{1 - 2\cos^2\theta} = \dfrac{\sec\theta - \sin\theta}{\tan\theta - 1}$

83. $\dfrac{\cos^2\theta - \sin^2\theta}{1 - \tan^2\theta} = \cos^2\theta$

84. $\dfrac{\cos\theta + \sin\theta - \sin^3\theta}{\sin\theta} = \cot\theta + \cos^2\theta$

85. $\dfrac{(2\cos^2\theta - 1)^2}{\cos^4\theta - \sin^4\theta} = 1 - 2\sin^2\theta$

86. $\dfrac{1 - 2\cos^2\theta}{\sin\theta\cos\theta} = \tan\theta - \cot\theta$

87. $\dfrac{1 + \sin\theta + \cos\theta}{1 + \sin\theta - \cos\theta} = \dfrac{1 + \cos\theta}{\sin\theta}$

88. $\dfrac{1 + \cos\theta + \sin\theta}{1 + \cos\theta - \sin\theta} = \sec\theta + \tan\theta$

89. $(a\sin\theta + b\cos\theta)^2 + (a\cos\theta - b\sin\theta)^2 = a^2 + b^2$

90. $(2a\sin\theta\cos\theta)^2 + a^2(\cos^2\theta - \sin^2\theta)^2 = a^2$

91. $\dfrac{\tan\alpha + \tan\beta}{\cot\alpha + \cot\beta} = \tan\alpha\tan\beta$

92. $(\tan\alpha + \tan\beta)(1 - \cot\alpha\cot\beta) + (\cot\alpha + \cot\beta)(1 - \tan\alpha\tan\beta) = 0$

93. $(\sin\alpha + \cos\beta)^2 + (\cos\beta + \sin\alpha)(\cos\beta - \sin\alpha) = 2\cos\beta(\sin\alpha + \cos\beta)$

94. $(\sin\alpha - \cos\beta)^2 + (\cos\beta + \sin\alpha)(\cos\beta - \sin\alpha) = -2\cos\beta(\sin\alpha - \cos\beta)$

95. $\ln|\sec\theta| = -\ln|\cos\theta|$

96. $\ln|\tan\theta| = \ln|\sin\theta| - \ln|\cos\theta|$

97. $\ln|1 + \cos\theta| + \ln|1 - \cos\theta| = 2\ln|\sin\theta|$

98. $\ln|\sec\theta + \tan\theta| + \ln|\sec\theta - \tan\theta| = 0$

99. Show that $\sec(\tan^{-1}v) = \sqrt{1 + v^2}$.

100. Show that $\tan(\sin^{-1}v) = \dfrac{v}{\sqrt{1 - v^2}}$.

101. Show that $\tan(\cos^{-1}v) = \dfrac{\sqrt{1 - v^2}}{v}$.

102. Show that $\sin(\cos^{-1}v) = \sqrt{1 - v^2}$.

103. Show that $\cos(\sin^{-1}v) = \sqrt{1 - v^2}$.

104. Show that $\cos(\tan^{-1}v) = \dfrac{1}{\sqrt{1 + v^2}}$.

Discussion and Writing

105. Write a few paragraphs outlining your strategy for establishing identities.

106. Write down the three Pythagorean Identities.

107. Why do you think it is usually preferable to start with the side containing the more complicated expression when establishing an identity?

108. Make up an identity that is not a Fundamental Identity.

'Are You Prepared?' Answers

1. True   **2.** True

6.4 Sum and Difference Formulas

PREPARING FOR THIS SECTION   *Before getting started, review the following:*

- Distance Formula (Section 1.1, p. 5)
- Values of the Trigonometric Functions (Section 5.2, pp. 374–382)

- Finding Exact Values Given the Value of a Trigonometric Function and the Quadrant of the Angle (Section 5.3, pp. 395–398)

 Now work the 'Are You Prepared?' problems on page 481.

OBJECTIVES   **1** Use Sum and Difference Formulas to Find Exact Values
   2 Use Sum and Difference Formulas to Establish Identities
   3 Use Sum and Difference Formulas Involving Inverse Trigonometric Functions

In this section, we continue our derivation of trigonometric identities by obtaining formulas that involve the sum or difference of two angles, such as $\cos(\alpha + \beta), \cos(\alpha - \beta),$ or $\sin(\alpha + \beta)$. These formulas are referred to as the **sum and difference formulas**. We begin with the formulas for $\cos(\alpha + \beta)$ and $\cos(\alpha - \beta)$.

Theorem

Sum and Difference Formulas for Cosines

$$\cos(\alpha + \beta) = \cos \alpha \cos \beta - \sin \alpha \sin \beta \qquad \textbf{(1)}$$
$$\cos(\alpha - \beta) = \cos \alpha \cos \beta + \sin \alpha \sin \beta \qquad \textbf{(2)}$$

In Words

Formula (1) states that the cosine of the sum of two angles equals the cosine of the first angle times the cosine of the second angle minus the sine of the first angle times the sine of the second angle.

Proof   We will prove formula (2) first. Although this formula is true for all numbers α and β, we shall assume in our proof that $0 < \beta < \alpha < 2\pi$. We begin with the unit circle and place the angles α and β in standard position, as shown in Figure 24(a). The point P_1 lies on the terminal side of β, so its coordinates are $(\cos \beta, \sin \beta)$; and the point P_2 lies on the terminal side of α, so its coordinates are $(\cos \alpha, \sin \alpha)$.

Figure 24

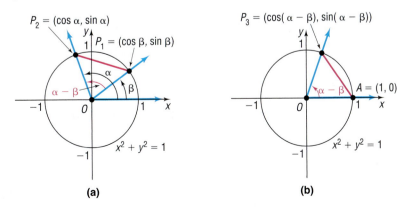

(a)   (b)

Now place the angle $\alpha - \beta$ in standard position, as shown in Figure 24(b). The point A has coordinates $(1, 0)$, and the point P_3 is on the terminal side of the angle $\alpha - \beta$, so its coordinates are $(\cos(\alpha - \beta), \sin(\alpha - \beta))$.

Looking at triangle OP_1P_2 in Figure 24(a) and triangle OAP_3 in Figure 24(b), we see that these triangles are congruent. (Do you see why? Two sides and the included angle, $\alpha - \beta$, are equal.) As a result, the unknown side of each triangle must be equal; that is,

$$d(A, P_3) = d(P_1, P_2)$$

Using the distance formula, we find that

$$\sqrt{[\cos(\alpha - \beta) - 1]^2 + [\sin(\alpha - \beta) - 0]^2} = \sqrt{(\cos \alpha - \cos \beta)^2 + (\sin \alpha - \sin \beta)^2} \qquad d(A, P_3) = d(P_1, P_2)$$

$$[\cos(\alpha - \beta) - 1]^2 + \sin^2(\alpha - \beta) = (\cos \alpha - \cos \beta)^2 + (\sin \alpha - \sin \beta)^2 \qquad \text{Square both sides.}$$

$$\cos^2(\alpha - \beta) - 2\cos(\alpha - \beta) + 1 + \sin^2(\alpha - \beta) = \cos^2 \alpha - 2\cos \alpha \cos \beta + \cos^2 \beta \qquad \text{Multiply out the squared terms.}$$
$$+ \sin^2 \alpha - 2 \sin \alpha \sin \beta + \sin^2 \beta$$

$$2 - 2\cos(\alpha - \beta) = 2 - 2\cos \alpha \cos \beta - 2 \sin \alpha \sin \beta \qquad \text{Apply a Pythagorean Identity (3 times).}$$

$$-2\cos(\alpha - \beta) = -2\cos \alpha \cos \beta - 2 \sin \alpha \sin \beta \qquad \text{Subtract 2 from each side.}$$

$$\cos(\alpha - \beta) = \cos \alpha \cos \beta + \sin \alpha \sin \beta \qquad \text{Divide each side by } -2.$$

This is formula (2). ∎

The proof of formula (1) follows from formula (2) and the Even–Odd Identities. We use the fact that $\alpha + \beta = \alpha - (-\beta)$. Then

$$\cos(\alpha + \beta) = \cos[\alpha - (-\beta)]$$

$$= \cos \alpha \cos(-\beta) + \sin \alpha \sin(-\beta) \qquad \text{Use formula (2).}$$

$$= \cos \alpha \cos \beta - \sin \alpha \sin \beta \qquad \text{Even–Odd Identities} \quad ∎$$

1 Use Sum and Difference Formulas to Find Exact Values

One use of formulas (1) and (2) is to obtain the exact value of the cosine of an angle that can be expressed as the sum or difference of angles whose sine and cosine are known exactly.

| **EXAMPLE 1** | **Using the Sum Formula to Find Exact Values** |

Find the exact value of $\cos 75°$.

Solution   Since $75° = 45° + 30°$, we use formula (1) to obtain

$$\cos 75° = \cos(45° + 30°) = \cos 45° \cos 30° - \sin 45° \sin 30°$$
$$\uparrow$$
$$\text{Formula (1)}$$

$$= \frac{\sqrt{2}}{2} \cdot \frac{\sqrt{3}}{2} - \frac{\sqrt{2}}{2} \cdot \frac{1}{2} = \frac{1}{4}\left(\sqrt{6} - \sqrt{2}\right)$$

◀

EXAMPLE 2	**Using the Difference Formula to Find Exact Values**

Find the exact value of $\cos\dfrac{\pi}{12}$.

Solution

$$\cos\frac{\pi}{12} = \cos\left(\frac{3\pi}{12} - \frac{2\pi}{12}\right) = \cos\left(\frac{\pi}{4} - \frac{\pi}{6}\right)$$

$$= \cos\frac{\pi}{4}\cos\frac{\pi}{6} + \sin\frac{\pi}{4}\sin\frac{\pi}{6} \qquad \text{Use formula (2).}$$

$$= \frac{\sqrt{2}}{2}\cdot\frac{\sqrt{3}}{2} + \frac{\sqrt{2}}{2}\cdot\frac{1}{2} = \frac{1}{4}\left(\sqrt{6} + \sqrt{2}\right) \qquad \blacktriangleleft$$

NOW WORK PROBLEM 11.

2 Use Sum and Difference Formula to Establish Identities

Another use of formulas (1) and (2) is to establish other identities. One important pair of identities is given next.

— Seeing the Concept —

Graph $Y_1 = \cos\left(\dfrac{\pi}{2} - \theta\right)$ and $Y_2 = \sin\theta$ on the same screen. Does this demonstrate the result 3(a)? How would you demonstrate the result 3(b)?

$$\cos\left(\frac{\pi}{2} - \theta\right) = \sin\theta \qquad\qquad \textbf{(3a)}$$

$$\sin\left(\frac{\pi}{2} - \theta\right) = \cos\theta \qquad\qquad \textbf{(3b)}$$

Proof To prove formula (3a), we use the formula for $\cos(\alpha - \beta)$ with $\alpha = \dfrac{\pi}{2}$ and $\beta = \theta$.

$$\cos\left(\frac{\pi}{2} - \theta\right) = \cos\frac{\pi}{2}\cos\theta + \sin\frac{\pi}{2}\sin\theta$$

$$= 0\cdot\cos\theta + 1\cdot\sin\theta$$

$$= \sin\theta \qquad\qquad \blacksquare$$

To prove formula (3b), we make use of the identity (3a) just established.

$$\sin\left(\frac{\pi}{2} - \theta\right) = \cos\left[\frac{\pi}{2} - \left(\frac{\pi}{2} - \theta\right)\right] = \cos\theta$$

$$\underset{\text{Use (3a).}}{\uparrow}$$

$$\blacksquare$$

Also, since

$$\cos\left(\frac{\pi}{2} - \theta\right) = \cos\left[-\left(\theta - \frac{\pi}{2}\right)\right] = \cos\left(\theta - \frac{\pi}{2}\right)$$

$$\underset{\text{Even Property of Cosine}}{\uparrow}$$

and since

$$\cos\left(\frac{\pi}{2} - \theta\right) = \sin\theta$$

$$\underset{3(a)}{\uparrow}$$

it follows that $\cos\left(\theta - \dfrac{\pi}{2}\right) = \sin\theta$. The graphs of $y = \cos\left(\theta - \dfrac{\pi}{2}\right)$ and $y = \sin\theta$ are identical, a fact that we conjectured earlier in Section 5.4.

Having established the identities in formulas (3a) and (3b), we now can derive the sum and difference formulas for $\sin(\alpha + \beta)$ and $\sin(\alpha - \beta)$.

Proof $\quad \sin(\alpha + \beta) = \cos\left[\dfrac{\pi}{2} - (\alpha + \beta)\right]$ Formula (3a)

$$= \cos\left[\left(\dfrac{\pi}{2} - \alpha\right) - \beta\right]$$

$$= \cos\left(\dfrac{\pi}{2} - \alpha\right)\cos\beta + \sin\left(\dfrac{\pi}{2} - \alpha\right)\sin\beta \quad \text{Formula (2)}$$

$$= \sin\alpha\cos\beta + \cos\alpha\sin\beta \quad \text{Formulas (3a) and (3b)}$$

■

$$\sin(\alpha - \beta) = \sin[\alpha + (-\beta)]$$

$$= \sin\alpha\cos(-\beta) + \cos\alpha\sin(-\beta) \quad \begin{array}{l}\text{Use the sum formula for}\\ \text{sine just obtained.}\end{array}$$

$$= \sin\alpha\cos\beta + \cos\alpha(-\sin\beta) \quad \text{Even–Odd Identities.}$$

$$= \sin\alpha\cos\beta - \cos\alpha\sin\beta$$

■

Theorem

Sum and Difference Formulas for Sines

In Words

Formula (4) states that the sine of the sum of two angles equals the sine of the first angle times the cosine of the second angle plus the cosine of the first angle times the sine of the second angle.

$$\sin(\alpha + \beta) = \sin\alpha\cos\beta + \cos\alpha\sin\beta \qquad (4)$$

$$\sin(\alpha - \beta) = \sin\alpha\cos\beta - \cos\alpha\sin\beta \qquad (5)$$

EXAMPLE 3   **Using the Sum Formula to Find Exact Values**

Find the exact value of $\sin\dfrac{7\pi}{12}$.

Solution $\sin\dfrac{7\pi}{12} = \sin\left(\dfrac{3\pi}{12} + \dfrac{4\pi}{12}\right) = \sin\left(\dfrac{\pi}{4} + \dfrac{\pi}{3}\right)$

$$= \sin\dfrac{\pi}{4}\cos\dfrac{\pi}{3} + \cos\dfrac{\pi}{4}\sin\dfrac{\pi}{3} \quad \text{Formula (4)}$$

$$= \dfrac{\sqrt{2}}{2}\cdot\dfrac{1}{2} + \dfrac{\sqrt{2}}{2}\cdot\dfrac{\sqrt{3}}{2} = \dfrac{1}{4}\left(\sqrt{2} + \sqrt{6}\right)$$

◀

NOW WORK PROBLEM 17.

EXAMPLE 4 **Using the Difference Formula to Find Exact Values**

Find the exact value of $\sin 80° \cos 20° - \cos 80° \sin 20°$.

Solution The form of the expression $\sin 80° \cos 20° - \cos 80° \sin 20°$ is that of the right side of the formula (5) for $\sin(\alpha - \beta)$ with $\alpha = 80°$ and $\beta = 20°$. That is,

$$\sin 80° \cos 20° - \cos 80° \sin 20° = \sin(80° - 20°) = \sin 60° = \frac{\sqrt{3}}{2}$$ ◀

 NOW WORK PROBLEMS 23 AND 27.

EXAMPLE 5 **Finding Exact Values**

If it is known that $\sin \alpha = \dfrac{4}{5}, \dfrac{\pi}{2} < \alpha < \pi$, and that $\sin \beta = -\dfrac{2}{\sqrt{5}} = -\dfrac{2\sqrt{5}}{5}$, $\pi < \beta < \dfrac{3\pi}{2}$, find the exact value of

(a) $\cos \alpha$ (b) $\cos \beta$ (c) $\cos(\alpha + \beta)$ (d) $\sin(\alpha + \beta)$

Solution (a) Since $\sin \alpha = \dfrac{4}{5} = \dfrac{y}{r}$ and $\dfrac{\pi}{2} < \alpha < \pi$, we let $y = 4$ and $r = 5$ and place α in quadrant II. The point $P = (x, y) = (x, 4), x < 0$, is on a circle of radius 5, namely, $x^2 + y^2 = 25$. See Figure 25. Then,

$$x^2 + y^2 = 25,$$
$$x^2 + 16 = 25 \qquad y = 4$$
$$x^2 = 25 - 16 = 9$$
$$x = -3 \qquad x < 0$$

Then,

$$\cos \alpha = \frac{x}{r} = -\frac{3}{5}$$

Alternatively, we can find $\cos \alpha$ using identities, as follows:

$$\cos \alpha = -\sqrt{1 - \sin^2 \alpha} = -\sqrt{1 - \frac{16}{25}} = -\sqrt{\frac{9}{25}} = -\frac{3}{5}$$

α in quadrant II,
$\cos \alpha < 0$

(b) Since $\sin \beta = \dfrac{-2}{\sqrt{5}} = \dfrac{y}{r}$ and $\pi < \beta < \dfrac{3\pi}{2}$, we let $y = -2$ and $r = \sqrt{5}$ and place β in quadrant III. The point $P = (x, y) = (x, -2), x < 0$, is on a circle of radius $\sqrt{5}$, namely, $x^2 + y^2 = 5$. See Figure 26. Then,

$$x^2 + y^2 = 5,$$
$$x^2 + 4 = 5 \qquad y = -2$$
$$x^2 = 1$$
$$x = -1 \qquad x < 0$$

Figure 25

$\sin \alpha = \dfrac{4}{5}, \dfrac{\pi}{2} < \alpha < \pi$

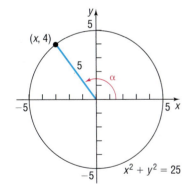

Figure 26

Given $\sin \beta = \dfrac{-2}{\sqrt{5}}, \pi < \beta < \dfrac{3\pi}{2}$

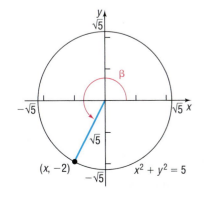

Then,

$$\cos \beta = \frac{x}{r} = \frac{-1}{\sqrt{5}} = -\frac{\sqrt{5}}{5}$$

Alternatively, we can find $\cos \beta$ using identities, as follows:

$$\cos \beta = -\sqrt{1 - \sin^2 \beta} = -\sqrt{1 - \frac{4}{5}} = -\sqrt{\frac{1}{5}} = -\frac{\sqrt{5}}{5}$$

(c) Using the results found in parts (a) and (b) and formula (1), we have

$$\cos(\alpha + \beta) = \cos \alpha \cos \beta - \sin \alpha \sin \beta$$

$$= -\frac{3}{5}\left(-\frac{\sqrt{5}}{5}\right) - \frac{4}{5}\left(-\frac{2\sqrt{5}}{5}\right) = \frac{11\sqrt{5}}{25}$$

(d) $\sin(\alpha + \beta) = \sin \alpha \cos \beta + \cos \alpha \sin \beta$

$$= \frac{4}{5}\left(-\frac{\sqrt{5}}{5}\right) + \left(-\frac{3}{5}\right)\left(-\frac{2\sqrt{5}}{5}\right) = \frac{2\sqrt{5}}{25}$$

◀

🖉 **NOW WORK PROBLEMS 31(a), (b), AND (c).**

EXAMPLE 6	**Establishing an Identity**

Establish the identity: $\dfrac{\cos(\alpha - \beta)}{\sin \alpha \sin \beta} = \cot \alpha \cot \beta + 1$

Solution
$$\frac{\cos(\alpha - \beta)}{\sin \alpha \sin \beta} = \frac{\cos \alpha \cos \beta + \sin \alpha \sin \beta}{\sin \alpha \sin \beta}$$

$$= \frac{\cos \alpha \cos \beta}{\sin \alpha \sin \beta} + \frac{\sin \alpha \sin \beta}{\sin \alpha \sin \beta}$$

$$= \frac{\cos \alpha}{\sin \alpha} \cdot \frac{\cos \beta}{\sin \beta} + 1$$

$$= \cot \alpha \cot \beta + 1$$

◀

🖉 **NOW WORK PROBLEMS 39 AND 51.**

We use the identity $\tan \theta = \dfrac{\sin \theta}{\cos \theta}$ and the sum formulas for $\sin(\alpha + \beta)$ and $\cos(\alpha + \beta)$ to derive a formula for $\tan(\alpha + \beta)$.

Proof $\tan(\alpha + \beta) = \dfrac{\sin(\alpha + \beta)}{\cos(\alpha + \beta)} = \dfrac{\sin \alpha \cos \beta + \cos \alpha \sin \beta}{\cos \alpha \cos \beta - \sin \alpha \sin \beta}$

Now we divide the numerator and denominator by $\cos \alpha \cos \beta$.

$$\tan(\alpha + \beta) = \frac{\dfrac{\sin\alpha\cos\beta + \cos\alpha\sin\beta}{\cos\alpha\cos\beta}}{\dfrac{\cos\alpha\cos\beta - \sin\alpha\sin\beta}{\cos\alpha\cos\beta}} = \frac{\dfrac{\sin\alpha\cos\beta}{\cos\alpha\cos\beta} + \dfrac{\cos\alpha\sin\beta}{\cos\alpha\cos\beta}}{\dfrac{\cos\alpha\cos\beta}{\cos\alpha\cos\beta} - \dfrac{\sin\alpha\sin\beta}{\cos\alpha\cos\beta}}$$

$$= \frac{\dfrac{\sin\alpha}{\cos\alpha} + \dfrac{\sin\beta}{\cos\beta}}{1 - \dfrac{\sin\alpha\sin\beta}{\cos\alpha\cos\beta}} = \frac{\tan\alpha + \tan\beta}{1 - \tan\alpha\tan\beta}$$ ∎

Proof We use the sum formula for $\tan(\alpha + \beta)$ and even–odd properties to get the difference formula.

$$\tan(\alpha - \beta) = \tan[\alpha + (-\beta)] = \frac{\tan\alpha + \tan(-\beta)}{1 - \tan\alpha\tan(-\beta)} = \frac{\tan\alpha - \tan\beta}{1 + \tan\alpha\tan\beta}$$ ∎

We have proved the following results:

Theorem

In Words
Formula (6) states that the tangent of the sum of two angles equals the tangent of the first angle plus the tangent of the second angle, all divided by 1 minus their product.

Sum and Difference Formulas for Tangents

$$\tan(\alpha + \beta) = \frac{\tan\alpha + \tan\beta}{1 - \tan\alpha\tan\beta} \tag{6}$$

$$\tan(\alpha - \beta) = \frac{\tan\alpha - \tan\beta}{1 + \tan\alpha\tan\beta} \tag{7}$$

NOW WORK PROBLEM 31(d).

EXAMPLE 7 **Establishing an Identity**

Prove the identity: $\tan(\theta + \pi) = \tan\theta$

Solution $\tan(\theta + \pi) = \dfrac{\tan\theta + \tan\pi}{1 - \tan\theta\tan\pi} = \dfrac{\tan\theta + 0}{1 - \tan\theta \cdot 0} = \tan\theta$ ◄

The result obtained in Example 7 verifies that the tangent function is periodic with period π, a fact that we mentioned earlier.

EXAMPLE 8 **Establishing an Identity**

Prove the identity: $\tan\left(\theta + \dfrac{\pi}{2}\right) = -\cot\theta$

Solution We cannot use formula (6), since $\tan\dfrac{\pi}{2}$ is not defined. Instead, we proceed as follows:

WARNING
Be careful when using formulas (6) and (7). These formulas can be used only for angles α and β for which $\tan\alpha$ and $\tan\beta$ are defined, that is, all angles except odd multiples of $\dfrac{\pi}{2}$. ∎

$$\tan\left(\theta + \frac{\pi}{2}\right) = \frac{\sin\left(\theta + \dfrac{\pi}{2}\right)}{\cos\left(\theta + \dfrac{\pi}{2}\right)} = \frac{\sin\theta\cos\dfrac{\pi}{2} + \cos\theta\sin\dfrac{\pi}{2}}{\cos\theta\cos\dfrac{\pi}{2} - \sin\theta\sin\dfrac{\pi}{2}}$$

$$= \frac{(\sin\theta)(0) + (\cos\theta)(1)}{(\cos\theta)(0) - (\sin\theta)(1)} = \frac{\cos\theta}{-\sin\theta} = -\cot\theta$$ ◄

Use Sum and Difference Formulas Involving Inverse Trigonometric Functions

---

EXAMPLE 9

Finding the Exact Value of an Expression Involving Inverse Trigonometric Functions

Find the exact value of: $\sin\left(\cos^{-1}\dfrac{1}{2} + \sin^{-1}\dfrac{3}{5}\right)$

Solution

We seek the sine of the sum of two angles, $\alpha = \cos^{-1}\dfrac{1}{2}$ and $\beta = \sin^{-1}\dfrac{3}{5}$. Then

$$\cos\alpha = \frac{1}{2},\quad 0 \le \alpha \le \pi,\quad\text{and}\quad \sin\beta = \frac{3}{5},\quad -\frac{\pi}{2} \le \beta \le \frac{\pi}{2}$$

NOTE

In Example 9, we could also find $\sin\alpha$ by using $\cos\alpha = \dfrac{1}{2} = \dfrac{x}{r}$, so $x = 1$ and $r = 2$. Then, $y = \sqrt{3}$ and $\sin\alpha = \dfrac{y}{r} = \dfrac{\sqrt{3}}{2}$. We could find $\cos\beta$ in a similar fashion. ∎

We use Pythagorean Identities to obtain $\sin\alpha$ and $\cos\beta$. Since $\sin\alpha \ge 0$ and $\cos\beta \ge 0$ (do you know why?), we find

$$\sin\alpha = \sqrt{1 - \cos^2\alpha} = \sqrt{1 - \frac{1}{4}} = \sqrt{\frac{3}{4}} = \frac{\sqrt{3}}{2}$$

$$\cos\beta = \sqrt{1 - \sin^2\beta} = \sqrt{1 - \frac{9}{25}} = \sqrt{\frac{16}{25}} = \frac{4}{5}$$

As a result,

$$\sin\left(\cos^{-1}\frac{1}{2} + \sin^{-1}\frac{3}{5}\right) = \sin(\alpha + \beta) = \sin\alpha\cos\beta + \cos\alpha\sin\beta$$

$$= \frac{\sqrt{3}}{2}\cdot\frac{4}{5} + \frac{1}{2}\cdot\frac{3}{5} = \frac{4\sqrt{3}+3}{10} \quad◀$$

 NOW WORK PROBLEM 67.

EXAMPLE 10

Writing a Trigonometric Expression as an Algebraic Expression

Write $\sin(\sin^{-1}u + \cos^{-1}v)$ as an algebraic expression containing u and v (that is, without any trigonometric functions).

Solution

Let $\alpha = \sin^{-1}u$ and $\beta = \cos^{-1}v$. Then

$$\sin\alpha = u,\quad -\frac{\pi}{2} \le \alpha \le \frac{\pi}{2},\quad\text{and}\quad \cos\beta = v,\quad 0 \le \beta \le \pi$$

Since $-\dfrac{\pi}{2} \le \alpha \le \dfrac{\pi}{2}$, we know that $\cos\alpha \ge 0$. As a result,

$$\cos\alpha = \sqrt{1 - \sin^2\alpha} = \sqrt{1 - u^2}$$

Similarly, since $0 \le \beta \le \pi$, we know that $\sin\beta \ge 0$. Then

$$\sin\beta = \sqrt{1 - \cos^2\beta} = \sqrt{1 - v^2}$$

Now

$$\sin(\sin^{-1}u + \cos^{-1}v) = \sin(\alpha + \beta) = \sin\alpha\cos\beta + \cos\alpha\sin\beta$$

$$= uv + \sqrt{1 - u^2}\,\sqrt{1 - v^2} \quad◀$$

NOW WORK PROBLEM 77.

Summary

Sum and Difference Formulas

$$\cos(\alpha + \beta) = \cos \alpha \cos \beta - \sin \alpha \sin \beta \qquad \cos(\alpha - \beta) = \cos \alpha \cos \beta + \sin \alpha \sin \beta$$

$$\sin(\alpha + \beta) = \sin \alpha \cos \beta + \cos \alpha \sin \beta \qquad \sin(\alpha - \beta) = \sin \alpha \cos \beta - \cos \alpha \sin \beta$$

$$\tan(\alpha + \beta) = \frac{\tan \alpha + \tan \beta}{1 - \tan \alpha \tan \beta} \qquad \tan(\alpha - \beta) = \frac{\tan \alpha - \tan \beta}{1 + \tan \alpha \tan \beta}$$

6.4 Assess Your Understanding

'Are You Prepared?'

Answers are given at the end of these exercises. If you get a wrong answer, read the pages listed in red.

1. The distance d from the point $(2, -3)$ to the point $(5, 1)$ is _____. (p. 5)

2. If $\sin \theta = \dfrac{4}{5}$ and θ is in quadrant II, then $\cos \theta =$ _____.
(pp. 395–398)

3. (a) $\sin \dfrac{\pi}{4} \cdot \cos \dfrac{\pi}{3} =$ _____. (p. 379)

　　(b) $\tan \dfrac{\pi}{4} - \sin \dfrac{\pi}{6} =$ _____. (p. 379)

Concepts and Vocabulary

4. $\cos(\alpha + \beta) = \cos \alpha \cos \beta$ ____ $\sin \alpha \sin \beta$.

5. $\sin(\alpha - \beta) = \sin \alpha \cos \beta$ ____ $\cos \alpha \sin \beta$.

6. *True or False:* $\sin(\alpha + \beta) = \sin \alpha + \sin \beta + 2 \sin \alpha \sin \beta$

7. *True or False:* $\tan 75° = \tan 30° + \tan 45°$

8. *True or False:* $\cos\left(\dfrac{\pi}{2} - \theta\right) = \cos \theta$

Skill Building

In Problems 9–20, find the exact value of each trigonometric function.

9. $\sin \dfrac{5\pi}{12}$

10. $\sin \dfrac{\pi}{12}$

11. $\cos \dfrac{7\pi}{12}$

12. $\tan \dfrac{7\pi}{12}$

13. $\cos 165°$

14. $\sin 105°$

15. $\tan 15°$

16. $\tan 195°$

17. $\sin \dfrac{17\pi}{12}$

18. $\tan \dfrac{19\pi}{12}$

19. $\sec\left(-\dfrac{\pi}{12}\right)$

20. $\cot\left(-\dfrac{5\pi}{12}\right)$

In Problems 21–30, find the exact value of each expression.

21. $\sin 20° \cos 10° + \cos 20° \sin 10°$

22. $\sin 20° \cos 80° - \cos 20° \sin 80°$

23. $\cos 70° \cos 20° - \sin 70° \sin 20°$

24. $\cos 40° \cos 10° + \sin 40° \sin 10°$

25. $\dfrac{\tan 20° + \tan 25°}{1 - \tan 20° \tan 25°}$

26. $\dfrac{\tan 40° - \tan 10°}{1 + \tan 40° \tan 10°}$

27. $\sin \dfrac{\pi}{12} \cos \dfrac{7\pi}{12} - \cos \dfrac{\pi}{12} \sin \dfrac{7\pi}{12}$

28. $\cos \dfrac{5\pi}{12} \cos \dfrac{7\pi}{12} - \sin \dfrac{5\pi}{12} \sin \dfrac{7\pi}{12}$

29. $\cos \dfrac{\pi}{12} \cos \dfrac{5\pi}{12} + \sin \dfrac{5\pi}{12} \sin \dfrac{\pi}{12}$

30. $\sin \dfrac{\pi}{18} \cos \dfrac{5\pi}{18} + \cos \dfrac{\pi}{18} \sin \dfrac{5\pi}{18}$

In Problems 31–36, find the exact value of each of the following under the given conditions:

(a) $\sin(\alpha + \beta)$　　(b) $\cos(\alpha + \beta)$　　(c) $\sin(\alpha - \beta)$　　(d) $\tan(\alpha - \beta)$

31. $\sin \alpha = \dfrac{3}{5}, 0 < \alpha < \dfrac{\pi}{2}; \quad \cos \beta = \dfrac{2\sqrt{5}}{5}, -\dfrac{\pi}{2} < \beta < 0$

32. $\cos \alpha = \dfrac{\sqrt{5}}{5}, 0 < \alpha < \dfrac{\pi}{2}; \quad \sin \beta = -\dfrac{4}{5}, -\dfrac{\pi}{2} < \beta < 0$

33. $\tan \alpha = -\dfrac{4}{3}, \dfrac{\pi}{2} < \alpha < \pi; \quad \cos \beta = \dfrac{1}{2}, 0 < \beta < \dfrac{\pi}{2}$

34. $\tan \alpha = \dfrac{5}{12}, \pi < \alpha < \dfrac{3\pi}{2}; \quad \sin \beta = -\dfrac{1}{2}, \pi < \beta < \dfrac{3\pi}{2}$

35. $\sin \alpha = \dfrac{5}{13}, -\dfrac{3\pi}{2} < \alpha < -\pi;$ $\tan \beta = -\sqrt{3}, \dfrac{\pi}{2} < \beta < \pi$ **36.** $\cos \alpha = \dfrac{1}{2}, -\dfrac{\pi}{2} < \alpha < 0;$ $\sin \beta = \dfrac{1}{3}, 0 < \beta < \dfrac{\pi}{2}$

37. If $\sin \theta = \dfrac{1}{3}, \theta$ in quadrant II, find the exact value of:

     (a) $\cos \theta$ (b) $\sin\left(\theta + \dfrac{\pi}{6}\right)$ (c) $\cos\left(\theta - \dfrac{\pi}{3}\right)$ (d) $\tan\left(\theta + \dfrac{\pi}{4}\right)$

38. If $\cos \theta = \dfrac{1}{4}, \theta$ in quadrant IV, find the exact value of:

     (a) $\sin \theta$ (b) $\sin\left(\theta - \dfrac{\pi}{6}\right)$ (c) $\cos\left(\theta + \dfrac{\pi}{3}\right)$ (d) $\tan\left(\theta - \dfrac{\pi}{4}\right)$

In Problems 39–64, establish each identity.

39. $\sin\left(\dfrac{\pi}{2} + \theta\right) = \cos \theta$ **40.** $\cos\left(\dfrac{\pi}{2} + \theta\right) = -\sin \theta$ **41.** $\sin(\pi - \theta) = \sin \theta$

42. $\cos(\pi - \theta) = -\cos \theta$ **43.** $\sin(\pi + \theta) = -\sin \theta$ **44.** $\cos(\pi + \theta) = -\cos \theta$

45. $\tan(\pi - \theta) = -\tan \theta$ **46.** $\tan(2\pi - \theta) = -\tan \theta$ **47.** $\sin\left(\dfrac{3\pi}{2} + \theta\right) = -\cos \theta$

48. $\cos\left(\dfrac{3\pi}{2} + \theta\right) = \sin \theta$ **49.** $\sin(\alpha + \beta) + \sin(\alpha - \beta) = 2 \sin \alpha \cos \beta$

50. $\cos(\alpha + \beta) + \cos(\alpha - \beta) = 2 \cos \alpha \cos \beta$ **51.** $\dfrac{\sin(\alpha + \beta)}{\sin \alpha \cos \beta} = 1 + \cot \alpha \tan \beta$

52. $\dfrac{\sin(\alpha + \beta)}{\cos \alpha \cos \beta} = \tan \alpha + \tan \beta$ **53.** $\dfrac{\cos(\alpha + \beta)}{\cos \alpha \cos \beta} = 1 - \tan \alpha \tan \beta$

54. $\dfrac{\cos(\alpha - \beta)}{\sin \alpha \cos \beta} = \cot \alpha + \tan \beta$ **55.** $\dfrac{\sin(\alpha + \beta)}{\sin(\alpha - \beta)} = \dfrac{\tan \alpha + \tan \beta}{\tan \alpha - \tan \beta}$

56. $\dfrac{\cos(\alpha + \beta)}{\cos(\alpha - \beta)} = \dfrac{1 - \tan \alpha \tan \beta}{1 + \tan \alpha \tan \beta}$ **57.** $\cot(\alpha + \beta) = \dfrac{\cot \alpha \cot \beta - 1}{\cot \beta + \cot \alpha}$

58. $\cot(\alpha - \beta) = \dfrac{\cot \alpha \cot \beta + 1}{\cot \beta - \cot \alpha}$ **59.** $\sec(\alpha + \beta) = \dfrac{\csc \alpha \csc \beta}{\cot \alpha \cot \beta - 1}$

60. $\sec(\alpha - \beta) = \dfrac{\sec \alpha \sec \beta}{1 + \tan \alpha \tan \beta}$ **61.** $\sin(\alpha - \beta) \sin(\alpha + \beta) = \sin^2 \alpha - \sin^2 \beta$

62. $\cos(\alpha - \beta) \cos(\alpha + \beta) = \cos^2 \alpha - \sin^2 \beta$ **63.** $\sin(\theta + k\pi) = (-1)^k \sin \theta, k$ any integer

64. $\cos(\theta + k\pi) = (-1)^k \cos \theta, k$ any integer

In Problems 65–76, find the exact value of each expression.

65. $\sin\left(\sin^{-1}\dfrac{1}{2} + \cos^{-1} 0\right)$ **66.** $\sin\left(\sin^{-1}\dfrac{\sqrt{3}}{2} + \cos^{-1} 1\right)$ **67.** $\sin\left[\sin^{-1}\dfrac{3}{5} - \cos^{-1}\left(-\dfrac{4}{5}\right)\right]$ **68.** $\sin\left[\sin^{-1}\left(-\dfrac{4}{5}\right) - \tan^{-1}\dfrac{3}{4}\right]$

69. $\cos\left(\tan^{-1}\dfrac{4}{3} + \cos^{-1}\dfrac{5}{13}\right)$ **70.** $\cos\left[\tan^{-1}\dfrac{5}{12} - \sin^{-1}\left(-\dfrac{3}{5}\right)\right]$ **71.** $\cos\left(\sin^{-1}\dfrac{5}{13} - \tan^{-1}\dfrac{3}{4}\right)$ **72.** $\cos\left(\tan^{-1}\dfrac{4}{3} + \cos^{-1}\dfrac{12}{13}\right)$

73. $\tan\left(\sin^{-1}\dfrac{3}{5} + \dfrac{\pi}{6}\right)$ **74.** $\tan\left(\dfrac{\pi}{4} - \cos^{-1}\dfrac{3}{5}\right)$ **75.** $\tan\left(\sin^{-1}\dfrac{4}{5} + \cos^{-1} 1\right)$ **76.** $\tan\left(\cos^{-1}\dfrac{4}{5} + \sin^{-1} 1\right)$

In Problems 77–82, write each trigonometric expression as an algebraic expression containing u and v.

77. $\cos(\cos^{-1} u + \sin^{-1} v)$ **78.** $\sin(\sin^{-1} u - \cos^{-1} v)$ **79.** $\sin(\tan^{-1} u - \sin^{-1} v)$

80. $\cos(\tan^{-1} u + \tan^{-1} v)$ **81.** $\tan(\sin^{-1} u - \cos^{-1} v)$ **82.** $\sec(\tan^{-1} u + \cos^{-1} v)$

Applications and Extensions

83. Show that $\sin^{-1} v + \cos^{-1} v = \dfrac{\pi}{2}$.   **84.** Show that $\tan^{-1} v + \cot^{-1} v = \dfrac{\pi}{2}$.

85. Show that $\tan^{-1}\left(\dfrac{1}{v}\right) = \dfrac{\pi}{2} - \tan^{-1} v$, if $v > 0$.   **86.** Show that $\cot^{-1} e^v = \tan^{-1} e^{-v}$.

87. Show that $\sin(\sin^{-1} v + \cos^{-1} v) = 1$.   **88.** Show that $\cos(\sin^{-1} v + \cos^{-1} v) = 0$.

89. Calculus Show that the difference quotient for $f(x) = \sin x$ is given by

$$\frac{f(x + h) - f(x)}{h} = \frac{\sin(x + h) - \sin x}{h}$$

$$= \cos x \cdot \frac{\sin h}{h} - \sin x \cdot \frac{1 - \cos h}{h}$$

90. Calculus Show that the difference quotient for $f(x) = \cos x$ is given by

$$\frac{f(x + h) - f(x)}{h} = \frac{\cos(x + h) - \cos x}{h}$$

$$= -\sin x \cdot \frac{\sin h}{h} - \cos x \cdot \frac{1 - \cos h}{h}$$

91. Explain why formula (7) cannot be used to show that

$$\tan\left(\frac{\pi}{2} - \theta\right) = \cot \theta$$

Establish this identity by using formulas (3a) and (3b).

92. If $\tan \alpha = x + 1$ and $\tan \beta = x - 1$, show that
$$2 \cot(\alpha - \beta) = x^2$$

93. Geometry: Angle Between Two Lines Let L_1 and L_2 denote two nonvertical intersecting lines, and let θ denote the acute angle between L_1 and L_2 (see the figure). Show that

$$\tan \theta = \frac{m_2 - m_1}{1 + m_1 m_2}$$

where m_1 and m_2 are the slopes of L_1 and L_2, respectively. [**Hint:** Use the facts that $\tan \theta_1 = m_1$ and $\tan \theta_2 = m_2$.]

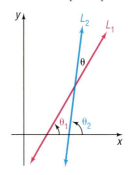

94. If $\alpha + \beta + \gamma = 180°$ and
$$\cot \theta = \cot \alpha + \cot \beta + \cot \gamma, \quad 0 < \theta < 90°$$
show that
$$\sin^3 \theta = \sin(\alpha - \theta) \sin(\beta - \theta) \sin(\gamma - \theta)$$

Discussion and Writing

95. Discuss the following derivation:

$$\tan\left(\theta + \frac{\pi}{2}\right) = \frac{\tan \theta + \tan\dfrac{\pi}{2}}{1 - \tan \theta \tan\dfrac{\pi}{2}} = \frac{\dfrac{\tan \theta}{\tan\dfrac{\pi}{2}} + 1}{\dfrac{1}{\tan\dfrac{\pi}{2}} - \tan \theta} = \frac{0 + 1}{0 - \tan \theta} = \frac{1}{-\tan \theta} = -\cot \theta$$

Can you justify each step?

'Are You Prepared?' Answers

1. 5   **2.** $-\dfrac{3}{5}$ **3.** (a) $\dfrac{\sqrt{2}}{4}$ (b) $\dfrac{1}{2}$

6.5 Double-angle and Half-angle Formulas

OBJECTIVES **1** Use Double-angle Formulas to Find Exact Values
2 Use Double-angle Formulas to Establish Identities
3 Use Half-angle Formulas to Find Exact Values

In this section we derive formulas for $\sin(2\theta)$, $\cos(2\theta)$, $\sin\left(\dfrac{1}{2}\theta\right)$, and $\cos\left(\dfrac{1}{2}\theta\right)$ in terms of $\sin \theta$ and $\cos \theta$. They are derived using the sum formulas.

In the sum formulas for $\sin(\alpha + \beta)$ and $\cos(\alpha + \beta)$, let $\alpha = \beta = \theta$. Then

$$\sin(\alpha + \beta) = \sin \alpha \cos \beta + \cos \alpha \sin \beta$$
$$\sin(\theta + \theta) = \sin \theta \cos \theta + \cos \theta \sin \theta$$
$$\sin(2\theta) = 2 \sin \theta \cos \theta$$

and

$$\cos(\alpha + \beta) = \cos \alpha \cos \beta - \sin \alpha \sin \beta$$
$$\cos(\theta + \theta) = \cos \theta \cos \theta - \sin \theta \sin \theta$$
$$\cos(2\theta) = \cos^2 \theta - \sin^2 \theta$$

An application of the Pythagorean Identity $\sin^2 \theta + \cos^2 \theta = 1$ results in two other ways to express $\cos(2\theta)$.

$$\cos(2\theta) = \cos^2 \theta - \sin^2 \theta = (1 - \sin^2 \theta) - \sin^2 \theta = 1 - 2 \sin^2 \theta$$

and

$$\cos(2\theta) = \cos^2 \theta - \sin^2 \theta = \cos^2 \theta - (1 - \cos^2 \theta) = 2 \cos^2 \theta - 1$$

We have established the following **Double-angle Formulas:**

Theorem

Double-angle Formulas

$$\sin(2\theta) = 2 \sin \theta \cos \theta \qquad \text{(1)}$$
$$\cos(2\theta) = \cos^2 \theta - \sin^2 \theta \qquad \text{(2)}$$
$$\cos(2\theta) = 1 - 2 \sin^2 \theta \qquad \text{(3)}$$
$$\cos(2\theta) = 2 \cos^2 \theta - 1 \qquad \text{(4)}$$

1 Use Double-angle Formulas to Find Exact Values

EXAMPLE 1 **Finding Exact Values Using the Double-angle Formula**

If $\sin \theta = \dfrac{3}{5}$, $\dfrac{\pi}{2} < \theta < \pi$, find the exact value of:

(a) $\sin(2\theta)$ (b) $\cos(2\theta)$

Solution

(a) Because $\sin(2\theta) = 2 \sin \theta \cos \theta$ and we already know that $\sin \theta = \dfrac{3}{5}$, we only need to find $\cos \theta$. Since $\sin \theta = \dfrac{3}{5} = \dfrac{y}{r}$, $\dfrac{\pi}{2} < \theta < \pi$, we let $y = 3$ and $r = 5$ and place θ in quadrant II. The point $P = (x, y) = (x, 3)$, $x < 0$, is on a circle of radius 5, namely, $x^2 + y^2 = 25$. See Figure 27. Then

$$x^2 + y^2 = 25,$$
$$x^2 + 9 = 25 \qquad \textcolor{blue}{y = 3}$$
$$x^2 = 25 - 9 = 16$$
$$x = -4 \qquad \textcolor{blue}{x < 0}$$

We find that $\cos \theta = \dfrac{x}{r} = \dfrac{-4}{5}$. Now we use formula (1) to obtain

$$\sin(2\theta) = 2 \sin \theta \cos \theta = 2\left(\dfrac{3}{5}\right)\left(-\dfrac{4}{5}\right) = -\dfrac{24}{25}$$

Figure 27

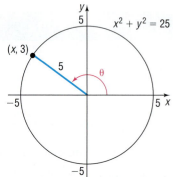

(b) Because we are given $\sin\theta = \dfrac{3}{5}$, it is easiest to use formula (4b) to get $\cos(2\theta)$.

$$\cos(2\theta) = 1 - 2\sin^2\theta = 1 - 2\left(\frac{9}{25}\right) = 1 - \frac{18}{25} = \frac{7}{25}$$ ◀

WARNING In finding $\cos(2\theta)$ in Example 1(b), we chose to use a version of the Double-angle Formula, formula (3). Note that we are unable to use the Pythagorean Identity $\cos(2\theta) = \pm\sqrt{1 - \sin^2(2\theta)}$, with $\sin(2\theta) = -\dfrac{24}{25}$, because we have no way of knowing which sign to choose. ■

NOW WORK PROBLEMS 7(a) AND (b).

2 Use Double-angle Formulas to Establish Identities

EXAMPLE 2 **Establishing Identities**

(a) Develop a formula for $\tan(2\theta)$ in terms of $\tan\theta$.
(b) Develop a formula for $\sin(3\theta)$ in terms of $\sin\theta$ and $\cos\theta$.

Solution (a) In the sum formula for $\tan(\alpha + \beta)$, let $\alpha = \beta = \theta$. Then

$$\tan(\alpha + \beta) = \frac{\tan\alpha + \tan\beta}{1 - \tan\alpha\tan\beta}$$

$$\tan(\theta + \theta) = \frac{\tan\theta + \tan\theta}{1 - \tan\theta\tan\theta}$$

$$\tan(2\theta) = \frac{2\tan\theta}{1 - \tan^2\theta} \qquad (5)$$

(b) To get a formula for $\sin(3\theta)$, we use the sum formula and write 3θ as $2\theta + \theta$.

$$\sin(3\theta) = \sin(2\theta + \theta) = \sin(2\theta)\cos\theta + \cos(2\theta)\sin\theta$$

Now use the Double-angle Formulas to get

$$\sin(3\theta) = (2\sin\theta\cos\theta)(\cos\theta) + (\cos^2\theta - \sin^2\theta)(\sin\theta)$$
$$= 2\sin\theta\cos^2\theta + \sin\theta\cos^2\theta - \sin^3\theta$$
$$= 3\sin\theta\cos^2\theta - \sin^3\theta$$ ◀

The formula obtained in Example 2(b) can also be written as

$$\sin(3\theta) = 3\sin\theta\cos^2\theta - \sin^3\theta = 3\sin\theta(1 - \sin^2\theta) - \sin^3\theta$$
$$= 3\sin\theta - 4\sin^3\theta$$

That is, $\sin(3\theta)$ is a third-degree polynomial in the variable $\sin\theta$. In fact, $\sin(n\theta)$, n a positive odd integer, can always be written as a polynomial of degree n in the variable $\sin\theta$.[*]

NOW WORK PROBLEM 53.

By rearranging the Double-angle Formulas (3) and (4), we obtain other formulas that we will use later in this section.

[*]Due to the work done by P.L. Chebyshëv, these polynomials are sometimes called *Chebyshëv polynomials*.

We begin with formula (3) and proceed to solve for $\sin^2\theta$.

$$\cos(2\theta) = 1 - 2\sin^2\theta$$
$$2\sin^2\theta = 1 - \cos(2\theta)$$

$$\sin^2\theta = \frac{1 - \cos(2\theta)}{2} \tag{6}$$

Similarly, using formula (4), we proceed to solve for $\cos^2\theta$.

$$\cos(2\theta) = 2\cos^2\theta - 1$$
$$2\cos^2\theta = 1 + \cos(2\theta)$$

$$\cos^2\theta = \frac{1 + \cos(2\theta)}{2} \tag{7}$$

Formulas (6) and (7) can be used to develop a formula for $\tan^2\theta$.

$$\tan^2\theta = \frac{\sin^2\theta}{\cos^2\theta} = \frac{\dfrac{1 - \cos(2\theta)}{2}}{\dfrac{1 + \cos(2\theta)}{2}}$$

$$\tan^2\theta = \frac{1 - \cos(2\theta)}{1 + \cos(2\theta)} \tag{8}$$

Formulas (6) through (8) do not have to be memorized since their derivations are so straightforward.

Formulas (6) and (7) are important in calculus. The next example illustrates a problem that arises in calculus requiring the use of formula (7).

EXAMPLE 3   **Establishing an Identity**

Write an equivalent expression for $\cos^4\theta$ that does not involve any powers of sine or cosine greater than 1.

Solution   The idea here is to apply formula (7) twice.

$$\cos^4\theta = (\cos^2\theta)^2 = \left(\frac{1 + \cos(2\theta)}{2}\right)^2 \qquad \text{Formula (7)}$$

$$= \frac{1}{4}[1 + 2\cos(2\theta) + \cos^2(2\theta)]$$

$$= \frac{1}{4} + \frac{1}{2}\cos(2\theta) + \frac{1}{4}\cos^2(2\theta)$$

$$= \frac{1}{4} + \frac{1}{2}\cos(2\theta) + \frac{1}{4}\left\{\frac{1 + \cos[2(2\theta)]}{2}\right\} \qquad \text{Formula (7)}$$

$$= \frac{1}{4} + \frac{1}{2}\cos(2\theta) + \frac{1}{8}[1 + \cos(4\theta)]$$

$$= \frac{3}{8} + \frac{1}{2}\cos(2\theta) + \frac{1}{8}\cos(4\theta) \qquad \blacktriangleleft$$

 NOW WORK PROBLEM 29.

Identities, such as the Double-angle Formulas, can sometimes be used to rewrite expressions in a more suitable form. Let's look at an example.

| EXAMPLE 4 | **Projectile Motion** |

Figure 28

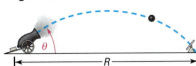

An object is propelled upward at an angle θ to the horizontal with an initial velocity of v_0 feet per second. See Figure 28. If air resistance is ignored, the **range** R, the horizontal distance that the object travels, is given by

$$R = \frac{1}{16} v_0^2 \sin \theta \cos \theta$$

(a) Show that $R = \frac{1}{32} v_0^2 \sin(2\theta)$.

(b) Find the angle θ for which R is a maximum.

Solution   (a) We rewrite the given expression for the range using the Double-angle Formula $\sin(2\theta) = 2 \sin \theta \cos \theta$. Then

$$R = \frac{1}{16} v_0^2 \sin \theta \cos \theta = \frac{1}{16} v_0^2 \frac{2 \sin \theta \cos \theta}{2} = \frac{1}{32} v_0^2 \sin(2\theta)$$

(b) In this form, the largest value for the range R can be found. For a fixed initial speed v_0, the angle θ of inclination to the horizontal determines the value of R. Since the largest value of a sine function is 1, occurring when the argument 2θ is $90°$, it follows that for maximum R we must have

$$2\theta = 90°$$
$$\theta = 45°$$

An inclination to the horizontal of $45°$ results in the maximum range. ◀

3 Use Half-angle Formulas to Find Exact Values

Another important use of formulas (6) through (8) is to prove the *Half-angle Formulas*. In formulas (6) through (8), let $\theta = \dfrac{\alpha}{2}$. Then

$$\sin^2 \frac{\alpha}{2} = \frac{1 - \cos \alpha}{2} \qquad \cos^2 \frac{\alpha}{2} = \frac{1 + \cos \alpha}{2} \qquad \tan^2 \frac{\alpha}{2} = \frac{1 - \cos \alpha}{1 + \cos \alpha} \quad \textbf{(9)}$$

 The identities in box (9) will prove useful in integral calculus.

If we solve for the trigonometric functions on the left sides of equations (9), we obtain the Half-angle Formulas.

Theorem   **Half-angle Formulas**

$$\sin \frac{\alpha}{2} = \pm \sqrt{\frac{1 - \cos \alpha}{2}} \qquad \textbf{(10a)}$$

$$\cos \frac{\alpha}{2} = \pm \sqrt{\frac{1 + \cos \alpha}{2}} \qquad \textbf{(10b)}$$

$$\tan \frac{\alpha}{2} = \pm \sqrt{\frac{1 - \cos \alpha}{1 + \cos \alpha}} \qquad \textbf{(10c)}$$

where the $+$ or $-$ sign is determined by the quadrant of the angle $\dfrac{\alpha}{2}$.

EXAMPLE 5 **Finding Exact Values Using Half-angle Formulas**

Use a Half-angle Formula find the exact value of:

(a) $\cos 15°$ (b) $\sin(-15°)$

Solution (a) Because $15° = \dfrac{30°}{2}$, we can use the Half-angle Formula for $\cos \dfrac{\alpha}{2}$ with $\alpha = 30°$.

Also, because $15°$ is in quadrant I, $\cos 15° > 0$, we choose the $+$ sign in using formula (10b):

$$\cos 15° = \cos \frac{30°}{2} = \sqrt{\frac{1 + \cos 30°}{2}}$$

$$= \sqrt{\frac{1 + \sqrt{3}/2}{2}} = \sqrt{\frac{2 + \sqrt{3}}{4}} = \frac{\sqrt{2 + \sqrt{3}}}{2}$$

(b) We use the fact that $\sin(-15°) = -\sin 15°$ and then apply formula (10a).

$$\sin(-15°) = -\sin \frac{30°}{2} = -\sqrt{\frac{1 - \cos 30°}{2}}$$

$$= -\sqrt{\frac{1 - \sqrt{3}/2}{2}} = -\sqrt{\frac{2 - \sqrt{3}}{4}} = -\frac{\sqrt{2 - \sqrt{3}}}{2} \quad \blacktriangleleft$$

It is interesting to compare the answer found in Example 5(a) with the answer to Example 2 of Section 6.4. There we calculated

$$\cos \frac{\pi}{12} = \cos 15° = \frac{1}{4}\left(\sqrt{6} + \sqrt{2}\right)$$

Based on this and the result of Example 5(a), we conclude that

$$\frac{1}{4}\left(\sqrt{6} + \sqrt{2}\right) \quad \text{and} \quad \frac{\sqrt{2 + \sqrt{3}}}{2}$$

are equal. (Since each expression is positive, you can verify this equality by squaring each expression.) Two very different looking, yet correct, answers can be obtained, depending on the approach taken to solve a problem.

✏ **NOW WORK PROBLEM 19.**

EXAMPLE 6 **Finding Exact Values Using Half-angle Formulas**

If $\cos \alpha = -\dfrac{3}{5}$, $\pi < \alpha < \dfrac{3\pi}{2}$, find the exact value of:

(a) $\sin \dfrac{\alpha}{2}$ (b) $\cos \dfrac{\alpha}{2}$ (c) $\tan \dfrac{\alpha}{2}$

Solution First, we observe that if $\pi < \alpha < \dfrac{3\pi}{2}$ then $\dfrac{\pi}{2} < \dfrac{\alpha}{2} < \dfrac{3\pi}{4}$. As a result, $\dfrac{\alpha}{2}$ lies in quadrant II.

(a) Because $\dfrac{\alpha}{2}$ lies in quadrant II, $\sin \dfrac{\alpha}{2} > 0$, so we use the $+$ sign in formula (10a) to get

$$\sin \frac{\alpha}{2} = \sqrt{\frac{1 - \cos \alpha}{2}} = \sqrt{\frac{1 - \left(-\dfrac{3}{5}\right)}{2}}$$

$$= \sqrt{\frac{\dfrac{8}{5}}{2}} = \sqrt{\frac{4}{5}} = \frac{2}{\sqrt{5}} = \frac{2\sqrt{5}}{5}$$

(b) Because $\dfrac{\alpha}{2}$ lies in quadrant II, $\cos \dfrac{\alpha}{2} < 0$, so we use the $-$ sign in formula (10b) to get

$$\cos \frac{\alpha}{2} = -\sqrt{\frac{1 + \cos \alpha}{2}} = -\sqrt{\frac{1 + \left(-\dfrac{3}{5}\right)}{2}}$$

$$= -\sqrt{\frac{\dfrac{2}{5}}{2}} = -\frac{1}{\sqrt{5}} = -\frac{\sqrt{5}}{5}$$

(c) Because $\dfrac{\alpha}{2}$ lies in quadrant II, $\tan \dfrac{\alpha}{2} < 0$, so we use the $-$ sign in formula (10c) to get

$$\tan \frac{\alpha}{2} = -\sqrt{\frac{1 - \cos \alpha}{1 + \cos \alpha}} = -\sqrt{\frac{1 - \left(-\dfrac{3}{5}\right)}{1 + \left(-\dfrac{3}{5}\right)}} = -\sqrt{\frac{\dfrac{8}{5}}{\dfrac{2}{5}}} = -2 \qquad \blacktriangleleft$$

Another way to solve Example 6(c) is to use the solutions found in parts (a) and (b).

$$\tan \frac{\alpha}{2} = \frac{\sin \dfrac{\alpha}{2}}{\cos \dfrac{\alpha}{2}} = \frac{\dfrac{2\sqrt{5}}{5}}{-\dfrac{\sqrt{5}}{5}} = -2$$

NOW WORK PROBLEMS 7(c) AND (d).

There is a formula for $\tan \dfrac{\alpha}{2}$ that does not contain $+$ and $-$ signs, making it more useful than Formula 10(c). To derive it, we use the formulas

$$1 - \cos \alpha = 2 \sin^2 \frac{\alpha}{2} \qquad \textit{Formula (9)}$$

and

$$\sin \alpha = \sin\left[2\left(\frac{\alpha}{2}\right)\right] = 2 \sin \frac{\alpha}{2} \cos \frac{\alpha}{2} \qquad \textit{Double-angle Formula}$$

Then

$$\frac{1 - \cos \alpha}{\sin \alpha} = \frac{2 \sin^2 \dfrac{\alpha}{2}}{2 \sin \dfrac{\alpha}{2} \cos \dfrac{\alpha}{2}} = \frac{\sin \dfrac{\alpha}{2}}{\cos \dfrac{\alpha}{2}} = \tan \frac{\alpha}{2}$$

Since it also can be shown that

$$\frac{1 - \cos \alpha}{\sin \alpha} = \frac{\sin \alpha}{1 + \cos \alpha}$$

we have the following two Half-angle Formulas:

Half-angle Formulas for $\tan \dfrac{\alpha}{2}$

$$\tan \frac{\alpha}{2} = \frac{1 - \cos \alpha}{\sin \alpha} = \frac{\sin \alpha}{1 + \cos \alpha} \tag{11}$$

With this formula, the solution to Example 6(c) can be obtained as follows

$$\cos \alpha = -\frac{3}{5}, \pi < \alpha < \frac{3\pi}{2}$$

$$\sin \alpha = -\sqrt{1 - \cos^2 \alpha} = -\sqrt{1 - \frac{9}{25}} = -\sqrt{\frac{16}{25}} = -\frac{4}{5}$$

Then, by equation (11),

$$\tan \frac{\alpha}{2} = \frac{1 - \cos \alpha}{\sin \alpha} = \frac{1 - \left(-\dfrac{3}{5}\right)}{-\dfrac{4}{5}} = \frac{\dfrac{8}{5}}{-\dfrac{4}{5}} = -2$$

6.5 Assess Your Understanding

Concepts and Vocabulary

1. $\cos(2\theta) = \cos^2 \theta - \underline{\quad} = \underline{\quad} - 1 = 1 - \underline{\quad}$.

2. $\sin^2 \dfrac{\theta}{2} = \dfrac{\underline{\quad}}{2}$.

3. $\tan \dfrac{\theta}{2} = \dfrac{1 - \cos \theta}{\underline{\quad}}$.

4. *True or False:* $\cos(2\theta)$ has three equivalent forms:
$$\cos^2 \theta - \sin^2 \theta, \quad 1 - 2 \sin^2 \theta, \quad 2 \cos^2 \theta - 1$$

5. *True or False:* $\sin(2\theta)$ has two equivalent forms:
$$2 \sin \theta \cos \theta \quad \text{and} \quad \sin^2 \theta - \cos^2 \theta$$

6. *True or False:* $\tan(2\theta) + \tan(2\theta) = \tan(4\theta)$

Skill Building

In Problems 7–18, use the information given about the angle θ, $0 \le \theta \le 2\pi$, to find the exact value of

(a) $\sin(2\theta)$ (b) $\cos(2\theta)$ (c) $\sin \dfrac{\theta}{2}$ (d) $\cos \dfrac{\theta}{2}$

7. $\sin \theta = \dfrac{3}{5}, \quad 0 < \theta < \dfrac{\pi}{2}$

8. $\cos \theta = \dfrac{3}{5}, \quad 0 < \theta < \dfrac{\pi}{2}$

9. $\tan \theta = \dfrac{4}{3}, \quad \pi < \theta < \dfrac{3\pi}{2}$

10. $\tan \theta = \dfrac{1}{2}, \quad \pi < \theta < \dfrac{3\pi}{2}$

11. $\cos \theta = -\dfrac{\sqrt{6}}{3}, \quad \dfrac{\pi}{2} < \theta < \pi$

12. $\sin \theta = -\dfrac{\sqrt{3}}{3}, \quad \dfrac{3\pi}{2} < \theta < 2\pi$

13. $\sec \theta = 3, \quad \sin \theta > 0$ **14.** $\csc \theta = -\sqrt{5}, \quad \cos \theta < 0$ **15.** $\cot \theta = -2, \quad \sec \theta < 0$

16. $\sec \theta = 2, \quad \csc \theta < 0$ **17.** $\tan \theta = -3, \quad \sin \theta < 0$ **18.** $\cot \theta = 3, \quad \cos \theta < 0$

In Problems 19–28, use the Half-angle Formulas to find the exact value of each trigonometric function.

19. $\sin 22.5°$ **20.** $\cos 22.5°$ **21.** $\tan \dfrac{7\pi}{8}$ **22.** $\tan \dfrac{9\pi}{8}$ **23.** $\cos 165°$

24. $\sin 195°$ **25.** $\sec \dfrac{15\pi}{8}$ **26.** $\csc \dfrac{7\pi}{8}$ **27.** $\sin\left(-\dfrac{\pi}{8}\right)$ **28.** $\cos\left(-\dfrac{3\pi}{8}\right)$

29. Show that $\sin^4 \theta = \dfrac{3}{8} - \dfrac{1}{2}\cos(2\theta) + \dfrac{1}{8}\cos(4\theta)$.

30. Show that $\sin(4\theta) = (\cos \theta)(4\sin \theta - 8\sin^3 \theta)$.

31. Develop a formula for $\cos(3\theta)$ as a third-degree polynomial in the variable $\cos \theta$.

32. Develop a formula for $\cos(4\theta)$ as a fourth-degree polynomial in the variable $\cos \theta$.

33. Find an expression for $\sin(5\theta)$ as a fifth-degree polynomial in the variable $\sin \theta$.

34. Find an expression for $\cos(5\theta)$ as a fifth-degree polynomial in the variable $\cos \theta$.

In Problems 35–56, establish each identity.

35. $\cos^4 \theta - \sin^4 \theta = \cos(2\theta)$

36. $\dfrac{\cot \theta - \tan \theta}{\cot \theta + \tan \theta} = \cos(2\theta)$

37. $\cot(2\theta) = \dfrac{\cot^2 \theta - 1}{2\cot \theta}$

38. $\cot(2\theta) = \dfrac{1}{2}(\cot \theta - \tan \theta)$

39. $\sec(2\theta) = \dfrac{\sec^2 \theta}{2 - \sec^2 \theta}$

40. $\csc(2\theta) = \dfrac{1}{2}\sec \theta \csc \theta$

41. $\cos^2(2\theta) - \sin^2(2\theta) = \cos(4\theta)$

42. $(4\sin \theta \cos \theta)(1 - 2\sin^2 \theta) = \sin(4\theta)$

43. $\dfrac{\cos(2\theta)}{1 + \sin(2\theta)} = \dfrac{\cot \theta - 1}{\cot \theta + 1}$

44. $\sin^2 \theta \cos^2 \theta = \dfrac{1}{8}[1 - \cos(4\theta)]$

45. $\sec^2 \dfrac{\theta}{2} = \dfrac{2}{1 + \cos \theta}$

46. $\csc^2 \dfrac{\theta}{2} = \dfrac{2}{1 - \cos \theta}$

47. $\cot^2 \dfrac{\theta}{2} = \dfrac{\sec \theta + 1}{\sec \theta - 1}$

48. $\tan \dfrac{\theta}{2} = \csc \theta - \cot \theta$

49. $\cos \theta = \dfrac{1 - \tan^2 \dfrac{\theta}{2}}{1 + \tan^2 \dfrac{\theta}{2}}$

50. $1 - \dfrac{1}{2}\sin(2\theta) = \dfrac{\sin^3 \theta + \cos^3 \theta}{\sin \theta + \cos \theta}$

51. $\dfrac{\sin(3\theta)}{\sin \theta} - \dfrac{\cos(3\theta)}{\cos \theta} = 2$

52. $\dfrac{\cos \theta + \sin \theta}{\cos \theta - \sin \theta} - \dfrac{\cos \theta - \sin \theta}{\cos \theta + \sin \theta} = 2\tan(2\theta)$

53. $\tan(3\theta) = \dfrac{3\tan \theta - \tan^3 \theta}{1 - 3\tan^2 \theta}$

54. $\tan \theta + \tan(\theta + 120°) + \tan(\theta + 240°) = 3\tan(3\theta)$

55. $\ln|\sin \theta| = \dfrac{1}{2}(\ln|1 - \cos(2\theta)| - \ln 2)$

56. $\ln|\cos \theta| = \dfrac{1}{2}(\ln|1 + \cos(2\theta)| - \ln 2)$

In Problems 57–68, find the exact value of each expression.

57. $\sin\left(2\sin^{-1}\dfrac{1}{2}\right)$ **58.** $\sin\left[2\sin^{-1}\dfrac{\sqrt{3}}{2}\right]$ **59.** $\cos\left(2\sin^{-1}\dfrac{3}{5}\right)$ **60.** $\cos\left(2\cos^{-1}\dfrac{4}{5}\right)$

61. $\tan\left[2\cos^{-1}\left(-\dfrac{3}{5}\right)\right]$ **62.** $\tan\left(2\tan^{-1}\dfrac{3}{4}\right)$ **63.** $\sin\left(2\cos^{-1}\dfrac{4}{5}\right)$ **64.** $\cos\left[2\tan^{-1}\left(-\dfrac{4}{3}\right)\right]$

65. $\sin^2\left(\dfrac{1}{2}\cos^{-1}\dfrac{3}{5}\right)$ **66.** $\cos^2\left(\dfrac{1}{2}\sin^{-1}\dfrac{3}{5}\right)$ **67.** $\sec\left(2\tan^{-1}\dfrac{3}{4}\right)$ **68.** $\csc\left[2\sin^{-1}\left(-\dfrac{3}{5}\right)\right]$

Applications and Extensions

69. If $x = 2\tan \theta$, express $\sin(2\theta)$ as a function of x.

70. If $x = 2\tan \theta$, express $\cos(2\theta)$ as a function of x.

71. Find the value of the number C:

$$\frac{1}{2}\sin^2 x + C = -\frac{1}{4}\cos(2x)$$

72. Find the value of the number C:

$$\frac{1}{2}\cos^2 x + C = \frac{1}{4}\cos(2x)$$

73. If $z = \tan\dfrac{\alpha}{2}$, show that $\sin\alpha = \dfrac{2z}{1+z^2}$.

74. If $z = \tan\dfrac{\alpha}{2}$, show that $\cos\alpha = \dfrac{1-z^2}{1+z^2}$.

---

75. Area of an Isosceles Triangle Show that the area A of an isosceles triangle whose equal sides are of length s and θ is the angle between them is

$$\frac{1}{2}s^2\sin\theta$$

[**Hint:** See the illustration. The height h bisects the angle θ and is the perpendicular bisector of the base.]

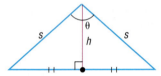

76. Geometry A rectangle is inscribed in a semicircle of radius 1. See the illustration.

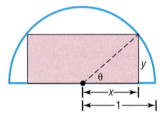

(a) Express the area A of the rectangle as a function of the angle θ shown in the illustration.
(b) Show that $A = \sin(2\theta)$.
(c) Find the angle θ that results in the largest area A.
(d) Find the dimensions of this largest rectangle.

77. Graph $f(x) = \sin^2 x = \dfrac{1-\cos(2x)}{2}$ for $0 \le x \le 2\pi$ by using transformations.

78. Repeat Problem 77 for $g(x) = \cos^2 x$.

79. Use the fact that

$$\cos\frac{\pi}{12} = \frac{1}{4}\left(\sqrt{6} + \sqrt{2}\right)$$

to find $\sin\dfrac{\pi}{24}$ and $\cos\dfrac{\pi}{24}$.

80. Show that

$$\cos\frac{\pi}{8} = \frac{\sqrt{2+\sqrt{2}}}{2}$$

and use it to find $\sin\dfrac{\pi}{16}$ and $\cos\dfrac{\pi}{16}$.

81. Show that

$$\sin^3\theta + \sin^3(\theta + 120°) + \sin^3(\theta + 240°) = -\frac{3}{4}\sin(3\theta)$$

82. If $\tan\theta = a\tan\dfrac{\theta}{3}$, express $\tan\dfrac{\theta}{3}$ in terms of a.

83. Projectile Motion An object is propelled upward at an angle θ, $45° < \theta < 90°$, to the horizontal with an initial velocity of v_0 feet per second from the base of a plane that makes an angle of $45°$ with the horizontal. See the illustration. If air resistance is ignored, the distance R that it travels up the inclined plane is given by

$$R = \frac{v_0^2\sqrt{2}}{16}\cos\theta(\sin\theta - \cos\theta)$$

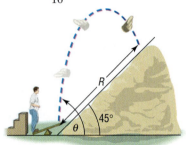

(a) Show that

$$R = \frac{v_0^2\sqrt{2}}{32}[\sin(2\theta) - \cos(2\theta) - 1]$$

(b) Graph $R = R(\theta)$. (Use $v_0 = 32$ feet per second.)
(c) What value of θ makes R the largest? (Use $v_0 = 32$ feet per second.)

84. Sawtooth Curve An oscilloscope often displays a sawtooth curve. This curve can be approximated by sinusoidal curves of varying periods and amplitudes. A first approximation to the sawtooth curve is given by

$$y = \frac{1}{2}\sin(2\pi x) + \frac{1}{4}\sin(4\pi x)$$

Show that $y = \sin(2\pi x)\cos^2(\pi x)$.

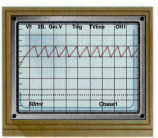

Discussion and Writing

85. Go to the library and research Chebyshëv polynomials. Write a report on your findings.

6.6 Product-to-Sum and Sum-to-Product Formulas

OBJECTIVES   1 Express Products as Sums
2 Express Sums as Products

1 Express Products as Sums

Sum and difference formulas can be used to derive formulas for writing the products of sines and/or cosines as sums or differences. These identities are usually called the **Product-to-Sum Formulas**.

Theorem

Product-to-Sum Formulas

$$\sin \alpha \sin \beta = \frac{1}{2}[\cos(\alpha - \beta) - \cos(\alpha + \beta)] \tag{1}$$

$$\cos \alpha \cos \beta = \frac{1}{2}[\cos(\alpha - \beta) + \cos(\alpha + \beta)] \tag{2}$$

$$\sin \alpha \cos \beta = \frac{1}{2}[\sin(\alpha + \beta) + \sin(\alpha - \beta)] \tag{3}$$

These formulas do not have to be memorized. Instead, you should remember how they are derived. Then, when you want to use them, either look them up or derive them, as needed.

To derive formulas (1) and (2), write down the sum and difference formulas for the cosine:

$$\cos(\alpha - \beta) = \cos \alpha \cos \beta + \sin \alpha \sin \beta \tag{4}$$

$$\cos(\alpha + \beta) = \cos \alpha \cos \beta - \sin \alpha \sin \beta \tag{5}$$

Subtract equation (5) from equation (4) to get

$$\cos(\alpha - \beta) - \cos(\alpha + \beta) = 2 \sin \alpha \sin \beta$$

from which

$$\sin \alpha \sin \beta = \frac{1}{2}[\cos(\alpha - \beta) - \cos(\alpha + \beta)]$$

Now add equations (4) and (5) to get

$$\cos(\alpha - \beta) + \cos(\alpha + \beta) = 2 \cos \alpha \cos \beta$$

from which

$$\cos \alpha \cos \beta = \frac{1}{2}[\cos(\alpha - \beta) + \cos(\alpha + \beta)]$$

To derive Product-to-Sum Formula (3), use the sum and difference formulas for sine in a similar way. (You are asked to do this in Problem 41.)

EXAMPLE 1 **Expressing Products as Sums**

Express each of the following products as a sum containing only sines or cosines.

(a) $\sin(6\theta) \sin(4\theta)$ (b) $\cos(3\theta) \cos\theta$ (c) $\sin(3\theta) \cos(5\theta)$

Solution (a) We use formula (1) to get

$$\sin(6\theta) \sin(4\theta) = \frac{1}{2}[\cos(6\theta - 4\theta) - \cos(6\theta + 4\theta)]$$

$$= \frac{1}{2}[\cos(2\theta) - \cos(10\theta)]$$

(b) We use formula (2) to get

$$\cos(3\theta) \cos\theta = \frac{1}{2}[\cos(3\theta - \theta) + \cos(3\theta + \theta)]$$

$$= \frac{1}{2}[\cos(2\theta) + \cos(4\theta)]$$

(c) We use formula (3) to get

$$\sin(3\theta) \cos(5\theta) = \frac{1}{2}[\sin(3\theta + 5\theta) + \sin(3\theta - 5\theta)]$$

$$= \frac{1}{2}[\sin(8\theta) + \sin(-2\theta)] = \frac{1}{2}[\sin(8\theta) - \sin(2\theta)]$$ ◀

━ **NOW WORK PROBLEM 1.**

2 **Express Sums as Products**

The **Sum-to-Product Formulas** are given next.

Theorem **Sum-to-Product Formulas**

$$\sin\alpha + \sin\beta = 2 \sin\frac{\alpha + \beta}{2} \cos\frac{\alpha - \beta}{2} \tag{6}$$

$$\sin\alpha - \sin\beta = 2 \sin\frac{\alpha - \beta}{2} \cos\frac{\alpha + \beta}{2} \tag{7}$$

$$\cos\alpha + \cos\beta = 2 \cos\frac{\alpha + \beta}{2} \cos\frac{\alpha - \beta}{2} \tag{8}$$

$$\cos\alpha - \cos\beta = -2 \sin\frac{\alpha + \beta}{2} \sin\frac{\alpha - \beta}{2} \tag{9}$$

We will derive formula (6) and leave the derivations of formulas (7) through (9) as exercises (see Problems 42 through 44).

Proof

$$2 \sin\frac{\alpha + \beta}{2} \cos\frac{\alpha - \beta}{2} = 2 \cdot \frac{1}{2}\left[\sin\left(\frac{\alpha + \beta}{2} + \frac{\alpha - \beta}{2}\right) + \sin\left(\frac{\alpha + \beta}{2} - \frac{\alpha - \beta}{2}\right)\right]$$

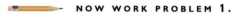
↑
Product-to-Sum Formula (3)

$$= \sin\frac{2\alpha}{2} + \sin\frac{2\beta}{2} = \sin\alpha + \sin\beta$$ ∎

| EXAMPLE 2 | **Expressing Sums (or Differences) as a Product** |

Express each sum or difference as a product of sines and/or cosines.

(a) $\sin(5\theta) - \sin(3\theta)$ (b) $\cos(3\theta) + \cos(2\theta)$

Solution   (a) We use formula (7) to get

$$\sin(5\theta) - \sin(3\theta) = 2 \sin \frac{5\theta - 3\theta}{2} \cos \frac{5\theta + 3\theta}{2}$$

$$= 2 \sin \theta \cos(4\theta)$$

(b) $\cos(3\theta) + \cos(2\theta) = 2 \cos \dfrac{3\theta + 2\theta}{2} \cos \dfrac{3\theta - 2\theta}{2}$ Formula (8)

$$= 2 \cos \frac{5\theta}{2} \cos \frac{\theta}{2}$$

◄

NOW WORK PROBLEM 11.

6.6 Assess Your Understanding

Skill Building

In Problems 1–10, express each product as a sum containing only sines or cosines.

1. $\sin(4\theta) \sin(2\theta)$ 2. $\cos(4\theta) \cos(2\theta)$ 3. $\sin(4\theta) \cos(2\theta)$ 4. $\sin(3\theta) \sin(5\theta)$ 5. $\cos(3\theta) \cos(5\theta)$

6. $\sin(4\theta) \cos(6\theta)$ 7. $\sin \theta \sin(2\theta)$ 8. $\cos(3\theta) \cos(4\theta)$ 9. $\sin \dfrac{3\theta}{2} \cos \dfrac{\theta}{2}$ 10. $\sin \dfrac{\theta}{2} \cos \dfrac{5\theta}{2}$

In Problems 11–18, express each sum or difference as a product of sines and/or cosines.

11. $\sin(4\theta) - \sin(2\theta)$ 12. $\sin(4\theta) + \sin(2\theta)$ 13. $\cos(2\theta) + \cos(4\theta)$ 14. $\cos(5\theta) - \cos(3\theta)$

15. $\sin \theta + \sin(3\theta)$ 16. $\cos \theta + \cos(3\theta)$ 17. $\cos \dfrac{\theta}{2} - \cos \dfrac{3\theta}{2}$ 18. $\sin \dfrac{\theta}{2} - \sin \dfrac{3\theta}{2}$

In Problems 19–36, establish each identity.

19. $\dfrac{\sin \theta + \sin(3\theta)}{2 \sin(2\theta)} = \cos \theta$ 20. $\dfrac{\cos \theta + \cos(3\theta)}{2 \cos(2\theta)} = \cos \theta$ 21. $\dfrac{\sin(4\theta) + \sin(2\theta)}{\cos(4\theta) + \cos(2\theta)} = \tan(3\theta)$

22. $\dfrac{\cos \theta - \cos(3\theta)}{\sin(3\theta) - \sin \theta} = \tan(2\theta)$ 23. $\dfrac{\cos \theta - \cos(3\theta)}{\sin \theta + \sin(3\theta)} = \tan \theta$ 24. $\dfrac{\cos \theta - \cos(5\theta)}{\sin \theta + \sin(5\theta)} = \tan(2\theta)$

25. $\sin \theta [\sin \theta + \sin(3\theta)] = \cos \theta [\cos \theta - \cos(3\theta)]$ 26. $\sin \theta [\sin(3\theta) + \sin(5\theta)] = \cos \theta [\cos(3\theta) - \cos(5\theta)]$

27. $\dfrac{\sin(4\theta) + \sin(8\theta)}{\cos(4\theta) + \cos(8\theta)} = \tan(6\theta)$ 28. $\dfrac{\sin(4\theta) - \sin(8\theta)}{\cos(4\theta) - \cos(8\theta)} = -\cot(6\theta)$

29. $\dfrac{\sin(4\theta) + \sin(8\theta)}{\sin(4\theta) - \sin(8\theta)} = -\dfrac{\tan(6\theta)}{\tan(2\theta)}$ 30. $\dfrac{\cos(4\theta) - \cos(8\theta)}{\cos(4\theta) + \cos(8\theta)} = \tan(2\theta) \tan(6\theta)$

31. $\dfrac{\sin \alpha + \sin \beta}{\sin \alpha - \sin \beta} = \tan \dfrac{\alpha + \beta}{2} \cot \dfrac{\alpha - \beta}{2}$ 32. $\dfrac{\cos \alpha + \cos \beta}{\cos \alpha - \cos \beta} = -\cot \dfrac{\alpha + \beta}{2} \cot \dfrac{\alpha - \beta}{2}$

33. $\dfrac{\sin \alpha + \sin \beta}{\cos \alpha + \cos \beta} = \tan \dfrac{\alpha + \beta}{2}$ 34. $\dfrac{\sin \alpha - \sin \beta}{\cos \alpha - \cos \beta} = -\cot \dfrac{\alpha + \beta}{2}$

35. $1 + \cos(2\theta) + \cos(4\theta) + \cos(6\theta) = 4 \cos \theta \cos(2\theta) \cos(3\theta)$

36. $1 - \cos(2\theta) + \cos(4\theta) - \cos(6\theta) = 4 \sin \theta \cos(2\theta) \sin(3\theta)$

Applications and Extensions

37. Touch-Tone Phones On a Touch-Tone phone, each button produces a unique sound. The sound produced is the sum of two tones, given by

$$y = \sin(2\pi l t) \quad \text{and} \quad y = \sin(2\pi h t)$$

where l and h are the low and high frequencies (cycles per second) shown on the illustration. For example, if you touch 7, the low frequency is $l = 852$ cycles per second and the high frequency is $h = 1209$ cycles per second. The sound emitted by touching 7 is

$$y = \sin[2\pi(852)t] + \sin[2\pi(1209)t]$$

Touch-Tone phone

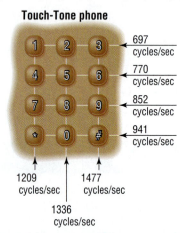

697 cycles/sec
770 cycles/sec
852 cycles/sec
941 cycles/sec

1209 cycles/sec   1477 cycles/sec
1336 cycles/sec

(a) Write this sound as a product of sines and/or cosines.
(b) Determine the maximum value of y.
(c) Graph the sound emitted by touching 7.

38. Touch-Tone Phones
(a) Write the sound emitted by touching the # key as a product of sines and/or cosines.
(b) Determine the maximum value of y.
(c) Graph the sound emitted by touching the # key.

39. If $\alpha + \beta + \gamma = \pi$, show that

$$\sin(2\alpha) + \sin(2\beta) + \sin(2\gamma) = 4 \sin \alpha \sin \beta \sin \gamma$$

40. If $\alpha + \beta + \gamma = \pi$, show that

$$\tan \alpha + \tan \beta + \tan \gamma = \tan \alpha \tan \beta \tan \gamma$$

41. Derive formula (3).

42. Derive formula (7).

43. Derive formula (8).

44. Derive formula (9).

6.7 Trigonometric Equations (I)

PREPARING FOR THIS SECTION *Before getting started, review the following:*

- Solving Equations (Appendix, Section A.5, pp. 692–703)
- Values of the Trigonometric Functions of Angles (Section 5.2, pp. 374–382)

 Now work the 'Are You Prepared?' problems on page 500.

OBJECTIVE 1 Solve Equations Involving a Single Trigonometric Function

1 Solve Equations Involving a Single Trigonometric Function

The previous four sections of this chapter were devoted to trigonometric identities, that is, equations involving trigonometric functions that are satisfied by every value in the domain of the variable. In the remaining two sections, we discuss **trigonometric equations**, that is, equations involving trigonometric functions that are satisfied only by some values of the variable (or, possibly, are not satisfied by any values of the variable). The values that satisfy the equation are called **solutions** of the equation.

EXAMPLE 1 | **Checking Whether a Given Number Is a Solution of a Trigonometric Equation**

Determine whether $\theta = \dfrac{\pi}{4}$ is a solution of the equation $\sin \theta = \dfrac{1}{2}$. Is $\theta = \dfrac{\pi}{6}$ a solution?

Solution Replace θ by $\dfrac{\pi}{4}$ in the given equation. The result is

$$\sin\frac{\pi}{4} = \frac{\sqrt{2}}{2} \neq \frac{1}{2}$$

We conclude that $\dfrac{\pi}{4}$ is not a solution.

Next replace θ by $\dfrac{\pi}{6}$ in the equation. The result is

$$\sin\frac{\pi}{6} = \frac{1}{2}$$

We conclude that $\dfrac{\pi}{6}$ is a solution of the given equation. ◀

The equation given in Example 1 has other solutions besides $\theta = \dfrac{\pi}{6}$. For example, $\theta = \dfrac{5\pi}{6}$ is also a solution, as is $\theta = \dfrac{13\pi}{6}$. (You should check this for yourself.) In fact, the equation has an infinite number of solutions due to the periodicity of the sine function. See Figure 29.

As before, our practice will be to solve equations, whenever possible, by finding exact solutions. In such cases, we will also verify the solution obtained by using a graphing utility. When traditional methods cannot be used, approximate solutions will be obtained using a graphing utility. The reader is encouraged to pay particular attention to the form of equations for which exact solutions are possible.

Unless the domain of the variable is restricted, we need to find *all* the solutions of a trigonometric equation. As the next example illustrates, finding all the solutions can be accomplished by first finding solutions over an interval whose length equals the period of the function and then adding multiples of that period to the solutions found. Let's look at some examples.

Figure 29

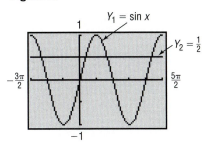

| EXAMPLE 2 | **Finding All the Solutions of a Trigonometric Equation** |

Figure 30

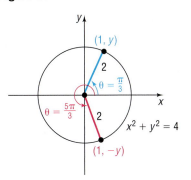

Solve the equation: $\cos\theta = \dfrac{1}{2}$

Give a general formula for all the solutions. List eight of the solutions.

Solution The period of the cosine function is 2π. In the interval $[0, 2\pi)$, there are two angles θ for which $\cos\theta = \dfrac{1}{2}$: $\theta = \dfrac{\pi}{3}$ and $\theta = \dfrac{5\pi}{3}$. See Figure 30. Because the cosine function has period 2π, all the solutions of $\cos\theta = \dfrac{1}{2}$ may be given by the general formula

$$\theta = \frac{\pi}{3} + 2k\pi \quad \text{or} \quad \theta = \frac{5\pi}{3} + 2k\pi \qquad \textcolor{blue}{k \text{ any integer}}$$

Eight of the solutions are

$$\underbrace{-\frac{5\pi}{3},}_{\textcolor{blue}{k = -1}} \quad \underbrace{-\frac{\pi}{3},}_{\textcolor{blue}{k = 0}} \quad \underbrace{\frac{\pi}{3}, \quad \frac{5\pi}{3},}_{} \quad \underbrace{\frac{7\pi}{3}, \quad \frac{11\pi}{3},}_{\textcolor{blue}{k = 1}} \quad \underbrace{\frac{13\pi}{3}, \quad \frac{17\pi}{3}}_{\textcolor{blue}{k = 2}}$$

Figure 31

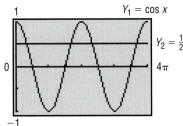

✔ **CHECK:** We can verify the solutions by graphing $Y_1 = \cos x$ and $Y_2 = \dfrac{1}{2}$ to determine where the graphs intersect. (Be sure to graph in radian mode.) See Figure 31.

The graph of Y_1 intersects the graph of Y_2 at $x = 1.05\left(\approx\dfrac{\pi}{3}\right), 5.24\left(\approx\dfrac{5\pi}{3}\right),$ $7.33\left(\approx\dfrac{7\pi}{3}\right),$ and $11.52\left(\approx\dfrac{11\pi}{3}\right),$ rounded to two decimal places. ◀

✏️ **NOW WORK PROBLEM 31.**

In most of our work, we shall be interested only in finding solutions of trigonometric equations for $0 \le \theta < 2\pi$.

EXAMPLE 3   **Solving a Linear Trigonometric Equation**

Solve the equation: $2 \sin \theta + \sqrt{3} = 0, \quad 0 \le \theta < 2\pi$

Solution   We solve the equation for $\sin \theta$.

$$2 \sin \theta + \sqrt{3} = 0$$
$$2 \sin \theta = -\sqrt{3} \qquad \text{Subtract } \sqrt{3} \text{ from both sides.}$$
$$\sin \theta = -\frac{\sqrt{3}}{2} \qquad \text{Divide both sides by 2.}$$

The period of the sine function is 2π. In the interval $[0, 2\pi)$, there are two angles θ for which $\sin \theta = -\frac{\sqrt{3}}{2}$: $\theta = \frac{4\pi}{3}$ and $\theta = \frac{5\pi}{3}$.   ◀

🖉➤ **NOW WORK PROBLEM 7.**

EXAMPLE 4   **Solving a Trigonometric Equation**

Solve the equation: $\sin(2\theta) = \frac{1}{2}, \quad 0 \le \theta < 2\pi$

Solution   The period of the sine function is 2π. In the interval $[0, 2\pi)$, the sine function has a value $\frac{1}{2}$ at $\frac{\pi}{6}$ and $\frac{5\pi}{6}$. See Figure 32. Because the argument is 2θ in the equation $\sin(2\theta) = \frac{1}{2}$, we have

Figure 32

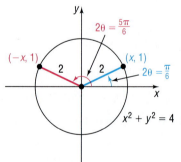

$$2\theta = \frac{\pi}{6} + 2k\pi \quad \text{or} \quad 2\theta = \frac{5\pi}{6} + 2k\pi \qquad k \text{ any integer}$$
$$\theta = \frac{\pi}{12} + k\pi \qquad\qquad \theta = \frac{5\pi}{12} + k\pi \qquad \text{Divide by 2.}$$

Then

$$\theta = \frac{\pi}{12} + (-1)\pi = \frac{-11\pi}{12} \qquad k = -1 \qquad \theta = \frac{5\pi}{12} + (-1)\pi = \frac{-7\pi}{12}$$

$$\theta = \frac{\pi}{12} + (0)\pi = \frac{\pi}{12} \qquad k = 0 \qquad \theta = \frac{5\pi}{12} + (0)\pi = \frac{5\pi}{12}$$

$$\theta = \frac{\pi}{12} + (1)\pi = \frac{13\pi}{12} \qquad k = 1 \qquad \theta = \frac{5\pi}{12} + (1)\pi = \frac{17\pi}{12}$$

$$\theta = \frac{\pi}{12} + (2)\pi = \frac{25\pi}{12} \qquad k = 2 \qquad \theta = \frac{5\pi}{12} + (2)\pi = \frac{29\pi}{12}$$

In the interval $[0, 2\pi)$, the solutions of $\sin(2\theta) = \frac{1}{2}$ are $\theta = \frac{\pi}{12}, \theta = \frac{5\pi}{12}, \theta = \frac{13\pi}{12}$, and $\theta = \frac{17\pi}{12}$.

✔ **CHECK:** Verify these solutions by graphing $Y_1 = \sin(2x)$ and $Y_2 = \frac{1}{2}$ for $0 \le x \le 2\pi$.   ◀

WARNING In solving a trigonometric equation for $\theta, 0 \leq \theta < 2\pi$, in which the argument is not θ (as in Example 4), you must write down all the solutions first and then list those that are in the interval $[0, 2\pi)$. Otherwise, solutions may be lost. For example, in solving $\sin(2\theta) = \dfrac{1}{2}$, if you merely write the solutions $2\theta = \dfrac{\pi}{6}$ and $2\theta = \dfrac{5\pi}{6}$, you will find only $\theta = \dfrac{\pi}{12}$ and $\theta = \dfrac{5\pi}{12}$ and miss the other solutions. ■

 **NOW WORK PROBLEM 13.**

| **EXAMPLE 5** | **Solving a Trigonometric Equation** |

Solve the equation: $\tan\left(\theta - \dfrac{\pi}{2}\right) = 1, \quad 0 \leq \theta < 2\pi$

Solution The period of the tangent function is π. In the interval $[0, \pi)$, the tangent function has the value 1 when the argument is $\dfrac{\pi}{4}$. Because the argument is $\theta - \dfrac{\pi}{2}$ in the given equation, we have

$$\theta - \frac{\pi}{2} = \frac{\pi}{4} + k\pi \qquad \textcolor{teal}{k \text{ any integer}}$$

$$\theta = \frac{3\pi}{4} + k\pi$$

In the interval $[0, 2\pi)$, $\theta = \dfrac{3\pi}{4}$ and $\theta = \dfrac{3\pi}{4} + \pi = \dfrac{7\pi}{4}$ are the only solutions.

✔ **CHECK:** Verify these solutions using a graphing utility. ◀

The next example illustrates how to solve trigonometric equations using a calculator. Remember that the function keys on a calculator will only give values consistent with the definition of the function.

| **EXAMPLE 6** | **Solving a Trigonometric Equation with a Calculator** |

Use a calculator to solve the equation: $\sin \theta = 0.3, 0 \leq \theta < 2\pi$ Express any solutions in radians, rounded to two decimal places.

Solution To solve $\sin \theta = 0.3$ on a calculator, first set the mode to radians. Then use the $\boxed{\sin^{-1}}$ key to obtain

$$\theta = \sin^{-1}(0.3) \approx 0.3046927$$

Figure 33

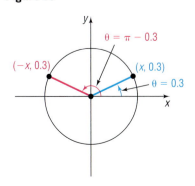

Rounded to two decimal places, $\theta = \sin^{-1}(0.3) = 0.30$ radian. Because of the definition of $y = \sin^{-1} x$, the angle θ that we obtain is the angle $-\dfrac{\pi}{2} \leq \theta \leq \dfrac{\pi}{2}$ for which $\sin \theta = 0.3$. Another angle for which $\sin \theta = 0.3$ is $\pi - 0.30$. See Figure 33. The angle $\pi - 0.30$ is the angle in quadrant II, where $\sin \theta = 0.3$. The solutions for $\sin \theta = 0.3, 0 \leq \theta < 2\pi$, are

$$\theta = 0.30 \text{ radian} \quad \text{and} \quad \theta = \pi - 0.30 \approx 2.84 \text{ radians} \qquad ◀$$

A second method for solving $\sin \theta = 0.3, 0 \le \theta < 2\pi$, would be to graph $Y_1 = \sin x$ and $Y_2 = 0.3$ for $0 \le x < 2\pi$ and find their point(s) of intersection. Try this method for yourself to verify the results obtained in Example 6.

WARNING   Example 6 illustrates that caution must be exercised when solving trigonometric equations on a calculator. Remember that the calculator supplies an angle only within the restrictions of the definition of the inverse trigonometric function. To find the remaining solutions, you must identify other quadrants, if any, in which a solution may be located. ■

NOW WORK PROBLEM 41.

6.7 Assess Your Understanding

'Are You Prepared?'

Answers are given at the end of these exercises. If you get a wrong answer, read the pages listed in red.

1. Solve: $3x - 5 = -x + 1$. (p. 694)

2. $\sin\left(\dfrac{\pi}{4}\right) = $ _____; $\cos\left(\dfrac{8\pi}{3}\right) = $ _____. (p. 379 and pp. 380–381)

Concepts and Vocabulary

3. Two solutions of the equation $\sin \theta = \dfrac{1}{2}$ are _____ and _____.

4. All the solutions of the equation $\sin \theta = \dfrac{1}{2}$ are _____.

5. *True or False:* Most trigonometric equations have unique solutions.

6. *True or False:* The equation $\sin \theta = 2$ has a real solution that can be found using a graphing calculator.

Skill Building

In Problems 7–30, solve each equation on the interval $0 \le \theta < 2\pi$.

7. $2 \sin \theta + 3 = 2$

8. $1 - \cos \theta = \dfrac{1}{2}$

9. $4 \cos^2 \theta = 1$

10. $\tan^2 \theta = \dfrac{1}{3}$

11. $2 \sin^2 \theta - 1 = 0$

12. $4 \cos^2 \theta - 3 = 0$

13. $\sin(3\theta) = -1$

14. $\tan \dfrac{\theta}{2} = \sqrt{3}$

15. $\cos(2\theta) = -\dfrac{1}{2}$

16. $\tan(2\theta) = -1$

17. $\sec \dfrac{3\theta}{2} = -2$

18. $\cot \dfrac{2\theta}{3} = -\sqrt{3}$

19. $2 \sin \theta + 1 = 0$

20. $\cos \theta + 1 = 0$

21. $\tan \theta + 1 = 0$

22. $\sqrt{3} \cot \theta + 1 = 0$

23. $4 \sec \theta + 6 = -2$

24. $5 \csc \theta - 3 = 2$

25. $3\sqrt{2} \cos \theta + 2 = -1$

26. $4 \sin \theta + 3\sqrt{3} = \sqrt{3}$

27. $\cos\left(2\theta - \dfrac{\pi}{2}\right) = -1$

28. $\sin\left(3\theta + \dfrac{\pi}{18}\right) = 1$

29. $\tan\left(\dfrac{\theta}{2} + \dfrac{\pi}{3}\right) = 1$

30. $\cos\left(\dfrac{\theta}{3} - \dfrac{\pi}{4}\right) = \dfrac{1}{2}$

In Problems 31–40, solve each equation. Give a general formula for all the solutions. List six solutions.

31. $\sin \theta = \dfrac{1}{2}$

32. $\tan \theta = 1$

33. $\tan \theta = -\dfrac{\sqrt{3}}{3}$

34. $\cos \theta = -\dfrac{\sqrt{3}}{2}$

35. $\cos \theta = 0$

36. $\sin \theta = \dfrac{\sqrt{2}}{2}$ **37.** $\cos(2\theta) = -\dfrac{1}{2}$ **38.** $\sin(2\theta) = -1$ **39.** $\sin \dfrac{\theta}{2} = -\dfrac{\sqrt{3}}{2}$ **40.** $\tan \dfrac{\theta}{2} = -1$

In Problems 41–52, use a calculator to solve each equation on the interval $0 \le \theta < 2\pi$. Round answers to two decimal places.

41. $\sin \theta = 0.4$ **42.** $\cos \theta = 0.6$ **43.** $\tan \theta = 5$ **44.** $\cot \theta = 2$

45. $\cos \theta = -0.9$ **46.** $\sin \theta = -0.2$ **47.** $\sec \theta = -4$ **48.** $\csc \theta = -3$

49. $5 \tan \theta + 9 = 0$ **50.** $4 \cot \theta = -5$ **51.** $3 \sin \theta - 2 = 0$ **52.** $4 \cos \theta + 3 = 0$

Applications and Extensions

53. Suppose that $f(x) = 3 \sin x$.

(a) Solve $f(x) = \dfrac{3}{2}$.

(b) For what values of x is $f(x) > \dfrac{3}{2}$ on the interval $[0, 2\pi)$?

54. Suppose that $f(x) = 2 \cos x$.

(a) Solve $f(x) = -\sqrt{3}$.

(b) For what values of x is $f(x) < -\sqrt{3}$ on the interval $[0, 2\pi)$?

55. Suppose that $f(x) = 4 \tan x$.

(a) Solve $f(x) = -4$.

(b) For what values of x is $f(x) < -4$ on the interval $\left(-\dfrac{\pi}{2}, \dfrac{\pi}{2}\right)$?

56. Suppose that $f(x) = \cot x$.

(a) Solve $f(x) = -\sqrt{3}$.

(b) For what values of x is $f(x) > -\sqrt{3}$ on the interval $(0, \pi)$?

57. The Ferris Wheel In 1893, George Ferris engineered the Ferris Wheel. It was 250 feet in diameter. If the wheel makes 1 revolution every 40 seconds, then

$$h(t) = 125 \sin\left(0.157t - \dfrac{\pi}{2}\right) + 125$$

represents the height h, in feet, of a seat on the wheel as a function of time t, where t is measured in seconds. The ride begins when $t = 0$.

(a) During the first 40 seconds of the ride, at what time t is an individual on the Ferris Wheel exactly 125 feet above the ground?

(b) During the first 80 seconds of the ride, at what time t is an individual on the Ferris Wheel exactly 250 feet above the ground?

(c) During the first 40 seconds of the ride, over what interval of time t is an individual on the Ferris Wheel more than 125 feet above the ground?

58. Tire Rotation The P215/65R15 Cobra Radial G/T tire has a diameter of exactly 26 inches. Suppose that a car's wheel is

making 2 revolutions per second (the car is traveling a little less than 10 miles per hour). Then $h(t) = 13 \sin\left(4\pi t - \dfrac{\pi}{2}\right) + 13$ represents the height h (in inches) of a point on the tire as a function of time t (in seconds). The car starts to move when $t = 0$.

(a) During the first second that the car is moving, at what time t is the point on the tire exactly 13 inches above the ground?

(b) During the first second that the car is moving, at what time t is the point on the tire exactly 6.5 inches above the ground?

(c) During the first second that the car is moving, at what time t is the point on the tire more than 13 inches above the ground?

SOURCE: Cobra Tire

59. Holding Pattern Suppose that an airplane is asked to stay within a holding pattern near Chicago's O'Hare International Airport. The function $d(x) = 70 \sin(0.65x) + 150$ represents the distance d, in miles, that the airplane is from the airport at time x, in minutes.

(a) When the plane enters the holding pattern, $x = 0$, how far is it from O'Hare?

(b) During the first 20 minutes after the plane enters the holding pattern, at what time x is the plane exactly 100 miles from the airport?

(c) During the first 20 minutes after the plane enters the holding pattern, at what time x is the plane more than 100 miles from the airport?

(d) While the plane is in the holding pattern, will it ever be within 70 miles of the airport? Why?

60. Projectile Motion A golfer hits a golf ball with an initial velocity of 100 miles per hour. The range R of the ball as a function of the angle θ to the horizontal is given by $R(\theta) = 672 \sin(2\theta)$, where R is measured in feet.

(a) At what angle θ should the ball be hit if the golfer wants the ball to travel 450 feet (150 yards)?

(b) At what angle θ should the ball be hit if the golfer wants the ball to travel 540 feet (180 yards)?

(c) At what angle θ should the ball be hit if the golfer wants the ball to travel at least 480 feet (160 yards)?

(d) Can the golfer hit the ball 720 feet (240 yards)?

The following discussion of Snell's Law of Refraction (named after Willebrord Snell, 1580–1626) is needed for Problems 61–67. Light, sound, and other waves travel at different speeds, depending on the media (air, water, wood, and so on) through which they pass. Suppose that light travels from a point A in one medium, where its speed is v_1, to a point B in another medium, where its speed is v_2. Refer to the figure, where the angle θ_1 is called the **angle of incidence** and the angle θ_2 is the **angle of refraction**. Snell's Law,[*] which can be proved using calculus, states that

$$\frac{\sin \theta_1}{\sin \theta_2} = \frac{v_1}{v_2}$$

The ratio $\dfrac{v_1}{v_2}$ is called the **index of refraction**. Some values are given in the following table.

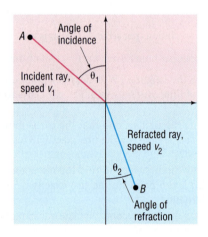

SOME INDEXES OF REFRACTION	
Medium	**Index of Refraction**[†]
Water	1.33
Ethyl alcohol	1.36
Carbon disulfide	1.63
Air (1 atm and 20°C)	1.0003
Methylene iodide	1.74
Fused quartz	1.46
Glass, crown	1.52
Glass, dense flint	1.66
Sodium chloride	1.54

[†]For light of wavelength 589 nanometers, measured with respect to a vacuum. The index with respect to air is negligibly different in most cases.

61. The index of refraction of light in passing from a vacuum into water is 1.33. If the angle of incidence is 40°, determine the angle of refraction.

62. The index of refraction of light in passing from a vacuum into dense glass is 1.66. If the angle of incidence is 50°, determine the angle of refraction.

63. Ptolemy, who lived in the city of Alexandria in Egypt during the second century AD, gave the measured values in the table below for the angle of incidence θ_1 and the angle of refraction θ_2 for a light beam passing from air into water. Do these values agree with Snell's Law? If so, what index of refraction results? (These data are interesting as the oldest recorded physical measurements.)[†]

θ_1	θ_2	θ_1	θ_2
10°	7°45′	50°	35°0′
20°	15°30′	60°	40°30′
30°	22°30′	70°	45°30′
40°	29°0′	80°	50°0′

64. The speed of yellow sodium light (wavelength of 589 nanometers) in a certain liquid is measured to be 1.92×10^8 meters per second. What is the index of refraction of this liquid, with respect to air, for sodium light?[†]

[**Hint:** The speed of light in air is approximately 2.997×10^8 meters per second.]

65. A beam of light with a wavelength of 589 nanometers traveling in air makes an angle of incidence of 40° on a slab of transparent material, and the refracted beam makes an angle of refraction of 26°. Find the index of refraction of the material.[†]

66. A light ray with a wavelength of 589 nanometers (produced by a sodium lamp) traveling through air makes an angle of incidence of 30° on a smooth, flat slab of crown glass. Find the angle of refraction.[†]

67. A light beam passes through a thick slab of material whose index of refraction is n_2. Show that the emerging beam is parallel to the incident beam.[‡]

[*]Because this law was also deduced by René Descartes in France, it is also known as Descartes's Law.
[†]Adapted from Halliday and Resnick, *Physics*, Part 1 & 2, 3rd ed. New York: Wiley.
[‡]*Physics for Scientists & Engineers* 3/E by Serway. © 1990. Reprinted with permission of Brooks/Cole, a division of Thomson Learning.

Discussion and Writing

68. Explain in your own words how you would use your calculator to solve the equation $\sin x = 0.3, 0 \leq x < 2\pi$. How would you modify your approach in order to solve the equation $\cot x = 5, 0 < x < 2\pi$?

'Are You Prepared?' Answers

1. $\left\{\dfrac{3}{2}\right\}$ **2.** $\dfrac{\sqrt{2}}{2}; -\dfrac{1}{2}$

6.8 Trigonometric Equations (II)

PREPARING FOR THIS SECTION   *Before getting started, review the following:*

- Solving Quadratic Equations by Factoring (Appendix, Section A.5, p. 697)
- Solving Equations in One Variable Using a Graphing Utility (Section 1.3, pp. 24–26)
- The Quadratic Formula (Appendix, Section A.5, pp. 700–703)
- Solve Equations Quadratic in Form (Appendix, Section A.5, pp. 703–704)

 Now work the 'Are You Prepared?' problems on page 508.

OBJECTIVES **1** Solve Trigonometric Equations Quadratic in Form
2 Solve Trigonometric Equations Using Identities
3 Solve Trigonometric Equations Linear in Sine and Cosine
4 Solve Trigonometric Equations Using a Graphing Utility

1 **Solve Trigonometric Equations Quadratic in From**

In this section we continue our study of trigonometric equations. Many trigonometric equations can be solved by applying techniques that we already know, such as applying the quadratic formula (if the equation is a second-degree polynomial) or factoring.

EXAMPLE 1   **Solving a Trigonometric Equation Quadratic in Form**

Solve the equation: $2 \sin^2 \theta - 3 \sin \theta + 1 = 0, \quad 0 \leq \theta < 2\pi$

Solution   The equation that we wish to solve is a quadratic equation (in $\sin \theta$) that can be factored.

$$2 \sin^2 \theta - 3 \sin \theta + 1 = 0 \qquad \color{teal}{2x^2 - 3x + 1 = 0, \quad x = \sin \theta}$$
$$(2 \sin \theta - 1)(\sin \theta - 1) = 0 \qquad \color{teal}{(2x - 1)(x - 1) = 0}$$
$$2 \sin \theta - 1 = 0 \quad \text{or} \quad \sin \theta - 1 = 0 \qquad \color{teal}{\text{Zero-product Property.}}$$
$$\sin \theta = \frac{1}{2} \quad \text{or} \qquad \sin \theta = 1$$

Solving each equation in the interval $[0, 2\pi)$, we obtain

$$\theta = \frac{\pi}{6}, \qquad \theta = \frac{5\pi}{6}, \qquad \theta = \frac{\pi}{2}$$

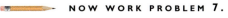 **NOW WORK PROBLEM 7.**

2 Solve Trigonometric Equations Using Identities

When a trigonometric equation contains more than one trigonometric function, identities sometimes can be used to obtain an equivalent equation that contains only one trigonometric function.

EXAMPLE 2 | **Solving a Trigonometric Equation Using Identities**

Solve the equation: $3 \cos \theta + 3 = 2 \sin^2 \theta, \quad 0 \le \theta < 2\pi$

Solution   The equation in its present form contains sines and cosines. However, a form of the Pythagorean Identity can be used to transform the equation into an equivalent expression containing only cosines.

$$3 \cos \theta + 3 = 2 \sin^2 \theta$$
$$3 \cos \theta + 3 = 2(1 - \cos^2 \theta) \qquad \qquad sin^2\,\theta = 1 - cos^2\,\theta$$
$$3 \cos \theta + 3 = 2 - 2 \cos^2 \theta$$
$$2 \cos^2 \theta + 3 \cos \theta + 1 = 0 \qquad \qquad \textit{Quadratic in cos θ}$$
$$(2 \cos \theta + 1)(\cos \theta + 1) = 0 \qquad \qquad \textit{Factor.}$$
$$2 \cos \theta + 1 = 0 \quad \text{or} \quad \cos \theta + 1 = 0 \qquad \textit{Zero-product Property.}$$
$$\cos \theta = -\frac{1}{2} \quad \text{or} \qquad \cos \theta = -1$$

Solving each equation in the interval $[0, 2\pi)$, we obtain

$$\theta = \frac{2\pi}{3}, \qquad \theta = \frac{4\pi}{3}, \qquad \theta = \pi$$

✔ **CHECK:** Graph $Y_1 = 3 \cos x + 3$ and $Y_2 = 2 \sin^2 x, 0 \le x \le 2\pi$, and find the points of intersection. How close are your approximate solutions to the exact ones found in this example? ◀

EXAMPLE 3 | **Solving a Trigonometric Equation Using Identities**

Solve the equation: $\cos(2\theta) + 3 = 5 \cos \theta, \quad 0 \le \theta < 2\pi$

Solution   First, we observe that the given equation contains two cosine functions, but with different arguments, θ and 2θ. We use the Double-angle Formula $\cos(2\theta) = 2 \cos^2 \theta - 1$ to obtain an equivalent equation containing only $\cos \theta$.

$$\cos(2\theta) + 3 = 5 \cos \theta$$
$$(2 \cos^2 \theta - 1) + 3 = 5 \cos \theta \qquad \qquad cos(2\theta) = 2\,cos^2\,\theta - 1$$
$$2 \cos^2 \theta - 5 \cos \theta + 2 = 0 \qquad \qquad \textit{Place in standard form.}$$
$$(\cos \theta - 2)(2 \cos \theta - 1) = 0 \qquad \qquad \textit{Factor.}$$
$$\cos \theta = 2 \quad \text{or} \quad \cos \theta = \frac{1}{2}$$

For any angle θ, $-1 \le \cos\theta \le 1$; therefore, the equation $\cos\theta = 2$ has no solution. The solutions of $\cos\theta = \dfrac{1}{2}, 0 \le \theta < 2\pi$, are

$$\theta = \frac{\pi}{3}, \qquad \theta = \frac{5\pi}{3}$$

✔ **CHECK:** Graph $Y_1 = \cos(2x) + 3$ and $Y_2 = 5\cos x, 0 \le x \le 2\pi$, and find the points of intersection. Compare your results with those of Example 3. ◀

▬▬▬ **NOW WORK PROBLEM 23.**

| **EXAMPLE 4** | **Solving a Trigonometric Equation Using Identities** |

Solve the equation: $\cos^2\theta + \sin\theta = 2$, $0 \le \theta < 2\pi$

Solution   This equation involves two trigonometric functions, sine and cosine. We use a form of the Pythagorean Identity, $\sin^2\theta + \cos^2\theta = 1$ to rewrite the equation in terms of $\sin\theta$.

Figure 34

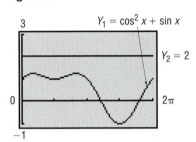

$$\cos^2\theta + \sin\theta = 2$$
$$(1 - \sin^2\theta) + \sin\theta = 2 \qquad \cos^2\theta = 1 - \sin^2\theta$$
$$\sin^2\theta - \sin\theta + 1 = 0$$

This is a quadratic equation in $\sin\theta$. The discriminant is $b^2 - 4ac = 1 - 4 = -3 < 0$. Therefore, the equation has no real solution.

✔ **CHECK:** Graph $Y_1 = \cos^2 x + \sin x$ and $Y_2 = 2$. See Figure 34. The two graphs do not intersect, so the equation $Y_1 = Y_2$ has no real solution. ◀

| **EXAMPLE 5** | **Solving a Trigonometric Equation Using Identities** |

Solve the equation: $\sin\theta\cos\theta = -\dfrac{1}{2}$, $0 \le \theta < 2\pi$

Solution   The left side of the given equation is in the form of the Double-angle Formula $2\sin\theta\cos\theta = \sin(2\theta)$, except for a factor of 2. We multiply each side by 2.

$$\sin\theta\cos\theta = -\frac{1}{2}$$
$$2\sin\theta\cos\theta = -1 \qquad \text{Multiply each side by 2.}$$
$$\sin(2\theta) = -1 \qquad \text{Double-angle Formula}$$

The argument here is 2θ. So we need to write all the solutions of this equation and then list those that are in the interval $[0, 2\pi)$. Because $\sin\left(\dfrac{3\pi}{2} + 2\pi k\right) = -1$, for any integer k we have

$$2\theta = \frac{3\pi}{2} + 2k\pi \qquad \text{k any integer}$$
$$\theta = \frac{3\pi}{4} + k\pi$$

$$\underset{\substack{\uparrow \\ k=-1}}{\theta = \frac{3\pi}{4} + (-1)\pi = -\frac{\pi}{4}}, \quad \underset{\substack{\uparrow \\ k=0}}{\theta = \frac{3\pi}{4} + (0)\pi = \frac{3\pi}{4}}, \quad \underset{\substack{\uparrow \\ k=1}}{\theta = \frac{3\pi}{4} + (1)\pi = \frac{7\pi}{4}}, \quad \underset{\substack{\uparrow \\ k=2}}{\theta = \frac{3\pi}{4} + (2)\pi = \frac{11\pi}{4}}$$

The solutions in the interval $[0, 2\pi)$ are

$$\theta = \frac{3\pi}{4}, \qquad \theta = \frac{7\pi}{4}$$

◀

3 Solve Trigonometric Equations Linear in Sine and Cosine

Sometimes it is necessary to square both sides of an equation to obtain expressions that allow the use of identities. Remember, squaring both sides of an equation may introduce extraneous solutions. As a result, apparent solutions must be checked.

EXAMPLE 6   **Solving a Trigonometric Equation Linear in Sine and Cosine**

Solve the equation: $\sin \theta + \cos \theta = 1, \quad 0 \le \theta < 2\pi$

Solution A   Attempts to use available identities do not lead to equations that are easy to solve. (Try it yourself.) Given the form of this equation, we decide to square each side.

$$\sin \theta + \cos \theta = 1$$
$$(\sin \theta + \cos \theta)^2 = 1 \qquad \textit{Square each side.}$$
$$\sin^2 \theta + 2 \sin \theta \cos \theta + \cos^2 \theta = 1 \qquad \textit{Remove parentheses.}$$
$$2 \sin \theta \cos \theta = 0 \qquad \textit{sin}^2\,\theta + \cos^2 \theta = 1$$
$$\sin \theta \cos \theta = 0$$

Setting each factor equal to zero, we obtain

$$\sin \theta = 0 \quad \text{or} \quad \cos \theta = 0$$

The apparent solutions are

$$\theta = 0, \qquad \theta = \pi, \qquad \theta = \frac{\pi}{2}, \qquad \theta = \frac{3\pi}{2}$$

Because we squared both sides of the original equation, we must check these apparent solutions to see if any are extraneous.

$$\theta = 0: \quad \sin 0 + \cos 0 = 0 + 1 = 1 \qquad \textit{A solution}$$
$$\theta = \pi: \quad \sin \pi + \cos \pi = 0 + (-1) = -1 \qquad \textit{Not a solution}$$
$$\theta = \frac{\pi}{2}: \quad \sin \frac{\pi}{2} + \cos \frac{\pi}{2} = 1 + 0 = 1 \qquad \textit{A solution}$$
$$\theta = \frac{3\pi}{2}: \quad \sin \frac{3\pi}{2} + \cos \frac{3\pi}{2} = -1 + 0 = -1 \qquad \textit{Not a solution}$$

The values $\theta = \pi$ and $\theta = \frac{3\pi}{2}$ are extraneous. The solution set is $\left\{0, \frac{\pi}{2}\right\}$.   ◀

Solution B   We start with the equation

$$\sin \theta + \cos \theta = 1$$

and divide each side by $\sqrt{2}$. (The reason for this choice will become apparent shortly.) Then

$$\frac{1}{\sqrt{2}}\sin\theta + \frac{1}{\sqrt{2}}\cos\theta = \frac{1}{\sqrt{2}}$$

The left side now resembles the formula for the sine of the sum of two angles, one of which is θ. The other angle is unknown (call it ϕ.) Then

$$\sin(\theta + \phi) = \sin\theta\cos\phi + \cos\theta\sin\phi = \frac{1}{\sqrt{2}} = \frac{\sqrt{2}}{2} \qquad \textbf{(1)}$$

where

$$\cos\phi = \frac{1}{\sqrt{2}} = \frac{\sqrt{2}}{2}, \qquad \sin\phi = \frac{1}{\sqrt{2}} = \frac{\sqrt{2}}{2}, \qquad 0 \le \phi < 2\pi$$

The angle ϕ is therefore $\dfrac{\pi}{4}$. As a result, equation (1) becomes

$$\sin\left(\theta + \frac{\pi}{4}\right) = \frac{\sqrt{2}}{2}$$

Figure 35

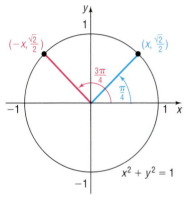

In the interval $[0, 2\pi)$, there are two angles whose sine is $\dfrac{\sqrt{2}}{2}$: $\dfrac{\pi}{4}$ and $\dfrac{3\pi}{4}$. See Figure 35. As a result,

$$\theta + \frac{\pi}{4} = \frac{\pi}{4} \quad \text{or} \quad \theta + \frac{\pi}{4} = \frac{3\pi}{4}$$

$$\theta = 0 \quad \text{or} \quad \theta = \frac{\pi}{2}$$

The solution set is $\left\{ 0, \dfrac{\pi}{2} \right\}$. ◄

This second method of solution can be used to solve any linear equation in the variables $\sin\theta$ and $\cos\theta$.

EXAMPLE 7 **Solving a Trigonometric Equation Linear in Sin θ and Cos θ**

Solve:
$$a\sin\theta + b\cos\theta = c, \qquad 0 \le \theta < 2\pi \qquad \textbf{(2)}$$

where a, b, and c are constants and either $a \ne 0$ or $b \ne 0$.

Solution We divide each side of equation (2) by $\sqrt{a^2 + b^2}$. Then

$$\frac{a}{\sqrt{a^2 + b^2}}\sin\theta + \frac{b}{\sqrt{a^2 + b^2}}\cos\theta = \frac{c}{\sqrt{a^2 + b^2}} \qquad \textbf{(3)}$$

There is a unique angle ϕ, $0 \le \phi < 2\pi$, for which

Figure 36

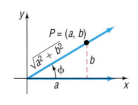

$$\cos\phi = \frac{a}{\sqrt{a^2 + b^2}} \quad \text{and} \quad \sin\phi = \frac{b}{\sqrt{a^2 + b^2}} \qquad \textbf{(4)}$$

See Figure 36. Equation (3) may be written as

$$\sin\theta\cos\phi + \cos\theta\sin\phi = \frac{c}{\sqrt{a^2 + b^2}}$$

or, equivalently,

$$\sin(\theta + \phi) = \frac{c}{\sqrt{a^2 + b^2}} \tag{5}$$

where ϕ satisfies equation (4).

If $|c| > \sqrt{a^2 + b^2}$, then $\sin(\theta + \phi) > 1$ or $\sin(\theta + \phi) < -1$, and equation (5) has no solution.

If $|c| \leq \sqrt{a^2 + b^2}$, then the solutions of equation (5) are

$$\theta + \phi = \sin^{-1}\frac{c}{\sqrt{a^2 + b^2}} \quad \text{or} \quad \theta + \phi = \pi - \sin^{-1}\frac{c}{\sqrt{a^2 + b^2}}$$

Because the angle ϕ is determined by equations (4), these are the solutions to equation (2).   ◀

 NOW WORK PROBLEM 41.

4   Solve Trigonometric Equations Using a Graphing Utility

The techniques introduced in this section apply only to certain types of trigonometric equations. Solutions for other types are usually studied in calculus, using numerical methods. In the next example, we show how a graphing utility may be used to obtain solutions.

EXAMPLE 8   **Solving Trigonometric Equations Using a Graphing Utility**

Solve: $5 \sin x + x = 3$

Express the solution(s) rounded to two decimal places.

Solution

Figure 37

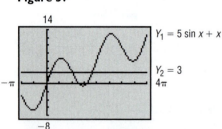

This type of trigonometric equation cannot be solved by previous methods. A graphing utility, though, can be used here. The solution(s) of this equation is the same as the points of intersection of the graphs of $Y_1 = 5 \sin x + x$ and $Y_2 = 3$. See Figure 37.
  There are three points of intersection; the x-coordinates are the solutions that we seek. Using INTERSECT, we find

$$x = 0.52, \qquad x = 3.18, \qquad x = 5.71$$ ◀

 NOW WORK PROBLEM 53.

6.8 Assess Your Understanding

'Are You Prepared?'

Answers are given at the end of these exercises. If you get a wrong answer, read the pages listed in red.

1. Find the real solutions of $4x^2 - x - 5 = 0$. (p. 697)

2. Find the real solutions of $x^2 - x - 1 = 0$. (pp. 700–703)

3. Find the real solutions of $(2x - 1)^2 - 3(2x - 1) - 4 = 0$ (pp. 703–704)

4. Use a graphing utility to solve $5x^3 - 2 = x - x^2$. Round answers to two decimal places. (pp. 24–26)

Skill Building

In Problems 5–46, solve each equation on the interval $0 \le \theta < 2\pi$.

5. $2\cos^2\theta + \cos\theta = 0$

6. $\sin^2\theta - 1 = 0$

7. $2\sin^2\theta - \sin\theta - 1 = 0$

8. $2\cos^2\theta + \cos\theta - 1 = 0$

9. $(\tan\theta - 1)(\sec\theta - 1) = 0$

10. $(\cot\theta + 1)\left(\csc\theta - \dfrac{1}{2}\right) = 0$

11. $\sin^2\theta - \cos^2\theta = 1 + \cos\theta$

12. $\cos^2\theta - \sin^2\theta + \sin\theta = 0$

13. $\sin^2\theta = 6(\cos\theta + 1)$

14. $2\sin^2\theta = 3(1 - \cos\theta)$

15. $\cos(2\theta) + 6\sin^2\theta = 4$

16. $\cos(2\theta) = 2 - 2\sin^2\theta$

17. $\cos\theta = \sin\theta$

18. $\cos\theta + \sin\theta = 0$

19. $\tan\theta = 2\sin\theta$

20. $\sin(2\theta) = \cos\theta$

21. $\sin\theta = \csc\theta$

22. $\tan\theta = \cot\theta$

23. $\cos(2\theta) = \cos\theta$

24. $\sin(2\theta)\sin\theta = \cos\theta$

25. $\sin(2\theta) + \sin(4\theta) = 0$

26. $\cos(2\theta) + \cos(4\theta) = 0$

27. $\cos(4\theta) - \cos(6\theta) = 0$

28. $\sin(4\theta) - \sin(6\theta) = 0$

29. $1 + \sin\theta = 2\cos^2\theta$

30. $\sin^2\theta = 2\cos\theta + 2$

31. $2\sin^2\theta - 5\sin\theta + 3 = 0$

32. $2\cos^2\theta - 7\cos\theta - 4 = 0$

33. $3(1 - \cos\theta) = \sin^2\theta$

34. $4(1 + \sin\theta) = \cos^2\theta$

35. $\tan^2\theta = \dfrac{3}{2}\sec\theta$

36. $\csc^2\theta = \cot\theta + 1$

37. $3 - \sin\theta = \cos(2\theta)$

38. $\cos(2\theta) + 5\cos\theta + 3 = 0$

39. $\sec^2\theta + \tan\theta = 0$

40. $\sec\theta = \tan\theta + \cot\theta$

41. $\sin\theta - \sqrt{3}\cos\theta = 1$

42. $\sqrt{3}\sin\theta + \cos\theta = 1$

43. $\tan(2\theta) + 2\sin\theta = 0$

44. $\tan(2\theta) + 2\cos\theta = 0$

45. $\sin\theta + \cos\theta = \sqrt{2}$

46. $\sin\theta + \cos\theta = -\sqrt{2}$

In Problems 47–52, solve each equation for x, $-\pi \le x \le \pi$. Express the solution(s) rounded to two decimal places.

47. Solve the equation $\cos x = e^x$ by graphing $Y_1 = \cos x$ and $Y_2 = e^x$ and finding their point(s) of intersection.

48. Solve the equation $\cos x = e^x$ by graphing $Y_1 = \cos x - e^x$ and finding the x-intercept(s).

49. Solve the equation $2\sin x = 0.7x$ by graphing $Y_1 = 2\sin x$ and $Y_2 = 0.7x$ and finding their point(s) of intersection.

50. Solve the equation $2\sin x = 0.7x$ by graphing $Y_1 = 2\sin x - 0.7x$ and finding the x-intercept(s).

51. Solve the equation $\cos x = x^2$ by graphing $Y_1 = \cos x$ and $Y_2 = x^2$ and finding their point(s) of intersection.

52. Solve the equation $\cos x = x^2$ by graphing $Y_1 = \cos x - x^2$ and finding the x-intercept(s).

In Problems 53–64, use a graphing utility to solve each equation. Express the solution(s) rounded to two decimal places.

53. $x + 5\cos x = 0$

54. $x - 4\sin x = 0$

55. $22x - 17\sin x = 3$

56. $19x + 8\cos x = 2$

57. $\sin x + \cos x = x$

58. $\sin x - \cos x = x$

59. $x^2 - 2\cos x = 0$

60. $x^2 + 3\sin x = 0$

61. $x^2 - 2\sin(2x) = 3x$

62. $x^2 = x + 3\cos(2x)$

63. $6\sin x - e^x = 2, \quad x > 0$

64. $4\cos(3x) - e^x = 1, \quad x > 0$

Applications and Extensions

65. Constructing a Rain Gutter   A rain gutter is to be constructed of aluminum sheets 12 inches wide. After marking off a length of 4 inches from each edge, this length is bent up at an angle θ. See the illustration. The area A of the opening as a function of θ is given by

$$A(\theta) = 16\sin\theta(\cos\theta + 1) \quad 0° < \theta < 90°$$

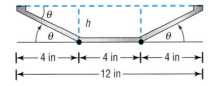

(a) In calculus, you will be asked to find the angle θ that maximizes A by solving the equation

$$\cos(2\theta) + \cos\theta = 0, \quad 0° < \theta < 90°$$

Solve this equation for θ by using the Double-angle Formula.

(b) Solve the equation for θ by writing the sum of the two cosines as a product.

(c) What is the maximum area A of the opening?

(d) Graph $A = A(\theta)$, $0° \le \theta \le 90°$, and find the angle θ that maximizes the area A. Also find the maximum area. Compare the results to the answers found earlier.

66. Projectile Motion An object is propelled upward at an angle θ, $45° < \theta < 90°$, to the horizontal with an initial velocity of v_0 feet per second from the base of a plane that makes an angle of $45°$ with the horizontal. See the illustration. If air resistance is ignored, the distance R that it travels up the inclined plane is given by

$$R(\theta) = \frac{v_0^2 \sqrt{2}}{32} [\sin(2\theta) - \cos(2\theta) - 1]$$

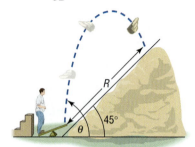

(a) In calculus, you will be asked to find the angle θ that maximizes R by solving the equation

$$\sin(2\theta) + \cos(2\theta) = 0$$

Solve this equation for θ using the method of Example 7.
(b) Solve this equation for θ by dividing each side by $\cos(2\theta)$.
(c) What is the maximum distance R if $v_0 = 32$ feet per second?
(d) Graph $R = R(\theta)$, $45° \le \theta \le 90°$, and find the angle θ that maximizes the distance R. Also find the maximum distance. Use $v_0 = 32$ feet per second. Compare the results with the answers found earlier.

67. Heat Transfer In the study of heat transfer, the equation $x + \tan x = 0$ occurs. Graph $Y_1 = -x$ and $Y_2 = \tan x$ for $x \ge 0$. Conclude that there are an infinite number of points of intersection of these two graphs. Now find the first two positive solutions of $x + \tan x = 0$ rounded to two decimal places.

68. Carrying a Ladder Around a Corner Two hallways, one of width 3 feet, the other of width 4 feet, meet at a right angle. See the illustration.

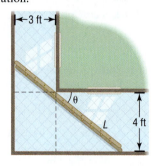

(a) Express the length L of the line segment shown as a function of θ.
(b) In calculus, you will be asked to find the length of the longest ladder that can turn the corner by solving the equation

$$3 \sec \theta \tan \theta - 4 \csc \theta \cot \theta = 0, \quad 0° < \theta < 90°$$

Solve this equation for θ.
(c) What is the length of the longest ladder that can be carried around the corner?
(d) Graph $L = L(\theta)$, $0° \le \theta \le 90°$, and find the angle θ that minimizes the length L.
(e) Compare the result with the one found in part (b). Explain why the two answers are the same.

69. Projectile Motion The horizontal distance that a projectile will travel in the air (ignoring air resistance) is given by the equation

$$R(\theta) = \frac{v_0^2 \sin(2\theta)}{g}$$

where v_0 is the initial velocity of the projectile, θ is the angle of elevation, and g is acceleration due to gravity (9.8 meters per second squared).
(a) If you can throw a baseball with an initial speed of 34.8 meters per second, at what angle of elevation θ should you direct the throw so that the ball travels a distance of 107 meters before striking the ground?
(b) Determine the maximum distance that you can throw the ball.
(c) Graph $R = R(\theta)$, with $v_0 = 34.8$ meters per second.
(d) Verify the results obtained in parts (a) and (b) using a graphing utility.

70. Projectile Motion Refer to Problem 69.
(a) If you can throw a baseball with an initial speed of 40 meters per second, at what angle of elevation θ should you direct the throw so that the ball travels a distance of 110 meters before striking the ground?
(b) Determine the maximum distance that you can throw the ball.
(c) Graph $R = R(\theta)$, with $v_0 = 40$ meters per second.
(d) Verify the results obtained in parts (a) and (b) using a graphing utility.

'Are You Prepared?' Answers

1. $\left\{-1, \dfrac{5}{4}\right\}$ **2.** $\left\{\dfrac{1 - \sqrt{5}}{2}, \dfrac{1 + \sqrt{5}}{2}\right\}$ **3.** $\left\{0, \dfrac{5}{2}\right\}$ **4.** $\{0.76\}$

Chapter Review

Things to Know

Definitions of the six inverse trigonometric functions

$y = \sin^{-1} x$ means $x = \sin y$ where $-1 \le x \le 1$, $-\dfrac{\pi}{2} \le y \le \dfrac{\pi}{2}$ (p. 449)

$y = \cos^{-1} x$ means $x = \cos y$ where $-1 \le x \le 1$, $0 \le y \le \pi$ (p. 453)

$y = \tan^{-1} x$ means $x = \tan y$ where $-\infty < x < \infty$, $-\dfrac{\pi}{2} < y < \dfrac{\pi}{2}$ (p. 456)

$y = \sec^{-1} x$ means $x = \sec y$ where $|x| \ge 1$, $0 \le y \le \pi$, $y \ne \dfrac{\pi}{2}$ (p. 462)

$y = \csc^{-1} x$ means $x = \csc y$ where $|x| \ge 1$, $-\dfrac{\pi}{2} \le y \le \dfrac{\pi}{2}$, $y \ne 0$ (p. 462)

$y = \cot^{-1} x$ means $x = \cot y$ where $-\infty < x < \infty$, $0 < y < \pi$ (p. 462)

Sum and Difference Formulas (pp. 473, 476, and 479)

$$\cos(\alpha + \beta) = \cos \alpha \cos \beta - \sin \alpha \sin \beta \qquad \cos(\alpha - \beta) = \cos \alpha \cos \beta + \sin \alpha \sin \beta$$

$$\sin(\alpha + \beta) = \sin \alpha \cos \beta + \cos \alpha \sin \beta \qquad \sin(\alpha - \beta) = \sin \alpha \cos \beta - \cos \alpha \sin \beta$$

$$\tan(\alpha + \beta) = \frac{\tan \alpha + \tan \beta}{1 - \tan \alpha \tan \beta} \qquad \tan(\alpha - \beta) = \frac{\tan \alpha - \tan \beta}{1 + \tan \alpha \tan \beta}$$

Double-angle Formulas (pp. 484 and 485)

$$\sin(2\theta) = 2 \sin \theta \cos \theta \qquad \cos(2\theta) = \cos^2 \theta - \sin^2 \theta \qquad \tan(2\theta) = \frac{2 \tan \theta}{1 - \tan^2 \theta}$$

$$\cos(2\theta) = 2 \cos^2 \theta - 1 \qquad \cos(2\theta) = 1 - 2 \sin^2 \theta$$

Half-angle Formulas (pp. 487 and 490)

$$\sin^2 \frac{\alpha}{2} = \frac{1 - \cos \alpha}{2} \qquad \cos^2 \frac{\alpha}{2} = \frac{1 + \cos \alpha}{2} \qquad \tan^2 \frac{\alpha}{2} = \frac{1 - \cos \alpha}{1 + \cos \alpha}$$

$$\sin \frac{\alpha}{2} = \pm\sqrt{\frac{1 - \cos \alpha}{2}} \qquad \cos \frac{\alpha}{2} = \pm\sqrt{\frac{1 + \cos \alpha}{2}} \qquad \tan \frac{\alpha}{2} = \pm\sqrt{\frac{1 - \cos \alpha}{1 + \cos \alpha}} = \frac{1 - \cos \alpha}{\sin \alpha} = \frac{\sin \alpha}{1 + \cos \alpha}$$

where the $+$ or $-$ is determined by the quadrant of $\dfrac{\alpha}{2}$

Product-to-Sum Formulas (p. 493)

$$\sin \alpha \sin \beta = \frac{1}{2}[\cos(\alpha - \beta) - \cos(\alpha + \beta)]$$

$$\cos \alpha \cos \beta = \frac{1}{2}[\cos(\alpha - \beta) + \cos(\alpha + \beta)]$$

$$\sin \alpha \cos \beta = \frac{1}{2}[\sin(\alpha + \beta) + \sin(\alpha - \beta)]$$

Sum-to-Product Formulas (p. 494)

$$\sin \alpha + \sin \beta = 2 \sin \frac{\alpha + \beta}{2} \cos \frac{\alpha - \beta}{2} \qquad \sin \alpha - \sin \beta = 2 \sin \frac{\alpha - \beta}{2} \cos \frac{\alpha + \beta}{2}$$

$$\cos \alpha + \cos \beta = 2 \cos \frac{\alpha + \beta}{2} \cos \frac{\alpha - \beta}{2} \qquad \cos \alpha - \cos \beta = -2 \sin \frac{\alpha + \beta}{2} \sin \frac{\alpha - \beta}{2}$$

Objectives

Review Exercises

In Problems 1–20, find the exact value of each expression. Do not use a calculator.

1. $\sin^{-1} 1$

2. $\cos^{-1} 0$

3. $\tan^{-1} 1$

4. $\sin^{-1}\left(-\dfrac{1}{2}\right)$

5. $\cos^{-1}\left(-\dfrac{\sqrt{3}}{2}\right)$

6. $\tan^{-1}(-\sqrt{3})$

7. $\sec^{-1}\sqrt{2}$

8. $\cot^{-1}(-1)$

9. $\tan\left[\sin^{-1}\left(-\dfrac{\sqrt{3}}{2}\right)\right]$

10. $\tan\left[\cos^{-1}\left(-\dfrac{1}{2}\right)\right]$

11. $\sec\left(\tan^{-1}\dfrac{\sqrt{3}}{3}\right)$

12. $\csc\left(\sin^{-1}\dfrac{\sqrt{3}}{2}\right)$

13. $\sin\left(\tan^{-1}\dfrac{3}{4}\right)$

14. $\cos\left(\sin^{-1}\dfrac{3}{5}\right)$

15. $\tan\left[\sin^{-1}\left(-\dfrac{4}{5}\right)\right]$

16. $\tan\left[\cos^{-1}\left(-\dfrac{3}{5}\right)\right]$

17. $\sin^{-1}\left(\cos\dfrac{2\pi}{3}\right)$

18. $\cos^{-1}\left(\tan\dfrac{3\pi}{4}\right)$

19. $\tan^{-1}\left(\tan\dfrac{7\pi}{4}\right)$

20. $\cos^{-1}\left(\cos\dfrac{7\pi}{6}\right)$

In Problems 21–52, establish each identity.

21. $\tan\theta\cot\theta - \sin^2\theta = \cos^2\theta$

22. $\sin\theta\csc\theta - \sin^2\theta = \cos^2\theta$

23. $\cos^2\theta(1 + \tan^2\theta) = 1$

24. $(1 - \cos^2\theta)(1 + \cot^2\theta) = 1$

25. $4\cos^2\theta + 3\sin^2\theta = 3 + \cos^2\theta$

26. $4\sin^2\theta + 2\cos^2\theta = 4 - 2\cos^2\theta$

27. $\dfrac{1 - \cos\theta}{\sin\theta} + \dfrac{\sin\theta}{1 - \cos\theta} = 2\csc\theta$

28. $\dfrac{\sin\theta}{1 + \cos\theta} + \dfrac{1 + \cos\theta}{\sin\theta} = 2\csc\theta$

29. $\dfrac{\cos\theta}{\cos\theta - \sin\theta} = \dfrac{1}{1 - \tan\theta}$

30. $1 - \dfrac{\cos^2\theta}{1 + \sin\theta} = \sin\theta$

31. $\dfrac{\csc\theta}{1 + \csc\theta} = \dfrac{1 - \sin\theta}{\cos^2\theta}$

32. $\dfrac{1 + \sec\theta}{\sec\theta} = \dfrac{\sin^2\theta}{1 - \cos\theta}$

33. $\csc\theta - \sin\theta = \cos\theta\cot\theta$

34. $\dfrac{\csc\theta}{1 - \cos\theta} = \dfrac{1 + \cos\theta}{\sin^3\theta}$

35. $\dfrac{1 - \sin\theta}{\sec\theta} = \dfrac{\cos^3\theta}{1 + \sin\theta}$

36. $\dfrac{1 - \cos\theta}{1 + \cos\theta} = (\csc\theta - \cot\theta)^2$

37. $\dfrac{1 - 2\sin^2\theta}{\sin\theta\cos\theta} = \cot\theta - \tan\theta$

38. $\dfrac{(2\sin^2\theta - 1)^2}{\sin^4\theta - \cos^4\theta} = 1 - 2\cos^2\theta$

39. $\dfrac{\cos(\alpha + \beta)}{\cos\alpha\sin\beta} = \cot\beta - \tan\alpha$

40. $\dfrac{\sin(\alpha - \beta)}{\sin\alpha\cos\beta} = 1 - \cot\alpha\tan\beta$

41. $\dfrac{\cos(\alpha - \beta)}{\cos\alpha\cos\beta} = 1 + \tan\alpha\tan\beta$

42. $\dfrac{\cos(\alpha + \beta)}{\sin\alpha\cos\beta} = \cot\alpha - \tan\beta$

43. $(1 + \cos\theta)\tan\dfrac{\theta}{2} = \sin\theta$

44. $\sin\theta\tan\dfrac{\theta}{2} = 1 - \cos\theta$

45. $2\cot\theta\cot(2\theta) = \cot^2\theta - 1$

46. $2\sin(2\theta)(1 - 2\sin^2\theta) = \sin(4\theta)$

47. $1 - 8\sin^2\theta\cos^2\theta = \cos(4\theta)$

48. $\dfrac{\sin(3\theta)\cos\theta - \sin\theta\cos(3\theta)}{\sin(2\theta)} = 1$

49. $\dfrac{\sin(2\theta) + \sin(4\theta)}{\cos(2\theta) + \cos(4\theta)} = \tan(3\theta)$

50. $\dfrac{\sin(2\theta) + \sin(4\theta)}{\sin(2\theta) - \sin(4\theta)} + \dfrac{\tan(3\theta)}{\tan\theta} = 0$

51. $\dfrac{\cos(2\theta) - \cos(4\theta)}{\cos(2\theta) + \cos(4\theta)} - \tan\theta\tan(3\theta) = 0$

52. $\cos(2\theta) - \cos(10\theta) = \tan(4\theta)[\sin(2\theta) + \sin(10\theta)]$

In Problems 53–60, find the exact value of each expression.

53. $\sin 165°$

54. $\tan 105°$

55. $\cos\dfrac{5\pi}{12}$

56. $\sin\left(-\dfrac{\pi}{12}\right)$

57. $\cos 80°\cos 20° + \sin 80°\sin 20°$

58. $\sin 70°\cos 40° - \cos 70°\sin 40°$

59. $\tan\dfrac{\pi}{8}$

60. $\sin\dfrac{5\pi}{8}$

In Problems 61–70, use the information given about the angles α and β to find the exact value of:

(a) $\sin(\alpha + \beta)$ (b) $\cos(\alpha + \beta)$ (c) $\sin(\alpha - \beta)$ (d) $\tan(\alpha + \beta)$

(e) $\sin(2\alpha)$ (f) $\cos(2\beta)$ (g) $\sin\dfrac{\beta}{2}$ (h) $\cos\dfrac{\alpha}{2}$

61. $\sin\alpha = \dfrac{4}{5}, 0 < \alpha < \dfrac{\pi}{2}; \sin\beta = \dfrac{5}{13}, \dfrac{\pi}{2} < \beta < \pi$

62. $\cos\alpha = \dfrac{4}{5}, 0 < \alpha < \dfrac{\pi}{2}; \cos\beta = \dfrac{5}{13}, -\dfrac{\pi}{2} < \beta < 0$

63. $\sin\alpha = -\dfrac{3}{5}, \pi < \alpha < \dfrac{3\pi}{2}; \cos\beta = \dfrac{12}{13}, \dfrac{3\pi}{2} < \beta < 2\pi$

64. $\sin\alpha = -\dfrac{4}{5}, -\dfrac{\pi}{2} < \alpha < 0; \cos\beta = -\dfrac{5}{13}, \dfrac{\pi}{2} < \beta < \pi$

65. $\tan\alpha = \dfrac{3}{4}, \pi < \alpha < \dfrac{3\pi}{2}; \tan\beta = \dfrac{12}{5}, 0 < \beta < \dfrac{\pi}{2}$

66. $\tan\alpha = -\dfrac{4}{3}, \dfrac{\pi}{2} < \alpha < \pi; \cot\beta = \dfrac{12}{5}, \pi < \beta < \dfrac{3\pi}{2}$

67. $\sec\alpha = 2, -\dfrac{\pi}{2} < \alpha < 0; \sec\beta = 3, \dfrac{3\pi}{2} < \beta < 2\pi$

68. $\csc\alpha = 2, \dfrac{\pi}{2} < \alpha < \pi; \sec\beta = -3, \dfrac{\pi}{2} < \beta < \pi$

69. $\sin\alpha = -\dfrac{2}{3}, \pi < \alpha < \dfrac{3\pi}{2}; \cos\beta = -\dfrac{2}{3}, \pi < \beta < \dfrac{3\pi}{2}$

70. $\tan\alpha = -2, \dfrac{\pi}{2} < \alpha < \pi; \cot\beta = -2, \dfrac{\pi}{2} < \beta < \pi$

In Problems 71–76, find the exact value of each expression.

71. $\cos\left(\sin^{-1}\dfrac{3}{5} - \cos^{-1}\dfrac{1}{2}\right)$

72. $\sin\left(\cos^{-1}\dfrac{5}{13} - \cos^{-1}\dfrac{4}{5}\right)$

73. $\tan\left[\sin^{-1}\left(-\dfrac{1}{2}\right) - \tan^{-1}\dfrac{3}{4}\right]$

74. $\cos\left[\tan^{-1}(-1) + \cos^{-1}\left(-\dfrac{4}{5}\right)\right]$

75. $\sin\left[2\cos^{-1}\left(-\dfrac{3}{5}\right)\right]$

76. $\cos\left(2\tan^{-1}\dfrac{4}{3}\right)$

In Problems 77–100, solve each equation on the interval $0 \le \theta < 2\pi$.

77. $\cos\theta = \dfrac{1}{2}$

78. $\sin\theta = -\dfrac{\sqrt{3}}{2}$

79. $2\cos\theta + \sqrt{2} = 0$

80. $\tan\theta + \sqrt{3} = 0$

81. $\sin(2\theta) + 1 = 0$

82. $\cos(2\theta) = 0$

83. $\tan(2\theta) = 0$

84. $\sin(3\theta) = 1$

85. $\sec^2\theta = 4$

86. $\csc^2\theta = 1$

87. $\sin\theta = \tan\theta$

88. $\cos\theta = \sec\theta$

89. $\sin\theta + \sin(2\theta) = 0$

90. $\cos(2\theta) = \sin\theta$

91. $\sin(2\theta) - \cos\theta - 2\sin\theta + 1 = 0$

92. $\sin(2\theta) - \sin\theta - 2\cos\theta + 1 = 0$

93. $2\sin^2\theta - 3\sin\theta + 1 = 0$

94. $2\cos^2\theta + \cos\theta - 1 = 0$

95. $4 \sin^2 \theta = 1 + 4 \cos \theta$

96. $8 - 12 \sin^2 \theta = 4 \cos^2 \theta$

97. $\sin(2\theta) = \sqrt{2} \cos \theta$

98. $1 + \sqrt{3} \cos \theta + \cos(2\theta) = 0$

99. $\sin \theta - \cos \theta = 1$

100. $\sin \theta - \sqrt{3} \cos \theta = 2$

In Problems 101–106, use a calculator to find an approximate value for each expression, rounded to two decimal places.

101. $\sin^{-1} 0.7$

102. $\cos^{-1} \dfrac{4}{5}$

103. $\tan^{-1}(-2)$

104. $\cos^{-1}(-0.2)$

105. $\sec^{-1} 3$

106. $\cot^{-1}(-4)$

In Problems 107–112, use a graphing utility to solve each equation on the interval $0 \le x \le 2\pi$. Approximate any solutions rounded to two decimal places.

107. $2x = 5 \cos x$

108. $2x = 5 \sin x$

109. $2 \sin x + 3 \cos x = 4x$

110. $3 \cos x + x = \sin x$

111. $\sin x = \ln x$

112. $\sin x = e^{-x}$

113. Use a Half-angle formula to find the exact value of sin 15°. Then use a difference formula to find the exact value of sin 15°. Show that the answers found are the same.

114. If you are given the value of $\cos \theta$ and want the exact value of $\cos(2\theta)$, what form of the Double-angle Formula for $\cos(2\theta)$ is most efficient to use?

Chapter Test

In Problems 1–6, find the exact value of each expression. Express all angles in radians.

1. $\sec^{-1}\left(\dfrac{2}{\sqrt{3}}\right)$

2. $\sin^{-1}\left(-\dfrac{\sqrt{2}}{2}\right)$

3. $\cos^{-1}\left(\sin \dfrac{11\pi}{6}\right)$

4. $\sin\left(\tan^{-1} \dfrac{7}{3}\right)$

5. $\cot\left(\csc^{-1} \sqrt{10}\right)$

6. $\sec\left(\cos^{-1}\left(-\dfrac{3}{4}\right)\right)$

In Problems 7–10, use a calculator to evaluate each espression. Express angles in radians.

7. $\sin^{-1} 0.382$

8. $\sec^{-1} 1.4$

9. $\tan^{-1} 3$

10. $\cot^{-1} 5$

In Problems 11–16 establish each identity.

11. $\dfrac{\csc \theta + \cot \theta}{\sec \theta + \tan \theta} = \dfrac{\sec \theta - \tan \theta}{\csc \theta - \cot \theta}$

12. $\sin \theta \tan \theta + \cos \theta = \sec \theta$

13. $\tan \theta + \cot \theta = 2 \csc(2\theta)$

14. $\dfrac{\sin(\alpha + \beta)}{\tan \alpha + \tan \beta} = \cos \alpha \cos \beta$

15. $\sin(3\theta) = 3 \sin \theta - 4 \sin^3 \theta$

16. $\dfrac{\tan \theta - \cot \theta}{\tan \theta + \cot \theta} = 1 - 2 \cos^2 \theta$

In Problems 17–24 use sum, difference, product or half-angle formulas to find the exact value of each expression.

17. $\cos 15°$

18. $\tan 75°$

19. $\sin\left(\dfrac{1}{2} \cos^{-1} \dfrac{3}{5}\right)$

20. $\tan\left(2 \sin^{-1} \dfrac{6}{11}\right)$

21. $\cos\left(\sin^{-1} \dfrac{2}{3} + \tan^{-1} \dfrac{3}{2}\right)$

22. $\sin 75° \cos 15°$

23. $\sin 75° + \sin 15°$

24. $\cos 65° \cos 20° + \sin 65° \sin 20°$

In Problems 25–29, solve each equation on $0 \le \theta < 2\pi$.

25. $4 \sin^2 \theta - 3 = 0$

26. $-3 \cos\left(\dfrac{\pi}{2} - \theta\right) = \tan \theta$

27. $\cos^2 \theta + 2 \sin \theta \cos \theta - \sin^2 \theta = 0$

28. $\sin(\theta + 1) = \cos \theta$

29. $4 \sin^2 \theta + 7 \sin \theta = 2$

30. Stage 16 of the 2004 Tour de France was a time trial from Bourg d'Oisans to L'Alpe d'Huez. The average grade (slope as a percent) for most of the 15 km mountainous trek was 7.9%. What was the change in elevation from the beginning to the end of the route?

Chapter Projects

1. **Waves**   A stretched string that is attached at both ends, pulled in a direction perpendicular to the string, and released has motion that is described as wave motion. If we assume no friction and a length such that there are no "echoes" (that is, the wave doesn't bounce back), the transverse motion (motion perpendicular to the string) can be described by the equation

$$y = y_m \sin(kx - \omega t)$$

where y_m is the amplitude measured in meters and k and ω are constants. The height of the sound wave depends on the distance x from one endpoint of the string and on the time t, so a typical wave has horizontal and vertical motion over time.

(a)   What is the amplitude of the wave
$y = 0.00421 \sin(68.3x - 2.68t)$?

(b)   The value of ω is the angular frequency measured in radians per second. What is the angular frequency of the wave given in part (a)?

(c)   The frequency f is the number of vibrations per second (hertz) made by the wave as it passes a certain point. Its value is found using the formula $f = \dfrac{\omega}{2\pi}$.
What is the frequency of the wave given in part (a)?

(d)   The wavelength, λ, of a wave is the shortest distance at which the wave pattern repeats itself for a constant t.
Thus, $\lambda = \dfrac{2\pi}{k}$. What is the wavelength of the wave given in part (a)?

(e)   Graph the height of the string a distance $x = 1$ meter from an endpoint.

(f)   If two waves travel simultaneously along the same stretched string, the vertical displacement of the string when both waves act is $y = y_1 + y_2$, where y_1 is the vertical displacement of the first wave and y_2 is the vertical displacement of the second wave. This result is called the Principle of Superposition and was analyzed by the French mathematician Jean Baptiste Fourier (1768–1830). When two waves travel along the same string, one wave will differ from the other wave by a phase constant ϕ. That is,

$$y_1 = y_m \sin(kx - \omega t)$$
$$y_2 = y_m \sin(kx - \omega t + \phi)$$

assuming that each wave has the same amplitude. Write $y_1 + y_2$ as a product using the Sum-to-Product Formulas.

(g)   Suppose that two waves are moving in the same direction along a stretched string. The amplitude of each wave is 0.0045 meter, and the phase difference between them is 2.5 radians. The wavelength, λ, of each wave is 0.09 meter and the frequency, f, is 2.3 hertz. Find y_1, y_2, and $y_1 + y_2$.

(h)   Using a graphing utility, graph y_1, y_2, and $y_1 + y_2$ on the same viewing window.

(i)   Redo parts (g) and (h) with the phase difference between the waves being 0.4 radian.

(j)   What effect does the phase difference have on the amplitude of $y_1 + y_2$?

The following projects are available on the Instructor's Resource Center (IRC):

2. **Project at Motorola**   *Sending Pictures Wirelessly*

3. **Jacob's Field**

4. **Calculus of Differences**

Cumulative Review

1. Find the real solutions, if any, of the equation $3x^2 + x - 1 = 0$.

2. Find an equation for the line containing the points $(-2, 5)$ and $(4, -1)$. What is the distance between these points? What is their midpoint?

3. Test the equation $3x + y^2 = 9$ for symmetry with respect to the x-axis, y-axis, and origin. List the intercepts.

4. Use transformations to graph the equation $y = |x - 3| + 2$.

5. Use transformations to graph the equation $y = 3e^x - 2$.

6. Use transformations to graph the equation $y = \cos\left(x - \dfrac{\pi}{2}\right) - 1$.

7. Sketch a graph of each of the following functions. Label at least three points on each graph. Name the inverse function of each and show its graph.
   (a) $y = x^3$
   (b) $y = e^x$
   (c) $y = \sin x$, $-\dfrac{\pi}{2} \le x \le \dfrac{\pi}{2}$
   (d) $y = \cos x$, $0 \le x \le \pi$

8. If $\sin\theta = -\dfrac{1}{3}$ and $\pi < \theta < \dfrac{3\pi}{2}$, find the exact value of:
   (a) $\cos\theta$ (b) $\tan\theta$ (c) $\sin(2\theta)$
   (d) $\cos(2\theta)$ (e) $\sin\left(\dfrac{1}{2}\theta\right)$ (f) $\cos\left(\dfrac{1}{2}\theta\right)$

9. Find the exact value of $\cos(\tan^{-1} 2)$.

10. If $\sin\alpha = \dfrac{1}{3}$, $\dfrac{\pi}{2} < \alpha < \pi$, and $\cos\beta = -\dfrac{1}{3}$, $\pi < \beta < \dfrac{3\pi}{2}$, find the exact value of:
   (a) $\cos\alpha$ (b) $\sin\beta$ (c) $\cos(2\alpha)$
   (d) $\cos(\alpha + \beta)$ (e) $\sin\dfrac{\beta}{2}$

11. For the function
$$f(x) = 2x^5 - x^4 - 4x^3 + 2x^2 + 2x - 1:$$
   (a) Find the real zeros and their multiplicity.
   (b) Find the intercepts.
   (c) Find the power function that the graph of f resembles for large $|x|$.
   (d) Graph f using a graphing utility.
   (e) Approximate the turning points, if any exist.
   (f) Use the information obtained in parts (a)–(e) to sketch a graph of f by hand.
   (g) Identify the intervals on which f is increasing, decreasing, or constant.

12. If $f(x) = 2x^2 + 3x + 1$ and $g(x) = x^2 + 3x + 2$, solve:
   (a) $f(x) = 0$
   (b) $f(x) = g(x)$
   (c) $f(x) > 0$
   (d) $f(x) \ge g(x)$

Applications of Trigonometric Functions

7

Chapter 7 in *Precalculus Essentials: Enhanced with Graphing Utilities*, 4e includes only the highlighted Sections 7.1, 7.2, and 7.3.

A LOOK BACK   In Chapter 5 we defined the six trigonometric functions using the unit circle and then extended this definition to include circles of radius *r*. In particular, we learned to evaluate the trigonometric functions. We also learned how to graph sinusoidal functions.

A LOOK AHEAD   In this chapter, we define the trigonometric functions using right triangles and use the trigonometric functions to solve applied problems. The first four sections deal with applications involving right triangles and *oblique triangles*, triangles that do not have a right angle. To solve problems involving oblique triangles, we will develop the Law of Sines and the Law of Cosines. We will also develop formulas for finding the area of a triangle.

The final section deals with applications of sinusoidal functions involving simple harmonic motion and damped motion.

OUTLINE

7.1 Right Triangle Trigonometry; Applications

PREPARING FOR THIS SECTION *Before getting started, review the following:*

- Pythagorean Theorem (Appendix A, Section A.2, pp. 669–670)
- Trigonometric Equations (I) (Section 6.7, pp. 496–500)

Now work the 'Are You Prepared?' problems on page 526.

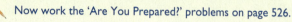

OBJECTIVES
1. Find the Value of Trigonometric Functions of Acute Angles
2. Use the Complementary Angle Theorem
3. Solve Right Triangles
4. Solve Applied Problems

---

1 Find the Value of Trigonometric Functions of Acute Angles

A triangle in which one angle is a right angle (90°) is called a **right triangle.** Recall that the side opposite the right angle is called the **hypotenuse,** and the remaining two sides are called the **legs** of the triangle. In Figure 1(a), we have labeled the hypotenuse as c to indicate that its length is c units, and, in a like manner, we have labeled the legs as a and b. Because the triangle is a right triangle, the Pythagorean Theorem tells us that

$$a^2 + b^2 = c^2$$

In Figure 1(a), we also show the angle θ. The angle θ is an **acute angle:** that is, $0° < \theta < 90°$ for θ measured in degrees and $0 < \theta < \dfrac{\pi}{2}$ for θ measured in radians.

Place θ in standard position, as shown in Figure 1(b). Then the coordinates of the point P are (a, b). Also, P is a point on the terminal side of θ that is on the circle $x^2 + y^2 = c^2$. (Do you see why?)

Figure 1

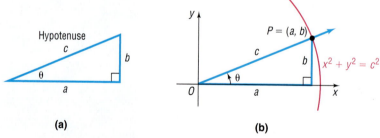

(a)   (b)

Now use the theorem on page 382. By referring to the lengths of the sides of the triangle by the names hypotenuse (c), opposite (b), and adjacent (a), as indicated in Figure 2, we can express the trigonometric functions of θ as ratios of the sides of a right triangle.

Figure 2

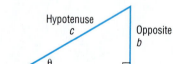

$$
\begin{aligned}
\sin \theta &= \frac{\text{Opposite}}{\text{Hypotenuse}} = \frac{b}{c} & \csc \theta &= \frac{\text{Hypotenuse}}{\text{Opposite}} = \frac{c}{b} \\
\cos \theta &= \frac{\text{Adjacent}}{\text{Hypotenuse}} = \frac{a}{c} & \sec \theta &= \frac{\text{Hypotenuse}}{\text{Adjacent}} = \frac{c}{a} \\
\tan \theta &= \frac{\text{Opposite}}{\text{Adjacent}} = \frac{b}{a} & \cot \theta &= \frac{\text{Adjacent}}{\text{Opposite}} = \frac{a}{b}
\end{aligned}
\tag{1}
$$

Notice that each of the trigonometric functions of the acute angle θ is positive.

EXAMPLE 1	**Finding the Value of Trigonometric Functions from a Right Triangle**

Find the exact value of the six trigonometric functions of the angle θ in Figure 3.

Solution   We see in Figure 3 that the two given sides of the triangle are

Figure 3

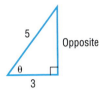

$$c = \text{Hypotenuse} = 5, \quad a = \text{Adjacent} = 3$$

To find the length of the opposite side, we use the Pythagorean Theorem.

$$(\text{Adjacent})^2 + (\text{Opposite})^2 = (\text{Hypotenuse})^2$$
$$3^2 + (\text{Opposite})^2 = 5^2$$
$$(\text{Opposite})^2 = 25 - 9 = 16$$
$$\text{Opposite} = 4$$

Now that we know the lengths of the three sides, we use the ratios in equations (1) to find the value of each of the six trigonometric functions.

$$\sin \theta = \frac{\text{Opposite}}{\text{Hypotenuse}} = \frac{4}{5} \quad \cos \theta = \frac{\text{Adjacent}}{\text{Hypotenuse}} = \frac{3}{5} \quad \tan \theta = \frac{\text{Opposite}}{\text{Adjacent}} = \frac{4}{3}$$

$$\csc \theta = \frac{\text{Hypotenuse}}{\text{Opposite}} = \frac{5}{4} \quad \sec \theta = \frac{\text{Hypotenuse}}{\text{Adjacent}} = \frac{5}{3} \quad \cot \theta = \frac{\text{Adjacent}}{\text{Opposite}} = \frac{3}{4} \quad ◀$$

NOW WORK PROBLEM 9.

The values of the trigonometric functions of an acute angle are ratios of the lengths of the sides of a right triangle. This way of viewing the trigonometric functions leads to many applications and, in fact, was the point of view used by early mathematicians (before calculus) in studying the subject of trigonometry.
We look at one such application next.

EXAMPLE 2	**Constructing a Rain Gutter**

Figure 4(a)

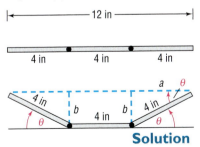

A rain gutter is to be constructed of aluminum sheets 12 inches wide. After marking off a length of 4 inches from each edge, this length is bent up at an angle θ. See Figure 4(a).

(a) Express the area A of the opening as a function of θ.

   [**Hint:** Let b denote the vertical height of the bend.]

(b) Graph $A = A(\theta)$. Find the angle θ that makes A largest. (This bend will allow the most water to flow through the gutter.)

Solution   (a) Look again at Figure 4(a). The area A of the opening is the sum of the areas of two congruent right triangles and one rectangle. Look at Figure 4(b), showing the triangle in Figure 4(a) redrawn. We see that

Figure 4(b)

$$\cos \theta = \frac{a}{4} \quad \text{so} \quad a = 4 \cos \theta \qquad \sin \theta = \frac{b}{4} \quad \text{so} \quad b = 4 \sin \theta$$

The area of the triangle is

$$\text{area} = \frac{1}{2}(\text{base})(\text{height}) = \frac{1}{2}ab = \frac{1}{2}(4\cos\theta)(4\sin\theta) = 8\sin\theta\cos\theta$$

So the area of the two triangles is $16\sin\theta\cos\theta$.

The rectangle has length 4 and height b, so its area is

$$4b = 4(4\sin\theta) = 16\sin\theta$$

The area A of the opening is

$$A = \text{area of the two triangles} + \text{area of the rectangle}$$
$$A(\theta) = 16\sin\theta\cos\theta + 16\sin\theta = 16\sin\theta(\cos\theta + 1)$$

(b) Figure 5 shows the graph of $A = A(\theta)$. Using MAXIMUM, the angle θ that makes A largest is 60°. ◀

Figure 5

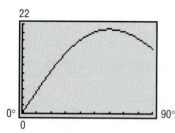

2 Use the Complementary Angle Theorem

Two acute angles are called **complementary** if their sum is a right angle. Because the sum of the angles of any triangle is 180°, it follows that, for a right triangle, the two acute angles are complementary.

Refer now to Figure 6. We have labeled the angle opposite side b as β and the angle opposite side a as α. Notice that side b is adjacent to angle α and side a is adjacent to angle β. As a result,

Figure 6

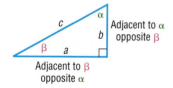

$$\sin\beta = \frac{b}{c} = \cos\alpha \qquad \cos\beta = \frac{a}{c} = \sin\alpha \qquad \tan\beta = \frac{b}{a} = \cot\alpha$$
$$\csc\beta = \frac{c}{b} = \sec\alpha \qquad \sec\beta = \frac{c}{a} = \csc\alpha \qquad \cot\beta = \frac{a}{b} = \tan\alpha$$
$$\text{(2)}$$

Because of these relationships, the functions sine and cosine, tangent and cotangent, and secant and cosecant are called **cofunctions** of each other. The identities (2) may be expressed in words as follows:

Complementary Angle Theorem

Cofunctions of complementary angles are equal.

Examples of this theorem are given next:

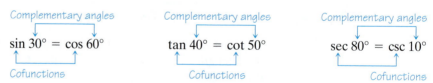

EXAMPLE 3   Using the Complementary Angle Theorem

(a) $\sin 62° = \cos(90° - 62°) = \cos 28°$

(b) $\tan\dfrac{\pi}{12} = \cot\left(\dfrac{\pi}{2} - \dfrac{\pi}{12}\right) = \cot\dfrac{5\pi}{12}$

(c) $\sin^2 40° + \sin^2 50° = \sin^2 40° + \cos^2 40° = 1$

$$\uparrow$$
$$\sin 50° = \cos 40°$$

◀

NOW WORK PROBLEM 19.

3 Solve Right Triangles

In the discussion that follows, we will always label a right triangle so that side a is opposite angle α, side b is opposite angle β, and side c is the hypotenuse, as shown in Figure 7. **To solve a right triangle** means to find the missing lengths of its sides and the measurements of its angles. We shall follow the practice of expressing the lengths of the sides rounded to two decimal places and expressing angles in degrees rounded to one decimal place. (Be sure that your calculator is in degree mode.)

To solve a right triangle, we need to know one of the acute angles α or β and a side, or else two sides. Then we make use of the Pythagorean Theorem and the fact that the sum of the angles of a triangle is 180°. The sum of the angles α and β in a right triangle is therefore 90°.

Figure 7

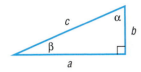

For the right triangle shown in Figure 7, we have

$$c^2 = a^2 + b^2, \qquad \alpha + \beta = 90°$$

EXAMPLE 4	**Solving a Right Triangle**

Use Figure 8. If $b = 2$ and $\alpha = 40°$, find a, c, and β.

Figure 8

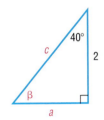

Solution   Since $\alpha = 40°$ and $\alpha + \beta = 90°$, we find that $\beta = 50°$. To find the sides a and c, we use the facts that

$$\tan 40° = \frac{a}{2} \quad \text{and} \quad \cos 40° = \frac{2}{c}$$

Now solve for a and c.

$$a = 2 \tan 40° \approx 1.68 \quad \text{and} \quad c = \frac{2}{\cos 40°} \approx 2.61 \qquad \blacktriangleleft$$

 NOW WORK PROBLEM 29.

EXAMPLE 5	**Solving a Right Triangle**

Use Figure 9. If $a = 3$ and $b = 2$, find c, α, and β.

Figure 9

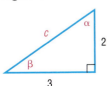

Solution   Since $a = 3$ and $b = 2$, then, by the Pythagorean Theorem, we have

$$c^2 = a^2 + b^2 = 3^2 + 2^2 = 9 + 4 = 13$$

$$c = \sqrt{13} \approx 3.61$$

To find angle α, we use the fact that

$$\tan \alpha = \frac{3}{2} \quad \text{so} \quad \alpha = \tan^{-1} \frac{3}{2}$$

Set the mode on your calculator to degrees. Then, rounded to one decimal place, we find that $\alpha = 56.3°$. Since $\alpha + \beta = 90°$, we find that $\beta = 33.7°$. ◄

NOTE To avoid round-off errors when using a calculator, we will store unrounded values in memory for use in subsequent calculations. ∎

 NOW WORK PROBLEM 39.

4 Solve Applied Problems

One common use for trigonometry is to measure heights and distances that are either awkward or impossible to measure by ordinary means.

EXAMPLE 6	**Finding the Width of a River**

A surveyor can measure the width of a river by setting up a transit[*] at a point C on one side of the river and taking a sighting of a point A on the other side. Refer to Figure 10. After turning through an angle of $90°$ at C, the surveyor walks a distance of 200 meters to point B. Using the transit at B, the angle β is measured and found to be $20°$. What is the width of the river rounded to the nearest meter?

Figure 10

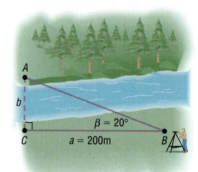

Solution   We seek the length of side b. We know a and β, so we use the fact that

$$\tan \beta = \frac{b}{a}$$

to get

$$\tan 20° = \frac{b}{200}$$

$$b = 200 \tan 20° \approx 72.79 \text{ meters}$$

The width of the river is 73 meters, rounded to the nearest meter. ◄

 NOW WORK PROBLEM 49.

[*]An instrument used in surveying to measure angles.

| EXAMPLE 7 | Finding the Inclination of a Mountain Trail |

A straight trail leads from the Alpine Hotel, elevation 8000 feet, to a scenic overlook, elevation 11,100 feet. The length of the trail is 14,100 feet. What is the inclination (grade) of the trail? That is, what is the angle β in Figure 11?

Solution   As Figure 11 illustrates, the angle β obeys the equation

Figure 11

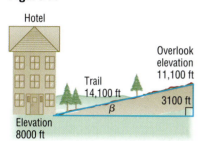

$$\sin \beta = \frac{3100}{14,100}$$

Using a calculator,

$$\beta = \sin^{-1} \frac{3100}{14,100} \approx 12.7°$$

The inclination (grade) of the trail is approximately 12.7°.   ◄

Vertical heights can sometimes be measured using either the *angle of elevation* or the *angle of depression*. If a person is looking up at an object, the acute angle measured from the horizontal to a line-of-sight observation of the object is called the **angle of elevation.** See Figure 12(a).

If a person is looking down at an object, the acute angle made by the line-of-sight observation of the object and the horizontal is called the **angle of depression.** See Figure 12(b).

Figure 12

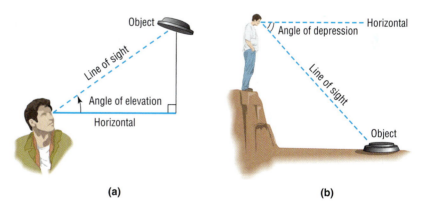

(a)   (b)

| EXAMPLE 8 | Finding the Height of a Cloud |

Meteorologists find the height of a cloud using an instrument called a **ceilometer.** A ceilometer consists of a **light projector** that directs a vertical light beam up to the cloud base and a **light detector** that scans the cloud to detect the light beam. See Figure 13(a). On December 1, 2004, at Midway Airport in Chicago, a ceilometer with a base of 300 feet was employed to find the height of the cloud cover. If the angle of elevation of the light detector is 75°, what is the height of the cloud cover? Round the answer to the nearest foot.

Figure 13

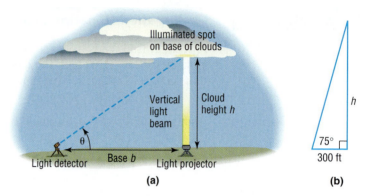

(a)

(b)

Solution   Figure 13(b) illustrates the situation. To find the height h, we use the fact that $\tan 75° = \dfrac{h}{300}$, so

$$h = 300 \tan 75° \approx 1120 \text{ feet}$$

The ceiling (height to the base of the cloud cover) is approximately 1120 feet.   ◄

⟶ **NOW WORK PROBLEM 51.**

The idea behind Example 8 can also be used to find the height of an object with a base that is not accessible to the horizontal.

EXAMPLE 9   **Finding the Height of a Statue on a Building**

Adorning the top of the Board of Trade building in Chicago is a statue of Ceres, the Roman goddess of wheat. From street level, two observations are taken 400 feet from the center of the building. The angle of elevation to the base of the statue is found to be 55.1°; the angle of elevation to the top of the statue is 56.5°. See Figure 14(a). What is the height of the statue? Round the answer to the nearest foot.

Figure 14

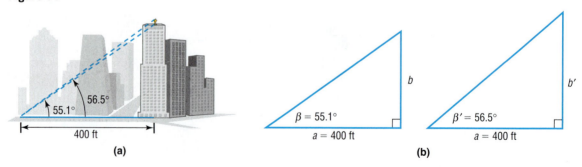

(a)

(b)

Solution   Figure 14(b) shows two triangles that replicate Figure 14(a). The height of the statue of Ceres will be $b' - b$. To find b and b', we refer to Figure 14(b).

$$\tan 55.1° = \frac{b}{400} \qquad\qquad \tan 56.5° = \frac{b'}{400}$$

$$b = 400 \tan 55.1° \approx 573 \qquad b' = 400 \tan 56.5° \approx 604$$

The height of the statue is approximately $604 - 573 = 31$ feet.   ◄

⟶ **NOW WORK PROBLEM 59.**

| EXAMPLE 10 | The Gibb's Hill Lighthouse, Southampton, Bermuda |

In operation since 1846, the Gibb's Hill Lighthouse stands 117 feet high on a hill 245 feet high, so its beam of light is 362 feet above sea level. A brochure states that the light can be seen on the horizon about 26 miles distant. Verify the accuracy of this statement.

Solution

Figure 15

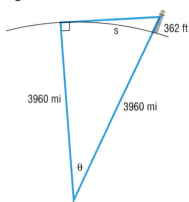

Figure 15 illustrates the situation. The central angle θ, positioned at the center of Earth, radius 3960 miles, obeys the equation

$$\cos \theta = \frac{3960}{3960 + \dfrac{362}{5280}} \approx 0.999982687 \qquad \text{1 mile = 5280 feet}$$

Solving for θ, we find

$$\theta \approx 0.33715° \approx 20.23'$$

The brochure does not indicate whether the distance is measured in nautical miles or statute miles. Let's calculate both distances.

The distance s in nautical miles (refer to Problem 114, p. 369) is the measure of angle θ in minutes, so $s = 20.23$ nautical miles.

The distance s in statute miles is given by the formula $s = r\theta$, where θ is measured in radians. Since

$$\theta = 20.23' \approx 0.33715° \approx 0.00588 \text{ radian}$$

$$1' = \frac{1}{60}° \qquad 1° = \frac{\pi}{180} \text{radian}$$

we find that

$$s = r\theta = (3960)(0.00588) = 23.3 \text{ miles}$$

Figure 16

In either case, it would seem that the brochure overstated the distance somewhat. ◄

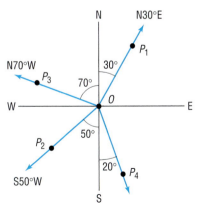

In navigation and surveying, the **direction** or **bearing** from a point O to a point P equals the acute angle θ between the ray OP and the vertical line through O, the north–south line.

Figure 16 illustrates some bearings. Notice that the bearing from O to P_1 is denoted by the symbolism N30°E, indicating that the bearing is 30° east of north. In writing the bearing from O to P, the direction north or south always appears first, followed by an acute angle, followed by east or west. In Figure 16, the bearing from O to P_2 is S50°W, and from O to P_3 it is N70°W.

| EXAMPLE 11 | Finding the Bearing of an Object |

In Figure 16, what is the bearing from O to an object at P_4?

Solution

The acute angle between the ray OP_4 and the north–south line through O is given as 20°. The bearing from O to P_4 is S20°E. ◄

EXAMPLE 12	**Finding the Bearing of an Airplane**

A Boeing 777 aircraft takes off from O'Hare Airport on runway 2 LEFT, which has a bearing of N20°E.* After flying for 1 mile, the pilot of the aircraft requests permission to turn 90° and head toward the northwest. The request is granted. After the plane goes 2 miles in this direction, what bearing should the control tower use to locate the aircraft?

Solution   Figure 17 illustrates the situation. After flying 1 mile from the airport O (the control tower), the aircraft is at P. After turning 90° toward the northwest and flying 2 miles, the aircraft is at the point Q. In triangle OPQ, the angle θ obeys the equation

$$\tan \theta = \frac{2}{1} = 2 \quad \text{so} \quad \theta = \tan^{-1} 2 \approx 63.4°$$

Figure 17

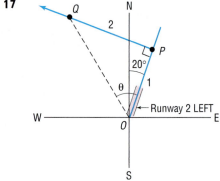

The acute angle between north and the ray OQ is $63.4° - 20° = 43.4°$. The bearing of the aircraft from O to Q is N43.4°W.   ◄

   **NOW WORK PROBLEM 67.**

---

*In air navigation, the term **azimuth** is employed to denote the positive angle measured clockwise from the north (N) to a ray OP. In Figure 16, the azimuth from O to P_1 is 30°; the azimuth from O to P_2 is 230°; the azimuth from O to P_3 is 290°. In naming runways, the units digit is left off the azimuth. Runway 2 LEFT means the left runway with a direction of azimuth 20° (bearing N20°E). Runway 23 is the runway with azimuth 230° and bearing S50°W.

7.1 Assess Your Understanding

'Are You Prepared?'

Answers are given at the end of these exercises. If you get a wrong answer, read the pages listed in red.

1. In a right triangle, if the length of the hypotenuse is 5 and the length of one of the other sides is 3, what is the length of the third side? (p. 670)

2. If θ is an angle, $0° < \theta < 90°$, solve the equation $\tan \theta = \frac{1}{2}$. (pp. 496–500)

Concepts and Vocabulary

3. *True or False:* The angles 52° and 48° are complementary.

4. *True or False:* In a right triangle, one of the angles is 90° and the sum of the other two angles is 90°.

5. When you look up at an object, the acute angle measured from the horizontal to a line-of-sight observation of the object is called the _____ _____ _____.

6. When you look down at an object, the acute angle described in Problem 5 is called the _____ _____ _____.

7. *True or False:* In a right triangle, if two sides are known, we can solve the triangle.

8. *True or False:* In a right triangle, if we know the two acute angles, we can solve the triangle.

Skill Building

In Problems 9–18, find the exact value of the six trigonometric functions of the angle θ in each figure.

9.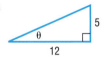

10.

11.

12.

13.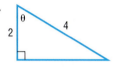

14.

15.

16.

17.

18.

In Problems 19–28, find the exact value of each expression. Do not use a calculator.

19. $\sin 38° - \cos 52°$

20. $\tan 12° - \cot 78°$

21. $\dfrac{\cos 10°}{\sin 80°}$

22. $\dfrac{\cos 40°}{\sin 50°}$

23. $1 - \cos^2 20° - \cos^2 70°$

24. $1 + \tan^2 5° - \csc^2 85°$

25. $\tan 20° - \dfrac{\cos 70°}{\cos 20°}$

26. $\cot 40° - \dfrac{\sin 50°}{\sin 40°}$

27. $\cos 35° \sin 55° + \sin 35° \cos 55°$

28. $\sec 35° \csc 55° - \tan 35° \cot 55°$

In Problems 29–42, use the right triangle shown in the margin. Then, using the given information, solve the triangle.

29. $b = 5$, $\beta = 20°$; find $a, c,$ and α

30. $b = 4$, $\beta = 10°$; find $a, c,$ and α

31. $a = 6$, $\beta = 40°$; find $b, c,$ and α

32. $a = 7$, $\beta = 50°$; find $b, c,$ and α

33. $b = 4$, $\alpha = 10°$; find $a, c,$ and β

34. $b = 6$, $\alpha = 20°$; find $a, c,$ and β

35. $a = 5$, $\alpha = 25°$; find $b, c,$ and β

36. $a = 6$, $\alpha = 40°$; find $b, c,$ and β

37. $c = 9$, $\beta = 20°$; find $b, a,$ and α

38. $c = 10$, $\alpha = 40°$; find $b, a,$ and β

39. $a = 5$, $b = 3$; find $c, \alpha,$ and β

40. $a = 2$, $b = 8$; find $c, \alpha,$ and β

41. $a = 2$, $c = 5$; find $b, \alpha,$ and β

42. $b = 4$, $c = 6$; find $a, \alpha,$ and β

Applications and Extensions

43. Geometry A right triangle has a hypotenuse of length 8 inches. If one angle is 35°, find the length of each leg.

44. Geometry A right triangle has a hypotenuse of length 10 centimeters. If one angle is 40°, find the length of each leg.

45. Geometry A right triangle contains a 25° angle. If one leg is of length 5 inches, what is the length of the hypotenuse? [**Hint:** Two answers are possible.]

46. Geometry A right triangle contains an angle of $\dfrac{\pi}{8}$ radian. If one leg is of length 3 meters, what is the length of the hypotenuse? [**Hint:** Two answers are possible.]

47. Geometry The hypotenuse of a right triangle is 5 inches. If one leg is 2 inches, find the degree measure of each angle.

48. Geometry The hypotenuse of a right triangle is 3 feet. If one leg is 1 foot, find the degree measure of each angle.

49. Finding the Width of a Gorge Find the distance from A to C across the gorge illustrated in the figure.

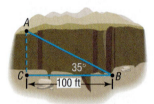

50. Finding the Distance across a Pond Find the distance from A to C across the pond illustrated in the figure.

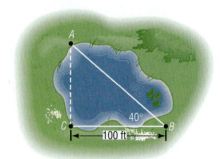

51. The Eiffel Tower The tallest tower built before the era of television masts, the Eiffel Tower was completed on March 31, 1889. Find the height of the Eiffel Tower (before a television mast was added to the top) using the information given in the illustration.

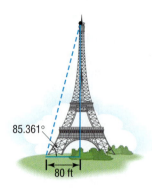

52. Finding the Distance of a Ship from Shore A ship, offshore from a vertical cliff known to be 100 feet in height, takes a sighting of the top of the cliff. If the angle of elevation is found to be 25°, how far offshore is the ship?

53. Finding the Distance to a Plateau Suppose that you are headed toward a plateau 50 meters high. If the angle of elevation to the top of the plateau is 20°, how far are you from the base of the plateau?

54. Statue of Liberty A ship is just offshore of New York City. A sighting is taken of the Statue of Liberty, which is about 305 feet tall. If the angle of elevation to the top of the statue is 20°, how far is the ship from the base of the statue?

55. Finding the Reach of a Ladder A 22-foot extension ladder leaning against a building makes a 70° angle with the ground. How far up the building does the ladder touch?

56. Finding the Height of a Building To measure the height of a building, two sightings are taken a distance of 50 feet apart. If the first angle of elevation is 40° and the second is 32°, what is the height of the building?

57. Finding the Distance between Two Objects A blimp, suspended in the air at a height of 500 feet, lies directly over a line from Soldier Field to the Adler Planetarium on Lake Michigan (see the figure). If the angle of depression from the blimp to the stadium is 32° and from the blimp to the planetarium is 23°, find the distance between Soldier Field and the Adler Planetarium.

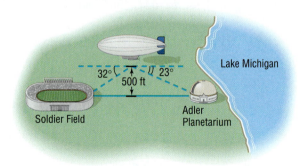

58. Finding the Angle of Elevation of the Sun At 10 AM on April 26, 2005, a building 300 feet high casts a shadow 50 feet long. What is the angle of elevation of the Sun?

59. Mt. Rushmore To measure the height of Lincoln's caricature on Mt. Rushmore, two sightings 800 feet from the base of the mountain are taken. If the angle of elevation to the bottom of Lincoln's face is 32° and the angle of elevation to the top is 35°, what is the height of Lincoln's face?

60. Directing a Laser Beam A laser beam is to be directed through a small hole in the center of a circle of radius 10 feet. The origin of the beam is 35 feet from the circle (see the figure). At what angle of elevation should the beam be aimed to ensure that it goes through the hole?

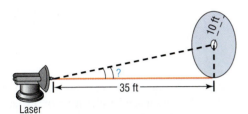

61. Finding the Length of a Guy Wire A radio transmission tower is 200 feet high. How long should a guy wire be if it is to be attached to the tower 10 feet from the top and is to make an angle of 21° with the ground?

62. Finding the Height of a Tower A guy wire 80 feet long is attached to the top of a radio transmission tower, making an angle of 25° with the ground. How high is the tower?

63. Washington Monument The angle of elevation of the Sun is 35.1° at the instant it casts a shadow 789 feet long of the Washington Monument. Use this information to calculate the height of the monument.

64. Finding the Length of a Mountain Trail A straight trail with an inclination of 17° leads from a hotel at an elevation of 9000 feet to a mountain lake at an elevation of 11,200 feet. What is the length of the trail?

65. Finding the Speed of a Truck A state trooper is hidden 30 feet from a highway. One second after a truck passes, the angle θ between the highway and the line of observation from the patrol car to the truck is measured. See the illustration.

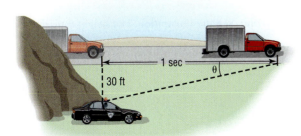

(a) If the angle measures 15°, how fast is the truck traveling? Express the answer in feet per second and in miles per hour.
(b) If the angle measures 20°, how fast is the truck traveling? Express the answer in feet per second and in miles per hour.
(c) If the speed limit is 55 miles per hour and a speeding ticket is issued for speeds of 5 miles per hour or more over the limit, for what angles should the trooper issue a ticket?

66. Security A security camera in a neighborhood bank is mounted on a wall 9 feet above the floor. What angle of depression should be used if the camera is to be directed to a spot 6 feet above the floor and 12 feet from the wall?

67. Finding the Bearing of an Aircraft A DC-9 aircraft leaves Midway Airport from runway 4 RIGHT, whose bearing is N40°E. After flying for $\frac{1}{2}$ mile, the pilot requests permission to turn 90° and head toward the southeast. The permission is granted. After the airplane goes 1 mile in this direction, what bearing should the control tower use to locate the aircraft?

68. Finding the Bearing of a Ship A ship leaves the port of Miami with a bearing of S80°E and a speed of 15 knots. After 1 hour, the ship turns 90° toward the south. After 2 hours, maintaining the same speed, what is the bearing to the ship from port?

69. Finding the Pitch of a Roof A carpenter is preparing to put a roof on a garage that is 20 feet by 40 feet by 20 feet. A steel support beam 46 feet in height is positioned in the center of the garage. To support the roof, another beam will be attached to the top of the center beam (see the figure). At what angle of elevation is the new beam? In other words, what is the pitch of the roof?

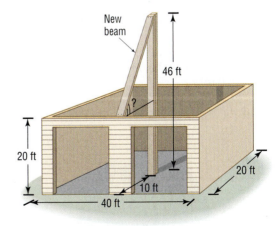

70. Shooting Free Throws in Basketball The eyes of a basketball player are 6 feet above the floor. The player is at the free-throw line, which is 15 feet from the center of the basket rim.

See the figure. What is the angle of elevation from the player's eyes to the center of the rim?
[**Hint:** The rim is 10 feet above the floor.]

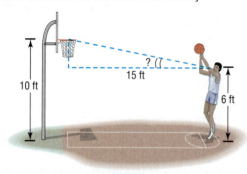

71. Geometry Find the value of the angle θ in degrees rounded to the nearest tenth of a degree.

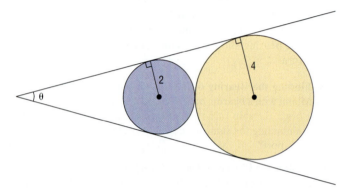

72. Surveillance Satellites A surveillance satellite circles Earth at a height of h miles above the surface. Suppose that d is the distance, in miles, on the surface of Earth that can be observed from the satellite. See the illustration.

(a) Find an equation that relates the central angle θ to the height h.
(b) Find an equation that relates the observable distance d and θ.
(c) Find an equation that relates d and h.
(d) If d is to be 2500 miles, how high must the satellite orbit above Earth?
(e) If the satellite orbits at a height of 300 miles, what distance d on the surface can be observed?

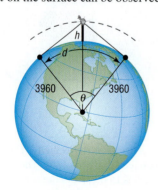

73. Photography A camera is mounted on a tripod 4 feet high at a distance of 10 feet from George, who is 6 feet tall. See the illustration. If the camera lens has an angle of depression and an angle of elevation of 20°, will George's feet and head be seen by the lens? If not, how far back will the camera need to be moved to include George's feet and head?

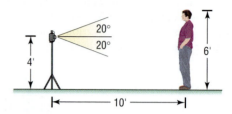

74. Construction A ramp for wheelchair accessibility is to be constructed with an angle of elevation of 15° and a final height of 5 feet. How long is the ramp?

75. Geometry A rectangle is inscribed in a semicircle of radius 1. See the illustration.

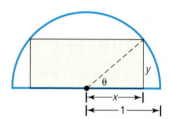

(a) Express the area A of the rectangle as a function of the angle θ shown in the illustration.
(b) Show that $A = \sin(2\theta)$.
(c) Find the angle θ that results in the largest area A.
(d) Find the dimensions of this largest rectangle.

76. Area of an Isosceles Triangle Show that the area A of an isosceles triangle, whose equal sides are of length s and the angle between them is θ, is

$$A = \frac{1}{2}s^2 \sin \theta$$

[**Hint:** See the illustration. The height h bisects the angle θ and is the perpendicular bisector of the base.]

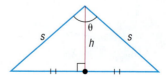

Discussion and Writing

77. The Gibb's Hill Lighthouse, Southampton, Bermuda In operation since 1846, the Gibb's Hill Lighthouse stands 117 feet high on a hill 245 feet high, so its beam of light is 362 feet above sea level. A brochure states that ships 40 miles away can see the light and planes flying at 10,000 feet can see it 120 miles away. Verify the accuracy of these statements. What assumption did the brochure make about the height of the ship?

78. Explain how you would measure the width of the Grand Canyon from a point on its ridge.

79. Explain how you would measure the height of a TV tower that is on the roof of a tall building.

'Are You Prepared?' Answers

1. 4   **2.** 26.6°

7.2 The Law of Sines

PREPARING FOR THIS SECTION   *Before getting started, review the following:*

- Trigonometric Equations (I) (Section 6.7, pp. 496–500)
- Difference Formula for Sines (Section 6.4, p. 476)

✏️ Now work the 'Are You Prepared?' problems on page 538.

OBJECTIVES **1** Solve SAA or ASA Triangles
  2 Solve SSA Triangles
  3 Solve Applied Problems

If none of the angles of a triangle is a right angle, the triangle is called **oblique**. An oblique triangle will have either three acute angles or two acute angles and one obtuse angle (an angle between 90° and 180°). See Figure 18.

Figure 18

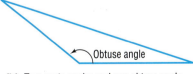

(a) All angles are acute   **(b)** Two acute angles and one obtuse angle

Figure 19

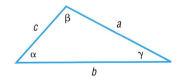

In the discussion that follows, we will always label an oblique triangle so that side a is opposite angle α, side b is opposite angle β, and side c is opposite angle γ, as shown in Figure 19.

To **solve an oblique triangle** means to find the lengths of its sides and the measurements of its angles. To do this, we shall need to know the length of one side* along with (i) two angles; (ii) one angle and one other side; or (iii) the other two sides. There are four possibilities to consider:

---

*The reason we need to know the length of one side is that, if we only know the angles, this will result in a family of *similar triangles*.

CASE 1: One side and two angles are known (ASA or SAA).

CASE 2: Two sides and the angle opposite one of them are known (SSA).

CASE 3: Two sides and the included angle are known (SAS).

CASE 4: Three sides are known (SSS).

Figure 20 illustrates the four cases.

Figure 20

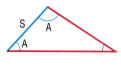
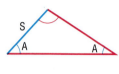
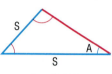
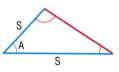
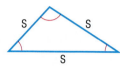

Case 1: ASA   Case 1: SAA   Case 2: SSA   Case 3: SAS   Case 4: SSS

WARNING

Oblique triangles cannot be solved using the methods of Section 7.1. Do you know why? ■

The **Law of Sines** is used to solve triangles for which Case 1 or 2 holds. Cases 3 and 4 are considered when we study the Law of Cosines in the next section.

Theorem

Law of Sines

For a triangle with sides a, b, c and opposite angles α, β, γ, respectively,

$$\frac{\sin \alpha}{a} = \frac{\sin \beta}{b} = \frac{\sin \gamma}{c} \qquad \textbf{(1)}$$

A proof of the Law of Sines is given at the end of this section.

In applying the Law of Sines to solve triangles, we use the fact that the sum of the angles of any triangle equals 180°; that is,

$$\alpha + \beta + \gamma = 180° \qquad \textbf{(2)}$$

✔ Solve SAA or ASA Triangles

Our first two examples show how to solve a triangle when one side and two angles are known (Case 1: SAA or ASA).

EXAMPLE 1 | **Using the Law of Sines to Solve a SAA Triangle**

Solve the triangle: $\alpha = 40°, \beta = 60°, a = 4$

Figure 21

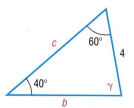

Solution Figure 21 shows the triangle that we want to solve. The third angle γ is found using equation (2).

$$\alpha + \beta + \gamma = 180°$$
$$40° + 60° + \gamma = 180°$$
$$\gamma = 80°$$

Now we use the Law of Sines (twice) to find the unknown sides b and c.

$$\frac{\sin \alpha}{a} = \frac{\sin \beta}{b} \qquad \frac{\sin \alpha}{a} = \frac{\sin \gamma}{c}$$

NOTE
Although not a check, we can verify the reasonableness of our answer by determining if the longest side is opposite the largest angle and the shortest side is opposite the smallest angle. ■

Because $a = 4$, $\alpha = 40°$, $\beta = 60°$, and $\gamma = 80°$, we have

$$\frac{\sin 40°}{4} = \frac{\sin 60°}{b} \qquad \frac{\sin 40°}{4} = \frac{\sin 80°}{c}$$

Solving for b and c, we find that

$$b = \frac{4 \sin 60°}{\sin 40°} \approx 5.39 \qquad c = \frac{4 \sin 80°}{\sin 40°} \approx 6.13$$ ◀

Notice in Example 1 that we found b and c by working with the given side a. This is better than finding b first and working with a rounded value of b to find c.

NOW WORK PROBLEM 9.

EXAMPLE 2

Using the Law of Sines to Solve an ASA Triangle

Solve the triangle: $\alpha = 35°$, $\beta = 15°$, $c = 5$

Solution

Figure 22

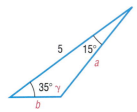

Figure 22 illustrates the triangle that we want to solve. Because we know two angles ($\alpha = 35°$ and $\beta = 15°$), we find the third angle using equation (2).

$$\alpha + \beta + \gamma = 180°$$
$$35° + 15° + \gamma = 180°$$
$$\gamma = 130°$$

Now we know the three angles and one side ($c = 5$) of the triangle. To find the remaining two sides a and b, we use the Law of Sines (twice).

$$\frac{\sin \alpha}{a} = \frac{\sin \gamma}{c} \qquad\qquad \frac{\sin \beta}{b} = \frac{\sin \gamma}{c}$$

$$\frac{\sin 35°}{a} = \frac{\sin 130°}{5} \qquad\qquad \frac{\sin 15°}{b} = \frac{\sin 130°}{5}$$

$$a = \frac{5 \sin 35°}{\sin 130°} \approx 3.74 \qquad b = \frac{5 \sin 15°}{\sin 130°} \approx 1.69$$ ◀

NOW WORK PROBLEM 23.

2 Solve SSA Triangles

Case 2 (SSA), which applies to triangles for which two sides and the angle opposite one of them are known, is referred to as the **ambiguous case**, because the known information may result in one triangle, two triangles, or no triangle at all. Suppose that we are given sides a and b and angle α, as illustrated in Figure 23. The key to determining the possible triangles, if any, that may be formed from the given information lies primarily with the height h and the fact that $h = b \sin \alpha$.

Figure 23

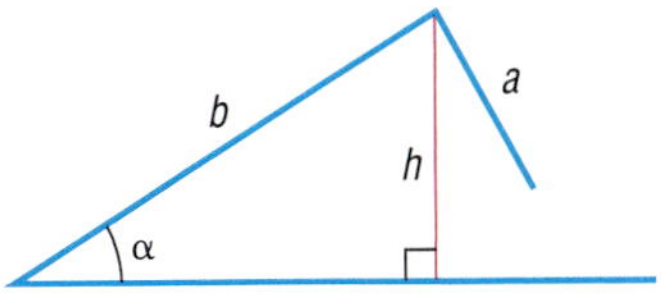

No Triangle If $a < h = b \sin \alpha$, then side a is not sufficiently long to form a triangle. See Figure 24.

One Right Triangle If $a = h = b \sin \alpha$, then side a is just long enough to form a right triangle. See Figure 25.

Figure 24
$a < h = b \sin \alpha$

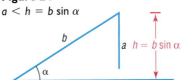

Figure 25
$a = b \sin \alpha$

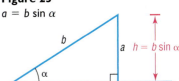

Two Triangles   If $a < b$ and $h = b \sin \alpha < a$, then two distinct triangles can be formed from the given information. See Figure 26.

One Triangle   If $a \geq b$, then only one triangle can be formed. See Figure 27.

Figure 26
$b \sin \alpha < a$ and $a < b$

Figure 27
$a \geq b$

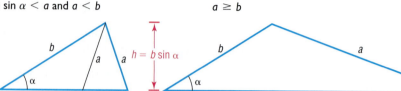

Fortunately, we do not have to rely on an illustration to draw the correct conclusion in the ambiguous case. The Law of Sines will lead us to the correct determination. Let's see how.

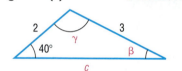

EXAMPLE 3

Using the Law of Sines to Solve a SSA Triangle (One Solution)

Solve the triangle: $a = 3, b = 2, \alpha = 40°$

Figure 28(a)

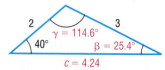

Solution   See Figure 28(a). Because $a = 3, b = 2$, and $\alpha = 40°$ are known, we use the Law of Sines to find the angle β.

$$\frac{\sin \alpha}{a} = \frac{\sin \beta}{b}$$

Then

$$\frac{\sin 40°}{3} = \frac{\sin \beta}{2}$$

$$\sin \beta = \frac{2 \sin 40°}{3} \approx 0.43$$

NOTE

Here we computed β_1 by determining the value of $\sin^{-1}\left(\frac{2 \sin 40°}{3}\right)$.

If you use the rounded value and evaluate $\sin^{-1}(0.43)$, you will obtain a slightly different result. ■

There are two angles $\beta, 0° < \beta < 180°$, for which $\sin \beta \approx 0.43$.

$$\beta_1 \approx 25.4° \quad \text{and} \quad \beta_2 \approx 180° - 25.4° = 154.6°$$

The second possibility, $\beta_2 \approx 154.6°$, is ruled out, because $\alpha = 40°$, making $\alpha + \beta_2 \approx 194.6° > 180°$. Now, using $\beta_1 \approx 25.4°$, we find that

$$\gamma = 180° - \alpha - \beta_1 \approx 180° - 40° - 25.4° = 114.6°$$

The third side c may now be determined using the Law of Sines.

$$\frac{\sin \alpha}{a} = \frac{\sin \gamma}{c}$$

$$\frac{\sin 40°}{3} = \frac{\sin 114.6°}{c}$$

$$c = \frac{3 \sin 114.6°}{\sin 40°} \approx 4.24$$

Figure 28(b)

Figure 28(b) illustrates the solved triangle.

EXAMPLE 4	**Using the Law of Sines to Solve a SSA Triangle (Two Solutions)**

Solve the triangle: $a = 6, b = 8, \alpha = 35°$

Figure 29(a)

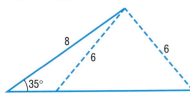

Solution   See Figure 29(a). Because $a = 6, b = 8$, and $\alpha = 35°$ are known, we use the Law of Sines to find the angle β.

$$\frac{\sin \alpha}{a} = \frac{\sin \beta}{b}$$

Then

$$\frac{\sin 35°}{6} = \frac{\sin \beta}{8}$$

$$\sin \beta = \frac{8 \sin 35°}{6} \approx 0.76$$

$$\beta_1 \approx 49.9° \quad \text{or} \quad \beta_2 \approx 180° - 49.9° = 130.1°$$

For both choices of β, we have $\alpha + \beta < 180°$. There are two triangles, one containing the angle $\beta_1 \approx 49.9°$ and the other containing the angle $\beta_2 \approx 130.1°$. The third angle γ is either

$$\gamma_1 = 180° - \alpha - \beta_1 \approx 95.1° \quad \text{or} \quad \gamma_2 = 180° - \alpha - \beta_2 \approx 14.9°$$

$$\uparrow \qquad\qquad\qquad\qquad\qquad \uparrow$$
$$\alpha = 35° \qquad\qquad\qquad\qquad \alpha = 35°$$
$$\beta_1 = 49.9° \qquad\qquad\qquad \beta_2 = 130.1°$$

The third side c obeys the Law of Sines, so we have

Figure 29(b)

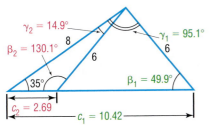

$$\frac{\sin \alpha}{a} = \frac{\sin \gamma_1}{c_1} \qquad\qquad \frac{\sin \alpha}{a} = \frac{\sin \gamma_2}{c_2}$$

$$\frac{\sin 35°}{6} = \frac{\sin 95.1°}{c_1} \qquad\qquad \frac{\sin 35°}{6} = \frac{\sin 14.9°}{c_2}$$

$$c_1 = \frac{6 \sin 95.1°}{\sin 35°} \approx 10.42 \qquad c_2 = \frac{6 \sin 14.9°}{\sin 35°} \approx 2.69$$

The two solved triangles are illustrated in Figure 29(b).   ◀

EXAMPLE 5	**Using the Law of Sines to Solve a SSA Triangle (No Solution)**

Solve the triangle: $a = 2, c = 1, \gamma = 50°$

Solution   Because $a = 2, c = 1$, and $\gamma = 50°$ are known, we use the Law of Sines to find the angle α.

$$\frac{\sin \alpha}{a} = \frac{\sin \gamma}{c}$$

$$\frac{\sin \alpha}{2} = \frac{\sin 50°}{1}$$

$$\sin \alpha = 2 \sin 50° \approx 1.53$$

Figure 30

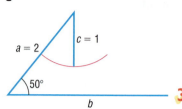

Since there is no angle α for which $\sin \alpha > 1$, there can be no triangle with the given measurements. Figure 30 illustrates the measurements given. Notice that, no matter how we attempt to position side c, it will never touch side b to form a triangle. ◀

✏ **NOW WORK PROBLEMS 25 AND 31.**

3 **Solve Applied Problems**

The Law of Sines is particularly useful for solving certain applied problems.

EXAMPLE 6 | **Finding the Height of a Mountain**

To measure the height of a mountain, a surveyor takes two sightings of the peak at a distance 900 meters apart on a direct line to the mountain.* See Figure 31(a). The first observation results in an angle of elevation of $47°$ and the second results in an angle of elevation of $35°$. If the transit is 2 meters high, what is the height h of the mountain?

Figure 31

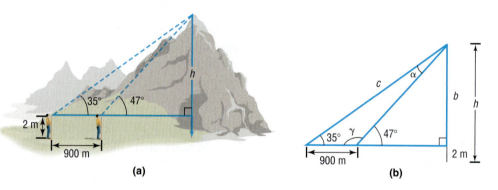

(a)   (b)

Solution Figure 31(b) shows the triangles that replicate the illustration in Figure 31(a). Since $\gamma + 47° = 180°$, we find that $\gamma = 133°$. Also, since $\alpha + \gamma + 35° = 180°$, we find that $\alpha = 180° - 35° - \gamma = 145° - 133° = 12°$. We use the Law of Sines to find c.

$$\frac{\sin \alpha}{a} = \frac{\sin \gamma}{c} \qquad \textcolor{blue}{\alpha = 12°, \gamma = 133°, a = 900}$$

$$c = \frac{900 \sin 133°}{\sin 12°} \approx 3165.86$$

Using the larger right triangle, we have

$$\sin 35° = \frac{b}{c} \qquad \textcolor{blue}{c = 3165.86}$$

$$b = 3165.86 \sin 35° \approx 1815.86 \approx 1816 \text{ meters}$$

The height of the peak from ground level is approximately $1816 + 2 = 1818$ meters. ◀

✏ **NOW WORK PROBLEM 39.**

---

*For simplicity, we assume that these sightings are at the same level.

| EXAMPLE 7 | **Rescue at Sea** |

Coast Guard Station Zulu is located 120 miles due west of Station X-ray. A ship at sea sends an SOS call that is received by each station. The call to Station Zulu indicates that the bearing of the ship from Zulu is N40°E (40° east of north). The call to Station X-ray indicates that the bearing of the ship from X-ray is N30°W (30° west of north).

(a) How far is each station from the ship?

(b) If a helicopter capable of flying 200 miles per hour is dispatched from the nearest station to the ship, how long will it take to reach the ship?

Figure 32

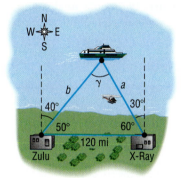

Solution   (a) Figure 32 illustrates the situation. The angle γ is found to be

$$\gamma = 180° - 50° - 60° = 70°$$

The Law of Sines can now be used to find the two distances a and b that we seek.

$$\frac{\sin 50°}{a} = \frac{\sin 70°}{120}$$

$$a = \frac{120 \sin 50°}{\sin 70°} \approx 97.82 \text{ miles}$$

$$\frac{\sin 60°}{b} = \frac{\sin 70°}{120}$$

$$b = \frac{120 \sin 60°}{\sin 70°} \approx 110.59 \text{ miles}$$

Station Zulu is about 111 miles from the ship, and Station X-ray is about 98 miles from the ship.

(b) The time t needed for the helicopter to reach the ship from Station X-ray is found by using the formula

$$(\text{Velocity, } v)(\text{Time, } t) = \text{Distance, } a$$

Then

$$t = \frac{a}{v} = \frac{97.82}{200} \approx 0.49 \text{ hour} \approx 29 \text{ minutes}$$

Figure 33

It will take about 29 minutes for the helicopter to reach the ship.   ◀

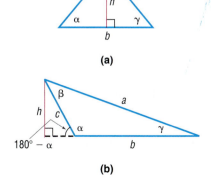

NOW WORK PROBLEM 37.

Proof of the Law of Sines   To prove the Law of Sines, we construct an altitude of length h from one of the vertices of a triangle. Figure 33(a) shows h for a triangle with three acute angles, and Figure 33(b) shows h for a triangle with an obtuse angle. In each case, the altitude is drawn from the vertex at β. Using either illustration, we have

$$\sin \gamma = \frac{h}{a}$$

from which

$$h = a \sin \gamma \qquad (3)$$

From Figure 33(a), it also follows that

$$\sin \alpha = \frac{h}{c}$$

from which

$$h = c \sin \alpha \qquad \textbf{(4)}$$

From Figure 33(b), it follows that

$$\sin(180° - \alpha) = \sin \alpha = \frac{h}{c}$$

$$\sin(180° - \alpha) = \sin 180° \cos \alpha - \cos 180° \sin \alpha = \sin \alpha$$

which again gives

$$h = c \sin \alpha$$

So, whether the triangle has three acute angles or has two acute angles and one obtuse angle, equations (3) and (4) hold. As a result, we may equate the expressions for h in equations (3) and (4) to get

$$a \sin \gamma = c \sin \alpha$$

from which

$$\frac{\sin \alpha}{a} = \frac{\sin \gamma}{c} \qquad \textbf{(5)}$$

In a similar manner, by constructing the altitude h' from the vertex of angle α as shown in Figure 34, we can show that

$$\sin \beta = \frac{h'}{c} \quad \text{and} \quad \sin \gamma = \frac{h'}{b}$$

Equating the expressions for h', we find that

$$h' = c \sin \beta = b \sin \gamma$$

from which

$$\frac{\sin \beta}{b} = \frac{\sin \gamma}{c} \qquad \textbf{(6)}$$

When equations (5) and (6) are combined, we have equation (1), the Law of Sines. ■

Figure 34

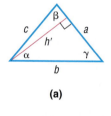

(a)

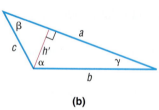

(b)

7.2 Assess Your Understanding

'Are You Prepared?'

Answers are given at the end of these exercises. If you get a wrong answer, read the pages listed in red.

1. The difference formula for sine is $\sin(\alpha - \beta) = $ _____. (p. 476)

2. If θ is an acute angle, solve the equation $\cos \theta = \frac{\sqrt{3}}{2}$. (pp. 497–498)

3. If θ is an acute angle, solve the equation $\sin \theta = 2$. (pp. 496–500)

Concepts and Vocabulary

4. If none of the angles of a triangle is a right angle, the triangle is called _____.

5. For a triangle with sides a, b, c and opposite angles α, β, γ, the Law of Sines states that _____.

6. *True or False:* An oblique triangle in which two sides and an angle are given always results in at least one triangle.

7. *True or False:* The sum of the angles of any triangle equals 180°.

8. *True or False:* The ambiguous case refers to the fact that, when two sides and the angle opposite one of them are given, sometimes the Law of Sines cannot be used.

Skill Building

In Problems 9–16, solve each triangle.

9.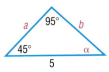

10.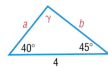

11.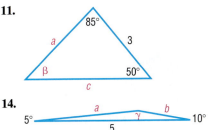

12.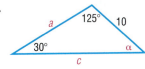

13.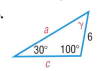

14.

15.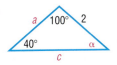

16.

In Problems 17–24, solve each triangle.

17. $\alpha = 40°$, $\beta = 20°$, $a = 2$

18. $\alpha = 50°$, $\gamma = 20°$, $a = 3$

19. $\beta = 70°$, $\gamma = 10°$, $b = 5$

20. $\alpha = 70°$, $\beta = 60°$, $c = 4$

21. $\alpha = 110°$, $\gamma = 30°$, $c = 3$

22. $\beta = 10°$, $\gamma = 100°$, $b = 2$

23. $\alpha = 40°$, $\beta = 40°$, $c = 2$

24. $\beta = 20°$, $\gamma = 70°$, $a = 1$

In Problems 25–36, two sides and an angle are given. Determine whether the given information results in one triangle, two triangles, or no triangle at all. Solve any triangle(s) that results.

25. $a = 3$, $b = 2$, $\alpha = 50°$

26. $b = 4$, $c = 3$, $\beta = 40°$

27. $b = 5$, $c = 3$, $\beta = 100°$

28. $a = 2$, $c = 1$, $\alpha = 120°$

29. $a = 4$, $b = 5$, $\alpha = 60°$

30. $b = 2$, $c = 3$, $\beta = 40°$

31. $b = 4$, $c = 6$, $\beta = 20°$

32. $a = 3$, $b = 7$, $\alpha = 70°$

33. $a = 2$, $c = 1$, $\gamma = 100°$

34. $b = 4$, $c = 5$, $\beta = 95°$

35. $a = 2$, $c = 1$, $\gamma = 25°$

36. $b = 4$, $c = 5$, $\beta = 40°$

Applications and Extensions

37. Rescue at Sea Coast Guard Station Able is located 150 miles due south of Station Baker. A ship at sea sends an SOS call that is received by each station. The call to Station Able indicates that the ship is located N55°E; the call to Station Baker indicates that the ship is located S60°E.
(a) How far is each station from the ship?
(b) If a helicopter capable of flying 200 miles per hour is dispatched from the station nearest to the ship, how long will it take to reach the ship?

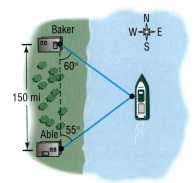

38. Surveying Consult the figure. To find the distance from the house at A to the house at B, a surveyor measures the angle BAC to be 40° and then walks off a distance of 100 feet to C and measures the angle ACB to be 50°. What is the distance from A to B?

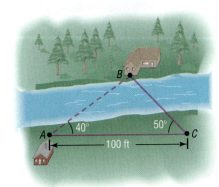

39. Finding the Length of a Ski Lift Consult the figure. To find the length of the span of a proposed ski lift from A to B, a surveyor measures the angle DAB to be $25°$ and then walks off a distance of 1000 feet to C and measures the angle ACB to be $15°$. What is the distance from A to B?

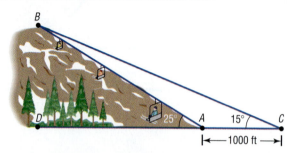

40. Finding the Height of a Mountain Use the illustration in Problem 39 to find the height BD of the mountain at B.

41. Finding the Height of an Airplane An aircraft is spotted by two observers who are 1000 feet apart. As the airplane passes over the line joining them, each observer takes a sighting of the angle of elevation to the plane, as indicated in the figure. How high is the airplane?

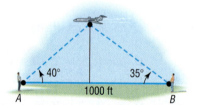

42. Finding the Height of the Bridge over the Royal Gorge The highest bridge in the world is the bridge over the Royal Gorge of the Arkansas River in Colorado. Sightings to the same point at water level directly under the bridge are taken from each side of the 880-foot-long bridge, as indicated in the figure. How high is the bridge?

SOURCE: *Guinness Book of World Records.*

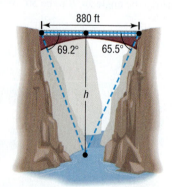

43. Navigation An airplane flies from city A to city B, a distance of 150 miles, and then turns through an angle of $40°$ and heads toward city C, as shown in the figure.

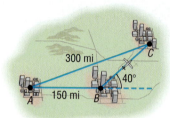

(a) If the distance between cities A and C is 300 miles, how far is it from city B to city C?
(b) Through what angle should the pilot turn at city C to return to city A?

44. Time Lost due to a Navigation Error In attempting to fly from city A to city B, an aircraft followed a course that was $10°$ in error, as indicated in the figure. After flying a distance of 50 miles, the pilot corrected the course by turning at point C and flying 70 miles farther. If the constant speed of the aircraft was 250 miles per hour, how much time was lost due to the error?

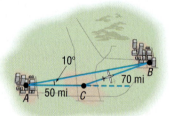

45. Finding the Lean of the Leaning Tower of Pisa The famous Leaning Tower of Pisa was originally 184.5 feet high.[*] At a distance of 123 feet from the base of the tower, the angle of elevation to the top of the tower is found to be $60°$. Find the angle CAB indicated in the figure. Also, find the perpendicular distance from C to AB.

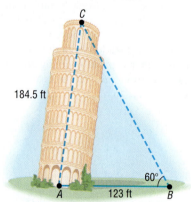

[*]In their 1986 report on the fragile seven-century-old bell tower, scientists in Pisa, Italy, said that the Leaning Tower of Pisa had increased its famous lean by 1 millimeter, or 0.04 inch. This is about the annual average, although the tilting had slowed to about half that much in the previous 2 years. (*Source:* United Press International, June 29, 1986.)

PISA, ITALY. September 1995. The Leaning Tower of Pisa has suddenly shifted, jeopardizing years of preservation work to stabilize it, Italian newspapers said Sunday. The tower, built on shifting subsoil between 1174 and 1350 as a belfry for the nearby cathedral, recently moved 0.07 inch in one night.

Update The tower, which had been closed to tourists since 1990, was reopened in December, 2001, after reinforcement of its base.

46. Crankshafts on Cars On a certain automobile, the crankshaft is 3 inches long and the connecting rod is 9 inches long (see the figure). At the time when the angle OPA is 15°, how far is the piston (P) from the center (O) of the crankshaft?

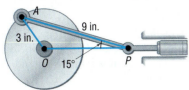

47. Constructing a Highway U.S. 41, a highway whose primary directions are north–south, is being constructed along the west coast of Florida. Near Naples, a bay obstructs the straight path of the road. Since the cost of a bridge is prohibitive, engineers decide to go around the bay. The illustration shows the path that they decide on and the measurements taken. What is the length of highway needed to go around the bay?

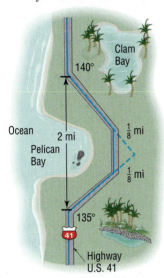

48. Calculating Distances at Sea The navigator of a ship at sea spots two lighthouses that she knows to be 3 miles apart along a straight seashore. She determines that the angles formed between two line-of-sight observations of the lighthouses and the line from the ship directly to shore are 15° and 35°. See the illustration.
(a) How far is the ship from lighthouse A?
(b) How far is the ship from lighthouse B?
(c) How far is the ship from shore?

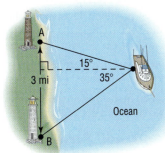

49. Designing an Awning An awning that covers a sliding glass door that is 88 inches tall forms an angle of 50° with the wall. The purpose of the awning is to prevent sunlight from entering the house when the angle of elevation of the sun is more than 65°. See the figure. Find the length L of the awning.

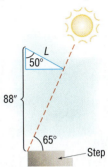

50. Finding Distances A forest ranger is walking on a path inclined at 5° to the horizontal directly toward a 100-foot-tall fire observation tower. The angle of elevation from the path to the top of the tower is 40°. How far is the ranger from the tower at this time?

51. Great Pyramid of Cheops One of the original Seven Wonders of the World, the Great Pyramid of Cheops was built about 2580 BC. Its original height was 480 feet 11 inches, but due to the loss of its topmost stones, it is now shorter. Find the current height of the Great Pyramid, using the information given in the illustration.
SOURCE: *Guinness Book of World Records.*

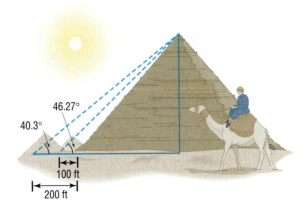

52. Determining the Height of an Aircraft Two sensors are spaced 700 feet apart along the approach to a small airport. When an aircraft is nearing the airport, the angle of elevation from the first sensor to the aircraft is 20°, and from the second sensor to the aircraft it is 15°. Determine how high the aircraft is at this time.

53. Mercury   The distance from the Sun to Earth is approximately 149,600,000 kilometers (km). The distance from the Sun to Mercury is approximately 57,910,000 km. The **elongation angle** α is the angle formed between the line of sight from Earth to the Sun and the line of sight from Earth to Mercury. See the figure. Suppose that the elongation angle for Mercury is 15°. Use this information to find the possible distances between Earth and Mercury.

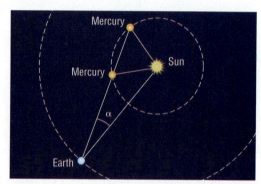

54. Venus   The distance from the Sun to Earth is approximately 149,600,000 km. The distance from the Sun to Venus is approximately 108,200,000 km. The elongation angle α is the angle formed between the line of sight from Earth to the Sun and the line of sight from Earth to Venus. Suppose that the elongation angle for Venus is 10°. Use this information to find the possible distances between Earth and Venus.

55. Landscaping   Pat needs to determine the height of a tree before cutting it down to be sure that it will not fall on a nearby fence. The angle of elevation of the tree from one position on a flat path from the tree is 30°, and from a second position 40 feet farther along this path it is 20°. What is the height of the tree?

56. Construction   A loading ramp 10 feet long that makes an angle of 18° with the horizontal is to be replaced by one that makes an angle of 12° with the horizontal. How long is the new ramp?

57. Finding the Height of a Helicopter   Two observers simultaneously measure the angle of elevation of a helicopter. One angle is measured as 25°, the other as 40° (see the figure). If the observers are 100 feet apart and the helicopter lies over the line joining them, how high is the helicopter?

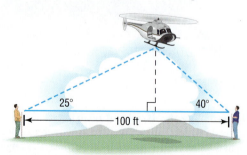

58. Mollweide's Formula   For any triangle, Mollweide's Formula (named after Karl Mollweide, 1774–1825) states that

$$\frac{a + b}{c} = \frac{\cos\left[\frac{1}{2}(\alpha - \beta)\right]}{\sin\left(\frac{1}{2}\gamma\right)}$$

Derive it.

[**Hint:** Use the Law of Sines and then a Sum-to-Product Formula. Notice that this formula involves all six parts of a triangle. As a result, it is sometimes used to check the solution of a triangle.]

59. Mollweide's Formula   Another form of Mollweide's Formula is

$$\frac{a - b}{c} = \frac{\sin\left[\frac{1}{2}(\alpha - \beta)\right]}{\cos\left(\frac{1}{2}\gamma\right)}$$

Derive it.

60. For any triangle, derive the formula

$$a = b\cos\gamma + c\cos\beta$$

[**Hint:** Use the fact that $\sin\alpha = \sin(180° - \beta - \gamma)$.]

61. Law of Tangents   For any triangle, derive the Law of Tangents.

$$\frac{a - b}{a + b} = \frac{\tan\left[\frac{1}{2}(\alpha - \beta)\right]}{\tan\left[\frac{1}{2}(\alpha + \beta)\right]}$$

[**Hint:** Use Mollweide's Formula.]

62. Circumscribing a Triangle   Show that

$$\frac{\sin\alpha}{a} = \frac{\sin\beta}{b} = \frac{\sin\gamma}{c} = \frac{1}{2r}$$

where r is the radius of the circle circumscribing the triangle ABC whose sides are a, b, and c, as shown in the figure.

[**Hint:** Draw the diameter AB'. Then β = angle ABC = angle $AB'C$, and angle ACB' = 90°.]

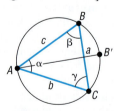

Discussion and Writing

63. Make up three problems involving oblique triangles. One should result in one triangle, the second in two triangles, and the third in no triangle.

64. What do you do first if you are asked to solve a triangle and are given one side and two angles?

65. What do you do first if you are asked to solve a triangle and are given two sides and the angle opposite one of them?

'Are You Prepared?' Answers

1. $\sin\alpha\cos\beta - \cos\alpha\sin\beta$ **2.** 30°   **3.** No solution

7.3 The Law of Cosines

PREPARING FOR THIS SECTION *Before getting started, review the following:*

- Trigonometric Equations (I) (Section 6.7, pp. 496–500)
- Distance Formula (Section 1.1, p. 5)

 Now work the 'Are You Prepared?' problems on page 546.

OBJECTIVES 1 Solve SAS Triangles
2 Solve SSS Triangles
3 Solve Applied Problems

In the previous section, we used the Law of Sines to solve Case 1 (SAA or ASA) and Case 2 (SSA) of an oblique triangle. In this section, we derive the Law of Cosines and use it to solve the remaining cases, 3 and 4.

Case 3: Two sides and the included angle are known (SAS).

Case 4: Three sides are known (SSS).

Theorem

Law of Cosines

For a triangle with sides a, b, c and opposite angles α, β, γ, respectively,

$$c^2 = a^2 + b^2 - 2ab \cos \gamma \tag{1}$$
$$b^2 = a^2 + c^2 - 2ac \cos \beta \tag{2}$$
$$a^2 = b^2 + c^2 - 2bc \cos \alpha \tag{3}$$

Figure 35

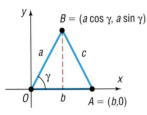

(a) Angle γ is acute

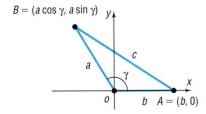

(b) Angle γ is obtuse

Proof We will prove only formula (1) here. Formulas (2) and (3) may be proved using the same argument.

We begin by strategically placing a triangle on a rectangular coordinate system so that the vertex of angle γ is at the origin and side b lies along the positive x-axis. Regardless of whether γ is acute, as in Figure 35(a), or obtuse, as in Figure 35(b), the vertex B has coordinates $(a \cos \gamma, a \sin \gamma)$. Vertex A has coordinates $(b, 0)$.

We can now use the distance formula to compute c^2.

$$c^2 = (b - a \cos \gamma)^2 + (0 - a \sin \gamma)^2$$
$$= b^2 - 2ab \cos \gamma + a^2 \cos^2 \gamma + a^2 \sin^2 \gamma$$
$$= b^2 - 2ab \cos \gamma + a^2(\cos^2 \gamma + \sin^2 \gamma)$$
$$= a^2 + b^2 - 2ab \cos \gamma \qquad \blacksquare$$

Each of formulas (1), (2), and (3) may be stated in words as follows:

Theorem

Law of Cosines

The square of one side of a triangle equals the sum of the squares of the other two sides minus twice their product times the cosine of their included angle.

Observe that if the triangle is a right triangle (so that, say, $\gamma = 90°$), then formula (1) becomes the familiar Pythagorean Theorem: $c^2 = a^2 + b^2$. The Pythagorean Theorem is a special case of the Law of Cosines!

1 Solve SAS Triangles

Let's see how to use the Law of Cosines to solve Case 3 (SAS), which applies to triangles for which two sides and the included angle are known.

EXAMPLE 1 | **Using the Law of Cosines to Solve a SAS Triangle**

Solve the triangle: $a = 2$, $b = 3$, $\gamma = 60°$

Figure 36

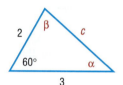

Solution   See Figure 36. The Law of Cosines makes it easy to find the third side, c.

$$c^2 = a^2 + b^2 - 2ab \cos \gamma$$
$$= 4 + 9 - 2 \cdot 2 \cdot 3 \cdot \cos 60°$$
$$= 13 - \left(12 \cdot \frac{1}{2} \right) = 7$$
$$c = \sqrt{7}$$

Side c is of length $\sqrt{7}$. To find the angles α and β, we may use either the Law of Sines or the Law of Cosines. It is preferable to use the Law of Cosines, since it will lead to an equation with one solution. Using the Law of Sines would lead to an equation with two solutions that would need to be checked to determine which solution fits the given data. We choose to use formulas (2) and (3) of the Law of Cosines to find α and β.

For α:

$$a^2 = b^2 + c^2 - 2bc \cos \alpha$$
$$2bc \cos \alpha = b^2 + c^2 - a^2$$
$$\cos \alpha = \frac{b^2 + c^2 - a^2}{2bc} = \frac{9 + 7 - 4}{2 \cdot 3\sqrt{7}} = \frac{12}{6\sqrt{7}} = \frac{2\sqrt{7}}{7}$$
$$\alpha \approx 40.9°$$

For β:

$$b^2 = a^2 + c^2 - 2ac \cos \beta$$
$$\cos \beta = \frac{a^2 + c^2 - b^2}{2ac} = \frac{4 + 7 - 9}{4\sqrt{7}} = \frac{1}{2\sqrt{7}} = \frac{\sqrt{7}}{14}$$
$$\beta \approx 79.1°$$

Notice that $\alpha + \beta + \gamma = 40.9° + 79.1° + 60° = 180°$, as required. ◀

━━━━ **NOW WORK PROBLEM 9.**

2 Solve SSS Triangles

The next example illustrates how the Law of Cosines is used when three sides of a triangle are known, Case 4 (SSS).

EXAMPLE 2 | **Using the Law of Cosines to Solve a SSS Triangle**

Solve the triangle: $a = 4, b = 3, c = 6$

Figure 37

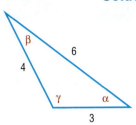

Solution See Figure 37. To find the angles α, β, and γ, we proceed as we did in the latter part of the solution to Example 1.

For α:

$$\cos \alpha = \frac{b^2 + c^2 - a^2}{2bc} = \frac{9 + 36 - 16}{2 \cdot 3 \cdot 6} = \frac{29}{36}$$

$$\alpha \approx 36.3°$$

For β:

$$\cos \beta = \frac{a^2 + c^2 - b^2}{2ac} = \frac{16 + 36 - 9}{2 \cdot 4 \cdot 6} = \frac{43}{48}$$

$$\beta \approx 26.4°$$

Since we know α and β,

$$\gamma = 180° - \alpha - \beta \approx 180° - 36.3° - 26.4° = 117.3° \quad \blacktriangleleft$$

 NOW WORK PROBLEM 15.

3 Solve Applied Problems

EXAMPLE 3 **Correcting a Navigational Error**

A motorized sailboat leaves Naples, Florida, bound for Key West, 150 miles away. Maintaining a constant speed of 15 miles per hour, but encountering heavy crosswinds and strong currents, the crew finds, after 4 hours, that the sailboat is off course by 20°.

(a) How far is the sailboat from Key West at this time?

(b) Through what angle should the sailboat turn to correct its course?

(c) How much time has been added to the trip because of this? (Assume that the speed remains at 15 miles per hour.)

Figure 38

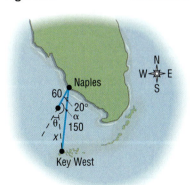

Solution See Figure 38. With a speed of 15 miles per hour, the sailboat has gone 60 miles after 4 hours. We seek the distance x of the sailboat from Key West. We also seek the angle θ that the sailboat should turn through to correct its course.

(a) To find x, we use the Law of Cosines, since we know two sides and the included angle.

$$x^2 = 150^2 + 60^2 - 2(150)(60) \cos 20° \approx 9186$$
$$x \approx 95.8$$

The sailboat is about 96 miles from Key West.

(b) We now know three sides of the triangle, so we can use the Law of Cosines again to find the angle α opposite the side of length 150 miles.

$$150^2 = 96^2 + 60^2 - 2(96)(60) \cos \alpha$$
$$9684 = -11,520 \cos \alpha$$
$$\cos \alpha \approx -0.8406$$
$$\alpha \approx 147.2°$$

The sailboat should turn through an angle of

$$\theta = 180° - \alpha \approx 180° - 147.2° = 32.8°$$

The sailboat should turn through an angle of about 33° to correct its course.

(c) The total length of the trip is now $60 + 96 = 156$ miles. The extra 6 miles will only require about 0.4 hour or 24 minutes more if the speed of 15 miles per hour is maintained. $\quad \blacktriangleleft$

 NOW WORK PROBLEM 35.

HISTORICAL FEATURE

The Law of Sines was known vaguely long before it was explicitly stated by Nasir Eddin (about AD 1250). Ptolemy (about AD 150) was aware of it in a form using a chord function instead of the sine function. But it was first clearly stated in Europe by Regiomontanus, writing in 1464.

The Law of Cosines appears first in Euclid's *Elements* (Book II), but in a well-disguised form in which squares built on the sides of triangles are added and a rectangle representing the cosine term is subtracted. It was thus known to all mathematicians because of their familiarity with Euclid's work. An early modern form of the Law of Cosines, that for finding the angle when the sides are known, was stated by François Viète (in 1593).

The Law of Tangents (see Problem 61 of Exercise 7.2) has become obsolete. In the past it was used in place of the Law of Cosines, because the Law of Cosines was very inconvenient for calculation with logarithms or slide rules. Mixing of addition and multiplication is now very easy on a calculator, however, and the Law of Tangents has been shelved along with the slide rule.

7.3 Assess Your Understanding

'Are You Prepared?'

Answers are given at the end of these exercises. If you get a wrong answer, read the pages listed in red.

1. Write the formula for the distance d from $P_1 = (x_1, y_1)$ to $P_2 = (x_2, y_2)$. (p. 5)

2. If θ is an acute angle, solve the equation $\cos\theta = \dfrac{\sqrt{2}}{2}$. (pp. 497–498)

Concepts and Vocabulary

3. If three sides of a triangle are given, the Law of _____ is used to solve the triangle.

4. If one side and two angles of a triangle are given, the Law of _____ is used to solve the triangle.

5. If two sides and the included angle of a triangle are given, the Law of _____ is used to solve the triangle.

6. *True or False:* Given only the three sides of a triangle, there is insufficient information to solve the triangle.

7. *True or False:* Given two sides and the included angle, the first thing to do to solve the triangle is to use the Law of Sines.

8. *True or False:* A special case of the Law of Cosines is the Pythagorean Theorem.

Skill Building

In Problems 9–16, solve each triangle.

9.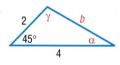

10.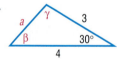

11.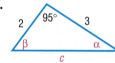

12.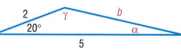

13.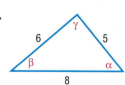

14.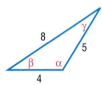

15.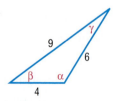

16.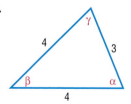

In Problems 17–32, solve each triangle.

17. $a = 3$, $b = 4$, $\gamma = 40°$

18. $a = 2$, $c = 1$, $\beta = 10°$

19. $b = 1$, $c = 3$, $\alpha = 80°$

20. $a = 6$, $b = 4$, $\gamma = 60°$

21. $a = 3$, $c = 2$, $\beta = 110°$

22. $b = 4$, $c = 1$, $\alpha = 120°$

23. $a = 2$, $b = 2$, $\gamma = 50°$ **24.** $a = 3$, $c = 2$, $\beta = 90°$ **25.** $a = 12$, $b = 13$, $c = 5$

26. $a = 4$, $b = 5$, $c = 3$ **27.** $a = 2$, $b = 2$, $c = 2$ **28.** $a = 3$, $b = 3$, $c = 2$

29. $a = 5$, $b = 8$, $c = 9$ **30.** $a = 4$, $b = 3$, $c = 6$ **31.** $a = 10$, $b = 8$, $c = 5$

32. $a = 9$, $b = 7$, $c = 10$

Applications and Extensions

33. Distance to the Green   A golfer hits an errant tee shot that lands in the rough. A marker in the center of the fairway is 150 yards from the center of the green. While standing on the marker and facing the green, the golfer turns 110° toward his ball. He then paces off 35 yards to his ball. See the figure. How far is the ball from the center of the green?

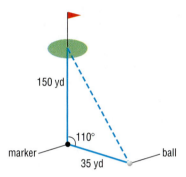

34. Navigation   An airplane flies from Ft. Myers to Sarasota, a distance of 150 miles, and then turns through an angle of 50° and flies to Orlando, a distance of 100 miles (see the figure).
(a) How far is it from Ft. Myers to Orlando?
(b) Through what angle should the pilot turn at Orlando to return to Ft. Myers?

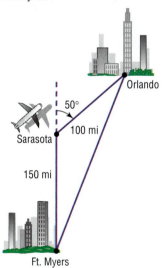

35. Avoiding a Tropical Storm   A cruise ship maintains an average speed of 15 knots in going from San Juan, Puerto Rico, to Barbados, West Indies, a distance of 600 nautical miles. To avoid a tropical storm, the captain heads out of San Juan in a direction of 20° off a direct heading to Barbados. The captain maintains the 15-knot speed for 10 hours, after which time the path to Barbados becomes clear of storms.

(a) Through what angle should the captain turn to head directly to Barbados?
(b) Once the turn is made, how long will it be before the ship reaches Barbados if the same 15-knot speed is maintained?

36. Revising a Flight Plan   In attempting to fly from Chicago to Louisville, a distance of 330 miles, a pilot inadvertently took a course that was 10° in error, as indicated in the figure.
(a) If the aircraft maintains an average speed of 220 miles per hour and if the error in direction is discovered after 15 minutes, through what angle should the pilot turn to head toward Louisville?
(b) What new average speed should the pilot maintain so that the total time of the trip is 90 minutes?

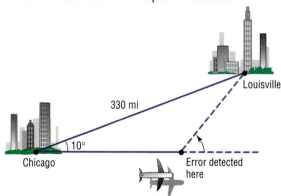

37. Major League Baseball Field   A Major League baseball diamond is actually a square 90 feet on a side. The pitching rubber is located 60.5 feet from home plate on a line joining home plate and second base.
(a) How far is it from the pitching rubber to first base?
(b) How far is it from the pitching rubber to second base?
(c) If a pitcher faces home plate, through what angle does he need to turn to face first base?

38. **Little League Baseball Field** According to Little League baseball official regulations, the diamond is a square 60 feet on a side. The pitching rubber is located 46 feet from home plate on a line joining home plate and second base.
    (a) How far is it from the pitching rubber to first base?
    (b) How far is it from the pitching rubber to second base?
    (c) If a pitcher faces home plate, through what angle does he need to turn to face first base?

39. **Finding the Length of a Guy Wire** The height of a radio tower is 500 feet, and the ground on one side of the tower slopes upward at an angle of 10° (see the figure).
    (a) How long should a guy wire be if it is to connect to the top of the tower and be secured at a point on the sloped side 100 feet from the base of the tower?
    (b) How long should a second guy wire be if it is to connect to the middle of the tower and be secured at a point 100 feet from the base on the flat side?

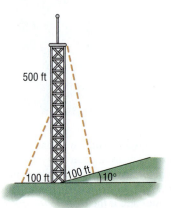

40. **Finding the Length of a Guy Wire** A radio tower 500 feet high is located on the side of a hill with an inclination to the horizontal of 5° (see the figure). How long should two guy wires be if they are to connect to the top of the tower and be secured at two points 100 feet directly above and directly below the base of the tower?

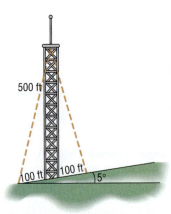

41. **Wrigley Field, Home of the Chicago Cubs** The distance from home plate to the fence in dead center in Wrigley Field

is 400 feet (see the figure). How far is it from the fence in dead center to third base?

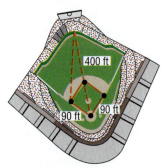

42. **Little League Baseball** The distance from home plate to the fence in dead center at the Oak Lawn Little League field is 280 feet. How far is it from the fence in dead center to third base?

    [**Hint:** The distance between the bases in Little League is 60 feet.]

43. **Rods and Pistons** Rod OA (see the figure) rotates about the fixed point O so that point A travels on a circle of radius r. Connected to point A is another rod AB of length $L > 2r$, and point B is connected to a piston. Show that the distance x between point O and point B is given by

    $$x = r \cos \theta + \sqrt{r^2 \cos^2 \theta + L^2 - r^2}$$

    where θ is the angle of rotation of rod OA.

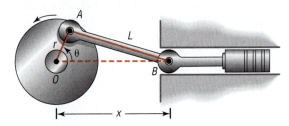

44. **Geometry** Show that the length d of a chord of a circle of radius r is given by the formula

    $$d = 2r \sin \frac{\theta}{2}$$

    where θ is the central angle formed by the radii to the ends of the chord (see the figure). Use this result to derive the fact that $\sin \theta < \theta$, where $\theta > 0$ is measured in radians.

45. For any triangle, show that

    $$\cos \frac{\gamma}{2} = \sqrt{\frac{s(s - c)}{ab}}$$

    where $s = \frac{1}{2}(a + b + c)$.

    [**Hint:** Use a Half-angle Formula and the Law of Cosines.]

46. For any triangle show that

$$\sin \frac{\gamma}{2} = \sqrt{\frac{(s-a)(s-b)}{ab}}$$

where $s = \frac{1}{2}(a + b + c)$.

47. Use the Law of Cosines to prove the identity

$$\frac{\cos \alpha}{a} + \frac{\cos \beta}{b} + \frac{\cos \gamma}{c} = \frac{a^2 + b^2 + c^2}{2abc}$$

Discussion and Writing

48. What do you do first if you are asked to solve a triangle and are given two sides and the included angle?

49. What do you do first if you are asked to solve a triangle and are given three sides?

50. Make up an applied problem that requires using the Law of Cosines.

51. Write down your strategy for solving an oblique triangle.

'Are You Prepared?' Answers

1. $d = \sqrt{(x_2 - x_1)^2 + (y_2 - y_1)^2}$ **2.** $\theta = 45°$

Polar Coordinates; Vectors

Chapter 8 in *Precalculus Essentials: Enhanced with Graphing Utilities*, 4e includes only the highlighted Sections 8.1 and 8.2.

8

A LOOK BACK, A LOOK AHEAD This chapter is in two parts: Polar Coordinates, Sections 8.1–8.3, and Vectors, Sections 8.4–8.7. They are independent of each other and may be covered in any order.

Sections 8.1–8.3: In Chapter 1 we introduced rectangular coordinates (x, y) and discussed the graph of an equation in two variables involving x and y. In Sections 8.1 and 8.2, we introduce an alternative to rectangular coordinates, polar coordinates, and discuss graphing equations that involve polar coordinates. In Section 4.3, we discussed raising a real number to a real power. In Section 8.3 we extend this idea by raising a complex number to a real power. As it turns out, polar coordinates are useful for the discussion.

Sections 8.4–8.7: We have seen in many chapters that often we are required to solve an equation to obtain a solution to applied problems. In the last four sections of this chapter, we develop the notion of a vector, and show how they can be used to solve certain types of applied problems, particularly in physics and engineering.

OUTLINE

8.1 Polar Coordinates

PREPARING FOR THIS SECTION *Before getting started, review the following:*

- Rectangular Coordinates (Section 1.1, pp. 2–5)
- Definitions of the Sine and Cosine Functions (Section 5.2, pp. 371–372)
- Inverse Tangent Function (Section 6.1, pp. 455–457)
- Completing the Square (Appendix, Section A.5, pp. 699–700)

 Now work the 'Are You Prepared?' problems on page 559.

OBJECTIVES 1 Plot Points Using Polar Coordinates
2 Convert from Polar Coordinates to Rectangular Coordinates
3 Convert from Rectangular Coordinates to Polar Coordinates

Figure 1

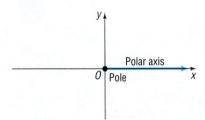

So far, we have always used a system of rectangular coordinates to plot points in the plane. Now we are ready to describe another system called *polar coordinates*. As we shall soon see, in many instances polar coordinates offer certain advantages over rectangular coordinates.

In a rectangular coordinate system, you will recall, a point in the plane is represented by an ordered pair of numbers (x, y), where x and y equal the signed distance of the point from the y-axis and x-axis, respectively. In a polar coordinate system, we select a point, called the **pole**, and then a ray with vertex at the pole, called the **polar axis**. See Figure 1. Comparing the rectangular and polar coordinate systems, we see that the origin in rectangular coordinates coincides with the pole in polar coordinates, and the positive x-axis in rectangular coordinates coincides with the polar axis in polar coordinates.

1 Plot Points Using Polar Coordinates

A point P in a polar coordinate system is represented by an ordered pair of numbers (r, θ). If $r > 0$, then r is the distance of the point from the pole; θ is an angle (in degrees or radians) formed by the polar axis and a ray from the pole through the point. We call the ordered pair (r, θ) the **polar coordinates** of the point. See Figure 2.

As an example, suppose that the polar coordinates of a point P are $\left(2, \dfrac{\pi}{4}\right)$.

We locate P by first drawing an angle of $\dfrac{\pi}{4}$ radian, placing its vertex at the pole and its initial side along the polar axis. Then we go out a distance of 2 units along the terminal side of the angle to reach the point P. See Figure 3.

Figure 2

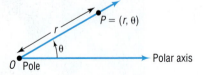

Figure 3

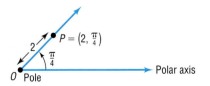

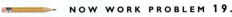 **NOW WORK PROBLEM 19.**

In using polar coordinates (r, θ), it is possible for the first entry r to be negative. When this happens, instead of the point being on the terminal side of θ, it is on the ray from the pole extending in the direction *opposite* the terminal side of θ at a distance $|r|$ units from the pole. See Figure 4 for an illustration.

For example, to plot the point $\left(-3, \dfrac{2\pi}{3}\right)$, we use the ray in the opposite direction of $\dfrac{2\pi}{3}$ and go out $|-3| = 3$ units along that ray. See Figure 5.

Figure 4

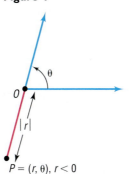

Figure 5

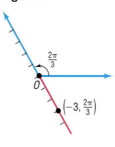

EXAMPLE 1

Plotting Points Using Polar Coordinates

Plot the points with the following polar coordinates:

(a) $\left(3, \dfrac{5\pi}{3}\right)$ (b) $\left(2, -\dfrac{\pi}{4}\right)$ (c) $(3, 0)$ (d) $\left(-2, \dfrac{\pi}{4}\right)$

Solution   Figure 6 shows the points.

Figure 6

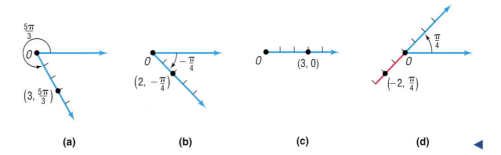

NOW WORK PROBLEMS 11 AND 27.

Recall that an angle measured counterclockwise is positive and an angle measured clockwise is negative. This convention has some interesting consequences relating to polar coordinates. Let's see what these consequences are.

| EXAMPLE 2 | **Finding Several Polar Coordinates of a Single Point** |

Consider again the point P with polar coordinates $\left(2, \dfrac{\pi}{4}\right)$, as shown in Figure 7(a). Because $\dfrac{\pi}{4}, \dfrac{9\pi}{4}$, and $-\dfrac{7\pi}{4}$ all have the same terminal side, we also could have located this point P by using the polar coordinates $\left(2, \dfrac{9\pi}{4}\right)$ or $\left(2, -\dfrac{7\pi}{4}\right)$, as shown in Figures 7(b) and (c). The point $\left(2, \dfrac{\pi}{4}\right)$ can also be represented by the polar coordinates $\left(-2, \dfrac{5\pi}{4}\right)$. See Figure 7(d).

Figure 7

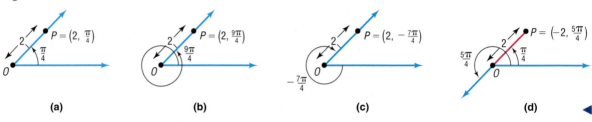

(a)   (b)   (c)   (d)

| EXAMPLE 3 | **Finding Other Polar Coordinates of a Given Point** |

Plot the point P with polar coordinates $\left(3, \dfrac{\pi}{6}\right)$, and find other polar coordinates (r, θ) of this same point for which:

(a) $r > 0, \quad 2\pi \le \theta < 4\pi$ (b) $r < 0, \quad 0 \le \theta < 2\pi$

(c) $r > 0, \quad -2\pi \le \theta < 0$

Figure 8

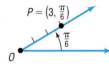

$P = \left(3, \dfrac{\pi}{6}\right)$

Solution The point $\left(3, \dfrac{\pi}{6}\right)$ is plotted in Figure 8.

(a) We add 1 revolution $(2\pi$ radians) to the angle $\dfrac{\pi}{6}$ to get $P = \left(3, \dfrac{\pi}{6} + 2\pi\right) = \left(3, \dfrac{13\pi}{6}\right)$. See Figure 9.

(b) We add $\dfrac{1}{2}$ revolution $(\pi$ radians) to the angle $\dfrac{\pi}{6}$ and replace 3 by -3 to get

$$P = \left(-3, \dfrac{\pi}{6} + \pi\right) = \left(-3, \dfrac{7\pi}{6}\right).$$ See Figure 10.

(c) We subtract 2π from the angle $\dfrac{\pi}{6}$ to get $P = \left(3, \dfrac{\pi}{6} - 2\pi\right) = \left(3, -\dfrac{11\pi}{6}\right)$.
See Figure 11.

Figure 9   **Figure 10**   **Figure 11**

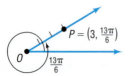   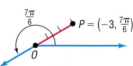   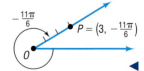

NOW WORK PROBLEM 31.

These examples show a major difference between rectangular coordinates and polar coordinates. In the former, each point has exactly one pair of rectangular coordinates; in the latter, a point can have infinitely many pairs of polar coordinates.

Summary

A point with polar coordinates (r, θ) also can be represented by either of the following:

$$(r, \theta + 2k\pi) \quad \text{or} \quad (-r, \theta + \pi + 2k\pi), \quad k \text{ any integer}$$

The polar coordinates of the pole are $(0, \theta)$, where θ can be any angle.

2 **Convert from Polar Coordinates to Rectangular Coordinates**

It is sometimes convenient and, indeed, necessary to be able to convert coordinates or equations in rectangular form to polar form, and vice versa. To do this, we recall that the origin in rectangular coordinates is the pole in polar coordinates and that the positive x-axis in rectangular coordinates is the polar axis in polar coordinates.

Theorem

Conversion from Polar Coordinates to Rectangular Coordinates

If P is a point with polar coordinates (r, θ), the rectangular coordinates (x, y) of P are given by

$$x = r \cos \theta \qquad y = r \sin \theta \qquad \text{(1)}$$

Figure 12

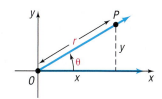

Proof Suppose that P has the polar coordinates (r, θ). We seek the rectangular coordinates (x, y) of P. Refer to Figure 12.

If $r = 0$, then, regardless of θ, the point P is the pole, for which the rectangular coordinates are $(0, 0)$. Formula (1) is valid for $r = 0$.

If $r > 0$, the point P is on the terminal side of θ, and $r = d(O, P) = \sqrt{x^2 + y^2}$. Since

$$\cos \theta = \frac{x}{r} \qquad \sin \theta = \frac{y}{r}$$

we have

$$x = r \cos \theta \qquad y = r \sin \theta$$

If $r < 0$, then the point $P = (r, \theta)$ can be represented as $(-r, \pi + \theta)$, where $-r > 0$. Since

$$\cos(\pi + \theta) = -\cos \theta = \frac{x}{-r} \qquad \sin(\pi + \theta) = -\sin \theta = \frac{y}{-r}$$

we have

$$x = r \cos \theta \qquad y = r \sin \theta \qquad \blacksquare$$

| EXAMPLE 4 | **Converting from Polar Coordinates to Rectangular Coordinates** |

Find the rectangular coordinates of the points with the following polar coordinates:

(a) $\left(6, \dfrac{\pi}{6}\right)$ (b) $\left(-4, -\dfrac{\pi}{4}\right)$

Solution We use formula (1): $x = r \cos\theta$ and $y = r \sin\theta$.

(a) Figure 13(a) shows $\left(6, \dfrac{\pi}{6}\right)$ plotted. Notice that $\left(6, \dfrac{\pi}{6}\right)$ lies in quadrant I of the rectangular coordinate system. So, we expect both the x-coordinate and the y-coordinate to be positive. With $r = 6$ and $\theta = \dfrac{\pi}{6}$, we have

$$x = r \cos\theta = 6 \cos\frac{\pi}{6} = 6 \cdot \frac{\sqrt{3}}{2} = 3\sqrt{3}$$

$$y = r \sin\theta = 6 \sin\frac{\pi}{6} = 6 \cdot \frac{1}{2} = 3$$

The rectangular coordinates of the point $\left(6, \dfrac{\pi}{6}\right)$ are $(3\sqrt{3}, 3)$, which lies in quadrant I, as expected.

(b) Figure 13(b) shows $\left(-4, -\dfrac{\pi}{4}\right)$ plotted. Notice that $\left(-4, -\dfrac{\pi}{4}\right)$ lies in quadrant II of the rectangular coordinate system. With $r = -4$ and $\theta = -\dfrac{\pi}{4}$, we have

$$x = r \cos\theta = -4 \cos\left(-\frac{\pi}{4}\right) = -4 \cdot \frac{\sqrt{2}}{2} = -2\sqrt{2}$$

$$y = r \sin\theta = -4 \sin\left(-\frac{\pi}{4}\right) = -4\left(-\frac{\sqrt{2}}{2}\right) = 2\sqrt{2}$$

The rectangular coordinates of the point $\left(-4, -\dfrac{\pi}{4}\right)$ are $(-2\sqrt{2}, 2\sqrt{2})$, which lies in quadrant II, as expected. ◀

Figure 13

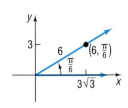

(a)

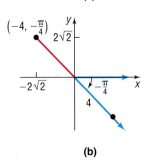

(b)

Figure 14

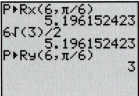

Most calculators have the capability of converting from polar coordinates to rectangular coordinates. Consult your owner's manual for the proper key strokes. Since in most cases this procedure is tedious, you will find that using formula (1) is faster.

Figure 14 verifies the result obtained in Example 4(a) using a TI-84 Plus. Note that the calculator is in radian mode.

━━━ **NOW WORK PROBLEMS 39 AND 51.**

3 **Convert from Rectangular Coordinates to Polar Coordinates**

Converting from rectangular coordinates (x, y) to polar coordinates (r, θ) is a little more complicated. Notice that we begin each example by plotting the given rectangular coordinates.

| EXAMPLE 5 | **Converting from Rectangular Coordinates to Polar Coordinates** |

Find polar coordinates of a point whose rectangular coordinates are $(0, 3)$.

Figure 15

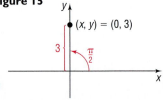

Solution See Figure 15. The point $(0, 3)$ lies on the y-axis a distance of 3 units from the origin (pole), so $r = 3$. A ray with vertex at the pole through $(0, 3)$ forms an angle $\theta = \dfrac{\pi}{2}$ with the polar axis. Polar coordinates for this point can be given by $\left(3, \dfrac{\pi}{2}\right)$. ◀

Figure 16

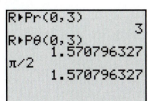

Most graphing calculators have the capability of converting from rectangular coordinates to polar coordinates. Consult you owner's manual for the proper keystrokes. Figure 16 verifies the results obtained in Example 5 using a TI-84 Plus. Note that the calculator is in radian mode.

Figure 17 shows polar coordinates of points that lie on either the x-axis or the y-axis. In each illustration, $a > 0$.

Figure 17

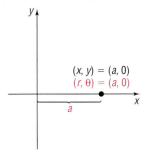 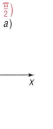 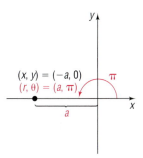 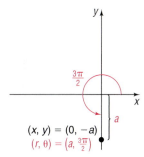

(a) $(x, y) = (a, 0)$, $a > 0$ (b) $(x, y) = (0, a)$, $a > 0$ (c) $(x, y) = (-a, 0)$, $a > 0$ (d) $(x, y) = (0, -a)$, $a > 0$

 NOW WORK PROBLEM 55.

EXAMPLE 6

Converting from Rectangular Coordinates to Polar Coordinates

Find polar coordinates of a point whose rectangular coordinates are:

(a) $(2, -2)$ (b) $\left(-1, -\sqrt{3}\right)$

Solution

(a) See Figure 18(a). The distance r from the origin to the point $(2, -2)$ is

$$r = \sqrt{x^2 + y^2} = \sqrt{(2)^2 + (-2)^2} = \sqrt{8} = 2\sqrt{2}$$

Figure 18

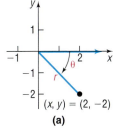

(a)

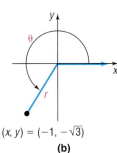

$(x, y) = (-1, -\sqrt{3})$

(b)

We find θ by recalling that $\tan\theta = \dfrac{y}{x}$, so $\theta = \tan^{-1}\dfrac{y}{x}$, $-\dfrac{\pi}{2} < \theta < \dfrac{\pi}{2}$.

Since $(2, -2)$ lies in quadrant IV, we know that $-\dfrac{\pi}{2} < \theta < 0$. As a result,

$$\theta = \tan^{-1}\frac{y}{x} = \tan^{-1}\left(\frac{-2}{2}\right) = \tan^{-1}(-1) = -\frac{\pi}{4}$$

A set of polar coordinates for this point is $\left(2\sqrt{2}, -\dfrac{\pi}{4}\right)$. Other possible representations include $\left(2\sqrt{2}, \dfrac{7\pi}{4}\right)$ and $\left(-2\sqrt{2}, \dfrac{3\pi}{4}\right)$.

(b) See Figure 18(b). The distance r from the origin to the point $\left(-1, -\sqrt{3}\right)$ is

$$r = \sqrt{(-1)^2 + \left(-\sqrt{3}\right)^2} = \sqrt{4} = 2$$

To find θ, we use $\theta = \tan^{-1}\dfrac{y}{x}$, $-\dfrac{\pi}{2} < \theta < \dfrac{\pi}{2}$. Since the point $\left(-1, -\sqrt{3}\right)$ lies in quadrant III and the inverse tangent function gives an angle in quadrant I, we add π to the result to obtain an angle in quadrant III.

$$\theta = \pi + \tan^{-1}\left(\frac{-\sqrt{3}}{-1}\right) = \pi + \tan^{-1}\sqrt{3} = \pi + \frac{\pi}{3} = \frac{4\pi}{3}$$

A set of polar coordinates for this point is $\left(2, \dfrac{4\pi}{3}\right)$. Other possible representations include $\left(-2, \dfrac{\pi}{3}\right)$ and $\left(2, -\dfrac{2\pi}{3}\right)$.

Figure 19

Figure 19 shows how to find polar coordinates of a point that lies in a quadrant when its rectangular coordinates (x, y) are given.

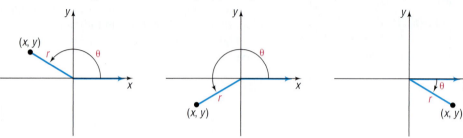

(a) $r = \sqrt{x^2 + y^2}$
$\theta = \tan^{-1}\dfrac{y}{x}$

(b) $r = \sqrt{x^2 + y^2}$
$\theta = \pi + \tan^{-1}\dfrac{y}{x}$

(c) $r = \sqrt{x^2 + y^2}$
$\theta = \pi + \tan^{-1}\dfrac{y}{x}$

(d) $r = \sqrt{x^2 + y^2}$
$\theta = \tan^{-1}\dfrac{y}{x}$

Based on the preceding discussion, we have the formulas

$$r^2 = x^2 + y^2 \qquad \tan\theta = \frac{y}{x} \qquad \text{if } x \neq 0 \qquad\qquad \textbf{(2)}$$

To use formula (2) effectively, follow these steps:

Steps for Converting from Rectangular to Polar Coordinates

STEP 1: Always plot the point (x, y) first, as we did in Examples 5 and 6.
STEP 2: If $x = 0$ or $y = 0$, use your illustration to find (r, θ). See Figure 7.
STEP 3: If $x \neq 0$ and $y \neq 0$, then $r = \sqrt{x^2 + y^2}$.
STEP 4: To find θ, first determine the quadrant that the point lies in.

Quadrant I: $\theta = \tan^{-1}\dfrac{y}{x}$ Quadrant II: $\theta = \pi + \tan^{-1}\dfrac{y}{x}$

Quadrant III: $\theta = \pi + \tan^{-1}\dfrac{y}{x}$ Quadrant IV: $\theta = \tan^{-1}\dfrac{y}{x}$

See Figure 19.

✏️   **NOW WORK PROBLEM 59.**

Formulas (1) and (2) may also be used to transform equations from polar form to rectangular form and vice-versa. Two common techniques for transforming an equation from polar form to rectangular form are (1) multiplying both sides of the equation by r and (2) squaring both sides of the equation.

EXAMPLE 7   **Transforming an Equation from Polar to Rectangular Form**

Transform the equation $r = 4\sin\theta$ from polar coordinates to rectangular coordinates, and identify the graph.

Solution   If we multiply each side by r, it will be easier to apply formulas (1) and (2).

$$r = 4\sin\theta$$
$$r^2 = 4r\sin\theta \qquad \text{Multiply each side by } r.$$
$$x^2 + y^2 = 4y \qquad r^2 = x^2 + y^2; y = r\sin\theta.$$

This is the equation of a circle; we proceed to complete the square to obtain the standard form of the equation.

$$x^2 + y^2 = 4y$$

$$x^2 + (y^2 - 4y) = 0 \qquad \text{\textcolor{blue}{General form}}$$

$$x^2 + (y^2 - 4y + 4) = 4 \qquad \text{\textcolor{blue}{Complete the square in y.}}$$

$$x^2 + (y - 2)^2 = 4 \qquad \text{\textcolor{blue}{Factor}}$$

This is the standard form of the equation of a circle with center $(0, 2)$ and radius 2. ◀

NOW WORK PROBLEM 75.

| EXAMPLE 8 | **Transforming an Equation from Rectangular to Polar Form** |

Transform the equation $4xy = 9$ from rectangular coordinates to polar coordinates.

Solution   We use formula (1): $x = r \cos \theta$ and $y = r \cos \theta$.

$$4xy = 9$$

$$4(r \cos \theta)(r \sin \theta) = 9 \qquad \text{\textcolor{blue}{$x = r \cos \theta, y = r \sin \theta$}}$$

$$4r^2 \cos \theta \sin \theta = 9$$

$$2r^2(2 \sin \theta \cos \theta) = 9 \qquad \text{\textcolor{blue}{Factor out $2r^2$.}}$$

$$2r^2 \sin(2\theta) = 9 \qquad \text{\textcolor{blue}{Double-angle Formula}} \qquad ◀$$

8.1 Assess Your Understanding

'Are You Prepared?'

Answers are given at the end of these exercises. If you get a wrong answer, read the pages listed in red.

1. Plot the point whose rectangular coordinates are $(3, -1)$. (pp. 2–5)

2. To complete the square of $x^2 + 6x$, add _____. (p. 699)

3. If $P = (x, y)$ is a point on a unit circle and on the terminal side of the angle θ, then $\sin \theta =$ _____. (p. 372)

4. $\tan^{-1}(-1) =$ _____ (pp. 455–457)

Concepts and Vocabulary

5. In polar coordinates, the origin is called the _____ and the positive x-axis is referred to as the _____ _____.

6. Another representation in polar coordinates for the point $\left(2, \dfrac{\pi}{3}\right)$ is $\left(\underline{\hspace{1cm}}, \dfrac{4\pi}{3}\right)$.

7. The polar coordinates $\left(-2, \dfrac{\pi}{6}\right)$ are represented in rectangular coordinates by (_____, _____).

8. *True or False:* The polar coordinates of a point are unique.

9. *True or False:* The rectangular coordinates of a point are unique.

10. *True or False:* In (r, θ), the number r can be negative.

Skill Building

In Problems 11–18, match each point in polar coordinates with either A, B, C, or D on the graph.

11. $\left(2, -\dfrac{11\pi}{6}\right)$ **12.** $\left(-2, -\dfrac{\pi}{6}\right)$ **13.** $\left(-2, \dfrac{\pi}{6}\right)$ **14.** $\left(2, \dfrac{7\pi}{6}\right)$

15. $\left(2, \dfrac{5\pi}{6}\right)$ **16.** $\left(-2, \dfrac{5\pi}{6}\right)$ **17.** $\left(-2, \dfrac{7\pi}{6}\right)$ **18.** $\left(2, \dfrac{11\pi}{6}\right)$

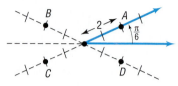

In Problems 19–30, plot each point given in polar coordinates.

19. $(3, 90°)$ **20.** $(4, 270°)$ **21.** $(-2, 0)$ **22.** $(-3, \pi)$

23. $\left(6, \dfrac{\pi}{6}\right)$ **24.** $\left(5, \dfrac{5\pi}{3}\right)$ **25.** $(-2, 135°)$ **26.** $(-3, 120°)$

27. $\left(-1, -\dfrac{\pi}{3}\right)$ **28.** $\left(-3, -\dfrac{3\pi}{4}\right)$ **29.** $(-2, -\pi)$ **30.** $\left(-3, -\dfrac{\pi}{2}\right)$

In Problems 31–38, plot each point given in polar coordinates, and find other polar coordinates (r, θ) of the point for which:
(a) $r > 0$, $-2\pi \le \theta < 0$ (b) $r < 0$, $0 \le \theta < 2\pi$ (c) $r > 0$, $2\pi \le \theta < 4\pi$

31. $\left(5, \dfrac{2\pi}{3}\right)$ **32.** $\left(4, \dfrac{3\pi}{4}\right)$ **33.** $(-2, 3\pi)$ **34.** $(-3, 4\pi)$

35. $\left(1, \dfrac{\pi}{2}\right)$ **36.** $(2, \pi)$ **37.** $\left(-3, -\dfrac{\pi}{4}\right)$ **38.** $\left(-2, -\dfrac{2\pi}{3}\right)$

In Problems 39–54, the polar coordinates of a point are given. Find the rectangular coordinates of each point. Verify your results using a graphing utility.

39. $\left(3, \dfrac{\pi}{2}\right)$ **40.** $\left(4, \dfrac{3\pi}{2}\right)$ **41.** $(-2, 0)$ **42.** $(-3, \pi)$

43. $(6, 150°)$ **44.** $(5, 300°)$ **45.** $\left(-2, \dfrac{3\pi}{4}\right)$ **46.** $\left(-2, \dfrac{2\pi}{3}\right)$

47. $\left(-1, -\dfrac{\pi}{3}\right)$ **48.** $\left(-3, -\dfrac{3\pi}{4}\right)$ **49.** $(-2, -180°)$ **50.** $(-3, -90°)$

51. $(7.5, 110°)$ **52.** $(-3.1, 182°)$ **53.** $(6.3, 3.8)$ **54.** $(8.1, 5.2)$

In Problems 55–66, the rectangular coordinates of a point are given. Find polar coordinates for each point. Verify your results using a graphing utility.

55. $(3, 0)$ **56.** $(0, 2)$ **57.** $(-1, 0)$ **58.** $(0, -2)$

59. $(1, -1)$ **60.** $(-3, 3)$ **61.** $\left(\sqrt{3}, 1\right)$ **62.** $\left(-2, -2\sqrt{3}\right)$

63. $(1.3, -2.1)$ **64.** $(-0.8, -2.1)$ **65.** $(8.3, 4.2)$ **66.** $(-2.3, 0.2)$

In Problems 67–74, the letters x and y represent rectangular coordinates. Write each equation using polar coordinates (r, θ).

67. $2x^2 + 2y^2 = 3$ **68.** $x^2 + y^2 = x$ **69.** $x^2 = 4y$ **70.** $y^2 = 2x$

71. $2xy = 1$ **72.** $4x^2 y = 1$ **73.** $x = 4$ **74.** $y = -3$

In Problems 75–82, the letters r and θ represent polar coordinates. Write each equation using rectangular coordinates (x, y).

75. $r = \cos \theta$ **76.** $r = \sin \theta + 1$ **77.** $r^2 = \cos \theta$ **78.** $r = \sin \theta - \cos \theta$

79. $r = 2$ **80.** $r = 4$ **81.** $r = \dfrac{4}{1 - \cos \theta}$ **82.** $r = \dfrac{3}{3 - \cos \theta}$

Applications and Extensions

83. Show that the formula for the distance d between two points $P_1 = (r_1, \theta_1)$ and $P_2 = (r_2, \theta_2)$ is

$$d = \sqrt{r_1^2 + r_2^2 - 2r_1 r_2 \cos(\theta_2 - \theta_1)}$$

Discussion and Writing

84. In converting from polar coordinates to rectangular coordinates, what formulas will you use?

85. Explain how you proceed to convert from rectangular coordinates to polar coordinates.

86. Is the street system in your town based on a rectangular coordinate system, a polar coordinate system, or some other system? Explain.

'Are You Prepared?' Answers

1. 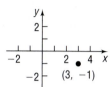   **2.** 9   **3.** y **4.** $-\dfrac{\pi}{4}$

8.2 Polar Equations and Graphs

PREPARING FOR THIS SECTION   *Before getting started, review the following:*

- Symmetry; (Section 1.2, pp. 17–19)
- Circles (Section 1.5, pp. 44–49)
- Even–Odd Properties of Trigonometric Functions (Section 5.3, pp. 398–399)

- Difference Formulas for Sine and Cosine (Section 6.4, pp. 473 and 476)
- Value of the Sine and Cosine Functions at Certain Angles (Section 5.2, pp. 374–381)

✎ Now work the 'Are You Prepared?' problems on page 577.

OBJECTIVES
1. Graph and Identify Polar Equations by Converting to Rectangular Equations
2. Graph Polar Equations Using a Graphing Utility
3. Test Polar Equations for Symmetry
4. Graph Polar Equations by Plotting Points

Just as a rectangular grid may be used to plot points given by rectangular coordinates, as in Figure 20(a), we can use a grid consisting of concentric circles (with centers at the pole) and rays (with vertices at the pole) to plot points given by polar coordinates, as shown in Figure 20(b). We shall use such **polar grids** to graph *polar equations*.

Figure 20

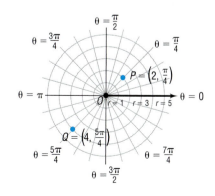

(a)   Rectangular grid

(b)   Polar grid

An equation whose variables are polar coordinates is called a **polar equation**. The **graph of a polar equation** consists of all points whose polar coordinates satisfy the equation.

1 Graph and Identify Polar Equations by Converting to Rectangular Equations

One method that we can use to graph a polar equation is to convert the equation to rectangular coordinates. In the discussion that follows, (x, y) represent the rectangular coordinates of a point P, and (r, θ) represent polar coordinates of the point P.

| EXAMPLE 1 | **Identifying and Graphing a Polar Equation By Hand (Circle)** |

Identify and graph the equation: $r = 3$

Solution   We convert the polar equation to a rectangular equation.

$$r = 3$$
$$r^2 = 9 \qquad \text{Square both sides.}$$
$$x^2 + y^2 = 9 \qquad r^2 = x^2 + y^2$$

The graph of $r = 3$ is a circle, with center at the pole and radius 3. See Figure 21.

Figure 21
$r = 3$ or $x^2 + y^2 = 9$

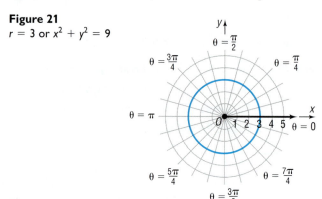

━━━━ **NOW WORK PROBLEM 13.**

| EXAMPLE 2 | **Identifying and Graphing a Polar Equation By Hand (Line)** |

Identify and graph the equation: $\theta = \dfrac{\pi}{4}$

Solution   We convert the polar equation to a rectangular equation.

$$\theta = \frac{\pi}{4}$$

$$\tan \theta = \tan \frac{\pi}{4} = 1$$

$$\frac{y}{x} = 1 \qquad \tan \theta = \frac{y}{x}$$

$$y = x$$

The graph of $\theta = \dfrac{\pi}{4}$ is a line passing through the pole making an angle of $\dfrac{\pi}{4}$ with the polar axis. See Figure 22.

Figure 22

$\theta = \dfrac{\pi}{4}$ or $y = x$

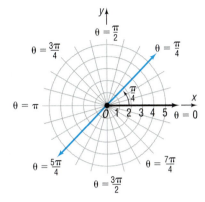

NOW WORK PROBLEM **15.**

EXAMPLE 3	**Identifying and Graphing a Polar Equation By Hand (Horizontal Line)**

Identify and graph the equation: $r \sin \theta = 2$

Solution   Since $y = r \sin \theta$, we can write the equation as

$$y = 2$$

Figure 23

$r \sin \theta = 2$ or $y = 2$

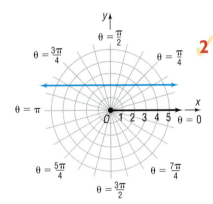

We conclude that the graph of $r \sin \theta = 2$ is a horizontal line 2 units above the pole. See Figure 23.

2   **Graph Polar Equations Using a Graphing Utility**

A second method we can use to graph a polar equation is to graph the equation using a graphing utility.

Most graphing utilities require the following steps to obtain the graph of an equation:

> **Graphing a Polar Equation Using a Graphing Utility**
>
> **STEP 1:** Solve the equation for r in terms of θ.
> **STEP 2:** Select the viewing window in POLar mode. In addition to setting Xmin, Xmax, Xscl, and so forth, the viewing window in polar mode requires setting minimum and maximum values for θ and an increment setting for θ (θstep). Finally, a square screen and radian measure should be used.
> **STEP 3:** Enter the expression involving θ that you found in Step 1. (Consult your manual for the correct way to enter the expression.)
> **STEP 4:** Press graph.

EXAMPLE 4	**Graphing a Polar Equation Using a Graphing Utility**

Use a graphing utility to graph the polar equation $r \sin \theta = 2$.

Solution   **STEP 1:**   We solve the equation for r in terms of θ.

$$r \sin \theta = 2$$

$$r = \dfrac{2}{\sin \theta}$$

STEP 2: From the polar mode, select a square viewing window. We will use the one given next.

$$\theta\text{min} = 0 \qquad X\text{min} = -9 \qquad Y\text{min} = -6$$
$$\theta\text{max} = 2\pi \qquad X\text{max} = 9 \qquad Y\text{max} = 6$$
$$\theta\text{step} = \frac{\pi}{24} \qquad X\text{scl} = 1 \qquad Y\text{scl} = 1$$

θstep determines the number of points the graphing utility will plot. For example, if θstep is $\dfrac{\pi}{24}$, then the graphing utility will evaluate r at $\theta = 0$ (θmin), $\dfrac{\pi}{24}, \dfrac{2\pi}{24}, \dfrac{3\pi}{24}$, and so forth, up to 2π(θmax).

Figure 24

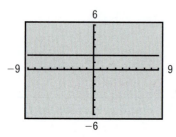

The smaller θstep, the more points the graphing utility will plot. The student is encouraged to experiment with different values for θmin, θmax, and θstep to see how the graph is affected.

STEP 3: Enter the expression $\dfrac{2}{\sin \theta}$ after the prompt $r = $.

STEP 4: Graph.

The graph is shown in Figure 24. ◀

EXAMPLE 5 **Identifying and Graphing a Polar Equation (Vertical Line)**

Identify and graph the equation: $r \cos \theta = -3$

Solution Since $x = r \cos \theta$, we can write the equation as

$$x = -3$$

We conclude that the graph of $r \cos \theta = -3$ is a vertical line 3 units to the left of the pole. Figure 25(a) shows the graph drawn by hand. Figure 25(b) shows the graph using a graphing utility with θmin $= 0$, θmax $= 2\pi$, and θstep $= \dfrac{\pi}{24}$.

Figure 25
$r \cos \theta = -3$ or $x = -3$

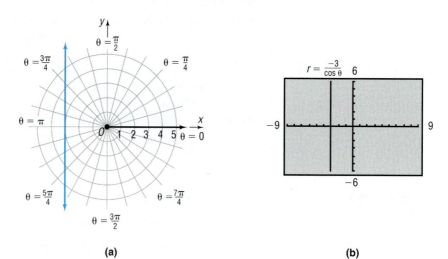

(a)   (b)   ◀

Based on Examples 3, 4, and 5, we are led to the following results. (The proofs are left as exercises.)

Theorem

Let a be a nonzero real number. Then the graph of the equation

$$r \sin \theta = a$$

is a horizontal line a units above the pole if $a > 0$ and $|a|$ units below the pole if $a < 0$.

The graph of the equation

$$r \cos \theta = a$$

is a vertical line a units to the right of the pole if $a > 0$ and $|a|$ units to the left of the pole if $a < 0$.

─ **NOW WORK PROBLEM 19.**

EXAMPLE 6 **Identifying and Graphing a Polar Equation (Circle)**

Identify and graph the equation: $r = 4 \sin \theta$

Solution To transform the equation to rectangular coordinates, we multiply each side by r.

$$r^2 = 4r \sin \theta$$

Now we use the facts that $r^2 = x^2 + y^2$ and $y = r \sin \theta$. Then

$$x^2 + y^2 = 4y$$

$$x^2 + (y^2 - 4y) = 0$$

$$x^2 + (y^2 - 4y + 4) = 4 \qquad \textcolor{teal}{\text{Complete the square in } y.}$$

$$x^2 + (y - 2)^2 = 4 \qquad \textcolor{teal}{\text{Factor.}}$$

This is the standard equation of a circle with center at $(0, 2)$ in rectangular coordinates and radius 2. Figure 26(a) shows the graph drawn by hand. Figure 26(b) shows the graph using a graphing utility with $\theta\min = 0$, $\theta\max = 2\pi$, and $\theta\text{step} = \dfrac{\pi}{24}$.

Figure 26
$r = 4 \sin \theta$ or $x^2 + (y - 2)^2 = 4$

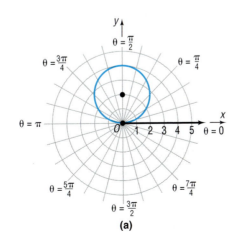

(a)

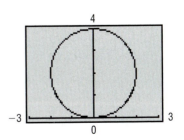

(b)

◀

| EXAMPLE 7 | Identifying and Graphing a Polar Equation (Circle) |

Identify and graph the equation: $r = -2 \cos \theta$

Solution

We proceed as in Example 6.

$$r^2 = -2r \cos \theta \qquad \text{Multiply both sides by } r.$$
$$x^2 + y^2 = -2x \qquad r^2 = x^2 + y^2; \quad x = r \cos \theta$$
$$x^2 + 2x + y^2 = 0$$
$$(x^2 + 2x + 1) + y^2 = 1 \qquad \text{Complete the square in } x.$$
$$(x + 1)^2 + y^2 = 1 \qquad \text{Factor.}$$

This is the standard equation of a circle with center at $(-1, 0)$ in rectangular coordinates and radius 1. Figure 27(a) shows the graph drawn by hand. Figure 27(b) shows the graph using a graphing utility with θmin $= 0$, θmax $= 2\pi$, and θstep $= \dfrac{\pi}{24}$.

Figure 27
$r = -2 \cos \theta$ or $(x + 1)^2 + y^2 = 1$

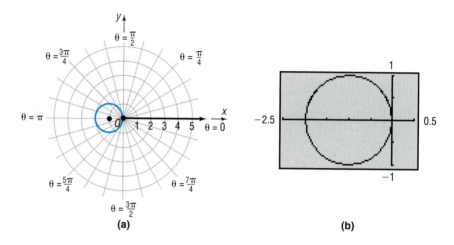

(a)

(b)

─── **Exploration** ───

Using a square screen, graph $r_1 = \sin \theta$, $r_2 = 2 \sin \theta$, and $r_3 = 3 \sin \theta$. Do you see the pattern? Clear the screen and graph $r_1 = -\sin \theta$, $r_2 = -2 \sin \theta$, and $r_3 = -3 \sin \theta$. Do you see the pattern? Clear the screen and graph $r_1 = \cos \theta$, $r_2 = 2 \cos \theta$, and $r_3 = 3 \cos \theta$. Do you see the pattern? Clear the screen and graph $r_1 = -\cos \theta$, $r_2 = -2 \cos \theta$, and $r_3 = -3 \cos \theta$. Do you see the pattern?

Based on Examples 6 and 7 and the preceding Exploration, we are led to the following results. (The proofs are left as exercises.)

Theorem

Let a be a positive real number. Then,

Equation **Description**

(a) $r = 2a \sin \theta$ Circle: radius a; center at $(0, a)$ in rectangular coordinates
(b) $r = -2a \sin \theta$ Circle: radius a; center at $(0, -a)$ in rectangular coordinates
(c) $r = 2a \cos \theta$ Circle: radius a; center at $(a, 0)$ in rectangular coordinates
(d) $r = -2a \cos \theta$ Circle: radius a; center at $(-a, 0)$ in rectangular coordinates

Each circle passes through the pole.

NOW WORK PROBLEM **21**.

The method of converting a polar equation to an identifiable rectangular equation to obtain the graph is not always helpful, nor is it always necessary. Usually, we set up a table that lists several points on the graph. By checking for symmetry, it may be possible to reduce the number of points needed to draw the graph.

3 Test Polar Equations for Symmetry

In polar coordinates, the points (r, θ) and $(r, -\theta)$ are symmetric with respect to the polar axis (and to the x-axis). See Figure 28(a). The points (r, θ) and $(r, \pi - \theta)$ are symmetric with respect to the line $\theta = \dfrac{\pi}{2}$ (the y-axis). See Figure 28(b). The points (r, θ) and $(-r, \theta)$ are symmetric with respect to the pole (the origin). See Figure 28(c).

Figure 28

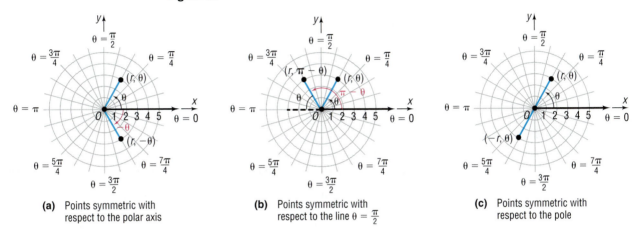

(a) Points symmetric with respect to the polar axis

(b) Points symmetric with respect to the line $\theta = \dfrac{\pi}{2}$

(c) Points symmetric with respect to the pole

The following tests are a consequence of these observations.

Theorem

Tests for Symmetry

Symmetry with Respect to the Polar Axis (x-Axis)

In a polar equation, replace θ by $-\theta$. If an equivalent equation results, the graph is symmetric with respect to the polar axis.

Symmetry with Respect to the Line $\theta = \dfrac{\pi}{2}$ (y-Axis)

In a polar equation, replace θ by $\pi - \theta$. If an equivalent equation results, the graph is symmetric with respect to the line $\theta = \dfrac{\pi}{2}$.

Symmetry with Respect to the Pole (Origin)

In a polar equation, replace r by $-r$. If an equivalent equation results, the graph is symmetric with respect to the pole.

The three tests for symmetry given here are *sufficient* conditions for symmetry, but they are not *necessary* conditions. That is, an equation may fail these tests and still have a graph that is symmetric with respect to the polar axis, the line $\theta = \dfrac{\pi}{2}$, or the pole. For example, the graph of $r = \sin(2\theta)$ turns out to be symmetric with respect to the polar axis, the line $\theta = \dfrac{\pi}{2}$, and the pole, but all three tests given here fail. See also Problems 87, 88, and 89.

4 Graph Polar Equations by Plotting Points

EXAMPLE 8	**Graphing a Polar Equation (Cardioid)**

Graph the equation: $r = 1 - \sin\theta$

Solution

We check for symmetry first.

Polar Axis: Replace θ by $-\theta$. The result is

$$r = 1 - \sin(-\theta) = 1 + \sin\theta$$

The test fails, so the graph may or may not be symmetric with respect to the polar axis.

The Line $\theta = \dfrac{\pi}{2}$: Replace θ by $\pi - \theta$. The result is

$$r = 1 - \sin(\pi - \theta) = 1 - (\sin\pi\cos\theta - \cos\pi\sin\theta)$$
$$= 1 - [0\cdot\cos\theta - (-1)\sin\theta] = 1 - \sin\theta$$

The test is satisfied, so the graph is symmetric with respect to the line $\theta = \dfrac{\pi}{2}$.

The Pole: Replace r by $-r$. Then the result is $-r = 1 - \sin\theta$, so $r = -1 + \sin\theta$. The test fails, so the graph may or may not be symmetric with respect to the pole.

Next, we identify points on the graph by assigning values to the angle θ and calculating the corresponding values of r. Due to the symmetry with respect to the line $\theta = \dfrac{\pi}{2}$, we only need to assign values to θ from $-\dfrac{\pi}{2}$ to $\dfrac{\pi}{2}$, as given in Table 1.

Now we plot the points (r, θ) from Table 1 and trace out the graph, beginning at the point $\left(2, -\dfrac{\pi}{2}\right)$ and ending at the point $\left(0, \dfrac{\pi}{2}\right)$. Then we reflect this portion of the graph about the line $\theta = \dfrac{\pi}{2}$ (the y-axis) to obtain the complete graph. Figure 29(a) shows the graph drawn by hand. Figure 29(b) shows the graph using a graphing utility with $\theta\min = 0$, $\theta\max = 2\pi$, and θstep $= \dfrac{\pi}{24}$.

Table 1

θ	$r = 1 - \sin\theta$
$-\dfrac{\pi}{2}$	$1 - (-1) = 2$
$-\dfrac{\pi}{3}$	$1 - \left(-\dfrac{\sqrt{3}}{2}\right) \approx 1.87$
$-\dfrac{\pi}{6}$	$1 - \left(-\dfrac{1}{2}\right) = \dfrac{3}{2}$
0	$1 - 0 = 1$
$\dfrac{\pi}{6}$	$1 - \dfrac{1}{2} = \dfrac{1}{2}$
$\dfrac{\pi}{3}$	$1 - \dfrac{\sqrt{3}}{2} \approx 0.13$
$\dfrac{\pi}{2}$	$1 - 1 = 0$

Figure 29

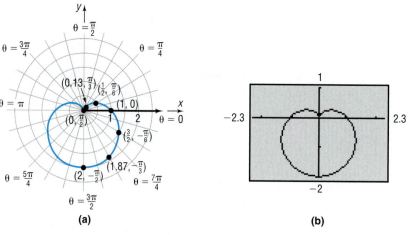

(a)   (b)

The curve in Figure 29 is an example of a *cardioid* (a heart-shaped curve).

--- **Exploration** ---

Graph $r_1 = 1 + \sin\theta$. Clear the screen and graph $r_1 = 1 - \cos\theta$. Clear the screen and graph $r_1 = 1 + \cos\theta$. Do you see a pattern?

Cardioids are characterized by equations of the form

$$r = a(1 + \cos\theta) \qquad r = a(1 + \sin\theta)$$
$$r = a(1 - \cos\theta) \qquad r = a(1 - \sin\theta)$$

where $a > 0$. The graph of a cardioid passes through the pole.

NOW WORK PROBLEM 43.

EXAMPLE 9 — Graphing a Polar Equation (Limaçon without Inner Loop)

Graph the equation: $r = 3 + 2\cos\theta$

Solution   We check for symmetry first.

Polar Axis: Replace θ by $-\theta$. The result is
$$r = 3 + 2\cos(-\theta) = 3 + 2\cos\theta$$
The test is satisfied, so the graph is symmetric with respect to the polar axis.

The Line $\theta = \dfrac{\pi}{2}$:   Replace θ by $\pi - \theta$. The result is

$$r = 3 + 2\cos(\pi - \theta) = 3 + 2(\cos\pi\cos\theta + \sin\pi\sin\theta)$$
$$= 3 - 2\cos\theta$$

The test fails, so the graph may or may not be symmetric with respect to the line $\theta = \dfrac{\pi}{2}$.

The Pole: Replace r by $-r$. The test fails, so the graph may or may not be symmetric with respect to the pole.

Next, we identify points on the graph by assigning values to the angle θ and calculating the corresponding values of r. Due to the symmetry with respect to the polar axis, we only need to assign values to θ from 0 to π, as given in Table 2.

Now we plot the points (r, θ) from Table 2 and trace out the graph, beginning at the point $(5, 0)$ and ending at the point $(1, \pi)$. Then we reflect this portion of the graph about the polar axis (the x-axis) to obtain the complete graph. Figure 30(a) shows the graph drawn by hand. Figure 30(b) shows the graph using a graphing utility with $\theta\min = 0$, $\theta\max = 2\pi$, and $\theta\text{step} = \dfrac{\pi}{24}$.

Table 2

θ	$r = 3 + 2\cos\theta$
0	$3 + 2(1) = 5$
$\dfrac{\pi}{6}$	$3 + 2\left(\dfrac{\sqrt{3}}{2}\right) \approx 4.73$
$\dfrac{\pi}{3}$	$3 + 2\left(\dfrac{1}{2}\right) = 4$
$\dfrac{\pi}{2}$	$3 + 2(0) = 3$
$\dfrac{2\pi}{3}$	$3 + 2\left(-\dfrac{1}{2}\right) = 2$
$\dfrac{5\pi}{6}$	$3 + 2\left(-\dfrac{\sqrt{3}}{2}\right) \approx 1.27$
π	$3 + 2(-1) = 1$

Figure 30

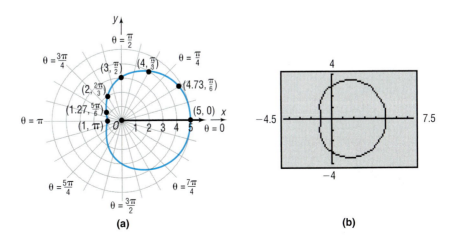

(a)   (b)

The curve in Figure 30 is an example of a *limaçon* (the French word for *snail*) *without an inner loop*.

Limaçons without an inner loop are characterized by equations of the form

$$r = a + b\cos\theta \qquad r = a + b\sin\theta$$
$$r = a - b\cos\theta \qquad r = a - b\sin\theta$$

where $a > 0$, $b > 0$, and $a > b$. The graph of a limaçon without an inner loop does not pass through the pole.

NOW WORK PROBLEM **49**.

EXAMPLE 10 | **Graphing a Polar Equation (Limaçon with Inner Loop)**

Graph the equation: $r = 1 + 2\cos\theta$

Solution First, we check for symmetry.

Polar Axis: Replace θ by $-\theta$. The result is

$$r = 1 + 2\cos(-\theta) = 1 + 2\cos\theta$$

The test is satisfied, so the graph is symmetric with respect to the polar axis.

The Line $\theta = \dfrac{\pi}{2}$: Replace θ by $\pi - \theta$. The result is

$$r = 1 + 2\cos(\pi - \theta) = 1 + 2(\cos\pi\cos\theta + \sin\pi\sin\theta)$$

$$= 1 - 2\cos\theta$$

The test fails, so the graph may or may not be symmetric with respect to the line $\theta = \dfrac{\pi}{2}$.

The Pole: Replace r by $-r$. The test fails, so the graph may or may not be symmetric with respect to the pole.

Next, we identify points on the graph of $r = 1 + 2\cos\theta$ by assigning values to the angle θ and calculating the corresponding values of r. Due to the symmetry with respect to the polar axis, we only need to assign values to θ from 0 to π, as given in Table 3.

Now we plot the points (r, θ) from Table 3, beginning at $(3, 0)$ and ending at $(-1, \pi)$. See Figure 31(a). Finally, we reflect this portion of the graph about the polar axis (the x-axis) to obtain the complete graph. See Figure 31(b). Figure 31(c) shows the graph using a graphing utility with θmin $= 0$, θmax $= 2\pi$, and θstep $= \dfrac{\pi}{24}$.

Table 3

θ	$r = 1 + 2\cos\theta$
0	$1 + 2(1) = 3$
$\dfrac{\pi}{6}$	$1 + 2\left(\dfrac{\sqrt{3}}{2}\right) \approx 2.73$
$\dfrac{\pi}{3}$	$1 + 2\left(\dfrac{1}{2}\right) = 2$
$\dfrac{\pi}{2}$	$1 + 2(0) = 1$
$\dfrac{2\pi}{3}$	$1 + 2\left(-\dfrac{1}{2}\right) = 0$
$\dfrac{5\pi}{6}$	$1 + 2\left(-\dfrac{\sqrt{3}}{2}\right) \approx -0.73$
π	$1 + 2(-1) = -1$

Figure 31

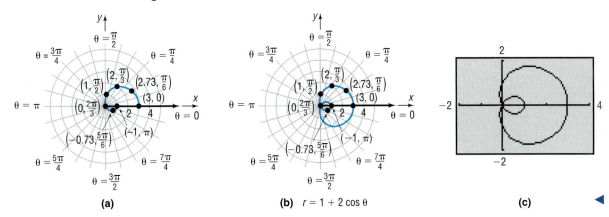

(a)   (b) $r = 1 + 2 \cos \theta$ (c)

—— **Exploration** ———

Graph $r_1 = 1 - 2 \cos \theta$. Clear the screen and graph $r_1 = 1 + 2 \sin \theta$. Clear the screen and graph $r_1 = 1 - 2 \sin \theta$. Do you see a pattern?

The curve in Figure 31(b) or 31(c) is an example of a *limaçon with an inner loop*.

Limaçons with an inner loop are characterized by equations of the form

$$r = a + b \cos \theta \qquad r = a + b \sin \theta$$
$$r = a - b \cos \theta \qquad r = a - b \sin \theta$$

where $a > 0, b > 0$, and $a < b$. The graph of a limaçon with an inner loop will pass through the pole twice.

NOW WORK PROBLEM 51.

EXAMPLE 11

Graphing a Polar Equation (Rose)

Graph the equation: $r = 2 \cos(2\theta)$

Solution   We check for symmetry.

Polar Axis: If we replace θ by $-\theta$, the result is

$$r = 2 \cos[2(-\theta)] = 2 \cos(2\theta)$$

The test is satisfied, so the graph is symmetric with respect to the polar axis.

The Line $\theta = \dfrac{\pi}{2}$:   If we replace θ by $\pi - \theta$, we obtain

$$r = 2 \cos[2(\pi - \theta)] = 2 \cos(2\pi - 2\theta) = 2 \cos(2\theta)$$

The test is satisfied, so the graph is symmetric with respect to the line $\theta = \dfrac{\pi}{2}$.

The Pole: Since the graph is symmetric with respect to both the polar axis and the line $\theta = \dfrac{\pi}{2}$, it must be symmetric with respect to the pole.

Table 4

θ	$r = 2\cos(2\theta)$
0	$2(1) = 2$
$\dfrac{\pi}{6}$	$2\left(\dfrac{1}{2}\right) = 1$
$\dfrac{\pi}{4}$	$2(0) = 0$
$\dfrac{\pi}{3}$	$2\left(-\dfrac{1}{2}\right) = -1$
$\dfrac{\pi}{2}$	$2(-1) = -2$

Next, we construct Table 4. Due to the symmetry with respect to the polar axis, the line $\theta = \dfrac{\pi}{2}$, and the pole, we consider only values of θ from 0 to $\dfrac{\pi}{2}$.

We plot and connect these points in Figure 32(a). Finally, because of symmetry, we reflect this portion of the graph first about the polar axis (the x-axis) and then about the line $\theta = \dfrac{\pi}{2}$ (the y-axis) to obtain the complete graph. See Figure 32(b). Figure 32(c) shows the graph using a graphing utility with $\theta\min = 0$, $\theta\max = 2\pi$, and $\theta\text{step} = \dfrac{\pi}{24}$.

Figure 32

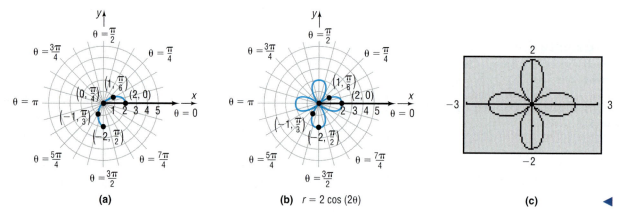

(a)　　　(b) $r = 2\cos(2\theta)$　　　(c)

──── **Exploration** ────

Graph $r_1 = 2\cos(4\theta)$; clear the screen and graph $r_1 = 2\cos(6\theta)$. How many petals did each of these graphs have?

Clear the screen and graph, in order, each on a clear screen, $r_1 = 2\cos(3\theta)$, $r_1 = 2\cos(5\theta)$, and $r_1 = 2\cos(7\theta)$. What do you notice about the number of petals?

The curve in Figure 32(b) or (c) is called a *rose* with four petals.

Rose curves are characterized by equations of the form

$$r = a\cos(n\theta), \qquad r = a\sin(n\theta), \qquad a \neq 0$$

and have graphs that are rose shaped. If $n \neq 0$ is even, the rose has $2n$ petals; if $n \neq \pm 1$ is odd, the rose has n petals.

NOW WORK PROBLEM 55.

EXAMPLE 12	**Graphing a Polar Equation (Lemniscate)**

Graph the equation: $r^2 = 4\sin(2\theta)$

Table 5

θ	$r^2 = 4\sin(2\theta)$	r
0	$4(0) = 0$	0
$\dfrac{\pi}{6}$	$4\left(\dfrac{\sqrt{3}}{2}\right) = 2\sqrt{3}$	± 1.9
$\dfrac{\pi}{4}$	$4(1) = 4$	± 2
$\dfrac{\pi}{3}$	$4\left(\dfrac{\sqrt{3}}{2}\right) = 2\sqrt{3}$	± 1.9
$\dfrac{\pi}{2}$	$4(0) = 0$	0

Solution We leave it to you to verify that the graph is symmetric with respect to the pole. Table 5 lists points on the graph for values of $\theta = 0$ through $\theta = \dfrac{\pi}{2}$. Note that there are no points on the graph for $\dfrac{\pi}{2} < \theta < \pi$ (quadrant II), since $\sin(2\theta) < 0$ for such values. The points from Table 5 where $r \geq 0$ are plotted in Figure 33(a). The remaining points on the graph may be obtained by using symmetry. Figure 33(b) shows the final graph drawn by hand. Figure 33(c) shows the graph using a graphing utility with $\theta\text{min} = 0$, $\theta\text{max} = 2\pi$, and $\theta\text{step} = \dfrac{\pi}{24}$.

Figure 33

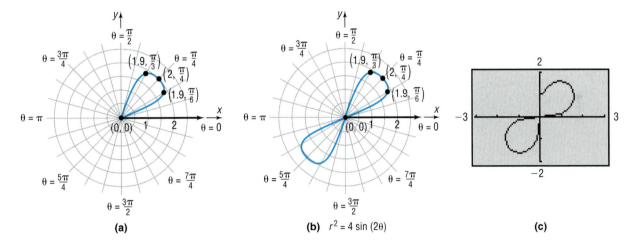

(a)   (b) $r^2 = 4\sin(2\theta)$ (c)

The curve in Figure 33(b) or (c) is an example of a *lemniscate* (from the Greek word *ribbon*).

Lemniscates are characterized by equations of the form

$$r^2 = a^2\sin(2\theta) \qquad r^2 = a^2\cos(2\theta)$$

where $a \neq 0$, and have graphs that are propeller shaped.

 NOW WORK PROBLEM 59.

EXAMPLE 13 **Graphing a Polar Equation (Spiral)**

Graph the equation: $r = e^{\theta/5}$

Solution The tests for symmetry with respect to the pole, the polar axis, and the line $\theta = \dfrac{\pi}{2}$ fail. Furthermore, there is no number θ for which $r = 0$, so the graph does not pass through the pole. We observe that r is positive for all θ, r increases as θ increases, $r \to 0$ as $\theta \to -\infty$, and $r \to \infty$ as $\theta \to \infty$. With the help of a calculator, we obtain

Table 6

θ	$r = e^{\theta/5}$
$-\dfrac{3\pi}{2}$	0.39
$-\pi$	0.53
$-\dfrac{\pi}{2}$	0.73
$-\dfrac{\pi}{4}$	0.85
0	1
$\dfrac{\pi}{4}$	1.17
$\dfrac{\pi}{2}$	1.37
π	1.87
$\dfrac{3\pi}{2}$	2.57
2π	3.51

the values in Table 6. See Figure 34(a) for the graph drawn by hand. Figure 34(b) shows the graph using a graphing utility with $\theta\min = -4\pi$, $\theta\max = 3\pi$, and $\theta\text{step} = \dfrac{\pi}{24}$.

Figure 34
$r = e^{\theta/5}$

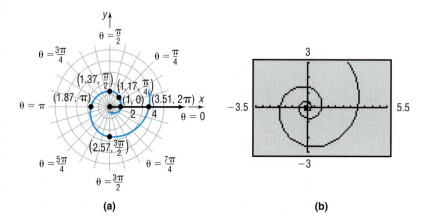

(a)   (b)

The curve in Figure 34 is called a **logarithmic spiral**, since its equation may be written as $\theta = 5 \ln r$ and it spirals infinitely both toward the pole and away from it.

Classification of Polar Equations

The equations of some lines and circles in polar coordinates and their corresponding equations in rectangular coordinates are given in Table 7. Also included are the names and the graphs of a few of the more frequently encountered polar equations.

Table 7

	Lines		
Description	Line passing through the pole making an angle α with the polar axis	Vertical line	Horizontal line
Rectangular equation	$y = (\tan \alpha)x$	$x = a$	$y = b$
Polar equation	$\theta = \alpha$	$r \cos \theta = a$	$r \sin \theta = b$
Typical graph			

	Circles		
Description	Center at the pole, radius a	Passing through the pole, tangent to the line $\theta = \dfrac{\pi}{2}$, center on the polar axis, radius a	Passing through the pole, tangent to the polar axis, center on the line $\theta = \dfrac{\pi}{2}$, radius a
Rectangular equation	$x^2 + y^2 = a^2$, $a > 0$	$x^2 + y^2 = \pm 2ax$, $a > 0$	$x^2 + y^2 = \pm 2ay$, $a > 0$
Polar equation	$r = a$, $a > 0$	$r = \pm 2a \cos \theta$, $a > 0$	$r = \pm 2a \sin \theta$, $a > 0$
Typical graph	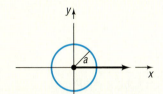	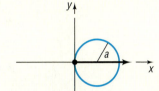	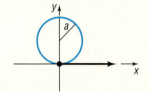

	Other Equations		
Name	Cardioid	Limaçon without inner loop	Limaçon with inner loop
Polar equations	$r = a \pm a \cos \theta$, $a > 0$ $r = a \pm a \sin \theta$, $a > 0$	$r = a \pm b \cos \theta$, $0 < b < a$ $r = a \pm b \sin \theta$, $0 < b < a$	$r = a \pm b \cos \theta$, $0 < a < b$ $r = a \pm b \sin \theta$, $0 < a < b$
Typical graph	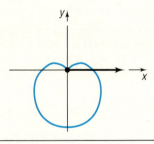	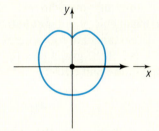	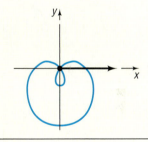
Name	Lemniscate	Rose with three petals	Rose with four petals
Polar equations	$r^2 = a^2 \cos(2\theta)$, $a > 0$ $r^2 = a^2 \sin(2\theta)$, $a > 0$	$r = a \sin(3\theta)$, $a > 0$ $r = a \cos(3\theta)$, $a > 0$	$r = a \sin(2\theta)$, $a > 0$ $r = a \cos(2\theta)$, $a > 0$
Typical graph	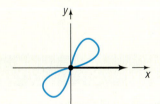	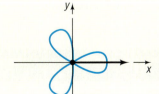	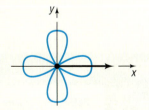

Sketching Quickly

If a polar equation involves only a sine (or cosine) function, you can quickly obtain a sketch of its graph by making use of Table 7, periodicity, and a short table.

EXAMPLE 14

Sketching the Graph of a Polar Equation Quickly by Hand

Graph the equation: $r = 2 + 2 \sin \theta$

Solution

We recognize the polar equation: Its graph is a cardioid. The period of $\sin \theta$ is 2π, so we form a table using $0 \le \theta \le 2\pi$, compute r, plot the points (r, θ), and sketch the graph of a cardioid as θ varies from 0 to 2π. See Table 8 and Figure 35.

Table 8

θ	$r = 2 + 2 \sin \theta$
0	$2 + 2(0) = 2$
$\dfrac{\pi}{2}$	$2 + 2(1) = 4$
π	$2 + 2(0) = 2$
$\dfrac{3\pi}{2}$	$2 + 2(-1) = 0$
2π	$2 + 2(0) = 2$

Figure 35

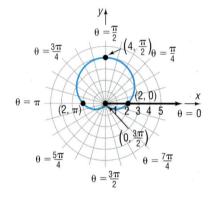

Calculus Comment

For those of you who are planning to study calculus, a comment about one important role of polar equations is in order.

In rectangular coordinates, the equation $x^2 + y^2 = 1$, whose graph is the unit circle, is not the graph of a function. In fact, it requires two functions to obtain the graph of the unit circle:

$$y_1 = \sqrt{1 - x^2} \quad \text{Upper semicircle} \qquad y_2 = -\sqrt{1 - x^2} \quad \text{Lower semicircle}$$

In polar coordinates, the equation $r = 1$, whose graph is also the unit circle, does define a function. That is, for each choice of θ there is only one corresponding value of r, that is, $r = 1$. Since many problems in calculus require the use of functions, the opportunity to express nonfunctions in rectangular coordinates as functions in polar coordinates becomes extremely useful.

Note also that the vertical-line test for functions is valid only for equations in rectangular coordinates.

HISTORICAL FEATURE

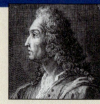

Jakob Bernoulli
(1654–1705)

Polar coordinates seem to have been invented by Jakob Bernoulli (1654–1705) in about 1691, although, as with most such ideas, earlier traces of the notion exist. Early users of calculus remained committed to rectangular coordinates, and polar coordinates did not become widely used until the early 1800s. Even then, it was mostly geometers who used them for describing odd curves. Finally, about the mid-1800s, applied mathematicians realized the tremendous simplification that polar coordinates make possible in the description of objects with circular or cylindrical symmetry. From then on their use became widespread.

8.2 Assess Your Understanding

'Are You Prepared?'

Answers are given at the end of these exercises. If you get a wrong answer, read the pages listed in red.

1. If the rectangular coordinates of a point are $(4, -6)$, the point symmetric to it with respect to the origin is _____. (pp. 17–19)

2. The difference formula for cosine is $\cos(\alpha - \beta) =$ _____. (p. 473)

3. The standard equation of a circle with center at $(-2, 5)$ and radius 3 is _____. (pp. 44–49)

4. Is the sine function even, odd, or neither? (pp. 398–399)

5. $\sin \dfrac{5\pi}{4} =$ _____. (pp. 380–381)

6. $\cos \dfrac{2\pi}{3} =$ _____. (pp. 380–381)

Concepts and Vocabulary

7. An equation whose variables are polar coordinates is called a _____ _____.

8. Using polar coordinates (r, θ), the circle $x^2 + y^2 = 2x$ takes the form _____.

9. A polar equation is symmetric with respect to the pole if an equivalent equation results when r is replaced by _____.

10. *True or False:* The tests for symmetry in polar coordinates are necessary, but not sufficient.

11. *True or False:* The graph of a cardioid never passes through the pole.

12. *True or False*: All polar equations have a symmetric feature.

Skill Building

In Problems 13–28, transform each polar equation to an equation in rectangular coordinates. Then identify and graph the equation. Verify your graph using a graphing utility.

13. $r = 4$

14. $r = 2$

15. $\theta = \dfrac{\pi}{3}$

16. $\theta = -\dfrac{\pi}{4}$

17. $r \sin \theta = 4$

18. $r \cos \theta = 4$

19. $r \cos \theta = -2$

20. $r \sin \theta = -2$

21. $r = 2 \cos \theta$

22. $r = 2 \sin \theta$

23. $r = -4 \sin \theta$

24. $r = -4 \cos \theta$

25. $r \sec \theta = 4$

26. $r \csc \theta = 8$

27. $r \csc \theta = -2$

28. $r \sec \theta = -4$

In Problems 29–36, match each of the graphs (A) through (H) to one of the following polar equations.

29. $r = 2$

30. $\theta = \dfrac{\pi}{4}$

31. $r = 2 \cos \theta$

32. $r \cos \theta = 2$

33. $r = 1 + \cos \theta$

34. $r = 2 \sin \theta$

35. $\theta = \dfrac{3\pi}{4}$

36. $r \sin \theta = 2$

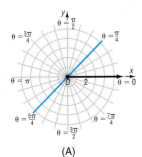

(A)

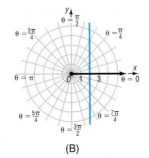

(B)

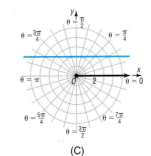

(C)

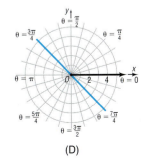

(D)

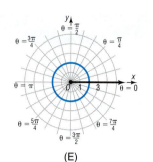

(E)

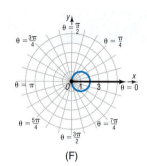

(F)

(G)

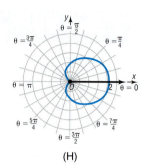

(H)

In Problems 37–42, match each of the graphs (A) through (F) to one of the following polar equations.

37. $r = 4$

38. $r = 3 \cos \theta$

39. $r = 3 \sin \theta$

40. $r \sin \theta = 3$

41. $r \cos \theta = 3$

42. $r = 2 + \sin \theta$

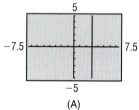

(A)

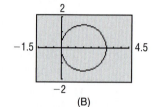

(B)

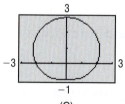

(C)

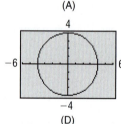

(D)

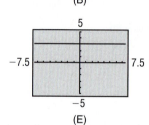

(E)

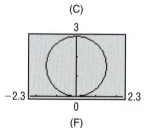

(F)

In Problems 43–66, identify and graph each polar equation. Verify your graph using a graphing utility.

43. $r = 2 + 2 \cos \theta$

44. $r = 1 + \sin \theta$

45. $r = 3 - 3 \sin \theta$

46. $r = 2 - 2 \cos \theta$

47. $r = 2 + \sin \theta$

48. $r = 2 - \cos \theta$

49. $r = 4 - 2 \cos \theta$

50. $r = 4 + 2 \sin \theta$

51. $r = 1 + 2 \sin \theta$

52. $r = 1 - 2 \sin \theta$

53. $r = 2 - 3 \cos \theta$

54. $r = 2 + 4 \cos \theta$

55. $r = 3 \cos(2\theta)$

56. $r = 2 \sin(3\theta)$

57. $r = 4 \sin(5\theta)$

58. $r = 3 \cos(4\theta)$

59. $r^2 = 9 \cos(2\theta)$

60. $r^2 = \sin(2\theta)$

61. $r = 2^\theta$

62. $r = 3^\theta$

63. $r = 1 - \cos \theta$

64. $r = 3 + \cos \theta$

65. $r = 1 - 3 \cos \theta$

66. $r = 4 \cos(3\theta)$

Applications and Extensions

In Problems 67–70, the polar equation for each graph is either $r = a + b \cos \theta$ or $r = a + b \sin \theta$, $a > 0, b > 0$. Select the correct equation and find the values of a and b.

67.

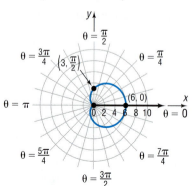

68.

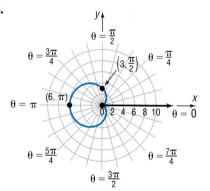

69.

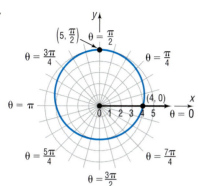

70.

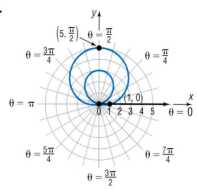

In Problems 71–80, graph each polar equation. Verify your graph using a graphing utility.

71. $r = \dfrac{2}{1 - \cos \theta}$ *(parabola)*

72. $r = \dfrac{2}{1 - 2 \cos \theta}$ *(hyperbola)*

73. $r = \dfrac{1}{3 - 2 \cos \theta}$ *(ellipse)*

74. $r = \dfrac{1}{1 - \cos \theta}$ *(parabola)*

75. $r = \theta, \quad \theta \geq 0$ *(spiral of Archimedes)*

76. $r = \dfrac{3}{\theta}$ *(reciprocal spiral)*

77. $r = \csc \theta - 2, \quad 0 < \theta < \pi$ *(conchoid)*

78. $r = \sin \theta \tan \theta$ *(cissoid)*

79. $r = \tan \theta, \quad -\dfrac{\pi}{2} < \theta < \dfrac{\pi}{2}$ *(kappa curve)*

80. $r = \cos \dfrac{\theta}{2}$

81. Show that the graph of the equation $r \sin \theta = a$ is a horizontal line a units above the pole if $a > 0$ and $|a|$ units below the pole if $a < 0$.

82. Show that the graph of the equation $r \cos \theta = a$ is a vertical line a units to the right of the pole if $a > 0$ and $|a|$ units to the left of the pole if $a < 0$.

83. Show that the graph of the equation $r = 2a \sin \theta, a > 0$, is a circle of radius a with center at $(0, a)$ in rectangular coordinates.

84. Show that the graph of the equation $r = -2a \sin \theta, a > 0$, is a circle of radius a with center at $(0, -a)$ in rectangular coordinates.

85. Show that the graph of the equation $r = 2a \cos \theta, a > 0$, is a circle of radius a with center at $(a, 0)$ in rectangular coordinates.

86. Show that the graph of the equation $r = -2a \cos \theta, a > 0$, is a circle of radius a with center at $(-a, 0)$ in rectangular coordinates.

Discussion and Writing

87. Explain why the following test for symmetry is valid: Replace r by $-r$ and θ by $-\theta$ in a polar equation. If an equivalent equation results, the graph is symmetric with respect to the line $\theta = \dfrac{\pi}{2}$ (*y*-axis).

(a) Show that the test on page 567 fails for $r^2 = \cos \theta$, yet this new test works.

(b) Show that the test on page 567 works for $r^2 = \sin \theta$, yet this new test fails.

88. Develop a new test for symmetry with respect to the pole.

(a) Find a polar equation for which this new test fails, yet the test on page 567 works.

(b) Find a polar equation for which the test on page 567 fails, yet the new test works.

89. Write down two different tests for symmetry with respect to the polar axis. Find examples in which one test works and the other fails. Which test do you prefer to use? Justify your answer.

'Are You Prepared?' Answers

1. $(-4, 6)$ **2.** $\cos \alpha \cos \beta + \sin \alpha \sin \beta$ **3.** $(x + 2)^2 + (y - 5)^2 = 9$ **4.** odd   **5.** $-\dfrac{\sqrt{2}}{2}$ **6.** $-\dfrac{1}{2}$

Analytic Geometry

9

Chapter 9 in *Precalculus Essentials: Enhanced with Graphing Utilities, 4e* includes only the highlighted Section 9.7.

A LOOK BACK In Chapter 1, we introduced rectangular coordinates and showed how geometry problems can be solved algebraically. We defined a circle geometrically and then used the distance formula and rectangular coordinates to obtain an equation for a circle.

A LOOK AHEAD In this chapter we give geometric definitions for the conics and use the distance formula and rectangular coordinates to obtain their equations.

Historically, Apollonius (200 BC) was among the first to study *conics* and discover some of their interesting properties. Today, conics are still studied because of their many uses. *Paraboloids of revolution* (parabolas rotated about their axes of symmetry) are used as signal collectors (the satellite dishes used with radar and cable TV, for example), as solar energy collectors, and as reflectors (telescopes, light projection, and so on). The planets circle the Sun in approximately *elliptical* orbits. Elliptical surfaces can be used to reflect signals such as light and sound from one place to another. And *hyperbolas* can be used to determine the location of lightning strikes.

The Greeks used the methods of Euclidean geometry to study conics. However, we shall use the more powerful methods of analytic geometry, bringing to bear both algebra and geometry, for our study of conics.

This chapter concludes with sections on equations of conics in polar coordinates and plane curves and parametric equations.

OUTLINE

9.7 Plane Curves and Parametric Equations

PREPARING FOR THIS SECTION   *Before getting started, review the following:*

• Amplitude and Period of Sinusoidal Graphs (Section 5.4, pp. 408–414)

Now work the 'Are You Prepared?' problem on page 593.

OBJECTIVES **1** Graph Parametric Equations by Hand
  2 Graph Parametric Equations Using a Graphing Utility
  3 Find a Rectangular Equation for a Curve Defined Parametrically
  4 Use Time as a Parameter in Parametric Equations
  5 Find Parametric Equations for Curves Defined by Rectangular Equations

Equations of the form $y = f(x)$, where f is a function, have graphs that are intersected no more than once by any vertical line. The graphs of many of the conics and certain other, more complicated, graphs do not have this characteristic. Yet each graph, like the graph of a function, is a collection of points (x, y) in the xy-plane; that is, each is a *plane curve*. In this section, we discuss another way of representing such graphs.

Let $x = f(t)$ and $y = g(t)$, where f and g are two functions whose common domain is some interval I. The collection of points defined by

$$(x, y) = (f(t), g(t))$$

is called a **plane curve**. The equations

$$x = f(t) \qquad y = g(t)$$

where t is in I, are called **parametric equations** of the curve. The variable t is called a **parameter**.

1 Graph Parametric Equations by Hand

Parametric equations are particularly useful in describing movement along a curve. Suppose that a curve is defined by the parametric equations

$$x = f(t), \qquad y = g(t), \qquad a \le t \le b$$

where f and g are each defined over the interval $a \le t \le b$. For a given value of t, we can find the value of $x = f(t)$ and $y = g(t)$, obtaining a point (x, y) on the curve. In fact, as t varies over the interval from $t = a$ to $t = b$, successive values of t give rise to a directed movement along the curve; that is, the curve is traced out in a certain direction by the corresponding succession of points (x, y). See Figure 59. The arrows show the direction, or **orientation**, along the curve as t varies from a to b.

Figure 59

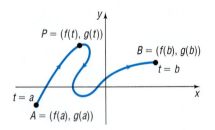

EXAMPLE 1

Discussing a Curve Defined by Parametric Equations

Discuss the curve defined by the parametric equations

$$x = 3t^2, \qquad y = 2t, \qquad -2 \le t \le 2$$

Solution   For each number t, $-2 \le t \le 2$, there corresponds a number x and a number y. For example, when $t = -2$, then $x = 3(-2)^2 = 12$ and $y = 2(-2) = -4$. When $t = 0$, then $x = 0$ and $y = 0$. Indeed, we can set up a table listing various choices of the parameter t and the corresponding values for x and y, as shown in Table 6. Plotting these points and connecting them with a smooth curve leads to Figure 60. The arrows in Figure 60 are used to indicate the orientation.

Table 6

t	x	y	(x, y)
−2	12	−4	(12, −4)
−1	3	−2	(3, −2)
0	0	0	(0, 0)
1	3	2	(3, 2)
2	12	4	(12, 4)

Figure 60

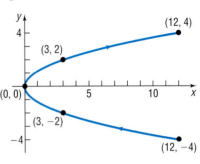

2 Graph Parametric Equations Using a Graphing Utility

Most graphing utilities have the capability of graphing parametric equations. The following steps are usually required to obtain the graph of parametric equations. Check your owner's manual to see how yours works.

Graphing Parametric Equations Using a Graphing Utility

STEP 1: Set the mode to PARametric, Enter $x(t)$ and $y(t)$.

STEP 2: Select the viewing window. In addition to setting Xmin, Xmax, Xscl, and so on, the viewing window in parametric mode requires setting minimum and maximum values for the parameter t and an increment setting for t(Tstep).

STEP 3: Graph.

EXAMPLE 2

Graphing a Curve Defined by Parametric Equations Using a Graphing Utility

Graph the curve defined by the parametric equations

$$x = 3t^2, \qquad y = 2t, \qquad -2 \leq t \leq 2 \qquad (1)$$

Solution **STEP 1:** Enter the equations $x(t) = 3t^2$, $y(t) = 2t$ with the graphing utility in PARametric mode.

STEP 2: Select the viewing window. The interval I is $-2 \leq t \leq 2$, so we select the following square viewing window:

$$T\text{min} = -2 \quad X\text{min} = 0 \quad Y\text{min} = -5$$
$$T\text{max} = 2 \quad X\text{max} = 15 \quad Y\text{max} = 5$$
$$T\text{step} = 0.1 \quad X\text{scl} = 1 \quad Y\text{scl} = 1$$

We choose Tmin $= -2$ and Tmax $= 2$ because $-2 \leq t \leq 2$. Finally, the choice for Tstep will determine the number of points the graphing utility will plot. For example, with Tstep at 0.1, the graphing utility will evaluate x

Figure 61

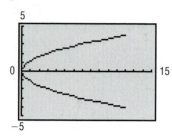

and y at $t = -2, -1.9, -1.8$, and so on. The smaller the Tstep, the more points the graphing utility will plot. The reader is encouraged to experiment with different values of Tstep to see how the graph is affected.

STEP 3: Graph. Notice the direction the graph is drawn in. This direction shows the orientation of the curve.

The graph shown in Figure 61 is complete. ◄

------ Exploration ------

Graph the following parametric equations using a graphing utility with Xmin = 0, Xmax = 15, Ymin = −5, Ymax = 5, and Tstep = 0.1:

1. $x = \dfrac{3t^2}{4}, y = t, -4 \le t \le 4$

2. $x = 3t^2 + 12t + 12, y = 2t + 4, -4 \le t \le 0$

3. $x = 3t^{\frac{2}{3}}, y = 2\sqrt[3]{t}, -8 \le t \le 8$

Compare these graphs to the graph in Figure 61. Conclude that parametric equations defining a curve are not unique; that is, different parametric equations can represent the same graph.

3 Find a Rectangular Equation for a Curve Defined Parametrically

The curve given in Examples 1 and 2 should be familiar. To identify it accurately, we find the corresponding rectangular equation by eliminating the parameter t from the parametric equations given in Example 1:

$$x = 3t^2, \qquad y = 2t, \qquad -2 \le t \le 2$$

Noting that we can readily solve for t in $y = 2t$, obtaining $t = \dfrac{y}{2}$, we substitute this expression in the other equation.

$$x = 3t^2 = 3\left(\frac{y}{2}\right)^2 = \frac{3y^2}{4}$$

$$t = \frac{y}{2}$$

This equation, $x = \dfrac{3y^2}{4}$, is the equation of a parabola with vertex at $(0, 0)$ and axis of symmetry along the x-axis.

------ Exploration ------

In FUNCtion mode, graph $x = \dfrac{3y^2}{4} \left(Y_1 = \sqrt{\dfrac{4x}{3}} \text{ and } Y_2 = -\sqrt{\dfrac{4x}{3}} \right)$ with Xmin = 0, Xmax = 15, Ymin = −5, Ymax = 5. Compare this graph with Figure 61. Why do the graphs differ?

Note that the parameterized curve defined by equation (1) and shown in Figure 60 (or 61) is only a part of the parabola $x = \dfrac{3y^2}{4}$. The graph of the rectangular equation obtained by eliminating the parameter will, in general, contain more points than the original parameterized curve. Care must therefore be taken when a parameterized curve is sketched by hand after eliminating the parameter. Even so, the process of eliminating the parameter t of a parameterized curve to identify it accurately is sometimes a better approach than merely plotting points. However, the elimination process sometimes requires a little ingenuity.

EXAMPLE 3	**Finding the Rectangular Equation of a Curve Defined Parametrically**

Find the rectangular equation of the curve whose parametric equations are

$$x = a \cos t \qquad y = a \sin t$$

where $a > 0$ is a constant. By hand, graph this curve, indicating its orientation.

Solution The presence of sines and cosines in the parametric equations suggests that we use a Pythagorean Identity. In fact, since

$$\cos t = \frac{x}{a} \qquad \sin t = \frac{y}{a}$$

we find that

$$\cos^2 t + \sin^2 t = 1$$
$$\left(\frac{x}{a}\right)^2 + \left(\frac{y}{a}\right)^2 = 1$$
$$x^2 + y^2 = a^2$$

Figure 62

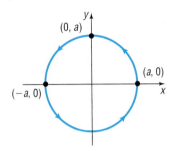

The curve is a circle with center at $(0, 0)$ and radius a. As the parameter t increases, say from $t = 0$ [the point $(a, 0)$] to $t = \dfrac{\pi}{2}$ [the point $(0, a)$] to $t = \pi$ [the point $(-a, 0)$], we see that the corresponding points are traced in a counterclockwise direction around the circle. The orientation is as indicated in Figure 62. ◀

═══ **NOW WORK PROBLEMS 7 AND 19.**

Let's discuss the curve in Example 3 further. The domain of each parametric equation is $-\infty < t < \infty$. Thus, the graph in Figure 62 is actually being repeated each time that t increases by 2π.

If we wanted the curve to consist of exactly 1 revolution in the counterclockwise direction, we could write

$$x = a \cos t, \qquad y = a \sin t, \qquad 0 \leq t \leq 2\pi$$

This curve starts at $t = 0$ [the point $(a, 0)$] and, proceeding counterclockwise around the circle, ends at $t = 2\pi$ [also the point $(a, 0)$].

If we wanted the curve to consist of exactly three revolutions in the counterclockwise direction, we could write

$$x = a \cos t, \qquad y = a \sin t, \qquad -2\pi \leq t \leq 4\pi$$

or

$$x = a \cos t, \qquad y = a \sin t, \qquad 0 \leq t \leq 6\pi$$

or

$$x = a \cos t, \qquad y = a \sin t, \qquad 2\pi \leq t \leq 8\pi$$

EXAMPLE 4	**Describing Parametric Equations**

Find rectangular equations for the following curves defined by parametric equations. Graph each curve.

(a) $x = a \cos t, \quad y = a \sin t, \quad 0 \leq t \leq \pi, \quad a > 0$

(b) $x = -a \sin t, \quad y = -a \cos t, \quad 0 \leq t \leq \pi, \quad a > 0$

Solution   (a) We eliminate the parameter t using a Pythagorean Identity.

$$\cos^2 t + \sin^2 t = 1$$

$$\left(\frac{x}{a}\right)^2 + \left(\frac{y}{a}\right)^2 = 1$$

$$x^2 + y^2 = a^2$$

The curve defined by these parametric equations is a circle, with radius a and center at $(0, 0)$. The circle begins at the point $(a, 0)$, $t = 0$; passes through the point $(0, a)$, $t = \dfrac{\pi}{2}$; and ends at the point $(-a, 0)$, $t = \pi$.

The parametric equations define an upper semicircle of radius a with a counterclockwise orientation. See Figure 63. The rectangular equation is

$$y = \sqrt{a^2 - x^2}, \qquad -a \le x \le a$$

Figure 63

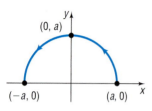

(b) We eliminate the parameter t using a Pythagorean Identity.

$$\sin^2 t + \cos^2 t = 1$$

$$\left(\frac{x}{-a}\right)^2 + \left(\frac{y}{-a}\right)^2 = 1$$

$$x^2 + y^2 = a^2$$

The curve defined by these parametric equations is a circle, with radius a and center at $(0, 0)$. The circle begins at the point $(0, -a)$, $t = 0$; passes through the point $(-a, 0)$, $t = \dfrac{\pi}{2}$; and ends at the point $(0, a)$, $t = \pi$. The parametric equations define a left semicircle of radius a with a clockwise orientation. See Figure 64. The rectangular equation is

$$x = -\sqrt{a^2 - y^2}, \qquad -a \le y \le a$$

Figure 64

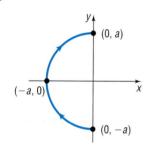

— **Seeing the Concept** —

Graph $x = \cos t, y = \sin t$ for $0 \le t \le 2\pi$. Compare to Figure 62.
Graph $x = \cos t, y = \sin t$ for $0 \le t \le \pi$. Compare to Figure 63.
Graph $x = -\sin t, y = -\cos t$ for $0 \le t \le \pi$. Compare to Figure 64.

Example 4 illustrates the versatility of parametric equations for replacing complicated rectangular equations, while providing additional information about orientation. These characteristics make parametric equations very useful in applications, such as projectile motion.

4   Use Time as a Parameter in Parametric Equations

If we think of the parameter t as time, then the parametric equations $x = f(t)$ and $y = g(t)$ of a curve C specify how the x- and y-coordinates of a moving point vary with time.

For example, we can use parametric equations to describe the motion of an object, sometimes referred to as **curvilinear motion**. Using parametric equations, we can specify not only where the object travels, that is, its location (x, y), but also when it gets there, that is, the time t.

When an object is propelled upward at an inclination θ to the horizontal with initial speed v_0, the resulting motion is called **projectile motion**. See Figure 65(a).

 In calculus it is shown that the parametric equations of the path of a projectile fired at an inclination θ to the horizontal, with an initial speed v_0, from a height h above the horizontal are

$$x = (v_0 \cos \theta)t \qquad y = -\frac{1}{2}gt^2 + (v_0 \sin \theta)t + h \tag{2}$$

where t is the time and g is the constant acceleration due to gravity (approximately 32 ft/sec/sec or 9.8 m/sec/sec). See Figure 65(b).

Figure 65

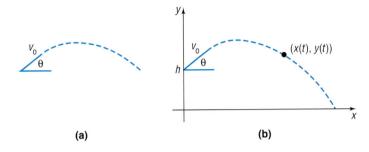

(a)   (b)

EXAMPLE 5

Figure 66

Projectile Motion

Suppose that Jim hit a golf ball with an initial velocity of 150 feet per second at an angle of 30° to the horizontal. See Figure 66.

(a) Find parametric equations that describe the position of the ball as a function of time.

(b) How long is the golf ball in the air?

(c) When is the ball at its maximum height? Determine the maximum height of the ball.

(d) Determine the horizontal distance that the ball traveled.

(e) Using a graphing utility, simulate the motion of the golf ball by simultaneously graphing the equations found in part (a).

Solution

(a) We have $v_0 = 150$ ft/sec, $\theta = 30°$, $h = 0$ (the ball is on the ground), and $g = 32$ (since the units are in feet and seconds). Substituting these values into equations (2), we find that

$$x = (v_0 \cos \theta)t = (150 \cos 30°)t = 75\sqrt{3}\,t$$

$$y = -\frac{1}{2}gt^2 + (v_0 \sin \theta)t + h = -\frac{1}{2}(32)t^2 + (150 \sin 30°)t + 0$$

$$= -16t^2 + 75t$$

(b) To determine the length of time that the ball is in the air, we solve the equation $y = 0$.

$$-16t^2 + 75t = 0$$
$$t(-16t + 75) = 0$$
$$t = 0 \sec \quad \text{or} \quad t = \frac{75}{16} = 4.6875 \sec$$

The ball will strike the ground after 4.6875 seconds.

(c) Notice that the height y of the ball is a quadratic function of t, so the maximum height of the ball can be found by determining the vertex of $y = -16t^2 + 75t$. The value of t at the vertex is

$$t = \frac{-b}{2a} = \frac{-75}{-32} = 2.34375 \sec$$

Figure 67

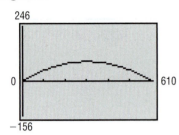

The ball is at its maximum height after 2.34375 seconds. The maximum height of the ball is found by evaluating the function y at $t = 2.34375$ seconds.

$$\text{Maximum height} = -16(2.34375)^2 + (75)2.34375 \approx 87.89 \text{ feet}$$

(d) Since the ball is in the air for 4.6875 seconds, the horizontal distance that the ball travels is

$$x = \left(75\sqrt{3}\right)4.6875 \approx 608.92 \text{ feet}$$

(e) We enter the equations from part (a) into a graphing utility with $T\text{min} = 0$, $T\text{max} = 4.7$, and $T\text{step} = 0.1$. We use ZOOM-SQUARE to avoid any distortion to the angle of elevation. See Figure 67. ◀

─── **Exploration** ───

Simulate the motion of a ball thrown straight up with an initial speed of 100 feet per second from a height of 5 feet above the ground. Use PARametric mode with $T\text{min} = 0$, $T\text{max} = 6.5$, $T\text{step} = 0.1$, $X\text{min} = 0$, $X\text{max} = 5$, $Y\text{min} = 0$, and $Y\text{max} = 180$. What happens to the speed with which the graph is drawn as the ball goes up and then comes back down? How do you interpret this physically? Repeat the experiment using other values for $T\text{step}$. How does this affect the experiment?

[**Hint:** In the projectile motion equations, let $\theta = 90°$, $v_0 = 100$, $h = 5$, and $g = 32$. Use $x = 3$ instead of $x = 0$ to see the vertical motion better.]

Result See Figure 68. In Figure 68(a) the ball is going up. In Figure 68(b) the ball is near its highest point. Finally, in Figure 68(c) the ball is coming back down.

Figure 68

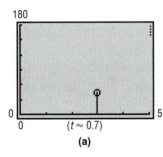 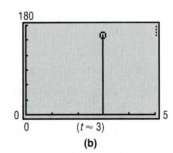 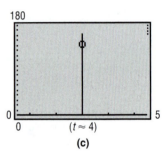

| (a) | (b) | (c) |
| $(t \approx 0.7)$ | $(t \approx 3)$ | $(t \approx 4)$ |

Notice that, as the ball goes up, its speed decreases, until at the highest point it is zero. Then the speed increases as the ball comes back down.

─── **NOW WORK PROBLEM 49.**

A graphing utility can be used to simulate other kinds of motion as well. Let's work again Example 4 from the Appendix, Section A.7.

EXAMPLE 6 **Simulating Motion**

Tanya, who is a long distance runner, runs at an average velocity of 8 miles per hour. Two hours after Tanya leaves your house, you leave in your Honda and follow the same route. If your average velocity is 40 miles per hour, how long will it be before you catch up to Tanya? See Figure 69. Use a simulation of the two motions to verify the answer.

Figure 69

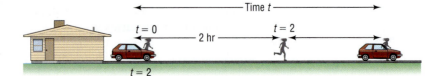

Solution

We begin with two sets of parametric equations: one to describe Tanya's motion, the other to describe the motion of the Honda. We choose time $t = 0$ to be when Tanya leaves the house. If we choose $y_1 = 2$ as Tanya's path, then we can use $y_2 = 4$ as the parallel path of the Honda. The horizontal distances traversed in time t (Distance = Velocity $\times$ Time) are

$$\text{Tanya:} \quad x_1 = 8t \qquad \text{Honda:} \quad x_2 = 40(t - 2)$$

The Honda catches up to Tanya when $x_1 = x_2$.

$$8t = 40(t - 2)$$
$$8t = 40t - 80$$
$$-32t = -80$$
$$t = \frac{-80}{-32} = 2.5$$

The Honda catches up to Tanya 2.5 hours after Tanya leaves the house. In PARametric mode with Tstep = 0.01, we simultaneously graph

$$\text{Tanya:} \quad x_1 = 8t \qquad \text{Honda:} \quad x_2 = 40(t - 2)$$
$$y_1 = 2 \qquad\qquad\qquad y_2 = 4$$

for $0 \le t \le 3$.

Figure 70 shows the relative position of Tanya and the Honda for $t = 0, t = 2, t = 2.25, t = 2.5,$ and $t = 2.75$.

Figure 70

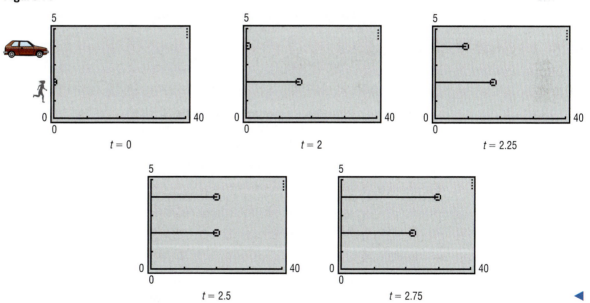

$t = 0$ $\qquad\qquad$ $t = 2$ $\qquad\qquad$ $t = 2.25$

$t = 2.5$ $\qquad\qquad$ $t = 2.75$

◀

5 **Find Parametric Equations for Curves Defined by Rectangular Equations**

We now take up the question of how to find parametric equations of a given curve.

If a curve is defined by the equation $y = f(x)$, where f is a function, one way of finding parametric equations is to let $x = t$. Then $y = f(t)$ and

$$x = t, \quad y = f(t), \qquad t \text{ in the domain of } f$$

are parametric equations of the curve.

EXAMPLE 7	**Finding Parametric Equations for a Curve Defined by a Rectangular Equation**

Find parametric equations for the equation $y = x^2 - 4$.

Solution   Let $x = t$. Then the parametric equations are

$$x = t, \quad y = t^2 - 4, \quad -\infty < t < \infty \qquad \blacktriangleleft$$

Another less obvious approach to Example 7 is to let $x = t^3$. Then the parametric equations become

$$x = t^3, \quad y = t^6 - 4, \quad -\infty < t < \infty$$

Care must be taken when using this approach, since the substitution for x must be a function that allows x to take on all the values stipulated by the domain of f. For example, letting $x = t^2$ so that $y = t^4 - 4$ does not result in equivalent parametric equations for $y = x^2 - 4$, since only points for which $x \geq 0$ are obtained.

 NOW WORK PROBLEM 33.

EXAMPLE 8	**Finding Parametric Equations for an Object in Motion**

Find parametric equations for the ellipse

$$x^2 + \frac{y^2}{9} = 1$$

where the parameter t is time (in seconds) and

(a) The motion around the ellipse is clockwise, begins at the point $(0, 3)$, and requires 1 second for a complete revolution.

(b) The motion around the ellipse is counterclockwise, begins at the point $(1, 0)$, and requires 2 seconds for a complete revolution.

Figure 71

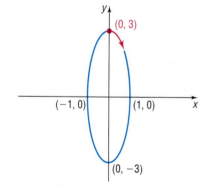

Solution   (a) See Figure 71. Since the motion begins at the point $(0, 3)$, we want $x = 0$ and $y = 3$ when $t = 0$. Furthermore, since the given equation is an ellipse, we begin by letting

$$x = \sin(\omega t) \qquad \frac{y}{3} = \cos(\omega t)$$

for some constant ω. These parametric equations satisfy the equation of the ellipse. Furthermore, with this choice, when $t = 0$, we have $x = 0$ and $y = 3$.

For the motion to be clockwise, the motion will have to begin with the value of x increasing and y decreasing as t increases. This requires that $\omega > 0$. [Do you know why? If $\omega > 0$, then $x = \sin(\omega t)$ is increasing when $t > 0$ is near zero and $y = 3 \cos(\omega t)$ is decreasing when $t > 0$ is near zero.] See the red part of the graph in Figure 71.

Finally, since 1 revolution requires 1 second, the period $\dfrac{2\pi}{\omega} = 1$, so $\omega = 2\pi$.

Parametric equations that satisfy the conditions stipulated are

$$x = \sin(2\pi t), \qquad y = 3\cos(2\pi t), \qquad 0 \leq t \leq 1 \qquad \textbf{(3)}$$

Figure 72

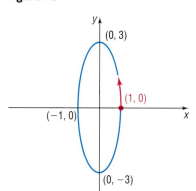

(b) See Figure 72. Since the motion begins at the point $(1, 0)$, we want $x = 1$ and $y = 0$ when $t = 0$. Furthermore, since the given equation is an ellipse, we begin by letting

$$x = \cos(\omega t) \qquad \frac{y}{3} = \sin(\omega t)$$

for some constant ω. These parametric equations satisfy the equation of the ellipse. Furthermore, with this choice, when $t = 0$, we have $x = 1$ and $y = 0$.

For the motion to be counterclockwise, the motion will have to begin with the value of x decreasing and y increasing as t increases. This requires that $\omega > 0$. [Do you know why?] Finally, since 1 revolution requires 2 seconds, the period is $\dfrac{2\pi}{\omega} = 2$, so $\omega = \pi$. The parametric equations that satisfy the conditions stipulated are

$$x = \cos(\pi t), \qquad y = 3\sin(\pi t), \qquad 0 \le t \le 2 \qquad \textbf{(4)} \ \blacktriangleleft$$

Either of equations (3) or (4) can serve as parametric equations for the ellipse $x^2 + \dfrac{y^2}{9} = 1$ given in Example 8. The direction of the motion, the beginning point, and the time for 1 revolution merely serve to help us arrive at a particular parametric representation.

 NOW WORK PROBLEM 39.

The Cycloid

Suppose that a circle of radius a rolls along a horizontal line without slipping. As the circle rolls along the line, a point P on the circle will trace out a curve called a **cycloid** (see Figure 73). We now seek parametric equations[*] for a cycloid.

Figure 73

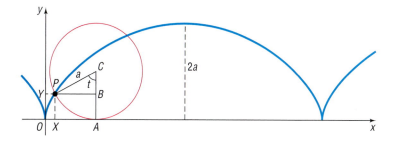

We begin with a circle of radius a and take the fixed line on which the circle rolls as the x-axis. Let the origin be one of the points at which the point P comes in contact with the x-axis. Figure 73 illustrates the position of this point P after the circle has rolled somewhat. The angle t (in radians) measures the angle through which the circle has rolled.

Since we require no slippage, it follows that

$$\text{Arc } AP = d(O, A)$$

---

[*]Any attempt to derive the rectangular equation of a cycloid would soon demonstrate how complicated the task is.

The length of the arc AP is given by $s = r\theta$, where $r = a$ and $\theta = t$ radians. Then

$$at = d(O, A) \qquad {\color{blue} s = r\theta, \text{ where } r = a \text{ and } \theta = t}$$

The x-coordinate of the point P is

$$d(O, X) = d(O, A) - d(X, A) = at - a \sin t = a(t - \sin t)$$

── **Exploration** ──────

Graph $x = t - \sin t, y = 1 - \cos t,$

$0 \le t \le 3\pi,$ using your graphing utility

with $T\text{step} = \dfrac{\pi}{36}$ and a square screen.

Compare your results with Figure 73.

The y-coordinate of the point P is equal to

$$d(O, Y) = d(A, C) - d(B, C) = a - a \cos t = a(1 - \cos t)$$

The parametric equations of the cycloid are

$$x = a(t - \sin t) \qquad y = a(1 - \cos t) \tag{5}$$

Applications to Mechanics

If a is negative in equation (5), we obtain an inverted cycloid, as shown in Figure 74(a). The inverted cycloid occurs as a result of some remarkable applications in the field of mechanics. We shall mention two of them: the *brachistochrone* and the *tautochrone*.[*]

Figure 74

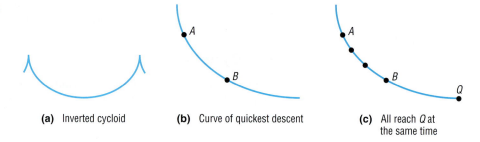

(a) Inverted cycloid   (b) Curve of quickest descent   (c) All reach Q at the same time

Figure 75

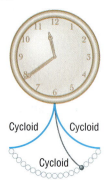

Cycloid   Cycloid

Cycloid

The **brachistochrone** is the curve of quickest descent. If a particle is constrained to follow some path from one point A to a lower point B (not on the same vertical line) and is acted on only by gravity, the time needed to make the descent is least if the path is an inverted cycloid. See Figure 74(b). This remarkable discovery, which is attributed to many famous mathematicians (including Johann Bernoulli and Blaise Pascal), was a significant step in creating the branch of mathematics known as the *calculus of variations*.

To define the **tautochrone**, let Q be the lowest point on an inverted cycloid. If several particles placed at various positions on an inverted cycloid simultaneously begin to slide down the cycloid, they will reach the point Q at the same time, as indicated in Figure 74(c). The tautochrone property of the cycloid was used by Christiaan Huygens (1629–1695), the Dutch mathematician, physicist, and astronomer, to construct a pendulum clock with a bob that swings along a cycloid (see Figure 75). In Huygen's clock, the bob was made to swing along a cycloid by suspending the bob on a thin wire constrained by two plates shaped like cycloids. In a clock of this design, the period of the pendulum is independent of its amplitude.

---

[*]In Greek, *brachistochrone* means "the shortest time" and *tautochrone* means "equal time."

9.7 Assess Your Understanding

'Are You Prepared?'

Answers are given at the end of these exercises. If you get a wrong answer, read the pages listed in red.

1. The function $f(x) = 3\sin(4x)$ has amplitude _____ and period _____. (p. 410)

Concepts and Vocabulary

2. Let $x = f(t)$ and $y = g(t)$, where f and g are two functions whose common domain is some interval I. The collection of points defined by $(x, y) = (f(t), g(t))$ is called a(n) _____ _____. The variable t is called a(n) _____.

3. The parametric equations $x = 2\sin t$, $y = 3\cos t$ define a(n) _____.

4. If a circle rolls along a horizontal line without slippage, a point P on the circle will trace out a curve called a(n) _____.

5. *True or False:* Parametric equations defining a curve are unique.

6. *True or False:* Curves defined using parametric equations have an orientation.

Skill Building

In Problems 7–26, graph the curve whose parametric equations are given by hand and show its orientation. Find the rectangular equation of each curve. Verify your graph using a graphing utility.

7. $x = 3t + 2$, $y = t + 1$; $0 \le t \le 4$

8. $x = t - 3$, $y = 2t + 4$; $0 \le t \le 2$

9. $x = t + 2$, $y = \sqrt{t}$; $t \ge 0$

10. $x = \sqrt{2t}$, $y = 4t$; $t \ge 0$

11. $x = t^2 + 4$, $y = t^2 - 4$; $-\infty < t < \infty$

12. $x = \sqrt{t} + 4$, $y = \sqrt{t} - 4$; $t \ge 0$

13. $x = 3t^2$, $y = t + 1$; $-\infty < t < \infty$

14. $x = 2t - 4$, $y = 4t^2$; $-\infty < t < \infty$

15. $x = 2e^t$, $y = 1 + e^t$; $t \ge 0$

16. $x = e^t$, $y = e^{-t}$; $t \ge 0$

17. $x = \sqrt{t}$, $y = t^{3/2}$; $t \ge 0$

18. $x = t^{3/2} + 1$, $y = \sqrt{t}$; $t \ge 0$

19. $x = 2\cos t$, $y = 3\sin t$; $0 \le t \le 2\pi$

20. $x = 2\cos t$, $y = 3\sin t$; $0 \le t \le \pi$

21. $x = 2\cos t$, $y = 3\sin t$; $-\pi \le t \le 0$

22. $x = 2\cos t$, $y = \sin t$; $0 \le t \le \dfrac{\pi}{2}$

23. $x = \sec t$, $y = \tan t$; $0 \le t \le \dfrac{\pi}{4}$

24. $x = \csc t$, $y = \cot t$; $\dfrac{\pi}{4} \le t \le \dfrac{\pi}{2}$

25. $x = \sin^2 t$, $y = \cos^2 t$; $0 \le t \le 2\pi$

26. $x = t^2$, $y = \ln t$; $t > 0$

In Problems 27–34, find two different parametric equations for each rectangular equation.

27. $y = 4x - 1$

28. $y = -8x + 3$

29. $y = x^2 + 1$

30. $y = -2x^2 + 1$

31. $y = x^3$

32. $y = x^4 + 1$

33. $x = y^{3/2}$

34. $x = \sqrt{y}$

In Problems 35–38, find parametric equations that define the curve shown.

35.

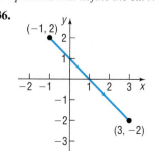

36.

37.

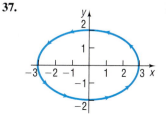

38.

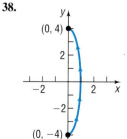

In Problems 39–42, find parametric equations for an object that moves along the ellipse $\dfrac{x^2}{4} + \dfrac{y^2}{9} = 1$ *with the motion described.*

39. The motion begins at $(2, 0)$, is clockwise, and requires 2 seconds for a complete revolution.

40. The motion begins at $(0, 3)$, is counterclockwise, and requires 1 second for a complete revolution.

41. The motion begins at $(0, 3)$, is clockwise, and requires 1 second for a complete revolution.

42. The motion begins at $(2, 0)$, is counterclockwise, and requires 3 seconds for a complete revolution.

In Problems 43 and 44, the parametric equations of four curves are given. Graph each of them, indicating the orientation.

43. C_1: $x = t$, $y = t^2$; $-4 \le t \le 4$

C_2: $x = \cos t$, $y = 1 - \sin^2 t$; $0 \le t \le \pi$

C_3: $x = e^t$, $y = e^{2t}$; $0 \le t \le \ln 4$

C_4: $x = \sqrt{t}$, $y = t$; $0 \le t \le 16$

44. C_1: $x = t$, $y = \sqrt{1 - t^2}$; $-1 \le t \le 1$

C_2: $x = \sin t$, $y = \cos t$; $0 \le t \le 2\pi$

C_3: $x = \cos t$, $y = \sin t$; $0 \le t \le 2\pi$

C_4: $x = \sqrt{1 - t^2}$, $y = t$; $-1 \le t \le 1$

In Problems 45–48, use a graphing utility to graph the curve defined by the given parametric equations.

45. $x = t \sin t$, $y = t \cos t$, $t > 0$

46. $x = \sin t + \cos t$, $y = \sin t - \cos t$

47. $x = 4 \sin t - 2 \sin(2t)$
$y = 4 \cos t - 2 \cos(2t)$

48. $x = 4 \sin t + 2 \sin(2t)$
$y = 4 \cos t + 2 \cos(2t)$

Applications and Extensions

49. Projectile Motion Bob throws a ball straight up with an initial speed of 50 feet per second from a height of 6 feet.
 (a) Find parametric equations that describe the motion of the ball as a function of time.
 (b) How long is the ball in the air?
 (c) When is the ball at its maximum height? Determine the maximum height of the ball.
 (d) Simulate the motion of the ball by graphing the equations found in part (a).

50. Projectile Motion Alice throws a ball straight up with an initial speed of 40 feet per second from a height of 5 feet.
 (a) Find parametric equations that describe the motion of the ball as a function of time.
 (b) How long is the ball in the air?
 (c) When is the ball at its maximum height? Determine the maximum height of the ball.
 (d) Simulate the motion of the ball by graphing the equations found in part (a).

51. Catching a Train Bill's train leaves at 8:06 AM and accelerates at the rate of 2 meters per second per second. Bill, who can run 5 meters per second, arrives at the train station 5 seconds after the train has left.
 (a) Find parametric equations that describe the motion of the train and Bill as a function of time.

 [**Hint:** The position s at time t of an object having acceleration a is $s = \dfrac{1}{2} a t^2$.]

 (b) Determine algebraically whether Bill will catch the train. If so, when?
 (c) Simulate the motion of the train and Bill by simultaneously graphing the equations found in part (a).

52. Catching a Bus Jodi's bus leaves at 5:30 PM and accelerates at the rate of 3 meters per second per second. Jodi, who can run 5 meters per second, arrives at the bus station 2 seconds after the bus has left.
 (a) Find parametric equations that describe the motion of the bus and Jodi as a function of time.

 [**Hint:** The position s at time t of an object having acceleration a is $s = \dfrac{1}{2} a t^2$.]

 (b) Determine algebraically whether Jodi will catch the bus. If so, when?
 (c) Simulate the motion of the bus and Jodi by simultaneously graphing the equations found in part (a).

53. Projectile Motion Ichiro throws a baseball with an initial speed of 145 feet per second at an angle of $20°$ to the horizontal. The ball leaves Ichiro's hand at a height of 5 feet.
 (a) Find parametric equations that describe the position of the ball as a function of time.
 (b) How long is the ball in the air?
 (c) When is the ball at its maximum height? Determine the maximum height of the ball.
 (d) Determine the horizontal distance that the ball traveled.
 (e) Using a graphing utility, simultaneously graph the equations found in part (a).

54. Projectile Motion Barry Bonds hit a baseball with an initial speed of 125 feet per second at an angle of $40°$ to the horizontal. The ball was hit at a height of 3 feet off the ground.
 (a) Find parametric equations that describe the position of the ball as a function of time.

(b) How long is the ball in the air?

(c) When is the ball at its maximum height? Determine the maximum height of the ball.

(d) Determine the horizontal distance that the ball traveled.

(e) Using a graphing utility, simultaneously graph the equations found in part (a).

55. Projectile Motion Suppose that Adam throws a tennis ball off a cliff 300 meters high with an initial speed of 40 meters per second at an angle of 45° to the horizontal.

(a) Find parametric equations that describe the position of the ball as a function of time.

(b) How long is the ball in the air?

(c) When is the ball at its maximum height? Determine the maximum height of the ball.

(d) Determine the horizontal distance that the ball traveled.

(e) Using a graphing utility, simultaneously graph the equations found in part (a).

56. Projectile Motion Suppose that Adam throws a tennis ball off a cliff 300 meters high with an initial speed of 40 meters per second at an angle of 45° to the horizontal on the Moon (gravity on the Moon is one-sixth of that on Earth).

(a) Find parametric equations that describe the position of the ball as a function of time.

(b) How long is the ball in the air?

(c) When is the ball at its maximum height? Determine the maximum height of the ball.

(d) Determine the horizontal distance that the ball traveled.

(e) Using a graphing utility, simultaneously graph the equations found in part (a).

57. Uniform Motion A Toyota Paseo (traveling east at 40 mph) and a Pontiac Bonneville (traveling north at 30 mph) are heading toward the same intersection. The Paseo is 5 miles from the intersection when the Bonneville is 4 miles from the intersection. See the figure.

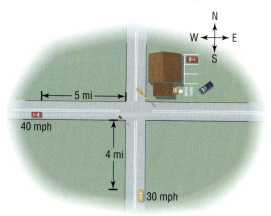

(a) Find parametric equations that describe the motion of the Paseo and Bonneville.

(b) Find a formula for the distance between the cars as a function of time.

(c) Graph the function in part (b) using a graphing utility.

(d) What is the minimum distance between the cars? When are the cars closest?

(e) Simulate the motion of the cars by simultaneously graphing the equations found in part (a).

58. Uniform Motion A Cessna (heading south at 120 mph) and a Boeing 747 (heading west at 600 mph) are flying toward the same point at the same altitude. The Cessna is 100 miles from the point where the flight patterns intersect, and the 747 is 550 miles from this intersection point. See the figure.

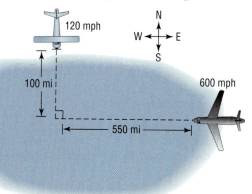

(a) Find parametric equations that describe the motion of the Cessna and 747.

(b) Find a formula for the distance between the planes as a function of time.

(c) Graph the function in part (b) using a graphing utility.

(d) What is the minimum distance between the planes? When are the planes closest?

(e) Simulate the motion of the planes by simultaneously graphing the equations found in part (a).

59. Show that the parametric equations for a line passing through the points (x_1, y_1) and (x_2, y_2) are

$$x = (x_2 - x_1)t + x_1$$
$$y = (y_2 - y_1)t + y_1, \quad -\infty < t < \infty$$

What is the orientation of this line?

60. Projectile Motion The position of a projectile fired with an initial velocity v_0 feet per second and at an angle θ to the horizontal at the end of t seconds is given by the parametric equations

$$x = (v_0 \cos \theta)t \qquad y = (v_0 \sin \theta)t - 16t^2$$

See the following illustration.

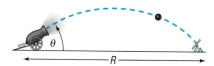

(a) Obtain the rectangular equation of the trajectory and identify the curve.

(b) Show that the projectile hits the ground ($y = 0$) when $t = \dfrac{1}{16} v_0 \sin \theta$.

(c) How far has the projectile traveled (horizontally) when it strikes the ground? In other words, find the range R.

(d) Find the time t when $x = y$. Then find the horizontal distance x and the vertical distance y traveled by the projectile in this time. Then compute $\sqrt{x^2 + y^2}$. This is the distance R, the range, that the projectile travels up a plane inclined at 45° to the horizontal $(x = y)$. See the following illustration. (See also Problem 83 in Exercise 6.5.)

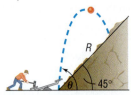

61. Hypocycloid The hypocycloid is a curve defined by the parametric equations

$$x(t) = \cos^3 t, \quad y(t) = \sin^3 t, \quad 0 \le t \le 2\pi$$

(a) Graph the hypocycloid using a graphing utility.
(b) Find rectangular equations of the hypocycloid.

Discussion and Writing

62. In Problem 61, we graphed the hypocycloid. Now graph the rectangular equations of the hypocycloid. Did you obtain a complete graph? If not, experiment until you do.

63. Look up the curves called *hypocycloid* and *epicycloid*. Write a report on what you find. Be sure to draw comparisons with the cycloid.

'Are You Prepared?' Answers

1. $3; \dfrac{\pi}{2}$

Systems of Equations and Inequalities

10

Chapter 10 in *Precalculus Essentials: Enhanced with Graphing Utilities*, 4e includes only the highlighted Sections 10.1 and 10.7.

A LOOK BACK In the Appendix, Sections A.5 and A.8, and Chapters 1, 3, and 4, we solved equations and inequalities involving a single variable.

A LOOK AHEAD In this chapter we take up the problem of solving equations and inequalities containing two or more variables. There are various ways to solve such problems:

The *method of substitution* for solving equations in several unknowns goes back to ancient times.

The *method of elimination*, although it had existed for centuries, was put into systematic order by Karl Friedrich Gauss (1777–1855) and by Camille Jordan (1838–1922).

The theory of *matrices* was developed in 1857 by Arthur Cayley (1821–1895), although only later were matrices used as we use them in this chapter. Matrices have become a very flexible instrument, useful in almost all areas of mathematics.

The method of *determinants* was invented by Takakazu Seki Kôwa (1642–1708) in 1683 in Japan and by Gottfried Wilhelm von Leibniz (1646–1716) in 1693 in Germany. *Cramer's Rule* is named after Gabriel Cramer (1704–1752) of Switzerland, who popularized the use of determinants for solving linear systems.

Section 10.5, on *partial fraction decomposition*, provides an application of systems of equations. This particular application is one that is used in integral calculus.

Section 10.8 introduces *linear programming*, a modern application of linear inequalities. This topic is particularly useful for students interested in operations research.

OUTLINE

10.1 Systems of Linear Equations: Substitution and Elimination

PREPARING FOR THIS SECTION *Before getting started, review the following:*

• Linear Equations (Appendix, Section A.5, pp. 694–695)   • Lines (Section 1.4, pp 27–38)

 Now work the 'Are You Prepared?' problems on page 609.

OBJECTIVES 1 Solve Systems of Equations by Substitution
                2 Solve Systems of Equations by Elimination
                3 Identify Inconsistent Systems of Equations Containing Two Variables
                4 Express the Solution of a System of Dependent Equations Containing Two Variables
                5 Solve Systems of Three Equations Containing Three Variables
                6 Identify Inconsistent Systems of Equations Containing Three Variables
                7 Express the Solution of a System of Dependent Equations Containing Three Variables

We begin with an example.

EXAMPLE 1   **Movie Theater Ticket Sales**

A movie theater sells tickets for $8.00 each, with seniors receiving a discount of $2.00. One evening the theater took in $3580 in revenue. If x represents the number of tickets sold at $8.00 and y the number of tickets sold at the discounted price of $6.00, write an equation that relates these variables.

Solution   Each nondiscounted ticket brings in $8.00, so x tickets will bring in $8x$ dollars. Similarly, y discounted tickets bring in $6y$ dollars. Since the total brought in is $3580, we must have

$$8x + 6y = 3580 \qquad \blacktriangleleft$$

In Example 1, suppose that we also know that 525 tickets were sold that evening. Then we have another equation relating the variables x and y.

$$x + y = 525$$

The two equations

$$8x + 6y = 3580$$
$$x + y = 525$$

form a *system* of equations.

In general, a **system of equations** is a collection of two or more equations, each containing one or more variables. Example 2 gives some illustrations of systems of equations.

| EXAMPLE 2 | **Examples of Systems of Equations** |

(a) $\begin{cases} 2x + y = 5 \\ -4x + 6y = -2 \end{cases}$ (1) Two equations containing two variables, x and y
  (2)

(b) $\begin{cases} x + y^2 = 5 \\ 2x + y = 4 \end{cases}$ (1) Two equations containing two variables, x and y
  (2)

(c) $\begin{cases} x + y + z = 6 \\ 3x - 2y + 4z = 9 \\ x - y - z = 0 \end{cases}$ (1) Three equations containing three variables, x, y, and z
  (2)
  (3)

(d) $\begin{cases} x + y + z = 5 \\ x - y = 2 \end{cases}$ (1) Two equations containing three variables, x, y, and z
  (2)

(e) $\begin{cases} x + y + z = 6 \\ 2x + 2z = 4 \\ y + z = 2 \\ x = 4 \end{cases}$ (1) Four equations containing three variables, x, y, and z
  (2)
  (3)
  (4)

◀

We use a brace, as shown, to remind us that we are dealing with a system of equations. We also will find it convenient to number each equation in the system.

A **solution** of a system of equations consists of values for the variables that are solutions of each equation of the system. To **solve** a system of equations means to find all solutions of the system.

For example, $x = 2$, $y = 1$ is a solution of the system in Example 2(a), because

$$\begin{cases} 2x + y = 5 \quad (1) \\ -4x + 6y = -2 \quad (2) \end{cases} \qquad \begin{cases} 2(2) + 1 = 4 + 1 = 5 \\ -4(2) + 6(1) = -8 + 6 = -2 \end{cases}$$

A solution of the system in Example 2(b) is $x = 1$, $y = 2$, because

$$\begin{cases} x + y^2 = 5 \quad (1) \\ 2x + y = 4 \quad (2) \end{cases} \qquad \begin{cases} 1 + 2^2 = 1 + 4 = 5 \\ 2(1) + 2 = 2 + 2 = 4 \end{cases}$$

Another solution of the system in Example 2(b) is $x = \dfrac{11}{4}$, $y = -\dfrac{3}{2}$, which you can check for yourself.

A solution of the system in Example 2(c) is $x = 3$, $y = 2$, $z = 1$, because

$$\begin{cases} x + y + z = 6 \quad (1) \\ 3x - 2y + 4z = 9 \quad (2) \\ x - y - z = 0 \quad (3) \end{cases} \qquad \begin{cases} 3 + 2 + 1 = 6 \quad\quad\quad\quad (1) \\ 3(3) - 2(2) + 4(1) = 9 - 4 + 4 = 9 \quad (2) \\ 3 - 2 - 1 = 0 \quad\quad\quad\quad (3) \end{cases}$$

Note that $x = 3$, $y = 3$, $z = 0$ is not a solution of the system in Example 2(c).

$$\begin{cases} x + y + z = 6 \quad (1) \\ 3x - 2y + 4z = 9 \quad (2) \\ x - y - z = 0 \quad (3) \end{cases} \qquad \begin{cases} 3 + 3 + 0 = 6 \quad\quad\quad\quad (1) \\ 3(3) - 2(3) + 4(0) = 3 \neq 9 \quad (2) \\ 3 - 3 - 0 = 0 \quad\quad\quad\quad (3) \end{cases}$$

Although these values satisfy equations (1) and (3), they do not satisfy equation (2). Any solution of the system must satisfy *each* equation of the system.

NOW WORK PROBLEM **9.**

When a system of equations has at least one solution, it is said to be **consistent**; otherwise, it is called **inconsistent**.

An equation in *n* variables is said to be **linear** if it is equivalent to an equation of the form

$$a_1x_1 + a_2x_2 + \cdots + a_nx_n = b$$

where $x_1, x_2, \ldots, x_n$ are *n* distinct variables, $a_1, a_2, \ldots, a_n, b$ are constants, and at least one of the *a*'s is not 0.

Some examples of linear equations are

$$2x + 3y = 2 \qquad 5x - 2y + 3z = 10 \qquad 8x + 8y - 2z + 5w = 0$$

If each equation in a system of equations is linear, then we have a **system of linear equations**. The systems in Examples 2(a), (c), (d), and (e) are linear, whereas the system in Example 2(b) is nonlinear. In this chapter we shall solve linear systems in Sections 10.1 to 10.3. We discuss nonlinear systems in Section 10.6.

We begin by discussing a system of two linear equations containing two variables. We can view the problem of solving such a system as a geometry problem. The graph of each equation in such a system is a line. So a system of two equations containing two variables represents a pair of lines. The lines either (1) intersect or (2) are parallel or (3) are **coincident** (that is, identical).

1. If the lines intersect, then the system of equations has one solution, given by the point of intersection. The system is **consistent** and the equations are **independent**. See Figure 1(a).

2. If the lines are parallel, then the system of equations has no solution, because the lines never intersect. The system is **inconsistent**. See Figure 1(b).

3. If the lines are coincident, then the system of equations has infinitely many solutions, represented by the totality of points on the line. The system is **consistent** and the equations are **dependent**. See Figure 1(c).

Figure 1

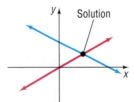
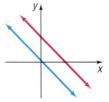
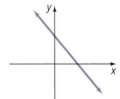

(a) Intersecting lines; system has one solution

(b) Parallel lines; system has no solution

(c) Coincident lines; system has infinitely many solutions

| EXAMPLE 3 | **Solving a System of Linear Equations Using a Graphing Utility** |

Solve: $\begin{cases} 2x + y = 5 & (1) \\ -4x + 6y = 12 & (2) \end{cases}$

Figure 2

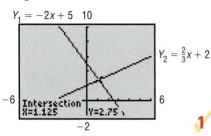

Solution First, we solve each equation for y. This is equivalent to writing each equation in slope–intercept form. Equation (1) in slope–intercept form is $Y_1 = -2x + 5$. Equation (2) in slope–intercept form is $Y_2 = \frac{2}{3}x + 2$. Figure 2 shows the graphs using a graphing utility. From the graph in Figure 2, we see that the lines intersect, so the system is consistent and the equations are independent. Using INTERSECT, we obtain the solution $(1.125, 2.75)$. ◄

1 Solve Systems of Equations by Substitution

Sometimes to obtain exact solutions, we must use algebraic methods. A number of methods are available to us for solving systems of linear equations algebraically. In this section, we introduce two methods: *substitution* and *elimination*. We illustrate the **method of substitution** by solving the system given in Example 3.

EXAMPLE 4

Solving a System of Linear Equations by Substitution

Solve: $\begin{cases} 2x + y = 5 & (1) \\ -4x + 6y = 12 & (2) \end{cases}$

Solution We solve the first equation for y, obtaining

$$2x + y = 5 \qquad (1)$$

$$y = -2x + 5 \qquad \text{Subtract 2x from each side of (1).}$$

We substitute this result for y into the second equation. The result is an equation containing just the variable x, which we can then solve.

$$-4x + 6y = 12 \qquad (2)$$

$$-4x + 6(-2x + 5) = 12 \qquad \text{Substitute } y = -2x + 5 \text{ in (2).}$$

$$-4x - 12x + 30 = 12 \qquad \text{Remove parentheses.}$$

$$-16x = -18 \qquad \text{Combine like terms and subtract 30 from both sides.}$$

$$x = \frac{-18}{-16} = \frac{9}{8} \qquad \text{Divide each side by } -16.$$

Once we know that $x = \frac{9}{8}$, we can easily find the value of y by **back-substitution**, that is, by substituting $\frac{9}{8}$ for x in one of the original equations. We will use the first equation.

$$2x + y = 5 \qquad (1)$$

$$y = -2x + 5 \qquad \text{Subtract 2x from each side.}$$

$$y = -2\left(\frac{9}{8}\right) + 5 \qquad \text{Substitute } x = \frac{9}{8} \text{ in (1).}$$

$$= \frac{-9}{4} + \frac{20}{4} = \frac{11}{4}$$

The solution of the system is $x = \frac{9}{8} = 1.125$, $y = \frac{11}{4} = 2.75$.

$$\checkmark \text{ CHECK:} \quad \begin{cases} 2x + y = 5: & 2\left(\dfrac{9}{8}\right) + \dfrac{11}{4} = \dfrac{9}{4} + \dfrac{11}{4} = \dfrac{20}{4} = 5 \\ -4x + 6y = 12: & -4\left(\dfrac{9}{8}\right) + 6\left(\dfrac{11}{4}\right) = -\dfrac{9}{2} + \dfrac{33}{2} = \dfrac{24}{2} = 12 \end{cases}$$

The method used to solve the system in Example 4 is called **substitution**. The steps to be used are outlined next.

Steps for Solving by Substitution

STEP 1: Pick one of the equations and solve for one of the variables in terms of the remaining variables.

STEP 2: Substitute the result into the remaining equations.

STEP 3: If one equation in one variable results, solve this equation. Otherwise repeat Steps 1 and 2 until a single equation with one variable remains.

STEP 4: Find the values of the remaining variables by back-substitution.

STEP 5: Check the solution found.

NOW USE SUBSTITUTION TO WORK PROBLEM 19.

2 Solve Systems of Equations by Elimination

A second method for solving a system of linear equations is the *method of elimination*. This method is usually preferred over substitution if substitution leads to fractions or if the system contains more than two variables. Elimination also provides the necessary motivation for solving systems using matrices (the subject of Section 10.2).

The idea behind the method of elimination is to replace the original system of equations by an equivalent system so that adding two of the equations eliminates a variable. The rules for obtaining equivalent equations are the same as those studied earlier. However, we may also interchange any two equations of the system and/or replace any equation in the system by the sum (or difference) of that equation and a nonzero multiple of any other equation in the system.

In Words

When using elimination, we want to get the coefficients of one of the variables to be negatives of one another.

Rules for Obtaining an Equivalent System of Equations

1. Interchange any two equations of the system.

2. Multiply (or divide) each side of an equation by the same nonzero constant.

3. Replace any equation in the system by the sum (or difference) of that equation and a nonzero multiple of any other equation in the system.

An example will give you the idea. As you work through the example, pay particular attention to the pattern being followed.

EXAMPLE 5 **Solving a System of Linear Equations by Elimination**

Solve: $\begin{cases} 2x + 3y = 1 & (1) \\ -x + y = -3 & (2) \end{cases}$

Solution   We multiply each side of equation (2) by 2 so that the coefficients of x in the two equations are negatives of one another. The result is the equivalent system

$$\begin{cases} 2x + 3y = 1 & (1) \\ -2x + 2y = -6 & (2) \end{cases}$$

If we add equations (1) and (2) we obtain an equation containing just the variable y, which we can solve.

$$\begin{cases} 2x + 3y = 1 & (1) \\ \underline{-2x + 2y = -6} & (2) \\ 5y = -5 & \text{Add (1) and (2).} \\ y = -1 & \text{Solve for } y. \end{cases}$$

We back-substitute this value for y in equation (1) and simplify.

$$\begin{aligned} 2x + 3y &= 1 & (1) \\ 2x + 3(-1) &= 1 & \text{Substitute } y = -1 \text{ in (1).} \\ 2x &= 4 & \text{Simplify.} \\ x &= 2 & \text{Solve for } x. \end{aligned}$$

The solution of the original system is $x = 2$, $y = -1$. We leave it to you to check the solution. ◄

The procedure used in Example 5 is called the **method of elimination**. Notice the pattern of the solution. First, we eliminated the variable x from the second equation. Then we back-substituted; that is, we substituted the value found for y back into the first equation to find x.

NOW USE ELIMINATION TO WORK PROBLEM 19.

Let's return to the movie theater example Example 1.

EXAMPLE 6	**Movie Theater Ticket Sales**

A movie theater sells tickets for $8.00 each, with seniors receiving a discount of $2.00. One evening the theater sold 525 tickets and took in $3580 in revenue. How many of each type of ticket were sold?

Solution   If x represents the number of tickets sold at $8.00 and y the number of tickets sold at the discounted price of $6.00, then the given information results in the system of equations

$$\begin{cases} 8x + 6y = 3580 & (1) \\ x + y = 525 & (2) \end{cases}$$

We use the method of elimination. First, multiply the second equation by -6, and then add the equations.

$$\begin{cases} 8x + 6y = 3580 \\ \underline{-6x - 6y = -3150} \\ 2x = 430 \qquad \text{Add the equations.} \\ x = 215 \end{cases}$$

Since $x + y = 525$, then $y = 525 - x = 525 - 215 = 310$. So 215 nondiscounted tickets and 310 senior discount tickets were sold. ◄

3 Identify Inconsistent Systems of Equations Containing Two Variables

The previous examples dealt with consistent systems of equations that had a single solution. The next two examples deal with two other possibilities that may occur, the first being a system that has no solution.

EXAMPLE 7 | **An Inconsistent System of Linear Equations**

Solve: $\begin{cases} 2x + y = 5 & (1) \\ 4x + 2y = 8 & (2) \end{cases}$

Solution We choose to use the method of substitution and solve equation (1) for y.

$$2x + y = 5 \qquad (1)$$
$$y = -2x + 5 \qquad \text{Subtract 2x from each side.}$$

Now substitute $y = -2x + 5$ for y in equation (2) and solve for x.

$$4x + 2y = 8 \qquad (2)$$
$$4x + 2(-2x + 5) = 8 \qquad \text{Substitute } y = -2x + 5 \text{ in (2).}$$
$$4x - 4x + 10 = 8 \qquad \text{Remove parentheses.}$$
$$0 \cdot x = -2 \qquad \text{Subtract 10 from both sides.}$$

This equation has no solution. We conclude that the system itself has no solution and is therefore inconsistent. ◀

Figure 3

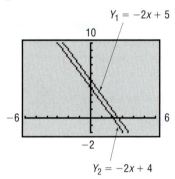

$Y_1 = -2x + 5$

$Y_2 = -2x + 4$

Figure 3 illustrates the pair of lines whose equations form the system in Example 7. Notice that the graphs of the two equations are lines, each with slope -2; one has a y-intercept of 5, the other a y-intercept of 4. The lines are parallel and have no point of intersection. This geometric statement is equivalent to the algebraic statement that the system has no solution.

4 Express the Solution of a System of Dependent Equations Containing Two Variables

EXAMPLE 8 | **Solving a System of Dependent Equations**

Solve: $\begin{cases} 2x + y = 4 & (1) \\ -6x - 3y = -12 & (2) \end{cases}$

Solution We choose to use the method of elimination.

$$\begin{cases} 2x + y = 4 & (1) \\ -6x - 3y = -12 & (2) \end{cases}$$

$$\begin{cases} 6x + 3y = 12 & (1) \quad \text{Multiply each side of equation (1) by 3.} \\ -6x - 3y = -12 & (2) \end{cases}$$

$$\begin{cases} 6x + 3y = 12 & (1) \\ 0 = 0 & (2) \quad \text{Replace equation (2) by the sum of equations (1) and (2).} \end{cases}$$

The original system is equivalent to a system containing one equation, so the equations are dependent. This means that any values of x and y for which $6x + 3y = 12$ or, equivalently, $2x + y = 4$ are solutions. For example, $x = 2, y = 0$; $x = 0, y = 4$; $x = -2, y = 8$; $x = 4, y = -4$; and so on, are solutions. There are, in fact, infinitely many values of x and y for which $2x + y = 4$, so the original system has infinitely many solutions. We will write the solution of the original system either as

$$y = -2x + 4$$

where x can be any real number, or as

$$x = -\frac{1}{2}y + 2$$

where y can be any real number. ◀

Figure 4

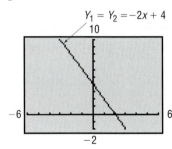

$Y_1 = Y_2 = -2x + 4$

Figure 4 illustrates the situation presented in Example 8. Notice that the graphs of the two equations are lines, each with slope -2 and each with y-intercept 4. The lines are coincident. Notice also that equation (2) in the original system is -3 times equation (1), indicating that the two equations are dependent.

For the system in Example 8, we can write down some of the infinite number of solutions by assigning values to x and then finding $y = -2x + 4$.

If $x = -2$, then $y = -2(-2) + 4 = 8$.

If $x = 0$, then $y = 4$.

If $x = 2$, then $y = 0$.

The ordered pairs $(-2, 8)$, $(0, 4)$, and $(2, 0)$ are three of the points on the line in Figure 4.

✏ **NOW WORK PROBLEMS 25 AND 29.**

5 Solve Systems of Three Equations Containing Three Variables

Just as with a system of two linear equations containing two variables, a system of three linear equations containing three variables also has either (1) exactly one solution (a consistent system with independent equations), or (2) no solution (an inconsistent system), or (3) infinitely many solutions (a consistent system with dependent equations).

We can view the problem of solving a system of three linear equations containing three variables as a geometry problem. The graph of each equation in such a system is a plane in space. A system of three linear equations containing three variables represents three planes in space. Figure 5 illustrates some of the possibilities.

Figure 5

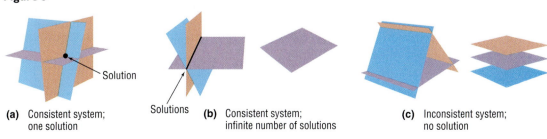

(a) Consistent system; one solution

Solution

Solutions

(b) Consistent system; infinite number of solutions

(c) Inconsistent system; no solution

Recall that a **solution** to a system of equations consists of values for the variables that are solutions of each equation of the system. For example, $x = 3, y = -1, z = -5$ is a solution to the system of equations

$$\begin{cases} x + y + z = -3 & \text{(1)} \\ 2x - 3y + 6z = -21 & \text{(2)} \\ -3x + 5y = -14 & \text{(3)} \end{cases} \qquad \begin{aligned} & 3 + (-1) + (-5) = -3 \\ & 2(3) - 3(-1) + 6(-5) = 6 + 3 - 30 = -21 \\ & -3(3) + 5(-1) = -9 - 5 = -14 \end{aligned}$$

because these values of the variables are solutions of each equation.

Typically, when solving a system of three linear equations containing three variables, we use the method of elimination. Recall that the idea behind the method of elimination is to form equivalent equations so that adding two of the equations eliminates a variable.

Let's see how elimination works on a system of three equations containing three variables.

EXAMPLE 9 **Solving a System of Three Linear Equations with Three Variables**

Use the method of elimination to solve the system of equations.

$$\begin{cases} x + y - z = -1 & \text{(1)} \\ 4x - 3y + 2z = 16 & \text{(2)} \\ 2x - 2y - 3z = 5 & \text{(3)} \end{cases}$$

Solution For a system of three equations, we attempt to eliminate one variable at a time, using pairs of equations until an equation with a single variable remains. Our plan of attack on this system will be to use equation (1) to eliminate the variable x from equations (2) and (3).

We begin by multiplying each side of equation (1) by -4 and adding the result to equation (2). (Do you see why? The coefficients of x are now negatives of one another.) We also multiply equation (1) by -2 and add the result to equation (3). Notice that these two procedures result in the removal of the variable x from equations (2) and (3).

$$\begin{aligned} x + y - z &= -1 \ \text{(1)} \quad \text{Multiply by }-4 \\ 4x - 3y + 2z &= 16 \ \text{(2)} \end{aligned} \qquad \begin{aligned} -4x - 4y + 4z &= 4 \ \text{(1)} \\ 4x - 3y + 2z &= 16 \ \text{(2)} \\ \hline -7y + 6z &= 20 \quad \text{Add.} \end{aligned}$$

$$\begin{aligned} x + y - z &= -1 \ \text{(1)} \quad \text{Multiply by }-2 \\ 2x - 2y - 3z &= 5 \ \text{(3)} \end{aligned} \qquad \begin{aligned} -2x - 2y + 2z &= 2 \ \text{(1)} \\ 2x - 2y - 3z &= 5 \ \text{(3)} \\ \hline -4y - z &= 7 \quad \text{Add.} \end{aligned}$$

$$\begin{cases} x + y - z = -1 & \text{(1)} \\ -7y + 6z = 20 & \text{(2)} \\ -4y - z = 7 & \text{(3)} \end{cases}$$

We now concentrate on equations (2) and (3), treating them as a system of two equations containing two variables. It is easiest to eliminate z. We multiply each side of equation (3) by 6 and add equations (2) and (3). The result is the new equation (3).

$$\begin{aligned} -7y + 6z &= 20 \ \text{(2)} \\ -4y - z &= 7 \ \text{(3)} \quad \text{Multiply by 6} \end{aligned} \qquad \begin{aligned} -7y + 6z &= 20 \ \text{(2)} \\ -24y - 6z &= 42 \ \text{(3)} \\ \hline -31y &= 62 \quad \text{Add.} \end{aligned} \qquad \begin{cases} x + y - z = -1 & \text{(1)} \\ -7y + 6z = 20 & \text{(2)} \\ -31y = 62 & \text{(3)} \end{cases}$$

We now solve equation (3) for y by dividing both sides of the equation by -31.

$$\begin{cases} x + y - z = -1 & (1) \\ -7y + 6z = 20 & (2) \\ y = -2 & (3) \end{cases}$$

Back-substitute $y = -2$ in equation (2) and solve for z.

$$\begin{aligned} -7y + 6z &= 20 && (2) \\ -7(-2) + 6z &= 20 && \text{Substitute } y = -2 \text{ in (2).} \\ 6z &= 6 && \text{Subtract 14 from both sides of the equation.} \\ z &= 1 && \text{Divide both sides of the equation by 6.} \end{aligned}$$

Finally, we back-substitute $y = -2$ and $z = 1$ in equation (1) and solve for x.

$$\begin{aligned} x + y - z &= -1 && (1) \\ x + (-2) - 1 &= -1 && \text{Substitute } y = -2 \text{ and } z = 1 \text{ in (1).} \\ x - 3 &= -1 && \text{Simplify.} \\ x &= 2 && \text{Add 3 to both sides.} \end{aligned}$$

The solution of the original system is $x = 2$, $y = -2$, $z = 1$. You should verify this solution. ◀

Look back over the solution given in Example 9. Note the pattern of removing one of the variables from two of the equations, followed by solving this system of two equations and two unknowns. Although which variables to remove is your choice, the methodology remains the same for all systems.

🖉 **NOW WORK PROBLEM 43.**

6 **Identify Inconsistent Systems of Equations Containing Three Variables**

EXAMPLE 10	**An Inconsistent System of Linear Equations**

Solve: $\begin{cases} 2x + y - z = -2 & (1) \\ x + 2y - z = -9 & (2) \\ x - 4y + z = 1 & (3) \end{cases}$

Solution   Our plan of attack is the same as in Example 9. However, in this system, it seems easiest to eliminate the variable z first. Do you see why?

Multiply each side of equation (1) by -1 and add the result to equation (2). Add equations (2) and (3).

$$\begin{array}{ll} -2x - y + z = 2 & (1) \text{ Multiply by } -1. \\ \underline{x + 2y - z = -9} & (2) \\ -x + y = -7 & \text{Add.} \end{array}$$

$$\begin{array}{ll} x + 2y - z = -9 & (2) \\ \underline{x - 4y + z = 1} & (3) \\ 2x - 2y = -8 & \text{Add.} \end{array}$$

$$\begin{cases} 2x + y - z = -2 & (1) \\ -x + y = -7 & (2) \\ 2x - 2y = -8 & (3) \end{cases}$$

We now concentrate on equations (2) and (3), treating them as a system of two equations containing two variables. Multiply each side of equation (2) by 2 and add the result to equation (3).

$$-x + y = -7 \quad \text{(2)} \quad \text{Multiply by 2.}$$
$$2x - 2y = -8 \quad \text{(3)}$$

$$\begin{array}{r} -2x + 2y = -14 \quad \text{(2)} \\ 2x - 2y = -8 \quad \text{(3)} \\ \hline 0 = -22 \quad \text{Add.} \end{array}$$

$$\begin{cases} 2x + y - z = -2 \quad \text{(1)} \\ -x + y \phantom{{}- z} = -7 \quad \text{(2)} \\ 0 = -22 \quad \text{(3)} \end{cases}$$

Equation (3) has no solution and the system is inconsistent. ◀

7 **Express the Solution of a System of Dependent Equations Containing Three Variables**

EXAMPLE 11 **Solving a System of Dependent Equations**

Solve: $\begin{cases} x - 2y - z = 8 \quad \text{(1)} \\ 2x - 3y + z = 23 \quad \text{(2)} \\ 4x - 5y + 5z = 53 \quad \text{(3)} \end{cases}$

Solution Our plan is to eliminate x from equations (2) and (3). Multiply each side of equation (1) by -2 and add the result to equation (2). Also, multiply each side of equation (1) by -4 and add the result to equation (3).

$$\begin{array}{l} x - 2y - z = 8 \quad \text{(1)} \quad \text{Multiply by } -2. \\ 2x - 3y + z = 23 \quad \text{(2)} \end{array}$$

$$\begin{array}{r} -2x + 4y + 2z = -16 \quad \text{(1)} \\ 2x - 3y + z = 23 \quad \text{(2)} \\ \hline y + 3z = 7 \quad \text{Add.} \end{array}$$

$$\begin{array}{l} x - 2y - z = 8 \quad \text{(1)} \quad \text{Multiply by } -4. \\ 4x - 5y + 5z = 53 \quad \text{(3)} \end{array}$$

$$\begin{array}{r} -4x + 8y + 4z = -32 \quad \text{(1)} \\ 4x - 5y + 5z = 53 \quad \text{(2)} \\ \hline 3y + 9z = 21 \quad \text{Add.} \end{array}$$

$$\begin{cases} x - 2y - z = 8 \quad \text{(1)} \\ y + 3z = 7 \quad \text{(2)} \\ 3y + 9z = 21 \quad \text{(3)} \end{cases}$$

Treat equations (2) and (3) as a system of two equations containing two variables, and eliminate the variable y by multiplying both sides of equation (2) by -3 and adding the result to equation (3).

$$\begin{array}{l} y + 3z = 7 \quad \text{Multiply by } -3. \\ 3y + 9z = 21 \end{array}$$

$$\begin{array}{r} -3y - 9z = -21 \\ 3y + 9z = 21 \quad \text{Add.} \\ \hline 0 = 0 \end{array}$$

$$\begin{cases} x - 2y - z = 8 \quad \text{(1)} \\ y + 3z = 7 \quad \text{(2)} \\ 0 = 0 \quad \text{(3)} \end{cases}$$

The original system is equivalent to a system containing two equations, so the equations are dependent and the system has infinitely many solutions. If we solve equation (2) for y, we can express y in terms of z as $y = -3z + 7$. Substitute this expression into equation (1) to determine x in terms of z.

$$\begin{array}{ll} x - 2y - z = 8 & \text{(1)} \\ x - 2(-3z + 7) - z = 8 & \text{Substitute } y = -3z + 7 \text{ in (1).} \\ x + 6z - 14 - z = 8 & \text{Remove parentheses.} \\ x + 5z = 22 & \text{Combine like terms.} \\ x = -5z + 22 & \text{Solve for } x. \end{array}$$

We will write the solution to the system as

$$\begin{cases} x = -5z + 22 \\ y = -3z + 7 \end{cases}$$

where z can be any real number.

This way of writing the solution makes it easier to find specific solutions of the system. To find specific solutions, choose any value of z and use the equations $x = -5z + 22$ and $y = -3z + 7$ to determine x and y. For example, if $z = 0$, then $x = 22$ and $y = 7$, and if $z = 1$, then $x = 17$ and $y = 4$. ◀

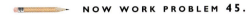 **NOW WORK PROBLEM 45.**

Two points in the Cartesian plane determine a unique line. Given three noncollinear points, we can find the (unique) quadratic function whose graph contains these three points.

EXAMPLE 12	**Curve Fitting**

Find real numbers a, b, and c so that the graph of the quadratic function $y = ax^2 + bx + c$ contains the points $(-1, -4)$, $(1, 6)$, and $(3, 0)$.

Solution

We require that the three points satisfy the equation $y = ax^2 + bx + c$.

For the point $(-1, -4)$ we have:	$-4 = a(-1)^2 + b(-1) + c$	$-4 = a - b + c$
For the point $(1, 6)$ we have:	$6 = a(1)^2 + b(1) + c$	$6 = a + b + c$
For the point $(3, 0)$ we have:	$0 = a(3)^2 + b(3) + c$	$0 = 9a + 3b + c$

We wish to determine a, b, and c so that each equation is satisfied. That is, we want to solve the following system of three equations containing three variables:

$$\begin{cases} a - b + c = -4 & (1) \\ a + b + c = 6 & (2) \\ 9a + 3b + c = 0 & (3) \end{cases}$$

Figure 6

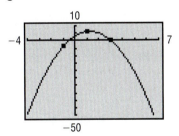

Solving this system of equations, we obtain $a = -2$, $b = 5$, and $c = 3$. So the quadratic function whose graph contains the points $(-1, -4)$, $(1, 6)$, and $(3, 0)$ is

$$y = -2x^2 + 5x + 3 \qquad y = ax^2 + bx + c, \quad a = -2, b = 5, c = 3$$

Figure 6 shows the graph of the function along with the three points. ◀

10.1 Assess Your Understanding

'Are You Prepared?'

Answers are given at the end of these exercises. If you get a wrong answer, read the pages listed in red.

1. Solve the equation: $3x + 4 = 8 - x$. (p. 694)

2. (a) Graph the line: $3x + 4y = 12$. (pp. 34–36)

(b) What is the slope of a line parallel to this line? (pp. 36–38)

Concepts and Vocabulary

3. If a system of equations has no solution, it is said to be _____.

4. If a system of equations has one or more solutions, the system is said to be _____.

5. *True or False:* A system of two linear equations containing two variables always has at least one solution.

6. *True or False:* A solution of a system of equations consists of values for the variables that are solutions of each equation of the system.

Skill Building

In Problems 7–16, verify that the values of the variables listed are solutions of the system of equations.

7. $\begin{cases} 2x - y = 5 \\ 5x + 2y = 8 \end{cases}$
$x = 2, y = -1$

8. $\begin{cases} 3x + 2y = 2 \\ x - 7y = -30 \end{cases}$
$x = -2, y = 4$

9. $\begin{cases} 3x - 4y = 4 \\ \dfrac{1}{2}x - 3y = -\dfrac{1}{2} \end{cases}$
$x = 2, y = \dfrac{1}{2}$

10. $\begin{cases} 2x + \dfrac{1}{2}y = 0 \\ 3x - 4y = -\dfrac{19}{2} \end{cases}$
$x = -\dfrac{1}{2}, y = 2$

11. $\begin{cases} x - y = 3 \\ \dfrac{1}{2}x + y = 3 \end{cases}$
$x = 4, y = 1$

12. $\begin{cases} x - y = 3 \\ -3x + y = 1 \end{cases}$
$x = -2, y = -5$

13. $\begin{cases} 3x + 3y + 2z = 4 \\ x - y - z = 0 \\ 2y - 3z = -8 \end{cases}$
$x = 1, y = -1, z = 2$

14. $\begin{cases} 4x - z = 7 \\ 8x + 5y - z = 0 \\ -x - y + 5z = 6 \end{cases}$
$x = 2, y = -3, z = 1$

15. $\begin{cases} 3x + 3y + 2z = 4 \\ x - 3y + z = 10 \\ 5x - 2y - 3z = 8 \end{cases}$
$x = 2, y = -2, z = 2$

16. $\begin{cases} 4x - 5z = 6 \\ 5y - z = -17 \\ -x - 6y + 5z = 24 \end{cases}$
$x = 4, y = -3, z = 2$

In Problems 17–54, solve each system of equations. If the system has no solution, say that it is inconsistent.

17. $\begin{cases} x + y = 8 \\ x - y = 4 \end{cases}$

18. $\begin{cases} x + 2y = 5 \\ x + y = 3 \end{cases}$

19. $\begin{cases} 5x - y = 13 \\ 2x + 3y = 12 \end{cases}$

20. $\begin{cases} x + 3y = 5 \\ 2x - 3y = -8 \end{cases}$

21. $\begin{cases} 3x = 24 \\ x + 2y = 0 \end{cases}$

22. $\begin{cases} 4x + 5y = -3 \\ -2y = -4 \end{cases}$

23. $\begin{cases} 3x - 6y = 2 \\ 5x + 4y = 1 \end{cases}$

24. $\begin{cases} 2x + 4y = \dfrac{2}{3} \\ 3x - 5y = -10 \end{cases}$

25. $\begin{cases} 2x + y = 1 \\ 4x + 2y = 3 \end{cases}$

26. $\begin{cases} x - y = 5 \\ -3x + 3y = 2 \end{cases}$

27. $\begin{cases} 2x - y = 0 \\ 3x + 2y = 7 \end{cases}$

28. $\begin{cases} 3x + 3y = -1 \\ 4x + y = \dfrac{8}{3} \end{cases}$

29. $\begin{cases} x + 2y = 4 \\ 2x + 4y = 8 \end{cases}$

30. $\begin{cases} 3x - y = 7 \\ 9x - 3y = 21 \end{cases}$

31. $\begin{cases} 2x - 3y = -1 \\ 10x + y = 11 \end{cases}$

32. $\begin{cases} 3x - 2y = 0 \\ 5x + 10y = 4 \end{cases}$

33. $\begin{cases} 2x + 3y = 6 \\ x - y = \dfrac{1}{2} \end{cases}$

34. $\begin{cases} \dfrac{1}{2}x + y = -2 \\ x - 2y = 8 \end{cases}$

35. $\begin{cases} \dfrac{1}{2}x + \dfrac{1}{3}y = 3 \\ \dfrac{1}{4}x - \dfrac{2}{3}y = -1 \end{cases}$

36. $\begin{cases} \dfrac{1}{3}x - \dfrac{3}{2}y = -5 \\ \dfrac{3}{4}x + \dfrac{1}{3}y = 11 \end{cases}$

37. $\begin{cases} 3x - 5y = 3 \\ 15x + 5y = 21 \end{cases}$

38. $\begin{cases} 2x - y = -1 \\ x + \dfrac{1}{2}y = \dfrac{3}{2} \end{cases}$

39. $\begin{cases} \dfrac{1}{x} + \dfrac{1}{y} = 8 \\ \dfrac{3}{x} - \dfrac{5}{y} = 0 \end{cases}$

40. $\begin{cases} \dfrac{4}{x} - \dfrac{3}{y} = 0 \\ \dfrac{6}{x} + \dfrac{3}{2y} = 2 \end{cases}$

[**Hint:** Let $u = \dfrac{1}{x}$ and $v = \dfrac{1}{y}$, and solve for u and v.
Then $x = \dfrac{1}{u}$ and $y = \dfrac{1}{v}$.]

41. $\begin{cases} x - y = 6 \\ 2x - 3z = 16 \\ 2y + z = 4 \end{cases}$

42. $\begin{cases} 2x + y = -4 \\ -2y + 4z = 0 \\ 3x - 2z = -11 \end{cases}$

43. $\begin{cases} x - 2y + 3z = 7 \\ 2x + y + z = 4 \\ -3x + 2y - 2z = -10 \end{cases}$

44. $\begin{cases} 2x + y - 3z = 0 \\ -2x + 2y + z = -7 \\ 3x - 4y - 3z = 7 \end{cases}$ **45.** $\begin{cases} x - y - z = 1 \\ 2x + 3y + z = 2 \\ 3x + 2y = 0 \end{cases}$ **46.** $\begin{cases} 2x - 3y - z = 0 \\ -x + 2y + z = 5 \\ 3x - 4y - z = 1 \end{cases}$ **47.** $\begin{cases} x - y - z = 1 \\ -x + 2y - 3z = -4 \\ 3x - 2y - 7z = 0 \end{cases}$

48. $\begin{cases} 2x - 3y - z = 0 \\ 3x + 2y + 2z = 2 \\ x + 5y + 3z = 2 \end{cases}$ **49.** $\begin{cases} 2x - 2y + 3z = 6 \\ 4x - 3y + 2z = 0 \\ -2x + 3y - 7z = 1 \end{cases}$ **50.** $\begin{cases} 3x - 2y + 2z = 6 \\ 7x - 3y + 2z = -1 \\ 2x - 3y + 4z = 0 \end{cases}$ **51.** $\begin{cases} x + y - z = 6 \\ 3x - 2y + z = -5 \\ x + 3y - 2z = 14 \end{cases}$

52. $\begin{cases} x - y + z = -4 \\ 2x - 3y + 4z = -15 \\ 5x + y - 2z = 12 \end{cases}$ **53.** $\begin{cases} x + 2y - z = -3 \\ 2x - 4y + z = -7 \\ -2x + 2y - 3z = 4 \end{cases}$ **54.** $\begin{cases} x + 4y - 3z = -8 \\ 3x - y + 3z = 12 \\ x + y + 6z = 1 \end{cases}$

Applications and Extensions

55. The perimeter of a rectangular floor is 90 feet. Find the dimensions of the floor if the length is twice the width.

56. The length of fence required to enclose a rectangular field is 3000 meters. What are the dimensions of the field if it is known that the difference between its length and width is 50 meters?

57. Cost of Fast Food   Four large cheeseburgers and two chocolate shakes cost a total of $7.90. Two shakes cost 15¢ more than one cheeseburger. What is the cost of a cheeseburger? A shake?

58. Movie Theater Tickets   A movie theater charges $9.00 for adults and $7.00 for senior citizens. On a day when 325 people paid an admission, the total receipts were $2495. How many who paid were adults? How many were seniors?

59. Mixing Nuts   A store sells cashews for $5.00 per pound and peanuts for $1.50 per pound. The manager decides to mix 30 pounds of peanuts with some cashews and sell the mixture for $3.00 per pound. How many pounds of cashews should be mixed with the peanuts so that the mixture will produce the same revenue as would selling the nuts separately?

60. Financial Planning   A recently retired couple need $12,000 per year to supplement their Social Security. They have $150,000 to invest to obtain this income. They have decided on two investment options: AA bonds yielding 10% per annum and a Bank Certificate yielding 5%.
  (a) How much should be invested in each to realize exactly $12,000?
  (b) If, after two years, the couple requires $14,000 per year in income, how should they reallocate their investment to achieve the new amount?

61. Computing Wind Speed   With a tail wind, a small Piper aircraft can fly 600 miles in 3 hours. Against this same wind, the Piper can fly the same distance in 4 hours. Find the average wind speed and the average airspeed of the Piper.

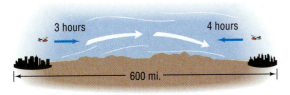

62. Computing Wind Speed   The average airspeed of a single-engine aircraft is 150 miles per hour. If the aircraft flew the same distance in 2 hours with the wind as it flew in 3 hours against the wind, what was the wind speed?

63. Restaurant Management   A restaurant manager wants to purchase 200 sets of dishes. One design costs $25 per set, while another costs $45 per set. If she only has $7400 to spend, how many of each design should be ordered?

64. Cost of Fast Food   One group of people purchased 10 hot dogs and 5 soft drinks at a cost of $12.50. A second bought 7 hot dogs and 4 soft drinks at a cost of $9.00. What is the cost of a single hot dog? A single soft drink?

We paid $12.50.
How much is one hot dog?
How much is one cola?

We paid $9.00.
How much is one hot dog?
How much is one cola?

65. Computing a Refund   The grocery store we use does not mark prices on its goods. My wife went to this store, bought three 1-pound packages of bacon and two cartons of eggs, and paid a total of $7.45. Not knowing that she went to the store, I also went to the same store, purchased two 1-pound packages of bacon and three cartons of eggs, and paid a total of $6.45. Now we want to return two 1-pound packages of bacon and two cartons of eggs. How much will be refunded?

66. Finding the Current of a Stream   Pamela requires 3 hours to swim 15 miles downstream on the Illinois River. The return trip upstream takes 5 hours. Find Pamela's average speed in still water. How fast is the current? (Assume that Pamela's speed is the same in each direction.)

67. Pharmacy A doctor's prescription calls for a daily intake containing 40 mg of vitamin C and 30 mg of vitamin D. Your pharmacy stocks two liquids that can be used: one contains 20% vitamin C and 30% vitamin D, the other 40% vitamin C and 20% vitamin D. How many milligrams of each compound should be mixed to fill the prescription?

68. Pharmacy A doctor's prescription calls for the creation of pills that contain 12 units of vitamin B_{12} and 12 units of vitamin E. Your pharmacy stocks two powders that can be used to make these pills: one contains 20% vitamin B_{12} and 30% vitamin E, the other 40% vitamin B_{12} and 20% vitamin E. How many units of each powder should be mixed in each pill?

69. Curve Fitting Find real numbers a, b, and c so that the graph of the function $y = ax^2 + bx + c$ contains the points $(-1, 4)$, $(2, 3)$, and $(0, 1)$.

70. Curve Fitting Find real numbers a, b, and c so that the graph of the function $y = ax^2 + bx + c$ contains the points $(-1, -2)$, $(1, -4)$, and $(2, 4)$.

71. IS-LM Model in Economics In economics, the IS curve is a linear equation that represents all combinations of income Y and interest rates r that maintain an equilibrium in the market for goods in the economy. The LM curve is a linear equation that represents all combinations of income Y and interest rates r that maintain an equilibrium in the market for money in the economy. In an economy, suppose the equilibrium level of income (in millions of dollars) and interest rates satisfy the system of equations

$$\begin{cases} 0.06Y - 5000r = 240 \\ 0.06Y + 6000r = 900 \end{cases}$$

Find the equilibrium level of income and interest rates.

72. IS-LM Model in Economics In economics, the IS curve is a linear equation that represents all combinations of income Y and interest rates r that maintain an equilibrium in the market for goods in the economy. The LM curve is a linear equation that represents all combinations of income Y and interest rates r that maintain an equilibrium in the market for money in the economy. In an economy, suppose the equilibrium level of income (in millions of dollars) and interest rates satisfy the system of equations

$$\begin{cases} 0.05Y - 1000r = 10 \\ 0.05Y + 800r = 100 \end{cases}$$

Find the equilibrium level of income and interest rates.

73. Electricity: Kirchhoff's Rules An application of Kirchhoff's Rules to the circuit shown results in the following system of equations:

$$\begin{cases} I_2 = I_1 + I_3 \\ 5 - 3I_1 - 5I_2 = 0 \\ 10 - 5I_2 - 7I_3 = 0 \end{cases}$$

Find the currents I_1, I_2, and I_3.

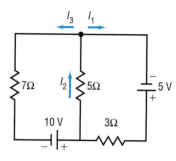

SOURCE: *Physics for Scientists & Engineers* 3/E by Serway. © 1990. Reprinted with permission of Brooks/Cole, a division of Thomson Learning.

74. Electricity: Kirchhoff's Rules An application of Kirchhoff's Rules to the circuit shown results in the following system of equations:

$$\begin{cases} I_3 = I_1 + I_2 \\ 8 = 4I_3 + 6I_2 \\ 8I_1 = 4 + 6I_2 \end{cases}$$

Find the currents I_1, I_2, and I_3.

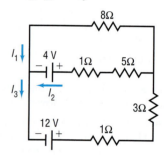

SOURCE: *Physics for Scientists & Engineers* 3/E by Serway. © 1990. Reprinted with permission of Brooks/Cole, a division of Thomson Learning.

75. Theater Revenues A Broadway theater has 500 seats, divided into orchestra, main, and balcony seating. Orchestra seats sell for $50, main seats for $35, and balcony seats for $25. If all the seats are sold, the gross revenue to the theater is $17,100. If all the main and balcony seats are sold, but only half the orchestra seats are sold, the gross revenue is $14,600. How many are there of each kind of seat?

76. Theater Revenues A movie theater charges $8.00 for adults, $4.50 for children, and $6.00 for senior citizens. One day the theater sold 405 tickets and collected $2320 in receipts. There were twice as many children's tickets sold as adult tickets. How many adults, children, and senior citizens went to the theater that day?

77. Nutrition A dietitian wishes a patient to have a meal that has 66 grams of protein, 94.5 grams of carbohydrates, and 910 milligrams of calcium. The hospital food service tells the dietitian that the dinner for today is chicken, corn, and 2% milk. Each serving of chicken has 30 grams of protein, 35 grams of carbohydrates, and 200 milligrams of calcium. Each serving of corn has 3 grams of protein, 16 grams of carbohydrates, and 10 milligrams of calcium. Each glass of 2% milk has 9 grams of protein, 13 grams of carbohydrates, and 300 milligrams of calcium. How many servings of each food should the dietitian provide for the patient?

78. **Investments** Kelly has $20,000 to invest. As her financial planner, you recommend that she diversify into three investments: Treasury bills that yield 5% simple interest, Treasury bonds that yield 7% simple interest, and corporate bonds that yield 10% simple interest. Kelly wishes to earn $1390 per year in income. Also, Kelly wants her investment in Treasury bills to be $3000 more than her investment in corporate bonds. How much money should Kelly place in each investment?

79. **Prices of Fast Food** One group of customers bought 8 deluxe hamburgers, 6 orders of large fries, and 6 large colas for $26.10. A second group ordered 10 deluxe hamburgers, 6 large fries, and 8 large colas and paid $31.60. Is there sufficient information to determine the price of each food item? If not, construct a table showing the various possibilities. Assume that the hamburgers cost between $1.75 and $2.25, the fries between $0.75 and $1.00, and the colas between $0.60 and $0.90.

80. **Prices of Fast Food** Use the information given in Problem 79. Suppose that a third group purchased 3 deluxe

hamburgers, 2 large fries, and 4 large colas for $10.95. Now is there sufficient information to determine the price of each food item? If so, determine each price.

81. **Painting a House** Three painters, Beth, Bill, and Edie, working together, can paint the exterior of a home in 10 hours. Bill and Edie together have painted a similar house in 15 hours. One day, all three worked on this same kind of house for 4 hours, after which Edie left. Beth and Bill required 8 more hours to finish. Assuming no gain or loss in efficiency, how long should it take each person to complete such a job alone?

Discussion and Writing

82. Make up a system of three linear equations containing three variables that has:
(a) No solution
(b) Exactly one solution
(c) Infinitely many solutions
Give the three systems to a friend to solve and critique.

83. Write a brief paragraph outlining your strategy for solving a system of two linear equations containing two variables.

84. Do you prefer the method of substitution or the method of elimination for solving a system of two linear equations containing two variables? Give reasons.

'Are You Prepared?' Answers

1. {1} 2. (a) (b) $-\dfrac{3}{4}$

10.7 Systems of Inequalities

PREPARING FOR THIS SECTION *Before getting started, review the following:*
• Solving Inequalities (Appendix, Section A.8, pp. 732–733)
• Lines (Section 1.4, pp. 27–38)
• Circles (Section 1.5, pp. 44–49)
• Graphing Techniques: Transformation (Section 2.6, pp. 118–126)

✎ Now work the 'Are You Prepared?' problems on page 622.

OBJECTIVES 1 Graph an Inequality by Hand
2 Graph an Inequality Using a Graphing Utility
3 Graph a System of Inequalities

In the Appendix, Section A.8, we discussed inequalities in one variable. In this section, we discuss inequalities in two variables.

| EXAMPLE 1 | **Examples of Inequalities in Two Variables** |

    (a) $3x + y \leq 6$ (b) $x^2 + y^2 < 4$ (c) $y^2 \leq x$ ◀

 Graph an Inequality by Hand

An inequality in two variables x and y is **satisfied** by an ordered pair (a, b) if, when x is replaced by a and y by b, a true statement results. The **graph of an inequality in two variables** x and y consists of all points (x, y) whose coordinates satisfy the inequality. Let's look at an example.

| EXAMPLE 2 | **Graphing an Inequality by Hand** |

Graph the linear inequality: $3x + y \leq 6$

Solution   We begin by graphing the equation

$$3x + y = 6$$

formed by replacing (for now) the $\leq$ symbol with an $=$ sign. The graph of the equation is a line. See Figure 26(a). This line is part of the graph of the inequality that we seek because the inequality is nonstrict. (Do you see why? We are seeking points for which $3x + y$ is less than *or equal to* 6.)

Now let's test a few randomly selected points to see whether they belong to the graph of the inequality.

	$3x + y \leq 6$	**Conclusion**
$(4, -1)$	$3(4) + (-1) = 11 > 6$	Does not belong to graph
$(5, 5)$	$3(5) + 5 = 20 > 6$	Does not belong to graph
$(-1, 2)$	$3(-1) + 2 = -1 \leq 6$	Belongs to graph
$(-2, -2)$	$3(-2) + (-2) = -8 \leq 6$	Belongs to graph

Look again at Figure 26(a). Notice that the two points that belong to the graph both lie on the same side of the line, and the two points that do not belong to the graph lie on the opposite side. As it turns out, this is always the case. The graph we seek consists of all points that lie on the same side of the line as $(-1, 2)$ and $(-2, -2)$ and is shown as the shaded region in Figure 26(b).

Figure 26

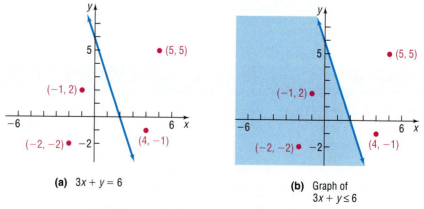

(a) $3x + y = 6$ **(b)** Graph of
                                    $3x + y \leq 6$ ◀

                        NOW WORK PROBLEM 15 BY HAND.

NOTE

The strict inequalities are $<$ or $>$. The nonstrict inequalities are $\leq$ or $\geq$. ■

The graph of any inequality in two variables may be obtained in a like way. First, the equation corresponding to the inequality is graphed, using dashes if the inequality is strict and solid marks if it is nonstrict. This graph, in almost every case, will separate the *xy*-plane into two or more regions. In each region, either all points satisfy the inequality or no points satisfy the inequality. The use of a single test point in each region is all that is required to determine whether the points of that region are part of the graph. The steps to follow are given next.

Steps for Graphing an Inequality by Hand

STEP 1: Replace the inequality symbol by an equal sign and graph the resulting equation. If the inequality is strict, use dashes; if it is nonstrict, use a solid mark. This graph separates the *xy*-plane into two or more regions.

STEP 2: In each region, select a test point *P*.
  (a) If the coordinates of *P* satisfy the inequality, then so do all the points in that region. Indicate this by shading the region.
  (b) If the coordinates of *P* do not satisfy the inequality, then none of the points in that region do.

EXAMPLE 3   **Graphing an Inequality by Hand**

Graph: $x^2 + y^2 \leq 4$

Solution   First, we graph the equation $x^2 + y^2 = 4$, a circle of radius 2, center at the origin. A solid circle will be used because the inequality is not strict. We use two test points, one inside the circle, the other outside.

Inside $(0, 0)$: $x^2 + y^2 = 0^2 + 0^2 = 0 \leq 4$ *Belongs to the graph*
Outside $(4, 0)$: $x^2 + y^2 = 4^2 + 0^2 = 16 > 4$ *Does not belong to the graph*

All the points inside and on the circle satisfy the inequality. See Figure 27.

Figure 27

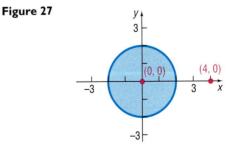

NOW WORK PROBLEM 17 BY HAND.

2 **Graph an Inequality Using a Graphing Utility**

Graphing utilities can also be used to graph inequalities. The steps to follow are given next.

Steps for Graphing an Inequality Using a Graphing Utility

STEP 1: Replace the inequality symbol by an equal sign and graph the resulting equation. This graph separates the xy-plane into two or more regions.

STEP 2: Select a test point P in each region.
  (a) Use a graphing utility to determine if the test point P satisfies the inequality. If the test point satisfies the inequality, then so do all the points in this region. Indicate this by using the graphing utility to shade the region.
  (b) If the coordinates of P do not satisfy the inequality, then none of the points in that region do.

EXAMPLE 4   Graphing an Inequality Using a Graphing Utility

Use a graphing utility to graph $3x + y \leq 6$.

Solution   **STEP 1:** We begin by graphing the equation $3x + y = 6$ ($Y_1 = -3x + 6$). See Figure 28.

Figure 28

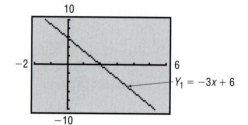

STEP 2: Select a test point in one of the regions and determine whether it satisfies the inequality. To test the point $(-1, 2)$, for example, enter $3(-1) + 2 \leq 6$. See Figure 29(a). The 1 that appears indicates that the statement entered (the inequality) is true. When the point $(5, 5)$ is tested, a 0 appears, indicating that the statement entered is false. So $(-1, 2)$ is a part of the graph of the inequality and $(5, 5)$ is not. We shade the region containing the point $(-1, 2)$ that is below Y_1. Figure 29(b) shows the graph of the inequality on a TI-84 Plus.

Figure 29

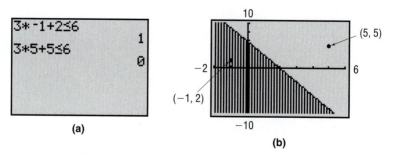

(a)   (b)

NOW WORK PROBLEM **15** USING A GRAPHING UTILITY.

Figure 30

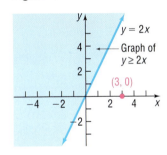

Linear inequalities are inequalities in one of the forms

$$Ax + By < C \qquad Ax + By > C \qquad Ax + By \leq C \qquad Ax + By \geq C$$

where A and B are not both zero.

The graph of the corresponding equation of a linear inequality is a line that separates the xy-plane into two regions, called **half-planes**. See Figure 30.

As shown, $Ax + By = C$ is the equation of the boundary line and it divides the plane into two half-planes: one for which $Ax + By < C$ and the other for which $Ax + By > C$. Because of this, for linear inequalities, only one test point is required.

EXAMPLE 5   **Graphing Linear Inequalities**

Graph: (a) $y < 2$ (b) $y \geq 2x$

Solution   (a) The graph of the equation $y = 2$ is a horizontal line and is not part of the graph of the inequality. Since $(0, 0)$ satisfies the inequality, the graph consists of the half-plane below the line $y = 2$. See Figure 31.

(b) The graph of the equation $y = 2x$ is a line and is part of the graph of the inequality. Using $(3, 0)$ as a test point, we find it does not satisfy the inequality $[0 < 2 \cdot 3]$. Points in the half-plane on the opposite side of $(3, 0)$ satisfy the inequality. See Figure 32.

Figure 31

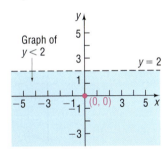

Figure 32

NOW WORK PROBLEM 13.

---

3   **Graph a System of Inequalities**

The **graph of a system of inequalities** in two variables x and y is the set of all points (x, y) that simultaneously satisfy *each* of the inequalities in the system. The graph of a system of inequalities can be obtained by graphing each inequality individually and then determining where, if at all, they intersect.

EXAMPLE 6   **Graphing a System of Linear Inequalities by Hand**

Graph the system: $\begin{cases} x + y \geq 2 \\ 2x - y \leq 4 \end{cases}$

Solution   First, we graph the inequality $x + y \geq 2$ as the shaded region in Figure 33(a). Next, we graph the inequality $2x - y \leq 4$ as the shaded region in Figure 33(b). Now, superimpose the two graphs, as shown in Figure 33(c). The points that are in both shaded regions [the overlapping, darker region in Figure 33(c)] are the solutions we seek to the system, because they simultaneously satisfy each linear inequality.

Figure 33

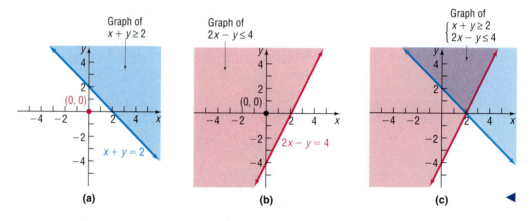

(a)   (b)   (c)

EXAMPLE 7   **Graphing a System of Linear Inequalities Using a Graphing Utility**

Graph the system: $\begin{cases} x + y \geq 2 \\ 2x - y \leq 4 \end{cases}$

Solution   First, we graph the lines $x + y = 2$ $(Y_1 = -x + 2)$ and $2x - y = 4$ $(Y_2 = 2x - 4)$. See Figure 34.

Figure 34

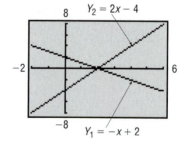

Notice that the graphs divide the viewing window into four regions. We select a test point for each region and determine whether the point makes *both* inequalities true. We choose to test $(0, 0)$, $(2, 3)$, $(4, 0)$, and $(2, -2)$. Figure 35(a) shows that $(2, 3)$ is the only point for which both inequalities are true. We obtain the graph shown in Figure 35(b).

Figure 35

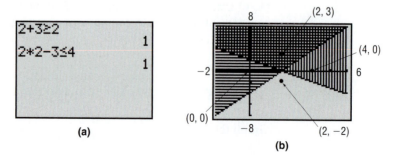

(a)   (b)

Rather than testing four points, we could just test the point $(0, 0)$ on each inequality. For example, $(0, 0)$ does not satisfy $x + y \geq 2$, so we shade above the line $x + y = 2$. In addition, $(0, 0)$ does satisfy $2x - y \leq 4$, so we shade above the line $2x - y = 4$. The intersection of the shaded regions gives us the result presented in Figure 35(b).

> **NOW WORK PROBLEM 23.**

| **EXAMPLE 8** | **Graphing a System of Linear Inequalities by Hand** |

Graph the system: $\begin{cases} x + y \leq 2 \\ x + y \geq 0 \end{cases}$

Solution   See Figure 36. The overlapping purple-shaded region between the two boundary lines is the graph of the system.

Figure 36

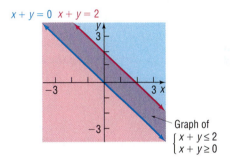

> **NOW WORK PROBLEM 29.**

| **EXAMPLE 9** | **Graphing a System of Linear Inequalities by Hand** |

Graph the system: $\begin{cases} 2x - y \geq 0 \\ 2x - y \geq 2 \end{cases}$

Solution   See Figure 37. The overlapping purple-shaded region is the graph of the system. Note that the graph of the system is identical to the graph of the single inequality $2x - y \geq 2$.

Figure 37

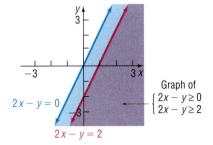

EXAMPLE 10	Graphing a System of Linear Inequalities by Hand

Graph the system: $\begin{cases} x + 2y \leq 2 \\ x + 2y \geq 6 \end{cases}$

Solution   See Figure 38. Because no overlapping region results, there are no points in the xy-plane that simultaneously satisfy each inequality. Hence, the system has no solution.

Figure 38

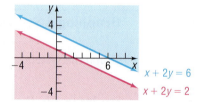

EXAMPLE 11	Graphing a System of Nonlinear Inequalities

Graph the region below the graph of $x + y = 2$ and above the graph of $y = x^2 - 4$ by graphing the system:

$$\begin{cases} y \geq x^2 - 4 \\ x + y \leq 2 \end{cases}$$

Label all points of intersection.

Solution   Figure 39 shows the graph of the region above the graph of the parabola $y = x^2 - 4$ and below the graph of the line $x + y = 2$. The points of intersection are found by solving the system of equations

Figure 39

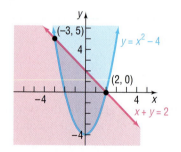

$$\begin{cases} y = x^2 - 4 \\ x + y = 2 \end{cases}$$

Using substitution, we find

$$x + (x^2 - 4) = 2$$

$$x^2 + x - 6 = 0$$

$$(x + 3)(x - 2) = 0$$

$$x = -3 \qquad x = 2$$

The two points of intersection are $(-3, 5)$ and $(2, 0)$.

NOW WORK PROBLEM 37.

EXAMPLE 12 — Graphing a System of Four Linear Inequalities by Hand

Figure 40

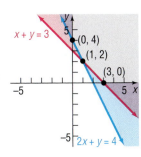

Graph the system:

$$\begin{cases} x + y \geq 3 \\ 2x + y \geq 4 \\ x \geq 0 \\ y \geq 0 \end{cases}$$

Solution   The two inequalities $x \geq 0$ and $y \geq 0$ require the graph of the system to be in quadrant I. We concentrate on the remaining two inequalities. The intersection of the graphs of these two inequalities and quadrant I, shown in light gray in Figure 40, is the graph of the system.   ◀

EXAMPLE 13 — Financial Planning

A retired couple has up to $25,000 to invest. As their financial adviser, you recommend that they place at least $15,000 in Treasury bills yielding 6% and at most $5000 in corporate bonds yielding 9%.

(a) Using x to denote the amount of money invested in Treasury bills and y the amount invested in corporate bonds, write a system of linear inequalities that describes the possible amounts of each investment. We shall assume that x and y are in thousands of dollars.

(b) Graph the system.

Solution

(a) The system of linear inequalities is

$$\begin{cases} x \geq 0 & \text{\textit{x} and \textit{y} are nonnegative variables since they represent} \\ y \geq 0 & \text{money invested in thousands of dollars.} \\ x + y \leq 25 & \text{The total of the two investments, \textit{x} + \textit{y}, cannot} \\ & \text{exceed \$25,000.} \\ x \geq 15 & \text{At least \$15,000 in Treasury bills.} \\ y \leq 5 & \text{At most \$5000 in corporate bonds.} \end{cases}$$

Figure 41

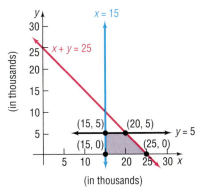

(b) See the shaded region in Figure 41. Note that the inequalities $x \geq 0$ and $y \geq 0$ again require that the graph of the system be in quadrant I.   ◀

The graph of the system of linear inequalities in Figure 41 is said to be **bounded**, because it can be contained within some circle of sufficiently large radius. A graph that cannot be contained in any circle is said to be **unbounded**. For example, the graph of the system of linear inequalities in Figure 40 is unbounded, since it extends indefinitely in a particular direction.

Notice in Figures 40 and 41 that those points belonging to the graph that are also points of intersection of boundary lines have been plotted. Such points are referred to as **vertices** or **corner points** of the graph. The system graphed in Figure 31 has three corner points: $(0, 4)$, $(1, 2)$, and $(3, 0)$. The system graphed in Figure 32 has four corner points: $(15, 0)$, $(25, 0)$, $(20, 5)$, and $(15, 5)$.

These ideas will be used in the next section in developing a method for solving linear programming problems, an important application of linear inequalities.

NOW WORK PROBLEM **45.**

10.7 Assess Your Understanding

'Are You Prepared?'

Answers are given at the end of these exercises. If you get a wrong answer, read the pages listed in red.

1. Solve the inequality: $3x + 4 < 8 - x$. (pp. 732–733)

2. Graph the equation: $3x - 2y = 6$. (pp. 34–36)

3. Graph the equation: $x^2 + y^2 = 9$. (pp. 44–47)

4. Graph the equation: $y = x^2 + 4$. (pp. 118–119)

5. *True or False:* The lines $2x + y = 4$ and $4x + 2y = 0$ are parallel. (pp. 36–38)

6. The graph of $y = (x - 2)^2$ may be obtained by shifting the graph of _____ to the (left/right) a distance of _____ units. (pp. 118–120)

Concepts and Vocabulary

7. An inequality in two variables x and y is _____ by an ordered pair (a, b) if, when x is replaced by a and y by b, a true statement results.

8. The graph of a linear inequality is called a(n) _____.

9. *True or False:* The graph of a linear inequality is a line.

10. *True or False:* The graph of a system of linear inequalities is sometimes unbounded.

Skill Building

In Problems 11–22, graph each inequality (a) by hand and (b) by using a graphing utility.

11. $x \geq 0$

12. $y \geq 0$

13. $x \geq 4$

14. $y \leq 2$

15. $2x + y \geq 6$

16. $3x + 2y \leq 6$

17. $x^2 + y^2 > 1$

18. $x^2 + y^2 \leq 9$

19. $y \leq x^2 - 1$

20. $y > x^2 + 2$

21. $xy \geq 4$

22. $xy \leq 1$

In Problems 23–42, graph each system of inequalities by hand.

23. $\begin{cases} x + y \leq 2 \\ 2x + y \geq 4 \end{cases}$

24. $\begin{cases} 3x - y \geq 6 \\ x + 2y \leq 2 \end{cases}$

25. $\begin{cases} 2x - y \leq 4 \\ 3x + 2y \geq -6 \end{cases}$

26. $\begin{cases} 4x - 5y \leq 0 \\ 2x - y \geq 2 \end{cases}$

27. $\begin{cases} 2x - 3y \leq 0 \\ 3x + 2y \leq 6 \end{cases}$

28. $\begin{cases} 4x - y \geq 2 \\ x + 2y \geq 2 \end{cases}$

29. $\begin{cases} x - 2y \leq 6 \\ 2x - 4y \geq 0 \end{cases}$

30. $\begin{cases} x + 4y \leq 8 \\ x + 4y \geq 4 \end{cases}$

31. $\begin{cases} 2x + y \geq -2 \\ 2x + y \geq 2 \end{cases}$

32. $\begin{cases} x - 4y \leq 4 \\ x - 4y \geq 0 \end{cases}$

33. $\begin{cases} 2x + 3y \geq 6 \\ 2x + 3y \leq 0 \end{cases}$

34. $\begin{cases} 2x + y \geq 0 \\ 2x + y \geq 2 \end{cases}$

35. $\begin{cases} x^2 + y^2 \leq 9 \\ x + y \geq 3 \end{cases}$

36. $\begin{cases} x^2 + y^2 \geq 9 \\ x + y \leq 3 \end{cases}$

37. $\begin{cases} y \geq x^2 - 4 \\ y \leq x - 2 \end{cases}$

38. $\begin{cases} y^2 \leq x \\ y \geq x \end{cases}$

39. $\begin{cases} x^2 + y^2 \leq 16 \\ y \geq x^2 - 4 \end{cases}$

40. $\begin{cases} x^2 + y^2 \leq 25 \\ y \leq x^2 - 5 \end{cases}$

41. $\begin{cases} xy \geq 4 \\ y \geq x^2 + 1 \end{cases}$

42. $\begin{cases} y + x^2 \leq 1 \\ y \geq x^2 - 1 \end{cases}$

In Problems 43–52, graph each system of linear inequalities. Tell whether the graph is bounded or unbounded, and label the corner points.

43. $\begin{cases} x \geq 0 \\ y \geq 0 \\ 2x + y \leq 6 \\ x + 2y \leq 6 \end{cases}$

44. $\begin{cases} x \geq 0 \\ y \geq 0 \\ x + y \geq 4 \\ 2x + 3y \geq 6 \end{cases}$

45. $\begin{cases} x \geq 0 \\ y \geq 0 \\ x + y \geq 2 \\ 2x + y \geq 4 \end{cases}$

46. $\begin{cases} x \geq 0 \\ y \geq 0 \\ 3x + y \leq 6 \\ 2x + y \leq 2 \end{cases}$

47. $\begin{cases} x \geq 0 \\ y \geq 0 \\ x + y \geq 2 \\ 2x + 3y \leq 12 \\ 3x + y \leq 12 \end{cases}$

48. $\begin{cases} x \geq 0 \\ y \geq 0 \\ x + y \geq 2 \\ x + y \leq 10 \\ 2x + y \leq 3 \end{cases}$
49. $\begin{cases} x \geq 0 \\ y \geq 0 \\ x + y \geq 2 \\ x + y \leq 8 \\ 2x + y \leq 10 \end{cases}$
50. $\begin{cases} x \geq 0 \\ y \geq 0 \\ x + y \geq 2 \\ x + y \leq 8 \\ x + 2y \geq 1 \end{cases}$
51. $\begin{cases} x \geq 0 \\ y \geq 0 \\ x + 2y \geq 1 \\ x + 2y \leq 10 \end{cases}$
52. $\begin{cases} x \geq 0 \\ y \geq 0 \\ x + 2y \geq 1 \\ x + 2y \leq 10 \\ x + y \geq 2 \\ x + y \leq 8 \end{cases}$

In Problems 53–56, write a system of linear inequalities that has the given graph.

53.

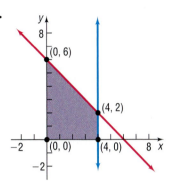

54.

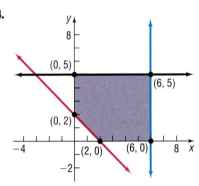

55.

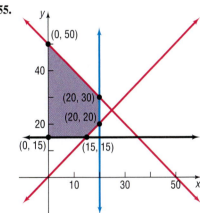

56.

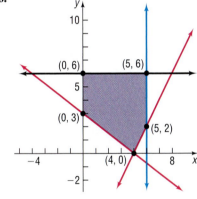

Applications and Extensions

57. Financial Planning A retired couple has up to $50,000 to invest. As their financial adviser, you recommend that they place at least $35,000 in Treasury bills yielding 7% and at most $10,000 in corporate bonds yielding 10%.
 (a) Using x to denote the amount of money invested in Treasury bills and y the amount invested in corporate bonds, write a system of linear inequalities that describes the possible amounts of each investment.
 (b) Graph the system and label the corner points.

58. Manufacturing Trucks Mike's Toy Truck Company manufacturers two models of toy trucks, a standard model and a deluxe model. Each standard model requires 2 hours for painting and 3 hours for detail work; each deluxe model requires 3 hours for painting and 4 hours for detail work. Two painters and three detail workers are employed by the company, and each works 40 hours per week.

 (a) Using x to denote the number of standard model trucks and y to denote the number of deluxe model trucks, write a system of linear inequalities that describes the possible number of each model of truck that can be manufactured in a week.
 (b) Graph the system and label the corner points.

59. Blending Coffee Bill's Coffee House, a store that specializes in coffee, has available 75 pounds of *A* grade coffee and 120 pounds of *B* grade coffee. These will be blended into 1 pound packages as follows: An economy blend that contains 4 ounces of *A* grade coffee and 12 ounces of *B* grade coffee and a superior blend that contains 8 ounces of *A* grade coffee and 8 ounces of *B* grade coffee.

(a) Using *x* to denote the number of packages of the economy blend and *y* to denote the number of packages of the superior blend, write a system of linear inequalities that describes the possible number of packages of each kind of blend.

(b) Graph the system and label the corner points.

60. Mixed Nuts Nola's Nuts, a store that specializes in selling nuts, has available 90 pounds of cashews and 120 pounds of peanuts. These are to be mixed in 12-ounce packages as follows: a lower-priced package containing 8 ounces of peanuts and 4 ounces of cashews and a quality package containing 6 ounces of peanuts and 6 ounces of cashews.

(a) Use *x* to denote the number of lower-priced packages and use *y* to denote the number of quality packages. Write a system of linear inequalities that describes the possible number of each kind of package.

(b) Graph the system and label the corner points.

61. Transporting Goods A small truck can carry no more than 1600 pounds of cargo nor more than 150 cubic feet of cargo. A printer weighs 20 pounds and occupies 3 cubic feet of space. A microwave oven weighs 30 pounds and occupies 2 cubic feet of space.

(a) Using *x* to represent the number of microwave ovens and *y* to represent the number of printers, write a system of linear inequalities that describes the number of ovens and printers that can be hauled by the truck.

(b) Graph the system and label the corner points.

'Are You Prepared?' Answers

1. $\{x \mid x < 1\}$

2.

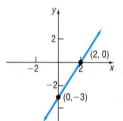

3.

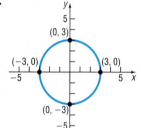

4.

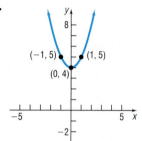

5. True **6.** $y = x^2$; right; 2

Sequences; Induction; the Binomial Theorem

11

The Future of the World Population

Chapter 11 in *Precalculus Essentials: Enhanced with Graphing Utilities*, 4e includes only the highlighted Sections 11.1, 11.2, and 11.3.

A LOOK BACK, A LOOK AHEAD This chapter may be divided into three independent parts: Sections 11.1–11.3, Section 11.4, and Section 11.5.

In Chapter 2, we defined a function and its domain, which was usually some set of real numbers. In Sections 11.1–11.3, we discuss sequences, which are functions whose domain is the set of positive integers.

Throughout this text, where it seemed appropriate, we have given proofs of many of the results. In Section 11.4, a technique for proving theorems involving natural numbers is discussed.

In the Appendix, Section A.3, we gave formulas for expanding $(x + a)^2$ and $(x + a)^3$. In Section 11.5, we discuss the Binomial Theorem, a formula for the expansion of $(x + a)^n$, where n is a positive integer.

The topics introduced in this chapter are covered in more detail in courses titled *Discrete Mathematics*. Applications of these topics can be found in the fields of computer science, engineering, business and economics, the social sciences, and the physical and biological sciences.

OUTLINE

11.1 Sequences

PREPARING FOR THIS SECTION   *Before getting started, review the following concept:*

• Functions (Section 2.1, pp. 56–63)   • Compound Interest (Section 4.7, pp. 315–322)

 Now work the 'Are You Prepared?' problems on page 635.

OBJECTIVES   1 Write the First Several Terms of a Sequence
            2 Write the Terms of a Sequence Defined by a Recursive Formula
            3 Use Summation Notation
            4 Find the Sum of a Sequence Algebraically and Using a Graphing Utility
            5 Solve Annuity and Amortization Problems

A **sequence** is a function whose domain is the set of positive integers.

Because a sequence is a function, it will have a graph. In Figure 1(a), we have the graph of the function $f(x) = \dfrac{1}{x}, x > 0$. If all the points on this graph were removed except those whose x-coordinates are positive integers, that is, if all points were removed except $(1, 1), \left(2, \dfrac{1}{2}\right), \left(3, \dfrac{1}{3}\right)$, and so on, the remaining points would be the graph of the sequence $f(n) = \dfrac{1}{n}$, as shown in Figure 1(b). Notice that we use n to represent the independent variable in a sequence. This serves to remind us that n is a positive integer.

Figure 1

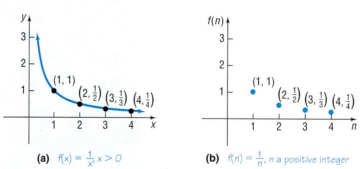

(a) $f(x) = \frac{1}{x}, x > 0$ (b) $f(n) = \frac{1}{n}, n$ a positive integer

1 Write the First Several Terms of a Sequence

A sequence is usually represented by listing its values in order. For example, the sequence whose graph is given in Figure 1(b) might be represented as

$$f(1), f(2), f(3), f(4), \ldots \quad \text{or} \quad 1, \frac{1}{2}, \frac{1}{3}, \frac{1}{4}, \ldots$$

The list never ends, as the ellipsis indicates. The numbers in this ordered list are called the **terms** of the sequence.

In dealing with sequences, we usually use subscripted letters, such as a_1, to represent the first term, a_2 for the second term, a_3 for the third term, and so on. For the sequence $f(n) = \dfrac{1}{n}$, we write

$$a_1 = f(1) = 1, \quad a_2 = f(2) = \frac{1}{2}, \quad a_3 = f(3) = \frac{1}{3}, \quad a_4 = f(4) = \frac{1}{4}, \ldots, \quad a_n = f(n) = \frac{1}{n}, \ldots$$

In other words, we usually do not use the traditional function notation $f(n)$ for sequences. For this particular sequence, we have a rule for the nth term, which is $a_n = \frac{1}{n}$, so it is easy to find any term of the sequence.

When a formula for the nth term (sometimes called the **general term**) of a sequence is known, rather than write out the terms of the sequence, we usually represent the entire sequence by placing braces around the formula for the nth term. For example, the sequence whose nth term is $b_n = \left(\frac{1}{2}\right)^n$ may be represented as

$$\{b_n\} = \left\{\left(\frac{1}{2}\right)^n\right\}$$

or by

$$b_1 = \frac{1}{2}, \quad b_2 = \frac{1}{4}, \quad b_3 = \frac{1}{8}, \ldots, \quad b_n = \left(\frac{1}{2}\right)^n, \ldots$$

EXAMPLE 1 Writing the First Several Terms of a Sequence

Write down the first six terms of the following sequence and graph it.

$$\{a_n\} = \left\{\frac{n-1}{n}\right\}$$

Algebraic Solution

The first six terms of the sequence are

$$a_1 = 0, \quad a_2 = \frac{1}{2}, \quad a_3 = \frac{2}{3},$$
$$a_4 = \frac{3}{4}, \quad a_5 = \frac{4}{5}, \quad a_6 = \frac{5}{6}$$

See Figure 2 for the graph.

Figure 2

a_n
1.0
0.8
0.6
0.4
0.2
(1, 0)
(2, $\frac{1}{2}$)
(3, $\frac{2}{3}$)
(4, $\frac{3}{4}$)
(5, $\frac{4}{5}$)
(6, $\frac{5}{6}$)
1 2 3 4 5 6
n

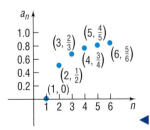

Graphing Solution

Figure 3 shows the sequence generated on a TI-84 Plus graphing calculator. We can see the first few terms of the sequence on the screen. You need to press the right arrow key to scroll right to see the remaining terms of the sequence.

Figure 3

seq((X-1)/X,X,1,
6,1)
{0 .5 .66666666...
Ans▸Frac
{0 1/2 2/3 3/4 ...

We could also obtain the terms of the sequence using the TABLE feature. First, put the graphing utility in SEQuence mode. Using $Y=$, enter the formula for the sequence into the graphing utility. See Figure 4. Set up the table with TblStart = 1 and ΔTbl = 1. See Table 1. Finally, we can graph the sequence. See Figure 5. Notice that the first term of the sequence is not visible since it lies on the x-axis. TRACEing the graph will allow you to determine the terms of the sequence.

Figure 4

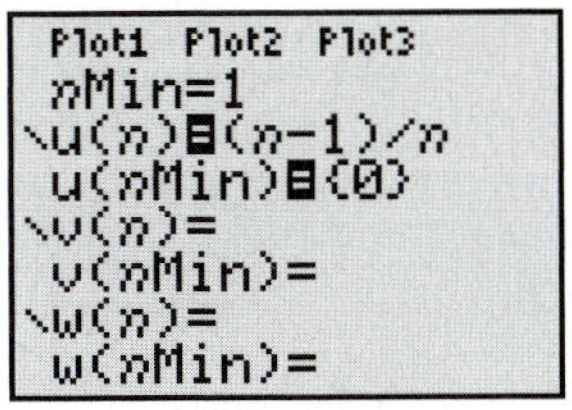

Plot1 Plot2 Plot3
nMin=1
\u(n)■(n-1)/n
u(nMin)■{0}
\v(n)=
v(nMin)=
\w(n)=
w(nMin)=

Table 1

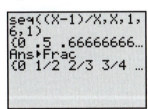

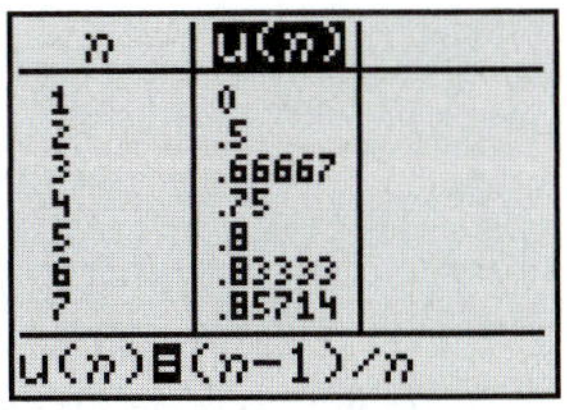

n	u(n)
1	0
2	.5
3	.66667
4	.75
5	.8
6	.83333
7	.85714

u(n)■(n-1)/n

Figure 5

1
0
0
7

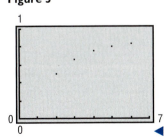

NOW WORK PROBLEM 19.

We will usually provide solutions done by hand. The reader is encouraged to verify solutions using a graphing utility.

Figure 6

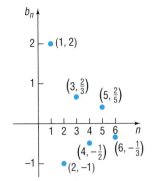

| EXAMPLE 2 | **Writing the First Several Terms of a Sequence** |

Write down the first six terms of the following sequence and graph it.

$$\{b_n\} = \left\{(-1)^{n+1}\left(\frac{2}{n}\right)\right\}$$

Solution   The first six terms of the sequence are

$$b_1 = 2, \quad b_2 = -1, \quad b_3 = \frac{2}{3}, \quad b_4 = -\frac{1}{2}, \quad b_5 = \frac{2}{5}, \quad b_6 = -\frac{1}{3}$$

See Figure 6 for the graph.   ◀

Notice in the sequence $\{b_n\}$ in Example 2 that the signs of the terms **alternate**. When this occurs, we use factors such as $(-1)^{n+1}$, which equals 1 if n is odd and -1 if n is even, or $(-1)^n$, which equals -1 if n is odd and 1 if n is even.

Figure 7

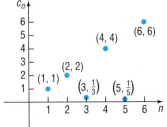

| EXAMPLE 3 | **Writing the First Several Terms of a Sequence** |

Write down the first six terms of the following sequence and graph it.

$$\{c_n\} = \begin{cases} n & \text{if } n \text{ is even} \\ \dfrac{1}{n} & \text{if } n \text{ is odd} \end{cases}$$

Solution   The first six terms of the sequence are

$$c_1 = 1, \quad c_2 = 2, \quad c_3 = \frac{1}{3}, \quad c_4 = 4, \quad c_5 = \frac{1}{5}, \quad c_6 = 6$$

See Figure 7 for the graph.   ◀

✏ **NOW WORK PROBLEM 21.**

Sometimes a sequence is indicated by an observed pattern in the first few terms that makes it possible to infer the makeup of the nth term. In the example that follows, a sufficient number of terms of the sequence is given so that a natural choice for the nth term is suggested.

| EXAMPLE 4 | **Determining a Sequence from a Pattern** |

(a) $e, \dfrac{e^2}{2}, \dfrac{e^3}{3}, \dfrac{e^4}{4}, \dots$ $\qquad a_n = \dfrac{e^n}{n}$

(b) $1, \dfrac{1}{3}, \dfrac{1}{9}, \dfrac{1}{27}, \dots$ $\qquad b_n = \dfrac{1}{3^{n-1}}$

(c) $1, 3, 5, 7, \dots$ $\qquad c_n = 2n - 1$

(d) $1, 4, 9, 16, 25, \dots$ $\qquad d_n = n^2$

(e) $1, -\dfrac{1}{2}, \dfrac{1}{3}, -\dfrac{1}{4}, \dfrac{1}{5}, \dots$ $\qquad e_n = (-1)^{n+1}\left(\dfrac{1}{n}\right)$

◀

✏ **NOW WORK PROBLEM 29.**

The Factorial Symbol

If $n \geq 0$ is an integer, the **factorial symbol** $n!$ is defined as follows:

$$0! = 1 \qquad 1! = 1$$
$$n! = n(n - 1) \cdot \ldots \cdot 3 \cdot 2 \cdot 1 \qquad \text{if } n \geq 2$$

For example, $2! = 2 \cdot 1 = 2$, $3! = 3 \cdot 2 \cdot 1 = 6$, $4! = 4 \cdot 3 \cdot 2 \cdot 1 = 24$, and so on. Table 2 lists the values of $n!$ for $0 \leq n \leq 6$.

Because

Table 2

n	0	1	2	3	4	5	6
$n!$	1	1	2	6	24	120	720

$$n! = \underbrace{n(n-1)(n-2) \cdot \ldots \cdot 3 \cdot 2 \cdot 1}_{(n-1)!}$$

we can use the formula

$$n! = n(n-1)!$$

NOTE

Your calculator has a factorial key. Use it to see how fast factorials increase in value. Find the value of 69!. What happens when you try to find 70!? In fact, 70! is larger than 10^{100} (a **googol**), the largest number most calculators can display. ∎

to find successive factorials. For example, because $6! = 720$, we have

$$7! = 7 \cdot 6! = 7(720) = 5040$$

and

$$8! = 8 \cdot 7! = 8(5040) = 40{,}320$$

2 Write the Terms of a Sequence Defined by a Recursive Formula

A second way of defining a sequence is to assign a value to the first (or the first few) term(s) and specify the nth term by a formula or equation that involves one or more of the terms preceding it. Sequences defined this way are said to be defined **recursively**, and the rule or formula is called a **recursive formula**.

EXAMPLE 5   **Writing the Terms of a Recursively Defined Sequence**

Write down the first five terms of the following recursively defined sequence.

$$s_1 = 1, \qquad s_n = n s_{n-1}$$

Algebraic Solution

The first term is given as $s_1 = 1$. To get the second term, we use $n = 2$ in the formula $s_n = n s_{n-1}$ to get $s_2 = 2s_1 = 2 \cdot 1 = 2$. To get the third term, we use $n = 3$ in the formula to get $s_3 = 3s_2 = 3 \cdot 2 = 6$. To get a new term requires that we know the value of the preceding term. The first five terms are

$$s_1 = 1$$
$$s_2 = 2 \cdot 1 = 2$$
$$s_3 = 3 \cdot 2 = 6$$
$$s_4 = 4 \cdot 6 = 24$$
$$s_5 = 5 \cdot 24 = 120$$

Do you recognize this sequence? $s_n = n!$ ◀

Graphing Solution

First, put the graphing utility into SEQuence mode. Using $Y =$, enter the recursive formula into the graphing utility. See Figure 8(a). Next, set up the viewing window to generate the desired sequence. Finally, graph the recursion relation and use TRACE to determine the terms in the sequence. See Figure 8(b). For example, we see that the fourth term of the sequence is 24. Table 3 also shows the terms of the sequence.

Figure 8

(a)

Table 3

n	$u(n)$
1	1
2	2
3	6
4	24
5	120
6	720
7	5040

$u(n) = nu(n-1)$

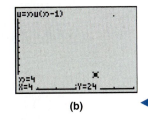

(b)

EXAMPLE 6	**Writing the Terms of a Recursively Defined Sequence**

Write down the first five terms of the following recursively defined sequence.

$$u_1 = 1, \qquad u_2 = 1, \qquad u_{n+2} = u_n + u_{n+1}$$

Solution We are given the first two terms. To get the third term requires that we know each of the previous two terms. That is,

$$u_1 = 1$$
$$u_2 = 1$$
$$u_3 = u_1 + u_2 = 1 + 1 = 2$$
$$u_4 = u_2 + u_3 = 1 + 2 = 3$$
$$u_5 = u_3 + u_4 = 2 + 3 = 5 \qquad \blacktriangleleft$$

The sequence defined in Example 6 is called a **Fibonacci sequence**, and the terms of this sequence are called **Fibonacci numbers**. These numbers appear in a wide variety of applications (see Problems 91–94).

> ✎ **NOW WORK PROBLEMS 37 AND 45.**

3 Use Summation Notation

It is often important to be able to find the sum of the first n terms of a sequence $\{a_n\}$, that is,

$$a_1 + a_2 + a_3 + \cdots + a_n$$

Rather than write down all these terms, we introduce a more concise way to express the sum, called **summation notation**. Using summation notation, we would write the sum as

$$a_1 + a_2 + a_3 + \cdots + a_n = \sum_{k=1}^{n} a_k$$

The symbol Σ (the Greek letter sigma, which is an S in our alphabet) is simply an instruction to sum, or add up, the terms. The integer k is called the **index** of the sum; it tells you where to start the sum and where to end it. The expression

$$\sum_{k=1}^{n} a_k$$

is an instruction to add the terms a_k of the sequence $\{a_n\}$ starting with $k = 1$ and ending with $k = n$. We read the expression as "the sum of a_k from $k = 1$ to $k = n$."

EXAMPLE 7	**Expanding Summation Notation**

Write out each sum.

(a) $\displaystyle\sum_{k=1}^{n} \frac{1}{k}$

(b) $\displaystyle\sum_{k=1}^{n} k!$

Solution (a) $\displaystyle\sum_{k=1}^{n} \frac{1}{k} = 1 + \frac{1}{2} + \frac{1}{3} + \cdots + \frac{1}{n}$

(b) $\displaystyle\sum_{k=1}^{n} k! = 1! + 2! + \cdots + n!$ $\qquad \blacktriangleleft$

EXAMPLE 8	**Writing a Sum in Summation Notation**

Express each sum using summation notation.

(a) $1^2 + 2^2 + 3^2 + \cdots + 9^2$

(b) $1 + \dfrac{1}{2} + \dfrac{1}{4} + \dfrac{1}{8} + \cdots + \dfrac{1}{2^{n-1}}$

Solution   (a) The sum $1^2 + 2^2 + 3^2 + \cdots + 9^2$ has 9 terms, each of the form k^2, and starts at $k = 1$ and ends at $k = 9$:

$$1^2 + 2^2 + 3^2 + \cdots + 9^2 = \sum_{k=1}^{9} k^2$$

(b) The sum

$$1 + \frac{1}{2} + \frac{1}{4} + \frac{1}{8} + \cdots + \frac{1}{2^{n-1}}$$

has n terms, each of the form $\dfrac{1}{2^{k-1}}$, and starts at $k = 1$ and ends at $k = n$:

$$1 + \frac{1}{2} + \frac{1}{4} + \frac{1}{8} + \cdots + \frac{1}{2^{n-1}} = \sum_{k=1}^{n} \frac{1}{2^{k-1}}$$ ◀

The index of summation need not always begin at 1 or end at n; for example, we could have expressed the sum in Example 8(b) as

$$\sum_{k=0}^{n-1} \frac{1}{2^k} = 1 + \frac{1}{2} + \frac{1}{4} + \cdots + \frac{1}{2^{n-1}}$$

Letters other than k may be used as the index. For example,

$$\sum_{j=1}^{n} j! \quad \text{and} \quad \sum_{i=1}^{n} i!$$

each represent the same sum as the one given in Example 7(b).

NOW WORK PROBLEMS 53 AND 63.

4   Find the Sum of a Sequence Algebraically and Using a Graphing Utility

Next we list some properties of sequences using summation notation. These properties are useful for adding the terms of a sequence algebraically.

Theorem   **Properties of Sequences**

If $\{a_n\}$ and $\{b_n\}$ are two sequences and c is a real number, then:

$$\sum_{k=1}^{n} c = \underbrace{c + c + \cdots + c}_{n \text{ terms}} = cn \tag{1}$$

$$\sum_{k=1}^{n} (ca_k) = ca_1 + ca_2 + \cdots + ca_n = c(a_1 + a_2 + \cdots + a_n) = c\sum_{k=1}^{n} a_k \tag{2}$$

$$\sum_{k=1}^{n} (a_k + b_k) = \sum_{k=1}^{n} a_k + \sum_{k=1}^{n} b_k \tag{3}$$

$$\sum_{k=1}^{n} (a_k - b_k) = \sum_{k=1}^{n} a_k - \sum_{k=1}^{n} b_k \tag{4}$$

$$\sum_{k=1}^{n} a_k = \sum_{k=1}^{j} a_k + \sum_{k=j+1}^{n} a_k, \quad \text{where } 0 < j < n \tag{5}$$

$$\sum_{k=1}^{n} k = 1 + 2 + 3 + \cdots + n = \frac{n(n+1)}{2} \tag{6}$$

$$\sum_{k=1}^{n} k^2 = 1^2 + 2^2 + 3^2 + \cdots + n^2 = \frac{n(n+1)(2n+1)}{6} \tag{7}$$

$$\sum_{k=1}^{n} k^3 = 1^3 + 2^3 + 3^3 + \cdots + n^3 = \left[\frac{n(n+1)}{2}\right]^2 \tag{8}$$

We shall not prove these properties. The proofs of (1) through (5) are based on properties of real numbers; the proofs of (7) and (8) require mathematical induction, which is discussed in Section 11.4. See Problem 97 for a derivation of (6).

EXAMPLE 9   **Finding the Sum of a Sequence**

Find the sum of each sequence.

(a) $\displaystyle\sum_{k=1}^{5}(3k)$ (b) $\displaystyle\sum_{k=1}^{3}(k^3+1)$ (c) $\displaystyle\sum_{k=1}^{4}(k^2-7k+2)$

Algebraic Solution

(a) $\displaystyle\sum_{k=1}^{5}(3k) = 3\sum_{k=1}^{5}k$ Property (2)

$= 3\left(\dfrac{5(5+1)}{2}\right)$ Property (6)

$= 3(15)$

$= 45$

(b) $\displaystyle\sum_{k=1}^{3}(k^3+1) = \sum_{k=1}^{3}k^3 + \sum_{k=1}^{3}1$ Property (3)

$= \left(\dfrac{3(3+1)}{2}\right)^2 + 1(3)$ Properties (1) and (8)

$= 36 + 3$

$= 39$

(c) $\displaystyle\sum_{k=1}^{4}(k^2-7k+2) = \sum_{k=1}^{4}k^2 - \sum_{k=1}^{4}(7k) + \sum_{k=1}^{4}2$ Properties (3), (4)

$= \displaystyle\sum_{k=1}^{4}k^2 - 7\sum_{k=1}^{4}k + \sum_{k=1}^{4}2$ Property (2)

$= \dfrac{4(4+1)(2\cdot4+1)}{6} - 7\left(\dfrac{4(4+1)}{2}\right) + 2(4)$ Properties (1), (6), (7)

$= 30 - 70 + 8$

$= -32$

Graphing Solution

(a) Figure 9 shows the solution using a TI-84 Plus graphing calculator.

Figure 9

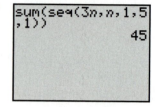

So, $\displaystyle\sum_{k=1}^{5}(3k) = 45$

(b) Figure 10 shows the solution using a TI-84 Plus graphing calculator.

Figure 10

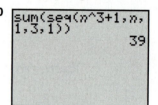

So, $\displaystyle\sum_{k=1}^{3}(k^3+1) = 39$

(c) Figure 11 shows the solution using a TI-84 Plus graphing calculator.

Figure 11

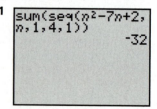

So, $\displaystyle\sum_{k=1}^{4}(k^2-7k+2) = -32$

◀

NOW WORK PROBLEM 75.

5   Solve Annuity and Amortization Problems

In Section 4.7 we developed the compound interest formula, which gives the future value when a fixed amount of money is deposited in an account that pays interest compounded periodically. Often, though, money is invested in small amounts at periodic intervals. An **annuity** is a sequence of equal periodic deposits. The periodic deposits may be made annually, quarterly, monthly, or daily.

When deposits are made at the same time that the interest is credited, the annuity is called **ordinary**. We will only deal with ordinary annuities here. The **amount of an annuity** is the sum of all deposits made plus all interest paid.

Suppose that the initial amount deposited in an annuity is $\$M$, the periodic deposit is $\$P$, and the per annum rate of interest is $r\%$ (expressed as a decimal) compounded N times per year. The periodic deposit is made at the same time that the interest is credited, so N deposits are made per year. The amount A_n of the annuity after n deposits will equal A_{n-1}, the amount of the annuity after $n - 1$ deposits, plus the interest earned on this amount, plus P, the periodic deposit. That is,

$$A_n = A_{n-1} + \frac{r}{N} A_{n-1} + P = \left(1 + \frac{r}{N}\right) A_{n-1} + P$$

$$\uparrow \qquad \uparrow \qquad \uparrow \qquad \uparrow$$

Amount   Amount   Interest   Periodic
after   in previous   earned   deposit
n deposits   period

We have established the following result:

Theorem   **Annuity Formula**

If $A_0 = M$ represents the initial amount deposited in an annuity that earns $r\%$ per annum compounded N times per year, and if P is the periodic deposit made at each payment period, then the amount A_n of the annuity after n deposits is given by the recursive sequence

$$A_0 = M, \qquad A_n = \left(1 + \frac{r}{N}\right) A_{n-1} + P, \qquad n \geq 1 \qquad \textbf{(9)}$$

Formula (9) may be explained as follows: the money in the account initially, A_0, is $\$M$; the money in the account after $n - 1$ payments, A_{n-1}, earns interest $\dfrac{r}{N}$ during the nth period; so when the periodic payment of P dollars is added, the amount after n payments, A_n, is obtained.

EXAMPLE 10   **Saving for Spring Break**

A trip to Cancun during spring break will cost $450 and full payment is due March 2. To have the money, a student, on September 1, deposits $100 in a savings account that pays 4% per annum compounded monthly. On the first of each month, the student deposits $50 in this account.

(a) Find a recursive sequence that explains how much is in the account after n months.

(b) Use the TABLE feature to list the amounts of the annuity for the first 6 months.

(c) After the deposit on March 1 is made, is there enough in the account to pay for the Cancun trip?

(d) If the student deposits $60 each month, will there be enough for the trip after the March 1 deposit?

Solution   (a) The initial amount deposited in the account is $A_0 = \$100$. The monthly deposit is $P = \$50$, and the per annum rate of interest is $r = 0.04$ compounded $N = 12$

Table 4

n	$u(n)$
0	100
1	150.33
2	200.83
3	251.5
4	302.34
5	353.35
6	404.53

$u(n)\blacksquare(1+.04/12)...$

Table 5

n	$u(n)$
0	100
1	160.33
2	220.87
3	281.6
4	342.54
5	403.68
6	465.03

$u(n)\blacksquare(1+.04/12)...$

times per year. The amount A_n in the account after n monthly deposits is given by the recursive sequence

$$A_0 = 100, \qquad A_n = \left(1 + \frac{r}{N}\right)A_{n-1} + P = \left(1 + \frac{0.04}{12}\right)A_{n-1} + 50$$

(b) In SEQuence mode on a TI-84 Plus, enter the sequence $\{A_n\}$ and create Table 4. On September 1 ($n = 0$), there is $100 in the account. After the first payment on October 1, the value of the account is $150.33. After the second payment on November 1, the value of the account is $200.83. After the third payment on December 1, the value of the account is $251.50, and so on.

(c) On March 1 ($n = 6$), there is only $404.53, not enough to pay for the trip to Cancun.

(d) If the periodic deposit, P, is $60, then on March 1, there is $465.03 in the account, enough for the trip. See Table 5. ◀

Recursive sequences can also be used to compute information about loans. When equal periodic payments are made to pay off a loan, the loan is said to be **amortized**.

Theorem

Amortization Formula

If $B is borrowed at an interest rate of r% (expressed as a decimal) per annum compounded monthly, the balance A_n due after n monthly payments of $P is given by the recursive sequence

$$A_0 = B, \qquad A_n = \left(1 + \frac{r}{12}\right)A_{n-1} - P, \qquad n \geq 1 \qquad \textbf{(10)}$$

Formula (10) may be explained as follows: The initial loan balance is $B. The balance due A_n after n payments will equal the balance due previously, A_{n-1}, plus the interest charged on that amount, reduced by the periodic payment P.

EXAMPLE 11 **Mortgage Payments**

Table 6

n	$u(n)$
0	180000
1	179852
2	179704
3	179555
4	179405
5	179254
6	179102

$u(n)\blacksquare(1+.07/12)...$

Table 7

n	$u(n)$
52	171067
53	170868
54	170667
55	170465
56	170262
57	170057
58	169852

$u(n)\blacksquare(1+.07/12)...$

John and Wanda borrowed $180,000 at 7% per annum compounded monthly for 30 years to purchase a home. Their monthly payment is determined to be $1197.54.

(a) Find a recursive formula that represents their balance after each payment of $1197.54 has been made.

(b) Determine their balance after the first payment is made.

(c) When will their balance be below $170,000?

Solution

(a) We use formula (10) with $A_0 = 180,000$, $r = 0.07$, and $P = $1197.54. Then

$$A_0 = 180,000 \qquad A_n = \left(1 + \frac{0.07}{12}\right)A_{n-1} - 1197.54$$

(b) In SEQuence mode on a TI-84 Plus, enter the sequence $\{A_n\}$ and create Table 6. After the first payment is made, the balance is $A_1 = $179,852.

(c) Scroll down until the balance is below $170,000. See Table 7. After the fifty-eighth payment is made ($n = 58$), the balance is below $170,000. ◀

NOW WORK PROBLEM 83.

11.1 Assess Your Understanding

'Are You Prepared?'

Answers are given at the end of these exercises. If you get a wrong answer, read the pages listed in red.

1. For the function $f(x) = \dfrac{x-1}{x}$, find $f(2)$ and $f(3)$. (pp. 61–63)

2. *True or False:* A function is a relation between two sets D and R so that each element x in the first set D is related to exactly one element y in the second set R. (pp. 56–61)

3. If \$1000 is invested at 4% per annum compounded semi-annually, how much is in the account after 2 years? (pp. 315–322)

4. How much do you need to invest now at 5% per annum compounded monthly so that in 1 year you will have \$10,000? (pp. 315–322)

Concepts and Vocabulary

5. A(n) _____ is a function whose domain is the set of positive integers.

6. For the sequence $\{s_n\} = \{4n - 1\}$, the first term is $s_1 = $ _____ and the fourth term is $s_4 = $ _____.

7. $\displaystyle\sum_{k=1}^{4}(2k) = $ _____.

8. *True or False:* Sequences are sometimes defined recursively.

9. *True or False:* A sequence is a function.

10. *True or False:* $\displaystyle\sum_{k=1}^{2}k = 3$

Skill Building

In Problems 11–16, evaluate each factorial expression. Verify your results using a graphing utility.

11. $10!$

12. $9!$

13. $\dfrac{9!}{6!}$

14. $\dfrac{12!}{10!}$

15. $\dfrac{3!7!}{4!}$

16. $\dfrac{5!8!}{3!}$

In Problems 17–28, write down the first five terms of each sequence.

17. $\{n\}$

18. $\{n^2 + 1\}$

19. $\left\{\dfrac{n}{n+2}\right\}$

20. $\left\{\dfrac{2n+1}{2n}\right\}$

21. $\{(-1)^{n+1}n^2\}$

22. $\left\{(-1)^{n-1}\left(\dfrac{n}{2n-1}\right)\right\}$

23. $\left\{\dfrac{2^n}{3^n+1}\right\}$

24. $\left\{\left(\dfrac{4}{3}\right)^n\right\}$

25. $\left\{\dfrac{(-1)^n}{(n+1)(n+2)}\right\}$

26. $\left\{\dfrac{3^n}{n}\right\}$

27. $\left\{\dfrac{n}{e^n}\right\}$

28. $\left\{\dfrac{n^2}{2^n}\right\}$

In Problems 29–36, the given pattern continues. Write down the nth term of each sequence suggested by the pattern.

29. $\dfrac{1}{2}, \dfrac{2}{3}, \dfrac{3}{4}, \dfrac{4}{5}, \dots$

30. $\dfrac{1}{1\cdot 2}, \dfrac{1}{2\cdot 3}, \dfrac{1}{3\cdot 4}, \dfrac{1}{4\cdot 5}, \dots$

31. $1, \dfrac{1}{2}, \dfrac{1}{4}, \dfrac{1}{8}, \dots$

32. $\dfrac{2}{3}, \dfrac{4}{9}, \dfrac{8}{27}, \dfrac{16}{81}, \dots$

33. $1, -1, 1, -1, 1, -1, \dots$

34. $1, \dfrac{1}{2}, 3, \dfrac{1}{4}, 5, \dfrac{1}{6}, 7, \dfrac{1}{8}, \dots$

35. $1, -2, 3, -4, 5, -6, \dots$

36. $2, -4, 6, -8, 10, \dots$

In Problems 37–50, a sequence is defined recursively. Write the first five terms.

37. $a_1 = 2$; $a_n = 3 + a_{n-1}$

38. $a_1 = 3$; $a_n = 4 - a_{n-1}$

39. $a_1 = -2$; $a_n = n + a_{n-1}$

40. $a_1 = 1$; $a_n = n - a_{n-1}$

41. $a_1 = 5$; $a_n = 2a_{n-1}$

42. $a_1 = 2$; $a_n = -a_{n-1}$

43. $a_1 = 3$; $a_n = \dfrac{a_{n-1}}{n}$

44. $a_1 = -2$; $a_n = n + 3a_{n-1}$

45. $a_1 = 1$; $a_2 = 2$; $a_n = a_{n-1} \cdot a_{n-2}$

46. $a_1 = -1$; $a_2 = 1$; $a_n = a_{n-2} + na_{n-1}$

47. $a_1 = A$; $a_n = a_{n-1} + d$

48. $a_1 = A$; $a_n = ra_{n-1}$, $r \neq 0$

49. $a_1 = \sqrt{2}$; $a_n = \sqrt{2 + a_{n-1}}$

50. $a_1 = \sqrt{2}$; $a_n = \sqrt{\dfrac{a_{n-1}}{2}}$

In Problems 51–60, write out each sum.

51. $\displaystyle\sum_{k=1}^{n}(k+2)$

52. $\displaystyle\sum_{k=1}^{n}(2k+1)$

53. $\displaystyle\sum_{k=1}^{n}\dfrac{k^2}{2}$

54. $\displaystyle\sum_{k=1}^{n}(k+1)^2$

55. $\displaystyle\sum_{k=0}^{n}\dfrac{1}{3^k}$

56. $\sum_{k=0}^{n} \left(\frac{3}{2}\right)^k$ **57.** $\sum_{k=0}^{n-1} \frac{1}{3^{k+1}}$ **58.** $\sum_{k=0}^{n-1} (2k + 1)$ **59.** $\sum_{k=2}^{n} (-1)^k \ln k$ **60.** $\sum_{k=3}^{n} (-1)^{k+1} 2^k$

In Problems 61–70, express each sum using summation notation.

61. $1 + 2 + 3 + \cdots + 20$

62. $1^3 + 2^3 + 3^3 + \cdots + 8^3$

63. $\frac{1}{2} + \frac{2}{3} + \frac{3}{4} + \cdots + \frac{13}{13 + 1}$

64. $1 + 3 + 5 + 7 + \cdots + [2(12) - 1]$

65. $1 - \frac{1}{3} + \frac{1}{9} - \frac{1}{27} + \cdots + (-1)^6 \left(\frac{1}{3^6}\right)$

66. $\frac{2}{3} - \frac{4}{9} + \frac{8}{27} - \cdots + (-1)^{11+1} \left(\frac{2}{3}\right)^{11}$

67. $3 + \frac{3^2}{2} + \frac{3^3}{3} + \cdots + \frac{3^n}{n}$

68. $\frac{1}{e} + \frac{2}{e^2} + \frac{3}{e^3} + \cdots + \frac{n}{e^n}$

69. $a + (a + d) + (a + 2d) + \cdots + (a + nd)$

70. $a + ar + ar^2 + \cdots + ar^{n-1}$

In Problems 71–82, find the sum of each sequence (a) algebraically and (b) using a graphing utility.

71. $\sum_{k=1}^{10} 5$ **72.** $\sum_{k=1}^{20} 8$ **73.** $\sum_{k=1}^{6} k$ **74.** $\sum_{k=1}^{4} (-k)$

75. $\sum_{k=1}^{5} (5k + 3)$ **76.** $\sum_{k=1}^{6} (3k - 7)$ **77.** $\sum_{k=1}^{3} (k^2 + 4)$ **78.** $\sum_{k=0}^{4} (k^2 - 4)$

79. $\sum_{k=1}^{6} (-1)^k 2^k$ **80.** $\sum_{k=1}^{4} (-1)^k 3^k$ **81.** $\sum_{k=1}^{4} (k^3 - 1)$ **82.** $\sum_{k=0}^{3} (k^3 + 2)$

Applications and Extensions

83. Credit Card Debt John has a balance of $3000 on his Discover card that charges 1% interest per month on any unpaid balance. John can afford to pay $100 toward the balance each month. His balance each month after making a $100 payment is given by the recursively defined sequence

$$B_0 = \$3000, \qquad B_n = 1.01B_{n-1} - 100$$

(a) Determine John's balance after making the first payment. That is, determine B_1.
(b) Using a graphing utility, determine when John's balance will be below $2000. How many payments of $100 have been made?
(c) Using a graphing utility, determine when John will pay off the balance. What is the total of all the payments?
(d) What was John's interest expense?

84. Car Loans Phil bought a car by taking out a loan for $18,500 at 0.5% interest per month. Phil's normal monthly payment is $434.47 per month, but he decides that he can afford to pay $100 extra toward the balance each month. His balance each month is given by the recursively defined sequence

$$B_0 = \$18,500, \qquad B_n = 1.005B_{n-1} - 534.47$$

(a) Determine Phil's balance after making the first payment. That is, determine B_1.
(b) Using a graphing utility, determine when Phil's balance will be below $10,000. How many payments of $534.47 have been made?
(c) Using a graphing utility, determine when Phil will pay off the balance. What is the total of all the payments?
(d) What was Phil's interest expense?

85. Trout Population A pond currently has 2000 trout in it. A fish hatchery decides to add an additional 20 trout each month. In addition, it is known that the trout population is growing 3% per month. The size of the population after n months is given by the recursively defined sequence

$$p_0 = 2000, \qquad p_n = 1.03p_{n-1} + 20$$

(a) How many trout are in the pond at the end of the second month? That is, what is p_2?
(b) Using a graphing utility, determine how long it will be before the trout population reaches 5000.

86. Environmental Control The Environmental Protection Agency (EPA) determines that Maple Lake has 250 tons of pollutants as a result of industrial waste and that 10% of the pollutant present is neutralized by solar oxidation every year. The EPA imposes new pollution control laws that result in 15 tons of new pollutant entering the lake each year. The amount of pollutant in the lake at the end of each year is given by the recursively defined sequence

$$p_0 = 250, \qquad p_n = 0.9p_{n-1} + 15$$

(a) Determine the amount of pollutant in the lake at the end of the second year. That is, determine p_2.
(b) Using a graphing utility, provide pollutant amounts for the next 20 years.
(c) What is the equilibrium level of pollution in Maple Lake? That is, what is $\lim_{n \to \infty} p_n$?

87. Roth IRA On January 1, 1999, Bob decided to place $500 at the end of each quarter into a Roth Individual Retirement Account.

(a) Find a recursive formula that represents Bob's balance at the end of each quarter if the rate of return is assumed to be 8% per annum compounded quarterly.
(b) How long will it be before the value of the account exceeds $100,000?
(c) What will be the value of the account in 25 years when Bob retires?

88. **Education IRA**   On January 1, 1999, John's parents decided to place $45 at the end of each month into an Education IRA.
(a) Find a recursive formula that represents the balance at the end of each month if the rate of return is assumed to be 6% per annum compounded monthly.
(b) How long will it be before the value of the account exceeds $4000?
(c) What will be the value of the account in 16 years when John goes to college?

89. **Home Loan**   Bill and Laura borrowed $150,000 at 6% per annum compounded monthly for 30 years to purchase a home. Their monthly payment is determined to be $899.33.
(a) Find a recursive formula for their balance after each monthly payment has been made.
(b) Determine Bill and Laura's balance after the first payment.
(c) Using a graphing utility, create a table showing Bill and Laura's balance after each monthly payment.
(d) Using a graphing utility, determine when Bill and Laura's balance will be below $140,000.
(e) Using a graphing utility, determine when Bill and Laura will pay off the balance.
(f) Determine Bill and Laura's interest expense when the loan is paid.
(g) Suppose that Bill and Laura decide to pay an additional $100 each month on their loan. Answer parts (a) to (f) under this scenario.
(h) Is it worthwhile for Bill and Laura to pay the additional $100? Explain.

90. **Home Loan**   Jodi and Jeff borrowed $120,000 at 6.5% per annum compounded monthly for 30 years to purchase a home. Their monthly payment is determined to be $758.48.
(a) Find a recursive formula for their balance after each monthly payment has been made.
(b) Determine Jodi and Jeff's balance after the first payment.
(c) Using a graphing utility, create a table showing Jodi and Jeff's balance after each monthly payment.
(d) Using a graphing utility, determine when Jodi and Jeff's balance will be below $100,000.
(e) Using a graphing utility, determine when Jodi and Jeff will pay off the balance.
(f) Determine Jodi and Jeff's interest expense when the loan is paid.
(g) Suppose that Jodi and Jeff decide to pay an additional $100 each month on their loan. Answer parts (a) to (f) under this scenario.
(h) Is it worthwhile for Jodi and Jeff to pay the additional $100? Explain.

91. **Growth of a Rabbit Colony**   A colony of rabbits begins with one pair of mature rabbits, which will produce a pair of offspring (one male, one female) each month. Assume that all rabbits mature in 1 month and produce a pair of offspring (one male, one female) after 2 months. If no rabbits ever die, how many pairs of mature rabbits are there after 7 months?
[**Hint:** A Fibonacci sequence models this colony. Do you see why?]

1 mature pair	
1 mature pair	
2 mature pairs	
3 mature pairs	

92. **Fibonacci Sequence**   Let

$$u_n = \frac{\left(1 + \sqrt{5}\right)^n - \left(1 - \sqrt{5}\right)^n}{2^n \sqrt{5}}$$

define the nth term of a sequence.
(a) Show that $u_1 = 1$ and $u_2 = 1$.
(b) Show that $u_{n+2} = u_{n+1} + u_n$.
(c) Draw the conclusion that $\{u_n\}$ is a Fibonacci sequence.

93. **Pascal's Triangle**   Divide the triangular array shown (called Pascal's triangle) using diagonal lines as indicated. Find the sum of the numbers in each of these diagonal rows. Do you recognize this sequence?

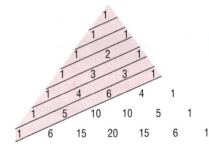

94. **Fibonacci Sequence**   Use the result of Problem 92 to do the following problems:
(a) Write the first 10 terms of the Fibonacci sequence.
(b) Compute the ratio $\dfrac{u_{n+1}}{u_n}$ for the first 10 terms.
(c) As n gets large, what number does the ratio approach? This number is referred to as the **golden ratio**. Rectangles whose sides are in this ratio were considered pleasing to the eye by the Greeks. For example, the facade of the Parthenon was constructed using the golden ratio.
(d) Compute the ratio $\dfrac{u_n}{u_{n+1}}$ for the first 10 terms.
(e) As n gets large, what number does the ratio approach? This number is also referred to as the **golden ratio**. This ratio is believed to have been used in the construction of the Great Pyramid in Egypt. The ratio equals the sum of the areas of the four face triangles divided by the total surface area of the Great Pyramid.

 **95. Approximating e^x**   In calculus, it can be shown that

$$e^x = \sum_{k=0}^{\infty} \frac{x^k}{k!}$$

We can approximate the value of e^x for any x using the following sum

$$e^x \approx \sum_{k=0}^{n} \frac{x^k}{k!}$$

for some n.
(a) Approximate $e^{1.3}$ with $n = 4$.
(b) Approximate $e^{1.3}$ with $n = 7$.
(c) Use a calculator to approximate the value of $e^{1.3}$.
(d) Using trial and error along with a graphing utility's SEQuence mode, determine the value of n required to approximate $e^{1.3}$ correct to eight decimal places.

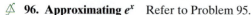 **96. Approximating e^x**   Refer to Problem 95.
(a) Approximate $e^{-2.4}$ with $n = 3$.
(b) Approximate $e^{-2.4}$ with $n = 6$.

(c) Use a calculator to approximate the value of $e^{-2.4}$.
(d) Using trial and error along with a graphing utility's SEQuence mode, determine the value of n required to approximate $e^{-2.4}$ correct to eight decimal places.

97. Show that

$$1 + 2 + \cdots + (n - 1) + n = \frac{n(n + 1)}{2}$$

[Hint: Let

$$S = 1 + 2 + \cdots + (n - 1) + n$$
$$S = n + (n - 1) + (n - 2) + \cdots + 1$$

Add these equations. Then

$$2S = \underbrace{[1 + n] + [2 + (n - 1)] + \cdots + [n + 1]}_{n \text{ terms in brackets}}$$

Now complete the derivation.]

Discussion and Writing

98. Investigate various applications that lead to a Fibonacci sequence, such as art, architecture, or financial markets. Write an essay on these applications.

'Are You Prepared?' Answers

1. $f(2) = \dfrac{1}{2}; f(3) = \dfrac{2}{3}$ **2.** True   **3.** $1082.43   **4.** $9513.28

11.2 Arithmetic Sequences

> **OBJECTIVES**   **1** Determine If a Sequence Is Arithmetic
> **2** Find a Formula for an Arithmetic Sequence
> **3** Find the Sum of an Arithmetic Sequence

1 Determine If a Sequence Is Arithmetic

When the difference between successive terms of a sequence is always the same number, the sequence is called **arithmetic**. An **arithmetic sequence*** may be defined recursively as $a_1 = a$, $a_n - a_{n-1} = d$, or as

$$a_1 = a, \qquad a_n = a_{n-1} + d \tag{1}$$

where $a = a_1$ and d are real numbers. The number a is the first term, and the number d is called the **common difference**.

The terms of an arithmetic sequence with first term a and common difference d follow the pattern

$$a, \ a + d, \ a + 2d, \ a + 3d, \ldots$$

---
*Sometimes called an **arithmetic progression**.

| EXAMPLE 1 | **Determining If a Sequence Is Arithmetic** |

The sequence
$$4, \quad 7, \quad 10, \quad 13, \ldots$$
is arithmetic since the difference of successive terms is 3. The first term is 4, and the common difference is 3. ◄

| EXAMPLE 2 | **Determining If a Sequence Is Arithmetic** |

Show that the following sequence is arithmetic. Find the first term and the common difference.
$$\{s_n\} = \{3n + 5\}$$

Solution The first term is $s_1 = 3 \cdot 1 + 5 = 8$. The nth and $(n - 1)$st terms of the sequence $\{s_n\}$ are
$$s_n = 3n + 5 \quad \text{and} \quad s_{n-1} = 3(n - 1) + 5 = 3n + 2$$

Their difference is
$$s_n - s_{n-1} = (3n + 5) - (3n + 2) = 5 - 2 = 3$$

Since the difference of two successive terms is constant, the sequence is arithmetic and the common difference is 3. ◄

| EXAMPLE 3 | **Determining If a Sequence Is Arithmetic** |

Show that the sequence $\{t_n\} = \{4 - n\}$ is arithmetic. Find the first term and the common difference.

Solution The first term is $t_1 = 4 - 1 = 3$. The nth and $(n - 1)$st terms are
$$t_n = 4 - n \quad \text{and} \quad t_{n-1} = 4 - (n - 1) = 5 - n$$

Their difference is
$$t_n - t_{n-1} = (4 - n) - (5 - n) = 4 - 5 = -1$$

Since the difference of two successive terms is constant, $\{t_n\}$ is an arithmetic sequence whose common difference is -1. ◄

NOW WORK PROBLEM 5.

2 Find a Formula for an Arithmetic Sequence

Suppose that a is the first term of an arithmetic sequence whose common difference is d. We seek a formula for the nth term, a_n. To see the pattern, we write down the first few terms.

$$a_1 = a$$
$$a_2 = a_1 + d = a + 1 \cdot d$$
$$a_3 = a_2 + d = (a + d) + d = a + 2 \cdot d$$
$$a_4 = a_3 + d = (a + 2 \cdot d) + d = a + 3 \cdot d$$
$$a_5 = a_4 + d = (a + 3 \cdot d) + d = a + 4 \cdot d$$
$$\vdots$$
$$a_n = a_{n-1} + d = [a + (n - 2)d] + d = a + (n - 1)d$$

We are led to the following result:

Theorem

nth Term of an Arithmetic Sequence

For an arithmetic sequence $\{a_n\}$ whose first term is a and whose common difference is d, the nth term is determined by the formula

$$a_n = a + (n-1)d \qquad \textbf{(2)}$$

EXAMPLE 4

Finding a Particular Term of an Arithmetic Sequence

Find the thirteenth term of the arithmetic sequence: $2, 6, 10, 14, 18, \ldots$

Solution

The first term of this arithmetic sequence is $a = 2$, and the common difference is 4. By formula (2), the nth term is

$$a_n = 2 + (n-1)4$$

Hence, the thirteenth term is

$$a_{13} = 2 + 12 \cdot 4 = 50 \qquad \blacktriangleleft$$

— **Exploration** —

Use a graphing utility to find the thirteenth term of the sequence given in Example 4. Use it to find the twentieth and fiftieth terms.

EXAMPLE 5

Finding a Recursive Formula for an Arithmetic Sequence

The eighth term of an arithmetic sequence is 75, and the twentieth term is 39. Find the first term and the common difference. Give a recursive formula for the sequence. What is the nth term of the sequence?

Solution

By formula (2), we know that $a_n = a + (n-1)d$. As a result,

$$\begin{cases} a_8 = a + 7d = 75 \\ a_{20} = a + 19d = 39 \end{cases}$$

This is a system of two linear equations containing two variables, a and d, which we can solve by elimination. Subtracting the second equation from the first equation, we get

$$-12d = 36$$

$$d = -3$$

With $d = -3$, we use $a + 7d = 75$ and find that $a = 75 - 7d = 75 - 7(-3) = 96$. The first term is $a = 96$, and the common difference is $d = -3$.

Using formula (1), a recursive formula for this sequence is

$$a_1 = 96, \qquad a_n = a_{n-1} - 3$$

Using formula (2), a formula for the nth term of the sequence $\{a_n\}$ is

$$a_n = a + (n-1)d = 96 + (n-1)(-3) = 99 - 3n \qquad \blacktriangleleft$$

— **Exploration** —

Graph the recursive formula from Example 5, $a_1 = 96$, $a_n = a_{n-1} - 3$, using a graphing utility. Conclude that the graph of the recursive formula behaves like the graph of a linear function. How is d, the common difference, related to m, the slope of a line?

NOW WORK PROBLEMS 21 AND 27.

3 Find the Sum of an Arithmetic Sequence

The next result gives a formula for finding the sum of the first n terms of an arithmetic sequence.

Theorem

Sum of n Terms of an Arithmetic Sequence

Let $\{a_n\}$ be an arithmetic sequence with first term a and common difference d. The sum S_n of the first n terms of $\{a_n\}$ is

$$S_n = \frac{n}{2}[2a + (n-1)d] = \frac{n}{2}(a + a_n) \qquad (3)$$

Proof

$$
\begin{aligned}
S_n &= a_1 + a_2 + a_3 + \cdots + a_n && \text{Sum of first } n \text{ terms}\\
&= a + (a+d) + (a+2d) + \cdots + [a+(n-1)d] && \text{Formula (2)}\\
&= \underbrace{(a + a + \cdots + a)}_{n \text{ terms}} + [d + 2d + \cdots + (n-1)d] && \text{Rearrange terms}\\
&= na + d[1 + 2 + \cdots + (n-1)]\\
&= na + d\left[\frac{(n-1)n}{2}\right] && \text{Property 6, Section 11.1}\\
&= na + \frac{n}{2}(n-1)d\\
&= \frac{n}{2}[2a + (n-1)d] && \text{Factor out } \frac{n}{2} && (4)\\
&= \frac{n}{2}[a + a + (n-1)d]\\
&= \frac{n}{2}(a + a_n) && \text{Formula (2)} && (5)
\end{aligned}
$$

∎

Formula (3) provides two ways to find the sum of the first n terms of an arithmetic sequence. Notice that formula (4) involves the first term and common difference, whereas formula (5) involves the first term and the nth term. Use whichever form is easier.

EXAMPLE 6

Finding the Sum of n Terms of an Arithmetic Sequence

Find the sum S_n of the first n terms of the sequence $\{3n + 5\}$; that is, find

$$8 + 11 + 14 + \cdots + (3n + 5)$$

Solution

The sequence $\{3n + 5\}$ is an arithmetic sequence with first term $a = 8$ and the nth term $(3n + 5)$. To find the sum S_n, we use formula (5).

$$S_n = \frac{n}{2}(a + a_n) = \frac{n}{2}[8 + (3n+5)] = \frac{n}{2}(3n + 13)$$

◀

NOW WORK PROBLEM 35.

Using a Graphing Utility to Find the Sum of 20 Terms of an Arithmetic Sequence

Figure 12

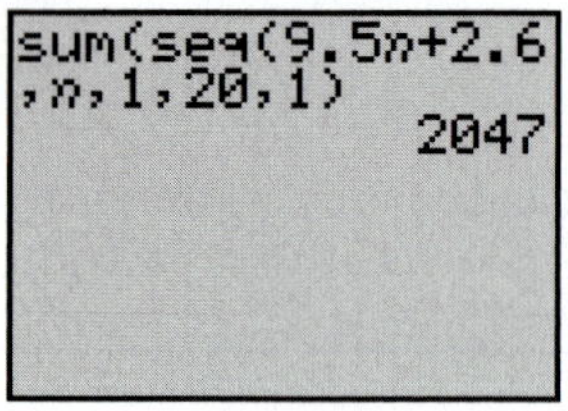

```
sum(seq(9.5n+2.6
,n,1,20,1)
            2047
```

Use a graphing utility to find the sum of the first 20 terms of the sequence $\{9.5n + 2.6\}$.

Solution   Figure 12 shows the results obtained using a TI-84 Plus graphing calculator.

The sum of the first 20 terms of the sequence $\{9.5n + 2.6\}$ is 2047.   ◄

✏ WORK EXAMPLE 7 USING FORMULA (3).

✏ NOW WORK PROBLEM 43.

EXAMPLE 8

Creating a Floor Design

A ceramic tile floor is designed in the shape of a trapezoid 20 feet wide at the base and 10 feet wide at the top. See Figure 13. The tiles, 12 inches by 12 inches, are to be placed so that each successive row contains one less tile than the preceding row. How many tiles will be required?

Figure 13

Solution   The bottom row requires 20 tiles and the top row, 10 tiles. Since each successive row requires one less tile, the total number of tiles required is

$$S = 20 + 19 + 18 + \cdots + 11 + 10$$

This is the sum of an arithmetic sequence; the common difference is -1. The number of terms to be added is $n = 11$, with the first term $a = 20$ and the last term $a_{11} = 10$. The sum S is

$$S = \frac{n}{2}(a + a_{11}) = \frac{11}{2}(20 + 10) = 165$$

In all, 165 tiles will be required.   ◄

11.2 Assess Your Understanding

Concepts and Vocabulary

1. In a(n) _____ sequence, the difference between successive terms is a constant.

2. *True or False:* In an arithmetic sequence the sum of the first and last terms equals twice the sum of all the terms.

Skill Building

In Problems 3–12, show that each sequence is arithmetic. Find the common difference and write out the first four terms.

3. $\{n + 4\}$ **4.** $\{n - 5\}$ **5.** $\{2n - 5\}$ **6.** $\{3n + 1\}$ **7.** $\{6 - 2n\}$

8. $\{4 - 2n\}$ **9.** $\left\{\dfrac{1}{2} - \dfrac{1}{3}n\right\}$ **10.** $\left\{\dfrac{2}{3} + \dfrac{n}{4}\right\}$ **11.** $\{\ln 3^n\}$ **12.** $\{e^{\ln n}\}$

In Problems 13–20, find the nth term of the arithmetic sequence whose initial term a and common difference d are given. What is the fifth term?

13. $a = 2;\quad d = 3$ **14.** $a = -2;\quad d = 4$ **15.** $a = 5;\quad d = -3$ **16.** $a = 6;\quad d = -2$

17. $a = 0;\quad d = \dfrac{1}{2}$ **18.** $a = 1;\quad d = -\dfrac{1}{3}$ **19.** $a = \sqrt{2};\quad d = \sqrt{2}$ **20.** $a = 0;\quad d = \pi$

In Problems 21–26, find the indicated term in each arithmetic sequence.

21. 12th term of $2, 4, 6, \ldots$ **22.** 8th term of $-1, 1, 3, \ldots$

23. 10th term of $1, -2, -5, \ldots$ **24.** 9th term of $5, 0, -5, \ldots$

25. 8th term of $a, a + b, a + 2b, \ldots$ **26.** 7th term of $2\sqrt{5}, 4\sqrt{5}, 6\sqrt{5}, \ldots$

In Problems 27–34, find the first term and the common difference of the arithmetic sequence described. Give a recursive formula for the sequence. Find a formula for the nth term.

27. 8th term is 8; 20th term is 44   **28.** 4th term is 3; 20th term is 35   **29.** 9th term is -5; 15th term is 31

30. 8th term is 4; 18th term is -96 **31.** 15th term is 0; 40th term is -50 **32.** 5th term is -2; 13th term is 30

33. 14th term is -1; 18th term is -9 **34.** 12th term is 4; 18th term is 28

In Problems 35–42, find each sum.

35. $1 + 3 + 5 + \cdots + (2n - 1)$ **36.** $2 + 4 + 6 + \cdots + 2n$ **37.** $7 + 12 + 17 + \cdots + (2 + 5n)$

38. $-1 + 3 + 7 + \cdots + (4n - 5)$ **39.** $2 + 4 + 6 + \cdots + 70$ **40.** $1 + 3 + 5 + \cdots + 59$

41. $5 + 9 + 13 + \cdots + 49$ **42.** $2 + 5 + 8 + \cdots + 41$

For Problems 43–48, use a graphing utility to find the sum of each sequence.

43. $\{3.45n + 4.12\}, \quad n = 20$ **44.** $\{2.67n - 1.23\}, \quad n = 25$ **45.** $2.8 + 5.2 + 7.6 + \cdots + 36.4$

46. $5.4 + 7.3 + 9.2 + \cdots + 32$ **47.** $4.9 + 7.48 + 10.06 + \cdots + 66.82$ **48.** $3.71 + 6.9 + 10.09 + \cdots + 80.27$

Applications and Extensions

49. Find x so that $x + 3, 2x + 1$, and $5x + 2$ are consecutive terms of an arithmetic sequence.

50. Find x so that $2x, 3x + 2$, and $5x + 3$ are consecutive terms of an arithmetic sequence.

51. Drury Lane Theater The Drury Lane Theater has 25 seats in the first row and 30 rows in all. Each successive row contains one additional seat. How many seats are in the theater?

52. Football Stadium The corner section of a football stadium has 15 seats in the first row and 40 rows in all. Each successive row contains two additional seats. How many seats are in this section?

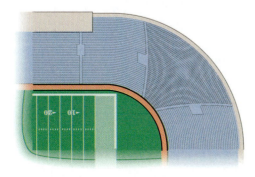

53. Creating a Mosaic A mosaic is designed in the shape of an equilateral triangle, 20 feet on each side. Each tile in the mosaic is in the shape of an equilateral triangle, 12 inches to a side. The tiles are to alternate in color as shown in the illustration. How many tiles of each color will be required?

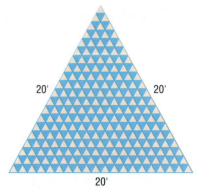

20'   20'

20'

54. Constructing a Brick Staircase A brick staircase has a total of 30 steps. The bottom step requires 100 bricks. Each successive step requires two less bricks than the prior step.
(a) How many bricks are required for the top step?
(b) How many bricks are required to build the staircase?

55. Stadium Construction How many rows are in the corner section of a stadium containing 2040 seats if the first row has 10 seats and each successive row has 4 additional seats?

56. Salary Suppose that you just received a job offer with a starting salary of $35,000 per year and a guaranteed raise of $1400 per year. How many years will it take before your aggregate salary is $280,000?

[**Hint:** Your aggregate salary after 2 years is $35,000 + ($35,000 + $1400).]

Discussion and Writing

57. Make up an arithmetic sequence. Give it to a friend and ask for its twentieth term.

58. Describe the similarities and differences between arithmetic sequences and linear functions.

11.3 Geometric Sequences; Geometric Series

OBJECTIVES 1 Determine If a Sequence Is Geometric
2 Find a Formula for a Geometric Sequence
3 Find the Sum of a Geometric Sequence
4 Find the Sum of a Geometric Series

1 Determine If a Sequence Is Geometric

When the ratio of successive terms of a sequence is always the same nonzero number, the sequence is called **geometric**. A **geometric sequence**[*] may be defined recursively as $a_1 = a$, $\dfrac{a_n}{a_{n-1}} = r$, or as

$$a_1 = a, \qquad a_n = ra_{n-1} \tag{1}$$

where $a_1 = a$ and $r \neq 0$ are real numbers. The number a is the first term, and the nonzero number r is called the **common ratio**.

The terms of a geometric sequence with first term a and common ratio r follow the pattern

$$a, \quad ar, \quad ar^2, \quad ar^3, \ldots$$

EXAMPLE 1 **Determining If a Sequence Is Geometric**

The sequence

$$2, \ 6, \ 18, \ 54, \ 162, \ldots$$

is geometric since the ratio of successive terms is 3 $\left(\dfrac{6}{2} = \dfrac{18}{6} = \dfrac{54}{18} = \cdots = 3 \right)$. The first term is 2, and the common ratio is 3. ◄

EXAMPLE 2 **Determining If a Sequence Is Geometric**

Show that the following sequence is geometric.

$$\{s_n\} = 2^{-n}$$

Find the first term and the common ratio.

---

[*]Sometimes called a **geometric progression**.

Solution   The first term is $s_1 = 2^{-1} = \dfrac{1}{2}$. The nth and $(n-1)$st terms of the sequence $\{s_n\}$ are

$$s_n = 2^{-n} \quad \text{and} \quad s_{n-1} = 2^{-(n-1)}$$

Their ratio is

$$\frac{s_n}{s_{n-1}} = \frac{2^{-n}}{2^{-(n-1)}} = 2^{-n+(n-1)} = 2^{-1} = \frac{1}{2}$$

Because the ratio of successive terms is a nonzero constant, the sequence $\{s_n\}$ is geometric with common ratio $\dfrac{1}{2}$.   ◄

EXAMPLE 3   **Determining If a Sequence Is Geometric**

Show that the following sequence is geometric.

$$\{t_n\} = \{4^n\}$$

Find the first term and the common ratio.

Solution   The first term is $t_1 = 4^1 = 4$. The nth and $(n-1)$st terms are

$$t_n = 4^n \quad \text{and} \quad t_{n-1} = 4^{n-1}$$

Their ratio is

$$\frac{t_n}{t_{n-1}} = \frac{4^n}{4^{n-1}} = 4^{n-(n-1)} = 4$$

The sequence, $\{t_n\}$, is a geometric sequence with common ratio 4.   ◄

NOW WORK PROBLEM 7.

2   **Find a Formula for a Geometric Sequence**

Suppose that a is the first term of a geometric sequence with common ratio $r \neq 0$. We seek a formula for the nth term a_n. To see the pattern, we write down the first few terms:

$$a_1 = a = a \cdot 1 = ar^0$$
$$a_2 = ra_1 = ar^1$$
$$a_3 = ra_2 = r(ar) = ar^2$$
$$a_4 = ra_3 = r(ar^2) = ar^3$$
$$a_5 = ra_4 = r(ar^3) = ar^4$$
$$\vdots$$
$$a_n = ra_{n-1} = r(ar^{n-2}) = ar^{n-1}$$

We are led to the following result:

Theorem

nth Term of a Geometric Sequence

For a geometric sequence $\{a_n\}$ whose first term is a and whose common ratio is r, the nth term is determined by the formula

$$a_n = ar^{n-1}, \qquad r \neq 0 \tag{2}$$

EXAMPLE 4

Finding a Particular Term of a Geometric Sequence

(a) Find the ninth term of the geometric sequence: $10,\ 9,\ \dfrac{81}{10},\ \dfrac{729}{100}\cdots$

(b) Find a recursive formula for this sequence.

Solution

(a) The first term of this geometric sequence is $a = 10$ and the common ratio is $\dfrac{9}{10}$.

── **Exploration** ──

Use a graphing utility to find the ninth term of the sequence given in Example 4. Use it to find the twentieth and fiftieth terms. Now use a graphing utility to graph the recursive formula found in Example 4(b). Conclude that the graph of the recursive formula behaves like the graph of an exponential function. How is r, the common ratio, related to a, the base of the exponential function $y = a^x$?

(Use $\dfrac{9}{10}$, or $\dfrac{\frac{81}{10}}{9} = \dfrac{9}{10}$, or any two successive terms.) By formula (2), the nth term is

$$a_n = 10\left(\frac{9}{10}\right)^{n-1}$$

The ninth term is

$$a_9 = 10\left(\frac{9}{10}\right)^{9-1} = 10\left(\frac{9}{10}\right)^8 \approx 4.3046721$$

(b) The first term in the sequence is 10 and the common ratio is $r = \dfrac{9}{10}$. Using formula (1), the recursive formula is $a_1 = 10,\ a_n = \dfrac{9}{10}a_{n-1}$.   ◄

 NOW WORK PROBLEMS 29 AND 37.

3 **Find the Sum of a Geometric Sequence**

The next result gives us a formula for finding the sum of the first n terms of a geometric sequence.

Theorem

Sum of n Terms of a Geometric Sequence

Let $\{a_n\}$ be a geometric sequence with first term a and common ratio r, where $r \neq 0, r \neq 1$. The sum S_n of the first n terms of $\{a_n\}$ is

$$S_n = a \cdot \frac{1 - r^n}{1 - r}, \qquad r \neq 0, 1 \tag{3}$$

Proof   The sum S_n of the first n terms of $\{a_n\} = \{ar^{n-1}\}$ is

$$S_n = a + ar + \cdots + ar^{n-1} \tag{4}$$

Multiply each side by r to obtain

$$rS_n = ar + ar^2 + \cdots + ar^n \tag{5}$$

Now, subtract (5) from (4). The result is

$$S_n - rS_n = a - ar^n$$
$$(1 - r)S_n = a(1 - r^n)$$

Since $r \neq 1$, we can solve for S_n.

$$S_n = a \cdot \frac{1 - r^n}{1 - r}$$

∎

| EXAMPLE 5 | **Finding the Sum of *n* Terms of a Geometric Sequence** |

Find the sum S_n of the first n terms of the sequence $\left\{\left(\frac{1}{2}\right)^n\right\}$; that is, find

$$\frac{1}{2} + \frac{1}{4} + \frac{1}{8} + \cdots + \left(\frac{1}{2}\right)^n$$

Solution The sequence $\left\{\left(\frac{1}{2}\right)^n\right\}$ is a geometric sequence with $a = \frac{1}{2}$ and $r = \frac{1}{2}$. The sum S_n that we seek is the sum of the first n terms of the sequence, so we use formula (3) to get

$$S_n = \sum_{k=1}^{n} \left(\frac{1}{2}\right)^k = \frac{1}{2} + \frac{1}{4} + \frac{1}{8} + \cdots + \left(\frac{1}{2}\right)^n$$

$$= \frac{1}{2}\left[\frac{1 - \left(\frac{1}{2}\right)^n}{1 - \frac{1}{2}}\right] \qquad \text{Formula (3)}$$

$$= \frac{1}{2}\left[\frac{1 - \left(\frac{1}{2}\right)^n}{\frac{1}{2}}\right]$$

$$= 1 - \left(\frac{1}{2}\right)^n$$

◀

━━━━ **NOW WORK PROBLEM 43.**

| EXAMPLE 6 | **Using a Graphing Utility to Find the Sum of a Geometric Sequence** |

Use a graphing utility to find the sum of the first 15 terms of the sequence $\left\{\left(\frac{1}{3}\right)^n\right\}$; that is, find

$$\frac{1}{3} + \frac{1}{9} + \frac{1}{27} + \cdots + \left(\frac{1}{3}\right)^{15}$$

Figure 14

```
sum(seq((1/3)^n,
n,1,15,1))
       .4999999652
```

Solution Figure 14 shows the result obtained using a TI-84 Plus graphing calculator. The sum of the first 15 terms of the sequence $\left\{\left(\frac{1}{3}\right)^n\right\}$ is 0.4999999652.

◀

━━━━ **NOW WORK PROBLEM 49.**

4 **Find the Sum of a Geometric Series**

An infinite sum of the form

$$a + ar + ar^2 + \cdots + ar^{n-1} + \cdots$$

with first term a and common ratio r, is called an **infinite geometric series** and is denoted by

$$\sum_{k=1}^{\infty} ar^{k-1}$$

Based on formula (3), the sum S_n of the first n terms of a geometric series is

$$S_n = a \cdot \frac{1 - r^n}{1 - r} = \frac{a}{1 - r} - \frac{ar^n}{1 - r} \qquad (6)$$

If this finite sum S_n approaches a number L as $n \to \infty$, then we say the infinite geometric series $\sum_{k=1}^{\infty} ar^{k-1}$ **converges**. We call L the **sum of the infinite geometric series**, and we write

$$L = \sum_{k=1}^{\infty} ar^{k-1}$$

If a series does not converge, it is called a **divergent series**.

Theorem

Sum of an Infinite Geometric Series

If $|r| < 1$, the sum of the infinite geometric series $\sum_{k=1}^{\infty} ar^{k-1}$ is

$$\sum_{k=1}^{\infty} ar^{k-1} = \frac{a}{1 - r} \qquad (7)$$

Intuitive Proof Since $|r| < 1$, it follows that $|r^n|$ approaches 0 as $n \to \infty$. Then, based on formula (6), the sum S_n approaches $\dfrac{a}{1 - r}$ as $n \to \infty$. ∎

EXAMPLE 7

Finding the Sum of a Geometric Series

Find the sum of the geometric series: $2 + \dfrac{4}{3} + \dfrac{8}{9} + \cdots$

Solution The first term is $a = 2$, and the common ratio is

$$r = \frac{\dfrac{4}{3}}{2} = \frac{4}{6} = \frac{2}{3}$$

Since $|r| < 1$, we use formula (7) to find that

$$2 + \frac{4}{3} + \frac{8}{9} + \cdots = \frac{2}{1 - \dfrac{2}{3}} = 6$$

— Exploration —

Use a graphing utility to graph $U_n = 2\left(\dfrac{2}{3}\right)^{n-1}$ in sequence mode. TRACE the graph for large values of n. What happens to the value of U_n as n increases without bound? What can you conclude about $\sum_{n=1}^{\infty} 2\left(\dfrac{2}{3}\right)^{n-1}$?

NOW WORK PROBLEM 55.

EXAMPLE 8	**Repeating Decimals**

Show that the repeating decimal $0.999\ldots$ equals 1.

Solution $0.999\ldots = \dfrac{9}{10} + \dfrac{9}{100} + \dfrac{9}{1000} + \cdots$

The decimal $0.999\ldots$ is a geometric series with first term $\dfrac{9}{10}$ and common ratio $\dfrac{1}{10}$. Using formula (7), we find

$$0.999\ldots = \frac{\dfrac{9}{10}}{1 - \dfrac{1}{10}} = \frac{\dfrac{9}{10}}{\dfrac{9}{10}} = 1$$

◄

EXAMPLE 9	**Pendulum Swings**

Figure 15

Initially, a pendulum swings through an arc of 18 inches. See Figure 15. On each successive swing, the length of the arc is 0.98 of the previous length.

(a) What is the length of the arc of the 10th swing?

(b) On which swing is the length of the arc first less than 12 inches?

(c) After 15 swings, what total distance will the pendulum have swung?

(d) When it stops, what total distance will the pendulum have swung?

Solution (a) The length of the first swing is 18 inches.

The length of the second swing is $0.98(18)$ inches.

The length of the third swing is $0.98(0.98)(18) = 0.98^2(18)$ inches.

The length of the arc of the 10th swing is

$$(0.98)^9(18) \approx 15.007 \text{ inches}$$

(b) The length of the arc of the nth swing is $(0.98)^{n-1}(18)$. For this to be exactly 12 inches requires that

$$(0.98)^{n-1}(18) = 12$$

$$(0.98)^{n-1} = \frac{12}{18} = \frac{2}{3} \qquad \text{\color{teal}{Divide both sides by 18.}}$$

$$n - 1 = \log_{0.98}\left(\frac{2}{3}\right) \qquad \text{\color{teal}{Express as a logarithm.}}$$

$$n = 1 + \frac{\ln\left(\dfrac{2}{3}\right)}{\ln 0.98} \approx 1 + 20.07 = 21.07 \qquad \text{\color{teal}{Solve for n; use the Change of Base Formula.}}$$

The length of the arc of the pendulum exceeds 12 inches on the 21st swing and is first less than 12 inches on the 22nd swing.

(c) After 15 swings, the pendulum will have swung the following total distance L:

$$L = \underset{\text{1st}}{18} + \underset{\text{2nd}}{0.98(18)} + \underset{\text{3rd}}{(0.98)^2(18)} + \underset{\text{4th}}{(0.98)^3(18)} + \cdots + \underset{\text{15th.}}{(0.98)^{14}(18)}$$

This is the sum of a geometric sequence. The common ratio is 0.98; the first term is 18. The sum has 15 terms, so

$$L = 18 \cdot \frac{1 - 0.98^{15}}{1 - 0.98} \approx 18(13.07) \approx 235.3 \text{ inches}$$

The pendulum will have swung through 235.3 inches after 15 swings.

(d) When the pendulum stops, it will have swung the following total distance T:

$$T = 18 + 0.98(18) + (0.98)^2(18) + (0.98)^3(18) + \cdots$$

This is the sum of a geometric series. The common ratio is $r = 0.98$; the first term is $a = 18$. The sum is

$$T = \frac{a}{1 - r} = \frac{18}{1 - 0.98} = 900$$

The pendulum will have swung a total of 900 inches when it finally stops. ◀

✎ **NOW WORK PROBLEM 69.**

HISTORICAL FEATURE

Fibonacci

Sequences are among the oldest objects of mathematical investigation, having been studied for over 3500 years. After the initial steps, however, little progress was made until about 1600.

Arithmetic and geometric sequences appear in the Rhind papyrus, a mathematical text containing 85 problems copied around 1650 BC by the Egyptian scribe Ahmes from an earlier work (see Historical Problem 1). Fibonacci (AD 1220) wrote about problems similar to those found in the Rhind papyrus, leading one to suspect that Fibonacci may have had material available that is now lost. This material would have been in the non-Euclidean Greek tradition of

Heron (about AD 75) and Diophantus (about AD 250). One problem, again modified slightly, is still with us in the familiar puzzle rhyme "As I was going to St. Ives …" (see Historical Problem 2).

The Rhind papyrus indicates that the Egyptians knew how to add up the terms of an arithmetic or geometric sequence, as did the Babylonians. The rule for summing up a geometric sequence is found in Euclid's *Elements* (Book IX, 35, 36), where, like all Euclid's algebra, it is presented in a geometric form.

Investigations of other kinds of sequences began in the 1500s, when algebra became sufficiently developed to handle the more complicated problems. The development of calculus in the 1600s added a powerful new tool, especially for finding the sum of infinite series, and the subject continues to flourish today.

Historical Problems

1. *Arithmetic sequence problem from the Rhind papyrus (statement modified slightly for clarity)* One hundred loaves of bread are to be divided among five people so that the amounts that they receive form an arithmetic sequence. The first two together receive one-seventh of what the last three receive. How many loaves does each receive?
   [*Partial answer:* First person receives $1\frac{2}{3}$ loaves.]

2. The following old English children's rhyme resembles one of the Rhind papyrus problems.

   As I was going to St. Ives
   I met a man with seven wives

   Each wife had seven sacks
   Each sack had seven cats
   Each cat had seven kits [kittens]
   Kits, cats, sacks, wives
   How many were going to St. Ives?

   (a) Assuming that the speaker and the cat fanciers met by traveling in opposite directions, what is the answer?
   (b) How many kittens are being transported?
   (c) Kits, cats, sacks, wives; how many?
   [**Hint**: It is easier to include the man, find the sum with the formula, and then subtract 1 for the man.]

11.3 Assess Your Understanding

Concepts and Vocabulary

1. In a(n) _____ sequence, the ratio of successive terms is a constant.

2. If $|r| < 1$, the sum of the geometric series $\sum_{k=1}^{\infty} ar^{k-1}$ is _____.

3. *True or False:* A geometric sequence may be defined recursively.

4. *True or False:* In a geometric sequence, the common ratio is always a positive number.

Skill Building

In Problems 5–14, show that each sequence is geometric. Find the common ratio and write out the first four terms.

5. $\{3^n\}$ **6.** $\{(-5)^n\}$ **7.** $\left\{-3\left(\frac{1}{2}\right)^n\right\}$ **8.** $\left\{\left(\frac{5}{2}\right)^n\right\}$ **9.** $\left\{\frac{2^{n-1}}{4}\right\}$

10. $\left\{\frac{3^n}{9}\right\}$ **11.** $\{2^{n/3}\}$ **12.** $\{3^{2n}\}$ **13.** $\left\{\frac{3^{n-1}}{2^n}\right\}$ **14.** $\left\{\frac{2^n}{3^{n-1}}\right\}$

In Problems 15–28, determine whether the given sequence is arithmetic, geometric, or neither. If the sequence is arithmetic, find the common difference; if it is geometric, find the common ratio.

15. $\{n + 2\}$ **16.** $\{2n - 5\}$ **17.** $\{4n^2\}$ **18.** $\{5n^2 + 1\}$ **19.** $\left\{3 - \frac{2}{3}n\right\}$

20. $\left\{8 - \frac{3}{4}n\right\}$ **21.** $1, 3, 6, 10, \ldots$ **22.** $2, 4, 6, 8, \ldots$ **23.** $\left\{\left(\frac{2}{3}\right)^n\right\}$ **24.** $\left\{\left(\frac{5}{4}\right)^n\right\}$

25. $-1, -2, -4, -8, \ldots$ **26.** $1, 1, 2, 3, 5, 8, \ldots$ **27.** $\{3^{n/2}\}$ **28.** $\{(-1)^n\}$

In Problems 29–36, find the fifth term and the nth term of the geometric sequence whose initial term a and common ratio r are given.

29. $a = 2;\quad r = 3$ **30.** $a = -2;\quad r = 4$ **31.** $a = 5;\quad r = -1$ **32.** $a = 6;\quad r = -2$

33. $a = 0;\quad r = \frac{1}{2}$ **34.** $a = 1;\quad r = -\frac{1}{3}$ **35.** $a = \sqrt{2};\quad r = \sqrt{2}$ **36.** $a = 0;\quad r = \frac{1}{\pi}$

In Problems 37–42, find the indicated term of each geometric sequence.

37. 7th term of $1, \frac{1}{2}, \frac{1}{4}, \ldots$ **38.** 8th term of $1, 3, 9, \ldots$ **39.** 9th term of $1, -1, 1, \ldots$

40. 10th term of $-1, 2, -4, \ldots$ **41.** 8th term of $0.4, 0.04, 0.004, \ldots$ **42.** 7th term of $0.1, 1.0, 10.0, \ldots$

In Problems 43–48, find each sum.

43. $\frac{1}{4} + \frac{2}{4} + \frac{2^2}{4} + \frac{2^3}{4} + \cdots + \frac{2^{n-1}}{4}$ **44.** $\frac{3}{9} + \frac{3^2}{9} + \frac{3^3}{9} + \cdots + \frac{3^n}{9}$ **45.** $\sum_{k=1}^{n} \left(\frac{2}{3}\right)^k$

46. $\sum_{k=1}^{n} 4 \cdot 3^{k-1}$ **47.** $-1 - 2 - 4 - 8 - \cdots - (2^{n-1})$ **48.** $2 + \frac{6}{5} + \frac{18}{25} + \cdots + 2\left(\frac{3}{5}\right)^{n-1}$

For Problems 49–54, use a graphing utility to find the sum of each geometric sequence.

49. $\frac{1}{4} + \frac{2}{4} + \frac{2^2}{4} + \frac{2^3}{4} + \cdots + \frac{2^{14}}{4}$ **50.** $\frac{3}{9} + \frac{3^2}{9} + \frac{3^3}{9} + \cdots + \frac{3^{15}}{9}$ **51.** $\sum_{n=1}^{15} \left(\frac{2}{3}\right)^n$

52. $\sum_{n=1}^{15} 4 \cdot 3^{n-1}$ **53.** $-1 - 2 - 4 - 8 - \cdots - 2^{14}$ **54.** $2 + \frac{6}{5} + \frac{18}{25} + \cdots + 2\left(\frac{3}{5}\right)^{15}$

In Problems 55–64, find the sum of each infinite geometric series.

55. $1 + \frac{1}{3} + \frac{1}{9} + \cdots$ **56.** $2 + \frac{4}{3} + \frac{8}{9} + \cdots$ **57.** $8 + 4 + 2 + \cdots$ **58.** $6 + 2 + \frac{2}{3} + \cdots$

59. $2 - \frac{1}{2} + \frac{1}{8} - \frac{1}{32} + \cdots$ **60.** $1 - \frac{3}{4} + \frac{9}{16} - \frac{27}{64} + \cdots$ **61.** $\sum_{k=1}^{\infty} 5\left(\frac{1}{4}\right)^{k-1}$ **62.** $\sum_{k=1}^{\infty} 8\left(\frac{1}{3}\right)^{k-1}$

63. $\sum_{k=1}^{\infty} 6\left(-\frac{2}{3}\right)^{k-1}$ **64.** $\sum_{k=1}^{\infty} 4\left(-\frac{1}{2}\right)^{k-1}$

Applications and Extensions

65. Find x so that x, $x + 2$, and $x + 3$ are consecutive terms of a geometric sequence.

66. Find x so that $x - 1$, x, and $x + 2$ are consecutive terms of a geometric sequence.

67. Salary Increases   Suppose that you have just been hired at an annual salary of $18,000 and expect to receive annual increases of 5%. What will your salary be when you begin your fifth year?

68. Equipment Depreciation   A new piece of equipment cost a company $15,000. Each year, for tax purposes, the company depreciates the value by 15%. What value should the company give the equipment after 5 years?

69. Pendulum Swings   Initially, a pendulum swings through an arc of 2 feet. On each successive swing, the length of the arc is 0.9 of the previous length.
  (a) What is the length of the arc of the 10th swing?

(b) On which swing is the length of the arc first less than 1 foot?

(c) After 15 swings, what total length will the pendulum have swung?

(d) When it stops, what total length will the pendulum have swung?

70. **Bouncing Balls** A ball is dropped from a height of 30 feet. Each time it strikes the ground, it bounces up to 0.8 of the previous height.

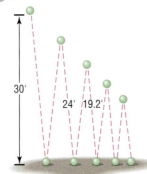

(a) What height will the ball bounce up to after it strikes the ground for the third time?

(b) How high will it bounce after it strikes the ground for the nth time?

(c) How many times does the ball need to strike the ground before its bounce is less than 6 inches?

(d) What total distance does the ball travel before it stops bouncing?

71. **Grains of Wheat on a Chess Board** In an old fable, a commoner who had saved the king's life was told he could ask the king for any just reward. Being a shrewd man, the commoner said, "A simple wish, sire. Place one grain of wheat on the first square of a chessboard, two grains on the second square, four grains on the third square, continuing until you have filled the board. This is all I seek." Compute the total number of grains needed to do this to see why the request, seemingly simple, could not be granted. (A chessboard consists of $8 \times 8 = 64$ squares.)

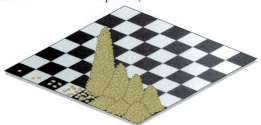

72. Look at the figure below. What fraction of the square is eventually shaded if the indicated shading process continues indefinitely?

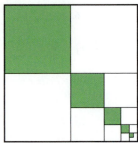

73. **Multiplier** Suppose that, throughout the U.S. economy, individuals spend 90% of every additional dollar that they earn. Economists would say that an individual's **marginal propensity to consume** is 0.90. For example, if Jane earns an additional dollar, she will spend $0.9(1) = \$0.90$ of it. The individual that earns \$0.90 (from Jane) will spend 90% of it or \$0.81. This process of spending continues and results in an infinite geometric series as follows:

$$1, 0.90, 0.90^2, 0.90^3, 0.90^4, \ldots.$$

The sum of this infinite geometric series is called the **multiplier**. What is the multiplier if individuals spend 90% of every additional dollar that they earn?

74. **Multiplier** Refer to Problem 73. Suppose that the marginal propensity to consume throughout the U.S. economy is 0.95. What is the multiplier for the U.S. economy?

75. **Stock Price** One method of pricing a stock is to discount the stream of future dividends of the stock. Suppose that a stock pays $\$P$ per year in dividends and, historically, the dividend has been increased $i\%$ per year. If you desire an annual rate of return of $r\%$, this method of pricing a stock states that the price that you should pay is the present value of an infinite stream of payments:

$$\text{Price} = P + P\frac{1+i}{1+r} + P\left(\frac{1+i}{1+r}\right)^2 + P\left(\frac{1+i}{1+r}\right)^3 + \cdots$$

The price of the stock is the sum of an infinite geometric series. Suppose that a stock pays an annual dividend of \$4.00 and, historically, the dividend has been increased 3% per year. You desire an annual rate of return of 9%. What is the most you should pay for the stock?

76. **Stock Price** Refer to Problem 75. Suppose that a stock pays an annual dividend of \$2.50 and, historically, the dividend has increased 4% per year. You desire an annual rate of return of 11%. What is the most that you should pay for the stock?

Discussion and Writing

77. **A Rich Man's Promise** A rich man promises to give you \$1000 on September 1, 2001. Each day thereafter he will give you $\dfrac{9}{10}$ of what he gave you the previous day. What is the first date on which the amount you receive is less than 1¢? How much have you received when this happens?

78. **Critical Thinking** You are interviewing for a job and receive two offers:

A: \$20,000 to start, with guaranteed annual increases of 6% for the first 5 years

B: \$22,000 to start, with guaranteed annual increases of 3% for the first 5 years

Which offer is best if your goal is to be making as much as possible after 5 years? Which is best if your goal is to make as much money as possible over the contract (5 years)?

79. Critical Thinking   Which of the following choices, *A* or *B*, results in more money?

   A: To receive $1000 on day 1, $999 on day 2, $998 on day 3, with the process to end after 1000 days

   B: To receive $1 on day 1, $2 on day 2, $4 on day 3, for 19 days

80. Critical Thinking   You have just signed a 7-year professional football league contract with a beginning salary of $2,000,000 per year. Management gives you the following options with regard to your salary over the 7 years.

   1. A bonus of $100,000 each year

   2. An annual increase of 4.5% per year beginning after 1 year

   3. An annual increase of $95,000 per year beginning after 1 year

Which option provides the most money over the 7-year period? Which the least? Which would you choose? Why?

81. Can a sequence be both arithmetic and geometric? Give reasons for your answer.

82. Make up a geometric sequence. Give it to a friend and ask for its 20th term.

83. Make up two infinite geometric series, one that has a sum and one that does not. Give them to a friend and ask for the sum of each series.

84. Describe the similarities and differences between geometric sequences and exponential functions.

Counting and Probability

12

The Two-children Problem

Chapter 12 is not included in *Precalculus Essentials: Enhanced with Graphing Utilities*, 4e.

A Preview of Calculus: The Limit, Derivative, and Integral of a Function

13

Chapter 13 is not included in *Precalculus Essentials: Enhanced with Graphing Utilities*, 4e.

A LOOK BACK In this book we have discussed a variety of functions: polynomial functions (including linear and quadratic functions), rational functions, exponential and logarithmic functions, trigonometric functions, and the inverse trigonometric functions. For each of these, we found their domain and range, intercepts, symmetry, if any, and asymptotes, if any, and we graphed each. We also discussed whether these functions were even, odd, or neither and determined on what intervals they were increasing and decreasing. We also discussed the idea of average rates of change.

A LOOK AHEAD In calculus, other properties are discussed, such as finding limits of functions, determining where functions are continuous, finding the derivative of functions, and finding the integral of functions. In this chapter, we give an introduction to these properties. By completing this chapter you will be well prepared for a first course in calculus.

OUTLINE

Review   Appendix

Outline

A.1   Algebra Review

PREPARING FOR THIS BOOK   *Before getting started, read "To the Student" at the beginning of this book on page xxiv.*

OBJECTIVES
1. Evaluate Algebraic Expressions
2. Determine the Domain of a Variable
3. Graph Inequalities
4. Find Distance on the Real Number Line
5. Use the Laws of Exponents
6. Evaluate Square Roots

Sets

When we want to treat a collection of similar but distinct objects as a whole, we use the idea of a **set**. For example, the set of *digits* consists of the collection of numbers $0, 1, 2, 3, 4, 5, 6, 7, 8$, and 9. If we use the symbol D to denote the set of digits, then we can write

$$D = \{0, 1, 2, 3, 4, 5, 6, 7, 8, 9\}$$

In this notation, the braces $\{\ \}$ are used to enclose the objects, or **elements**, in the set. This method of denoting a set is called the **roster method**. A second way to denote a set is to use **set-builder notation**, where the set D of digits is written as

$$D = \{ \quad x \quad | \quad x \text{ is a digit}\}$$

Read as "D is the set of all x such that x is a digit."

| EXAMPLE 1 | **Using Set-builder Notation and the Roster Method** |

(a) $E = \{x \mid x \text{ is an even digit}\} = \{0, 2, 4, 6, 8\}$

(b) $O = \{x \mid x \text{ is an odd digit}\} = \{1, 3, 5, 7, 9\}$ ◀

In listing the elements of a set, we do not list an element more than once because the elements of a set are distinct. Also, the order in which the elements are listed is not relevant. For example, $\{2, 3\}$ and $\{3, 2\}$ both represent the same set.

If every element of a set A is also an element of a set B, then we say that A is a **subset** of B. If two sets A and B have the same elements, then we say that A **equals** B. For example, $\{1, 2, 3\}$ is a subset of $\{1, 2, 3, 4, 5\}$; and $\{1, 2, 3\}$ equals $\{2, 3, 1\}$.

Finally, if a set has no elements, it is called the **empty set**, or the **null set**, and it is denoted by the symbol $\varnothing$.

Real Numbers

Real numbers are represented by symbols such as

$$25, \quad 0, \quad -3, \quad \frac{1}{2}, \quad -\frac{5}{4}, \quad 0.125, \quad \sqrt{2}, \quad \pi, \quad \sqrt[3]{-2}, \quad 0.666\ldots$$

The set of **counting numbers**, or **natural numbers**, contains the numbers in the set $\{1, 2, 3, 4, \ldots\}$. (The three dots, called an **ellipsis**, indicate that the pattern continues indefinitely.) The set of **integers** contains the numbers in the set $\{\ldots, -3, -2, -1, 0, 1, 2, 3, \ldots\}$. A **rational number** is a number that can be expressed as a *quotient* $\dfrac{a}{b}$ of two integers, where the integer b cannot be 0. Examples of rational numbers are $\dfrac{3}{4}, \dfrac{5}{2}, \dfrac{0}{4}$, and $-\dfrac{2}{3}$. Since $\dfrac{a}{1} = a$ for any integer a, every integer is also a rational number. Real numbers that are not rational are called **irrational**. Examples of irrational numbers are $\sqrt{2}$ and π (the Greek letter pi), which equals the constant ratio of the circumference to the diameter of a circle. See Figure 1.

Real numbers can be represented as **decimals**. Rational real numbers have decimal representations that either **terminate** or are non terminating with **repeating** blocks of digits. For example, $\dfrac{3}{4} = 0.75$, which terminates; and $\dfrac{2}{3} = 0.666\ldots$, in which the digit 6 repeats indefinitely. Irrational real numbers have decimal representations that neither repeat nor terminate. For example, $\sqrt{2} = 1.414213\ldots$ and $\pi = 3.14159\ldots$. In practice, the decimal representation of an irrational number is given as an approximation. We use the symbol $\approx$ (read as "approximately equal to") to write $\sqrt{2} \approx 1.4142$ and $\pi \approx 3.1416$.

Two properties of real numbers that we shall use often are given next.

Suppose that a, b, and c are real numbers.

Figure 1 $\pi = \dfrac{C}{d}$

Distributive Property

$$a \cdot (b + c) = ab + ac$$

Zero-Product Property

If $ab = 0$, then either $a = 0$ or $b = 0$ or both equal 0.

The Distributive Property can be used to remove parentheses:
$2(x + 3) = 2x + 2 \cdot 3 = 2x + 6$.

The Zero-Product Property will be used to solve equations (Section A.5). If $2x = 0$, then $2 = 0$ or $x = 0$. Since $2 \neq 0$, it follows that $x = 0$.

Constants and Variables

In algebra we use letters to represent numbers. If the letter used is to represent *any* number from a given set of numbers, it is called a **variable**. A **constant** is either a fixed number, such as 5 or $\sqrt{3}$, or a letter that represents a fixed (possibly unspecified) number.

Constants and variables are combined using the operations of addition, subtraction, multiplication, and division to form *algebraic expressions*. Examples of algebraic expressions include

$$x + 3 \qquad \frac{3}{1 - t} \qquad 7x - 2y$$

1 Evaluate Algebraic Expressions

To evaluate an algebraic expression, substitute for each variable its numerical value.

EXAMPLE 2 **Evaluating an Algebraic Expression**

Evaluate each expression if $x = 3$ and $y = -1$.

(a) $x + 3y$ 　　　　(b) $5xy$ 　　　　(c) $\dfrac{3y}{2 - 2x}$

Solution (a) Substitute 3 for x and -1 for y in the expression $x + 3y$.

$$x + 3y = 3 + 3(-1) = 3 + (-3) = 0$$
$$\uparrow$$
$$x = 3, y = -1$$

(b) If $x = 3$ and $y = -1$, then

$$5xy = 5(3)(-1) = -15$$

(c) If $x = 3$ and $y = -1$, then

$$\frac{3y}{2 - 2x} = \frac{3(-1)}{2 - 2(3)} = \frac{-3}{2 - 6} = \frac{-3}{-4} = \frac{3}{4}$$ ◀

NOW WORK PROBLEM 9.

2 Determine the Domain of a Variable

In working with expressions or formulas involving variables, the variables may be allowed to take on values from only a certain set of numbers. For example, in the formula for the area A of a circle of radius r, $A = \pi r^2$, the variable r is necessarily restricted to the positive real numbers. In the expression $\dfrac{1}{x}$, the variable x cannot take on the value 0, since division by 0 is not defined.

The set of values that a variable in an expression may assume is called the **domain of the variable**.

| EXAMPLE 3 | Finding the Domain of a Variable |

The domain of the variable x in the expression

$$\frac{5}{x-2}$$

is $\{x \mid x \neq 2\}$, since, if $x = 2$, the denominator becomes 0, which is not defined. ◀

| EXAMPLE 4 | Circumference of a Circle |

In the formula for the circumference C of a circle of radius r,

$$C = 2\pi r$$

the domain of the variable r, representing the radius of the circle, is the set of positive real numbers. The domain of the variable C, representing the circumference of the circle, is also the set of positive real numbers. ◀

In describing the domain of a variable, we may use either set notation or words, whichever is more convenient.

 NOW WORK PROBLEM 17.

The Real Number Line

The real numbers can be represented by points on a line called the **real number line**. There is a one-to-one correspondence between real numbers and points on a line. That is, every real number corresponds to a point on the line, and each point on the line has a unique real number associated with it.

Pick a point on the line somewhere in the center, and label it O. This point, called the **origin**, corresponds to the real number 0. See Figure 2. The point 1 unit to the right of O corresponds to the number 1. The distance between 0 and 1 determines the scale of the number line. For example, the point associated with the number 2 is twice as far from O as 1 is. Notice that an arrowhead on the right end of the line indicates the direction in which the numbers increase. Figure 2 also shows the points associated with the irrational numbers $\sqrt{2}$ and π. Points to the left of the origin correspond to the real numbers -1, -2, and so on.

Figure 2
Real number line

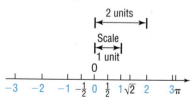

The real number associated with a point P is called the **coordinate** of P, and the line whose points have been assigned coordinates is called the **real number line**.

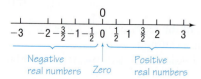 **NOW WORK PROBLEM 29.**

The real number line consists of three classes of real numbers, as shown in Figure 3.

Figure 3

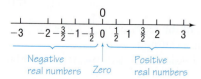

1. The **negative real numbers** are the coordinates of points to the left of the origin O.
2. The real number **zero** is the coordinate of the origin O.
3. The **positive real numbers** are the coordinates of points to the right of the origin O.

3 Graph Inequalities

An important property of the real number line follows from the fact that, given two numbers (points) a and b, either a is to the left of b, a is at the same location as b, or a is to the right of b. See Figure 4.

Figure 4

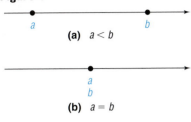

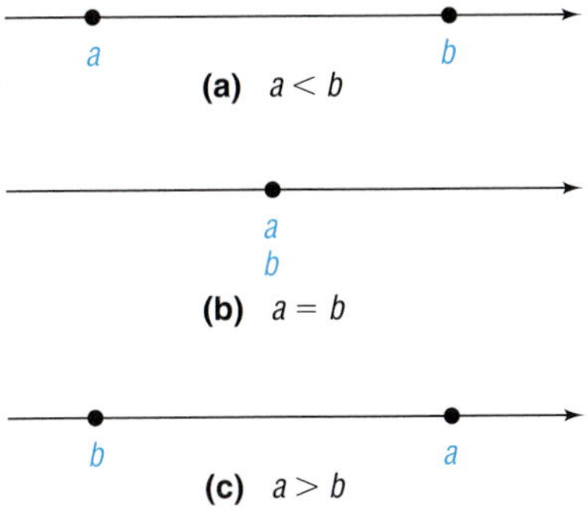

(a) $a < b$

(b) $a = b$

(c) $a > b$

If a is to the left of b, we say that "a is less than b" and write $a < b$. If a is to the right of b, we say that "a is greater than b" and write $a > b$. If a is at the same location as b, then $a = b$. If a is either less than or equal to b, we write $a \leq b$. Similarly, $a \geq b$ means that a is either greater than or equal to b. Collectively, the symbols $<$, $>$, $\leq$, and $\geq$ are called **inequality symbols**.

Note that $a < b$ and $b > a$ mean the same thing. It does not matter whether we write $2 < 3$ or $3 > 2$.

Furthermore, if $a < b$ or if $b > a$, then the difference $b - a$ is positive. Do you see why?

An **inequality** is a statement in which two expressions are related by an inequality symbol. The expressions are referred to as the **sides** of the inequality. Statements of the form $a < b$ or $b > a$ are called **strict inequalities**, while statements of the form $a \leq b$ or $b \geq a$ are called **nonstrict inequalities**.

Based on the discussion thus far, we conclude that

$$a > 0 \quad \text{is equivalent to} \quad a \text{ is positive}$$
$$a < 0 \quad \text{is equivalent to} \quad a \text{ is negative}$$

We sometimes read $a > 0$ by saying that "a is positive." If $a \geq 0$, then either $a > 0$ or $a = 0$, and we may read this as "a is nonnegative."

 NOW WORK PROBLEMS 33 AND 43.

We shall find it useful in later work to graph inequalities on the real number line.

EXAMPLE 5	**Graphing Inequalities**

(a) On the real number line, graph all numbers x for which $x > 4$.

(b) On the real number line, graph all numbers x for which $x \leq 5$.

Solution

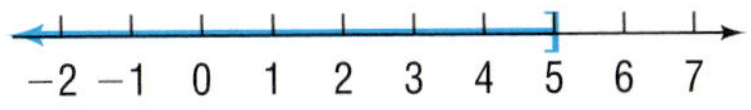

(a) See Figure 5. Notice that we use a left parenthesis to indicate that the number 4 is not part of the graph.

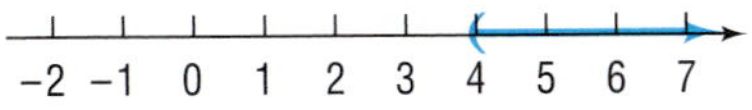

(b) See Figure 6. Notice that we use a right bracket to indicate that the number 5 is part of the graph. ◀

Figure 5

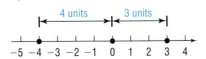

Figure 6

 NOW WORK PROBLEM 49.

Absolute Value

Figure 7

The *absolute value* of a number a is the distance from 0 to a on the number line. For example, -4 is 4 units from 0; and 3 is 3 units from 0. See Figure 7. Thus, the absolute value of -4 is 4, and the absolute value of 3 is 3.

A more formal definition of absolute value is given next.

The **absolute value** of a real number a, denoted by the symbol $|a|$, is defined by the rules

$$|a| = a \quad \text{if } a \geq 0 \qquad \text{and} \qquad |a| = -a \quad \text{if } a < 0$$

For example, since $-4 < 0$, the second rule must be used to get $|-4| = -(-4) = 4$.

EXAMPLE 6	**Computing Absolute Value**

(a) $|8| = 8$ (b) $|0| = 0$ (c) $|-15| = -(-15) = 15$ ◀

NOW WORK PROBLEM 51.

4 Find Distance on the Real Number Line

Look again at Figure 7. The distance from -4 to 3 is 7 units. This distance is the difference $3 - (-4)$, obtained by subtracting the smaller coordinate from the larger. However, since $|3 - (-4)| = |7| = 7$ and $|-4 - 3| = |-7| = 7$, we can use absolute value to calculate the distance between two points without being concerned about which is smaller.

If P and Q are two points on a real number line with coordinates a and b, respectively, the **distance between P and Q**, denoted by $d(P, Q)$, is

$$d(P, Q) = |b - a|$$

Since $|b - a| = |a - b|$, it follows that $d(P, Q) = d(Q, P)$.

EXAMPLE 7	**Finding Distance on a Number Line**

Let P, Q, and R be points on a real number line with coordinates $-5, 7$, and -3, respectively. Find the distance

(a) between P and Q (b) between Q and R

Solution   See Figure 8.

Figure 8

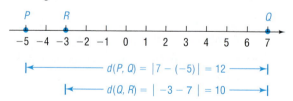

(a) $d(P, Q) = |7 - (-5)| = |12| = 12$

(b) $d(Q, R) = |-3 - 7| = |-10| = 10$ ◀

NOW WORK PROBLEM 65.

5 Use the Laws of Exponents

Integer exponents provide a shorthand device for representing repeated multiplications of a real number.

If a is a real number and n is a positive integer, then the symbol a^n represents the product of n factors of a. That is,

$$a^n = \underbrace{a \cdot a \cdot \ldots \cdot a}_{n \text{ factors}}$$

where it is understood that $a^1 = a$. Then, $a^2 = a \cdot a$, $a^3 = a \cdot a \cdot a$, and so on. In the expression a^n, a is called the **base** and n is called the **exponent**, or **power**. We read a^n as "a raised to the power n" or as "a to the nth power." We usually read a^2 as "a squared" and a^3 as "a cubed."

In working with exponents, the operation of *raising to a power* is performed before any other operation. For example,

$$4 \cdot 3^2 + 5 = 4 \cdot 9 + 5 = 36 + 5 = 41 \qquad -2^4 = -16 \qquad 2^2 + 3^2 = 4 + 9 = 13$$

Parentheses are used to indicate operations to be performed first. For example,

$$(-2)^4 = (-2)(-2)(-2)(-2) = 16 \qquad (2 + 3)^2 = 5^2 = 25$$

If $a \neq 0$, we define

$$a^0 = 1 \qquad \text{if } a \neq 0$$

If $a \neq 0$ and if n is a positive integer, then we define

$$a^{-n} = \frac{1}{a^n} \qquad \text{if } a \neq 0$$

With these definitions, the symbol a^n is defined for any integer n.

The following properties, called the **laws of exponents**, can be proved using the preceding definitions. In the list, a and b are real numbers, and m and n are integers.

Laws of Exponents

$$a^m a^n = a^{m+n} \qquad (a^m)^n = a^{mn} \qquad (ab)^n = a^n b^n$$

$$\frac{a^m}{a^n} = a^{m-n} = \frac{1}{a^{n-m}}, \quad \text{if } a \neq 0 \qquad \left(\frac{a}{b}\right)^n = \frac{a^n}{b^n}, \quad \text{if } b \neq 0$$

EXAMPLE 8   **Using the Laws of Exponents**

Write each expression so that all exponents are positive.

(a) $\dfrac{x^5 y^{-2}}{x^3 y}$, $x \neq 0$, $y \neq 0$ (b) $\left(\dfrac{x^{-3}}{3y^{-1}}\right)^{-2}$, $x \neq 0$, $y \neq 0$

Solution   (a) $\dfrac{x^5 y^{-2}}{x^3 y} = \dfrac{x^5}{x^3} \cdot \dfrac{y^{-2}}{y} = x^{5-3} \cdot y^{-2-1} = x^2 y^{-3} = x^2 \cdot \dfrac{1}{y^3} = \dfrac{x^2}{y^3}$

(b) $\left(\dfrac{x^{-3}}{3y^{-1}}\right)^{-2} = \dfrac{(x^{-3})^{-2}}{(3y^{-1})^{-2}} = \dfrac{x^6}{3^{-2}(y^{-1})^{-2}} = \dfrac{x^6}{\dfrac{1}{9}y^2} = \dfrac{9x^6}{y^2}$

◀

NOW WORK PROBLEMS 71 AND 81.

6 Evaluate Square Roots

A real number is squared when it is raised to the power 2. The inverse of squaring is finding a **square root**. For example, since $6^2 = 36$ and $(-6)^2 = 36$, the numbers 6 and -6 are square roots of 36.

The symbol $\sqrt{}$, called a **radical sign**, is used to denote the **principal**, or nonnegative, square root. Thus, $\sqrt{36} = 6$.

> In general, if a is a nonnegative real number, the nonnegative number b such that $b^2 = a$ is **the principal square root** of a and is denoted by $b = \sqrt{a}$.

The following comments are noteworthy:

1. Negative numbers do not have square roots (in the real number system), because the square of any real number is *nonnegative*. For example, $\sqrt{-4}$ is not a real number, because there is no real number whose square is -4.

2. The principal square root of 0 is 0, since $0^2 = 0$. That is, $\sqrt{0} = 0$.

3. The principal square root of a positive number is positive.

4. If $c \geq 0$, then $(\sqrt{c})^2 = c$. For example, $(\sqrt{2})^2 = 2$ and $(\sqrt{3})^2 = 3$.

EXAMPLE 9   **Evaluating Square Roots**

(a) $\sqrt{64} = 8$ (b) $\sqrt{\dfrac{1}{16}} = \dfrac{1}{4}$ (c) $(\sqrt{1.4})^2 = 1.4$ (d) $\sqrt{(-3)^2} = |-3| = 3$ ◀

Examples 9(a) and (b) are examples of square roots of perfect squares, since $64 = 8^2$ and $\dfrac{1}{16} = \left(\dfrac{1}{4}\right)^2$.

Notice the need for the absolute value in Example 9(d). Since $a^2 \geq 0$, the principal square root of a^2 is defined whether $a > 0$ or $a < 0$. However, since the principal square root is nonnegative, we need the absolute value to ensure the nonnegative result.

In general, we have

$$\sqrt{a^2} = |a| \tag{1}$$

EXAMPLE 10   **Using Equation (1)**

(a) $\sqrt{(2.3)^2} = |2.3| = 2.3$ (b) $\sqrt{(-2.3)^2} = |-2.3| = 2.3$ (c) $\sqrt{x^2} = |x|$ ◀

NOW WORK PROBLEM 77.

Calculators

Calculators are finite machines. As a result, they are incapable of displaying decimals that contain a large number of digits. For example, some calculators are capable of displaying only eight digits. When a number requires more than eight digits, the calculator either truncates or rounds. To see how your calculator handles decimals, divide 2 by 3. How many digits do you see? Is the last digit a 6 or a 7? If it is a 6, your calculator truncates; if it is a 7, your calculator rounds.

There are different kinds of calculators. An **arithmetic** calculator can only add, subtract, multiply, and divide numbers; therefore, this type is not adequate for this course. **Scientific** calculators have all the capabilities of arithmetic calculators and also contain **function keys** labeled ln, log, sin, cos, tan, x^y, inv, and so on. As you proceed through this text, you will discover how to use many of the function keys. **Graphing** calculators have all the capabilities of scientific calculators and contain a screen on which graphs can be displayed.

A.1 Assess Your Understanding

Concepts and Vocabulary

1. A _____ is a letter used in algebra to represent any number from a given set of numbers.

2. On the real number line, the real number zero is the coordinate of the _____.

3. An inequality of the form $a > b$ is called a(n) _____ inequality.

4. In the expression 2^4, the number 2 is called the _____ and 4 is called the _____.

5. $\sqrt{(-5)^2} = $ _____.

6. *True or False:* The distance between two points on the real number line is always greater than zero.

7. *True or False:* The absolute value of a real number is always greater than zero.

8. *True or False:* To multiply two expressions having the same base, retain the base and multiply the exponents.

Skill Building

In Problems 9–16, find the value of each expression if $x = -2$ and $y = 3$.

9. $x + 2y$

10. $3x + y$

11. $5xy + 2$

12. $-2x + xy$

13. $\dfrac{2x}{x - y}$

14. $\dfrac{x + y}{x - y}$

15. $\dfrac{3x + 2y}{2 + y}$

16. $\dfrac{2x - 3}{y}$

In Problems 17–24, determine which of the value(s) given below, if any, must be excluded from the domain of the variable in each expression.

(a) $x = 3$

(b) $x = 1$

(c) $x = 0$

(d) $x = -1$

17. $\dfrac{x^2 - 1}{x}$

18. $\dfrac{x^2 + 1}{x}$

19. $\dfrac{x}{x^2 - 9}$

20. $\dfrac{x}{x^2 + 9}$

21. $\dfrac{x^2}{x^2 + 1}$

22. $\dfrac{x^3}{x^2 - 1}$

23. $\dfrac{x^2 + 5x - 10}{x^3 - x}$

24. $\dfrac{-9x^2 - x + 1}{x^3 + x}$

In Problems 25–28, determine the domain of the variable x in each expression.

25. $\dfrac{4}{x - 5}$

26. $\dfrac{-6}{x + 4}$

27. $\dfrac{x}{x + 4}$

28. $\dfrac{x - 2}{x - 6}$

29. On the real number line, label the points with coordinates 0, $1, -1, \dfrac{5}{2}, -2.5, \dfrac{3}{4}$, and 0.25.

30. Repeat Problem 29 for the coordinates 0, $-2, 2, -1.5, \dfrac{3}{2}, \dfrac{1}{3}$, and $\dfrac{2}{3}$.

In Problems 31–40, replace the question mark by <, >, or =, whichever is correct.

31. $\frac{1}{2}$? 0   **32.** 5 ? 6   **33.** −1 ? −2   **34.** −3 ? −$\frac{5}{2}$ **35.** π ? 3.14

36. $\sqrt{2}$? 1.41   **37.** $\frac{1}{2}$? 0.5   **38.** $\frac{1}{3}$? 0.33   **39.** $\frac{2}{3}$? 0.67   **40.** $\frac{1}{4}$? 0.25

In Problems 41–46, write each statement as an inequality.

41. x is positive.   **42.** z is negative.   **43.** x is less than 2.

44. y is greater than −5.   **45.** x is less than or equal to 1.   **46.** x is greater than or equal to 2.

In Problems 47–50, graph the numbers x on the real number line.

47. $x \geq -2$ **48.** $x < 4$ **49.** $x > -1$ **50.** $x \leq 7$

In Problems 51–60, find the value of each expression if x = 3 and y = −2.

51. $|x + y|$ **52.** $|x - y|$ **53.** $|x| + |y|$ **54.** $|x| - |y|$ **55.** $\frac{|x|}{x}$

56. $\frac{|y|}{y}$ **57.** $|4x - 5y|$ **58.** $|3x + 2y|$ **59.** $||4x| - |5y||$ **60.** $3|x| + 2|y|$

In Problems 61–66, use the real number line below to compute each distance.

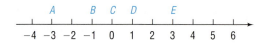

61. $d(C, D)$ **62.** $d(C, A)$ **63.** $d(D, E)$ **64.** $d(C, E)$ **65.** $d(A, E)$ **66.** $d(D, B)$

In Problems 67–78, simplify each expression.

67. $(-4)^2$ **68.** -4^2 **69.** 4^{-2} **70.** -4^{-2} **71.** $3^{-6} \cdot 3^4$ **72.** $4^{-2} \cdot 4^3$

73. $(3^{-2})^{-1}$ **74.** $(2^{-1})^{-3}$ **75.** $\sqrt{25}$ **76.** $\sqrt{36}$ **77.** $\sqrt{(-4)^2}$ **78.** $\sqrt{(-3)^2}$

In Problems 79–88, simplify each expression. Express the answer so that all exponents are positive. Whenever an exponent is 0 or negative, we assume that the base is not 0.

79. $(8x^3)^{-2}$ **80.** $(-4x^2)^{-1}$ **81.** $(x^2y^{-1})^2$ **82.** $(x^{-1}y)^3$ **83.** $\frac{x^{-2}y^3}{xy^4}$

84. $\frac{x^{-2}y}{xy^2}$ **85.** $\frac{(-2)^3x^4(yz)^2}{3^2xy^3z}$ **86.** $\frac{4x^{-2}(yz)^{-1}}{2^3x^4y}$ **87.** $\left(\frac{3x^{-1}}{4y^{-1}}\right)^{-2}$ **88.** $\left(\frac{5x^{-2}}{6y^{-2}}\right)^{-3}$

Applications and Extensions

In Problems 89–100, express each statement as an equation involving the indicated variables.

89. Area of a Rectangle   The area A of a rectangle is the product of its length l and its width w.

90. Perimeter of a Rectangle   The perimeter P of a rectangle is twice the sum of its length l and its width w.

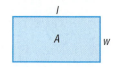

91. Circumference of a Circle   The circumference C of a circle is the product of π and its diameter d.

92. Area of a Triangle   The area A of a triangle is one-half the product of its base b and its height h.

93. Area of an Equilateral Triangle   The area A of an equilateral triangle is $\frac{\sqrt{3}}{4}$ times the square of the length x of one side.

94. Perimeter of an Equilateral Triangle The perimeter P of an equilateral triangle is 3 times the length x of one side.

95. Volume of a Sphere The volume V of a sphere is $\frac{4}{3}$ times π times the cube of the radius r.

96. Surface Area of a Sphere The surface area S of a sphere is 4 times π times the square of the radius r.

97. Volume of a Cube The volume V of a cube is the cube of the length x of a side.

98. Surface Area of a Cube The surface area S of a cube is 6 times the square of the length x of a side.

99. U.S. Voltage In the United States, normal household voltage is 115 volts. It is acceptable for the actual voltage x to differ from normal by at most 5 volts. A formula that describes this is

$$|x - 115| \le 5$$

(a) Show that a voltage of 113 volts is acceptable.
(b) Show that a voltage of 109 volts is not acceptable.

100. Foreign Voltage In countries other than the United States, normal household voltage is 220 volts. It is acceptable for the actual voltage x to differ from normal by at most 8 volts. A formula that describes this is

$$|x - 220| \le 8$$

(a) Show that a voltage of 214 volts is acceptable.
(b) Show that a voltage of 209 volts is not acceptable.

101. Making Precision Ball Bearings The FireBall Company manufactures ball bearings for precision equipment. One of their products is a ball bearing with a stated radius of 3 centimeters (cm). Only ball bearings with a radius within 0.01 cm of this stated radius are acceptable. If x is the radius of a ball bearing, a formula describing this situation is

$$|x - 3| \le 0.01$$

(a) Is a ball bearing of radius $x = 2.999$ acceptable?
(b) Is a ball bearing of radius $x = 2.89$ acceptable?

102. Body Temperature Normal human body temperature is 98.6°F. A temperature x that differs from normal by at least 1.5°F is considered unhealthy. A formula that describes this is

$$|x - 98.6| \ge 1.5$$

(a) Show that a temperature of 97°F is unhealthy.
(b) Show that a temperature of 100°F is not unhealthy.

103. Does $\frac{1}{3}$ equal 0.333? If not, which is larger? By how much?

104. Does $\frac{2}{3}$ equal 0.666? If not, which is larger? By how much?

Discussion and Writing

105. Is there a positive real number "closest" to 0?

106. I'm thinking of a number! It lies between 1 and 10; its square is rational and lies between 1 and 10. The number is larger than π. Correct to two decimal places, name the number. Now think of your own number, describe it, and challenge a fellow student to name it.

107. Write a brief paragraph that illustrates the similarities and differences between "less than" ($<$) and "less than or equal" ($\le$).

A.2 Geometry Review

> **OBJECTIVES** **1** Use the Pythagorean Theorem and Its Converse
>
> **2** Know Geometry Formulas

In this section we review some topics studied in geometry that we shall need for our study of algebra.

1 Use the Pythagorean Theorem and Its Converse

The *Pythagorean Theorem* is a statement about *right triangles*. A **right triangle** is one that contains a **right angle**, that is, an angle of 90°. The side of the triangle opposite the 90° angle is called the **hypotenuse**; the remaining two sides are called **legs**. In Figure 9 we have used c to represent the length of the hypotenuse and a and b to represent the lengths of the legs. Notice the use of the symbol ⌐ to show the 90° angle. We now state the Pythagorean Theorem.

Figure 9

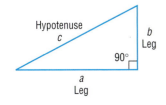

Pythagorean Theorem

In a right triangle, the square of the length of the hypotenuse is equal to the sum of the squares of the lengths of the legs. That is, in the right triangle shown in Figure 9,

$$c^2 = a^2 + b^2 \tag{1}$$

EXAMPLE 1   **Finding the Hypotenuse of a Right Triangle**

In a right triangle, one leg is of length 4 and the other is of length 3. What is the length of the hypotenuse?

Solution   Since the triangle is a right triangle, we use the Pythagorean Theorem with $a = 4$ and $b = 3$ to find the length c of the hypotenuse. From equation (1), we have

$$c^2 = a^2 + b^2$$
$$c^2 = 4^2 + 3^2 = 16 + 9 = 25$$
$$c = \sqrt{25} = 5$$

◀

   NOW WORK PROBLEM 9.

The converse of the Pythagorean Theorem is also true.

Converse of the Pythagorean Theorem

In a triangle, if the square of the length of one side equals the sum of the squares of the lengths of the other two sides, then the triangle is a right triangle. The 90° angle is opposite the longest side.

EXAMPLE 2   **Verifying That a Triangle Is a Right Triangle**

Show that a triangle whose sides are of lengths 5, 12, and 13 is a right triangle. Identify the hypotenuse.

Solution   We square the lengths of the sides.

$$5^2 = 25, \qquad 12^2 = 144, \qquad 13^2 = 169$$

Figure 10

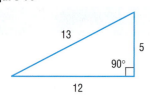

Notice that the sum of the first two squares (25 and 144) equals the third square (169). Hence, the triangle is a right triangle. The longest side, 13, is the hypotenuse. See Figure 10.

◀

NOW WORK PROBLEM 17.

EXAMPLE 3   **Applying the Pythagorean Theorem**

The tallest inhabited building in the world is the Sears Tower in Chicago. If the observation tower is 1450 feet above ground level, how far can a person standing in the observation tower see (with the aid of a telescope)? Use 3960 miles for the radius of Earth. See Figure 11.

SOURCE: Council on Tall Buildings and Urban Habitat (1997); Sears Tower No. 1 for tallest roof (1450 ft) and tallest occupied floor (1431 ft).

Figure 11

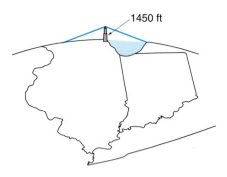

1450 ft

Solution From the center of Earth, draw two radii: one through the Sears Tower and the other to the farthest point a person can see from the tower. See Figure 12. Apply the Pythagorean Theorem to the right triangle.

Since 1 mile = 5280 feet, then 1450 feet = $\dfrac{1450}{5280}$ miles. So we have

$$d^2 + (3960)^2 = \left(3960 + \frac{1450}{5280}\right)^2$$

$$d^2 = \left(3960 + \frac{1450}{5280}\right)^2 - (3960)^2 \approx 2175.08$$

$$d \approx 46.64$$

Figure 12

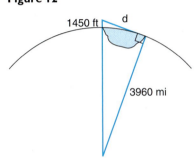

1450 ft d

3960 mi

A person can see about 47 miles from the observation tower. ◀

 NOW WORK PROBLEM 43.

2 Know Geometry Formulas

Certain formulas from geometry are useful in solving algebra problems. We list some of these formulas next.

For a rectangle of length l and width w,

w

l

Area = lw Perimeter = $2l + 2w$

For a triangle with base b and altitude h,

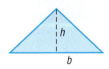

h

b

Area = $\dfrac{1}{2}bh$

For a circle of radius r (diameter $d = 2r$),

d r

Area = πr^2 Circumference = $2\pi r = \pi d$

For a closed rectangular box of length l, width w, and height h,

h

w

l

Volume = lwh Surface area = $2lh + 2wh + 2lw$

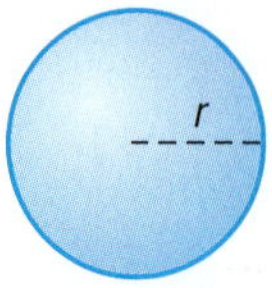

For a sphere of radius r,

$$\text{Volume} = \frac{4}{3}\pi r^3 \qquad \text{Surface area} = 4\pi r^2$$

For a right circular cylinder of height h and radius r,

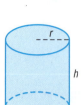

$$\text{Volume} = \pi r^2 h \qquad \text{Surface area} = 2\pi r^2 + 2\pi rh$$

NOW WORK PROBLEM 25.

EXAMPLE 4 **Using Geometry Formulas**

A Christmas tree ornament is in the shape of a semicircle on top of a triangle. How many square centimeters (cm) of copper is required to make the ornament if the height of the triangle is 6 cm and the base is 4 cm?

Solution

See Figure 13. The amount of copper required equals the shaded area. This area is the sum of the area of the triangle and the semicircle. The triangle has height $h = 6$ and base $b = 4$. The semicircle has diameter $d = 4$, so its radius is $r = 2$.

$$\text{Area} = \text{Area of triangle} + \text{Area of semicircle}$$

Figure 13

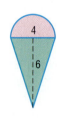

$$= \frac{1}{2}bh + \frac{1}{2}\pi r^2 = \frac{1}{2}(4)(6) + \frac{1}{2}\pi \cdot 2^2 \qquad \textcolor{blue}{b = 4; h = 6; r = 2}$$

$$= 12 + 2\pi \approx 18.28 \text{ cm}^2$$

About 18.28 cm^2 of copper is required. ◀

NOW WORK PROBLEM 39.

A.2 Assess Your Understanding

Concepts and Vocabulary

1. A _____ triangle is one that contains an angle of 90 degrees. The longest side is called the _____.

2. For a triangle with base b and altitude h, a formula for the area A is _____.

3. The formula for the circumference C of a circle of radius r is _____.

4. *True or False:* In a right triangle, the square of the length of the longest side equals the sum of the squares of the lengths of the other two sides.

5. *True or False:* The triangle with sides of length 6, 8, and 10 is a right triangle.

6. *True or False:* The volume of a sphere of radius r is $\frac{4}{3}\pi r^2$.

Skill Building

In Problems 7–12, the lengths of the legs of a right triangle are given. Find the hypotenuse.

7. $a = 5$, $b = 12$

8. $a = 6$, $b = 8$

9. $a = 10$, $b = 24$

10. $a = 4$, $b = 3$

11. $a = 7$, $b = 24$

12. $a = 14$, $b = 48$

In Problems 13–20, the lengths of the sides of a triangle are given. Determine which are right triangles. For those that are, identify the hypotenuse.

13. 3, 4, 5   **14.** 6, 8, 10   **15.** 4, 5, 6   **16.** 2, 2, 3

17. 7, 24, 25   **18.** 10, 24, 26   **19.** 6, 4, 3   **20.** 5, 4, 7

21. Find the area A of a rectangle with length 4 inches and width 2 inches.

22. Find the area A of a rectangle with length 9 centimeters and width 4 centimeters.

23. Find the area A of a triangle with height 4 inches and base 2 inches.

24. Find the area A of a triangle with height 9 centimeters and base 4 centimeters.

25. Find the area A and circumference C of a circle of radius 5 meters.

26. Find the area A and circumference C of a circle of radius 2 feet.

27. Find the volume V and surface area S of a rectangular box with length 8 feet, width 4 feet, and height 7 feet.

28. Find the volume V and surface area S of a rectangular box with length 9 inches, width 4 inches, and height 8 inches.

29. Find the volume V and surface area S of a sphere of radius 4 centimeters.

30. Find the volume V and surface area S of a sphere of radius 3 feet.

31. Find the volume V and surface area S of a right circular cylinder with radius 9 inches and height 8 inches.

32. Find the volume V and surface area S of a right circular cylinder with radius 8 inches and height 9 inches.

In Problems 33–36, find the area of the shaded region.

33.

34.

35.

36.

Applications and Extensions

37. How many feet does a wheel with a diameter of 16 inches travel after four revolutions?

38. How many revolutions will a circular disk with a diameter of 4 feet have completed after it has rolled 20 feet?

39. In the figure shown, $ABCD$ is a square, with each side of length 6 feet. The width of the border (shaded portion) between the outer square $EFGH$ and $ABCD$ is 2 feet. Find the area of the border.

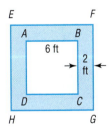

40. Refer to the figure. Square $ABCD$ has an area of 100 square feet; square $BEFG$ has an area of 16 square feet. What is the area of the triangle CGF?

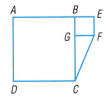

41. Architecture   A Norman window consists of a rectangle surmounted by a semicircle. Find the area of the Norman window shown in the illustration. How much wood frame is needed to enclose the window?

42. Construction   A circular swimming pool, 20 feet in diameter, is enclosed by a wooden deck that is 3 feet wide. What is the area of the deck? How much fence is required to enclose the deck?

In Problems 43–45, use the facts that the radius of Earth is 3960 miles and 1 mile = 5280 feet.

43. **How Far Can You See?** The conning tower of the U.S.S. *Silversides*, a World War II submarine now permanently stationed in Muskegon, Michigan, is approximately 20 feet above sea level. How far can you see from the conning tower?

44. **How Far Can You See?** A person who is 6 feet tall is standing on the beach in Fort Lauderdale, Florida, and looks out onto the Atlantic Ocean. Suddenly, a ship appears on the horizon. How far is the ship from shore?

45. **How Far Can You See?** The deck of a destroyer is 100 feet above sea level. How far can a person see from the deck? How far can a person see from the bridge, which is 150 feet above sea level?

46. Suppose that m and n are positive integers with $m > n$. If $a = m^2 - n^2$, $b = 2mn$, and $c = m^2 + n^2$, show that a, b, and c are the lengths of the sides of a right triangle. (This formula can be used to find the sides of a right triangle that are integers, such as 3, 4, 5; 5, 12, 13; and so on. Such triplets of integers are called **Pythagorean triples.**)

Discussion and Writing

47. You have 1000 feet of flexible pool siding and wish to construct a swimming pool. Experiment with rectangular-shaped pools with perimeters of 1000 feet. How do their areas vary? What is the shape of the rectangle with the largest area? Now compute the area enclosed by a circular pool with a perimeter (circumference) of 1000 feet. What would be your choice of shape for the pool? If rectangular, what is your preference for dimensions? Justify your choice. If your only consideration is to have a pool that encloses the most area, what shape should you use?

48. **The Gibb's Hill Lighthouse, Southampton, Bermuda,** in operation since 1846, stands 117 feet high on a hill 245 feet high, so its beam of light is 362 feet above sea level. A brochure states that the light itself can be seen on the horizon about 26 miles distant. Verify the correctness of this information. The brochure further states that ships 40 miles away can see the light and planes flying at 10,000 feet can see it 120 miles away. Verify the accuracy of these statements. What assumption did the brochure make about the height of the ship?

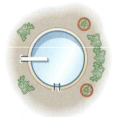

A.3 Polynomials and Rational Expressions

OBJECTIVES 1 Recognize Special Products
2 Factor Polynomials
3 Simplify Rational Expressions
4 Use the LCM to Add Rational Expressions

As we said earlier, in algebra we use letters to represent real numbers. We shall use the letters at the end of the alphabet, such as x, y, and z, to represent variables and the letters at the beginning of the alphabet, such as a, b, and c, to represent constants. In the expressions $3x + 5$ and $ax + b$, it is understood that x is a variable and that a and b are constants, even though the constants a and b are unspecified. As you will find out, the context usually makes the intended meaning clear.

Now we introduce some basic vocabulary.

A **monomial** in one variable is the product of a constant and a variable raised to a nonnegative integer power. Thus, a monomial is of the form

$$ax^k$$

where a is a constant, x is a variable, and $k \geq 0$ is an integer. The constant a is called the **coefficient** of the monomial. If $a \neq 0$, then k is called the **degree** of the monomial.

Examples of monomials follow:

Monomial	Coefficient	Degree	
$6x^2$	6	2	
$-\sqrt{2}x^3$	$-\sqrt{2}$	3	
3	3	0	Since $3 = 3 \cdot 1 = 3x^0$
$-5x$	-5	1	Since $-5x = -5x^1$
x^4	1	4	Since $x^4 = 1 \cdot x^4$

Two monomials ax^k and bx^k with the same degree and the same variable are called **like terms**. Such monomials when added or subtracted can be combined into a single monomial by using the distributive property. For example,

$$2x^2 + 5x^2 = (2 + 5)x^2 = 7x^2 \quad \text{and} \quad 8x^3 - 5x^3 = (8 - 5)x^3 = 3x^3$$

The sum or difference of two monomials having different degrees is called a **binomial**. The sum or difference of three monomials with three different degrees is called a **trinomial**. For example,

$x^2 - 2$ is a binomial.

$x^3 - 3x + 5$ is a trinomial.

$2x^2 + 5x^2 + 2 = 7x^2 + 2$ is a binomial.

In Words

A polynomial is a sum of monomials.

A **polynomial** in one variable is an algebraic expression of the form

$$a_n x^n + a_{n-1} x^{n-1} + \cdots + a_1 x + a_0 \tag{1}$$

where $a_n, a_{n-1}, \ldots, a_1, a_0$ are constants[*] called the **coefficients** of the polynomial, $n \geq 0$ is an integer, and x is a variable. If $a_n \neq 0$, it is called the **leading coefficient**, and n is called the **degree** of the polynomial.

The monomials that make up a polynomial are called its **terms**. If all the coefficients are 0, the polynomial is called the **zero polynomial**, which has no degree.

---

[*]The notation a_n is read as "a sub n." The number n is called a **subscript** and should not be confused with an exponent. We use subscripts in order to distinguish one constant from another when a large or undetermined number of constants is required.

Polynomials are usually written in **standard form**, beginning with the nonzero term of highest degree and continuing with terms in descending order according to degree. If a power of x is missing, it is because its coefficient is zero. Examples of polynomials follow:

Polynomial	Coefficients	Degree
$3x^2 - 5 = 3x^2 + 0 \cdot x + (-5)$	$3, 0, -5$	2
$8 - 2x + x^2 = 1 \cdot x^2 - 2x + 8$	$1, -2, 8$	2
$5x + \sqrt{2} = 5x^1 + \sqrt{2}$	$5, \sqrt{2}$	1
$3 = 3 \cdot 1 = 3 \cdot x^0$	3	0
0	0	No degree

Although we have been using x to represent the variable, letters such as y or z are also commonly used.

$3x^4 - x^2 + 2$ is a polynomial (in x) of degree 4.

$9y^3 - 2y^2 + y - 3$ is a polynomial (in y) of degree 3.

$z^5 + \pi$ is a polynomial (in z) of degree 5.

Algebraic expressions such as

$$\frac{1}{x} \quad \text{and} \quad \frac{x^2 + 1}{x + 5}$$

are not polynomials. The first is not a polynomial because $\dfrac{1}{x} = x^{-1}$ has an exponent that is not a nonnegative integer. Although the second expression is the quotient of two polynomials, the polynomial in the denominator has degree greater than 0, so the expression cannot be a polynomial.

1 Recognize Special Products

Certain products, which we call **special products**, occur frequently in algebra. In the list that follows, x, a, b, c, and d are real numbers.

Difference of Two Squares

$$(x - a)(x + a) = x^2 - a^2 \tag{2}$$

Squares of Binomials, or Perfect Squares

$$(x + a)^2 = x^2 + 2ax + a^2 \tag{3a}$$

$$(x - a)^2 = x^2 - 2ax + a^2 \tag{3b}$$

Miscellaneous Trinomials

$$(x + a)(x + b) = x^2 + (a + b)x + ab \tag{4a}$$

$$(ax + b)(cx + d) = acx^2 + (ad + bc)x + bd \tag{4b}$$

Cubes of Binomials, or Perfect Cubes

$$(x + a)^3 = x^3 + 3ax^2 + 3a^2x + a^3 \qquad \textbf{(5a)}$$

$$(x - a)^3 = x^3 - 3ax^2 + 3a^2x - a^3 \qquad \textbf{(5b)}$$

Difference of Two Cubes

$$(x - a)(x^2 + ax + a^2) = x^3 - a^3 \qquad \textbf{(6)}$$

Sum of Two Cubes

$$(x + a)(x^2 - ax + a^2) = x^3 + a^3 \qquad \textbf{(7)}$$

The special product formulas in equations (2) through (7) are used often, and their patterns should be committed to memory. But if you forget one or are unsure of its form, you should be able to derive it as needed.

EXAMPLE 1   **Using Special Formulas**

(a) $(x - 4)(x + 4) = x^2 - 4^2 = x^2 - 16$

(b) $(2x + 5)(3x - 1) = 6x^2 - 2x + 15x - 5 = 6x^2 + 13x - 5$

(c) $(x - 2)^3 = x^3 - 3(2)x^2 + 3(2)^2x - (2)^3 = x^3 - 6x^2 + 12x - 8$ ◀

NOW WORK PROBLEM 17.

2   Factor Polynomials

Consider the following product:

$$(2x + 3)(x - 4) = 2x^2 - 5x - 12$$

The two polynomials on the left are called **factors** of the polynomial on the right. Expressing a given polynomial as a product of other polynomials, that is, finding the factors of a polynomial, is called **factoring**.

We shall restrict our discussion here to factoring polynomials in one variable into products of polynomials in one variable, where all coefficients are integers. We call this **factoring over the integers**.

Any polynomial can be written as the product of 1 times itself or as −1 times its additive inverse. If a polynomial cannot be written as the product of two other polynomials (excluding 1 and −1), then the polynomial is said to be **prime**. When a polynomial has been written as a product consisting only of prime factors, it is said to be **factored completely**. Examples of prime polynomials are

$$2, \quad 3, \quad 5, \quad x, \quad x + 1, \quad x - 1, \quad 3x + 4$$

The first factor to look for in a factoring problem is a common monomial factor present in each term of the polynomial. If one is present, use the distributive property to factor it out. For example,

Polynomial	Common Monomial Factor	Remaining Factor	Factored Form
$2x + 4$	2	$x + 2$	$2x + 4 = 2(x + 2)$
$3x - 6$	3	$x - 2$	$3x - 6 = 3(x - 2)$
$2x^2 - 4x + 8$	2	$x^2 - 2x + 4$	$2x^2 - 4x + 8 = 2(x^2 - 2x + 4)$
$8x - 12$	4	$2x - 3$	$8x - 12 = 4(2x - 3)$
$x^2 + x$	x	$x + 1$	$x^2 + x = x(x + 1)$
$x^3 - 3x^2$	x^2	$x - 3$	$x^3 - 3x^2 = x^2(x - 3)$
$6x^2 + 9x$	$3x$	$2x + 3$	$6x^2 + 9x = 3x(2x + 3)$

The list of special products (2) through (7) given earlier provides a list of factoring formulas when the equations are read from right to left. For example, equation (2) states that if the polynomial is the difference of two squares, $x^2 - a^2$, it can be factored into $(x - a)(x + a)$. The following example illustrates several factoring techniques.

EXAMPLE 2 **Factoring Polynomials**

Factor completely each polynomial.

(a) $x^4 - 16$ (b) $x^3 - 1$ (c) $9x^2 - 6x + 1$

(d) $x^2 + 4x - 12$ (e) $3x^2 + 10x - 8$ (f) $x^3 - 4x^2 + 2x - 8$

Solution

(a) $x^4 - 16 = (x^2 - 4)(x^2 + 4) = (x - 2)(x + 2)(x^2 + 4)$

      Difference of squares   Difference of squares

(b) $x^3 - 1 = (x - 1)(x^2 + x + 1)$

      Difference of cubes

(c) $9x^2 - 6x + 1 = (3x - 1)^2$

      Perfect square

(d) $x^2 + 4x - 12 = (x + 6)(x - 2)$

    The product of 6 and -2 is -12, and the sum of 6 and -2 is 4.

    $12x - 2x = 10x$

(e) $3x^2 + 10x - 8 = (3x - 2)(x + 4)$

    $3x^2$ -8

(f) $x^3 - 4x^2 + 2x - 8 = (x^3 - 4x^2) + (2x - 8)$

      Regroup

    $= x^2(x - 4) + 2(x - 4) = (x^2 + 2)(x - 4)$

      Distributive property   Distributive property ◀

The technique used in Example 2(f) is called **factoring by grouping**.

NOW WORK PROBLEMS 31, 47, AND 79.

3 Simplify Rational Expressions

If we form the quotient of two polynomials, the result is called a **rational expression.** Some examples of rational expressions are

$$\text{(a)} \ \frac{x^3 + 1}{x} \qquad \text{(b)} \ \frac{3x^3 + x - 2}{x^5 + 5} \qquad \text{(c)} \ \frac{x}{x^2 - 1} \qquad \text{(d)} \ \frac{xy^2}{(x - y)^2}$$

Expressions (a), (b), and (c) are rational expressions in one variable, x, whereas (d) is a rational expression in two variables, x and y.

Rational expressions are described in the same manner as rational numbers. Thus, in expression (a), the polynomial $x^3 + 1$ is called the **numerator**, and x is called the **denominator**. When the numerator and denominator of a rational expression contain no common factors (except 1 and -1), we say that the rational expression is **reduced to lowest terms**, or **simplified**.

A rational expression is reduced to lowest terms by completely factoring the numerator and the denominator and canceling any common factors by using the cancellation property.

$$\frac{ac}{bc} = \frac{a}{b}, \qquad b \neq 0, \ \ c \neq 0$$

We shall follow the common practice of using a slash mark to indicate cancellation. For example,

$$\frac{x^2 - 1}{x^2 - 2x - 3} = \frac{(x - 1)\cancel{(x + 1)}}{(x - 3)\cancel{(x + 1)}} = \frac{x - 1}{x - 3}$$

| **EXAMPLE 3** | **Simplifying Rational Expressions** |

Reduce each rational expression to lowest terms.

$$\text{(a)} \ \frac{x^2 + 4x + 4}{x^2 + 3x + 2} \qquad\qquad \text{(b)} \ \frac{x^3 - 8}{x^3 - 2x^2} \qquad\qquad \text{(c)} \ \frac{8 - 2x}{x^2 - x - 12}$$

Solution (a) $\dfrac{x^2 + 4x + 4}{x^2 + 3x + 2} = \dfrac{\cancel{(x + 2)}(x + 2)}{\cancel{(x + 2)}(x + 1)} = \dfrac{x + 2}{x + 1}, \qquad x \neq -2, -1$

(b) $\dfrac{x^3 - 8}{x^3 - 2x^2} = \dfrac{\cancel{(x - 2)}(x^2 + 2x + 4)}{x^2\cancel{(x - 2)}} = \dfrac{x^2 + 2x + 4}{x^2}, \qquad x \neq 0, 2$

(c) $\dfrac{8 - 2x}{x^2 - x - 12} = \dfrac{2(4 - x)}{(x - 4)(x + 3)} = \dfrac{2(-1)\cancel{(x - 4)}}{\cancel{(x - 4)}(x + 3)} = \dfrac{-2}{x + 3}, \quad x \neq -3, 4$ ◀

NOW WORK PROBLEM 89.

The rules for multiplying and dividing rational expressions are the same as the rules for multiplying and dividing rational numbers.

$$\frac{a}{b} \cdot \frac{c}{d} = \frac{ac}{bd}, \qquad \text{if } b \neq 0, d \neq 0 \tag{8}$$

$$\frac{\dfrac{a}{b}}{\dfrac{c}{d}} = \frac{a}{b} \cdot \frac{d}{c} = \frac{ad}{bc}, \qquad \text{if } b \neq 0, c \neq 0, d \neq 0 \tag{9}$$

In using equations (8) and (9) with rational expressions, be sure first to factor each polynomial completely so that common factors can be canceled. We shall follow the practice of leaving our answers in factored form.

| EXAMPLE 4 | **Finding Products and Quotients of Rational Expressions** |

Perform the indicated operation and simplify the result. Leave your answer in factored form.

(a) $\dfrac{x^2 - 2x + 1}{x^3 + x} \cdot \dfrac{4x^2 + 4}{x^2 + x - 2}$

(b) $\dfrac{\dfrac{x + 3}{x^2 - 4}}{\dfrac{x^2 - x - 12}{x^3 - 8}}$

Solution

(a) $\dfrac{x^2 - 2x + 1}{x^3 + x} \cdot \dfrac{4x^2 + 4}{x^2 + x - 2} = \dfrac{(x - 1)^2}{x(x^2 + 1)} \cdot \dfrac{4(x^2 + 1)}{(x + 2)(x - 1)}$

$\qquad = \dfrac{(x - 1)^2(4)\cancel{(x^2 + 1)}}{x\cancel{(x^2 + 1)}(x + 2)\cancel{(x - 1)}} = \dfrac{4(x - 1)}{x(x + 2)}, \quad x \neq -2, 0, 1$

(b) $\dfrac{\dfrac{x + 3}{x^2 - 4}}{\dfrac{x^2 - x - 12}{x^3 - 8}} = \dfrac{x + 3}{x^2 - 4} \cdot \dfrac{x^3 - 8}{x^2 - x - 12}$

$\qquad = \dfrac{x + 3}{(x - 2)(x + 2)} \cdot \dfrac{(x - 2)(x^2 + 2x + 4)}{(x - 4)(x + 3)}$

$\qquad = \dfrac{\cancel{(x + 3)}\cancel{(x - 2)}(x^2 + 2x + 4)}{\cancel{(x - 2)}(x + 2)(x - 4)\cancel{(x + 3)}} = \dfrac{x^2 + 2x + 4}{(x + 2)(x - 4)}, x \neq -3, -2, 2, 4$ ◄

NOTE Slanting the cancellation marks in different directions for different factors, as in Example 4, is a good practice to follow, since it will help in checking for errors. ∎

✎— **NOW WORK PROBLEM 67.**

In Words

Keep the common denominator and add (subtract) the numerators.

If the denominators of two rational expressions to be added (or subtracted) are the same, we add (or subtract) the numerators and keep the common denominator. That is, if $\dfrac{a}{b}$ and $\dfrac{c}{b}$ are two rational expressions, then

$$\frac{a}{b} + \frac{c}{b} = \frac{a + c}{b} \qquad \frac{a}{b} - \frac{c}{b} = \frac{a - c}{b}, \qquad \text{if } b \neq 0 \tag{10}$$

EXAMPLE 5   **Finding the Sum of Two Rational Expressions**

Perform the indicated operation and simplify the result. Leave your answer in factored form.

$$\frac{2x^2 - 4}{2x + 5} + \frac{x + 3}{2x + 5} \qquad x \neq -\frac{5}{2}$$

Solution $\dfrac{2x^2 - 4}{2x + 5} + \dfrac{x + 3}{2x + 5} = \dfrac{(2x^2 - 4) + (x + 3)}{2x + 5}$

$$= \frac{2x^2 + x - 1}{2x + 5} = \frac{(2x - 1)(x + 1)}{2x + 5} \qquad \blacktriangleleft$$

If the denominators of two rational expressions to be added or subtracted are not the same, we can use the general formulas for adding and subtracting quotients.

$$\frac{a}{b} + \frac{c}{d} = \frac{a \cdot d}{b \cdot d} + \frac{b \cdot c}{b \cdot d} = \frac{ad + bc}{bd}, \qquad \text{if } b \neq 0, d \neq 0$$

$$\frac{a}{b} - \frac{c}{d} = \frac{a \cdot d}{b \cdot d} - \frac{b \cdot c}{b \cdot d} = \frac{ad - bc}{bd}, \qquad \text{if } b \neq 0, d \neq 0$$

(11)

EXAMPLE 6   **Finding the Difference of Two Rational Expressions**

Perform the indicated operation and simplify the result. Leave your answer in factored form.

$$\frac{x^2}{x^2 - 4} - \frac{1}{x} \qquad x \neq -2, 0, 2$$

Solution $\dfrac{x^2}{x^2 - 4} - \dfrac{1}{x} = \dfrac{x^2(x) - (x^2 - 4)(1)}{(x^2 - 4)(x)} = \dfrac{x^3 - x^2 + 4}{(x - 2)(x + 2)(x)} \qquad \blacktriangleleft$

 Use the Least Common Multiple (LCM) to Add Rational Expressions

If the denominators of two rational expressions to be added (or subtracted) have common factors, we usually do not use the general rules given by equation (11), since, in doing so, we make the problem more complicated than it needs to be. Instead, just as with fractions, we apply the **least common multiple (LCM) method** by using the polynomial of least degree that contains each denominator polynomial as a factor. Then we rewrite each rational expression using the LCM as the common denominator and use equation (10) to do the addition (or subtraction).

To find the least common multiple of two or more polynomials, first factor completely each polynomial. The LCM is the product of the different prime factors of each polynomial, each factor appearing the greatest number of times it occurs in each polynomial. The next example will give you the idea.

| EXAMPLE 7 | **Finding the Least Common Multiple** |

Find the least common multiple of the following pair of polynomials:

$$x(x - 1)^2(x + 1) \quad \text{and} \quad 4(x - 1)(x + 1)^3$$

Solution The polynomials are already factored completely as

$$x(x - 1)^2(x + 1) \quad \text{and} \quad 4(x - 1)(x + 1)^3$$

Start by writing the factors of the left-hand polynomial. (Alternatively, you could start with the one on the right.)

$$x(x - 1)^2(x + 1)$$

Now look at the right-hand polynomial. Its first factor, 4, does not appear in our list, so we insert it:

$$4x(x - 1)^2(x + 1)$$

The next factor, $x - 1$, is already in our list, so no change is necessary. The final factor is $(x + 1)^3$. Since our list has $x + 1$ to the first power only, we replace $x + 1$ in the list by $(x + 1)^3$. The LCM is

$$4x(x - 1)^2(x + 1)^3$$

Notice that the LCM is, in fact, the polynomial of least degree that contains $x(x - 1)^2(x + 1)$ and $4(x - 1)(x + 1)^3$ as factors. ◀

The next example illustrates how the LCM is used for adding and subtracting rational expressions.

| EXAMPLE 8 | **Using the LCM to Add Rational Expressions** |

Perform the indicated operation and simplify the result. Leave your answer in factored form.

$$\frac{x}{x^2 + 3x + 2} + \frac{2x - 3}{x^2 - 1}, \quad x \neq -2, -1, 1$$

Solution First, we find the LCM of the denominators.

$$x^2 + 3x + 2 = (x + 2)(x + 1)$$
$$x^2 - 1 = (x - 1)(x + 1)$$

The LCM is $(x + 2)(x + 1)(x - 1)$. Next, we rewrite each rational expression using the LCM as the denominator.

$$\frac{x}{x^2 + 3x + 2} = \frac{x}{(x + 2)(x + 1)} = \frac{x(x - 1)}{(x + 2)(x + 1)(x - 1)}$$

↑ Multiply numerator and denominator by $x - 1$ to get the LCM in the denominator.

$$\frac{2x - 3}{x^2 - 1} = \frac{2x - 3}{(x - 1)(x + 1)} = \frac{(2x - 3)(x + 2)}{(x - 1)(x + 1)(x + 2)}$$

↑ Multiply numerator and denominator by $x + 2$ to get the LCM in the denominator.

Now we can add using equation (10).

$$\frac{x}{x^2 + 3x + 2} + \frac{2x - 3}{x^2 - 1} = \frac{x(x - 1)}{(x + 2)(x + 1)(x - 1)} + \frac{(2x - 3)(x + 2)}{(x + 2)(x + 1)(x - 1)}$$

$$= \frac{(x^2 - x) + (2x^2 + x - 6)}{(x + 2)(x + 1)(x - 1)}$$

$$= \frac{3x^2 - 6}{(x + 2)(x + 1)(x - 1)} = \frac{3(x^2 - 2)}{(x + 2)(x + 1)(x - 1)} \quad \blacktriangleleft$$

If we had not used the LCM technique to add the quotients in Example 8, but decided instead to use the general rule of equation (11), we would have obtained a more complicated expression, as follows:

$$\frac{x}{x^2 + 3x + 2} + \frac{2x - 3}{x^2 - 1} = \frac{x(x^2 - 1) + (x^2 + 3x + 2)(2x - 3)}{(x^2 + 3x + 2)(x^2 - 1)}$$

$$= \frac{3x^3 + 3x^2 - 6x - 6}{(x^2 + 3x + 2)(x^2 - 1)} = \frac{3(x^3 + x^2 - 2x - 2)}{(x^2 + 3x + 2)(x^2 - 1)}$$

Now we are faced with a more complicated problem of expressing this quotient in lowest terms. It is always best to first look for common factors in the denominators of expressions to be added or subtracted and to use the LCM if any common factors are found.

 **NOW WORK PROBLEM 71.**

A.3 Assess Your Understanding

Concepts and Vocabulary

1. The polynomial $3x^4 - 2x^3 + 13x^2 - 5$ is of degree _____. The leading coefficient is _____.

2. $(x^2 - 4)(x^2 + 4) = $ _____.

3. $(x - 2)(x^2 + 2x + 4) = $ _____.

4. *True or False:* $4x^{-2}$ is a monomial of degree -2.

5. *True or False:* The degree of the product of two nonzero polynomials equals the sum of their degrees.

6. *True or False:* $(x + a)(x^2 + ax + a) = x^3 + a^3$.

7. If factored completely, $3x^3 - 12x = $ _____.

8. If a polynomial cannot be written as the product of two other polynomials (excluding 1 and -1), then the polynomial is said to be _____.

9. *True or False:* The polynomial $x^2 + 4$ is prime.

10. *True or False:* $3x^3 - 2x^2 - 6x + 4 = (3x - 2)(x^3 + 2)$.

11. When the numerator and denominator of a rational expression contain no common factors (except 1 and -1), the rational expression is _____.

12. LCM is an abbreviation for _____ _____ _____.

13. *True or False:* The rational expression $\dfrac{2x^3 - 4x}{x - 2}$ is reduced to lowest terms.

14. *True or False:* The LCM of $2x^3 + 6x^2$ and $6x^4 + 4x^3$ is $4x^3(x + 1)$.

Skill Building

In Problems 15–24, perform the indicated operations. Express each answer as a polynomial written in standard form.

15. $(10x^5 - 8x^2) + (3x^3 - 2x^2 + 6)$

16. $3(x^2 - 3x + 1) + 2(3x^2 + x - 4)$

17. $(x + a)^2 - x^2$

18. $(x - a)^2 - x^2$

19. $(x + 8)(2x + 1)$

20. $(2x - 1)(x + 2)$

21. $(x^2 + x - 1)(x^2 - x + 1)$

22. $(x^2 + 2x + 1)(x^2 - 3x + 4)$

23. $(x + 1)^3 - (x - 1)^3$

24. $(x + 1)^3 - (x + 2)^3$

In Problems 25–66, factor completely each polynomial. If the polynomial cannot be factored, say it is prime.

25. $x^2 - 36$

26. $x^2 - 9$

27. $1 - 4x^2$

28. $1 - 9x^2$

29. $x^2 + 7x + 10$

30. $x^2 + 5x + 4$

31. $x^2 - 2x + 8$

32. $x^2 - 4x + 5$

33. $x^2 + 4x + 16$

34. $x^2 + 12x + 36$

35. $15 + 2x - x^2$

36. $14 + 6x - x^2$

37. $3x^2 - 12x - 36$

38. $x^3 + 8x^2 - 20x$

39. $y^4 + 11y^3 + 30y^2$

40. $3y^3 - 18y^2 - 48y$

41. $4x^2 + 12x + 9$

42. $9x^2 - 12x + 4$

43. $3x^2 + 4x + 1$

44. $4x^2 + 3x - 1$

45. $x^4 - 81$

46. $x^4 - 1$

47. $x^6 - 2x^3 + 1$

48. $x^6 + 2x^3 + 1$

49. $x^7 - x^5$

50. $x^8 - x^5$

51. $5 + 16x - 16x^2$

52. $5 + 11x - 16x^2$

53. $4y^2 - 16y + 15$

54. $9y^2 + 9y - 4$

55. $1 - 8x^2 - 9x^4$

56. $4 - 14x^2 - 8x^4$

57. $x(x + 3) - 6(x + 3)$

58. $5(3x - 7) + x(3x - 7)$

59. $(x + 2)^2 - 5(x + 2)$

60. $(x - 1)^2 - 2(x - 1)$

61. $6x(2 - x)^4 - 9x^2(2 - x)^3$

62. $6x(1 - x^2)^4 - 24x^3(1 - x^2)^3$

63. $x^3 + 2x^2 - x - 2$

64. $x^3 - 3x^2 - x + 3$

65. $x^4 - x^3 + x - 1$

66. $x^4 + x^3 + x + 1$

In Problems 67–78, perform the indicated operation and simplify the result. Leave your answer in factored form.

67. $\dfrac{3x - 6}{5x} \cdot \dfrac{x^2 - x - 6}{x^2 - 4}$

68. $\dfrac{9x^2 - 25}{2x - 2} \cdot \dfrac{1 - x^2}{6x - 10}$

69. $\dfrac{4x^2 - 1}{x^2 - 16} \cdot \dfrac{x^2 - 4x}{2x + 1}$

70. $\dfrac{12}{x^2 - x} \cdot \dfrac{x^2 - 1}{4x - 2}$

71. $\dfrac{x}{x^2 - 7x + 6} - \dfrac{x}{x^2 - 2x - 24}$

72. $\dfrac{x}{x - 3} - \dfrac{x + 1}{x^2 + 5x - 24}$

73. $\dfrac{4}{x^2 - 4} - \dfrac{2}{x^2 + x - 6}$

74. $\dfrac{3}{x - 1} - \dfrac{x - 4}{x^2 - 2x + 1}$

75. $\dfrac{1}{x} - \dfrac{2}{x^2 + x} + \dfrac{3}{x^3 - x^2}$

76. $\dfrac{x}{(x - 1)^2} + \dfrac{2}{x} - \dfrac{x + 1}{x^3 - x^2}$

77. $\dfrac{1}{h}\left(\dfrac{1}{x + h} - \dfrac{1}{x}\right)$

78. $\dfrac{1}{h}\left[\dfrac{1}{(x + h)^2} - \dfrac{1}{x^2}\right]$

In Problems 79–88, expressions that occur in calculus are given. Factor completely each expression.

79. $2(3x + 4)^2 + (2x + 3) \cdot 2(3x + 4) \cdot 3$

80. $5(2x + 1)^2 + (5x - 6) \cdot 2(2x + 1) \cdot 2$

81. $2x(2x + 5) + x^2 \cdot 2$

82. $3x^2(8x - 3) + x^3 \cdot 8$

83. $2(x + 3)(x - 2)^3 + (x + 3)^2 \cdot 3(x - 2)^2$

84. $4(x + 5)^3(x - 1)^2 + (x + 5)^4 \cdot 2(x - 1)$

85. $(4x - 3)^2 + x \cdot 2(4x - 3) \cdot 4$

86. $3x^2(3x + 4)^2 + x^3 \cdot 2(3x + 4) \cdot 3$

87. $2(3x - 5) \cdot 3(2x + 1)^3 + (3x - 5)^2 \cdot 3(2x + 1)^2 \cdot 2$

88. $3(4x + 5)^2 \cdot 4(5x + 1)^2 + (4x + 5)^3 \cdot 2(5x + 1) \cdot 5$

In Problems 89–96, expressions that occur in calculus are given. Reduce each expression to lowest terms.

89. $\dfrac{(2x + 3) \cdot 3 - (3x - 5) \cdot 2}{(3x - 5)^2}$

90. $\dfrac{(4x + 1) \cdot 5 - (5x - 2) \cdot 4}{(5x - 2)^2}$

91. $\dfrac{x \cdot 2x - (x^2 + 1) \cdot 1}{(x^2 + 1)^2}$

92. $\dfrac{x \cdot 2x - (x^2 - 4) \cdot 1}{(x^2 - 4)^2}$

93. $\dfrac{(3x + 1) \cdot 2x - x^2 \cdot 3}{(3x + 1)^2}$

94. $\dfrac{(2x - 5) \cdot 3x^2 - x^3 \cdot 2}{(2x - 5)^2}$

95. $\dfrac{(x^2 + 1) \cdot 3 - (3x + 4) \cdot 2x}{(x^2 + 1)^2}$

96. $\dfrac{(x^2 + 9) \cdot 2 - (2x - 5) \cdot 2x}{(x^2 + 9)^2}$

A.4 Polynomial Division; Synthetic Division

OBJECTIVES 1 Divide Polynomials Using Long Division

2 Divide Polynomials Using Synthetic Division

1 Divide Polynomials Using Long Division

The procedure for dividing two polynomials is similar to the procedure for dividing two integers.

EXAMPLE 1 **Dividing Two Integers**

Divide 842 by 15.

Solution

$$
\begin{array}{r}
56 \quad \leftarrow Quotient \\
Divisor \rightarrow \; 15\overline{)842} \quad \leftarrow Dividend \\
75 \quad \leftarrow 5 \cdot 15 \;(Subtract) \\
\overline{92} \\
90 \quad \leftarrow 6 \cdot 15 \;(Subtract) \\
\overline{2} \quad \leftarrow Remainder
\end{array}
$$

So, $\dfrac{842}{15} = 56 + \dfrac{2}{15}$. ◀

In the long division process detailed in Example 1, the number 15 is called the **divisor**, the number 842 is called the **dividend**, the number 56 is called the **quotient**, and the number 2 is called the **remainder**.

To check the answer obtained in a division problem, multiply the quotient by the divisor and add the remainder. The answer should be the dividend.

$$(\text{Quotient})(\text{Divisor}) + \text{Remainder} = \text{Dividend}$$

For example, we can check the results obtained in Example 1 as follows:

$$(56)(15) + 2 = 840 + 2 = 842$$

To divide two polynomials, we first must write each polynomial in standard form. The process then follows a pattern similar to that of Example 1. The next example illustrates the procedure.

EXAMPLE 2 **Dividing Two Polynomials**

Find the quotient and the remainder when

$$3x^3 + 4x^2 + x + 7 \quad \text{is divided by} \quad x^2 + 1$$

Solution Each polynomial is in standard form. The dividend is $3x^3 + 4x^2 + x + 7$, and the divisor is $x^2 + 1$.

STEP 1: Divide the leading term of the dividend, $3x^3$, by the leading term of the divisor, x^2. Enter the result, $3x$, over the term $3x^3$, as follows:

$$\frac{3x}{x^2 + 1\overline{)3x^3 + 4x^2 + x + 7}}$$

STEP 2: Multiply $3x$ by $x^2 + 1$ and enter the result below the dividend.

$$
\begin{array}{r}
3x \\
x^2 + 1\overline{)3x^3 + 4x^2 + x + 7} \\
\underline{3x^3 + 3x} \\
\end{array}
$$

$\leftarrow 3x \cdot (x^2 + 1) = 3x^3 + 3x$

Notice that we align the $3x$ term under the x to make the next step easier.

STEP 3: Subtract and bring down the remaining terms.

$$
\begin{array}{r}
3x \\
x^2 + 1\overline{)3x^3 + 4x^2 + x + 7} \\
\underline{3x^3 + 3x} \\
4x^2 - 2x + 7 \\
\end{array}
$$

$\leftarrow$ Subtract (change the signs and add)
$\leftarrow$ Bring down the $4x^2$ and the 7.

STEP 4: Repeat Steps 1–3 using $4x^2 - 2x + 7$ as the dividend.

$$
\begin{array}{r}
3x + 4 \\
x^2 + 1\overline{)3x^3 + 4x^2 + x + 7} \\
\underline{3x^3 + 3x} \\
4x^2 - 2x + 7 \\
\underline{4x^2 + 4} \\
-2x + 3 \\
\end{array}
$$

$\leftarrow$ Divide $4x^2$ by x^2 to get 4.
$\leftarrow$ Multiply $x^2 + 1$ by 4; subtract.

Since x^2 does not divide $-2x$ evenly (that is, the result is not a monomial), the process ends. The quotient is $3x + 4$, and the remainder is $-2x + 3$.

✔ **CHECK:** (Quotient)(Divisor) + Remainder

$$= (3x + 4)(x^2 + 1) + (-2x + 3)$$
$$= 3x^3 + 4x^2 + 3x + 4 + (-2x + 3)$$
$$= 3x^3 + 4x^2 + x + 7$$
$$= \text{Dividend}$$

Then

$$\frac{3x^3 + 4x^2 + x + 7}{x^2 + 1} = 3x + 4 + \frac{-2x + 3}{x^2 + 1}$$

◄

The next example combines the steps involved in long division.

| **EXAMPLE 3** | **Dividing Two Polynomials** |

Find the quotient and the remainder when

$$x^4 - 3x^3 + 2x - 5 \quad \text{is divided by} \quad x^2 - x + 1$$

Solution In setting up this division problem, it is necessary to leave a space for the missing x^2 term in the dividend.

$$
\begin{array}{r}
x^2 - 2x \; - \; 3 \qquad \leftarrow \text{Quotient} \\
x^2 - x + 1 \overline{)\; x^4 - 3x^3 \qquad\quad + 2x - 5\;} \qquad \leftarrow \text{Dividend} \\
\underline{x^4 - \; x^3 + \; x^2 \qquad\qquad} \\
-2x^3 - \; x^2 + 2x - 5 \\
\underline{-2x^3 + 2x^2 - 2x \qquad} \\
-3x^2 + 4x - 5 \\
\underline{-3x^2 + 3x - 3} \\
x - 2 \qquad \leftarrow \text{Remainder}
\end{array}
$$

Divisor $\rightarrow$; Subtract $\rightarrow$; Subtract $\rightarrow$; Subtract $\rightarrow$

✔ **CHECK:** (Quotient)(Divisor) + Remainder

$$
\begin{aligned}
&= (x^2 - 2x - 3)(x^2 - x + 1) + x - 2 \\
&= x^4 - x^3 + x^2 - 2x^3 + 2x^2 - 2x - 3x^2 + 3x - 3 + x - 2 \\
&= x^4 - 3x^3 + 2x - 5 \\
&= \text{Dividend}
\end{aligned}
$$

As a result,

$$
\frac{x^4 - 3x^3 + 2x - 5}{x^2 - x + 1} = x^2 - 2x - 3 + \frac{x - 2}{x^2 - x + 1} \qquad \blacktriangleleft
$$

The process of dividing two polynomials leads to the following result:

Theorem Let Q be a polynomial of positive degree and let P be a polynomial whose degree is greater than the degree of Q. The remainder after dividing P by Q is either the zero polynomial or a polynomial whose degree is less than the degree of the divisor Q.

NOW WORK PROBLEM 9.

2 **Divide Polynomials Using Synthetic Division**

To find the quotient as well as the remainder when a polynomial of degree 1 or higher is divided by $x - c$, a shortened version of long division, called **synthetic division**, makes the task simpler.

To see how synthetic division works, we will use long division to divide the polynomial $2x^3 - x^2 + 3$ by $x - 3$.

$$
\begin{array}{r}
2x^2 + 5x \; + 15 \qquad \leftarrow \text{Quotient} \\
x - 3 \overline{)\; 2x^3 - \; x^2 \qquad\quad + \; 3\;} \\
\underline{2x^3 - 6x^2 \qquad\qquad} \\
5x^2 \qquad + \; 3 \\
\underline{5x^2 - 15x \qquad} \\
15x + \; 3 \\
\underline{15x - 45} \\
48 \qquad \leftarrow \text{Remainder}
\end{array}
$$

✔ **CHECK:** (Divisor) $\cdot$ (Quotient) + Remainder

$$
\begin{aligned}
&= (x - 3)(2x^2 + 5x + 15) + 48 \\
&= 2x^3 + 5x^2 + 15x - 6x^2 - 15x - 45 + 48 \\
&= 2x^3 - x^2 + 3
\end{aligned}
$$

The process of synthetic division arises from rewriting long division in a more compact form, using simpler notation. For example, in the long division on p. 979, the terms in blue are not really necessary because they are identical to the terms directly above them. With these terms removed, we have

$$
\begin{array}{r}
2x^2 + 5x + 15 \\
x - 3\overline{)2x^3 - x^2 \qquad\qquad + 3} \\
- 6x^2 \\
\overline{5x^2} \\
- 15x \\
\overline{15x} \\
- 45 \\
\overline{48}
\end{array}
$$

Most of the x's that appear in this process can also be removed, provided that we are careful about positioning each coefficient. In this regard, we will need to use 0 as the coefficient of x in the dividend, because that power of x is missing. Now we have

$$
\begin{array}{r}
2x^2 + 5x + 15 \\
x - 3\overline{)2 \quad - 1 \qquad 0 \qquad 3} \\
- 6 \\
\overline{5} \\
- 15 \\
\overline{15} \\
- 45 \\
\overline{48}
\end{array}
$$

We can make this display more compact by moving the lines up until the numbers in color align horizontally.

$$
\begin{array}{rl}
2x^2 + 5x + 15 & \text{Row 1} \\
x - 3\overline{)2 \quad - 1 \qquad 0 \qquad 3} & \text{Row 2} \\
- 6 \;\; - 15 \; - 45 & \text{Row 3} \\
\overline{\bigcirc \quad 5 \qquad 15 \quad 48} & \text{Row 4}
\end{array}
$$

Because the leading coefficient of the divisor is always 1, we know that the leading coefficient of the dividend will also be the leading coefficient of the quotient. So we place the leading coefficient of the quotient, 2, in the circled position. Now, the first three numbers in row 4 are precisely the coefficients of the quotient, and the last number in row 4 is the remainder. Thus, row 1 is not really needed, so we can compress the process to three rows, where the bottom row contains both the coefficients of the quotient and the remainder.

$$
\begin{array}{rl}
x - 3\overline{)2 \quad - 1 \qquad 0 \qquad 3} & \text{Row 1} \\
- 6 - 15 - 45 & \text{Row 2 (subtract)} \\
\overline{2 \quad 5 \quad 15 \quad 48} & \text{Row 3}
\end{array}
$$

Recall that the entries in row 3 are obtained by subtracting the entries in row 2 from those in row 1. Rather than subtracting the entries in row 2, we can change the sign of each entry and add. With this modification, our display will look like this:

$$x - 3\overline{)2 \quad -1 \quad 0 \quad 3} \quad \text{Row 1}$$
$$\underline{ 6 \quad 15 \quad 45} \quad \text{Row 2 (add)}$$
$$2 \quad 5 \quad 15 \quad 48 \quad \text{Row 3}$$

Notice that the entries in row 2 are three times the prior entries in row 3. Our last modification to the display replaces the $x - 3$ by 3. The entries in row 3 give the quotient and the remainder, as shown next.

$$3\overline{)2 \quad -1 \quad 0 \quad 3} \quad \text{Row 1}$$
$$\underline{ 6 \quad 15 \quad 45} \quad \text{Row 2 (add)}$$
$$2 \quad 5 \quad 15 \quad 48 \quad \text{Row 3}$$

$$\underbrace{}_{\text{Quotient}} \qquad \overset{\text{Remainder}}{}$$

$$2x^2 + 5x + 15 \quad 48$$

Let's go through an example step by step.

EXAMPLE 4   **Using Synthetic Division to Find the Quotient and Remainder**

Use synthetic division to find the quotient and remainder when

$$x^3 - 4x^2 - 5 \quad \text{is divided by} \quad x - 3$$

Solution   **STEP 1:** Write the dividend in descending powers of x. Then copy the coefficients, remembering to insert a 0 for any missing powers of x.

$$1 \quad -4 \quad 0 \quad -5 \quad \text{Row 1}$$

STEP 2: Insert the usual division symbol. In synthetic division, the divisor is of the form $x - c$, and c is the number placed to the left of the division symbol. Here, since the divisor is $x - 3$, we insert 3 to the left of the division symbol.

$$3\overline{)1 \quad -4 \quad 0 \quad -5} \quad \text{Row 1}$$

STEP 3: Bring the 1 down two rows, and enter it in row 3.

$$3\overline{)1 \quad -4 \quad 0 \quad -5} \quad \text{Row 1}$$
$$\big\downarrow \quad \text{Row 2}$$
$$1 \quad \text{Row 3}$$

STEP 4: Multiply the latest entry in row 3 by 3, and place the result in row 2, one column over to the right.

$$3\overline{)1 \quad -4 \quad 0 \quad -5} \quad \text{Row 1}$$
$$\underline{ 3 } \quad \text{Row 2}$$
$$1 \quad \text{Row 3}$$

STEP 5: Add the entry in row 2 to the entry above it in row 1, and enter the sum in row 3.

$$3\overline{)1 \quad -4 \quad 0 \quad -5} \quad \text{Row 1}$$
$$\underline{ 3 } \quad \text{Row 2}$$
$$1 \quad -1 \quad \text{Row 3}$$

STEP 6: Repeat Steps 4 and 5 until no more entries are available in row 1.

$$
\begin{array}{r}
3\overline{)1 \quad -4 \quad 0 \quad -5} \quad \text{Row 1}\\
\underline{3 \quad -3 \quad -9} \quad \text{Row 2}\\
1 \quad -1 \quad -3 \quad -14 \quad \text{Row 3}
\end{array}
$$

STEP 7: The final entry in row 3, the -14, is the remainder; the other entries in row 3, the 1, -1, and -3, are the coefficients (in descending order) of a polynomial whose degree is 1 less than that of the dividend. This is the quotient. Thus,

$$\text{Quotient} = x^2 - x - 3 \quad \text{Remainder} = -14$$

✔ **CHECK:** (Divisor)(Quotient) + Remainder

$$
\begin{aligned}
&= (x - 3)(x^2 - x - 3) + (-14)\\
&= (x^3 - x^2 - 3x - 3x^2 + 3x + 9) + (-14)\\
&= x^3 - 4x^2 - 5 = \text{Dividend} \qquad \blacktriangleleft
\end{aligned}
$$

Let's do an example in which all seven steps are combined.

EXAMPLE 5 **Using Synthetic Division to Verify a Factor**

Use synthetic division to show that $x + 3$ is a factor of

$$2x^5 + 5x^4 - 2x^3 + 2x^2 - 2x + 3$$

Solution The divisor is $x + 3 = x - (-3)$, so we place -3 to the left of the division symbol. Then the row 3 entries will be multiplied by -3, entered in row 2, and added to row 1.

$$
\begin{array}{r}
-3\overline{)2 \quad 5 \quad -2 \quad 2 \quad -2 \quad 3} \quad \text{Row 1}\\
\underline{-6 \quad 3 \quad -3 \quad 3 \quad -3} \quad \text{Row 2}\\
2 \quad -1 \quad 1 \quad -1 \quad 1 \quad 0 \quad \text{Row 3}
\end{array}
$$

Because the remainder is 0, we have

(Divisor)(Quotient) + Remainder

$$= (x + 3)(2x^4 - x^3 + x^2 - x + 1) = 2x^5 + 5x^4 - 2x^3 + 2x^2 - 2x + 3$$

As we see, $x + 3$ is a factor of $2x^5 + 5x^4 - 2x^3 + 2x^2 - 2x + 3$. ◀

As Example 5 illustrates, the remainder after division gives information about whether the divisor is, or is not, a factor.

NOW WORK PROBLEMS 23 AND 33.

A.4 Assess Your Understanding

Concepts and Vocabulary

1. To check division, the expression that is being divided, the dividend, should equal the product of the _____ and the _____ plus the _____.

2. To divide $2x^3 - 5x + 1$ by $x + 3$ using synthetic division, the first step is to write $\underline{}\overline{)}$.

3. *True or False:* In using synthetic division, the divisor is always a polynomial of degree 1, whose leading coefficient is 1.

4. *True or False:* $-2{\overline{\smash{\big)}\,5\quad 3\quad 2\quad 1}}$ means $\dfrac{5x^3 + 3x^2 + 2x + 1}{x + 2} = 5x^2 - 7x + 16 + \dfrac{-31}{x + 2}.$

$$\begin{array}{r} -2{\overline{\smash{\big)}\,5\quad\ 3\quad\ 2\quad\ 1\ }}\\ \underline{-10\ \ 14\ -32}\\ 5\ \ -7\ \ 16\ -31 \end{array}$$

Skill Building

In Problems 5–20, find the quotient and the remainder. Check your work by verifying that

$$(Quotient)(Divisor) + Remainder = Dividend$$

5. $4x^3 - 3x^2 + x + 1$ divided by $x + 2$

6. $3x^3 - x^2 + x - 2$ divided by $x + 2$

7. $4x^3 - 3x^2 + x + 1$ divided by x^2

8. $3x^3 - x^2 + x - 2$ divided by x^2

9. $5x^4 - 3x^2 + x + 1$ divided by $x^2 + 2$

10. $5x^4 - x^2 + x - 2$ divided by $x^2 + 2$

11. $4x^5 - 3x^2 + x + 1$ divided by $2x^3 - 1$

12. $3x^5 - x^2 + x - 2$ divided by $3x^3 - 1$

13. $2x^4 - 3x^3 + x + 1$ divided by $2x^2 + x + 1$

14. $3x^4 - x^3 + x - 2$ divided by $3x^2 + x + 1$

15. $-4x^3 + x^2 - 4$ divided by $x - 1$

16. $-3x^4 - 2x - 1$ divided by $x - 1$

17. $1 - x^2 + x^4$ divided by $x^2 + x + 1$

18. $1 - x^2 + x^4$ divided by $x^2 - x + 1$

19. $x^3 - a^3$ divided by $x - a$

20. $x^5 - a^5$ divided by $x - a$

In Problems 21–32, use synthetic division to find the quotient and remainder.

21. $x^3 - x^2 + 2x + 4$ divided by $x - 2$

22. $x^3 + 2x^2 - 3x + 1$ divided by $x + 1$

23. $3x^3 + 2x^2 - x + 3$ divided by $x - 3$

24. $-4x^3 + 2x^2 - x + 1$ divided by $x + 2$

25. $x^5 - 4x^3 + x$ divided by $x + 3$

26. $x^4 + x^2 + 2$ divided by $x - 2$

27. $4x^6 - 3x^4 + x^2 + 5$ divided by $x - 1$

28. $x^5 + 5x^3 - 10$ divided by $x + 1$

29. $0.1x^3 + 0.2x$ divided by $x + 1.1$

30. $0.1x^2 - 0.2$ divided by $x + 2.1$

31. $x^5 - 1$ divided by $x - 1$

32. $x^5 + 1$ divided by $x + 1$

In Problems 33–42, use synthetic division to determine whether $x - c$ is a factor of the given polynomial.

33. $4x^3 - 3x^2 - 8x + 4$; $x - 2$

34. $-4x^3 + 5x^2 + 8$; $x + 3$

35. $3x^4 - 6x^3 - 5x + 10$; $x - 2$

36. $4x^4 - 15x^2 - 4$; $x - 2$

37. $3x^6 + 82x^3 + 27$; $x + 3$

38. $2x^6 - 18x^4 + x^2 - 9$; $x + 3$

39. $4x^6 - 64x^4 + x^2 - 15$; $x + 4$

40. $x^6 - 16x^4 + x^2 - 16$; $x + 4$

41. $2x^4 - x^3 + 2x - 1$; $x - \dfrac{1}{2}$

42. $3x^4 + x^3 - 3x + 1$; $x + \dfrac{1}{3}$

43. Find the sum of $a, b, c,$ and d if

$$\frac{x^3 - 2x^2 + 3x + 5}{x + 2} = ax^2 + bx + c + \frac{d}{x + 2}$$

Discussion and Writing

44. When dividing a polynomial by $x - c$, do you prefer to use long division or synthetic division? Does the value of c make a difference to you in choosing? Give reasons.

A.5 Solving Equations

PREPARING FOR THIS SECTION *Before getting started, review the following:*

- Factoring Polynomials (Appendix, Section A.3, pp. 677–678)
- Zero-Product Property (Appendix, Section A.1, p. 660)
- Square Roots (Appendix, Section A.1, p. 666)
- Absolute Value (Appendix, Section A.1, pp. 663–664)
- Rational Expressions (Appendix, Section A.3, pp. 679–683)

 Now work the 'Are You Prepared?' problems on page 705.

OBJECTIVES
1. Solve Linear Equations
2. Solve Rational Equations
3. Solve Quadratic Equations by Factoring
4. Solve Quadratic Equations Using the Square Root Method
5. Solve Quadratic Equations by Completing the Square
6. Solve Quadratic Equations Using the Quadratic Formula
7. Solve Equations Quadratic in Form
8. Solve Absolute Value Equations
9. Solve Equations by Factoring

An **equation in one variable** is a statement in which two expressions, at least one containing the variable, are equal. The expressions are called the **sides** of the equation. Since an equation is a statement, it may be true or false, depending on the value of the variable. Unless otherwise restricted, the admissible values of the variable are those in the domain of the variable. Those admissible values of the variable, if any, that result in a true statement are called **solutions**, or **roots**, of the equation. To **solve an equation** means to find all the solutions of the equation.

For example, the following are all equations in one variable, x:

$$x + 5 = 9 \qquad x^2 + 5x = 2x - 2 \qquad \frac{x^2 - 4}{x + 1} = 0 \qquad \sqrt{x^2 + 9} = 5$$

The first of these statements, $x + 5 = 9$, is true when $x = 4$ and false for any other choice of x. Thus, 4 is a solution of the equation $x + 5 = 9$. We also say that 4 **satisfies** the equation $x + 5 = 9$, because, when we substitute 4 for x, a true statement results.

Sometimes an equation will have more than one solution. For example, the equation

$$\frac{x^2 - 4}{x + 1} = 0$$

has $x = -2$ and $x = 2$ as solutions.

Usually, we will write the solution of an equation in set notation. This set is called the **solution set** of the equation. For example, the solution set of the equation $x^2 - 9 = 0$ is $\{-3, 3\}$.

Some equations have no real solution. For example, $x^2 + 9 = 5$ has no real solution, because there is no real number whose square when added to 9 equals 5.

An equation that is satisfied for every choice of the variable for which both sides are defined is called an **identity**. For example, the equation

$$3x + 5 = x + 3 + 2x + 2$$

is an identity, because this statement is true for any real number x.

Solving Equations Algebraically

One method for solving equations algebraically requires that a series of *equivalent equations* be developed from the original equation until an obvious solution results.

> Two or more equations that have precisely the same solutions are called **equivalent equations**.

For example, all the following equations are equivalent, because each has only the solution $x = 5$:

$$2x + 3 = 13$$
$$2x = 10$$
$$x = 5$$

The question, though, is "How do I obtain an equivalent equation?" In general, there are five ways to do so.

Procedures That Result in Equivalent Equations

1. Interchange the two sides of the equation:

Replace $3 = x$ by $x = 3$

2. Simplify the sides of the equation by combining like terms, eliminating parentheses, and so on:

Replace $x + 2 + 6 = 2x + 3(x + 1)$

by $x + 8 = 5x + 3$

3. Add or subtract the same expression on both sides of the equation:

Replace $3x - 5 = 4$

by $(3x - 5) + 5 = 4 + 5$

4. Multiply or divide both sides of the equation by the same nonzero expression:

Replace $\dfrac{3x}{x - 1} = \dfrac{6}{x - 1}$ $x \neq 1$

by $\dfrac{3x}{x - 1} \cdot (x - 1) = \dfrac{6}{x - 1} \cdot (x - 1)$

5. If one side of the equation is 0 and the other side can be factored, then we may use the Zero-Product Property and set each factor equal to 0:

Replace $x(x - 3) = 0$

by $x = 0$ or $x - 3 = 0$

WARNING

Squaring both sides of an equation does not necessarily lead to an equivalent equation. ■

Whenever it is possible to solve an equation in your head, do so. For example:

The solution of $2x = 8$ is $x = 4$.

The solution of $3x - 15 = 0$ is $x = 5$.

NOW WORK PROBLEM 13.

We now introduce specific types of equations that can be solved algebraically to obtain exact solutions. We start with **linear equations**.

1 Solve Linear Equations

Linear equations are equations such as

$$3x + 12 = 0, \qquad \frac{3}{4}x - \frac{1}{5} = 0, \qquad 0.62x - 0.3 = 0$$

A **linear equation in one variable** is equivalent to an equation of the form

$$ax + b = 0$$

where a and b are real numbers and $a \neq 0$.

Sometimes a linear equation is called a **first-degree equation**, because the left side is a polynomial in x of degree 1.

EXAMPLE 1	**Solving a Linear Equation**

Solve the equation: $3(x - 2) = 5(x - 1)$

Solution

$$
\begin{aligned}
3(x - 2) &= 5(x - 1) & & \text{} \\
3x - 6 &= 5x - 5 & & \text{Use the Distributive Property.} \\
3x - 6 - 5x &= 5x - 5 - 5x & & \text{Subtract 5x from each side.} \\
-2x - 6 &= -5 & & \text{Simplify.} \\
-2x - 6 + 6 &= -5 + 6 & & \text{Add 6 to each side.} \\
-2x &= 1 & & \text{Simplify.} \\
\frac{-2x}{-2} &= \frac{1}{-2} & & \text{Divide each side by } -2. \\
x &= -\frac{1}{2} & & \text{Simplify.}
\end{aligned}
$$

✔ **CHECK:**   Let $x = -\dfrac{1}{2}$ in the expression in x on the left side of the equation and simplify. Let $x = -\dfrac{1}{2}$ in the expression in x on the right side of the equation and simplify. If the two expressions are equal, the solution checks.

$$3(x - 2) = 3\left(-\frac{1}{2} - 2\right) = 3\left(-\frac{5}{2}\right) = -\frac{15}{2}$$

$$5(x - 1) = 5\left(-\frac{1}{2} - 1\right) = 5\left(-\frac{3}{2}\right) = -\frac{15}{2}$$

Since the two expressions are equal, the solution $x = -\dfrac{1}{2}$ checks.

The solution set is $\left\{-\dfrac{1}{2}\right\}$.

◀

NOW WORK PROBLEM 23.

The next example illustrates the solution of an equation that does not appear to be linear, but leads to a linear equation upon simplification.

| EXAMPLE 2 | **Solving an Equation That Leads to a Linear Equation** |

Solve the equation: $(2x - 1)(x - 1) = (x - 5)(2x - 5)$

Solution

$$(2x - 1)(x - 1) = (x - 5)(2x - 5)$$

$2x^2 - 3x + 1 = 2x^2 - 15x + 25$	Multiply and combine like terms.
$2x^2 - 3x + 1 - 2x^2 = 2x^2 - 15x + 25 - 2x^2$	Subtract $2x^2$ from each side.
$-3x + 1 = -15x + 25$	Simplify.
$-3x + 1 - 1 = -15x + 25 - 1$	Subtract 1 from each side.
$-3x = -15x + 24$	Simplify.
$-3x + 15x = -15x + 24 + 15x$	Add 15x to each side.
$12x = 24$	Simplify.
$\dfrac{12x}{12} = \dfrac{24}{12}$	Divide each side by 12.
$x = 2$	Simplify.

✔ CHECK: $(2x - 1)(x - 1) = (2 \cdot 2 - 1)(2 - 1) = (3)(1) = 3$

$(x - 5)(2x - 5) = (2 - 5)(2 \cdot 2 - 5) = (-3)(-1) = 3$

Since the two expressions are equal, the solution checks. The solution set is $\{2\}$. ◀

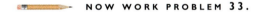 **NOW WORK PROBLEM 33.**

2 Solve Rational Equations

We now introduce another type of equation, the *rational equation*. A **rational equation** is an equation that contains a rational expression. Examples of rational equations are

$$\frac{3}{x + 1} = \frac{2}{x - 1} + 7 \quad \text{and} \quad \frac{x - 5}{x - 4} = \frac{3}{x + 2}$$

To solve a rational equation, multiply both sides of the equation by the least common multiple of the denominators of the rational expressions that make up the rational equation.

| EXAMPLE 3 | **Solving a Rational Equation** |

Solve the equation: $\dfrac{3}{x - 2} = \dfrac{1}{x - 1} + \dfrac{7}{(x - 1)(x - 2)}$

Solution

First, we note that the domain of the variable is $\{x \mid x \neq 1, x \neq 2\}$. We clear the equation of rational expressions by multiplying both sides by the least common multiple of the denominators of the three rational expressions, $(x - 1)(x - 2)$.

$$\frac{3}{x - 2} = \frac{1}{x - 1} + \frac{7}{(x - 1)(x - 2)}$$

$(x - 1)(x - 2)\dfrac{3}{x - 2} = (x - 1)(x - 2)\left[\dfrac{1}{x - 1} + \dfrac{7}{(x - 1)(x - 2)}\right]$
  Multiply both sides by $(x - 1)(x - 2)$. Cancel on the left.

$3x - 3 = (x - 1)(x - 2)\dfrac{1}{x - 1} + (x - 1)(x - 2)\dfrac{7}{(x - 1)(x - 2)}$
  Use the Distributive Property on each side; cancel on the right.

$3x - 3 = (x - 2) + 7$

$3x - 3 = x + 5$ *Combine like terms.*

$2x = 8$ *Add 3 to each side; subtract x from each side.*

$x = 4$ *Divide by 2.*

✔ **CHECK:** $\dfrac{3}{x - 2} = \dfrac{3}{4 - 2} = \dfrac{3}{2}$

$\dfrac{1}{x - 1} + \dfrac{7}{(x - 1)(x - 2)} = \dfrac{1}{4 - 1} + \dfrac{7}{(4 - 1)(4 - 2)} = \dfrac{1}{3} + \dfrac{7}{3 \cdot 2} = \dfrac{2}{6} + \dfrac{7}{6} = \dfrac{9}{6} = \dfrac{3}{2}$

Since the two expressions are equal, the solution $x = 4$ checks.

The solution set is $\{4\}$. ◀

NOW WORK PROBLEM 45.

Quadratic Equations

Quadratic equations are equations such as

$$2x^2 + x + 8 = 0, \qquad 3x^2 - 5x = 0, \qquad x^2 - 9 = 0$$

A general definition is given next.

A **quadratic equation** is an equation equivalent to one of the form

$$ax^2 + bx + c = 0 \qquad\qquad \textbf{(1)}$$

where a, b, and c are real numbers and $a \neq 0$.

A quadratic equation written in the form $ax^2 + bx + c = 0$ is said to be in **standard form**.

Sometimes, a quadratic equation is called a **second-degree equation**, because the left side is a polynomial of degree 2. We shall discuss three algebraic ways of solving quadratic equations: by factoring, by completing the square, and by using the quadratic formula.

3 Solve Quadratic Equations by Factoring

When a quadratic equation is written in standard form, $ax^2 + bx + c = 0$, it may be possible to factor the expression on the left side as the product of two first-degree polynomials. Then, by setting each factor equal to 0 and solving the resulting linear equations, we obtain the *exact* solutions of the quadratic equation. This approach leads us to a basic premise in mathematics. Whenever a problem is encountered, use techniques that reduce the problem to one you already know how to solve. In this instance, we are reducing quadratic equations to linear equations using the technique of factoring.

Let's look at an example.

EXAMPLE 4   **Solving a Quadratic Equation by Factoring**

Solve the equation: $x^2 = 12 - x$

Solution   We put the equation in standard form by adding $x - 12$ to each side:

$$x^2 = 12 - x$$
$$x^2 + x - 12 = 0$$

The left side of the equation may now be factored as

$$(x + 4)(x - 3) = 0$$

Then, by the Zero-Product Property, we have

$$x + 4 = 0 \quad \text{or} \quad x - 3 = 0$$
$$x = -4 \qquad\qquad x = 3$$

The solution set is $\{-4, 3\}$.   ◀

━━━ **NOW WORK PROBLEM 67.**

When the left side factors into two linear equations with the same solution, the quadratic equation is said to have a **repeated solution**. We also call this solution a **root of multiplicity 2**, or a **double root**.

EXAMPLE 5   **Solving a Quadratic Equation by Factoring**

Solve the equation: $x^2 - 6x + 9 = 0$

Solution   This equation is already in standard form, and the left side can be factored:

$$x^2 - 6x + 9 = 0$$
$$(x - 3)(x - 3) = 0$$

so

$$x = 3 \quad \text{or} \quad x = 3$$

This equation has only the repeated solution 3. The solution set is $\{3\}$.   ◀

4 Solve Quadratic Equations Using the Square Root Method

Suppose that we wish to solve the quadratic equation

$$x^2 = p \qquad\qquad (2)$$

where $p \geq 0$ is a nonnegative number. We proceed as in the earlier examples:

$$x^2 - p = 0 \qquad \text{Put in standard form.}$$
$$(x - \sqrt{p})(x + \sqrt{p}) = 0 \qquad \text{Factor (over the real numbers).}$$
$$x = \sqrt{p} \quad \text{or} \quad x = -\sqrt{p} \qquad \text{Solve.}$$

We have the following result:

If $x^2 = p$ and $p \geq 0$, then $x = \sqrt{p}$ or $x = -\sqrt{p}$.   **(3)**

When statement (3) is used, it is called the **Square Root Method**. In statement (3), note that if $p > 0$ the equation $x^2 = p$ has two solutions, $x = \sqrt{p}$ and $x = -\sqrt{p}$. We usually abbreviate these solutions as $x = \pm\sqrt{p}$, read as "x equals plus or minus the square root of p." For example, the two solutions of the equation

$$x^2 = 4$$

are

$$x = \pm\sqrt{4}$$

and, since $\sqrt{4} = 2$, we have

$$x = \pm 2$$

The solution set is $\{-2, 2\}$.

EXAMPLE 6	**Solving Quadratic Equations by Using the Square Root Method**

Solve each equation.

(a) $x^2 = 5$ (b) $(x - 2)^2 = 16$

Solution   (a) $x^2 = 5$

$$x = \pm\sqrt{5} \qquad \text{Use the Square Root Method.}$$
$$x = \sqrt{5} \quad \text{or} \quad x = -\sqrt{5}$$

The solution set is $\{-\sqrt{5}, \sqrt{5}\}$.

(b) $(x - 2)^2 = 16$

$$x - 2 = \pm\sqrt{16} \qquad \text{Use the Square Root Method.}$$
$$x - 2 = \sqrt{16} \quad \text{or} \quad x - 2 = -\sqrt{16}$$
$$x - 2 = 4 \qquad\qquad\quad x - 2 = -4$$
$$x = 6 \qquad\qquad\qquad x = -2$$

The solution set is $\{-2, 6\}$. ◀

NOW WORK PROBLEM 95.

5 Solve Quadratic Equations by Completing the Square

We now introduce the method of **completing the square**. The idea behind this method is to "adjust" the left side of a quadratic equation, $ax^2 + bx + c = 0$, so that it becomes a perfect square, that is, the square of a first-degree polynomial. For example, $x^2 + 6x + 9$ and $x^2 - 4x + 4$ are perfect squares because

$$x^2 + 6x + 9 = (x + 3)^2 \quad \text{and} \quad x^2 - 4x + 4 = (x - 2)^2$$

How do we "adjust" the left side? We do it by adding the appropriate number to create a perfect square. For example, to make $x^2 + 6x$ a perfect square, we add 9.

Let's look at several examples of completing the square when the coefficient of x^2 is 1.

Start	Add	Result
$x^2 + 4x$	4	$x^2 + 4x + 4 = (x + 2)^2$
$x^2 + 12x$	36	$x^2 + 12x + 36 = (x + 6)^2$
$x^2 - 6x$	9	$x^2 - 6x + 9 = (x - 3)^2$
$x^2 + x$	$\dfrac{1}{4}$	$x^2 + x + \dfrac{1}{4} = \left(x + \dfrac{1}{2}\right)^2$

Do you see the pattern? Provided that the coefficient of x^2 is 1, we complete the square by adding the square of one-half of the coefficient of x.

Start	Add	Result
$x^2 + mx$	$\left(\dfrac{m}{2}\right)^2$	$x^2 + mx + \left(\dfrac{m}{2}\right)^2 = \left(x + \dfrac{m}{2}\right)^2$

 NOW WORK PROBLEM 99.

The next example illustrates how the procedure of completing the square can be used to solve a quadratic equation.

EXAMPLE 7

Solving a Quadratic Equation by Completing the Square

Solve by completing the square: $x^2 + 5x + 4 = 0$

Solution

We always begin this procedure by rearranging the equation so that the constant is on the right side.

$$x^2 + 5x + 4 = 0$$
$$x^2 + 5x = -4$$

Since the coefficient of x^2 is 1, we can complete the square on the left side by adding $\left(\dfrac{1}{2} \cdot 5\right)^2 = \dfrac{25}{4}$. Of course, in an equation, whatever we add to the left side also must be added to the right side. We add $\dfrac{25}{4}$ to *both* sides.

$$x^2 + 5x + \frac{25}{4} = -4 + \frac{25}{4} \qquad \text{Add } \frac{25}{4} \text{ to both sides.}$$

$$\left(x + \frac{5}{2}\right)^2 = \frac{9}{4} \qquad \text{Factor; simplify.}$$

$$x + \frac{5}{2} = \pm\sqrt{\frac{9}{4}} \qquad \text{Use the Square Root Method.}$$

$$x + \frac{5}{2} = \pm\frac{3}{2}$$

$$x = -\frac{5}{2} \pm \frac{3}{2}$$

$$x = -\frac{5}{2} + \frac{3}{2} = -1 \quad \text{or} \quad x = -\frac{5}{2} - \frac{3}{2} = -4$$

The solution set is $\{-4, -1\}$. ◀

The solution of the equation in Example 7 can also be obtained by factoring. Rework Example 7 using factoring.

NOW WORK PROBLEM 105.

6 Solve Quadratic Equations Using The Quadratic Formula

We can use the method of completing the square to obtain a general formula for solving the quadratic equation

$$ax^2 + bx + c = 0, \qquad a > 0$$

As in Example 7, we begin by rearranging the terms as

$$ax^2 + bx = -c$$

Since $a > 0$, we can divide both sides by a to get

$$x^2 + \frac{b}{a}x = -\frac{c}{a}$$

Now the coefficient of x^2 is 1. To complete the square on the left side, add the square of $\frac{1}{2}$ of the coefficient of x; that is, add

$$\left(\frac{1}{2} \cdot \frac{b}{a}\right)^2 = \frac{b^2}{4a^2}$$

to each side. Then

$$x^2 + \frac{b}{a}x + \frac{b^2}{4a^2} = \frac{b^2}{4a^2} - \frac{c}{a}$$

$$\left(x + \frac{b}{2a}\right)^2 = \frac{b^2 - 4ac}{4a^2} \qquad \frac{b^2}{4a^2} - \frac{c}{a} = \frac{b^2}{4a^2} - \frac{4ac}{4a^2} = \frac{b^2 - 4ac}{4a^2} \qquad \textbf{(4)}$$

Provided that $b^2 - 4ac \geq 0$, we now can use the Square Root Method to get

$$x + \frac{b}{2a} = \pm\sqrt{\frac{b^2 - 4ac}{4a^2}}$$

$$x + \frac{b}{2a} = \frac{\pm\sqrt{b^2 - 4ac}}{2a} \qquad \text{The square root of a quotient equals the quotient of the square roots. Also, } \sqrt{4a^2} = 2a \text{ since } a > 0.$$

$$x = -\frac{b}{2a} \pm \frac{\sqrt{b^2 - 4ac}}{2a} \qquad \text{Add } -\frac{b}{2a} \text{ to both sides.}$$

$$= \frac{-b \pm \sqrt{b^2 - 4ac}}{2a} \qquad \text{Combine the quotients on the right.}$$

What if $b^2 - 4ac$ is negative? Then equation (4) states that the left expression (a real number squared) equals the right expression (a negative number). Since this occurrence is impossible for real numbers, we conclude that if $b^2 - 4ac < 0$ the quadratic equation has no *real* solution.*

We now state the *quadratic formula*.

Theorem

Quadratic Formula

Consider the quadratic equation

$$ax^2 + bx + c = 0 \qquad a \neq 0$$

If $b^2 - 4ac < 0$, this equation has no real solution.

If $b^2 - 4ac \geq 0$, the real solution(s) of this equation is (are) given by the **quadratic formula**.

$$x = \frac{-b \pm \sqrt{b^2 - 4ac}}{2a}$$

The quantity $b^2 - 4ac$ is called the **discriminant** of the quadratic equation, because its value tells us whether the equation has real solutions. In fact, it also tells us how many solutions to expect.

Discriminant of a Quadratic Equation

For a quadratic equation $ax^2 + bx + c = 0$:

1. If $b^2 - 4ac > 0$, there are two unequal real solutions.

2. If $b^2 - 4ac = 0$, there is a repeated real solution, a root of multiplicity 2.

3. If $b^2 - 4ac < 0$, there is no real solution.

When asked to find the real solutions, if any, of a quadratic equation, always evaluate the discriminant first to see how many real solutions there are.

---

*We consider quadratic equations where $b^2 - 4ac$ is negative in the next section.

EXAMPLE 8	**Solving a Quadratic Equation by Using the Quadratic Formula**

Find the real solutions, if any, of the equation $3x^2 - 5x + 1 = 0$.

Solution The equation is in standard form, so we compare it to $ax^2 + bx + c = 0$ to find $a, b,$ and c.

$$3x^2 - 5x + 1 = 0$$
$$ax^2 + bx + c = 0, \quad a = 3, b = -5, c = 1$$

With $a = 3, b = -5,$ and $c = 1$, we evaluate the discriminant $b^2 - 4ac$.

$$b^2 - 4ac = (-5)^2 - 4(3)(1) = 25 - 12 = 13$$

Since $b^2 - 4ac > 0$, there are two real solutions.
We use the quadratic formula with $a = 3, b = -5, c = 1,$ and $b^2 - 4ac = 13$.

$$x = \frac{-b \pm \sqrt{b^2 - 4ac}}{2a} = \frac{-(-5) \pm \sqrt{13}}{2(3)} = \frac{5 \pm \sqrt{13}}{6}$$

The solution set is $\left\{ \dfrac{5 - \sqrt{13}}{6}, \dfrac{5 + \sqrt{13}}{6} \right\}$. ◄

NOW WORK PROBLEM **111.**

EXAMPLE 9	**Solving a Quadratic Equation by Using the Quadratic Formula**

Find the real solutions, if any, of the equation

$$3x^2 + 2 = 4x$$

Solution The equation, as given, is not in standard form.

$$3x^2 + 2 = 4x$$

$$3x^2 - 4x + 2 = 0 \qquad \text{Subtract 4x from both sides to put the equation in standard form.}$$

$$ax^2 + bx + c = 0 \qquad \text{Compare to standard form.}$$

With $a = 3, b = -4,$ and $c = 2$, we find that

$$b^2 - 4ac = (-4)^2 - 4(3)(2)$$
$$= 16 - 24$$
$$= -8$$

Since $b^2 - 4ac < 0$, the equation has no real solution. ◄

NOW WORK PROBLEM **117.**

Summary

Procedure for Solving a Quadratic Equation Algebraically

To solve a quadratic equation, first put it in standard form:

$$ax^2 + bx + c = 0$$

Then:

STEP 1: Identify a, b, and c.

STEP 2: Evaluate the discriminant, $b^2 - 4ac$.

STEP 3: (a) If the discriminant is negative, the equation has no real solution.
(b) If the discriminant is nonnegative, determine whether the left side can be factored. If you can easily spot factors, use the factoring method to solve the equation. Otherwise, use the quadratic formula or the method of completing the square.

 Solve Equations Quadratic in Form

The equation $x^4 + x^2 - 12 = 0$ is not quadratic in x, but it is quadratic in x^2. That is, if we let $u = x^2$, we get $u^2 + u - 12 = 0$, a quadratic equation. This equation can be solved for u and, in turn, by using $u = x^2$, we can find the solutions x of the original equation.

In general, if an appropriate substitution u transforms an equation into one of the form

$$au^2 + bu + c = 0, \quad a \neq 0$$

then the original equation is called an **equation of the quadratic type** or an **equation quadratic in form**.

The difficulty of solving such an equation lies in the determination that the equation is, in fact, quadratic in form. After you are told an equation is quadratic in form, it is easy enough to see it, but some practice is needed to enable you to recognize them on your own.

EXAMPLE 10   **Solving Equations That Are Quadratic in Form**

Find the real solutions of the equation: $(x + 2)^2 + 11(x + 2) - 12 = 0$

Solution   For this equation, let $u = x + 2$. Then $u^2 = (x + 2)^2$, and the original equation,

$$(x + 2)^2 + 11(x + 2) - 12 = 0$$

becomes

$$u^2 + 11u - 12 = 0 \quad \text{Let } u = x + 2.$$
$$(u + 12)(u - 1) = 0 \quad \text{Factor.}$$
$$u = -12 \quad \text{or} \quad u = 1 \quad \text{Solve.}$$

But we want to solve for x. Because $u = x + 2$, we have

$$x + 2 = -12 \quad \text{or} \quad x + 2 = 1$$
$$x = -14 \qquad\qquad x = -1$$

✔ **CHECK:** $x = -14$: $(-14 + 2)^2 + 11(-14 + 2) - 12$
$$= (-12)^2 + 11(-12) - 12 = 144 - 132 - 12 = 0$$

$x = -1$: $(-1 + 2)^2 + 11(-1 + 2) - 12 = 1 + 11 - 12 = 0$

The original equation has the solution set $\{-14, -1\}$. ◀

The idea should now be clear. If an equation contains an expression and that same expression squared, make a substitution for the expression. You may get a quadratic equation.

✏ **NOW WORK PROBLEM 85.**

8 Solve Absolute Value Equations

Recall that, on the real number line, the absolute value of a equals the distance from the origin to the point whose coordinate is a. For example, there are two points whose distance from the origin is 5 units, -5 and 5. Thus the equation $|x| = 5$ will have the solution set $\{-5, 5\}$. This leads to the following result:

> **Equations Involving Absolute Value**
>
> If a is a positive real number and if u is any algebraic expression, then
>
> $$|u| = a \quad \text{is equivalent to} \quad u = a \text{ or } u = -a \qquad \textbf{(5)}$$

EXAMPLE 11 **Solving an Equation Involving Absolute Value**

Solve the equation $|x + 4| = 13$.

Solution This follows the form of equation (5), where $u = x + 4$. There are two possibilities.

$$x + 4 = 13 \quad \text{or} \quad x + 4 = -13$$
$$x = 9 \qquad\qquad x = -17$$

The solution set is $\{-17, 9\}$. ◀

✏ **NOW WORK PROBLEM 49.**

9 Solve Equations by Factoring

We have already solved certain quadratic equations using factoring. Let's look at examples of other kinds of equations that can be solved by factoring.

| EXAMPLE 12 | **Solving Equations by Factoring** |

Solve the equation: $x^3 - x^2 - 4x + 4 = 0$

Solution   Do you recall the method of factoring by grouping? (If not, review Example 2(f) on p. 680.) We group the terms of $x^3 - x^2 - 4x + 4 = 0$ as follows:

$$(x^3 - x^2) - (4x - 4) = 0$$

Factor out x^2 from the first grouping and 4 from the second.

$$x^2(x - 1) - 4(x - 1) = 0$$

This reveals the common factor $(x - 1)$, so we have

$$(x^2 - 4)(x - 1) = 0$$
$$(x - 2)(x + 2)(x - 1) = 0 \qquad \text{\color{blue}Factor again.}$$
$$x - 2 = 0 \quad \text{or} \quad x + 2 = 0 \quad \text{or} \quad x - 1 = 0 \qquad \text{\color{blue}Set each factor equal to 0.}$$
$$x = 2 \qquad\qquad x = -2 \qquad\qquad x = 1 \qquad \text{\color{blue}Solve.}$$

The solution set is $\{-2, 1, 2\}$.

✔ **CHECK:**

$x = -2$: $(-2)^3 - (-2)^2 - 4(-2) + 4 = -8 - 4 + 8 + 4 = 0$ \color{blue}-2 is a solution.

$x = 1$: $1^3 - 1^2 - 4(1) + 4 = 1 - 1 - 4 + 4 = 0$ \color{blue}1 is a solution.

$x = 2$: $2^3 - 2^2 - 4(2) + 4 = 8 - 4 - 8 + 4 = 0$ \color{blue}2 is a solution. ◀

 NOW WORK PROBLEM 89.

A.5 Assess Your Understanding

'Are You Prepared?'

Answers are given at the end of these exercises. If you get a wrong answer, read the pages listed in \color{red}red.

1. Find the least common denominator of $\dfrac{3}{x^2 - 4}$ and $\dfrac{5}{x^2 - 3x + 2}$. (pp. 681–683)

2. Factor $2x^2 - x - 3$. (pp. 677–678)

3. The solution set of the equation $(x - 3)(3x + 5) = 0$ is _____. (p. 660)

4. *True or False:* $\sqrt{x^2} = |x|$. (pp. 666)

Concepts and Vocabulary

5. Two equations that have the same solution set are called _____.

6. An equation that is satisfied for every choice of the variable for which both sides are defined is called a(n) _____.

7. *True or False:* The solution of the equation $3x - 8 = 0$ is $\dfrac{3}{8}$.

8. *True or False:* Some equations have no solution.

9. To complete the square of the expression $x^2 + 5x$, you would _____ the number _____.

10. The quantity $b^2 - 4ac$ is called the _____ of a quadratic equation. If it is _____, the equation has no real solution.

11. *True or False:* Quadratic equations always have two real solutions.

12. *True or False:* If the discriminant of a quadratic equation is positive, then the equation has two solutions that are negatives of one another.

Skill Building

In Problems 13–92, solve each equation.

13. $3x = 21$ | **14.** $3x = -24$ | **15.** $5x + 15 = 0$ | **16.** $3x + 18 = 0$

17. $2x - 3 = 5$ | **18.** $3x + 4 = -8$ | **19.** $\dfrac{1}{3}x = \dfrac{5}{12}$ | **20.** $\dfrac{2}{3}x = \dfrac{9}{2}$

21. $6 - x = 2x + 9$ | **22.** $3 - 2x = 2 - x$ | **23.** $2(3 + 2x) = 3(x - 4)$ | **24.** $3(2 - x) = 2x - 1$

25. $8x - (2x + 1) = 3x - 10$ | **26.** $5 - (2x - 1) = 10$ | **27.** $\dfrac{1}{2}x - 4 = \dfrac{3}{4}x$ | **28.** $1 - \dfrac{1}{2}x = 5$

29. $0.9t = 0.4 + 0.1t$ | **30.** $0.9t = 1 + t$ | **31.** $\dfrac{2}{y} + \dfrac{4}{y} = 3$ | **32.** $\dfrac{4}{y} - 5 = \dfrac{5}{2y}$

33. $(x + 7)(x - 1) = (x + 1)^2$ | **34.** $(x + 2)(x - 3) = (x - 3)^2$ | **35.** $z(z^2 + 1) = 3 + z^3$

36. $w(4 - w^2) = 8 - w^3$ | **37.** $x^2 = 9x$ | **38.** $x^3 = x^2$ | **39.** $t^3 - 9t^2 = 0$

40. $4z^3 - 8z^2 = 0$ | **41.** $\dfrac{3}{2x - 3} = \dfrac{2}{x + 5}$ | **42.** $\dfrac{-2}{x + 4} = \dfrac{-3}{x + 1}$ | **43.** $(x + 2)(3x) = (x + 2)(6)$

44. $(x - 5)(2x) = (x - 5)(4)$ | **45.** $\dfrac{2}{x - 2} = \dfrac{3}{x + 5} + \dfrac{10}{(x + 5)(x - 2)}$ | **46.** $\dfrac{1}{2x + 3} + \dfrac{1}{x - 1} = \dfrac{1}{(2x + 3)(x - 1)}$

47. $|2x| = 6$ | **48.** $|3x| = 12$ | **49.** $|2x + 3| = 5$ | **50.** $|3x - 1| = 2$

51. $|1 - 4t| = 5$ | **52.** $|1 - 2z| = 3$ | **53.** $|-2x| = 8$ | **54.** $|-x| = 1$

55. $|-2|x = 4$ | **56.** $|3|x = 9$ | **57.** $|x - 2| = -\dfrac{1}{2}$ | **58.** $|2 - x| = -1$

59. $|x^2 - 4| = 0$ | **60.** $|x^2 - 9| = 0$ | **61.** $|x^2 - 2x| = 3$ | **62.** $|x^2 + x| = 12$

63. $|x^2 + x - 1| = 1$ | **64.** $|x^2 + 3x - 2| = 2$ | **65.** $x^2 = 4x$ | **66.** $x^2 = -8x$

67. $z^2 + 4z - 12 = 0$ | **68.** $v^2 + 7v + 12 = 0$ | **69.** $2x^2 - 5x - 3 = 0$ | **70.** $3x^2 + 5x + 2 = 0$

71. $x(x - 7) + 12 = 0$ | **72.** $x(x + 1) = 12$ | **73.** $4x^2 + 9 = 12x$ | **74.** $25x^2 + 16 = 40x$

75. $6x - 5 = \dfrac{6}{x}$ | **76.** $x + \dfrac{12}{x} = 7$ | **77.** $\dfrac{4(x - 2)}{x - 3} + \dfrac{3}{x} = \dfrac{-3}{x(x - 3)}$ | **78.** $\dfrac{5}{x + 4} = 4 + \dfrac{3}{x - 2}$

79. $\dfrac{x}{x^2 - 1} - \dfrac{x + 3}{x^2 - x} = \dfrac{-3}{x^2 + x}$ | **80.** $\dfrac{x + 1}{x^2 + 2x} - \dfrac{x + 4}{x^2 + x} = \dfrac{-3}{x^2 + 3x + 2}$ | **81.** $x^4 - 5x^2 + 4 = 0$

82. $x^4 - 10x^2 + 25 = 0$ | **83.** $(x + 2)^2 + 7(x + 2) + 12 = 0$ | **84.** $(2x + 5)^2 - (2x + 5) - 6 = 0$

85. $2(s + 1)^2 - 5(s + 1) = 3$ | **86.** $3(1 - y)^2 + 5(1 - y) + 2 = 0$ | **87.** $x^3 + x^2 - 20x = 0$

88. $x^3 + 6x^2 - 7x = 0$ | **89.** $x^3 + x^2 - x - 1 = 0$ | **90.** $x^3 + 4x^2 - x - 4 = 0$

91. $2x^3 + 4 = x^2 + 8x$ | **92.** $3x^3 + 4x^2 = 27x + 36$

In Problems 93–98, solve each equation by the Square Root Method.

93. $x^2 = 25$ | **94.** $x^2 = 36$ | **95.** $(x - 1)^2 = 4$

96. $(x + 2)^2 = 1$ | **97.** $(2x + 3)^2 = 9$ | **98.** $(3x - 2)^2 = 4$

In Problems 99–104, what number should be added to complete the square of each expression?

99. $x^2 + 8x$ | **100.** $x^2 - 4x$ | **101.** $x^2 + \dfrac{1}{2}x$

102. $x^2 - \dfrac{1}{3}x$ | **103.** $x^2 - \dfrac{2}{3}x$ | **104.** $x^2 - \dfrac{2}{5}x$

In Problems 105–110, solve each equation by completing the square.

105. $x^2 + 4x = 21$ | **106.** $x^2 - 6x = 13$ | **107.** $x^2 - \dfrac{1}{2}x - \dfrac{3}{16} = 0$

108. $x^2 + \dfrac{2}{3}x - \dfrac{1}{3} = 0$ | **109.** $3x^2 + x - \dfrac{1}{2} = 0$ | **110.** $2x^2 - 3x - 1 = 0$

In Problems 111–122, find the real solutions, if any, of each equation. Use the quadratic formula.

111. $x^2 - 4x + 2 = 0$

112. $x^2 + 4x + 2 = 0$

113. $x^2 - 5x - 1 = 0$

114. $x^2 + 5x + 3 = 0$

115. $2x^2 - 5x + 3 = 0$

116. $2x^2 + 5x + 3 = 0$

117. $4y^2 - y + 2 = 0$

118. $4t^2 + t + 1 = 0$

119. $4x^2 = 1 - 2x$

120. $2x^2 = 1 - 2x$

121. $x^2 + \sqrt{3}x - 3 = 0$

122. $x^2 + \sqrt{2}x - 2 = 0$

In Problems 123–128, use the discriminant to determine whether each quadratic equation has two unequal real solutions, a repeated real solution, or no real solution without solving the equation.

123. $x^2 - 5x + 7 = 0$

124. $x^2 + 5x + 7 = 0$

125. $9x^2 - 30x + 25 = 0$

126. $25x^2 - 20x + 4 = 0$

127. $3x^2 + 5x - 8 = 0$

128. $2x^2 - 3x - 4 = 0$

Problems 129–134 list some formulas that occur in applications. Solve each formula for the indicated variable.

129. Electricity $\dfrac{1}{R} = \dfrac{1}{R_1} + \dfrac{1}{R_2}$ for R

130. Finance $A = P(1 + rt)$ for r

131. Mechanics $F = \dfrac{mv^2}{R}$ for R

132. Chemistry $PV = nRT$ for T

133. Mathematics $S = \dfrac{a}{1 - r}$ for r

134. Mechanics $v = -gt + v_0$ for t

---

135. Show that the sum of the roots of a quadratic equation is $-\dfrac{b}{a}$.

136. Show that the product of the roots of a quadratic equation is $\dfrac{c}{a}$.

137. Find k so that the equation $kx^2 + x + k = 0$ has a repeated real solution.

138. Find k so that the equation $x^2 - kx + 4 = 0$ has a repeated real solution.

139. Show that the real solutions of the equation $ax^2 + bx + c = 0$ are the negatives of the real solutions of the equation $ax^2 - bx + c = 0$. Assume that $b^2 - 4ac \geq 0$.

140. Show that the real solutions of the equation $ax^2 + bx + c = 0$ are the reciprocals of the real solutions of the equation $cx^2 + bx + a = 0$. Assume that $b^2 - 4ac \geq 0$.

Discussion and Writing

141. Which of the following pairs of equations are equivalent? Explain.

(a) $x^2 = 9; \quad x = 3$

(b) $x = \sqrt{9}; \quad x = 3$

(c) $(x - 1)(x - 2) = (x - 1)^2; \quad x - 2 = x - 1$

142. The equation

$$\frac{5}{x + 3} + 3 = \frac{8 + x}{x + 3}$$

has no solution, yet when we go through the process of solving it we obtain $x = -3$. Write a brief paragraph to explain what causes this to happen.

143. Make up an equation that has no solution and give it to a fellow student to solve. Ask the fellow student to write a critique of your equation.

144. Describe three ways you might solve a quadratic equation. State your preferred method; explain why you chose it.

145. Explain the benefits of evaluating the discriminant of a quadratic equation before attempting to solve it.

146. Make up three quadratic equations: one having two distinct solutions, one having no real solution, and one having exactly one real solution.

147. The word *quadratic* seems to imply four (*quad*), yet a quadratic equation is an equation that involves a polynomial of degree 2. Investigate the origin of the term *quadratic* as it is used in the expression *quadratic equation*. Write a brief essay on your findings.

148. The equation $|x| = -2$ has no real solution. Why?

'Are You Prepared?' Answers

1. $(x - 2)(x + 2)(x - 1)$ **2.** $(2x - 3)(x + 1)$ **3.** $\left\{-\dfrac{5}{3}, 3\right\}$ **4.** True

A.6 Complex Numbers; Quadratic Equations in the Complex Number System

One property of a real number is that its square is nonnegative (greater than or equal to 0). For example, there is no real number x for which

$$x^2 = -1$$

To remedy this situation, we introduce a number called the **imaginary unit**, which we denote by i and whose square is -1; that is,

$$i^2 = -1$$

This should not surprise you. If our universe were to consist only of integers, there would be no number x for which $2x = 1$. This unfortunate circumstance was remedied by introducing numbers such as $\frac{1}{2}$ and $\frac{2}{3}$, the *rational numbers*. If our universe were to consist only of rational numbers, there would be no x whose square equals 2. That is, there would be no number x for which $x^2 = 2$. To remedy this, we introduced numbers such as $\sqrt{2}$ and $\sqrt[3]{5}$, the *irrational numbers*. The *real numbers*, you will recall, consist of the rational numbers and the irrational numbers. Now, if our universe were to consist only of real numbers, then there would be no number x whose square is -1. To remedy this, we introduce a number i, whose square is -1.

In the progression outlined, each time that we encountered a situation that was unsuitable, we introduced a new number system to remedy this situation. And each new number system contained the earlier number system as a subset. The number system that results from introducing the number i is called the **complex number system**.

Complex numbers are numbers of the form $a + bi$, where a and b are real numbers. The real number a is called the **real part** of the number $a + bi$; the real number b is called the **imaginary part** of $a + bi$.

For example, the complex number $-5 + 6i$ has the real part -5 and the imaginary part 6.

When a complex number is written in the form $a + bi$, where a and b are real numbers, we say it is in **standard form**. However, if the imaginary part of a complex number is negative, such as in the complex number $3 + (-2)i$, we agree to write it instead in the form $3 - 2i$.

Also, the complex number $a + 0i$ is usually written merely as a. This serves to remind us that the real numbers are a subset of the complex numbers. The complex number $0 + bi$ is usually written as bi. Sometimes the complex number bi is called a **pure imaginary number**.

Add, Subtract, Multiply, and Divide Complex Numbers

Equality, addition, subtraction, and multiplication of complex numbers are defined so as to preserve the familiar rules of algebra for real numbers. Thus, two complex numbers are equal if and only if their real parts are equal and their imaginary parts are equal. That is,

Equality of Complex Numbers

$$a + bi = c + di \quad \text{if and only if} \quad a = c \text{ and } b = d \qquad \textbf{(1)}$$

Two complex numbers are added by forming the complex number whose real part is the sum of the real parts and whose imaginary part is the sum of the imaginary parts. That is,

Sum of Complex Numbers

$$(a + bi) + (c + di) = (a + c) + (b + d)i \qquad \textbf{(2)}$$

To subtract two complex numbers, we use this rule:

Difference of Complex Numbers

$$(a + bi) - (c + di) = (a - c) + (b - d)i \qquad \textbf{(3)}$$

EXAMPLE 1 **Adding and Subtracting Complex Numbers**

(a) $(3 + 5i) + (-2 + 3i) = [3 + (-2)] + (5 + 3)i = 1 + 8i$

(b) $(6 + 4i) - (3 + 6i) = (6 - 3) + (4 - 6)i = 3 + (-2)i = 3 - 2i$ ◄

Figure 14

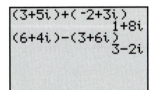

Some graphing calculators have the capability of handling complex numbers. For example, Figure 14 shows the results of Example 1 using a TI-84 Plus graphing calculator.

 NOW WORK PROBLEM 13.

Products of complex numbers are calculated as illustrated in Example 2.

EXAMPLE 2 **Multiplying Complex Numbers**

$$\begin{aligned}
(5 + 3i) \cdot (2 + 7i) &= 5 \cdot (2 + 7i) + 3i(2 + 7i) && \text{Distributive Property} \\
&= 10 + 35i + 6i + 21i^2 && \text{Distributive Property} \\
&= 10 + 41i + 21(-1) && i^2 = -1 \\
&= -11 + 41i && \blacktriangleleft
\end{aligned}$$

Figure 15

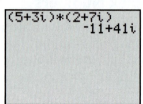

Graphing calculators may also be used to multiply complex numbers. Figure 15 shows the result obtained in Example 2 using a TI-84 Plus graphing calculator.

Based on the procedure of Example 2, we define the **product** of two complex numbers by the following formula:

Product of Complex Numbers

$$(a + bi) \cdot (c + di) = (ac - bd) + (ad + bc)i \qquad (4)$$

Do not bother to memorize formula (4). Instead, whenever it is necessary to multiply two complex numbers, follow the usual rules for multiplying two binomials, as in Example 2, remembering that $i^2 = -1$. For example,

$$(2i)(2i) = 4i^2 = -4$$
$$(2 + i)(1 - i) = 2 - 2i + i - i^2 = 3 - i$$

⟶ **NOW WORK PROBLEM 19.**

Algebraic properties for addition and multiplication, such as the Commutative, Associative, and Distributive Properties, hold for complex numbers. However, the property that every nonzero complex number has a multiplicative inverse, or reciprocal, requires a closer look.

Conjugates

If $z = a + bi$ is a complex number, then its **conjugate**, denoted by $\bar{z}$, is defined as

$$\bar{z} = \overline{a + bi} = a - bi$$

For example, $\overline{2 + 3i} = 2 - 3i$ and $\overline{-6 - 2i} = -6 + 2i$.

EXAMPLE 3 | **Multiplying a Complex Number by Its Conjugate**

Find the product of the complex number $z = 3 + 4i$ and its conjugate $\bar{z}$.

Solution Since $\bar{z} = 3 - 4i$, we have

$$z\bar{z} = (3 + 4i)(3 - 4i) = 9 - 12i + 12i - 16i^2 = 9 + 16 = 25 \qquad ◀$$

The result obtained in Example 3 has an important generalization.

Theorem The product of a complex number and its conjugate is a nonnegative real number. That is, if $z = a + bi$, then

$$z\bar{z} = a^2 + b^2 \qquad (5)$$

Proof If $z = a + bi$, then

$$z\bar{z} = (a + bi)(a - bi) = a^2 - (bi)^2 = a^2 - b^2i^2 = a^2 + b^2 \qquad ■$$

To express the reciprocal of a nonzero complex number z in standard form, multiply the numerator and denominator of $\dfrac{1}{z}$ by its conjugate $\bar{z}$. That is, if $z = a + bi$ is a nonzero complex number, then

$$\frac{1}{a + bi} = \frac{1}{z} = \frac{1}{z} \cdot \frac{\bar{z}}{\bar{z}} = \frac{\bar{z}}{z\bar{z}} \underset{\substack{\uparrow \\ \text{Use (5).}}}{=} \frac{a - bi}{a^2 + b^2} = \frac{a}{a^2 + b^2} - \frac{b}{a^2 + b^2}i$$

EXAMPLE 4	**Writing the Reciprocal of a Complex Number in Standard Form**

Write $\dfrac{1}{3 + 4i}$ in standard form $a + bi$; that is, find the reciprocal of $3 + 4i$.

Figure 16

$$\boxed{\begin{aligned} &1/(3+4i)\blacktriangleright Frac \\ &\quad\quad 3/25-4/25i \end{aligned}}$$

Solution The idea is to multiply the numerator and denominator by the conjugate of $3 + 4i$, that is, the complex number $3 - 4i$. The result is

$$\frac{1}{3 + 4i} = \frac{1}{3 + 4i} \cdot \frac{3 - 4i}{3 - 4i} = \frac{3 - 4i}{9 + 16} = \frac{3}{25} - \frac{4}{25}i \qquad \blacktriangleleft$$

A graphing calculator can be used to verify the result of Example 4. See Figure 16.

To express the quotient of two complex numbers in standard form, we multiply the numerator and denominator of the quotient by the conjugate of the denominator.

EXAMPLE 5	**Writing the Quotient of Complex Numbers in Standard Form**

Write each of the following in standard form.

(a) $\dfrac{1 + 4i}{5 - 12i}$ (b) $\dfrac{2 - 3i}{4 - 3i}$

Solution (a) $\dfrac{1 + 4i}{5 - 12i} = \dfrac{1 + 4i}{5 - 12i} \cdot \dfrac{5 + 12i}{5 + 12i} = \dfrac{5 + 12i + 20i + 48i^2}{25 + 144}$

$\qquad\qquad = \dfrac{-43 + 32i}{169} = \dfrac{-43}{169} + \dfrac{32}{169}i$

(b) $\dfrac{2 - 3i}{4 - 3i} = \dfrac{2 - 3i}{4 - 3i} \cdot \dfrac{4 + 3i}{4 + 3i} = \dfrac{8 + 6i - 12i - 9i^2}{16 + 9}$

$\qquad\qquad = \dfrac{17 - 6i}{25} = \dfrac{17}{25} - \dfrac{6}{25}i \qquad \blacktriangleleft$

 NOW WORK PROBLEM 27.

EXAMPLE 6	**Writing Other Expressions in Standard Form**

If $z = 2 - 3i$ and $w = 5 + 2i$, write each of the following expressions in standard form.

(a) $\dfrac{z}{w}$ (b) $\overline{z + w}$ (c) $z + \bar{z}$

Solution (a) $\dfrac{z}{w} = \dfrac{z \cdot \bar{w}}{w \cdot \bar{w}} = \dfrac{(2 - 3i)(5 - 2i)}{(5 + 2i)(5 - 2i)} = \dfrac{10 - 4i - 15i + 6i^2}{25 + 4}$

$\qquad\qquad\qquad = \dfrac{4 - 19i}{29} = \dfrac{4}{29} - \dfrac{19}{29}i$

(b) $\overline{z + w} = \overline{(2 - 3i) + (5 + 2i)} = \overline{7 - i} = 7 + i$

(c) $z + \overline{z} = (2 - 3i) + (2 + 3i) = 4$ ◀

The conjugate of a complex number has certain general properties that we shall find useful later.

For a real number $a = a + 0i$, the conjugate is $\overline{a} = \overline{a + 0i} = a - 0i = a$. That is,

Theorem The conjugate of a real number is the real number itself.

Other properties that are direct consequences of the definition of the conjugate are given next. In each statement, z and w represent complex numbers.

Theorem The conjugate of the conjugate of a complex number is the complex number itself.

$$(\overline{\overline{z}}) = z \qquad (6)$$

The conjugate of the sum of two complex numbers equals the sum of their conjugates.

$$\overline{z + w} = \overline{z} + \overline{w} \qquad (7)$$

The conjugate of the product of two complex numbers equals the product of their conjugates.

$$\overline{z \cdot w} = \overline{z} \cdot \overline{w} \qquad (8)$$

We leave the proofs of equations (6), (7), and (8) as exercises. See Problems 86–88.

Powers of i

The powers of i follow a pattern that is useful to know.

$$i^1 = i \qquad\qquad i^5 = i^4 \cdot i = 1 \cdot i = i$$
$$i^2 = -1 \qquad\qquad i^6 = i^4 \cdot i^2 = -1$$
$$i^3 = i^2 \cdot i = -i \qquad\qquad i^7 = i^4 \cdot i^3 = -i$$
$$i^4 = i^2 \cdot i^2 = (-1)(-1) = 1 \qquad i^8 = i^4 \cdot i^4 = 1$$

And so on. The powers of i repeat with every fourth power.

EXAMPLE 7 **Evaluating Powers of i**

(a) $i^{27} = i^{24} \cdot i^3 = (i^4)^6 \cdot i^3 = 1^6 \cdot i^3 = -i$

(b) $i^{101} = i^{100} \cdot i^1 = (i^4)^{25} \cdot i = 1^{25} \cdot i = i$ ◀

EXAMPLE 8	**Writing the Power of a Complex Number in Standard Form**

Write $(2 + i)^3$ in standard form.

Solution   We use the special product formula for $(x + a)^3$.

$$(x + a)^3 = x^3 + 3ax^2 + 3a^2x + a^3$$

Using this special product formula,

$$(2 + i)^3 = 2^3 + 3 \cdot i \cdot 2^2 + 3 \cdot i^2 \cdot 2 + i^3$$
$$= 8 + 12i + 6(-1) + (-i)$$
$$= 2 + 11i$$

NOW WORK PROBLEMS **33** AND **41**.

2   Solve Quadratic Equations with a Negative Discriminant

Quadratic equations with a negative discriminant have no real number solution. However, if we extend our number system to allow complex numbers, quadratic equations will always have a solution. Since the solution to a quadratic equation involves the square root of the discriminant, we begin with a discussion of square roots of negative numbers.

> If N is a positive real number, we define the **principal square root of** $-N$, denoted by $\sqrt{-N}$, as
>
> $$\sqrt{-N} = \sqrt{N}i$$
>
> where i is the imaginary unit and $i^2 = -1$.

EXAMPLE 9	**Evaluating the Square Root of a Negative Number**

(a) $\sqrt{-1} = \sqrt{1}i = i$ (b) $\sqrt{-4} = \sqrt{4}i = 2i$

(c) $\sqrt{-8} = \sqrt{8}i = 2\sqrt{2}i$

NOW WORK PROBLEM **49**.

EXAMPLE 10	**Solving Equations**

WARNING
When working with square roots of negative numbers, do not set the square root of a product equal to the product of the square roots (which can be done with positive numbers). To see why, look at this calculation: We know that $\sqrt{100} = 10$. However, it is also true that $100 = (-25)(-4)$, so

$10 = \sqrt{100}$
$ = \sqrt{(-25)(-4)}$
$ \neq \sqrt{-25}\sqrt{-4}$
because $\sqrt{-25} \cdot \sqrt{-4}$
$ = \left(\sqrt{25}i\right)\left(\sqrt{4}i\right)$
$ = (5i)(2i)$
$ = 10i^2 = -10$

Solve each equation in the complex number system.

(a) $x^2 = 4$ (b) $x^2 = -9$

Solution   (a) $x^2 = 4$

$$x = \pm\sqrt{4} = \pm 2$$

The equation has the solution set $\{-2, 2\}$.

(b) $x^2 = -9$

$$x = \pm\sqrt{-9} = \pm\sqrt{9}i = \pm 3i$$

The equation has the solution set $\{-3i, 3i\}$.

NOW WORK PROBLEM **53**.

Because we have defined the square root of a negative number, we can now re-state the quadratic formula without restriction.

Theorem

In the complex number system, the solutions of the quadratic equation $ax^2 + bx + c = 0$, where a, b, and c are real numbers and $a \neq 0$, are given by the formula

$$x = \frac{-b \pm \sqrt{b^2 - 4ac}}{2a} \qquad \textbf{(9)}$$

EXAMPLE 11

Solving Quadratic Equations in the Complex Number System

Solve the equation $x^2 - 4x + 8 = 0$ in the complex number system.

Solution

Here $a = 1$, $b = -4$, $c = 8$, and $b^2 - 4ac = 16 - 4(1)(8) = -16$. Using equation (9), we find that

$$x = \frac{-(-4) \pm \sqrt{-16}}{2(1)} = \frac{4 \pm \sqrt{16}i}{2} = \frac{4 \pm 4i}{2} = 2 \pm 2i$$

The equation has the solution set $\{2 - 2i, 2 + 2i\}$.

✔ **CHECK:**

Figure 17

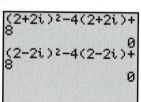

$$2 + 2i: \quad (2 + 2i)^2 - 4(2 + 2i) + 8 = 4 + 8i + 4i^2 - 8 - 8i + 8$$
$$= 4 - 4 = 0$$

$$2 - 2i: \quad (2 - 2i)^2 - 4(2 - 2i) + 8 = 4 - 8i + 4i^2 - 8 + 8i + 8$$
$$= 4 - 4 = 0 \qquad ◀$$

Figure 17 shows the check of the solution using a TI-84 Plus graphing calculator.

NOW WORK PROBLEM 59.

The discriminant $b^2 - 4ac$ of a quadratic equation still serves as a way to determine the character of the solutions.

Character of the Solutions of a Quadratic Equation

In the complex number system, consider a quadratic equation $ax^2 + bx + c = 0$ with real coefficients.

1. If $b^2 - 4ac > 0$, the equation has two unequal real solutions.
2. If $b^2 - 4ac = 0$, the equation has a repeated real solution, a double root.
3. If $b^2 - 4ac < 0$, the equation has two complex solutions that are not real. The solutions are conjugates of each other.

The third conclusion in the display is a consequence of the fact that if $b^2 - 4ac = -N < 0$ then, by the quadratic formula, the solutions are

$$x = \frac{-b + \sqrt{b^2 - 4ac}}{2a} = \frac{-b + \sqrt{-N}}{2a} = \frac{-b + \sqrt{N}i}{2a} = \frac{-b}{2a} + \frac{\sqrt{N}}{2a}i$$

and

$$x = \frac{-b - \sqrt{b^2 - 4ac}}{2a} = \frac{-b - \sqrt{-N}}{2a} = \frac{-b - \sqrt{N}i}{2a} = \frac{-b}{2a} - \frac{\sqrt{N}}{2a}i$$

which are conjugates of each other.

EXAMPLE 12   **Determining the Character of the Solutions of a Quadratic Equation**

Without solving, determine the character of the solutions of each equation.

(a) $3x^2 + 4x + 5 = 0$ (b) $2x^2 + 4x + 1 = 0$

(c) $9x^2 - 6x + 1 = 0$

Solution   (a) Here $a = 3, b = 4,$ and $c = 5,$ so $b^2 - 4ac = 16 - 4(3)(5) = -44.$ The solutions are two complex numbers that are not real and are conjugates of each other.

(b) Here $a = 2, b = 4,$ and $c = 1,$ so $b^2 - 4ac = 16 - 8 = 8.$ The solutions are two unequal real numbers.

(c) Here $a = 9, b = -6,$ and $c = 1,$ so $b^2 - 4ac = 36 - 4(9)(1) = 0.$ The solution is a repeated real number, that is, a double root.   ◀

NOW WORK PROBLEM 73.

A.6 Assess Your Understanding

Concepts and Vocabulary

1. *True or False:* $i = \sqrt{-1}.$

2. $(2 + i)(2 - i) = $ _____.

3. *True or False:* In the complex number system, a quadratic equation has four solutions.

4. In the complex number $5 + 2i,$ the number 5 is called the _____ part; the number 2 is called the _____ part; the number i is called the _____ _____.

5. The equation $x^2 = -4$ has the solution set _____.

6. *True or False:* The conjugate of $2 + 5i$ is $-2 - 5i.$

7. *True or False:* All real numbers are complex numbers.

8. *True or False:* If $2 - 3i$ is a solution of a quadratic equation with real coefficients, then $-2 + 3i$ is also a solution.

Skill Building

In Problems 9–46, write each expression in the standard form a + bi.

9. $(2 - 3i) + (6 + 8i)$ **10.** $(4 + 5i) + (-8 + 2i)$ **11.** $(-3 + 2i) - (4 - 4i)$ **12.** $(3 - 4i) - (-3 - 4i)$

13. $(2 - 5i) - (8 + 6i)$ **14.** $(-8 + 4i) - (2 - 2i)$ **15.** $3(2 - 6i)$ **16.** $-4(2 + 8i)$

17. $2i(2 - 3i)$ **18.** $3i(-3 + 4i)$ **19.** $(3 - 4i)(2 + i)$ **20.** $(5 + 3i)(2 - i)$

21. $(-6 + i)(-6 - i)$ **22.** $(-3 + i)(3 + i)$ **23.** $\dfrac{10}{3 - 4i}$ **24.** $\dfrac{13}{5 - 12i}$

25. $\dfrac{2 + i}{i}$ **26.** $\dfrac{2 - i}{-2i}$ **27.** $\dfrac{6 - i}{1 + i}$ **28.** $\dfrac{2 + 3i}{1 - i}$

29. $\left(\dfrac{1}{2} + \dfrac{\sqrt{3}}{2}i\right)^2$ **30.** $\left(\dfrac{\sqrt{3}}{2} - \dfrac{1}{2}i\right)^2$ **31.** $(1 + i)^2$ **32.** $(1 - i)^2$

33. i^{23} **34.** i^{14} **35.** i^{-15} **36.** i^{-23}

37. $i^6 - 5$ **38.** $4 + i^3$ **39.** $6i^3 - 4i^5$ **40.** $4i^3 - 2i^2 + 1$

41. $(1 + i)^3$ **42.** $(3i)^4 + 1$ **43.** $i^7(1 + i^2)$ **44.** $2i^4(1 + i^2)$

45. $i^6 + i^4 + i^2 + 1$ **46.** $i^7 + i^5 + i^3 + i$

In Problems 47–52, perform the indicated operations and express your answer in the form a + bi.

47. $\sqrt{-4}$ **48.** $\sqrt{-9}$ **49.** $\sqrt{-25}$

50. $\sqrt{-64}$ **51.** $\sqrt{(3 + 4i)(4i - 3)}$ **52.** $\sqrt{(4 + 3i)(3i - 4)}$

In Problems 53–72, solve each equation in the complex number system.

53. $x^2 + 4 = 0$ **54.** $x^2 - 4 = 0$ **55.** $x^2 - 16 = 0$ **56.** $x^2 + 25 = 0$

57. $x^2 - 6x + 13 = 0$ **58.** $x^2 + 4x + 8 = 0$ **59.** $x^2 - 6x + 10 = 0$ **60.** $x^2 - 2x + 5 = 0$

61. $8x^2 - 4x + 1 = 0$ **62.** $10x^2 + 6x + 1 = 0$ **63.** $5x^2 + 1 = 2x$ **64.** $13x^2 + 1 = 6x$

65. $x^2 + x + 1 = 0$ **66.** $x^2 - x + 1 = 0$ **67.** $x^3 - 8 = 0$ **68.** $x^3 + 27 = 0$

69. $x^4 = 16$ **70.** $x^4 = 1$ **71.** $x^4 + 13x^2 + 36 = 0$ **72.** $x^4 + 3x^2 - 4 = 0$

In Problems 73–78, without solving, determine the character of the solutions of each equation in the complex number system.

73. $3x^2 - 3x + 4 = 0$ **74.** $2x^2 - 4x + 1 = 0$ **75.** $2x^2 + 3x = 4$

76. $x^2 + 6 = 2x$ **77.** $9x^2 - 12x + 4 = 0$ **78.** $4x^2 + 12x + 9 = 0$

79. $2 + 3i$ is a solution of a quadratic equation with real coefficients. Find the other solution.

80. $4 - i$ is a solution of a quadratic equation with real coefficients. Find the other solution.

In Problems 81–84, z = 3 − 4i and w = 8 + 3i. Write each expression in the standard form a + bi.

81. $z + \bar{z}$ **82.** $w - \bar{w}$ **83.** $z\bar{z}$ **84.** $\overline{z - w}$

85. Use $z = a + bi$ to show that $z + \bar{z} = 2a$ and $z - \bar{z} = 2bi$.

86. Use $z = a + bi$ to show that $\bar{\bar{z}} = z$.

87. Use $z = a + bi$ and $w = c + di$ to show that $\overline{z + w} = \bar{z} + \bar{w}$.

88. Use $z = a + bi$ and $w = c + di$ to show that $\overline{z \cdot w} = \bar{z} \cdot \bar{w}$.

Discussion and Writing

89. Explain to a friend how you would add two complex numbers and how you would multiply two complex numbers. Explain any differences in the two explanations.

A.7   Problem Solving

Applied (word) problems do not come in the form "Solve the equation. . . ." Instead, they supply information using words, a verbal description of the real problem. So, to solve applied problems, we must be able to translate the verbal description into the language of mathematics. We do this by using variables to represent unknown quantities and then finding relationships (such as equations) that involve these variables. The process of doing all this is called **mathematical modeling**.

Any solution to the mathematical problem must be checked against the mathematical problem, the verbal description, and the real problem. See Figure 18 for an illustration of the **modeling process**.

Figure 18

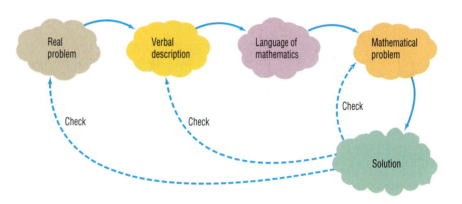

1   Translate Verbal Descriptions into Mathematical Expressions

Let's look at a few examples that will help you to translate certain words into mathematical symbols.

EXAMPLE 1   **Translating Verbal Descriptions into Mathematical Expressions**

(a) The area of a rectangle is the product of its length and its width.

   Translation: If A is used to represent the area, l the length, and w the width, then $A = lw$.

(b) For uniform motion, the velocity of an object equals the distance traveled divided by the time required.

   Translation: If v is the velocity, s the distance, and t the time, then $v = \dfrac{s}{t}$.

(c) A total of $5000 is invested, some in stocks and some in bonds. If the amount invested in stocks is x, express the amount invested in bonds in terms of x.

Translation: If x is the amount invested in stocks, then the amount invested in bonds is $5000 - x$, since their sum is $x + (5000 - x) = 5000$.

(d) Let x denote a number.

The number 5 times as large as x is $5x$.

The number 3 less than x is $x - 3$.

The number that exceeds x by 4 is $x + 4$.

The number that, when added to x, gives 5 is $5 - x$. ◀

NOW WORK PROBLEM 7.

Always check the units used to measure the variables of an applied problem. In Example 1(a), if l is measured in feet, then w also must be expressed in feet, and A will be expressed in square feet. In Example 1(b), if v is measured in miles per hour, then the distance s must be expressed in miles and the time t must be expressed in hours. It is a good practice to check units to be sure that they are consistent and make sense.

Although each situation has unique features, we can provide an outline of the steps to follow in setting up applied problems.

Steps for Setting Up Applied Problems

STEP 1: Read the problem carefully, perhaps two or three times. Pay particular attention to the question being asked in order to identify what you are looking for. If you can, determine realistic possibilities for the answer.

STEP 2: Assign a letter (variable) to represent what you are looking for, and, if necessary, express any remaining unknown quantities in terms of this variable.

STEP 3: Make a list of all the known facts, and translate them into mathematical expressions. These may take the form of an equation (or, later, an inequality) involving the variable. If possible, draw an appropriately labeled diagram to assist you. Sometimes a table or chart helps.

STEP 4: Solve the equation for the variable, and then answer the question, usually using a complete sentence.

STEP 5: Check the answer with the facts in the problem. If it agrees, congratulations! If it does not agree, try again.

---

2 Solve Interest Problems

Interest is money paid for the use of money. The total amount borrowed (whether by an individual from a bank in the form of a loan or by a bank from an individual in the form of a savings account) is called the **principal**. The **rate of interest**, expressed as a percent, is the amount charged for the use of the principal for a given period of time, usually on a yearly (that is, per annum) basis.

Simple Interest Formula

If a principal of P dollars is borrowed for a period of t years at a per annum interest rate r, expressed as a decimal, the interest I charged is

$$I = Prt \qquad\qquad (1)$$

Interest charged according to formula (1) is called **simple interest**. In using formula (1), be sure to express r as a decimal.

EXAMPLE 2

Financial Planning

Candy has $70,000 to invest and requires an overall rate of return of 9%. She can invest in a safe, government-insured certificate of deposit, but it only pays 8%. To obtain 9%, she agrees to invest some of her money in noninsured corporate bonds paying 12%. How much should be placed in each investment to achieve her goal?

Solution   STEP 1: The question is asking for two dollar amounts: the principal to invest in the corporate bonds and the principal to invest in the certificate of deposit.

STEP 2: We let x represent the amount (in dollars) to be invested in the bonds. Then $70,000 - x$ is the amount that will be invested in the certificate. (Do you see why?)

STEP 3: We set up a table:

	Principal ($)	Rate	Time (yr)	Interest ($)
Bonds	x	12% = 0.12	1	$0.12x$
Certificate	$70,000 - x$	8% = 0.08	1	$0.08(70,000 - x)$
Total	70,000	9% = 0.09	1	$0.09(70,000) = 6300$

Since the total interest from the investments is equal to $0.09(70,000) = 6300$, we must have the equation

$$0.12x + 0.08(70,000 - x) = 6300$$

(Note that the units are consistent: the unit is dollars on each side.)

STEP 4: $0.12x + 5600 - 0.08x = 6300$

$$0.04x = 700$$

$$x = 17,500$$

Candy should place $17,500 in the bonds and $70,000 - \$17,500 = \$52,500$ in the certificate.

STEP 5: The interest on the bonds after 1 year is $0.12(\$17,500) = \$2100$; the interest on the certificate after 1 year is $0.08(\$52,500) = \$4200$. The total annual interest is $6300, the required amount. ◀

NOW WORK PROBLEMS 17 AND 23.

3   Solve Mixture Problems

Oil refineries sometimes produce gasoline that is a blend of two or more types of fuel; bakeries occasionally blend two or more types of flour for their bread. These problems are referred to as **mixture problems** because they combine two or more quantities to form a mixture.

| EXAMPLE 3 |

Blending Coffees

The manager of a Starbucks store decides to experiment with a new blend of coffee. She will mix some B grade Colombian coffee that sells for $5 per pound with some A grade Arabica coffee that sells for $10 per pound to get 100 pounds of the new blend. The selling price of the new blend is to be $7 per pound, and there is to be no difference in revenue from selling the new blend versus selling the other types. How many pounds of the B grade Colombian and A grade Arabica coffees are required?

Solution   Let x represent the number of pounds of the B grade Colombian coffee. Then $100 - x$ equals the number of pounds of the A grade Arabica coffee. See Figure 19.

Figure 19

$5 per pound $10 per pound $7 per pound

B Grade   A Grade   Blend
Colombian   Arabica
x pounds   + $100 - x$ pounds   =   100 pounds

Since there is to be no difference in revenue between selling the A and B grades separately versus the blend, we have

$$\left\{\begin{array}{c}\text{Price per pound}\\ \text{of B grade}\end{array}\right\}\left\{\begin{array}{c}\text{Pounds}\\ \text{of B grade}\end{array}\right\} + \left\{\begin{array}{c}\text{Price per pound}\\ \text{of A grade}\end{array}\right\}\left\{\begin{array}{c}\text{Pounds}\\ \text{of A grade}\end{array}\right\} = \left\{\begin{array}{c}\text{Price per pound}\\ \text{of blend}\end{array}\right\}\left\{\begin{array}{c}\text{Pounds}\\ \text{of blend}\end{array}\right\}$$

$$\$5 \quad \cdot \quad x \quad + \quad \$10 \quad \cdot (100 - x) \quad = \quad \$7 \quad \cdot \quad 100$$

We have the equation

$$5x + 10(100 - x) = 700$$
$$5x + 1000 - 10x = 700$$
$$-5x = -300$$
$$x = 60$$

The manager should blend 60 pounds of B grade Colombian coffee with $100 - 60 = 40$ pounds of A grade Arabica coffee to get the desired blend.

✔ CHECK: The 60 pounds of B grade coffee would sell for $(\$5)(60) = \$300$, and the 40 pounds of A grade coffee would sell for $(\$10)(40) = \$400$; the total revenue, $700, equals the revenue obtained from selling the blend, as desired.   ◄

NOW WORK PROBLEM **27.**

4 **Solve Uniform Motion Problems**

Objects that move at a constant velocity are said to be in **uniform motion**. When the average velocity of an object is known, it can be interpreted as its constant velocity. For example, a bicyclist traveling at an average velocity of 25 miles per hour is in uniform motion.

> **Uniform Motion Formula**
>
> If an object moves at an average velocity v, the distance s covered in time t is given by the formula
>
> $$s = vt \tag{2}$$

That is, Distance = Velocity · Time.

EXAMPLE 4 **Physics: Uniform Motion**

Tanya, who is a long-distance runner, runs at an average velocity of 8 miles per hour (mi/hr). Two hours after Tanya leaves your house, you leave in your Honda and follow the same route. If your average velocity is 40 mi/hr, how long will it be before you catch up to Tanya? How far will each of you be from your home?

Solution   Refer to Figure 20. We use t to represent the time (in hours) that it takes the Honda to catch up to Tanya. When this occurs, the total time elapsed for Tanya is $t + 2$ hours.

Figure 20

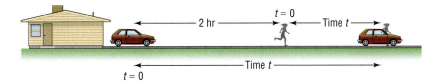

Set up the following table:

	Velocity mi/hr	Time hr	Distance mi
Tanya	8	$t + 2$	$8(t + 2)$
Honda	40	t	$40t$

Since the distance traveled is the same, we are led to the following equation:

$$8(t + 2) = 40t$$
$$8t + 16 = 40t$$
$$32t = 16$$
$$t = \frac{1}{2} \text{ hour}$$

It will take the Honda $\frac{1}{2}$ hour to catch up to Tanya. Each of you will have gone 20 miles.

✔ **CHECK:**   In 2.5 hours, Tanya travels a distance of $(2.5)(8) = 20$ miles. In $\frac{1}{2}$ hour, the Honda travels a distance of $\left(\frac{1}{2}\right)(40) = 20$ miles.   ◀

| **EXAMPLE 5** | **Physics: Uniform Motion** |

A motorboat heads upstream a distance of 24 miles on the Illinois River, whose current is running at 3 miles per hour (mi/hr). The trip up and back takes 6 hours. Assuming that the motorboat maintained a constant speed relative to the water, what was its speed?

Solution   See Figure 21. We use v to represent the constant speed of the motorboat relative to the water. Then the true speed going upstream is $v - 3$ mi/hr, and the true speed going downstream is $v + 3$ mi/hr. Since Distance = Velocity × Time, then Time $= \dfrac{\text{Distance}}{\text{Velocity}}$. We set up a table.

Figure 21

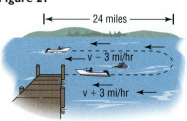

	Velocity **(mi/hr)**	**Distance** **(mi)**	**Time** $=$ $\dfrac{\text{Distance}}{\text{Velocity}}$ **(hr)**
Upstream	$v - 3$	24	$\dfrac{24}{v-3}$
Downstream	$v + 3$	24	$\dfrac{24}{v+3}$

Since the total time up and back is 6 hours, we have

$$\frac{24}{v-3} + \frac{24}{v+3} = 6$$

$$\frac{24(v+3) + 24(v-3)}{(v-3)(v+3)} = 6 \qquad \text{Add the quotients on the left.}$$

$$\frac{48v}{v^2 - 9} = 6 \qquad \text{Simplify.}$$

$$48v = 6(v^2 - 9) \qquad \text{Clear fractions.}$$

$$8v = v^2 - 9 \qquad \text{Divide each side by 6.}$$

$$v^2 - 8v - 9 = 0 \qquad \text{Place in standard form.}$$

$$(v - 9)(v + 1) = 0 \qquad \text{Factor.}$$

$$v = 9 \quad \text{or} \quad v = -1 \qquad \text{Apply the Zero-Product Property and solve.}$$

We discard the solution $v = -1$ mi/hr, so the speed of the motorboat relative to the water is 9 mi/hr.   ◀

NOW WORK PROBLEM 31.

5 Solve Constant Rate Job Problems

This section involves jobs that are performed at a **constant rate**. Our assumption is that, if a job can be done in t units of time, $\dfrac{1}{t}$ of the job is done in 1 unit of time. Let's look at an example.

EXAMPLE 6   ### Working Together to Do a Job

At 10 AM Danny is asked by his father to weed the garden. From past experience, Danny knows that this will take him 4 hours, working alone. His older brother, Mike, when it is his turn to do this job, requires 6 hours. Since Mike wants to go golfing with Danny and has a reservation for 1 PM, he agrees to help Danny. Assuming no gain or loss of efficiency, when will they finish if they work together? Can they make the golf date?

Solution   We set up Table 1. In 1 hour, Danny does $\dfrac{1}{4}$ of the job, and in 1 hour, Mike does $\dfrac{1}{6}$ of the job. Let t be the time (in hours) that it takes them to do the job together. In 1 hour, then, $\dfrac{1}{t}$ of the job is completed. We reason as follows:

$$\begin{pmatrix} \text{Part done by Danny} \\ \text{in 1 hour} \end{pmatrix} + \begin{pmatrix} \text{Part done by Mike} \\ \text{in 1 hour} \end{pmatrix} = \begin{pmatrix} \text{Part done together} \\ \text{in 1 hour} \end{pmatrix}$$

Table 1

	Hours to Do Job	Part of Job Done in 1 Hour
Danny	4	$\dfrac{1}{4}$
Mike	6	$\dfrac{1}{6}$
Together	t	$\dfrac{1}{t}$

From Table 1,

$$\frac{1}{4} + \frac{1}{6} = \frac{1}{t}$$

$$\frac{3 + 2}{12} = \frac{1}{t}$$

$$\frac{5}{12} = \frac{1}{t}$$

$$5t = 12$$

$$t = \frac{12}{5}$$

Working together, the job can be done in $\dfrac{12}{5}$ hours, or 2 hours, 24 minutes. They should make the golf date, since they will finish at 12:24 PM.   ◀

 NOW WORK PROBLEM 35.

   Here is an applied problem that you will probably see again in a slightly different form if you study calculus.

EXAMPLE 7   ### Constructing a Box

From each corner of a square piece of sheet metal, remove a square of side 9 centimeters. Turn up the edges to form an open box. If the box is to hold 144 cubic centimeters (cm^3), what should be the dimensions of the piece of sheet metal?

Solution   We use Figure 22 as a guide. We have labeled by x the length of a side of the square piece of sheet metal. The box will be of height 9 centimeters, and its square base will have $x - 18$ as the length of a side. The volume (Length × Width × Height) of the box is therefore

$$9(x - 18)(x - 18) = 9(x - 18)^2$$

Figure 22

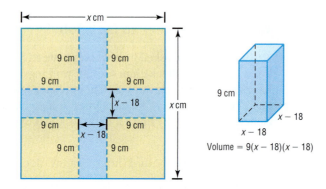

Since the volume of the box is to be 144 cm³, we have

$$9(x - 18)^2 = 144$$
$$(x - 18)^2 = 16 \qquad \text{Divide each side by 9.}$$
$$x - 18 = \pm 4 \qquad \text{Use the Square Root Method.}$$
$$x = 18 \pm 4$$
$$x = 22 \quad \text{or} \quad x = 14$$

We discard the solution $x = 14$ (do you see why?) and conclude that the sheet metal should be 22 centimeters by 22 centimeters.   ◄

✔ **Check:**   If we begin with a piece of sheet metal 22 centimeters by 22 centimeters, cut out a 9 centimeter square from each corner, and fold up the edges, we get a box whose dimensions are 9 by 4 by 4, with volume $9 \times 4 \times 4 = 144$ cm³, as required.

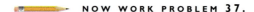   **NOW WORK PROBLEM 37.**

A.7 Assess Your Understanding

Concepts and Vocabulary

1. In applied problems, we translate a verbal description into the language of mathematics. Then we use variables to represent unknown quantities and find relationships that involve these variables. This process is referred to as _____ _____.

2. The money paid for the use of money is _____.

3. Objects that move at a constant velocity are said to be in _____ _____.

4. *True or False:* The amount charged for the use of principal for a given period of time is called the rate of interest.

5. *True or False:* If an object moves at an average velocity v, the distance s covered in time t is given by the formula $s = vt$.

6. Suppose that you want to mix two coffees in order to obtain 100 pounds of the blend. If x represents the number of pounds of coffee A, write an algebraic expression that represents the number of pounds of coffee B.

Applications and Extensions

In Problems 7–16, translate each sentence into a mathematical equation. Be sure to identify the meaning of all symbols.

7. Geometry   The area of a circle is the product of the number π and the square of the radius.

8. Geometry   The circumference of a circle is the product of the number π and twice the radius.

9. Geometry   The area of a square is the square of the length of a side.

10. Geometry   The perimeter of a square is four times the length of a side.

11. Physics   Force equals the product of mass and acceleration.

12. Physics   Pressure is force per unit area.

13. Physics   Work equals force times distance.

14. Physics   Kinetic energy is one-half the product of the mass and the square of the velocity.

15. Business   The total variable cost of manufacturing x dishwashers is $150 per dishwasher times the number of dishwashers manufactured.

16. Business   The total revenue derived from selling x dishwashers is $250 per dishwasher times the number of dishwashers sold.

17. Finance   A total of $20,000 is to be invested, some in bonds and some in certificates of deposit (CDs). If the amount invested in bonds is to exceed that in CDs by $3000, how much will be invested in each type of investment?

18. Finance   A total of $10,000 is to be divided between Sean and George, with George to receive $3000 less than Sean. How much will each receive?

19. Computing Grades   Going into the final exam, which will count as two tests, Brooke has test scores of 80, 83, 71, 61, and 95. What score does Brooke need on the final in order to have an average score of 80?

20. Computing Grades   Going into the final exam, which will count as two-thirds of the final grade, Mike has test scores of 86, 80, 84, and 90. What score does Mike need on the final in order to earn a B, which requires an average score of 80? What does he need to earn an A, which requires an average of 90?

21. Geometry   The perimeter of a rectangle is 60 feet. Find its length and width if the length is 8 feet longer than the width.

22. Geometry   The perimeter of a rectangle is 42 meters. Find its length and width if the length is twice the width.

23. Financial Planning   Betsy, a recent retiree, requires $6000 per year in extra income. She has $50,000 to invest and can invest in B-rated bonds paying 15% per year or in a certificate of deposit (CD) paying 7% per year. How much money should be invested in each to realize exactly $6000 in interest per year?

24. Financial Planning   After 2 years, Betsy (see Problem 23) finds that she will now require $7000 per year. Assuming that the remaining information is the same, how should the money be reinvested?

25. Banking   A bank loaned out $12,000, part of it at the rate of 8% per year and the rest at the rate of 18% per year. If the interest received totaled $1000, how much was loaned at 8%?

26. Banking   Wendy, a loan officer at a bank, has $1,000,000 to lend and is required to obtain an average return of 18% per year. If she can lend at the rate of 19% or at the rate of 16%, how much can she lend at the 16% rate and still meet her requirement?

27. Business: Blending Teas   The manager of a store that specializes in selling tea decides to experiment with a new blend. She will mix some Earl Grey tea that sells for $5 per pound with some Orange Pekoe tea that sells for $3 per pound to get 100 pounds of the new blend. The selling price of the new blend is to be $4.50 per pound, and there is to be no difference in revenue from selling the new blend versus selling the other types. How many pounds of the Earl Grey tea and Orange Pekoe tea are required?

28. Business: Blending Coffee   A coffee manufacturer wants to market a new blend of coffee that sells for $3.90 per pound by mixing two coffees that sell for $2.75 and $5 per pound, respectively. What amounts of each coffee should be blended to obtain the desired mixture?

[**Hint:** Assume that the total weight of the desired blend is 100 pounds.]

29. Business: Mixing Nuts   A nut store normally sells cashews for $4.00 per pound and peanuts for $1.50 per pound. But at the end of the month the peanuts had not sold well, so, in order to sell 60 pounds of peanuts, the manager decided to mix the 60 pounds of peanuts with some cashews and sell the mixture for $2.50 per pound. How many pounds of cashews should be mixed with the peanuts to ensure no change in the profit?

30. Business: Mixing Candy   A candy store sells boxes of candy containing caramels and cremes. Each box sells for $12.50 and holds 30 pieces of candy (all pieces are the same size). If the caramels cost $0.25 to produce and the cremes cost $0.45 to produce, how many of each should be in a box to make a profit of $3?

31. Physics: Uniform Motion   A motorboat can maintain a constant speed of 16 miles per hour relative to the water. The boat makes a trip upstream to a certain point in 20 minutes; the return trip takes 15 minutes. What is the speed of the current? (See the figure.)

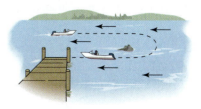

32. Physics: Uniform Motion   A motorboat heads upstream on a river that has a current of 3 miles per hour. The trip upstream takes 5 hours, and the return trip takes 2.5 hours. What is the speed of the motorboat? (Assume that the motorboat maintains a constant speed relative to the water.)

33. Physics: Uniform Motion A Metra commuter train leaves Union Station in Chicago at 12 noon. Two hours later, an Amtrak train leaves on the same track, traveling at an average speed that is 50 miles per hour faster than the Metra train. At 3 PM the Amtrak train is 10 miles behind the commuter train. How fast is each going?

34. Physics: Uniform Motion Two cars enter the Florida Turnpike at Commercial Boulevard at 8:00 AM, each heading for Wildwood. One car's average speed is 10 miles per hour more than the other's. The faster car arrives at Wildwood at 11:00 AM, $\frac{1}{2}$ hour before the other car. What is the average speed of each car? How far did each travel?

35. Working Together on a Job Trent can deliver his newspapers in 30 minutes. It takes Lois 20 minutes to do the same route. How long would it take them to deliver the newspapers if they work together?

36. Working Together on a Job Patrice, by himself, can paint four rooms in 10 hours. If he hires April to help, they can do the same job together in 6 hours. If he lets April work alone, how long will it take her to paint four rooms?

37. Constructing a Box An open box is to be constructed from a square piece of sheet metal by removing a square of side 1 foot from each corner and turning up the edges. If the box is to hold 4 cubic feet, what should be the dimensions of the sheet metal?

38. Constructing a Box Rework Problem 37 if the piece of sheet metal is a rectangle whose length is twice its width.

39. Physics: Uniform Motion A motorboat maintained a constant speed of 15 miles per hour relative to the water in going 10 miles upstream and then returning. The total time for the trip was 1.5 hours. Use this information to find the speed of the current.

40. Dimensions of a Patio A contractor orders 8 cubic yards of premixed cement, all of which is to be used to pour a rectangular patio that will be 4 inches thick. If the length of the patio is specified to be twice the width, what will be the patio dimensions? (1 cubic yard = 27 cubic feet)

41. Dimensions of a Window The area of the opening of a rectangular window is to be 143 square feet. If the length is to be 2 feet more than the width, what are the dimensions?

42. Dimensions of a Window The area of a rectangular window is to be 306 square centimeters. If the length exceeds the width by 1 centimeter, what are the dimensions?

43. Geometry Find the dimensions of a rectangle whose perimeter is 26 meters and whose area is 40 square meters.

44. Watering a Field An adjustable water sprinkler that sprays water in a circular pattern is placed at the center of a square field whose area is 1250 square feet (see the figure). What is the shortest radius setting that can be used if the field is to be completely enclosed within the circle?

45. Physics A ball is thrown vertically upward from the top of a building 96 feet tall with an initial velocity of 80 feet per second. The distance s (in feet) of the ball from the ground after t seconds is $s = 96 + 80t - 16t^2$.
(a) After how many seconds does the ball strike the ground?
(b) After how many seconds will the ball pass the top of the building on its way down?

46. Physics An object is propelled vertically upward with an initial velocity of 20 meters per second. The distance s (in meters) of the object from the ground after t seconds is $s = -4.9t^2 + 20t$.
(a) When will the object be 15 meters above the ground?
(b) When will it strike the ground?
(c) Will the object reach a height of 100 meters?
(d) What is the maximum height?

47. Enclosing a Garden A gardener has 46 feet of fencing to be used to enclose a rectangular garden that has a border 2 feet wide surrounding it (see the figure).
(a) If the length of the garden is to be twice its width, what will be the dimensions of the garden?
(b) What is the area of the garden?
(c) If the length and width of the garden are to be the same, what would be the dimensions of the garden?
(d) What would be the area of the square garden?

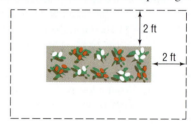

48. Construction A pond is enclosed by a wooden deck that is 3 feet wide. The fence surrounding the deck is 100 feet long.
(a) If the pond is square, what are its dimensions?
(b) If the pond is rectangular and the length of the pond is to be three times its width, what are its dimensions?
(c) If the pond is circular, what is its diameter?
(d) Which pond has the most area?

49. Constructing a Border around a Pool A pool in the shape of a circle measures 10 feet across. One cubic yard of concrete is to be used to create a circular border of uniform width around the pool. If the border is to have a depth of 3 inches, how wide will the border be? (1 cubic yard = 27 cubic feet)

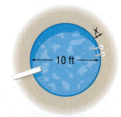

50. Constructing a Border around a Pool   Rework Problem 49 if the depth of the border is 4 inches.

51. Constructing a Border around a Garden   A landscaper, who just completed a rectangular flower garden measuring 6 feet by 10 feet, orders 1 cubic yard of premixed cement, all of which is to be used to create a border of uniform width around the garden. If the border is to have a depth of 3 inches, how wide will the border be? (1 cubic yard = 27 cubic feet)

52. Constructing a Coffee Can   A 39 ounce can of Hill Bros.® coffee requires 188.5 square inches of aluminum. If its height is 7 inches, what is its radius? (The surface area S of a right cylinder is $S = 2\pi r^2 + 2\pi rh$, where r is the radius and h is the height.)

53. Mixing Water and Antifreeze   How much water should be added to 1 gallon of pure antifreeze to obtain a solution that is 60% antifreeze?

54. Mixing Water and Antifreeze   The cooling system of a certain foreign-made car has a capacity of 15 liters. If the system is filled with a mixture that is 40% antifreeze, how much of this mixture should be drained and replaced by pure antifreeze so that the system is filled with a solution that is 60% antifreeze?

55. Chemistry: Salt Solutions   How much water must be evaporated from 32 ounces of a 4% salt solution to make a 6% salt solution?

56. Chemistry: Salt Solutions   How much water must be evaporated from 240 gallons of a 3% salt solution to produce a 5% salt solution?

57. Purity of Gold   The purity of gold is measured in karats, with pure gold being 24 karats. Other purities of gold are expressed as proportional parts of pure gold. Thus 18 karat gold is $\frac{18}{24}$, or 75% pure gold; 12 karat gold is $\frac{12}{24}$, or 50% pure gold; and so on. How much 12 karat gold should be mixed with pure gold to obtain 60 grams of 16 karat gold?

58. Chemistry: Sugar Molecules   A sugar molecule has twice as many atoms of hydrogen as it does oxygen and one more atom of carbon than oxygen. If a sugar molecule has a total of 45 atoms, how many are oxygen? How many are hydrogen?

59. Running a Race   Mike can run the mile in 6 minutes, and Dan can run the mile in 9 minutes. If Mike gives Dan a head start of 1 minute, how far from the start will Mike pass Dan? (See the figure.) How long does it take?

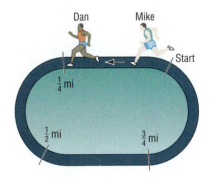

60. Range of an Airplane   An air rescue plane averages 300 miles per hour in still air. It carries enough fuel for 5 hours of flying time. If, upon takeoff, it encounters a head wind of 30 mi/hr, how far can it fly and return safely? (Assume that the wind remains constant.)

61. Emptying Oil Tankers   An oil tanker can be emptied by the main pump in 4 hours. An auxiliary pump can empty the tanker in 9 hours. If the main pump is started at 9 AM, when should the auxiliary pump be started so that the tanker is emptied by noon?

62. Cement Mix   A 20-pound bag of Economy brand cement mix contains 25% cement and 75% sand. How much pure cement must be added to produce a cement mix that is 40% cement?

63. Filling a Tub   A bathroom tub will fill in 15 minutes with both faucets open and the stopper in place. With both faucets closed and the stopper removed, the tub will empty in 20 minutes. How long will it take for the tub to fill if both faucets are open and the stopper is removed?

64. Using Two Pumps   A 5 horsepower (hp) pump can empty a pool in 5 hours. A smaller, 2 hp pump empties the same pool in 8 hours. The pumps are used together to begin emptying this pool. After two hours, the 2 hp pump breaks down. How long will it take the larger pump to empty the pool?

65. Comparing Olympic Heroes   In the 1984 Olympics, Carl Lewis of the United States won the gold medal in the 100 meter race with a time of 9.99 seconds. In the 1896 Olympics, Thomas Burke, also of the United States, won the gold medal in the 100 meter race in 12.0 seconds. If they ran in the same race repeating their respective times, by how many meters would Lewis beat Burke?

66. Football   A tight end can run the 100 yard dash in 12 seconds. A defensive back can do it in 10 seconds. The tight end catches a pass at his own 20 yard line with the defensive back at the 15 yard line. (See the figure.) If no other players are nearby, at what yard line will the defensive back catch up to the tight end?

[**Hint:** At time $t = 0$, the defensive back is 5 yards behind the tight end.]

Discussion and Writing

71. Computing Average Speed In going from Chicago to Atlanta, a car averages 45 miles per hour, and in going from Atlanta to Miami, it averages 55 miles per hour. If Atlanta is halfway between Chicago and Miami, what is the average speed from Chicago to Miami? Discuss an intuitive solution. Write a paragraph defending your intuitive solution. Then solve the problem algebraically. Is your intuitive solution the same as the algebraic one? If not, find the flaw.

72. Speed of a Plane On a recent flight from Phoenix to Kansas City, a distance of 919 nautical miles, the plane arrived 20 minutes early. On leaving the aircraft, I asked the captain, "What was our tail wind?" He replied, "I don't know, but our ground speed was 550 knots." How can you determine if enough information is provided to find the tail wind? If possible, find the tail wind. (1 knot = 1 nautical mile per hour)

67. Computing Business Expense Therese, an outside salesperson, uses her car for both business and pleasure. Last year, she traveled 30,000 miles, using 900 gallons of gasoline. Her car gets 40 miles per gallon on the highway and 25 in the city. She can deduct all highway travel, but no city travel, on her taxes. How many miles should Therese be allowed as a business expense?

68. Summing Consecutive Integers The sum of the consecutive integers $1, 2, 3, \ldots, n$ is given by the formula $\frac{1}{2}n(n + 1)$. How many consecutive integers, starting with 1, must be added to get a sum of 666?

69. Geometry If a polygon of n sides has $\frac{1}{2}n(n - 3)$ diagonals, how many sides will a polygon with 65 diagonals have? Is there a polygon with 80 diagonals?

70. Geometry A right triangle has legs x and $x + 1$. The hypotenuse of the triangle is $2x - 1$. What is the length of the hypotenuse?

73. Critical Thinking You are the manager of a clothing store and have just purchased 100 dress shirts for $20.00 each. After 1 month of selling the shirts at the regular price, you plan to have a sale giving 40% off the original selling price. However, you still want to make a profit of $4 on each shirt at the sale price. What should you price the shirts at initially to ensure this? If, instead of 40% off at the sale, you give 50% off, by how much is your profit reduced?

74. Critical Thinking Make up a word problem that requires solving a linear equation as part of its solution. Exchange problems with a friend. Write a critique of your friend's problem.

75. Critical Thinking Without solving, explain what is wrong with the following mixture problem: How many liters of 25% ethanol should be added to 20 liters of 48% ethanol to obtain a solution of 58% ethanol? Now go through an algebraic solution. What happens?

A.8 Interval Notation; Solving Inequalities

PREPARING FOR THIS SECTION *Before getting started, review the following:*

- Inequalities (Appendix, Section A.1, p. 663)
- Absolute Value (Appendix, Section A.1, pp. 663–664)

Now work the 'Are You Prepared?' problems on page 736.

OBJECTIVES
1. Use Interval Notation
2. Use Properties of Inequalities
3. Solve Linear Inequalities
4. Solve Combined Inequalities
5. Solve Absolute Value Inequalities

Suppose that a and b are two real numbers and $a < b$. We shall use the notation $a < x < b$ to mean that x is a number *between* a and b. Thus, the expression $a < x < b$ is equivalent to the two inequalities $a < x$ and $x < b$. Similarly, the expression $a \leq x \leq b$ is equivalent to the two inequalities $a \leq x$ and $x \leq b$. The remaining two possibilities, $a \leq x < b$ and $a < x \leq b$, are defined similarly.

Although it is acceptable to write $3 \geq x \geq 2$, it is preferable to reverse the inequality symbols and write instead $2 \leq x \leq 3$ so that, as you read from left to right, the values go from smaller to larger.

A statement such as $2 \leq x \leq 1$ is false because there is no number x for which $2 \leq x$ and $x \leq 1$. Finally, we never mix inequality symbols, as in $2 \leq x \geq 3$.

1 Use Interval Notation

Let a and b represent two real numbers with $a < b$:

A **closed interval**, denoted by $[a, b]$, consists of all real numbers x for which $a \leq x \leq b$.
An **open interval**, denoted by (a, b), consists of all real numbers x for which $a < x < b$.
The **half-open**, or **half-closed**, **intervals** are $(a, b]$, consisting of all real numbers x for which $a < x \leq b$, and $[a, b)$, consisting of all real numbers x for which $a \leq x < b$.

In each of these definitions, a is called the **left endpoint** and b the **right endpoint** of the interval.

The symbol ∞ (read as "infinity") is not a real number but a notational device used to indicate unboundedness in the positive direction. The symbol $-\infty$ (read as "minus infinity" or "negative infinity") also is not a real number, but a notational device used to indicate unboundedness in the negative direction. Using the symbols ∞ and $-\infty$, we can define five other kinds of intervals:

$[a, \infty)$	consists of all real numbers x for which $x \geq a$
(a, ∞)	consists of all real numbers x for which $x > a$
$(-\infty, a]$	consists of all real numbers x for which $x \leq a$
$(-\infty, a)$	consists of all real numbers x for which $x < a$
$(-\infty, \infty)$	consists of all real numbers x

Note that ∞ and $-\infty$ are never included as endpoints since they are not real numbers.

Table 2 summarizes interval notation, corresponding inequality notation, and their graphs.

Table 2

Interval	Inequality	Graph
The open interval (a, b)	$a < x < b$	
The closed interval $[a, b]$	$a \leq x \leq b$	
The half-open interval $[a, b)$	$a \leq x < b$	
The half-open interval $(a, b]$	$a < x \leq b$	
The interval $[a, \infty)$	$x \geq a$	
The interval (a, ∞)	$x > a$	
The interval $(-\infty, a]$	$x \leq a$	
The interval $(-\infty, a)$	$x < a$	
The interval $(-\infty, \infty)$	All real numbers	

| EXAMPLE 1 | **Writing Inequalities Using Interval Notation** |

Write each inequality using interval notation.

(a) $1 \leq x \leq 3$ (b) $-4 < x < 0$ (c) $x > 5$ (d) $x \leq 1$

Solution
(a) $1 \leq x \leq 3$ describes all numbers x between 1 and 3, inclusive. In interval notation, we write $[1, 3]$.

(b) In interval notation, $-4 < x < 0$ is written $(-4, 0)$.

(c) $x > 5$ consists of all numbers x greater than 5. In interval notation, we write $(5, \infty)$.

(d) In interval notation, $x \leq 1$ is written $(-\infty, 1]$. ◄

| EXAMPLE 2 | **Writing Intervals Using Inequality Notation** |

Write each interval as an inequality involving x.

(a) $[1, 4)$ (b) $(2, \infty)$ (c) $[2, 3]$ (d) $(-\infty, -3]$

Solution
(a) $[1, 4)$ consists of all numbers x for which $1 \leq x < 4$.

(b) $(2, \infty)$ consists of all numbers x for which $x > 2$.

(c) $[2, 3]$ consists of all numbers x for which $2 \leq x \leq 3$.

(d) $(-\infty, -3]$ consists of all numbers x for which $x \leq -3$. ◄

NOW WORK PROBLEMS 11, 23, AND 31.

2 Use Properties of Inequalities

The product of two positive real numbers is positive, the product of two negative real numbers is positive, and the product of 0 and 0 is 0. For any real number a, the value of a^2 is 0 or positive; that is, a^2 is nonnegative. This is called the **nonnegative property**.

In Words
The square of a real number is never negative.

Nonnegative Property

For any real number a,

$$a^2 \geq 0 \qquad \text{(1)}$$

If we add the same number to both sides of an inequality, we obtain an equivalent inequality. For example, since $3 < 5$, then $3 + 4 < 5 + 4$ or $7 < 9$. This is called the **addition property** of inequalities.

Addition Property of Inequalities

For real numbers $a, b,$ and c,

$$\text{if } a < b, \text{ then } a + c < b + c \qquad \text{(2a)}$$
$$\text{if } a > b, \text{ then } a + c > b + c \qquad \text{(2b)}$$

The addition property states that the sense, or direction, of an inequality remains unchanged if the same number is added to each side. Figure 23 illustrates the addition property (2a). In Figure 23(a), we see that a lies to the left of b. If c is positive, then $a + c$ and $b + c$ each lie c units to the right of a and b, respectively. Consequently, $a + c$ must lie to the left of $b + c$; that is, $a + c < b + c$. Figure 23(b) illustrates the situation if c is negative.

Figure 23

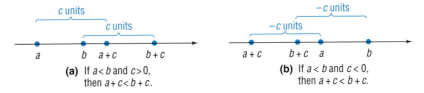

(a) If $a < b$ and $c > 0$,
then $a + c < b + c$.

(b) If $a < b$ and $c < 0$,
then $a + c < b + c$.

Draw an illustration similar to Figure 23 that illustrates the addition property (2b).

| **EXAMPLE 3** | **Addition Property of Inequalities** |

(a) If $x < -5$, then $x + 5 < -5 + 5$ or $x + 5 < 0$.
(b) If $x > 2$, then $x + (-2) > 2 + (-2)$ or $x - 2 > 0$. ◄

NOW WORK PROBLEM 39.

We will use two examples to arrive at our next property.

| **EXAMPLE 4** | **Multiplying an Inequality by a Positive Number** |

Express as an inequality the result of multiplying each side of the inequality $3 < 7$ by 2.

Solution   We begin with

$$3 < 7$$

Multiplying each side by 2 yields the numbers 6 and 14, so we have

$$6 < 14$$ ◄

| **EXAMPLE 5** | **Multiplying an Inequality by a Negative Number** |

Express as an inequality the result of multiplying each side of the inequality $9 > 2$ by -4.

Solution   We begin with

$$9 > 2$$

Multiplying each side by -4 yields the numbers -36 and -8, so we have

$$-36 < -8$$ ◄

Note that the effect of multiplying both sides of $9 > 2$ by the negative number -4 is that the direction of the inequality symbol is reversed.

Examples 4 and 5 illustrate the following general **multiplication properties** for inequalities:

Multiplication Properties for Inequalities

For real numbers a, b, and c,

if $a < b$ and if $c > 0$, then $ac < bc$.	
if $a < b$ and if $c < 0$, then $ac > bc$.	**(3a)**
if $a > b$ and if $c > 0$, then $ac > bc$.	
if $a > b$ and if $c < 0$, then $ac < bc$.	**(3b)**

The multiplication properties state that the sense, or direction, of an inequality *remains the same* if each side is multiplied by a *positive* real number, whereas the direction is *reversed* if each side is multiplied by a *negative* real number.

EXAMPLE 6

Multiplication Property of Inequalities

(a) If $2x < 6$, then $\dfrac{1}{2}(2x) < \dfrac{1}{2}(6)$ or $x < 3$.

(b) If $\dfrac{x}{-3} > 12$, then $-3\left(\dfrac{x}{-3}\right) < -3(12)$ or $x < -36$.

(c) If $-4x > -8$, then $\dfrac{-4x}{-4} < \dfrac{-8}{-4}$ or $x < 2$.

(d) If $-x < 8$, then $(-1)(-x) > (-1)(8)$ or $x > -8$. ◀

NOW WORK PROBLEM 45.

3 **Solve Linear Inequalities**

An **inequality in one variable** is a statement involving two expressions, at least one containing the variable, separated by one of the inequality symbols, $<$, $\leq$, $>$, or $\geq$. To **solve an inequality** means to find all values of the variable for which the statement is true. These values are called **solutions** of the inequality.

For example, the following are all inequalities involving one variable, x:

$$x + 5 < 8, \qquad 2x - 3 \geq 4, \qquad x^2 - 1 \leq 3, \qquad \frac{x+1}{x-2} > 0$$

Two inequalities having exactly the same solution set are called **equivalent inequalities**. As with equations, one method for solving an inequality is to replace it by a series of equivalent inequalities until an inequality with an obvious solution, such as $x < 3$, is obtained. We obtain equivalent inequalities by applying some of the same operations as those used to find equivalent equations. The addition property and the multiplication properties form the basis for the following procedures.

Procedures That Leave the Inequality Symbol Unchanged

1. Simplify both sides of the inequality by combining like terms and eliminating parentheses:

   Replace $\quad x + 2 + 6 > 2x + 5(x + 1)$

   by $\qquad x + 8 > 7x + 5$

2. Add or subtract the same expression on both sides of the inequality:

   Replace $\qquad 3x - 5 < 4$

   by $\quad (3x - 5) + 5 < 4 + 5$

3. Multiply or divide both sides of the inequality by the same *positive* expression:

   Replace $\quad 4x > 16$ by $\dfrac{4x}{4} > \dfrac{16}{4}$

Procedures That Reverse the Sense or Direction of the Inequality Symbol

1. Interchange the two sides of the inequality:

   Replace $\quad 3 < x$ by $x > 3$

2. Multiply or divide both sides of the inequality by the same *negative* expression:

   Replace $\quad -2x > 6$ by $\dfrac{-2x}{-2} < \dfrac{6}{-2}$

As the examples that follow illustrate, we solve inequalities using many of the same steps that we would use to solve equations. In writing the solution of an inequality, we may use either set notation or interval notation, whichever is more convenient.

EXAMPLE 7 **Solving an Inequality**

Solve the inequality $4x + 7 \geq 2x - 3$, and graph the solution set.

Solution

$$4x + 7 \geq 2x - 3$$

$$4x + 7 - 7 \geq 2x - 3 - 7 \qquad \text{Subtract 7 from both sides.}$$

$$4x \geq 2x - 10 \qquad \text{Simplify.}$$

$$4x - 2x \geq 2x - 10 - 2x \qquad \text{Subtract 2x from both sides.}$$

$$2x \geq -10 \qquad \text{Simplify.}$$

$$\frac{2x}{2} \geq \frac{-10}{2} \qquad \text{Divide both sides by 2. (The direction of the inequality symbol is unchanged.)}$$

$$x \geq -5 \qquad \text{Simplify.}$$

Figure 24

The solution set is $\{x \mid x \geq -5\}$ or, using interval notation, all numbers in the interval $[-5, \infty)$.

See Figure 24 for the graph of the solution set. ◀

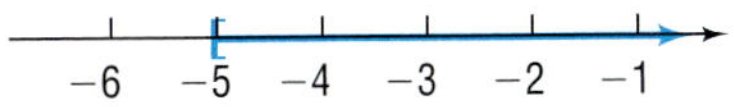

NOW WORK PROBLEM **53.**

4 Solve Combined Inequalities

Now let's look at how to solve combined inequalities.

EXAMPLE 8 **Solving Combined Inequalities**

Solve the inequality $-5 < 3x - 2 < 1$ and draw a graph to illustrate the solution.

Solution   Recall that the inequality

$$-5 < 3x - 2 < 1$$

is equivalent to the two inequalities

$$-5 < 3x - 2 \quad \text{and} \quad 3x - 2 < 1$$

We will solve each of these inequalities separately.

$-5 < 3x - 2$		$3x - 2 < 1$
$-5 + 2 < 3x - 2 + 2$ Add 2 to both sides.		$3x - 2 + 2 < 1 + 2$
$-3 < 3x$ Simplify.		$3x < 3$
$\dfrac{-3}{3} < \dfrac{3x}{3}$ Divide both sides by 3.		$\dfrac{3x}{3} < \dfrac{3}{3}$
$-1 < x$ Simplify.		$x < 1$

The solution set of the original pair of inequalities consists of all x for which

$$-1 < x \quad \text{and} \quad x < 1$$

This may be written more compactly as $\{x \mid -1 < x < 1\}$. In interval notation, the solution is $(-1, 1)$.

See Figure 25 for the graph of the solution set.   ◀

Figure 25

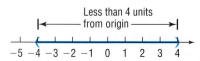

We observe in the solution of Example 8 that the two inequalities that were solved required exactly the same steps. A shortcut to solving the original inequality is to deal with the two inequalities at the same time, as follows:

$$-5 < \qquad 3x - 2 < 1$$
$$-5 + 2 < 3x - 2 + 2 < 1 + 2 \qquad \text{Add 2 to each part.}$$
$$-3 < \qquad 3x \qquad < 3 \qquad \text{Simplify.}$$
$$\frac{-3}{3} < \qquad \frac{3x}{3} \qquad < \frac{3}{3} \qquad \text{Divide each part by 3.}$$
$$-1 < \qquad x \qquad < 1 \qquad \text{Simplify.}$$

NOW WORK PROBLEM 73.

---

5 Solve Absolute Value Inequalities

Let's look at an inequality involving absolute value.

EXAMPLE 9 **Solving an Inequality Involving Absolute Value**

Figure 26

Solve the inequality: $|x| < 4$

Solution   We are looking for all points whose coordinate x is a distance less than 4 units from the origin. See Figure 26 for an illustration.

Because any x between -4 and 4 satisfies the condition $|x| < 4$, the solution set consists of all numbers x for which $-4 < x < 4$, that is, all x in the interval $(-4, 4)$. ◀

We are led to the following results:

Inequalities Involving Absolute Value

If a is any positive number and if u is any algebraic expression, then

$\|u\| < a$	is equivalent to	$-a < u < a$	**(4)**
$\|u\| \le a$	is equivalent to	$-a \le u \le a$	**(5)**

In other words, $|u| < a$ is equivalent to $-a < u$ and $u < a$.

EXAMPLE 10 | **Solving an Inequality Involving Absolute Value**

Solve the inequality $|2x + 4| \le 3$, and graph the solution set.

Solution

$$|2x + 4| \le 3$$
This follows the form of statement (5); the expression $u = 2x + 4$ is inside the absolute value bars.

$$-3 \le 2x + 4 \le 3$$ Apply statement (5).

$$-3 - 4 \le 2x + 4 - 4 \le 3 - 4$$ Subtract 4 from each part.

$$-7 \le 2x \le -1$$ Simplify.

$$\frac{-7}{2} \le \frac{2x}{2} \le \frac{-1}{2}$$ Divide each part by 2.

$$-\frac{7}{2} \le x \le -\frac{1}{2}$$ Simplify.

Figure 27

The solution set is $\left\{ x \mid -\frac{7}{2} \le x \le -\frac{1}{2} \right\}$, that is, all x in the interval $\left[-\frac{7}{2}, -\frac{1}{2} \right]$. See Figure 27 for a graph of the solution set. ◀

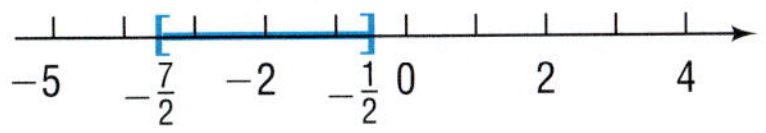

NOW WORK PROBLEM 91.

EXAMPLE 11 | **Solving an Inequality Involving Absolute Value**

Solve the inequality $|x| > 3$.

Solution
We are looking for all points whose coordinate x is a distance greater than 3 units from the origin. Figure 28 illustrates the situation.

Figure 28

We conclude that any x less than -3 or greater than 3 satisfies the condition $|x| > 3$. Consequently, the solution set consists of all numbers x for which $x < -3$ or $x > 3$, that is, all x in the intervals $(-\infty, -3)$ or $(3, \infty)$. ◀

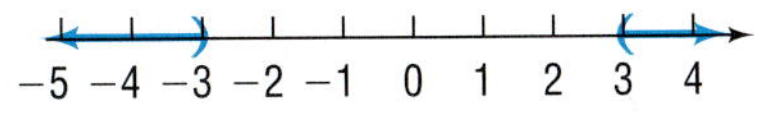

WARNING

A common error to be avoided is to attempt to write the solution $x < 1$ or $x > 4$ as $1 > x > 4$, which is incorrect, since there are no numbers x for which $1 > x$ and $x > 4$. Another common error is to "mix" the symbols and write $1 < x > 4$, which makes no sense. ■

Inequalities Involving Absolute Value

If a is any positive number and u is any algebraic expression, then

$\lvert u \rvert > a$	is equivalent to	$u < -a$ or $u > a$ **(6)**
$\lvert u \rvert \geq a$	is equivalent to	$u \leq -a$ or $u \geq a$ **(7)**

EXAMPLE 12   **Solving an Inequality Involving Absolute Value**

Solve the inequality $\lvert 2x - 5 \rvert > 3$, and graph the solution set.

Solution

$$\lvert 2x - 5 \rvert > 3 \qquad \text{This follows the form of statement (6); the expression } u = 2x - 5 \text{ is inside the absolute value bars.}$$

$2x - 5 < -3$	or	$2x - 5 > 3$	Apply statement (6).
$2x - 5 + 5 < -3 + 5$	or	$2x - 5 + 5 > 3 + 5$	Add 5 to each part.
$2x < 2$	or	$2x > 8$	Simplify.
$\dfrac{2x}{2} < \dfrac{2}{2}$	or	$\dfrac{2x}{2} > \dfrac{8}{2}$	Divide each part by 2.
$x < 1$	or	$x > 4$	Simplify.

Figure 29

The solution set is $\{x \mid x < 1 \text{ or } x > 4\}$, that is, all x in the intervals $(-\infty, 1)$ or $(4, \infty)$.

See Figure 29 for a graph of the solution set. ◀

✏ **NOW WORK PROBLEM 95.**

A.8 Assess Your Understanding

'Are You Prepared?'

Answers are given at the end of these exercises. If you get a wrong answer, read the pages listed in red.

1. Graph the inequality: $x \geq -2$. (p. 663)

2. *True or False:* The absolute value of a negative number is positive. (p. 663-664)

Concepts and Vocabulary

3. If each side of an inequality is multiplied by a(n) _____ number, then the sense of the inequality symbol is reversed.

4. A(n) _____ _____, denoted $[a, b]$, consists of all real numbers x for which $a \leq x \leq b$.

5. The _____ _____ state that the sense, or direction, of an inequality remains the same if each side is multiplied by a positive number, while the direction is reversed if each side is multiplied by a negative number.

In Problems 6–9, determine if the statement is True or False if $a < b$ and $c < 0$.

6. $a + c < b + c$ **7.** $a - c < b - c$ **8.** $ac > bc$ **9.** $\dfrac{a}{c} < \dfrac{b}{c}$

10. *True or False:* The square of any real number is always nonnegative.

Skill Building

In Problems 11–16, express the graph shown in color using interval notation. Also express each graph as an inequality involving x.

11.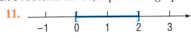

12.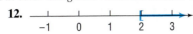

13.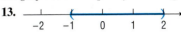

14.

15.

16.

In Problems 17–22, an inequality is given. Write the inequality obtained by:
(a) *Adding 3 to each side of the given inequality.*   (b) *Subtracting 5 from each side of the given inequality.*
(c) *Multiplying each side of the given inequality by 3.*   (d) *Multiplying each side of the given inequality by −2.*

17. $3 < 5$

18. $2 > 1$

19. $4 > -3$

20. $-3 > -5$

21. $2x + 1 < 2$

22. $1 - 2x > 5$

In Problems 23–30, write each inequality using interval notation, and illustrate each inequality using the real number line.

23. $0 \leq x \leq 4$

24. $-1 < x < 5$

25. $4 \leq x < 6$

26. $-2 < x < 0$

27. $x \geq 4$

28. $x \leq 5$

29. $x < -4$

30. $x > 1$

In Problems 31–38, write each interval as an inequality involving x, and illustrate each inequality using the real number line.

31. $[2, 5]$

32. $(1, 2)$

33. $(-3, -2)$

34. $[0, 1)$

35. $[4, \infty)$

36. $(-\infty, 2]$

37. $(-\infty, -3)$

38. $(-8, \infty)$

In Problems 39–52, fill in the blank with the correct inequality symbol.

39. If $x < 5$, then $x - 5$ _____ 0.

40. If $x < -4$, then $x + 4$ _____ 0.

41. If $x > -4$, then $x + 4$ _____ 0.

42. If $x > 6$, then $x - 6$ _____ 0.

43. If $x \geq -4$, then $3x$ _____ -12.

44. If $x \leq 3$, then $2x$ _____ 6.

45. If $x > 6$, then $-2x$ _____ -12.

46. If $x > -2$, then $-4x$ _____ 8.

47. If $x \geq 5$, then $-4x$ _____ -20.

48. If $x \leq -4$, then $-3x$ _____ 12.

49. If $2x < 6$, then x _____ 3.

50. If $3x \leq 12$, then x _____ 4.

51. If $-\dfrac{1}{2}x \leq 3$, then x _____ -6.

52. If $-\dfrac{1}{4}x > 1$, then x _____ -4.

In Problems 53–106, solve each inequality. Express your answer using set notation or interval notation. Graph the solution set.

53. $x + 1 < 5$

54. $x - 6 < 1$

55. $1 - 2x \leq 3$

56. $2 - 3x \leq 5$

57. $3x - 7 > 2$

58. $2x + 5 > 1$

59. $3x - 1 \geq 3 + x$

60. $2x - 2 \geq 3 + x$

61. $-2(x + 3) < 8$

62. $-3(1 - x) < 12$

63. $4 - 3(1 - x) \leq 3$

64. $8 - 4(2 - x) \leq -2x$

65. $\dfrac{1}{2}(x - 4) > x + 8$

66. $3x + 4 > \dfrac{1}{3}(x - 2)$

67. $\dfrac{x}{2} \geq 1 - \dfrac{x}{4}$

68. $\dfrac{x}{3} \geq 2 + \dfrac{x}{6}$

69. $0 \leq 2x - 6 \leq 4$

70. $4 \leq 2x + 2 \leq 10$

71. $-5 \leq 4 - 3x \leq 2$

72. $-3 \leq 3 - 2x \leq 9$

73. $-3 < \dfrac{2x - 1}{4} < 0$

74. $0 < \dfrac{3x + 2}{2} < 4$

75. $1 < 1 - \dfrac{1}{2}x < 4$

76. $0 < 1 - \dfrac{1}{3}x < 1$

77. $(x + 2)(x - 3) > (x - 1)(x + 1)$

78. $(x - 1)(x + 1) > (x - 3)(x + 4)$

79. $x(4x + 3) \le (2x + 1)^2$

80. $x(9x - 5) \le (3x - 1)^2$

81. $\dfrac{1}{2} \le \dfrac{x + 1}{3} < \dfrac{3}{4}$

82. $\dfrac{1}{3} < \dfrac{x + 1}{2} \le \dfrac{2}{3}$

83. $|x| < 6$

84. $|x| < 9$

85. $|x| > 4$

86. $|x| > 1$

87. $|2x| < 8$

88. $|3x| < 15$

89. $|3x| > 12$

90. $|2x| > 6$

91. $|x - 2| + 2 < 3$

92. $|x + 4| + 3 < 5$

93. $|3t - 2| \le 4$

94. $|2u + 5| \le 7$

95. $|x - 3| \ge 2$

96. $|x + 4| \ge 2$

97. $|1 - 4x| - 7 < -2$

98. $|1 - 2x| - 4 < -1$

99. $|1 - 2x| > |-3|$

100. $|2 - 3x| > |-1|$

101. $|2x + 1| < -1$

102. $|3x - 4| \ge 0$

103. $-3 < x + 5 < 2x$

104. $2 < x - 3 < 2x$

105. $x + 2 < 2x - 1 < 5x$

106. $2x - 1 < 3x + 5 < 5x - 7$

Applications and Extensions

107. Express the fact that x differs from 2 by less than $\dfrac{1}{2}$ as an inequality involving an absolute value. Solve for x.

108. Express the fact that x differs from -1 by less than 1 as an inequality involving an absolute value. Solve for x.

109. Express the fact that x differs from -3 by more than 2 as an inequality involving an absolute value. Solve for x.

110. Express the fact that x differs from 2 by more than 3 as an inequality involving an absolute value. Solve for x.

111. A young adult may be defined as someone older than 21, but less than 30 years of age. Express this statement using inequalities.

112. Middle-aged may be defined as being 40 or more and less than 60. Express this statement using inequalities.

113. Body Temperature Normal human body temperature is 98.6°F. If a temperature x that differs from normal by at least 1.5° is considered unhealthy, write the condition for an unhealthy temperature x as an inequality involving an absolute value, and solve for x.

114. Household Voltage In the United States, normal household voltage is 115 volts. However, it is not uncommon for actual voltage to differ from normal voltage by at most 5 volts. Express this situation as an inequality involving an absolute value. Use x as the actual voltage and solve for x.

115. Life Expectancy Metropolitan Life Insurance Co. reported that an average 25-year-old male in 2000 could expect to live at least 50.6 more years, and an average 25-year-old female in 2000 could expect to live at least 55.4 more years.

(a) To what age can an average 25-year-old male expect to live? Express your answer as an inequality.
(b) To what age can an average 25-year-old female expect to live? Express your answer as an inequality.
(c) Who can expect to live longer, a male or a female? By how many years?

116. General Chemistry For a certain ideal gas, the volume V (in cubic centimeters) equals 20 times the temperature T in kelvins (K). If the temperature varies from 353 to 393 K, inclusive, what is the corresponding range of the volume of the gas?

117. Real Estate A real estate agent agrees to sell a large apartment complex according to the following commission schedule: $45,000 plus 25% of the selling price in excess of $900,000. Assuming that the complex will sell at some price between $900,000 and $1,100,000, inclusive, over what range does the agent's commission vary? How does the commission vary as a percent of selling price?

118. Sales Commission A used car salesperson is paid a commission of $25 plus 40% of the selling price in excess of owner's cost. The owner claims that used cars typically sell for at least owner's cost plus $70 and at most owner's cost plus $300. For each sale made, over what range can the salesperson expect the commission to vary?

119. Federal Tax Withholding   The percentage method of withholding for federal income tax (2004) states that a single person whose weekly wages, after subtracting withholding allowances, are over $592, but not over $1317, shall have $74.35 plus 25% of the excess over $592 withheld. Over what range does the amount withheld vary if the weekly wages vary from $600 to $700, inclusive?
SOURCE: *Employer's Tax Guide*. Department of the Treasury, Internal Revenue Service, 2004.

120. Federal Tax Withholding   Rework Problem 119 if the weekly wages vary from $800 to $900, inclusive.

121. Electricity Rates   Commonwealth Edison Company's summer charge for electricity is 8.275¢ per kilowatt-hour. In addition, each monthly bill contains a customer charge of $10.07. If last summer's bills ranged from a low of $65.96 to a high of $217.02, over what range did usage vary (in kilowatt-hours)?
SOURCE: Commonwealth Edison Co., Chicago, Illinois, 2004.

122. Water Bills   The Village of Oak Lawn charges homeowners $27.18 per quarter-year plus $1.90 per 1000 gallons for water usage in excess of 12,000 gallons. In 2004, one homeowner's quarterly bill ranged from a high of $43.33 to a low of $30.03. Over what range did water usage vary?
SOURCE: Village of Oak Lawn, Illinois, 2004.

123. Markup of a New Car   The markup over dealer's cost of a new car ranges from 12% to 18%. If the sticker price is $8800, over what range will the dealer's cost vary?

124. IQ Tests   A standard intelligence test has an average score of 100. According to statistical theory, of the people who take the test, the 2.5% with the highest scores will have scores of more than 1.96σ above the average, where σ (sigma, a number called the *standard deviation*) depends on the nature of the test. If $\sigma = 12$ for this test and there is (in principle) no upper limit to the score possible on the test, write the interval of possible test scores of the people in the top 2.5%.

125. Computing Grades   In your Economics 101 class, you have scores of 68, 82, 87, and 89 on the first four of five tests. To get a grade of B, the average of the first five test scores must

be greater than or equal to 80 and less than 90. Solve an inequality to find the range of the score that you need on the last test to get a B.

What do I need to get a B?

126. Computing Grades   Repeat Problem 125 if the fifth test counts double.

127. A car that averages 25 miles per gallon has a tank that holds 20 gallons of gasoline. After a trip that covered at least 300 miles, the car ran out of gasoline. What is the range of the amount of gasoline (in gallons) that was in the tank at the start of the trip?

128. Repeat Problem 127 if the same car runs out of gasoline after a trip of no more than 250 miles.

129. Arithmetic Mean   If $a < b$, show that $a < \dfrac{a + b}{2} < b$. The number $\dfrac{a + b}{2}$ is called the **arithmetic mean** of a and b.

130. Refer to Problem 129. Show that the arithmetic mean of a and b is equidistant from a and b.

131. Geometric Mean   If $0 < a < b$, show that $a < \sqrt{ab} < b$. The number $\sqrt{ab}$ is called the **geometric mean** of a and b.

132. Refer to Problems 129 and 131. Show that the geometric mean of a and b is less than the arithmetic mean of a and b.

133. Harmonic Mean   For $0 < a < b$, let h be defined by

$$\frac{1}{h} = \frac{1}{2}\left(\frac{1}{a} + \frac{1}{b}\right)$$

Show that $a < h < b$. The number h is called the **harmonic mean** of a and b.

134. Refer to Problems 129, 131, and 133. Show that the harmonic mean of a and b equals the geometric mean squared divided by the arithmetic mean.

Discussion and Writing

135. Make up an inequality that has no solution. Make up one that has exactly one solution.

136. How would you explain to a fellow student the underlying reason for the multiplication properties for inequalities (page 1024); that is, the sense or direction of an inequality remains the same if each side is multiplied by a positive real number, while the direction is reversed if each side is multiplied by a negative real number.

137. The inequality $x^2 + 1 < -5$ has no solution. Explain why.

138. Do you prefer to use inequality notation or interval notation to express the solution to an inequality? Give your reasons. Are there particular circumstances when you prefer one to the other? Cite examples.

'Are You Prepared?' Answers

1. -4 -2 0 **2.** True

A.9 *nth* Roots; Rational Exponents; Radical Equations

PREPARING FOR THIS SECTION *Before getting started, review the following:*

• Exponents, Square Roots (Appendix A, Section A.1, pp. 664–666)

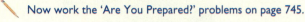 Now work the 'Are You Prepared?' problems on page 745.

OBJECTIVES 1 Work with *nth* Roots
2 Simplify Radicals
3 Rationalize Denominators
4 Solve Radical Equations
5 Simplify Expressions with Rational Exponents

1 Work with *nth* Roots

The **principal *nth* root of a number *a*,** symbolized by $\sqrt[n]{a}$, where $n \geq 2$ is an integer, is defined as follows:

$$\sqrt[n]{a} = b \quad \text{means} \quad a = b^n$$

where $a \geq 0$ and $b \geq 0$ if $n \geq 2$ is even, and a, b are any real numbers if $n \geq 3$ is odd.

Notice that if *a* is negative and *n* is even then $\sqrt[n]{a}$ is not defined. When it is defined, the principal *nth* root of a number is unique.

The symbol $\sqrt[n]{a}$ for the principal *nth* root of *a* is sometimes called a **radical**; the integer *n* is called the **index**, and *a* is called the **radicand**. If the index of a radical is 2, we call $\sqrt[n]{a}$ the **square root** of *a* and omit the index 2 by simply writing $\sqrt{a}$. If the index is 3, we call $\sqrt[3]{a}$ the **cube root** of *a*.

EXAMPLE 1 | **Evaluating Principal *nth* Roots**

(a) $\sqrt[3]{8} = \sqrt[3]{2^3} = 2$ (b) $\sqrt[3]{-64} = \sqrt[3]{(-4)^3} = -4$

(c) $\sqrt[4]{\dfrac{1}{16}} = \sqrt[4]{\left(\dfrac{1}{2}\right)^4} = \dfrac{1}{2}$ (d) $\sqrt[6]{(-2)^6} = |-2| = 2$ ◀

These are examples of **perfect roots**, since each simplifies to a rational number. Notice the absolute value in Example 1(d). If *n* is even, the principal *nth* root must be nonnegative.

In general, if $n \geq 2$ is a positive integer and *a* is a real number, we have

$$\sqrt[n]{a^n} = a, \quad \text{if } n \geq 3 \text{ is odd} \tag{1a}$$

$$\sqrt[n]{a^n} = |a|, \quad \text{if } n \geq 2 \text{ is even} \tag{1b}$$

NOW WORK PROBLEM **7.**

2 Simplify Radicals

Let $n \geq 2$ and $m \geq 2$ denote positive integers, and let a and b represent real numbers. Assuming that all radicals are defined, we have the following properties:

$$\sqrt[n]{ab} = \sqrt[n]{a}\,\sqrt[n]{b} \tag{2a}$$

$$\sqrt[n]{\frac{a}{b}} = \frac{\sqrt[n]{a}}{\sqrt[n]{b}} \tag{2b}$$

$$\sqrt[n]{a^m} = (\sqrt[n]{a})^m \tag{2c}$$

When used in reference to radicals, the direction to "simplify" will mean to remove from the radicals any perfect roots that occur as factors. Let's look at some examples of how the preceding rules are applied to simplify radicals.

EXAMPLE 2   **Simplifying Radicals**

(a) $\sqrt{32} = \sqrt{16 \cdot 2} = \sqrt{16} \cdot \sqrt{2} = 4\sqrt{2}$

      ↑
    16 is a perfect square.

(b) $\sqrt[3]{16} = \sqrt[3]{8 \cdot 2} = \sqrt[3]{8} \cdot \sqrt[3]{2} = 2\sqrt[3]{2}$

      ↑   ↑
8 is a perfect cube.   (2a)

(c) $\sqrt[3]{-16x^4} = \sqrt[3]{-8 \cdot 2 \cdot x^3 \cdot x} = \sqrt[3]{(-8x^3)(2x)}$

      ↑   ↑
Factor perfect cubes   Combine perfect cubes.
inside radical.

$$= \sqrt[3]{(-2x)^3 \cdot 2x} = \sqrt[3]{(-2x)^3} \cdot \sqrt[3]{2x}$$

                ↑
             (2a)

$$= -2x\sqrt[3]{2x} \qquad ◄$$

🖉── **NOW WORK PROBLEM 13.**

EXAMPLE 3   **Combining Like Radicals**

(a) $-8\sqrt{12} + \sqrt{3} = -8\sqrt{4 \cdot 3} + \sqrt{3} = -8 \cdot \sqrt{4}\,\sqrt{3} + \sqrt{3}$

          $= -16\sqrt{3} + \sqrt{3} = -15\sqrt{3}$

(b) $\sqrt[3]{8x^4} + \sqrt[3]{-x} + 4\sqrt[3]{27x} = \sqrt[3]{2^3 x^3 x} + \sqrt[3]{-1 \cdot x} + 4\sqrt[3]{3^3 x}$

        $= \sqrt[3]{(2x)^3} \cdot \sqrt[3]{x} + \sqrt[3]{-1} \cdot \sqrt[3]{x} + 4\sqrt[3]{3^3} \cdot \sqrt[3]{x}$

        $= 2x\sqrt[3]{x} - 1 \cdot \sqrt[3]{x} + 12\sqrt[3]{x}$

        $= (2x + 11)\sqrt[3]{x} \qquad ◄$

🖉── **NOW WORK PROBLEM 31.**

3 Rationalize Denominators

When radicals occur in quotients, it is customary to rewrite the quotient so that the denominator contains no square roots. This process is referred to as **rationalizing the denominator**.

The idea is to multiply by an appropriate expression so that the new denominator contains no radicals. For example:

If Denominator Contains the Factor	Multiply by	To Obtain Denominator Free of Radicals
$\sqrt{3}$	$\sqrt{3}$	$\left(\sqrt{3}\right)^2 = 3$
$\sqrt{3} + 1$	$\sqrt{3} - 1$	$\left(\sqrt{3}\right)^2 - 1^2 = 3 - 1 = 2$
$\sqrt{2} - 3$	$\sqrt{2} + 3$	$\left(\sqrt{2}\right)^2 - 3^2 = 2 - 9 = -7$
$\sqrt{5} - \sqrt{3}$	$\sqrt{5} + \sqrt{3}$	$\left(\sqrt{5}\right)^2 - \left(\sqrt{3}\right)^2 = 5 - 3 = 2$
$\sqrt[3]{4}$	$\sqrt[3]{2}$	$\sqrt[3]{4} \cdot \sqrt[3]{2} = \sqrt[3]{8} = 2$

In rationalizing the denominator of a quotient, be sure to multiply both the numerator and the denominator by the expression.

EXAMPLE 4	**Rationalizing Denominators**

Rationalize the denominator of each expression.

(a) $\dfrac{4}{\sqrt{2}}$ (b) $\dfrac{\sqrt{3}}{\sqrt[3]{2}}$ (c) $\dfrac{\sqrt{x} - 2}{\sqrt{x} + 2}, \quad x \geq 0$

Solution

(a) $\dfrac{4}{\sqrt{2}} = \dfrac{4}{\sqrt{2}} \cdot \dfrac{\sqrt{2}}{\sqrt{2}} = \dfrac{4\sqrt{2}}{\left(\sqrt{2}\right)^2} = \dfrac{4\sqrt{2}}{2} = 2\sqrt{2}$

Multiply by $\dfrac{\sqrt{2}}{\sqrt{2}}$.

(b) $\dfrac{\sqrt{3}}{\sqrt[3]{2}} = \dfrac{\sqrt{3}}{\sqrt[3]{2}} \cdot \dfrac{\sqrt[3]{4}}{\sqrt[3]{4}} = \dfrac{\sqrt{3} \, \sqrt[3]{4}}{\sqrt[3]{8}} = \dfrac{\sqrt{3} \, \sqrt[3]{4}}{2}$

Multiply by $\dfrac{\sqrt[3]{4}}{\sqrt[3]{4}}$.

(c) $\dfrac{\sqrt{x} - 2}{\sqrt{x} + 2} = \dfrac{\sqrt{x} - 2}{\sqrt{x} + 2} \cdot \dfrac{\sqrt{x} - 2}{\sqrt{x} - 2} = \dfrac{(\sqrt{x} - 2)^2}{(\sqrt{x})^2 - 2^2}$

$= \dfrac{(\sqrt{x})^2 - 4\sqrt{x} + 4}{x - 4} = \dfrac{x - 4\sqrt{x} + 4}{x - 4}$ ◀

NOW WORK PROBLEM 39.

4 Solve Radical Equations

When the variable in an equation occurs in a square root, cube root, and so on, that is, when it occurs under a radical, the equation is called a **radical equation**. Sometimes a suitable operation will change a radical equation to one that is linear or quadratic. The most commonly used procedure is to isolate the most complicated radical on one side of the equation and then eliminate it by raising each side to a power equal to the index of the radical. Care must be taken, because extraneous solutions may result. Thus, when working with radical equations, we always check apparent solutions. Let's look at an example.

EXAMPLE 5

Solving Radical Equations

Solve the equation: $\sqrt[3]{2x - 4} - 2 = 0$

Solution

The equation contains a radical whose index is 3. We isolate it on the left side.

$$\sqrt[3]{2x - 4} - 2 = 0$$
$$\sqrt[3]{2x - 4} = 2$$

Now raise each side to the third power (since the index of the radical is 3) and solve.

$$\left(\sqrt[3]{2x - 4}\right)^3 = 2^3 \qquad \text{Raise each side to the 3rd power.}$$
$$2x - 4 = 8 \qquad \text{Simplify.}$$
$$2x = 12 \qquad \text{Solve for } x.$$
$$x = 6$$

✔ **CHECK:** $\sqrt[3]{2(6) - 4} - 2 = \sqrt[3]{12 - 4} - 2 = \sqrt[3]{8} - 2 = 2 - 2 = 0.$

The solution is $x = 6$. ◄

━ **NOW WORK PROBLEM 47.**

5 **Simplify Expressions with Rational Exponents**

Radicals are used to define rational exponents.

> If a is a real number and $n \geq 2$ is an integer, then
>
> $$a^{1/n} = \sqrt[n]{a} \qquad (3)$$
>
> provided that $\sqrt[n]{a}$ exists.

Note that if n is even and $a < 0$, then $\sqrt[n]{a}$ and $a^{1/n}$ do not exist.

EXAMPLE 6

Using Equation (3)

(a) $4^{1/2} = \sqrt{4} = 2$ (b) $(-27)^{1/3} = \sqrt[3]{-27} = -3$

(c) $8^{1/2} = \sqrt{8} = 2\sqrt{2}$ (d) $16^{1/3} = \sqrt[3]{16} = 2\sqrt[3]{2}$ ◄

> If a is a real number and m and n are integers containing no common factors with $n \geq 2$, then
>
> $$a^{m/n} = \sqrt[n]{a^m} = \left(\sqrt[n]{a}\right)^m \qquad (4)$$
>
> provided that $\sqrt[n]{a}$ exists.

We have two comments about equation (4):

1. The exponent $\dfrac{m}{n}$ must be in lowest terms and n must be positive.

2. In simplifying $a^{m/n}$, either $\sqrt[n]{a^m}$ or $(\sqrt[n]{a})^m$ may be used. Generally, taking the root first, as in $(\sqrt[n]{a})^m$, is easier.

EXAMPLE 7 **Using Equation (4)**

(a) $4^{3/2} = \left(\sqrt{4}\right)^3 = 2^3 = 8$ (b) $(-8)^{4/3} = \left(\sqrt[3]{-8}\right)^4 = (-2)^4 = 16$

(c) $(32)^{-2/5} = \left(\sqrt[5]{32}\right)^{-2} = 2^{-2} = \dfrac{1}{4}$ ◀

NOW WORK PROBLEMS 51 AND 57.

It can be shown that the laws of exponents hold for rational exponents.

EXAMPLE 8 **Simplifying Expressions with Rational Exponents**

Simplify each expression. Express your answer so that only positive exponents occur. Assume that the variables are positive.

(a) $\left(\dfrac{2x^{1/3}}{y^{2/3}}\right)^{-3}$ (b) $(x^{2/3}y)(x^{-2}y)^{1/2}$

Solution (a) $\left(\dfrac{2x^{1/3}}{y^{2/3}}\right)^{-3} = \left(\dfrac{y^{2/3}}{2x^{1/3}}\right)^3 = \dfrac{(y^{2/3})^3}{(2x^{1/3})^3} = \dfrac{y^2}{2^3(x^{1/3})^3} = \dfrac{y^2}{8x}$

(b) $(x^{2/3}y)(x^{-2}y)^{1/2} = (x^{2/3}y)[(x^{-2})^{1/2}y^{1/2}]$

$\qquad = x^{2/3}yx^{-1}y^{1/2} = (x^{2/3}x^{-1})(y \cdot y^{1/2})$

$\qquad = x^{-1/3}y^{3/2} = \dfrac{y^{3/2}}{x^{1/3}}$ ◀

NOW WORK PROBLEM 67.

 The next two examples illustrate some algebra that you will need to know for certain calculus problems.

EXAMPLE 9 **Writing an Expression as a Single Quotient**

Write the following expression as a single quotient in which only positive exponents appear.

$$(x^2 + 1)^{1/2} + x \cdot \dfrac{1}{2}(x^2 + 1)^{-1/2} \cdot 2x$$

Solution

$$(x^2 + 1)^{1/2} + x \cdot \frac{1}{2}(x^2 + 1)^{-1/2} \cdot 2x = (x^2 + 1)^{1/2} + \frac{x^2}{(x^2 + 1)^{1/2}}$$

$$= \frac{(x^2 + 1)^{1/2}(x^2 + 1)^{1/2} + x^2}{(x^2 + 1)^{1/2}}$$

$$= \frac{(x^2 + 1) + x^2}{(x^2 + 1)^{1/2}}$$

$$= \frac{2x^2 + 1}{(x^2 + 1)^{1/2}}$$

NOW WORK PROBLEM 73.

EXAMPLE 10

Factoring an Expression Containing Rational Exponents

Factor: $4x^{1/3}(2x + 1) + 2x^{4/3}$

Solution

We begin by looking for factors that are common to the two terms. Notice that 2 and $x^{1/3}$ are common factors. Then

$$4x^{1/3}(2x + 1) + 2x^{4/3} = 2x^{1/3}[2(2x + 1) + x]$$
$$= 2x^{1/3}(5x + 2)$$

A.9 Assess Your Understanding

'Are You Prepared?'

Answers are given at the end of these exercises. If you get a wrong answer, read the pages listed in red.

1. *True or False:* One of the Laws of Exponents states that $a^n + b^n = (a + b)^n$ (pp. 664–665)

2. $\sqrt{(-4)^2} =$ _____ . (p. 666)

Concepts and Vocabulary

3. In the symbol $\sqrt[n]{a}$, the integer n is called the _____ .

4. We call $\sqrt[3]{a}$ the _____ _____ of a.

5. *True or False:* $\sqrt[5]{-32} = -2$

6. *True or False:* $\sqrt[4]{(-3)^4} = -3$

Skill Building

In Problems 7–34, simplify each expression. Assume that all variables are positive when they appear.

7. $\sqrt[3]{27}$

8. $\sqrt[4]{16}$

9. $\sqrt[3]{-8}$

10. $\sqrt[3]{-1}$

11. $\sqrt{8}$

12. $\sqrt[3]{54}$

13. $\sqrt[3]{-8x^4}$

14. $\sqrt[4]{48x^5}$

15. $\sqrt[4]{x^{12}y^8}$

16. $\sqrt[5]{x^{10}y^5}$

17. $\sqrt[4]{\dfrac{x^9 y^7}{xy^3}}$

18. $\sqrt[3]{\dfrac{3xy^2}{81x^4y^2}}$

19. $\sqrt{36x}$

20. $\sqrt{9x^5}$

21. $\sqrt{3x^2}\sqrt{12x}$

22. $\sqrt{5x}\sqrt{20x^3}$

23. $\left(\sqrt{5}\sqrt[3]{9}\right)^2$

24. $\left(\sqrt[3]{3}\sqrt{10}\right)^4$

25. $\left(3\sqrt{6}\right)\left(2\sqrt{2}\right)$

26. $\left(5\sqrt{8}\right)\left(-3\sqrt{3}\right)$

27. $\left(\sqrt{3} + 3\right)\left(\sqrt{3} - 1\right)$

28. $\left(\sqrt{5} - 2\right)\left(\sqrt{5} + 3\right)$

29. $\left(\sqrt{x} - 1\right)^2$

30. $\left(\sqrt{x} + \sqrt{5}\right)^2$

31. $3\sqrt{2} - 4\sqrt{8}$

32. $\sqrt[3]{-x^4} + \sqrt[3]{8x}$

33. $\sqrt[3]{16x^4} - \sqrt[3]{2x}$

34. $\sqrt[4]{32x} + \sqrt[4]{2x^5}$

In Problems 35–46, rationalize the denominator of each expression. Assume that all variables are positive when they appear.

35. $\dfrac{1}{\sqrt{2}}$

36. $\dfrac{6}{\sqrt[3]{4}}$

37. $\dfrac{-\sqrt{3}}{\sqrt{5}}$

38. $\dfrac{-\sqrt[3]{3}}{\sqrt[3]{8}}$

39. $\dfrac{\sqrt{3}}{5-\sqrt{2}}$

40. $\dfrac{\sqrt{2}}{\sqrt{7}+2}$

41. $\dfrac{2-\sqrt{5}}{2+3\sqrt{5}}$

42. $\dfrac{\sqrt{3}-1}{2\sqrt{3}+3}$

43. $\dfrac{5}{\sqrt[3]{2}}$

44. $\dfrac{-2}{\sqrt[3]{9}}$

45. $\dfrac{\sqrt{x+h}-\sqrt{x}}{\sqrt{x+h}+\sqrt{x}}$

46. $\dfrac{\sqrt{x+h}+\sqrt{x-h}}{\sqrt{x+h}-\sqrt{x-h}}$

In Problems 47–50, solve each equation.

47. $\sqrt[3]{2t-1}=2$

48. $\sqrt[3]{3t+1}=-2$

49. $\sqrt{15-2x}=x$

50. $\sqrt{12-x}=x$

In Problems 51–62, simplify each expression.

51. $8^{2/3}$

52. $4^{3/2}$

53. $(-27)^{1/3}$

54. $16^{3/4}$

55. $16^{3/2}$

56. $64^{3/2}$

57. $9^{-3/2}$

58. $25^{-5/2}$

59. $\left(\dfrac{9}{8}\right)^{3/2}$

60. $\left(\dfrac{27}{8}\right)^{2/3}$

61. $\left(\dfrac{8}{9}\right)^{-3/2}$

62. $\left(\dfrac{8}{27}\right)^{-2/3}$

In Problems 63–70, simplify each expression. Express your answer so that only positive exponents occur. Assume that the variables are positive.

63. $x^{3/4}x^{1/3}x^{-1/2}$

64. $x^{2/3}x^{1/2}x^{-1/4}$

65. $(x^3y^6)^{1/3}$

66. $(x^4y^8)^{3/4}$

67. $(x^2y)^{1/3}(xy^2)^{2/3}$

68. $(xy)^{1/4}(x^2y^2)^{1/2}$

69. $(16x^2y^{-1/3})^{3/4}$

70. $(4x^{-1}y^{1/3})^{3/2}$

In Problems 71–84, expressions that occur in calculus are given. Write each expression as a single quotient in which only positive exponents and/or radicals appear.

71. $\dfrac{x}{(1+x)^{1/2}}+2(1+x)^{1/2}, \quad x>-1$

72. $\dfrac{1+x}{2x^{1/2}}+x^{1/2}, \quad x>0$

73. $2x(x^2+1)^{1/2}+x^2\cdot\dfrac{1}{2}(x^2+1)^{-1/2}\cdot 2x$

74. $(x+1)^{1/3}+x\cdot\dfrac{1}{3}(x+1)^{-2/3}, \quad x\neq-1$

75. $\sqrt{4x+3}\cdot\dfrac{1}{2\sqrt{x-5}}+\sqrt{x-5}\cdot\dfrac{1}{5\sqrt{4x+3}}, \quad x>5$

76. $\dfrac{\sqrt[3]{8x+1}}{3\sqrt[3]{(x-2)^2}}+\dfrac{\sqrt[3]{x-2}}{24\sqrt[3]{(8x+1)^2}}, \quad x\neq 2, x\neq -\dfrac{1}{8}$

77. $\dfrac{\sqrt{1+x}-x\cdot\dfrac{1}{2\sqrt{1+x}}}{1+x}, \quad x>-1$

78. $\dfrac{\sqrt{x^2+1}-x\cdot\dfrac{2x}{2\sqrt{x^2+1}}}{x^2+1}$

79. $\dfrac{(x+4)^{1/2}-2x(x+4)^{-1/2}}{x+4}, \quad x>-4$

80. $\dfrac{(9-x^2)^{1/2}+x^2(9-x^2)^{-1/2}}{9-x^2}, \quad -3<x<3$

81. $\dfrac{\dfrac{x^2}{(x^2-1)^{1/2}}-(x^2-1)^{1/2}}{x^2}, \quad x<-1 \text{ or } x>1$

82. $\dfrac{(x^2+4)^{1/2}-x^2(x^2+4)^{-1/2}}{x^2+4}$

83. $\dfrac{\dfrac{1+x^2}{2\sqrt{x}}-2x\sqrt{x}}{(1+x^2)^2}, \quad x>0$

84. $\dfrac{2x(1-x^2)^{1/3}+\dfrac{2}{3}x^3(1-x^2)^{-2/3}}{(1-x^2)^{2/3}}, \quad x\neq -1, x\neq 1$

△ *In Problems 85–96, expressions that occur in calculus are given. Factor each expression, express your answer so that only positive exponents occur.*

85. $(x + 1)^{3/2} + x \cdot \dfrac{3}{2}(x + 1)^{1/2}, \quad x \geq -1$

86. $(x^2 + 4)^{4/3} + x \cdot \dfrac{4}{3}(x^2 + 4)^{1/3} \cdot 2x$

87. $6x^{1/2}(x^2 + x) - 8x^{3/2} - 8x^{1/2}, \quad x \geq 0$

88. $6x^{1/2}(2x + 3) + x^{3/2} \cdot 8, \quad x \geq 0$

89. $3(x^2 + 4)^{4/3} + x \cdot 4(x^2 + 4)^{1/3} \cdot 2x$

90. $2x(3x + 4)^{4/3} + x^2 \cdot 4(3x + 4)^{1/3}$

91. $4(3x + 5)^{1/3}(2x + 3)^{3/2} + 3(3x + 5)^{4/3}(2x + 3)^{1/2}, \quad x \geq -\dfrac{3}{2}$

92. $6(6x + 1)^{1/3}(4x - 3)^{3/2} + 6(6x + 1)^{4/3}(4x - 3)^{1/2}, \quad x \geq \dfrac{3}{4}$

93. $3x^{-1/2} + \dfrac{3}{2}x^{1/2}, \quad x > 0$

94. $8x^{1/3} - 4x^{-2/3}, \quad x \neq 0$

95. $x\left(\dfrac{1}{2}\right)(8 - x^2)^{-1/2}(-2x) + (8 - x^2)^{1/2}$

96. $2x(1 - x^2)^{3/2} + x^2\left(\dfrac{3}{2}\right)(1 - x^2)^{1/2}(-2x)$

Discussion and Writing

97. Write a brief paragraph that compares the method used to rationalize the denominator of a rational expression and the method used to write the quotient of two complex numbers in standard form.

'Are You Prepared?' Answers

1. False   **2.** 4

ANSWERS

CHAPTER 1 Graphs

1.1 Assess Your Understanding *(page 8)*

5. abscissa; ordinate **6.** quadrants **7.** midpoint **8.** F **9.** F **10.** T

11. (a) Quadrant II **(b)** Positive x-axis
(c) Quadrant III **(d)** Quadrant I
(e) Negative y-axis **(f)** Quadrant IV

13. The points will be on a vertical line that is 2 units to the right of the y-axis.

15. $(-1, 4)$; Quadrant II
17. $(3, 1)$; Quadrant I
19. $X\min = -11$, $X\max = 5$, $X\text{scl} = 1$, $Y\min = -3$, $Y\max = 6$, $Y\text{scl} = 1$
21. $X\min = -30$, $X\max = 50$, $X\text{scl} = 10$, $Y\min = -90$, $Y\max = 50$, $Y\text{scl} = 10$

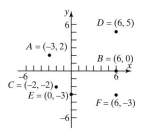

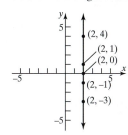

23. $X\min = -10$, $X\max = 110$, $X\text{scl} = 10$, $Y\min = -10$, $Y\max = 160$, $Y\text{scl} = 10$ **25.** $X\min = -6$, $X\max = 6$, $X\text{scl} = 2$, $Y\min = -4$, $Y\max = 4$, $Y\text{scl} = 2$ **27.** $X\min = -6$, $X\max = 6$, $X\text{scl} = 2$, $Y\min = -1$, $Y\max = 3$, $Y\text{scl} = 1$ **29.** $X\min = 3$, $X\max = 9$, $X\text{scl} = 1$, $Y\min = 2$, $Y\max = 10$, $Y\text{scl} = 2$ **31.** $\sqrt{5}$ **33.** $\sqrt{10}$ **35.** $2\sqrt{17}$ **37.** $\sqrt{85}$ **39.** $\sqrt{53}$ **41.** $\sqrt{6.89} \approx 2.62$ **43.** $\sqrt{a^2 + b^2}$ **45.** $4\sqrt{10}$ **47.** $2\sqrt{65}$

49. $d(A, B) = \sqrt{13}$
$d(B, C) = \sqrt{13}$
$d(A, C) = \sqrt{26}$
$(\sqrt{13})^2 + (\sqrt{13})^2 = (\sqrt{26})^2$
Area $= \dfrac{13}{2}$ square units

51. $d(A, B) = \sqrt{130}$
$d(B, C) = \sqrt{26}$
$d(A, C) = 2\sqrt{26}$
$(\sqrt{26})^2 + (2\sqrt{26})^2 = (\sqrt{130})^2$
Area $= 26$ square units

53. $d(A, B) = 4$
$d(A, C) = 5$
$d(B, C) = \sqrt{41}$
$4^2 + 5^2 = 16 + 25 = (\sqrt{41})^2$
Area $= 10$ square units

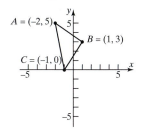

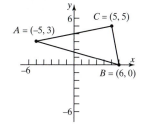

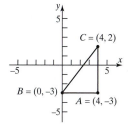

55. $(4, 0)$ **57.** $\left(\dfrac{3}{2}, 1\right)$ **59.** $(5, -1)$ **61.** $(1.05, 0.7)$ **63.** $\left(\dfrac{a}{2}, \dfrac{b}{2}\right)$ **65.** $(2, 2)$; $(2, -4)$ **67.** $(0, 0)$; $(8, 0)$ **69.** $\sqrt{17}$; $2\sqrt{5}$; $\sqrt{29}$ **71.** $\left(\dfrac{s}{2}, \dfrac{s}{2}\right)$
73. $d(P_1, P_2) = 6$; $d(P_2, P_3) = 4$; $d(P_1, P_3) = 2\sqrt{13}$; right triangle **75.** $d(P_1, P_2) = 2\sqrt{17}$; $d(P_2, P_3) = \sqrt{34}$; $d(P_1, P_3) = \sqrt{34}$; isosceles right triangle **77.** $90\sqrt{2} \approx 127.28$ ft **79. (a)** $(90, 0)$, $(90, 90)$, $(0, 90)$ **(b)** $5\sqrt{2161} \approx 232.43$ ft **(c)** $30\sqrt{149} \approx 366.20$ ft **81.** $d = 50t$

1.2 Assess Your Understanding *(page 21)*

3. intercepts **4.** zeros; roots **5.** y-axis **6.** 4 **7.** $(-3, 4)$ **8.** T **9.** F **10.** F

11.

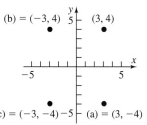

13.

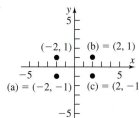

15.

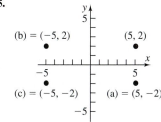

17.

(a) = (−3, 4)   (c) = (3, 4)

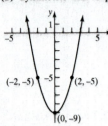

(−3, −4)   (b) = (3, −4)

19.

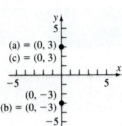

(a) = (0, 3)
(c) = (0, 3)

(0, −3)
(b) = (0, −3)

21. (0, 0) is on the graph.
23. (0, 3) is on the graph.
25. (0, 2) and $(\sqrt{2}, \sqrt{2})$ are on the graph.

27. (a) $(−1, 0), (1, 0)$

 (b) Symmetric with respect to the x-axis, the y-axis, and the origin.

29. (a) $\left(−\dfrac{\pi}{2}, 0\right), (0, 1), \left(\dfrac{\pi}{2}, 0\right)$

 (b) Symmetric with respect to the y-axis

31. (a) $(0, 0)$

 (b) Symmetric with respect to the x-axis

33. (a) $(−2, 0), (0, 0), (2, 0)$

 (b) Symmetric with respect to the origin

35.

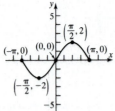

(−2, −5)   (2, −5)

(0, −9)

37.

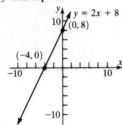

$\left(\dfrac{\pi}{2}, 2\right)$

$(−\pi, 0)$ $(0, 0)$ $(\pi, 0)$

$\left(−\dfrac{\pi}{2}, −2\right)$

39. x-intercept: −2;
 y-intercept: 2

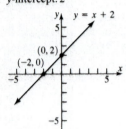

$y = x + 2$

(0, 2)
(−2, 0)

41. x-intercept: −4;
 y-intercept: 8

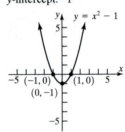

$y = 2x + 8$
(0, 8)

(−4, 0)

43. x-intercepts: −1, 1;
 y-intercept: −1

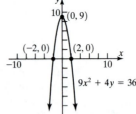

$y = x^2 − 1$

(−1, 0)   (1, 0)
(0, −1)

45. x-intercepts: −2, 2;
 y-intercept: 4

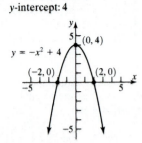

$y = −x^2 + 4$ (0, 4)

(−2, 0)   (2, 0)

47. x-intercept: 3;
 y-intercept: 2

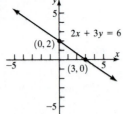

$2x + 3y = 6$
(0, 2)

(3, 0)

49. x-intercepts: −2, 2;
 y-intercept: 9

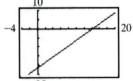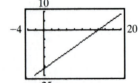

(0, 9)

(−2, 0)   (2, 0)

$9x^2 + 4y = 36$

51. x-intercept: 6.5;
 y-intercept: −13

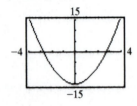

10

−5   10

−20

53. x-intercepts: −2.74, 2.74;
 y-intercept: −15

15

−4   4

−15

55. x-intercept: 14.33;
 y-intercept: −21.5

10

−4   20

−25

57. x-intercepts: −2.72, 2.72;
 y-intercept: 12.33

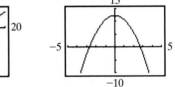

15

−5   5

−10

59. $(-4,0), (0,-2), (0,2)$; symmetric with respect to the x-axis   **61.** $(0,0)$; symmetric with respect to the origin   **63.** $(0,9), (3,0), (-3,0)$; symmetric with respect to the y-axis   **65.** $(-2,0), (2,0), (0,-3), (0,3)$; symmetric with respect to the x-axis, y-axis, and origin   **67.** $(0,-27), (3,0)$; no symmetry   **69.** $(0,-4), (4,0), (-1,0)$; no symmetry   **71.** $(0,0)$; symmetric with respect to the origin   **73.** $(0,0)$; symmetric with respect to the origin

75. 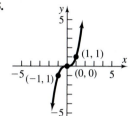   **77.** 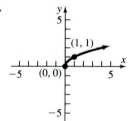   **79.** 13   **81.** -4 or 1

83. (a) $y = \sqrt{x^2}$ and $y = |x|$ have the same graph.   **(b)** $\sqrt{x^2} = |x|$ **(c)** $x \geq 0$ for $y = (\sqrt{x})^2$, while x can be any real number for $y = x$.   **(d)** $y \geq 0$ for $y = \sqrt{x^2}$

1.3 Assess Your Understanding (page 26)

3. ZERO   **4.** F   **5.** $\{-2.21, 0.54, 1.68\}$ **7.** $\{-1.55, 1.15\}$ **9.** $\{-1.12, 0.36\}$ **11.** $\{-2.69, -0.49, 1.51\}$ **13.** $\{-2.86, -1.34, 0.20, 1.00\}$
15. No real solutions   **17.** $\{-18\}$ **19.** $\{-4\}$ **21.** $\left\{\dfrac{46}{5}\right\}$ **23.** $\{3\}$ **25.** $\{2\}$ **27.** $\{-4, 7\}$ **29.** $\left\{-\dfrac{2}{3}, 2\right\}$ **31.** $\{-2, -1, 2\}$ **33.** $\{15\}$
35. $\left\{-4, -\dfrac{1}{8}\right\}$

1.4 Assess Your Understanding (page 40)

1. undefined; zero   **2.** $m_1 = m_2$; y-intercepts; $m_1 = -\dfrac{1}{m_2}$ **3.** $y = b$; y-intercept   **4.** T   **5.** F   **6.** F

7. (a) $\dfrac{1}{2}$ **(b)** If x increases by 2 units, y will increase by 1 unit.   **9. (a)** $-\dfrac{1}{3}$ **(b)** If x increases by 3 units, y will decrease by 1 unit.

11. Slope $= -\dfrac{3}{2}$ **13.** Slope $= -\dfrac{1}{2}$ **15.** Slope $= 0$ **17.** Slope undefined

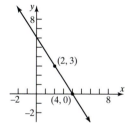   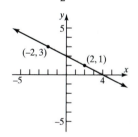   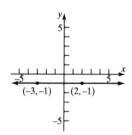   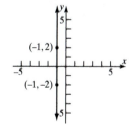

19. 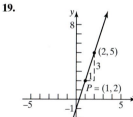   **21.** 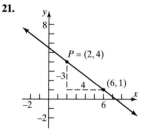   **23.** 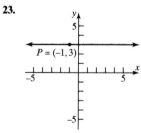   **25.**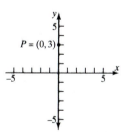

27. $(2,6); (3,10); (4,14)$ **29.** $(4,-7); (6,-10); (8,-13)$ **31.** $(-1,-5); (0,-7); (1,-9)$ **33.** $x - 2y = 0$ or $y = \dfrac{1}{2}x$ **35.** $x + y = 2$ or $y = -x + 2$
37. $2x - y = 3$ or $y = 2x - 3$ **39.** $x + 2y = 5$ or $y = -\dfrac{1}{2}x + \dfrac{5}{2}$ **41.** $3x - y = -9$ or $y = 3x + 9$ **43.** $2x + 3y = -1$ or $y = -\dfrac{2}{3}x - \dfrac{1}{3}$
45. $x - 2y = -5$ or $y = \dfrac{1}{2}x + \dfrac{5}{2}$ **47.** $3x + y = 3$ or $y = -3x + 3$ **49.** $x - 2y = 2$ or $y = \dfrac{1}{2}x - 1$ **51.** $x = 2$; no slope–intercept form
53. $2x - y = -4$ or $y = 2x + 4$ **55.** $2x - y = 0$ or $y = 2x$ **57.** $x = 4$; no slope–intercept form   **59.** $2x + y = 0$ or $y = -2x$
61. $x - 2y = -3$ or $y = \dfrac{1}{2}x + \dfrac{3}{2}$ **63.** $y = 4$

65. Slope $= 2$; y-intercept $= 3$

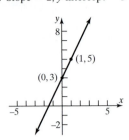

67. $y = 2x - 2$; slope $= 2$; y-intercept $= -2$

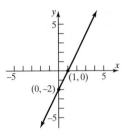

69. Slope $= \dfrac{1}{2}$; y-intercept $= 2$

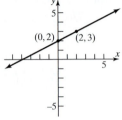

71. $y = -\dfrac{1}{2}x + 2$; slope $= -\dfrac{1}{2}$; y-intercept $= 2$

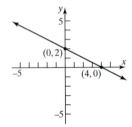

73. $y = \dfrac{2}{3}x - 2$; slope $= \dfrac{2}{3}$; y-intercept $= -2$

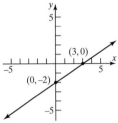

75. $y = -x + 1$; slope $= -1$; y-intercept $= 1$

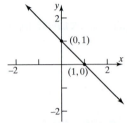

77. Slope undefined; no y-intercept

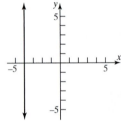

79. Slope $= 0$; y-intercept $= 5$

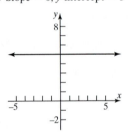

81. $y = x$; slope $= 1$; y-intercept $= 0$

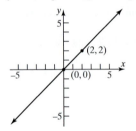

83. $y = \dfrac{3}{2}x$; slope $= \dfrac{3}{2}$; y-intercept $= 0$

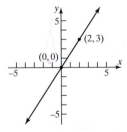

85. **(a)** x-intercept: 3; y-intercept: 2

(b)

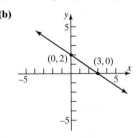

87. **(a)** x-intercept: -10; y-intercept: 8

(b)

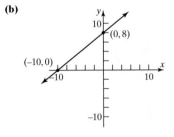

89. **(a)** x-intercept: 3; y-intercept: $\dfrac{21}{2}$

(b)

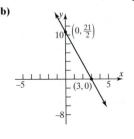

91. **(a)** x-intercept: 2; y-intercept: 3

(b)

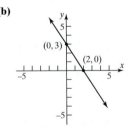

93. **(a)** x-intercept: 5; y-intercept: -2

(b)

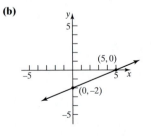

95. $y = 0$ **97.** (b)   **99.** (d)   **101.** $x - y = -2$ or $y = x + 2$ **103.** $x + 3y = 3$ or $y = -\dfrac{1}{3}x + 1$

105. $C = 0.07x + 29$; $36.70; $45.10   **107.** $C = 0.53x + 1,070,000$

109. (a) $C = 0.08275x + 7.58, 0 \le x \le 400$ **(b)**   **(c)** $15.86   **(d)** $32.41
(e) Each additional kW-h used adds $0.08275 to the bill.

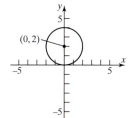

111. $°C = \dfrac{5}{9}(°F - 32)$; approximately 21°C   **113. (a)** $A = \dfrac{1}{5}(x - 100,000) + 40,000$ **(b)** $80,000

(c) Each additional box sold requires an additional $0.20 in advertising.

115. All have the same slope, 2; the lines are parallel.

117. (b), (c), (e), (g)
119. (c)
125. No; no
127. They are the same line.
129. Yes

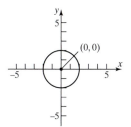

1.5 Assess Your Understanding *(page 49)*

1. F   **2.** radius   **3.** T   **4.** F   **5.** Center $(2, 1)$; radius 2; $(x - 2)^2 + (y - 1)^2 = 4$ **7.** Center $\left(\dfrac{5}{2}, 2\right)$; radius $\dfrac{3}{2}$; $\left(x - \dfrac{5}{2}\right)^2 + (y - 2)^2 = \dfrac{9}{4}$

9. $x^2 + y^2 = 4$;
$x^2 + y^2 - 4 = 0$

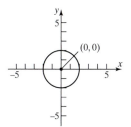

11. $x^2 + (y - 2)^2 = 4$;
$x^2 + y^2 - 4y = 0$

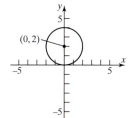

13. $(x - 4)^2 + (y + 3)^2 = 25$;
$x^2 + y^2 - 8x + 6y = 0$

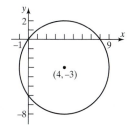

15. $(x + 2)^2 + (y - 1)^2 = 16$;
$x^2 + y^2 + 4x - 2y - 11 = 0$

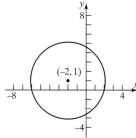

17. $\left(x - \dfrac{1}{2}\right)^2 + y^2 = \dfrac{1}{4}$;
$x^2 + y^2 - x = 0$

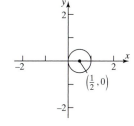

19. (a) $(h, k) = (0, 0); r = 2$

(b)

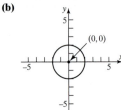

(c) $(\pm 2, 0); (0, \pm 2)$

21. (a) $(h, k) = (3, 0); r = 2$

(b)

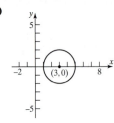

(c) $(1, 0); (5, 0)$

23. (a) $(h, k) = (1, 2); r = 3$

(b)

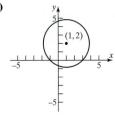

(c) $(1 \pm \sqrt{5}, 0); (0, 2 \pm 2\sqrt{2})$

25. (a) $(h, k) = (-2, 2); r = 3$

(b)

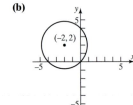

(c) $(-2 \pm \sqrt{5}, 0); (0, 2 \pm \sqrt{5})$

27. (a) $(h, k) = \left(\frac{1}{2}, -1\right); r = \frac{1}{2}$

(b)

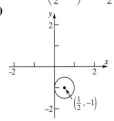

(c) $(0, -1)$

29. (a) $(h, k) = (3, -2); r = 5$

(b)

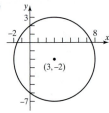

(c) $(3 \pm \sqrt{21}, 0); (0, -6), (0, 2)$

31. $x^2 + y^2 - 13 = 0$ **33.** $x^2 + y^2 - 4x - 6y + 4 = 0$ **35.** $x^2 + y^2 + 2x - 6y + 5 = 0$ **37. (c)**   **39. (b)**
41. $(x + 3)^2 + (y - 1)^2 = 16$ **43.** $(x - 2)^2 + (y - 2)^2 = 9$ **45.** $x^2 + y^2 + 2x + 4y - 4168.16 = 0$ **47.** $\sqrt{2}x + 4y - 9\sqrt{2} = 0$
49. $(1, 0)$ **51.** $y = 2$ **53.** (b), (e), (g)

Review Exercises *(page 52)*

1. (a) $2\sqrt{5}$ **(b)** $(2, 1)$ **(c)** $\frac{1}{2}$ **(d)** For each run of 2, there is a rise of 1.   **3. (a)** 5   **(b)** $\left(-\frac{1}{2}, 1\right)$ **(c)** $-\frac{4}{3}$ **(d)** For each run of 3, there is a rise

of -4.   **5. (a)** 12   **(b)** $(4, 2)$ **(c)** Undefined   **(d)** No change in x **7.** $(-4, 0), (0, 2), (0, 0), (0, -2), (2, 0)$

9.

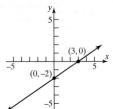

11.

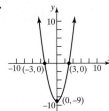

13.

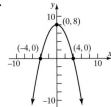

15. x-axis   **17.** x-axis, y-axis, origin
19. y-axis   **21.** no symmetry

23.

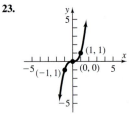

25. $\{-2.49, 0.66, 1.83\}$
27. $\{-1.14, 1.64\}$

29. $2x + y = 5$ or $y = -2x + 5$ **31.** $x = -3$; no slope-intercept form

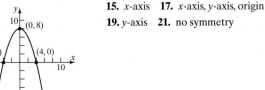

33. $x + 5y = -10$ or $y = -\frac{1}{5}x - 2$

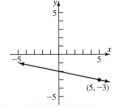

35. $2x - 3y = -19$ or $y = \frac{2}{3}x + \frac{19}{3}$

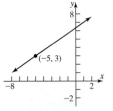

37. $-x + y = -7$ or $y = x - 7$

39. Slope $= -\dfrac{2}{3}$, y-intercept $= 6$ **41.** Slope $= -\dfrac{1}{5}$, y-intercept $= 4$ **43.** $(x + 2)^2 + (y - 3)^2 = 16$ **45.** $(x + 1)^2 + (y + 2)^2 = 1$

47. Center $(1, -2)$; radius $= 3$ **49.** Center $(1, -2)$; radius $= \sqrt{5}$ **51.** $d(A, B) = \sqrt{13}$; $d(B, C) = \sqrt{13}$ **53.** $m_{AB} = -1$; $m_{BC} = -1$

55. Center $(1, -2)$; radius $= 4\sqrt{2}$; $x^2 + y^2 - 2x + 4y - 27 = 0$

57.

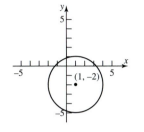

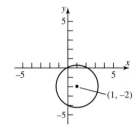

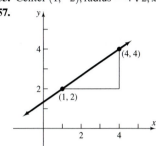

Chapter Test *(page 54)*

1. (a) $d = 10$ **(b)** $M = (1, 1)$ **(c)** $(x - 1)^2 + (y - 1)^2 = 25$ **2.**   **3.**

(d)

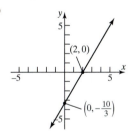

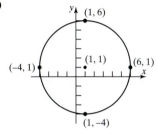

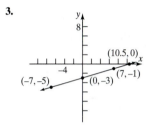

4. (a) $m = \dfrac{5}{3}$ **(b)** The x–intercept is 2. The y-intercept is $-\dfrac{10}{3}$

(c)

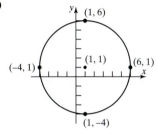

(d) $y = -\dfrac{3}{5}x + \dfrac{22}{5}$ **(e)** $y = \dfrac{5}{3}x + \dfrac{20}{3}$

5. $\{-1, 0.5, 1\}$
6. $\{-2.50, 2.50\}$
7. $\{-2.46, -0.24, 1.70\}$
8. x-axis

C H A P T E R   2   Functions and Their Graphs

2.1 Assess Your Understanding *(page 68)*

5. independent; dependent   **6.** range   **7.** $[0, 5]$ **8.** $\ne$; f; g **9.** $(g - f)(x)$ **10.** F   **11.** T   **12.** T   **13.** F   **14.** F
15. Function; Domain: {Elvis, Colleen, Kaleigh, Marissa}, Range: {January 8, March 15, September 17}   **17.** Not a function   **19.** Not a function
21. Function; Domain: $\{1, 2, 3, 4\}$; Range: $\{3\}$ **23.** Not a function   **25.** Function; Domain: $\{-2, -1, 0, 1\}$, Range: $\{0, 1, 4\}$ **27.** Function
29. Function   **31.** Not a function   **33.** Not a function   **35.** Function   **37.** Not a function   **39. (a)** -4 **(b)** 1   **(c)** -3 **(d)** $3x^2 - 2x - 4$
(e) $-3x^2 - 2x + 4$ **(f)** $3x^2 + 8x + 1$ **(g)** $12x^2 + 4x - 4$ **(h)** $3x^2 + 6xh + 3h^2 + 2x + 2h - 4$

41. (a) 0   **(b)** $\dfrac{1}{2}$ **(c)** $-\dfrac{1}{2}$ **(d)** $\dfrac{-x}{x^2 + 1}$ **(e)** $\dfrac{-x}{x^2 + 1}$ **(f)** $\dfrac{x + 1}{x^2 + 2x + 2}$ **(g)** $\dfrac{2x}{4x^2 + 1}$ **(h)** $\dfrac{x + h}{x^2 + 2xh + h^2 + 1}$

43. (a) 4   **(b)** 5   **(c)** 5   **(d)** $|x| + 4$ **(e)** $-|x| - 4$ **(f)** $|x + 1| + 4$ **(g)** $2|x| + 4$ **(h)** $|x + h| + 4$

45. (a) $-\dfrac{1}{5}$ **(b)** $-\dfrac{3}{2}$ **(c)** $\dfrac{1}{8}$ **(d)** $\dfrac{-2x + 1}{-3x - 5}$ **(e)** $\dfrac{-2x - 1}{3x - 5}$ **(f)** $\dfrac{2x + 3}{3x - 2}$ **(g)** $\dfrac{4x + 1}{6x - 5}$ **(h)** $\dfrac{2x + 2h + 1}{3x + 3h - 5}$

47. All real numbers   **49.** All real numbers   **51.** $\{x | x \ne -4, x \ne 4\}$ **53.** $\{x | x \ne 0\}$ **55.** $\{x | x \ge 4\}$ **57.** $\{x | x > 9\}$ **59.** $\{x | x > 1\}$
61. (a) $(f + g)(x) = 5x + 1$; All real numbers   **(b)** $(f - g)(x) = x + 7$; All real numbers

(c) $(f \cdot g)(x) = 6x^2 - x - 12$; All real numbers   **(d)** $\left(\dfrac{f}{g}\right)(x) = \dfrac{3x + 4}{2x - 3}$; $\left\{ x \middle| x \ne \dfrac{3}{2} \right\}$

63. (a) $(f + g)(x) = 2x^2 + x - 1$; All real numbers   **(b)** $(f - g)(x) = -2x^2 + x - 1$; All real numbers

(c) $(f \cdot g)(x) = 2x^3 - 2x^2$; All real numbers   **(d)** $\left(\dfrac{f}{g}\right)(x) = \dfrac{x - 1}{2x^2}$; $\{x | x \neq 0\}$ **65. (a)** $(f + g)(x) = \sqrt{x} + 3x - 5$; $\{x | x \geq 0\}$

(b) $(f - g)(x) = \sqrt{x} - 3x + 5$; $\{x | x \geq 0\}$ **(c)** $(f \cdot g)(x) = 3x\sqrt{x} - 5\sqrt{x}$; $\{x | x \geq 0\}$ **(d)** $\left(\dfrac{f}{g}\right)(x) = \dfrac{\sqrt{x}}{3x - 5}$; $\left\{x | x \geq 0, x \neq \dfrac{5}{3}\right\}$

67. (a) $(f + g)(x) = 1 + \dfrac{2}{x}$; $\{x | x \neq 0\}$ **(b)** $(f - g)(x) = 1$; $\{x | x \neq 0\}$ **(c)** $(f \cdot g)(x) = \dfrac{1}{x} + \dfrac{1}{x^2}$; $\{x | x \neq 0\}$

(d) $\left(\dfrac{f}{g}\right)(x) = x + 1$; $\{x | x \neq 0\}$ **69. (a)** $(f + g)(x) = \dfrac{6x + 3}{3x - 2}$; $\left\{x \middle| x \neq \dfrac{2}{3}\right\}$ **(b)** $(f - g)(x) = \dfrac{-2x + 3}{3x - 2}$; $\left\{x \middle| x \neq \dfrac{2}{3}\right\}$

(c) $(f \cdot g)(x) = \dfrac{8x^2 + 12x}{(3x - 2)^2}$; $\left\{x \middle| x \neq \dfrac{2}{3}\right\}$ **(d)** $\left(\dfrac{f}{g}\right)(x) = \dfrac{2x + 3}{4x}$; $\left\{x \middle| x \neq 0, x \neq \dfrac{2}{3}\right\}$ **71.** $g(x) = 5 - \dfrac{7}{2}x$

73. 4   **75.** $2x + h - 1$ **77.** $3x^2 + 3xh + h^2$ **79.** $A = -\dfrac{7}{2}$ **81.** $A = -4$ **83.** $A = 8$; undefined at $x = 3$ **85.** $A(x) = \dfrac{1}{2}x^2$ **87.** $G(x) = 10x$

89. (a) 15.1 m, 14.07 m, 12.94 m, 11.72 m   **(b)** 1.01 sec, 1.43 sec, 1.75 sec   **(c)** 2.02 sec   **91. (a)** \$222   **(b)** \$225   **(c)** \$220   **(d)** \$230

93. $R(x) = \dfrac{L(x)}{P(x)}$ **95.** $H(x) = P(x) \cdot I(x)$ **97.** Only $h(x) = 2x$

2.2 Assess Your Understanding *(page 75)*

3. vertical   **4.** 5; −3   **5.** $a = -2$ **6.** F   **7.** F   **8.** T

9. (a) $f(0) = 3$; $f(-6) = -3$ **(b)** $f(6) = 0$; $f(11) = 1$ **(c)** Positive   **(d)** Negative   **(e)** −3, 6, and 10   **(f)** $-3 < x < 6$; $10 < x \leq 11$

(g) $\{x | -6 \leq x \leq 11\}$ **(h)** $\{y | -3 \leq y \leq 4\}$ **(i)** −3, 6, 10   **(j)** 3   **(k)** 3 times   **(l)** Once   **(m)** 0, 4   **(n)** −5, 8

11. Not a function   **13.** Function **(a)** Domain: $\{x | -\pi \leq x \leq \pi\}$; Range: $\{y | -1 \leq y \leq 1\}$ **(b)** $\left(-\dfrac{\pi}{2}, 0\right), \left(\dfrac{\pi}{2}, 0\right), (0, 1)$ **(c)** y-axis

15. Not a function   **17.** Function   **(a)** Domain: $\{x | x > 0\}$; Range: all real numbers   **(b)** $(1, 0)$ **(c)** None

19. Function   **(a)** Domain: all real numbers; Range: $\{y | y \leq 2\}$ **(b)** $(-3, 0), (3, 0), (0, 2)$ **(c)** y-axis

21. Function   **(a)** Domain: all real numbers; Range: $\{y | y \geq -3\}$ **(b)** $(1, 0), (3, 0), (0, 9)$ **(c)** None

23. (a) Yes   **(b)** $f(-2) = 9$; $(-2, 9)$ **(c)** $0, \dfrac{1}{2}$; $(0, -1), \left(\dfrac{1}{2}, -1\right)$ **(d)** All real numbers   **(e)** $-\dfrac{1}{2}, 1$ **(f)** −1

25. (a) No   **(b)** $f(4) = -3$; $(4, -3)$ **(c)** 14; $(14, 2)$ **(d)** $\{x | x \neq 6\}$ **(e)** −2   **(f)** $-\dfrac{1}{3}$

27. (a) Yes   **(b)** $f(2) = \dfrac{8}{17}$; $\left(2, \dfrac{8}{17}\right)$ **(c)** $-1, 1$; $(-1, 1), (1, 1)$ **(d)** All real numbers   **(e)** 0   **(f)** 0

29. (a) About 81.07 ft   **(b)** About 129.59 ft   **31. (a)** 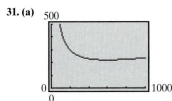   **(b)**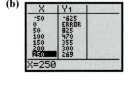

(c) About 26.63 ft   **(d)** About 528.13 ft

(e)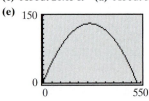

(c) 600 mi/hr

(f) 115.07 ft and 413.05 ft

(g) 275 ft; maximum height shown in the table is 131.8 ft

(h) 264 ft

35. There is at most one y-intercept.   **37. (a)** III   **(b)** IV   **(c)** I   **(d)** V   **(e)** II

39.

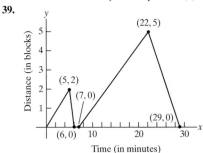

41. (a) 2 hr elapsed during which Kevin was between 0 and 3 mi from home.   **(b)** 0.5 hr elapsed during which Kevin was 3 mi from home.   **(c)** 0.3 hr elapsed during which Kevin was between 0 and 3 mi from home.   **(d)** 0.2 hr elapsed during which Kevin was 0 mi from home.

(e) 0.9 hr elapsed during which Kevin was between 0 and 2.8 mi from home.   **(f)** 0.3 hr elapsed during which Kevin was 2.8 mi from home.   **(g)** 1.1 hr elapsed during which Kevin was between 0 and 2.8 mi from home.   **(h)** 3 mi   **(i)** 2 times

43. No points whose x-coordinate is 5 or whose y-coordinate is 0 can be on the graph.

2.3 Assess Your Understanding *(page 88)*

6. increasing **7.** even; odd **8.** T **9.** T **10.** F **11.** Yes **13.** No **15.** $(-8, -2); (0, 2); (5, \infty)$

17. Yes; 10 **19.** $-2, 2; 6, 10$ **21. (a)** $(-2, 0), (0, 3), (2, 0)$ **(b)** Domain: $\{x | -4 \le x \le 4\}$ or $[-4, 4]$; Range: $\{y | 0 \le y \le 3\}$ or $[0, 3]$

(c) Increasing on $(-2, 0)$ and $(2, 4)$; Decreasing on $(-4, -2)$ and $(0, 2)$ **(d)** Even

23. (a) $(0, 1)$ **(b)** Domain: all real numbers; Range: $\{y | y > 0\}$ or $(0, \infty)$ **(c)** Increasing on $(-\infty, \infty)$ **(d)** Neither

25. (a) $(-\pi, 0), (0, 0), (\pi, 0)$ **(b)** Domain: $\{x | -\pi \le x \le \pi\}$ or $[-\pi, \pi]$; Range: $\{y | -1 \le y \le 1\}$ or $[-1, 1]$

(c) Increasing on $\left(-\dfrac{\pi}{2}, \dfrac{\pi}{2}\right)$; Decreasing on $\left(-\pi, -\dfrac{\pi}{2}\right)$ and $\left(\dfrac{\pi}{2}, \pi\right)$ **(d)** Odd **27. (a)** $\left(0, \dfrac{1}{2}\right), \left(\dfrac{1}{2}, 0\right), \left(\dfrac{5}{2}, 0\right)$

(b) Domain: $\{x | -3 \le x \le 3\}$ or $[-3, 3]$; Range: $\{y | -1 \le y \le 2\}$ or $[-1, 2]$ **(c)** Increasing on $(2, 3)$; Decreasing on $(-1, 1)$;

Constant on $(-3, -1)$ and $(1, 2)$ **(d)** Neither **29. (a)** $0; 3$ **(b)** $-2, 2; 0, 0$

31. (a) $\dfrac{\pi}{2}; 1$ **(b)** $-\dfrac{\pi}{2}; -1$ **33.** Odd **35.** Even **37.** Odd **39.** Neither **41.** Even **43.** Odd

45.

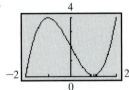

Increasing: $(-2, -1), (1, 2)$
Decreasing: $(-1, 1)$
Local maximum: $(-1, 4)$
Local minimum: $(1, 0)$

47.

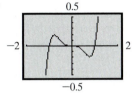

Increasing: $(-2, -0.77), (0.77, 2)$
Decreasing: $(-0.77, 0.77)$
Local maximum: $(-0.77, 0.19)$
Local minimum: $(0.77, -0.19)$

49.

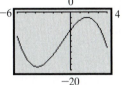

Increasing: $(-3.77, 1.77)$
Decreasing: $(-6, -3.77), (1.77, 4)$
Local maximum: $(1.77, -1.91)$
Local minimum: $(-3.77, -18.89)$

51.

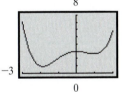

Increasing: $(-1.87, 0), (0.97, 2)$
Decreasing: $(-3, -1.87), (0, 0.97)$
Local maximum: $(0, 3)$
Local minima: $(-1.87, 0.95), (0.97, 2.65)$

53. (a) -4 **(b)** -8 **(c)** -10 **55. (a)** 17 **(b)** -1 **(c)** 11

57. (a) 5

(b) 5; the slope of the line joining $(1, f(1))$ and $(3, f(3))$ is 5

(c) $y = 5x - 2$

(d)

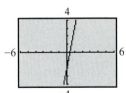

59. (a) $x - 2$

(b) -1; the slope of the line joining $(-2, g(-2))$ and $(1, g(1))$ is -1.

(c) $y = -x$

(d)

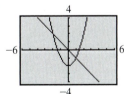

61. (a) x

(b) 4; the slope of the line joining $(2, h(2))$ and $(4, h(4))$ is 4.

(c) $y = 4x - 8$

(d)

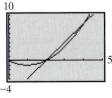

63. (a) $V(x) = x(24 - 2x)^2$

(b) 972 in.^{3} **(c)** 160 in.^{3}

(c) 160 in.^{3}

(d)

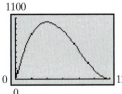

65. (a)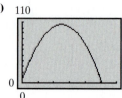

110

0
0   6

(b) 2.5 sec
(c) 106 ft

67. (a)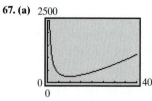

2500

0   40
0

(b) 10 riding lawn mowers/hr
(c) $239/mower

69. (a), (b), (e)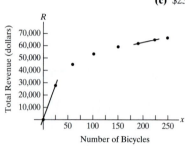

(c) 1120 dollars/bicycle
(d) For each additional bicycle sold between 0 and 25 bicycles, total
    revenue increases, on average, by $1120.
(f) 75 dollars/bicycle
(g) For each additional bicycle sold between 190 and 223 bicycles, total revenue
    increases, on average, by $75.
(h) The average rate of change is decreasing as the number of
    bicycles increase.

71. (a) 1   **(b)** 0.5   **(c)** 0.1   **(d)** 0.01   **(e)** 0.00

(f)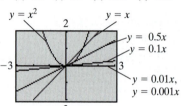

(g) They are getting closer to the tangent line at $(0, 0)$.
(h) They are getting closer to 0.

73. (a) 2   **(b)** 2; 2; 2; 2   **(c)** $y = 2x + 5$

(d)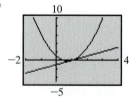

75. (a) $2x + h + 2$ **(b)** 4.5; 4.1; 4.01; 4   **(c)** $y = 4.01x - 1.01$

(d)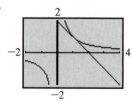

77. (a) $4x + 2h - 3$ **(b)** 2; 1.2; 1.02; 1

(c) $y = 1.02x - 1.02$

(d)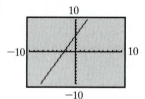

79. (a) $-\dfrac{1}{(x + h)x}$ **(b)** $-\dfrac{2}{3}; -\dfrac{10}{11}; -\dfrac{100}{101}; -1$

(c) $y = -\dfrac{100}{101}x + \dfrac{201}{101}$

(d)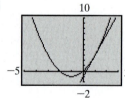

83. At most one   **85.** Yes; the function $f(x) = 0$ is both even and odd.

2.4 Assess Your Understanding *(page 101)*

5. slope; y-intercept   **6.** scatter diagram   **7.** $y = kx$ **8.** T   **9.** T   **10.** T
11. 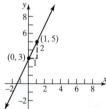   ; 2   **13.** 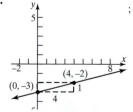   ; −3   **15.** 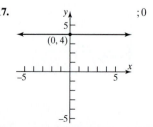   ; $\dfrac{1}{4}$ **17.**   ; 0

19. Linear relation   **21.** Linear relation   **23.** Nonlinear relation

25. (a) 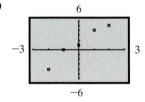   **(b)** Answers will vary. Using $(4, 6)$ and $(8, 14)$: $y = 2x - 2$

(c) 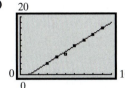   **(d)** $y = 2.0357x - 2.3571$ **(e)**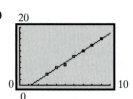

27. (a) 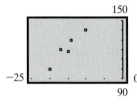   **(b)** Answers will vary. Using $(-2, -4)$ and $(1, 4)$: $y = \dfrac{8}{3}x + \dfrac{4}{3}$

(c) 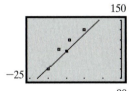   **(d)** $y = 2.2x + 1.2$ **(e)**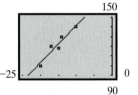

29. (a) 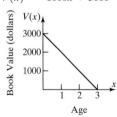   **(b)** Answers will vary. Using $(-20, 100)$ and $(-15, 118)$: $y = \dfrac{18}{5}x + 172$

(c) 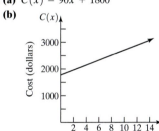   **(d)** $y = 3.8613x + 180.2920$ **(e)**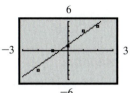

31. (a) \$45   **(b)** 180 mi   **(c)** 259 mi   **33. (a)** \$778.22   **(b)** 2006   **(c)** 2012   **35. (a)** $p = \$16, q = 600$ T-shirts   **(b)** $0 < p < \$16$

37. (a) $x = 5000$ **(b)** $x > 5000$

39. (a) $V(x) = -1000x + 3000$

(b)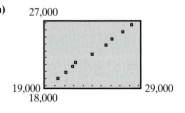

(c) \$1000   **(d)** After 1 year

41. (a) $C(x) = 90x + 1800$

(b)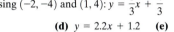

(c) \$3060   **(d)** 22 bicycles

43. (a) $C(x) = 0.07x + 29$ **(b)** \$36.70; \$45.10   **45.** $p(b) = 0.00649B$; \$941.05   **47.** $R(g) = 1.95g$; \$20.48   **49.** 5 gallons

51. (a) *[graph]*

(b) $C(I) = 0.9241I + 479.6584$
(c) If disposable income increases by \$1, consumption increases by \$0.92.
(d) \$27,048
(e) \$28,590

53. (a)

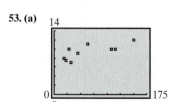

(b) $L(G) = 0.0261G + 7.8738$

(c) If gestation period increases by 1 day, then life expectancy increases by about 0.0261 years.

(d) About 10.2 years

55. (a) No **(b)**

(c) $D = -1.3355p + 86.1974$

(d) If the price increases \$1, the quantity sold per day decreases by about 1.34 pairs of jeans.

(e) $D(p) = -1.3355p + 86.1974$

(f) $\{p \mid p > 0\}$

(g) $D(28) \approx 48.80$; about 49 pairs

57.

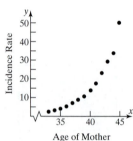

Age of Mother

The data do not follow a linear pattern.

59. When the y-intercept is 0. Yes, if the slope is 0.   **61.** No linear relation

2.5 Assess Your Understanding *(page 114)*

4. less   **5.** piecewise-defined   **6.** T   **7.** F   **8.** F   **9.** C   **11.** E   **13.** B   **15.** F

17.

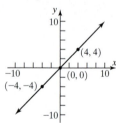

19.

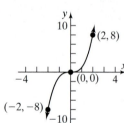

21.

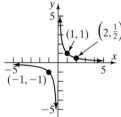

23.

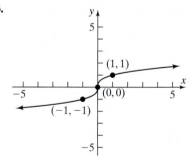

25. (a) 4   **(b)** 2   **(c)** 5   **27. (a)** 2   **(b)** 3   **(c)** −4

29. (a) All real numbers

(b) $(0, 1)$

(c)

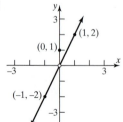

(d) $\{y \mid y \neq 0\}; (-\infty, 0)$ or $(0, \infty)$

31. (a) All real numbers

(b) $(0, 3)$

(c)

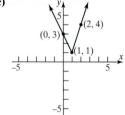

(d) $\{y \mid y \geq 1\}; [1, \infty)$

33. (a) $\{x \mid x \geq -2\}; [-2, \infty)$

(b) $(0, 3), (2, 0)$

(c)

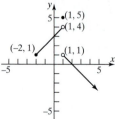

(d) $\{y \mid y < 4, y = 5\}; (-\infty, 4)$ and $\{5\}$

35. (a) All real numbers
(b) $(-1, 0), (0, 0)$
(c)

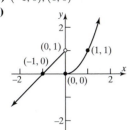

(d) All real numbers

37. (a) $\{x | x \geq -2\}; [-2, \infty)$
(b) $(0, 1)$
(c)

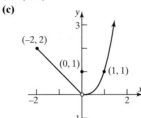

(d) $\{y | y > 0\}; (0, \infty)$

39. (a) All real numbers
(b) $(x, 0)$ for $0 \leq x < 1$
(c)

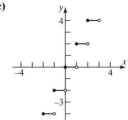

(d) Set of even integers

41. $f(x) = \begin{cases} -x & \text{if } -1 \leq x \leq 0 \\ \frac{1}{2}x & \text{if } 0 < x \leq 2 \end{cases}$ (Other answers are possible.)

43. $f(x) = \begin{cases} -x & \text{if } x \leq 0 \\ -x + 2 & \text{if } 0 < x \leq 2 \end{cases}$ (Other answers are possible.)

45. (a) \$35   **(b)** \$61   **(c)** \$35.40

47. (a) \$59.33   **(b)** \$396.04
(c) $C = \begin{cases} 0.99755x + 9.45 & \text{if } 0 \leq x \leq 50 \\ 0.74825x + 21.915 & \text{if } x > 50 \end{cases}$
(d)

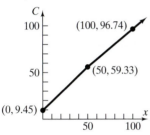

49. For schedule X: $f(x) = \begin{cases} 0.10x & \text{if } & 0 < x \leq 7150 \\ 715 + 0.15(x - 7150) & \text{if } & 7150 < x \leq 29{,}050 \\ 4000 + 0.25(x - 29{,}050) & \text{if } & 29{,}050 < x \leq 70{,}350 \\ 14{,}325 + 0.28(x - 70{,}350) & \text{if } & 70{,}350 < x \leq 146{,}750 \\ 35{,}717 + 0.33(x - 146{,}750) & \text{if } & 146{,}750 < x \leq 319{,}100 \\ 92{,}592.50 + 0.35(x - 319{,}100) & \text{if } & x > 319{,}100 \end{cases}$

51. (a)

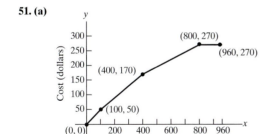

(b) $C = 50 + 0.4(x - 100)$
(c) $C = 170 + 0.25(x - 400)$

53. $f(x) = \begin{cases} x & \text{if } & 0 \leq x < 10 \\ 10 & \text{if } & 10 \leq x < 500 \\ 30 & \text{if } & 500 \leq x < 1000 \\ 50 & \text{if } & 1000 \leq x < 1500 \\ 70 & \text{if } & 1500 \leq x \end{cases}$

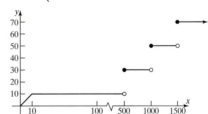

55. (a) 10°C   **(b)** 4°C   **(c)** −3°C   **(d)** −4°C   **(e)** The wind chill is equal to the air temperature.   **(f)** At wind speed greater than 20 m/sec, the wind chill factor depends only on the air temperature.   **57.** Each graph is that of $y = x^2$, but shifted vertically. If $y = x^2 + k, k > 0$, the shift is up k units; if $y = x^2 - k, k > 0$, the shift is down k units.   **59.** Each graph is that of $y = |x|$, but either compressed or stretched. If $y = k|x|$ and $k > 1$, the graph is stretched vertically; if $y = k|x|, 0 < k < 1$, the graph is compressed vertically.   **61.** The graph of $y = f(-x)$ is the reflection about the y-axis of the graph of $y = f(x)$.   **63.** They are all ∪-shaped and open upward. All three go through the points $(-1, 1), (0, 0)$ and $(1, 1)$. As the exponent increases, the steepness of the curve increases (except near $x = 0$).

2.6 Assess Your Understanding *(page 126)*

1. horizontal; right   **2.** y **3.** $-5; -2; 2$ **4.** T   **5.** F   **6.** T   **7.** B   **9.** H   **11.** I   **13.** L   **15.** F   **17.** G   **19.** $y = (x - 4)^3$
21. $y = x^3 + 4$ **23.** $y = -x^3$ **25.** $y = 4x^3$ **27.** (1) $y = \sqrt{x} + 2$; (2) $y = -(\sqrt{x} + 2)$; (3) $y = -(\sqrt{-x} + 2)$
29. (1) $y = -\sqrt{x}$; (2) $y = -\sqrt{x} + 2$; (3) $y = -\sqrt{x + 3} + 2$ **31.** (c)   **33.** (c)

35.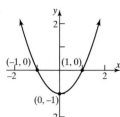

37.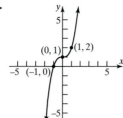

39.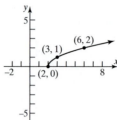

41.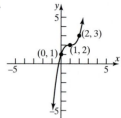

43.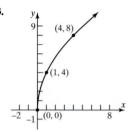

45.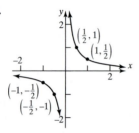

47.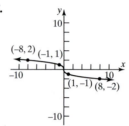

49.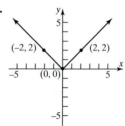

51.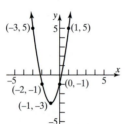

53.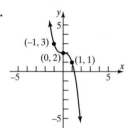

55.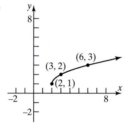

57.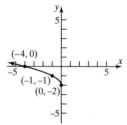

59.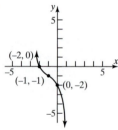

61.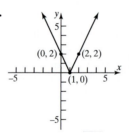

63.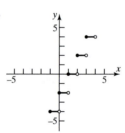

65. (a) $F(x) = f(x) + 3$ **(b)** $G(x) = f(x + 2)$ **(c)** $P(x) = -f(x)$ **(d)** $H(x) = f(x + 1) - 2$

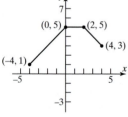 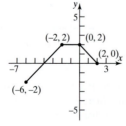 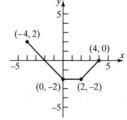 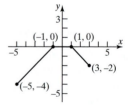

(e) $Q(x) = \frac{1}{2}f(x)$ **(f)** $g(x) = f(-x)$ **(g)** $h(x) = f(2x)$

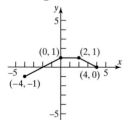 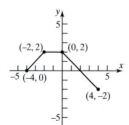 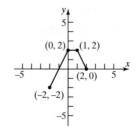

67. (a) $F(x) = f(x) + 3$

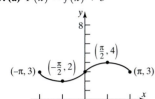

(b) $G(x) = f(x + 2)$

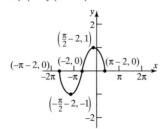

(c) $P(x) = -f(x)$

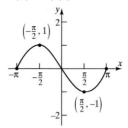

(d) $H(x) = f(x + 1) - 2$

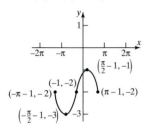

(e) $Q(x) = \dfrac{1}{2}f(x)$

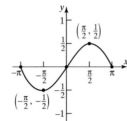

(f) $g(x) = f(-x)$

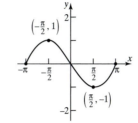

(g) $h(x) = f(2x)$

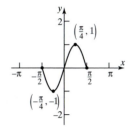

69. (a) -7 and 1 **(b)** -3 and 5 **(c)** -5 and 3 **(d)** -3 and 5 **71. (a)** $(-3, 3)$ **(b)** $(4, 10)$ **(c)** Decreasing on $(-1, 5)$ **(d)** Decreasing on $(-5, 1)$

73. (a)

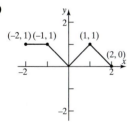

(b)

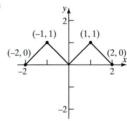

75. $f(x) = (x + 1)^2 - 1$

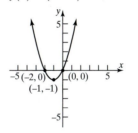

77. $f(x) = (x - 4)^2 - 15$

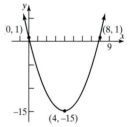

79. $f(x) = \left(x + \dfrac{1}{2}\right)^2 + \dfrac{3}{4}$

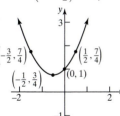

81. $f(x) = 2(x - 3)^2 + 1$

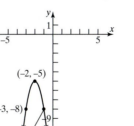

83. $f(x) = -3(x + 2)^2 - 5$

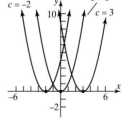

85.

87.

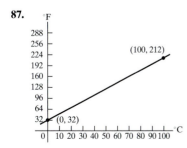

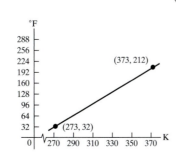

89. (a)

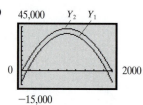

(b) 10% tax

(c) Y_1 is the graph of $p(x)$ shifted down vertically 10,000 units.
  Y_2 is the graph of $p(x)$ vertically compressed by a factor of 0.9.

(d) 10% tax

91. (a) $y = |x + 1|$

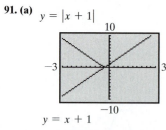

$y = x + 1$

(b)

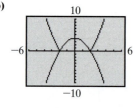

(c)

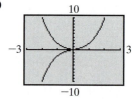

(d) Any part of the graph of $y = f(x)$ that lies below the x-axis is reflected about the x-axis to obtain the graph of $y = |f(x)|$.

2.7 Assess Your Understanding *(page 134)*

1. $V(r) = 2\pi r^3$

3. (a) $R(x) = -\dfrac{1}{6}x^2 + 100x$

(b) \$13,333.33

(c) 15,000

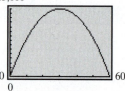

(d) 300; \$15,000   **(e)** \$50

5. (a) $R(x) = -\dfrac{1}{5}x^2 + 20x$

(b) \$255

(c) 500

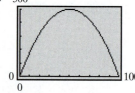

(d) 50; \$500   **(e)** \$10

7. (a) $A(x) = -x^2 + 200x$

(b) $0 < x < 200$

(c) A is largest when $x = 100$ yd.

10,000

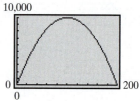

9. (a) $d(x) = \sqrt{x^4 - 15x^2 + 64}$

(b) $d(0) = 8$

(c) $d(1) = \sqrt{50} \approx 7.07$

(d)

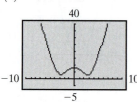

(e) d is smallest when $x \approx -2.74$
  or $x \approx 2.74$.

11. (a) $d(x) = \sqrt{x^2 - x + 1}$

(b)

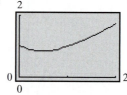

(c) d is smallest when $x = \dfrac{1}{2}$.

13. $A(x) = \dfrac{1}{2}x^4$

15. (a) $A(x) = x(16 - x^2)$

(b) Domain: $\{x \mid 0 < x < 4\}$

(c) The area is largest when $x \approx 2.31$.

30

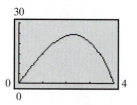

17. (a) $A(x) = 4x\sqrt{4 - x^2}$ **(c)** A is largest when $x \approx 1.41$.   **(d)** p is largest when $x \approx 1.41$.
(b) $p(x) = 4x + 4\sqrt{4 - x^2}$

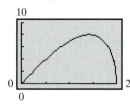

19. (a) $A(x) = x^2 + \dfrac{25 - 20x + 4x^2}{\pi}$ **21. (a)** $C(x) = x$ **(b)** $A(x) = \dfrac{x^2}{4\pi}$

(b) Domain: $\{x | 0 < x < 2.5\}$ **23. (a)** $A(r) = 2r^2$ **(b)** $p(r) = 6r$

(c) A is smallest when $x \approx 1.40$ m.   **25.** $A(x) = \left(\dfrac{\pi}{3} - \dfrac{\sqrt{3}}{4} \right)x^2$

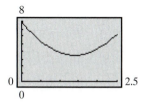

27. (a) $d(t) = \sqrt{2500t^2 - 360t + 13}$ **29.** $V(r) = \dfrac{\pi H(R - r)r^2}{R}$ **31. (a)** $T(x) = \dfrac{12 - x}{5} + \dfrac{\sqrt{x^2 + 4}}{3}$
(b) d is smallest when $t \approx 0.07$ hr.   **(b)** $\{x | 0 \le x \le 12\}$
(c) 3.09 hr
(d) 3.55 hr

Review Exercises *(page 140)*

1. Function; domain $\{-1, 2, 4\}$, range $\{0, 3\}$

3. (a) 2 **(b)** -2 **(c)** $-\dfrac{3x}{x^2 - 1}$ **(d)** $-\dfrac{3x}{x^2 - 1}$ **(e)** $\dfrac{3(x - 2)}{x^2 - 4x + 3}$ **(f)** $\dfrac{6x}{4x^2 - 1}$ **5. (a)** 0 **(b)** 0 **(c)** $\sqrt{x^2 - 4}$ **(d)** $-\sqrt{x^2 - 4}$ **(e)** $\sqrt{x^2 - 4x}$

(f) $2\sqrt{x^2 - 1}$ **7. (a)** 0 **(b)** 0 **(c)** $\dfrac{x^2 - 4}{x^2}$ **(d)** $-\dfrac{x^2 - 4}{x^2}$ **(e)** $\dfrac{x(x - 4)}{(x - 2)^2}$ **(f)** $\dfrac{x^2 - 1}{x^2}$ **9.** $\{x | x \ne -3, x \ne 3\}$ **11.** $\{x | x \le 2\}$ **13.** $\{x | x > 0\}$

15. $\{x | x \ne -3, x \ne 1\}$

17. $(f + g)(x) = 2x + 3$; Domain: all real numbers   **19.** $(f + g)(x) = 3x^2 + 4x + 1$; Domain: all real numbers
$(f - g)(x) = -4x + 1$; Domain: all real numbers $(f - g)(x) = 3x^2 - 2x + 1$; Domain: all real numbers
$(f \cdot g)(x) = -3x^2 + 5x + 2$; Domain: all real numbers $(f \cdot g)(x) = 9x^3 + 3x^2 + 3x$; Domain: all real numbers
$\left(\dfrac{f}{g}\right)(x) = \dfrac{2 - x}{3x + 1}$; Domain: $\left\{x \middle| x \ne -\dfrac{1}{3}\right\}$ $\left(\dfrac{f}{g}\right)(x) = \dfrac{3x^2 + x + 1}{3x}$; Domain: $\{x | x \ne 0\}$

21. $(f + g)(x) = \dfrac{x^2 + 2x - 1}{x(x - 1)}$; Domain: $\{x | x \ne 0, 1\}$ **23.** $-4x + 1 - 2h$

$(f - g)(x) = \dfrac{x^2 + 1}{x(x - 1)}$; Domain: $\{x | x \ne 0, 1\}$

$(f \cdot g)(x) = \dfrac{x + 1}{x(x - 1)}$; Domain: $\{x | x \ne 0, 1\}$

$\left(\dfrac{f}{g}\right)(x) = \dfrac{x(x + 1)}{x - 1}$; Domain: $\{x | x \ne 0, 1\}$

25. (a) Domain: $\{x|-4 \le x \le 3\}$; Range: $\{y|-3 \le y \le 3\}$ **(b)** $(0,0)$ **(c)** -1 **(d)** -4

(e) $\{x|0 < x \le 3\}$ **(f)**   **(g)**   **(h)**

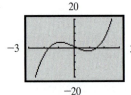   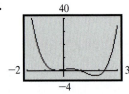

27. (a) Domain: $\{x|-4 \le x \le 4\}$ or $[-4, 4]$
    Range: $\{y|-3 \le y \le 1\}$ or $[-3, 1]$
(b) Increasing on $(-4, -1)$ and $(3, 4)$;
    Decreasing on $(-1, 3)$
(c) Local maximum is 1 and occurs at $x = -1$;
    Local minimum is -3 and occurs at $x = 3$
(d) No symmetry
(e) Neither
(f) x-intercepts: $-2, 0, 4$; y-intercept: 0

29. Odd
31. Even
33. Neither
35. Odd

37.

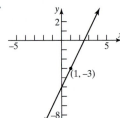

Local maximum: $(-0.91, 4.04)$
Local minimum: $(0.91, -2.04)$
Increasing: $(-3, -0.91)$; $(0.91, 3)$
Decreasing: $(-0.91, 0.91)$

39.

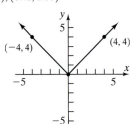

Local maximum: $(0.41, 1.53)$
Local minimum: $(-0.34, 0.54)$; $(1.80, -3.56)$
Increasing: $(-0.34, 0.41)$; $(1.80, 3)$
Decreasing: $(-2, -0.34)$; $(0.41, 1.80)$

41. (a) 23   **(b)** 7   **(c)** 47
43. -5
45. $-4x - 5$
47. (b)

49.   **51.**   **53.**

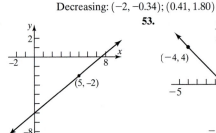

55.   **57.**   **59.**   **61.**

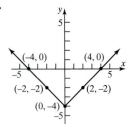   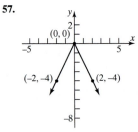   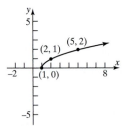   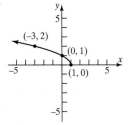

Intercepts: $(-4, 0), (4, 0), (0, -4)$
Domain: all real numbers
Range: $\{y|y \ge -4\}$ or $[-4, \infty)$

Intercept: $(0, 0)$
Domain: all real numbers
Range: $\{y|y \le 0\}$ or $(-\infty, 0]$

Intercept: $(1, 0)$
Domain: $\{x|x \ge 1\}$ or $[1, \infty)$
Range: $\{y|y \ge 0\}$ or $[0, \infty)$

Intercepts: $(0, 1), (1, 0)$
Domain: $\{x|\ x \le 1\}$ or $(-\infty, 1]$
Range: $\{y|y \ge 0\}$ or $[0, \infty)$

63.

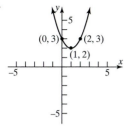

Intercept: $(0, 3)$

Domain: all real numbers

Range: $\{y | y \geq 2\}$ or $[2, \infty)$

65.

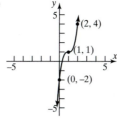

Intercepts: $(0, -2)$,

$\left(1 - \dfrac{\sqrt[3]{9}}{3}, 0\right)$ or about $(0.3, 0)$

Domain: all real numbers

Range: all real numbers

67. (a) $\{x | x > -2\}; (-2, \infty)$

   (b) $(0, 0)$

   (c)

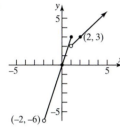

   (d) $\{y | y > -6\}; (-6, \infty)$

69. (a) $\{x | x \geq -4\}; [-4, \infty)$

   (b) $(0, 1)$

   (c)

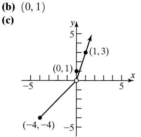

   (d) $\{y | y \geq -4\}; [-4, \infty)$

71. $f(x) = -2x + 3$ **73.** $A = 11$ **75.** $T(h) = -0.0025h + 30, 0 \leq x \leq 10{,}000$

77. If the radius doubles, the volume of the new sphere is 8 times as large as the original sphere, and the surface area of the new sphere is 4 times as large as the original sphere.

79. $S(x) = kx(36 - x^2)^{3/2}$; Domain: $\{x | 0 < x < 6\}$

81. (a) Yes

   (b)

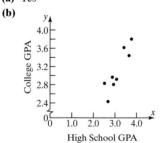

83. $R(g) = 2.14g$; 23.97

85. 2 by 8 ft

87. (a) 63   **(b)** 151.90

89. $(2, 3)$

   (c) $G = 0.964x + 0.072$

   (d) As high school GPA increases 1 point, college GPA increases by 0.964 point.

   (e) $G(x) = 0.964x + 0.072$

   (f) $\{x | 0 \leq x \leq 4\}$

   (g) 3.19

91. (a) $A(x) = 2x^2 + \dfrac{40}{x}$ **(b)** 42 ft^2

   (c) 28 ft^2 **(d)**

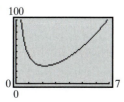

A is smallest when $x \approx 2.15$ ft.

Chapter Test *(page 145)*

1. (a) Function; domain: $\{2, 4, 6, 8\}$; range: $\{5, 6, 7, 8\}$ **(b)** Not a function **(c)** Not a function

(d) Function; domain: all real numbers; range: $\{y | y \geq 2\}$ **2.** Domain: $\left\{x | x \leq \dfrac{4}{5}\right\}$; $f(-1) = 3$ **3.** Domain: $\{x | x \neq -2\}$; $g(-1) = 1$

4. Domain: $\{x | x \neq -9, x \neq 4\}$; $h(-1) = \dfrac{1}{8}$ **5. (a)** Domain: $\{x | x -5 \leq x \leq 5\}$; range: $\{y | -3 \leq y \leq 3\}$ **(b)** $(0, 2), (-2, 0)$, and $(2, 0)$

(c) $f(1) = 3$ **(d)** $x = -5$ and $x = 3$ **(e)** $\{x | -5 \leq x < -2 \text{ or } 2 < x \leq 5\}; [-5, -2) \cup (2, 5]$

6. Local maxima: $f(-0.85) \approx -0.86$
$$f(2.35) \approx 15.55$$
Local minima: $f(0) = -2$
The function is increasing on the intervals $(-5, -0.85)$ and
$(0, 2.35)$ and decreasing on the intervals $(-0.85, 0)$ and $(2.35, 5)$.

7. (a)

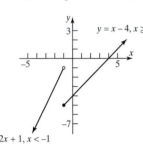

(b) $(0, -4), (4, 0)$
(c) $g(-5) = -9$
(d) $g(2) = -2$

8. $\dfrac{f(x) - f(3)}{x - 3} = 3x + 7, x \neq 3$ **9. (a)** $(f - g) = 2x^2 - 3x + 3$ **(b)** $(f \cdot g) = 6x^3 - 4x^2 + 3x - 2$ **(c)** $f(x + h) - f(x) = 4xh + 2h^2$

10. (a)

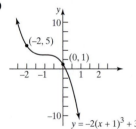

(b)

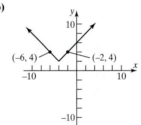

11. (a) Set B is more linear.   **(b)** $y = 2.02x - 5.33$

12. (a) 8.67% occurring in 1997 $(x \approx 5)$ **(b)** The model predicts that the interest rate will be -10.343%. This is not reasonable.

13. (a) $V(x) = \dfrac{x^2}{8} - \dfrac{5x}{4} + \dfrac{\pi x^2}{64}$ **(b)** 1297.61 ft³

Cumulative Review *(page 147)*

1. $\left\{\dfrac{4}{5}\right\}$ **2.** $\{3, 4\}$ **3.** $\left\{-\dfrac{1}{3}, 2\right\}$ **4.** $\left\{-\dfrac{1}{2}\right\}$ **5.** No real solution   **6.** $\{-7\}$ **7.** $\{-31\}$ **8.** $\left\{\dfrac{1}{3}, 1\right\}$ **9.** $\left\{\dfrac{1 - \sqrt{15}i}{4}, \dfrac{1 + \sqrt{15}i}{4}\right\}$

10. $\{x | 1 < x < 4\}$

11.

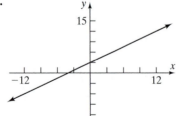

12.

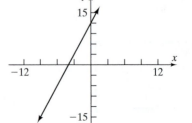

13.

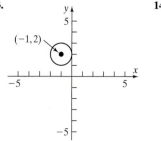

14.

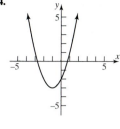

15. (a) Domain: $\{x|-4 \leq x \leq 4\}$; Range: $\{y|-1 \leq y \leq 3\}$ **(b)** $(-1, 0), (0, -1), (1, 0)$ **(c)** y-axis   **(d)** 1   **(e)** -4 and 4

(f) $\{x|-1 < x < 1\}$ **(g)**

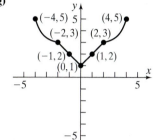

(h)

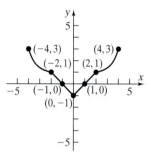

(i)

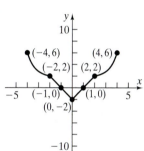

(j) even   **(k)** $(0, 4)$ **(l)** $(-4, 0)$ **(m)** f has a local minimum of -1 at $x = 0$ **(n)** 1

16. $5\sqrt{2}$ **17.** $(-2, -1)$ and $(2, 3)$ are on the graph.   **18.** x-intercepts: $-5, \dfrac{1}{3}$; y-intercept: -5 **19.** $-1.10, 0.26, 1.48, 2.36$

20. $\quad y = -\dfrac{1}{2}x + \dfrac{13}{2}$

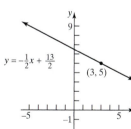

21. Yes, a function

22. (a) -3

(b) $x^2 - 4x - 2$

(c) $x^2 + 4x + 1$

(d) $-x^2 + 4x - 1$

(e) $x^2 - 3$

(f) $2x + h - 4$

23. $\{z|z \neq -1, z \neq 7\}$

24. Yes, a function

25. (a) No

(b) -1; $(-2, -1)$ is on the graph.

(c) -8; $(-8, 2)$ is on the graph.

C H A P T E R   3   Polynomial and Rational Functions

3.1 Assess Your Understanding (page 163)

5. parabola **6.** axis or axis of symmetry **7.** $-\dfrac{b}{2a}$ **8.** T **9.** T **10.** T **11.** C **13.** F **15.** G **17.** H

19.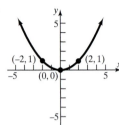

21.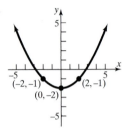

23.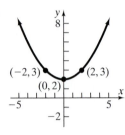

25.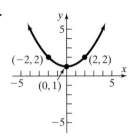

27. $f(x) = (x + 2)^2 - 2$

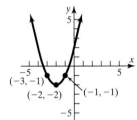

29. $f(x) = 2(x - 1)^2 - 1$

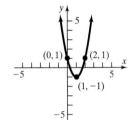

31. $(x) = -(x + 1)^2 + 1$

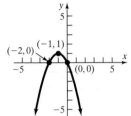

33. $f(x) = \dfrac{1}{2}(x + 1)^2 - \dfrac{3}{2}$

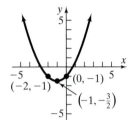

35.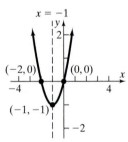

Domain: $(-\infty, \infty)$
Range $[-1, \infty)$
Decreasing on $(-\infty, -1)$
Increasing on $(-1, \infty)$

37.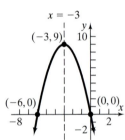

Domain: $(-\infty, \infty)$
Range $(-\infty, 9]$
Increasing on $(-\infty, -3)$
Decreasing on $(-3, \infty)$

39.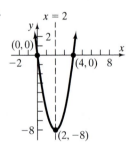

Domain: $(-\infty, \infty)$
Range $[-8, \infty)$
Decreasing on $(-\infty, 2)$
Increasing on $(2, \infty)$

41.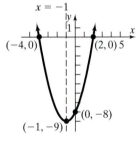

Domain: $(-\infty, \infty)$
Range $[-9, \infty)$
Decreasing on $(-\infty, -1)$
Increasing on $(-1, \infty)$

43.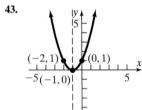

Domain: $(-\infty, \infty)$

Range $[0, \infty)$

Decreasing on $(-\infty, -1)$

Increasing on $(-1, \infty)$

45.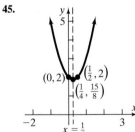

Domain: $(-\infty, \infty)$

Range $\left[\dfrac{15}{8}, \infty\right)$

Decreasing on $\left(-\infty, \dfrac{1}{4}\right)$

Increasing on $\left(\dfrac{1}{4}, \infty\right)$

47.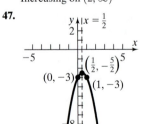

Domain: $(-\infty, \infty)$

Range $\left(-\infty, -\dfrac{5}{2}\right]$

Decreasing on $\left(\dfrac{1}{2}, \infty\right)$

Increasing on $\left(-\infty, \dfrac{1}{2}\right)$

49.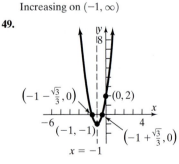

Domain: $(-\infty, \infty)$

Range $[-1, \infty)$

Decreasing on $(-\infty, -1)$

Increasing on $(-1, \infty)$

51.

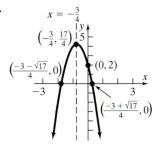

$x = -\frac{3}{4}$

$\left(-\frac{3}{4}, \frac{17}{4}\right)$

$\left(\frac{-3 - \sqrt{17}}{4}, 0\right)$ $(0, 2)$

$\left(\frac{-3 + \sqrt{17}}{4}, 0\right)$

Domain: $(-\infty, \infty)$

Range $\left(-\infty, \frac{17}{4}\right]$

Decreasing on $\left(-\frac{3}{4}, \infty\right)$

Increasing on $\left(-\infty, -\frac{3}{4}\right)$

53. $f(x) = (x + 1)^2 - 2 = x^2 + 2x - 1$
55. $f(x) = -(x + 3)^2 + 5 = -x^2 - 6x - 4$
57. $f(x) = 2(x - 1)^2 - 3 = 2x^2 - 4x - 1$
59. Minimum value; -18
61. Minimum value; -21
63. Maximum value; 21
65. Maximum value; 13
67. $a = 6, b = 0, c = 2$

69. (a) $a = 1: f(x) = (x + 3)(x - 1) = x^2 + 2x - 3$
    $a = 2: f(x) = 2(x + 3)(x - 1) = 2x^2 + 4x - 6$
    $a = -2: f(x) = -2(x + 3)(x - 1) = -2x^2 - 4x + 6$
    $a = 5: f(x) = 5(x + 3)(x - 1) = 5x^2 + 10x - 15$
(b) The value of a does not affect the x-intercepts
    but it changes the y-intercept by a factor of a.

(c) The value of a does not affect the axis of symmetry. It is $x = -1$ for all
    values of a.
(d) The value of a does not affect the x-coordinate of the vertex.
    However, the y-coordinate of the vertex is multiplied by a.
(e) The midpoint of the x-intercepts is the x-coordinate of the vertex.

71. $\$500; \$1,000,000$ **73. (a)** $R(x) = -\frac{1}{6}x^2 + 100x$ **(b)** $\$13,333.33$ **(c)** $300; \$15,000$ **(d)** $\$50$

75. (a) $R(x) = -\frac{1}{5}x^2 + 20x$ **(b)** $\$255$ **(c)** $50; \$500$ **(d)** $\$10$

77. (a) $A(w) = -w^2 + 200w$ **(b)** A is largest when $w = 100$ yd.   **(c)** 10,000 sq yd   **79.** 2,000,000 m²

81. (a) $\frac{625}{16} \approx 39$ ft   **(b)** 219.5 ft   **(c)** 170 ft

(d)

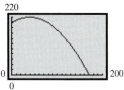

220

0   200
0

(e) When the height is 100 ft, the projectile is
    135.7 ft from the cliff.

83. 18.75 m   **85.** 3 in.

87. $\frac{750}{\pi}$ by 375 m   **89.** $x = \frac{a}{2}$

91. $\frac{38}{3}$ **93.** $\frac{248}{3}$ **95.** 25 square units

97. (a) $\$56,600$; 3685 hunters
(b)

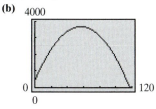

4000

0   120
0
Increasing

99. (a) 1795   **(b)** 28 years old
(c)

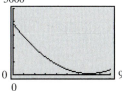

5000

0   90
0

(d) The number of victims initially
    decreases, then begins to increase.

101. (a) Quadratic, $a < 0$

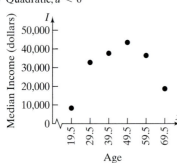

Age

(b) $I(x) = -42.67x^2 + 3998.27x - 51{,}873.20$
(c) 46.9 yr old
(d) \$41,788
(e) 50,000

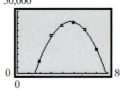

103. (a) Quadratic, $a < 0$

(b) $h(x) = -0.0037x^2 + 1.0318x + 5.6667$

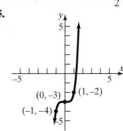

Distance

(c) 139.4 ft
(d) 77.6 ft
(e) 80

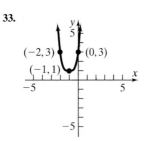

3.2 Assess Your Understanding *(page 182)*

5. smooth; continuous   **6.** zero or root   **7.** touches   **8.** T   **9.** F   **10.** F   **11.** Yes; degree 3   **13.** Yes; degree 2

15. No; x is raised to the -1 power.   **17.** No; x is raised to the $\dfrac{3}{2}$ power.   **19.** Yes; degree 4   **21.** Yes; degree 4

23.

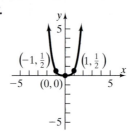

25.

27.

29.

31.

33.

35.

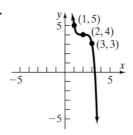

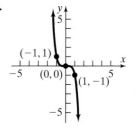

37. $f(x) = x^3 - 3x^2 - x + 3$ for $a = 1$ **39.** $f(x) = x^3 - x^2 - 12x$ for $a = 1$ **41.** $f(x) = x^4 - 15x^2 + 10x + 24$ for $a = 1$

43. $f(x) = x^3 - 5x^2 + 3x + 9$ for $a = 1$

45. (a) 7, multiplicity 1; -3, multiplicity 2   **(b)** graph touches the x-axis at -3 and crosses it at 7   **(c)** $y = 3x^3$

47. (a) 2, multiplicity 3   **(b)** graph crosses the x-axis at 2   **(c)** $y = 4x^5$

49. (a) $-\dfrac{1}{2}$, multiplicity 2   **(b)** graph touches the x-axis at $-\dfrac{1}{2}$ **(c)** $y = -2x^6$

51. (a) 5, multiplicity 3; -4, multiplicity 2   **(b)** graph touches the x-axis at -4 and crosses it at 5   **(c)** $y = x^5$

53. (a) no real zeros   **(b)** graph neither crosses nor touches the x-axis   **(c)** $y = 3x^6$

55. (a) 0, multiplicity 2; $-\sqrt{2}$, $\sqrt{2}$, multiplicity 1   **(b)** graph touches the x-axis at 0 and crosses at $-\sqrt{2}$ and $\sqrt{2}$ **(c)** $y = -2x^4$

57. (a) Degree 2; $y = x^2$
(b) x-intercept: 1; y-intercept: 1
(c) 1: Touches
(d)
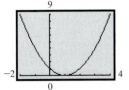
(e) Local minimum at $(1, 0)$
(f)
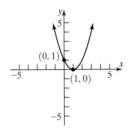
(g) Domain: $(-\infty, \infty)$
Range: $[0, \infty)$
(h) Increasing on $(1, \infty)$
Decreasing on $(-\infty, 1)$

59. (a) Degree 3; $y = x^3$
(b) x-intercepts: 0, 3; y-intercept: 0
(c) 0: Touches; 3: Crosses
(d)
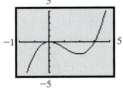
(e) Local minimum at $(2, -4)$
Local maximum at $(0, 0)$
(f)
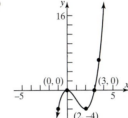
(g) Domain: $(-\infty, \infty)$
Range: $(-\infty, \infty)$
(h) Increasing on $(-\infty, 0)$ and $(2, \infty)$
Decreasing on $(0, 2)$

61. (a) Degree 3; $y = x^3$
(b) x-intercepts: -4, 2; y-intercept: 16
(c) -4: Crosses; 2: Touches
(d)
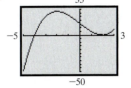
(e) Local maximum at $(-2, 32)$
Local minimum at $(2, 0)$
(f)
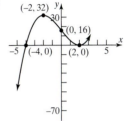
(g) Domain: $(-\infty, \infty)$
Range: $(-\infty, \infty)$
(h) Increasing on $(-\infty, -2)$ and $(2, \infty)$
Decreasing on $(-2, 2)$

63. (a) Degree 4; $y = -2x^4$
(b) x-intercepts: -2, 2
y-intercept: 32
(c) -2: Crosses; 2: Crosses
(d)
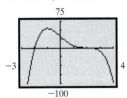
(e) Local maximum at $(-1, 54)$
(f)
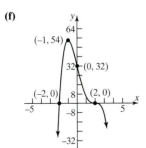
(g) Domain: $(-\infty, \infty)$
Range: $(-\infty, 54]$
(h) Increasing on $(-\infty, -1)$
Decreasing on $(-1, \infty)$

65. (a) Degree 3; $y = x^3$
(b) x-intercepts: -4, -1, 2
y-intercept: -8
(c) -4, -1, 2: Crosses
(d)
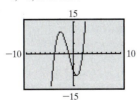
(e) Local maximum at $(-2.73, 10.39)$
Local minimum at $(0.73, -10.39)$
(f)
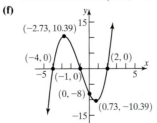
(g) Domain: $(-\infty, \infty)$
Range: $(-\infty, \infty)$
(h) Increasing on $(-\infty, -2.73)$ and $(0.73, \infty)$
Decreasing on $(-2.73, 0.73)$

67. (a) Degree 3; $y = -x^3$
(b) x-intercepts: -2, 0, 2
y-intercept: 0
(c) -2, 0, 2: Crosses
(d)
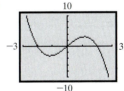
(e) Local minimum at $(-1.15, -3.08)$
Local maximum at $(1.15, 3.08)$
(f)
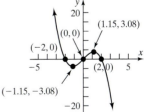
(g) Domain: $(-\infty, \infty)$
Range: $(-\infty, \infty)$
(h) Increasing on $(-1.15, 1.15)$
Decreasing on $(-\infty, -1.15)$
and $(1.15, \infty)$

69. (a) Degree 4; $y = x^4$

(b) x-intercepts: $-2, 0, 2$
  y-intercept: 0

(c) $-2, 2$: Crosses; 0: Touches

(d)

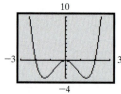

(e) Local minima at $(-1.41, -4)$, $(1.41, -4)$; Local maximum at $(0, 0)$

(f)

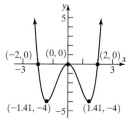

(g) Domain: $(-\infty, \infty)$
  Range: $[-4, \infty)$

(h) Increasing on $(-1.41, 0)$ and $(1.41, \infty)$
  Decreasing on $(-\infty, -1.41)$ and $(0, 1.41)$

71. (a) Degree 4; $y = x^4$

(b) x-intercepts: $-1, 2$
  y-intercept: 4

(c) $-1, 2$: Touches

(d)

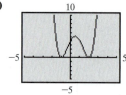

(e) Local minima at $(-1, 0)$, $(2, 0)$
  Local maximum at $(0.5, 5.06)$

(f)

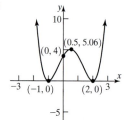

(g) Domain: $(-\infty, \infty)$
  Range: $[0, \infty)$

(h) Increasing on $(-1, 0.5)$ and $(2, \infty)$
  Decreasing on $(-\infty, -1)$ and $(0.5, 2)$

73. (a) Degree 4; $y = x^4$

(b) x-intercepts: $-1, 0, 3$;
  y-intercept: 0

(c) $-1, 3$: Crosses; 0: Touches

(d)

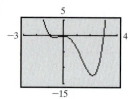

(e) Local minima at $(-0.69, -0.54)$, $(2.19, -12.39)$; Local maximum at $(0, 0)$

(f)

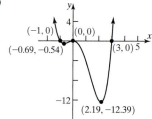

(g) Domain: $(-\infty, \infty)$
  Range: $[-12.39, \infty)$

(h) Increasing on $(-0.69, 0)$ and $(2.19, \infty)$
  Decreasing on $(-\infty, -0.69)$ and $(0, 2.19)$

75. (a) Degree 4; $y = x^4$

(b) x-intercepts: $-2, 4$; y-intercept: 64

(c) $-2, 4$: Touches

(d)

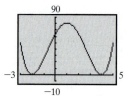

(e) Local minima at $(-2, 0)$, $(4, 0)$
  Local maximum at $(1, 81)$

(f)

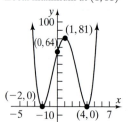

(g) Domain: $(-\infty, \infty)$
  Range: $(0, \infty)$

(h) Increasing on $(-2, 1)$ and $(4, \infty)$
  Decreasing on $(-\infty, -2)$ and $(1, 4)$

77. (a) Degree 5; $y = x^5$

(b) x-intercepts: $0, 2$; y-intercept: 0

(c) 0: Touches; 2: Crosses

(d)

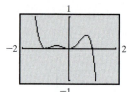

(e) Local maximum at $(0, 0)$
  Local minimum at $(1.48, -5.91)$

(f)

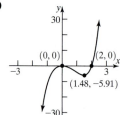

(g) Domain: $(-\infty, \infty)$
  Range: $(-\infty, \infty)$

(h) Increasing on $(-\infty, 0)$ and $(1.48, \infty)$
  Decreasing on $(0, 1.48)$

79. (a) (a) Degree 5; $y = -x^5$

(b) x-intercepts: $-1, 0, 1$; y-intercept: 0

(c) $-1, 0$: Touches; 1: Crosses

(d)

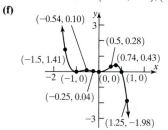

(e) Local minima at $(-1, 0)$, $(0, 0)$
  Local maxima at $(-0.54, 0.10)$, $(0.74, 0.43)$

(f)

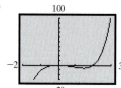

(g) Domain: $(-\infty, \infty)$
  Range: $(-\infty, \infty)$

(h) Increasing on $(-1, -0.54)$ and $(0, 0.74)$
  Decreasing on $(-\infty, -1)$, $(-0.54, 0)$,
  and $(0.74, \infty)$

81. (a) Degree 3; $y = x^3$

(b)

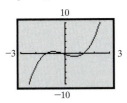

(c) x-intercepts: $-1.26, -0.20, 1.26$
y-intercept: -0.31752

(d)

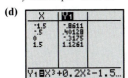

(e) Local maximum at $(-0.80, 0.57)$
Local minimum at $(0.66, -0.99)$

(f)

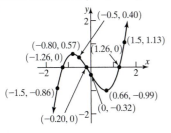

(g) Domain: $(-\infty, \infty)$; Range: $(-\infty, \infty)$
(h) Increasing on $(-\infty, -0.80)$ and $(0.66, \infty)$
Decreasing on $(-0.80, 0.66)$

83. (a) Degree 3; $y = x^3$

(b)

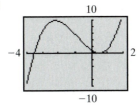

(c) x-intercepts: $-3.56, 0.50$
y-intercept: 0.89

(d)

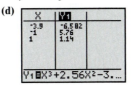

(e) Local maximum at $(-2.21, 9.91)$
Local minimum at $(0.50, 0)$

(f)

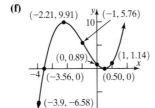

(g) Domain: $(-\infty, \infty)$; Range: $(-\infty, \infty)$
(h) Increasing on $(-\infty, -2.21)$ and $(0.50, \infty)$
Decreasing on $(-2.21, 0.50)$

85. (a) Degree 4; $y = x^4$

(b)

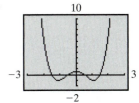

(c) x-intercepts: $-1.5, -0.5, 0.5, 1.5$
y-intercept: 0.5625

(d)

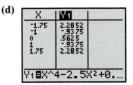

(e) Local minima at $(-1.12, -1), (1.12, -1)$
Local maximum at $(0, 0.56)$

(f)

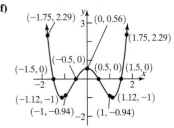

(g) Domain: $(-\infty, \infty)$; Range: $[-1, \infty]$
(h) Increasing on $(-1.12, 0)$ and $(1.12, \infty)$
Decreasing on $(-\infty, -1.12)$ and $(0, 1.12)$

87. (a) Degree 4; $y = 2x^4$

(b)

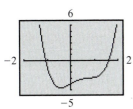

(c) x-intercepts: $-1.07, 1.62$
y-intercept: -4

(d)

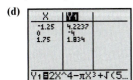

(e) Local minimum at $(-0.42, -4.64)$

(f)

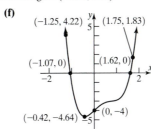

(g) Domain: $(-\infty, \infty)$
Range: $[-4.64, \infty)$
(h) Increasing on $(-0.42, \infty)$
Decreasing on $(-\infty, -0.42)$

89. (a) Degree 5; $y = -2x^5$

(b)

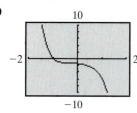

(c) x-intercept: -0.98

y-intercept: $-\sqrt{2}$

(d)

(e) None

(f)

(g) Domain: $(-\infty, \infty)$

Range: $(-\infty, \infty)$

(h) Decreasing on $(-\infty, \infty)$

91. Answers will vary. One possibility is $f(x) = x(x - 1)(x - 2)$.

93. Answers will vary. One possibility is $f(x) = -\dfrac{1}{2}(x + 1)(x - 1)(x - 2)$.

95. (a) Cubic, $a > 0$ **(b)** \$7000 per car   **(c)** \$20,000 per car

(d) $C(x) = 0.2156x^3 - 2.3473x^2 + 14.3275x + 10.2238$

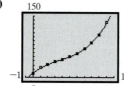

(e)

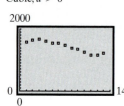

(f) About \$171,000

(g) Fixed costs of about \$10,200

97. (a) Cubic, $a > 0$ **(b)** $T(x) = 1.2582x^3 - 28.6146x^2 + 139.5808x + 1453.2098$

(c)

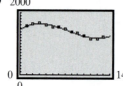

(d) About 1,738,000 thefts

99. No; yes   **103.** $f(x) = \text{int}(x); g(x) = |x|$

3.3 Assess Your Understanding *(page 195)*

5. $y = 1$ **6.** $x = -1$ **7.** proper   **8.** F   **9.** T   **10.** T   **11.** All real numbers except 3; $\{x|x \neq 3\}$

13. All real numbers except 2 and -4; $\{x|x \neq 2, x \neq -4\}$ **15.** All real numbers except $-\dfrac{1}{2}$ and 3; $\left\{x\middle|x \neq -\dfrac{1}{2}, x \neq 3\right\}$

17. All real numbers except 2; $\{x|x \neq 2\}$ **19.** All real numbers   **21.** All real numbers except -3 and 3; $\{x|x \neq -3, x \neq 3\}$

23. (a) Domain: $\{x|x \neq 2\}$; Range: $\{y|y \neq 1\}$ **(b)** $(0,0)$ **(c)** $y = 1$ **(d)** $x = 2$ **(e)** None

25. (a) Domain: $\{x|x \neq 0\}$; Range: all real numbers   **(b)** $(-1,0), (1,0)$ **(c)** None   **(d)** $x = 0$ **(e)** $y = 2x$

27. (a) Domain: $\{x|x \neq -2, x \neq 2\}$; Range: $\{y|y \leq 0, y > 1\}$ **(b)** $(0,0)$ **(c)** $y = 1$ **(d)** $x = -2, x = 2$ **(e)** None

29.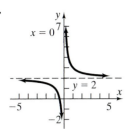

31.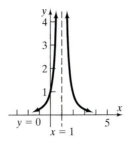

33.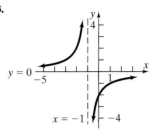

35.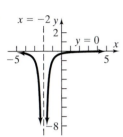

37.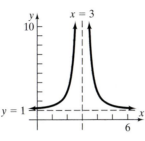

39.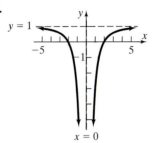

41. Horizontal asymptote: $y = 3$; vertical asymptote: $x = -4$ **43.** Vertical asymptote: $x = 2$; oblique asymptote: $y = x + 5$
45. Horizontal asymptote: $y = 0$; vertical asymptotes: $x = 1, x = -1$ **47.** Horizontal asymptote: $y = 0$; vertical asymptote: $x = 0$
49. Oblique asymptote: $y = 3x$; vertical asymptote: $x = 0$ **51.** Oblique asymptote: $y = -(x + 1)$; vertical asymptote: $x = 0$
53. (a) 9.8208 m/sec² **(b)** 9.8195 m/sec² **(c)** 9.7936 m/sec² **(d)** h-axis

3.4 Assess Your Understanding *(page 207)*

3. in lowest terms   **4.** F   **5.** F   **6.** T
7. 1. Domain: $\{x \mid x \neq 0, x \neq -4\}$

  2. $R(x)$ is in lowest terms.
  3. x-intercept: -1; no y-intercept
  4. No symmetry
  5. Vertical asymptotes: $x = 0, x = -4$
  6. Horizontal asymptote: $y = 0$, intersected at $(-1, 0)$

7.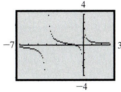

8.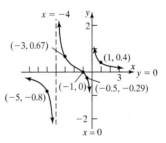

9. 1. Domain: $\{x \mid x \neq -2\}$

  2. $R(x) = \dfrac{3(x + 1)}{2(x + 2)}$

  3. x-intercept: -1; y-intercept: $\dfrac{3}{4}$

  4. No symmetry
  5. Vertical asymptote: $x = -2$

  6. Horizontal asymptote: $y = \dfrac{3}{2}$, not intersected

7.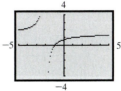

8.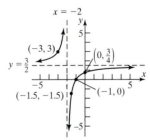

11. 1. Domain: $\{x \mid x \neq -2, x \neq 2\}$

  2. $R(x) = \dfrac{3}{(x + 2)(x - 2)}$

  3. No x-intercept; y-intercept: $-\dfrac{3}{4}$

  4. Symmetric with respect to the y-axis
  5. Vertical asymptotes: $x = 2, x = -2$
  6. Horizontal asymptote: $y = 0$, not intersected

7.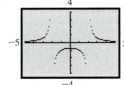

8.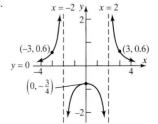

13. 1. Domain: $\{x \mid x \neq -1, x \neq 1\}$

2. $P(x) = \dfrac{x^4 + x^2 + 1}{(x + 1)(x - 1)}$

3. No x-intercept; y-intercept: -1

4. Symmetric with respect to the y-axis

5. Vertical asymptotes: $x = -1, x = 1$

6. No horizontal or oblique asymptotes

7.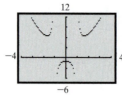

8.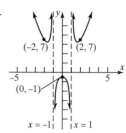

15. 1. Domain: $\{x \mid x \neq -3, x \neq 3\}$

2. $H(x) = \dfrac{(x - 1)(x^2 + x + 1)}{(x + 3)(x - 3)}$

3. x-intercept: 1; y-intercept: $\dfrac{1}{9}$

4. No symmetry

5. Vertical asymptotes: $x = 3, x = -3$

6. Oblique asymptote: $y = x$, intersected at $\left(\dfrac{1}{9}, \dfrac{1}{9}\right)$

7.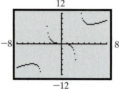

8.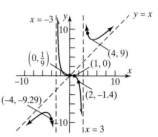

17. 1. Domain: $\{x \neq -3, x \neq 2\}$

2. $R(x) = \dfrac{x^2}{(x + 3)(x - 2)}$

3. Intercept: $(0, 0)$

4. No symmetry

5. Vertical asymptotes: $x = 2, x = -3$

6. Horizontal asymptote: $y = 1$, intersected at $(6, 1)$

7.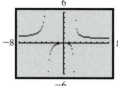

8.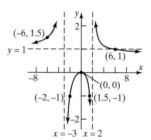

19. 1. Domain: $\{x \mid x \neq -2, x \neq 2\}$

2. $G(x) = \dfrac{x}{(x + 2)(x - 2)}$

3. Intercept: $(0, 0)$

4. Symmetry with respect to the origin

5. Vertical asymptotes: $x = -2, x = 2$

6. Horizontal asymptote: $y = 0$, intersected at $(0, 0)$

7.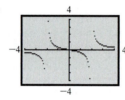

8.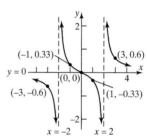

21. 1. Domain: $\{x \mid x \neq 1, x \neq -2, x \neq 2\}$

2. $R(x) = \dfrac{3}{(x - 1)(x - 2)(x + 2)}$

3. No x-intercept; y-intercept: $\dfrac{3}{4}$

4. No symmetry

5. Vertical asymptotes: $x = -2, x = 1, x = 2$

6. Horizontal asymptote: $y = 0$, not intersected

7.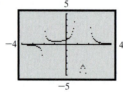

8.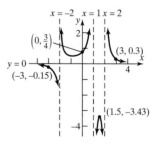

23. 1. Domain: $\{x \mid x \neq -2, x \neq 2\}$

2. $H(x) = \dfrac{4(x + 1)(x - 1)}{(x^2 + 4)(x + 2)(x - 2)}$

3. x-intercepts: $-1, 1$; y-intercept: $\dfrac{1}{4}$

4. Symmetry with respect to the y-axis

5. Vertical asymptotes: $x = -2, x = 2$

6. Horizontal asymptote: $y = 0$,
    intersected at $(-1, 0)$ and $(1, 0)$

7.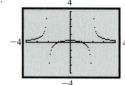

8.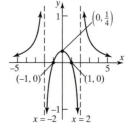

25. 1. Domain: $\{x | x \neq -2\}$

2. $F(x) = \dfrac{(x - 4)(x + 1)}{x + 2}$

3. x-intercepts: $-1, 4$; y-intercept: -2

4. No symmetry

5. Vertical asymptote: $x = -2$

6. Oblique asymptote: $y = x - 5$, not intersected

7.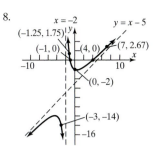

8.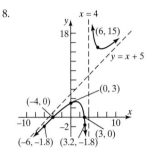

27. 1. Domain: $\{x | x \neq 4\}$

2. $R(x) = \dfrac{(x + 4)(x - 3)}{x - 4}$

3. x-intercepts: $-4, 3$; y-intercept: 3

4. No symmetry

5. Vertical asymptote: $x = 4$

6. Oblique asymptote: $y = x + 5$, not intersected

7.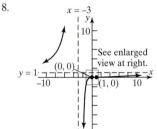

8.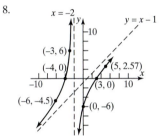

29. 1. Domain: $\{x | x \neq -2\}$

2. $F(x) = \dfrac{(x + 4)(x - 3)}{x + 2}$

3. x-intercepts: $-4, 3$; y-intercept: -6

4. No symmetry

5. Vertical asymptote: $x = -2$

6. Oblique asymptote: $y = x - 1$, not intersected

7.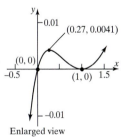

8.

31. 1. Domain: $\{x | x \neq -3\}$

2. $R(x)$ is in lowest terms.

3. x-intercepts: $0, 1$; y-intercept: 0

4. No symmetry

5. Vertical asymptote: $x = -3$

6. Horizontal asymptote: $y = 1$, not intersected

7.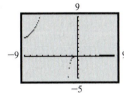

8.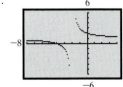

33. 1. Domain: $\{x | x \neq -2, x \neq 3\}$

2. $R(x) = \dfrac{x + 4}{x + 2}$

3. x-intercept: -4; y-intercept: 2

4. No symmetry

5. Vertical asymptote: $x = -2$; hole at $\left(3, \dfrac{7}{5}\right)$

6. Horizontal asymptote: $y = 1$, not intersected

7.

8.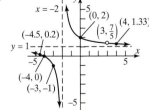

35. 1. Domain: $\left\{x \mid x \neq \dfrac{3}{2}, x \neq 2\right\}$

2. $R(x) = \dfrac{3x + 1}{x - 2}$

3. x-intercept: $-\dfrac{1}{3}$; y-intercept: $-\dfrac{1}{2}$

4. No symmetry

5. Vertical asymptote: $x = 2$; hole at $\left(\dfrac{3}{2}, -11\right)$

6. Horizontal asymptote: $y = 3$, not intersected

7.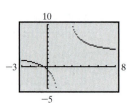

8.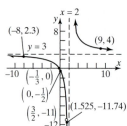

37. 1. Domain: $\{x \mid x \neq -3\}$

2. $R(x) = x + 2$

3. x-intercept: -2; y-intercept: 2

4. No symmetry

5. Vertical asymptote: none; hole at $(-3, -1)$

6. Oblique asymptote: $y = x + 2$

7.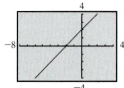

8.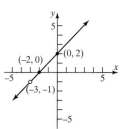

39. 1. Domain: $\{x \mid x \neq 0\}$

2. $f(x) = \dfrac{x^2 + 1}{x}$

3. No x-intercepts; no y-intercept

4. Symmetric with respect to the origin

5. Vertical asymptote: $x = 0$

6. Oblique asymptote: $y = x$, not intersected

7.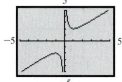

8.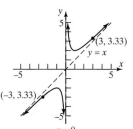

41. 1. Domain: $\{x \mid x \neq 0\}$

2. $f(x) = \dfrac{x^3 + 1}{x} = \dfrac{(x + 1)(x^2 - x + 1)}{x}$

3. x-intercept: -1; no y-intercept

4. No symmetry

5. Vertical asymptote: $x = 0$

6. No horizontal or oblique asymptotes

7.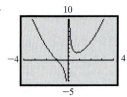

8.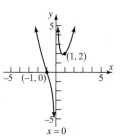

43. 1. Domain: $\{x \mid x \neq 0\}$

2. $f(x) = \dfrac{x^4 + 1}{x^3}$

3. No x-intercepts; no y-intercept

4. Symmetric with respect to the origin

5. Vertical asymptote: $x = 0$

6. Oblique asymptote: $y = x$, not intersected

7.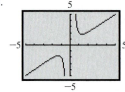

8.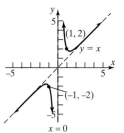

45. Minimum value: 2.00 at $x = 1.00$

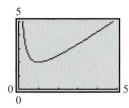

47. Minimum value: 1.89 at $x = 0.79$

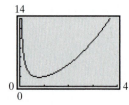

49. Minimum value: 1.75 at $x = 1.32$

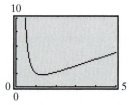

51. One possibility: $R(x) = \dfrac{x^2}{x^2 - 4}$ **53.** One possibility: $R(x) = \dfrac{(x - 1)(x - 3)(x^2 + \frac{4}{3})}{(x + 1)^2(x - 2)^2}$

55. (a) t-axis; $C(t) \to 0$

(b)

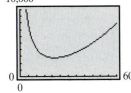

(c) 0.71 hr after injection

57. (a) $\overline{C}(x) = \dfrac{0.2x^3 - 2.3x^2 + 14.3x + 10.2}{x}$

(b) \$9400
(c) $\approx$ \$10,933
(d)

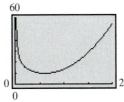

(e) 6
(f) \$9400

59. (a) $S(x) = 2x^2 + \dfrac{40{,}000}{x}$

(b)

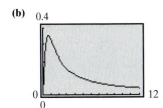

(c) 2784.95 sq in.
(d) 21.54 in. $\times$ 21.54 in. $\times$ 21.54 in.
(e) To minimize the cost of material needed for construction

61. (a) $C(x) = 0.12\pi r^2 + \dfrac{40}{r}$

(b)

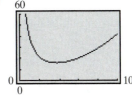

The cost is smallest when $r \approx 3.76$ cm.

63. (a) $D(p) = \dfrac{429}{p}$ **(b)** 143   **65.** 450 cm³   **67.** 124.76 pounds   **69.** $V = \pi r^2 h$ **71.** $\sqrt[3]{6} \approx 1.82$ in.   **73.** 900 foot-lb   **75.** 384 psi

77. No. Each function is a quotient of polynomials, but it is not written in lowest terms. Each function is undefined for $x = 1$;
each graph has a hole at $x = 1$.

3.5 Assess Your Understanding (page 217)

2. T   **3.** $\{x|-2 < x < 5\}$; $(-2, 5)$ **5.** $\{x|x < 0 \text{ or } x > 4\}$; $(-\infty, 0)$ or $(4, \infty)$ **7.** $\{x|-3 < x < 3\}$; $(-3, 3)$

9. $\{x|x \leq -4 \text{ or } x \geq 3\}$; $(-\infty, -4]$ or $[3, \infty)$ **11.** $\left\{x\middle|x < -\dfrac{1}{2} \text{ or } x > 3\right\}$; $\left(-\infty, -\dfrac{1}{2}\right)$ or $(3, \infty)$ **13.** $\{x|-1 < x < 8\}$; $(-1, 8)$

15. No real solution   **17.** $\left\{x\middle|x \leq -\dfrac{2}{3} \text{ or } x \geq \dfrac{3}{2}\right\}$; $\left(-\infty, -\dfrac{2}{3}\right]$ or $\left[\dfrac{3}{2}, \infty\right)$ **19.** $\{x|x > 1\}$; $(1, \infty)$ **21.** $\{x|x < 1 \text{ or } 2 < x < 3\}$; $(-\infty, 1)$ or $(2, 3)$

23. $\{x|-1 \leq x \leq 0 \text{ or } x \geq 3\}$; $[-1, 0]$ or $[3, \infty)$ **25.** $\{x|x < -1 \text{ or } x > 1\}$; $(-\infty, -1)$ or $(1, \infty)$ **27.** $\{x|x > 1\}$; $(1, \infty)$

29. $\{x|x < -1 \text{ or } x > 1\}$; $(-\infty, -1)$ or $(1, \infty)$ **31.** $\{x|-1 < x < 8\}$; $(-1, 8)$ **33.** $\{x|x < -2 \text{ or } x > 2\}$; $(-\infty, -2)$ or $(2, \infty)$

35. $\{x|x < -1 \text{ or } x > 1\}$; $(-\infty, -1)$ or $(1, \infty)$ **37.** $\{x|x < -1 \text{ or } 0 < x < 1\}$; $(-\infty, -1)$ or $(0, 1)$ **39.** $\{x|x < -1 \text{ or } x > 1\}$; $(-\infty, -1)$ or $(1, \infty)$

41. $\left\{x\middle|x < -\dfrac{2}{3} \text{ or } 0 < x < \dfrac{3}{2}\right\}$; $\left(-\infty, -\dfrac{2}{3}\right)$ or $\left(0, \dfrac{3}{2}\right)$ **43.** $\{x|x < 2\}$; $(-\infty, 2)$ **45.** $\{x|-2 < x \leq 9\}$; $(-2, 9]$

47. $\{x|x < 2 \text{ or } 3 < x < 5\}$; $(-\infty, 2)$ or $(3, 5)$ **49.** $\{x|x < -3 \text{ or } -1 < x < 1 \text{ or } x > 2\}$; $(-\infty, -3)$ or $(-1, 1)$ or $(2, \infty)$

51. $\{x|x < -5 \text{ or } -4 < x < -3 \text{ or } x > 1\}$; $(-\infty, -5)$ or $(-4, -3)$ or $(1, \infty)$ **53.** $\left\{x\middle|x \leq -\dfrac{1}{2} \text{ or } 1 \leq x < 4\right\}$; $\left(-\infty, -\dfrac{1}{2}\right]$ or $[1, 4)$

55. $\left\{x\middle|\dfrac{-3 - \sqrt{13}}{2} < x < -3 \text{ or } x > \dfrac{-3 + \sqrt{13}}{2}\right\}$; $\left(\dfrac{-3 - \sqrt{13}}{2}, -3\right)$ or $\left(\dfrac{-3 + \sqrt{13}}{2}, \infty\right)$ **57.** $\{x|x > 4\}$; $(4, \infty)$

59. $\{x|x \leq -4 \text{ or } x \geq 4\}$; $(-\infty, -4]$ or $[4, \infty)$ **61.** $\{x|x < -4 \text{ or } x \geq 2\}$; $(-\infty, -4)$ or $[2, \infty)$

63. (a) The ball is more than 96 feet above the ground from time t between 2 and 3 seconds, $2 < t < 3$.

(b)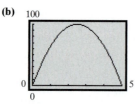

(c) 100 ft **(d)** 2.5 sec

65. (a) For a profit of at least \$50, between 8 and 32 watches must be sold, $8 \le x \le 32$.

(b)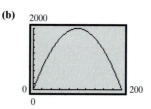

(c) \$2000 **(d)** 100

(e) [graph, 80, 40]

(f) \$80 **(g)** 20

67. Chevy can produce at most 8 Cavaliers in an hour, assuming that cars cannot be partially completed in an hour.

Historical Problems (page 230)

1.
$$\left(x - \frac{b}{3}\right)^3 + b\left(x - \frac{b}{3}\right)^2 + c\left(x - \frac{b}{3}\right) + d = 0$$
$$x^3 - bx^2 + \frac{b^2 x}{3} - \frac{b^3}{27} + bx^2 - \frac{2b^2 x}{3} + \frac{b^3}{9} + cx - \frac{bc}{3} + d = 0$$
$$x^3 + \left(c - \frac{b^2}{3}\right)x + \left(\frac{2b^3}{27} - \frac{bc}{3} + d\right) = 0$$

Let $p = c - \dfrac{b^2}{3}$ and $q = \dfrac{2b^3}{27} - \dfrac{bc}{3} + d$. Then $x^3 + px + q = 0$.

2.
$$(H + K)^3 + p(H + K) + q = 0$$
$$H^3 + 3H^2K + 3HK^2 + K^3 + pH + pK + q = 0$$
Let $3HK = -p$.
$$H^3 - pH - pK + K^3 + pH + pK + q = 0$$
$$H^3 + K^3 = -q$$

3.
$$3HK = -p$$
$$K = -\frac{p}{3H}$$
$$H^3 + \left(-\frac{p}{3H}\right)^3 = -q$$
$$H^3 - \frac{p^3}{27H^3} = -q$$
$$27H^6 - p^3 = -27qH^3$$
$$27H^6 + 27qH^3 - p^3 = 0$$
$$H^3 = \frac{-27q \pm \sqrt{(27q)^2 - 4(27)(-p^3)}}{2 \cdot 27}$$
$$H^3 = \frac{-q}{2} \pm \sqrt{\frac{27^2 q^2}{2^2(27^2)} + \frac{4(27)p^3}{2^2(27^2)}}$$
$$H^3 = \frac{-q}{2} \pm \sqrt{\frac{q^2}{4} + \frac{p^3}{27}}$$
$$H = \sqrt[3]{\frac{-q}{2} + \sqrt{\frac{q^2}{4} + \frac{p^3}{27}}}$$

Choose the positive root for now.

4.
$$H^3 + K^3 = -q$$
$$K^3 = -q - H^3$$
$$K^3 = -q - \left[\frac{-q}{2} + \sqrt{\frac{q^2}{4} + \frac{p^3}{27}}\right]$$
$$K^3 = \frac{-q}{2} - \sqrt{\frac{q^2}{4} + \frac{p^3}{27}}$$
$$K = \sqrt[3]{\frac{-q}{2} - \sqrt{\frac{q^2}{4} + \frac{p^3}{27}}}$$

5. $x = H + K$
$$x = \sqrt[3]{\frac{-q}{2} + \sqrt{\frac{q^2}{4} + \frac{p^3}{27}}} + \sqrt[3]{\frac{-q}{2} - \sqrt{\frac{q^2}{4} + \frac{p^3}{27}}}$$ (Note that if we had used the negative root in 3 the result would be the same.)

6. $x = 3$ **7.** $x = 2$ **8.** $x = 2$

3.6 Assess Your Understanding *(page 230)*

5. Remainder; dividend   **6.** $f(c)$ **7.** -4 **8.** F   **9.** F   **10.** T   **11.** No; $f(3) = 61$ **13.** No; $f(1) = 2$
15. Yes; $f(x) = (x + 2)(3x^5 - 6x^4 + 12x^3 - 22x^2 + 44x - 88)$ **17.** Yes; $f(x) = (x - 4)(4x^5 + 16x^4 + x + 4)$

19. No; $f\left(-\dfrac{1}{2}\right) = -\dfrac{7}{4}$ **21.** $4; \pm 1, \pm \dfrac{1}{3}$ **23.** $5; \pm 1, \pm 3$ **25.** $3; \pm 1, \pm 2, \pm \dfrac{1}{4}, \pm \dfrac{1}{2}$ **27.** $4; \pm 1, \pm 2, \pm \dfrac{1}{3}, \pm \dfrac{2}{3}$ **29.** $5; \pm 1, \pm 2, \pm 4, \pm \dfrac{1}{2}$

31. $4; \pm 1, \pm 2, \pm \dfrac{1}{6}, \pm \dfrac{1}{3}, \pm \dfrac{1}{2}, \pm \dfrac{2}{3}$

33. -1 and 1

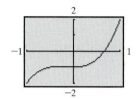

35. -12 and 12

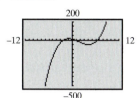

37. -10 and 10

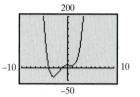

39. $-3, -1, 2; f(x) = (x + 3)(x + 1)(x - 2)$ **41.** $\dfrac{1}{2}, 3, 3; f(x) = (2x - 1)(x - 3)^2$ **43.** $-\dfrac{1}{3}; f(x) = (3x + 1)(x^2 + x + 1)$

45. $3, \dfrac{5 + \sqrt{17}}{2}, \dfrac{5 - \sqrt{17}}{2}; f(x) = (x - 3)\left(x - \left(\dfrac{5 + \sqrt{17}}{2}\right)\right)\left(x - \left(\dfrac{5 - \sqrt{17}}{2}\right)\right)$ **47.** $-2, -1, 1, 1; f(x) = (x + 2)(x + 1)(x - 1)^2$

49. $-5, -3, -\dfrac{3}{2}, 1; f(x) = (x + 5)(x + 3)(2x + 3)(x - 1)$ **51.** $-2, -\dfrac{3}{2}, 1, 4; f(x) = (x + 2)(2x + 3)(x - 1)(x - 4)$

53. $-\dfrac{1}{2}, \dfrac{1}{2}; f(x) = (2x + 1)(2x - 1)(x^2 + 2)$ **55.** $\dfrac{\sqrt{2}}{2}, -\dfrac{\sqrt{2}}{2}, 2; f(x) = (x - 2)(2x - \sqrt{2})(2x + \sqrt{2})\left(x^2 + \dfrac{1}{2}\right)$ **57.** $-5.9, -0.3, 3$

59. $-3.8, 4.5$ **61.** $-43.5, 1, 23$ **63.** $\{-1, 2\}$ **65.** $\left\{\dfrac{2}{3}, -1 + \sqrt{2}, -1 - \sqrt{2}\right\}$ **67.** $\left\{\dfrac{1}{3}, \sqrt{5}, -\sqrt{5}\right\}$ **69.** $\{-3, -2\}$ **71.** $\left\{-\dfrac{1}{3}\right\}$

73. $f(0) = -1; f(1) = 10$; Zero: 0.22 **75.** $f(-5) = -58; f(-4) = 2$; Zero: -4.05 **77.** $f(1.4) = -0.17536; f(1.5) = 1.40625$; Zero: 1.41
79. ≈ 27 Cavaliers   **81.** $k = 5$ **83.** -7 **85.** 5 **87.** 7 in.   **89.** If $f(x) = x^n - c^n$, then $f(c) = c^n - c^n = 0$, so $x - c$ is a factor of f.
91. All the potential rational zeros are integers. Hence, r is either an integer or is not a rational zero (and is therefore irrational).

3.7 Assess Your Understanding *(page 237)*

3. one   **4.** $3 - 4i$ **5.** T   **6.** F   **7.** $4 + i$ **9.** $-i, 1 - i$ **11.** $-i, -2i$ **13.** $-i$ **15.** $2 - i, -3 + i$
17. $f(x) = x^4 - 14x^3 + 77x^2 - 200x + 208; a = 1$ **19.** $f(x) = x^5 - 4x^4 + 7x^3 - 8x^2 + 6x - 4; a = 1$

21. $f(x) = x^4 - 6x^3 + 10x^2 - 6x + 9; a = 1$ **23.** $-2i, 4$ **25.** $2i, -3, \dfrac{1}{2}$ **27.** $3 + 2i, -2, 5$ **29.** $4i, -\sqrt{11}, \sqrt{11}, -\dfrac{2}{3}$

31. $1, -\dfrac{1}{2} - \dfrac{\sqrt{3}}{2}i, -\dfrac{1}{2} + \dfrac{\sqrt{3}}{2}i; f(x) = (x - 1)\left(x + \dfrac{1}{2} + \dfrac{\sqrt{3}}{2}i\right)\left(x + \dfrac{1}{2} - \dfrac{\sqrt{3}}{2}i\right)$

33. $2, 3 - 2i, 3 + 2i; f(x) = (x - 2)(x - 3 + 2i)(x - 3 - 2i)$ **35.** $-i, i, -2i, 2i; f(x) = (x + i)(x - i)(x + 2i)(x - 2i)$

37. $-5i, 5i, -3, 1; f(x) = (x + 5i)(x - 5i)(x + 3)(x - 1)$ **39.** $-4, \dfrac{1}{3}, 2 - 3i, 2 + 3i; f(x) = 3(x + 4)\left(x - \dfrac{1}{3}\right)(x - 2 + 3i)(x - 2 - 3i)$

41. Zeros that are complex numbers must occur in conjugate pairs; or a polynomial with real coefficients of odd degree must have at least one real zero.   **43.** If the remaining zero were a complex number, then its conjugate would also be a zero, creating a polynomial of degree 5.

Review Exercises *(page 240)*

1.

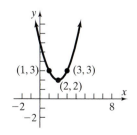

3.

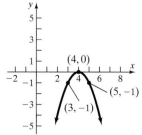

5.

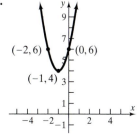

7.

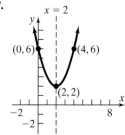

9.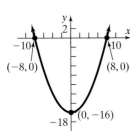

11.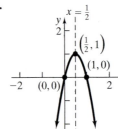

13.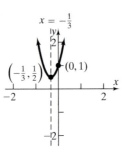

15.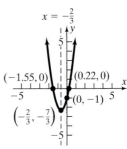

17. Minimum value; 1   **19.** Maximum value; 12   **21.** Maximum value; 16   **23.** Polynomial of degree 5   **25.** Not a polynomial

27.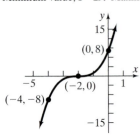

29.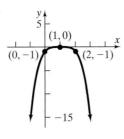

31.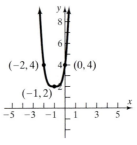

33. (a) x-intercepts: $-4, -2, 0$; y-intercept: 0 **(b)** $-4, -2, 0$: Crosses   **(c)** $y = x^3$

(d)

(e) $2; (-3.15, 3.08), (-0.85, -3.08)$

(f)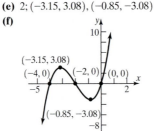

(g) Domain: $(-\infty, \infty)$
Range: $(-\infty, \infty)$

(h) Increasing on $(-\infty, -3.15)$ and $(-0.85, \infty)$
Decreasing on $(-3.15, -0.85)$

35. (a) x-intercepts: $-4, 2$;
y-intercept: 16
(b) -4: Crosses; 2: Touches
(c) $y = x^3$
(d)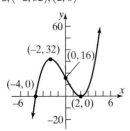

(e) $2; (-2, 32), (2, 0)$
(f)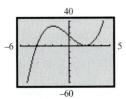

(g) Domain: $(-\infty, \infty)$
Range: $(-\infty, \infty)$
(h) Increasing on $(-\infty, -2)$ and $(2, \infty)$
Decreasing on $(-2, 2)$

37. $f(x) = x^3 - 4x^2 = x^2(x - 4)$
(a) x-intercepts: $0, 4$; y-intercept: 0
(b) 0: Touches; 4: Crosses
(c) $y = x^3$
(d)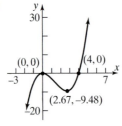

(e) $2; (0, 0), (2.67, -9.48)$
(f)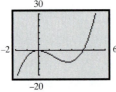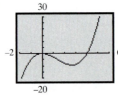

(g) Domain: $(-\infty, \infty)$
Range: $(-\infty, \infty)$
(h) Increasing on $(-\infty, 0)$ and $(2.67, \infty)$
Decreasing on $(0, 2.67)$

39. (a) x-intercepts: $-3, -1, 1$;
y-intercept: 3
(b) $-3, -1$: Crosses; 1: Touches
(c) $y = x^4$
(d)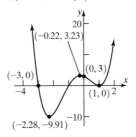

(e) $3; (-2.28, -9.91), (-0.22, 3.23), (1, 0)$
(f)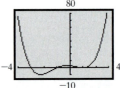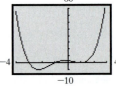

(g) Domain: $(-\infty, \infty)$
Range: $(-\infty, \infty)$
(h) Increasing on $(-2.28, -0.22)$ and $(1, \infty)$
Decreasing on $(-\infty, -2.28)$ and $(-0.22, 1)$

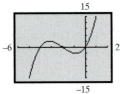

41. Domain: $\{x | x \neq -3, x \neq 3\}$; Horizontal asymptote: $y = 0$; Vertical asymptotes: $x = -3, x = 3$

43. Domain: $\{x | x \neq -2\}$; Horizontal asymptote: $y = 1$; Vertical asymptote: $x = -2$

45. 1. Domain: $\{x | x \neq 0\}$

    2. $R(x)$ is in lowest terms.

    3. x-intercept: 3; no y-intercept

    4. No symmetry

    5. Vertical asymptote: $x = 0$

    6. Horizontal asymptote: $y = 2$, not intersected

7.

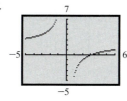

8.

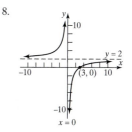

47. 1. Domain: $\{x | x \neq 0, x \neq 2\}$

    2. $H(x)$ is in lowest terms

    3. x-intercept: -2; no y-intercept

    4. No symmetry

    5. Vertical asymptotes: $x = 0, x = 2$

    6. Horizontal asymptote: $y = 0$, intersected at $(-2, 0)$

7.

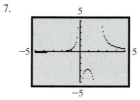

8.

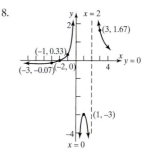

49. 1. Domain: $\{x | x \neq -2, x \neq 3\}$

    2. $R(x) = \dfrac{(x + 3)(x - 2)}{(x - 3)(x + 2)}$

    3. x-intercepts: $-3, 2$; y-intercept: 1

    4. No symmetry

    5. Vertical asymptotes: $x = -2, x = 3$

    6. Horizontal asymptote: $y = 1$, intersected at $(0, 1)$

7.

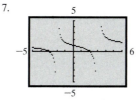

8.

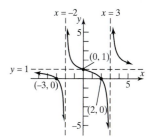

51. 1. Domain: $\{x | x \neq -2, x \neq 2\}$

    2. $F(x) = \dfrac{x^3}{(x + 2)(x - 2)}$

    3. Intercept: $(0, 0)$

    4. Symmetric with respect to the origin

    5. Vertical asymptotes: $x = -2, x = 2$

    6. Oblique asymptote: $y = x$, intersected at $(0, 0)$

7.

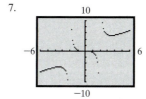

8.

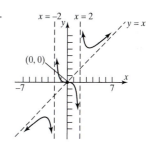

53. 1. Domain: $\{x | x \neq 1\}$

    2. $R(x)$ is in lowest terms

    3. Intercept: $(0, 0)$

    4. No symmetry

    5. Vertical asymptote: $x = 1$

    6. No oblique or horizontal asymptote

7.

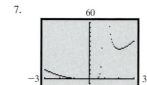

8.

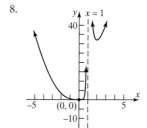

55. 1. Domain: $\{x|x \neq -1, x \neq 2\}$

2. $G(x) = \dfrac{x+2}{x+1}$

3. x-intercept: -2; y-intercept: 2

4. No symmetry

5. Vertical asymptote: $x = -1$; hole at $\left(2, \dfrac{4}{3}\right)$

6. Horizontal asymptote: $y = 1$, not intersected

7.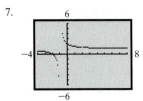

8.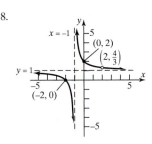

57. $\left\{x \middle| -4 < x < \dfrac{3}{2}\right\}; \left(-4, \dfrac{3}{2}\right)$ **59.** $\{x|-3 < x \leq 3\}; (-3, 3]$ **61.** $\{x|x < 1 \text{ or } x > 2\}; (-\infty, 1) \text{ or } (2, \infty)$

63. $\{x|1 < x < 2 \text{ or } x > 3\}; (1, 2) \text{ or } (3, \infty)$ **65.** $\{x|x < -4 \text{ or } 2 < x < 4 \text{ or } x > 6\}; (-\infty, -4) \text{ or } (2, 4) \text{ or } (6, \infty)$

67. $R = 10$; g is not a factor of f.   **69.** $R = 0$; g is a factor of f.   **71.** $f(4) = 47{,}105$ **73.** $8; \pm\dfrac{1}{2}, \pm 1, \pm\dfrac{3}{2}, \pm 3$ **75.** $-2, 1, 4$

77. $\dfrac{1}{2}$, multiplicity 2; -2 **79.** 2, multiplicity 2   **81.** $-2.5, 3.1, 5.32$ **83.** $-11.3, -0.6, 4, 9.33$ **85.** $\{-3, 2\}$ **87.** $\left\{-3, -1, -\dfrac{1}{2}, 1\right\}$

89. -5 and 5 **91.** $-\dfrac{37}{2}$ and $\dfrac{37}{2}$

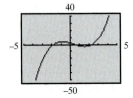   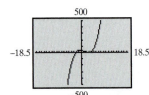

93. $f(0) = -1, f(1) = 1; 0.85$ **95.** $f(0) = -1; f(1) = 1; 0.94$ **97.** $4 - i; f(x) = x^3 - 14x^2 + 65x - 102$

99. $-i, 1 - i; f(x) = x^4 - 2x^3 + 3x^2 - 2x + 2$ **101.** $\left\{-\dfrac{1}{2} - \dfrac{\sqrt{3}}{2}i, -\dfrac{1}{2} + \dfrac{\sqrt{3}}{2}i\right\}$ **103.** $\left\{\dfrac{-1 - \sqrt{17}}{4}, \dfrac{-1 + \sqrt{17}}{4}\right\}$

105. $\left\{\dfrac{1}{2} - \dfrac{\sqrt{11}}{2}i, \dfrac{1}{2} + \dfrac{\sqrt{11}}{2}i\right\}$ **107.** $\left\{\dfrac{1}{2} - \dfrac{\sqrt{23}}{2}i, \dfrac{1}{2} + \dfrac{\sqrt{23}}{2}i\right\}$ **109.** $\{-\sqrt{2}, \sqrt{2}, -2i, 2i\}$ **111.** $\{-3, 2\}$ **113.** $\left\{\dfrac{1}{3}, 1, -i, i\right\}$ **115.** $(2, 2)$

117. $4{,}166{,}666.7$ m^2 **119.** The side with the semicircles should be $\dfrac{50}{\pi}$ ft; the other side should be 25 ft.   **121. (a)** 63 clubs **(b)** \$151.90

123. 199.9 pounds

125. (a) Quadratic, $a < 0$.

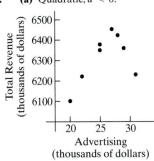

(b) $R(A) = -7.760A^2 + 411.875A + 942.721$
(c) About \$26.5 thousand
(d) \$6408 thousand
(e)

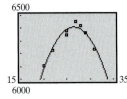

127. (a) $A(r) = 2\pi r^2 + \dfrac{500}{r}$
(b) 223.22 cm^2
(c) 257.08 cm^2
(d)

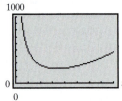

A is smallest when $r \approx 3.41$ cm.

Chapter Test *(page 244)*

1.

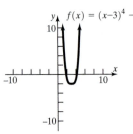

2. (a) The leading coefficient (i.e., the coefficient on x^2) is positive so the graph will open up. Thus, the graph has a minimum.

(b) $(2, -8)$ **(c)** $x = 2$

(d) $(0, 4), \left(\dfrac{6 - 2\sqrt{6}}{3}, 0\right), \left(\dfrac{6 + 2\sqrt{6}}{3}, 0\right)$.

(e)

3. (a) 3

(b) Every zero of g lies between -15 and 15.

(c) $\dfrac{p}{q}$: $\pm\dfrac{1}{2}, \pm 1, \pm\dfrac{3}{2}, \pm\dfrac{5}{2}, \pm 3, \pm 5, \pm\dfrac{15}{2}, \pm 15$

(d) $-5, -\dfrac{1}{2}, 3$; $g(x) = (x+5)(2x+1)(x-3)$

8.

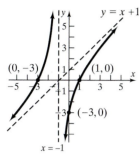

14. (a) $(0, -4), (-2, 0), (1, 0)$

(d) Because the power function has an odd degree, we expect the ends of the graph to go in opposite directions. The leading coefficient is negative, so the left side of the graph will go up and the right side will go down.

Reading the graph from left to right, we would expect to see it decreasing and cross the x-axis at $x = -2$. Somewhere between $x = -2$ and $x = 1$ the graph turns so that it is increasing, touches the x-axis at $x = 1$, turns around and decreases from that point on.

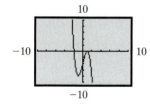

4. $4, -5i, 5i$

5. $\left\{ 1, \dfrac{5 - \sqrt{61}}{6}, \dfrac{5 + \sqrt{61}}{6} \right\}$

6. Domain: $\{x \mid x \neq -10, x \neq 4\}$

Asymptotes: $x = -10, y = 2$

7. Domain: $\{x \mid x \neq -1\}$;

Asymptotes: $x = -1, y = x + 1$

9. Answers may vary. One possibility is

$f(x) = x^4 - 4x^3 - 2x^2 + 20x$

10. Answers may vary. One possibility is $r(x) = \dfrac{2(x-9)(x-1)}{(x-4)(x-9)}$

11. $f(0) = 8$; $f(4) = -36$

Since $f(0) = 8 > 0$ and $f(4) = -36 < 0$, the Intermediate Value Theorem guarantees that there is at least one real zero between 0 and 4.

12. $\{x \mid -\infty < x \leq -3 \text{ or } 2 \leq x < \infty\}$, or $(-\infty, -3] \cup [2, \infty)$

13. $\{x \mid x < 3 \text{ or } x > 8\}$, or $(-\infty, 3) \cup (8, \infty)$

(b) Touches at 1 and crosses at -2.   **(c)** $y = -2x^3$

(e) $(-1, -8)$, and $(1, 0)$.

(f)

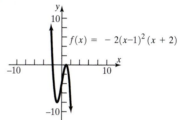

15. (a) $A(x) = -0.604x^2 + 6.038x + 25.350$

(b) The model predicts that the average 2004 home game attendance for the St. Louis Cardinals will be 30,768.

Cumulative Review *(page 246)*

1. $\sqrt{26}$

2. $\{x \mid x \leq 0 \text{ or } x \geq 1\}$ or $(-\infty, 0] \cup [1, \infty)$

3. $\{x \mid -1 < x < 4\}$ or $(-1, 4)$

4. $f(x) = -3x + 1$

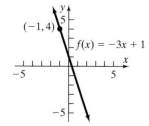

5. $y = 2x - 1$

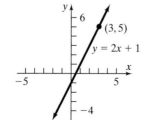

6.

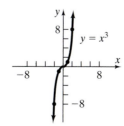

7. Not a function; 3 has two images   **8.** $\{0, 2, 4\}$ **9.** $\left\{x \middle| x \geq \dfrac{3}{2}\right\}; \left[\dfrac{3}{2}, \infty\right)$ **10.** Center: $(-2, 1)$; Radius: 3

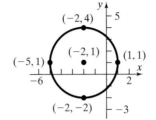

11. x-intercepts: $-3, 0, 3$
y-intercept: 0
Symmetric with respect to the origin

12. $y = -\dfrac{2}{3}x + \dfrac{17}{3}$

13. Not a function; it fails the Vertical Line Test.

14. (a) 22   **(b)** $x^2 - 5x - 2$ **(c)** $-x^2 - 5x + 2$ **(d)** $9x^2 + 15x - 2$ **(e)** $2x + h + 5$

15. (a) $\{x | x \neq 1\}$ **(b)** No, $(2, 7)$ is on the graph.   **(c)** $4; (3, 4)$ is on the graph.   **(d)** $\dfrac{7}{4}; \left(\dfrac{7}{4}, 9\right)$ is on the graph.

16.

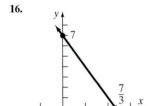

17.

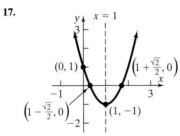

18. $x + 4; m_{\sec} = 6$

19. (a) x-intercepts: $-5, -1, 5$; y-intercept: -3
(b) No symmetry
(c) Neither
(d) Increasing: $(-\infty, -3)$ and $(2, \infty)$
Decreasing: $(-3, 2)$
(e) Local maximum is 5 and occurs at $x = -3$.
(f) Local minimum is -6 and occurs at $x = 2$.

20. Odd   **21. (a)** Domain: $\{x | -3 < x\}$ or $(-3, \infty)$

(b) x-intercept: $-\dfrac{1}{2}$; y-intercept: 1

(c)

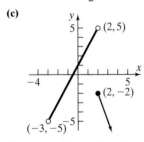

22.

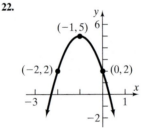

(d) Range: $\{y | y < 5\}$ or $(-\infty, 5)$

23. (a) $(f + g)(x) = x^2 - 9x - 6$; domain: all real numbers   **(b)** $\left(\dfrac{f}{g}\right)(x) = \dfrac{x^2 - 5x + 1}{-4x - 7}$; domain: $\left\{x \middle| x \neq -\dfrac{7}{4}\right\}$

24. (a) $R(x) = -\dfrac{1}{10}x^2 + 150x$ **(b)** \$14,000   **(c)** 750; \$56,250   **(d)** \$75

C H A P T E R   4   **Exponential and Logarithmic Functions**

4.1 Assess Your Understanding (page 253)

4. $(g \circ f)(x)$ **5.** F   **6.** F   **7. (a)** -1 **(b)** -1 **(c)** 8   **(d)** 0   **(e)** 8   **(f)** -7 **9. (a)** 4   **(b)** 5   **(c)** -1 **(d)** -2 **11. (a)** 98   **(b)** 49

(c) 4   **(d)** 4   **13. (a)** 97   **(b)** $-\dfrac{163}{2}$ **(c)** 1   **(d)** $-\dfrac{3}{2}$ **15. (a)** $2\sqrt{2}$ **(b)** $2\sqrt{2}$ **(c)** 1   **(d)** 0   **17. (a)** $\dfrac{1}{17}$ **(b)** $\dfrac{1}{5}$ **(c)** 1   **(d)** $\dfrac{1}{2}$

19. (a) $\dfrac{3}{\sqrt[3]{4} + 1}$ **(b)** 1   **(c)** $\dfrac{6}{5}$ **(d)** 0   **21.** $\{x | x \neq 0, x \neq 2\}$ **23.** $\{x | x \neq -4, x \neq 0\}$ **25.** $\left\{x \middle| x \geq -\dfrac{3}{2}\right\}$ **27.** $\{x | x \geq 1\}$

29. (a) $(f \circ g)(x) = 6x + 3$; All real numbers   **(b)** $(g \circ f)(x) = 6x + 9$; All real numbers   **(c)** $(f \circ f)(x) = 4x + 9$; All real numbers
(d) $(g \circ g)(x) = 9x$; All real numbers   **31. (a)** $(f \circ g)(x) = 3x^2 + 1$; All real numbers   **(b)** $(g \circ f)(x) = 9x^2 + 6x + 1$; All real numbers
(c) $(f \circ f)(x) = 9x + 4$; All real numbers   **(d)** $(g \circ g)(x) = x^4$; All real numbers   **33. (a)** $(f \circ g)(x) = x^4 + 8x^2 + 16$; All real numbers
(b) $(g \circ f)(x) = x^4 + 4$; All real numbers   **(c)** $(f \circ f)(x) = x^4$; All real numbers   **(d)** $(g \circ g)(x) = x^4 + 8x^2 + 20$; All real numbers

35. (a) $(f \circ g)(x) = \dfrac{3x}{2-x}; \{x | x \neq 0, x \neq 2\}$ **(b)** $(g \circ f)(x) = \dfrac{2(x-1)}{3}; \{x | x \neq 1\}$

(c) $(f \circ f)(x) = \dfrac{3(x-1)}{4-x}; \{x | x \neq 1, x \neq 4\}$ **(d)** $(g \circ g)(x) = x; \{x | x \neq 0\}$

37. (a) $(f \circ g)(x) = \dfrac{4}{4+x}; \{x | x \neq -4, x \neq 0\}$ **(b)** $(g \circ f)(x) = \dfrac{-4(x-1)}{x}; \{x | x \neq 0, x \neq 1\}$ **(c)** $(f \circ f)(x) = x; \{x | x \neq 1\}$

(d) $(g \circ g)(x) = x; \{x | x \neq 0\}$ **39. (a)** $(f \circ g)(x) = \sqrt{2x+3}; \left\{x \,\middle|\, x \geq -\dfrac{3}{2}\right\}$ **(b)** $(g \circ f)(x) = 2\sqrt{x} + 3; \{x | x \geq 0\}$

(c) $(f \circ f)(x) = \sqrt[4]{x}; \{x | x \geq 0\}$ **(d)** $(g \circ g)(x) = 4x + 9;$ All real numbers   **41. (a)** $(f \circ g)(x) = x; \{x | x \geq 1\}$

(b) $(g \circ f)(x) = |x|;$ All real numbers   **(c)** $(f \circ f)(x) = x^4 + 2x^2 + 2;$ All real numbers   **(d)** $(g \circ g)(x) = \sqrt{\sqrt{x-1}-1}; \{x | x \geq 2\}$

43. (a) $(f \circ g)(x) = acx + ad + b;$ All real numbers   **(b)** $(g \circ f)(x) = acx + bc + d;$ All real numbers

(c) $(f \circ f)(x) = a^2 x + ab + b;$ All real numbers   **(d)** $(g \circ g)(x) = c^2 x + cd + d;$ All real numbers

45. $(f \circ g)(x) = f(g(x)) = f\left(\dfrac{1}{2}x\right) = 2\left(\dfrac{1}{2}x\right) = x; (g \circ f)(x) = g(f(x)) = g(2x) = \dfrac{1}{2}(2x) = x$

47. $(f \circ g)(x) = f(g(x)) = f(\sqrt[3]{x}) = (\sqrt[3]{x})^3 = x; (g \circ f)(x) = g(f(x)) = g(x^3) = \sqrt[3]{x^3} = x$

49. $(f \circ g)(x) = f(g(x)) = f\left(\dfrac{1}{2}(x+6)\right) = 2\left[\dfrac{1}{2}(x+6)\right] - 6 = x + 6 - 6 = x;$

$(g \circ f)(x) = g(f(x)) = g(2x - 6) = \dfrac{1}{2}(2x - 6 + 6) = x$

51. $(f \circ g)(x) = f(g(x)) = f\left(\dfrac{1}{a}(x-b)\right) = a\left[\dfrac{1}{a}(x-b)\right] + b = x; (g \circ f)(x) = g(f(x)) = g(ax + b) = \dfrac{1}{a}(ax + b - b) = x$

53. $f(x) = x^4; g(x) = 2x + 3$ (Other answers are possible.)   **55.** $f(x) = \sqrt{x}; g(x) = x^2 + 1$ (Other answers are possible.)

57. $f(x) = |x|; g(x) = 2x + 1$ (Other answers are possible.)   **59.** $(f \circ g)(x) = 11; (g \circ f)(x) = 2$ **61.** $-3, 3$ **63.** $S(t) = \dfrac{16}{9}\pi t^6$

65. $C(t) = 15{,}000 + 800{,}000t - 40{,}000t^2$ **67.** $C(p) = \dfrac{2\sqrt{100-p}}{25} + 600, 0 \leq p \leq 100$ **69.** $V(r) = 2\pi r^3$

71. (a) $f(x) = 0.857118x$ **(b)** $g(x) = 128.6054x$ **(c)** $g(f(x)) = g(0.857118x) = 110.23x$ **(d)** 110,230.00 yen

73. f is an odd function, so $f(-x) = -f(x). g$ is an even function, so $g(-x) = g(x)$.

Then $(f \circ g)(-x) = f(g(-x)) = f(g(x)) = (f \circ g)(x).$ So $f \circ g$ is even.

Also, $(g \circ f)(-x) = g(f(-x)) = g(-f(x)) = g(f(x)) = (g \circ f)(x),$ so $g \circ f$ is even.

4.2 Assess Your Understanding   *(page 267)*

4. One-to-one   **5.** $y = x$ **6.** $[4, \infty)$ **7.** F   **8.** T

9. One-to-one   **11.** Not one-to-one   **13.** Not one-to-one   **15.** One-to-one

17. One-to-one   **19.** Not one-to-one   **21.** One-to-one

23.

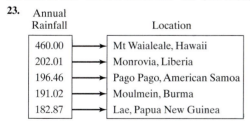

Domain: $\{460.00, 202.01, 196.46, 191.02; 182.87\}$

Range: $\{$Mt. Waialeale, Monrovia, Pago Pago, Moulmein, Lae$\}$

25.

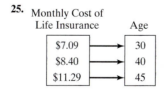

Domain: $\{\$7.09, \$8.04, \$11.29\}$

Range: $\{30, 40, 45\}$

27. $\{(5, -3), (9, -2), (2, -1), (11, 0), (-5, 1)\}$

Domain: $\{5, 9, 2, 11, -5\}$

Range: $\{-3, -2, -1, 0, 1\}$

29. $\{(1, -2), (2, -3), (0, -10), (9, 1), (4, 2)\}$

Domain: $\{1, 2, 0, 9, 4\}$

Range: $\{-2, -3, -10, 1, 2\}$

31.

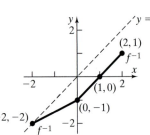

33.

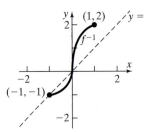

35.

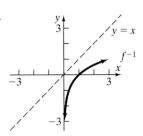

37. $f(g(x)) = f\left(\dfrac{1}{3}(x - 4)\right) = 3\left[\dfrac{1}{3}(x - 4)\right] + 4$

$\qquad = (x - 4) + 4 = x;$

$\quad g(f(x)) = g(3x + 4) = \dfrac{1}{3}[(3x + 4) - 4] = \dfrac{1}{3}(3x) = x$

39. $f(g(x)) = f\left(\dfrac{x}{4} + 2\right) = 4\left[\dfrac{x}{4} + 2\right] - 8 = (x + 8) - 8 = x;$

$\quad g(f(x)) = g(4x - 8) = \dfrac{4x - 8}{4} + 2 = (x - 2) + 2 = x$

41. $f(g(x)) = f(\sqrt[3]{x + 8}) = (\sqrt[3]{x + 8})^3 - 8 = (x + 8) - 8 = x;$

$\quad g(f(x)) = g(x^3 - 8) = \sqrt[3]{(x^3 - 8) + 8} = \sqrt[3]{x^3} = x$

43. $f(g(x)) = f\left(\dfrac{1}{x}\right) = \dfrac{1}{\left(\dfrac{1}{x}\right)} = x; g(f(x)) = g\left(\dfrac{1}{x}\right) = \dfrac{1}{\left(\dfrac{1}{x}\right)} = x$

45. $f(g(x)) = f\left(\dfrac{4x - 3}{2 - x}\right) = \dfrac{2\left(\dfrac{4x - 3}{2 - x}\right) + 3}{\dfrac{4x - 3}{2 - x} + 4}$

$\quad = \dfrac{2(4x - 3) + 3(2 - x)}{4x - 3 + 4(2 - x)} = \dfrac{5x}{5} = x;$

$\quad g(f(x)) = g\left(\dfrac{2x + 3}{x + 4}\right) = \dfrac{4\left(\dfrac{2x + 3}{x + 4}\right) - 3}{2 - \dfrac{2x + 3}{x + 4}}$

$\quad = \dfrac{4(2x + 3) - 3(x + 4)}{2(x + 4) - (2x + 3)} = \dfrac{5x}{5} = x$

47. $f^{-1}(x) = \dfrac{1}{3}x$

$\quad f(f^{-1}(x)) = f\left(\dfrac{1}{3}x\right) = 3\left(\dfrac{1}{3}x\right) = x$

$\quad f^{-1}(f(x)) = f^{-1}(3x) = \dfrac{1}{3}(3x) = x$

$\quad$ Domain f = Range f^{-1} = All real numbers

$\quad$ Range f = Domain f^{-1} = All real numbers

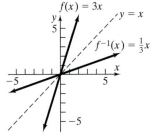

49. $f^{-1}(x) = \dfrac{x}{4} - \dfrac{1}{2}$

$\quad f(f^{-1}(x)) = f\left(\dfrac{x}{4} - \dfrac{1}{2}\right) = 4\left(\dfrac{x}{4} - \dfrac{1}{2}\right) + 2 = (x - 2) + 2 = x$

$\quad f^{-1}(f(x)) = f^{-1}(4x + 2) = \dfrac{4x + 2}{4} - \dfrac{1}{2} = \left(x + \dfrac{1}{2}\right) - \dfrac{1}{2} = x$

$\quad$ Domain f = Range f^{-1} = All real numbers

$\quad$ Range f = Domain f^{-1} = All real numbers

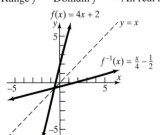

51. $f^{-1}(x) = \sqrt[3]{x + 1}$

$\quad f(f^{-1}(x)) = f(\sqrt[3]{x + 1}) = (\sqrt[3]{x + 1})^3 - 1 = x$

$\quad f^{-1}(f(x)) = f^{-1}(x^3 - 1) = \sqrt[3]{(x^3 - 1) + 1} = x$

$\quad$ Domain f = Range f^{-1} = All real numbers

$\quad$ Range f = Domain f^{-1} = All real numbers

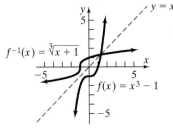

53. $f^{-1}(x) = \sqrt{x - 4}, x \geq 4$

$f(f^{-1}(x)) = f(\sqrt{x - 4}) = (\sqrt{x - 4})^2 + 4 = x$

$f^{-1}(f(x)) = f^{-1}(x^2 + 4) = \sqrt{(x^2 + 4) - 4} = \sqrt{x^2} = x, x \geq 0$

Domain f = Range f^{-1} = $\{x | x \geq 0\}$ or $[0, \infty)$

Range f = Domain f^{-1} = $\{x | x \geq 4\}$ or $[4, \infty)$

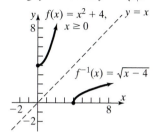

55. $f^{-1}(x) = \dfrac{4}{x}$

$f(f^{-1}(x)) = f\left(\dfrac{4}{x}\right) = \dfrac{4}{\left(\dfrac{4}{x}\right)} = x$

$f^{-1}(f(x)) = f^{-1}\left(\dfrac{4}{x}\right) = \dfrac{4}{\left(\dfrac{4}{x}\right)} = x$

Domain f = Range f^{-1} = All real numbers except 0

Range f = Domain f^{-1} = All real numbers except 0

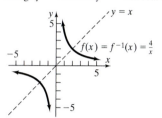

57. $f^{-1}(x) = \dfrac{2x + 1}{x}$

$f(f^{-1}(x)) = f\left(\dfrac{2x + 1}{x}\right) = \dfrac{1}{\dfrac{2x + 1}{x} - 2} = \dfrac{x}{(2x + 1) - 2x} = x$

$f^{-1}(f(x)) = f^{-1}\left(\dfrac{1}{x - 2}\right) = \dfrac{2\left(\dfrac{1}{x - 2}\right) + 1}{\dfrac{1}{x - 2}} = \dfrac{2 + (x - 2)}{1} = x$

Domain f = Range f^{-1} = All real numbers except 2

Range f = Domain f^{-1} = All real numbers except 0

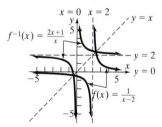

59. $f^{-1}(x) = \dfrac{2 - 3x}{x}$

$f(f^{-1}(x)) = f\left(\dfrac{2 - 3x}{x}\right) = \dfrac{2}{3 + \dfrac{2 - 3x}{x}} = \dfrac{2x}{3x + 2 - 3x} = \dfrac{2x}{2} = x$

$f^{-1}(f(x)) = f^{-1}\left(\dfrac{2}{3 + x}\right) = \dfrac{2 - 3\left(\dfrac{2}{3 + x}\right)}{\dfrac{2}{3 + x}} = \dfrac{2(3 + x) - 3 \cdot 2}{2}$

$= \dfrac{2x}{2} = x$

Domain f = All real numbers except -3

Range f = Domain f^{-1} = All real numbers except 0

61. $f^{-1}(x) = \dfrac{-2x}{x - 3}$

$f(f^{-1}(x)) = f\left(\dfrac{-2x}{x - 3}\right) = \dfrac{3\left(\dfrac{-2x}{x - 3}\right)}{\dfrac{-2x}{x - 3} + 2} = \dfrac{3(-2x)}{-2x + 2(x - 3)}$

$= \dfrac{-6x}{-6} = x$

$f^{-1}(f(x)) = f^{-1}\left(\dfrac{3x}{x + 2}\right) = \dfrac{-2\left(\dfrac{3x}{x + 2}\right)}{\dfrac{3x}{x + 2} - 3} = \dfrac{-2(3x)}{3x - 3(x + 2)}$

$= \dfrac{-6x}{-6} = x$

Domain f = All real numbers except -2

Range f = Domain f^{-1} = All real numbers except 3

63. $f^{-1}(x) = \dfrac{x}{3x - 2}$

$f(f^{-1}(x)) = f\left(\dfrac{x}{3x - 2}\right) = \dfrac{2\left(\dfrac{x}{3x - 2}\right)}{3\left(\dfrac{x}{3x - 2}\right) - 1} = \dfrac{2x}{3x - (3x - 2)}$

$= \dfrac{2x}{2} = x$

$f^{-1}(f(x)) = f^{-1}\left(\dfrac{2x}{3x - 1}\right) = \dfrac{\dfrac{2x}{3x - 1}}{3\left(\dfrac{2x}{3x - 1}\right) - 2} = \dfrac{2x}{6x - 2(3x - 1)}$

$= \dfrac{2x}{2} = x$

Domain f = All real numbers except $\dfrac{1}{3}$

Range f = Domain f^{-1} = All real numbers except $\dfrac{2}{3}$

65. $f^{-1}(x) = \dfrac{3x + 4}{2x - 3}$

$f(f^{-1}(x)) = f\left(\dfrac{3x + 4}{2x - 3}\right) = \dfrac{3\left(\dfrac{3x + 4}{2x - 3}\right) + 4}{2\left(\dfrac{3x + 4}{2x - 3}\right) - 3} = \dfrac{3(3x + 4) + 4(2x - 3)}{2(3x + 4) - 3(2x - 3)}$

$= \dfrac{17x}{17} = x$

$f^{-1}(f(x)) = f^{-1}\left(\dfrac{3x + 4}{2x - 3}\right) = \dfrac{3\left(\dfrac{3x + 4}{2x - 3}\right) + 4}{2\left(\dfrac{3x + 4}{2x - 3}\right) - 3} = \dfrac{3(3x + 4) + 4(2x - 3)}{2(3x + 4) - 3(2x - 3)}$

$= \dfrac{17x}{17} = x$

Domain f = All real numbers except $\dfrac{3}{2}$

Range f = Domain f^{-1} = All real numbers except $\dfrac{3}{2}$

67. $f^{-1}(x) = \dfrac{-2x + 3}{x - 2}$

$f(f^{-1}(x)) = f\left(\dfrac{-2x + 3}{x - 2}\right) = \dfrac{2\left(\dfrac{-2x + 3}{x - 2}\right) + 3}{\dfrac{-2x + 3}{x - 2} + 2}$

$= \dfrac{2(-2x + 3) + 3(x - 2)}{-2x + 3 + 2(x - 2)} = \dfrac{-x}{-1} = x$

$f^{-1}(f(x)) = f^{-1}\left(\dfrac{2x + 3}{x + 2}\right) = \dfrac{2\left(\dfrac{-2x + 3}{x - 2}\right) + 3}{\dfrac{-2x + 3}{x - 2} + 2}$

$= \dfrac{2(-2x + 3) + 3(x - 2)}{-2x + 3 + 2(x - 2)} = \dfrac{-x}{-1} = x$

Domain f = All real numbers except -2
Range f = Domain f^{-1} = All real numbers except 2

69. $f^{-1}(x) = \dfrac{2}{\sqrt{1 - 2x}}$

$f(f^{-1}(x)) = f\left(\dfrac{2}{\sqrt{x - 2x}}\right) = \dfrac{\dfrac{4}{1 - 2x} - 4}{2 \cdot \dfrac{4}{1 - 2x}} = \dfrac{4 - 4(1 - 2x)}{2 \cdot 4} = \dfrac{8x}{8} = x$

$f^{-1}(f(x)) = f^{-1}\left(\dfrac{x^2 - 4}{2x^2}\right) = \dfrac{2}{\sqrt{1 - 2\left(\dfrac{x^2 - 4}{2x^2}\right)}} = \dfrac{2}{\sqrt{\dfrac{4}{x^2}}} = \sqrt{x^2}$

$= x$, since $x > 0$.

Domain f = $\{x | x > 0\}$ or $(0, \infty)$

Range f = Domain f^{-1} = $\left\{x \,\middle|\, x < \dfrac{1}{2}\right\}$ or $\left(-\infty, \dfrac{1}{2}\right)$

71. (a) 0 **(b)** 2 **(c)** 0 **(d)** 1 **73.** $f^{-1}(x) = \dfrac{1}{m}(x - b)$,

$m \neq 0$ **75.** Quadrant I **77.** Possible answer: $f(x) = |x|$,

$x \geq 0$, is one-to-one; $f^{-1}(x) = x, x \geq 0$

79. (a) $C(H) = \dfrac{H + 10.53}{2.15}$ **(b)** 16.99 in.

81. $x(p) = \dfrac{300 - p}{50}, p \leq 300$

83. $f^{-1}(x) = \dfrac{-dx + b}{cx - a}; f = f^{-1}$ if $a = -d$

87. Yes, if the domain is $\{x | x \geq 0\}$.

89. On the line $y = x$; no; no.

4.3 Assess Your Understanding (page 282)

6. $\left(-1, \dfrac{1}{a}\right), (0, 1), (1, a)$ **7.** 1 **8.** 4 **9.** F **10.** F **11. (a)** 11.212 **(b)** 11.587 **(c)** 11.664 **(d)** 11.665 **13. (a)** 8.815 **(b)** 8.821 **(c)** 8.824

(d) 8.825 **15. (a)** 21.217 **(b)** 22.217 **(c)** 22.440 **(d)** 22.459 **17.** 3.320 **19.** 0.427 **21.** Not exponential **23.** Exponential; $a = 4$

25. Exponential; $a = 2$ **27.** Not exponential **29.** B **31.** D **33.** A **35.** E

37.

Domain: All real numbers
Range: $\{y | y > 1\}$ or $(1, \infty)$
Horizontal asymptote: $y = 1$

39.

Domain: All real numbers
Range: $\{y | y > -2\}$ or $(-2, \infty)$
Horizontal asymptote: $y = -2$

41.

Domain: All real numbers
Range: $\{y | y > 2\}$ or $(2, \infty)$
Horizontal asymptote: $y = 2$

43.

Domain: All real numbers
Range: $\{y | y > 2\}$ or $(2, \infty)$
Horizontal asymptote: $y = 2$

45.

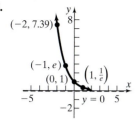

Domain: All real numbers
Range: $\{y | y > 0\}$ or $(0, \infty)$
Horizontal asymptote: $y = 0$

47.

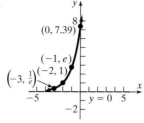

Domain: All real numbers
Range: $\{y | y > 0\}$ or $(0, \infty)$
Horizontal asymptote: $y = 0$

49.

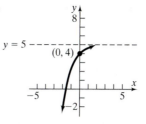

Domain: All real numbers
Range: $\{y | y < 5\}$ or $(-\infty, 5)$
Horizontal asymptote: $y = 5$

51.

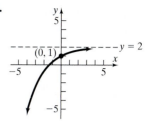

Domain: All real numbers
Range: $\{y | y < 2\}$ or $(-\infty, 2)$
Horizontal asymptote: $y = 2$

53. $\left\{\dfrac{1}{2}\right\}$ **55.** $\{-\sqrt{2}, 0, \sqrt{2}\}$ **57.** $\left\{1 - \dfrac{\sqrt{6}}{3}, 1 + \dfrac{\sqrt{6}}{3}\right\}$ **59.** $\{0\}$ **61.** $\{4\}$ **63.** $\left\{\dfrac{3}{2}\right\}$ **65.** $\{1, 2\}$ **67.** $\dfrac{1}{49}$ **69.** $\dfrac{1}{4}$ **71.** $f(x) = 3^x$

73. $f(x) = -6^x$ **75. (a)** 74%   **(b)** 47%   **77. (a)** \$12,123 **(b)** \$6443   **79.** 3.35 mg; 0.45 mg

81. (a) 0.63   **(b)** 0.98   **(c)** 1

(d)

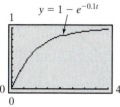

(e) About 7 min

83. (a) 0.0516   **(b)** 0.0888
85. (a) 70.95%   **(b)** 72.62%   **(c)** 100%

87. (a) 5.41 amp, 7.59 amp, 10.38 amp   **(b)** 12 amp   **(d)** 3.34 amp, 5.31 amp, 9.44 amp   **(e)** 24 amp

(c), (f)

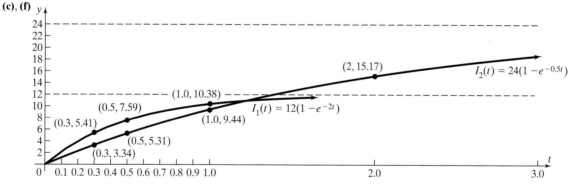

89. $n = 4: 2.7083; n = 6: 2.7181; n = 8: 2.7182788; n = 10: 2.7182818$

91. $\dfrac{f(x + h) - f(x)}{h} = \dfrac{a^{x+h} - a^x}{h} = \dfrac{a^x a^h - a^x}{h} = \dfrac{a^x(a^h - 1)}{h}$ **93.** $f(-x) = a^{-x} = \dfrac{1}{a^x} = \dfrac{1}{f(x)}$

95. (a) $f(-x) = \dfrac{1}{2}(e^{-x} - e^{-(-x)}) = \dfrac{1}{2}(e^{-x} - e^x)$

$= -\dfrac{1}{2}(e^x - e^{-x}) = -f(x)$

97. $f(1) = 5, f(2) = 17, f(3) = 257, f(4) = 65{,}537,$
$f(5) = 4{,}294{,}967{,}297 = 641 \times 6{,}700{,}417$

(b)

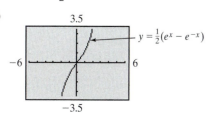

4.4 Assess Your Understanding *(page 296)*

4. $\{x|x > 0\}$ or $(0, \infty)$ **5.** $\left(\dfrac{1}{a}, -1\right), (1, 0), (a, 1)$ **6.** 1   **7.** F   **8.** T   **9.** $2 = \log_3 9$ **11.** $2 = \log_a 1.6$

13. $2 = \log_{1.1} M$ **15.** $x = \log_2 7.2$ **17.** $\sqrt{2} = \log_x \pi$ **19.** $x = \ln 8$ **21.** $2^3 = 8$ **23.** $a^6 = 3$ **25.** $3^x = 2$ **27.** $2^{1.3} = M$ **29.** $(\sqrt{2})^x = \pi$

31. $e^x = 4$ **33.** 0   **35.** 2   **37.** -4 **39.** $\dfrac{1}{2}$ **41.** 4   **43.** $\dfrac{1}{2}$ **45.** $\{x|x > 3\}; (3, \infty)$ **47.** All real numbers except 0; $\{x|x \neq 0\}$ **49.** $\{x|x > 0\}; (0, \infty)$

51. $\{x|x > -1\}; (-1, \infty)$ **53.** $\{x|x < -1 \text{ or } x > 0\}; (-\infty, -1) \text{ or } (0, \infty)$ **55.** $\{x|x \geq 1\}; [1, \infty)$ **57.** 0.511   **59.** 30.099   **61.** $\sqrt{2}$

63.

65.

67. B   **69.** D   **71.** A   **73.** E

75.

Domain: $(-4, \infty)$
Range: $(-\infty, \infty)$
Vertical asymptote: $x = -4$

77.

Domain: $(0, \infty)$
Range: $(-\infty, \infty)$
Vertical asymptote: $x = 0$

79.

Domain: $(0, \infty)$
Range: $(-\infty, \infty)$
Vertical asymptote: $x = 0$

81.

Domain: $(0, \infty)$
Range: $(-\infty, \infty)$
Vertical asymptote: $x = 0$

83.

Domain: $(4, \infty)$
Range: $(-\infty, \infty)$
Vertical asymptote: $x = 4$

85.

Domain: $(0, \infty)$
Range: $(-\infty, \infty)$
Vertical asymptote: $x = 0$

87.

Domain: $(0, \infty)$
Range: $(-\infty, \infty)$
Vertical asymptote: $x = 0$

89.

Domain: $(-2, \infty)$
Range: $(-\infty, \infty)$
Vertical asymptote: $x = -2$

91. $\{9\}$ **93.** $\left\{\dfrac{7}{2}\right\}$ **95.** $\{2\}$ **97.** $\{5\}$ **99.** $\{3\}$ **101.** $\{2\}$ **103.** $\left\{\dfrac{\ln 10}{3}\right\}$ **105.** $\left\{\dfrac{\ln 8 - 5}{2}\right\}$ **107.** $\{-2\sqrt{2}, 2\sqrt{2}\}$ **109.** $\{-1\}$

111. (a)

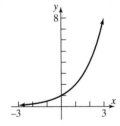

Domain: $(-\infty, \infty)$
Range: $(0, \infty)$
Horizontal asymptote: $y = 0$

(b) $f^{-1}(x) = \log_2 x$

(c)

Domain of f^{-1} = Range of $f = (0, \infty)$
Range of f^{-1} = Domain of $f = (-\infty, \infty)$
Vertical asymptote of f^{-1}: $x = 0$

113. (a)

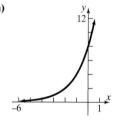

(b) $f^{-1}(x) = \log_2 x + 3$ **(c)**

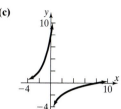

Domain: $(-\infty, \infty)$
Range: $(0, \infty)$
Horizontal asymptote: $y = 0$

Domain of f^{-1} = Range of $f = (0, \infty)$
Range of f^{-1} = Domain of $f = (-\infty, \infty)$
Vertical asymptote of f^{-1}: $x = 0$

115. (a) 1   **(b)** 2   **(c)** 3   **(d)** It increases.   **(e)** 0.000316   **(f)** 3.981×10^{-8} **117. (a)** 5.97 km   **(b)** 0.90 km   **119. (a)** 6.93 min
(b) 16.09 min   **(c)** No, since $F(t)$ can never equal 1   **121.** $h \approx 2.29$, so the time between injections is about 2 hr, 17 min

123. 0.2695 sec
0.8959 sec

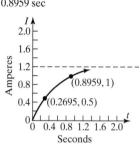

125. 50 decibels (dB)
127. 110 dB
129. 8.1
131. (a) $k = 20.07$ **(b)** 91%   **(c)** 0.175   **(d)** 0.08
133. Because $y = \log_1 x$ means $1^y = 1 = x$, which cannot be true for $x \neq 1$

4.5 Assess Your Understanding *(page 307)*

1. Sum   **2.** 7   **3.** $r \log_a M$ **4.** F   **5.** F   **6.** T   **7.** 71   **9.** -4 **11.** 7   **13.** 1   **15.** 1   **17.** 2   **19.** $\dfrac{5}{4}$ **21.** 4   **23.** $a + b$ **25.** $b - a$

27. $3a$ **29.** $\dfrac{1}{5}(a + b)$ **31.** $2 + \log_5 x$ **33.** $3 \log_2 z$ **35.** $1 + \ln x$ **37.** $\ln x + x$ **39.** $2 \log_a u + 3 \log_a v$ **41.** $2 \ln x + \dfrac{1}{2} \ln(1 - x)$

43. $3 \log_2 x - \log_2(x - 3)$ **45.** $\log x + \log(x + 2) - 2 \log(x + 3)$ **47.** $\dfrac{1}{3} \ln(x - 2) + \dfrac{1}{3} \ln(x + 1) - \dfrac{2}{3} \ln(x + 4)$

49. $\ln 5 + \ln x + \dfrac{1}{2} \ln(1 + 3x) - 3 \ln(x - 4)$ **51.** $\log_5 u^3 v^4$ **53.** $-\dfrac{5}{2} \log_3 x$ **55.** $\log_4 \left[\dfrac{x - 1}{(x + 1)^4}\right]$ **57.** $-2 \ln(x - 1)$

59. $\log_2[x(3x - 2)^4]$ **61.** $\log_a\left(\dfrac{25x^6}{\sqrt{2x + 3}}\right)$ **63.** $\log_2\left[\dfrac{(x + 1)^2}{(x + 3)(x - 1)}\right]$ **65.** 2.771   **67.** -3.880 **69.** 5.615   **71.** 0.874

73. $y = \dfrac{\log x}{\log 4}$

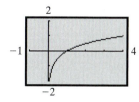

75. $y = \dfrac{\log(x + 2)}{\log 2}$

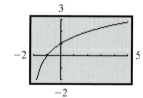

77. $y = \dfrac{\log(x + 1)}{\log(x - 1)}$

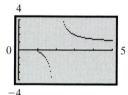

79. $y = Cx$ **81.** $y = Cx(x + 1)$ **83.** $y = Ce^{3x}$ **85.** $y = Ce^{-4x} + 3$ **87.** $y = \dfrac{\sqrt[3]{C}(2x + 1)^{1/6}}{(x + 4)^{1/9}}$ **89.** 3   **91.** 1

93. $\log_a(x + \sqrt{x^2 - 1}) + \log_a(x - \sqrt{x^2 - 1}) = \log_a[(x + \sqrt{x^2 - 1})(x - \sqrt{x^2 - 1})]$
$= \log_a[x^2 - (x^2 - 1)] = \log_a 1 = 0$

95. $\ln(1 + e^{2x}) = \ln[e^{2x}(e^{-2x} + 1)] = \ln e^{2x} + \ln(e^{-2x} + 1) = 2x + \ln(1 + e^{-2x})$

97. $y = f(x) = \log_a x; a^y = x$ implies $a^y = \left(\dfrac{1}{a}\right)^{-y} = x$, so $-y = \log_{1/a} x = -f(x)$.

99. $f(x) = \log_a x; f\left(\dfrac{1}{x}\right) = \log_a \dfrac{1}{x} = \log_a 1 - \log_a x = -f(x)$ **101.** $\log_a \dfrac{M}{N} = \log_a(M \cdot N^{-1}) = \log_a M + \log_a N^{-1} = \log_a M - \log_a N$,
since $a^{\log_a N^{-1}} = N^{-1}$ implies $a^{-\log_a N^{-1}} = N$; i.e., $\log_a N = -\log_a N^{-1}$.

4.6 Assess Your Understanding *(page 313)*

5. 6   **7.** 16   **9.** 8   **11.** 3   **13.** 5   **15.** $-1 + \sqrt{1 + e^4} \approx 6.456$ **17.** $\dfrac{\ln 3}{\ln 2} \approx 1.585$ **19.** 0   **21.** $\dfrac{\ln 10}{\ln 2} \approx 3.322$ **23.** $-\dfrac{\ln 1.2}{\ln 8} \approx -0.088$

25. $\dfrac{\ln 3}{2 \ln 3 + \ln 4} \approx 0.307$ **27.** $\dfrac{\ln 7}{\ln 0.6 + \ln 7} \approx 1.356$ **29.** 0   **31.** $\dfrac{\ln \pi}{1 + \ln \pi} \approx 0.534$ **33.** $\dfrac{\ln 1.6}{3 \ln 2} \approx 0.226$ **35.** $\dfrac{9}{2}$ **37.** 2   **39.** 1   **41.** 16

43. $-1, \dfrac{2}{3}$ **45.** 0   **47.** $\ln(2 + \sqrt{5}) \approx 1.444$ **49.** $\dfrac{1}{10^{\frac{1}{\log 5} + \frac{1}{\log 3}}}$ **51.** 2.79   **53.** -0.57 **55.** -0.70 **57.** 0.57   **59.** 0.39, 1.00   **61.** 1.32

63. 1.31   **65. (a)** Around the middle of the year 2006   **(b)** In the beginning of the year 2021
67. (a) After 2.4 years   **(b)** After 6.5 years   **(c)** After 10 years

4.7 Assess Your Understanding *(page 322)*

3. $108.29   **5.** $609.50   **7.** $697.09   **9.** $12.46   **11.** $125.23   **13.** $88.72   **15.** $860.72   **17.** $554.09   **19.** $59.71   **21.** $361.93   **23.** 5.35%

25. 26%   **27.** $6\frac{1}{4}$% compounded annually   **29.** 9% compounded monthly   **31.** 104.32 mo (about 8.7 yr); 103.97 mo (about 8.66 yr)

33. 61.02 mo; 60.82 mo   **35.** 15.27 yr or 15 yr, 3 mo   **37.** $104,335   **39.** $12,910.62   **41.** About $30.17 per share or $3017   **43.** 9.35%
45. Not quite. Jim will have $1057.60. The second bank gives a better deal, since Jim will have $1060.62 after 1 yr.
47. Will has $11,632.73; Henry has $10,947.89.   **49. (a)** Interest is $30,000   **(b)** Interest is $38,613.59   **(c)** Interest is $37,752.73. Simple
interest at 12% is best.   **51. (a)** $1364.62   **(b)** $1353.35   **53.** $4631.93

55. (a) 6.1 yr   **(b)** 18.45 yr   **(c)** $mP = P\left(1 + \dfrac{r}{n}\right)^{nt}$

$$m = \left(1 + \dfrac{r}{n}\right)^{nt}$$

$$\ln m = \ln\left(1 + \dfrac{r}{n}\right)^{nt} = nt \ln\left(1 + \dfrac{r}{n}\right)$$

$$t = \dfrac{\ln m}{n \ln\left(1 + \dfrac{r}{n}\right)}$$

4.8 Assess Your Understanding *(page 334)*

1. (a) 500 insects
(b) $0.02 = 2\%$
(c)

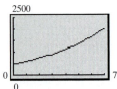

(d) About 611 insects
(e) After about 23.5 days
(f) After about 34.7 days

13. (a) 5:18 PM
(b)

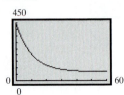

(c) About 14.3 min
(d) The temperature of the pizza
approaches 70°F.

3. (a) $-0.0244 = -2.44\%$
(b)

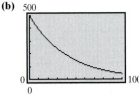

(c) About 391.7 g
(d) After about 9.1 yr
(e) 28.4 yr

15. (a) 18.63°C; 25.1°C
(b)

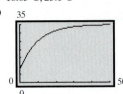

5. 5832; 3.9 days
7. 25,198
9. 9.797 g

17. 7.34 kg; 76.6 h
19. 26.6 days

11. (a) 9727 yr ago
(b)

(c) 5600 yr

21. (a) 90% **(b)** 12.86%

(c)

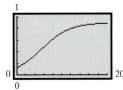

(d) 85.77%

(e) 1996

(f) About 5.6 yr

23. (a) 1000 g **(b)** 43.9% **(c)** 30 g

(d)

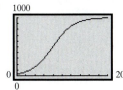

(e) 616.6 g

(f) After 9.85 h

(g) About 7.9 h

25. (a) 9.23×10^{-3}, or about 0

(b) 0.81, or about 1

(c) 5.01, or about 5

(d) 57.91°, 43.99°, 30.07°

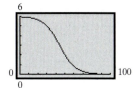

27. (a) In 1984, 91.8% of households did not own a personal computer.

(b)

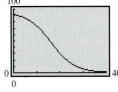

(c) 70.6% **(d)** During 2007

29. (a)

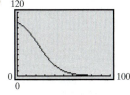

(b) 0.78 or 78%

(c) 50 people

4.9 Assess Your Understanding *(page 342)*

1. (a)

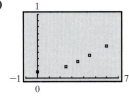

(b) $y = 0.0903(1.3384)^x$

(c) $N(t) = 0.0903e^{0.2915t}$

(d)

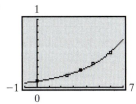

(e) 0.69 **(f)** After about 7.26 h

3. (a)

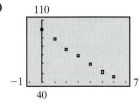

(b) $y = 100.326(0.8769)^x$

(c) $A = 100.326e^{-0.1314t}$

(d)

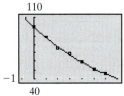

(e) 5.3 weeks **(f)** 0.14 g

(g) After about 12.3 weeks

5. (a)

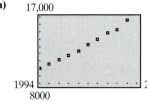

(b) $y = 2.2908 \times 10^{-44}(1.056554737)^x$

(c) 5.66%

(d) $44,230.54

(e) In 2023

7. (a)

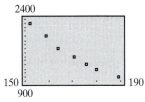

(b) $y = 32,741.02 - 6070.96 \ln x$

(c)

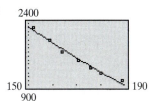

(d) Approximately 168 computers

9.(a) 290,000,000

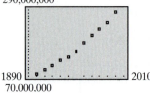

1890   2010
70.000.000

(b) $y = \dfrac{799{,}475{,}916.5}{1 + 1.56344 \times 10^{14} e^{-0.0160x}}$

(c) 290,000,000

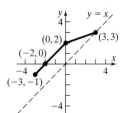

1890   2010
70,000,000

(d) 799,475,917

(e) Approximately 279,809,184   **(f)** 2011

11.(a) 12,500,000

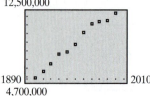

1890   2010
4,700,000

(b) $y = \dfrac{14{,}471{,}245.24}{1 + 3.860 \times 10^{20} e^{-0.0246x}}$

(c) 12,500,000

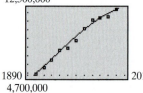

1890   2010
4,700,000

(d) 14,471,245

(e) Approximately 12,811,429

Review Exercises *(page 348)*

1. (a) −26   **(b)** −241   **(c)** 16   **(d)** −1   **3. (a)** $\sqrt{11}$ **(b)** 1   **(c)** $\sqrt{\sqrt{6}+2}$ **(d)** 19   **5. (a)** e^4 **(b)** $3e^{-2}-2$ **(c)** e^{e^4} **(d)** −17

7. $(f \circ g)(x) = 1 - 3x$, all real numbers; $(g \circ f)(x) = 7 - 3x$, all real numbers;
$(f \circ f)(x) = x$, all real numbers; $(g \circ g)(x) = 9x + 4$, all real numbers

9. $(f \circ g)(x) = 27x^2 + 3|x| + 1$, all real numbers; $(g \circ f)(x) = 3|3x^2 + x + 1|$, all real numbers;
$(f \circ f)(x) = 3(3x^2 + x + 1)^2 + 3x^2 + x + 2$, all real numbers; $(g \circ g)(x) = 9|x|$, all real numbers

11. $(f \circ g)(x) = \dfrac{1+x}{1-x}, \{x | x \neq 0, x \neq 1\}$; $(g \circ f)(x) = \dfrac{x-1}{x+1}, \{x | x \neq -1, x \neq 1\}$; $(f \circ f)(x) = x, \{x | x \neq 1\}$; $(g \circ g)(x) = x, \{x | x \neq 0\}$

13. (a) One-to-one   **(b)** $\{(2,1), (5,3), (8,5), (10,6)\}$

15.

17. $f^{-1}(x) = \dfrac{2x+3}{5x-2}$

$f(f^{-1}(x)) = \dfrac{2\left(\dfrac{2x+3}{5x-2}\right) + 3}{5\left(\dfrac{2x+3}{5x-2}\right) - 2} = x$

$f^{-1}(f(x)) = \dfrac{2\left(\dfrac{2x+3}{5x-2}\right) + 3}{5\left(\dfrac{2x+3}{5x-2}\right) - 2} = x$

Domain of f = Range of f^{-1} = all real numbers except $\dfrac{2}{5}$

Range of f = Domain of f^{-1} = all real numbers except $\dfrac{2}{5}$

19. $f^{-1}(x) = \dfrac{x+1}{x}$

$f(f^{-1}(x)) = \dfrac{1}{\dfrac{x+1}{x} - 1} = x$

$f^{-1}(f(x)) = \dfrac{\dfrac{1}{x-1} + 1}{\dfrac{1}{x-1}} = x$

Domain of f = Range of f^{-1} = all real numbers except 1

Range of f = Domain of f^{-1} = all real numbers except 0

21. $f^{-1}(x) = \dfrac{27}{x^3}$

$f(f^{-1}(x)) = \dfrac{3}{\left(\dfrac{27}{x^3}\right)^{1/3}} = x$

$f^{-1}(f(x)) = \dfrac{27}{\left(\dfrac{3}{x^{1/3}}\right)^3} = x$

Domain of f = Range of f^{-1} = all real numbers except 0
Range of f = Domain of f^{-1} = all real numbers except 0

23. (a) 81 **(b)** 2 **(c)** $\dfrac{1}{9}$ **(d)** -3

25. $\log_5 z = 2$ **27.** $5^{13} = u$

29. $\left\{x \,\middle|\, x > \dfrac{2}{3}\right\}; \left(\dfrac{2}{3}, \infty\right)$ **31.** $\{x \mid x < 1 \text{ or } x > 2\}; (-\infty, 1) \text{ or } (2, \infty)$

33. -3 **35.** $\sqrt{2}$ **37.** 0.4 **39.** $\log_3 u + 2\log_3 v - \log_3 w$ **41.** $2\log x + \dfrac{1}{2}\log(x^3 + 1)$ **43.** $\ln x + \dfrac{1}{3}\ln(x^2 + 1) - \ln(x - 3)$

45. $\dfrac{25}{4}\log_4 x$ **47.** $-2\ln(x + 1)$ **49.** $\log\left(\dfrac{4x^3}{[(x + 3)(x - 2)]^{1/2}}\right)$ **51.** 2.124

53.

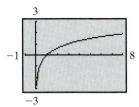

55.

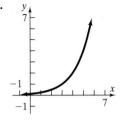

Domain: $(-\infty, \infty)$
Range: $(0, \infty)$
Horizontal asymptote: $y = 0$

57.

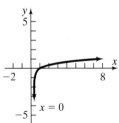

Domain: $(-\infty, \infty)$
Range: $(0, \infty)$
Horizontal asymptote: $y = 0$

59.

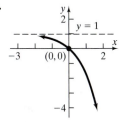

Domain: $(-\infty, \infty)$
Range: $(-\infty, 1)$
Horizontal asymptote: $y = 1$

61.

Domain: $(0, \infty)$
Range: $(-\infty, \infty)$
Vertical asymptote: $x = 0$

63.

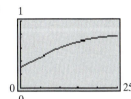

Domain: $(-\infty, \infty)$
Range: $(-\infty, 3)$
Horizontal asymptote: $y = 3$

65. $\left\{\dfrac{1}{4}\right\}$ **67.** $\left\{\dfrac{-1 - \sqrt{3}}{2}, \dfrac{-1 + \sqrt{3}}{2}\right\}$ **69.** $\left\{\dfrac{1}{4}\right\}$ **71.** $\left\{\dfrac{2\ln 3}{\ln 5 - \ln 3} \approx 4.301\right\}$

73. $\left\{\dfrac{12}{5}\right\}$ **75.** $\{83\}$ **77.** $\left\{\dfrac{1}{2}, -3\right\}$ **79.** $\{-1\}$ **81.** $\{1 - \ln 5 \approx -0.609\}$

83. $\left\{\dfrac{\ln 3}{3\ln 2 - 2\ln 3} \approx -9.327\right\}$ **85.** 3229.5 m **87. (a)** 37.3 W **(b)** 6.9 dB

89. (a) 9.85 yr **(b)** 4.27 yr **91.** \$41,668.97 **93.** 24,203 yr ago

95. 6,835,600,129

97. (a) 0.3
(b) 0.8
(c)

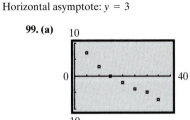

(d) In 2023

99. (a)

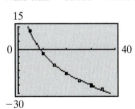

(b) Wind chill = $18.921 - 7.096\ln$ (wind speed)
(c)

15

0 ———————— 40

-30

(d) Approximately $-3°$F

Chapter Test *(page 352)*

1. (a) $f \circ g = \dfrac{2x + 7}{2x + 3}$; Domain: $\left\{ x \Big| x \neq -\dfrac{3}{2} \right\}$ **(b)** $(g \circ f)(-2) = 5$ **(c)** $(f \circ g)(-2) = -3$

2. (a) The function is not one-to-one.   **(b)** The function is one-to-one.

3. $f^{-1}(x) = \dfrac{2 + 5x}{3x}$

	Domain	Range
f	$x \neq \dfrac{5}{3}$	$y \neq 0$
f^{-1}	$x \neq 0$	$y \neq \dfrac{5}{3}$

4. The point $(-5, 3)$ must be on the graph of f^{-1}.   **5.** $x = 5$ **6.** $b = 4$ **7.** $x = 625$ **8.** $e^3 + 2 \approx 22.086$ **9.** $\log 20 \approx 1.301$

10. $\log_3 21 = \dfrac{\ln 21}{\ln 3} \approx 2.771$ **11.** $\ln 133 \approx 4.890$

12. Domain: $\{x | -\infty < x < \infty\}$
Range: $\{y | y > -2\}$
Asymptote: $y = -2$

13. Domain: $\{x | x > 2\}$
Range: $\{y | -\infty < y < \infty\}$
Asymptote: $x = 2$

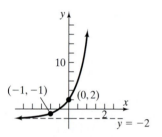

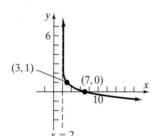

14. $x = 1$ **15.** $x = 91$ **16.** $x = -\ln 2 \approx -0.693$ **17.** $\dfrac{1 - \sqrt{13}}{2} \approx -1.303, \dfrac{1 + \sqrt{13}}{2} \approx 2.303$ **18.** $x = \dfrac{3 \ln 7}{1 - \ln 7} \approx -6.172$

19. $x = 2\sqrt{6} \approx 4.899$ **20.** $2 + 3 \log_2 x - \log_2 (x - 6) - \log_2 (x + 3)$ **21.** About 250.39 days   **22. (a)** \$35,298   **(b)** \$21,409

23. (a) About 83 decibles   **(b)** The pain threshold will be exceeded if 31,623 people shouted at the same time.

24. $y = \dfrac{213}{1 + 205.86 e^{-0.3564t}}$; 197 million U.S. cell phone subscribers

Cumulative Review *(page 353)*

1. Yes; no   **2. (a)** 10   **(b)** $2x^2 + 3x + 1$ **(c)** $2x^2 + 4xh + 2h^2 - 3x - 3h + 1$ **3.** $\left(\dfrac{1}{2}, \dfrac{\sqrt{3}}{2} \right)$ is on the graph   **4.** -26

5.

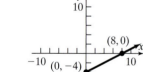

6. (a)

(b) $\{x | -\infty < x < \infty\}$

7. $f(x) = 2(x - 4)^2 - 8 = 2x^2 - 16x + 24$

8.

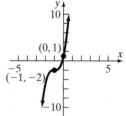

9. $f(g(x)) = \dfrac{4}{(x - 3)^2} + 2$; domain: $\{x | x \neq 3\}$; 3

10. (a) Zeros: $-4, -\dfrac{1}{4}, 2$

(b) x-intercepts: $-4, -\dfrac{1}{4}, 2$; y-intercept: -8

(c) Local maximum of 60.75 occurs at $x = -2.5$;
Local minimum of -25 occurs at $x = 1$

(d)

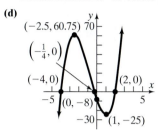

11. (a), (c)

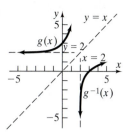

Domain g = Range $g^{-1} = (-\infty, \infty)$
Range g = Domain $g^{-1} = (2, \infty)$

(b) $g^{-1}(x) = \log_3(x - 2)$

12. $-\dfrac{3}{2}$ **13.** 2   **14. (a)** -1 **(b)** $\{x \mid x > -1\}$ or $(-1, \infty)$ **(c)** 25   **15. (a)**

(b) Answers will vary.

C H A P T E R 5   Trigonometric Functions

5.1 Assess Your Understanding *(page 366)*

3. Standard position   **4.** $r\theta$; $\dfrac{1}{2}r^2\theta$ **5.** $\dfrac{s}{t}$; $\dfrac{\theta}{t}$ **6.** F   **7.** T   **8.** T   **9.** T   **10.** F

11.

13.

15.

17.

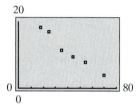

19.

21.

23. $40.17°$ **25.** $1.03°$ **27.** $9.15°$ **29.** $40°19'12''$ **31.** $18°15'18''$ **33.** $19°59'24''$ **35.** $\dfrac{\pi}{6}$ **37.** $\dfrac{4\pi}{3}$ **39.** $-\dfrac{\pi}{3}$ **41.** π **43.** $-\dfrac{3\pi}{4}$ **45.** $-\dfrac{\pi}{2}$

47. $60°$ **49.** $-225°$ **51.** $90°$ **53.** $15°$ **55.** $-90°$ **57.** $-30°$ **59.** 0.30 **61.** -0.70 **63.** 2.18 **65.** $179.91°$ **67.** $114.59°$ **69.** $362.11°$

71. 5 m   **73.** 6 ft   **75.** 0.6 radian   **77.** $\dfrac{\pi}{3} \approx 1.047$ in.   **79.** 25 m²   **81.** $2\sqrt{3} \approx 3.464$ ft   **83.** 0.24 radian   **85.** $\dfrac{\pi}{3} \approx 1.047$ in²

87. $s = 2.094$ ft; $A = 2.094$ ft²   **89.** $s = 14.661$ yd; $A = 87.965$ yd²   **91.** $3\pi \approx 9.4248$ in.; $5\pi \approx 15.7080$ in.   **93.** $2\pi \approx 6.28$ m²

95. $\dfrac{675\pi}{2} \approx 1060.29$ ft²   **97.** $\omega = \dfrac{1}{60}$ radian/sec; $v = \dfrac{1}{12}$ cm/sec   **99.** Approximately 452.5 rpm   **101.** Approximately 359 mi

103. Approximately 898 mi/hr   **105.** Approximately 2292 mi/hr   **107.** $\dfrac{3}{4}$ rpm   **109.** Approximately 2.86 mi/hr   **111.** Approximately 31.47 rpm

113. Approximately 1037 mi/hr   **115.** radius ≈ 3979 miles; circumference $\approx 25{,}000$ miles

5.2 Assess Your Understanding *(page 384)*

7. $\dfrac{3}{2}$ **8.** 0.91   **9.** T   **10.** F   **11.** $\sin t = \dfrac{1}{2}; \cos t = \dfrac{\sqrt{3}}{2}; \tan t = \dfrac{\sqrt{3}}{3}; \csc t = 2; \sec t = \dfrac{2\sqrt{3}}{3}; \cot t = \sqrt{3}$

13. $\sin t = \dfrac{\sqrt{21}}{5}; \cos t = -\dfrac{2}{5}; \tan t = -\dfrac{\sqrt{21}}{2}; \csc t = \dfrac{5\sqrt{21}}{21}; \sec t = -\dfrac{5}{2}; \cot t = -\dfrac{2\sqrt{21}}{21}$ **15.** $\sin t = \dfrac{\sqrt{2}}{2}; \cos t = -\dfrac{\sqrt{2}}{2}; \tan t = -1;$

$\csc t = \sqrt{2}; \sec t = -\sqrt{2}; \cot t = -1$ **17.** $\sin t = -\dfrac{1}{3}; \cos t = \dfrac{2\sqrt{2}}{3}; \tan t = -\dfrac{\sqrt{2}}{4}; \csc t = -3; \sec t = \dfrac{3\sqrt{2}}{4}; \cot t = -2\sqrt{2}$

19. -1 **21.** 0   **23.** -1 **25.** 0   **27.** -1 **29.** $\dfrac{1}{2}(\sqrt{2}+1)$ **31.** 2   **33.** $\dfrac{1}{2}$ **35.** $\sqrt{6}$ **37.** 4   **39.** 0   **41.** 0   **43.** $2\sqrt{2}+\dfrac{4\sqrt{3}}{3}$ **45.** -1 **47.** 1

49. $\sin\dfrac{2\pi}{3} = \dfrac{\sqrt{3}}{2}; \cos\dfrac{2\pi}{3} = -\dfrac{1}{2}; \tan\dfrac{2\pi}{3} = -\sqrt{3}; \csc\dfrac{2\pi}{3} = \dfrac{2\sqrt{3}}{3}; \sec\dfrac{2\pi}{3} = -2; \cot\dfrac{2\pi}{3} = -\dfrac{\sqrt{3}}{3}$

51. $\sin 210° = -\dfrac{1}{2}; \cos 210° = -\dfrac{\sqrt{3}}{2}; \tan 210° = \dfrac{\sqrt{3}}{3}; \csc 210° = -2; \sec 210° = -\dfrac{2\sqrt{3}}{3}; \cot 210° = \sqrt{3}$

53. $\sin\dfrac{3\pi}{4} = \dfrac{\sqrt{2}}{2}; \cos\dfrac{3\pi}{4} = -\dfrac{\sqrt{2}}{2}; \tan\dfrac{3\pi}{4} = -1; \csc\dfrac{3\pi}{4} = \sqrt{2}; \sec\dfrac{3\pi}{4} = -\sqrt{2}; \cot\dfrac{3\pi}{4} = -1$

55. $\sin\dfrac{8\pi}{3} = \dfrac{\sqrt{3}}{2}; \cos\dfrac{8\pi}{3} = -\dfrac{1}{2}; \tan\dfrac{8\pi}{3} = -\sqrt{3}; \csc\dfrac{8\pi}{3} = \dfrac{2\sqrt{3}}{3}; \sec\dfrac{8\pi}{3} = -2; \cot\dfrac{8\pi}{3} = -\dfrac{\sqrt{3}}{3}$

57. $\sin 405° = \dfrac{\sqrt{2}}{2}; \cos 405° = \dfrac{\sqrt{2}}{2}; \tan 405° = 1; \csc 405° = \sqrt{2}; \sec 405° = \sqrt{2}; \cot 405° = 1$

59. $\sin\left(-\dfrac{\pi}{6}\right) = -\dfrac{1}{2}; \cos\left(-\dfrac{\pi}{6}\right) = \dfrac{\sqrt{3}}{2}; \tan\left(-\dfrac{\pi}{6}\right) = -\dfrac{\sqrt{3}}{3}; \csc\left(-\dfrac{\pi}{6}\right) = -2; \sec\left(-\dfrac{\pi}{6}\right) = \dfrac{2\sqrt{3}}{3}; \cot\left(-\dfrac{\pi}{6}\right) = -\sqrt{3}$

61. $\sin(-45°) = -\dfrac{\sqrt{2}}{2}; \cos(-45°) = \dfrac{\sqrt{2}}{2}; \tan(-45°) = -1; \csc(-45°) = -\sqrt{2}; \sec(-45°) = \sqrt{2}; \cot(-45°) = -1$

63. $\sin\dfrac{5\pi}{2} = 1; \cos\dfrac{5\pi}{2} = 0; \tan\dfrac{5\pi}{2}$ is undefined; $\csc\dfrac{5\pi}{2} = 1; \sec\dfrac{5\pi}{2}$ is undefined; $\cot\dfrac{5\pi}{2} = 0$ **65.** $\sin 720° = 0; \cos 720° = 1; \tan 720° = 0;$

$\csc 720°$ is undefined; $\sec 720° = 1; \cot 720°$ is undefined   **67.** 0.47   **69.** 0.38   **71.** 1.33   **73.** 0.31

75. 3.73   **77.** 1.04   **79.** 0.84   **81.** 0.02   **83.** $\sin\theta = \dfrac{4}{5}; \cos\theta = -\dfrac{3}{5}; \tan\theta = -\dfrac{4}{3}; \csc\theta = \dfrac{5}{4}; \sec\theta = -\dfrac{5}{3}; \cot\theta = -\dfrac{3}{4}$

85. $\sin\theta = -\dfrac{3\sqrt{13}}{13}; \cos\theta = \dfrac{2\sqrt{13}}{13}; \tan\theta = -\dfrac{3}{2}; \csc\theta = -\dfrac{\sqrt{13}}{3}; \sec\theta = \dfrac{\sqrt{13}}{2}; \cot\theta = -\dfrac{2}{3}$

87. $\sin\theta = -\dfrac{\sqrt{2}}{2}; \cos\theta = -\dfrac{\sqrt{2}}{2}; \tan\theta = 1; \csc\theta = -\sqrt{2}; \sec\theta = -\sqrt{2}; \cot\theta = 1$

89. $\sin\theta = -\dfrac{2\sqrt{13}}{13}; \cos\theta = -\dfrac{3\sqrt{13}}{13}; \tan\theta = \dfrac{2}{3}; \csc\theta = -\dfrac{\sqrt{13}}{2}; \sec\theta = -\dfrac{\sqrt{13}}{3}; \cot\theta = \dfrac{3}{2}$

91. $\sin\theta = -\dfrac{3}{5}; \cos\theta = \dfrac{4}{5}; \tan\theta = -\dfrac{3}{4}; \csc\theta = -\dfrac{5}{3}; \sec\theta = \dfrac{5}{4}; \cot\theta = -\dfrac{4}{3}$

93. 0   **95.** -0.1 **97.** 3   **99.** 5   **101.** $\dfrac{\sqrt{3}}{2}$ **103.** $\dfrac{1}{2}$ **105.** $\dfrac{3}{4}$ **107.** $\dfrac{\sqrt{3}}{2}$ **109.** $\sqrt{3}$ **111.** $-\dfrac{\sqrt{3}}{2}$

113.

θ	0.5	0.4	0.2	0.1	0.01	0.001	0.0001	0.00001
$\sin\theta$	0.4794	0.3894	0.1987	0.0998	0.0100	0.0010	0.0001	0.00001
$\dfrac{\sin\theta}{\theta}$	0.9589	0.9735	0.9933	0.9983	1.0000	1.0000	1.0000	1.0000

$\dfrac{\sin\theta}{\theta}$ approaches 1 as θ approaches 0.

115. $R \approx 310.56$ ft; $H \approx 77.64$ ft   **117.** $R \approx 19{,}542$ m; $H \approx 2278$ m   **119. (a)** 1.20 s   **(b)** 1.12 s   **(c)** 1.20 s

121. (a) 1.9 hr; 0.57 hr   **(b)** 1.69 hr; 0.75 hr   **(c)** 1.63 hr; 0.86 hr   **(d)** 1.67 hr; $\tan 90°$ is undefined

123. (a) 16.56 ft   **(b)**   **(c)** 67.5°

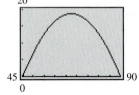

125. (a) values estimated to the nearest tenth: $\sin 1 \approx 0.8$; $\cos 1 \approx 0.5$; $\tan 1 \approx 1.6$; $\csc 1 \approx 1.3$; $\sec 1 \approx 2.0$; $\cot 1 \approx 0.6$; actual values to the nearest tenth: $\sin 1 \approx 0.8$; $\cos 1 \approx 0.5$; $\tan 1 \approx 1.6$; $\csc 1 \approx 1.2$; $\sec 1 \approx 1.9$; $\cot 1 \approx 0.6$ **(b)** values estimated to the nearest tenth: $\sin 5.1 \approx -0.9$; $\cos 5.1 \approx 0.4$; $\tan 5.1 \approx -2.3$; $\csc 5.1 \approx -1.1$; $\sec 5.1 \approx 2.5$; $\cot 5.1 \approx -0.4$; actual values to the nearest tenth: $\sin 5.1 \approx -0.9$; $\cos 5.1 \approx 0.4$; $\tan 5.1 \approx -2.4$; $\csc 5.1 \approx -1.1$; $\sec 5.1 \approx 2.6$; $\cot 5.1 \approx -0.4$ **(c)** values estimated to the nearest tenth: $\sin 2.4 \approx 0.7$; $\cos 2.4 \approx -0.7$; $\tan 2.4 \approx -1.0$; $\csc 2.4 \approx 1.4$; $\sec 2.4 \approx -1.4$; $\cot 2.4 \approx -1.0$; actual values to the nearest tenth: $\sin 2.4 \approx 0.7$; $\cos 2.4 \approx -0.7$; $\tan 2.4 \approx -0.9$; $\csc 2.4 \approx 1.5$; $\sec 2.4 \approx -1.4$; $\cot 2.4 \approx -1.1$

127. (a) values estimated to the nearest tenth: $\sin 1.5 \approx 1.0$; $\cos 1.5 \approx 0.1$; $\tan 1.5 \approx 10.0$; $\csc 1.5 \approx 1.0$; $\sec 1.5 \approx 10.0$; $\cot 1.5 \approx 0.1$; actual values to the nearest tenth: $\sin 1.5 \approx 1.0$; $\cos 1.5 \approx 0.1$; $\tan 1.5 \approx 14.1$; $\csc 1.5 \approx 1.0$; $\sec 1.5 \approx 14.1$; $\cot 1.5 \approx 0.1$ **(b)** values estimated to the nearest tenth: $\sin 4.3 \approx -0.9$; $\cos 4.3 \approx -0.4$; $\tan 4.3 \approx 2.3$; $\csc 4.3 \approx -1.1$; $\sec 4.3 \approx -2.5$; $\cot 4.3 \approx 0.4$; actual values to the nearest tenth: $\sin 4.3 \approx -0.9$; $\cos 4.3 \approx -0.4$; $\tan 4.3 \approx 2.3$; $\csc 4.3 \approx -1.1$; $\sec 4.3 \approx -2.5$; $\cot 4.3 \approx 0.4$ **(c)** values estimated to the nearest tenth: $\sin 5.3 \approx -0.8$; $\cos 5.3 \approx 0.6$; $\tan 5.3 \approx -1.3$; $\csc 5.3 \approx -1.3$; $\sec 5.3 \approx 1.7$; $\cot 5.3 = -0.8$; actual values to the nearest tenth: $\sin 5.3 \approx -0.8$; $\cos 5.3 \approx 0.6$; $\tan 5.3 \approx -1.5$; $\csc 5.3 \approx -1.2$; $\sec 5.3 \approx 1.8$; $\cot 5.3 \approx -0.7$

5.3 Assess Your Understanding (page 399)

5. 2π; π **6.** all real numbers except odd multiples of $\dfrac{\pi}{2}$ **7.** $[-1, 1]$ **8.** T   **9.** F   **10.** F   **11.** $\dfrac{\sqrt{2}}{2}$ **13.** 1   **15.** 1   **17.** $\sqrt{3}$

19. $\dfrac{\sqrt{2}}{2}$ **21.** 0   **23.** $\sqrt{2}$ **25.** $\dfrac{\sqrt{3}}{3}$ **27.** II   **29.** IV   **31.** IV   **33.** II   **35.** $\tan \theta = -\dfrac{3}{4}$; $\cot \theta = -\dfrac{4}{3}$; $\sec \theta = \dfrac{5}{4}$; $\csc \theta = -\dfrac{5}{3}$

37. $\tan \theta = 2$; $\cot \theta = \dfrac{1}{2}$; $\sec \theta = \sqrt{5}$; $\csc \theta = \dfrac{\sqrt{5}}{2}$ **39.** $\tan \theta = \dfrac{\sqrt{3}}{3}$; $\cot \theta = \sqrt{3}$; $\sec \theta = \dfrac{2\sqrt{3}}{3}$; $\csc \theta = 2$

41. $\tan \theta = -\dfrac{\sqrt{2}}{4}$; $\cot \theta = -2\sqrt{2}$; $\sec \theta = \dfrac{3\sqrt{2}}{4}$; $\csc \theta = -3$ **43.** $\cos \theta = -\dfrac{5}{13}$; $\tan \theta = -\dfrac{12}{5}$; $\csc \theta = \dfrac{13}{12}$; $\sec \theta = -\dfrac{13}{5}$; $\cot \theta = -\dfrac{5}{12}$

45. $\sin \theta = -\dfrac{3}{5}$; $\tan \theta = \dfrac{3}{4}$; $\csc \theta = -\dfrac{5}{3}$; $\sec \theta = -\dfrac{5}{4}$; $\cot \theta = \dfrac{4}{3}$ **47.** $\cos \theta = -\dfrac{12}{13}$; $\tan \theta = -\dfrac{5}{12}$; $\csc \theta = \dfrac{13}{5}$; $\sec \theta = -\dfrac{13}{12}$; $\cot \theta = -\dfrac{12}{5}$

49. $\sin \theta = \dfrac{2\sqrt{2}}{3}$; $\tan \theta = -2\sqrt{2}$; $\csc \theta = \dfrac{3\sqrt{2}}{4}$; $\sec \theta = -3$; $\cot \theta = -\dfrac{\sqrt{2}}{4}$ **51.** $\cos \theta = -\dfrac{\sqrt{5}}{3}$; $\tan \theta = -\dfrac{2\sqrt{5}}{5}$; $\csc \theta = \dfrac{3}{2}$; $\sec \theta = -\dfrac{3\sqrt{5}}{5}$; $\cot \theta = -\dfrac{\sqrt{5}}{2}$ **53.** $\sin \theta = -\dfrac{\sqrt{3}}{2}$; $\cos \theta = \dfrac{1}{2}$; $\tan \theta = -\sqrt{3}$; $\csc \theta = -\dfrac{2\sqrt{3}}{3}$; $\cot \theta = -\dfrac{\sqrt{3}}{3}$ **55.** $\sin \theta = -\dfrac{3}{5}$; $\cos \theta = -\dfrac{4}{5}$; $\csc \theta = -\dfrac{5}{3}$; $\sec \theta = -\dfrac{5}{4}$; $\cot \theta = \dfrac{4}{3}$ **57.** $\sin \theta = \dfrac{\sqrt{10}}{10}$; $\cos \theta = -\dfrac{3\sqrt{10}}{10}$; $\csc \theta = \sqrt{10}$; $\sec \theta = -\dfrac{\sqrt{10}}{3}$; $\cot \theta = -3$ **59.** $-\dfrac{\sqrt{3}}{2}$

61. $-\dfrac{\sqrt{3}}{3}$ **63.** 2   **65.** -1 **67.** -1 **69.** $\dfrac{\sqrt{2}}{2}$ **71.** 0   **73.** $-\sqrt{2}$ **75.** $\dfrac{2\sqrt{3}}{3}$ **77.** 1   **79.** 1   **81.** 0   **83.** 1   **85.** -1 **87.** 0

89. 0.9   **91.** 9   **93.** 0   **95.** All real numbers   **97.** Odd multiples of $\dfrac{\pi}{2}$ **99.** Odd multiples of $\dfrac{\pi}{2}$ **101.** $-1 \le y \le 1$

103. All real numbers   **105.** $|y| \ge 1$ **107.** Odd; yes; origin   **109.** Odd; yes; origin   **111.** Even; yes; y-axis   **113. (a)** $-\dfrac{1}{3}$ **(b)** 1

115. (a) -2 **(b)** 6   **117. (a)** -4 **(b)** -12 **119.** About 15.81 min

121. Let a be a real number and $P = (x, y)$ be the point on the unit circle that corresponds to t. Consider the equation $\tan t = \dfrac{y}{x} = a$. Then $y = ax$. But $x^2 + y^2 = 1$ so $x^2 + a^2x^2 = 1$. Thus, $x = \pm \dfrac{1}{\sqrt{1 + a^2}}$ and $y = \pm \dfrac{a}{\sqrt{1 + a^2}}$; that is, for any real number a, there is a point $P = (x, y)$ on the unit circle for which $\tan t = a$. In other words, the range of the tangent function is the set of all real numbers.

123. Suppose that there is a number p, $0 < p < 2\pi$, for which $\sin(\theta + p) = \sin \theta$ for all θ. If $\theta = 0$, then $\sin(0 + p) = \sin p = \sin 0 = 0$, so $p = \pi$. If $\theta = \dfrac{\pi}{2}$, then $\sin\left(\dfrac{\pi}{2} + p\right) = \sin\left(\dfrac{\pi}{2}\right)$. But $p = \pi$. Thus, $\sin\left(\dfrac{3\pi}{2}\right) = -1 = \sin\left(\dfrac{\pi}{2}\right) = 1$. This is impossible. Therefore, the smallest positive number p for which $\sin(\theta + p) = \sin \theta$ for all θ is 2π.

125. $\sec \theta = \dfrac{1}{\cos \theta}$; since $\cos \theta$ has period 2π, so does $\sec \theta$.

127. If $P = (a, b)$ is the point on the unit circle corresponding to θ, then $Q = (-a, -b)$ is the point on the unit circle corresponding to $\theta + \pi$. Thus, $\tan(\theta + \pi) = \dfrac{-b}{-a} = \dfrac{b}{a} = \tan \theta$. Suppose that there exists a number p, $0 < p < \pi$, for which $\tan(\theta + p) = \tan \theta$ for all θ. Then, if $\theta = 0$, then $\tan p = \tan 0 = 0$. But this means that p is a multiple of π. Since no multiple of π exists in the interval $(0, \pi)$, this is a contradiction. Therefore, the period of $f(\theta) = \tan \theta$ is π.

129. Let $P = (a, b)$ be the point on the unit circle corresponding to θ. Then $\csc \theta = \dfrac{1}{b} = \dfrac{1}{\sin \theta}$; $\sec \theta = \dfrac{1}{a} = \dfrac{1}{\cos \theta}$; $\cot \theta = \dfrac{a}{b} = \dfrac{1}{b/a} = \dfrac{1}{\tan \theta}$.

131. $(\sin \theta \cos \phi)^2 + (\sin \theta \sin \phi)^2 + \cos^2 \theta = \sin^2 \theta \cos^2 \phi + \sin^2 \theta \sin^2 \phi + \cos^2 \theta$
$$= \sin^2 \theta(\cos^2 \phi + \sin^2 \phi) + \cos^2 \theta = \sin^2 \theta + \cos^2 \theta = 1$$

5.4 Assess Your Understanding (page 414)

3. $1; \dfrac{\pi}{2} + 2\pi k, k$ any integer   **4.** $3; \pi$ **5.** $3; \dfrac{\pi}{3}$ **6.** T   **7.** F   **8.** T   **9.** 0   **11.** $-\dfrac{\pi}{2} \le x \le \dfrac{\pi}{2}$ **13.** 1

15. $0, \pi, 2\pi$ **17.** $\sin x = 1$ for $x = -\dfrac{3\pi}{2}, \dfrac{\pi}{2}; \sin x = -1$ for $x = -\dfrac{\pi}{2}, \dfrac{3\pi}{2}$ **19.** B, C, F

21. 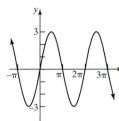   **23.** 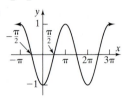   **25.** 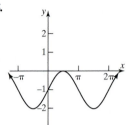   **27.**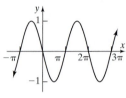

29. 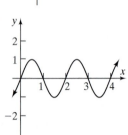   **31.** 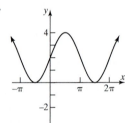   **33.** 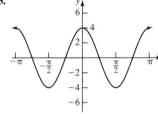   **35.**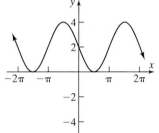

37. Amplitude $= 2$; Period $= 2\pi$ **39.** Amplitude $= 4$; Period $= \pi$ **41.** Amplitude $= 6$; Period $= 2$

43. Amplitude $= \dfrac{1}{2}$; Period $= \dfrac{4\pi}{3}$ **45.** Amplitude $= \dfrac{5}{3}$; Period $= 3$ **47.** F   **49.** A   **51.** H   **53.** C   **55.** J   **57.** A   **59.** B

61. 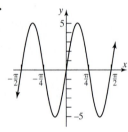   **63.** 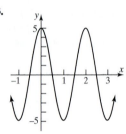   **65.** 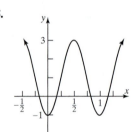   **67.**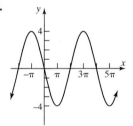

69. 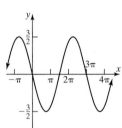   **71.** $y = \pm 3 \sin(2x)$ **73.** $y = \pm 3 \sin(\pi x)$

75. $y = 5 \cos\left(\dfrac{\pi}{4}x\right)$ **77.** $y = -3 \cos\left(\dfrac{1}{2}x\right)$

79. $y = \dfrac{3}{4} \sin(2\pi x)$ **81.** $y = -\sin\left(\dfrac{3}{2}x\right)$

83. $y = -\cos\left(\dfrac{4\pi}{3}x\right) + 1$ **85.** $y = 3 \sin\left(\dfrac{\pi}{2}x\right)$ **87.** $y = -4 \cos(3x)$

89. Period $= \dfrac{1}{30}$; Amplitude $= 220$

91. (a) Amplitude $= 220$; Period $= \dfrac{1}{60}$ **(c)** $I = 22 \sin(120\pi t)$

(b), (e)   **(d)** Amplitude $= 22$; Period $= \dfrac{1}{60}$

93. (a) $P = \dfrac{[V_0 \sin(2\pi ft)]^2}{R} = \dfrac{V_0^2}{R}\sin^2(2\pi ft)$

(b) Since the graph of P has amplitude $\dfrac{V_0^2}{2R}$ and period $\dfrac{1}{2f}$ and is of the form

$y = A\cos(\omega t) + B$, then $A = -\dfrac{V_0^2}{2R}$ and $B = \dfrac{V_0^2}{2R}$. Since $\dfrac{1}{2f} = \dfrac{2\pi}{\omega}$, then

$\omega = 4\pi f$. Therefore, $P = -\dfrac{V_0^2}{2R}\cos(4\pi ft) + \dfrac{V_0^2}{2R} = \dfrac{V_0^2}{2R}[1 - \cos(4\pi ft)]$.

95.

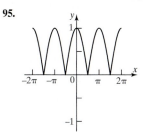

5.5 Assess Your Understanding *(page 423)*

3. origin; odd multiples of $\dfrac{\pi}{2}$ **4.** y-axis; odd multiples of $\dfrac{\pi}{2}$ **5.** $y = \cos x$ **6.** T **7.** 0 **9.** 1

11. $\sec x = 1$ for $x = -2\pi, 0, 2\pi$; $\sec x = -1$ for $x = -\pi, \pi$ **13.** $-\dfrac{3\pi}{2}, -\dfrac{\pi}{2}, \dfrac{\pi}{2}, \dfrac{3\pi}{2}$ **15.** $-\dfrac{3\pi}{2}, -\dfrac{\pi}{2}, \dfrac{\pi}{2}, \dfrac{3\pi}{2}$ **17.** D **19.** B

21.

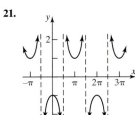

23.

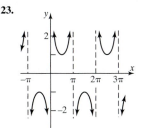

25.

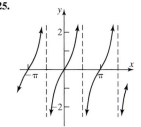

27.

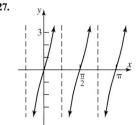

29.

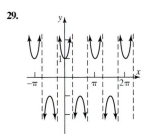

31.

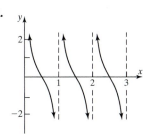

33.

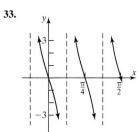

35.

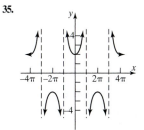

37.

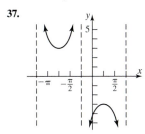

39.

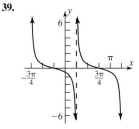

41.

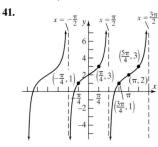

43.

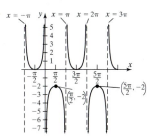

45. (a) $L(\theta) = \dfrac{3}{\cos\theta} + \dfrac{4}{\sin\theta} = 3\sec\theta + 4\csc\theta$

(b)

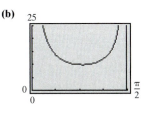

(c) 0.83
(d) 9.86 ft

47.

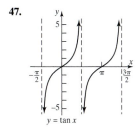
$y = \tan x$

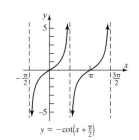
$y = -\cot\left(x + \dfrac{\pi}{2}\right)$

5.6 Assess Your Understanding *(page 434)*

1. phase shift **2.** F

3. Amplitude = 4
Period = π
Phase shift = $\dfrac{\pi}{2}$

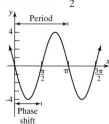

5. Amplitude = 2
Period = $\dfrac{2\pi}{3}$
Phase shift = $-\dfrac{\pi}{6}$

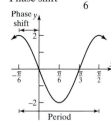

7. Amplitude = 3
Period = π
Phase shift = $-\dfrac{\pi}{4}$

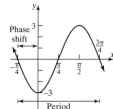

9. Amplitude = 4
Period = 2
Phase shift = $-\dfrac{2}{\pi}$

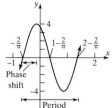

11. Amplitude = 3
Period = 2
Phase shift = $\dfrac{2}{\pi}$

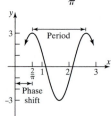

13. Amplitude = 3
Period = π
Phase shift = $\dfrac{\pi}{4}$

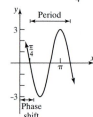

15. $y = 2\sin\left[2\left(x - \dfrac{1}{2}\right)\right]$ or $y = 2\sin(2x - 1)$

17. $y = 3\sin\left[\dfrac{2}{3}\left(x + \dfrac{1}{3}\right)\right]$ or $y = 3\sin\left(\dfrac{2}{3}x + \dfrac{2}{9}\right)$

19. Period = $\dfrac{1}{15}$; Amplitude = 120; Phase shift = $\dfrac{1}{90}$

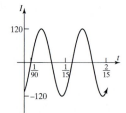

21. (a)

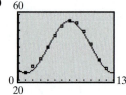

(b) $y = 15.9\sin\left[\dfrac{\pi}{6}(x - 4)\right] + 40.1$ or $y = 15.9\sin\left(\dfrac{\pi}{6}x - \dfrac{2\pi}{3}\right) + 40.1$

(c)

(d) $y = 15.62\sin(0.517x - 2.096) + 40.377$

(e)

23. (a)

(b) $y = 24.95\sin\left[\dfrac{\pi}{6}(x - 4)\right] + 50.45$ or $y = 24.95\sin\left(\dfrac{\pi}{6}x - \dfrac{2\pi}{3}\right) + 50.45$

(c)

(d) $y = 25.693\sin(0.476x - 1.814) + 49.854$

(e)

25. (a) 4:08 PM

(b) $y = 4.4\sin\left[\dfrac{4\pi}{25}(x - 0.5083)\right] + 3.8$ or

$y = 4.4\sin\left[\dfrac{4\pi}{25}x - 0.2555\right] + 3.8$

(c)

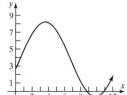

(d) 8.2 ft

27. (a) $y = 1.0835 \sin\left[\dfrac{2\pi}{365}(x - 80.75)\right] + 11.6665$ or

$y = 1.0835 \sin\left(\dfrac{2\pi}{365}x - 1.3900\right) + 11.6665$

(b) 11.86 hr   **(c)**
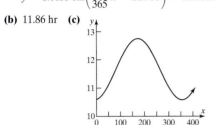

29. (a) $y = 5.3915 \sin\left[\dfrac{2\pi}{365}(x - 80.75)\right] + 10.8415$ or

$y = 5.3915 \sin\left(\dfrac{2\pi}{365}x - 1.3900\right) + 10.8415$

(b) 11.76 hr   **(c)**
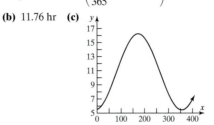

Review Exercises (page 440)

1. $\dfrac{3\pi}{4}$ **3.** $\dfrac{\pi}{10}$ **5.** $135°$ **7.** $-450°$ **9.** $\dfrac{1}{2}$ **11.** $\dfrac{3\sqrt{2}}{2} - \dfrac{4\sqrt{3}}{3}$ **13.** $-3\sqrt{2} - 2\sqrt{3}$ **15.** 3 **17.** 0 **19.** 0 **21.** 1 **23.** 1 **25.** 1 **27.** -1 **29.** 1

31. $\cos\theta = \dfrac{3}{5}; \tan\theta = \dfrac{4}{3}; \csc\theta = \dfrac{5}{4}; \sec\theta = \dfrac{5}{3}; \cot\theta = \dfrac{3}{4}$ **33.** $\sin\theta = -\dfrac{12}{13}; \cos\theta = -\dfrac{5}{13}; \csc\theta = -\dfrac{13}{12}; \sec\theta = -\dfrac{13}{5}; \cot\theta = \dfrac{5}{12}$

35. $\sin\theta = \dfrac{3}{5}; \cos\theta = -\dfrac{4}{5}; \tan\theta = -\dfrac{3}{4}; \csc\theta = \dfrac{5}{3}; \cot\theta = -\dfrac{4}{3}$ **37.** $\cos\theta = -\dfrac{5}{13}; \tan\theta = -\dfrac{12}{5}; \csc\theta = \dfrac{13}{12}; \sec\theta = -\dfrac{13}{5}; \cot\theta = -\dfrac{5}{12}$

39. $\cos\theta = \dfrac{12}{13}; \tan\theta = -\dfrac{5}{12}; \csc\theta = -\dfrac{13}{5}; \sec\theta = \dfrac{13}{12}; \cot\theta = -\dfrac{12}{5}$ **41.** $\sin\theta = -\dfrac{\sqrt{10}}{10}; \cos\theta = -\dfrac{3\sqrt{10}}{10}; \csc\theta = -\sqrt{10}; \sec\theta = -\dfrac{\sqrt{10}}{3};$

$\cot\theta = 3$ **43.** $\sin\theta = -\dfrac{2\sqrt{2}}{3}; \cos\theta = \dfrac{1}{3}; \tan\theta = -2\sqrt{2}; \csc\theta = -\dfrac{3\sqrt{2}}{4}; \cot\theta = -\dfrac{\sqrt{2}}{4}$ **45.** $\sin\theta = \dfrac{\sqrt{5}}{5}; \cos\theta = -\dfrac{2\sqrt{5}}{5}; \tan\theta = -\dfrac{1}{2};$

$\csc\theta = \sqrt{5}; \sec\theta = -\dfrac{\sqrt{5}}{2}$

47.
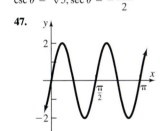

49.
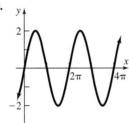

51.
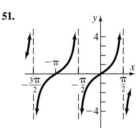

53.
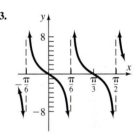

55.
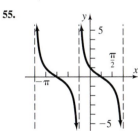

57.
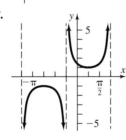

59. Amplitude = 4; period = 2π
61. Amplitude = 8; period = 4

63. Amplitude = 4

Period = $\dfrac{2\pi}{3}$

Phase shift = 0

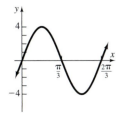

65. Amplitude = 2

Period = π

Phase Shift = $\dfrac{\pi}{2}$

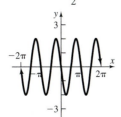

67. Amplitude = $\dfrac{1}{2}$

Period = $\dfrac{4\pi}{3}$

Phase shift = $\dfrac{2\pi}{3}$

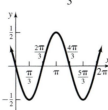

69. Amplitude = $\dfrac{2}{3}$

Period = 2

Phase shift = $\dfrac{6}{\pi}$

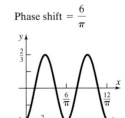

71. $y = 5 \cos \dfrac{x}{4}$ **73.** $y = -6 \cos\left(\dfrac{\pi}{4}x\right)$ **75.** 0.38   **77.** Sine, Cosine, Cosecant and Secant: Negative; Tangent and Cotangent: Positive

79. $\sin\theta = \dfrac{2\sqrt{2}}{3}$; $\cos\theta = -\dfrac{1}{3}$; $\tan\theta = -2\sqrt{2}$; $\csc\theta = \dfrac{3\sqrt{2}}{4}$; $\sec\theta = -3$; $\cot\theta = -\dfrac{\sqrt{2}}{4}$

81. Domain: $\left\{x \,\middle|\, x \neq \text{odd multiple of } \dfrac{\pi}{2}\right\}$; range: $\{y \,|\, |y| \geq 1\}$; period $= 2(\pi)$

83. $\dfrac{\pi}{3} \approx 1.05$ ft; $\dfrac{\pi}{3} \approx 1.05$ ft^2 **85.** Approximately 114.59 revolutions/hr   **87.** 0.1 revolution/s $= \dfrac{\pi}{5}$ radian/s

89. (a) 120 (b) $\dfrac{1}{60}$ (c)

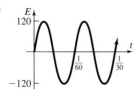

91. (a)

(b) $y = 19.5 \sin\left(\dfrac{\pi}{6}x - \dfrac{2\pi}{3}\right) + 70.5$

(c)

(d) $y = 19.518 \sin(0.541x - 2.283) + 71.014$

(e)

93. (a) $y = 1.85 \sin\left(\dfrac{2\pi}{365}x - \dfrac{357}{146}\pi\right) + 11.517$

(b)

(c) 11.83 hr

Chapter Test *(page 443)*

1. $\dfrac{13\pi}{9}$ **2.** $-\dfrac{20\pi}{9}$ **3.** $\dfrac{13\pi}{180}$ **4.** $-22.5°$ **5.** $810°$ **6.** $135°$ **7.** $\dfrac{1}{2}$ **8.** 0 **9.** $-\dfrac{1}{2}$ **10.** $-\dfrac{\sqrt{3}}{3}$ **11.** 2 **12.** $\dfrac{3(1-\sqrt{2})}{2}$ **13.** ≈ 0.292 **14.** ≈ 0.309

15. ≈ -1.524 **16.** ≈ 2.747 **17.**

	$\sin\theta$	$\cos\theta$	$\tan\theta$	$\sec\theta$	$\csc\theta$	$\cot\theta$
θ in QI	+	+	+	+	+	+
θ in QII	+	−	−	−	+	−
θ in QIII	−	−	+	−	−	+
θ in QIV	−	+	−	+	−	−

18. $-\dfrac{3}{5}$

19. $\cos\theta = -\dfrac{2\sqrt{6}}{7}$

$\tan\theta = -\dfrac{5}{2\sqrt{6}} = -\dfrac{5\sqrt{6}}{12}$

$\csc\theta = \dfrac{7}{5}$

$\sec\theta = -\dfrac{7}{2\sqrt{6}} = -\dfrac{7\sqrt{6}}{12}$

$\cot\theta = -\dfrac{2\sqrt{6}}{5}$

20. $\sin\theta = -\dfrac{\sqrt{5}}{3}$

$\tan\theta = -\dfrac{\sqrt{5}}{2}$

$\csc\theta = -\dfrac{3}{\sqrt{5}} = -\dfrac{3\sqrt{5}}{5}$

$\sec\theta = \dfrac{3}{2}$

$\cot\theta = -\dfrac{2}{\sqrt{5}} = -\dfrac{2\sqrt{5}}{5}$

21. $\sin\theta = \dfrac{12}{13}$

$\cos\theta = -\dfrac{5}{13}$

$\csc\theta = \dfrac{13}{12}$

$\sec\theta = -\dfrac{13}{5}$

$\cot\theta = -\dfrac{5}{12}$

22. $\dfrac{7\sqrt{53}}{53}$

23. $-\dfrac{5\sqrt{146}}{146}$

24. $-\dfrac{1}{2}$

25.

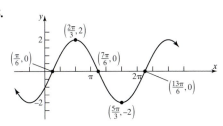

26.

$y = \tan\left(-x + \frac{\pi}{4}\right) + 2$

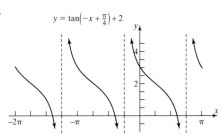

27. $y = -3\sin\left(3x + \frac{3\pi}{4}\right)$

28. 121.19 ft^2 **29.** 143.5 rpm   **30. (a)** 2633 ft   **(b)** 818 ft

Cumulative Review *(page 446)*

1. $\left\{-1, \frac{1}{2}\right\}$ **2.** $y - 5 = -3(x + 2)$ or $y = -3x - 1$ **3.** $x^2 + (y + 2)^2 = 16$

4. A line. Slope $\frac{2}{3}$; intercepts $(6, 0)$ and $(0, -4)$ **5.** A circle. Center $(1, -2)$; Radius 3

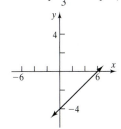

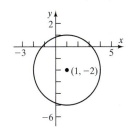

6.

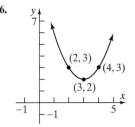

7. (a)

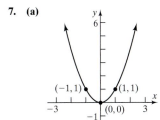

(b)

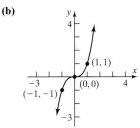

(c)

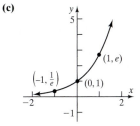

(d)

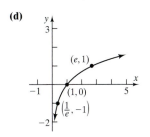

(e)

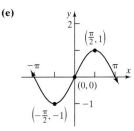

(f)

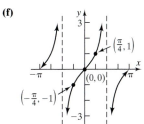

8. $f^{-1}(x) = \frac{1}{3}(x + 2)$ **9.** -2 **10.**

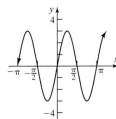

11. $3 - \frac{3\sqrt{3}}{2}$ **12.** $y = 2(3^x)$ **13.** $y = 3\cos\left(\frac{\pi}{6}x\right)$

14. (a) $f(x) = -3x - 3; m = -3; (-1, 0), (0, -3)$ **(b)** $f(x) = (x - 1)^2 - 6; (0, -5), (-\sqrt{6} + 1, 0); (\sqrt{6} + 1, 0)$

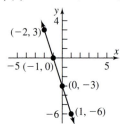 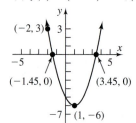

(c) We have that $y = 3$ when $x = -2$ and $y = -6$ when $x = 1$. Both points satisfy $y = ae^x$. Therefore, for $(-2, 3)$ we have $3 = ae^{-2}$ which implies that $a = 3e^2$. But for $(1, -6)$ we have $-6 = ae^1$, which implies that $a = -6e^{-1}$. Therefore, there is no exponential function $y = ae^x$ that contains $(-2, 3)$ and $(1, -6)$.

15. (a) $f(x) = \dfrac{1}{6}(x + 2)(x - 3)(x - 5)$ **(b)** $R(x) = -\dfrac{(x + 2)(x - 3)(x - 5)}{3(x - 2)}$

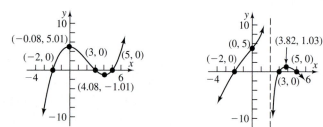 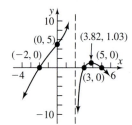

C H A P T E R 6   Analytic Trigonometry

6.1 Assess Your Understanding *(page 457)*

7. $x = \sin y$ **8.** 0   **9.** $\dfrac{\pi}{5}$ **10.** F   **11.** T   **12.** T   **13.** 0   **15.** $-\dfrac{\pi}{2}$ **17.** 0   **19.** $\dfrac{\pi}{4}$ **21.** $\dfrac{\pi}{3}$ **23.** $\dfrac{5\pi}{6}$ **25.** 0.10   **27.** 1.37   **29.** 0.51

31. -0.38 **33.** -0.12 **35.** 1.08   **37.** 0.54   **39.** $\dfrac{4\pi}{5}$ **41.** -3.5 **43.** $-\dfrac{3\pi}{7}$ **45.** Yes; $-\dfrac{\pi}{6}$ lies in the interval $\left[-\dfrac{\pi}{2}, \dfrac{\pi}{2}\right]$.

47. No; 2 is not in the domain of $\sin^{-1} x$.   **49.** No; $-\dfrac{\pi}{6}$ does not lie in the interval $[0, \pi]$.   **51.** Yes; $-\dfrac{1}{2}$ is in the domain of $\cos^{-1} x$.

53. Yes; $-\dfrac{\pi}{3}$ lies in the interval $\left(-\dfrac{\pi}{2}, \dfrac{\pi}{2}\right)$.   **55.** Yes; 2 is in the domain of $\tan^{-1} x$.   **57. (a)** 13.92 hr or 13 hr, 55 min   **(b)** 12 hr

(c) 13.85 hr or 13 hr, 51 min   **59. (a)** 13.3 hr or 13 hr, 18 min   **(b)** 12 hr   **(c)** 13.26 hr or 13 hr, 15 min   **61. (a)** 12 hr   **(b)** 12 hr

(c) 12 hr   **(d)** It's 12 hr.   **63.** 3.35 min   **65. (a)** $\dfrac{\pi}{3}$ square units   **(b)** $\dfrac{5\pi}{12}$ square units

6.2 Assess Your Understanding *(page 464)*

4. $x = \sec y; \geq 1; 0; \pi$ **5.** $\dfrac{\sqrt{2}}{2}$ **6.** F   **7.** T   **8.** T   **9.** $\dfrac{\sqrt{2}}{2}$ **11.** $-\dfrac{\sqrt{3}}{3}$ **13.** 2   **15.** $\sqrt{2}$ **17.** $-\dfrac{\sqrt{2}}{2}$ **19.** $\dfrac{2\sqrt{3}}{3}$ **21.** $\dfrac{3\pi}{4}$ **23.** $\dfrac{\pi}{6}$ **25.** $\dfrac{\sqrt{2}}{4}$ **27.** $\dfrac{\sqrt{5}}{2}$

29. $-\dfrac{\sqrt{14}}{2}$ **31.** $-\dfrac{3\sqrt{10}}{10}$ **33.** $\sqrt{5}$ **35.** $-\dfrac{\pi}{4}$ **37.** $\dfrac{\pi}{6}$ **39.** $-\dfrac{\pi}{2}$ **41.** $\dfrac{\pi}{6}$ **43.** $\dfrac{2\pi}{3}$ **45.** 1.32   **47.** 0.46   **49.** -0.34 **51.** 2.72   **53.** -0.73 **55.** 2.55

57. 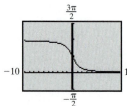   **59.**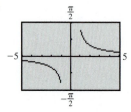

6.3 Assess Your Understanding (page 471)

3. identity; conditional   **4.** -1 **5.** 0 **6.** T   **7.** F   **8.** T   **9.** $\dfrac{1}{\cos\theta}$ **11.** $\dfrac{1+\sin\theta}{\cos\theta}$ **13.** $\dfrac{1}{\sin\theta\cos\theta}$ **15.** 2 **17.** $\dfrac{3\sin\theta+1}{\sin\theta+1}$

19. $\csc\theta\cdot\cos\theta = \dfrac{1}{\sin\theta}\cdot\cos\theta = \dfrac{\cos\theta}{\sin\theta} = \cot\theta$ **21.** $1+\tan^2(-\theta) = 1 + (-\tan\theta)^2 = 1 + \tan^2\theta = \sec^2\theta$

23. $\cos\theta(\tan\theta + \cot\theta) = \cos\theta\left(\dfrac{\sin\theta}{\cos\theta} + \dfrac{\cos\theta}{\sin\theta}\right) = \cos\theta\left(\dfrac{\sin^2\theta + \cos^2\theta}{\cos\theta\sin\theta}\right) = \cos\theta\left(\dfrac{1}{\cos\theta\sin\theta}\right) = \dfrac{1}{\sin\theta} = \csc\theta$

25. $\tan\theta\cot\theta - \cos^2\theta = \dfrac{\sin\theta}{\cos\theta}\cdot\dfrac{\cos\theta}{\sin\theta} - \cos^2\theta = 1 - \cos^2\theta = \sin^2\theta$ **27.** $(\sec\theta - 1)(\sec\theta + 1) = \sec^2\theta - 1 = \tan^2\theta$

29. $(\sec\theta + \tan\theta)(\sec\theta - \tan\theta) = \sec^2\theta - \tan^2\theta = 1$

31. $\cos^2\theta(1 + \tan^2\theta) = \cos^2\theta + \cos^2\theta\tan^2\theta = \cos^2\theta + \cos^2\theta\cdot\dfrac{\sin^2\theta}{\cos^2\theta} = \cos^2\theta + \sin^2\theta = 1$

33. $(\sin\theta + \cos\theta)^2 + (\sin\theta - \cos\theta)^2 = \sin^2\theta + 2\sin\theta\cos\theta + \cos^2\theta + \sin^2\theta - 2\sin\theta\cos\theta + \cos^2\theta = \sin^2\theta + \cos^2\theta + \sin^2\theta + \cos^2\theta$
$= 1 + 1 = 2$ **35.** $\sec^4\theta - \sec^2\theta = \sec^2\theta(\sec^2\theta - 1) = (1 + \tan^2\theta)\tan^2\theta = \tan^4\theta + \tan^2\theta$

37. $\sec\theta - \tan\theta = \dfrac{1}{\cos\theta} - \dfrac{\sin\theta}{\cos\theta} = \dfrac{1-\sin\theta}{\cos\theta}\cdot\dfrac{1+\sin\theta}{1+\sin\theta} = \dfrac{1-\sin^2\theta}{\cos\theta(1+\sin\theta)} = \dfrac{\cos^2\theta}{\cos\theta(1+\sin\theta)} = \dfrac{\cos\theta}{1+\sin\theta}$

39. $3\sin^2\theta + 4\cos^2\theta = 3\sin^2\theta + 3\cos^2\theta + \cos^2\theta = 3(\sin^2\theta + \cos^2\theta) + \cos^2\theta = 3 + \cos^2\theta$

41. $1 - \dfrac{\cos^2\theta}{1+\sin\theta} = 1 - \dfrac{1-\sin^2\theta}{1+\sin\theta} = 1 - \dfrac{(1+\sin\theta)(1-\sin\theta)}{1+\sin\theta} = 1 - (1-\sin\theta) = \sin\theta$

43. $\dfrac{1+\tan\theta}{1-\tan\theta} = \dfrac{1 + \dfrac{1}{\cot\theta}}{1 - \dfrac{1}{\cot\theta}} = \dfrac{\dfrac{\cot\theta+1}{\cot\theta}}{\dfrac{\cot\theta-1}{\cot\theta}} = \dfrac{\cot\theta+1}{\cot\theta-1}$ **45.** $\dfrac{\sec\theta}{\csc\theta} + \dfrac{\sin\theta}{\cos\theta} = \dfrac{\dfrac{1}{\cos\theta}}{\dfrac{1}{\sin\theta}} + \tan\theta = \dfrac{\sin\theta}{\cos\theta} + \tan\theta = \tan\theta + \tan\theta = 2\tan\theta$

47. $\dfrac{1+\sin\theta}{1-\sin\theta} = \dfrac{1 + \dfrac{1}{\csc\theta}}{1 - \dfrac{1}{\csc\theta}} = \dfrac{\dfrac{\csc\theta+1}{\csc\theta}}{\dfrac{\csc\theta-1}{\csc\theta}} = \dfrac{\csc\theta+1}{\csc\theta-1}$

49. $\dfrac{1-\sin\theta}{\cos\theta} + \dfrac{\cos\theta}{1-\sin\theta} = \dfrac{(1-\sin\theta)^2 + \cos^2\theta}{\cos\theta(1-\sin\theta)} = \dfrac{1 - 2\sin\theta + \sin^2\theta + \cos^2\theta}{\cos\theta(1-\sin\theta)} = \dfrac{2 - 2\sin\theta}{\cos\theta(1-\sin\theta)} = \dfrac{2(1-\sin\theta)}{\cos\theta(1-\sin\theta)} = \dfrac{2}{\cos\theta} = 2\sec\theta$

51. $\dfrac{\sin\theta}{\sin\theta - \cos\theta} = \dfrac{1}{\dfrac{\sin\theta - \cos\theta}{\sin\theta}} = \dfrac{1}{1 - \dfrac{\cos\theta}{\sin\theta}} = \dfrac{1}{1 - \cot\theta}$

53. $(\sec\theta - \tan\theta)^2 = \sec^2\theta - 2\sec\theta\tan\theta + \tan^2\theta = \dfrac{1}{\cos^2\theta} - \dfrac{2\sin\theta}{\cos^2\theta} + \dfrac{\sin^2\theta}{\cos^2\theta} = \dfrac{1 - 2\sin\theta + \sin^2\theta}{\cos^2\theta} = \dfrac{(1-\sin\theta)^2}{1-\sin^2\theta}$
$= \dfrac{(1-\sin\theta)^2}{(1-\sin\theta)(1+\sin\theta)} = \dfrac{1-\sin\theta}{1+\sin\theta}$

55. $\dfrac{\cos\theta}{1-\tan\theta} + \dfrac{\sin\theta}{1-\cot\theta} = \dfrac{\cos\theta}{1 - \dfrac{\sin\theta}{\cos\theta}} + \dfrac{\sin\theta}{1 - \dfrac{\cos\theta}{\sin\theta}} = \dfrac{\cos\theta}{\dfrac{\cos\theta - \sin\theta}{\cos\theta}} + \dfrac{\sin\theta}{\dfrac{\sin\theta - \cos\theta}{\sin\theta}} = \dfrac{\cos^2\theta}{\cos\theta - \sin\theta} + \dfrac{\sin^2\theta}{\sin\theta - \cos\theta}$
$= \dfrac{\cos^2\theta - \sin^2\theta}{\cos\theta - \sin\theta} = \dfrac{(\cos\theta - \sin\theta)(\cos\theta + \sin\theta)}{\cos\theta - \sin\theta} = \sin\theta + \cos\theta$

57. $\tan\theta + \dfrac{\cos\theta}{1+\sin\theta} = \dfrac{\sin\theta}{\cos\theta} + \dfrac{\cos\theta}{1+\sin\theta} = \dfrac{\sin\theta(1+\sin\theta) + \cos^2\theta}{\cos\theta(1+\sin\theta)} = \dfrac{\sin\theta + \sin^2\theta + \cos^2\theta}{\cos\theta(1+\sin\theta)} = \dfrac{\sin\theta + 1}{\cos\theta(1+\sin\theta)} = \dfrac{1}{\cos\theta} = \sec\theta$

59. $\dfrac{\tan\theta + \sec\theta - 1}{\tan\theta - \sec\theta + 1} = \dfrac{\tan\theta + (\sec\theta - 1)}{\tan\theta - (\sec\theta - 1)}\cdot\dfrac{\tan\theta + (\sec\theta - 1)}{\tan\theta + (\sec\theta - 1)} = \dfrac{\tan^2\theta + 2\tan\theta(\sec\theta - 1) + \sec^2\theta - 2\sec\theta + 1}{\tan^2\theta - (\sec^2\theta - 2\sec\theta + 1)}$
$= \dfrac{\sec^2\theta - 1 + 2\tan\theta(\sec\theta - 1) + \sec^2\theta - 2\sec\theta + 1}{\sec^2\theta - 1 - \sec^2\theta + 2\sec\theta - 1} = \dfrac{2\sec^2\theta - 2\sec\theta + 2\tan\theta(\sec\theta - 1)}{-2 + 2\sec\theta}$
$= \dfrac{2\sec\theta(\sec\theta - 1) + 2\tan\theta(\sec\theta - 1)}{2(\sec\theta - 1)} = \dfrac{2(\sec\theta - 1)(\sec\theta + \tan\theta)}{2(\sec\theta - 1)} = \tan\theta + \sec\theta$

61. $\dfrac{\tan\theta - \cot\theta}{\tan\theta + \cot\theta} = \dfrac{\dfrac{\sin\theta}{\cos\theta} - \dfrac{\cos\theta}{\sin\theta}}{\dfrac{\sin\theta}{\cos\theta} + \dfrac{\cos\theta}{\sin\theta}} = \dfrac{\dfrac{\sin^2\theta - \cos^2\theta}{\cos\theta\sin\theta}}{\dfrac{\sin^2\theta + \cos^2\theta}{\cos\theta\sin\theta}} = \dfrac{\sin^2\theta - \cos^2\theta}{1} = \sin^2\theta - \cos^2\theta$

63. $\dfrac{\tan\theta - \cot\theta}{\tan\theta + \cot\theta} + 1 = \dfrac{\dfrac{\sin\theta}{\cos\theta} - \dfrac{\cos\theta}{\sin\theta}}{\dfrac{\sin\theta}{\cos\theta} + \dfrac{\cos\theta}{\sin\theta}} + 1 = \dfrac{\dfrac{\sin^2\theta - \cos^2\theta}{\cos\theta\sin\theta}}{\dfrac{\sin^2\theta + \cos^2\theta}{\cos\theta\sin\theta}} + 1 = \sin^2\theta - \cos^2\theta + 1 = \sin^2\theta + (1 - \cos^2\theta) = 2\sin^2\theta$

65. $\dfrac{\sec\theta + \tan\theta}{\cot\theta + \cos\theta} = \dfrac{\dfrac{1}{\cos\theta} + \dfrac{\sin\theta}{\cos\theta}}{\dfrac{\cos\theta}{\sin\theta} + \cos\theta} = \dfrac{\dfrac{1 + \sin\theta}{\cos\theta}}{\dfrac{\cos\theta + \cos\theta\sin\theta}{\sin\theta}} = \dfrac{1 + \sin\theta}{\cos\theta} \cdot \dfrac{\sin\theta}{\cos\theta(1 + \sin\theta)} = \dfrac{\sin\theta}{\cos\theta} \cdot \dfrac{1}{\cos\theta} = \tan\theta\sec\theta$

67. $\dfrac{1 - \tan^2\theta}{1 + \tan^2\theta} + 1 = \dfrac{1 - \tan^2\theta}{\sec^2\theta} + 1 = \dfrac{1}{\sec^2\theta} - \dfrac{\tan^2\theta}{\sec^2\theta} + 1 = \cos^2\theta - \dfrac{\dfrac{\sin^2\theta}{\cos^2\theta}}{\dfrac{1}{\cos^2\theta}} + 1 = \cos^2\theta - \sin^2\theta + 1$

$= \cos^2\theta + (1 - \sin^2\theta) = 2\cos^2\theta$

69. $\dfrac{\sec\theta - \csc\theta}{\sec\theta\csc\theta} = \dfrac{\sec\theta}{\sec\theta\csc\theta} - \dfrac{\csc\theta}{\sec\theta\csc\theta} = \dfrac{1}{\csc\theta} - \dfrac{1}{\sec\theta} = \sin\theta - \cos\theta$

71. $\sec\theta - \cos\theta = \dfrac{1}{\cos\theta} - \cos\theta = \dfrac{1 - \cos^2\theta}{\cos\theta} = \dfrac{\sin^2\theta}{\cos\theta} = \sin\theta \cdot \dfrac{\sin\theta}{\cos\theta} = \sin\theta\tan\theta$

73. $\dfrac{1}{1 - \sin\theta} + \dfrac{1}{1 + \sin\theta} = \dfrac{1 + \sin\theta + 1 - \sin\theta}{(1 + \sin\theta)(1 - \sin\theta)} = \dfrac{2}{1 - \sin^2\theta} = \dfrac{2}{\cos^2\theta} = 2\sec^2\theta$

75. $\dfrac{\sec\theta}{1 - \sin\theta} = \dfrac{\sec\theta}{1 - \sin\theta} \cdot \dfrac{1 + \sin\theta}{1 + \sin\theta} = \dfrac{\sec\theta(1 + \sin\theta)}{1 - \sin^2\theta} = \dfrac{\sec\theta(1 + \sin\theta)}{\cos^2\theta} = \dfrac{1 + \sin\theta}{\cos^3\theta}$

77. $\dfrac{(\sec\theta - \tan\theta)^2 + 1}{\csc\theta(\sec\theta - \tan\theta)} = \dfrac{\sec^2\theta - 2\sec\theta\tan\theta + \tan^2\theta + 1}{\dfrac{1}{\sin\theta}\left(\dfrac{1}{\cos\theta} - \dfrac{\sin\theta}{\cos\theta}\right)} = \dfrac{2\sec^2\theta - 2\sec\theta\tan\theta}{\dfrac{1}{\sin\theta}\left(\dfrac{1 - \sin\theta}{\cos\theta}\right)} = \dfrac{\dfrac{2}{\cos^2\theta} - \dfrac{2\sin\theta}{\cos^2\theta}}{\dfrac{1 - \sin\theta}{\sin\theta\cos\theta}} = \dfrac{2 - 2\sin\theta}{\cos^2\theta} \cdot \dfrac{\sin\theta\cos\theta}{1 - \sin\theta}$

$= \dfrac{2(1 - \sin\theta)}{\cos\theta} \cdot \dfrac{\sin\theta}{1 - \sin\theta} = \dfrac{2\sin\theta}{\cos\theta} = 2\tan\theta$

79. $\dfrac{\sin\theta + \cos\theta}{\cos\theta} - \dfrac{\sin\theta - \cos\theta}{\sin\theta} = \dfrac{\sin\theta}{\cos\theta} + 1 - 1 + \dfrac{\cos\theta}{\sin\theta} = \dfrac{\sin^2\theta + \cos^2\theta}{\cos\theta\sin\theta} = \dfrac{1}{\cos\theta\sin\theta} = \sec\theta\csc\theta$

81. $\dfrac{\sin^3\theta + \cos^3\theta}{\sin\theta + \cos\theta} = \dfrac{(\sin\theta + \cos\theta)(\sin^2\theta - \sin\theta\cos\theta + \cos^2\theta)}{\sin\theta + \cos\theta} = \sin^2\theta + \cos^2\theta - \sin\theta\cos\theta = 1 - \sin\theta\cos\theta$

83. $\dfrac{\cos^2\theta - \sin^2\theta}{1 - \tan^2\theta} = \dfrac{\cos^2\theta - \sin^2\theta}{1 - \dfrac{\sin^2\theta}{\cos^2\theta}} = \dfrac{\cos^2\theta - \sin^2\theta}{\dfrac{\cos^2\theta - \sin^2\theta}{\cos^2\theta}} = \cos^2\theta$

85. $\dfrac{(2\cos^2\theta - 1)^2}{\cos^4\theta - \sin^4\theta} = \dfrac{[2\cos^2\theta - (\sin^2\theta + \cos^2\theta)]^2}{(\cos^2\theta - \sin^2\theta)(\cos^2\theta + \sin^2\theta)} = \dfrac{(\cos^2\theta - \sin^2\theta)^2}{\cos^2\theta - \sin^2\theta} = \cos^2\theta - \sin^2\theta = (1 - \sin^2\theta) - \sin^2\theta = 1 - 2\sin^2\theta$

87. $\dfrac{1 + \sin\theta + \cos\theta}{1 + \sin\theta - \cos\theta} = \dfrac{(1 + \sin\theta) + \cos\theta}{(1 + \sin\theta) - \cos\theta} \cdot \dfrac{(1 + \sin\theta) + \cos\theta}{(1 + \sin\theta) + \cos\theta} = \dfrac{1 + 2\sin\theta + \sin^2\theta + 2(1 + \sin\theta)\cos\theta + \cos^2\theta}{1 + 2\sin\theta + \sin^2\theta - \cos^2\theta}$

$= \dfrac{1 + 2\sin\theta + \sin^2\theta + 2(1 + \sin\theta)(\cos\theta) + (1 - \sin^2\theta)}{1 + 2\sin\theta + \sin^2\theta - (1 - \sin^2\theta)} = \dfrac{2 + 2\sin\theta + 2(1 + \sin\theta)(\cos\theta)}{2\sin\theta + 2\sin^2\theta}$

$= \dfrac{2(1 + \sin\theta) + 2(1 + \sin\theta)(\cos\theta)}{2\sin\theta(1 + \sin\theta)} = \dfrac{2(1 + \sin\theta)(1 + \cos\theta)}{2\sin\theta(1 + \sin\theta)} = \dfrac{1 + \cos\theta}{\sin\theta}$

89. $(a\sin\theta + b\cos\theta)^2 + (a\cos\theta - b\sin\theta)^2 = a^2\sin^2\theta + 2ab\sin\theta\cos\theta + b^2\cos^2\theta + a^2\cos^2\theta - 2ab\sin\theta\cos\theta + b^2\sin^2\theta$
$= a^2(\sin^2\theta + \cos^2\theta) + b^2(\cos^2\theta + \sin^2\theta) = a^2 + b^2$

91. $\dfrac{\tan\alpha + \tan\beta}{\cot\alpha + \cot\beta} = \dfrac{\tan\alpha + \tan\beta}{\dfrac{1}{\tan\alpha} + \dfrac{1}{\tan\beta}} = \dfrac{\tan\alpha + \tan\beta}{\dfrac{\tan\beta + \tan\alpha}{\tan\alpha\tan\beta}} = (\tan\alpha + \tan\beta) \cdot \dfrac{\tan\alpha\tan\beta}{\tan\alpha + \tan\beta} = \tan\alpha\tan\beta$

93. $(\sin \alpha + \cos \beta)^2 + (\cos \beta + \sin \alpha)(\cos \beta - \sin \alpha) = (\sin^2 \alpha + 2 \sin \alpha \cos \beta + \cos^2 \beta) + (\cos^2 \beta - \sin^2 \alpha)$
$= 2 \cos^2 \beta + 2 \sin \alpha \cos \beta = 2 \cos \beta (\cos \beta + \sin \alpha)$

95. $\ln|\sec \theta| = \ln|\cos \theta|^{-1} = -\ln|\cos \theta|$ **97.** $\ln|1 + \cos \theta| + \ln|1 - \cos \theta| = \ln(|1 + \cos \theta||1 - \cos \theta|) = \ln|1 - \cos^2 \theta| = \ln|\sin^2 \theta| = 2 \ln|\sin \theta|$

99. Let $\theta = \tan^{-1} v$. Then $\tan \theta = v, -\dfrac{\pi}{2} < \theta < \dfrac{\pi}{2}$. Now, $\sec \theta > 0$ and $\tan^2 \theta + 1 = \sec^2 \theta$. Thus $\sec(\tan^{-1} v) = \sec \theta = \sqrt{1 + v^2}$.

101. Let $\theta = \cos^{-1} v$. Then $\cos \theta = v, 0 \le \theta \le \pi$, and $\tan(\cos^{-1} v) = \tan \theta = \dfrac{\sin \theta}{\cos \theta} = \dfrac{\sqrt{1 - \cos^2 \theta}}{\cos \theta} = \dfrac{\sqrt{1 - v^2}}{v}$.

103. Let $\theta = \sin^{-1} v$. Then $\sin \theta = v, -\dfrac{\pi}{2} \le \theta \le \dfrac{\pi}{2}$, and $\cos(\sin^{-1} v) = \cos \theta = \sqrt{1 - \sin^2 \theta} = \sqrt{1 - v^2}$.

6.4 Assess Your Understanding (page 481)

4. − **5.** − **6.** F **7.** F **8.** F **9.** $\dfrac{1}{4}(\sqrt{6} + \sqrt{2})$ **11.** $\dfrac{1}{4}(\sqrt{2} - \sqrt{6})$ **13.** $-\dfrac{1}{4}(\sqrt{2} + \sqrt{6})$ **15.** $2 - \sqrt{3}$ **17.** $-\dfrac{1}{4}(\sqrt{6} + \sqrt{2})$

19. $\sqrt{6} - \sqrt{2}$ **21.** $\dfrac{1}{2}$ **23.** 0 **25.** 1 **27.** -1 **29.** $\dfrac{1}{2}$ **31. (a)** $\dfrac{2\sqrt{5}}{25}$ **(b)** $\dfrac{11\sqrt{5}}{25}$ **(c)** $\dfrac{2\sqrt{5}}{5}$ **(d)** 2 **33. (a)** $\dfrac{4 - 3\sqrt{3}}{10}$ **(b)** $\dfrac{-3 - 4\sqrt{3}}{10}$

(c) $\dfrac{4 + 3\sqrt{3}}{10}$ **(d)** $\dfrac{25\sqrt{3} + 48}{39}$ **35. (a)** $-\dfrac{5 + 12\sqrt{3}}{26}$ **(b)** $\dfrac{12 - 5\sqrt{3}}{26}$ **(c)** $\dfrac{-5 + 12\sqrt{3}}{26}$ **(d)** $\dfrac{-240 + 169\sqrt{3}}{69}$

37. (a) $-\dfrac{2\sqrt{2}}{3}$ **(b)** $\dfrac{-2\sqrt{2} + \sqrt{3}}{6}$ **(c)** $\dfrac{-2\sqrt{2} + \sqrt{3}}{6}$ **(d)** $\dfrac{9 - 4\sqrt{2}}{7}$ **39.** $\sin\left(\dfrac{\pi}{2} + \theta\right) = \sin \dfrac{\pi}{2} \cos \theta + \cos \dfrac{\pi}{2} \sin \theta = 1 \cdot \cos \theta + 0 \cdot \sin \theta = \cos \theta$

41. $\sin(\pi - \theta) = \sin \pi \cos \theta - \cos \pi \sin \theta = 0 \cdot \cos \theta - (-1)\sin \theta = \sin \theta$

43. $\sin(\pi + \theta) = \sin \pi \cos \theta + \cos \pi \sin \theta = 0 \cdot \cos \theta + (-1)\sin \theta = -\sin \theta$

45. $\tan(\pi - \theta) = \dfrac{\tan \pi - \tan \theta}{1 + \tan \pi \tan \theta} = \dfrac{0 - \tan \theta}{1 + 0 \cdot \tan \theta} = -\tan \theta$ **47.** $\sin\left(\dfrac{3\pi}{2} + \theta\right) = \sin \dfrac{3\pi}{2} \cos \theta + \cos \dfrac{3\pi}{2} \sin \theta = (-1)\cos \theta + 0 \cdot \sin \theta = -\cos \theta$

49. $\sin(\alpha + \beta) + \sin(\alpha - \beta) = \sin \alpha \cos \beta + \cos \alpha \sin \beta + \sin \alpha \cos \beta - \cos \alpha \sin \beta = 2 \sin \alpha \cos \beta$

51. $\dfrac{\sin(\alpha + \beta)}{\sin \alpha \cos \beta} = \dfrac{\sin \alpha \cos \beta + \cos \alpha \sin \beta}{\sin \alpha \cos \beta} = \dfrac{\sin \alpha \cos \beta}{\sin \alpha \cos \beta} + \dfrac{\cos \alpha \sin \beta}{\sin \alpha \cos \beta} = 1 + \cot \alpha \tan \beta$

53. $\dfrac{\cos(\alpha + \beta)}{\cos \alpha \cos \beta} = \dfrac{\cos \alpha \cos \beta - \sin \alpha \sin \beta}{\cos \alpha \cos \beta} = \dfrac{\cos \alpha \cos \beta}{\cos \alpha \cos \beta} - \dfrac{\sin \alpha \sin \beta}{\cos \alpha \cos \beta} = 1 - \tan \alpha \tan \beta$

55. $\dfrac{\sin(\alpha + \beta)}{\sin(\alpha - \beta)} = \dfrac{\sin \alpha \cos \beta + \cos \alpha \sin \beta}{\sin \alpha \cos \beta - \cos \alpha \sin \beta} = \dfrac{\dfrac{\sin \alpha \cos \beta + \cos \alpha \sin \beta}{\cos \alpha \cos \beta}}{\dfrac{\sin \alpha \cos \beta - \cos \alpha \sin \beta}{\cos \alpha \cos \beta}} = \dfrac{\dfrac{\sin \alpha \cos \beta}{\cos \alpha \cos \beta} + \dfrac{\cos \alpha \sin \beta}{\cos \alpha \cos \beta}}{\dfrac{\sin \alpha \cos \beta}{\cos \alpha \cos \beta} - \dfrac{\cos \alpha \sin \beta}{\cos \alpha \cos \beta}} = \dfrac{\tan \alpha + \tan \beta}{\tan \alpha - \tan \beta}$

57. $\cot(\alpha + \beta) = \dfrac{\cos(\alpha + \beta)}{\sin(\alpha + \beta)} = \dfrac{\cos \alpha \cos \beta - \sin \alpha \sin \beta}{\sin \alpha \cos \beta + \cos \alpha \sin \beta} = \dfrac{\dfrac{\cos \alpha \cos \beta - \sin \alpha \sin \beta}{\sin \alpha \sin \beta}}{\dfrac{\sin \alpha \cos \beta + \cos \alpha \sin \beta}{\sin \alpha \sin \beta}} = \dfrac{\dfrac{\cos \alpha \cos \beta}{\sin \alpha \sin \beta} - \dfrac{\sin \alpha \sin \beta}{\sin \alpha \sin \beta}}{\dfrac{\sin \alpha \cos \beta}{\sin \alpha \sin \beta} + \dfrac{\cos \alpha \sin \beta}{\sin \alpha \sin \beta}} = \dfrac{\cot \alpha \cot \beta - 1}{\cot \beta + \cot \alpha}$

59. $\sec(\alpha + \beta) = \dfrac{1}{\cos(\alpha + \beta)} = \dfrac{1}{\cos \alpha \cos \beta - \sin \alpha \sin \beta} = \dfrac{\dfrac{1}{\sin \alpha \sin \beta}}{\dfrac{\cos \alpha \cos \beta - \sin \alpha \sin \beta}{\sin \alpha \sin \beta}} = \dfrac{\dfrac{1}{\sin \alpha} \cdot \dfrac{1}{\sin \beta}}{\dfrac{\cos \alpha \cos \beta}{\sin \alpha \sin \beta} - \dfrac{\sin \alpha \sin \beta}{\sin \alpha \sin \beta}} = \dfrac{\csc \alpha \csc \beta}{\cot \alpha \cot \beta - 1}$

61. $\sin(\alpha - \beta) \sin(\alpha + \beta) = (\sin \alpha \cos \beta - \cos \alpha \sin \beta)(\sin \alpha \cos \beta + \cos \alpha \sin \beta) = \sin^2 \alpha \cos^2 \beta - \cos^2 \alpha \sin^2 \beta$
$= (\sin^2 \alpha)(1 - \sin^2 \beta) - (1 - \sin^2 \alpha)(\sin^2 \beta) = \sin^2 \alpha - \sin^2 \beta$

63. $\sin(\theta + k\pi) = \sin \theta \cos k\pi + \cos \theta \sin k\pi = (\sin \theta)(-1)^k + (\cos \theta)(0) = (-1)^k \sin \theta, k$ any integer

65. $\dfrac{\sqrt{3}}{2}$ **67.** $-\dfrac{24}{25}$ **69.** $-\dfrac{33}{65}$ **71.** $\dfrac{63}{65}$ **73.** $\dfrac{48 + 25\sqrt{3}}{39}$ **75.** $\dfrac{4}{3}$ **77.** $u\sqrt{1 - v^2} - v\sqrt{1 - u^2}$

79. $\dfrac{u\sqrt{1 - v^2} - v}{\sqrt{1 + u^2}}$ **81.** $\dfrac{uv - \sqrt{1 - u^2}\sqrt{1 - v^2}}{v\sqrt{1 - u^2} + u\sqrt{1 - v^2}}$

83. Let $\alpha = \sin^{-1} v$ and $\beta = \cos^{-1} v$. Then $\sin \alpha = \cos \beta = v$, and since $\sin \alpha = \cos\left(\dfrac{\pi}{2} - \alpha\right), \cos\left(\dfrac{\pi}{2} - \alpha\right) = \cos \beta$.

If $v \ge 0$, then $0 \le \alpha \le \dfrac{\pi}{2}$, so $\left(\dfrac{\pi}{2} - \alpha\right)$ and β both lie on $\left[0, \dfrac{\pi}{2}\right]$. If $v < 0$, then $-\dfrac{\pi}{2} \le \alpha < 0$, so $\left(\dfrac{\pi}{2} - \alpha\right)$ and β both lie on

$\left(\dfrac{\pi}{2}, \pi\right]$. Either way, $\cos\left(\dfrac{\pi}{2} - \alpha\right) = \cos \beta$ implies $\dfrac{\pi}{2} - \alpha = \beta$, or $\alpha + \beta = \dfrac{\pi}{2}$.

85. Let $\alpha = \tan^{-1}\dfrac{1}{v}$, and $\beta = \tan^{-1} v$. Because $v \neq 0$, $\alpha, \beta \neq 0$. Then $\tan \alpha = \dfrac{1}{v} = \dfrac{1}{\tan \beta} = \cot \beta$, and since

$\tan \alpha = \cot\left(\dfrac{\pi}{2} - \alpha\right)$, $\cot\left(\dfrac{\pi}{2} - \alpha\right) = \cot \beta$. Because $v > 0$, $0 < \alpha < \dfrac{\pi}{2}$, and so $\left(\dfrac{\pi}{2} - \alpha\right)$ and β both lie on $\left(0, \dfrac{\pi}{2}\right)$.

Then $\cot\left(\dfrac{\pi}{2} - \alpha\right) = \cot \beta$ implies $\dfrac{\pi}{2} - \alpha = \beta$, or $\alpha = \dfrac{\pi}{2} - \beta$.

87. $\sin(\sin^{-1} v + \cos^{-1} v) = \sin(\sin^{-1} v)\cos(\cos^{-1} v) + \cos(\sin^{-1} v)\sin(\cos^{-1} v) = (v)(v) + \sqrt{1 - v^2}\sqrt{1 - v^2} = v^2 + 1 - v^2 = 1$

89. $\dfrac{\sin(x + h) - \sin x}{h} = \dfrac{\sin x \cos h + \cos x \sin h - \sin x}{h} = \dfrac{\cos x \sin h - \sin x(1 - \cos h)}{h} = \cos x \cdot \dfrac{\sin h}{h} - \sin x \cdot \dfrac{1 - \cos h}{h}$

91. $\tan \dfrac{\pi}{2}$ is not defined; $\tan\left(\dfrac{\pi}{2} - \theta\right) = \dfrac{\sin\left(\dfrac{\pi}{2} - \theta\right)}{\cos\left(\dfrac{\pi}{2} - \theta\right)} = \dfrac{\cos \theta}{\sin \theta} = \cot \theta$

93. $\tan \theta = \tan(\theta_2 - \theta_1) = \dfrac{\tan \theta_2 - \tan \theta_1}{1 + \tan \theta_1 \tan \theta_2} = \dfrac{m_2 - m_1}{1 + m_1 m_2}$ **95.** No; $\tan \dfrac{\pi}{2}$ is undefined.

6.5 Assess Your Understanding (page 490)

1. $\sin^2 \theta; 2\cos^2 \theta; 2\sin^2 \theta$ **2.** $1 - \cos \theta$ **3.** $\sin \theta$ **4.** T **5.** F **6.** F

7. (a) $\dfrac{24}{25}$ **(b)** $\dfrac{7}{25}$ **(c)** $\dfrac{\sqrt{10}}{10}$ **(d)** $\dfrac{3\sqrt{10}}{10}$ **9. (a)** $\dfrac{24}{25}$ **(b)** $-\dfrac{7}{25}$ **(c)** $\dfrac{2\sqrt{5}}{5}$ **(d)** $-\dfrac{\sqrt{5}}{5}$ **11. (a)** $-\dfrac{2\sqrt{2}}{3}$ **(b)** $\dfrac{1}{3}$ **(c)** $\sqrt{\dfrac{3 + \sqrt{6}}{6}}$ **(d)** $\sqrt{\dfrac{3 - \sqrt{6}}{6}}$

13. (a) $\dfrac{4\sqrt{2}}{9}$ **(b)** $-\dfrac{7}{9}$ **(c)** $\dfrac{\sqrt{3}}{3}$ **(d)** $\dfrac{\sqrt{6}}{3}$ **15. (a)** $-\dfrac{4}{5}$ **(b)** $\dfrac{3}{5}$ **(c)** $\sqrt{\dfrac{5 + 2\sqrt{5}}{10}}$ **(d)** $\sqrt{\dfrac{5 - 2\sqrt{5}}{10}}$ **17. (a)** $-\dfrac{3}{5}$ **(b)** $-\dfrac{4}{5}$ **(c)** $\dfrac{1}{2}\sqrt{\dfrac{10 - \sqrt{10}}{5}}$

(d) $-\dfrac{1}{2}\sqrt{\dfrac{10 + \sqrt{10}}{5}}$ **19.** $\dfrac{\sqrt{2 - \sqrt{2}}}{2}$ **21.** $1 - \sqrt{2}$ **23.** $-\dfrac{\sqrt{2 + \sqrt{3}}}{2}$ **25.** $\dfrac{2}{\sqrt{2 + \sqrt{2}}} = (2 - \sqrt{2})\sqrt{2 + \sqrt{2}}$ **27.** $-\dfrac{\sqrt{2 - \sqrt{2}}}{2}$

29. $\sin^4 \theta = (\sin^2 \theta)^2 = \left(\dfrac{1 - \cos(2\theta)}{2}\right)^2 = \dfrac{1}{4}[1 - 2\cos(2\theta) + \cos^2(2\theta)] = \dfrac{1}{4} - \dfrac{1}{2}\cos(2\theta) + \dfrac{1}{4}\cos^2(2\theta)$

$= \dfrac{1}{4} - \dfrac{1}{2}\cos(2\theta) + \dfrac{1}{4}\left(\dfrac{1 + \cos(4\theta)}{2}\right) = \dfrac{1}{4} - \dfrac{1}{2}\cos(2\theta) + \dfrac{1}{8} + \dfrac{1}{8}\cos(4\theta) = \dfrac{3}{8} - \dfrac{1}{2}\cos(2\theta) + \dfrac{1}{8}\cos(4\theta)$

31. $\cos(3\theta) = 4\cos^3 \theta - 3\cos \theta$ **33.** $\sin(5\theta) = 16\sin^5 \theta - 20\sin^3 \theta + 5\sin \theta$ **35.** $\cos^4 \theta - \sin^4 \theta = (\cos^2 \theta + \sin^2 \theta)(\cos^2 \theta - \sin^2 \theta) = \cos(2\theta)$

37. $\cot(2\theta) = \dfrac{1}{\tan(2\theta)} = \dfrac{1 - \tan^2 \theta}{2\tan \theta} = \dfrac{1 - \dfrac{1}{\cot^2 \theta}}{2\left(\dfrac{1}{\cot \theta}\right)} = \dfrac{\dfrac{\cot^2 \theta - 1}{\cot^2 \theta}}{\dfrac{2}{\cot \theta}} = \dfrac{\cot^2 \theta - 1}{\cot^2 \theta} \cdot \dfrac{\cot \theta}{2} = \dfrac{\cot^2 \theta - 1}{2\cot \theta}$

39. $\sec(2\theta) = \dfrac{1}{\cos(2\theta)} = \dfrac{1}{2\cos^2 \theta - 1} = \dfrac{1}{\dfrac{2}{\sec^2 \theta} - 1} = \dfrac{1}{\dfrac{2 - \sec^2 \theta}{\sec^2 \theta}} = \dfrac{\sec^2 \theta}{2 - \sec^2 \theta}$ **41.** $\cos^2(2\theta) - \sin^2(2\theta) = \cos[2(2\theta)] = \cos(4\theta)$

43. $\dfrac{\cos(2\theta)}{1 + \sin(2\theta)} = \dfrac{\cos^2 \theta - \sin^2 \theta}{1 + 2\sin \theta \cos \theta} = \dfrac{(\cos \theta - \sin \theta)(\cos \theta + \sin \theta)}{\sin^2 \theta + \cos^2 \theta + 2\sin \theta \cos \theta} = \dfrac{(\cos \theta - \sin \theta)(\cos \theta + \sin \theta)}{(\sin \theta + \cos \theta)(\sin \theta + \cos \theta)} = \dfrac{\cos \theta - \sin \theta}{\cos \theta + \sin \theta}$

$= \dfrac{\dfrac{\cos \theta - \sin \theta}{\sin \theta}}{\dfrac{\cos \theta + \sin \theta}{\sin \theta}} = \dfrac{\dfrac{\cos \theta}{\sin \theta} - \dfrac{\sin \theta}{\sin \theta}}{\dfrac{\cos \theta}{\sin \theta} + \dfrac{\sin \theta}{\sin \theta}} = \dfrac{\cot \theta - 1}{\cot \theta + 1}$

45. $\sec^2 \dfrac{\theta}{2} = \dfrac{1}{\cos^2\left(\dfrac{\theta}{2}\right)} = \dfrac{1}{\dfrac{1 + \cos \theta}{2}} = \dfrac{2}{1 + \cos \theta}$

47. $\cot^2 \dfrac{\theta}{2} = \dfrac{1}{\tan^2\left(\dfrac{\theta}{2}\right)} = \dfrac{1}{\dfrac{1 - \cos \theta}{1 + \cos \theta}} = \dfrac{1 + \cos \theta}{1 - \cos \theta} = \dfrac{1 + \dfrac{1}{\sec \theta}}{1 - \dfrac{1}{\sec \theta}} = \dfrac{\dfrac{\sec \theta + 1}{\sec \theta}}{\dfrac{\sec \theta - 1}{\sec \theta}} = \dfrac{\sec \theta + 1}{\sec \theta} \cdot \dfrac{\sec \theta}{\sec \theta - 1} = \dfrac{\sec \theta + 1}{\sec \theta - 1}$

49. $\dfrac{1 - \tan^2\left(\dfrac{\theta}{2}\right)}{1 + \tan^2\left(\dfrac{\theta}{2}\right)} = \dfrac{1 - \dfrac{1 - \cos \theta}{1 + \cos \theta}}{1 + \dfrac{1 - \cos \theta}{1 + \cos \theta}} = \dfrac{\dfrac{1 + \cos \theta - (1 - \cos \theta)}{1 + \cos \theta}}{\dfrac{1 + \cos \theta + 1 - \cos \theta}{1 + \cos \theta}} = \dfrac{2\cos \theta}{1 + \cos \theta} \cdot \dfrac{1 + \cos \theta}{2} = \cos \theta$

51. $\dfrac{\sin(3\theta)}{\sin\theta} - \dfrac{\cos(3\theta)}{\cos\theta} = \dfrac{\sin(3\theta)\cos\theta - \cos(3\theta)\sin\theta}{\sin\theta\cos\theta} = \dfrac{\sin(3\theta - \theta)}{\dfrac{1}{2}(2\sin\theta\cos\theta)} = \dfrac{2\sin(2\theta)}{\sin(2\theta)} = 2$

53. $\tan(3\theta) = \tan(\theta + 2\theta) = \dfrac{\tan\theta + \tan(2\theta)}{1 - \tan\theta\tan(2\theta)} = \dfrac{\tan\theta + \dfrac{2\tan\theta}{1 - \tan^2\theta}}{1 - \dfrac{\tan\theta(2\tan\theta)}{1 - \tan^2\theta}} = \dfrac{\tan\theta - \tan^3\theta + 2\tan\theta}{1 - \tan^2\theta - 2\tan^2\theta} = \dfrac{3\tan\theta - \tan^3\theta}{1 - 3\tan^2\theta}$

55. $\dfrac{1}{2}(\ln|1 - \cos(2\theta)| - \ln 2) = \ln\left(\dfrac{|1 - \cos(2\theta)|}{2}\right)^{1/2} = \ln|\sin^2\theta|^{1/2} = \ln|\sin\theta|$

57. $\dfrac{\sqrt{3}}{2}$ **59.** $\dfrac{7}{25}$ **61.** $\dfrac{24}{7}$ **63.** $\dfrac{24}{25}$ **65.** $\dfrac{1}{5}$ **67.** $\dfrac{25}{7}$ **69.** $\sin(2\theta) = \dfrac{4x}{4 + x^2}$ **71.** $-\dfrac{1}{4}$

73. $\dfrac{2z}{1 + z^2} = \dfrac{2\tan\left(\dfrac{\alpha}{2}\right)}{1 + \tan^2\left(\dfrac{\alpha}{2}\right)} = \dfrac{2\tan\left(\dfrac{\alpha}{2}\right)}{\sec^2\left(\dfrac{\alpha}{2}\right)} = \dfrac{\dfrac{2\sin\left(\dfrac{\alpha}{2}\right)}{\cos\left(\dfrac{\alpha}{2}\right)}}{\dfrac{1}{\cos^2\left(\dfrac{\alpha}{2}\right)}} = 2\sin\left(\dfrac{\alpha}{2}\right)\cos\left(\dfrac{\alpha}{2}\right) = \sin\left(2\cdot\dfrac{\alpha}{2}\right) = \sin\alpha$

75. $A = \dfrac{1}{2}h(\text{base}) = h\left(\dfrac{1}{2}\text{base}\right) = s\cos\dfrac{\theta}{2}\cdot s\sin\dfrac{\theta}{2} = \dfrac{1}{2}s^2\sin\theta$

77.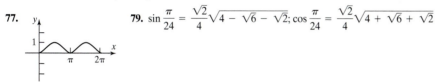

79. $\sin\dfrac{\pi}{24} = \dfrac{\sqrt{2}}{4}\sqrt{4 - \sqrt{6} - \sqrt{2}};\ \cos\dfrac{\pi}{24} = \dfrac{\sqrt{2}}{4}\sqrt{4 + \sqrt{6} + \sqrt{2}}$

81. $\sin^3\theta + \sin^3(\theta + 120°) + \sin^3(\theta + 240°) = \sin^3\theta + (\sin\theta\cos 120° + \cos\theta\sin 120°)^3 + (\sin\theta\cos 240° + \cos\theta\sin 240°)^3$

$= \sin^3\theta + \left(-\dfrac{1}{2}\sin\theta + \dfrac{\sqrt{3}}{2}\cos\theta\right)^3 + \left(-\dfrac{1}{2}\sin\theta - \dfrac{\sqrt{3}}{2}\cos\theta\right)^3$

$= \sin^3\theta + \dfrac{1}{8}(3\sqrt{3}\cos^3\theta - 9\cos^2\theta\sin\theta + 3\sqrt{3}\cos\theta\sin^2\theta - \sin^3\theta) - \dfrac{1}{8}(\sin^3\theta + 3\sqrt{3}\sin^2\theta\cos\theta + 9\sin\theta\cos^2\theta + 3\sqrt{3}\cos^3\theta)$

$= \dfrac{3}{4}\sin^3\theta - \dfrac{9}{4}\cos^2\theta\sin\theta = \dfrac{3}{4}[\sin^3\theta - 3\sin\theta(1 - \sin^2\theta)] = \dfrac{3}{4}(4\sin^3\theta - 3\sin\theta) = -\dfrac{3}{4}\sin(3\theta)$ (from Example 2)

83. (a) $R = \dfrac{v_0^2\sqrt{2}}{16}(\sin\theta\cos\theta - \cos^2\theta)$ **(b)** 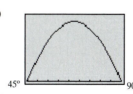   **(c)** $\theta = 67.5°$ makes R largest.

6.6 Assess Your Understanding *(page 495)*

1. $\dfrac{1}{2}[\cos(2\theta) - \cos(6\theta)]$ **3.** $\dfrac{1}{2}[\sin(6\theta) + \sin(2\theta)]$ **5.** $\dfrac{1}{2}[\cos(2\theta) + \cos(8\theta)]$ **7.** $\dfrac{1}{2}[\cos\theta - \cos(3\theta)]$ **9.** $\dfrac{1}{2}[\sin(2\theta) + \sin\theta]$

11. $2\sin\theta\cos(3\theta)$ **13.** $2\cos(3\theta)\cos\theta$ **15.** $2\sin(2\theta)\cos\theta$ **17.** $2\sin\theta\sin\dfrac{\theta}{2}$ **19.** $\dfrac{\sin\theta + \sin(3\theta)}{2\sin(2\theta)} = \dfrac{2\sin(2\theta)\cos\theta}{2\sin(2\theta)} = \cos\theta$

21. $\dfrac{\sin(4\theta) + \sin(2\theta)}{\cos(4\theta) + \cos(2\theta)} = \dfrac{2\sin(3\theta)\cos\theta}{2\cos(3\theta)\cos\theta} = \dfrac{\sin(3\theta)}{\cos(3\theta)} = \tan(3\theta)$ **23.** $\dfrac{\cos\theta - \cos(3\theta)}{\sin\theta + \sin(3\theta)} = \dfrac{2\sin(2\theta)\sin\theta}{2\sin(2\theta)\cos\theta} = \dfrac{\sin\theta}{\cos\theta} = \tan\theta$

25. $\sin\theta[\sin\theta + \sin(3\theta)] = \sin\theta[2\sin(2\theta)\cos\theta] = \cos\theta[2\sin(2\theta)\sin\theta] = \cos\theta\left[2\cdot\dfrac{1}{2}[\cos\theta - \cos(3\theta)]\right] = \cos\theta[\cos\theta - \cos(3\theta)]$

27. $\dfrac{\sin(4\theta) + \sin(8\theta)}{\cos(4\theta) + \cos(8\theta)} = \dfrac{2\sin(6\theta)\cos(2\theta)}{2\cos(6\theta)\cos(2\theta)} = \dfrac{\sin(6\theta)}{\cos(6\theta)} = \tan(6\theta)$

29. $\dfrac{\sin(4\theta) + \sin(8\theta)}{\sin(4\theta) - \sin(8\theta)} = \dfrac{2\sin(6\theta)\cos(-2\theta)}{2\sin(-2\theta)\cos(6\theta)} = \dfrac{\sin(6\theta)}{\cos(6\theta)}\cdot\dfrac{\cos(2\theta)}{-\sin(2\theta)} = \tan(6\theta)[-\cot(2\theta)] = -\dfrac{\tan(6\theta)}{\tan(2\theta)}$

31. $\dfrac{\sin\alpha + \sin\beta}{\sin\alpha - \sin\beta} = \dfrac{2\sin\dfrac{\alpha+\beta}{2}\cos\dfrac{\alpha-\beta}{2}}{2\sin\dfrac{\alpha-\beta}{2}\cos\dfrac{\alpha+\beta}{2}} = \dfrac{\sin\dfrac{\alpha+\beta}{2}}{\cos\dfrac{\alpha+\beta}{2}}\cdot\dfrac{\cos\dfrac{\alpha-\beta}{2}}{\sin\dfrac{\alpha-\beta}{2}} = \tan\dfrac{\alpha+\beta}{2}\cot\dfrac{\alpha-\beta}{2}$

33. $\dfrac{\sin\alpha + \sin\beta}{\cos\alpha + \cos\beta} = \dfrac{2\sin\dfrac{\alpha+\beta}{2}\cos\dfrac{\alpha-\beta}{2}}{2\cos\dfrac{\alpha+\beta}{2}\cos\dfrac{\alpha-\beta}{2}} = \dfrac{\sin\dfrac{\alpha+\beta}{2}}{\cos\dfrac{\alpha+\beta}{2}} = \tan\dfrac{\alpha+\beta}{2}$

35. $1 + \cos(2\theta) + \cos(4\theta) + \cos(6\theta) = [1 + \cos(6\theta)] + [\cos(2\theta) + \cos(4\theta)] = 2\cos^2(3\theta) + 2\cos(3\theta)\cos(-\theta)$
$= 2\cos(3\theta)[\cos(3\theta) + \cos\theta] = 2\cos(3\theta)[2\cos(2\theta)\cos\theta] = 4\cos\theta\cos(2\theta)\cos(3\theta)$

37. **(a)** $y = 2\sin(2061\pi t)\cos(357\pi t)$ **(b)** $y_{\max} = 2$
(c)

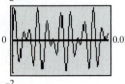

39. $\sin(2\alpha) + \sin(2\beta) + \sin(2\gamma) = 2\sin(\alpha + \beta)\cos(\alpha - \beta) + \sin(2\gamma) = 2\sin(\alpha + \beta)\cos(\alpha - \beta) + 2\sin\gamma\cos\gamma$
$= 2\sin(\pi - \gamma)\cos(\alpha - \beta) + 2\sin\gamma\cos\gamma = 2\sin\gamma\cos(\alpha - \beta) + 2\sin\gamma\cos\gamma = 2\sin\gamma[\cos(\alpha - \beta) + \cos\gamma]$
$= 2\sin\gamma\left(2\cos\dfrac{\alpha - \beta + \gamma}{2}\cos\dfrac{\alpha - \beta - \gamma}{2}\right) = 4\sin\gamma\cos\dfrac{\pi - 2\beta}{2}\cos\dfrac{2\alpha - \pi}{2} = 4\sin\gamma\cos\left(\dfrac{\pi}{2} - \beta\right)\cos\left(\alpha - \dfrac{\pi}{2}\right)$
$= 4\sin\gamma\sin\beta\sin\alpha$

41.
$$\sin(\alpha - \beta) = \sin\alpha\cos\beta - \cos\alpha\sin\beta$$
$$\sin(\alpha + \beta) = \sin\alpha\cos\beta + \cos\alpha\sin\beta$$
$$\sin(\alpha - \beta) + \sin(\alpha + \beta) = 2\sin\alpha\cos\beta$$
$$\sin\alpha\cos\beta = \frac{1}{2}[\sin(\alpha + \beta) + \sin(\alpha - \beta)]$$

43. $2\cos\dfrac{\alpha+\beta}{2}\cos\dfrac{\alpha-\beta}{2} = 2\cdot\dfrac{1}{2}\left[\cos\left(\dfrac{\alpha+\beta}{2} + \dfrac{\alpha-\beta}{2}\right) + \cos\left(\dfrac{\alpha+\beta}{2} - \dfrac{\alpha-\beta}{2}\right)\right] = \cos\dfrac{2\alpha}{2} + \cos\dfrac{2\beta}{2} = \cos\alpha + \cos\beta$

6.7 Assess Your Understanding *(page 500)*

3. $\left\{\dfrac{\pi}{6}, \dfrac{5\pi}{6}\right\}$ **4.** $\left\{\theta\middle| \theta = \dfrac{\pi}{6} + 2\pi k, \theta = \dfrac{5\pi}{6} + 2\pi k, k \text{ any integer}\right\}$ **5.** F   **6.** F

7. $\left\{\dfrac{7\pi}{6}, \dfrac{11\pi}{6}\right\}$ **9.** $\left\{\dfrac{\pi}{3}, \dfrac{2\pi}{3}, \dfrac{4\pi}{3}, \dfrac{5\pi}{3}\right\}$ **11.** $\left\{\dfrac{\pi}{4}, \dfrac{3\pi}{4}, \dfrac{5\pi}{4}, \dfrac{7\pi}{4}\right\}$ **13.** $\left\{\dfrac{\pi}{2}, \dfrac{7\pi}{6}, \dfrac{11\pi}{6}\right\}$ **15.** $\left\{\dfrac{\pi}{3}, \dfrac{2\pi}{3}, \dfrac{4\pi}{3}, \dfrac{5\pi}{3}\right\}$ **17.** $\left\{\dfrac{4\pi}{9}, \dfrac{8\pi}{9}, \dfrac{16\pi}{9}\right\}$

19. $\left\{\dfrac{7\pi}{6}, \dfrac{11\pi}{6}\right\}$ **21.** $\left\{\dfrac{3\pi}{4}, \dfrac{7\pi}{4}\right\}$ **23.** $\left\{\dfrac{2\pi}{3}, \dfrac{4\pi}{3}\right\}$ **25.** $\left\{\dfrac{3\pi}{4}, \dfrac{5\pi}{4}\right\}$ **27.** $\left\{\dfrac{3\pi}{4}, \dfrac{7\pi}{4}\right\}$ **29.** $\left\{\dfrac{11\pi}{6}\right\}$

31. $\left\{\theta\middle| \theta = \dfrac{\pi}{6} + 2k\pi, \theta = \dfrac{5\pi}{6} + 2k\pi\right\}; \dfrac{\pi}{6}, \dfrac{5\pi}{6}, \dfrac{13\pi}{6}, \dfrac{17\pi}{6}, \dfrac{25\pi}{6}, \dfrac{29\pi}{6}$ **33.** $\left\{\theta\middle| \theta = \dfrac{5\pi}{6} + k\pi\right\}; \dfrac{5\pi}{6}, \dfrac{11\pi}{6}, \dfrac{17\pi}{6}, \dfrac{23\pi}{6}, \dfrac{29\pi}{6}, \dfrac{35\pi}{6}$

35. $\left\{\theta\middle| \theta = \dfrac{\pi}{2} + 2k\pi, \theta = \dfrac{3\pi}{2} + 2k\pi\right\}; \dfrac{\pi}{2}, \dfrac{3\pi}{2}, \dfrac{5\pi}{2}, \dfrac{7\pi}{2}, \dfrac{9\pi}{2}, \dfrac{11\pi}{2}$ **37.** $\left\{\theta\middle| \theta = \dfrac{\pi}{3} + k\pi, \theta = \dfrac{2\pi}{3} + k\pi\right\}; \dfrac{\pi}{3}, \dfrac{2\pi}{3}, \dfrac{4\pi}{3}, \dfrac{5\pi}{3}, \dfrac{7\pi}{3}, \dfrac{8\pi}{3}$

39. $\left\{\theta\middle| \theta = \dfrac{8\pi}{3} + 4k\pi, \theta = \dfrac{10\pi}{3} + 4k\pi\right\}; \dfrac{8\pi}{3}, \dfrac{10\pi}{3}, \dfrac{20\pi}{3}, \dfrac{22\pi}{3}, \dfrac{32\pi}{3}, \dfrac{34\pi}{3}$ **41.** $\{0.41, 2.73\}$ **43.** $\{1.37, 4.51\}$ **45.** $\{2.69, 3.59\}$

47. $\{1.82, 4.46\}$ **49.** $\{2.08, 5.22\}$ **51.** $\{0.73, 2.41\}$ **53. (a)** $\left\{x\middle| x = \dfrac{\pi}{6} + 2\pi k, x = \dfrac{5\pi}{6} + 2\pi k\right\}$ **(b)** $\dfrac{\pi}{6} < x < \dfrac{5\pi}{6}$ or $\left(\dfrac{\pi}{6}, \dfrac{5\pi}{6}\right)$

55. (a) $\left\{x\middle| x = -\dfrac{\pi}{4} + k\pi\right\}$ **(b)** $-\dfrac{\pi}{2} < x < -\dfrac{\pi}{4}$ or $\left(-\dfrac{\pi}{2}, -\dfrac{\pi}{4}\right)$ **57. (a)** 10 sec; 30 sec   **(b)** 20 sec; 60 sec   **(c)** $10 < x < 30$ or $(10, 30)$

59. (a) 150 mi   **(b)** 6.06, 8.44, 15.72, 18.11 min   **(c)** Before 6.06 min, between 8.44 and 15.72 min, and after 18.11 min   **(d)** No

61. 28.90°   **63.** Yes; it varies from 1.28 to 1.34.   **65.** 1.47

67. If θ is the original angle of incidence and ϕ is the angle of refraction, then $\dfrac{\sin\theta}{\sin\phi} = n_2$. The angle of incidence of the emerging beam is also ϕ, and the index of refraction is $\dfrac{1}{n_2}$. Thus, θ is the angle of refraction of the emerging beam.

6.8 Assess Your Understanding *(page 508)*

5. $\dfrac{\pi}{2}, \dfrac{2\pi}{3}, \dfrac{4\pi}{3}, \dfrac{3\pi}{2}$ **7.** $\dfrac{\pi}{2}, \dfrac{7\pi}{6}, \dfrac{11\pi}{6}$ **9.** $0, \dfrac{\pi}{4}, \dfrac{5\pi}{4}$ **11.** $\dfrac{\pi}{2}, \dfrac{2\pi}{3}, \dfrac{4\pi}{3}, \dfrac{3\pi}{2}$ **13.** π **15.** $\dfrac{\pi}{3}, \dfrac{2\pi}{3}, \dfrac{4\pi}{3}, \dfrac{5\pi}{3}$ **17.** $\dfrac{\pi}{4}, \dfrac{5\pi}{4}$ **19.** $0, \dfrac{\pi}{3}, \pi, \dfrac{5\pi}{3}$ **21.** $\dfrac{\pi}{2}, \dfrac{3\pi}{2}$

23. $0, \dfrac{2\pi}{3}, \dfrac{4\pi}{3}$ **25.** $0, \dfrac{\pi}{3}, \dfrac{\pi}{2}, \dfrac{2\pi}{3}, \pi, \dfrac{4\pi}{3}, \dfrac{3\pi}{2}, \dfrac{5\pi}{3}$ **27.** $0, \dfrac{\pi}{5}, \dfrac{2\pi}{5}, \dfrac{3\pi}{5}, \dfrac{4\pi}{5}, \pi, \dfrac{6\pi}{5}, \dfrac{7\pi}{5}, \dfrac{8\pi}{5}, \dfrac{9\pi}{5}$ **29.** $\dfrac{\pi}{6}, \dfrac{5\pi}{6}, \dfrac{3\pi}{2}$ **31.** $\dfrac{\pi}{2}$

33. 0 **35.** $\dfrac{\pi}{3}, \dfrac{5\pi}{3}$ **37.** No real solutions **39.** No real solutions **41.** $\dfrac{\pi}{2}, \dfrac{7\pi}{6}$ **43.** $0, \dfrac{\pi}{3}, \pi, \dfrac{5\pi}{3}$ **45.** $\dfrac{\pi}{4}$

47. $-1.29, 0$ **49.** $-2.24, 0, 2.24$ **51.** $-0.82, 0.82$ **53.** $-1.31, 1.98, 3.84$
55. 0.52
57. 1.26

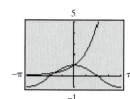   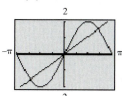   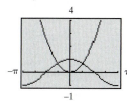

59. $-1.02, 1.02$
61. $0, 2.15$
63. $0.76, 1.35$

65. (a) $60°$ **(b)** $60°$ **67.** $2.03, 4.91$
(c) $A(60°) = 12\sqrt{3}$ in.²   **69. (a)** $30°, 60°$
(d)   **(b)** 123.6 m
                                       (d)

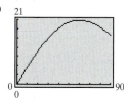

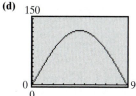

$\theta_{max} = 60°$
Maximum area $= 20.78$ in.²

Review Exercises *(page 512)*

1. $\dfrac{\pi}{2}$ **3.** $\dfrac{\pi}{4}$ **5.** $\dfrac{5\pi}{6}$ **7.** $\dfrac{\pi}{4}$ **9.** $-\sqrt{3}$ **11.** $\dfrac{2\sqrt{3}}{3}$ **13.** $\dfrac{3}{5}$ **15.** $-\dfrac{4}{3}$ **17.** $-\dfrac{\pi}{6}$ **19.** $-\dfrac{\pi}{4}$ **21.** $\tan\theta\cot\theta - \sin^2\theta = 1 - \sin^2\theta = \cos^2\theta$

23. $\cos^2\theta(1 + \tan^2\theta) = \cos^2\theta\sec^2\theta = 1$ **25.** $4\cos^2\theta + 3\sin^2\theta = \cos^2\theta + 3(\cos^2\theta + \sin^2\theta) = 3 + \cos^2\theta$

27. $\dfrac{1 - \cos\theta}{\sin\theta} + \dfrac{\sin\theta}{1 - \cos\theta} = \dfrac{(1 - \cos\theta)^2 + \sin^2\theta}{\sin\theta(1 - \cos\theta)} = \dfrac{1 - 2\cos\theta + \cos^2\theta + \sin^2\theta}{\sin\theta(1 - \cos\theta)} = \dfrac{2(1 - \cos\theta)}{\sin\theta(1 - \cos\theta)} = 2\csc\theta$

29. $\dfrac{\cos\theta}{\cos\theta - \sin\theta} = \dfrac{\dfrac{\cos\theta}{\cos\theta}}{\dfrac{\cos\theta - \sin\theta}{\cos\theta}} = \dfrac{1}{1 - \dfrac{\sin\theta}{\cos\theta}} = \dfrac{1}{1 - \tan\theta}$

31. $\dfrac{\csc\theta}{1 + \csc\theta} = \dfrac{\dfrac{1}{\sin\theta}}{1 + \dfrac{1}{\sin\theta}} = \dfrac{1}{1 + \sin\theta} = \dfrac{1}{1 + \sin\theta} \cdot \dfrac{1 - \sin\theta}{1 - \sin\theta} = \dfrac{1 - \sin\theta}{1 - \sin^2\theta} = \dfrac{1 - \sin\theta}{\cos^2\theta}$

33. $\csc\theta - \sin\theta = \dfrac{1}{\sin\theta} - \sin\theta = \dfrac{1 - \sin^2\theta}{\sin\theta} = \dfrac{\cos^2\theta}{\sin\theta} = \cos\theta \cdot \dfrac{\cos\theta}{\sin\theta} = \cos\theta\cot\theta$

35. $\dfrac{1 - \sin\theta}{\sec\theta} = \cos\theta(1 - \sin\theta) \cdot \dfrac{1 + \sin\theta}{1 + \sin\theta} = \dfrac{\cos\theta(1 - \sin^2\theta)}{1 + \sin\theta} = \dfrac{\cos^3\theta}{1 + \sin\theta}$

37. $\cot\theta - \tan\theta = \dfrac{\cos\theta}{\sin\theta} - \dfrac{\sin\theta}{\cos\theta} = \dfrac{\cos^2\theta - \sin^2\theta}{\sin\theta\cos\theta} = \dfrac{1 - 2\sin^2\theta}{\sin\theta\cos\theta}$

39. $\dfrac{\cos(\alpha + \beta)}{\cos\alpha\sin\beta} = \dfrac{\cos\alpha\cos\beta - \sin\alpha\sin\beta}{\cos\alpha\sin\beta} = \dfrac{\cos\alpha\cos\beta}{\cos\alpha\sin\beta} - \dfrac{\sin\alpha\sin\beta}{\cos\alpha\sin\beta} = \cot\beta - \tan\alpha$

41. $\dfrac{\cos(\alpha - \beta)}{\cos\alpha\cos\beta} = \dfrac{\cos\alpha\cos\beta + \sin\alpha\sin\beta}{\cos\alpha\cos\beta} = \dfrac{\cos\alpha\cos\beta}{\cos\alpha\cos\beta} + \dfrac{\sin\alpha\sin\beta}{\cos\alpha\cos\beta} = 1 + \tan\alpha\tan\beta$

43. $(1 + \cos\theta)\left(\tan\dfrac{\theta}{2}\right) = (1 + \cos\theta) \cdot \dfrac{\sin\theta}{1 + \cos\theta} = \sin\theta$

45. $2\cot\theta\cot 2\theta = 2\left(\dfrac{\cos\theta}{\sin\theta}\right)\left(\dfrac{\cos 2\theta}{\sin 2\theta}\right) = \dfrac{2\cos\theta(\cos^2\theta - \sin^2\theta)}{2\sin^2\theta\cos\theta} = \dfrac{\cos^2\theta - \sin^2\theta}{\sin^2\theta} = \cot^2\theta - 1$

47. $1 - 8\sin^2\theta\cos^2\theta = 1 - 2(2\sin\theta\cos\theta)^2 = 1 - 2\sin^2(2\theta) = \cos(4\theta)$ **49.** $\dfrac{\sin(2\theta) + \sin(4\theta)}{\cos(2\theta) + \cos(4\theta)} = \dfrac{2\sin(3\theta)\cos(-\theta)}{2\cos(3\theta)\cos(-\theta)} = \tan(3\theta)$

51. $\dfrac{\cos(2\theta) - \cos(4\theta)}{\cos(2\theta) + \cos(4\theta)} - \tan\theta\tan(3\theta) = \dfrac{-2\sin(3\theta)\sin(-\theta)}{2\cos(3\theta)\cos(-\theta)} - \tan\theta\tan(3\theta) = \tan(3\theta)\tan\theta - \tan\theta\tan(3\theta) = 0$

53. $\dfrac{1}{4}(\sqrt{6} - \sqrt{2})$ **55.** $\dfrac{1}{4}(\sqrt{6} - \sqrt{2})$ **57.** $\dfrac{1}{2}$ **59.** $\sqrt{2} - 1$

61. (a) $-\dfrac{33}{65}$ **(b)** $-\dfrac{56}{65}$ **(c)** $-\dfrac{63}{65}$ **(d)** $\dfrac{33}{56}$ **(e)** $\dfrac{24}{25}$ **(f)** $\dfrac{119}{169}$ **(g)** $\dfrac{5\sqrt{26}}{26}$ **(h)** $\dfrac{2\sqrt{5}}{5}$

63. (a) $-\dfrac{16}{65}$ **(b)** $-\dfrac{63}{65}$ **(c)** $-\dfrac{56}{65}$ **(d)** $\dfrac{16}{63}$ **(e)** $\dfrac{24}{25}$ **(f)** $\dfrac{119}{169}$ **(g)** $\dfrac{\sqrt{26}}{26}$ **(h)** $-\dfrac{\sqrt{10}}{10}$

65. (a) $-\dfrac{63}{65}$ **(b)** $\dfrac{16}{65}$ **(c)** $\dfrac{33}{65}$ **(d)** $-\dfrac{63}{16}$ **(e)** $\dfrac{24}{25}$ **(f)** $-\dfrac{119}{169}$ **(g)** $\dfrac{2\sqrt{13}}{13}$ **(h)** $-\dfrac{\sqrt{10}}{10}$

67. (a) $\dfrac{-\sqrt{3} - 2\sqrt{2}}{6}$ **(b)** $\dfrac{1 - 2\sqrt{6}}{6}$ **(c)** $\dfrac{-\sqrt{3} + 2\sqrt{2}}{6}$ **(d)** $\dfrac{8\sqrt{2} + 9\sqrt{3}}{23}$ **(e)** $-\dfrac{\sqrt{3}}{2}$ **(f)** $-\dfrac{7}{9}$ **(g)** $\dfrac{\sqrt{3}}{3}$ **(h)** $\dfrac{\sqrt{3}}{2}$

69. (a) 1 **(b)** 0 **(c)** $-\dfrac{1}{9}$ **(d)** Not defined **(e)** $\dfrac{4\sqrt{5}}{9}$ **(f)** $-\dfrac{1}{9}$ **(g)** $\dfrac{\sqrt{30}}{6}$ **(h)** $-\dfrac{\sqrt{6}\sqrt{3 - \sqrt{5}}}{6}$

71. $\dfrac{4 + 3\sqrt{3}}{10}$ **73.** $-\dfrac{48 + 25\sqrt{3}}{39}$ **75.** $-\dfrac{24}{25}$ **77.** $\left\{\dfrac{\pi}{3}, \dfrac{5\pi}{3}\right\}$ **79.** $\left\{\dfrac{3\pi}{4}, \dfrac{5\pi}{4}\right\}$ **81.** $\left\{\dfrac{3\pi}{4}, \dfrac{7\pi}{4}\right\}$ **83.** $\left\{0, \dfrac{\pi}{2}, \pi, \dfrac{3\pi}{2}\right\}$ **85.** $\left\{\dfrac{\pi}{3}, \dfrac{2\pi}{3}, \dfrac{4\pi}{3}, \dfrac{5\pi}{3}\right\}$

87. $\{0, \pi\}$ **89.** $\left\{0, \dfrac{2\pi}{3}, \pi, \dfrac{4\pi}{3}\right\}$ **91.** $\left\{0, \dfrac{\pi}{6}, \dfrac{5\pi}{6}\right\}$ **93.** $\left\{\dfrac{\pi}{6}, \dfrac{\pi}{2}, \dfrac{5\pi}{6}\right\}$ **95.** $\left\{\dfrac{\pi}{3}, \dfrac{5\pi}{3}\right\}$ **97.** $\left\{\dfrac{\pi}{4}, \dfrac{\pi}{2}, \dfrac{3\pi}{4}, \dfrac{3\pi}{2}\right\}$ **99.** $\left\{\dfrac{\pi}{2}, \pi\right\}$ **101.** 0.78 **103.** −1.11

105. 1.23 **107.** {1.11} **109.** {0.87} **111.** {2.22} **113.** $\sin\left(\dfrac{30°}{2}\right) = \dfrac{\sqrt{2 - \sqrt{3}}}{2}; \sin(45° - 30°) = \dfrac{1}{4}(\sqrt{6} - \sqrt{2})$

Chapter Test *(page 514)*

1. $\dfrac{\pi}{6}$ **2.** $-\dfrac{\pi}{4}$ **3.** $\dfrac{2\pi}{3}$ **4.** $\dfrac{7\sqrt{58}}{58}$ **5.** 3 **6.** $-\dfrac{4}{3}$ **7.** ≈0.392 **8.** ≈0.775 **9.** ≈1.249 **10.** ≈0.197

11. $\dfrac{(\csc\theta + \cot\theta)}{(\sec\theta + \tan\theta)} = \dfrac{(\csc\theta + \cot\theta)}{(\sec\theta + \tan\theta)} \cdot \dfrac{(\csc\theta - \cot\theta)}{(\csc\theta - \cot\theta)} = \dfrac{(\csc^2\theta - \cot^2\theta)}{(\sec\theta + \tan\theta)(\csc\theta - \cot\theta)} = \dfrac{1}{(\sec\theta + \tan\theta)(\csc\theta - \cot\theta)}$

$= \dfrac{1}{(\sec\theta + \tan\theta)(\csc\theta - \cot\theta)} \cdot \dfrac{\sec\theta - \tan\theta}{\sec\theta - \tan\theta} = \dfrac{\sec\theta - \tan\theta}{(\sec^2\theta - \tan^2\theta)(\csc\theta - \cot\theta)} = \dfrac{\sec\theta - \tan\theta}{\csc\theta - \cot\theta}$

12. $\sin\theta\tan\theta + \cos\theta = \sin\theta \cdot \dfrac{\sin\theta}{\cos\theta} + \cos\theta = \dfrac{\sin^2\theta}{\cos\theta} + \dfrac{\cos^2\theta}{\cos\theta} = \dfrac{\sin^2\theta + \cos^2\theta}{\cos\theta} = \dfrac{1}{\cos\theta} = \sec\theta$

13. $\tan\theta + \cot\theta = \dfrac{\sin\theta}{\cos\theta} + \dfrac{\cos\theta}{\sin\theta} = \dfrac{\sin^2\theta}{\sin\theta\cos\theta} + \dfrac{\cos^2\theta}{\sin\theta\cos\theta} = \dfrac{\sin^2\theta + \cos^2\theta}{\sin\theta\cos\theta} = \dfrac{1}{\sin\theta\cos\theta}$

$= \dfrac{2}{2\sin\theta\cos\theta} = \dfrac{2}{\sin(2\theta)} = 2\csc(2\theta)$

14. $\dfrac{\sin(\alpha + \beta)}{\tan\alpha + \tan\beta} = \dfrac{\sin\alpha\cos\beta + \cos\alpha\sin\beta}{\dfrac{\sin\alpha}{\cos\alpha} + \dfrac{\sin\beta}{\cos\beta}} = \dfrac{\sin\alpha\cos\beta + \cos\alpha\sin\beta}{\dfrac{\sin\alpha\cos\beta}{\cos\alpha\cos\beta} + \dfrac{\cos\alpha\sin\beta}{\cos\alpha\cos\beta}} = \dfrac{\sin\alpha\cos\beta + \cos\alpha\sin\beta}{\dfrac{\sin\alpha\cos\beta + \cos\alpha\sin\beta}{\cos\alpha\cos\beta}}$

$= \dfrac{\sin\alpha\cos\beta + \cos\alpha\sin\beta}{1} \cdot \dfrac{\cos\alpha\cos\beta}{\sin\alpha\cos\beta + \cos\alpha\sin\beta} = \cos\alpha\cos\beta$

15. $\sin(3\theta) = \sin(\theta + 2\theta) = \sin\theta\cos(2\theta) + \cos\theta\sin(2\theta) = \sin\theta \cdot (\cos^2\theta - \sin^2\theta) + \cos\theta \cdot 2\sin\theta\cos\theta = \sin\theta\cos^2\theta - \sin^3\theta + 2\sin\theta\cos^2\theta$
$= 3\sin\theta\cos^2\theta - \sin^3\theta = 3\sin\theta(1 - \sin^2\theta) - \sin^3\theta = 3\sin\theta - 3\sin^3\theta - \sin^3\theta = 3\sin\theta - 4\sin^3\theta$

16. $\dfrac{(\tan\theta - \cot\theta)}{(\tan\theta + \cot\theta)} = \dfrac{\dfrac{\sin\theta}{\cos\theta} - \dfrac{\cos\theta}{\sin\theta}}{\dfrac{\sin\theta}{\cos\theta} + \dfrac{\cos\theta}{\sin\theta}} = \dfrac{\dfrac{\sin^2\theta - \cos^2\theta}{\sin\theta\cos\theta}}{\dfrac{\sin^2\theta + \cos^2\theta}{\sin\theta\cos\theta}} = \dfrac{\sin^2\theta - \cos^2\theta}{\sin^2\theta + \cos^2\theta} = \dfrac{-\cos(2\theta)}{1} = -(2\cos^2\theta - 1) = 1 - 2\cos^2\theta$

17. $\dfrac{\sqrt{2}}{4}(\sqrt{3}+1)$ **18.** $2+\sqrt{3}$ **19.** $\dfrac{\sqrt{5}}{5}$ **20.** $\dfrac{12\sqrt{85}}{49}$ **21.** $\dfrac{2\sqrt{13}(\sqrt{5}-3)}{39}$ **22.** $\dfrac{2+\sqrt{3}}{4}$ **23.** $\dfrac{\sqrt{6}}{2}$ **24.** $\dfrac{\sqrt{2}}{2}$

25. $\left\{\dfrac{\pi}{3},\dfrac{2\pi}{3},\dfrac{4\pi}{3},\dfrac{5\pi}{3}\right\}$ **26.** $\{0,1.911,\pi,4.373\}$ **27.** $\left\{\dfrac{3\pi}{8},\dfrac{7\pi}{8},\dfrac{11\pi}{8},\dfrac{15\pi}{8}\right\}$ **28.** $\{0.285,3.427\}$ **29.** $\{0.253,2.889\}$

30. The change in elevation during the time trial was 1.185 kilometers.

Cumulative Review *(page 516)*

1. $\left\{\dfrac{-1-\sqrt{13}}{6},\dfrac{-1+\sqrt{13}}{6}\right\}$ **2.** $y+1=-1(x-4)$ or $x+y=3; 6\sqrt{2};(1,2)$ **3.** *x*-axis symmetry; $(0,-3),(0,3),(3,0)$

4. 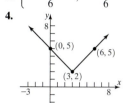   **5.** 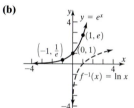   **6.**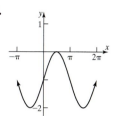

7. (a) 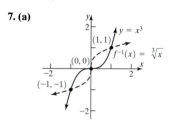   **(b)** 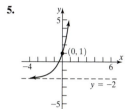   **(c)** 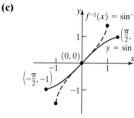   **(d)**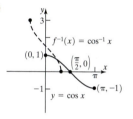

8. (a) $-\dfrac{2\sqrt{2}}{3}$ **(b)** $\dfrac{\sqrt{2}}{4}$ **(c)** $\dfrac{4\sqrt{2}}{9}$ **(d)** $\dfrac{7}{9}$ **(e)** $\sqrt{\dfrac{3+2\sqrt{2}}{6}}$ **(f)** $-\sqrt{\dfrac{3-2\sqrt{2}}{6}}$ **9.** $\dfrac{\sqrt{5}}{5}$

10. (a) $-\dfrac{2\sqrt{2}}{3}$ **(b)** $-\dfrac{2\sqrt{2}}{3}$ **(c)** $\dfrac{7}{9}$ **(d)** $\dfrac{4\sqrt{2}}{9}$ **(e)** $\dfrac{\sqrt{6}}{3}$

11. (a) $f(x)=(2x-1)(x-1)^2(x+1)^2; \dfrac{1}{2}$ multiplicity 1; 1 and -1 multiplicity 2

(b) $(0,-1); \left(\dfrac{1}{2},0\right);(-1,0);(1,0)$ **(c)** $y=2x^5$

(d)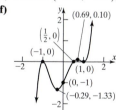

(e) Minima $(-0.29,-1.33),(1,0)$; maxima $(-1,0),(0.69,0.10)$

(f)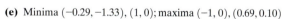

(g) Increasing: $(-\infty,-1),(-0.29,0.69),(1,\infty)$; Decreasing: $(-1,-0.29),(0.69,1)$

12. (a) $\left\{-1,-\dfrac{1}{2}\right\}$ **(b)** $\{-1,1\}$ **(c)** $(-\infty,-1)$ or $\left(-\dfrac{1}{2},\infty\right)$ **(d)** $(-\infty,-1],[1,\infty)$

C H A P T E R   7   Applications of Trigonometric Functions

7.1 Assess Your Understanding *(page 526)*

3. F   **4.** T   **5.** angle of elevation   **6.** angle of depression   **7.** T   **8.** F

9. $\sin\theta=\dfrac{5}{13}; \cos\theta=\dfrac{12}{13}; \tan\theta=\dfrac{5}{12}; \cot\theta=\dfrac{12}{5}; \sec\theta=\dfrac{13}{12}; \csc\theta=\dfrac{13}{5};$

11. $\sin\theta=\dfrac{2\sqrt{13}}{13}; \cos\theta=\dfrac{3\sqrt{13}}{13}; \tan\theta=\dfrac{2}{3}; \cot\theta=\dfrac{3}{2}; \sec\theta=\dfrac{\sqrt{13}}{3}; \csc\theta=\dfrac{\sqrt{13}}{2}$

13. $\sin\theta=\dfrac{\sqrt{3}}{2}; \cos\theta=\dfrac{1}{2}; \tan\theta=\sqrt{3}; \cot\theta=\dfrac{\sqrt{3}}{3}; \sec\theta=2; \csc\theta=\dfrac{2\sqrt{3}}{3}$

15. $\sin\theta = \dfrac{\sqrt{6}}{3}$; $\cos\theta = \dfrac{\sqrt{3}}{3}$; $\tan\theta = \sqrt{2}$; $\cot\theta = \dfrac{\sqrt{2}}{2}$; $\sec\theta = \sqrt{3}$; $\csc\theta = \dfrac{\sqrt{6}}{2}$

17. $\sin\theta = \dfrac{\sqrt{5}}{5}$; $\cos\theta = \dfrac{2\sqrt{5}}{5}$; $\tan\theta = \dfrac{1}{2}$; $\cot\theta = 2$; $\sec\theta = \dfrac{\sqrt{5}}{2}$; $\csc\theta = \sqrt{5}$

19. 0 **21.** 1 **23.** 0 **25.** 0 **27.** 1 **29.** $a \approx 13.74, c \approx 14.62, \alpha = 70°$ **31.** $b \approx 5.03, c \approx 7.83, \alpha = 50°$ **33.** $a \approx 0.71, c \approx 4.06, \beta = 80°$

35. $b \approx 10.72, c \approx 11.83, \beta = 65°$ **37.** $b \approx 3.08, a \approx 8.46, \alpha = 70°$ **39.** $c \approx 5.83, \alpha \approx 59.0°, \beta = 31.0°$ **41.** $b \approx 4.58, \alpha \approx 23.6°, \beta = 66.4°$

43. 4.59 in., 6.55 in. **45.** 5.52 in. or 11.83 in. **47.** 23.6° and 66.4° **49.** 70.02 ft **51.** 985.91 ft **53.** 137.37 m **55.** 20.67 ft **57.** 1978.09 ft

59. 60.27 ft **61.** 530.18 ft **63.** 554.52 ft **65. (a)** 111.96 ft/sec or 76.3 mi/hr **(b)** 82.42 ft/sec or 56.2 mi/hr **(c)** Under 18.8°

67. S76.6°E **69.** 69.0° **71.** 38.9° **73.** No; move the tripod back about 1 ft. **75. (a)** $A(\theta) = 2\sin\theta\cos\theta$ **(b)** From Double-angle Formula,

since $2\sin\theta\cos\theta = \sin(2\theta)$ **(c)** $\theta = 45°$ **(d)** $\dfrac{\sqrt{2}}{2}$ by $\sqrt{2}$

7.2 Assess Your Understanding *(page 538)*

4. oblique **5.** $\dfrac{\sin\alpha}{a} = \dfrac{\sin\beta}{b} = \dfrac{\sin\gamma}{c}$ **6.** F **7.** T **8.** F

9. $a \approx 3.23, b \approx 3.55, \alpha = 40°$ **11.** $a \approx 3.25, c \approx 4.23, \beta = 45°$ **13.** $\gamma = 95°, c \approx 9.86, a \approx 6.36$ **15.** $\alpha = 40°, a = 2, c \approx 3.06$

17. $\gamma = 120°, b \approx 1.06, c \approx 2.69$ **19.** $\alpha = 100°, a \approx 5.24, c \approx 0.92$ **21.** $\beta = 40°, a \approx 5.64, b \approx 3.86$ **23.** $\gamma = 100°, a \approx 1.31, b \approx 1.31$

25. One triangle; $\beta \approx 30.7°, \gamma \approx 99.3°, c \approx 3.86$ **27.** One triangle; $\gamma \approx 36.2°, \alpha \approx 43.8°, a \approx 3.51$ **29.** No triangle

31. Two triangles; $\gamma_1 \approx 30.9°, \alpha_1 \approx 129.1°, a_1 \approx 9.07$ or $\gamma_2 \approx 149.1°, \alpha_2 \approx 10.9°, a_2 \approx 2.20$ **33.** No triangle

35. Two triangles; $\alpha_1 \approx 57.7°, \beta_1 \approx 97.3°, b_1 \approx 2.35$ or $\alpha_2 \approx 122.3°, \beta_2 \approx 32.7°, b_2 \approx 1.28$

37. (a) Station Able is about 143.33 mi from the ship; Station Baker is about 135.58 mi from the ship. **(b)** Approximately 41 min

39. 1490.48 ft **41.** 381.69 ft **43. (a)** 169.18 mi **(b)** 161.3° **45.** 84.7°; 183.72 ft **47.** 2.64 mi **49.** 38.5 in. **51.** 449.36 ft

53. 187,600,000 km or 101,440,000 km **55.** 39.39 ft **57.** 29.97 ft

59. $\dfrac{a-b}{c} = \dfrac{a}{c} - \dfrac{b}{c} = \dfrac{\sin\alpha}{\sin\gamma} - \dfrac{\sin\beta}{\sin\gamma} = \dfrac{\sin\alpha - \sin\beta}{\sin\gamma} = \dfrac{2\sin\left(\dfrac{\alpha-\beta}{2}\right)\cos\left(\dfrac{\alpha+\beta}{2}\right)}{2\sin\dfrac{\gamma}{2}\cos\dfrac{\gamma}{2}} = \dfrac{\sin\left(\dfrac{\alpha-\beta}{2}\right)\cos\left(\dfrac{\pi}{2}-\dfrac{\gamma}{2}\right)}{\sin\dfrac{\gamma}{2}\cos\dfrac{\gamma}{2}} = \dfrac{\sin\left(\dfrac{\alpha-\beta}{2}\right)}{\cos\dfrac{\gamma}{2}}$

61. $\dfrac{a-b}{a+b} = \dfrac{\dfrac{a-b}{c}}{\dfrac{a+b}{c}} = \dfrac{\dfrac{\cos\dfrac{\gamma}{2}\,\sin\left[\dfrac{1}{2}(\alpha-\beta)\right]}{\cos\left[\dfrac{1}{2}(\alpha-\beta)\right]}}{\dfrac{\sin\dfrac{\gamma}{2}}{}} \cdot = \dfrac{\tan\left[\dfrac{1}{2}(\alpha-\beta)\right]}{\cot\dfrac{\gamma}{2}} = \dfrac{\tan\left[\dfrac{1}{2}(\alpha-\beta)\right]}{\tan\left(\dfrac{\pi}{2}-\dfrac{\gamma}{2}\right)} = \dfrac{\tan\left[\dfrac{1}{2}(\alpha-\beta)\right]}{\tan\left[\dfrac{1}{2}(\alpha+\beta)\right]}$

7.3 Assess Your Understanding *(page 546)*

3. Cosines **4.** Sines **5.** Cosines **6.** F **7.** F **8.** T **9.** $b \approx 2.95, \alpha \approx 28.7°, \gamma \approx 106.3°$

11. $c \approx 3.75, \alpha \approx 32.1°, \beta \approx 52.9°$ **13.** $\alpha \approx 48.5°, \beta \approx 38.6°, \gamma \approx 92.9°$ **15.** $\alpha \approx 127.2°, \beta \approx 32.1°, \gamma \approx 20.7°$ **17.** $c \approx 2.57, \alpha \approx 48.6°, \beta \approx 91.4°$

19. $a \approx 2.99, \beta \approx 19.2°, \gamma \approx 80.8°$ **21.** $b \approx 4.14, \alpha \approx 43.0°, \gamma \approx 27.0°$ **23.** $c \approx 1.69, \alpha \approx 65.0°, \beta \approx 65.0°$ **25.** $\alpha \approx 67.4°, \beta = 90°, \gamma \approx 22.6°$

27. $\alpha = 60°, \beta = 60°, \gamma = 60°$ **29.** $\alpha \approx 33.6°, \beta \approx 62.2°, \gamma \approx 84.3°$ **31.** $\alpha \approx 97.9°, \beta \approx 52.4°, \gamma \approx 29.7°$ **33.** 165 yd **35. (a)** 26.4° **(b)** 30.8 hr

37. (a) 63.7 ft **(b)** 66.8 ft **(c)** 92.8° **39. (a)** 492.6 ft **(b)** 269.3 ft **41.** 342.3 ft

43. Using the Law of Cosines:

$L^2 = x^2 + r^2 - 2rx\cos\theta$

$x^2 - 2rx\cos\theta + r^2 - L^2 = 0$

Then, using the quadratic formula:

$x = r\cos\theta + \sqrt{r^2\cos^2\theta + L^2 - r^2}$

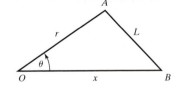

45. $\cos \dfrac{\gamma}{2} = \sqrt{\dfrac{1 + \cos \gamma}{2}} = \sqrt{\dfrac{1 + \dfrac{a^2 + b^2 - c^2}{2ab}}{2}} = \sqrt{\dfrac{2ab + a^2 + b^2 - c^2}{4ab}} = \sqrt{\dfrac{(a + b)^2 - c^2}{4ab}} = \sqrt{\dfrac{(a + b + c)(a + b - c)}{4ab}}$

$= \sqrt{\dfrac{2s(2s - 2c)}{4ab}} = \sqrt{\dfrac{s(s - c)}{ab}}$

47. $\dfrac{\cos \alpha}{a} + \dfrac{\cos \beta}{b} + \dfrac{\cos \gamma}{c} = \dfrac{b^2 + c^2 - a^2}{2abc} + \dfrac{a^2 + c^2 - b^2}{2abc} + \dfrac{a^2 + b^2 - c^2}{2abc} = \dfrac{b^2 + c^2 - a^2 + a^2 + c^2 - b^2 + a^2 + b^2 - c^2}{2abc} = \dfrac{a^2 + b^2 + c^2}{2abc}$

CHAPTER 8   Polar Coordinates; Vectors

8.1 Assess Your Understanding *(page 559)*

5. pole; polar axis   **6.** -2 **7.** $(-\sqrt{3}, -1)$ **8.** F   **9.** T   **10.** T   **11.** A   **13.** C   **15.** B   **17.** A

19.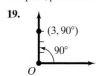

21.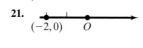

23.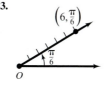

25.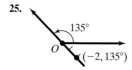

27.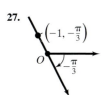

29.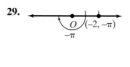

31.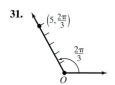

33.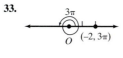

35.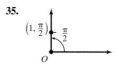

37.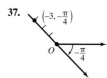

(a) $\left(5, -\dfrac{4\pi}{3}\right)$

(b) $\left(-5, \dfrac{5\pi}{3}\right)$

(c) $\left(5, \dfrac{8\pi}{3}\right)$

(a) $(2, -2\pi)$

(b) $(-2, \pi)$

(c) $(2, 2\pi)$

(a) $\left(1, -\dfrac{3\pi}{2}\right)$

(b) $\left(-1, \dfrac{3\pi}{2}\right)$

(c) $\left(1, \dfrac{5\pi}{2}\right)$

(a) $\left(3, -\dfrac{5\pi}{4}\right)$

(b) $\left(-3, \dfrac{7\pi}{4}\right)$

(c) $\left(3, \dfrac{11\pi}{4}\right)$

39. $(0, 3)$ **41.** $(-2, 0)$ **43.** $(-3\sqrt{3}, 3)$ **45.** $(\sqrt{2}, -\sqrt{2})$ **47.** $\left(-\dfrac{1}{2}, \dfrac{\sqrt{3}}{2}\right)$ **49.** $(2, 0)$ **51.** $(-2.57, 7.05)$ **53.** $(-4.98, -3.85)$ **55.** $(3, 0)$

57. $(1, \pi)$ **59.** $\left(\sqrt{2}, -\dfrac{\pi}{4}\right)$ **61.** $\left(2, \dfrac{\pi}{6}\right)$ **63.** $(2.47, -1.02)$ **65.** $(9.30, 0.47)$ **67.** $r^2 = \dfrac{3}{2}$ **69.** $r^2 \cos^2 \theta - 4r \sin \theta = 0$ **71.** $r^2 \sin 2\theta = 1$

73. $r \cos \theta = 4$ **75.** $x^2 + y^2 - x = 0$ or $\left(x - \dfrac{1}{2}\right)^2 + y^2 = \dfrac{1}{4}$ **77.** $(x^2 + y^2)^{3/2} - x = 0$ **79.** $x^2 + y^2 = 4$ **81.** $y^2 = 8(x + 2)$

83. $d = \sqrt{(r_2 \cos \theta_2 - r_1 \cos \theta_1)^2 + (r_2 \sin \theta_2 - r_1 \sin \theta_1)^2}$

$= \sqrt{(r_2^2 \cos^2 \theta_2 - 2r_2 \cos \theta_2 r_1 \cos \theta_1 + r_1^2 \cos^2 \theta_1) + (r_2^2 \sin^2 \theta_2 - 2r_2 \sin \theta_2 r_1 \sin \theta_1 + r_1^2 \sin^2 \theta_1)}$

$= \sqrt{r_1^2 + r_2^2 - 2r_1 r_2 (\cos \theta_2 \cos \theta_1 + \sin \theta_2 \sin \theta_1)}$

$= \sqrt{r_1^2 + r_2^2 - 2r_1 r_2 \cos(\theta_2 - \theta_1)}$

8.2 Assess Your Understanding *(page 577)*

7. polar equation   **8.** $r = 2\cos\theta$ **9.** $-r$ **10.** F   **11.** F   **12.** F

13. $x^2 + y^2 = 16$

Circle, radius 4, center at pole

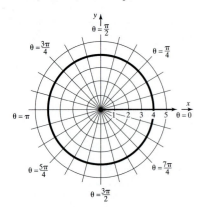

15. $y = \sqrt{3}x$; Line through pole, making an angle of $\dfrac{\pi}{3}$ with polar axis

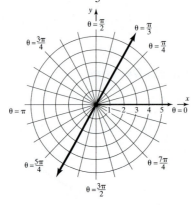

17. $y = 4$; Horizontal line 4 units above the pole

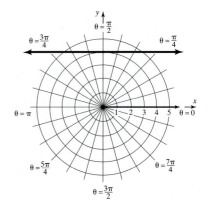

19. $x = -2$; Vertical line 2 units to the left of the pole

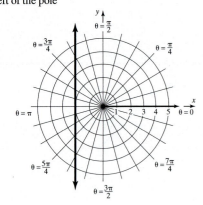

21. $(x - 1)^2 + y^2 = 1$; Circle, radius 1, center $(1, 0)$ in rectangular coordinates

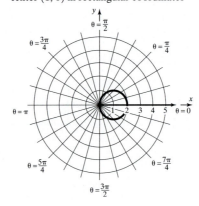

23. $x^2 + (y + 2)^2 = 4$

Circle, radius 2, center at $(0, -2)$ in rectangular coordinates

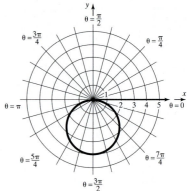

25. $(x - 2)^2 + y^2 = 4$

Circle, radius 2, center at $(2, 0)$ in rectangular coordinates

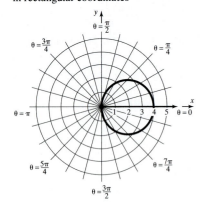

27. $x^2 + (y + 1)^2 = 1$

Circle, radius 1, center at $(0, -1)$ in rectangular coordinates

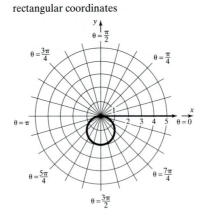

29. E   **31.** F   **33.** H   **35.** D   **37.** D   **39.** F   **41.** A

43. Cardioid

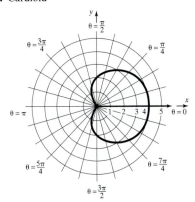

45. Cardioid

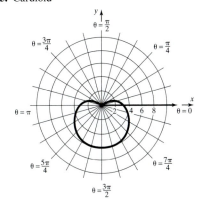

47. Limaçon without inner loop

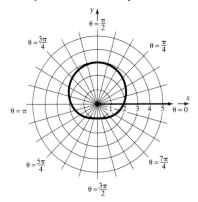

49. Limaçon without inner loop

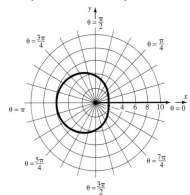

51. Limaçon with inner loop

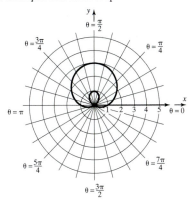

53. Limaçon with inner loop

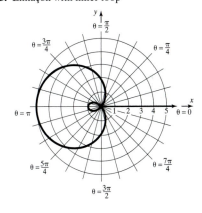

55. Rose

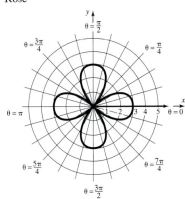

57. Rose

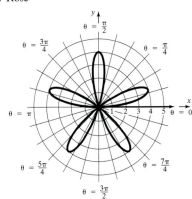

59. Lemniscate

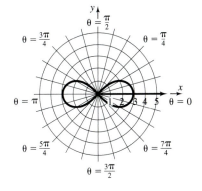

61. Spiral

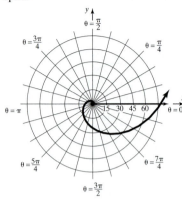

63. Cardioid

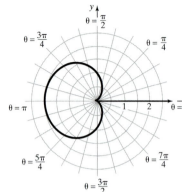

65. Limaçon with inner loop

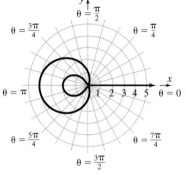

67. $r = 3 + 3 \cos \theta$

69. $r = 4 + \sin \theta$

71.

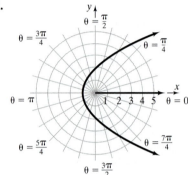

73.

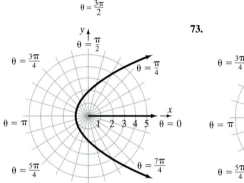

75.

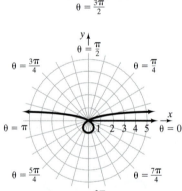

77.

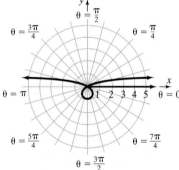

79.

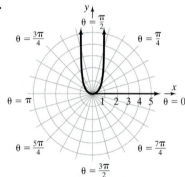

81. $r \sin \theta = a$

$y = a$

83.
$$r = 2a \sin \theta$$
$$r^2 = 2ar \sin \theta$$
$$x^2 + y^2 = 2ay$$
$$x^2 + y^2 - 2ay = 0$$
$$x^2 + (y - a)^2 = a^2$$
Circle, radius a, center at $(0, a)$
in rectangular coordinates

85.
$$r = 2a \cos \theta$$
$$r^2 = 2ar \cos \theta$$
$$x^2 + y^2 = 2ax$$
$$x^2 - 2ax + y^2 = 0$$
$$(x - a)^2 + y^2 = a^2$$
Circle, radius a, center at $(0, a)$
in rectangular coordinates

87. (a) $r^2 = \cos \theta$: $r^2 = \cos(\pi - \theta)$
$$r^2 = -\cos \theta$$
Not equivalent; test fails.
$$(-r)^2 = \cos(-\theta)$$
$$r^2 = \cos \theta$$
New test works.

(b) $r^2 = \sin \theta$: $r^2 = \sin(\pi - \theta)$
$$r^2 = \sin \theta$$
Test works.
$$(-r)^2 = \sin(-\theta)$$
$$r^2 = -\sin \theta$$
Not equivalent; new test fails.

C H A P T E R 9   Analytic Geometry

9.7 Assess Your Understanding (page 593)

2. plane curve; parameter **3.** ellipse **4.** cycloid **5.** F **6.** T

7.
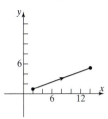
$x - 3y + 1 = 0$

9.
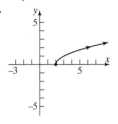
$y = \sqrt{x - 2}$

11.
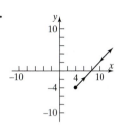
$x = y + 8$

13.
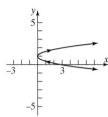
$x = 3(y - 1)^2$

15.
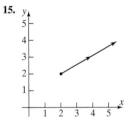
$2y = 2 + x$

17.
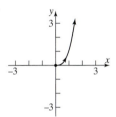
$y = x^3$

19.
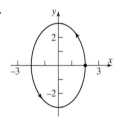
$\dfrac{x^2}{4} + \dfrac{y^2}{9} = 1$

21.
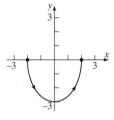
$\dfrac{x^2}{4} + \dfrac{y^2}{9} = 1$

23.
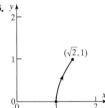
$x^2 - y^2 = 1$

25.
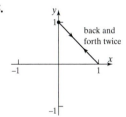
back and forth twice
$x + y = 1$

27. $x = t$ $x = \dfrac{t + 1}{4}$
$\quad\quad$ or
$y = 4t - 1$ $y = t$

29. $x = t$ $x = t^3$
$\quad\quad\quad$ or
$\quad y = t^2 + 1$ $y = t^6 + 1$

31. $x = t$ $x = \sqrt[3]{t}$
$\quad\quad\quad$ or
$\quad y = t^3$ $y = t$

33. $x = t$ $x = t^3$
$\quad\quad$ or
$y = t^{2/3}$ $y = t^2, t \geq 0$

35. $x = t + 2, y = t, 0 \leq t \leq 5$

37. $x = 3 \cos t, y = 2 \sin t, 0 \leq t \leq 2\pi$

39. $x = 2 \cos(\pi t), y = -3 \sin(\pi t), 0 \leq t \leq 2$ **41.** $x = 2 \sin(2\pi t), y = 3 \cos(2\pi t), 0 \leq t \leq 1$

43.
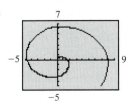
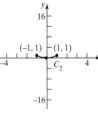
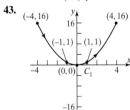
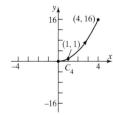

45.
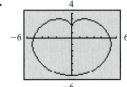

47.

49. (a) $x = 3$
$y = -16t^2 + 50t + 6$
(b) 3.24 sec
(c) 1.5625 sec; 45.0625 ft

51. (a) Train: $x_1 = t^2, y_1 = 1$;
Bill: $x_2 = 5(t - 5), y_2 = 3$
(b) Bill won't catch the train.
(c)

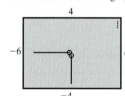

53. (a) $x = (145 \cos 20°)t$
$y = -16t^2 + (145 \sin 20°)t + 5$
(b) 3.197 sec
(c) 1.55 sec; 43.43 ft
(d) 435.61 ft
(e)

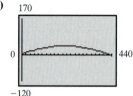

55. (a) $x = (40 \cos 45°)t$
$y = -4.9t^2 + (40 \sin 45°)t + 300$
(b) 11.226 sec
(c) 2.886 sec; 340.8 m
(d) 317.5 m
(e)

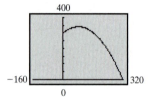

57. (a) Paseo: $x = 40t - 5, y = 0$;
Bonneville: $x = 0, y = 30t - 4$
(b) $d = \sqrt{(40t - 5)^2 + (30t - 4)^2}$
(c)

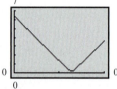

(d) 0.2 mi; 7.68 min
(e) Turn axes off to see the graph:

59. The orientation is from (x_1, y_1) to (x_2, y_2).
61. (a)

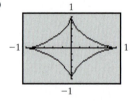

(b) $x^{2/3} + y^{2/3} = 1$

C H A P T E R   1 0   Systems of Equations and Inequalities

10.1 Assess Your Understanding *(page 609)*

3. inconsistent   **4.** consistent   **5.** F   **6.** T

7. $\begin{cases} 2(2) - (-1) = 5 \\ 5(2) + 2(-1) = 8 \end{cases}$

9. $\begin{cases} 3(2) - 4\left(\dfrac{1}{2}\right) = 4 \\ \dfrac{1}{2}(2) - 3\left(\dfrac{1}{2}\right) = -\dfrac{1}{2} \end{cases}$

11. $\begin{cases} 4 - 1 = 3 \\ \dfrac{1}{2}(4) + 1 = 3 \end{cases}$

13. $\begin{cases} 3(1) + 3(-1) + 2(2) = 4 \\ 1 - (-1) - 2 = 0 \\ 2(-1) - 3(2) = -8 \end{cases}$

15. $\begin{cases} 3(2) + 3(-2) + 2(2) = 4 \\ 2 - 3(-2) + 2 = 10 \\ 5(2) - 2(-2) - 3(2) = 8 \end{cases}$

17. $x = 6, y = 2$ **19.** $x = 3, y = 2$ **21.** $x = 8, y = -4$ **23.** $x = \dfrac{1}{3}, y = -\dfrac{1}{6}$ **25.** Inconsistent   **27.** $x = 1, y = 2$

29. $x = 4 - 2y, y$ is any real number   **31.** $x = 1, y = 1$ **33.** $x = \dfrac{3}{2}, y = 1$ **35.** $x = 4, y = 3$ **37.** $x = \dfrac{4}{3}, y = \dfrac{1}{5}$ **39.** $x = \dfrac{1}{5}, y = \dfrac{1}{3}$

41. $x = 8, y = 2, z = 0$ **43.** $x = 2, y = -1, z = 1$ **45.** Inconsistent   **47.** $x = 5z - 2, y = 4z - 3$, where z is any real number   **49.** Inconsistent

51. $x = 1, y = 3, z = -2$ **53.** $x = -3, y = \dfrac{1}{2}, z = 1$ **55.** Length 30 ft; width 15 ft   **57.** Cheeseburger \$1.55; shake \$0.85   **59.** 22.5 lb

61. Average wind speed 25 mph; average airspeed 175 mph   **63.** 80 \$25 sets and 120 \$45 sets   **65.** \$5.56   **67.** Mix 50 mg of first compound with 75 mg of second.

69. $a = \frac{4}{3}, b = -\frac{5}{3}, c = 1$ **71.** $y = 9000, r = 0.06$ **73.** $I_1 = \frac{10}{71}, I_2 = \frac{65}{71}, I_3 = \frac{55}{71}$

75. 100 orchestra, 210 main, and 190 balcony seats   **77.** 1.5 chicken, 1 corn, 2 milk

79. If x = price of hamburgers, y = price of fries, z = price of colas, then $x = 2.75 - z$, $y = \frac{41}{60} + \frac{1}{3}z$, $0.60 \le z \le 0.90$.

There is not sufficient information:

x	$2.13	$2.01	$1.86
y	$0.89	$0.93	$0.98
z	$0.62	$0.74	$0.89

81. It will take Beth 30 hr, Bill 24 hr, and Edie 40 hr.

10.7 Assess Your Understanding *(page 622)*

7. satisfied   **8.** half-plane   **9.** F   **10.** T

11.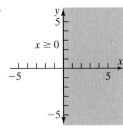

13.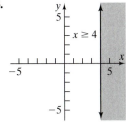

15.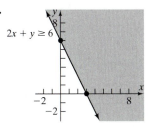

17.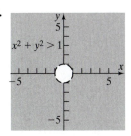

19.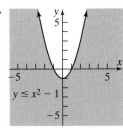

21.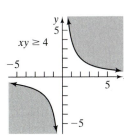

23.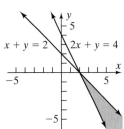

25.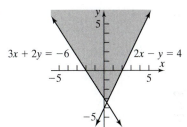

27.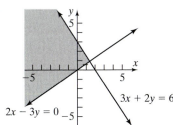

29.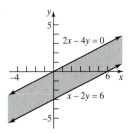

31.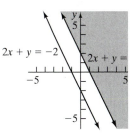

33. No solution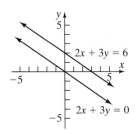

35.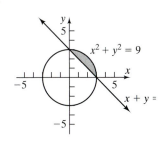

37.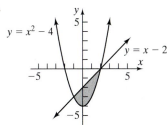

39.

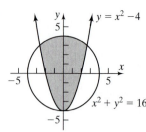

41.

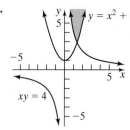

43. Bounded; corner points

$(0, 0), (3, 0), (2, 2), (0, 3)$

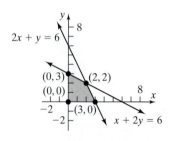

45. Unbounded; corner points

$(2, 0), (0, 4)$

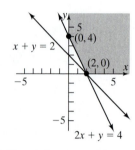

47. Bounded; corner points $(2, 0), (4, 0)$,

$\left(\dfrac{24}{7}, \dfrac{12}{7}\right), (0, 4), (0, 2)$

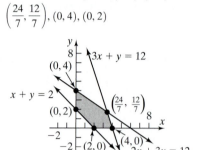

49. Bounded; corner points $(2, 0), (5, 0)$,

$(2, 6), (0, 8), (0, 2)$

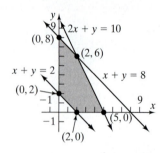

51. Bounded; corner points

$(1, 0), (10, 0), (0, 5), \left(0, \dfrac{1}{2}\right)$

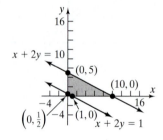

53. $\begin{cases} x \leq 4 \\ x + y \leq 6 \\ x \geq 0 \\ y \geq 0 \end{cases}$ **55.** $\begin{cases} x \leq 20 \\ y \geq 15 \\ x + y \leq 50 \\ x - y \leq 0 \\ x \geq 0 \end{cases}$

57. (a) $\begin{cases} x + y \leq 50{,}000 \\ x \geq 35{,}000 \\ y \leq 10{,}000 \\ x \geq 0 \\ y \geq 0 \end{cases}$

59. (a) $\begin{cases} x \geq 0 \\ y \geq 0 \\ x + 2y \leq 300 \\ 3x + 2y \leq 480 \end{cases}$

61. (a) $\begin{cases} 3x + 2y \leq 160 \\ 2x + 3y \leq 150 \\ x \geq 0 \\ y \geq 0 \end{cases}$

(b)

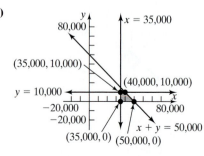

(b)

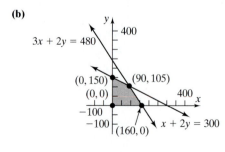

(b)

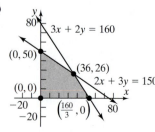

C H A P T E R   1 1   Sequences; Induction; the Binomial Theorem

11.1 Assess Your Understanding *(page 635)*

5. sequence   **6.** $3; 15$ **7.** 20 **8.** T   **9.** T   **10.** T   **11.** $3,628,800$ **13.** 504 **15.** 1260 **17.** $1, 2, 3, 4, 5$

19. $\dfrac{1}{3}, \dfrac{1}{2}, \dfrac{3}{5}, \dfrac{2}{3}, \dfrac{5}{7}$ **21.** $1, -4, 9, -16, 25$ **23.** $\dfrac{1}{2}, \dfrac{2}{5}, \dfrac{2}{7}, \dfrac{8}{41}, \dfrac{8}{61}$ **25.** $-\dfrac{1}{6}, \dfrac{1}{12}, -\dfrac{1}{20}, \dfrac{1}{30}, -\dfrac{1}{42}$ **27.** $\dfrac{1}{e}, \dfrac{2}{e^2}, \dfrac{3}{e^3}, \dfrac{4}{e^4}, \dfrac{5}{e^5}$ **29.** $a_n = \dfrac{n}{n+1}$ **31.** $a_n = \dfrac{1}{2^{n-1}}$

33. $a_n = (-1)^{n+1}$ **35.** $a_n = (-1)^{n+1}n$ **37.** $a_1 = 2, a_2 = 5, a_3 = 8, a_4 = 11, a_5 = 14$ **39.** $a_1 = -2, a_2 = 0, a_3 = 3, a_4 = 7, a_5 = 12$

41. $a_1 = 5, a_2 = 10, a_3 = 20, a_4 = 40, a_5 = 80$ **43.** $a_1 = 3, a_2 = \dfrac{3}{2}, a_3 = \dfrac{1}{2}, a_4 = \dfrac{1}{8}, a_5 = \dfrac{1}{40}$

45. $a_1 = 1, a_2 = 2, a_3 = 2, a_4 = 4, a_5 = 8$ **47.** $a_1 = A, a_2 = A + d, a_3 = A + 2d, a_4 = A + 3d, a_5 = A + 4d$

49. $a_1 = \sqrt{2}, a_2 = \sqrt{2 + \sqrt{2}}, a_3 = \sqrt{2 + \sqrt{2 + \sqrt{2}}}, a_4 = \sqrt{2 + \sqrt{2 + \sqrt{2 + \sqrt{2}}}}, a_5 = \sqrt{2 + \sqrt{2 + \sqrt{2 + \sqrt{2 + \sqrt{2}}}}}$

51. $3 + 4 + \cdots + (n + 2)$ **53.** $\dfrac{1}{2} + 2 + \dfrac{9}{2} + \cdots + \dfrac{n^2}{2}$ **55.** $1 + \dfrac{1}{3} + \dfrac{1}{9} + \cdots + \dfrac{1}{3^n}$ **57.** $\dfrac{1}{3} + \dfrac{1}{9} + \cdots + \dfrac{1}{3^n}$

59. $\ln 2 - \ln 3 + \ln 4 - \cdots + (-1)^n \ln n$ **61.** $\displaystyle\sum_{k=1}^{20} k$ **63.** $\displaystyle\sum_{k=1}^{13} \dfrac{k}{k+1}$ **65.** $\displaystyle\sum_{k=0}^{6} (-1)^k \left(\dfrac{1}{3^k}\right)$ **67.** $\displaystyle\sum_{k=1}^{n} \dfrac{3^k}{k}$ **69.** $\displaystyle\sum_{k=1}^{n} (a + kd)$ or $\displaystyle\sum_{k=1}^{n+1} [a + (k-1)d]$

71. 50 **73.** 21 **75.** 90 **77.** 26 **79.** 42 **81.** 96 **83. (a)** $\$2930$ **(b)** 14 payments have been made.   **(c)** 36 payments; $\$3584.62$ **(d)** $\$584.62$

85. (a) 2162 **(b)** After 26 months   **87. (a)** $a_0 = 0, a_n = (1.02)a_{n-1} + 500$ **(b)** After 82 quarters   **(c)** $\$156,116.15$

89. (a) $a_0 = 150,000, a_n = (1.005)a_{n-1} - 899.33$

   (b) $\$149,850.67$

   (c)

   (d) After 58 payments or 4 years and 10 months later

   (e) After 359 payments of $\$899.33$, plus last payment of $\$895.10$

   (f) $\$173,754.57$

   (g) (a) $a_0 = 150,000, a_n = (1.005)a_{n-1} - 999.33$

   (b) $\$149,750.67$ **(c)**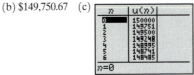

   (d) After 37 payments or 3 years and 1 month later

   (e) After 278 payments of $\$999.33$, plus last payment of $\$353.69(1.005) = \$355.46$

   (f) $\$128,169.20$

91. 21 **93.** A Fibonacci sequence   **95. (a)** 3.630170833 **(b)** 3.669060828 **(c)** 3.669296668 **(d)** 12

97. $2S = \underbrace{(1 + n) + (1 + n) + \cdots + (n + 1)}_{n \text{ terms}}$

$2S = n(n + 1)$

$S = \dfrac{1}{2}n(n + 1)$

11.2 Assess Your Understanding *(page 642)*

1. arithmetic   **2.** F   **3.** $t_n - t_{n-1} = (n + 4) - [(n - 1) + 4] = n + 4 - (n + 3) = n + 4 - n - 3 = 1$, a constant; $d = 1; 5, 6, 7, 8$

5. $t_n - t_{n-1} = (2n - 5) - [2(n - 1) - 5] = 2n - 5 - (2n - 2 - 5) = 2n - 5 - (2n - 7) = 2n - 5 - 2n + 7 = 2$, a constant;
$d = 2; -3, -1, 1, 3$

7. $t_n - t_{n-1} = (6 - 2n) - [6 - 2(n - 1)] = 6 - 2n - (6 - 2n + 2) = 6 - 2n - (8 - 2n) = 6 - 2n - 8 + 2n = -2$, a constant;
$d = -2; 4, 2, 0, -2$

9. $t_n - t_{n-1} = \left(\dfrac{1}{2} - \dfrac{1}{3}n\right) - \left[\dfrac{1}{2} - \dfrac{1}{3}(n - 1)\right] = \dfrac{1}{2} - \dfrac{1}{3}n - \left(\dfrac{1}{2} - \dfrac{1}{3}n + \dfrac{1}{3}\right) = \dfrac{1}{2} - \dfrac{1}{3}n - \left(\dfrac{5}{6} - \dfrac{1}{3}n\right) = \dfrac{1}{2} - \dfrac{1}{3}n - \dfrac{5}{6} + \dfrac{1}{3}n = -\dfrac{1}{3}$, a constant;
$d = -\dfrac{1}{3}; \dfrac{1}{6}, -\dfrac{1}{6}, -\dfrac{1}{2}, -\dfrac{5}{6}$

11. $t_n - t_{n-1} = \ln 3^n - \ln 3^{n-1} = n \ln 3 - (n - 1)\ln 3 = n \ln 3 - (n \ln 3 - \ln 3) = n \ln 3 - n \ln 3 + \ln 3 = \ln 3$, a constant;
$d = \ln 3; \ln 3, 2 \ln 3, 3 \ln 3, 4 \ln 3$

13. $a_n = 3n - 1; a_5 = 14$ **15.** $a_n = 8 - 3n; a_5 = -7$ **17.** $a_n = \dfrac{1}{2}(n - 1); a_5 = 2$ **19.** $a_n = \sqrt{2}n; a_5 = 5\sqrt{2}$ **21.** $a_{12} = 24$ **23.** $a_{10} = -26$

25. $a_8 = a + 7b$ **27.** $a_1 = -13; d = 3; a_n = a_{n-1} + 3; a_n = -16 + 3n$ **29.** $a_1 = -53; d = 6; a_n = a_{n-1} + 6; a_n = -59 + 6n$

30. $a_1 = 74; d = -10; a_n = a_{n-1} - 10; a_n = 84 - 10n$ **31.** $a_1 = 28; d = -2; a_n = a_{n-1} - 2; a_n = 30 - 2n$

33. $a_1 = 25; d = -2; a_n = a_{n-1} - 2; a_n = 27 - 2n$ **35.** n^2 **37.** $\dfrac{n}{2}(9 + 5n)$ **39.** 1260 **41.** 324

43.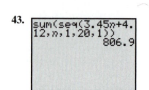

45.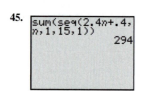

47.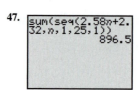

49. $-\dfrac{3}{2}$ **50.** 1   **51.** 1185 seats   **53.** 210 of beige and 190 blue   **55.** 30 rows

Historical Problems *(page 650)*

1. $1\dfrac{2}{3}$ loaves, $10\dfrac{5}{6}$ loaves, 20 loaves, $29\dfrac{1}{6}$ loaves, $38\dfrac{1}{3}$ loaves   **2. (a)** 1   **(b)** 2401   **(c)** 2800

11.3 Assess Your Understanding *(page 650)*

1. geometric   **2.** $\dfrac{a}{1-r}$ **3.** T   **4.** F   **5.** $\dfrac{S_n}{S_{n-1}} = \dfrac{3^n}{3^{n-1}} = 3^{n-(n-1)} = 3^1 = 3$, a nonzero constant; $r = 3$; 3, 9, 27, 81

7. $\dfrac{S_n}{S_{n-1}} = \dfrac{-3\left(\frac{1}{2}\right)^n}{-3\left(\frac{1}{2}\right)^{n-1}} = \left(\dfrac{1}{2}\right)^{n-(n-1)} = \left(\dfrac{1}{2}\right)^1 = \dfrac{1}{2}$, a nonzero constant; $r = \dfrac{1}{2}$; $-\dfrac{3}{2}, -\dfrac{3}{4}, -\dfrac{3}{8}, -\dfrac{3}{16}$

9 $\dfrac{S_n}{S_{n-1}} = \dfrac{\frac{2^{n-1}}{4}}{\frac{2^{(n-1)-1}}{4}} = 2^{(n-1)-[(n-1)-1]} = 2^1 = 2$, a nonzero constant; $r = 2$; $\dfrac{1}{4}, \dfrac{1}{2}, 1, 2$

11. $\dfrac{S_n}{S_{n-1}} = \dfrac{2^{n/3}}{2^{(n-1)/3}} = 2^{n/3-[(n-1)/3]} = 2^{1/3}$, a nonzero constant; $r = 2^{1/3}$; $2^{1/3}, 2^{2/3}, 2, 2^{4/3}$

13. $\dfrac{S_n}{S_{n-1}} = \dfrac{\frac{3^{n-1}}{2^n}}{\frac{3^{(n-1)-1}}{2^{n-1}}} = \dfrac{3^{n-1}}{2^n} \cdot \dfrac{2^{n-1}}{3^{n-2}} = \dfrac{3^{n-1} \cdot 2^{n-1}}{2^n \cdot 3^{n-2}} = 3^{n-1-(n-2)} \cdot 2^{n-1-n} = 3^1 \cdot 2^{-1} = \dfrac{3}{2}$, a nonzero constant; $r = \dfrac{3}{2}$; $\dfrac{1}{2}, \dfrac{3}{4}, \dfrac{9}{8}, \dfrac{27}{16}$

15. Arithmetic; $d = 1$ **17.** Neither   **19.** Arithmetic; $d = -\dfrac{2}{3}$ **21.** Neither   **23.** Geometric; $r = \dfrac{2}{3}$ **25.** Geometric; $r = 2$

27. Geometric; $r = 3^{1/2}$ **29.** $a_5 = 162$; $a_n = 2 \cdot 3^{n-1}$ **31.** $a_5 = 5$; $a_n = 5 \cdot (-1)^{n-1}$ **33.** $a_5 = 0$; $a_n = 0$ **35.** $a_5 = 4\sqrt{2}$; $a_n = (\sqrt{2})^n$

37. $a_7 = \dfrac{1}{64}$ **39.** $a_9 = 1$ **41.** $a_8 = 0.00000004$ **43.** $-\dfrac{1}{4}(1 - 2^n)$ **45.** $2\left[1 - \left(\dfrac{2}{3}\right)^n\right]$ **47.** $1 - 2^n$

49.

51.

53.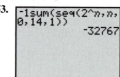

55. $\dfrac{3}{2}$ **57.** 16   **59.** $\dfrac{8}{5}$ **61.** $\dfrac{20}{3}$ **63.** $\dfrac{18}{5}$ **65.** -4 **67.** \$21,879.11   **69. (a)** 0.775 ft   **(b)** 8th   **(c)** 15.88 ft   **(d)** 20 ft   **71.** 1.845×10^{19}

73. 10   **75.** \$72.67 per share   **77.** December 20, 2001 (111 days); \$9999.92   **79.** Option B results in more money (\$524,287 versus \$500,500).

81. Yes. A constant sequence is both arithmetic and geometric. For example, 3, 3, 3, … is an arithmetic sequence with $a_1 = 3$ and $d = 0$ and is a geometric sequence with $a_1 = 3$ and $r = 1$.

CHAPTER 12 Counting and Probability

Chapter 12 is not included in *Precalculus Essentials: Enhanced with Graphing Utilities*, 4e.

CHAPTER 13 A Preview of Calculus: The Limit, Derivative, and Integral of a Function

Chapter 13 is not included in *Precalculus Essentials: Enhanced with Graphing Utilities*, 4e.

APPENDIX   Review

A.1 Assess Your Understanding *(page 667)*

1. variable   **2.** origin   **3.** strict   **4.** base; exponent or power   **5.** 5   **6.** F   **7.** F   **8.** F   **9.** 4   **11.** -28 **13.** $\frac{4}{5}$ **15.** 0   **17.** $x = 0$ **19.** $x = 3$
21. None   **23.** $x = 1, x = 0, x = -1$ **25.** $\{x | x \neq 5\}$ **27.** $\{x | x \neq -4\}$ **29.**   **31.** $>$ **33.** $>$ **35.** $>$

37. $=$ **39.** $<$ **41.** $x > 0$ **43.** $x < 2$ **45.** $x \leq 1$ **47.**   **49.**   **51.** 1   **53.** 5   **55.** 1   **57.** 22   **59.** 2

61. 1   **63.** 2   **65.** 6   **67.** 16   **69.** $\frac{1}{16}$ **71.** $\frac{1}{9}$ **73.** 9   **75.** 5   **77.** 4   **79.** $\frac{1}{64x^6}$ **81.** $\frac{x^4}{y^2}$ **83.** $\frac{1}{x^3 y}$ **85.** $-\frac{8x^3 z}{9y}$ **87.** $\frac{16x^2}{9y^2}$ **89.** $A = lw$ **91.** $C = \pi d$

93. $A = \frac{\sqrt{3}}{4}x^2$ **95.** $V = \frac{4}{3}\pi r^3$ **97.** $V = x^3$ **99.** **(a)** $2 \leq 5$ **(b)** $6 > 5$ **101.** **(a)** Yes   **(b)** No   **103.** No; $\frac{1}{3}$ is larger; 0.000333...   **105.** No

A.2 Assess Your Understanding *(page 672)*

1. right; hypotenuse   **2.** $A = \frac{1}{2}bh$ **3.** $C = 2\pi r$ **4.** T   **5.** T   **6.** F   **7.** 13   **9.** 26   **11.** 25   **13.** Right triangle; 5
15. Not a right triangle   **17.** Right triangle; 25   **19.** Not a right triangle   **21.** 8 in^2 **23.** 4 in^2 **25.** $A = 25\pi$ m^2; $C = 10\pi$ m
27. $V = 224$ ft^3; $S = 232$ ft^2 **29.** $V = \frac{256}{3}\pi$ cm^3; $S = 64\pi$ cm^2 **31.** $V = 648\pi$ in^3; $S = 306\pi$ in^2 **33.** π square units   **35.** 2π square units
37. About 16.8 ft   **39.** 64 ft^2 **41.** $24 + 2\pi \approx 30.28$ ft^2; $16 + 2\pi \approx 22.28$ ft^2 **43.** About 5.477 mi   **45.** From 100 ft: 12.247 mi; From 150 ft: 15.000 mi

A.3 Assess Your Understanding *(page 683)*

1. 4; 3   **2.** $x^4 - 16$ **3.** $x^3 - 8$ **4.** F   **5.** T   **6.** F   **7.** $3x(x - 2)(x + 2)$ **8.** prime   **9.** T   **10.** F   **11.** in lowest terms
12. least common multiple   **13.** T   **14.** F   **15.** $10x^5 + 3x^3 - 10x^2 + 6$ **17.** $2ax + a^2$ **19.** $2x^2 + 17x + 8$ **21.** $x^4 - x^2 + 2x - 1$
23. $6x^2 + 2$ **25.** $(x - 6)(x + 6)$ **27.** $(1 - 2x)(1 + 2x)$ **29.** $(x + 2)(x + 5)$ **31.** Prime   **33.** Prime   **35.** $(5 - x)(3 + x)$
37. $3(x + 2)(x - 6)$ **39.** $y^2(y + 5)(y + 6)$ **41.** $(2x + 3)^2$ **43.** $(3x + 1)(x + 1)$ **45.** $(x^2 + 9)(x - 3)(x + 3)$ **47.** $(x - 1)^2(x^2 + x + 1)^2$
49. $x^5(x - 1)(x + 1)$ **51.** $(5 - 4x)(1 + 4x)$ **53.** $(2y - 3)(2y - 5)$ **55.** $(1 + x^2)(1 - 3x)(1 + 3x)$ **57.** $(x - 6)(x + 3)$ **59.** $(x + 2)(x - 3)$
61. $3x(2 - x)^3(4 - 5x)$ **63.** $(x + 2)(x - 1)(x + 1)$ **65.** $(x - 1)(x + 1)(x^2 - x + 1)$ **67.** $\frac{3(x - 3)}{5x}$ **69.** $\frac{x(2x - 1)}{x + 4}$
71. $\frac{5x}{(x - 6)(x - 1)(x + 4)}$ **73.** $\frac{2(x + 4)}{(x - 2)(x + 2)(x + 3)}$ **75.** $\frac{x^3 - 2x^2 + 4x + 3}{x^2(x + 1)(x - 1)}$ **77.** $\frac{-1}{x(x + h)}$ **79.** $2(3x + 4)(9x + 13)$
81. $2x(3x + 5)$ **83.** $5(x + 3)(x - 2)^2(x + 1)$ **85.** $3(4x - 3)(4x - 1)$ **87.** $6(3x - 5)(2x + 1)^2(5x - 4)$ **89.** $\frac{19}{(3x - 5)^2}$
91. $\frac{x^2 - 1}{(x^2 + 1)^2}$ **93.** $\frac{x(3x + 2)}{(3x + 1)^2}$ **95.** $-\frac{3x^2 + 8x - 3}{(x^2 + 1)^2}$

A.4 Assess Your Understanding *(page 690)*

1. quotient; divisor; remainder   **2.** $-3)\overline{2\ 0\ -5\ 1}$ **3.** T   **4.** T   **5.** $4x^2 - 11x + 23$; remainder -45 **7.** $4x - 3$; remainder $x + 1$
9. $5x^2 - 13$; remainder $x + 27$ **11.** $2x^2$; remainder $-x^2 + x + 1$ **13.** $x^2 - 2x + \frac{1}{2}$; remainder $\frac{5}{2}x + \frac{1}{2}$ **15.** $-4x^2 - 3x - 3$; remainder -7
17. $x^2 - x - 1$; remainder $2x + 2$ **19.** $x^2 + ax + a^2$; remainder 0   **21.** $x^2 + x + 4$; remainder 12   **23.** $3x^2 + 11x + 32$; remainder 99
25. $x^4 - 3x^3 + 5x^2 - 15x + 46$; remainder -138 **27.** $4x^5 + 4x^4 + x^3 + x^2 + 2x + 2$; remainder 7   **29.** $0.1x^2 - 0.11x + 0.321$; remainder -0.3531
31. $x^4 + x^3 + x^2 + x + 1$; remainder 0   **33.** No   **35.** Yes   **37.** Yes   **39.** No   **41.** Yes   **43.** $a = 1, b = -4, c = 11, d = -17$; $a + b + c + d = -9$

A.5 Assess Your Understanding *(page 705)*

5. equivalent equations   **6.** identity   **7.** F   **8.** T   **9.** add; $\frac{25}{4}$ **10.** discriminant; negative   **11.** F   **12.** F   **13.** $\{7\}$ **15.** $\{-3\}$ **17.** $\{4\}$ **19.** $\left\{\frac{5}{4}\right\}$
21. $\{-1\}$ **23.** $\{-18\}$ **25.** $\{-3\}$ **27.** $\{-16\}$ **29.** $\{0.5\}$ **31.** $\{2\}$ **33.** $\{2\}$ **35.** $\{3\}$ **37.** $\{0, 9\}$ **39.** $\{0, 9\}$ **41.** $\{21\}$ **43.** $\{-2, 2\}$ **45.** $\{6\}$
47. $\{-3, 3\}$ **49.** $\{-4, 1\}$ **51.** $\left\{-1, \frac{3}{2}\right\}$ **53.** $\{-4, 4\}$ **55.** $\{2\}$ **57.** No real solution   **59.** $\{-2, 2\}$ **61.** $\{-1, 3\}$ **63.** $\{-2, -1, 0, 1\}$ **65.** $\{0, 4\}$
67. $\{-6, 2\}$ **69.** $\left\{-\frac{1}{2}, 3\right\}$ **71.** $\{3, 4\}$ **73.** $\left\{\frac{3}{2}\right\}$ **75.** $\left\{-\frac{2}{3}, \frac{3}{2}\right\}$ **77.** $\left\{-\frac{3}{4}, 2\right\}$ **79.** $\{-6\}$ **81.** $\{-2, -1, 1, 2\}$ **83.** $\{-6, -5\}$ **85.** $\left\{-\frac{3}{2}, 2\right\}$
87. $\{-5, 0, 4\}$ **89.** $\{-1, 1\}$ **91.** $\left\{-2, \frac{1}{2}, 2\right\}$ **93.** $\{-5, 5\}$ **95.** $\{-1, 3\}$ **97.** $\{-3, 0\}$ **99.** 16   **101.** $\frac{1}{16}$ **103.** $\frac{1}{9}$ **105.** $\{-7, 3\}$ **107.** $\left\{-\frac{1}{4}, \frac{3}{4}\right\}$

109. $\left\{\dfrac{-1 - \sqrt{7}}{6}, \dfrac{-1 + \sqrt{7}}{6}\right\}$ **111.** $\{2 - \sqrt{2}, 2 + \sqrt{2}\}$ **113.** $\left\{\dfrac{5 - \sqrt{29}}{2}, \dfrac{5 + \sqrt{29}}{2}\right\}$ **115.** $\left\{1, \dfrac{3}{2}\right\}$ **117.** No real solution

119. $\left\{\dfrac{-1 - \sqrt{5}}{4}, \dfrac{-1 + \sqrt{5}}{4}\right\}$ **121.** $\left\{\dfrac{-\sqrt{3} - \sqrt{15}}{2}, \dfrac{-\sqrt{3} + \sqrt{15}}{2}\right\}$ **123.** No real solution **125.** Repeated real solution **127.** Two unequal

real solutions **129.** $R = \dfrac{R_1 R_2}{R_1 + R_2}$ **131.** $R = \dfrac{mv^2}{F}$ **133.** $r = \dfrac{S - a}{S}$ **135.** $\dfrac{-b + \sqrt{b^2 - 4ac}}{2a} + \dfrac{-b - \sqrt{b^2 - 4ac}}{2a} = \dfrac{-2b}{2a} = \dfrac{-b}{a}$

137. $k = -\dfrac{1}{2} \text{ or } \dfrac{1}{2}$ **139.** The solutions of $ax^2 - bx + c = 0$ are $\dfrac{b + \sqrt{b^2 - 4ac}}{2a}$ and $\dfrac{b - \sqrt{b^2 - 4ac}}{2a}$. **141.** (b)

A.6 Assess Your Understanding *(page 715)*

1. F **2.** 5 **3.** F **4.** real; imaginary; imaginary unit **5.** $\{-2i, 2i\}$ **6.** F **7.** T **8.** F **9.** $8 + 5i$ **11.** $-7 + 6i$ **13.** $-6 - 11i$ **15.** $6 - 18i$
17. $6 + 4i$ **19.** $10 - 5i$ **21.** 37 **23.** $\dfrac{6}{5} + \dfrac{8}{5}i$ **25.** $1 - 2i$ **27.** $\dfrac{5}{2} - \dfrac{7}{2}i$ **29.** $-\dfrac{1}{2} + \dfrac{\sqrt{3}}{2}i$ **31.** $2i$ **33.** $-i$ **35.** i **37.** -6 **39.** $-10i$ **41.** $-2 + 2i$
43. 0 **45.** 0 **47.** $2i$ **49.** $5i$ **51.** $5i$ **53.** $\{-2i, 2i\}$ **55.** $\{-4, 4\}$ **57.** $\{3 - 2i, 3 + 2i\}$ **59.** $\{3 - i, 3 + i\}$ **61.** $\left\{\dfrac{1}{4} - \dfrac{1}{4}i, \dfrac{1}{4} + \dfrac{1}{4}i\right\}$
63. $\left\{\dfrac{1}{5} - \dfrac{2}{5}i, \dfrac{1}{5} + \dfrac{2}{5}i\right\}$ **65.** $\left\{-\dfrac{1}{2} - \dfrac{\sqrt{3}}{2}i, -\dfrac{1}{2} + \dfrac{\sqrt{3}}{2}i\right\}$ **67.** $\{2, -1 - \sqrt{3}i, -1 + \sqrt{3}i\}$ **69.** $\{-2, 2, -2i, 2i\}$ **71.** $\{-3i, -2i, 2i, 3i\}$
73. Two complex solutions that are conjugates of each other **75.** Two unequal real solutions **77.** A repeated real solution **79.** $2 - 3i$ **81.** 6 **83.** 25
85. $z + \overline{z} = a + bi + a - bi = 2a + (b - b)i = 2a$
$\quad\ z - \overline{z} = a + bi - a - bi = 0 + (b + b)i = 2bi$
87. $\overline{z + w} = \overline{(a + bi) + (c + di)} = \overline{(a + c) + (b + d)i} = (a + c) - (b + d)i$
$\quad\ \overline{z} + \overline{w} = (a - bi) + (c - di) = (a + c) + (-b - d)i = (a + c) - (b + d)i$

A.7 Assess Your Understanding *(page 724)*

1. mathematical modeling **2.** interest **3.** uniform motion **4.** T **5.** T **6.** $100 - x$ **7.** $A = \pi r^2; r = $ radius, $A = $ area
9. $A = s^2; A = $ area, $s = $ length of a side **11.** $F = ma; F = $ force, $m = $ mass, $a = $ acceleration **13.** $W = Fd; W = $ work, $F = $ force, $d = $ distance
15. $C = 150x; C = $ total variable cost, $x = $ number of dishwashers **17.** \$11,500 will be invested in bonds and \$8500 in CDs. **19.** Brooke needs a
score of 85. **21.** The length is 19 ft; the width is 11 ft. **23.** Invest \$31,250 in bonds and \$18,750 in CDs. **25.** \$11,600 was loaned out at 8%.
27. Mix 75 lb of Earl Grey tea with 25 lb of Orange Pekoe tea. **29.** Mix 40 lb of cashews with the peanuts. **31.** The speed of the current is 2.286 mi/hr.
33. The Metra commuter averages 30 mi/hr; the Amtrak averages 80 mi/hr. **35.** Working together, it takes 12 min. **37.** The dimensions should
be 4 ft by 4 ft. **39.** The speed of the current is 5 mph. **41.** The dimensions are 11 ft by 13 ft. **43.** The dimensions are 5 m by 8 m. **45. (a)** The ball
strikes the ground after 6 sec. **(b)** The ball passes the top of the building on its way down after 5 sec. **47. (a)** The dimensions are 10 ft by 5 ft.
(b) The area is 50 sq ft. **(c)** The dimensions would be 7.5 ft by 7.5 ft. **(d)** The area would be 56.25 sq ft. **49.** The border will be 2.71 ft wide.
51. The border will be 2.56 ft wide. **53.** Add $\dfrac{2}{3}$ gal of water. **55.** Evaporate 10.67 oz of water. **57.** 40 g of 12 karat gold should be mixed with
20 g of pure gold. **59.** Mike passes Dan $\dfrac{1}{3}$ mi from the start, 2 min from the time Mike started to race. **61.** Start the auxiliary pump at 9:45 AM.
63. The tub will fill in 1 hr. **65.** Lewis would beat Burke by 16.75 m. **67.** Therese should be allowed 20,000 mi as a business expense.
69. 13 sides; No **71.** The average speed is 49.5 mph. **73.** Set the original price at \$40. At 50% off, there will be no profit.

A.8 Assess Your Understanding *(page 736)*

3. negative **4.** closed interval **5.** multiplication properties **6.** T **7.** T **8.** T **9.** F **10.** T **11.** $[0, 2]; 0 \le x \le 2$
13. $(-1, 2); -1 < x < 2$ **15.** $[0, 3); 0 \le x < 3$ **17. (a)** $6 < 8$ **(b)** $-2 < 0$ **(c)** $9 < 15$ **(d)** $-6 > -10$
19. (a) $7 > 0$ **(b)** $-1 > -8$ **(c)** $12 > -9$ **(d)** $-8 < 6$ **21. (a)** $2x + 4 < 5$ **(b)** $2x - 4 < -3$ **(c)** $6x + 3 < 6$ **(d)** $-4x - 2 > -4$
23. $[0, 4]$ **25.** $[4, 6)$ **27.** $[4, \infty)$ **29.** $(-\infty, -4)$

31. $2 \le x \le 5$ **33.** $-3 < x < -2$ **35.** $x \ge 4$ **37.** $x < -3$

39. $<$ **41.** $>$ **43.** $\ge$ **45.** $<$ **47.** $\le$ **49.** $<$ **51.** $\ge$
53. $\{x | x < 4\}; (-\infty, 4)$ **55.** $\{x | x \ge -1\}; [-1, \infty)$ **57.** $\{x | x > 3\}; (3, \infty)$ **59.** $\{x | x \ge 2\}; [2, \infty)$

61. $\{x | x > -7\}; (-7, \infty)$ **63.** $\left\{x \middle| x \le \dfrac{2}{3}\right\}; \left(-\infty, \dfrac{2}{3}\right]$ **65.** $\{x | x < -20\}; (-\infty, -20)$ **67.** $\left\{x \middle| x \ge \dfrac{4}{3}\right\}; \left[\dfrac{4}{3}, \infty\right)$

69. $\{x|3 \le x \le 5\}; [3, 5]$ **71.** $\left\{x \Big| \dfrac{2}{3} \le x \le 3\right\}; \left[\dfrac{2}{3}, 3\right]$ **73.** $\left\{x \Big| -\dfrac{11}{2} < x < \dfrac{1}{2}\right\}; \left(-\dfrac{11}{2}, \dfrac{1}{2}\right)$ **75.** $\{x|-6 < x < 0\}; (-6, 0)$

77. $\{x|x < -5\}; (-\infty, -5)$ **79.** $\{x|x \ge -1\}; [-1, \infty)$ **81.** $\left\{x \Big| \dfrac{1}{2} \le x < \dfrac{5}{4}\right\}; \left[\dfrac{1}{2}, \dfrac{5}{4}\right)$ **83.** $\{x|-6 < x < 6\}; (-6, 6)$

85. $\{x|x < -4$ or $x > 4\};$
$(-\infty, -4)$ or $(4, \infty)$ **87.** $\{x|-4 < x < 4\}; (-4, 4)$ **89.** $\{x|x < -4$ or $x > 4\};$
$(-\infty, -4)$ or $(4, \infty)$ **91.** $\{x|1 < x < 3\}; (1, 3)$

93. $\left\{t \Big| -\dfrac{2}{3} \le t \le 2\right\}; \left[-\dfrac{2}{3}, 2\right]$ **95.** $\{x|x \le 1$ or $x \ge 5\};$
$(-\infty, 1]$ or $[5, \infty)$ **97.** $\left\{x \Big| -1 < x < \dfrac{3}{2}\right\}; \left(-1, \dfrac{3}{2}\right)$ **99.** $\{x|x < -1$ or $x > 2\};$
$(-\infty, -1)$ or $(2, \infty)$

101. No real solution; $\varnothing$ **103.** $\{x|x > 5\}; (5, \infty)$ **105.** $\{x|x > 3\}; (3, \infty)$

107. $\left|x - 2\right| < \dfrac{1}{2}; \left\{x \Big| \dfrac{3}{2} < x < \dfrac{5}{2}\right\}$ **109.** $|x + 3| > 2; \{x|x < -5$ or $x > -1\}$ **111.** $21 < $ age < 30
113. $|x - 98.6| \ge 1.5; \{x|x \le 97.1$ or $x \ge 100.1\}$ **115. (a)** Male ≥ 75.6 **(b)** Female ≥ 80.4 **(c)** A female can expect to live at least 4.8 years longer.
117. The agent's commission ranges from \$45,000 to \$95,000, inclusive. As a percent of selling price, the commission ranges from 5% to approximately
8.6%, inclusive.   **119.** The amount withheld varies from \$76.35 to \$101.35, inclusive.   **121.** The usage varies from approximately 675.41 to
2500.91 kilowatt-hours, inclusive.   **123.** The dealer's cost varies from \$7457.63 to \$7857.14, inclusive.   **125.** You need at least a 74 on the last test.
127. The amount of gasoline ranged from 12 to 20 gal, inclusive.
129. $\dfrac{a + b}{2} - a = \dfrac{a + b - 2a}{2} = \dfrac{b - a}{2} > 0;$ therefore, $a < \dfrac{a + b}{2}; \; b - \dfrac{a + b}{2} = \dfrac{2b - a - b}{2} = \dfrac{b - a}{2} > 0;$ therefore, $b > \dfrac{a + b}{2}$.
131. $(\sqrt{ab})^2 - a^2 = ab - a^2 = a(b - a) > 0;$ thus, $(\sqrt{ab})^2 > a^2$ and $\sqrt{ab} > a; \; b^2 - (\sqrt{ab})^2 = b^2 - ab = b(b - a) > 0;$
thus $b^2 > (\sqrt{ab})^2$ and $b > \sqrt{ab}$.   **133.** $0 < a < b$ so $0 < \dfrac{1}{b} < \dfrac{1}{a}; \; \dfrac{1}{b} < \dfrac{1}{2}\left(\dfrac{1}{b} + \dfrac{1}{a}\right) < \dfrac{1}{a}$ by Problem 129, so $\dfrac{1}{b} < \dfrac{1}{h} < \dfrac{1}{a}$; therefore $a < h < b$.

A.9 Assess Your Understanding *(page 745)*

3. index   **4.** cube root   **5.** T   **6.** F   **7.** 3   **9.** -2 **11.** $2\sqrt{2}$ **13.** $-2x\sqrt[3]{x}$ **15.** $x^3 y^2$ **17.** $x^2 y$ **19.** $6\sqrt{x}$ **21.** $6x\sqrt{x}$ **23.** $15\sqrt[3]{3}$ **25.** $12\sqrt{3}$
27. $2\sqrt{3}$ **29.** $x - 2\sqrt{x} + 1$ **31.** $-5\sqrt{2}$ **33.** $(2x - 1)\sqrt[3]{2x}$ **35.** $\dfrac{\sqrt{2}}{2}$ **37.** $-\dfrac{\sqrt{15}}{5}$ **39.** $\dfrac{\sqrt{3}(5 + \sqrt{2})}{23}$ **41.** $\dfrac{-19 + 8\sqrt{5}}{41}$ **43.** $\dfrac{5\sqrt[3]{4}}{2}$
45. $\dfrac{2x + h - 2\sqrt{x(x + h)}}{h}$ **47.** $\dfrac{9}{2}$ **49.** 3   **51.** 4   **53.** -3 **55.** 64   **57.** $\dfrac{1}{27}$ **59.** $\dfrac{27\sqrt{2}}{32}$ **61.** $\dfrac{27\sqrt{2}}{32}$ **63.** $x^{7/12}$ **65.** xy^2 **67.** $x^{4/3} y^{5/3}$ **69.** $\dfrac{8x^{3/2}}{y^{1/4}}$
71. $\dfrac{3x + 2}{(1 + x)^{1/2}}$ **73.** $\dfrac{x(3x^2 + 2)}{(x^2 + 1)^{1/2}}$ **75.** $\dfrac{22x + 5}{10\sqrt{x - 5}\sqrt{4x + 3}}$ **77.** $\dfrac{2 + x}{2(1 + x)^{3/2}}$ **79.** $\dfrac{4 - x}{(x + 4)^{3/2}}$ **81.** $\dfrac{1}{x^2(x^2 - 1)^{1/2}}$ **83.** $\dfrac{1 - 3x^2}{2\sqrt{x}(1 + x^2)^2}$
85. $\dfrac{1}{2}(5x + 2)(x + 1)^{1/2}$ **87.** $2x^{1/2}(3x - 4)(x + 1)$ **89.** $(x^2 + 4)^{1/3}(11x^2 + 12)$ **91.** $(3x + 5)^{1/3}(2x + 3)^{1/2}(17x + 27)$ **93.** $\dfrac{3(x + 2)}{2x^{1/2}}$
95. $\dfrac{2(2 - x)(2 + x)}{(8 - x^2)^{1/2}}$

Index

TRIGONOMETRIC FUNCTIONS

Let t be a real number and let $P = (x, y)$ be the point on the unit circle that corresponds to t.

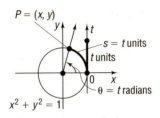

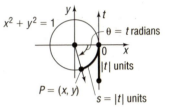

$$\sin t = y \qquad \cos t = x \qquad \tan t = \frac{y}{x}, \quad x \neq 0$$

$$\csc t = \frac{1}{y}, \quad y \neq 0 \qquad \sec t = \frac{1}{x}, \quad x \neq 0 \qquad \cot t = \frac{x}{y}, \quad y \neq 0$$

TRIGONOMETRIC IDENTITIES

Fundamental Identities

$$\tan\theta = \frac{\sin\theta}{\cos\theta} \qquad \cot\theta = \frac{\cos\theta}{\sin\theta}$$

$$\csc\theta = \frac{1}{\sin\theta} \qquad \sec\theta = \frac{1}{\cos\theta} \qquad \cot\theta = \frac{1}{\tan\theta}$$

$$\sin^2\theta + \cos^2\theta = 1$$
$$\tan^2\theta + 1 = \sec^2\theta$$
$$\cot^2\theta + 1 = \csc^2\theta$$

Even-Odd Identities

$$\sin(-\theta) = -\sin\theta \qquad \csc(-\theta) = -\csc\theta$$

$$\cos(-\theta) = \cos\theta \qquad \sec(-\theta) = \sec\theta$$

$$\tan(-\theta) = -\tan\theta \qquad \cot(-\theta) = -\cot\theta$$

Sum and Difference Formulas

$$\sin(\alpha + \beta) = \sin\alpha\cos\beta + \cos\alpha\sin\beta$$
$$\sin(\alpha - \beta) = \sin\alpha\cos\beta - \cos\alpha\sin\beta$$
$$\cos(\alpha + \beta) = \cos\alpha\cos\beta - \sin\alpha\sin\beta$$
$$\cos(\alpha - \beta) = \cos\alpha\cos\beta + \sin\alpha\sin\beta$$

$$\tan(\alpha + \beta) = \frac{\tan\alpha + \tan\beta}{1 - \tan\alpha\tan\beta}$$

$$\tan(\alpha - \beta) = \frac{\tan\alpha - \tan\beta}{1 + \tan\alpha\tan\beta}$$

Half-Angle Formulas

$$\sin\frac{\theta}{2} = \pm\sqrt{\frac{1 - \cos\theta}{2}}$$

$$\cos\frac{\theta}{2} = \pm\sqrt{\frac{1 + \cos\theta}{2}}$$

$$\tan\frac{\theta}{2} = \frac{1 - \cos\theta}{\sin\theta}$$

Double-Angle Formulas

$$\sin(2\theta) = 2\sin\theta\cos\theta$$
$$\cos(2\theta) = \cos^2\theta - \sin^2\theta$$
$$\cos(2\theta) = 2\cos^2\theta - 1$$
$$\cos(2\theta) = 1 - 2\sin^2\theta$$

$$\tan(2\theta) = \frac{2\tan\theta}{1 - \tan^2\theta}$$

Product-to-Sum Formulas

$$\sin\alpha\sin\beta = \tfrac{1}{2}\left[\cos(\alpha - \beta) - \cos(\alpha + \beta)\right]$$

$$\cos\alpha\cos\beta = \tfrac{1}{2}\left[\cos(\alpha - \beta) + \cos(\alpha + \beta)\right]$$

$$\sin\alpha\cos\beta = \tfrac{1}{2}\left[\sin(\alpha + \beta) + \sin(\alpha - \beta)\right]$$

Sum-to-Product Formulas

$$\sin\alpha + \sin\beta = 2\sin\frac{\alpha + \beta}{2}\cos\frac{\alpha - \beta}{2}$$

$$\sin\alpha - \sin\beta = 2\sin\frac{\alpha - \beta}{2}\cos\frac{\alpha + \beta}{2}$$

$$\cos\alpha + \cos\beta = 2\cos\frac{\alpha + \beta}{2}\cos\frac{\alpha - \beta}{2}$$

$$\cos\alpha - \cos\beta = -2\sin\frac{\alpha + \beta}{2}\sin\frac{\alpha - \beta}{2}$$

SOLVING TRIANGLES

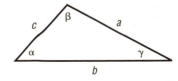

Law of Sines

$$\frac{\sin\alpha}{a} = \frac{\sin\beta}{b} = \frac{\sin\gamma}{c}$$

Law of Cosines

$$a^2 = b^2 + c^2 - 2bc\cos\alpha$$
$$b^2 = a^2 + c^2 - 2ac\cos\beta$$
$$c^2 = a^2 + b^2 - 2ab\cos\gamma$$

FORMULAS/EQUATIONS

Distance Formula
If $P_1 = (x_1, y_1)$ and $P_2 = (x_2, y_2)$, the distance from P_1 to P_2 is

$$d(P_1, P_2) = \sqrt{(x_2 - x_1)^2 + (y_2 - y_1)^2}$$

Standard Equation of a Circle
The standard equation of a circle of radius r with center at (h, k) is
$$(x - h)^2 + (y - k)^2 = r^2$$

Slope Formula
The slope m of the line containing the points $P_1 = (x_1, y_1)$ and $P_2 = (x_2, y_2)$ is

$$m = \frac{y_2 - y_1}{x_2 - x_1} \quad \text{if } x_1 \neq x_2$$

$$m \text{ is undefined} \quad \text{if } x_1 = x_2$$

Point–Slope Equation of a Line
The equation of a line with slope m containing the point (x_1, y_1) is
$$y - y_1 = m(x - x_1)$$

Slope–Intercept Equation of a Line
The equation of a line with slope m and y-intercept b is
$$y = mx + b$$

Quadratic Formula
The solutions of the equation $ax^2 + bx + c = 0, a \neq 0$, are

$$x = \frac{-b \pm \sqrt{b^2 - 4ac}}{2a}$$

If $b^2 - 4ac > 0$, there are two unequal real solutions.
If $b^2 - 4ac = 0$, there is a repeated real solution.
If $b^2 - 4ac < 0$, there are two complex solutions that are not real.

GEOMETRY FORMULAS

Circle

r = Radius, A = Area, C = Circumference
$A = \pi r^2 \quad C = 2\pi r$

Triangle

b = Base, h = Altitude (Height), A = area
$A = \frac{1}{2} bh$

Rectangle

l = Length, w = Width, A = area, P = perimeter
$A = lw \quad P = 2l + 2w$

Rectangular Box

l = Length, w = Width, h = Height, V = Volume, S = Surface area
$V = lwh \quad S = 2lw + 2lh + 2wh$

Sphere

r = Radius, V = Volume, S = Surface area
$V = \frac{4}{3} \pi r^3 \quad S = 4\pi r^2$

Right Circular Cylinder

r = Radius, h = Height, V = Volume, S = Surface area
$V = \pi r^2 h \quad S = 2\pi r^2 + 2\pi rh$